THE MOLECULAR AND CELLULAR BASIS OF NEURODEGENERATIVE DISEASES

THE MOLECULAR AND CELLULAR BASIS OF NEURODEGENERATIVE DISEASES

Underlying Mechanisms

Edited by

MICHAEL S. WOLFE

University of Kansas, Lawrence, KS, United States

Academic Press is an imprint of Elsevier
125 London Wall, London EC2Y 5AS, United Kingdom
525 B Street, Suite 1800, San Diego, CA 92101-4495, United States
50 Hampshire Street, 5th Floor, Cambridge, MA 02139, United States
The Boulevard, Langford Lane, Kidlington, Oxford OX5 1GB, United Kingdom

British Library Cataloguing-in-Publication Data
A catalogue record for this book is available from the British Library

Library of Congress Cataloging-in-Publication Data
A catalog record for this book is available from the Library of Congress

ISBN: 978-0-12-811304-2

For Information on all Academic Press publications
visit our website at https://www.elsevier.com/books-and-journals

Publisher: Nikki Levy
Acquisition Editor: Natalie Farra
Editorial Project Manager: Kathy Padilla
Production Project Manager: Mohana Natarajan
Cover Designer: Miles Hitchen

Typeset by MPS Limited, Chennai, India

Dedication

I dedicate this book to my dear friend Missy, for all her love and support.

Contents

6. Parkinson's Disease and Other Synucleinopathies

MARK R. COOKSON

7. Huntington's Disease and Other Polyglutamine Repeat Diseases: Molecular Mechanisms and Pathogenic Pathways

AUDREY S. DICKEY AND ALBERT R. LA SPADA

8. Prion-Like Propagation in Neurodegenerative Diseases

WOUTER PEELAERTS, VEERLE BAEKELANDT AND PATRIK BRUNDIN

9. Neurodegenerative Diseases as Protein Folding Disorders

JEREMY D. BAKER, JACK M. WEBSTER, LINDSEY B. SHELTON, JOHN KOREN, III, VLADIMIR N. UVERSKY, LAURA J. BLAIR AND CHAD A. DICKEY

10. Heat Shock Proteins and Protein Quality Control in Alzheimer's Disease

FRED W. VAN LEEUWEN AND HARM H. KAMPINGA

11. Neurodegenerative Diseases and Autophagy

ANGELEEN FLEMING, MARIELLA VICINANZA, MAURIZIO RENNA, CLAUDIA PURI, THOMAS RICKETTS, JENS FÜLLGRABE, ANA LOPEZ, SARAH M. DE JAGER, AVRAHAM ASHKENAZI, MARIANA PAVEL, FLORIANA LICITRA, ANDREA CARICASOLE, STEPHEN P. ANDREWS, JOHN SKIDMORE AND DAVID C. RUBINSZTEIN

12. Neurodegenerative Diseases and Axonal Transport

LAWRENCE S. GOLDSTEIN AND UTPAL DAS

13. Mitochondrial Function and Neurodegenerative Diseases

HEATHER M. WILKINS, IAN WEIDLING, SCOTT KOPPEL, XIAOWAN WANG, ALEX VON SCHULZE AND RUSSELL H. SWERDLOW

14. Non-cell Autonomous Degeneration: Role of Astrocytes in Neurodegenerative Diseases

SARAH E. SMITH AND AZAD BONNI

15. Neurodegenerative Diseases and RNA-Mediated Toxicity

TIFFANY W. TODD AND LEONARD PETRUCELLI

16. Neuroinflammation in Age-Related Neurodegenerative Diseases

KATHRYN P. MACPHERSON, MARIA E. DE SOUSA RODRIGUES, AMARALLYS F. CINTRON AND MALÚ G. TANSEY

17. Neurodegenerative Diseases and the Aging Brain

STEPHEN K. GODIN, JINSOO SEO AND LI-HUEI TSAI

List of Contributors

Stephen P. Andrews Alzheimer's Research UK Cambridge Drug Discovery Institute, University of Cambridge, Cambridge, United Kingdom

Avraham Ashkenazi Cambridge Institute for Medical Research, University of Cambridge, Cambridge Biomedical Campus, Cambridge, United Kingdom

Veerle Baekelandt KU Leuven, Leuven, Belgium

Jeremy D. Baker University of South Florida, Tampa, FL, United States

Konrad Beyreuther The University of Heidelberg, Heidelberg, Germany

Laura J. Blair University of South Florida, Tampa, FL, United States

Azad Bonni Department of Neuroscience, Washington University School of Medicine, St. Louis, MO, United States

Erin Bove-Fenderson Boston University School of Medicine, Boston, MA, United States

Patrik Brundin Van Andel Research Institute, Grand Rapids, MI, United States

Andrea Caricasole Alzheimer's Research UK Cambridge Drug Discovery Institute, University of Cambridge, Cambridge, United Kingdom

Amarallys F. Cintron Emory University School of Medicine, Atlanta, GA, United States

Mark R. Cookson National Institute on Aging, National Institutes of Health, Bethesda, MD, United States

Utpal Das University of California San Diego, School of Medicine, San Diego, CA, United States

Sarah M. de Jager Cambridge Institute for Medical Research, University of Cambridge, Cambridge Biomedical Campus, Cambridge, United Kingdom

Maria E. de Sousa Rodrigues Emory University School of Medicine, Atlanta, GA, United States

Audrey S. Dickey Duke University School of Medicine, Durham, NC, United States

Chad A. Dickey University of South Florida, Tampa, FL, United States

Cheng Fang Boston University School of Medicine, Boston, MA, United States

Angeleen Fleming Cambridge Institute for Medical Research, University of Cambridge, Cambridge Biomedical Campus, Cambridge, United Kingdom

Jens Füllgrabe Cambridge Institute for Medical Research, University of Cambridge, Cambridge Biomedical Campus, Cambridge, United Kingdom

Stephen K. Godin Massachusetts Institute of Technology, Cambridge, MA, United States

Michel Goedert MRC Laboratory of Molecular Biology, Cambridge, United Kingdom

Lawrence S. Goldstein University of California San Diego, School of Medicine, San Diego, CA, United States

Jorge Gomez-Deza King's College London, London, United Kingdom

Ben Gu The Florey Institute, The University of Melbourne, Melbourne, VIC, Australia

David A. Harris Boston University School of Medicine, Boston, MA, United States

Harm H. Kampinga University of Groningen, Groningen, The Netherlands

Scott Koppel University of Kansas School of Medicine, Kansas City, KS, United States

John Koren III University of South Florida, Tampa, FL, United States

Albert R. La Spada Duke University School of Medicine, Durham, NC, United States

Simon Laws Edith Cowan University, Joondalup, WA, Australia

Floriana Licitra Cambridge Institute for Medical Research, University of Cambridge, Cambridge Biomedical Campus, Cambridge, United Kingdom

Yen Y. Lim The Florey Institute, The University of Melbourne, Melbourne, VIC, Australia

Ana Lopez Cambridge Institute for Medical Research, University of Cambridge, Cambridge Biomedical Campus, Cambridge, United Kingdom

Kathryn P. MacPherson Emory University School of Medicine, Atlanta, GA, United States

Colin L. Masters The Florey Institute, The University of Melbourne, Melbourne, VIC, Australia

Alex J. McDonald Boston University School of Medicine, Boston, MA, United States

Robert C.C. Mercer Boston University School of Medicine, Boston, MA, United States

Mariana Pavel Cambridge Institute for Medical Research, University of Cambridge, Cambridge Biomedical Campus, Cambridge, United Kingdom

Wouter Peelaerts KU Leuven, Leuven, Belgium; Van Andel Research Institute, Grand Rapids, MI, United States

Leonard Petrucelli Mayo Clinic, Jacksonville, FL, United States

Claudia Puri Cambridge Institute for Medical Research, University of Cambridge, Cambridge Biomedical Campus, Cambridge, United Kingdom

Maurizio Renna Cambridge Institute for Medical Research, University of Cambridge, Cambridge Biomedical Campus, Cambridge, United Kingdom

Thomas Ricketts Cambridge Institute for Medical Research, University of Cambridge, Cambridge Biomedical Campus, Cambridge, United Kingdom

Blaine Roberts The Florey Institute, The University of Melbourne, Melbourne, VIC, Australia

David C. Rubinsztein Cambridge Institute for Medical Research, University of Cambridge, Cambridge Biomedical Campus, Cambridge, United Kingdom; UK Dementia Research Institute, Cambridge Biomedical Campus, Cambridge, United Kingdom

Jinsoo Seo Massachusetts Institute of Technology, Cambridge, MA, United States

Christopher E. Shaw King's College London, London, United Kingdom

Lindsey B. Shelton University of South Florida, Tampa, FL, United States

John Skidmore Alzheimer's Research UK Cambridge Drug Discovery Institute, University of Cambridge, Cambridge, United Kingdom

Sarah E. Smith Medical Scientist Training Program, Washington University School of Medicine, St. Louis, MO, United States

Russell H. Swerdlow University of Kansas School of Medicine, Kansas City, KS, United States

Malú G. Tansey Emory University School of Medicine, Atlanta, GA, United States

Tiffany W. Todd Mayo Clinic, Jacksonville, FL, United States

Li-Huei Tsai Massachusetts Institute of Technology, Cambridge, MA, United States

Vladimir N. Uversky University of South Florida, Tampa, FL, United States

Fred W. van Leeuwen University of Maastricht, Maastricht, The Netherlands

Mariella Vicinanza Cambridge Institute for Medical Research, University of Cambridge, Cambridge Biomedical Campus, Cambridge, United Kingdom

Victor L. Villemagne The Florey Institute, The University of Melbourne, Melbourne, VIC, Australia; Austin Health, Melbourne, VIC, Australia

Alex von Schulze University of Kansas School of Medicine, Kansas City, KS, United States

Xiaowan Wang University of Kansas School of Medicine, Kansas City, KS, United States

Jack M. Webster University of South Florida, Tampa, FL, United States

Ian Weidling University of Kansas School of Medicine, Kansas City, KS, United States

Heather M. Wilkins University of Kansas School of Medicine, Kansas City, KS, United States

Michael S. Wolfe University of Kansas, Lawrence, KS, United States

Bei Wu Boston University School of Medicine, Boston, MA, United States

Preface

Neurodegenerative diseases are among the most devastating of human ailments, slowly robbing a person of memories, reason, personality, or movement. All are ultimately fatal. The human cost goes far beyond the patient, as family and friends watch their loved one deteriorate and caregivers struggle to provide daily assistance in the face of relentless decline. Economic costs are exorbitant as well, not only because of medical expenses but also lost productivity and wages.

Despite the dire need, there are no medications or procedures that effectively slow or halt the neurodegenerative process. For Parkinson's disease, replacement therapy for lost dopamine provides effective relief from tremors and other motor symptoms. However, the loss of dopaminergic neurons in the substantia nigra continues, and ultimately replacement therapy becomes ineffective. For Alzheimer's disease, acetylcholinesterase inhibitors boost signaling at cholinergic synapses, but neurodegeneration continues in areas critical to memory and cognition, such as the hippocampus.

Solving these difficult problems in human health will require better understanding of the molecular and cellular mechanisms that lead to pathogenesis and progression. In this book, we put forward the leading ideas and evidence regarding these mechanisms as well as prospects for developing means of prevention and treatment of neurodegenerative diseases. Each chapter is authored by an international expert on their topic.

We begin with an overview chapter for this broad area of biomedical investigation, describing the major neurodegenerative diseases and their epidemiology, genetics, pathology, and molecular and cell biology, as well as animal models and the special challenges to developing therapeutics. The next set of chapters focuses on particular neurodegenerative diseases, beginning with prion diseases, as recent research suggests that many neurodegenerative diseases may involve similar mechanisms for the spread of molecular pathology. Specific chapters follow on Alzheimer's disease, tauopathies such as frontotemporal dementia, amyotrophic lateral sclerosis, Parkinson's disease, and Huntington's disease.

Subsequent chapters focus on general mechanisms that are apparently shared by many neurodegenerative diseases. This includes trans-synaptic propagation of prion-like proteins, problems with protein folding and quality control, disruption of axonal transport, mitochondrial dysfunction, RNA-mediated neurotoxicity, neuroinflammation, and aging. As this list suggests, there are many ways to cause neuronal dysfunction and death. This is due to the special biology of neurons with processes extending often quite far from the cell body that need to connect functionally with other neurons or effector cells.

The general aim of this book is to provide a resource that condenses and consolidates the overwhelming amount of information in the scientific literature. In doing so, we hope to give the researcher as well as the student a sense of the big picture and common themes and to allow the opportunity to make connections between seemingly disparate areas of investigation. We also hope to make clear the open questions that remain and to suggest where the field might be heading. Thus the broader goal is to stimulate new ideas and future research that will ultimately lead to effective means of prevention and treatment.

I would like to thank all contributing authors for devoting considerable time, thought, and effort in putting together superb chapters on challenging topics. More broadly, I am grateful to all those carrying out research to understand these difficult diseases and work toward prevention and treatment. I also thank my friend and colleague David Teplow at UCLA for his careful reading and critiquing of my introductory chapter. Finally, I thank Natalie Farra and Kathy Padilla and their colleagues at Elsevier for all their help in bringing the idea of this book into reality.

Michael S. Wolfe
University of Kansas, Lawrence, KS, United States
September, 2017

CHAPTER

1

Solving the Puzzle of Neurodegeneration

Michael S. Wolfe

University of Kansas, Lawrence, KS, United States

OUTLINE

INTRODUCTION: THE GENERAL PROBLEM OF NEURODEGENERATION

Neurodegenerative diseases are among the most difficult biomedical problems to solve. Despite intense efforts around the world by many laboratories, both academic and industrial, little can be done for the patient who contracts one of these debilitating and deadly disorders, which include Alzheimer's disease (AD), Parkinson's disease (PD), frontotemporal dementia (FTD), amyotrophic lateral sclerosis (ALS), Huntington's disease (HD), and prion diseases.

All approved therapeutics, at best, work at the symptomatic level; none slow or stop the inexorable loss of neurons and neuronal connections. Although tremendous progress has been made toward understanding the molecular and cellular basis of neurodegenerative diseases, this progress has yet to be translated into efficacious medicines. The failure of so many drug candidates in the clinic suggests that our understanding of disease mechanisms is still insufficient.

The need to solve these problems is dire. Over six million people in the United States, and perhaps over 50 million worldwide, have a neurodegenerative disease (Alzheimer's_Association,

The Molecular and Cellular Basis of Neurodegenerative Diseases
DOI: https://doi.org/10.1016/B978-0-12-811304-2.00001-8

2017; Parkinson's_Disease_Foundation, 2017; World_Alzheimer_Report, 2015). These diseases are invariably progressive, devastatingly debilitating, and ultimately lethal. As the victim becomes more and more disabled, the strain—emotional, physical, and financial—on patients, their families, and caregivers can become overwhelming. The healthcare costs become exorbitant, and as age is generally the greatest risk factor for acquiring a neurodegenerative disorder, demographic changes suggest societies will be overburdened in the decades to come.

A major part of the reason why neurodegenerative diseases have been so difficult to solve therapeutically is the special characteristics of neurons, which are postmitotic and generally not replaced once lost. Although neurogenesis does occur to a limited extent in the adult human brain (Bergmann, Spalding, & Frisen, 2015), the great majority of the approximately 100 billion neurons are in place around the time of birth (Johnson, 2001). Furthermore, neurons can be especially vulnerable to disease because their axons can extend large distances to connect with other neurons. Some, for example those of certain motor neurons, may extend a meter or more. Axons, as well as dendrites, require transport systems to convey needed biomolecules and organelles to distance synapses. The disruption and blocking of these systems can lead to synaptic failure, and the health of neurons depends on healthy synaptic connections (Morfini et al., 2009). Synaptic failure can lead to neuronal loss.

These characteristics—postmitotic, largely irreplaceable, long processes, and dependence on proper connections—make neurons particular vulnerable. There are many ways to kill neurons. One route is proteotoxicity, the buildup of toxic proteins due to overproduction, or inefficient clearance (Douglas & Dillin, 2010). In most neurodegenerative diseases, abnormal deposition of specific proteins in the brain is a defining feature (Table 1.1). The classic pathological description of AD is the deposition of the amyloid β-protein in extracellular plaques and the intracellular accumulation of neurofibrillary tangles formed by the protein tau (Vinters, 2015). Other diseases such as FTD are classified as tauopathies, with tau deposition similar to that seen in AD but without amyloid plaques (Lee, Goedert, & Trojanowski, 2001; Wang & Mandelkow, 2016). In PD, the membrane-associated protein α-synuclein aggregates within dopaminergic neurons of the substantia nigra (Kalia & Lang, 2015). ALS commonly displays deposition of the TAR DNA-binding protein 43 (TDP-43) in motor neurons (Neumann, 2009), and the huntingtin (Htt) protein can be found aggregated in neurons of the basal ganglia in HD (Walker, 2007). Prion diseases, such as Creutzfeldt–Jakob

TABLE 1.1 Primary Familial Genes and Molecular Pathology for Major Neurodegenerative Diseases

Neurodegenerative Disease	Mutant Genes	Protein Deposition
Alzheimer's disease	APP, presenilins	Aβ, tau
Frontotemporal dementia	Tau, progranulin, c9orf72	Tau, TDP-43
Amyotrophic lateral sclerosis	TDP-43, c9orf72	TDP-43
Parkinson's disease	α-Synuclein, LRRK2	α-Synuclein
Huntington's disease	Huntingtin	Huntingtin
Prion diseases	PrP	PrP

disease (CJD), display plaques composed of the prion protein (PrP) (Johnson, 2005).

The pathways to protein aggregation include overproduction of the disease-associated protein in rare genetic cases. Typically though, these proteins become misfolded, due to the failure of molecular chaperones to ensure proper protein folding. As a result, they are not cleared sufficiently, due to the inability of the ubiquitin-proteasome system or autophagic mechanisms to keep up (Yerbury et al., 2016). Other mechanisms by which neurons become dysfunctional or are destroyed involve RNA toxicity, in which mRNA may become enmeshed in RNA foci or specific mRNA aggregates, causing gain of neurotoxic function as well as loss of normal function (Wojciechowska & Krzyzosiak, 2011). Neuroinflammation provides yet more routes to neurotoxicity, through noncell autonomous effects of support cells such as astrocytes, or of the brain's immune cells, microglia (Ransohoff, 2016).

Why certain types of neurons or neuronal networks are particularly vulnerable to the abnormal buildup of certain proteins and RNA is unclear and remains a central problem in this field of investigation. What is clear is that this selective neurotoxicity leads to the manifestation of a specific disease. For instance, because neurons of the substantia nigra are unable to effectively clear misfolded or aggregated α-synuclein and are selectively vulnerable to this protein, the result is PD, as these neurons are important in controlling movement. Selective vulnerability of neurons to abnormal tau aggregates in the frontotemporal lobe lead to the specific cognitive and behavioral symptoms in FTD. Deciphering why specific neurons and neuronal networks are affected by certain molecular changes would help elucidate why these molecular changes cause-specific neurodegenerative diseases.

The hope is that elucidating the molecular and cellular basis of neurodegenerative diseases will reveal new therapeutic targets, provide critical information about how an ideal drug should interact with and affect its target, and suggest screening strategies for drug discovery. What follows is a general overview of the nature of these diseases, current hypotheses and evidence for disease mechanisms, important remaining questions, and potential avenues for solving these complex puzzles and developing effective therapeutics.

EPIDEMIOLOGY AND CLINICAL PRESENTATION

AD is the most common neurodegenerative disorder, affecting nearly six million people in the United States and over 35 million worldwide (Prince et al., 2013). The illness manifests itself primarily as a decline in memory and cognition, a consequence of degeneration of the hippocampus and the neocortex, and generally strikes those over age 65, although some 1%–2% of cases are early-onset genetic forms of the disease (Alves, Correia, Miguel, Alegria, & Bugalho, 2012). On average, the course of the disease is roughly 8 years from the onset of symptoms until death, although this can be as long as 20 years. The debilitating nature of the disease, combined with the slow decline and large numbers of people affected, make AD highly costly to society.

FTD is an umbrella term for a spectrum of related diseases that range from decline in language ability to movement disorders to dramatic personality changes and compulsive behaviors (Seelaar, Rohrer, Pijnenburg, Fox, & van Swieten, 2011). As the name suggests, the degeneration takes place primarily in the frontal and temporal lobes. Although not well appreciated by the public, FTD is the most common form of dementia in those under 65 years of age, typically between ages 45 and 64. At least 15% of cases are familial, following a Mendelian genetic pattern of inheritance, and a genetic cause may account for up to 40% of all cases of FTD.

PD is the most common neurodegenerative movement disorder, striking 1% of all people over the age of 65 (Sveinbjornsdottir, 2016). The disease manifests itself clinically with symptoms that include a resting tremor in the limbs and face, bradykinesia (slowness of movement), rigidity in the limbs and trunk, and postural instability. These symptoms are due to the destruction of neurons in the pars compacta of the substantia nigra in the midbrain, neurons that are integrated in circuits that control areas of the basal ganglia involved in voluntary movement. The neurons that are lost are dopaminergic, so treatment with the dopamine precursor L-DOPA has been a longstanding treatment of symptoms. However, this replacement therapy does not stop the underlying neurodegenerative process and ultimately becomes ineffective. Survival after diagnosis of PD is generally between 7 and 11 years (de Lau, Schipper, Hofman, Koudstaal, & Breteler, 2005).

ALS involves the progressive degeneration of motor neurons in the brain and spinal cord (Taylor, Brown, & Cleveland, 2016). The resulting reduced innervation of muscles leads to their wasting (thus "amyotrophic") and descending axons in the lateral spinal cord appear scarred (thus "lateral sclerosis"). ALS is a relatively rare disease, with some 20,000–30,000 in the United States affected at a given time. The disease progression is generally very rapid, with death typically coming 3–5 years after diagnosis, although some cases can very slowly progress over decades. The disease typically strikes in middle adulthood, with a mean age of onset of 55 years, with initial symptoms of subtle cramping or weakness in muscles of the limbs or those involved in speech and swallowing. Ultimately, the disease progresses to paralysis of most skeletal muscles.

HD is the only major neurodegenerative disease that is 100% hereditary (Walker, 2007). HD clinically presents with symptoms overlapping with those of AD, PD, and ALS. The disease is commonly considered a movement disorder, with its original name being Huntington's chorea, as involuntary movement of the limbs was thought to resemble a dance. However, other symptoms include diminished speech and difficulty swallowing as well as dementia and personality changes such as depression and irritability. Disease onset is typically in midlife, although it may strike at any age. The course of the illness runs 10–20 years and, like all the progressive neurodegenerative diseases, is ultimately fatal. Although neurodegeneration is widespread in the brain, areas affected early include a part of the basal ganglia called the striatum, involved in motor coordination, cognition, and reward and motivation. Other areas affected include the substantia nigra, parts of the cerebral cortex, and the hippocampus.

Prion diseases are rare neurodegenerative disorders that may present with different symptoms and neuropathology but that are all caused by a protein agent termed a prion. Prion disease can occur sporadically or through infection, as well as being inherited (Johnson, 2005). CJD, fatal familial insomnia (FFI), Gerstmann–Straussler–Scheinker syndrome (GSS), and kuru are examples of human prion diseases. The study of kuru led to the discovery of the infectious nature of these brain-wasting diseases. The disease was endemic in the Fore ethnic group in the highlands of Papua New Guinea and was found to be acquired as a result of ritual cannibalism involving contact with or ingestion of brains of deceased family members. Prion diseases all involve a long incubation period and rapid progression after onset. Postmortem analysis reveals a spongiform pathology and extensive plaque deposition of the PrP. Prion diseases also are found in cows (bovine spongiform encephalopathy or mad cow disease), sheep (scrapie), and deer and elk (chronic wasting disease).

MOLECULAR PATHOLOGY

The first clues to molecular mechanisms of these diseases were their postmortem pathological features. As mentioned earlier, AD is characterized by plaque deposition and neurofibrillary tangles (Vinters, 2015). Biochemical analysis identified the proteins Aβ and tau as the major components of the plaques and tangles, respectively. Aβ deposition appears to be the earliest pathological change, seen up to 25 years before the expected onset of symptoms. In contrast, tau deposition, although observed later, is more spatially and temporally correlated with the loss of neurons. Together these findings suggest that aberrant Aβ could be a pathogenic initiator, while downstream pathological changes in tau may be the more proximal cause of neurodegeneration. Other pathological events include neuroinflammation, which may be causative or exacerbating factors of neuronal dysfunction and loss.

PD is characterized by the presence of Lewy bodies in neuronal cell bodies and neuronal loss in the pars compacta region of the substantia nigra (Kalia & Lang, 2015). Lewy bodies are deposits of the protein α-synuclein, and the neurons that are lost are primarily dopaminergic, innervate the basal ganglia, and are critical to motor functions. Deposition of Aβ or tau is typically not seen in PD unless the subject also displayed dementia.

In contrast, neurofibrillary tangles composed of aggregated tau are found in nearly half of all FTD cases (Irwin et al., 2015). Interestingly though, Aβ deposition is not. A spectrum of neurodegenerative diseases collectively called tauopathies share this specific pathology (Ballatore, Lee, & Trojanowski, 2007; Wang & Mandelkow, 2016), leading to the idea that aberrant Aβ in AD is one of a number of means by which neurotoxic tau can be elicited.

In some 50% of other FTD cases, however, tau-negative, ubiquitin-positive protein neuronal deposits are observed (Irwin et al., 2015). In most of these cases, the deposits are composed of the RNA-binding protein TDP-43. Normally a nuclear protein, TDP-43 is translocated to the cytoplasm and aggregates in neurons in FTD. TDP-43 is also the main component in deposits within motor neurons found in most cases of ALS, and this and other evidence suggests that FTD and ALS may be related diseases on different ends of a spectrum of TDP-43 proteinopathies (Geser, Lee, & Trojanowski, 2010). Neuronal deposits of other proteins such as superoxide dismutase 1 (SOD1) or an RNA-binding protein related to TDP-43 called fused in sarcoma (FUS) can be found in rare genetic forms of ALS (Taylor et al., 2016). RNA foci are also found in many cases of ALS, composed of transcripts of a gene called c9orf72 that contains an expansion of a hexanucleotide repeat region (Haeusler, Donnelly, & Rothstein, 2016).

Cytoplasmic aggregates and nuclear inclusions are also found throughout the brain, but particularly in the basal ganglia, as a major and common pathological feature of HD (Labbadia & Morimoto, 2013). The primary component of these deposits is a mutated form of the Htt protein containing an expansion of a polyglutamate region at the N-terminus of the protein. Because the interactome of Htt is quite large, many other proteins become entrapped as well, including those involved in transcription and protein quality control.

Prion diseases display extensive plaque deposition along with a spongiform pathology. The primary component of these plaques is PrP, which can assume a variety of possible conformations and glycosylation patterns capable of aggregation and spreading throughout the brain (Collinge, 2016). These different conformations and glycosylation patterns are thought to lead to different strains of the infectious forms of PrP that affect different regions of the brain, result in different clinical presentations (e.g., CJD, FFI, and GSS), and affect the ability of PrP to infect different species. The ability of a protein alone to transmit disease and even encode different infectious strains was a paradigm-shifting discovery for which Dr. Stanley Prusiner was awarded the Nobel Prize in 1997.

Perhaps the most intriguing and common pathological finding in neurodegeneration in recent years has been the increasing appreciation of the spread of molecular pathology from neuron to neuron, in a networked manner through what is termed synaptic transmission (Guo & Lee, 2014). This process has been likened to the assembly and spreading of PrP pathology. It is critical to point out, however, that PrP is the only protein known to be infectious and that the prion-like character of other proteins involved in neurodegenerative diseases appears to be limited to the molecular and cellular levels. Nevertheless, prion-like assembly and synaptic transmission is an important emerging concept in the field with potential therapeutic implications. This issue will be discussed in further detail later.

GENETICS

Major clues to disease mechanisms have also come from genetics, particularly from the study of families with classical Mendelian inheritance patterns. For some neurodegenerative diseases, different genes may be mutated in different families, which taken together can suggest a particular cellular process, pathway, or function. For instance, three genes are sites of autosomal-dominant missense mutations that lead to familial early-onset AD: the amyloid β-protein precursor (*APP*), presenilin-1 (*PSEN1*) and presenilin-2 (*PSEN2*). APP is the precursor to the Aβ peptide that deposits in the AD brain, and presenilin is the catalytic component of γ-secretase, one of two proteases responsible for producing the Aβ peptide. Thus, the genetic evidence from familial cases, combined with pathological and biochemical evidence, strongly points to a role of Aβ in AD pathogenesis (Tanzi & Bertram, 2005).

A major risk factor for sporadic, late-onset AD is the apolipoprotein E (*APOE*) gene, which encodes a cholesterol-transporting protein (Cuyvers & Sleegers, 2016). The E4 variant of this gene increases risk (3–4-fold for one allele and 12–15-fold for both alleles), while the E2 variant decreases risk and the E3 variant is neutral. A rare missense mutation in the gene encoding an immune cell receptor, triggering receptor expressed on myeloid cells 2 (*TREM2*), also confers substantial risk of late-onset AD, providing an important clue to possible roles of neuroinflammation in AD pathogenesis.

The search for genetic causes of dominantly inherited FTD led to the discovery of mutations in the microtubule-associate protein tau (*MAPT*) gene (Goedert & Jakes, 2005; Wang & Mandelkow, 2016). These are mostly point mutations in the coding region that increase the tendency of the protein to aggregate, but some are silent or intronic mutations that shift pre-mRNA splicing toward a set of normal tau isoforms that are more prone to aggregation. The discovery of tau mutations causing FTD bolstered a role of tau in AD as well. Interestingly, other familial cases of FTD, without tau pathology but with ubiquitin-positive deposits, are caused by loss-of-function mutations in the progranulin (*PRG*) gene, located in the same region of chromosome 17 as the *MAPT* gene (Baker et al., 2006; Cruts et al., 2006). Another major site of dominantly inherited mutations leading to FTD is *c9orf72*, caused by expansion of a hexanucleotide repeat in the promoter of this gene (Haeusler et al., 2016), suggesting a neurotoxic role of the RNA transcript.

The *c9orf72* repeat expansion is also associated with ALS, accounting for 25% of familial cases and 10% of sporadic cases (Haeusler et al., 2016; Taylor et al., 2016). Thus, like TDP-43 pathology, the *c9orf72* mutations suggest that FTD and ALS are related diseases: mutations in the same genes can lead to either disease or a combination of the two. FTD and ALS are apparently on either end of a spectrum of possible disease states caused by common molecular changes (Geser et al., 2010). Missense mutations in *SOD1* are responsible for another 20% of familial ALS, and these mutations lead to misfolding and aggregation of the SOD1 protein. Other mutations associated with familial ALS include *TARDBP* (encoding TDP-43 itself) and *FUS*, encoding a related RNA-binding protein, suggesting disrupted RNA metabolism as a common mechanism. Genes encoding proteins involved in autophagy such as optineurin, ubiquilin-2, and sequestosome-1 are also mutated in familial ALS, pointing to problems with aberrant disposal of cellular waste. Other ALS-causing mutations, such as dynactin subunit 1 and tubulin alpha-4A chain, suggest that another pathway to dysfunctional and destroyed motor neurons is disruption of axonal transport.

Genes associated with familial PD include those encoding Parkin, an E3 ubiquitin ligase, and PTEN-induced putative kinase 1 (PINK1), a mitochondrial-associated kinase (Lubbe & Morris, 2014). These two proteins apparently work together to regulate mitophagy, the degradation of defective mitochondria via autophagy. PD-associated recessive PINK1 and Parkin mutations lead to reduced mitophagy, suggesting that neurons of the substantia nigra are particularly susceptible to mitochondrial dysfunction. Dominant mutations in the gene encoding α-synuclein, the protein that deposits in the characteristic Lewy bodies in PD, also cause familial PD. This discovery provided compelling evidence that misfolded or aggregated α-synuclein can trigger PD, especially as duplication or triplication of the wild-type α-synuclein gene also leads to familial PD. The gene encoding leucine-rich repeat kinase 2 (LRRK2) is also a major site of PD-associated mutations, not only for dominantly inherited familial PD but also for some 1%–2% of all sporadic cases of PD. LRRK2 is a large multidomain protein, and how the mutations alter function to cause disease is still unclear. Recessive mutations in the *DJ-1* gene, encoding a protein thought to be important to protect neurons from oxidative stress, are also associated with familial PD.

HD is the only major neurodegenerative disease that is 100% genetic—every case is due to expansion of a CAG trinucleotide repeat in exon 1 of the *Htt* gene that leads to polyglutamate expansion in the encoded protein (Labbadia & Morimoto, 2013). Normally, the number of repeats range from 16 to 20, and expansion beyond 35 repeats leads to HD. Htt is a high-molecular-weight protein with a large interactome. Polyglutamine expansion occurs in other genes associated with other neurodegenerative diseases, including a variety of ataxias, and leads to aggregation of the mutant protein (Polling, Hill, & Hatters, 2012). Although protein misfolding or aggregation can apparently result in neurotoxic entities, the mutant mRNA can also play a role in pathogenesis (Wojciechowska & Krzyzosiak, 2011). This is particularly true for certain rare neurodegenerative diseases in which nucleotide repeat expansion occurs in noncoding regions of a gene.

Prion diseases are unique in that they can be contracted via proteinaceous infectious particles (the origin of the term "prion") (Collinge, 2016; Johnson, 2005). Nevertheless, they can also be caused by genetic mutations in the *PrP* gene, which encodes PrP. These mutations lead to protein aggregation and plaque formation in the brain. Different mutations can lead to different patterns of protease-resistance, different brain regions with neurodegeneration, and different clinical presentations (e.g., CJD, FFI, and GSS). These specific effects with specific mutations are thought to be caused by particular conformational states and glycosylation patterns in the protein, related to the concept of PrP strains mentioned earlier.

MOLECULAR CLUES TO MECHANISMS OF PATHOGENESIS

The identification of proteins in pathological deposits and the discovery of genes associated with familial forms of neurodegenerative diseases opened the door to experiments to determine the normal functions and pathological roles of these proteins.

For prion diseases, the involvement of PrP is inescapable (Soto & Satani, 2011). The protein deposits in the brain in the form of amyloid plaques, dominant mutations in the PrP gene cause familial disease, introduction of PrP from diseased brains causes highly reproducible and specific disease in the recipient, and knockout of the PrP gene in the recipient prevents this "protein-only" disease.

Prion diseases have three causes: dominant inheritance of a *PrP* mutation, spontaneous or sporadic occurrence, and infection through environmental exposure (Colby & Prusiner, 2011). PrP is a membrane-anchored glycoprotein, the normal function of which remains unknown. The leading hypothesis for the pathogenic mechanism of PrP is that it can assume an ensemble of conformations, some of which are capable of propagation and templating the conversion of normal cellular PrP to a lethal form. The propagation of infectious forms can occur over a long incubation period that is essentially independent of PrP expression level.

Upon the leveling off of the prion titer, the conversion to lethal PrP and manifestation of disease is inversely proportional to PrP expression: disease onset occurs faster in PrP-overexpressing transgenic mice than in wild-type mice, and wild-type mice develop disease faster than PrP heterozygous knockout mice (Collinge, 2016). These kinetic findings demonstrate that prion propagation and toxicity can be uncoupled; that is, the infectious prion particle is distinct from the lethal form of the protein. Despite the clear progress and the understanding of the essential role of PrP in prion diseases, the identities of the infectious and lethal forms are unknown, as are the mechanisms of neurotoxicity.

For AD, the case for Aβ in some form as a pathogenic entity is not as air-tight as it is for PrP in prion diseases, but the evidence is nevertheless compelling (Selkoe & Hardy, 2016). As mentioned earlier, missense mutations that cause autosomal-dominant familial AD are found in and around the Aβ region of APP and in the catalytic component of the protease that produces Aβ. Aβ is produced from the single-pass membrane protein APP through two proteolytic events: cleavage outside the membrane by β-secretase, followed by cleavage of the C-terminal remnant inside its transmembrane domain by γ-secretase. Secreted Aβ ranges from 38 to 43 amino acids, with the 40-residue peptide (Aβ40) being the major form. The longer Aβ42 peptide (Aβ42), containing more of the hydrophobic transmembrane domain, is more aggregation-prone and is the major protein component of the amyloid plaques of AD.

Mutations in APP near the Aβ N-terminus, which cause FAD in midlife, increase cleavage by β-secretase, leading to increased Aβ production throughout life (Tanzi & Bertram, 2005). In contrast, a protective mutation, also near the Aβ N-terminus, that substantially reduces AD risk in old age decreases this same proteolytic event to lower Aβ production throughout life (Jonsson et al., 2012). APP mutations within the Aβ region increase the propensity of Aβ to aggregate. Finally, mutations near the C-terminus of the Aβ region in APP alter cleavage by γ-secretase to increase the proportion of aggregation-prone peptides (Tanzi & Bertram, 2005).

Mutation in the *PSEN1* and *PSEN2* genes likewise alter the proportion of aggregation-prone Aβ peptides (Tanzi & Bertram, 2005). Presenilin is the catalytic component of γ-secretase, a membrane protein complex that carries out hydrolysis within the hydrophobic environment of the lipid bilayer. AD-causing mutations in presenilins affect the proteolysis of the APP transmembrane domain by γ-secretase to generate longer, more aggregation-prone Aβ peptides.

As with prion diseases, however, the pathogenic entity in AD remains at large. The amyloid plaques per se do not correlate with neurodegeneration, and the current leading hypothesis is that oligomeric forms of Aβ42 are synaptotoxic (Benilova, Karran, & De Strooper, 2012). Oligomers from dimers and trimers to dodecamers and larger have been touted as the primary pathogenic species, but the high heterogeneity makes it challenging to deconvolute and identify a single molecular culprit. One possibility is that an "Aβ soup", with many toxic forms, is responsible. How pathogenic Aβ triggers pathological tau and neurofibrillary tangles also remains unknown. Roles for risk factors APOE and TREM2 are also unclear, although both of these proteins are apparently involved in Aβ clearance (Castellano et al., 2011; Wang et al., 2015).

In AD, tau pathology correlates better with degree of cognitive impairment than does Aβ pathology (Ballatore et al., 2007). This finding is consistent with Aβ being a pathogenic trigger, with tau being downstream in the process and temporally and spatially more proximal to neuronal cell death. While mutations that cause familial AD are found in the substrate and the enzyme that produce Aβ, AD-associated mutations in the tau gene (*MAPT*) have not been identified. However, mutations in tau do cause familial forms of FTD, demonstrating that altered tau protein alone can lead to tau pathology, neurodegeneration, and dementia (Wolfe, 2009). Such findings support a central role for tau in AD pathogenesis as well. Tau pathology is seen in a variety of neurodegenerative diseases, including chronic traumatic encephalopathy (Lee et al., 2001; McKee et al., 2009). Apparently, pathological tau is a common mediator of neurodegeneration that can be elicited by a variety of factors, including pathogenic Aβ.

As mentioned earlier, familial FTD can also result from dominant mutations in the *PRG* gene, located very near the *MAPT* gene. These mutations lead to truncated transcripts that undergo nonsense-mediated delay (Baker et al., 2006; Cruts et al., 2006). How heterozygous loss of PRG leads to FTD is unclear. These mutations are not associated with tau pathology, but rather with neuronal deposition of the RNA-binding protein TDP-43. TDP-43 deposits are seen in a spectrum of neurodegenerative diseases with FTD on one end and ALS on the other, and together these are referred to as TDP-43 proteinopathies (Geser et al., 2010). Pathological TDP-43 translocates from the nucleus to the cytoplasm, where it aggregates in a hyperphosphorylated, ubiquitinated, and truncated form (Neumann et al., 2006). However, TDP-43 is an RNA-binding protein, and aberrant RNA processing may be the main driver of pathogenesis, rather than toxic TDP-43 protein aggregates (Janssens & Van Broeckhoven, 2013).

FTD-ALS TDP-43 pathology is also seen with dominant mutations in the *TARPBP* gene, encoding TDP-43, as well as with repeat expansion of an intronic region of the *c9orf72* gene. Hypotheses for how *c9orf72* repeat expansion causes neurodegeneration include (1) haploinsufficiency, in which loss of expression from the mutant allele leads to reduction of function below a critical threshold; (2) aggregation of the expanded mRNA into inclusions that sequester proteins critical to RNA processing and function; (3) non-ATG-initiated translation of this repeat-expanded region into different aggregation-prone dipeptide-repeat proteins (Haeusler et al., 2016). Protein aggregation also occurs with ALS-mutant SOD1, and in general, ALS is associated with an inability to clear misfolded or aggregated proteins from motor neurons (Ruegsegger & Saxena, 2016).

Like motor neurons, dopaminergic neurons in the substantia nigra are also apparently highly vulnerable to a variety of molecular insults, as a number of different genes are associated with familial PD (Kalia & Lang, 2015; Lubbe & Morris, 2014). Misfolded and/or aggregated α-synuclein is a common feature and is elicited by mutation, duplication, or triplication of the α-synuclein gene. α-Synuclein, which normally is associated with synaptic vesicles, forms inclusions in sporadic PD as well. Other mutations lead to mitochondrial dysfunction. PD proteins Parkin and PINK1 normally work together to signal defective mitochondria that require disposal through mitophagy (Nguyen, Padman, & Lazarou, 2016). PINK1 accumulates on the surface of defective mitochondria, which leads to the recruitment and activation of Parkin, an E3 ubiquitin ligase. Parkin then ubiquitinates proteins in the outer mitochondrial membrane to trigger autophagy of the organelle. More generally, a number of gene products associated with PD play important roles in vesicle trafficking, particularly to lysosomes, and defects in these proteins can lead to the inability to clear misfolded and aggregated proteins as well as defective mitochondria (Abeliovich & Gitler, 2016).

Although HD is a 100% monogenetic neurodegenerative disease, its mechanisms of pathogenesis are quite complex (Labbadia & Morimoto, 2013). The Htt protein, in which polyglutamine expansion occurs, has a large interactome that couples it to many different cellular processes. The mutant form of Htt (mHtt) causes widespread changes in the transcriptome, and consistent with this finding, mHtt interacts with and disrupts a number of proteins involved in general transcription. Moreover, mHtt also disrupts the function of histone acetyltransferases, which are critical regulators of gene expression. The mutant protein also affects general proteostasis, preventing proper folding by interfering with chaperones and disrupting degradation via the proteasome. Thus, the general protein disposal system can become overwhelmed. Moreover, mHtt can cause impaired energy metabolism by interfering with mitochondrial function, biogenesis, and quality control by mitophagy. Why mHtt particularly affects medium spiny neurons of the striatum is unclear, although enhanced excitotoxicity of these glutaminergic neurons is a leading hypothesis.

RNA toxicity is also implicated as a contributor to HD pathogenesis (Marti, 2016). The CAG triplet repeat expansion can form hairpin structures that bind nuclear proteins, leading to altered RNA splicing, disrupted nuclear export, and nucleolar stress. The contributions to pathogenicity from the mHTT transcript can be difficult to separate from those elicited by the polyglutamine-expanded protein. However, the presence of triplet repeat expansion in noncoding regions of other genes can lead to degenerative diseases, including myotonic dystrophy 1, HD-like 2, and spinocerebellar ataxia 8, which are caused by CTG expansion in noncoding regions of *DMPK*, *JPH3*, and *ATXN8* genes, respectively. Moreover, expression of translated and untranslated mHtt exon 1 mRNA containing the expanded CAG tract in human neuronal cell lines demonstrated a purely mRNA-mediated neurotoxicity (Banez-Coronel et al., 2012; Sun et al., 2015). Evidence that folding of the expanded CAG regions in mRNA can contribute to disease include the finding that interruption of such CAG tracks with the synonymous CAA codon can mitigate neurodegeneration in *Drosophila* (Li, Yu, Teng, & Bonini, 2008).

COMMON THEMES AND CONTROVERSIES IN NEURODEGENERATION

Given the above findings on the pathology, genetics, and mechanisms of pathogenesis for the major neurodegenerative diseases, some general principles for the molecular and cellular basis of pathogenesis and progression can be appreciated (Fig. 1.1). Some of these concepts are still controversial, with robust debate in the field.

Perhaps the most important emerging concept is that of prion-like spread of pathogenic protein seeds in various neurodegenerative diseases (Collinge, 2016). Prion diseases are rare and work by a paradigm-shifting mechanism of protein-only transmission from organism to organism. An infectious protein seed containing misfolded and aggregated PrP serves as a template to convert normal cellular PrP into this same misfolded and aggregated conformation, increasing the titer of infectious particles in the brain. Ultimately, misfolded and/or aggregated PrP serves as a template to convert cellular PrP into a lethal form. As mentioned above, the kinetics of prion titer buildup and disease onset and progression demonstrate that the infectious and lethal forms of PrP can be dissociated.

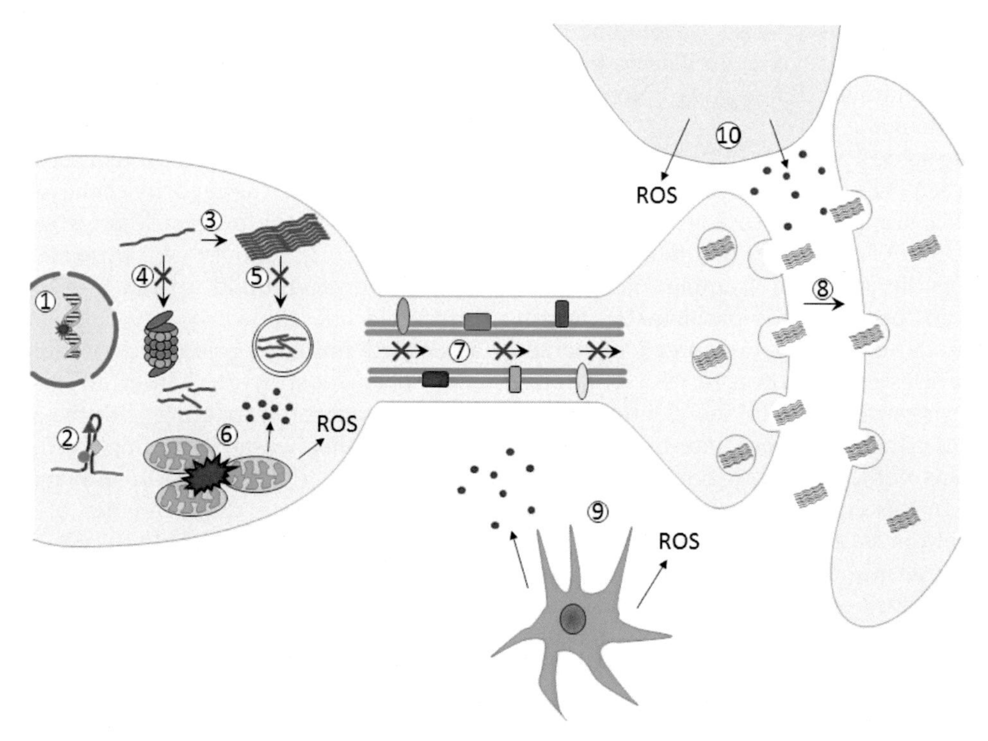

FIGURE 1.1 **Molecular and cellular mechanisms of neurodegeneration.** Multiple factors are known to contribute to the pathogenesis and progression of neurodegenerative diseases. Among these include the following: *(1)* Genetic mutations, either dominant or recessive, are associated with familial forms of all the major neurodegenerative diseases. *(2)* Huntington's disease and other nucleotide repeat disorders express mutant mRNAs that can form hairpins and other structures that sequester RNA-binding proteins and alter mRNA metabolism and function. *(3)* Disease-associated proteins can misfold and aggregate into neurotoxic forms. *(4)* Misfolded proteins are not adequately degraded through the ubiquitin-proteasome system. *(5)* Aggregated proteins are not adequately degraded via autophagy. *(6)* Mitochondrial dysfunction and inadequate mitophagy can result in release of apoptotic signals and formation of ROS and altered energy metabolism. *(7)* Axonal transport can be blocked by aggregated proteins or mutant transport proteins. *(8)* Synaptic transmission of pathogenic protein seeds can spread protein pathology from neuron to neuron. *(9)* Activated microglia or *(10)* astrocytes release neurotoxic signals and ROS. *ROS*, reactive oxygen species.

In other neurodegenerative diseases, some of the proteins implicated in pathogenesis and progression appear capable of propagating their misfolded and aggregated states through transsynaptic propagation, spreading potentially pathogenic seeds through a neuronal network (Guo & Lee, 2014). In this hypothesis, specific disease manifestation results from the anatomical location of the initiating pathogenic seed and how the neurons are integrated into neural circuits. This concept has major implications, not only for understanding disease mechanisms but also for developing new strategies for the discovery and development of potential therapeutic agents for disease prevention and treatment (Hasegawa, Nonaka, & Masuda-Suzukake, 2017).

Despite some apparent similarities, the analogy between PrP in prion diseases and proteins such as Aβ, tau, and α-synuclein and others in AD, FTD, PD, and other neurodegenerative diseases is limited, and framing these other proteins as "prion-like" is problematic (Walsh & Selkoe, 2016). To call the spread of pathogenic particles from neuron to neuron "transmission" is misleading to the degree that it alludes to a concept from infectious disease that means contagion from one organism to another. In this context, the spread from neuron to neuron is more akin to the concept of synaptic transmission in neuronal signaling. The term "trans-synaptic propagation" more accurately describes the process (Liu et al., 2012). Interestingly, pathogenic PrP itself does not clearly propagate from neuron to neuron in a network-like manner.

Evidence continues to build for the transsynaptic propagation of certain pathological proteins, especially for tau and α-synuclein (Guo & Lee, 2014). Even the concept of different strains is under consideration, with some of these pathological proteins apparently misfolding and aggregating into distinct states that are capable of propagation. Calling these different aggregated states "strains", however, may not be appropriate, as it is unclear whether these ultimately manifest themselves as distinct diseases, as clearly occurs with PrP and the prion diseases.

Regardless of the nature of the spread of the protein pathology, the accumulation of misfolded and/or aggregated proteins is responsible for it. But how does such accumulation occur? One reason is the inability of the protein folding machinery to handle the disease-associated protein (Lindberg et al., 2015). Indeed, overexpression of molecular chaperones and foldases can prevent protein aggregation and neurodegeneration in animal models of disease. The inability to properly fold disease-associated proteins and prevent them from aggregating can then overwhelm the protein disposal mechanisms. As misfolded protein accumulates inside the neuron, the capacity of the ubiquitin-proteasome system to degrade proteins is compromised (Ciechanover & Brundin, 2003). These in turn can build up and lead to general proteotoxicity. Similarly, as the disease-associated protein aggregates, the autophagic machinery must work harder to dispose of these aggregates, interfering with the ability to clear out other waste, including dysfunctional mitochondria (Nixon, 2013). As mentioned before, the unique characteristics of neurons make them highly vulnerable to cellular stress, and this includes proteotoxic stress. Protein aggregates can clog axons, blocking the transport of vital macromolecules and organelles between the cell body and synaptic termini (Morfini et al., 2009). Synapses may be directly affected by protein aggregates as well, and failure of synaptic function can ultimately lead to neurodegeneration (Haass & Selkoe, 2007).

Neurons also use considerable energy to carry out their critical functions and therefore need functional mitochondria. As just pointed out, overwhelming the autophagic machinery with aggregated protein can interfere with mitochondrial quality control through mitophagy (Yerbury et al., 2016). Moreover, axonal clogging by aggregated proteins can block proper transport of mitochondria to synapses (Correia, Perry, & Moreira, 2016). Mitochondria also have their own unfolded protein response system, and failure of this may lead to compromised energy metabolism that interferes with neuronal health and function (Jovaisaite, Mouchiroud, & Auwerx, 2014). Other mechanisms beyond protein misfolding and aggregation may adversely affect mitochondria. For instance, mitophagy can be disrupted by PD-associated mutations in Parkin and PINK1, two proteins critical to the regulation of mitophagy (Nguyen et al., 2016). Mitochondria also have their own DNA, and accumulated mutations with aging may lead to mitochondrial dysfunction and altered energy metabolism (Keogh & Chinnery, 2015).

Neurodegenerative disease is not all about rogue proteins: RNA-mediated mechanisms can apparently also be operative (Belzil, Gendron, & Petrucelli, 2013). RNA-binding proteins such as TDP-43 can aggregate in FTD and ALS, bringing along cognate mRNA into the resultant stress granules. In so doing, these mRNA fail to get processed and translated into functional proteins, leading to reduction or loss of function. The mRNA in triplet repeat diseases such as HD and certain ataxias abnormally fold into hairpins that can sequester critical nuclear proteins (Marti, 2016). Similarly, the mRNA of the hexanucleotide repeat-expanded c9orf72 gene in FTD and ALS aggregates in the nucleus and cytoplasm and can sequester proteins needed for proper RNA processing (Haeusler et al., 2016).

Neurodegenerative disease is also not all about neurons. Nonneural cells in the brain, including astrocytes, microglia, and oligodendrocytes, can play important—perhaps even major—roles in the dysfunction and destruction of neurons (Ransohoff, 2016). For example, selective deletion in astrocytes of an ALS-causing SOD1-mutant gene in transgenic mice inhibited microglial activation and slowed the loss of motor function phenotype (Yamanaka et al., 2008). A number of genes implicated in neuroinflammation, with overactivated microglia causing neuronal damage, have been identified as risk factors for neurodegenerative diseases, particularly for AD (Villegas-Llerena, Phillips, Garcia-Reitboeck, Hardy, & Pocock, 2016). For example, a rare allelic variant in the TREM2 receptor expressed in microglia and macrophages can increase the risk of sporadic AD as much as the ApoE4 variant (Guerreiro et al., 2013; Jonsson et al., 2013), although the latter is much more common and therefore implicated in many more cases of late-onset AD.

Perhaps the most common risk factor for neurodegeneration is age (see Chapter 17, Neurodegenerative Diseases and the Aging Brain). Damage to neurons can accrue with age, and the large majority of neurons are not replaced via neurogenesis. As neurons age, they accumulate more mutations, in both nuclear and mitochondrial DNA, as reactive oxygen species increase and DNA repair mechanisms fail to compensate. Such mutations result in altered gene expression, including of genes important for learning, memory, and neuronal survival. A decline in protein quality control machinery and waste disposal through the proteasome and autophagy also occurs with age and leads to buildup of toxic proteins and protein aggregates (Douglas & Dillin, 2010). Mitochondrial function and energy metabolism likewise lessen with age and interfere with neuronal health and function (Lane, Hilsabeck, & Rea, 2015).

ANIMAL MODELS

Good animal models are essential to better understanding of pathogenesis and progression of neurodegenerative diseases. Moreover, these models are critical for providing proof of concept for experimental therapeutics before they can advance into human trials. The identification of Mendelian genetic mutations that cause specific neurodegenerative diseases has presented the opportunity to develop transgenic, knock-in and knockout animals and examine the pathological changes as well as the physical and behavioral phenotypes that result.

For AD, transgenic mice expressing disease-causing mutations in human APP in the brain, with or without coexpression of mutant *PSEN1*, can result in age-dependent formation of amyloid plaques composed primarily in Aβ42 that are similar to what is seen in the postmortem human AD brain (Esquerda-Canals, Montoliu-Gaya, Guell-Bosch, & Villegas, 2017). Moreover, these mice can develop age-related learning and memory deficits reminiscent of what occurs in AD, with concurrent decreases in hippocampal long-term potentiation that is considered critical to learning and memory (Webster, Bachstetter, Nelson, Schmitt, & Van Eldik, 2014). Knockout of tau alleles in APP transgenic mice can rescue the cognitive deficits (Roberson et al., 2007). However, APP transgenic mice do not develop tau pathology, and little neurodegeneration is observed.

Transgenic expression of FTD-mutant tau, in contrast, does result in age-dependent tau pathology, neurodegeneration, and cognitive and motor deficits (Gotz et al., 2007). Furthermore, crossing mutant tau transgenic mice with mutant APP transgenic mice revealed that overexpression of APP exacerbates the tau pathology (Lewis et al., 2001), another clue that these two proteins may work in tandem in AD pathogenesis and progression. The concern, however, is that overexpression of aggregation-prone proteins such as Aβ or tau in the brain can lead to neural dysfunction through mechanisms that are not related to what actually occurs in AD. Efforts to address these limitations include the development of knock-in mice, in which the endogenous mouse gene is replaced with the disease-causing human mutant gene (see, e.g., Saito et al., 2014). Thus, the encoded protein would be expressed at normal levels and under normal physiological control, both spatially and temporally.

Mouse models of PD have suffered from similar limitations (Chesselet & Richter, 2011). For instance, some transgenic mice overexpressing mutant α-synuclein have developed deposits of this protein but without neurodegeneration (Fleming et al., 2004). Others have not developed α-synuclein deposits but have developed motor deficits (Lee et al., 2002). The latter, however, was due to pathology in the spinal cord, not the substantia nigra, the site of neurodegeneration in PD. Likewise, transgenic or knock-in mice expressing PD-mutant LRRK2 show little or neurodegeneration or α-synuclein deposits. However, these mice can develop age-dependent locomotor deficits as well as impaired striatal dopamine transmission (Li et al., 2009, 2010). Knockout of PD genes such as Parkin and PINK1 that are associated with recessive, loss-of-function mutations do not result in PD-like neuropathology, such as loss of dopaminergic neurons (Gispert et al., 2009; Goldberg et al., 2003), suggesting other factors are critical to disease pathogenesis in people with these mutations. Chemically induced rodent or primate models of parkinsonism, using toxins such as MPTP or 6-hydroxydopamine, lead to highly selective loss of dopamine neurons in the substantia nigra and PD-like motor deficits, but the degree to which the molecular mechanisms resemble what goes on in PD is unclear (Bove & Perier, 2012). Interestingly though, in monkey MPTP models, α-synuclein upregulation is observed and associated with loss of dopaminergic neurons (Chu & Kordower, 2007).

Transgenic expression of ALS genes has produced mouse models that recapitulate some of the features of the disease (Picher-Martel, Valdmanis, Gould, Julien, & Dupre, 2016). An early model expressed SOD1 G93A under control of the SOD1 promoter, and these mice developed progressive motor dysfunction, loss of muscle innervation, and motor neurodegeneration, along with gliosis (Gurney et al., 1994). Knockout of SOD1 does not lead to neuronal loss or an ALS-like phenotype, strongly suggesting that loss of function is not the major cause of the disease (Reaume et al., 1996). Rather, a gain of neurotoxic function (e.g., through protein misfolding and aggregation) is apparently responsible. Overexpression of wild-type and mutant TDP-43 in neurons can lead to age-dependent motor dysfunction and TDP-43 cytosolic inclusions (Swarup et al., 2011). However, in general, these mouse models display some but not all major histological and motor features of ALS. A number of c9orf72 repeat expansion mouse models have already been developed, even though this genetic association was only discovered a few years ago. Some of these models display neurodegeneration, motor dysfunction, RNA foci, dipeptide repeat formation, and TDP-43 neuronal cytosolic inclusions (Chew et al., 2015). As with SOD1, knockout of c9orf72 did not lead to an ALS phenotype, suggesting that loss of function is not sufficient to cause the disease (Picher-Martel et al., 2016).

Mouse models of HD are of three general types (Ferrante, 2009). The first involves transgenic overexpression of the Htt N-terminus with an expanded polyglutamine tract. Such models typically display an accelerated disease phenotype, developing progressive neurological abnormalities, including abnormal gait, loss of coordination, neuropathology, and premature death (e.g., Laforet et al., 2001; Mangiarini et al., 1996). However, such models do not address the possibility of contributions by the rest of the large Htt gene and protein. Moreover, expression of the N-terminally truncated mHtt is not under endogenous control. To address these issues, mice have been developed that either have the full-length mutant human Htt with the human promoter region inserted into the mouse genome (Gray et al., 2008) or that have the expanded CAG repeat inserted into the mouse Htt gene (Sathasivam et al., 2013). These mice show a much slower disease progression, and in some cases may be considered only as a predisease model. Nevertheless, they provide important tools to address questions of disease mechanisms, such as the role of proteolytic cleavage producing the N-terminal Htt. The inserted full-length mutant human Htt typically contains interspersed CAA (synonymous with CAG for glutamine). This has the advantage of avoiding the genomic instability cause by long CAG repeats; however, such disruption of the repeat sequence can interfere with RNA-mediated mechanisms of cytotoxicity. Knock-in mice are heterozygous for the mHtt, as in human patients, and contain uninterrupted CAG expansions, but this makes these mouse lines vulnerable to genomic instability.

While no animal model can perfectly recapitulate all the features of a human disease, the mouse models of prion diseases arguably are the best in the field of neurodegeneration (Brandner & Jaunmuktane, 2017; Telling, 2008). Transgenic human PrP mice infected with prions can develop the same pathology and phenotype as seen in their human counterparts. Key to this was the development of PrP null mice, which demonstrated that endogenous PrP is required for propagation of prions. The knockout mice were crossed with human PrP transgenic mice to allow expression of the human protein without the confounding effects of the endogenous protein. Depending on the level of expression of the human PrP transgene, such mice reliably exhibited disease phenotypes upon infection that were strain-specific and displayed incubation times that were remarkably reproducible. Low-expressing as well as knock-in lines often do not develop overt disease but do show preclinical pathological changes that allow these mice to be used to address hypotheses about disease mechanisms.

While mice have been primarily used for animal models of neurodegenerative diseases, other species have also been used to develop such models. Nematodes and fruit flies, with their rapid growth rates and short life cycles, have been employed particularly for the search for genetic modifiers but also for small molecule drug screening (Lu & Vogel, 2009; Sin, Michels, & Nollen, 2014). Examples include the use of such models to investigate the role of histone deacetylases, via both RNAi and chemical inhibition, in mitigating phenotypes caused by mHtt expression (Steffan et al., 2001). Finally, larger animals, including dogs and pigs, but especially nonhuman primates, have provided models that closely mimic human disease (Eaton & Wishart, 2017). While such models can be especially useful, the time and cost—as well as sociopolitical resistance—make widespread utility unrealistic.

PROSPECTS FOR THERAPEUTICS

At present, there are no effective disease-modifying therapies for any neurodegenerative disease. While some symptomatic treatments are available, none of these prevent inexorable neuronal loss and disease progression. One of the main reasons is the lack of clear mechanistic understanding of disease. For instance, considerable efforts have gone into the discovery of specific kinase inhibitors for AD and other tauopathies (Dolan & Johnson, 2010), as the pathological tau filaments have long been known to be hyperphosphorylated. However, the specific phosphorylation sites on tau and the responsible kinases that might be involved in pathogenicity are unclear. Multiple sites may be critical, and inhibiting any single kinase may not be sufficient. More concerning, it is not even clear if hyperphosphorylation is a cause or consequence of the buildup of tau protein aggregates in neurons. The uncertainty about disease mechanisms is true of every neurodegenerative disease, as the responsible pathogenic entity and the pathway to neuronal loss are unknown in all cases. Even with prion diseases and HD, the forms of the PrP or Htt that trigger neurotoxicity are still unknown, and for HD, RNA-mediated neurotoxicity may be involved (Marti, 2016).

Another major problem with treating neurodegenerative diseases is that the proteins and nucleic acids involved are endogenous human macromolecules that have critical functions. While knockout of genes encoding APP and PrP may be tolerated, loss of Htt and TDP-43 are lethal (Nasir et al., 1995; Wu et al., 2010). Thus, pursuing small molecule drugs, immunotherapies, or gene therapies that interfere with the normal function of these proteins would not be appropriate. Targeting γ-secretase to block the production of Aβ from APP provides a clear illustration of the perils of blocking the normal function of a vital macromolecule. γ-Secretase cleaves not only APP but dozens of other membrane proteins, most notably the Notch family of cell-surface receptors (Beel & Sanders, 2008). Ligand-induced proteolysis of Notch leads to release of its intracellular domain, with translocation to the nucleus and activation of the transcription of genes involved in cell fate determinations. Knockout of γ-secretase components (e.g., PSEN1) is embryonic lethal (Shen et al., 1997; Wong et al., 1997), and γ-secretase inhibitors that entered clinical trials for AD failed in large part due to adverse effects caused by blocking Notch signaling (Doody et al., 2014).

Neurodegenerative diseases present further difficulties by requiring entry of therapeutics into the central nervous system (Banks, 2016). The blood–brain barrier typically prevents access of compounds larger than 500 Da. Moreover, the P-glycoprotein efflux pump, located at the blood–brain barrier, effectively pumps small molecules of general lipophilic character back out of the brain. This presents a major challenge in keeping the molecular weight of drug candidates low and reducing their potential to act as *P*-glycoprotein substrates, while retaining potency and specificity for the target macromolecule. In the cases where the small molecule is intended to block protein–protein interactions (e.g., self-assembly of aggregation-prone protein), this can be especially daunting, as protein–protein interaction surfaces are often relatively large, typically requiring larger drug molecules for effective inhibition.

Early diagnosis will be critical for effective treatment of neurodegenerative diseases. It is now clear that these disease processes begin many years, even decades before the onset of symptoms, and by the time symptoms appear, considerable neurodegeneration has already taken place. Replacing lost neurons, with the appropriate neuronal subtypes and interneuronal connections, is not possible at this time, nor is it obvious how that might be realistically accomplished in the near future. Prevention would be preferable in any event. In AD, changes in CSF Aβ levels can occur 25 years prior to expected onset of symptoms and Aβ plaque deposition can occur 15 years prior (Bateman et al., 2012). Thus, there is justifiable concern that the pathogenic process becomes largely Aβ-independent and tau-driven by the time of clinical presentation. This may be the reason why all anti-Aβ approaches to treating AD have failed in clinical trials: those enrolled in those trials already had signs and symptoms of AD. The development of amyloid imaging agents has made it possible to identify those at high risk of developing AD before it happens (Mallik, Drzezga, & Minoshima, 2017). These imaging agents now allow for more focused trials, weeding out those with other forms of dementia and keeping only those with Aβ pathology. In this way, target engagement can be demonstrated for anti-Aβ agents (e.g., monoclonal antibodies) by showing that drug candidate treatment leads to clearance of amyloid deposits.

Even with such focused patient recruitment, however, clinical trials will likely need to be long and large. Neurodegenerative diseases typically progress very slowly. Variability in age of disease onset and rate of progression between at-risk individuals means that large numbers of subjects and long trial times will be needed in order to statistically demonstrate that drug treatment leads to disease modification. Biomarkers, including those in CSF and plasma, will be important surrogates for determining target engagement and effective drug dosing. The most suitable patients for initial trials may be those with familial disease, carrying mutations with known disease onset and progression and providing essential clues to pathogenic mechanisms. Such a personalized medicine approach is proving invaluable in treating other diseases, such as cancer, and should be possible with neurodegenerative diseases, given the current state of knowledge. Once efficacy is established in Mendelian genetic cases, trials can be conducted for those with sporadic disease with more confidence.

CONCLUSIONS AND PERSPECTIVE

Remarkable progress has been made in elucidating the mechanisms of many neurodegenerative diseases. Determination of the composition of pathological lesions along with advances in the genetics of these diseases has provided important clues that have empowered investigators to develop and test specific hypotheses regarding the molecular and cellular processes responsible for neurodegeneration. Common themes that have emerged include protein misfolding and aggregation, disrupted protein disposal, dysfunctional mitochondria and energy metabolism, clogged axonal transport, neuroinflammation, and RNA-mediated toxicity. The spread of protein pathology in a prion-like manner is also emerging as a common theme and may open new avenues to the development of therapeutics. Connecting all these themes is aging, as disease incidence increases with age. Neurons are mostly postmitotic and particularly vulnerable due to their unique morphology and function, and with age neurons become less able to prevent the buildup of misfolded proteins and maintain overall homeostasis. The development of a wide variety of animal models that recapitulate certain molecular, pathological, and behavioral disease phenotypes has provided critical tools to address hypotheses about disease mechanisms and test candidate therapeutic agents. Although no disease-modifying drugs are yet approved for any neurodegenerative disorder, continued advances in the understanding of the underlying causes of disease, and the development of new tools and approaches for drug discovery, offer reasons for optimism that effective therapeutics can be identified and developed in the near future.

References

Abeliovich, A., & Gitler, A. D. (2016). Defects in trafficking bridge Parkinson's disease pathology and genetics. *Nature*, *539*, 207–216.

Alves, L., Correia, A. S., Miguel, R., Alegria, P., & Bugalho, P. (2012). Alzheimer's disease: A clinical practice-oriented review. *Frontiers in Neurology*, *3*, 63.

Alzheimer's_Association (2017). Alzheimer's disease 2017 facts and figures. http://www.alz.org/facts/overview.asp.

Baker, M., Mackenzie, I. R., Pickering-Brown, S. M., Gass, J., Rademakers, R., Lindholm, C., ... Rollinson, S. (2006). Mutations in progranulin cause tau-negative frontotemporal dementia linked to chromosome 17. *Nature*, *442*, 916–919.

Ballatore, C., Lee, V. M., & Trojanowski, J. Q. (2007). Tau-mediated neurodegeneration in Alzheimer's disease and related disorders. *Nature Reviews Neuroscience*, *8*, 663–672.

Banez-Coronel, M., Porta, S., Kagerbauer, B., Mateu-Huertas, E., Pantano, L., Ferrer, I., ... Marti, E. (2012). A pathogenic mechanism in Huntington's disease involves small CAG-repeated RNAs with neurotoxic activity. *PLoS Genetics*, *8*, e1002481.

Banks, W. A. (2016). From blood–brain barrier to blood–brain interface: New opportunities for CNS drug delivery. *Nature Reviews Drug Discovery*, *15*, 275–292.

Bateman, R. J., Xiong, C., Benzinger, T. L., Fagan, A. M., Goate, A., Fox, N. C., ... Blazey, T. M. (2012). Clinical and biomarker changes in dominantly inherited Alzheimer's disease. *The New England Journal of Medicine*, *367*, 795–804.

Beel, A. J., & Sanders, C. R. (2008). Substrate specificity of gamma-secretase and other intramembrane proteases. *Cellular and Molecular Life Sciences: CMLS*, *65*, 1311–1334.

Belzil, V. V., Gendron, T. F., & Petrucelli, L. (2013). RNA-mediated toxicity in neurodegenerative disease. *Molecular and Cellular Neurosciences*, *56*, 406–419.

Benilova, I., Karran, E., & De Strooper, B. (2012). The toxic Abeta oligomer and Alzheimer's disease: An emperor in need of clothes. *Nature Neuroscience*, *15*, 349–357.

Bergmann, O., Spalding, K. L., & Frisen, J. (2015). Adult neurogenesis in humans. *Cold Spring Harbor Perspectives in Biology*, *7*, a018994.

Bove, J., & Perier, C. (2012). Neurotoxin-based models of Parkinson's disease. *Neuroscience*, *211*, 51–76.

Brandner, S., & Jaunmuktane, Z. (2017). Prion disease: Experimental models and reality. *Acta neuropathologica*, *133*, 197–222.

Castellano, J. M., Kim, J., Stewart, F. R., Jiang, H., DeMattos, R. B., Patterson, B. W., ... Cruchaga, C. (2011). Human apoE isoforms differentially regulate brain amyloid-beta peptide clearance. *Science Translational Medicine*, *3*, 89ra57.

Chesselet, M. F., & Richter, F. (2011). Modelling of Parkinson's disease in mice. *The Lancet Neurology*, *10*, 1108–1118.

Chew, J., Gendron, T. F., Prudencio, M., Sasaguri, H., Zhang, Y. J., Castanedes-Casey, M., ... Murray, M. E. (2015). Neurodegeneration. C9ORF72 repeat expansions in mice cause TDP-43 pathology, neuronal loss, and behavioral deficits. *Science (New York, NY)*, *348*, 1151–1154.

Chu, Y., & Kordower, J. H. (2007). Age-associated increases of alpha-synuclein in monkeys and humans are associated with nigrostriatal dopamine depletion: Is this the target for Parkinson's disease?. *Neurobiology of Disease*, *25*, 134–149.

Ciechanover, A., & Brundin, P. (2003). The ubiquitin proteasome system in neurodegenerative diseases: Sometimes the chicken, sometimes the egg. *Neuron*, *40*, 427–446.

Colby, D. W., & Prusiner, S. B. (2011). Prions. *Cold Spring Harbor Perspectives in Biology*, *3*, a006833.

Collinge, J. (2016). Mammalian prions and their wider relevance in neurodegenerative diseases. *Nature*, *539*, 217–226.

Correia, S. C., Perry, G., & Moreira, P. I. (2016). Mitochondrial traffic jams in Alzheimer's disease—pinpointing the roadblocks. *Biochimica et Biophysica Acta*, *1862*, 1909–1917.

Cruts, M., Gijselinck, I., van der Zee, J., Engelborghs, S., Wils, H., Pirici, D., ... Martin, J. J. (2006). Null mutations in progranulin cause ubiquitin-positive frontotemporal dementia linked to chromosome 17q21. *Nature, 442*, 920–924.

Cuyvers, E., & Sleegers, K. (2016). Genetic variations underlying Alzheimer's disease: Evidence from genome-wide association studies and beyond. *The Lancet Neurology, 15*, 857–868.

de Lau, L. M., Schipper, C. M., Hofman, A., Koudstaal, P. J., & Breteler, M. M. (2005). Prognosis of Parkinson disease: Risk of dementia and mortality: The Rotterdam Study. *Archives of Neurology, 62*, 1265–1269.

Dolan, P. J., & Johnson, G. V. (2010). The role of tau kinases in Alzheimer's disease. *Current Opinion in Drug Discovery & Development, 13*, 595–603.

Doody, R. S., Thomas, R. G., Farlow, M., Iwatsubo, T., Vellas, B., Joffe, S., ... Aisen, P. S. (2014). Phase 3 trials of solanezumab for mild-to-moderate Alzheimer's disease. *The New England Journal of Medicine, 370*, 311–321.

Douglas, P. M., & Dillin, A. (2010). Protein homeostasis and aging in neurodegeneration. *The Journal of Cell Biology, 190*, 719–729.

Eaton, S. L., & Wishart, T. M. (2017). Bridging the gap: Large animal models in neurodegenerative research. *Mammalian Genome: Official Journal of the International Mammalian Genome Society, 238*, 247–256.

Esquerda-Canals, G., Montoliu-Gaya, L., Guell-Bosch, J., & Villegas, S. (2017). Mouse models of Alzheimer's disease. *Journal of Alzheimer's Disease: JAD, 57*, 1171–1183.

Ferrante, R. J. (2009). Mouse models of Huntington's disease and methodological considerations for therapeutic trials. *Biochimica et Biophysica Acta, 1792*, 506–520.

Fleming, S. M., Salcedo, J., Fernagut, P. O., Rockenstein, E., Masliah, E., Levine, M. S., & Chesselet, M. F. (2004). Early and progressive sensorimotor anomalies in mice overexpressing wild-type human alpha-synuclein. *The Journal of Neuroscience: The Official Journal of the Society for Neuroscience, 24*, 9434–9440.

Geser, F., Lee, V. M., & Trojanowski, J. Q. (2010). Amyotrophic lateral sclerosis and frontotemporal lobar degeneration: A spectrum of TDP-43 proteinopathies. *Neuropathology: Official Journal of the Japanese Society of Neuropathology, 30*, 103–112.

Gispert, S., Ricciardi, F., Kurz, A., Azizov, M., Hoepken, H. H., Becker, D., ... Kudin, A. P. (2009). Parkinson phenotype in aged PINK1-deficient mice is accompanied by progressive mitochondrial dysfunction in absence of neurodegeneration. *PLoS ONE, 4*, e5777.

Goedert, M., & Jakes, R. (2005). Mutations causing neurodegenerative tauopathies. *Biochimica et Biophysica Acta, 1739*, 240–250.

Goldberg, M. S., Fleming, S. M., Palacino, J. J., Cepeda, C., Lam, H. A., Bhatnagar, A., ... Klapstein, G. J. (2003). Parkin-deficient mice exhibit nigrostriatal deficits but not loss of dopaminergic neurons. *The Journal of Biological Chemistry, 278*, 43628–43635.

Gotz, J., Deters, N., Doldissen, A., Bokhari, L., Ke, Y., Wiesner, A., ... Ittner, L. M. (2007). A decade of tau transgenic animal models and beyond. *Brain Pathology (Zurich, Switzerland), 17*, 91–103.

Gray, M., Shirasaki, D. I., Cepeda, C., Andre, V. M., Wilburn, B., Lu, X. H., ... Sun, Y. E. (2008). Full-length human mutant huntingtin with a stable polyglutamine repeat can elicit progressive and selective neuropathogenesis in BACHD mice. *The Journal of Neuroscience: the Official Journal of the Society for Neuroscience, 28*, 6182–6195.

Guerreiro, R., Wojtas, A., Bras, J., Carrasquillo, M., Rogaeva, E., Majounie, E., ... Younkin, S. (2013). TREM2 variants in Alzheimer's disease. *The New England Journal of Medicine, 368*, 117–127.

Guo, J. L., & Lee, V. M. (2014). Cell-to-cell transmission of pathogenic proteins in neurodegenerative diseases. *Nature Medicine, 20*, 130–138.

Gurney, M. E., Pu, H., Chiu, A. Y., Dal Canto, M. C., Polchow, C. Y., Alexander, D. D., ... Deng, H. X. (1994). Motor neuron degeneration in mice that express a human Cu, Zn superoxide dismutase mutation. *Science (New York, NY), 264*, 1772–1775.

Haass, C., & Selkoe, D. J. (2007). Soluble protein oligomers in neurodegeneration: Lessons from the Alzheimer's amyloid beta-peptide. *Nature Reviews Molecular Cell Biology, 8*, 101–112.

Haeusler, A. R., Donnelly, C. J., & Rothstein, J. D. (2016). The expanding biology of the C9orf72 nucleotide repeat expansion in neurodegenerative disease. *Nature Reviews Neuroscience, 17*, 383–395.

Hasegawa, M., Nonaka, T., & Masuda-Suzukake, M. (2017). Prion-like mechanisms and potential therapeutic targets in neurodegenerative disorders. *Pharmacology & Therapeutics, 172*, 22–33.

Irwin, D. J., Cairns, N. J., Grossman, M., McMillan, C. T., Lee, E. B., Van Deerlin, V. M., ... Trojanowski, J. Q. (2015). Frontotemporal lobar degeneration: Defining phenotypic diversity through personalized medicine. *Acta Neuropathologica, 129*, 469–491.

Janssens, J., & Van Broeckhoven, C. (2013). Pathological mechanisms underlying TDP-43 driven neurodegeneration in FTLD–ALS spectrum disorders. *Human Molecular Genetics, 22*, R77–R87.

Johnson, M. H. (2001). Functional brain development in humans. *Nature Reviews Neuroscience, 2*, 475–483.

Johnson, R. T. (2005). Prion diseases. *The Lancet Neurology, 4*, 635–642.

Jonsson, T., Atwal, J. K., Steinberg, S., Snaedal, J., Jonsson, P. V., Bjornsson, S., ... Maloney, J. (2012). A mutation in APP protects against Alzheimer's disease and age-related cognitive decline. *Nature, 488*, 96–99.

Jonsson, T., Stefansson, H., Steinberg, S., Jonsdottir, I., Jonsson, P. V., Snaedal, J., ... Lah, J. J. (2013). Variant of TREM2 associated with the risk of Alzheimer's disease. *The New England Journal of Medicine, 368*, 107–116.

Jovaisaite, V., Mouchiroud, L., & Auwerx, J. (2014). The mitochondrial unfolded protein response, a conserved stress response pathway with implications in health and disease. *The Journal of Experimental Biology, 217*, 137–143.

Kalia, L. V., & Lang, A. E. (2015). Parkinson's disease. *Lancet (London, England), 386*, 896–912.

Keogh, M. J., & Chinnery, P. F. (2015). Mitochondrial DNA mutations in neurodegeneration. *Biochimica et Biophysica Acta, 1847*, 1401–1411.

Labbadia, J., & Morimoto, R. I. (2013). Huntington's disease: Underlying molecular mechanisms and emerging concepts. *Trends in Biochemical Sciences, 38*, 378–385.

Laforet, G. A., Sapp, E., Chase, K., McIntyre, C., Boyce, F. M., Campbell, M., ... Reddy, P. H. (2001). Changes in cortical and striatal neurons predict behavioral and electrophysiological abnormalities in a transgenic murine model of Huntington's disease. *The Journal of Neuroscience: The Official Journal of the Society for Neuroscience, 21*, 9112–9123.

Lane, R. K., Hilsabeck, T., & Rea, S. L. (2015). The role of mitochondrial dysfunction in age-related diseases. *Biochimica et Biophysica Acta, 1847*, 1387–1400.

Lee, M. K., Stirling, W., Xu, Y., Xu, X., Qui, D., Mandir, A. S., ... Price, D. L. (2002). Human alpha-synuclein-harboring familial Parkinson's disease-linked Ala-53 → Thr mutation causes neurodegenerative disease with alpha-synuclein aggregation in transgenic mice. *Proceedings of the National Academy of Sciences of the United States of America, 99*, 8968–8973.

Lee, V. M., Goedert, M., & Trojanowski, J. Q. (2001). Neurodegenerative tauopathies. *Annual Review of Neuroscience, 24*, 1121–1159.

Lewis, J., Dickson, D. W., Lin, W. L., Chisholm, L., Corral, A., Jones, G., ... Yager, D. (2001). Enhanced neurofibrillary degeneration in transgenic mice expressing mutant tau and APP. *Science (New York, NY), 293*, 1487–1491.

Li, L. B., Yu, Z., Teng, X., & Bonini, N. M. (2008). RNA toxicity is a component of ataxin-3 degeneration in *Drosophila*. *Nature, 453*, 1107–1111.

Li, X., Patel, J. C., Wang, J., Avshalumov, M. V., Nicholson, C., Buxbaum, J. D., ... Yue, Z. (2010). Enhanced striatal dopamine transmission and motor performance with LRRK2 overexpression in mice is eliminated by familial Parkinson's disease mutation G2019S. *The Journal of Neuroscience: The Official Journal of the Society for Neuroscience, 30*, 1788–1797.

Li, Y., Liu, W., Oo, T. F., Wang, L., Tang, Y., Jackson-Lewis, V., ... Przedborski, S. (2009). Mutant LRRK2(R1441G) BAC transgenic mice recapitulate cardinal features of Parkinson's disease. *Nature Neuroscience, 12*, 826–828.

Lindberg, I., Shorter, J., Wiseman, R. L., Chiti, F., Dickey, C. A., & McLean, P. J. (2015). Chaperones in neurodegeneration. *Journal of Neuroscience, 35*, 13853–13859.

Liu, L., Drouet, V., Wu, J. W., Witter, M. P., Small, S. A., Clelland, C., & Duff, K. (2012). Trans-synaptic spread of tau pathology in vivo. *PLoS ONE, 7*, e31302.

Lu, B., & Vogel, H. (2009). *Drosophila* models of neurodegenerative diseases. *Annual Review of Pathology, 4*, 315–342.

Lubbe, S., & Morris, H. R. (2014). Recent advances in Parkinson's disease genetics. *Journal of Neurology, 261*, 259–266.

Mallik, A., Drzezga, A., & Minoshima, S. (2017). Clinical amyloid imaging. *Seminars in Nuclear Medicine, 47*, 31–43.

Mangiarini, L., Sathasivam, K., Seller, M., Cozens, B., Harper, A., Hetherington, C., ... Davies, S. W. (1996). Exon 1 of the HD gene with an expanded CAG repeat is sufficient to cause a progressive neurological phenotype in transgenic mice. *Cell, 87*, 493–506.

Marti, E. (2016). RNA toxicity induced by expanded CAG repeats in Huntington's disease. *Brain Pathology (Zurich, Switzerland), 26*, 779–786.

McKee, A. C., Cantu, R. C., Nowinski, C. J., Hedley-Whyte, E. T., Gavett, B. E., Budson, A. E., ... Stern, R. A. (2009). Chronic traumatic encephalopathy in athletes: Progressive tauopathy after repetitive head injury. *Journal of Neuropathology and Experimental Neurology, 68*, 709–735.

Morfini, G. A., Burns, M., Binder, L. I., Kanaan, N. M., LaPointe, N., Bosco, D. A., ... Hayward, L. (2009). Axonal transport defects in neurodegenerative diseases. *The Journal of Neuroscience: The Official Journal of the Society for Neuroscience, 29*, 12776–12786.

Nasir, J., Floresco, S. B., O'Kusky, J. R., Diewert, V. M., Richman, J. M., Zeisler, J., ... Hayden, M. R. (1995). Targeted disruption of the Huntington's disease gene results in embryonic lethality and behavioral and morphological changes in heterozygotes. *Cell, 81*, 811–823.

Neumann, M. (2009). Molecular neuropathology of TDP-43 proteinopathies. *International Journal of Molecular Sciences, 10*, 232–246.

Neumann, M., Sampathu, D. M., Kwong, L. K., Truax, A. C., Micsenyi, M. C., Chou, T. T., ... Clark, C. M. (2006). Ubiquitinated TDP-43 in frontotemporal lobar degeneration and amyotrophic lateral sclerosis. *Science (New York, NY), 314*, 130–133.

Nguyen, T. N., Padman, B. S., & Lazarou, M. (2016). Deciphering the molecular signals of PINK1/parkin mitophagy. *Trends in Cell Biology, 26*, 733–744.

Nixon, R. A. (2013). The role of autophagy in neurodegenerative disease. *Nature Medicine, 19*, 983–997.

Parkinson's_Disease_Foundation (2017). http://parkinson.org/Understanding-Parkinsons/Causes-and-Statistics/Statistics. Parkinson's Foundation: Statistics.

Picher-Martel, V., Valdmanis, P. N., Gould, P. V., Julien, J. P., & Dupre, N. (2016). From animal models to human disease: A genetic approach for personalized medicine in ALS. *Acta Neuropathologica Communications, 4*, 70.

Polling, S., Hill, A. F., & Hatters, D. M. (2012). Polyglutamine aggregation in Huntington and related diseases. *Advances in Experimental Medicine and Biology, 769*, 125–140.

Prince, M., Bryce, R., Albanese, E., Wimo, A., Ribeiro, W., & Ferri, C. P. (2013). The global prevalence of dementia: A systematic review and metaanalysis. *Alzheimer's & Dementia: The Journal of the Alzheimer's Association, 9*, 63-75.e62.

Ransohoff, R. M. (2016). How neuroinflammation contributes to neurodegeneration. *Science (New York, NY), 353*, 777–783.

Reaume, A. G., Elliott, J. L., Hoffman, E. K., Kowall, N. W., Ferrante, R. J., Siwek, D. F., ... Brown, R. H., Jr. (1996). Motor neurons in Cu/Zn superoxide dismutase-deficient mice develop normally but exhibit enhanced cell death after axonal injury. *Nature Genetics, 13*, 43–47.

Roberson, E. D., Scearce-Levie, K., Palop, J. J., Yan, F., Cheng, I. H., Wu, T., ... Mucke, L. (2007). Reducing endogenous tau ameliorates amyloid beta-induced deficits in an Alzheimer's disease mouse model. *Science (New York, NY), 316*, 750–754.

Ruegsegger, C., & Saxena, S. (2016). Proteostasis impairment in ALS. *Brain Research, 1648*, 571–579.

Saito, T., Matsuba, Y., Mihira, N., Takano, J., Nilsson, P., Itohara, S., ... Saido, T. C. (2014). Single App knock-in mouse models of Alzheimer's disease. *Nature Neuroscience, 17*, 661–663.

Sathasivam, K., Neueder, A., Gipson, T. A., Landles, C., Benjamin, A. C., Bondulich, M. K., ... Howland, D. (2013). Aberrant splicing of HTT generates the pathogenic exon 1 protein in Huntington disease. *Proceedings of the National Academy of Sciences of the United States of America, 110*, 2366–2370.

Seelaar, H., Rohrer, J. D., Pijnenburg, Y. A., Fox, N. C., & van Swieten, J. C. (2011). Clinical, genetic and pathological heterogeneity of frontotemporal dementia: A review. *Journal of Neurology, Neurosurgery, and Psychiatry, 82*, 476–486.

Selkoe, D. J., & Hardy, J. (2016). The amyloid hypothesis of Alzheimer's disease at 25 years. *EMBO Molecular Medicine, 8*, 595–608.

Shen, J., Bronson, R. T., Chen, D. F., Xia, W., Selkoe, D. J., & Tonegawa, S. (1997). Skeletal and CNS defects in Presenilin-1-deficient mice. *Cell, 89*, 629–639.

Sin, O., Michels, H., & Nollen, E. A. (2014). Genetic screens in *Caenorhabditis* elegans models for neurodegenerative diseases. *Biochimica et Biophysica Acta, 1842*, 1951–1959.

Soto, C., & Satani, N. (2011). The intricate mechanisms of neurodegeneration in prion diseases. *Trends in Molecular Medicine, 17*, 14–24.

Steffan, J. S., Bodai, L., Pallos, J., Poelman, M., McCampbell, A., Apostol, B. L., ... Greenwald, M. (2001). Histone deacetylase inhibitors arrest polyglutamine-dependent neurodegeneration in *Drosophila*. *Nature, 413*, 739–743.

Sun, X., Li, P. P., Zhu, S., Cohen, R., Marque, L. O., Ross, C. A., ... Rudnicki, D. D. (2015). Nuclear retention of full-length HTT RNA is mediated by splicing factors MBNL1 and U2AF65. *Scientific Reports, 5*, 12521.

Sveinbjornsdottir, S. (2016). The clinical symptoms of Parkinson's disease. *Journal of Neurochemistry, 139*(Suppl 1), 318–324.

Swarup, V., Phaneuf, D., Bareil, C., Robertson, J., Rouleau, G. A., Kriz, J., & Julien, J. P. (2011). Pathological hallmarks of amyotrophic lateral sclerosis/frontotemporal lobar degeneration in transgenic mice produced with TDP-43 genomic fragments. *Brain: A Journal of Neurology, 134*, 2610–2626.

Tanzi, R. E., & Bertram, L. (2005). Twenty years of the Alzheimer's disease amyloid hypothesis: A genetic perspective. *Cell, 120*, 545–555.

Taylor, J. P., Brown, R. H., Jr., & Cleveland, D. W. (2016). Decoding ALS: From genes to mechanism. *Nature, 539*, 197–206.

Telling, G. C. (2008). Transgenic mouse models of prion diseases. *Methods in Molecular Biology (Clifton, NJ), 459*, 249–263.

Villegas-Llerena, C., Phillips, A., Garcia-Reitboeck, P., Hardy, J., & Pocock, J. M. (2016). Microglial genes regulating neuroinflammation in the progression of Alzheimer's disease. *Current Opinion in Neurobiology, 36*, 74–81.

Vinters, H. V. (2015). Emerging concepts in Alzheimer's disease. *Annual Review of Pathology, 10*, 291–319.

Walker, F. O. (2007). Huntington's disease. *Lancet (London, England), 369*, 218–228.

Walsh, D. M., & Selkoe, D. J. (2016). A critical appraisal of the pathogenic protein spread hypothesis of neurodegeneration. *Nature Reviews Neuroscience, 17*, 251–260.

Wang, Y., Cella, M., Mallinson, K., Ulrich, J. D., Young, K. L., Robinette, M. L., ... Zinselmeyer, B. H. (2015). TREM2 lipid sensing sustains the microglial response in an Alzheimer's disease model. *Cell, 160*, 1061–1071.

Wang, Y., & Mandelkow, E. (2016). Tau in physiology and pathology. *Nature Reviews Neuroscience, 17*, 5–21.

Webster, S. J., Bachstetter, A. D., Nelson, P. T., Schmitt, F. A., & Van Eldik, L. J. (2014). Using mice to model Alzheimer's dementia: An overview of the clinical disease and the preclinical behavioral changes in 10 mouse models. *Frontiers in Genetics, 5*, 88.

Wojciechowska, M., & Krzyzosiak, W. J. (2011). Cellular toxicity of expanded RNA repeats: Focus on RNA foci. *Human Molecular Genetics, 20*, 3811–3821.

Wolfe, M. S. (2009). Tau mutations in neurodegenerative diseases. *The Journal of Biological Chemistry, 284*, 6021–6025.

Wong, P. C., Zheng, H., Chen, H., Becher, M. W., Sirinathsinghji, D. J., Trumbauer, M. E., ... Sisodia, S. S. (1997). Presenilin 1 is required for Notch1 and Dll1 expression in the paraxial mesoderm. *Nature, 387*, 288–292.

World_Alzheimer_Report (2015). World Alzheimer's report 2015: The global impact of Dementia. https://www.alz.co.uk/research/WorldAlzheimerReport2015-sheet.pdf.

Wu, L. S., Cheng, W. C., Hou, S. C., Yan, Y. T., Jiang, S. T., & Shen, C. K. (2010). TDP-43, a neuro-pathosignature factor, is essential for early mouse embryogenesis. *Genesis (New York, NY : 2000), 48*, 56–62.

Yamanaka, K., Chun, S. J., Boillee, S., Fujimori-Tonou, N., Yamashita, H., Gutmann, D. H., ... Cleveland, D. W. (2008). Astrocytes as determinants of disease progression in inherited amyotrophic lateral sclerosis. *Nature Neuroscience, 11*, 251–253.

Yerbury, J. J., Ooi, L., Dillin, A., Saunders, D. N., Hatters, D. M., Beart, P. M., ... Ecroyd, H. (2016). Walking the tightrope: Proteostasis and neurodegenerative disease. *Journal of Neurochemistry, 137*, 489–505.

CHAPTER

2

Prion Diseases

Robert C.C. Mercer, Alex J. McDonald*, Erin Bove-Fenderson, Cheng Fang, Bei Wu and David A. Harris*

Boston University School of Medicine, Boston, MA, United States

OUTLINE

INTRODUCTION AND HISTORICAL PERSPECTIVE

Prion diseases are invariably fatal neurodegenerative disease of humans and animals. They are unprecedented in the history of medicine because the infectious agent that causes them is a protein molecule that is devoid of a nucleic acid genome. The earliest recorded prion disease is scrapie, first described in the 18th century in sheep and goats (McGowan, 1922), and subsequently recognized to be infectious by laboratory transmission to rodents (Chandler, 1961). Creutzfeldt–Jakob disease (CJD), a progressive neurodegenerative disorder of humans, was originally recognized in the 1920s

* These two authors contributed equally.

The Molecular and Cellular Basis of Neurodegenerative Diseases
DOI: https://doi.org/10.1016/B978-0-12-811304-2.00002-X

(Creutzfeldt, 1920). In the 1950s, Gajdusek and Zigas (1957) described kuru, a "shaking disease" of the Fore people of Papua New Guinea that was transmitted by ritual cannibalism. It was soon appreciated that CJD and kuru shared neuropathological similarities, and that these features resembled those found in the brains of scrapie-infected sheep (Hadlow, 1959; Klatzo, Gajdusek, & Zigas, 1959). This recognition led to the demonstration that kuru and CJD were infectious, as demonstrated by experimental transmission to primates (Gajdusek, Gibbs, & Alpers, 1966; Gibbs et al., 1968). Subsequent work on the nature of the infectious agent determined that it was highly resistant to ultraviolet and ionizing radiation (procedures known to destroy nucleic acid), suggesting that it might represent a previously unknown class of pathogen (Alper & Cramp, 1967; Alper, Haig, & Clarke, 1966; Gibbs, Gajdusek, & Latarjet, 1978; Latarjet, Muel, Haig, Clarke, & Alper, 1970). Initial theories postulated that it was a "slow virus," based on the long incubation times it produced (Eklund, Kennedy, & Hadlow, 1967).

In order to define further the nature of the infectious agent, it was necessary to develop a workable bioassay and to purify the agent. An important experimental advance was adaptation of the scrapie agent to replication in rodents (Fig. 2.1). Although bioassays in mice took almost half a year to complete (Chandler, 1961), they were much more practical than investigations using sheep, which had incubation times of 1–2 years. The use of Syrian hamsters as a recipient further shortened the time required for assay to about 60 days, allowing for rapid measurement of infectivity in biochemically fractionated samples (Marsh & Kimberlin, 1975). By employing the hamster bioassay in an incubation time format, in which infectious titer was inversely related to incubation time, it became feasible to purify and characterize the infectious agent.

Using a scheme involving detergent extraction, centrifugation, and digestion with proteinase K, highly infectious samples were purified from the brains of infected hamsters (Bolton, McKinley, & Prusiner, 1982). These samples were found to contain a single kind of protein molecule and appeared to be devoid of DNA or RNA. It was observed that treatment with protein denaturants such as sodium dodecyl sulfate (SDS), guanidinium thiosulfate, and phenol destroyed infectivity, while treatments that destroyed nucleic acid had no effect, leading to the conclusion that the long-sought infectious agent is composed solely of a protein (Prusiner, 1982). It was at this point that the term prion was first coined as a portmanteau of *pro(teinaceous)* and *in(fectious particle)* (Prusiner, 1982). Following sequencing of the purified protein by Edman degradation, the N-terminal sequence was used to create degenerate oligonucleotide primers that hybridized to mRNA and genomic DNA in both infected and noninfected animals, leading to the discovery of a corresponding protein expressed by healthy individuals (Basler et al., 1986; Kretzschmar, Prusiner, Stowring, & DeArmond, 1986; Oesch et al., 1985).

MOLECULAR MECHANISM OF PRION PROPAGATION

Based on a great deal of biochemical and molecular genetic work over the past 25 years, it is now recognized that the prion protein (PrP) can exist in two distinct conformations: PrP^{Sc}, the infectious form, and PrP^{C}, the normal cellular form. The two forms have the same primary amino acid sequence and post-translational modifications, but differ in their conformations, with PrP^{Sc} having a much higher content of β-sheet than PrP^{C} (Nguyen, Baldwin, Cohen, & Prusiner, 1995; Pan et al., 1993; Prusiner, 1982; Requena & Wille, 2014; Wille et al., 2002). The essential event in prion propagation is the conformational conversion of PrP^{C} into PrP^{Sc}. In this process, PrP^{Sc} serves as a molecular template, which physically

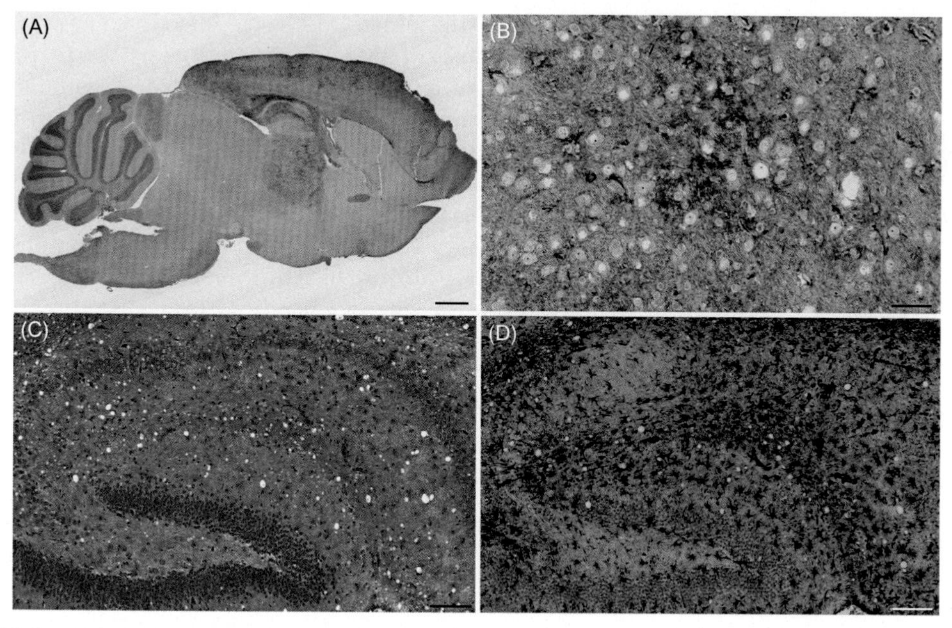

FIGURE 2.1 **Pathological hallmarks of prion disease.** An FVB/N mouse at the terminal stage following intracerebral inoculation with the rodent-adapted scrapie strain, Rocky Mountain Laboratories (RML). Sections were fixed in 10% phosphate buffered formalin and embedded in paraffin. **(A)**, deparaffinized sections were treated with formic acid and guanidine thiocyanate to enhance detection of PrP^{Sc} using the anti-PrP antibody SAF83. Scale bar = 1 mm. **(B)** Magnified view of the thalamus. The diffuse distribution of PrP^{Sc} shown here is referred to as "synaptic-like" to distinguish it from plaque-like deposits seen in some cases of prion disease. Scale bar = 500 μm. (C) A magnified view of the hippocampus is shown following staining with hematoxylin and eosin to visualize spongiform change typical of some but not all prion diseases. Scale bar = 150 μm. (D) Reactive gliosis is demonstrated by immunostaining for glial acidic fibrillary protein, GFAP. The hippocampus is shown. Scale bar = 150 μm. *PrP*, prion protein. *These data provided courtesy of Dr. Nathalie Daude and Ms. Hristina Gapeshina from the Centre for Prions and Protein Folding Diseases at the University of Alberta.*

interacts with PrP^{C}, converting the latter into PrP^{Sc} in an autocatalytic fashion (Fig. 2.2). The two conformers also differ biochemically, with PrP^{Sc} being aggregated and protease-resistant, in contrast to PrP^{C}, which is monomeric and protease-sensitive. These features are often used to distinguish the two conformations in experimental and diagnostic settings.

Although it is clear that the conformations of PrP^{C} and PrP^{Sc} differ, it has proven difficult to determine an atomic-scale structure of PrP^{Sc} using high-resolution structural techniques such as NMR and X-ray crystallography, due to mainly the aggregated and heterogeneous nature of the infectious protein. Recently, cryoelectron microscopy was used to create a three-dimensional reconstruction of infectious prion fibrils. This work demonstrated that these fibrils were composed of two protofilaments with a four-rung β-solenoid as the underlying subunit (Vázquez-Fernández et al., 2016). Other models have been proposed, most notably structures comprising parallel, in-register β-sheets, but these are based on noninfectious forms of PrP amyloid (Cobb, Sonnichsen, McHaourab, & Surewicz, 2007; Groveman et al., 2014). A complete, atomic-level structure of infectious PrP^{Sc} remains an important goal for the field (Diaz-Espinoza & Soto, 2012; Zweckstetter, Requena, & Wille, 2017).

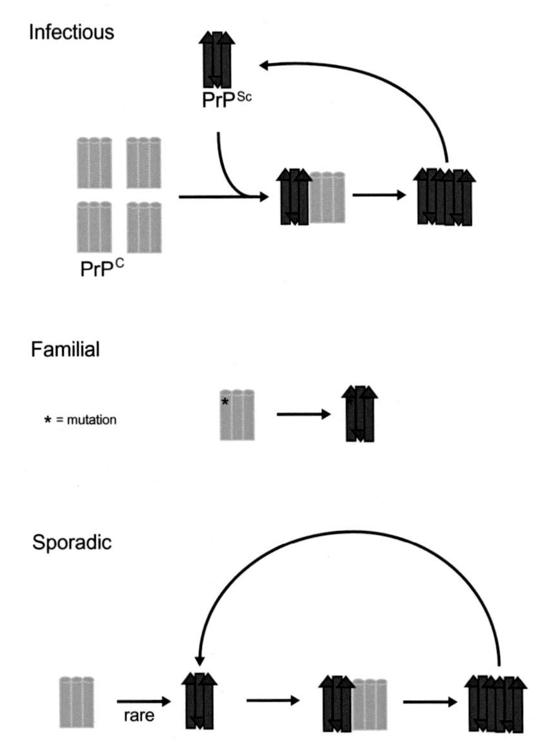

FIGURE 2.2 **Mechanisms of prion propagation.** Three schematics of prion propagation are presented, corresponding to infectious, familial, and sporadic manifestations of prion disease. When initiated by infection, exogenous PrP^{Sc} interacts with endogenous PrP^{C}, inducing a structural transformation that then propagates itself through a self-templating mechanism. Familial prion diseases are due to one of several germline mutations (represented here by an asterisk) in the *PRNP* gene, which are thought to make the spontaneous transition from PrP^{C} to PrP^{Sc} more energetically favorable. Sporadic prion disease is depicted here as a spontaneous refolding of PrP^{C} to PrP^{Sc} (a rare event) which then propagates itself through the same self-templating mechanism as described above. Another possibility is that a somatic mutation in *PRNP* could initiate the misfolding cascade. It has also been hypothesized that PrP^{Sc} is always present at undetectable levels, and a rare stabilizing event allows for its propagation. *PrP*, prion protein.

The prion hypothesis is now supported by an extensive network of experimental findings. Historically, perhaps the most persuasive piece of evidence was the discovery that mice in which the gene encoding PrP^{C} (*Prnp*) is ablated are completely resistant to prion disease, demonstrating that endogenous PrP^{C} is an essential substrate for prion propagation (Büeler et al., 1993). Moreover, it was shown that the species tropism of prions could be manipulated by the introduction of different PrP-encoding transgenes into mice (Scott et al., 1989). Additionally, mutations of *PRNP* (the gene encoding PrP^{C} in humans) were found in patients with hereditary prion diseases, and expression of these mutants in mice caused spontaneous disease (Bouybayoune et al., 2015; Chiesa, Piccardo, Ghetti, & Harris, 1998; Hsiao et al., 1989, 1990, 1991; Medori et al., 1992; Yang et al., 2009; Mercer et al., 2018). Finally, in recent years, prions have been created in vitro using PrP from brain or recombinant sources (Wang, Wang, Yuan, & Ma, 2010). The evidence for the proteinaceous nature of the infectious agent is now overwhelming, although there remain a few investigators who still advocate for a cryptic virus as the causative agent (Manuelidis, Yu, Barquero, & Mullins, 2007).

Prions exemplify a mechanism for propagation of biological information that is independent of nucleic acid, and that has been found to occur in other organisms, including yeast, filamentous fungi, and bacteria (Wickner, 2016; Yuan & Hochschild, 2017). Self-templating changes in protein conformation have recently been shown to participate in a number of physiological phenomena, including innate immunity (Hou et al., 2011), formation of membraneless cellular organelles (Kato et al., 2012), and possibly even memory formation in the brain (Fioriti et al., 2015). As will be discussed below, a prion-like mechanism may also underlie the spread within the central nervous system (CNS) of protein aggregates involved in other neurodegenerative diseases, including Alzheimer's disease (AD), Parkinson's disease (PD), and tauopathies.

THE CELLULAR PRION PROTEIN: STRUCTURE AND PROTEOLYTIC PROCESSING

Mammalian PrP^C is a highly conserved, protein of approximately 250 amino acids, which is expressed in a variety of tissues throughout the body but primarily in the CNS (Ford, Burton, Morris, & Hall, 2002; Oesch et al., 1985). Paralogues of mammalian PrP are also found in fish, avian, and reptilian species (Rivera-Milla, Stuermer, & Málaga-Trillo, 2003; Simonic et al., 2000; Wopfner et al., 1999). After removal of the N-terminal signal sequence and conjugation of a C-terminal glycosylphosphatidylinositol (GPI) anchor, the mature protein is delivered to the cell surface, where it is attached to the outer leaflet of the lipid bilayer via its GPI anchor and is localized in lipid rafts (Stahl, Borchelt, Hsiao, & Prusiner, 1987).

While the structure of PrP^{Sc} has been difficult to determine, the three-dimensional structure of PrP^C was worked out over 20 years ago using recombinant forms of the protein. PrP^C consists of a structured C-terminal domain containing three α-helices and two short β-strands arranged in a β1-α1-β2-α2-α3 orientation, and an intrinsically disordered N-terminus (Riek et al., 1996; Riek, Hornemann, Wider, Glockshuber, & Wüthrich, 1997). Connecting the N- and C-terminal domains is a hydrophobic linker region that is thought to contribute to the core of PrP^{Sc} (Norstrom & Mastrianni, 2005). The C-terminal domain of PrP^C contains two cysteine residues, which form a structurally critical disulfide bond (Turk, Teplow, Hood, & Prusiner, 1988), as well as two asparagine residues, N181 and N197, which can be N-glycosylated in the secretory pathway (Haraguchi et al., 1989). Mature PrP^C is found at the membrane surface in unglycosylated, monoglycosylated (at N181 or N197), and diglycosylated forms (Fig. 2.3).

PrP is subject to posttranslational proteolytic processing (Altmeppen et al., 2012; Glatzel et al., 2015; McDonald, Dibble, Evans, & Millhauser, 2014), including α-cleavage, which occurs between residues K109 and H110, and β-cleavage, which occurs within or just C-terminal to the octapeptide repeats (near Q90). Both of these processing events leave the structured C-terminal domain attached to the outer leaflet of the plasma membrane. PrP^C is also cleaved adjacent to its GPI anchor resulting in the shedding of the protein from the membrane surface (Fig. 2.3A). The physiological function of these cleavage events and the resulting PrP fragments are uncertain. α-cleavage is thought to be carried out by members of the ADAM (a disintegrin and metalloproteinase) family including ADAM 8, 10, and 17 (Liang et al., 2012; Vincent et al., 2001). Each of these transmembrane proteinases processes a number of other membrane-attached proteins (Saftig & Lichtenthaler, 2015). β-Cleavage may occur through two or more separate pathways. In normal brain tissue, ADAM 8 is capable of cleaving PrP^C within the octarepeat domain as part of the protein's normal processing (McDonald et al., 2014). However, in scrapie-infected tissue, cathepsins are partly responsible for the digestion of PrP^{Sc}, trimming it to a tightly misfolded core by cleavage around G89 and Q90 (Dron et al., 2010).

PHYSIOLOGICAL FUNCTION OF PRP^C

The normal, physiological function of PrP^C has remained elusive. *Prnp* knockout mice are resistant to prion infection but display no gross anatomical or development defects (Büeler et al., 1993). However, a number of subtle phenotypes have been described in mice or cells lacking PrP^C expression, leading to a plethora of purported functions (for a comprehensive review see Wulf, Senatore, & Aguzzi, 2017). An important caveat to the interpretation of phenotypes in knockout animals is that some may be attributable to

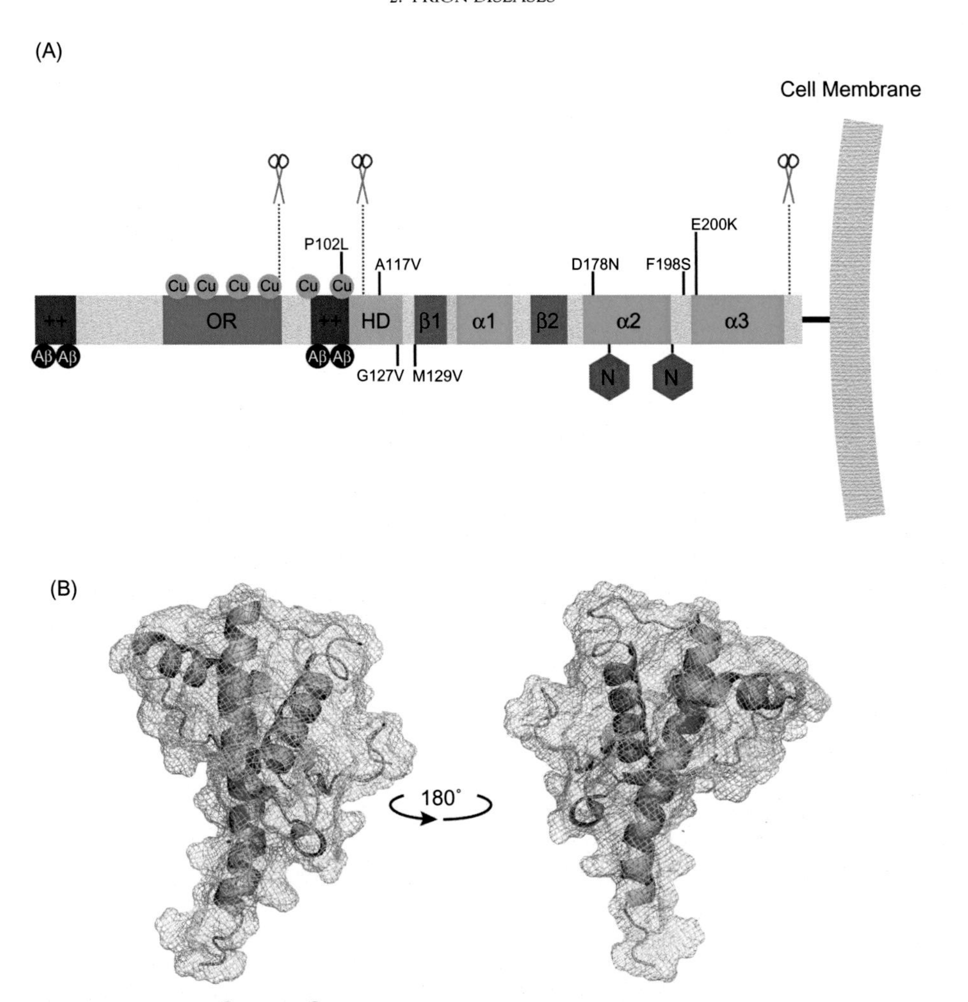

FIGURE 2.3 **Structure of PrP^{C}.** (A) PrP^{C} consists of a flexibly disordered N-terminal domain and a structured C-terminal domain. The N-terminal domain contains two positively charged patches indicated by '++' (which are the sites of Aβ binding), as well as four OR. At saturation, one Cu^{2+} ion can bind to each of the OR, and two additional Cu^{2+} ions can bind to histidine residues in the second positively charged patch. The C-terminal domain is formed by three α-helices as well as two short β strands flanking the first helix. A disulfide bond links helices 2 and 3, and the two N-linked glycosylation sites are indicated by hexagons. The C-terminal end of the protein is anchored to the outer leaflet of the plasma membrane via a GPI anchor. Linking the N- and C-terminal domains is a palindromic HD. Sites of physiologically relevant proteolytic cleavage are indicated by scissors and dashed lines. Pathogenic mutations and polymorphisms mentioned in the text are indicated above and below the line diagram, respectively. There are many more known mutations and polymorphisms in the human population (Lloyd, Mead, & Collinge, 2013). (B) A crystallographic structure (Gossert, Bonjour, Lysek, Fiorito, & Wuthrich, 2005) (1XYX) of the C-terminal domain of murine PrP^{C}. *HD*, hydrophobic domain; *OR*, octapeptide repeats; *PrP*, prion protein.

the effects of genetic loci outside of the *Prnp* gene, which have been altered in certain lines of knockout mice (Nuvolone et al., 2016).

One set of potential roles for PrP^{C} concerns synaptic transmission. One of the original studies on the physiology of *Prnp* knockout mice reported defective synaptic transmission due to an impairment of gamma-aminobutyric acid (GABA) receptor-mediated fast inhibition and long-term potentiation in CA1 pyramidal neurons of the hippocampus, deficits that could be rescued by the reintroduction of PrP^C (Collinge et al., 1994; Whittington et al., 1995). Although this finding has been disputed (Lledo, Tremblay, Dearmond, Prusiner, & Nicoll, 1996), there is other evidence that PrP^C plays a role in synaptic biology (Curtis, Errington, Bliss, Voss, & MacLeod, 2003; Maglio, Perez, Martins, Brentani, & Ramirez, 2004; Mallucci et al., 2002). Hippocampal brain slices from *Prnp* knockout animals show disrupted Ca^{2+}-activated K^+ currents and reorganization of mossy fibers (Colling, Collinge, & Jefferys, 1996; Colling, Khana, Collinge, & Jefferys, 1997). Mice devoid of PrP^C have an increased latency period when finding a food source and cannot discriminate between odors, defects that have been attributed to abnormal dendrodendritic synaptic transmission between olfactory bulb granule and mitral cells (Le Pichon et al., 2009). Alterations in circadian rhythm have been noted in *Prnp* knockout mice, which may be related to changes in synaptic transmission (Tobler, Deboer, & Fischer, 1997; Tobler et al., 1996).

Other studies have implicated PrP^C in excitotoxic phenomena. It has been reported that *Prnp* knockout mice are more susceptible to drug-induced seizures than their wild-type counterparts when exposed to various convulsive agents (Walz et al., 1999). A similar finding has also been reported in zebrafish (*Danio rerio*) harboring a zinc finger-targeted disruption of the *Prnp* paralogue, *prp2* (Fleisch et al., 2013). PrP^C appears to play a protective role during stroke, which may be related to a dampening effect on excitotoxicity. Following an ischemic event, PrP^C levels in plasma and the tissue surrounding the insult increase (McLennan et al., 2004; Mitsios et al., 2007; Weise et al., 2004), and animals lacking PrP^C have larger infarct sizes (Spudich et al., 2005).

The involvement of PrP^C in neuronal excitability is further supported by data demonstrating modulatory interactions with various ion channels and receptors. PrP^C has been shown to interact with AMPA (Watt et al., 2012) and NMDA receptors (Khosravani et al., 2008; You et al., 2012), as well as with the α7 nicotinic acetylcholine receptor through interaction with stress-inducible protein 1 (Beraldo et al., 2010; Zanata et al., 2002). PrP^C has also been shown to modulate the activity of the voltage-gated K^+ channel, Kv4.2, as well as voltage-gated Ca^{2+} channels, through interaction with DPP6 and $\alpha2\delta$-1, their respective auxiliary subunits (Mercer et al., 2013; Senatore et al., 2012). The metabotropic glutamate receptor, mGluR5, has also been identified as a binding partner of PrP^C (Um et al., 2013).

There is evidence that PrP^C is required for the maintenance of myelin, a function that requires neuronal expression of the protein as well as its normal proteolytic processing. When the PrP gene was ablated or proteolytic processing impaired by mutation of the cleavage sites in PrP, a chronic demyelinating polyneuropathy was produced (Bremer et al., 2010; Nishida et al., 1999). Recent work suggests that neuronal PrP^C promotes myelin homeostasis through transactivation of the G protein-coupled receptor, Gpr126, which is present on Schwann cells (Küffer et al., 2016). The flexible, N-terminal tail of PrP^C is proposed to be the region that interacts with Gpr126.

Several lines of evidence suggest that PrP^C could play a role in neural cell adhesion and neurite outgrowth. An in vivo chemical crosslinking study to identify potential PrP^C interactors recovered neural cell adhesion molecule (NCAM), a regulator of neurite outgrowth, as one candidate (Schmitt-Ulms et al., 2001, 2004). NCAM was subsequently shown to interact with PrP^C using superresolution nanoscopy (Slapsak et al., 2016). It had been shown previously that PrP^C recruits NCAM to lipid rafts and activates fyn kinase to induce neuritogenesis (Santuccione, Sytnyk, Leshchyns'ka, & Schachner, 2005), likely through NCAM's fibronectin domains (Slapsak et al., 2016). There is evidence that PrP^C regulates polysialylation of NCAM, although it is not clear if this requires direct interaction with NCAM (Mehrabian et al., 2015). Experiments in zebrafish also support a role for PrP^C in cell adhesion via activation of a fyn kinase pathway (Kaiser et al., 2012; Malaga-Trillo et al., 2009; Sempou, Biasini, Pinzon-Olejua, Harris, & Malaga-Trillo, 2016; Solis et al., 2013). However, the original studies employed morpholino knockdown of zebrafish PrP homologs, and subsequent studies using zinc-finger nucleases to achieve genomic knockout have failed to reproduce some of the phenotypes seen with morpholinos (Fleisch et al., 2013).

Finally, the ability of PrP^C to coordinate Cu^{2+} and Zn^{2+} suggests a possible functional role in metal ion homeostasis. The unstructured N-terminal domain of PrP^C contains a highly conserved octomeric peptide repeat, $(PHGGGWGQ)_4$, that binds up to four equivalents of Cu^{2+} in three distinct coordination modes (Fig. 2.3A) (Brown et al., 1997; Burns et al., 2003; Chattopadhyay et al., 2005; Hornshaw, McDermott, & Candy, 1995). Two additional stoichiometric equivalents of copper can be bound by histidine residues at positions 96 and 111. The only other physiologically relevant metal species that can coordinate to the octarepeat domain is Zn^{2+}. Although Cu^{2+} is found in the CNS at lower concentrations than Zn^{2+}, PrP^C has a significantly stronger affinity for Cu^{2+} (nM) than Zn^{2+} (μM) (Walter, Chattopadhyay, & Millhauser, 2006; Walter, Stevens, Visconte, & Millhauser, 2007).

Recent work by Glenn Millhauser's group has shed light on structural changes in PrP induced by metal ions (Evans, Pushie, Markham, Lee, & Millhauser, 2016; Spevacek et al., 2013). Their data indicate that chelation of Cu^{2+} or Zn^{2+} by the N-terminal domain of PrP^C induces an intramolecular interaction between this region and the C-terminal domain. Moreover, pathogenic mutants such as P102L, D178N, and E200K (Fig. 2.3A) show a systematic weakening of this interaction compared with wild-type PrP^C, suggesting that a loss of metal ion-induced conformational changes could underlie some genetic prion diseases.

Cu^{2+} also plays an important role in PrP^C trafficking. Exposure of cells expressing PrP^C to exogenous Cu^{2+} causes the protein to undergo clathrin-mediated endocytosis (Pauly & Harris, 1998) via interaction with the low-density lipoprotein receptor-related protein 1 (Taylor & Hooper, 2007). While a portion of the endocytosed PrP^C is degraded via lysosomal pathways, much of the protein is returned to the cell surface from early endosomes, suggesting the possibility that PrP^C could serve as an endocytic recycling receptor for cellular uptake of Cu^{2+}. However, attempts to demonstrate such a role by correlating overall Cu^{2+} uptake with PrP^C expression levels have been unsuccessful (unpublished data). Moreover, expressing mammalian PrP^C in yeast does not alter any of several well-characterized copper-related phenotypes (Li, Dong, & Harris, 2004). PrP^C may also play an important role in protecting the cell against harmful oxidative stress caused by reactive oxygen species, possibly via its ability to coordinate Cu^{2+} (Roucou, Gains, & LeBlanc, 2004; Vassallo & Herms, 2003; Watt et al., 2012). Nigel Hooper's research group has shown that PrP^C acts to modulate Zn^{2+} transport into the cell, possibly via an interaction with the α-amino-3-hydroxy-5-methyl-4-isoxazolepropionate receptor (Watt et al., 2012).

MECHANISMS OF PRP^{SC} TOXICITY: THE N-TERMINAL DOMAIN OF PRP^C POSSESS A TOXIC EFFECTOR ACTIVITY

Although a great deal is now known about the properties of prions and how they propagate, the mechanisms by which they cause neurodegeneration have remained mysterious.

Are prion diseases caused by a toxic gain of function of misfolded PrP^{Sc}, or a loss of the normal function of PrP^{C}? Arguing against a loss of function mechanism is the observation that *Prnp* knockout mice develop normally and do not display any features of prion disease (Büeler et al., 1993). In contrast, a considerable body of literature suggests that alterations in the functional activity of PrP^{C} may produce toxic effects. Importantly, there is strong evidence that PrP^{C} may act as a cell surface receptor, which, upon binding PrP^{Sc}, generates a neurotoxic signal (see Biasini, Turnbaugh, Unterberger, & Harris, 2012 for a review). Neurons which lack endogenous expression of PrP^{C} are resistant to the toxic effects of PrP^{Sc} delivered extracellularly (Brandner et al., 1996; Mallucci et al., 2003).

Important clues to PrP^{C}-mediated gain of toxicity have emerged from the study of transgenic mice that express PrP molecules harboring deletions that span the central linker region connecting the flexible N-terminal domain to the structured C-terminal domain. Surprisingly, these mice display dramatic neurodegenerative phenotypes, including neuronal loss and white matter lesions in the brain (Shmerling et al., 1998). Interestingly, PrP constructs lacking residues 32–121 or 32–134 (PrP Δ32–121 or Δ32–134) were toxic while constructs with shorter deletions were not. Additionally, PrP Δ105–125 (referred to as ΔCR) was found to be even more toxic than PrP Δ32–121 or Δ32–134 (Li et al., 2007). For each deletion construct, coexpression of wild-type PrP^{C} rescued toxicity. In the case of PrP ΔCR, a fivefold excess of PrP^{C} expression was required, suggesting that the more toxic the deletion, the more wild-type PrP^{C} is required for rescue. Interestingly, the first nine amino acids of mature PrP, which constitute a polybasic region (^{23}KKRPKPGGW^{31}), are required for the toxicity of the N-terminal deletion constructs (Westergard, Turnbaugh, & Harris, 2011). The fact that these phenotypes could be rescued by wild-type PrP^{C} suggests that the observed toxicity may be related in some way to normal PrP^{C} functionality. These experiments raise the possibility that in prion diseases alterations in the physiological activity of the N-terminal domain of PrP^{C} could lead to a toxic gain of function.

Further investigation of PrP ΔCR toxicity through patch-clamping experiments revealed spontaneous ionic currents in cells expressing PrP ΔCR (Solomon, Huettner, & Harris, 2010; Solomon et al., 2011). These currents could be completely suppressed by coexpression of wild-type PrP^{C}, paralleling rescue experiments in transgenic mice. Likewise, deleting the same nine amino acids (^{23}KKRPKPGGW^{31}) in the PrP ΔCR construct also prevented spontaneous currents. These spontaneous ionic currents may account for the neurotoxic activity of ΔCR and related PrP deletion mutants, and they reveal an otherwise hidden action of PrP^{C} driven by the N-terminal tail of the protein.

Important new insights on the mechanism by which of PrP^{C} acquires a toxic function have emerged from a recently published study from the Harris laboratory (Wu et al., 2017). In that work, it was shown that several different ligands binding to the N-terminal domain, including pentosan sulfate, Cu^{2+}, and antibodies, abolish the spontaneous ionic currents associated with ΔCR PrP (Fig. 2.4A). Amazingly, the isolated N-terminal domain-induced currents when expressed in the absence of the C-terminal domain (Fig. 2.4B). Finally, anti-PrP antibodies targeting epitopes in the C-terminal domain induced currents (Fig. 2.4C and D) and caused massive degeneration of dendrites on murine hippocampal neurons (Fig. 2.4E–H). These latter effects were entirely dependent on the presence of the N-terminal domain. Taken together, these results suggested that that the flexible, N-terminal domain of PrP^{C} functions as a powerful toxicity-transducing effector whose activity is tightly regulated in *cis* by the globular C-terminal domain. Consistent with this model, NMR experiments demonstrate intramolecular

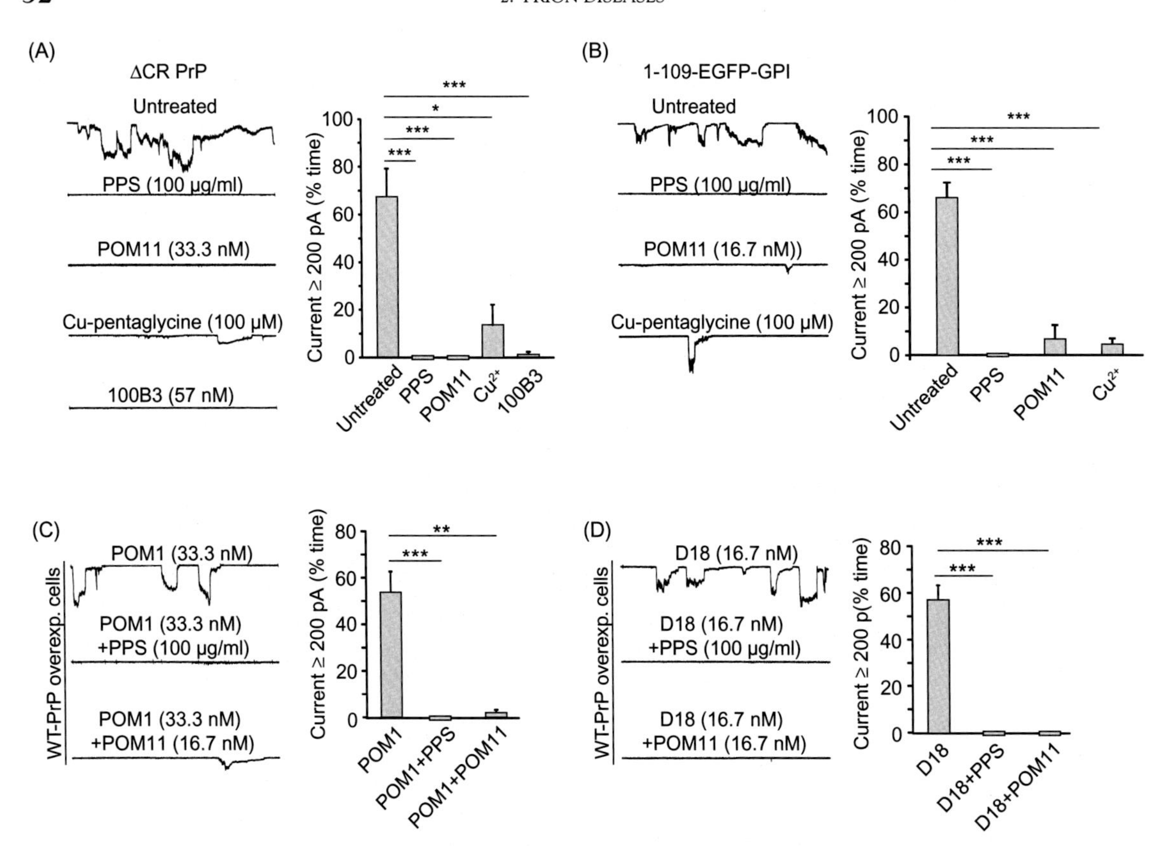

FIGURE 2.4 **The N-terminal domain of PrP^C is a toxic effector whose activity is regulated by the C-terminal domain.** Ligands binding to the N-terminal domain of PrP^C block spontaneous currents and the loss of dendritic spines induced by mutant PrP (Δ105–125, referred to as ΔCR), PrP1-109-GFP-GPI, or antiprion antibodies. (A) Left, representative traces of currents recorded from N2a cells expressing ΔCR in the absence or presence of PPS (100 μg/mL), POM11 (33.3 nM), 100B3 (57 nM), or Cu-pentaglycine (100 μM). Right, quantitation of the currents, plotted as the percentage of the total time the cells exhibited inward current ≥200 pA ($n = 10$). (B) Left, representative traces of currents recorded from N2a cells expressing PrP1-109-GFP-GPI in the absence or presence of PPS (100 μg/ml), POM11 (33.3 nM), or Cu-pentaglycine (100 μM). Right, quantitation of the currents, plotted as the percentage of the total time the cells exhibited inward current ≥200 pA ($n = 10$). (C) Left, representative traces of currents recorded from cells expressing WT PrP in the presence of POM1, POM1 + PPS, or POM1 + POM11. Antibody POM1 binds to the outer surface of helix 1 in the C-terminal domain of PrP^C, and POM11 binds to the octapeptide repeats in the N-terminal domain. Right, quantitative analysis of the currents ($n = 10$). (D) Left, representative traces of currents recorded from cells expressing WT PrP in the presence of D18, D18 + PPS, or D18 + POM11. Antibody D18 binds to the outer surface of helix 1, similar to POM1. Right, quantitative analysis of the currents ($n = 10$). (E) Top, representative images showing dendrite morphology of cultured hippocampal neurons from Tga20 mice (which overexpress WT PrP^C), WT mice, or *Prnp*$^{-/-}$ mice after treatment for 48 h with D18 (16.7 nM), POM1 (33.3 nM), or nonspecific IgG (33.3 nM). The cells were stained with an antibody to MAP2 to visualize dendrites. Boxed areas are enlarged below each image. Scale bar = 10 μm. Bottom, quantitation of dendritic degeneration, expressed as the length of beaded dendrite segments as a percentage of total dendrite length, from 10 images in 3 independent cultures for each experimental condition. (F–H) Quantitation of dendritic beading following treatment with IgG, D18 alone, N-terminal ligand (PPS, 100B3, or POM11) alone, or D18 together with the N-terminal ligand. Data represent mean ± SEM. *$P < 0.05$; **$P < 0.01$; ***$P < 0.005$. *PrP*, prion protein; *WT*, wild-type. *Adapted from Wu, B., McDonald, A.J., Markham, K., Rich, C.B., Mchugh, K.P., Tatzelt, J., et al. (2017). The N-terminus of the prion protein is a toxic effector regulated by the C-terminus. eLife, 6, e23473.*

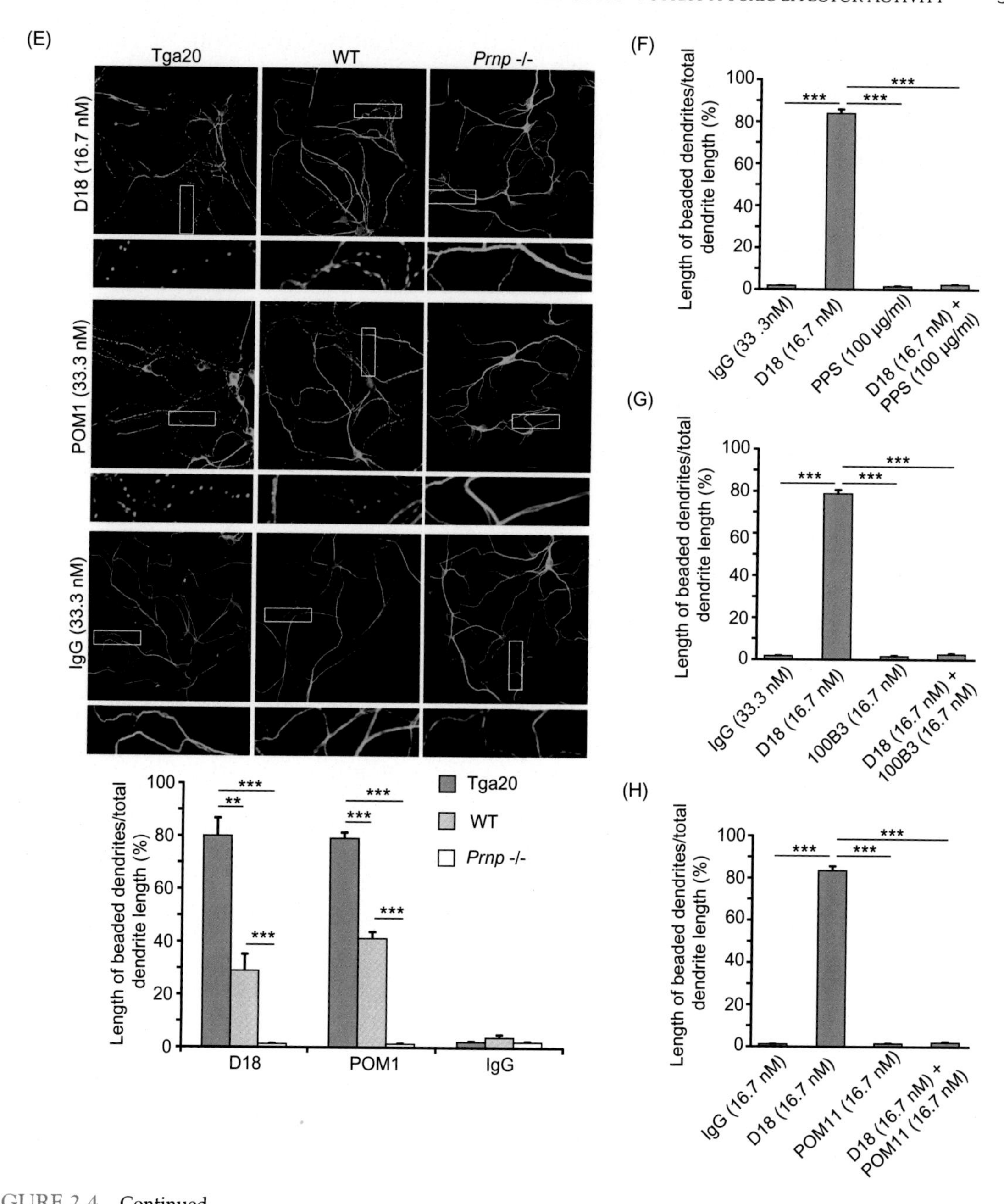

FIGURE 2.4 Continued.

docking between N- and C-terminal domains of PrP^{C}. Thus, a novel autoinhibitory mechanism regulates the functional activity of PrP^{C}, and loss of this inhibitory brake leads to variety of toxic consequences. It is possible that the neurotoxicities of PrP^{Sc} and amyloid-β (Aβ) oligomers, both of which interact with the N-terminal domain of PrP^C, act by disrupting the autoinhibitory loop.

Work from Aguzzi and colleagues also supports such a model (Sonati et al., 2013). They found that when the antibody POM1, which recognizes an epitope on helices 1 and 3 of PrP^C, was incubated with cerebellar organotypic cultured slices (COCS), there was marked death of cerebellar granule cells. Additionally, when POM1 was injected into the cerebella of mice overexpressing PrP^C, brain lesions were observed within 72 hours. However, when COCS expressing PrP deleted for residues 32–93 were incubated with POM1, no loss of cerebellar granule cells was observed, implicating the N-terminal domain of PrP as being essential for toxicity. POM1 toxicity was also suppressed by coincubation with the N-terminal antibody POM2. Based on these results, they concluded that the N-terminal domain of PrP^C acts as a toxic effector.

It remains to be determined how the N-terminus of PrP^C exerts a toxic activity when freed of C-terminal regulation. The observation that ΔCR PrP induces spontaneous ionic currents, as does binding of C-terminal antibodies, suggests that the N-terminus of PrP^C could interact directly with the lipid bilayer to induce transient pores through which ions could permeate. In fact, the N-terminal polybasic region of the protein (^{23}KKRPKPGGW^{31}) is chemically similar to protein transduction domains that are known to cause membrane disruption (Wadia, Schaller, Williamson, & Dowdy, 2008). Alternatively, the N-terminal domain might interact with other membrane proteins to cause toxic effects.

Whether the PrP^{Sc}–PrP^C interaction elicits toxic signaling via dysregulation of the N-terminal domain is the subject of ongoing investigation. To dissect the cellular mechanisms underlying PrP^{Sc} neurotoxicity, the Harris laboratory has studied the effect of PrP^{Sc} on cultures of primary hippocampal neurons (Fang, Imberdis, Garza, Wille, & Harris, 2016). Treatment of cultures with PrP^{Sc}, either in the form of brain homogenate or purified protein, causes a rapid (within 24 hours) and dramatic retraction of dendritic spines. Dendritic spines are the sites of synaptic contact on excitatory neurons, and changes in spine morphology and function are thought to underlie learning and memory. Other than retraction of their spines, the neurons remain morphologically normal, suggesting that this system is revealing one of the earliest deleterious actions of infectious prions, occurring well before neuronal death ensues. Importantly, neuronal cultures derived from *Prnp* knockout mice do not exhibit dendritic spine retraction, suggesting that PrP^C on the target neurons is likely interacting with added PrP^{Sc} to produce a toxic signal (Fig. 2.5A–L). Remarkably, neuronal cultures derived from mice expressing PrP missing residues 23–31 (the same nine amino acids that are necessary for other PrP^C-mediated toxic activities mentioned above) are completely resistant to PrP^{Sc}-induced dendritic retraction (Fig. 2.5M–R). These findings strongly implicate the N-terminal domain of PrP^C in the toxic effects of PrP^{Sc}. Ongoing work using this neuronal culture system is aimed at dissecting the signal transduction mechanisms that lead from PrP^{Sc}–PrP^C binding on the cell surface to the cytoskeletal changes that underlie dendritic spine retraction.

HUMAN PRION DISEASES

Human prion diseases exist in sporadic, genetic, or infectious forms, whose origins can be rationalized in terms of the molecular theory of prion propagation (Fig. 2.2). The most common prion disease of humans, sporadic CJD (sCJD), has a worldwide occurrence of one case per million people per year (Masters et al., 1979). sCJD is characterized by a rapid progressive dementia and myoclonus. Once symptoms appear, patients typically succumb within 1 year. Neuropathologically, sCJD is characterized by abundant spongiform change in the brain, with minimal accumulation of PrP-containing amyloid plaques.

There are several inherited prion diseases of humans. Genetic CJD (gCJD) is caused by any of a number of mutations in the *PRNP* gene (Fig. 2.3A), with E200K being the most common. Interestingly, the identity of the polymorphic residue at position 129 plays an important role in determining the phenotype of some genetic prion diseases. In particular, individuals with a valine in *cis* to the D178N mutation present with gCJD while those with a methionine in *cis* to this mutation present with fatal familial insomnia (FFI). FFI is characterized by a progressive and profound inability to sleep, leading to hallucinations and eventual death (Goldfarb et al., 1992; Lugaresi et al., 1986; Medori et al., 1992). Fatal insomnia has also been found to occur sporadically, with very similar clinical findings (Blase et al., 2014; Mastrianni et al., 1999; Parchi et al., 1999). The third genetic form of prion disease, Gerstmann–Sträussler–Scheinker disease (GSS), presents with a primarily ataxic phenotype and is typically characterized by the accumulation of abundant amyloid plaques composed of internal fragments of PrP. GSS is associated with a number of PRNP mutations, with P102L, A117V, and F198S being the most common (Doh-ura, Tateishi, Sasaki, Kitamoto, & Sakaki, 1989; Hsiao et al., 1989, 1992). Octarepeat expansions have also been shown to cause GSS (Laplanche et al., 1999) and gCJD (Owen et al., 1989).

Perhaps the best known example of a human prion disease acquired via the infectious route is kuru. Kuru, a prion disease endemic to the Fore linguistic group of the northern highlands of Papua New Guinea, was caused by the consumption of brain tissue during ritualistic mortuary feasts (Collinge et al., 2006; Gajdusek & Zigas, 1957). Recently, a novel *PRNP* mutation (G127V), which prevens infection with kuru, was observed in the Fore (Asante et al., 2015; Mead et al., 2009). Astonishingly, transgenic mice expressing this sequence were completely resistant to sCJD, iatrogenic CJD (iCJD), and variant CJD (vCJD) (Asante et al., 2015).

iCJD, another example of the infectious origin of a prion disease, is acquired through exposure to improperly sterilized surgical tools or contaminated dura mater grafts. Children who received human growth hormone purified from the pituitaries of cadavers infected with sCJD have also developed iCJD (Brown et al., 2000). The use of cadaverically derived growth hormone was discontinued in 1985 in favor of recombinant hormone.

vCJD is caused by the transmission of bovine spongiform encephalopathy (BSE), colloquially known as "mad cow disease," to humans through the consumption of contaminated beef. It has been determined that vCJD can be transmitted through blood transfusion, prompting strict policies regarding donor exclusion for blood products (Bruce et al., 1997; Collinge, Sidle, Meads, Ironside, & Hill, 1996; Llewelyn et al., 2004; Peden, Head, Ritchie, Bell, & Ironside, 2004; Scott et al., 1999; Wroe et al., 2006). The polymorphism at position 129 also plays a role in the manifestation of vCJD (Wadsworth et al., 2004). Until recently, only individuals homozygous for methionine at this position had presented with the disease, but an individual with heterozygosity at position 129 has now been reported (Mok et al., 2017). An analysis of surgically removed appendices from the UK population revealed that 1 in 2000 samples tested positive for prions, and that a higher proportion belonged to carriers of the valine polymorphism than was to be expected from its prevalence in the normal population. This observation suggests that a larger proportion of exposed individuals may have been infected than previously appreciated (Gill et al., 2013).

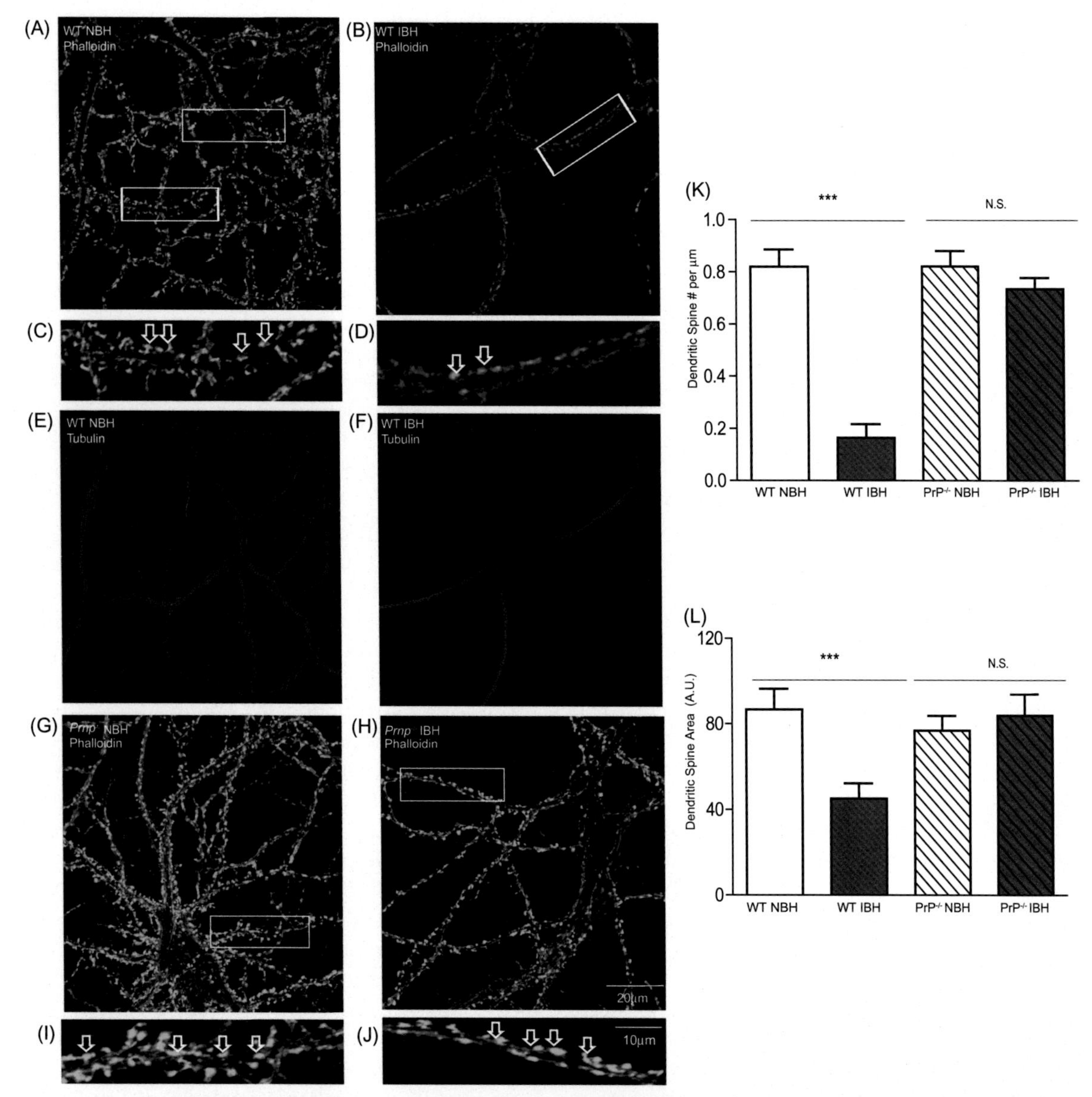

FIGURE 2.5 **PrP^{Sc} causes PrP^C-dependent retraction of dendritic spines, and the N-terminal domain of PrP^C is essential for this effect.** Primary hippocampal neurons from WT mice (A–F) or *Prnp* knockout (*$Prnp^{0/0}$*) mice (G–J) were treated for 24 h with brain homogenate [0.16% (w/v) final concentration] prepared from either normal mice normal brain homogenate (NBH) (A, C, E, G, I) or from terminally ill, scrapie-infected mice infected brain homogenate (IBH) (B, D, F, H, J). Neurons were then fixed and stained with Alexa 488-phalloidin (green) (A-D, G-J) to visualize F-actin, which is enriched in dendritic spines; and with antitubulin (red) (E and F) to visualize overall dendritic morphology. The boxed regions in panels (A), (B), (G), and (H) are shown at higher magnification in panels (C), (D), (I), and (J), respectively. Arrows in panels (C), (I), and (J) point to dendritic spines, and arrows in panel (D) indicate the positions of spines that have retracted. Pooled measurements of spine number (K) and area (L) were collected from 16 neurons from 4 independent experiments for each genotype and each treatment. Spine number is expressed per μm length of dendrite, and spine

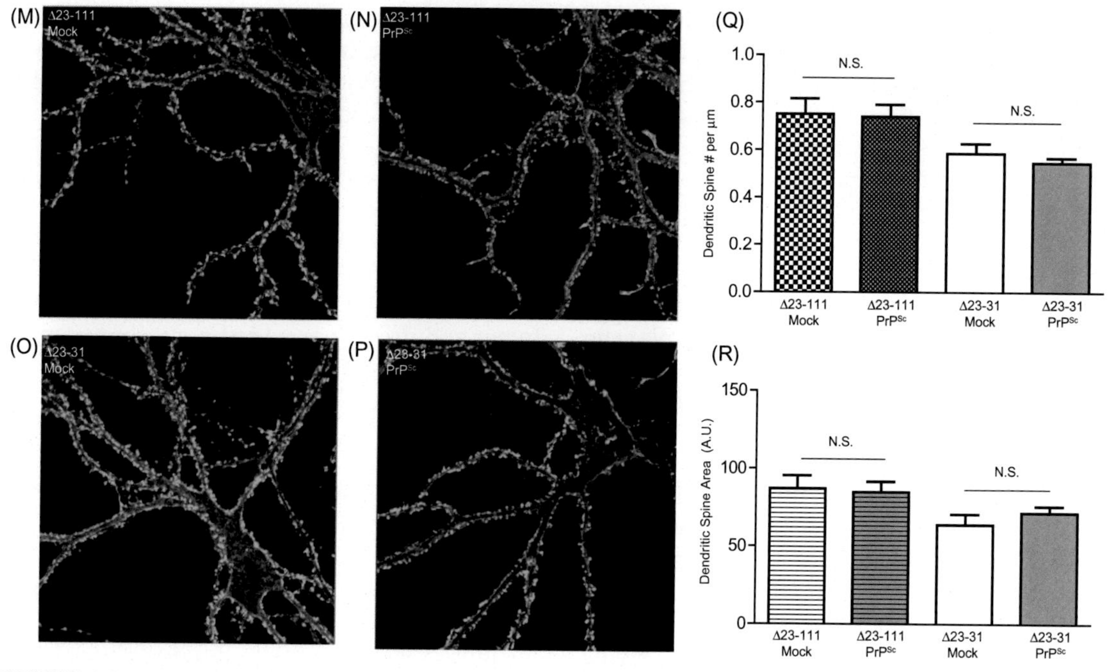

FIGURE 2.5 Continued.

area as the average area of an individual spine in AU. *** $P < 0.001$ by Student's *t*-test; *N.S.*, not significantly different. The decrease in spine area (L) reflects the fact that spines gradually shrink prior to completely disappearing; the magnitude of this effect is typically less than the reduction in the number of spines (K). Hippocampal neurons from T_g (Δ23–111) mice (M and N) and T_g(Δ23–31) mice (O and P) (both on the *Prnp*$^{0/0}$ background) were treated for 24 h with 4.4 μg/ml of purified PrP^{Sc} (N and P), or with an equivalent amount of mock-purified material from uninfected brains (M and O). Neurons were then fixed and stained with Alexa 488-phalloidin. Pooled measurements of spine number (Q) and area (R) were collected from 20 to 24 cells from four independent experiments. *N.S.*, not significantly different by Student's *t*-test. Scale bar in panel (H) = 20 μm [applicable to panels (A), (B), (E), (F), (G), (M)–(P)]; scale bar in panel J = 10 μm [applicable to panels (C), (D), and (I)]. Data represent mean ± SEM. *AU*, arbitrary units; *PrP*, prion protein; *WT*, wild-type. *Adapted from Fang, C., Imberdis, T., Garza, M.C., Wille, H., & Harris, D.A. (2016). A neuronal culture system to detect prion synaptotoxicity. PLoS Pathogens, 12(5), e1005623.*

ANIMAL PRION DISEASES

As mentioned previously, scrapie in sheep and goats was first recognized almost 250 years ago (McGowan, 1922). The name is derived from the tendency of affected animals to rub their fleece against rocks, trees, or fences. The mode of transmission of scrapie in the wild is still unclear, although it is known that susceptibility is strongly influenced by the polymorphisms of the sheep PrP^C gene (Clouscard et al., 1995; Goldmann, Hunter, Smith, Foster, & Hope, 1994; Westaway et al., 1994).

The first case of BSE was recognized in 1986, and the disease subsequently reached epidemic proportions in the cattle population of the United Kingdom with a peak in 1991. There have been over 180,000 cases of BSE in the United Kingdom, and the disease has been found in other European countries as well as Japan and

North America (Donnelly, 2000; Dudas et al., 2010; Giles, 2001). BSE may have originated as a sporadic disease but was then was spread among the cattle population because of the use of improperly processed meat and bone meal as a feed supplement (Wilesmith et al., 1988).

Both sheep and cattle are also affected by atypical forms of prion disease that are differentiated from "classical" forms by their distinct clinical, pathological, and biochemical profiles (Benestad et al., 2003; Biacabe, Laplanche, Ryder, & Baron, 2004; Casalone et al., 2004). These are thought to represent distinct strains of the infectious agent.

Chronic wasting disease (CWD) is found in free-ranging as well as captive cervid populations. CWD primarily affects deer (both mule and white-tail) and elk (Williams, 2005). Experimental transmission of CWD to other cervids, namely, moose and caribou, has been accomplished (Kreeger, Montgomery, Jewell, Schultz, & Williams, 2006; Mitchell et al., 2012). Until recently, CWD had only been detected in North America and Korea, but recent reports from Norway of infected reindeer and moose has raised the specter that it is also present in Europe (Benestad, Mitchell, Simmons, Ytrehus, & Vikøren, 2016; Herbst, Velásquez, Triscott, Aiken, & McKenzie, 2017). Because CWD is capable of horizontal transmission, this situation will have to be monitored very closely. There is currently no published experimental or epidemiological evidence to suggest that CWD is transmissible to humans, although squirrel monkeys appear to be susceptible (Belay, 2004; Kong et al., 2005; Marsh, Kincaid, Bessen, & Bartz, 2005; Sandberg et al., 2010).

Transmissible mink encephalopathy (TME), first described in the mid-1960s, affects farm-raised mink, although an outbreak has not been documented for more than 30 years (Hartsough & Burger, 1965; Marsh, Bessen, Lehmann, & Hartsough, 1991; Marsh, Burger, Eckroade, Rhein, & Hanson, 1969). Two well-studied strains have been isolated after adaptation of TME to Syrian golden hamsters: hyper, characterized by hyper excitability and a short incubation period; and drowsy, which has a long incubation period with progressive lethargy. Proteinase K treatment of these agents revealed a difference in the molecular weights of the resistant fragments and biophysical studies of PrP demonstrated distinct proportions of β-sheet secondary structure in the two strains. These comparative analyses of hyper and drowsy provided early clues toward understanding the molecular basis of prion strain diversity (Bessen & Marsh, 1992, 1994).

PRION STRAINS AND SPECIES BARRIERS

By definition, prion strains cause distinctive disease phenotypes in a common host, such as an inbred strain of mice. Besides clinical outcomes, distribution of PrP^{Sc} in the brain, profile of spongiosis, and incubation period are typical parameters used to define a strain. In bacteria and viruses, strains result from nucleotide sequence variations; since prions lack nucleic acid genomes, it is hypothesized that the propagation of distinct conformations of PrP^{Sc} from identical PrP^{C} precursor molecules is the underlying mechanism of prion strain diversity (Bartz, 2016). These variant conformations are reflected in biochemical differences in the corresponding PrP^{Sc} molecules, such as the size or glycosylation profiles of proteinase K resistant fragments (Bessen & Marsh, 1994; Parchi et al., 1996). A conformation-dependent immunoassay, which quantifies the binding of antibodies to epitopes that are differentially accessible on native PrP^{Sc}, has been used to probe the structural heterogeneity of strains (Safar et al., 1998). Strain discrimination can also be achieved through analysis of the spectral shifts induced by the binding of luminescent conjugated polymers to various tertiary structures of PrP^{Sc} (Sigurdson et al., 2007). The existence of prion strains clearly illustrates how biological information can be enciphered in different protein conformations.

Species barrier effects of prion infection are closely related to the concept of strains. The barrier to interspecies transmission of prions is due to the inefficient conversion of PrP^C by PrP^{Sc} with an unmatched primary sequence. The result is that not all infected hosts succumb to disease and may only do so after protracted incubation times. This barrier can sometimes be overcome by successive passage into the new host by a process known as adaptation (Collinge & Clarke, 2007).

METHODS FOR PROPAGATION AND DETECTION OF PRIONS

Besides bioassay in living animals, which remains the gold standard for detection, prions can be propagated in cell culture or in cell-free systems. The most widely utilized cell culture system is the murine neuroblastoma cell line, N2a, which can maintain chronic prion infection without overt signs of toxicity (Butler et al., 1988). These "ScN2a" cells have been used extensively to understand the cell biology of PrP^{Sc} and in the search for antiprion compounds.

In recent years, new cell-free technologies for propagation of prions have emerged, which are sensitive enough to be used for antemortem diagnostic purposes. One of these methods, protein misfolding cyclic amplification (PMCA), is a protein-based analog of PCR; it utilizes repeated steps of sonication followed by periods of incubation to propagate PrP^{Sc} in vitro. The levels of PrP^{Sc} in the reaction are monitored through proteinase K digestion of aliquots from every cycle (Saborio, Permanne, & Soto, 2001). Real-time quaking-induced conversion, unlike PMCA, does not produce infectious material but is as sensitive as standard animal bioassays (Wilham et al., 2010). In this method, amplification of misfolded PrP is monitored in real time through thioflavin-T binding. This method can detect prions from accessible human samples such as blood, CSF, and nasal brushings, making it the most likely candidate to be used as a clinical diagnostic (Orru et al., 2011, 2014, 2015).

THERAPEUTIC APPROACHES

The hunt for therapeutics for prion diseases has been fraught with failure. Although some antiprion compounds prolong scrapie incubation times in animal models, none of them completely prevents the disease, and their efficacy is usually diminished when administered in the symptomatic phase. There have been very few clinical trials of existing antiprion agents, and those that have been undertaken have proven disappointing. An exhaustive summary of prion therapeutics is beyond the scope of this chapter, but a brief discussion is nonetheless warranted. For comprehensive reviews, see Giles, Olson, and Prusiner (2017) and Sim (2012).

It is known from studies of *Prnp* knockout mice that PrP^C is required for the propagation and toxicity of PrP^{Sc} (Büeler et al., 1993). Thus, one obvious therapeutic strategy is the elimination or suppression of endogenous PrP^C expression. The use of siRNA targeting *Prnp* mRNA clears infected cells of prions (Daude, Marella, & Chabry, 2003). Conditional ablation of the *Prnp* gene in neurons using cre/lox technology prevents neuronal death and reverses spongiosis following prion inoculation in vivo, despite the continued accumulation of PrP^{Sc} by nonneuronal cells (Mallucci et al., 2003). Reduction of PrP^C levels using lentivirally encoded shRNAs introduced directly into the brain has been shown to prolong the incubation period of scrapie-infected mice (White et al., 2008). Taken together, these studies represent proofs of the principle that therapies aiming to reduce neuronal PrP^C levels have merit (Daude et al., 2003; Mallucci et al., 2003).

However, these approaches pose obvious challenges of agent delivery to the brain and would only be applicable to end-stage patients.

Another major therapeutic approach is the use of small organic molecules or other synthetic compounds. Polyanions such as pentosan polysulfate were among the first molecules used to treat prion disease (Ehlers & Diringer, 1984). While the effects in vivo were modest, these molecules (and others such as Congo red) were shown to have prion-clearing effects in cell culture and were used to gain insight into the mechanisms of prion accumulation (Caughey & Race, 1992). Quinacrine, which had long been used in humans due to its antimalarial properties, was also found to have prion clearing activity in cells (Korth, May, Cohen, & Prusiner, 2001). Due to its known safety in humans, clinical trials were quickly initiated, but no significant improvements were observed (Collinge et al., 2009; Geschwind et al., 2013; Haik et al., 2004). Doxycycline was another molecule preapproved for human use that showed promise as a prion therapeutic (De Luigi et al., 2008; Forloni et al., 2002; Tagliavini et al., 2000). Unfortunately, a trial using CJD patients was unsuccessful (Haik et al., 2014).

Several new lead compounds, discovered by high-throughput screening methods, are capable of significantly prolonging the incubation period of intracerebrally inoculated animals. A phenylhydrazine derivative, compound B (compB), is one such molecule (Kawasaki et al., 2007). Unfortunately, compB is only effective against rodent adapted strains, and one of its potential metabolites is a carcinogen (Lu et al., 2013). The restricted specificity against rodent-adapted strains is also true of the 2-aminothiazole compounds, IND24 and IND81 (Berry et al., 2013; Silber et al., 2013). Anle138b is another promising compound which, unlike compB, IND24, and IND81, is active against human prion strains, although it has been suggested that it too may produce toxic metabolites (Berry et al., 2013; Wagner et al., 2013).

In the search for clinically useful compounds, it was realized that prion strains respond to a drug-induced "selection pressure." For example, it was found that continuous treatment of chronically infected cells with quinacrine conferred an advantage upon a previously undetectable strain that emerged from within the prion population (Ghaemmaghami et al., 2009). This phenomenon has been observed for other drugs, notably swainsonine and IND24, suggesting that combination therapy may be the ultimate answer in the search for efficacious prion therapeutics (Berry et al., 2013; Li, Browning, Mahal, Oelschlegel, & Weissmann, 2010).

Recently, the Harris laboratory has discovered a novel class of antiprion compounds, phenethyl piperidines, using a cellular assay for PrP^{C} functional activity. In addition to clearing cells of infection, these compounds are capable of inhibiting PrP^{Sc}-induced synatotoxicity in primary cultures of hippocampal neurons (Fang et al., 2016; Imberdis et al., 2016). Current efforts are aimed at identifying the molecular target of these compounds.

Some antiprion compounds bind to PrP^{C}, and may act by sterically occluding critical interactions, or by stabilizing the structure of the folded, C-terminal domain. Other compounds appear to act by binding to PrP^{Sc}, destabilizing or otherwise altering its structure, and possibly enhancing its clearance. Importantly, there is evidence that many antiprion compounds discovered in high-throughput screens of ScN2a cells do not interact with either PrP^{C} or PrP^{Sc}, and presumably target non-PrP molecules (Poncet-Montange et al., 2011). Identifying these non-PrP targets may provide further insight into the cellular mechanisms of PrP^{Sc} formation and may motivate additional therapeutic approaches.

One exciting non-PrP molecular target is the translation initiation factor eIF2α. ER stress leads to the accumulation of misfolded proteins and triggers the unfolded protein response (UPR). The UPR results in the upregulation of multiple chaperones and, if the response is overwhelmed, cell death by apoptosis. PrP^{Sc} can induce ER stress in vivo, and UPR-regulated chaperones are more abundant in the brains of CJD patients (Hetz, Russelakis-Carneiro, Maundrell, Castilla, & Soto, 2003). It has been demonstrated recently that the UPR is activated during prion disease via activation of PERK, resulting in translational repression of protein synthesis through the phosphorylation of eIF2α (Moreno et al., 2012). Modulation of the phosphorylation level of eIF2α was later found to correlate with disease progression (Moreno et al., 2013). Encouragingly, small molecules that inhibit eIF2α-P-mediated translational attenuation were found to block prion-induced neurodegeneration (Halliday et al., 2015). Targeting downstream neurotoxic pathways may represent a promising therapeutic strategy that could synergize with interventions that reduce levels of PrP^{Sc}.

PRP^C AND THE ALZHEIMER'S Aβ PEPTIDE

A fascinating and unexpected discovery has been the role of PrP^C as a receptor for the Aβ peptide in AD. Aβ is a 38–42 amino acid peptide that is produced by the sequential cleavage of the transmembrane amyloid precursor protein (APP) by β and γ secretases (Haass & Selkoe, 2007). Aβ accumulates in the form of amyloid plaques in the brains of AD patients, and these deposits are considered pathognomonic of the disease. According to the amyloid cascade hypothesis, Aβ is the primary instigator of AD pathology, and its accumulation leads to other downstream effects including the appearance of tau-containing neurofibrillary tangles, synaptic loss, and clinical symptoms (Hardy & Higgins, 1992). A great deal of evidence now supports the idea that soluble, oligomeric forms of Aβ, rather than large amyloid fibrils, are the primary neurotoxic culprit in AD, and that these forms act at the level of synapses to disrupt key aspects of neurotransmission (Lacor et al., 2007; Shankar et al., 2008; Walsh & Selkoe, 2007).

It has long been assumed that Aβ oligomers require cell-surface receptors to transmit toxic signals to neurons, and a number of candidate Aβ receptors have been reported (Xia, Cheng, Yi, Gao, & Xiong, 2016). Unexpectedly, a search using expression cloning to identify cellular receptors for Aβ revealed that PrP^C is a prominent cell-surface binding partner for Aβ oligomers (Laurén, Gimbel, Nygaard, Gilbert, & Strittmatter, 2009). Aβ oligomers were shown to bind to postsynaptic sites in wild-type mouse hippocampal neuronal cultures, and binding was significantly diminished in cultures derived from *Prnp* knockout mice. Suppression of long-term potentiation, a functional correlate of Aβ synaptotoxicity, was also observed to be dependent on PrP^C expression. In follow-up studies, APPswe/PSen1ΔE9 transgenic mice, which model certain aspects of AD, were found to display significantly improved learning and memory when crossed onto the *Prnp* knockout background (Gimbel et al., 2010). This result was independent of Aβ levels in the brain. Other studies have shown that PrP^C is required for loss of dendritic spines (Um et al., 2012) and increased long-term depression in response to Aβ oligomers (Hu et al., 2014). Overall, strong evidence indicates that PrP^C plays an important role in the early synaptotoxicity associated with Aβ oligomers. While some studies have refuted the importance of PrP^C in Aβ synaptotoxicity (Calella et al., 2010; Cisse et al., 2011; Kessels, Nguyen, Nabavi, & Malinow, 2010), these discrepancies may be attributable to the use of different experimental paradigms, or the involvement of multiple neurotoxic pathways.

Structural studies have revealed two key Aβ binding sites on PrP^C, both of which are in the unstructured N-terminal domain. In initial studies, anti-PrP antibodies 6D11 (epitope 93-109) and 8G8 (epitope 95-110) were found to significantly decrease PrP-Aβ binding, as did deleting PrP residues 23-95 (Laurén et al., 2009). ICSM35, an antibody that recognizes an epitope similar to 6D11 also blocked binding of Aβ to PrP^C (Freir et al., 2011). Surprisingly, the antibody ICSM18, which recognizes helix 1 in the C-terminal domain of PrP^C, also inhibited Aβ binding, a result that was attributed to steric hindrance (Freir et al., 2011). The identity of the Aβ binding sites on PrP^C were further refined by testing of a series of deletion mutants; two polybasic regions in the unstructured N-terminal domain of PrP^C, consisting of residues 23-27 and 95-110, were found to be necessary for full binding affinity (Chen et al., 2010; Fluharty et al., 2013) (Figure 2.3A).

The affinity of PrP^C for Aβ oligomers is higher than for monomer. Dissociation constants for oligomers have been reported in the range of 50–100 nM (based on monomer-equivalents) (Chen, Yadav, & Surewicz, 2010; Freir et al., 2011; Laurén et al., 2009). In contrast, Aβ monomers bind to PrP^C with micromolar affinity (Laurén et al., 2009). Only a few studies have examined binding interactions between PrP^C and Aβ fibrils, although Nicoll et al. showed that PrP^C interacts with Aβ protofibrils 20–200 nm in length (Nicoll et al., 2013). There is evidence that PrP^C binds to mature fibrils of Aβ (Nieznanski, Surewicz, Chen, Nieznanska, & Surewicz, 2014), and that it can inhibit the process of Aβ polymerization at substoichiometric ratios (Nieznanski, Choi, Chen, Surewicz, & Surewicz, 2012; Nieznanski et al., 2014; Younan, Sarell, Davies, Brown, & Viles, 2013). However, the interaction between PrP^C and Aβ oligomers is considered of special significance due to the neurotoxic nature of these aggregates. The high affinity of PrP^C for Aβ oligomers supports a role for this interaction in an early toxic signaling cascade leading to AD.

A recent study from the Harris laboratory (Bove-Fenderson, Urano, Straub, & Harris, 2017) has reexamined the interaction of PrP^C with Aβ, uncovering an interesting and unexpected mechanism. They demonstrate that PrP^C binds with high affinity to the ends of growing fibrils, thereby specifically inhibiting the elongation step of fibril growth. This inhibitory effect required the globular C-terminal domain of PrP^C, which had not been previously implicated in interactions with Aβ. These results suggests that PrP^C specifically recognizes structural features that are common to both Aβ oligomers and fibril ends, and that this interaction could contribute to the neurotoxic effect of Aβ aggregates.

Mechanistic studies of Aβ oligomer toxicity have implicated several downstream cellular pathways, including those involving NMDA receptors (Li et al., 2011; Shankar et al., 2007; Um et al., 2012). Um et al. (2012) studied the relationship between these proposed downstream mechanisms and PrP^C–Aβ interactions, revealing that Aβ initiates a fyn kinase-dependent change in NMDAR localization at the synapse upon binding to PrP^C. As PrP^C is a GPI-anchored cell-surface protein, the mechanism of cytoplasmic fyn kinase activation presumably required a transmembrane linker protein; the metabotropic glutamate receptor, mGluR5, was found to be one such linker (Um et al., 2013).

An important outcome of understanding the interactions between PrP^C and Aβ is the potential for new therapeutic targets for AD. The flexible, N-terminal domain of PrP^C does not present well-defined binding sites for small molecules, although antibodies against this region have been have been shown to block synaptotoxicity by preventing interaction with Aβ (Freir et al., 2011). The structured, C-terminal domain may represent a more druggable molecular target. Antibodies against this domain have been shown to block the synaptotoxic effects of soluble Aβ in rats (Klyubin et al., 2014), and the recent results of Bove-Fenderson et al (2017) demonstrate the importance of this domain in PrP^C interactions with Aβ fibril ends. Recently, Chicago Sky blue, a Congo red-like organic molecule, was reported to block the Aβ–PrP^C interaction, although its precise binding site on PrP^C was not determined (Risse et al., 2015). Exogenously administered recombinant forms of PrP^C, in particular the isolated N-terminal domain, have been shown to protect against Aβ oligomer toxicity, possibly by acting as a "sink" to bind Aβ oligomers in the extracellular fluid (Balducci et al., 2010; Fluharty et al., 2013; Nieznanski et al., 2012). Interestingly, this fragment is produced endogenously by normal proteolytic processing of PrP^C (see above). Inhibitors of signaling processes downstream of Aβ–PrP^C binding, such as those involving mGluR5 or fyn, may also have therapeutic potential (Hamilton et al., 2016; Kaufman et al., 2015).

PRION-LIKE PROPAGATION OF MISFOLDED PROTEINS IN OTHER NEURODEGENERATIVE DISEASES

There has been a great deal of interest over the past 10 years in the idea that misfolded proteins in other neurodegenerative diseases may propagate through the CNS by a prion-like, self-templating process. This concept originated in the observation that grafts of fetal dopaminergic neurons into Parkinson's patients were sometimes found to contain accumulations of aggregated α-synuclein (Kordower, Chu, Hauser, Freeman, & Olanow, 2008; Li et al., 2008). Subsequent studies have shown that aggregated forms of α-synuclein, Aβ, tau, Htt, TDP-43, and SOD-1 can be propagated between cells in culture (reviewed in Stopschinski & Diamond, 2017; Walker & Jucker, 2015; see other chapters in this volume). In some cases, inoculation of mice with fibrillar or aggregated forms of these proteins has been shown to initiate misfolding of the corresponding endogenous proteins and propagation of the aggregates through the CNS, in a subset of these cases accompanied by neuropathological changes and clinical symptoms (Clavaguera et al., 2013; Luk et al., 2012; Meyer-Luehmann et al., 2006). In addition, misfolded forms of Aβ, tau and α-synuclein have been shown to form distinct "strains," which, like authentic prion strains, are associated with characteristic clinical or pathological phenotypes (Walker, 2016). Taken together, these observations have led to the proposal that perhaps all neurological diseases caused by protein misfolding should be considered prion diseases. Although there is very little evidence that AD, PD and other common neurodegenerative diseases are normally transmissible between individuals, it has been reported that some patients who were administered cadaverically derived growth hormone and who subsequently developed iCJD also showed accumulation of Aβ in their brains, possibly due to seeding by traces of Aβ in these preparations (Jaunmuktane et al., 2015; Ritchie et al., 2017).

In one sense, it should not be a surprise that misfolded proteins underlying other neurodegenerative disorders display prion-like spread, since all of these proteins are capable of undergoing a seeded polymerization reaction in vitro. These proteins are natively unstructured but can undergo spontaneous conversion to β-rich, amyloid forms, which can then seed further conversion of the unstructured molecules in a self-templating reaction (Eisenberg & Jucker, 2012). Particularly if large amounts of fibrillar protein are directly inoculated into the brain, it is not unexpected that the same seeding process occurring in vitro would also occur in brain tissue. Moreover, this process might easily produce diverse, prion-like strains, since it is well known that the structures of amyloid polymorphs are determined by the conformation of the seeds that initiate their polymerization (Eisenberg & Sawaya, 2017).

Although it is conceptually attractive to apply the unifying concept of prion diseases to other neurodegenerative conditions, a number of caveats must be kept in mind. First, most of the evidence for the prion-like features of these disorders is derived from cell culture models and from the artificial experimental situation in which preformed amyloid fibrils are injected into the brains of mice. What is lacking is compelling evidence that a prion-like propagation process actually occurs during the natural course of these diseases in animals and humans. Admittedly, this kind of evidence is challenging to obtain. One approach has been to employ cell type-specific promoters to express mutant forms of tau or other proteins in particular areas of the brain and then monitor the spread of misfolded protein to other areas (de Calignon et al., 2012; Liu et al., 2012). Another tactic is to use antibodies or other inhibitors to block the spread of misfolded protein during the disease process in experimental animals (Yanamandra et al., 2013). A third strategy is to image the spread of aggregates during disease progression using ligands that can be detected in vivo (Maruyama et al., 2013).

It has been pointed out that pathological changes in AD, PD, and tauopathies often start in a defined locus within the brain, and then spread to adjacent or anatomically connected areas (Jucker & Walker, 2013). Indeed, this observation is often cited to support the prion-like nature of these disorders. However, it must be kept in mind that observations of postmortem pathology are descriptive in nature and do not provide an underlying mechanism. Further work is needed to perturb experimentally the spread of protein aggregates that originate endogenously in the brain and to demonstrate that this dissemination involves a prion conversion mechanism. It will also be important to measure directly levels of misfolded protein seeds early in the course of the disease and correlate these with the eventual development of pathology (Kaufman, Thomas, Del Tredici, Braak, & Diamond, 2017).

It should be recognized that, even for authentic prion diseases, the mechanisms that account for development of specific patterns of neuropathology remain poorly understood. For example, it is uncertain how prions propagate along axonal tracts. It is unclear if this occurs by a "domino" effect in which PrP^{Sc} on the axonal membrane converts adjacent molecules of PrP^{C} (Aguzzi, 2003), or by an intracellular transport process in which vesicles containing PrP^{Sc} move along the axon by a motor-driven process (Encalada, Szpankowski, Xia, & Goldstein, 2011). How PrP^{Sc} traverses synapses is also unknown. Moreover, it is clear that factors other than spread along neuroanatomical connections must play a role in the development of prion diseases, since different prion strains inoculated in the same brain region produce distinct patterns of neuropathology. Moreover, different genetic prion diseases display unique pathological profiles, even though the mutant PrP is expressed in neurons throughout the brain. Additional factors that might dictate the evolution of prion pathology include selective targeting, uptake or propagation of PrP^{Sc}, or selective neuronal vulnerability to the toxic effects of PrP^{Sc}.

Another area that requires further work is the cell biology underlying the spread of protein aggregates. PrP^{Sc} is cell-surface protein, so it is easy to imagine how it can be released from a neuron and catalyze conversion of PrP^{C} on adjacent neurons. Indeed, when transgenic mice expressing a nonmembrane-anchored version of PrP^{C} are inoculated with prions, PrP^{Sc} is seen to accumulate widely in dense plaques around blood vessels, and even in peripheral tissues (Chesebro et al., 2005). Since Aβ is released extracellularly from transmembrane APP, it may spread in a similar fashion. In contrast, α-synuclein, tau, Htt, TDP-43, and SOD-1 are all cytoplasmic proteins. The mechanisms by which these proteins escape from cells and are then taken up by other cells to convert substrate molecules in their cytoplasm remain poorly understood.

Finally, the mechanisms by which misfolded proteins are neurotoxic require additional exploration. Many of the studies documenting prion-like spread of misfolded proteins in AD, PD, and other disorders do not directly address how the misfolded proteins interact with neurons and their synapses to cause dysfunction and pathology. In the prion field, it is clear that the infectious and neurotoxic properties of prions are distinct (Chiesa & Harris, 2001; Sandberg, Al-Doujaily, Sharps, Clarke, & Collinge, 2011). Indeed, there is considerable evidence for the existence of toxic forms of PrP that are noninfectious, and for infectious forms that are nontoxic (Chiesa et al., 2003; Sandberg et al., 2014). Thus, the fact that a misfolded protein can propagate itself within the CNS (i.e., be infectious) does not necessarily explain how it can cause neuronal damage. To understand fully the origin of neurodegenerative disease will require knowledge of both features of prion-like molecules.

Perhaps the most important implication of the prion hypothesis for common neurodegenerative diseases lies in the therapeutic realm. If misfolded forms of normally intracellular proteins like α-synuclein, tau, Htt, TDP-43, and SOD-1 exhibit an extracellular phase that is responsible for their spread throughout the CNS, antibodies and other therapeutic ligands would have direct access to them. Restricting the dissemination of these aggregates may be sufficient to produce significant therapeutic benefit (Yanamandra et al., 2013).

CONCLUDING REMARKS

Much progress has been made in the prion field over the past 50 years, in terms of demonstrating the protein nature of the infectious agent, and elucidating the mechanisms by which it propagates. Nevertheless, there are still many important questions that remain to be answered. One important goal will be a deeper understanding of how prions propagate in a cellular context. Although prions can be produced in vitro from purified components, this process is often inefficient, suggesting that cellular cofactors play an important role. Despite its conservation across species, the physiological function of PrP^C also remains enigmatic. Understanding this function may provide important clues to how interactions between PrP^C and PrP^{Sc} result in activation of neurotoxic signaling cascades, a subject that has only recently become the subject of investigation. While the existence of prion strains is widely accepted, their structural basis is poorly understood, since a complete, atomic-level structure of PrP^{Sc} has proven elusive. Finally, therapeutics and diagnostics are critical areas for future investigation. Despite the identification of compounds that inhibit PrP^{Sc} formation in cellular systems, none of them has proven effective in humans. It will be important to identify preclinical biomarkers if we hope to monitor the progression of prion disease in response to new treatment modalities. Although there are gaps in our knowledge of classical prion diseases, they have provided an important paradigm for understanding other, more common neurodegenerative disorders due to protein misfolding, including AD, PD, and tauopathies. It is our hope that the continued application of the once heretical prion concept will provide a deeper understanding of these devastating maladies, and will spark new insights into the role of self-propagating protein aggregation in other biological phenomena.

References

Aguzzi, A. (2003). Prions and the immune system: A journey through gut, spleen, and nerves. *Advances in Immunology*, *81*, 123–171.

Alper, T., & Cramp, W. (1967). Does the agent of scrapie replicate without nucleic acid? *Nature*, *214*, 764–766.

Alper, T., Haig, D., & Clarke, M. (1966). The exceptionally small size of the scrapie agent. *Biochemical and Biophysical Research Communications*, *22*(3), 278–284.

Altmeppen, H. C., Puig, B., Dohler, F., Thurm, D. K., Falker, C., Krasemann, S., & Glatzel, M. (2012). Proteolytic processing of the prion protein in health and disease. *American Journal of Neurodegenerative Disease*, *1*(1), 15–31.

Asante, E. A., Smidak, M., Grimshaw, A., Houghton, R., Tomlinson, A., Jeelani, A., ... Collinge, J. (2015). A naturally occurring variant of the human prion protein completely prevents prion disease. *Nature*, *522*(7557), 478–481.

Balducci, C., Beeg, M., Stravalaci, M., Bastone, A., Sclip, A., Biasini, E., ... Forloni, G. (2010). Synthetic amyloid-β oligomers impair long-term memory independently of cellular prion protein. *Proceedings of the National Academy of Sciences of the United States of America*, *107*(5), 2295–2300.

Bartz, J. C. (2016). Prion strain diversity. *Cold Spring Harbor Perspectives in Medicine*, *6*(12), a024349.

Basler, K., Oesch, B., Scott, M., Westaway, D., Walchli, M., Groth, D. F., ... Weissmann, C. (1986). Scrapie and cellular PrP isoforms are encoded by the same chromosomal gene. *Cell*, *46*(3), 417–428.

Belay, E. D. (2004). Chronic wasting disease and potential transmission to humans. *Emerging Infectious Disease Journal*, *10*(6), 977–984.

Benestad, S., Sarradin, P., Thu, B., Schönheit, J., Tranulis, M., & Bratberg, B. (2003). Cases of scrapie with unusual features in Norway and designation of a new type, Nor98. *The Veterinary Record*, *153*(7), 202–208.

Benestad, S. L., Mitchell, G., Simmons, M., Ytrehus, B., & Vikøren, T. (2016). First case of chronic wasting disease in Europe in a Norwegian free-ranging reindeer. *Veterinary Research*, *47*(1), 88.

Beraldo, F. H., Arantes, C. P., Santos, T. G., Queiroz, N. G., Young, K., Rylett, R. J., ... Martins, V. R. (2010). Role of α7 nicotinic acetylcholine receptor in calcium signaling induced by prion protein interaction with stress-inducible protein 1. *Journal of Biological Chemistry*, *285* (47), 36542–36550.

Berry, D. B., Lu, D., Geva, M., Watts, J. C., Bhardwaj, S., Oehler, A., ... Giles, K. (2013). Drug resistance confounding prion therapeutics. *Proceedings of the National Academy of Sciences of the United States of America*, *110* (44), E4160–E4169.

Bessen, R. A., & Marsh, R. F. (1992). Biochemical and physical properties of the prion protein from two strains of the transmissible mink encephalopathy agent. *Journal of Virology*, *66*(4), 2096–2101.

Bessen, R. A., & Marsh, R. F. (1994). Distinct PrP properties suggest the molecular basis of strain variation in transmissible mink encephalopathy. *Journal of Virology*, *68*(12), 7859–7868.

Biacabe, A. G., Laplanche, J. L., Ryder, S., & Baron, T. (2004). Distinct molecular phenotypes in bovine prion diseases. *EMBO Reports*, *5*(1), 110–115.

Biasini, E., Turnbaugh, J. A., Unterberger, U., & Harris, D. A. (2012). Prion protein at the crossroads of physiology and disease. *Trends in Neurosciences*, *35*(2), 92–103.

Blase, J. L., Cracco, L., Schonberger, L. B., Maddox, R. A., Cohen, Y., Cali, I., & Belay, E. D. (2014). Sporadic fatal insomnia in an adolescent. *Pediatrics*, *133*(3), e766–e770.

Bolton, D. C., McKinley, M. P., & Prusiner, S. B. (1982). Identification of a protein that purifies with the scrapie prion. *Science*, *218*(4579), 1309–1311.

Bouybayoune, I., Mantovani, S., Del Gallo, F., Bertani, I., Restelli, E., Comerio, L., ... Mangieri, M. (2015). Transgenic fatal familial insomnia mice indicate prion infectivity-independent mechanisms of pathogenesis and phenotypic expression of disease. *PLoS Pathogens*, *11*(4), e1004796.

Bove-Fenderson, E., Urano, R., Straub, J. E., & Harris, D. A. (2017). Cellular prion protein targets amyloid-β fibril ends via its C-terminal domain to prevent elongation. *Journal of Biological Chemistry*, *292*(41), 16858–16871.

Brandner, S., Isenmann, S., Raeber, A., Fischer, M., Sailer, A., Kobayashi, Y., ... Aguzzi, A. (1996). Normal host prion protein necessary for scrapie-induced neurotoxicity. *Nature*, *379*, 339–343.

Bremer, J., Baumann, F., Tiberi, C., Wessig, C., Fischer, H., Schwarz, P., ... Aguzzi, A. (2010). Axonal prion protein is required for peripheral myelin maintenance. *Nature Neuroscience*, *13*, 310–318.

Brown, D. R., Qin, K. F., Herms, J. W., Madlung, A., Manson, J., Strome, R., ... Kretzschmar, H. (1997). The cellular prion protein binds copper in vivo. *Nature*, *390* (6661), 684–687.

Brown, P., Preece, M., Brandel, J. P., Sato, T., McShane, L., Zerr, I., ... Collins, S. J. (2000). Iatrogenic Creutzfeldt–Jakob disease at the millennium. *Neurology*, *55*(8), 1075–1081.

Bruce, M. E., Will, R. G., Ironside, J. W., McConnell, I., Drummond, D., Suttie, A., ... Bostock, C. J. (1997). Transmissions to mice indicate that 'new variant' CJD is caused by the BSE agent. *Nature*, *389*(6650), 498–501.

Büeler, H., Aguzzi, A., Sailer, A., Greiner, R. A., Autenried, P., Aguet, M., & Weissmann, C. (1993). Mice devoid of PrP are resistant to scrapie. *Cell*, *73*(7), 1339–1347.

Burns, C. S., Aronoff-Spencer, E., Legname, G., Prusiner, S. B., Antholine, W. E., Gerfen, G. J., ... Millhauser, G. L. (2003). Copper coordination in the full-length, recombinant prion protein. *Biochemistry*, *42*(22), 6794–6803.

Butler, D. A., Scott, M. R. D., Bockman, J. M., Borchelt, D. R., Taraboulos, A., Hsiao, K. K., ... Prusiner, S. B. (1988). Scrapie-infected murine neuroblastoma cells produce protease-resistant prion proteins. *Journal of Virology, 62*, 1558–1564.

Calella, A. M., Farinelli, M., Nuvolone, M., Mirante, O., Moos, R., Falsig, J., ... Aguzzi, A. (2010). Prion protein and Aβ-related synaptic toxicity impairment. *EMBO Molecular Medicine, 2*(8), 306–314.

Casalone, C., Zanusso, G., Acutis, P., Ferrari, S., Capucci, L., Tagliavini, F., ... Caramelli, M. (2004). Identification of a second bovine amyloidotic spongiform encephalopathy: Molecular similarities with sporadic Creutzfeldt–Jakob disease. *Proceedings of the National Academy of Sciences of the United States of America, 101*(9), 3065–3070.

Caughey, B., & Race, R. E. (1992). Potent inhibition of scrapie-associated PrP accumulation by Congo red. *Journal of Neurochemistry, 59*(2), 768–771.

Chandler, R. L. (1961). Encephalopathy in mice produced by inoculation with scrapie brain material. *The Lancet, 277*(7191), 1378–1379.

Chattopadhyay, M., Walter, E. D., Newell, D. J., Jackson, P. J., Aronoff-Spencer, E., Peisach, J., ... Millhauser, G. L. (2005). The octarepeat domain of the prion protein binds Cu(II) with three distinct coordination modes at pH 7.4. *Journal of the American Chemical Society, 127*(36), 12647–12656.

Chen, S., Yadav, S. P., & Surewicz, W. K. (2010). Interaction between human prion protein and amyloid-β (Aβ) oligomers: The role of N-terminal residues. *Journal of Biological Chemistry, 285*, 26377–26383.

Chesebro, B., Trifilo, M., Race, R., Meade-White, K., Teng, C., LaCasse, R., ... Oldstone, M. (2005). Anchorless prion protein results in infectious amyloid disease without clinical scrapie. *Science, 308*(5727), 1435–1439.

Chiesa, R., & Harris, D. A. (2001). Prion diseases: What is the neurotoxic molecule? *Neurobiology of Disease, 8*(5), 743–763.

Chiesa, R., Piccardo, P., Ghetti, B., & Harris, D. A. (1998). Neurological illness in transgenic mice expressing a prion protein with an insertional mutation. *Neuron, 21*, 1339–1351.

Chiesa, R., Piccardo, P., Quaglio, E., Drisaldi, B., Si-Hoe, S. L., Takao, M., ... Harris, D. A. (2003). Molecular distinction between pathogenic and infectious properties of the prion protein. *Journal of Virology, 77*(13), 7611–7622.

Cisse, M., Sanchez, P. E., Kim, D. H., Ho, K., Yu, G. Q., & Mucke, L. (2011). Ablation of cellular prion protein does not ameliorate abnormal neural network activity or cognitive dysfunction in the J20 line of human amyloid precursor protein transgenic mice. *Journal of Neuroscience, 31*(29), 10427–10431.

Clavaguera, F., Akatsu, H., Fraser, G., Crowther, R. A., Frank, S., Hench, J., ... Tolnay, M. (2013). Brain homogenates from human tauopathies induce tau inclusions in mouse brain. *Proceedings of the National Academy of Sciences of the United States of America, 110*(23), 9535–9540.

Clouscard, C., Beaudry, P., Elsen, J., Milan, D., Dussaucy, M., Bounneau, C., ... Laplanche, J. (1995). Different allelic effects of the codons 136 and 171 of the prion protein gene in sheep with natural scrapie. *Journal of General Virology, 76*(8), 2097–2101.

Cobb, N. J., Sonnichsen, F. D., McHaourab, H., & Surewicz, W. K. (2007). Molecular architecture of human prion protein amyloid: A parallel, in-register beta-structure. *Proceedings of the National Academy of Sciences of the United States of America, 104*(48), 18946–18951. Available from https://doi.org/10.1073/pnas.0706522104.

Colling, S. B., Collinge, J., & Jefferys, J. G. R. (1996). Hippocampal slices from prion protein null mice: Disrupted Ca^{2+}-activated K^{+} currents. *Neuroscience Letters, 209*(1), 49–52.

Colling, S. B., Khana, M., Collinge, J., & Jefferys, J. G. R. (1997). Mossy fibre reorganization in the hippocampus of prion protein null mice. *Brain Research, 755*(1), 28–35.

Collinge, J., & Clarke, A. R. (2007). A general model of prion strains and their pathogenicity. *Science, 318*(5852), 930–936.

Collinge, J., Gorham, M., Hudson, F., Kennedy, A., Keogh, G., Pal, S., ... Darbyshire, J. (2009). Safety and efficacy of quinacrine in human prion disease (PRION-1 study): A patient-preference trial. *Lancet Neurol., 8*(4), 334–344.

Collinge, J., Sidle, K. C. L., Meads, J., Ironside, J., & Hill, A. F. (1996). Molecular analysis of prion strain variation and the aetiology of 'new variant' CJD. *Nature, 383*, 685–690.

Collinge, J., Whitfield, J., McKintosh, E., Beck, J., Mead, S., Thomas, D. J., & Alpers, M. P. (2006). Kuru in the 21st century—An acquired human prion disease with very long incubation periods. *The Lancet, 367*(9528), 2068–2074.

Collinge, J., Whittington, M. A., Sidle, K. C., Smith, C. J., Palmer, M. S., Clarke, A. R., & Jefferys, J. G. (1994). Prion protein is necessary for normal synaptic function. *Nature, 370*(6487), 295–297.

Creutzfeldt, H. G. (1920). Über eine eigenartige herdförmige Erkrankung des Zentralnervensystems (vorläufige Mitteilung). *Zeitschrift für die gesamte Neurologie und Psychiatrie, 57*(1), 1–18.

Curtis, J., Errington, M., Bliss, T., Voss, K., & MacLeod, N. (2003). Age-dependent loss of PTP and LTP in the hippocampus of PrP-null mice. *Neurobiology of Disease, 13*(1), 55–62.

Daude, N., Marella, M., & Chabry, J. (2003). Specific inhibition of pathological prion protein accumulation by small interfering RNAs. *Journal of Cell Science*, *116*(Pt 13), 2775–2779.

de Calignon, A., Polydoro, M., Suarez-Calvet, M., William, C., Adamowicz, D. H., Kopeikina, K. J., ... Hyman, B. T. (2012). Propagation of tau pathology in a model of early Alzheimer's disease. *Neuron*, *73*(4), 685–697.

De Luigi, A., Colombo, L., Diomede, L., Capobianco, R., Mangieri, M., Miccolo, C., ... Salmona, M. (2008). The efficacy of tetracyclines in peripheral and intracerebral prion infection. *PLoS ONE*, *3*(3), e1888.

Diaz-Espinoza, R., & Soto, C. (2012). High-resolution structure of infectious prion protein: The final frontier. *Nat Struct Mol Biol*, *19*(4), 370–377.

Doh-ura, K., Tateishi, J., Sasaki, H., Kitamoto, T., & Sakaki, Y. (1989). Pro→Leu change at position 102 of prion protein is the most common but not the sole mutation related to Gerstmann–Sträussler syndrome. *Biochemical and Biophysical Research Communications*, *163*(2), 974–979.

Donnelly, C. A. (2000). Likely size of the French BSE epidemic. *Nature*, *408*(6814), 787–788.

Dron, M., Moudjou, M., Chapuis, J., Salamat, M. K., Bernard, J., Cronier, S., ... Laude, H. (2010). Endogenous proteolytic cleavage of disease-associated prion protein to produce C2 fragments is strongly cell- and tissue-dependent. *Journal of Biological Chemistry*, *285* (14), 10252–10264.

Dudas, S., Yang, J., Graham, C., Czub, M., McAllister, T. A., Coulthart, M. B., & Czub, S. (2010). Molecular, biochemical and genetic characteristics of BSE in Canada. *PLoS ONE*, *5*(5), e10638.

Ehlers, B., & Diringer, H. (1984). Dextran sulphate 500 delays and prevents mouse scrapie by impairment of agent replication in spleen. *Journal of General Virology*, *65*(8), 1325–1330.

Eisenberg, D., & Jucker, M. (2012). The amyloid state of proteins in human diseases. *Cell*, *148*(6), 1188–1203.

Eisenberg, D. S., & Sawaya, M. R. (2017). Structural studies of amyloid proteins at the molecular level. *Annual Review of Biochemistry*. Available from https://doi.org/10.1146/annurev-biochem-061516-045104.

Eklund, C., Kennedy, R. C., & Hadlow, W. (1967). Pathogenesis of scrapie virus infection in the mouse. *The Journal of Infectious Diseases*, *177*(1), 15–22.

Encalada, S. E., Szpankowski, L., Xia, C. H., & Goldstein, L. S. (2011). Stable kinesin and dynein assemblies drive the axonal transport of mammalian prion protein vesicles. *Cell*, *144*(4), 551–565.

Evans, E. G., Pushie, M. J., Markham, K. A., Lee, H. W., & Millhauser, G. L. (2016). Interaction between prion protein's copper-bound octarepeat domain and a charged C-terminal pocket suggests a mechanism for N-terminal regulation. *Structure*, *24*(7), 1057–1067.

Fang, C., Imberdis, T., Garza, M. C., Wille, H., & Harris, D. A. (2016). A neuronal culture system to detect prion synaptotoxicity. *PLoS Pathogens*, *12*(5), e1005623.

Fioriti, L., Myers, C., Huang, Y.-Y., Li, X., Stephan, J. S., Trifilieff, P., ... Pavlopoulos, E. (2015). The persistence of hippocampal-based memory requires protein synthesis mediated by the prion-like protein CPEB3. *Neuron*, *86*(6), 1433–1448.

Fleisch, V. C., Leighton, P. L., Wang, H., Pillay, L. M., Ritzel, R. G., Bhinder, G., ... Waskiewicz, A. J. (2013). Targeted mutation of the gene encoding prion protein in zebrafish reveals a conserved role in neuron excitability. *Neurobiology of Disease*, *55*, 11–25.

Fluharty, B. R., Biasini, E., Stravalaci, M., Sclip, A., Diomede, L., Balducci, C., ... Harris, D. A. (2013). An N-terminal fragment of the prion protein binds to amyloid-β oligomers and inhibits their neurotoxicity *in vivo*. *Journal of Biological Chemistry*, *288*, 7857–7866.

Ford, M. J., Burton, L. J., Morris, R. J., & Hall, S. M. (2002). Selective expression of prion protein in peripheral tissues of the adult mouse. *Neuroscience*, *113*(1), 177–192.

Forloni, G., Iussich, S., Awan, T., Colombo, L., Angeretti, N., Girola, L., ... Bruzzone, M. G. (2002). Tetracyclines affect prion infectivity. *Proceedings of the National Academy of Sciences of the United States of America*, *99*(16), 10849–10854.

Freir, D. B., Nicoll, A. J., Klyubin, I., Panico, S., Mc Donald, J. M., Risse, E., ... Collinge, J. (2011). Interaction between prion protein and toxic amyloid β assemblies can be therapeutically targeted at multiple sites. *Nature Communications*, *2*, 336.

Gajdusek, D. C., Gibbs, C. J., & Alpers, M. (1966). Experimental transmission of a kuru-like syndrome to chimpanzee. *Nature*, *209*, 794–796.

Gajdusek, D. C., & Zigas, V. (1957). Degenerative disease of the central nervous system in New Guinea: The endemic occurrence of "kuru" in the native population. *New England Journal of Medicine*, *257*, 974–978.

Geschwind, M. D., Kuo, A. L., Wong, K. S., Haman, A., Devereux, G., Raudabaugh, B. J., ... Garcia, P. (2013). Quinacrine treatment trial for sporadic Creutzfeldt–Jakob disease. *Neurology*, *81*(23), 2015–2023.

Ghaemmaghami, S., Ahn, M., Lessard, P., Giles, K., Legname, G., DeArmond, S. J., & Prusiner, S. B. (2009). Continuous quinacrine treatment results in the formation of drug-resistant prions. *PLoS Pathogens*, *5*(11), e1000673.

Gibbs, C. J., Gajdusek, D. C., Asher, D., Alpers, M., Beck, E., Daniel, P., & Matthews, W. B. (1968). Creutzfeldt–Jakob disease (spongiform encephalopathy): Transmission to the chimpanzee. *Science*, *161* (3839), 388–389.

Gibbs, C. J., Gajdusek, D. C., & Latarjet, R. (1978). Unusual resistance to ionizing radiation of the viruses of kuru, Creutzfeldt–Jakob disease, and scrapie. *Proceedings of the National Academy of Sciences of the United States of America*, *75*(12), 6268–6270.

Giles, J. (2001). Mad cow disease comes to Japan. *Nature*, *413*(6853), 240.

Giles, K., Olson, S. H., & Prusiner, S. B. (2017). Developing Therapeutics for PrP Prion Diseases. *Cold Spring Harbor Perspectives in Medicine*, *7*(4), a023747.

Gill, O. N., Spencer, Y., Richard-Loendt, A., Kelly, C., Dabaghian, R., Boyes, L., ... Bellerby, P. (2013). Prevalent abnormal prion protein in human appendixes after bovine spongiform encephalopathy epizootic: Large scale survey. *BMJ*, *347*, f5675.

Gimbel, D. A., Nygaard, H. B., Coffey, E. E., Gunther, E. C., Lauren, J., Gimbel, Z. A., & Strittmatter, S. M. (2010). Memory impairment in transgenic Alzheimer mice requires cellular prion protein. *Journal of Neuroscience*, *30*(18), 6367–6374.

Glatzel, M., Linsenmeier, L., Dohler, F., Krasemann, S., Puig, B., & Altmeppen, H. C. (2015). Shedding light on prion disease. *Prion*, *9*(4), 244–256.

Goldfarb, L. G., Petersen, R. B., Tabaton, M., Brown, P., LeBlanc, A. C., Montagna, P., ... Gambetti, P. (1992). Fatal familial insomnia and familial Creutzfeldt–Jakob disease: Disease phenotype determined by a DNA polymorphism. *Science*, *258*, 806–808.

Goldmann, W., Hunter, N., Smith, G., Foster, J., & Hope, J. (1994). PrP genotype and agent effects in scrapie: Change in allelic interaction with different isolates of agent in sheep, a natural host of scrapie. *Journal of General Virology*, *75*(5), 989–995.

Gossert, A. D., Bonjour, S., Lysek, D. A., Fiorito, F., & Wuthrich, K. (2005). Prion protein NMR structures of elk and of mouse/elk hybrids. *Proceedings of the National Academy of Sciences of the United States of America*, *102*(3), 646–650.

Groveman, B. R., Dolan, M. A., Taubner, L. M., Kraus, A., Wickner, R. B., & Caughey, B. (2014). Parallel in-register intermolecular beta-sheet architectures for prion-seeded prion protein (PrP) amyloids. *Journal of Biological Chemistry*, *289*(35), 24129–24142.

Haass, C., & Selkoe, D. J. (2007). Soluble protein oligomers in neurodegeneration: Lessons from the Alzheimer's amyloid beta-peptide. *Nature Reviews Molecular Cell Biology*, *8*(2), 101–112.

Hadlow, W. J. (1959). Scrapie and kuru. *The Lancet*, *2*, 289–290.

Haik, S., Brandel, J., Salomon, D., Sazdovitch, V., Delasnerie-Laupretre, N., Laplanche, J., ... Belorgey, C. (2004). Compassionate use of quinacrine in Creutzfeldt–Jakob disease fails to show significant effects. *Neurology*, *63*(12), 2413–2415.

Haik, S., Marcon, G., Mallet, A., Tettamanti, M., Welaratne, A., Giaccone, G., ... Tagliavini, F. (2014). Doxycycline in Creutzfeldt–Jakob disease: A phase 2, randomised, double-blind, placebo-controlled trial. *The Lancet Neurology*, *13*(2), 150–158.

Halliday, M., Radford, H., Sekine, Y., Moreno, J., Verity, N., Le Quesne, J., ... Fischer, P. (2015). Partial restoration of protein synthesis rates by the small molecule ISRIB prevents neurodegeneration without pancreatic toxicity. *Cell Death & Disease*, *6*(3), e1672.

Hamilton, A., Vasefi, M., Vander Tuin, C., McQuaid, R. J., Anisman, H., & Ferguson, S. S. (2016). Chronic pharmacological mGluR5 inhibition prevents cognitive impairment and reduces pathogenesis in an Alzheimer disease mouse model. *Cell Rep*, *15*(9), 1859–1865.

Haraguchi, T., Fisher, S., Olofsson, S., Endo, T., Groth, D., Tarentino, A., ... Prusiner, S. B. (1989). Asparagine-linked glycosylation of the scrapie and cellular prion proteins. *Archives of Biochemistry and Biophysics*, *274*, 1–13.

Hardy, J. A., & Higgins, G. A. (1992). Alzheimer's disease: The amyloid cascade hypothesis. *Science*, *256*(5054), 184–185.

Hartsough, G., & Burger, D. (1965). Encephalopathy of mink: I. Epizootiologic and clinical observations. *Journal of Infectious Diseases*, *115*(4), 387–392.

Herbst, A., Velásquez, C. D., Triscott, E., Aiken, J. M., & McKenzie, D. (2017). Chronic wasting disease prion strain emergence and host range expansion. *Emerging Infectious Diseases*, *23*(9), 1598–1600.

Hetz, C., Russelakis-Carneiro, M., Maundrell, K., Castilla, J., & Soto, C. (2003). Caspase-12 and endoplasmic reticulum stress mediate neurotoxicity of pathological prion protein. *The EMBO Journal*, *22*(20), 5435–5445.

Hornshaw, M. P., McDermott, J. R., & Candy, J. M. (1995). Copper binding to the N-terminal tandem repeat regions of mammalian and avian prion protein. *Biochemical and Biophysical Research Communications*, *207* (2), 621–629.

Hou, F., Sun, L., Zheng, H., Skaug, B., Jiang, Q.-X., & Chen, Z. J. (2011). MAVS forms functional prion-like aggregates to activate and propagate antiviral innate immune response. *Cell*, *146*(3), 448–461.

Hsiao, K., Baker, H., Crow, T. J., Poulter, M., Owen, F., Terwilliger, J. D., ... Prusiner, S. B. (1989). Linkage of a prion protein missense variant to Gerstmann–Sträussler syndrome. *Nature*, *338*, 342–345.

Hsiao, K., Dlouhy, S. R., Farlow, M. R., Cass, C., Da Costa, M., Conneally, P. M., ... Prusiner, S. B. (1992). Mutant prion proteins in Gerstmann–Sträussler–Scheinker disease with neurofibrillary tangles. *Nature Genetics*, *1*, 68–71.

Hsiao, K., Meiner, Z., Kahana, E., Cass, C., Kahana, I., Avrahami, D., ... Gabizon, R. (1991). Mutation of the prion protein in Libyan Jews with Creutzfeldt–Jakob

disease. *New England Journal of Medicine, 324*(16), 1091–1097.
Hsiao, K. K., Scott, M., Foster, D., Groth, D. F., DeArmond, S. J., & Prusiner, S. B. (1990). Spontaneous neurodegeneration in transgenic mice with mutant prion protein. *Science, 250*, 1587–1590.
Hu, N. W., Nicoll, A. J., Zhang, D., Mably, A. J., O'Malley, T., Purro, S. A., ... Rowan, M. J. (2014). mGlu5 receptors and cellular prion protein mediate amyloid-β-facilitated synaptic long-term depression *in vivo*. *Nature Communications, 5*, 3374.
Imberdis, T., Heeres, J. T., Yueh, H., Fang, C., Zhen, J., Rich, C. B., ... Harris, D. A. (2016). Identification of anti-prion compounds using a novel cellular assay. *Journal of Biological Chemistry, 291*(50), 26164–26176.
Jaunmuktane, Z., Mead, S., Ellis, M., Wadsworth, J. D., Nicoll, A. J., Kenny, J., ... Brandner, S. (2015). Evidence for human transmission of amyloid-beta pathology and cerebral amyloid angiopathy. *Nature, 525*(7568), 247–250.
Jucker, M., & Walker, L. C. (2013). Self-propagation of pathogenic protein aggregates in neurodegenerative diseases. *Nature, 501*(7465), 45–51.
Kaiser, D. M., Acharya, M., Leighton, P. L., Wang, H., Daude, N., Wohlgemuth, S., ... Allison, W. T. (2012). Amyloid beta precursor protein and prion protein have a conserved interaction affecting cell adhesion and CNS development. *PLoS ONE, 7*(12), e51305.
Kato, M., Han, T. W., Xie, S., Shi, K., Du, X., Wu, L. C., ... Pei, J. (2012). Cell-free formation of RNA granules: Low complexity sequence domains form dynamic fibers within hydrogels. *Cell, 149*(4), 753–767.
Kaufman, A. C., Salazar, S. V., Haas, L. T., Yang, J., Kostylev, M. A., Jeng, A. T., ... Strittmatter, S. M. (2015). Fyn inhibition rescues established memory and synapse loss in Alzheimer mice. *Annals of Neurology, 77*(6), 953–971.
Kaufman, S. K., Thomas, T. L., Del Tredici, K., Braak, H., & Diamond, M. I. (2017). Characterization of tau prion seeding activity and strains from formaldehyde-fixed tissue. *Acta Neuropathologica Communications, 5*(1), 41.
Kawasaki, Y., Kawagoe, K., Chen, C.-J., Teruya, K., Sakasegawa, Y., & Doh-ura, K. (2007). Orally administered amyloidophilic compound is effective in prolonging the incubation periods of animals cerebrally infected with prion diseases in a prion strain-dependent manner. *Journal of Virology, 81*(23), 12889–12898.
Kessels, H. W., Nguyen, L. N., Nabavi, S., & Malinow, R. (2010). The prion protein as a receptor for amyloid-β. *Nature, 466*(7308), E3–4.
Khosravani, H., Zhang, Y., Tsutsui, S., Hameed, S., Altier, C., Hamid, J., ... Zamponi, G. W. (2008). Prion protein attenuates excitotoxicity by inhibiting NMDA receptors. *Journal of Cell Biology, 181*(3), 551–565.
Klatzo, I., Gajdusek, D. C., & Zigas, V. (1959). Pathology of Kuru. *Laboratory Investigation; A Journal of Technical Methods and Pathology, 8*(4), 799.
Klyubin, I., Nicoll, A. J., Khalili-Shirazi, A., Farmer, M., Canning, S., Mably, A., ... Collinge, J. (2014). Peripheral administration of a humanized anti-PrP antibody blocks Alzheimer's disease Aβ synaptotoxicity. *Journal of Neuroscience, 34*(18), 6140–6145.
Kong, Q., Huang, S., Zou, W., Vanegas, D., Wang, M., Wu, D., ... Deng, H. (2005). Chronic wasting disease of elk: Transmissibility to humans examined by transgenic mouse models. *Journal of Neuroscience, 25*(35), 7944–7949.
Kordower, J. H., Chu, Y., Hauser, R. A., Freeman, T. B., & Olanow, C. W. (2008). Lewy body-like pathology in long-term embryonic nigral transplants in Parkinson's disease. *Nature Medicine, 14*(5), 504–506.
Korth, C., May, B. C., Cohen, F. E., & Prusiner, S. B. (2001). Acridine and phenothiazine derivatives as pharmacotherapeutics for prion disease. *Proceedings of the National Academy of Sciences of the United States of America, 98*(17), 9836–9841.
Kreeger, T. J., Montgomery, D., Jewell, J. E., Schultz, W., & Williams, E. S. (2006). Oral transmission of chronic wasting disease in captive Shira's moose. *Journal of Wildlife Diseases, 42*(3), 640–645.
Kretzschmar, H. A., Prusiner, S. B., Stowring, L. E., & DeArmond, S. J. (1986). Scrapie prion proteins are synthesized in neurons. *The American Journal of Pathology, 122*(1), 1–5.
Küffer, A., Lakkaraju, A. K., Mogha, A., Petersen, S. C., Airich, K., Doucerain, C., ... Monnard, A. (2016). The prion protein is an agonistic ligand of the G protein-coupled receptor Adgrg6. *Nature, 536*(7617), 464–468.
Lacor, P. N., Buniel, M. C., Furlow, P., Clemente, A. S., Velasco, P. T., Wood, M., ... Klein, W. L. (2007). Aβ oligomer-induced aberrations in synapse composition, shape, and density provide a molecular basis for loss of connectivity in Alzheimer's disease. *Journal of Neuroscience, 27*(4), 796–807.
Laplanche, J.-L., El Hachimi, K. H., Durieux, I., Thuillet, P., Defebvre, L., Delasnerie-Lauprêtre, N., ... Destée, A. (1999). Prominent psychiatric features and early onset in an inherited prion disease with a new insertional mutation in the prion protein gene. *Brain, 122*(12), 2375–2386.
Latarjet, R., Muel, B., Haig, D., Clarke, M., & Alper, T. (1970). Inactivation of the scrapie agent by near monochromatic ultraviolet light. *Nature, 227*(5265), 1341–1343.
Laurén, J., Gimbel, D. A., Nygaard, H. B., Gilbert, J. W., & Strittmatter, S. M. (2009). Cellular prion protein

mediates impairment of synaptic plasticity by amyloid-β oligomers. *Nature*, *457*(7233), 1128–1132.

Le Pichon, C. E., Valley, M. T., Polymenidou, M., Chesler, A. T., Sagdullaev, B. T., Aguzzi, A., & Firestein, S. (2009). Olfactory behavior and physiology are disrupted in prion protein knockout mice. *Nature Neuroscience*, *12* (1), 60–69.

Li, A., Christensen, H. M., Stewart, L. R., Roth, K. A., Chiesa, R., & Harris, D. A. (2007). Neonatal lethality in transgenic mice expressing prion protein with a deletion of residues 105–125. *EMBO Journal*, *26*, 548–558.

Li, A., Dong, J., & Harris, D. A. (2004). Cell surface expression of the prion protein in yeast does not alter copper utilization phenotypes. *Journal of Biological Chemistry*, *279*(28), 29469–29477.

Li, J., Browning, S., Mahal, S. P., Oelschlegel, A. M., & Weissmann, C. (2010). Darwinian evolution of prions in cell culture. *Science*, *327*(5967), 869–872.

Li, J. Y., Englund, E., Holton, J. L., Soulet, D., Hagell, P., Lees, A. J., … Brundin, P. (2008). Lewy bodies in grafted neurons in subjects with Parkinson's disease suggest host-to-graft disease propagation. *Nature Medicine*, *14*(5), 501–503.

Li, S., Jin, M., Koeglsperger, T., Shepardson, N. E., Shankar, G. M., & Selkoe, D. J. (2011). Soluble Abeta oligomers inhibit long-term potentiation through a mechanism involving excessive activation of extrasynaptic NR2B-containing NMDA receptors. *Journal of Neuroscience*, *31* (18), 6627–6638.

Liang, J., Wang, W., Sorensen, D., Medina, S., Ilchenko, S., Kiselar, J., … Kong, Q. (2012). Cellular prion protein regulates its own alpha-cleavage through ADAM8 in skeletal muscle. *Journal of Biological Chemistry*, *287*(20), 16510–16520.

Liu, L., Drouet, V., Wu, J. W., Witter, M. P., Small, S. A., Clelland, C., & Duff, K. (2012). Trans-synaptic spread of tau pathology *in vivo*. *PLoS ONE*, *7*(2), e31302.

Lledo, P. M., Tremblay, P., Dearmond, S. J., Prusiner, S. B., & Nicoll, R. A. (1996). Mice deficient for prion protein exhibit normal neuronal excitability and synaptic transmission in the hippocampus. *Proceedings of the National Academy of Sciences of the United States of America*, *93*(6), 2403–2407.

Llewelyn, C., Hewitt, P., Knight, R., Amar, K., Cousens, S., Mackenzie, J., & Will, R. G. (2004). Possible transmission of variant Creutzfeldt–Jakob disease by blood transfusion. *The Lancet*, *363*(9407), 417–421.

Lloyd, S. E., Mead, S., & Collinge, J. (2013). Genetics of prion diseases. *Current Opinion in Genetics and Development*, *23*(3), 345–351.

Lu, D., Giles, K., Li, Z., Rao, S., Dolghih, E., Gever, J. R., … Bryant, C. (2013). Biaryl amides and hydrazones as therapeutics for prion disease in transgenic mice. *Journal of Pharmacology and Experimental Therapeutics*, *347*(2), 325–338.

Lugaresi, E., Medori, R., Montagna, P., Baruzzi, A., Cortelli, P., Lugaresi, A., … Gambetti, P. (1986). Fatal familial insomnia and dysautonomia with selective degeneration of thalamic nuclei. *New England Journal of Medicine*, *315*(16), 997–1003.

Luk, K. C., Kehm, V., Carroll, J., Zhang, B., O'Brien, P., Trojanowski, J. Q., & Lee, V. M. (2012). Pathological α-synuclein transmission initiates Parkinson-like neurodegeneration in nontransgenic mice. *Science*, *338*(6109), 949–953.

Maglio, L. E., Perez, M. F., Martins, V. R., Brentani, R. R., & Ramirez, O. A. (2004). Hippocampal synaptic plasticity in mice devoid of cellular prion protein. *Brain Research. Molecular Brain Research*, *131*(1-2), 58–64.

Malaga-Trillo, E., Solis, G. P., Schrock, Y., Geiss, C., Luncz, L., Thomanetz, V., & Stuermer, C. A. (2009). Regulation of embryonic cell adhesion by the prion protein. *PLoS Biology*, *7*(3), e55.

Mallucci, G., Dickinson, A., Linehan, J., Klohn, P. C., Brandner, S., & Collinge, J. (2003). Depleting neuronal PrP in prion infection prevents disease and reverses spongiosis. *Science*, *302*(5646), 871–874.

Mallucci, G. R., Ratte, S., Asante, E. A., Linehan, J., Gowland, I., Jefferys, J. G., & Collinge, J. (2002). Post-natal knockout of prion protein alters hippocampal CA1 properties, but does not result in neurodegeneration. *EMBO Journal*, *21*(3), 202–210.

Manuelidis, L., Yu, Z.-X., Barquero, N., & Mullins, B. (2007). Cells infected with scrapie and Creutzfeldt–Jakob disease agents produce intracellular 25-nm virus-like particles. *Proceedings of the National Academy of Sciences of the United States of America*, *104*(6), 1965–1970.

Marsh, R., Bessen, R. A., Lehmann, S., & Hartsough, G. (1991). Epidemiological and experimental studies on a new incident of transmissible mink encephalopathy. *Journal of General Virology*, *72*(3), 589–594.

Marsh, R., Burger, D., Eckroade, R., Rhein, G. Z., & Hanson, R. (1969). A preliminary report on the experimental host range of the transmissible mink encephalopathy agent. *The Journal of Infectious Diseases*, 713–719.

Marsh, R., & Kimberlin, R. (1975). Comparison of scrapie and transmissible mink encephalopathy in hamsters. II. Clinical signs, pathology, and pathogenesis. *Journal of Infectious Diseases*, *131*(2), 104–110.

Marsh, R. F., Kincaid, A. E., Bessen, R. A., & Bartz, J. C. (2005). Interspecies transmission of chronic wasting disease prions to squirrel monkeys (*Saimiri sciureus*). *Journal of Virology*, *79*(21), 13794–13796.

Maruyama, M., Shimada, H., Suhara, T., Shinotoh, H., Ji, B., Maeda, J., … Higuchi, M. (2013). Imaging of tau pathology in a tauopathy mouse model and in Alzheimer patients compared to normal controls. *Neuron, 79*(6), 1094–1108.

Masters, C. L., Harris, J. O., Gajdusek, D. C., Gibbs, C. J., Bernoulli, C., & Asher, D. M. (1979). Creutzfeldt–Jakob disease: Patterns of worldwide occurrence and the significance of familial and sporadic clustering. *Annals of Neurology, 5*(2), 177–188.

Mastrianni, J. A., Nixon, R., Layzer, R., Telling, G. C., Han, D., DeArmond, S. J., & Prusiner, S. B. (1999). Prion protein conformation in a patient with sporadic fatal insomnia. *New England Journal of Medicine, 340*(21), 1630–1638.

McDonald, A. J., Dibble, J. P., Evans, E. G., & Millhauser, G. L. (2014). A new paradigm for enzymatic control of α-cleavage and β-cleavage of the prion protein. *Journal of Biological Chemistry, 289*(2), 803–813.

McGowan, J. (1922). Scrapie in sheep. *Scottish Journal of Agriculture, 5*, 365–375.

McLennan, N. F., Brennan, P. M., McNeill, A., Davies, I., Fotheringham, A., Rennison, K. A., … Bell, J. E. (2004). Prion protein accumulation and neuroprotection in hypoxic brain damage. *American Journal of Pathology, 165*(1), 227–235.

Mead, S., Whitfield, J., Poulter, M., Shah, P., Uphill, J., Campbell, T., … Mein, C. A. (2009). A novel protective prion protein variant that colocalizes with kuru exposure. *New England Journal of Medicine, 361*(21), 2056–2065.

Medori, R., Tritschler, H.-J., LeBlanc, A., Villare, F., Manetto, V., Chen, H. Y., … Cortelli, P. (1992). Fatal familial insomnia, a prion disease with a mutation at codon 178 of the prion protein gene. *New England Journal of Medicine, 326*(7), 444–449.

Mehrabian, M., Brethour, D., Wang, H., Xi, Z., Rogaeva, E., & Schmitt-Ulms, G. (2015). The prion protein controls polysialylation of neural cell adhesion molecule 1 during cellular morphogenesis. *PLoS ONE, 10*(8), e0133741.

Mercer, R. C., Daude, N., Dorosh, L., Fu, Z. L., Mays, C. E., Gapeshina, H., … Westaway, D. (2018). A novel Gerstmann-Sträussler-Scheinker disease mutation defines a precursor for amyloidogenic 8 kDa PrP fragments and reveals N-terminal structural changes shared by other GSS alleles. *PLoS Pathogens, 14*(1), e1006826.

Mercer, R. C., Ma, L., Watts, J. C., Strome, R., Wohlgemuth, S., Yang, J., … Jhamandas, J. H. (2013). The prion protein modulates A-type K + currents mediated by Kv4. 2 complexes through dipeptidyl aminopeptidase-like protein 6. *Journal of Biological Chemistry, 288*(52), 37241–37255.

Meyer-Luehmann, M., Coomaraswamy, J., Bolmont, T., Kaeser, S., Schaefer, C., Kilger, E., … Jucker, M. (2006). Exogenous induction of cerebral β-amyloidogenesis is governed by agent and host. *Science, 313*(5794), 1781–1784.

Mitchell, G. B., Sigurdson, C. J., O'Rourke, K. I., Algire, J., Harrington, N. P., Walther, I., … Balachandran, A. (2012). Experimental oral transmission of chronic wasting disease to reindeer (*Rangifer tarandus tarandus*). *PLoS ONE, 7*(6), e39055.

Mitsios, N., Saka, M., Krupinski, J., Pennucci, R., Sanfeliu, C., Miguel Turu, M., … Slevin, M. (2007). Cellular prion protein is increased in the plasma and peri-infarcted brain tissue after acute stroke. *Journal of Neuroscience Research, 85*(3), 602–611.

Mok, T., Jaunmuktane, Z., Joiner, S., Campbell, T., Morgan, C., Wakerley, B., … Jäger, H. R. (2017). Variant Creutzfeldt–Jakob disease in a patient with heterozygosity at PRNP codon 129. *New England Journal of Medicine, 376*(3), 292–294.

Moreno, J. A., Halliday, M., Molloy, C., Radford, H., Verity, N., Axten, J. M., … Barrett, D. A. (2013). Oral treatment targeting the unfolded protein response prevents neurodegeneration and clinical disease in prion-infected mice. *Science Translational Medicine, 5*(206), 206ra138.

Moreno, J. A., Radford, H., Peretti, D., Steinert, J. R., Verity, N., Martin, M. G., … Mallucci, G. R. (2012). Sustained translational repression by eIF2alpha-P mediates prion neurodegeneration. *Nature, 485*(7399), 507–511.

Nguyen, J., Baldwin, M. A., Cohen, F. E., & Prusiner, S. B. (1995). Prion protein peptides induce α-helix to β-sheet conformational transitions. *Biochemistry, 34*, 4186–4192.

Nicoll, A. J., Panico, S., Freir, D. B., Wright, D., Terry, C., Risse, E., … Collinge, J. (2013). Amyloid-β nanotubes are associated with prion protein-dependent synaptotoxicity. *Nature Communications, 4*, 2416.

Nieznanski, K., Choi, J. K., Chen, S., Surewicz, K., & Surewicz, W. K. (2012). Soluble prion protein inhibits amyloid-β (Aβ) fibrillization and toxicity. *Journal of Biological Chemistry, 287*(40), 33104–33108.

Nieznanski, K., Surewicz, K., Chen, S., Nieznanska, H., & Surewicz, W. K. (2014). Interaction between prion protein and Abeta amyloid fibrils revisited. *ACS Chemical Neuroscience, 5*(5), 340–345.

Nishida, N., Tremblay, P., Sugimoto, T., Shigematsu, K., Shirabe, S., Petromilli, C., … Katamine, S. (1999). A mouse prion protein transgene rescues mice deficient for the prion protein gene from Purkinje cell degeneration and demyelination. *Laboratory Investigation, 79*(6), 689–697.

Norstrom, E. M., & Mastrianni, J. A. (2005). The AGAAAAGA palindrome in PrP is required to generate a productive PrP^{Sc}–PrP^{C} complex that leads to prion propagation. *Journal of Biological Chemistry, 280*(29), 27236–27243.

Nuvolone, M., Hermann, M., Sorce, S., Russo, G., Tiberi, C., Schwarz, P., … Aguzzi, A. (2016). Strictly co-isogenic C57BL/6J-Prnp$^{-/-}$ mice: A rigorous resource for prion science. *Journal of Experimental Medicine, JEM., 213*(3), 313–327.

Oesch, B., Westaway, D., Walchli, M., McKinley, M. P., Kent, S. B., Aebersold, R., … Weissmann, C. (1985). A cellular gene encodes scrapie PrP 27–30 protein. *Cell, 40*(4), 735–746.

Orru, C. D., Bongianni, M., Tonoli, G., Ferrari, S., Hughson, A. G., Groveman, B. R., … Zanusso, G. (2014). A test for Creutzfeldt–Jakob disease using nasal brushings. *New England Journal of Medicine, 371*(6), 519–529.

Orru, C. D., Groveman, B. R., Hughson, A. G., Zanusso, G., Coulthart, M. B., & Caughey, B. (2015). Rapid and sensitive RT-QuIC detection of human Creutzfeldt–Jakob disease using cerebrospinal fluid. *MBio, 6*(1), e02451–14.

Orru, C. D., Wilham, J. M., Raymond, L. D., Kuhn, F., Schroeder, B., Raeber, A. J., & Caughey, B. (2011). Prion disease blood test using immunoprecipitation and improved quaking-induced conversion. *MBio, 2*(3), e00078–00011. Available from https://doi.org/10.1128/mBio.00078-11.

Owen, F., Lofthouse, R., Crow, T., Baker, H., Poulter, M., Collinge, J., … Prusiner, S. (1989). Insertion in prion protein gene in familial Creutzfeldt–Jakob disease. *The Lancet, 333*(8628), 51–52.

Pan, K.-M., Baldwin, M., Nguyen, J., Gasset, M., Serban, A., Groth, D., … Prusiner, S. B. (1993). Conversion of α-helices into β-sheets features in the formation of the scrapie prion proteins. *Proceedings of the National Academy of Sciences of the United States of America, 90*, 10962–10966.

Parchi, P., Capellari, S., Chin, S., Schwarz, H., Schecter, N., Butts, J., … Gambetti, P. (1999). A subtype of sporadic prion disease mimicking fatal familial insomnia. *Neurology, 52*(9), 1757-1757.

Parchi, P., Castellani, R., Capellari, S., Ghetti, B., Young, K., Chen, S. G., … Gambetti, P. (1996). Molecular basis of phenotypic variability in sporadic Creutzfeldt–Jakob disease. *Annals of Neurology, 39*, 767–778.

Pauly, P. C., & Harris, D. A. (1998). Copper stimulates endocytosis of the prion protein. *Journal of Biological Chemistry, 273*, 33107–33110.

Peden, A. H., Head, M. W., Ritchie, D. L., Bell, J. E., & Ironside, J. W. (2004). Preclinical vCJD after blood transfusion in a PRNP codon 129 heterozygous patient. *The Lancet, 364*(9433), 527–529.

Poncet-Montange, G., St Martin, S. J., Bogatova, O. V., Prusiner, S. B., Shoichet, B. K., & Ghaemmaghami, S. (2011). A survey of antiprion compounds reveals the prevalence of non-PrP molecular targets. *Journal of Biological Chemistry, 286*(31), 27718–27728.

Prusiner, S. B. (1982). Novel proteinaceous infectious particles cause scrapie. *Science, 216*, 136–144.

Requena, J. R., & Wille, H. (2014). The structure of the infectious prion protein: Experimental data and molecular models. *Prion, 8*(1), 60–66.

Riek, R., Hornemann, S., Wider, G., Billeter, M., Glockshuber, R., & Wüthrich, K. (1996). NMR structure of the mouse prion protein domain PrP(121–231). *Nature, 382*, 180–182.

Riek, R., Hornemann, S., Wider, G., Glockshuber, R., & Wüthrich, K. (1997). NMR characterization of the full-length recombinant murine prion protein, mPrP (23–231). *FEBS Letters, 413*(2), 282–288.

Risse, E., Nicoll, A. J., Taylor, W. A., Wright, D., Badoni, M., Yang, X., … Collinge, J. (2015). Identification of a compound which disrupts binding of amyloid-beta to the prion protein using a novel fluorescence-based assay. *Journal of Biological Chemistry, 290*(27), 17020–17028.

Ritchie, D. L., Adlard, P., Peden, A. H., Lowrie, S., Le Grice, M., Burns, K., … Ironside, J. W. (2017). Amyloid-beta accumulation in the CNS in human growth hormone recipients in the UK. *Acta Neuropathologica, 134*(2), 221–240.

Rivera-Milla, E., Stuermer, C. A., & Málaga-Trillo, E. (2003). An evolutionary basis for scrapie disease: Identification of a fish prion mRNA. *Trends in Genetics, 19*(2), 72–75.

Roucou, X., Gains, M., & LeBlanc, A. C. (2004). Neuroprotective functions of prion protein. *Journal of Neuroscience Research, 75*(2), 153–161.

Saborio, G. P., Permanne, B., & Soto, C. (2001). Sensitive detection of pathological prion protein by cyclic amplification of protein misfolding. *Nature, 411*(6839), 810–813.

Safar, J., Wille, H., Itrri, V., Groth, D., Serban, H., Torchia, M., … Prusiner, S. B. (1998). Eight prion strains have PrP^{Sc} molecules with different conformations. *Nature Medicine, 4*(10), 1157–1165.

Saftig, P., & Lichtenthaler, S. F. (2015). The alpha secretase ADAM10: A metalloprotease with multiple functions in the brain. *Progress in Neurobiology, 135*, 1–20.

Sandberg, M. K., Al-Doujaily, H., Sharps, B., Clarke, A. R., & Collinge, J. (2011). Prion propagation and toxicity in vivo occur in two distinct mechanistic phases. *Nature, 470*(7335), 540–542.

Sandberg, M. K., Al-Doujaily, H., Sharps, B., De Oliveira, M. W., Schmidt, C., Richard-Londt, A., … Collinge, J. (2014). Prion neuropathology follows the accumulation of alternate prion protein isoforms after infective titre has peaked. *Nature Communications, 5*, 4347.

Sandberg, M. K., Al-Doujaily, H., Sigurdson, C. J., Glatzel, M., O'Malley, C., Powell, C., … Wadsworth, J. D.

(2010). Chronic wasting disease prions are not transmissible to transgenic mice overexpressing human prion protein. *Journal of General Virology*, *91*(10), 2651–2657.

Santuccione, A., Sytnyk, V., Leshchyns'ka, I., & Schachner, M. (2005). Prion protein recruits its neuronal receptor NCAM to lipid rafts to activate p59fyn and to enhance neurite outgrowth. *Journal of Cell Biology*, *169*(2), 341–354.

Schmitt-Ulms, G., Hansen, K., Liu, J., Cowdrey, C., Yang, J., DeArmond, S. J., ... Baldwin, M. A. (2004). Time-controlled transcardiac perfusion cross-linking for the study of protein interactions in complex tissues. *Nature Biotechnology*, *22*(6), 724–731.

Schmitt-Ulms, G., Legname, G., Baldwin, M. A., Ball, H. L., Bradon, N., Bosque, P. J., ... Prusiner, S. B. (2001). Binding of neural cell adhesion molecules (N-CAMs) to the cellular prion protein. *Journal of Molecular Biology*, *314*(5), 1209–1225.

Scott, M., Foster, D., Mirenda, C., Serban, D., Coufal, F., Wälchli, M., ... Prusiner, S. B. (1989). Transgenic mice expressing hamster prion protein produce species-specific scrapie infectivity and amyloid plaques. *Cell*, *59*, 847–857.

Scott, M. R., Will, R., Ironside, J., Nguyen, H. O., Tremblay, P., DeArmond, S. J., & Prusiner, S. B. (1999). Compelling transgenetic evidence for transmission of bovine spongiform encephalopathy prions to humans. *Proceedings of the National Academy of Sciences of the United States of America*, *96*(26), 15137–15142.

Sempou, E., Biasini, E., Pinzon-Olejua, A., Harris, D. A., & Malaga-Trillo, E. (2016). Activation of zebrafish Src family kinases by the prion protein is an amyloid-β-sensitive signal that prevents the endocytosis and degradation of E-cadherin/β-catenin complexes in vivo. *Molecular Neurodegeneration*, *11*, 18.

Senatore, A., Colleoni, S., Verderio, C., Restelli, E., Morini, R., Condliffe, S. B., ... Chiesa, R. (2012). Mutant PrP suppresses glutamatergic neurotransmission in cerebellar granule neurons by impairing membrane delivery of VGCC $\alpha_2\delta$-1 subunit. *Neuron*, *74*(2), 300–313.

Shankar, G. M., Bloodgood, B. L., Townsend, M., Walsh, D. M., Selkoe, D. J., & Sabatini, B. L. (2007). Natural oligomers of the Alzheimer amyloid-beta protein induce reversible synapse loss by modulating an NMDA-type glutamate receptor-dependent signaling pathway. *Journal of Neuroscience*, *27*(11), 2866–2875.

Shankar, G. M., Li, S., Mehta, T. H., Garcia-Munoz, A., Shepardson, N. E., Smith, I., ... Selkoe, D. J. (2008). Amyloid-beta protein dimers isolated directly from Alzheimer's brains impair synaptic plasticity and memory. *Nature Medicine*, *14*(8), 837–842.

Shmerling, D., Hegyi, I., Fischer, M., Blättler, T., Brandner, S., Götz, J., ... Weissmann, C. (1998). Expression of amino-terminally truncated PrP in the mouse leading to ataxia and specific cerebellar lesions. *Cell*, *93*, 203–214.

Sigurdson, C. J., Nilsson, K. P. R., Hornemann, S., Manco, G., Polymenidou, M., Schwarz, P., ... Aguzzi, A. (2007). Prion strain discrimination using luminescent conjugated polymers. *Nature Methods*, *4*(12), 1023–1030.

Silber, B. M., Rao, S., Fife, K. L., Gallardo-Godoy, A., Renslo, A. R., Dalvie, D. K., ... Gever, J. R. (2013). Pharmacokinetics and metabolism of 2-aminothiazoles with antiprion activity in mice. *Pharmaceutical Research*, *30*(4), 932–950.

Sim, V. L. (2012). Prion disease: Chemotherapeutic strategies. *Infectious Disorders Drug Targets*, *12*(2), 144–160.

Simonic, T., Duga, S., Strumbo, B., Asselta, R., Ceciliani, F., & Ronchi, S. (2000). cDNA cloning of turtle prion protein. *FEBS Letters*, *469*(1), 33–38.

Slapsak, U., Salzano, G., Amin, L., Abskharon, R. N., Ilc, G., Zupancic, B., ... Legname, G. (2016). The N terminus of the prion protein mediates functional interactions with the neuronal cell adhesion molecule (NCAM) fibronectin domain. *Journal of Biological Chemistry*, *291* (42), 21857–21868.

Solis, G. P., Radon, Y., Sempou, E., Jechow, K., Stuermer, C. A., & Málaga-Trillo, E. (2013). Conserved roles of the prion protein domains on subcellular localization and cell–cell adhesion. *PLoS ONE*, *8*(7), e70327.

Solomon, I. H., Huettner, J. E., & Harris, D. A. (2010). Neurotoxic mutants of the prion protein induce spontaneous ionic currents in cultured cells. *Journal of Biological Chemistry*, *285*, 26719–26726.

Solomon, I. H., Khatri, N., Biasini, E., Massignan, T., Huettner, J. E., & Harris, D. A. (2011). An N-terminal polybasic domain and cell surface localization are required for mutant prion protein toxicity. *Journal of Biological Chemistry*, *286*(16), 14724–14736.

Sonati, T., Reimann, R. R., Falsig, J., Baral, P. K., O'Connor, T., Hornemann, S., ... Aguzzi, A. (2013). The toxicity of antiprion antibodies is mediated by the flexible tail of the prion protein. *Nature*, *501*(7465), 102–106.

Spevacek, A. R., Evans, E. G., Miller, J. L., Meyer, H. C., Pelton, J. G., & Millhauser, G. L. (2013). Zinc drives a tertiary fold in the prion protein with familial disease mutation sites at the interface. *Structure*, *21*(2), 236–246.

Spudich, A., Frigg, R., Kilic, E., Kilic, U., Oesch, B., Raeber, A., ... Hermann, D. M. (2005). Aggravation of ischemic brain injury by prion protein deficiency: Role of ERK-1/-2 and STAT-1. *Neurobiology of Disease*, *20*, 442–449.

Stahl, N., Borchelt, D. R., Hsiao, K., & Prusiner, S. B. (1987). Scrapie prion protein contains a phosphatidylinositol glycolipid. *Cell*, *51*, 229–249.

Stopschinski, B. E., & Diamond, M. I. (2017). The prion model for progression and diversity of neurodegenerative diseases. *The Lancet Neurology, 16*(4), 323–332.

Tagliavini, F., Forloni, G., Colombo, L., Rossi, G., Girola, L., Canciani, B., ... Salmona, M. (2000). Tetracycline affects abnormal properties of synthetic PrP peptides and PrP^{Sc} in vitro. *Journal of Molecular Biology, 300*(5), 1309–1322.

Taylor, D. R., & Hooper, N. M. (2007). The low-density lipoprotein receptor-related protein 1 (LRP1) mediates the endocytosis of the cellular prion protein. *Biochemical Journal, 402*(1), 17–23.

Tobler, I., Deboer, T., & Fischer, M. (1997). Sleep and sleep regulation in normal and prion protein-deficient mice. *Journal of Neuroscience, 17*(5), 1869–1879.

Tobler, I., Gaus, S. E., Deboer, T., Achermann, P., Fischer, M., Rulicke, T., ... Manson, J. C. (1996). Altered circadian activity rhythms and sleep in mice devoid of prion protein. *Nature, 380*(6575), 639–642.

Turk, E., Teplow, D. B., Hood, L. E., & Prusiner, S. B. (1988). Purification and properties of the cellular and scrapie hamster prion proteins. *European Journal of Biochemistry, 176*, 21–30.

Um, J. W., Kaufman, A. C., Kostylev, M., Heiss, J. K., Stagi, M., Takahashi, H., ... Strittmatter, S. M. (2013). Metabotropic glutamate receptor 5 is a coreceptor for Alzheimer Aβ oligomer bound to cellular prion protein. *Neuron, 79*(5), 887–902. Available from https://doi.org/10.1016/j.neuron.2013.06.036.

Um, J. W., Nygaard, H. B., Heiss, J. K., Kostylev, M. A., Stagi, M., Vortmeyer, A., ... Strittmatter, S. M. (2012). Alzheimer amyloid-β oligomer bound to postsynaptic prion protein activates Fyn to impair neurons. *Nature Neuroscience, 15*(9), 1227–1235.

Vassallo, N., & Herms, J. (2003). Cellular prion protein function in copper homeostasis and redox signalling at the synapse. *Journal of Neurochemistry, 86*(3), 538–544.

Vázquez-Fernández, E., Vos, M. R., Afanasyev, P., Cebey, L., Sevillano, A. M., Vidal, E., ... Peters, P. J. (2016). The structural architecture of an infectious mammalian prion using electron cryomicroscopy. *PLoS Pathogens, 12*(9), e1005835.

Vincent, B., Paitel, E., Saftig, P., Frobert, Y., Hartmann, D., De Strooper, B., ... Checler, F. (2001). The disintegrins ADAM10 and TACE contribute to the constitutive and phorbol ester-regulated normal cleavage of the cellular prion protein. *Journal of Biological Chemistry, 276*(41), 37743–37746.

Wadia, J. S., Schaller, M., Williamson, R. A., & Dowdy, S. F. (2008). Pathologic prion protein infects cells by lipid-raft dependent macropinocytosis. *PLoS ONE, 3*, e3314.

Wadsworth, J. D., Asante, E. A., Desbruslais, M., Linehan, J. M., Joiner, S., Gowland, I., ... Hill, A. F. (2004). Human prion protein with valine 129 prevents expression of variant CJD phenotype. *Science, 306*(5702), 1793–1796.

Wagner, J., Ryazanov, S., Leonov, A., Levin, J., Shi, S., Schmidt, F., ... Mitteregger-Kretzschmar, G. (2013). Anle138b: A novel oligomer modulator for disease-modifying therapy of neurodegenerative diseases such as prion and Parkinson's disease. *Acta Neuropathologica, 125*(6), 795–813.

Walker, L. C. (2016). Proteopathic strains and the heterogeneity of neurodegenerative diseases. *Annual Review of Genetics, 50*, 329–346. Available from https://doi.org/10.1146/annurev-genet-120215-034943.

Walker, L. C., & Jucker, M. (2015). Neurodegenerative diseases: Expanding the prion concept. *Annual Review of Neuroscience, 38*, 87–103. Available from https://doi.org/10.1146/annurev-neuro-071714-033828.

Walsh, D. M., & Selkoe, D. J. (2007). Aβ oligomers—A decade of discovery. *Journal of Neurochemistry, 101*(5), 1172–1184.

Walter, E. D., Chattopadhyay, M., & Millhauser, G. L. (2006). The affinity of copper binding to the prion protein octarepeat domain: Evidence for negative cooperativity. *Biochemistry, 45*(43), 13083–13092.

Walter, E. D., Stevens, D. J., Visconte, M. P., & Millhauser, G. L. (2007). The prion protein is a combined zinc and copper binding protein: Zn^{2+} alters the distribution of Cu^{2+} coordination modes. *Journal of the American Chemical Society, 129*(50), 15440–15441.

Walz, R., Amaral, O. B., Rockenbach, I. C., Roesler, R., Izquierdo, I., Cavalheiro, E. A., ... Brentani, R. R. (1999). Increased sensitivity to seizures in mice lacking cellular prion protein. *Epilepsia, 40*(12), 1679–1682.

Wang, F., Wang, X., Yuan, C. G., & Ma, J. (2010). Generating a prion with bacterially expressed recombinant prion protein. *Science, 327*(5969), 1132–1135.

Watt, N. T., Taylor, D. R., Kerrigan, T. L., Griffiths, H. H., Rushworth, J. V., Whitehouse, I. J., & Hooper, N. M. (2012). Prion protein facilitates uptake of zinc into neuronal cells. *Nature Communications, 3*, 1134.

Weise, J., Crome, O., Sandau, R., Schulz-Schaeffer, W., Bahr, M., & Zerr, I. (2004). Upregulation of cellular prion protein (PrP^{C}) after focal cerebral ischemia and influence of lesion severity. *Neuroscience Letters, 372*(1-2), 146–150.

Westaway, D., Zuliani, V., Cooper, C. M., Da Costa, M., Neuman, S., Jenny, A. L., ... Prusiner, S. B. (1994). Homozygosity for prion protein alleles encoding glutamine-171 renders sheep susceptible to natural scrapie. *Genes and Development, 8*(8), 959–969.

Westergard, L., Turnbaugh, J. A., & Harris, D. A. (2011). A nine amino acid domain is essential for mutant prion protein toxicity. *Journal of Neuroscience, 31*(39), 14005–14017.

White, M. D., Farmer, M., Mirabile, I., Brandner, S., Collinge, J., & Mallucci, G. R. (2008). Single treatment with RNAi against prion protein rescues early neuronal dysfunction and prolongs survival in mice with prion disease. *Proceedings of the National Academy of Sciences of the United States of America*, *105*(29), 10238–10243.

Whittington, M. A., Sidle, K. C., Gowland, I., Meads, J., Hill, A. F., Palmer, M. S., ... Collinge, J. (1995). Rescue of neurophysiological phenotype seen in PrP null mice by transgene encoding human prion protein. *Nature Genetics*, *9*(2), 197–201.

Wickner, R. B. (2016). Yeast and fungal prions. *Cold Spring Harbor Perspectives in Biology*, *8*(9), a023531.

Wilesmith, J. W., Wells, G., Cranwell, M. P., & Ryan, J. (1988). Bovine spongiform encephalopathy: Epidemiological studies. *The Veterinary Record*, *123*(25), 638–644.

Wilham, J. M., Orru, C. D., Bessen, R. A., Atarashi, R., Sano, K., Race, B., ... Caughey, B. (2010). Rapid end-point quantitation of prion seeding activity with sensitivity comparable to bioassays. *PLoS Pathogens*, *6*(12), e1001217.

Wille, H., Michelitsch, M. D., Guénebaut, V., Supattapone, S., Serban, A., Cohen, F. E., ... Prusiner, S. B. (2002). Structural studies of the scrapie prion protein by electron crystallography. *Proceedings of the National Academy of Sciences of the United States of America*, *99*(6), 3563–3568.

Williams, E. (2005). Chronic wasting disease. *Veterinary Pathology*, *42*(5), 530–549.

Wopfner, F., Weidenhöfer, G., Schneider, R., von Brunn, A., Gilch, S., Schwarz, T. F., ... Schätzl, H. M. (1999). Analysis of 27 mammalian and 9 avian PrPs reveals high conservation of flexible regions of the prion protein. *Journal of Molecular Biology*, *289*(5), 1163–1178.

Wroe, S. J., Pal, S., Siddique, D., Hyare, H., Macfarlane, R., Joiner, S., ... Hewitt, P. (2006). Clinical presentation and pre-mortem diagnosis of variant Creutzfeldt–Jakob disease associated with blood transfusion: A case report. *The Lancet*, *368*(9552), 2061–2067.

Wu, B., McDonald, A. J., Markham, K., Rich, C. B., Mchugh, K. P., Tatzelt, J., ... Harris, D. A. (2017). The N-terminus of the prion protein is a toxic effector regulated by the C-terminus. *eLife*, *6*, e23473.

Wulf, M.-A., Senatore, A., & Aguzzi, A. (2017). The biological function of the cellular prion protein: An update. *BMC Biology*, *15*(1), 34.

Xia, M., Cheng, X., Yi, R., Gao, D., & Xiong, J. (2016). The binding receptors of Aβ: An alternative therapeutic target for Alzheimer's disease. *Molecular Neurobiology*, *53*(1), 455–471.

Yanamandra, K., Kfoury, N., Jiang, H., Mahan, T. E., Ma, S., Maloney, S. E., ... Holtzman, D. M. (2013). Anti-tau antibodies that block tau aggregate seeding in vitro markedly decrease pathology and improve cognition *in vivo*. *Neuron*, *80*(2), 402–414.

Yang, W., Cook, J., Rassbach, B., Lemus, A., DeArmond, S. J., & Mastrianni, J. A. (2009). A new transgenic mouse model of Gerstmann–Straussler–Scheinker syndrome caused by the A117V mutation of PRNP. *Journal of Neuroscience*, *29*(32), 10072–10080.

You, H., Tsutsui, S., Hameed, S., Kannanayakal, T. J., Chen, L., Xia, P., ... Zamponi, G. W. (2012). Aβ neurotoxicity depends on interactions between copper ions, prion protein, and *N*-methyl-D-aspartate receptors. *Proceedings of the National Academy of Sciences of the United States of America*, *109*(5), 1737–1742.

Younan, N. D., Sarell, C. J., Davies, P., Brown, D. R., & Viles, J. H. (2013). The cellular prion protein traps Alzheimer's Aβ in an oligomeric form and disassembles amyloid fibers. *FASEB Journal*, *27*(5), 1847–1858.

Yuan, A. H., & Hochschild, A. (2017). A bacterial global regulator forms a prion. *Science*, *355*(6321), 198–201.

Zanata, S. M., Lopes, M. H., Mercadante, A. F., Hajj, G. N., Chiarini, L. B., Nomizo, R., ... Martins, V. R. (2002). Stress-inducible protein 1 is a cell surface ligand for cellular prion that triggers neuroprotection. *EMBO Journal*, *21*(13), 3307–3316.

Zweckstetter, M., Requena, J. R., & Wille, H. (2017). Elucidating the structure of an infectious protein. *PLoS Pathogens*, *13*(4), e1006229.

CHAPTER

3

Alzheimer's Disease: Toward a Quantitative Biological Approach in Describing its Natural History and Underlying Mechanisms

Colin L. Masters[1], *Ben Gu*[1], *Simon Laws*[2], *Yen Y. Lim*[1], *Blaine Roberts*[1], *Victor L. Villemagne*[1,3] *and Konrad Beyreuther*[4]

[1]The Florey Institute, The University of Melbourne, Melbourne, VIC, Australia
[2]Edith Cowan University, Joondalup, WA, Australia [3]Austin Health, Melbourne, VIC, Australia
[4]The University of Heidelberg, Heidelberg, Germany

"Hypotheses non fingo" (I contrive no hypotheses) ***Isaac Newton, Principia, 1713***

OUTLINE

The Molecular and Cellular Basis of Neurodegenerative Diseases
DOI: https://doi.org/10.1016/B978-0-12-811304-2.00003-1

Alzheimer's disease (AD) is a debilitating, progressive, unremitting chronic degenerative illness of the brain. It is common in the elderly population (mean age at onset 75–80 years), striking apparently at random, with anecdotal reports of increased genetic risk within nuclear families. Comorbidities, especially cerebrovascular disease, are also common in this population. In contrast, onset in young or middle-aged persons is rare, but is often associated with a strong pattern of autosomal-dominant inheritance, in which pathogenic mutations in three genes have been identified (mean age at onset 45 years). Pathologically, the AD-affected brain is characterized by plaque deposition of an amyloid peptide (Aβ) within the parenchyma and around the outer margins of small vessels (mainly arterioles) and the occurrence of tangled skeins of neurofilamentous material composed of aggregated tau within the soma of larger neurons and in neurites in close proximity to the Aβ amyloid deposits. Reactive changes (astrocytic and microglial proliferation and activation) and synaptic and neuronal loss occur to varying degrees. Recent advances in molecular genetics and in detecting the accumulation of Aβ and tau during life have enabled a fresh approach to the nosologic classification of AD into preclinical, prodromal, and clinical stages. Quantitative estimates of the relative risks associated with heredity, environment, and replication of somatic DNA are now beginning to emerge. There is general agreement in the field that the majority of AD cases (sporadic) are the result of the failure of clearance of Aβ from the aging brain. The mechanisms underlying this failure are only just now becoming clear.

The above description of AD is entirely qualitative and is more or less the approach familiar to most clinicians working since the time of Alzheimer (Alzheimer, 1907). When the nature of the amyloid became known in the period 1983–1985 (Allsop, Landon, & Kidd, 1983; Glenner & Wong, 1984a, 1984b; Masters et al., 1985), and its proteolytic origin from a neuronal precursor was discovered in 1987 (Kang et al., 1987), it then became possible to describe the pathogenesis of AD in more quantitative terms. Coupled with the characterization of the microtubule-associated protein tau as a component of neurofibrillary tangles (NFT) and neurites (Brion, van den Bosch de Aguilar, & Flament-Durand, 1985), the path was set toward a quantitative biological approach to the natural history of AD.

What might a fully quantitative biological approach to AD look like? It should begin with a precise analytical epidemiological approach to basic demographics: age, sex, race, incidence, and prevalence. Next would be an analysis of cut points and change over time: rates of production and accumulation of Aβ-amyloid (and tau, α-synuclein, TAR DNA-binding protein 43 (TDP-43), etc., if such data were available); brain structural change (atrophy) and synaptic loss; and finally, the rates of cognitive change and the confounding effects of comorbidities. These quantitative parameters would then be used to discover and develop disease-modifying strategies, including primary (if possible) and secondary prevention strategies. In this chapter, we begin the process of drawing together some of the more robust quantitative aspects of AD. For the purposes of simplicity and clarity, we will not incorporate the full sets of confidence intervals and estimates of variance in these numbers in this review; these are available from the primary references quoted in the text.

We can then start to address the following questions:

1. How much Aβ accumulates in the AD brain?
2. How long does it take to accumulate?
3. How much is the normal clearance mechanism of Aβ impaired in AD?
4. Can real-time biomarker read-outs be used in clinical trial design to monitor drug efficacy?

Over the past 40 years, much has been made of the "amyloid hypothesis" for the causation of AD (Beyreuther & Masters, 1991; Hardy & Higgins, 1992; Hardy & Selkoe, 2002). Like Newton, quoted above, we would argue that hypotheses are not of overriding importance: rather, it will eventually be the numbers which speak for themselves (see Table 3.1).

QUANTITATIVE APPROACH TO BASIC AD DEMOGRAPHICS

Surprisingly, despite the availability of validated biomarkers (measuring Aβ accumulation in vivo using position emission tomography (PET) and cerebrospinal fluid (CSF) assays), there is still a lack of population-based analytical epidemiology on the incidence, prevalence, and disability-adjusted life-years of AD (GBD 2013 DALYs and HALE Collaborators, 2015). Current estimates are based on clinical criteria, which carry an accuracy of approximately 80% and do not adequately account for comorbidities such as cardio/cerebrovascular disease.

A starting point would be an overall estimate of annual incidence (and mortality) of 1 in 300 (3300 per million) of the total population, or 1%–2% of the population aged more than 65 years (Bachman et al., 1993; Brookmeyer, Gray, & Kawas, 1998; Evans et al., 2003; Herbert et al., 1995; Kawas, Gray, Brookmeyer, Fozard, & Zonderman, 2000). From being a rare occurrence (<1%) under the age of 65 years, the incidence approaches 10% in the 80–90-year age group. What happens after the age of 90 years is still uncertain and remains a fundamental unanswered question: a continued increase, a leveling off, or even a decrease in incidence would have major implications for understanding the underlying pathogenesis. Further unanswered questions relate to prevalence. In the absence of reliable data on the duration of the clinical dementia phase of AD and its effect on the age-at-onset, the best estimates of prevalence of "dementia" based purely on clinical surveys lie in the 6%–7% range for populations aged >60 years (Prince et al., 2015). This figure is much lower than the emerging-specific biomarker data based on Aβ-PET in which the prevalence of preclinical AD is 30%–35% in the 60–80-year age group (Jack et al., 2017; Rowe et al., 2010; Sperling, Mormino, & Johnson, 2014). The reasons for this must lie in the long (20 years) preclinical and prodromal phases. The prevalence of combined preclinical, prodromal, and clinical AD in the >90-year age group remains to be determined but, as with incidence mentioned above, remains a fundamental issue relating to the question of the inevitability of AD in the aging brain. Once having moved beyond the peak age of crude mortality (approximately 85 years), does the human brain progressively escape the risk of AD?

The large excess of female deaths attributable to "dementia" in most Western countries (66% female, 34% male) appears to reflect the differential disease burdens of the other common causes of death. The female:male imbalance largely disappears when an age-adjustment is made (Buckley et al., submitted). Other risk factors for "dementia" [viz. type 2 diabetes, midlife hypertension, midlife obesity, physical inactivity, depression, smoking, and low-educational attainment (the seven deadly sins of our advanced economies)] are, of course, well known and are also validated risk factors for atherosclerotic vascular disease. This major comorbidity alone may account for most of the population-attributable risk in conjunction with the major AD hereditable genetic age-at-onset risk imparted by the common polymorphic alleles of the apolipoprotein E (*APOE*) gene. Combined with the other minor AD genetic risk association factors [such as *CD33*, triggering receptor expressed on myeloid cells-2 (*TREM2*), ATP-binding cassette transporter A7 (*ABCA7*), *CLU*, *Ms4A4A*, *BIN1*, and *PICALM*], these genetic risks point toward a deficit in the clearance pathways that normally remove Aβ from the aging brain (Huang et al., 2017). Educational attainment has a significant association with Aβ

TABLE 3.1 "Sporadic" AD by the Numbers

DEMOGRAPHICS				
Incidence	2% population >65 years			
Prevalence	6% population >60 years			
Mean age onset	$\varepsilon4^{+/+}$ 68 years, $\varepsilon4^{+/-}$ 76 years, $\varepsilon4^{-/-}$ 84 years			
Mean age at death	88 years			
Mean duration	10 years			
Aβ PRODUCTION^{+}				
$[A\beta]_{CSF}$	581 ng/h			
$[A\beta_{42}]_{CSF}$	57 ng/h			
$[A\beta]_{CSF}$ turnover	13 h control, 19 h AD			
$[A\beta]_{CSF}$ $t^{1/2}$	3.8 h young, 9.4 h old			
$[A\beta]_{CSF}$ exchange rate constant	0.006 pools/h Aβ, 0.049 pools/h Aβ			
Aβ ACCUMULATION				
Aβ burden by PET	0.048 SUVR/year over 20 years			
Aβ threshold	1.4 SUVR (5 μg/g)			
Aβ dementia	2.33 SUVR (11.2 μg/g)			
Total brain mass Aβ	1.7-mg control, 6.5-mg AD			
Estimated rate of accumulation	28 ng/h			
7×10^{6} plaques in AD Brain	200 molecules of Aβ added to each plaque/sec			
BRAIN ATROPHY (SYNAPTIC LOSS)				
	CN Low Aβ-PET	**Preclinical AD High Aβ-PET**	**MCI Low Aβ-PET**	**Prodromal AD High Aβ-PET**
Hippocampal atrophy (cm^3/year)	− 0.031	− 0.052	− 0.043	− 0.093
Gray matter atrophy (cm^3/year)	− 1.6	− 2.9	− 0.6	− 2.6
COGNITIVE LOSS (PRECLINICAL AND PRODROMAL AD, SD UNITS)				
	CN Low Aβ-PET	**Preclinical AD High Aβ-PET**	**MCI Low Aβ-PET**	**Prodromal AD High Aβ-PET**
Episodic memory	0.04	− 0.40	0.37	− 0.50
Executive function	− 0.10	− 0.01	0.14	− 0.22
Language	− 0.03	− 0.15	− 0.17	− 0.57
Attention	− 0.11	− 0.15	0.01	− 0.34
Global (e.g., PACC)	0.20	− 0.26	*n/a	*n/a

AD, Alzheimer's disease; *PACC*, Preclinical Alzheimer's Cognitive Composite.
*n/a because the PACC has been designed for the preclinical AD population, and its utility in the prodromal AD population is yet to be determined.
[+]Data from Bateman Laboratory *Lucey, B. P., Mawuenyega, K. G., Patterson, B. W., Elbert, D. L., Ovod, V., Kasten, T., et al., (2017) Associations between β-amyloid kinetics and β-amyloid diurnal pattern in the central nervous system, JAMA Neurology, 74, 207–215.*

positivity in cognitively normal and mild cognitive impairment (MCI) subjects. The prevalence of Aβ deposition was found to be 5% higher in individuals with an education above the median than in those with education below the median regardless of cognitive status, age, and *APOE*-ε4 carrier status (Jansen et al., 2015). This may be due to higher synaptic density in higher education, which is usually summarized as cognitive reserve. In due course, a quantitative algorithm will be devised which takes into account the various demographic factors discussed above, in the form of {Age X Education X Aβ burden X Genetic Risk Score X Vascular Risk Score}.

Change Over Time: (i) Biogenesis, Production, and Turnover of Aβ

APP, the parent molecule for Aβ (Kang et al., 1987), is highly expressed in neurons and is abundant in the brain. The half-life of APP in tissue culture is approximately 30 minutes (Gersbacher et al., 2013; Weidemann et al., 1989), and there are about 10^6 copies of APP per neuron (Bahmanyar et al., 1987; Weidemann et al., 1989). Our discovery of the proteolytic origin of Aβ from APP led to the eventual characterization of the α-, β-, and γ- cleavages that generate Aβ/p3 (Kang et al., 1987). With the development of the stable isotope labeling kinetic (SILK) technique, Bateman et al. (2006) have been able to characterize the CSF pools of Aβ in the living human. Although we estimate that less than 15% of total Aβ production in the brain is delivered into the extracellular interstitial fluid (ISF) compartment, and that the total soluble pool of Aβ in the brain is <1% (Roberts et al., 2017), the kinetics of Aβ in the measurable CSF pool has provided important insights into its production, turnover, and clearance (Huang et al., 2012; Lucey et al., 2017; Mawuenyega et al., 2010; Patterson et al., 2015; Potter et al., 2013).

Using SILK, it has been shown that $[A\beta]_{CSF}$ is rapidly produced and cleared (Bateman et al., 2006). 8% of the total $[A\beta]_{CSF}$ is produced and cleared (i.e., turned over) per hour. The half-life of $[A\beta]_{CSF}$ is 9 hours. In control subjects (mean age 48 years), total $[A\beta_{42+40+38}]_{CSF}$ production is 581 ng/h ($A\beta_{38}$ 18%; $A\beta_{40}$ 72%; $A\beta_{42}$ 10%) (Potter et al., 2013). Newly formed $[A\beta]_{CSF}$ is in exchange equilibrium with preexisting pools of Aβ (oligomers, aggregates, fibrils, complexes with chaperones/lipids). The amounts of $[A\beta]_{CSF}$ show diurnal/circadian variations with increased levels during sleeping hours (Huang et al., 2012; Lucey et al., 2017).

In sporadic AD, the turnover rates of $[A\beta_{42}]_{CSF}$ and $[A\beta_{40}]_{CSF}$ are 7%/hour (Mawuenyega et al., 2010). The clearance rates, however, are 49% slower ($[A\beta]_{CSF}$ turnover 5.3% in AD compared to 7.6% in controls). This equates to a 13-hour complete turnover of the $[A\beta]_{CSF}$ pool in controls compared to a 19-hour complete turnover in AD. There is therefore a 42% impairment in the production:clearance ratio (Mawuenyega et al., 2010).

In contrast, in autosomal-dominant AD (with *PSEN1* or *PSEN2* pathogenic mutations), there is an 18% increased production of $[A\beta_{42}]_{CSF}$ with no change in $[A\beta_{40,38}]_{CSF}$ (Potter et al., 2013). There is a 24% increase in the $[A\beta_{42:40}]_{CSF}$ production ratios, and there is a 65% increase in the $[A\beta_{42}]_{CSF}$ relative to $[A\beta_{40}]_{CSF}$ fractional turnover rates (FTRs), consistent with an increased removal of both $A\beta_{42}$ and $A\beta_{40}$ through extracellular deposition.

With the clinical onset of sporadic AD, there is a loss of the normal linear kinetics of production/clearance and diurnal variation in the $[A\beta]_{CSF}$ pool (Huang et al., 2012). The mean differences of the SD of residuals ($A\beta_{40}$ 7.42 pm, $A\beta_{42}$ 3.72 pm) are altered in subjects with Aβ deposited (above normal thresholds) in their brains ($A\beta^+$ subjects). These $A\beta^+$ subjects have diminished diurnal amplitudes and the linear increase of $A\beta_{42}$, but not $A\beta_{40}$. Increasing age also diminished the amplitude of the diurnal variation of both $A\beta_{40}$ and $A\beta_{42}$ (Lucey et al., 2017).

Increasing age also correlates with a slower $[A\beta]_{CSF}$ turnover (Patterson et al., 2015). There is a 2.5-fold longer half-life over five decades (30–80 years), with $t_{1/2}$ increasing from 3.8 to 9.4 hours, affecting all Aβ isoforms in a similar fashion. The FTRs slow approximately 60% between ages 30 (FTR 0.184/hours) and 80 years (0.074/hours). The FTR of $A\beta_{38}$, $A\beta_{40}$, and $A\beta_{42}$ are each highly correlated with age (Patterson et al., 2015).

Once Aβ is deposited above threshold levels ($A\beta^+$ subjects), then in CSF there is >50% irreversible loss of soluble $A\beta_{42}$ and a 10-fold higher $A\beta_{42}$ reversible exchange rate. In $A\beta^+$ subjects, $A\beta_{38,40}$ SILK time courses are similar, but $A\beta_{42}$ peaks earlier (Patterson et al., 2015). The $A\beta_{38:40}$ isotope enrichment ratio is close to one in both $A\beta^-$ and $A\beta^+$ subjects. But in $A\beta^+$ subjects, the $A\beta_{42}$:$A\beta_{40}$ ratio is >1 in the rise to peak, and <1 after the peak. This would indicate a faster turnover in $A\beta_{42}$ kinetics in $A\beta^+$ subjects (also seen in AD subjects—Potter & Wisniewski, 2012), consistent with a specific disturbance of soluble $A\beta_{42}$ kinetics in $A\beta^+$ subjects.

$A\beta_{42}$ is in an exchange process in the $A\beta^+$ group (exchange rate constant 0.049 pools/hour, about half the magnitude of the $A\beta_{42}$ FTR) but is nearly absent in the $A\beta^-$ group (0.006 pools/hour). This is consistent with the $A\beta_{42}$ brain pool acting as a major sink for newly produced $A\beta_{42}$ species.

The FTR of all Aβ isoforms ($A\beta_{38}$, $A\beta_{40}$, $A\beta_{42}$) correlated with both cognition and biomarkers (Aβ-PET/CSF) (Patterson et al., 2015). In cognitively normal $A\beta^+$ and prodromal $A\beta^+$ subjects, the FTR's are faster compared to cognitively normal $A\beta^-$ subjects ($A\beta_{42}$ 57%, $A\beta_{40}$ 17%, and $A\beta_{38}$ 22% faster). The FTR ($A\beta_{42}$) correlates positively with the rate of change of Aβ-PET.

Change Over Time: (ii) The Accumulation, Spread, Propagation of Aβ

The progressive accumulation, topographic spread, and propagation of the key morphological (plaques, perivascular Aβ amyloid deposits, drusige Entartung, neurofibrillary tangles, neuritic changes) and molecular determinants (Aβ amyloid, tau, α-synuclein, TDP-43) have been inferred ever since these markers were first described. Based on cross-sectional postmortem studies, it has always been clear that change over time was an essential feature of all types of neurodegenerative diseases (Davies et al., 1988). But until the advent of real-time biomarker assays (using molecular PET imaging or biochemical CSF assays), it was impossible to construct the curves which portray the actual rates of change in individuals during the evolution of the degenerative process. This has now changed with the assembly of longitudinal observational cohorts such as AIBL (the Australian Imaging, Biomarker and Lifestyle) study (Ellis et al., 2014), the Alzheimer's Disease Neuroimaging Initiative (Tosun et al., 2017), the Adult Children's Study (Xiong et al., 2016), the Mayo Clinic Study of Aging (Jack et al., 2014), and the Harvard Aging Brain Study (Mormino et al., 2014a, 2014b). Using data derived from three or more time points (e.g., baseline, 18 and 36 months and beyond), we have been able to reconstruct the mean rates of change of Aβ accumulation across the preclinical, prodromal, and clinical phases of AD.

There are several Aβ-amyloid imaging PET tracers, possessing different pharmacokinetics and optimal analytical methods that are reflected in a wide range of different results. Therefore, a method has recently been developed to convert each Aβ-amyloid imaging tracer value into a universal unit of measurement that is called the Centiloid (Klunk et al., 2015). That conversion allows comparing and compiling Aβ-amyloid imaging results obtained with different PET tracers.

Fig. 3.1 shows the overall sigmoidal shape of the rate of Aβ accumulation over a 50-year interval. This rate is calculated by measuring the change in Aβ PET-SUVR (standard uptake value ratio) at three of more time points during

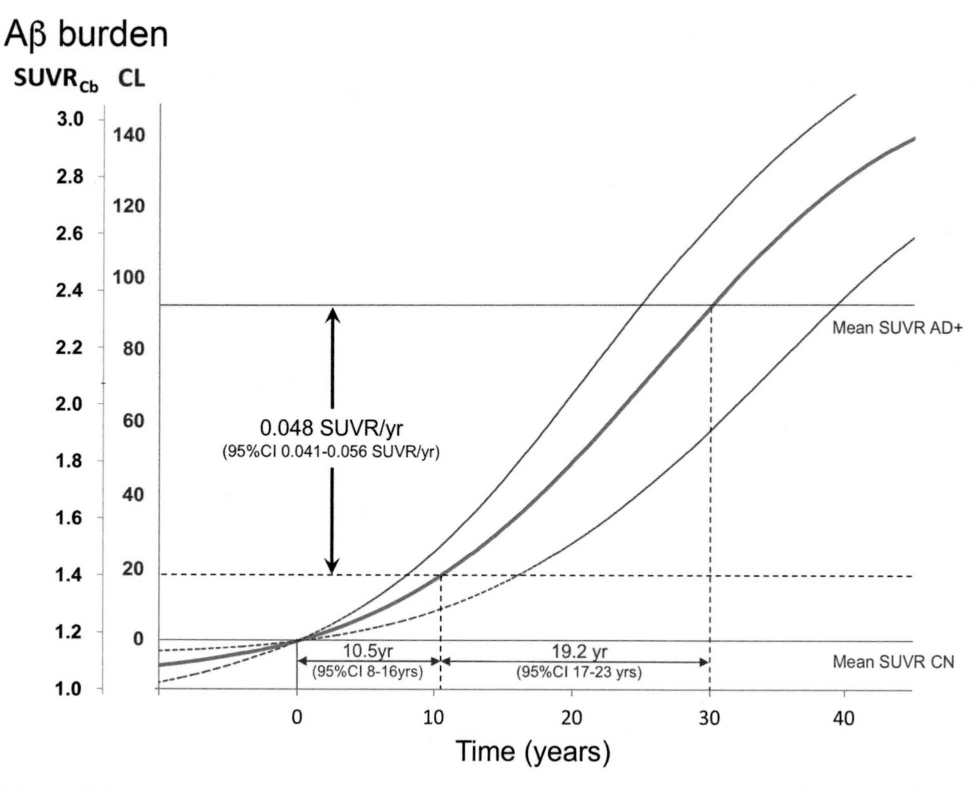

FIGURE 3.1 **Global Aβ accumulation expressed in both $SUVR_{Cb}$ and Centiloids (CL) for ^{11}C-PiB and ^{18}F-NAV4694.** The time required to transition from the levels observed in CN- to the threshold of abnormality is 10 years, requiring another 20 years to reach the levels observed in mild AD. *^{11}C-PiB*, carbon-11-labeled Pittsburgh compound B; *$SUVR_{Cb}$*, standard uptake value ratio generated with the cerebellar cortex as reference region; *CL*, *Centiloids*; *CN*, cognitively normal; *95% CI*, 95% confidence interval. *Adapted from Villemagne, V. L., Burnham, S., Bourgeat, P., Brown, B., Ellis, K. A., Salvado, O., et al. (2013). Amyloid β deposition, neurodegeneration, and cognitive decline in sporadic Alzheimer's disease: a prospective cohort study. Lancet Neurology, 12, 357–367.*

the natural history of Aβ evolution. The SUVR is a measure of the global cortical binding of the Aβ-PET ligand compound to a reference region. The Aβ-PET global SUVR lower cut point of 1.4 marks the current boundary below which changes in episodic memory are difficult to detect using standard neuropsychological instruments. The upper Aβ-PET SUVR cut point is set at a value of 2.33, the mean value for cases at the beginning of the AD dementia phase. This value of 2.33 indicates that the amount of Aβ in the clinical AD brain is 2.33 times increased over the control value (usually set to a reference obtained from the cerebellar cortex). Between these 1.4 and 2.33 cut points, the rate of Aβ accumulation is linear at 0.048 SUVR/year. The time needed to transition from 1.4 to 2.33 is 19 years. Since the mean age at clinical AD onset is approximately 75–80 years, we can estimate that the preclinical phase of AD starts at approximately 55–60 years. There is a 10-year window below the lower cut point where we estimate that the SUVR moves from 1.2 to 1.3 to 1.4. We have designated the 1.3–1.4 band (a 5-year window from approximate age 55 to 60 years) as a phase of "pre-AD" in which subjects would be suitable for primary prevention interventional studies, assuming we

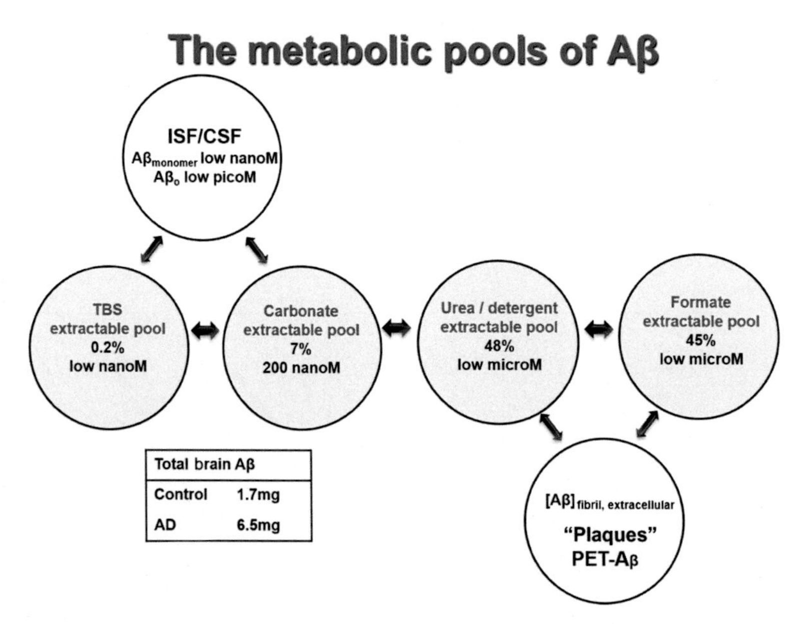

FIGURE 3.2 **Exchange between biochemically defined pools, CSF, and plaque.** The main location of Aβ is in the lipid/membrane pool that is represented by the urea/detergent and carbonate pools. After cleavage, Aβ predominately remains associated with the lipid/membrane with a decreasing concentration gradient to the extra cellular/cytosolic soluble pool represented by the TBS extract. This gradient continues to the ISF and CSF. The trapping and accumulation of $A\beta_{42}$ species as Aβ plaques and perivascular deposits by mass action would shift the equilibrium, which is consistent with decreased levels of $A\beta_{42}$ species in the CSF of AD cases. *AD*, Alzheimer's disease; *ISF*, interstitial fluid. *Adapted from Roberts, B. R., Lind, M., Wagen, A. Z., Rembach, A., Frugier, T., Li, Q. X., et al. (2017). Biochemically-defined pools of amyloid-β in sporadic Alzheimer's disease: Correlation with amyloid PET. Brain, 140; 1486–1498.*

can identify these subjects based on the {Age X Education X Aβ burden X Genetic Risk Score X Vascular Risk Score} algorithm.

While these timeframes reflect the global average of Aβ accumulation in several cortical regions in the brain, namely, the frontal, temporal, parietal cortices, as well as the posterior and anterior cingulate gyri, each of these regions have their idiosyncratic onset of Aβ deposition as well as their particular rate of Aβ accumulation. As we reported before (Villemagne et al., 2011), while early Aβ deposition is observed in the posterior cingulate, orbitofrontal and supramarginal gyrus regions of the brain, continuous and progressive accumulation is observed in the temporal cortex, even after reaching the AD stage where other regions are already approaching a plateau. These particular features, regional onset of Aβ accumulation and regional rates of change over time, should be considered in the design of anti-Aβ therapeutic trials.

Having established the rates of Aβ accumulation based on PET results, we next determined the values of these PET units in terms of the mass of Aβ. First, we reestablished quantitative measurements of Aβ in the brain tissue of AD and control subjects (Roberts et al., 2017) (Fig. 3.2). By determining the amounts of Aβ in the four major biochemically defined pools of homogenized gray matter (Tris-buffered saline (TBS), carbonate, urea/detergent, and formate extractable pools), adding these together, we could show that the average AD brain has a total mass of 6.5 mg of Aβ and an aged-matched control brain has 1.7 mg of Aβ (a ratio of 3.8, a similar order of magnitude as found in the Aβ-PET SUVR itself). Across these pools, the concentration of Aβ ranges from low nanomolar (TBS) to low micromolar (the bulk of Aβ resident in the urea/detergent and formate extractable pools) (Fig. 3.2). The carbonate pool showed the largest proportional increased value in the AD group (a 7.7-fold increase over controls). $A\beta_{42}$ (especially the N-terminally truncated $A\beta_{4-42}$) were the predominantly increased species of Aβ in the AD brain, compared with the $A\beta_{40}$ species. In control brains, there was a good correlation between the $A\beta_{40}$ and $A\beta_{42}$ species in the four major fractions, but these correlations were lost for $A\beta_{42}$ in the AD cases. Previous studies of the total Aβ burden in the AD brain have yielded widely disparate figures, although a mean value of 8 mg of $A\beta_{42}$ was established (Cohen et al., 2013).

In passing, we speculate that the Aβ-PET signals reflect the urea/detergent pool more than the formate pool, even though these two pools correlate closely with the Aβ-PET SUVR. This is based on the presumption that an intravenously-administered ligand, with only 60 minutes equilibration time, would not have the capacity to enter the solid fibrillar core of the center of the Aβ-amyloid plaque. In contrast, the TBS and carbonate pools show no or little correlation with the Aβ-PET, and we speculate that these will reflect more the concentrations of Aβ in the CSF and ISF spaces of the brain. The kinetics of the equilibria between the four major biochemical pools remains to be determined. Since the urea/detergent and formate pools constitute more than 90% of the Aβ accumulating in the brain, it is likely that these two pools will be the major determinants of the bulk flow of Aβ.

From a subset of cases in whom an Aβ-PET scan had been conducted before death, we were then able to correlate the Aβ-PET SUVR with the mass of Aβ in the cortical gray matter (Fig. 3.3). This demonstrated that the lower PET SUVR cutoff of 1.4 equates to 5 μg of Aβ/g wet weight of gray matter, and the mean clinical AD upper Aβ-PET SUVR cutoff of 2.33 equates to 11.2 μg of Aβ/g wet weight of gray matter. Since we had previously shown that it takes approximately 19 years to move from the

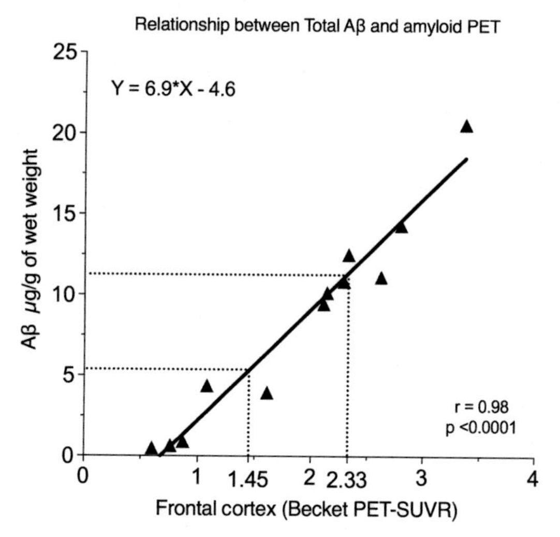

FIGURE 3.3 **Translation of PET SUVR into Aβ mass units.** The lower threshold equates to 5.0 μg and the dementia level 11.2 μg Aβ/g of gray matter; estimated rate of accumulation of Aβ from SUVR 1.4 to 2.3 over 19 years is 28 ng/h (a 2%–5% decrease in clearance rates). *SUVR*, standard uptake value ratio. *Adapted from Roberts, B. R., Lind, M., Wagen, A. Z., Rembach, A., Frugier, T., Li, Q. X., et al. (2017). Biochemically-defined pools of amyloid-β in sporadic Alzheimer's disease: Correlation with amyloid PET. Brain, 140; 1486–1498.*

Aβ-PET SUVR 1.4 to 2.33, we can estimate that Aβ is accumulating at a rate of 28 ng/hour over these two decades. Allowing for a rather wide confidence intervals and uncertainties associated with these calculations, we would state that a first approximation lies within a 2–100-ng/hour order of magnitude. Based on clearance data from Aβ-CSF in vivo studies (see below), we estimate that these values represent approximately a 2%–5% decrease or deficit in the normal clearance rate of Aβ over this time interval. The advent of high performance assays of $A\beta_{42}$ in plasma now present us with an opportunity to explore the kinetics of the equilibria between the central and peripheral pools of $A\beta_{42}$ (Nakamura et al., 2018).

Change Over Time: (iii) The Growth and Accumulation of Plaque Cores

How many years are required for the formation of mature amyloid-Aβ plaques? One way to answer this question is to use racemization of the amino acids aspartic acid and asparagine as a biological clock of proteins. On the assumption that approximately 0.15% D-aspartate are formed per year, the 4.57% D-aspartate of amyloid of plaque cores correspond to approximately 30 years, the time it takes to fully develop these lesions (Müller-Hill & Beyreuther, 1989). As noted above, using carbon-11-labeled Pittsburgh compound B (^{11}C-PiB) PET scans, the accumulation of amyloid-Aβ in sporadic AD was suggested to occur over 20–30 years (Villemagne et al., 2013). The FTR for irreversible loss of $A\beta_{42}$ observed in individuals with Aβ deposition is also consistent with a 15–20-year estimate (Patterson et al., 2015). Using a rate of deposition of $A\beta_{42}$ into plaques of 28 (Roberts et al., 2017) or 31.2 ng/hour (Patterson et al., 2015) yields after 15–20 years 3.7–4.9 or 4.1–5.5 mg $A\beta_{42}$ accumulation per brain, respectively. Accumulation of 4.8 mg $A\beta_{42}$ over 19 years with a rate of 28 ng/hour suggests that approximately 5% of the hourly produced 570 ng $A\beta_{42}$ associates with insoluble amyloid-β-plaques, but this corresponds to only 0.5% of the hourly generated 5800 ng of total Aβ ($A\beta_{38}$, $A\beta_{40}$, $A\beta n_{42}$). That Aβ-plaques consist mainly of $A\beta_{42}$ is not only consistent with a selective FTR for irreversible loss of $A\beta_{42}$ and no corresponding loss of $A\beta_{40}$ or $A\beta_{38}$ in $A\beta^{+}$ individuals (Patterson et al., 2015), but also with $A\beta_{42}$ folding and aggregation (Schmidt et al., 2015). Such $A\beta_{42}$ folding excludes mixed oligomer formation with $A\beta_{40}$ or other Aβ species (Schmidt et al., 2015).

The estimated Aβ accumulation of 28 ng/hour for $A\beta_{42}$ equates to an addition rate of 7.8 pg/second or that approximately 1.2 billion molecules of $A\beta_{42}$ associate with preexisting Aβ deposits each second. Using 10 μm as the diameter of crystalline Aβ-plaques yields a plaque mass of approximately 0.72 ng that correspond to approximately 108 billion $A\beta_{42}$ molecules. The hourly addition of 28 ng or 7.8 pg per second of $A\beta_{42}$ equals a mass of approximately 39 mature plaques, and that approximately every 100 seconds an amount of aggregating $A\beta_{42}$ is retained that corresponds to the mass of a single plaque. The growth rate of a total plaque mass of 4–6 mg divided by single plaque mass yields an approximate estimate of the total average plaque load as being 5.6–8.3 million plaques in a patient with AD. Using an intermediate value of 6.9 million plaques and a plaque buildup time of 20 or 30 years, the simple calculation estimates that approximately 174–261 (range 145–214 and 261–392, respectively) molecules are added to each plaque/second. These simple estimations do not account for the potentially exponential growth in the early phase of Aβ-plaque growth observed by PiB-PET (Villemagne et al., 2013), nor does it account for Aβ degradation and the predicted plateau in plaque load as noted previously (Villemagne et al., 2013) (Fig. 3.1). However, it does suggest that, ideally, disease-modifying nonmedical and medicinal observational and interventional studies in AD should be started before or early in the estimated 15–20 years of Aβ accumulation.

Change Over Time: (iv) Brain Atrophy

Hippocampal and cortical gray matter atrophy, reflecting dendritic tree and synaptic loss, along with ventricular enlargement are typical MRI findings in AD (Frisoni, Fox, Jack, Scheltens, & Thompson, 2010), and it has been postulated that the rates of hippocampal atrophy might predict conversion from MCI to AD (Jack et al., 2008). Interestingly, no association has been observed between hippocampal Aβ burden and hippocampal atrophy at any stage of the disease spectrum. When comparing $A\beta^-$ to $A\beta^+$ nondemented individuals, contrasting results have been reported. On the one hand, hippocampal atrophy was reported in the elderly with high global Aβ burden (Storandt, Mintun, Head, & Morris, 2009), while this was not found in other studies (Dickerson et al., 2009). Along the same line of contradictory findings, significant correlations between hippocampal atrophy and global Aβ burden were reported in some studies of normal controls and MCI subjects (Mormino et al., 2009; Rowe et al., 2010), but not in others (Dickerson et al., 2009). Voxelwise studies assessing the association between regional Aβ deposition and local atrophy reported a significant relationship between atrophy in the posterior and anterior cingulate gyri and the regional Aβ deposition (Chetelat et al., 2010), a relationship that was only found in elderly with subjective cognitive impairment but not in the MCI or in the AD patients. However, by studying in vivo tau, Aβ and gray matter profiles, tau and Aβ-deposits were reported to be associated with distinctive spatial patterns of brain tissue loss. The latter revealed a network of interdigitations of tau and Aβ in the cortical mantle (Sepulcre et al., 2016).

When looking at the rates of atrophy, cognitively normal controls with high global Aβ burden in the brain have shown significantly faster rates of atrophy than those controls with low Aβ burden in the brain (Chetelat et al., 2012; Villemagne et al., 2013). Specifically, rates of -1.6 and $-2.9\ cm^3/year$ (for cortical gray matter atrophy) and $-0.03\ cm^3/year$ (for hippocampal atrophy) and $-0.05\ cm^3/year$ were reported in cognitively normal controls with low and high global Aβ burden in the brain, respectively (Villemagne et al., 2013). In AD patients with high global Aβ burden, the rates of atrophy were even faster than both control groups, with rates of $-4.76\ cm^3/year$ and $-0.098\ cm^3/year$ for cortical gray matter and hippocampal atrophy, respectively (Villemagne et al., 2013).

The rates of hippocampal atrophy are not associated with the rates of Aβ accumulation (Jack et al., 2014). This, together with the aforementioned controversial relationship between Aβ and hippocampal atrophy, might explain the time lag between Aβ becoming abnormal 15–17 years before reaching a Clinical Dementia Rating (CDR) score of 1.0 (dementia), while hippocampal atrophy becomes abnormal just 5 years before reaching a CDR of 1.0.

When implementing a criterion of two markers, Jack et al. (2012) found that about 43% of the cognitively unimpaired elderly had no positive marker of Aβ deposition or neurodegeneration, with 16% presenting high Aβ burden, 12% both high Aβ deposition and neurodegeneration, and, interestingly, 23% were classified with neurodegeneration without evidence of Aβ deposition which was termed "suspected non-AD pathophysiology" (SNAP) (Jack et al., 2012). In the resulting efforts to elucidate the short- and long-term clinical, cognitive, and volumetric trajectories of these four groups, the overwhelming majority showed that, in contrast with those on the AD pathway, those classified as SNAP did not decline over time and were indistinguishable from those elderly controls with no evidence of Aβ or neurodegeneration, suggesting a different and non-AD-underlying pathogenic mechanism (Burnham et al., 2016; Mormino et al., 2014a), most likely vascular or "cognitive frailty" (an ill-defined concept of loss of "cognitive reserve" or the ability of the normal aging brain to withstand the effects of aging per se) (Woods, Cohen, & Pahor, 2013).

The lack of a clear-cut association between Aβ deposition and measures of cognition, synaptic loss, and neurodegeneration in AD suggests that Aβ is an early and necessary, though not sufficient, cause for cognitive decline in AD (Rabinovici & Roberson, 2010; Villemagne et al., 2008), indicating the involvement of other downstream mechanisms, likely triggered by Aβ, such as NFT formation, synaptic failure, and eventual neuronal loss.

Change Over Time: (v) Cognition

In cognitively normal older adults, when cognition is considered at a single assessment, the presence of high levels of Aβ does not manifest as impairment in any aspect of cognitive function (Lim et al., 2013b), although we and others have now shown that when Aβ levels are considered as a continuous rather than a categorical measure at baseline, higher levels of Aβ are moderately associated with worse visual and verbal episodic memory, but only in carriers of the *APOE* ε4 allele (Kantarci et al., 2012; Lim et al., 2013a). When studied prospectively over several years, cognitively normal older adults with a baseline classification of high Aβ show significant and relentless decline on measures of visual and verbal episodic memory when compared to cognitively normal older adults with low Aβ, who showed no decline, with the magnitude of this decline estimated to be approximately 0.1–0.4 standard deviations per year (Baker et al., 2016; Doraiswamy et al., 2014; Hedden, Oh, Younger, & Patel, 2013; Lim et al., 2014a) (see Table 3.1 and Figs 3.4, 3.5). This $Aβ^+$-related decline in episodic memory is also associated with higher rates of progression to a clinical diagnosis of MCI and clinically classified AD dementia itself (Lim et al., 2014a; Rowe et al., 2013). This suggests that the combination of decline in memory with parameters from Aβ imaging may be useful for the identification of AD processes in individuals who do not yet meet any clinical criteria for cognitive impairment.

Despite the reliable decline in episodic memory that can be detected in $Aβ^+$ cognitively normal older adults, there remains much variability in the rate at which this memory decline occurs. We and others have now shown that individual characteristics can act to reduce the neurotoxic effects of $Aβ^+$ on memory decline and brain volume loss. Some researchers have proposed that cognitive reserve, as defined by higher levels of education, occupational complexity, and cognitive stimulation or activity, may increase the brain's ability to withstand neuronal injury or insult before any clinical or cognitive symptoms become evident. These studies have observed that in some older adults, higher levels of cognitive reserve are associated with later onset of clinically diagnosed AD, and even some protection against $Aβ^+$-related memory impairment (Kemppainen et al., 2008; Rentz et al., 2010) and memory decline (Yaffe et al., 2011). However, despite this, the parameters that reflect cognitive reserve have not been defined reliably, with most definitions focusing on indirect indicators such as premorbid intelligence, education levels, or socioeconomic status.

Recently, we have argued that a concept of genetic neural reserve provides a more parsimonious framework for considering risk and protection against $Aβ^+$ neurotoxicity. Within this framework, we and others have found that the presence of even one copy of the *APOE* ε4 allele can increase the rate of memory decline in $Aβ^+$ cognitively normal older adults (Lim et al., 2015b, 2016b; Mormino et al., 2014b). This

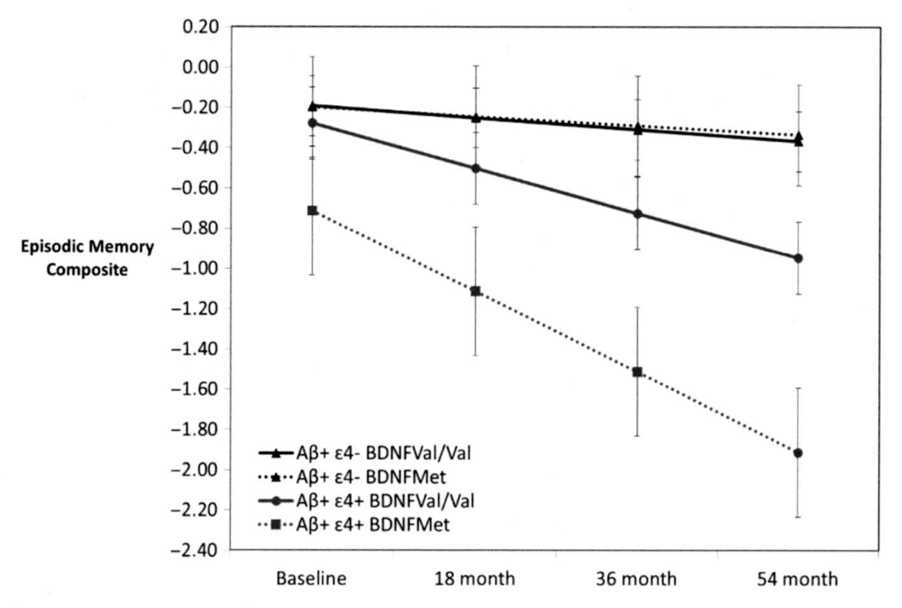

FIGURE 3.4 **54-month data on Aβ$^+$ cognitively normal older adults from AIBL, showing the interactive effects of APOE and BDNF on memory decline.** No effect of *BDNF* was observed in Aβ$^+$ ε4 noncarriers (black lines). Aβ$^+$ ε4 Met66 carriers showed substantially faster decline compared to Aβ$^+$ ε4 Val66 homozygotes. *AIBL*, the Australian Imaging, Biomarker and Lifestyle; *APOE*, apolipoprotein E; *BDNF*, brain-derived neurotrophic factor. *Adapted from Lim, Y. Y., Villemagne, V. L., Laws, S. M., Pietrzak, R. H., Snyder, P. J., Ames, D., et al. (2015a). APOE and BDNF polymorphisms moderate amyloid β-related cognitive decline in preclinical Alzheimer's disease. Molecular Psychiatry, 20, 1322–1328.*

accords with data from epidemiological studies showing that carriage of an *APOE* ε4 allele lowers the age of onset for AD dementia in a gene dose-dependent manner, with the mean age of onset of AD dementia in ε4 homozygotes 68 years, compared to 76 years for ε4 heterozygotes and 84 years for ε4 noncarriers (Raber, Huang, & Ashford, 2004). In addition to *APOE*, a genetic polymorphism related to synaptic excitation and neuronal plasticity in the central nervous system, the brain-derived neurotrophic factor (*BDNF*) Val66Met (rs6265) polymorphism (Fahnestock, 2011; Lu, Nagappan, Guan, Nathan, & Wren, 2013), has also been shown to moderate the rate of memory decline and brain volume loss, in both Aβ$^+$ cognitively normal older adults and MCI patients (Boots et al., 2017; Lim et al., 2013c, 2014b, 2015a). Specifically, we and others have found that Aβ$^+$ Met66 carriers show increased rates of memory decline and brain volume loss when compared to Aβ$^+$ Val66 homozygotes, despite equivalent Aβ levels at baseline, and equivalent rates of Aβ accumulation. This was similarly observed in young mutation carriers with autosomal-dominant AD (pathogenic mutations in *APP*, *PSEN1*, and *PSEN2*), where mutation carriers who were also Met66 carriers showed worse memory performance and hippocampal function than mutation carriers who were Val66 homozygotes, despite equivalent Aβ levels at baseline (Lim et al., 2016a). Intriguingly, mutation carriers who were Met66 carriers also showed higher levels of tau and phosphorylated tau in the cerebrospinal fluid than mutation carriers who were Val66 homozygotes (Lim et al., 2016a), suggesting that the increased rates of memory decline and brain volume loss observed in Met66 carriers may be a result of an interaction between BDNF and tau, although

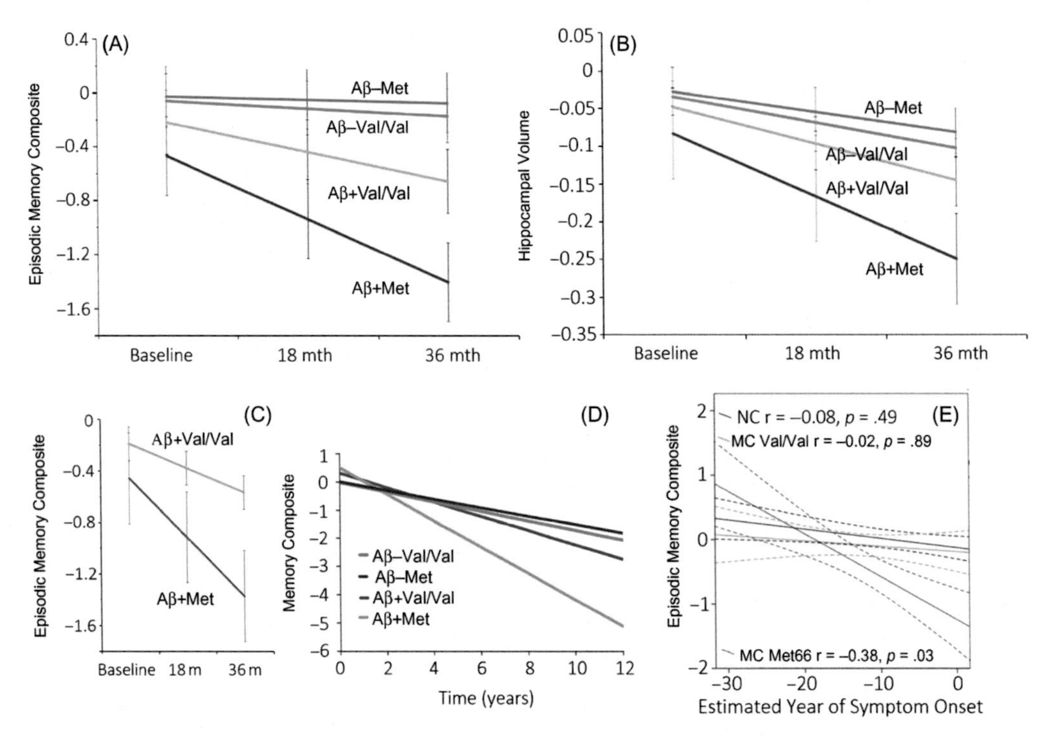

FIGURE 3.5 **Change in cognition over time**. In $A\beta^+$ cognitively normal older adults, *BDNF* Met66 carriers show faster decline on an episodic memory composite (A) and hippocampal volume (B) over 3 years; similarly, *BDNF* Met66 carriers show faster memory decline in episodic memory composite (C), in prodromal AD (aged 60–95), and (D) $A\beta^+$ middle-aged adults (aged 47–73). (E) Baseline data from adults with autosomal dominant AD (aged 18–55) show a strong relationship between memory and estimated year of symptom onset in Met66 carriers (red line), while no such relation was observed in Val66 homozygotes (green line), or nonmutation carriers (black line). *AD*, Alzheimer's disease; *BDNF*, brain-derived neurotrophic factor. *Adapted from Lim, Y. Y., Villemagne, V. L., Laws, S. M., Ames, D., Pietrzak, R. H., Ellis, K. A., et al. (2013c). BDNF Val66Met, Aβ amyloid and cognitive decline in preclinical Alzheimer's disease. Neurobiology of Ageing, 34, 2457–2464 (**A, B**); Lim, Y.Y., Villemagne, V. L., Laws, S. M., Ames, D., Pietrzak, R. H., Ellis, K. A., et al., (2014b) BDNF Val66Met moderates Aβ-related memory decline and hippocampal atrophy in prodromal Alzheimer's disease: A preliminary study. PLoS One, 9, e86498 (**C**); Boots, E. A., Schultz, S. A., Clark, L. R., Racine, A. M., Darst, B. F., Koscik, R. L., et al. (2017) BDNF polymorphism predicts cognitive decline in the Wisconsin Registry for Alzheimer's Prevention. Neurology , 88, 2098–2106 (**D**); and Lim, Y. Y., Hassenstab, J., Cruchaga C., Goate, A., Fagan, A. M., Benzinger T. L. S., et al. (2016a). BDNF Val66Met moderates memory impairment, hippocampal function and tau in preclinical autosomal dominant Alzheimer's disease. Brain, 139, 2766–2777.*

this hypothesis requires confirmation from further study (Figs. 3.4 and 3.5).

CLEARANCE MECHANISMS AND IMPAIRED PHAGOCYTOSIS

There is strong evidence for an age-dependent failure of clearance pathways of hydrophobic Aβ, which is naturally prone to aggregation (Masters et al., 2015). Increasingly, genome-wide association studies (GWAS) have identified loci within the endosomal, lysosomal, autophagic, and phagocytic pathways in sporadic AD. These include ABCA7, CD33, TREM2, and complement receptor 1 (Chan et al., 2015; Crehan, Hardy, & Pocock, 2013; Guerreiro et al., 2013; Vasquez, Fardo, & Estus, 2013). Experimental mouse models of AD also point to a role for innate immune pathways in which monocytic/macrophagic and microglial cells participate in the clearance of Aβ

(Fassbender et al., 2004; Kim et al., 2013; Wang et al., 2015). We have also found an age-dependent decline of phagocytic function in peripheral blood monocytes, brain microglia, and retinal macrophages from C57BL/6 mice (Vessey et al., 2017). Recent progress of antibody-mediated Aβ clearance in human clinical trials (gantenerumab, solanezumab, and aducanumab) also supports this line of investigation (Boche et al., 2008; Ostrowitzki et al., 2012; Sevigny et al., 2016; Siemers et al., 2016).

Since it is difficult to study monocytic/macrophage and microglial activity in the living human brain, we sought to investigate whether these central functions could be assessed in the peripheral pool of monocytes/macrophages in the blood, some of which are proposed to be involved in the normal innate immune pathways of the CNS (Durafourt et al., 2012). Using a standardized real-time tricolor flow cytometry method, we measured the phagocytic function of human peripheral blood monocyte subsets, including nonclassic $CD14^{dim}CD16^{+}$, intermediate $CD14^{+}CD16^{+}$, and classic $CD14^{+}CD16^{-}$ monocytes from preclinical, prodromal, and clinical AD, matched with cognitively normal control subjects (Fig. 3.6A and B). Basal levels of phagocytosis in all three subsets of monocytes were similar between controls and AD patients, while a significant 20%–25% increase of basal phagocytosis was found in subjects with high Aβ burden as assessed by PET scans (Fig. 3.6C), and this alteration was independent of age and *APOE* status (Gu et al., 2016).

Importantly, pretreating cells with glatiramer acetate (Copaxone, a peptide drug found to stimulate phagocytosis) or ATP (an inhibitor of P2X7-mediated phagocytosis) showed a differential response depending on clinical or Aβ-burden status, indicating a relative functional deficit of phagocytosis in relation to Aβ burden (Gu et al., 2016).

Our results therefore demonstrated evidence for a perturbed innate phagocytic function in AD that correlates with brain Aβ burden. The results not only offer insight into the possible causes of failure of clearance of Aβ in sporadic AD but also offer a pathway toward diagnostic biomarkers and treatment strategies targeting innate phagocytosis.

CALCULATING POLYGENIC RISK SCORES

After age, genetic factors are arguably the next greatest contributor to the risk of developing AD, with the *APOE* ε4 allele the most clearly established genetic risk factor (Liu, Kanekiyo, Xu, & Bu, 2013). Whilst the presence of the *APOE* ε4 allele alone has been shown not to have a major effect on the rates of clinical progression, we have shown that it plays a significant role in moderating Aβ-related cognitive decline (Thai et al., 2015). We have also reported that combining genetic risk factors can further account for some of the variability in Aβ-mediated cognitive decline. Initially, we reported carriage of both *APOE* ε4 and *BDNF* met66 significantly influenced rate of cognitive decline such that cognitively normal older adults, with abnormal levels of Aβ, declined at a clinically significant rate (1.5 SD) that would meet criteria for memory impairment after 3 years (Lim et al., 2013a, 2015a).

While such studies, for example, in cognitive decline, have important prognostic impact for the management of preclinical AD and design of secondary prevention clinical trials, *BDNF* and *APOE* are only two genetic factors that may influence this decline. Profiles or algorithms that incorporate a larger number of genetic factors may be able to account for more of the variation in rates of change and therefore be applicable to larger percentages of the disease-at-risk population. Such profiles predicting rates of cognitive change, let alone rates of accumulation of the pathological features, remain to be clearly defined in the early stages of AD. GWAS have identified AD risk and protective loci, and

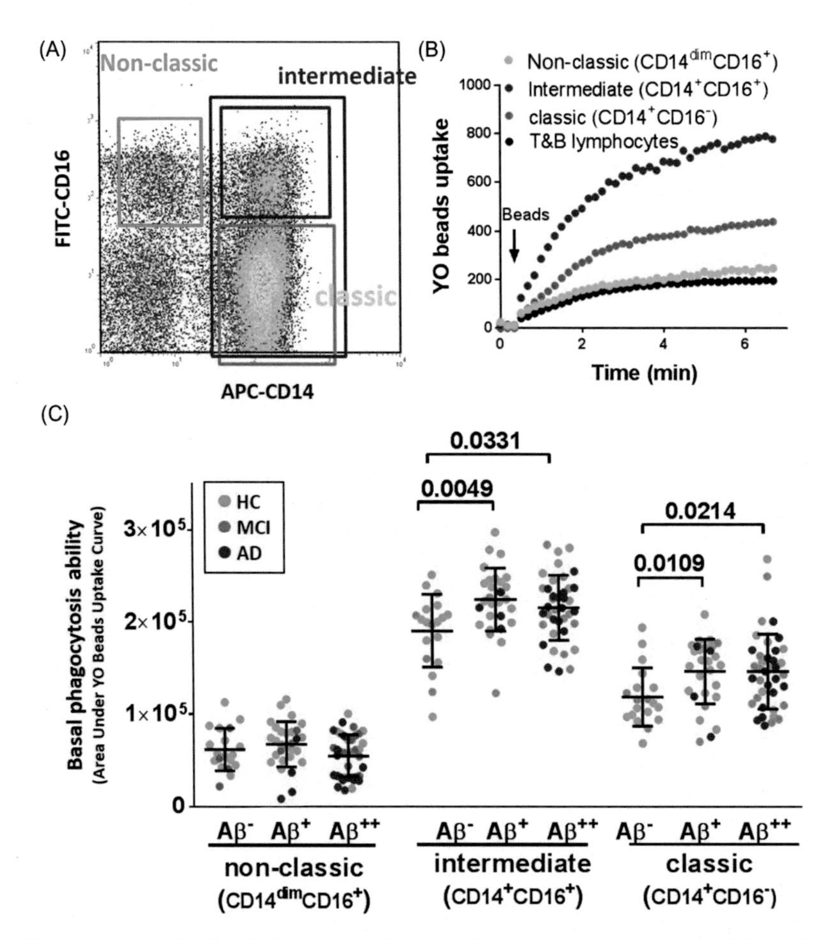

FIGURE 3.6 **Phagocytosis of YO beads by subpopulations of monocytes is altered in AD.** Fresh human peripheral blood mononuclear cells were labeled with allophycocyanin-conjugated CD14 and FITC-conjugated CD16 before the addition of 1 μm YO beads. The YO beads fluorescence intensity was analyzed by real time flow cytometry. (A) A typical example of density plot in CD14 versus CD16 is shown. Monocytes were gated first by forward and side scatter. (B) Typical real-time flow cytometry YO beads uptake curves by $CD14^{dim}CD16^{+}$ (green), $CD14^{+}CD16^{+}$ (red), and $CD14^{+}CD16^{-}$ (brown) monocytes. (C) Basal phagocytic ability of human monocytes subsets as assayed by quantified YO beads uptake. Monocytes from $A\beta^{-}$ ($n = 19$), $A\beta^{+}$ ($n = 29$), and $A\beta^{++}$ ($n = 39$) subjects. Phagocytosis ability was quantitated as arbitrary units of area under bead uptake curve in the first 6 min after the addition of YO beads. *AD*, Alzheimer's disease; *YO*, yellow–orange. *Adapted from Gu, B. J., Huang, X., Ou, A., Rembach, A., Fowler, C., Avula, P. K., et al. (2016). Innate phagocytosis by peripheral blood monocytes is altered in Alzheimer's disease. Acta Neuropathologica, 132, 377–389.*

using complex trait analysis has shown that 33% of total phenotypic variance is explained by all common SNPs (Lambert et al., 2013). Such large GWAS have also provided the initial gene variants used for generating a priori "AD-risk" polygenic risk scores (PRS).

These PRS that incorporate lead AD risk genes have been generally implemented using a weighting by odds ratios or effect sizes (Che & Motsinger-Reif, 2013) derived from those reported in the large GWAS metaanalysis of the IGAP consortium (Lambert et al., 2013) or those from the AlzGene database (Bertram, McQueen, Mullin, Blacker, & Tanzi, 2007). These approaches, using core selection genetic variants, reported significant increases in association with overall risk for AD, an earlier age at onset, cross-sectional CSF biomarkers (Sleegers et al., 2015) and cross-sectional cognitive performance, CSF, and imaging biomarkers (Darst et al., 2017). Likewise, higher genetic risk was associated with faster rates of longitudinal cognitive decline (Andrews, Das, Cherbuin, Anstey, & Easteal, 2016; Marden et al., 2016). However, by focusing on genome-wide significant SNPs, these PRS exclude a significant level of genetic variance explained beyond these GWAS-confirmed variants. More expansive PRS that employ a more liberal level of genetic association (threshold of $P = 0.01$) have provided a greater level of significance across multiple modalities, particularly with respect to longitudinal cognitive decline and rates of clinical progression (Mormino et al., 2016). PRS that take a different approach to simple effect size weighting provide further compelling evidence for the value of PRS in risk stratification and utilization in therapeutic trial design. For example, the use of survival model frameworks that integrate AD-risk SNPs with population-based incidence rates have shown promising utility for deriving age-specific genetic risk for AD on a per-individual basis (Desikan et al., 2017).

Overall, further development of PRS is warranted, and future scores may focus on leveraging data from longitudinal cohort studies. The use of PRS developed from a priori "AD risk" gene candidates could be perceived as biased, as they might exclude variants that, whilst not associated with the greatest level AD risk, may be associated with rates of change in AD-related phenotypes. To overcome this limitation, other approaches could be taken for the generation of cognitive-related gene profiles or rate-of-change specific PGS. The latter approach would require cohorts with significant longitudinal data but would provide an unbiased discovery approach to identify individual genetic variant contributions to rates of change. Using a similar approach to that applied for AD-risk weighted PRS (Che & Motsinger-Reif, 2013), this information could be used to define a true rate of change PRS that could have important clinical utilization.

DISEASE-MODIFYING STRATEGIES: MODELS OF Aβ ACCUMULATION IN ALZHEIMER'S DISEASE—IMPLICATIONS FOR Aβ AMYLOID-TARGETING THERAPIES

In a healthy brain, $A\beta_{42}$, $A\beta_{40}$, and $A\beta_{38}$ move freely through the brain parenchyma. As noted above, its production, concentration, and clearance can be determined by the SILKs method developed by Bateman et al. (2006) and Patterson et al. (2015). The method revealed that in the absence of Aβ accumulation, in CSF, the production and clearance rates of Aβ are 418 ± 83 ng/hour for $A\beta_{40}$, 57 ng/hour for $A\beta_{42}$, and 108 ng/hour for $A\beta_{38}$ (Potter et al., 2013). Thus, approximately 72% of newly produced and cleared Aβ is $A\beta_{40}$, 9.8% is $A\beta_{42}$, and 18.2% is $A\beta_{38}$. In the presence of Aβ plaque accumulation, there is an increased irreversible loss of soluble $A\beta_{42}$ in CSF and a concomitant 10-fold increase in the reversible rate of soluble CSF $A\beta_{42}$ with parenchymal pools of $A\beta_{42}$ (Potter et al., 2013). There is no change in the kinetics of CSF $A\beta_{38}$ and $A\beta_{40}$ with regard to the presence of Aβ plaques. The irreversible loss of soluble $A\beta_{42}$ was >5%, corresponding to at least 28.5 ng of the hourly produced approximately 570 ng of total cerebral $A\beta_{42}$. This strongly suggests that cerebral $A\beta_{42}$ kinetics are altered with Aβ accumulation. These species of newly formed Aβ at 28.5 ng/hour seem to bind to Aβ aggregates of $A\beta_{42}$ or $A\beta_{42}$ fibrils of Aβ-plaques and to drive the progression of fibrillogenesis. The consequence of this is that therapies directed toward lowering of total Aβ production should aim to lower $A\beta_{42}$ production by more than 95% if treatment is started after a large amount of Aβ has accumulated. A 95% lowering is not enough to modify disease progression, since Aβ accumulation in AD is expected to grow by 28.5 ng/hour, a value that is in agreement with the rate of Aβ growth in Aβ-positive participants of the AIBL cohort (Roberts et al., 2017; Villemagne et al., 2013). Accordingly, for disease-modifying preventative treatments, a reduction in total brain Aβ production by 95% would result in a residual $A\beta_{42}$ production of approximately 14 ng/hour. This amount should translate to an approximate 50% reduction of the rate of disease progression, corresponding to a half of the rate of cognitive decline and a doubling of the time it takes for onset of disease. Likewise, at 97.5% reduction of Aβ production, the residual newly produced hourly amount of 7.0 ng of $A\beta_{42}$ should lead to a 80% slowing in the rate of cognitive decline, corresponding to a delay of onset of disease from 7 to 28 years. This means that reduction of Aβ production by inhibition of secretases may have to be performed with doses of inhibitors that are able to inhibit at least 95% of total Aβ generation and not 50% as in previous or current trials. Modeling of this type of dose–response relationship will need to be conducted for each stage of the preclinical and prodromal phases of AD evolution.

The same holds true for antibody-based anti-Aβ immunotherapies. How much anti-Aβ monoclonal antibody is needed to reach a 95% lowering of newly generated soluble $A\beta_{42}$ depends on brain levels of the corresponding anti-Aβ antibody. Upon intravenous infusion, the brain concentration of monoclonal anti-Aβ antibodies may only reach levels of 0.1%–1% of a given dose in mg/kg body weight (Goure, Krafft, Jerecic, & Hefti, 2014). For antibodies binding Aβ monomers, an approximate molar excess of 100–1000 times of antibody to Aβ would be required to remove newly synthesized soluble brain Aβ. Since the human monoclonal anti-Aβ antibodies have a half-life of about 28 days in the human peripheral circulation, removal of the hourly newly produced brain Aβ of approximately 5800 ng would require between 7520 and 75,200 mg of antibody (calculated for brain levels of 1 and 0.1% of total monthly dose, respectively). Theoretically, 10 times less would be required for $A\beta_{42}$-specific antibodies and even less for antibodies with specificity for oligomeric Aβ or Aβ plaques since both contain mainly $A\beta_{42}$ (i.e., a removal of newly synthesized Aβ such that the total cerebral Aβ production falls from 5800 to 580 ng/h and of $A\beta_{42}$ falls from 570 to 57 ng/h). The apparent reduction to baseline levels of Aβ burden by the autoantibody aducanumab over a 12-month interval would be an excellent starting point for validating this quantitative approach (Sevigny et al., 2016).

References

Allsop, D., Landon, M., & Kidd, M. (1983). The isolation and amino acid composition of senile plaque core protein. *Brain Research*, *259*, 348–352.

Alzheimer, A. (1907). Über eine eigenartige erkrankung der hirnrinde. *Allgemeine Zeitschrift Psychiatrie Psychischgerichtliche Medizin*, *64*, 146–148.

Andrews, S. J., Das, D., Cherbuin, N., Anstey, K. J., & Easteal, S. (2016). Association of genetic risk factors with cognitive decline: The PATH through life project. *Neurobiology of Aging*, *41*, 150–158.

Bachman, D. L., Wolf, P. A., Linn, R. T., Knoefel, J. E., Cobb, J. L., Belanger, A. J., ... D'Agostino, R. B. (1993). Incidence of dementia and probable Alzheimer's disease in a general population: The Framingham study. *Neurology*, *43*, 515–519.

Bahmanyar, S., Higgins, G. A., Goldgaber, D., Lewis, D. A., Morrison, J. H., Wilson, M. C., ... Gajdusek, D. C. (1987). Localization of amyloid β protein messenger RNA in brains from patients with Alzheimer's disease. *Science*, *237*, 77–80.

Baker, J. E., Lim, Y. Y., Pietrzak, R. H., Hassenstab, J., Snyder, P. J., Masters, C. L., & Maruff, P. (2016). Cognitive impairment and decline in cognitively normal older adults with high amyloid-β: A meta-analysis. *Alzheimer's and Dementia: Diagnosis, Assessment and Disease Monitoring*, *6*, 108–121.

Bateman, R. J., Munsell, L. Y., Morris, J. C., Swarm, R., Yarasheski, K. E., & Holtzman, D. M. (2006). Human amyloid-β synthesis and clearance rates as measured in cerebrospinal fluid in vivo. *Nature Medicine*, *12*, 856–861.

Bertram, L., McQueen, M. B., Mullin, K., Blacker, D., & Tanzi, R. E. (2007). Systematic meta-analyses of Alzheimer disease genetic association studies: The AlzGene database. *Nature Genetics*, *39*, 17–23.

Beyreuther, K., & Masters, C. L. (1991). Amyloid precursor protein (APP) and βA4 amyloid in the etiology of Alzheimer's disease: Precursor–product relationships in the derangement of neuronal function. *Brain Pathology*, *1*, 241–251.

Boche, D., Zotova, E., Weller, R. O., Love, S., Neal, J. W., Pickering, R. M., ... Nicoll, J. A. (2008). Consequence of Aβ immunization on the vasculature of human Alzheimer's disease brain. *Brain*, *131*, 3299–3310.

Boots, E.A., Schultz, S.A., Clark, L.R., Racine, A.M., Darst, B.F., Koscik, R.L., . . . Okonkwo, O.C. BDNF Val66Met predicts cognitive decline in the Wisconsin Registry for Alzheimer's Prevention. Neurology **88**, 2017, 2098-2106.

Brion, J. P., van den Bosch de Aguilar, P., & Flament-Durand, J. (1985). Senile dementia of the Alzheimer type: Morphological and immunocytochemical studies. In J. Traber, & W. H. Gispen (Eds.), *Senile dementia of the Alzheimer type: Early diagnosis, neuropathology and animal models* (pp. 164–174). Berlin: Springer-Verlag.

Brookmeyer, R., Gray, S., & Kawas, C. (1998). Projections of Alzheimer's disease in the United States and the public health impact of delaying disease onset. *American Journal of Public Health*, *88*, 1337–1342.

Burnham, S. C., Bourgeat, P., Dore, V., Savage, G., Brown, B., Laws, S., ... the AIBL Research Group. (2016). Clinical and cognitive trajectories in cognitively healthy elderly individuals with suspected non-Alzheimer's disease pathophysiology (SNAP) or Alzheimer's disease pathology: A longitudinal study. *The Lancet Neurology*, *15*, 1044–1053.

Chan, G., White, C. C., Winn, P. A., Cimpean, M., Replogle, J. M., Glick, L. R., ... De Jager, P. L. (2015). CD33 modulates TREM2: Convergence of Alzheimer loci. *Nature Neuroscience*, *18*, 1556–1558.

Che, R., & Motsinger-Reif, A. A. (2013). Evaluation of genetic risk score models in the presence of interaction and linkage disequilibrium. *Frontiers in Genetics*, *4*, 138.

Chetelat, G., Villemagne, V. L., Bourgeat, P., Pike, K. E., Jones, G., Ames, D., ... Rowe, C. C. (2010). Relationship between atrophy and β-amyloid deposition in Alzheimer disease. *Annals of Neurology*, *67*, 317–324.

Chetelat, G., Villemagne, V. L., Villain, N., Jones, G., Ellis, K. A., Ames, D., . . . Rowe, C. C. (2012). Accelerated cortical atrophy in cognitively normal elderly with high beta-amyloid deposition. *Neurology*, *78*, 477–484.

Cohen, S. I., Linse, S., Luheshi, L. M., Hellstrand, E., White, D. A., Rajah, L., ... Knowles, T. P. (2013). Proliferation of amyloid-β42 aggregates occurs through a secondary nucleation mechanism. *Proceedings of the National Academy of Science of the United States of America*, *110*, 9758–9763.

Crehan, H., Hardy, J., & Pocock, J. (2013). Blockage of CR1 prevents activation of rodent microglia. *Neurobiology of Disease*, *54*, 139–149.

Darst, B. F., Koscik, R. L., Racine, A. M., Oh, J. M., Krause, R. A., Carlsson, C. M., ... Engelman, C. D. (2017). Pathway-specific polygenic risk scores as predictors of amyloid-β deposition and cognitive function in a sample at increased risk for Alzheimer's disease. *Journal of Alzheimer's Disease*, *55*, 473–484.

Davies, L., Wolska, B., Hilbich, C., Multhaup, G., Martins, R., Simms, G., ... Masters, C. L. (1988). A4 amyloid protein deposition and the diagnosis of Alzheimer's disease: Prevalence in aged brains determined by immunocytochemistry compared with conventional neuropathologic techniques. *Neurology*, *38*, 1688–1693.

Desikan, R. S., Fan, C. C., Wang, Y., Schork, A. J., Cabral, H. J., Cupples, L. A., ... Dale, A. M. (2017). Genetic assessment of age-associated Alzheimer disease risk: Development and validation of a polygenic hazard score. *PLoS Medicine*, *14*(3), e1002258.

Dickerson, B. C., Bakkour, A., Salat, D. H., Feczko, E., Pacheco, J., Greve, D. N., ... Buckner, R. L. (2009). The cortical signature of Alzheimer's disease: Regionally specific cortical thinning relates to symptom severity in very mild to mild AD dementia and is detectable in asymptomatic amyloid-positive individuals. *Cerebral Cortex*, *19*, 497–510.

Doraiswamy, P. M., Sperling, R. A., Johnson, K., Reiman, E. M., Wong, T. Z., Sabbagh, M. N., ... the Av45-A11 Study Group. (2014). Florbetapir F 18 amyloid PET and 36-month cognitive decline: A prospective multicenter study. *Molecular Psychiatry*, *19*, 1044–1051.

Durafourt, B. A., Moore, C. S., Zammit, D. A., Johnson, T. A., Zaguia, F., Guiot, M. C., ... Antel, J. P. (2012). Comparison of polarization properties of human adult microglia and blood-derived macrophages. *Glia*, *60*, 717–727.

Ellis, K. A., Szoeke, C., Bush, A. I., Darby, D., Graham, P. L., Lautenschlager, N. T., ... Ames, D. (2014). Rates of diagnostic transition and cognitive change at 18-month follow-up among 1,112 participants in the Australian Imaging, Biomarkers and Lifestyle Flagship Study of Ageing (AIBL). *International Psychogeriatrics*, *26*, 543–554.

Evans, D. A., Bennett, D. A., Wilson, R. S., Bienias, J. L., Morris, M. C., Scherr, P. A., ... Schneider, J. (2003). Incidence of Alzheimer disease in a biracial urban community: Relation to apolipoprotein E allele status. *Archives of Neurology*, *6*, 185–189.

Fahnestock, M. (2011). Brain-derived neurotrophic factor: The link between amyloid-β and memory loss. *Future Neurology*, *6*, 627–639.

Fassbender, K., Walter, S., Kühl, S., Landmann, R., Ishii, K., Bertsch, T., ... Beyreuther, K. (2004). The LPS receptor (CD14) links innate immunity with Alzheimer's disease. *FASEB Journal, 18*(1), 203–205.

Frisoni, G. B., Fox, N. C., Jack, C. R., Scheltens, P., & Thompson, P. M. (2010). The clinical use of structural MRI in Alzheimer disease. *Nature Reviews Neurology, 6*, 67–77.

GBD 2013 DALYs and HALE Collaborators. (2015). Global, regional, and national disability-adjusted life years (DALYs) for 306 diseases and injuries and healthy life expectancy (HALE) for 188 countries, 1990–2013: Quantifying the epidemiological transition. *The Lancet, 386*, 2145–2191.

Gersbacher, M. T., Goodger, Z. V., Trutzel, A., Bundschuh, D., Nitsch, R. M., & Konietzko, U. (2013). Turnover of amyloid precursor protein family members determines their nuclear signaling capability. *PLoS ONE, 8*(7), e69363. Available from https://doi.org/10.1371/journal.pone.0069363.

Glenner, G. G., & Wong, C. W. (1984a). Alzheimer's disease: Initial report of the purification and characterization of a novel cerebrovascular amyloid protein. *Biochemical and Biophysical Research Communications, 120*, 885–890.

Glenner, G. G., & Wong, C. W. (1984b). Alzheimer's disease and Down's syndrome: Sharing of a unique cerebrovascular amyloid fibril protein. *Biochemical and Biophysical Research Communications, 122*, 1131–1135.

Goure, W. F., Krafft, G. A., Jerecic, J., & Hefti, F. (2014). Targeting the proper amyloid-beta neuronal toxins: A path forward for Alzheimer's disease immunotherapeutics. *Alzheimer's Research and Therapy, 6*, 42.

Gu, B. J., Huang, X., Ou, A., Rembach, A., Fowler, C., Avula, P. K., ... Masters, C. L. (2016). Innate phagocytosis by peripheral blood monocytes is altered in Alzheimer's disease. *Acta Neuropathologica, 132*, 377–389.

Guerreiro, R., Wojtas, A., Bras, J., Carrasquillo, M., Rogaeva, E., Majounie, E., ... Alzheimer Genetic Analysis Group. (2013). TREM2 variants in Alzheimer's disease. *New England Journal of Medicine, 368*, 117–127.

Hardy, J. A., & Higgins, G. A. (1992). Alzheimer's disease: The amyloid cascade hypothesis. *Science, 256*, 184–185.

Hardy, J. A., & Selkoe, D. J. (2002). The amyloid hypothesis of Alzheimer's disease: Progress and problems on the road to therapeutics. *Science, 297*, 353–356.

Hebert, L. E., Scherr, P. A., Beckett, L. A., Albert, M. S., Pilgrim, D. M., Chown, M. J., ... Evans, D. A. (1995). Age-specific incidence of Alzheimer's disease in a community population. *Journal of the American Medical Association, 273*, 1354–1359.

Hedden, T., Oh, H., Younger, A. P., & Patel, T. A. (2013). Meta-analysis of amyloid-cognition relations in cognitively normal older adults. *Neurology, 80*, 1341–1348.

Huang, K.-l, Marcora, E., Pimenova, A., Di Narzo, A., Kapoor, M., Jin, S. C., ... Goate, A. (2017). A common haplotype lowers SPI1 (PU.1) expression in myeloid cells and delays age at onset for Alzheimer's disease. *bioRxiv*. Available from https://doi.org/10.1101/110957.

Huang, Y., Potter, R., Sigurdson, W., Santacruz, A., Shih, S., Ju, Y. E., ... Bateman, R. J. (2012). Effects of age and amyloid deposition on Aβ dynamics in the human central nervous system. *Archives of Neurology, 69*, 51–58.

Jack, C. R., Lowe, V. J., Senjem, M. L., Weigand, S. D., Kemp, B. J., Shiung, M. M., ... Petersen, R. C. (2008). C^{11} PiB and structural MRI provide complementary information in imaging of Alzheimer's disease and amnestic mild cognitive impairment. *Brain, 131*, 665–680.

Jack, C. R., Vemuri, P., Wiste, H. J., Weigand, S. D., Lesnick, T. G., Lowe, V., ... Knopman, D. S. (2012). Shapes of the trajectories of 5 major biomarkers of Alzheimer disease. *Archives of Neurology, 69*, 856–867.

Jack, C. R., Wiste, H. J., Weigand, S. D., Rocca, W. A., Knopman, D. S., Mielke, M. M., ... Petersen, R. C. (2014). Age-specific population frequencies of cerebral β-amyloidosis and neurodegeneration among people with normal cognitive function aged 50–89 years: A cross-sectional study. *The Lancet Neurology, 13*, 997–1005.

Jack, C. R., Wiste, H. J., Weigand, S. D., Therneau, T. M., Knopman, D. S., Lowe, V., ... Petersen, R. C. (2017). Age-specific and sex-specific prevalence of cerebral β-amyloidosis, tauopathy, and neurodegeneration in cognitively unimpaired individuals aged 50–95 years: A cross-sectional study. *The Lancet Neurology, 16*, 435–444.

Jansen, W. J., Ossenkoppele, R., Knol, D. L., Tijms, B. M., Scheltens, P., Verhey, F. R., ... Aalten, P. (2015). Prevalence of cerebral amyloid pathology in persons without dementia: A meta-analysis. *Journal of the American Medical Association, 313*, 1924–1938.

Kang, J., Lemaire, H., Unterbeck, A., Salbaum, J. M., Masters, C. L., Grzeschik, K., ... Müller-Hill, B. (1987). The precursor of Alzheimer's disease amyloid A4 protein resembles a cell-surface receptor. *Nature, 325*, 733–736.

Kantarci, K., Lowe, V., Przybelski, S. A., Weigand, S. D., Senjem, M. L., Ivnik, R. J., ... Jack, C. R. (2012). *APOE* modifies the association between Aβ load and cognition in cognitively normal older adults. *Neurology, 78*, 232–240.

Kawas, C., Gray, S., Brookmeyer, R., Fozard, J., & Zonderman, A. (2000). Age-specific incidence rates of Alzheimer's disease: The Baltimore Longitudinal Study of Aging. *Neurology, 54*, 2072–2077.

Kemppainen, N. M., Aalto, S., Karrasch, M., Någren, K., Savisto, N., Oikonen, V., ... Rinne, J. O. (2008). Cognitive reserve hypothesis: Pittsburgh compound B and fluorodeoxyglucose positron emission tomography in relation to education in mild Alzheimer's disease. *Annals of Neurology, 63*, 112–118.

Kim, W. S., Hongyun, L., Ruberu, K., Chan, S., Elliott, D. A., Low, J. K., ... Garner, B. (2013). Deletion of ABCA7 increases cerebral amyloid-β accumulation in the J20 mouse model of Alzheimer's disease. *Journal of Neuroscience, 33*, 4387–4394.

Klunk, W. E., Koeppe, R. A., Price, J. C., Benzinger, T. L., Devous, M. D., Sr., Jagust, W. J., ... Mintun, M. A. (2015). The Centiloid project: Standardizing quantitative amyloid plaque estimation by PET. *Alzheimer's and Dementia, 11*, 1–15.

Lambert, J. C., Ibrahim-Verbaas, C. A., Harold, D., Naj, A. C., Sims, R., Bellenguez, C., DeStafano, A. L., et al. (2013). Meta-analysis of 74,046 individuals identifies 11 new susceptibility loci for Alzheimer's disease. *Nature Genetics, 45*, 1452–1458.

Lim, Y. Y., Ellis, K. A., Ames, D., Darby, D., Harrington, K., Martins, R. N., ... the AIBL Research Group. (2013a). Aβ amyloid, cognition and *APOE* genotype in healthy older adults. *Alzheimer's and Dementia, 9*, 538–545.

Lim, Y. Y., Ellis, K. A., Harrington, K., Kamer, A., Pietrzak, R. H., Bush, A. I., ... the AIBL Research Group. (2013b). Cognitive consequences of high Aβ amyloid in mild cognitive impairment and healthy older adults: Implications for early detection of Alzheimer's disease. *Neuropsychology, 27*, 322–332.

Lim, Y. Y., Hassenstab, J., Cruchaga, C., Goate, A., Fagan, A. M., Benzinger, T. L. S., ... the Dominantly Inherited Alzheimer's Network. (2016a). BDNF Val66Met moderates memory impairment, hippocampal function and tau in preclinical autosomal dominant Alzheimer's disease. *Brain, 139*, 2766–2777.

Lim, Y. Y., Laws, S. M., Villemagne, V. L., Pietrzak, R. H., Porter, T., Ames, D., ... Maruff, P. (2016b). Aβ-related memory decline in APOE ε4 non-carriers: Implications for Alzheimer's disease. *Neurology, 86*, 1635–1642.

Lim, Y. Y., Maruff, P., Pietrzak, R. H., Ames, D., Ellis, K. A., Harrington, K., ... the AIBL Research Group. (2014a). Effect of amyloid on memory and non-memory decline from preclinical to clinical Alzheimer's disease. *Brain, 137*, 221–231.

Lim, Y. Y., Villemagne, V. L., Laws, S. M., Ames, D., Pietrzak, R. H., Ellis, K. A., ... the AIBL Research Group. (2013c). BDNF Val66Met, Aβ amyloid and cognitive decline in preclinical Alzheimer's disease. *Neurobiology of Aging, 34*, 2457–2464.

Lim, Y. Y., Villemagne, V. L., Laws, S. M., Ames, D., Pietrzak, R. H., Ellis, K. A., ... Maruff, P. (2014b). BDNF Val66Met moderates Aβ-related memory decline and hippocampal atrophy in prodromal Alzheimer's disease: A preliminary study. *PLoS ONE, 9*, e86498.

Lim, Y. Y., Villemagne, V. L., Laws, S. M., Pietrzak, R. H., Snyder, P. J., Ames, D., ... Maruff, P. (2015a). APOE and BDNF polymorphisms moderate amyloid β-related cognitive decline in preclinical Alzheimer's disease. *Molecular Psychiatry, 20*, 1322–1328.

Lim, Y. Y., Villemagne, V. L., Pietrzak, R. H., Ames, D., Ellis, K. A., Harrington, K., ... Maruff, P. (2015b). APOE ε4 moderates amyloid-related memory decline in preclinical Alzheimer's disease. *Neurobiology of Aging, 36*, 1239–1244.

Liu, C. C., Kanekiyo, T., Xu, H., & Bu, G. (2013). Apolipoprotein E and Alzheimer disease: Risk, mechanisms and therapy. *Nature Reviews Neurology, 9*, 106–118.

Lu, B., Nagappan, G., Guan, X., Nathan, P. J., & Wren, P. (2013). BDNF-based synaptic repair as a disease-modifying strategy for neurodegenerative diseases. *Nature Reviews Neuroscience, 14*, 401–416.

Lucey, B. P., Mawuenyega, K. G., Patterson, B. W., Elbert, D. L., Ovod, V., Kasten, T., ... Bateman, R. J. (2017). Associations between β-amyloid kinetics and the β-amyloid diurnal pattern in the central nervous system. *JAMA Neurology, 74*, 207–215.

Marden, J. R., Mayeda, E. R., Walter, S., Vivot, A., Tchetgen Tchetgen, E. J., Kawachi, I., & Glymour, M. M. (2016). Using an Alzheimer disease polygenic risk score to predict memory decline in black and white Americans over 14 years of follow-up. *Alzheimer Disease and Associated Disorders, 30*, 195–202.

Masters, C. L., Bateman, R., Blennow, K., Rowe, C. C., Sperling, R. A., & Cummings, J. L. (2015). Alzheimer's disease. *Nature Reviews Disease Primers, 1*, 15056. Available from https://doi.org/10.1038/nrdp.2015.56.

Masters, C. L., Simms, G., Weinman, N. A., Multhaup, G., McDonald, B. L., & Beyreuther, K. (1985). Amyloid plaque core protein in Alzheimer disease and Down syndrome. *Proceedings of the National Academy of Science of the United States of America, 82*, 4245–4249.

Mawuenyega, K. G., Sigurdson, W., Ovod, V., Munsell, L., Kasten, T., Morris, J. C., ... Bateman, R. J. (2010). Decreased clearance of CNS β-amyloid in Alzheimer's disease. *Science, 330*, 1774.

Mormino, E. C., Betensky, R. A., Hedden, T., Schultz, A. P., Amariglio, R. E., Rentz, D. M., ... Sperling, R. A. (2014a). Synergistic effect of β-amyloid and neurodegeneration on cognitive decline in clinically normal individuals. *JAMA Neurology, 71*, 1379–1385.

Mormino, E. C., Betensky, R. A., Hedden, T., Schultz, A. P., Ward, A., Huijbers, W., ... Sperling, R. A. (2014b). Amyloid and APOE E4 interact to influence short-term decline in preclinical Alzheimer's disease. *Neurology, 82*, 1760–1767.

Mormino, E. C., Kluth, J. T., Madison, C. M., Rabinovici, G. D., Baker, S. L., Miller, B. L., ... Alzheimer's Dis Neuroimaging Initiative. (2009). Episodic memory loss is related to hippocampal-mediated β-amyloid deposition in elderly subjects. *Brain, 132*, 1310–1323.

Mormino, E. C., Sperling, R. A., Holmes, A. J., Buckner, R. L., De Jager, P. L., Smoller, J. W., ... Alzheimer's Disease Neuroimaging Initiative. (2016). Polygenic risk of Alzheimer disease is associated with early- and late-life processes. *Neurology, 87*, 481–488.

Müller-Hill, B., & Beyreuther, K. (1989). Molecular biology of Alzheimer's disease. *Annual Review of Biochemistry, 58*, 287–307.

Nakamura, A., Kaneko, N., Villemagne, V. L., Kato, T., Doecke, J., Doré, V., ... Yanagisawa, K. (2018). High performance plasma amyloid-β biomarkers for Alzheimer's disease. *Nature, 554*, 249–254.

Ostrowitzki, S., Deptula, D., Thurfjell, L., Barkhof, F., Bohrmann, B., Brooks, D. J., ... Santarelli, L. (2012). Mechanism of amyloid removal in patients with Alzheimer disease treated with gantenerumab. *Archives of Neurology, 69*, 198–207.

Patterson, B. W., Elbert, D. L., Mawuenyega, K. G., Kasten, T., Ovod, V., Ma, S., ... Bateman, R. J. (2015). Age and amyloid effects on human central nervous system amyloid-beta kinetics. *Annals of Neurology, 78*, 439–453.

Potter, H., & Wisniewski, T. (2012). Apolipoprotein E: Essential catalyst of the Alzheimer amyloid cascade. *International Journal of Alzheimer's Disease, 2012*. Available from https://doi.org/10.1155/2012/489428, 489428.

Potter, R., Patterson, B. W., Elbert, D. L., Ovod, V., Kasten, T., Sigurdson, W., ... Bateman, R. J. (2013). Increased in vivo amyloid-β42 production, exchange, and loss in presenilin mutation carriers. *Science Translational Medicine, 5*(189). Available from https://doi.org/10.1126/scitranslmed.3005615.

Prince, M., Wimo, A., Guerchet, M., Ali, G.-C., Wu, Y.-T., & Prina, M. (2015). *World Alzheimer's report 2015: The global impact of dementia, an analysis of prevalence, incidence, cost and trends*. London: Alzheimer's Disease International (ADI).

Raber, J., Huang, Y., & Ashford, J. W. (2004). *APOE* genotype accounts for the vast majority of AD risk and AD pathology. *Neurobiology of Aging, 25*, 641–650.

Rabinovici, G. D., & Roberson, E. D. (2010). Beyond diagnosis: What biomarkers are teaching us about the "bio"logy of Alzheimer disease. *Annals of Neurology, 67*, 283–285.

Rentz, D. M., Locascio, J. J., Becker, J. A., Moran, E. K., Eng, E., Buckner, R. L., ... Johnson, K. A. (2010). Cognition, reserve, and amyloid deposition in normal aging. *Annals of Neurology, 67*, 353–364.

Roberts, B. R., Lind, M., Wagen, A. Z., Rembach, A., Frugier, T., Li, Q. X., ... Masters, C. L. (2017). Biochemically-defined pools of amyloid-β in sporadic Alzheimer's disease: Correlation with amyloid PET. *Brain, 140*, 1486–1498.

Rowe, C. C., Bourgeat, P., Ellis, K. A., Brown, B., Lim, Y. Y., Mulligan, R., ... the AIBL Research Group. (2013). Predicting Alzheimer disease with β-amyloid imaging: Results from the Australian Imaging, Biomarkers, and Lifestyle study of ageing. *Annals of Neurology, 74*, 905–913.

Rowe, C. C., Ellis, K. A., Rimajova, M., Bourgeat, P., Pike, K. E., Jones, G., ... Villemagne, V. L. (2010). Amyloid imaging results from the Australian Imaging, Biomarkers and Lifestyle (AIBL) study of aging. *Neurobiology of Aging, 31*, 1275–1283.

Schmidt, M., Rohou, A., Lasker, K., Yadav, J. K., Schiene-Fischer, C., Fandrich, M., & Grigorieff, N. (2015). Peptide dimer structure in an Aβ(1-42) fibril visualized with cryo-EM. *Proceedings of the National Academy of Sciences of the United States of America, 112*, 11858–11863.

Sepulcre, J., Schultz, A. P., Sabuncu, M., Gomez-Isla, T., Chhatwal, J., Becker, A., ... Johnson, K. A. (2016). In vivo tau, amyloid, and gray matter profiles in the aging brain. *Journal of Neuroscience, 36*, 7364–7374.

Sevigny, J., Chiao, P., Bussiere, T., Weinreb, P. H., Williams, L., Maier, M., ... Sandrock, A. (2016). The antibody aducanumab reduces Aβ plaques in Alzheimer's disease. *Nature, 537*, 50–56.

Siemers, E. R., Sundell, K. L., Carlson, C., Case, M., Sethuraman, G., Liu-Seifert, H., ... Demattos, R. (2016). Phase 3 solanezumab trials: Secondary outcomes in mild Alzheimer's disease patients. *Alzheimer's and Dementia, 12*, 110–120.

Sleegers, K., Bettens, K., De Roeck, A., Van Cauwenberghe, C., Cuyvers, E., Verheijen, J., ... BELNEU consortium. (2015). A 22-single nucleotide polymorphism Alzheimer's disease risk score correlates with family history, onset age, and cerebrospinal fluid Aβ42. *Alzheimer's and Dementia, 11*, 1452–1460.

Sperling, R., Mormino, E., & Johnson, K. (2014). The evolution of preclinical Alzheimer's disease: Implications for prevention trials. *Neuron, 84*, 608–622.

Storandt, M., Mintun, M. A., Head, D., & Morris, J. C. (2009). Cognitive decline and brain volume loss as signatures of cerebral amyloid-β peptide deposition identified with Pittsburgh compound B: Cognitive decline associated with Abeta deposition. *Archives of Neurology, 66*, 1476–1481.

Thai, C., Lim, Y. Y., Villemagne, V. L., Laws, S. M., Ames, D., Ellis, K. A., … … Australian Imaging, Biomarkers and Lifestyle Research Group. (2015). Amyloid-related memory decline in preclinical Alzheimer's disease is dependent on apoe ε4 and is detectable over 18-months. *PLoS ONE*, *10*, e0139082.

Tosun, D., Landau, S., Aisen, P. S., Petersen, R. C., Mintun, M., Jagust, W., … … Alzheimer's Disease Neuroimaging Initiative. (2017). Association between tau deposition and antecedent amyloid-beta accumulation rates in normal and early symptomatic individuals. *Brain*, *140*, 1499–1512.

Vasquez, J. B., Fardo, D. W., & Estus, S. (2013). ABCA7 expression is associated with Alzheimer's disease polymorphism and disease status. *Neuroscience Letters*, *556*, 58–62.

Vessey, K. A., Gu, B. J., Jobling, A. I., Phipps, J. A., Greferath, U., Tran, M. X., … … Fletcher, E. L. (2017). Loss of function of P2X7 receptor scavenger activity produces an early age related macular degeneration phenotype in mice. *American Journal of Pathology*, *187*, 1670–1685.

Villemagne, V. L., Burnham, S., Bourgeat, P., Brown, B., Ellis, K. A., Salvado, O., … … Masters, C. L. (2013). Amyloid β deposition, neurodegeneration, and cognitive decline in sporadic Alzheimer's disease: A prospective cohort study. *The Lancet Neurology*, *12*, 357–367.

Villemagne, V. L., Pike, K. E., Chetelat, G., Ellis, K. A., Mulligan, R. S., Bourgeat, P., … … Rowe, C. C. (2011). Longitudinal assessment of A β and cognition in aging and Alzheimer disease. *Annals of Neurology*, *69*, 181–192.

Villemagne, V. L., Pike, K. E., Darby, D., Maruff, P., Savage, G., Ng, S., … … Rowe, C. C. (2008). Aβ deposits in older non-demented individuals with cognitive decline are indicative of preclinical Alzheimer's disease. *Neuropsychologia*, *46*, 1688–1697.

Wang, Y., Cella, M., Mallinson, K., Ulrich, J. D., Young, K. L., Robinette, M. L., … … Colonna, M. (2015). TREM2 lipid sensing sustains the microglial response in an Alzheimer's disease model. *Cell*, *160*, 1–11.

Weidemann, A., König, G., Bunke, D., Fischer, P., Salbaum, J. M., Masters, C. L., & Beyreuther, K. (1989). Identification, biogenesis, and localization of precursors of Alzheimer's disease A4 amyloid protein. *Cell*, *57*, 115–126.

Woods, A. J., Cohen, R. A., & Pahor, M. (2013). Cognitive frailty: Frontiers and challenges. *The Journal of Nutrition Health and Aging*, *17*, 741–743.

Xiong, C., Jasielec, M. S., Weng, H., Fagan, A. M., Benzinger, T. L., Head, D., … … Morris, J. C. (2016). Longitudinal relationships among biomarkers for Alzheimer disease in the Adult Children Study. *Neurology*, *86*, 1499–1506.

Yaffe, K., Weston, A., Graff-Radford, N. R., Satterfield, S., Simonsick, E. M., Younkin, S. G., … … Harris, T. B. (2011). Association of plasma β-amyloid level and cognitive reserve with subsequent cognitive decline. *Journal of the American Medical Association*, *305*, 261–266.

CHAPTER

4

Neurodegeneration and the Ordered Assembly of Tau

Michel Goedert

MRC Laboratory of Molecular Biology, Cambridge, United Kingdom

OUTLINE

INTRODUCTION

Common human neurodegenerative diseases, including Alzheimer's disease (AD) and Parkinson's disease (PD), are characterized by the presence of abundant inclusions with the properties of amyloid filaments (Goedert, 2015). Each inclusion has only one protein as its major component, with Aβ, Tau, and α-synuclein being the most commonly involved. Most neurodegenerative diseases are defined by a single type of inclusion. AD, the most prevalent, is characterized by two different inclusions, abundant extracellular Aβ deposits, and intraneuronal Tau aggregates.

Aβ, Tau, and α-synuclein undergo transformation from a soluble to an insoluble filamentous amyloid state, with several intermediates. Most cases of disease are sporadic, but a small percentage is inherited, often in a dominant manner. The latter are caused by mutations in the genes encoding the proteins that make up the inclusions, or proteins that increase their production, underscoring the importance of inclusion formation for neurodegeneration. The finding that the three genes, mutations in which cause familial AD [Aβ precursor protein (APP), presenilin-1, and presenilin-2], influence APP processing, and are involved in the production of Aβ peptides (Goate et al., 1991; Sherrington

The Molecular and Cellular Basis of Neurodegenerative Diseases
DOI: https://doi.org/10.1016/B978-0-12-811304-2.00004-3

et al., 1995), led to the amyloid cascade hypothesis (Hardy & Allsop, 1991). Mutations in *MAPT*, the microtubule-associated protein Tau gene, give rise to an inherited form of frontotemporal dementia and parkinsonism with abundant filamentous Tau inclusions in brain, in the absence of Aβ deposits (Hutton et al., 1998; Poorkaj et al., 1998; Spillantini et al., 1998). Although APP dysfunction may initiate a cascade of events in AD, Tau aggregation appears to be the efficient cause of neurodegeneration.

Until recently, cell autonomous mechanisms were believed to account for sporadic human neurodegenerative diseases, implying that the same aggregation events occur independently in many otherwise healthy brain cells, resulting in degeneration. At death, protein inclusions are present in thousands of postmitotic nerve cells. Alternatively, the first inclusions may form in a localized fashion, from where they propagate to normal cells through cell nonautonomous mechanisms, resulting in degeneration.

Propagation is consistent with staging schemes, which have postulated that Tau inclusions progress from a limited number of sites in a predictable manner [locus coeruleus and transentorhinal cortex in AD, cortical sulci in chronic traumatic encephalopathy (CTE), frontotemporal cortex in Pick's disease (PiD), substantia nigra and subthalamic nucleus in progressive supranuclear palsy (PSP), striatum in corticobasal degeneration (CBD), and ambient gyrus in argyrophilic grain disease (AGD)] (Braak & Braak, 1991; Braak & Del Tredici, 2011; Irwin et al., 2016; Ling et al., 2016; McKee et al., 2013; Saito et al., 2004; Yoshida et al., 2017). However, staging is also compatible with brain regions being affected sequentially, without aggregate transfer. Propagation is supported by the absence of Tau inclusions in the disconnected frontal cortex of an individual with AD who had undergone an operation to remove a meningioma 27 years earlier (Duyckaerts, Uchihara, Seilhean, He, & Hauw, 1997). Although there were abundant Tau inclusions in limbic and isocortical regions, there were none in the disconnected frontal cortex.

Experimental studies in mice have shown that the injection of Tau inclusions induces nerve cells to form intracellular inclusions at the injection sites, from where they can spread to distant brain regions (Clavaguera et al., 2009). Provided these findings are related to what happens in human brain, cell nonautonomous mechanisms and spreading must also be at work in human neurodegenerative diseases.

Propagation of pathology is often called prion-like, referring to the formation of ordered protein assemblies and their intercellular spread. The acronym "prion" stands for "proteinaceous infectious particle," encompassing the formation of ordered assemblies, intercellular propagation, and interorganismal transmission, but not necessarily amyloid formation (Prusiner, 1998). There is no evidence to suggest that human tauopathies can transfer between individuals, hence our use of prion-like. Propagation of aggregates requires their release into the extracellular space, uptake by connected cells, and seeded aggregation of soluble proteins. Studying the underlying mechanisms may lead to the identification of novel therapeutic targets.

TAU ISOFORMS

Tau is expressed predominantly in the central and peripheral nervous systems, where it is most abundant in nerve cell axons. It can be divided into an N-terminal projection domain, a proline-rich region, a repeat region, and a C-terminal domain.

Six Tau isoforms ranging from 352 to 441 amino acids are expressed in adult human brain, produced by alternative mRNA splicing of transcripts from *MAPT* on chromosome 17q21.31 (Fig. 4.1A) (Goedert, Spillantini, Jakes, Rutherford, & Crowther, 1989). They differ by the presence or absence of inserts of 29 or 58

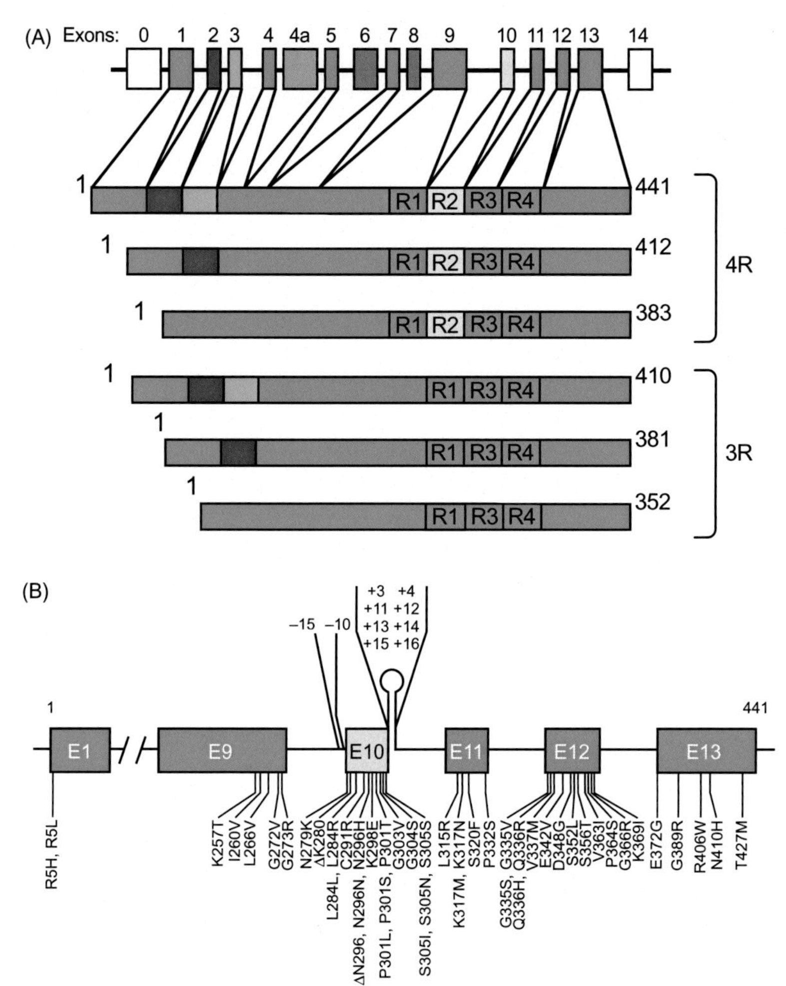

FIGURE 4.1 Human brain Tau isoforms and *MAPT* mutations. (A) *MAPT* and the six Tau isoforms expressed in adult human brain. *MAPT* consists of 16 exons (E). Alternative mRNA splicing of E2 (red), E3 (green), and E10 (yellow) gives rise to Tau isoforms ranging from 352 to 441 amino acids. Constitutively spliced exons (E1, E4, E5, E7, E9, E11, E12, E13) are shown in blue. E0, which is part of the promoter, and E14 are noncoding (white). E6 and E8 (violet) are not transcribed in human brain. E4a (orange) is only expressed in the peripheral nervous system. The repeats (R1–R4) are shown, with three isoforms having four repeats each (4R) and three isoforms having three repeats each (3R). The repeats are 31 or 32 amino acids in length. Exons and introns are not drawn to scale. (B) Mutations in *MAPT* in cases of FTDP-17T; 49 coding region mutations and 10 intronic mutations flanking E10 are shown. *FTDP-17*, frontotemporal dementia and parkinsonism linked to chromosome 17.

amino acids in the N-terminal half, and the inclusion, or not, of the 31 amino acid repeat encoded by exon 10 of *MAPT*, in the C-terminal half. Inclusion of exon 10 results in the production of three Tau isoforms with four repeats each (4R) and its exclusion in a further three isoforms with three repeats each (3R). Four repeats comprise residues 244–368 or 252–376, depending on the alignment, in the numbering of the 441 amino acid isoform. 4R Tau isoforms have R1, R2, R3, R4, whereas 3R Tau isoforms have R1, R3, R4. Together with some adjoining sequences, the repeats constitute the microtubule-binding domains of Tau (Kadavath et al., 2015).

Single-molecule tracking has revealed a kiss-and-hop mechanism, with a dwell time of Tau on individual microtubules of about 40 ms (Janning et al., 2014). Isoform differences did not influence this interaction (Niewidok et al., 2016). Despite these rapid dynamics, Tau promoted microtubule assembly, but it remains to be seen if microtubules were also stabilized. In brain, Tau is subject to posttranslational modifications, including phosphorylation, acetylation, methylation, glycation, isomerization, *O*-GlcNAcylation, nitration, sumoylation, ubiquitination, and truncation (Spillantini & Goedert, 2013). Big Tau, which carries an additional large exon in the N-terminal half, is expressed in the peripheral nervous system (Couchie et al., 1992; Goedert, Spillantini, & Crowther, 1992a). Tau is natively unfolded. However, this does not preclude some global order. Single-molecule studies have shown the presence of long-range contacts between N- and C-termini, as well as between both termini and the repeats, resulting in an S-shaped fold (Elbaum-Garfinkle & Rhoades, 2012).

Similar levels of 3R and 4R Tau are expressed in the cerebral cortex of adults (Goedert & Jakes, 1990). In developing human brain, only the shortest Tau isoform is present. 3R, 4R, and 5R Tau isoforms are found in the brains of adult chickens (Yoshida & Goedert, 2002), whereas most adult rodents express only 4R Tau (Götz et al., 1995). What is conserved is the expression of one hyperphosphorylated 3R Tau isoform lacking N-terminal inserts during vertebrate development. The genomes of *Caenorhabditis elegans* and *Drosophila melanogaster* each encode one protein with Tau-like repeats (Goedert et al., 1996a; Heidary & Fortini, 2001). Similar repeats are present in the high-molecular weight proteins MAP2 and MAP4 (Aizawa et al., 1990; Lewis, Wang, & Cowan, 1988). MAP4 may derive from an invertebrate ancestor, whereas Tau and MAP2 may have shared a more recent common ancestor (Sündermann, Fernandez, & Morgan, 2016).

TAU AGGREGATION

Full-length Tau assembles into filaments through its repeats, with most of the N-terminal half and the C-terminus forming the fuzzy coat (Goedert, Wischik, Crowther, Walker, & Klug, 1988; Wischik et al., 1988a, b). Tau filaments from human brain and those assembled from expressed protein have a cross-β structure characteristic of amyloid filaments (Berriman et al., 2003). The region that binds to microtubules overlaps with that forming the core of Tau filaments, suggesting that physiological function and pathological assembly may be mutually exclusive.

Phosphorylation of Tau negatively regulates its ability to interact with microtubules, and filamentous Tau is always hyperphosphorylated (Iqbal, Liu, & Gong, 2016). However, it remains to be proved that phosphorylation is a trigger for aggregation. Alternatively, a conformational change in Tau having to do with abnormal assembly may cause hyperphosphorylation. There is no strong evidence to suggest that the activities of protein kinases, phosphatases, or both, are changed in human Tauopathies. Other posttranslational changes may also be involved. Many publications incorrectly equate Tau phosphorylation with aggregation. Although aggregated Tau is heavily phosphorylated in human brain, not all phosphorylated Tau is aggregated or on its way to aggregation. For instance, highly phosphorylated Tau formed during hibernation, and this was not associated with filament formation (Arendt et al., 2003). Hyperphosphorylation was reversible upon arousal.

In AD, CTE, postencephalitic parkinsonism, and several other Tauopathies, all six isoforms of brain Tau are present in the disease filaments (Buée-Scherrer et al., 1997; Goedert, Spillantini, Cairns, & Crowther, 1992b; Schmidt, Zhukareva, Newell, Lee, & Trojanowski, 2001). Paired helical and straight filaments of AD are ultrastructural polymorphs, which consist of two identical protofilaments, each made of residues 306–378 of Tau (Fig. 4.2) (Fitzpatrick et al., 2017). They comprise R3 and R4, as well as 10 amino acids C-terminal to the repeats. Additional, weaker densities appear at the N-terminal region of the core. They may correspond to a mixture of peptides and/or a more dynamic or transiently

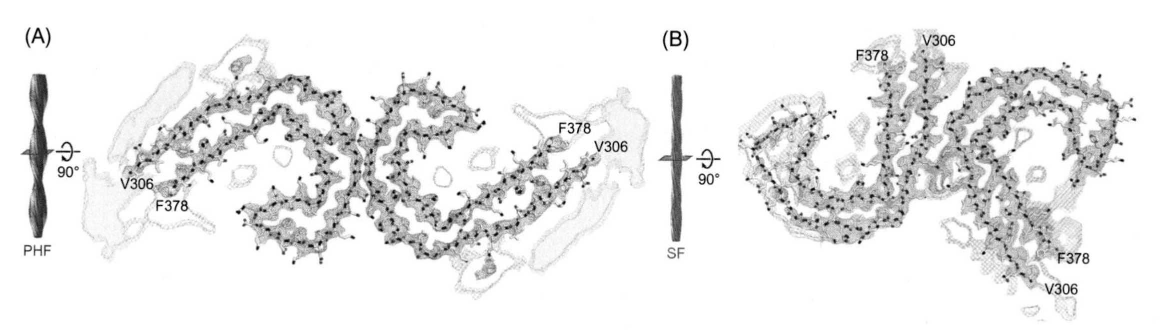

FIGURE 4.2 Cross-sections of the paired helical and straight Tau filament core structures from Alzheimer's disease brain, as determined by cryo-EM.
Cryo-EM density and atomic models of paired helical (A) and straight (B) filaments. Overviews of the helical reconstructions (left) show the orientation of the cross-sectional densities (right). Sharpened, high-resolution maps are shown in blue (paired helical filaments) and green (straight filaments). Paired helical and straight filaments are ultrastructural polymorphs. Their protofilament core corresponds to residues V306–F378 of Tau (R3 + R4 + 10 amino acids carboxy-terminal to the repeats). Each protofilament consists of eight β-strands, corresponding to two pairs of antiparallel β-sheets and a β-helix. Additional densities are in contact with the side-chains of K317, T319, and K321. Unsharpened, 4.5 Å low-pass filtered density is shown in gray. Unsharpened density highlighted with an orange background is reminiscent of a less-ordered β-sheet and could accommodate an additional 16 amino acids, which would correspond to a mixture of residues 259–274 (R1) from 3R Tau and residues 290–305 (R2) from 4R Tau. *cryo-EM*, cryoelectron microscopy.

occupied structure than the filament core. This weaker density can accommodate approximately 16 amino acids in a β-sheet-like conformation that may represent a mixture of the C-termini of R1 and R2. It explains the presence of all six Tau isoforms in the AD filaments. Another region of additional density is found interacting with the side-chains of K317, T319, and K321. We speculate that this density corresponds to ^{7}EFE^{9}, three residues that form part of the epitopes of antibodies Alz-50 and MC-1 (Jicha, Bowser, Kazam, & Davies, 1997). The ultrastructural polymorphism between paired helical and straight filaments is due to differences in lateral contacts between protofilaments (Fitzpatrick et al., 2017). In paired helical filaments, the two protofilaments are packed symmetrically, and the interaction is formed by antiparallel stacking of residues ^{332}PGGGQ^{336}. In straight filaments, the two protofilaments are packed asymmetrically. Their backbones are nearest each other between residues ^{321}KCGS^{324} of the first and ^{313}VDLSK^{317} of the second protofilament.

In other diseases—such as PSP, CBD, AGD, globular glial Tauopathy (GGT), and aging-related Tau astrogliopathy—only isoforms with 4R Tau are found in the filaments (Flament, Delacourte, Verny, Hauw, & Javoy-Agid, 1991; Kovacs et al., 2008, 2016; Ksiezak-Reding et al., 1994; Togo et al., 2002). In PiD, 3R Tau isoforms predominate in the inclusions (Delacourte et al., 1996). Unlike AD, these other diseases lack Aβ deposits. The morphologies of Tau filaments in different diseases vary, even when they are made of the same isoforms (Crowther & Goedert, 2000). Isoform differences can be detected by silver staining (Uchihara, Tsuchiya, Nakamura, & Akiyama, 2005). Inclusions made of all six Tau isoforms stain with Gallyas–Braak and Campbell–Switzer. Those made of 4R Tau are positive with Gallyas–Braak, whereas those made of 3R Tau stain only with Campbell–Switzer.

It has been suggested that patients with AD-type neurofibrillary degeneration restricted to hippocampus and medial temporal lobe, who lack Aβ deposits, suffer from primary age-related Tauopathy (PART), a condition that differs from AD (Crary et al., 2014). Tangle-only dementia (TD), a rare form of dementia (Ulrich, Spillantini, Goedert, Dukas, & Stähelin, 1992), may represent a severe form of PART. However, the view that PART is different from the pathological process of AD has been challenged (Duyckaerts et al., 2015). In AD, following the death of tangle-bearing cells, Tau filaments can remain in the extracellular space as ghost tangles, which consist largely of Tau repeats that have lost their fuzzy coat through proteolysis. In PiD, PSP, CBD, and most cases caused by *MAPT* mutations, Tau filaments do not accumulate to a significant extent in the extracellular space following the death of aggregate-bearing cells. The reason why Tau filaments from AD brains are less soluble remains to be established.

Overexpression of human wild-type Tau in mouse brain does not result in the formation of abundant filamentous inclusions (Götz et al., 1995; Probst et al., 2000). By contrast, overexpression of human Tau with disease-causing mutations, such as P301L and P301S, gives rise to models that reproduce the essential molecular and cellular characteristics of human Tauopathies, including the formation of abundant Tau filaments and extensive neurodegeneration (Allen et al., 2002; Lewis et al., 2001).

The interaction in vitro between unphosphorylated full-length, recombinant Tau, and some negatively charged compounds, such as sulphated glycosaminoglycans, results in filament assembly (Goedert et al., 1996b; Pérez, Valpuesta, Medina, De Garcini, & Avila, 1996). Heparin may induce dimerization of Tau, with filaments growing through monomer addition (Ramachandran & Udgaonkar, 2011). Filaments were decorated by antibodies directed against the N- and C-termini of Tau (BR133 and BR134), but not by an antibody against R3 (BR135, which was raised against residues 323–335 of Tau). These findings, which indicate that at least part of the repeat region is inaccessible to antibodies, are similar to those obtained in AD. Synthetic Tau filaments resemble those in human Tauopathies. They also appear to be characterized by the N-terminal region contacting residues in R3 of Tau (Bibow et al., 2011). These findings, in conjunction with the structural analysis of Tau filaments from AD brain (Fitzpatrick et al., 2017), indicate that intermolecular interactions give rise to the high-affinity epitopes of Alz-50 and MC-1, in apparent contrast to findings obtained using single-molecule studies (Elbaum-Garfinkle & Rhoades, 2012). Electron cryomicroscopy will tell how cores of recombinant Tau filaments may differ from those of AD. The mechanisms leading to filament formation inside brain cells in sporadic human Tauopathies remain to be identified. Heparin is probably not involved.

Hexapeptide sequences at the beginning of R2 (amino acids 275–280, VQIINK) and R3 (amino acids 306–311, VQIVYK) are essential for the induced assembly of recombinant Tau into filaments (Von Bergen et al., 2000, 2001). Residues 310–313 in Tau (YKPV) differ from the equivalent residues in MAP2 (TKKI). When the latter were changed to YKPV, MAP2c also assembled into filaments (Xie et al., 2015). Although VQIVYK is necessary for the assembly of Tau into filaments, it is not sufficient for the seeded aggregation of tagged four Tau repeats (residues 244–372) with mutations P301L and V337M (Stöhr et al., 2017).

GENETICS OF *MAPT*

The connection between Tau dysfunction and neurodegeneration was established through human genetics. In 1994, a dominantly inherited form of frontotemporal dementia and parkinsonism was linked to chromosome 17q21–22 (FTDP-17) (Wilhelmsen, Lynch, Pavlou, Higgins, & Nygaard, 1994). In June

1998, mutations in *MAPT* were reported in this and other families with FTDP-17 (Hutton et al., 1998; Poorkaj et al., 1998; Spillantini et al., 1998). Fifty-nine pathogenic *MAPT* mutations had been identified by August 2017 (Fig. 4.1B). Behavioral symptoms are the most common clinical sign. However, in some cases, *MAPT* mutations have been associated with parkinsonism. In addition, syndromes similar to PSP, CBD, PiD, GGT, and motor neuron disease have been described. The ages at disease onset are variable but can be as early as in the third decade. *MAPT* mutations are associated with abundant Tau inclusions predominantly in nerve cells (exons 9, 11, 12, and 13) or in both nerve cells and glial cells (exons 1 and 10, introns 9 and 10) (Ghetti et al., 2015). They can give rise to different clinicopathological phenotypes, even within a given family. Environmental and genetic factors other than the mutations themselves may play a role in determining phenotypic heterogeneity.

Mutations in *MAPT* account for about 5% of cases of frontotemporal dementia. They are concentrated in exons 9–12 (encoding R1–R4) and the introns flanking exon 10. Mutations can be divided into those with a primary effect at the protein level and those affecting the alternative splicing of Tau pre-mRNA. Mutations that act at the protein-level change or delete single amino acids, reducing the ability of Tau to interact with microtubules. Some mutations also promote the assembly of Tau into filaments. Mutations with a primary effect at the RNA level are intronic or exonic and increase the alternative mRNA splicing of exon 10 of *MAPT*. This affects the ratio of 3R to 4R isoforms, resulting in the relative overproduction of 4R Tau and its assembly into filaments. A single mutation (ΔK280) has been reported to cause the relative overexpression of 3R Tau and its assembly into filaments.

No mutations have been found in residues ^{306}VQIVYK^{311}, the sequence required for Tau aggregation at the N-terminus of R3. The same is true of residues ^{373}THKLTF^{378}, which form an antiparallel β-sheet with ^{306}VQIVYK^{311} in the protofilament structure from AD brain (Fitzpatrick et al., 2017). This contrasts with residues ^{275}VQIINK^{280} at the N-terminus of R2, where mutations N279K and ΔK280 have been described (Ghetti et al., 2015). It remains to be seen if these mutations act at the protein and/or the RNA level. Moreover, N279 is mostly deamidated to D279 in assembled Tau in AD (Dan et al., 2013). In the protofilament structure of Tau from AD brain, the N-terminus of R2 is located in the fuzzy coat (Fitzpatrick et al., 2017).

Aggregated Tau can show different isoform patterns and filament morphologies, depending on the mutations in *MAPT* (Ghetti et al., 2015). Mutations V337M in exon 12 and R406W in exon 13 give rise to insoluble Tau bands of 60, 64, and 68 kDa and a weaker band of 72 kDa. Following dephosphorylation, six bands are present that align with recombinant Tau, like what is seen in AD. By electron microscopy, paired helical and straight filaments are seen. The brains of many individuals with missense *MAPT* mutations in exons 9–13 (K257T, L266V, S305N, G272V, L315R, S320F, S320Y, P332S, Q336H, Q336R, K369I, E372G, and G389R) are characterized by abundant Pick bodies made predominantly of 3R Tau. As in sporadic PiD, insoluble Tau shows strong bands of 60 and 64 kDa. However, variable amounts of the 68- and 72-kDa bands are also observed. A third pattern is characteristic of *MAPT* mutations that affect the alternative mRNA splicing of exon 10, resulting in the relative overproduction of 4R Tau (intronic mutations and exonic mutations N279K, ΔK280, L284L, L284R, ΔN296, N296D, N296H, N296N, S305L, S305N, and S305S). Insoluble Tau runs as two strong bands of 64 and 68 kDa and a weaker band of 72 kDa; following dephosphorylation, three bands are present that align with recombinant 4R Tau (isoforms of 383, 412, and 441 amino acids). A similar pattern of pathological Tau bands is observed for mutations in exon 10, such as P301L and P301S, which have their primary effects at the protein level. Aggregation of 4R Tau has also been described for mutations I260V in exon 9, K317N in exon 11, E342V in exon 12, and N410H in exon 13, showing that it is possible to alter 3R and 4R Tau mRNAs through mutations located outside exon 10.

The effects of *MAPT* mutations can vary. Neighboring mutations in exon 12 (G335S, G335V, Q336H, Q336R, and V337M) give rise to structurally distinct filamentous Tau aggregates and exert different functional effects. They are located at the interface between the two protofilaments of paired helical filaments (Fitzpatrick et al., 2017). Mutation G335S is characterized by the presence of abundant filamentous Tau inclusions in nerve cells and glial cells, in the absence of Pick bodies (Spina et al., 2007). Mutations Q336H and Q336R give rise to what is a familial form of PiD, with abundant Pick bodies in nerve cells (Pickering-Brown et al., 2004; Tacik et al., 2015), whereas mutation V337M produces a neuronal filamentous Tau pathology indistinguishable from that of AD (Poorkaj et al., 1998; Spillantini, Crowther, & Goedert, 1996). These findings on *MAPT* mutations in three adjacent codons reinforce the view that the mechanisms resulting in the formation of neurofibrillary lesions and Pick bodies are closely related. Recombinant Tau with the G335S, G335V (Neumann et al., 2005), or V337M mutation shows a greatly reduced ability to promote microtubule assembly. By contrast, mutations Q336H and Q336R increase the ability of Tau to promote microtubule assembly. Mutations G335V and V337M fail to increase heparin-induced assembly into filaments significantly, whereas mutations Q336H and Q336R increase the assembly of 3R, but not 4R, Tau.

MAPT in populations of European descent is characterized by two haplotypes that result from a 900-kb inversion (H1) or noninversion (H2) polymorphism (Stefansson et al., 2005). Inheritance of the H1 haplotype is a risk factor for PSP (Baker et al., 1999; Conrad et al., 1997), CBD (Houlden et al., 2001), and PD (Pastor et al., 2000). This has been confirmed in genome-wide association studies (Höglinger et al., 2011; Kouri et al., 2015; Satake et al., 2009; Simón-Sánchez et al., 2009). The association with PD is particularly surprising, since PD is not characterized by Tau inclusions.

For PSP and CBD, an association with an allele at the *MOBP/SLC25A38* locus results in elevated levels of appoptosin, a protein that activates caspase-3, which can cleave Tau (Zhao et al., 2015). This may cause increased aggregation of 4R Tau. Additional loci were unique to PSP or CBD. The association of the H1 *MAPT* haplotype with PSP had a higher odds ratio than that between apolipoprotein E epsilon 4 (APOEε4) and AD (Höglinger et al., 2011). APOEε4 is the major risk factor allele for late-onset AD (Corder et al., 1993). The H2 haplotype is associated with increased expression of exon 3 of *MAPT* in gray matter, suggesting that inclusion of exon 3 is protective (Caffrey, Joachim, & Wade-Martins, 2008). Reduced expression of 1N4R has also been associated with the H2 haplotype (Valenca et al., 2016). Tau isoforms containing exons 2 and 10 promote aggregation, whereas exon 3-containing isoforms are inhibitory (Zhong, Condon, Nagaraja, & Kuret, 2012). Even though all six Tau isoforms give rise to the paired helical and straight filaments of AD, no known mutations in *MAPT* give rise to AD. Tau with an A152T substitution has been reported to be a risk factor for AD (Coppola et al., 2012), as well as for PSP, CBD, and unusual Tauopathies (Coppola et al., 2012; Kovacs et al., 2011; Kara et al., 2012, but see also Pastor et al., 2016).

PROPAGATION OF TAU AGGREGATES

Assembly of Tau into filaments can be initiated or accelerated through the addition of seeds. If mechanisms for the intercellular transfer of seeds exist, then human Tauopathies can propagate through the brain. Much evidence has been adduced to suggest that Tau assemblies, when applied extracellularly, can seed the formation of Tau aggregates, followed by their spreading (Fig. 4.3) (Clavaguera et al., 2009; Frost, Jacks, & Diamond, 2009). Because Tau is an intracellular protein, its propagation requires seeding, as well as aggregate uptake and release. Although monomeric Tau is taken up by cells, from which it can be released, it is probably not able to seed aggregation. Expressed Tau can only be seeded when it is aggregation competent. Deletion of amino acids 275–280 in R2 and 306–311 in R3 abolished the seeding activity of full-length aggregated Tau (Falcon et al., 2015). Aggregation inhibitors may thus be able to reduce Tau-induced seeding and spreading.

Uptake of ordered Tau assemblies requires heparan sulphate proteoglycans on the cell surface and may occur through macropinocytosis, at least in cultured cells (Holmes et al., 2013). Seeds probably escape from endosomal vesicles to induce the assembly of cytoplasmic Tau. Tau aggregates are then released through

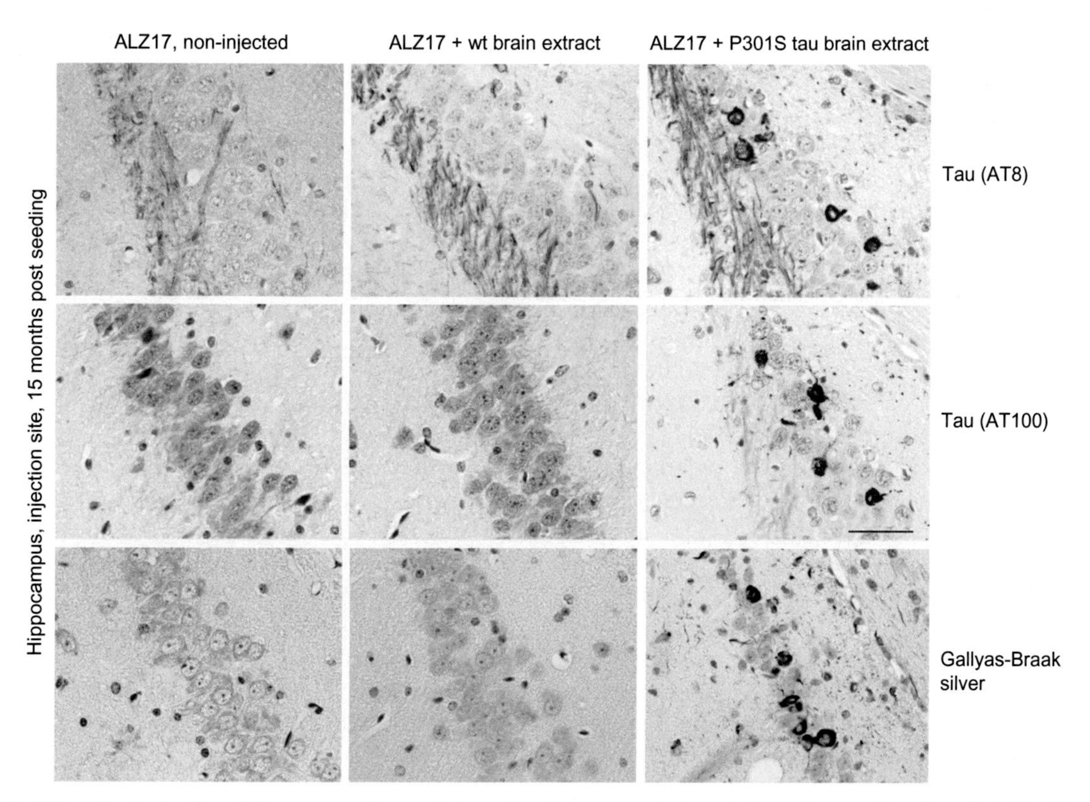

FIGURE 4.3 Induction of filamentous Tau pathology in mice transgenic for wild-type human Tau (line ALZ17) following injection with brain extracts from symptomatic mice transgenic for human mutant P301S Tau.
Staining of the hippocampal CA3 region of 18-month-old ALZ17 mice with anti-Tau antibodies AT8 and AT100 and Gallyas–Braak silver. Noninjected (left), 15 months after injection of brain extract from nontransgenic control mice (middle) and 15 months after injection with brain extract from 6-month-old mice transgenic for human P301S Tau (right). The sections were counterstained with hematoxylin.

ill-defined mechanisms. Intracellular Tau may transfer between cells through tunneling nanotubes. Alternatively, aggregated Tau could be released into the extracellular space, either freely or inside vesicles. It has been detected in the interstitial fluid of transgenic mice (Yamada et al., 2011).

Microglial cells may promote Tau propagation through exosome-dependent mechanisms (Asai et al., 2015). Another study concluded that Tau was released from cells through an exosome-independent pathway that required heat shock cognate 70, its cochaperone DnaJ and synaptosomal-associated protein 23 (Fontaine et al., 2016). Most of this work probably described the release of aggregation-incompetent soluble Tau. By contrast, optogenetic and chemogenetic approaches in transgenic mouse models have shown that an increase in neural activity can accelerate Tauopathy (Wu et al., 2016).

Phosphorylation of Tau seeds is not required, even though seeded Tau aggregates are hyperphosphorylated (Falcon et al., 2015). Filaments may grow by incorporating unphosphorylated Tau that then undergoes a conformational change and becomes hyperphosphorylated. Phosphorylation of Tau has been reported to influence seeded aggregation (Wu et al., 2016). It remains to be determined if the detailed morphological characteristics of Tau seeds can be replicated in seeded aggregates.

Intracerebral injection of brain extracts from mice expressing human P301S Tau with inclusions (Allen et al., 2002) into mice transgenic for wild-type human Tau lacking inclusions (line ALZ17) (Probst et al., 2000) induced the assembly of wild-type Tau into filaments and its spreading to distant brain regions (Clavaguera et al., 2009) (Fig. 4.3). No inclusions formed when Tau was immunodepleted from the extracts prior to injection. When presymptomatic, P301S Tau transgenic mice were intracerebrally injected with brain extracts from symptomatic animals, Tau inclusions formed rapidly at the injection sites. Contralateral and rostrocaudal propagation was seen, indicating that the spread of Tau inclusions was dependent on connectivity, not proximity.

Several groups studied the spreading of pathological Tau along the entorhinal cortex/hippocampal pathway. They used mouse models that apparently only expressed human P301L Tau in the entorhinal cortex (De Calignon et al., 2012; Harris et al., 2012; Liu et al., 2012, but see also Yetman, Lillehaug, Bjaalie, Leergaard, & Jankowsky, 2016). Several months after the appearance of Tau inclusions, hippocampal neurons also developed Tau pathology. Similarly, Tau aggregates were observed in dentate gyrus following adeno-associated virus-mediated expression of P301L Tau in entorhinal cortex. Over time, activated microglia and astrocytes accumulated in the medial entorhinal cortex, where nerve cell loss was observed at late time points (Fu et al., 2016). This is consistent with previous findings in mice transgenic for human P301S Tau, suggesting that glial activation and neuroinflammation are downstream of Tau aggregation (Bellucci et al., 2004). In mice transgenic for human mutant amyloid precursor protein, presenilin-1, and/or mutant Tau, neuroinflammatory changes must be secondary, even though they may influence the development of neurodegeneration. Primary microgliopathies, such as Nasu–Hakola disease (Paloneva et al., 2000) and hereditary diffuse leukoencephalopathy with spheroids (Rademakers et al., 2012), are not characterized by Tauopathy. However, genome-wide association studies have shown the involvement of innate immunity in the development of sporadic AD (Huang et al., 2017).

The intraperitoneal injection of brain extracts from symptomatic P301S Tau transgenic mice into presymptomatic mice promoted the formation of cerebral Tau inclusions (Clavaguera et al., 2014). Aggregated Tau can thus promote inclusion formation in the central nervous system of transgenic mice following peripheral administration. Similar findings have been reported for prions, assembled Aβ, and assembled α-synuclein.

Aggregated recombinant human Tau induced inclusion formation, but with a lower efficiency than aggregated Tau from transgenic mouse or AD brain (Clavaguera et al., 2013; Guo et al., 2016; Iba et al., 2013). Similar differences have been described for prions, Aβ, α-synuclein, and reactive serum amyloid A assemblies (Luk et al., 2012; Meyer-Luehmann et al., 2006; Prusiner, 2013; Recasens et al., 2014; Stöhr et al., 2012; Zhang et al., 2008). Recombinant Tau aggregates were more resistant to disaggregation by guanidine hydrochloride and digestion by proteinase K than Tau aggregates from transgenic mouse brain, consistent with the view that more stable aggregates possess lower seeding activity (Falcon et al., 2015). Recombinant aggregated Tau was also more stable than Tau filaments from AD brain.

Distinct conformations accounted for differences in seeding potency. Thus, Tau filaments formed from recombinant P301S Tau (in the absence of heparin) following seeding with aggregated Tau from transgenic mouse brain showed resistance to guanidine hydrochloride, which was similar to that of Tau seeds from the brains of mice transgenic for human P301S Tau (Falcon et al., 2015). The seeding potency of Tau filaments was like that of brain-derived aggregated Tau.

We dissected the molecular characteristics of seed-competent Tau from the brains of symptomatic P301S Tau transgenic mice (Jackson et al., 2016). Sucrose gradient fractions caused aggregation in transfected cells only when Tau aggregates larger than 10mers were present. They also induced the formation and spreading of filamentous Tau in presymptomatic P301S Tau transgenic mice, whereas fractions containing monomers and small aggregates were inactive. Immunoelectron microscopy showed that seed-competent sucrose fractions contained aggregated Tau species, including ring-like structures and short fibrils. It has been suggested that cells may interact with particles, not the monomeric proteins that make up those particles (Pieri, Madiona, Bousset, & Melki, 2012). Because Tau oligomers may be made of 50–100 and fibrils of a much larger number of monomers, at equal concentrations of monomer, many more oligomers than fibrils were injected. These findings were obtained in transgenic mice; it will be important to determine if similar species of aggregated Tau underlie seeding, spreading, and neurodegeneration in AD and other Tauopathies.

Provided that intracellular aggregation of Tau in human brain requires seeds moving from one cell to another, it follows that inhibition of seed uptake, seeding, and release provides novel therapeutic opportunities. Antibodies may be able to target what is an obligatory extracellular stage, provided that all Tau aggregates are not enclosed by membranous structures, such as exosomes. Anti-Tau antibodies have been shown to reduce the amount of hyperphosphorylated and aggregated Tau in transgenic mice (Chai et al., 2011; Yanamandra et al., 2013; Sankaranarayanan et al., 2015). Similarly, antisense oligonucleotides reduced Tau pathology (DeVos et al., 2017).

When Tau assemblies enter cells, they can be detected and neutralized via a danger response mediated by bound anti-Tau antibodies and the cytosolic Fc receptor tripartite motif protein 21 (TRIM21) (McEwan et al., 2017). Tau seeds are neutralized by TRIM21 through the activity of the proteasome and the AAA ATPase p97/valosin-containing protein, in a manner similar to viruses. Most studies have used transfected nonneuronal cells, some of which were of human origin, or primary nerve cells from rodents. However, human neurons derived from induced pluripotent stem cells also took up seeds and exhibited seed-induced Tau aggregation (Usenovic et al., 2015; Verheyen et al., 2015).

STRAINS OF AGGREGATED TAU

The intracerebral injection of brain homogenates from humans with pathologically confirmed Tauopathies led to the formation of neuronal and glial Tau inclusions in ALZ17 mice (Clavaguera et al., 2013). Inclusions formed after inoculation of brain homogenates from all cases of AD, TD, PiD, AGD, PSP, and CBD. Brain homogenates from patients with AGD, PSP, and CBD produced lesions similar to those of the human disorders. With the exception of PiD, the inclusions of the Tauopathies used were made of either 4R Tau (AGD, PSP, and CBD) or mixture of 3R and 4R Tau (AD and TD).

Injection of PSP homogenates into ALZ17 mice gave rise to silver-positive neuronal and glial Tau aggregates; the latter resembled tufted astrocytes, a hallmark lesion of PSP. The injection of CBD homogenates produced neuronal inclusions and silver-positive structures reminiscent of astrocytic plaques. With AGD homogenates, argyrophilic grains and silver-negative astrocytic Tau inclusions were seen, like in the human disease. With the exception of PiD, Tau inclusions propagated over time to connected brain regions. Similar, although fewer, inclusions formed after the intracerebral injection of brain homogenates from human Tauopathies into nontransgenic mice. A subsequent study showed that the intracerebral injection of Tau filaments purified from AD brain resulted in the formation of abundant Tau inclusions in nontransgenic mice (Guo et al., 2016). The electron microscopic appearance of these inclusions remains to be determined. Since adult wild-type mice only express 4R Tau, unlike AD filaments, which comprise 3R and 4R Tau isoforms, it follows that these inclusions were probably made of 4R Tau.

Induced Tau pathology propagated serially when brain homogenates from ALZ17 mice that had received bilateral injections of brain extracts from human P301S Tau transgenic mice 18 months earlier were injected into 3-month-old ALZ17 mice (Clavaguera et al., 2013). In a different experiment, homogenates were prepared from the brains of nontransgenic mice that had been injected bilaterally with AGD brain homogenates 18 months earlier. Twelve months after the intracerebral injection of these homogenates into ALZ17 mice, many neuropil threads and Tau aggregates were present at the injection sites.

Morphologically different Tau assemblies made of four Tau repeats formed in human embryonic kidney cells (Kaufman et al., 2016; Sanders et al., 2014). Inoculation of these assemblies into the hippocampus of young transgenic mice induced pathologies that were stable through serial transmission. When cells expressing four Tau repeats were seeded with homogenates from these brains, inclusions formed that were identical to those present initially.

These observations suggest that different strains of aggregated Tau may exist, but additional work is required. In particular, it will be important to see if a given Tau strain possesses unique features. High-resolution structures of the cores of Tau filaments from AD brain have been reported (Fitzpatrick et al., 2017). It remains to be seen if Tau filaments from other diseases possess different core structures.

Acknowledgments

M.G. is an Honorary Professor in the Department of Clinical Neurosciences of the University of Cambridge. Our work is supported by the UK Medical Research Council (MC_U105184291) and the European Union (Joint Programme-Neurodegeneration Research and Horizon 2020 IMPRiND).

References

Aizawa, H., Emori, Y., Murofushi, H., Kawasaki, H., Sakai, H., & Suzuki, K. (1990). Molecular cloning of a ubiquitously distributed microtubule-associated protein with M_r 190,000. *Journal of Biological Chemistry, 265,* 13849–13855.

Allen, B., Ingram, E., Takao, M., Smith, M. J., Jakes, R., Virdee, K., ... Goedert, M. (2002). Abundant tau filaments and nonapoptotic neurodegeneration in transgenic mice expressing human P301S tau protein. *Journal of Neuroscience, 22,* 9340–9351.

Arendt, T., Stieler, J., Strijkstra, A. M., Hut, R. A., Rüdiger, J., Van der Zee, E. A., ... Härtig, W. (2003). Reversible paired helical filament-like phosphorylation of tau is an adaptive process associated with neuronal plasticity in hibernating animals. *Journal of Neuroscience, 23,* 6972–6981.

Asai, H., Ikezu, S., Tsunoda, S., Medalla, M., Luebke, J., Haydar, T., ... Ikezu, T. (2015). Depletion of microglia and inhibition of exosome synthesis halt tau propagation. *Nature Neuroscience, 18,* 1584–1593.

Baker, M., Litvan, I., Houlden, H., Adamson, J., Dickson, D. W., Perez-Tur, J., ... Hutton, M. (1999). Association of an extended haplotype in the tau gene with progressive supranuclear palsy. *Human Molecular Genetics, 8,* 711–715.

Bellucci, A., Westwood, A. J., Ingram, E., Casamenti, F., Goedert, M., & Spillantini, M. G. (2004). Induction of inflammatory mediators and microglial activation in mice transgenic for mutant human P301S tau protein. *American Journal of Pathology, 165,* 1643–1652.

Berriman, J., Serpell, L. C., Oberg, K. A., Fink, A. L., Goedert, M., & Crowther, R. A. (2003). Tau filaments from human brain and from in vitro assembly of recombinant protein show cross-β structure. *Proceedings of the National Academy of Sciences of the United States of America, 100,* 9034–9038.

Bibow, S., Mukrasch, M. D., Chinnathambi, S., Biernat, J., Griesinger, C., Mandelkow, E., & Zweckstetter, M. (2011). The dynamic structure of filamentous tau. *Angewandte Chemie International Edition, 50,* 11520–11524.

Braak, H., & Braak, E. (1991). Neuropathological staging of Alzheimer-related changes. *Acta Neuropathologica, 82,* 239–259.

Braak, H., & Del Tredici, K. (2011). The pathological process underlying Alzheimer's disease in individuals under thirty. *Acta Neuropathologica, 121,* 171–181.

Buée-Scherrer, V., Buée, L., Leveugle, P., Perl, D. P., Vermersch, P., Hof, P. R., & Delacourte, A. (1997). Pathological tau proteins in postencephalitic parkinsonism: Comparison with Alzheimer's disease and other neurodegenerative disorders. *Annals of Neurology, 42,* 356–359.

Caffrey, T. M., Joachim, C., & Wade-Martins, R. (2008). Haplotype-specific expression of the N-terminal exon 2 and 3 at the human *MAPT* locus. *Neurobiology of Aging, 29,* 1923–1929.

Chai, X., Wu, S., Murray, T. K., Kinley, R., Vella, C. V., Sims, H., ... Citron, M. (2011). Passive immunization with anti-tau antibodies in two transgenic models. *Journal of Biological Chemistry, 286,* 34457–34467.

Clavaguera, F., Akatsu, H., Fraser, G., Crowther, R. A., Frank, S., Hench, J., ... Tolnay, M. (2013). Brain homogenates from human tauopathies induce tau inclusions in mouse brain. *Proceedings of the National Academy of Sciences of the United States of America, 110,* 9535–9540.

Clavaguera, F., Bolmont, T., Crowther, R. A., Abramowski, D., Frank, S., Probst, A., ... Tolnay, M. (2009). Transmission and spreading of tauopathy in transgenic mouse brain. *Nature Cell Biology, 11,* 909–913.

Clavaguera, F., Hench, J., Lavenir, I., Schweighauser, G., Frank, S., Goedert, M., & Tolnay, M. (2014). Peripheral administration of tau aggregates triggers intracerebral tauopathy in transgenic mice. *Acta Neuropathologica, 127,* 299–301.

Conrad, C., Andreadis, A., Trojanoewski, jQ., Dickson, D. W., Kang, D., Chen, X., ... Saitoh, T. (1997). Genetic evidence for the involvement of tau in progressive supranuclear palsy. *Annals of Neurology, 41,* 277–281.

Coppola, G., Chinnathambi, S., Lee, J. J., Dombroski, B. A., Baker, M. C., Soto-Ortolazo, A. L., ... Geschwind, D. H. (2012). Evidence for a role of the rare p.A152T variant in *MAPT* in increasing the risk for FTD-spectrum and Alzheimer's diseases. *Human Molecular Genetics, 21,* 3500–3512.

Corder, E. H., Saunders, A. M., Strittmatter, W. J., Schmechel, D. E., Gaskell, P. C., Small, G. W., ... Pericak-Vance, M. A. (1993). Gene dose of apolipoprotein E type 4 allele and the risk of Alzheimer's disease in late onset families. *Science, 261,* 921–923.

Couchie, D., Mavilia, C., Georgieff, I. S., Liem, R. K. H., Shelanski, M. L., & Nunez, J. (1992). Primary structure of high molecular weight tau present in the peripheral nervous system. *Proceedings of the National Academy of Sciences of the United States of America, 89,* 4378–4381.

Crary, J. F., Trojanowski, J. Q., Schneider, J. A., Abisambra, J. F., Abner, E. L., Alazuloff, I., ... Nelson, P. T. (2014). Primary age-related tauopathy (PART): A common pathology associated with human aging. *Acta Neuropathologica, 128,* 755–766.

Crowther, R. A., & Goedert, M. (2000). Abnormal tau-containing filaments in neurodegenerative diseases. *Journal of Structural Biology, 130,* 271–279.

Dan, A., Takahashi, M., Masuda-Suzukake, M., Kametani, F., Nonaka, T., Kondo, H., ... Hasegawa, M. (2013). Extensive deamidation at asparagine residue 279 accounts for weak immunoreactivity of tau with RD4 antibody in Alzheimer's disease brain. *Acta Neuropathologica Communications*, *1*, 54.

De Calignon, A., Polydoro, M., Suárez-Calvet, M., William, C., Adamowicz, D. H., Kopeikina, K. J., ... Hyman, B. T. (2012). Propagation of tau pathology in a model of early Alzheimer's disease. *Neuron*, *73*, 685–697.

Delacourte, A., Robitaille, Y., Sergeant, N., Buée, L., Hof, P. R., Wattez, A., ... Gauvreau, D. (1996). Specific pathological tau protein variants characterize Pick's disease. *Journal of Neuropathology and Experimental Neurology*, *55*, 159–168.

DeVos, S. L., Miller, R. L., Schoch, K. M., Holmes, B. B., Kebodeaux, C. S., Wegener, A. J., ... Miller, T. M. (2017). Tau reduction prevents neuronal loss and reverses pathological tau deposition and seeding in mice with tauopathy. *Science Translational Medicine*, *9*, eaag0481.

Duyckaerts, C., Uchihara, T., Seilhean, D., He, Y., & Hauw, J. J. (1997). Dissociation of Alzheimer type pathology in a disconnected piece of cortex. *Acta Neuropathologica*, *93*, 501–507.

Duyckaerts, C., Braak, H., Brion, J. P., Buée, L., Del Tredici, K., Goedert, M., ... Uchihara, T. (2015). PART is part of Alzheimer disease. *Acta Neuropathologica*, *129*, 749–756.

Elbaum-Garfinkle, S., & Rhoades, E. (2012). Identification of an aggregation-prone structure of tau. *Journal of the American Chemical Society*, *134*, 16607–16613.

Falcon, B., Cavallini, A., Angers, R., Glover, S., Murray, T. K., Barnham, L., ... Bose, S. (2015). Conformation determines the seeding potencies of native and recombinant tau aggregates. *Journal of Biological Chemistry*, *290*, 1049–1065.

Fitzpatrick, A. W. P., Falcon, B., He, S., Murzin, A. G., Murshudov, G., Garringer, H. J., ... Scheres, S. H. W. (2017). Cryo-EM structures of tau filaments from Alzheimer's disease. *Nature*, *547*, 185–190.

Flament, S., Delacourte, A., Verny, M., Hauw, J. J., & Javoy-Agid, F. (1991). Abnormal tau proteins in progressive supranuclear palsy. Similarities and differences with the neurofibrillary degeneration of the Alzheimer type. *Acta Neuropathologica*, *81*, 591–596.

Fontaine, S. N., Zheng, D., Sabbagh, J. J., Martin, M. D., Chaput, D., Darling, A., ... Dickey, C. A. (2016). DnaJ/Hsc70 chaperone complexes control the extracellular release of neurodegenerative-associated proteins. *EMBO J.*, *35*, 1537–1549.

Frost, B., Jacks, R. L., & Diamond, M. I. (2009). Propagation of tau misfolding from the outside to the inside of a cell. *Journal of Biological Chemistry*, *284*, 12845–12852.

Fu, H., Hussaini, S. A., Wegmann, S., Profaci, C., Daniels, J. D., Herman, M., ... Duff, K. E. (2016). 3D visualization of the temporal and spatial spread of tau pathology reveals extensive sites of tau accumulation associated with neuronal loss and recognition memory deficit in aged tau transgenic mice. *PLoS ONE*, *11*, e0159463.

Ghetti, B., Oblak, A. L., Boeve, B. F., Johnson, K. A., Dickerson, B. C., & Goedert, M. (2015). Frontotemporal dementia caused by microtubule-associated protein tau gene (*MAPT*) mutations: A chameleon for neuropathology and neuroimaging. *Neuropathology and Applied Neurobiology*, *41*, 24–46.

Goate, A. M., Chartier-Harlin, M. C., Mullan, M., Brown, J., Crawford, F., Fidani, L., ... Hardy, J. (1991). Segregation of a missense mutation in the amyloid precursor protein gene with familial Alzheimer's disease. *Nature*, *394*, 704–706.

Goedert, M., Wischik, C. M., Crowther, R. A., Walker, J. E., & Klug, A. (1988). Cloning and sequencing of the cDNA encoding a core protein of the paired helical filament of Alzheimer disease: Identification as the microtubule-associated protein tau. *Proceedings of the National Academy of Sciences of the United States of America*, *85*, 4051–4055.

Goedert, M., Spillantini, M. G., Jakes, R., Rutherford, D., & Crowther, R. A. (1989). Multiple isoforms of human microtubule-associated protein tau: Sequences and localization in neurofibrillary tangles of Alzheimer's disease. *Neuron*, *3*, 519–526.

Goedert, M., & Jakes, R. (1990). Expression of separate isoforms of human tau protein: Correlation with the tau pattern in brain and effects on tubulin polymerization. *EMBO Journal*, *9*, 4225–4230.

Goedert, M., Spillantini, M. G., & Crowther, R. A. (1992a). Cloning of a big tau microtubule-associated protein characteristic of the peripheral nervous system. *Proceedings of the National Academy of Sciences of the United States of America*, *89*, 1983–1987.

Goedert, M., Spillantini, M. G., Cairns, N. J., & Crowther, R. A. (1992b). Tau proteins of Alzheimer paired helical filaments: Abnormal phosphorylation of all six brain isoforms. *Neuron*, *8*, 159–168.

Goedert, M., Baur, C. P., Ahringer, J., Jakes, R., Hasegawa, M., Spillantini, M. G., ... Hill, F. (1996a). PTL-1, a microtubule-associated protein with tau-like repeats from the nematode *Caenorhabditis elegans*. *Journal of Cell Science*, *109*, 2661–2672.

Goedert, M., Jakes, R., Spillantini, M. G., Hasegawa, M., Smith, M. J., & Crowther, R. A. (1996b). Assembly of microtubule-associated protein tau into Alzheimer-like filaments induced by sulphated glycosaminoglycans. *Nature*, *383*, 550–553.

Goedert, M. (2015). Alzheimer's and Parkinson's diseases: The prion concept in relation to assembled Aβ, tau, and α-synuclein. *Science, 349*, 1255555.

Götz, J., Probst, A., Spillantini, M. G., Schäfer, T., Jakes, R., Bürki, K., & Goedert, M. (1995). Somatodendritic localisation and hyperphosphorylation of tau protein in transgenic mice expressing the longest human brain tau isoform. *EMBO Journal, 14*, 1304–1313.

Guo, J. L., Narasimhan, S., Changolkar, L., He, Z., Stieber, A., Zhang, B., ... Lee, V. M. Y. (2016). Unique pathological tau conformers from Alzheimer's brains transmit tau pathology in nontransgenic mice. *Journal of Experimental Medicine, 213*, 2635–2654.

Hardy, J., & Allsop, D. (1991). Amyloid deposition as the central event in the aetiology of Alzheimer's disease. *Trends in Pharmacological Sciences, 12*, 383–388.

Harris, J. A., Koyama, A., Maeda, S., Ho, K., Devidze, N., Dubal, D. B., ... Mucke, L. (2012). Human P301L-mutant tau expression in mouse entorhinal-hippocampal network causes tau aggregation and presynaptic pathology but no cognitive defects. *PLoS ONE, 7*, e45881. (2012).

Heidary, G., & Fortini, M. E. (2001). Identification and characterization of the *Drosophila tau* homolog. *Mechanisms of Development, 108*, 171–178.

Holmes, B. B., DeVos, S. L., Kfoury, N., Li, M., Jacks, R., Yanamandra, K., ... Diamond, M. I. (2013). Heparan sulphate proteoglycans mediate internalization and propagation of specific proteopathic seeds. *Proceedings of the National Academy of Sciences of the United States of America, 110*, E3138–E3147.

Houlden, H., Baker, M., Morris, H. R., MacDonald, N., Pickering-Brown, S., Adamson, J., ... Hutton, M. (2001). Corticobasal degeneration and progressive supranuclear palsy share a common tau haplotype. *Neurology, 56*, 1702–1706.

Huang, K. O., Marcora, E., Pimenova, A. A., Di Narzo, A. F., Kapoor, M., Jin, S. C., ... Goate, A. M. (2017). A common haplotype lowers PU.1 expression in myeloid cells and delays onset of Alzheimer's disease. *Nature Neuroscience, 20*, 1052–1061.

Hutton, M., Lendon, C. L., Rizzu, P., Baker, M., Froelich, S., Houlden, H., ... Heutink, P. (1998). Association of missense and 5′-splice site mutations in *tau* with the inherited dementia FTDP-17. *Nature, 393*, 702–705.

Höglinger, G. U., Melhem, N. M., Dickson, D. W., Sleiman, P. M. A., Wang, L. S., Klei, L., ... Schellenberg, G. D. (2011). Identification of common variants influencing risk of the tauopathy progressive supranuclear paly. *Nature Genetics, 43*, 699–705. (2011).

Iba, M., Guo, J. L., McBride, J. D., Zhang, B., Trojanowski, J. Q., & Lee, V. M. Y. (2013). Synthetic tau fibrils mediate transmission of neurofibrillary tangles in a transgenic mouse model of Alzheimer's-like tauopathy. *Journal of Neuroscience, 33*, 1024–1037.

Iqbal, K., Liu, F., & Gong, C. X. (2016). Tau and neurodegenerative disease: The story so far. *Nature Reviews Neurology, 12*, 15–27.

Irwin, D. J., Brettschneider, J., McMillan, C. T., Cooper, F., Olm, C., Arnold, S. E., ... Trojanowski, J. Q. (2016). Deep clinical and neuropathological phenotyping of Pick disease. *Annals of Neurology, 79*, 272–287.

Jackson, S. J., Kerridge, C., Cooper, J., Cavallini, A., Faslcon, B., Cella, C. V., ... Bose, S. (2016). Short fibrils constitute the major species of seed-competent tau in the brains of mice transgenic for human P301S tau. *Journal of Neuroscience, 36*, 762–772.

Janning, D., Igaev, M., Sündermann, F., Brühmann, J., Beutel, O., Heinisch, J. J., ... Brandt, R. (2014). Single-molecule tracking of tau reveals fast kiss-and-hop interaction with microtubules in living neurons. *Molecular Biology of the Cell, 25*, 3541–3551.

Jicha, G. A., Bowser, R., Kazam, I. G., & Davies, P. (1997). Alz-50 and MC-1, a new monoclonal antibody raised to paired helical filaments, recognize conformational epitopes on recombinant tau. *Journal of Neuroscience Research, 48*, 128–132.

Kadavath, H., Jaremko, M., Jaremko, L., Biernat, J., Mandelkow, E., & Zweckstetter, M. (2015). Folding of the tau protein on microtubules. *Angewandte Chemie International Edition, 54*, 10347–10351.

Kara, E., Ling, H., Pittman, A. M., Shaw, L., de Silva, R., Simone, R., ... Revesz, T. (2012). The MAPT p. A152T variant is a risk factor associated with tauopathies with atypical clinical and neuropathological features. *Neurobiology of Aging, 33*, 2231e7–2231e14.

Kaufman, S. D. K., Sanders, D. W., Thomas, T. L., Ruchinskas, A. J., Vaquer-Alicea, J., Sharma, A. M., ... Diamond, M. I. (2016). Tau prion strains dictate patterns of cell pathology, progression rate, and regional vulnerability in vivo. *Neuron, 92*, 796–812.

Kouri, N., Ross, O. A., Dombroski, B., Younkin, C. S., Serie, D. J., Soto-Ortolaza, A., ... Dickson, D. W. (2015). Genome-wide association study of corticobasal degeneration identifies risk variants shared with progressive supranuclear palsy. *Nature Communications, 6*, 7247.

Kovacs, G. G., Majtenyi, K., Spina, S., Murrell, J. R., Gelpi, E., Hoftberger, R., ... Ghetti, B. (2008). White matter tauopathy with globular glial inclusions: A distinct sporadic frontotemporal lobar degeneration. *Journal of Neuropathology and Experimental Neurology, 67*, 963–975.

Kovacs, G. G., Wöhrer, A., Ströbel, T., Botond, G., Attems, J., & Budka, H. (2011). Unclassifiable tauopathy associated with an A152T variation in *MAPT* exon 7. *Clinical Neuropathology, 30*, 3–10.

Kovacs, G. G., Ferrer, I., Grinberg, L. T., Alazuloff, I., Attems, J., Budka, H., ... Dickson, D. W. (2016). Aging-related tau astrogliopathy (ARTAG): Harmonized evaluation strategy. *Acta Neuropathologica, 131*, 87–102.

Ksiezak-Reding, H., Morgan, K., Mattiace, L. A., Davies, P., Liu, W. K., Yen, S. H., ... Dickson, D. W. (1994). Ultrastructure and biochemical composition of paired helical filaments in corticobasal degeneration. *American Journal of Pathology, 145*, 1496–1508.

Lewis, S. A., Wang, D., & Cowan, N. J. (1988). Microtubule-associated protein MAP2 shares a microtubule binding motif with tau protein. *Science, 242*, 936–939.

Lewis, J., Dickson, D. W., Lin, W. L., Chisholm, L., Corral, A., Jones, G., ... McGowan, E. (2001). Enhanced neurofibrillary degeneration in transgenic mice expressing mutant tau and APP. *Science, 293*, 1487–1491.

Ling, H., Kovacs, G. G., Vonsattel, J. P. G., Davey, K., Mok, K. Y., Hardy, J., ... Revesz, T. (2016). Astrogliopathy predominates the earliest stage of corticobasal degeneration pathology. *Brain, 139*, 3237–3252.

Liu, L., Drouet, V., Wu, J. W., Witter, M. P., Small, S. A., & Clelland, C. (2012). Trans-synaptic spread of tau pathology *in vivo*. *PLoS ONE, 7*, e31302.

Luk, K. C., Kehm, V. M., Zhang, B., O'Brien, P., Trojanowski, J. Q., & Lee, V. M. Y. (2012). Intracerebral inoculation of pathological α-synuclein initiates a rapidly progressive neurodegenerative α-synucleinopathy in mice. *Journal of Experimental Medicine, 209*, 975–986.

McEwan, W. A., Falcon, B., Vaysburd, M., Clift, D., Oblak, A. L., Ghetti, B., ... James, L. C. (2017). Cytosolic Fc receptor TRIM21 inhibits seeded tau aggregation. *Proceedings of the National Academy of Sciences of the United States of America, 114*, 574–579.

McKee, A. C., Stein, T. D., Nowinski, C. J., Stern, R. A., Daneshvar, D. H., Alvarez, V. E., ... Cantu, R. C. (2013). The spectrum of disease in chronic traumatic encephalopathy. *Brain, 136*, 43–64.

Meyer-Luehmann, M., Coomaraswamy, J., Bolmont, T., Kaeser, S., Schäfer, C., Kilger, E., ... Jucker, M. (2006). Exogenous induction of cerebral β-amyloidogenesis is governed by agent and host. *Science, 313*, 1781–1784.

Neumann, M., Diekmann, S., Bertsch, U., Vanmassenhove, B., Bogerts, B., & Kretzschmar, H. A. (2005). Novel G335V mutation in the *tau* gene associated with early onset familial frontotemporal dementia. *Neurogenetics, 6*, 91–95.

Niewidok, B., Igaev, M., Sündermann, F., Janning, D., Bakota, L., & Brandt, R. (2016). Presence of a carboxy-terminal pseudorepeat and disease-like pseudophosphorylation critically influence tau's interaction with microtubules in axon-like processes. *Molecular Biology of the Cell, 27*, 3537–3549.

Paloneva, J., Kestilä, M., Wu, J., Salminen, A., Böhling, T., Ruotsalainen, V., ... Peltonen, L. (2000). Loss-of-function mutations in TYROBP (DAP12) result in a presenile dementia with bone cysts. *Nature Genetics, 25*, 357–361.

Pastor, P., Ezquerra, M., Munoz, E., Marti, M. J., Blesa, R., Tolosa, E., & Oliva, R. (2000). Significant association between the tau gene A0/A0 genotype and Parkinson's disease. *Annals of Neurology, 47*, 242–245.

Pastor, P., Moreno, F., Clarimón, J., Ruiz, A., Combarros, O., Calero, M., ... Sánchez-Juan, P. (2016). *MAPT* H1 haplotype is associated with late-onset Alzheimer's disease risk in *APOE* ε4 noncarriers: Results from the dementia genetics Spanish consortium. *Journal of Alzheimer's disease, 49*, 343–352.

Pérez, M., Valpuesta, J. M., Medina, M., De Garcini, E. M., & Avila, J. (1996). Polymerization of tau into filaments in the presence of heparin: The minimal sequence required for tau-tau interaction. *Journal of Neurochemistry, 67*, 1183–1190.

Pickering-Brown, S. M., Baker, M., Nonaka, T., Ikeda, K., Sharma, S., Mackenzie, J., ... Mann, D. M. A. (2004). Frontotemporal dementia with Pick-type histology associated with Q336R mutation in the *tau* gene. *Brain, 127*, 1415–1426.

Pieri, L., Madiona, K., Bousset, L., & Melki, R. (2012). Fibrillar α-synuclein and huntingtin exon 1 assemblies are toxic to the cells. *Biophysical Journal, 102*, 2894–2905.

Poorkaj, P., Bird, T. D., Wijsman, E., Nemens, E., Garruto, R. M., Anderson, L., ... Schellenberg, G. D. (1998). Tau is a candidate gene for chromosome 17 frontotemporal dementia. *Annals of Neurology, 43*, 815–825.

Probst, A., Götz, J., Wiederhold, K. H., Tolnay, M., Mistl, C., Jaton, A. L., ... Goedert, M. (2000). Axonopathy and amyotrophy in mice transgenic for human four-repeat tau protein. *Acta Neuropathologica, 99*, 469–481.

Prusiner, S. B. (1998). Prions. *Proceedings of the National Academy of Sciences of the United States of America, 95*, 13363–13383.

Prusiner, S. B. (2013). Biology and genetics of prions causing neurodegeneration. *Annual Review of Genetics, 47*, 601–623.

Rademakers, R., Baker, M., Nicholson, A. M., Rutherford, N. J., Finch, N., Soto-Ortolaza, A., ... Wszolek, Z. K. (2012). Mutations in the colony stimulating factor 1 receptor (*CSF1R*) gene cause hereditary diffuse leukoencephalopathy with spheroids. *Nature Genetics, 44*, 200–205.

Ramachandran, G., & Udgaonkar, J. (2011). Understanding the kinetic roles of the inducer heparin and of rod-like protofibrils during amyloid fibril formation by tau protein. *Journal of Biological Chemistry, 286*, 38948–38959.

Recasens, A., Dehay, B., Bové, J., Carballo-Carbajal, I., Dovero, S., Pérez-Villalba, A., ... Vila, M. (2014). Lewy body extracts from Parkinson disease brains trigger α-synuclein pathology and neurodegeneration in mice and monkeys. *Annals of Neurology, 75*, 351–362.

Saito, Y., Ruberu, N. N., Sawabe, M., Arai, T., Tanaka, N., Kakuta, Y., ... Murayama, S. (2004). Staging of argyrophilic grains: An age-associated tauopathy. *Journal of Neuropathology and Experimental Neurology, 63*, 911–918.

Sanders, D. W., Kaufman, S. K., DeVos, S. L., Sharma, A. M., Mirhaba, H., Li, A., ... Diamond, M. I. (2014). Distinct tau prion strains propagate in cells and mice and define different tauopathies. *Neuron, 82*, 1271–1288.

Sankaranarayanan, S., Barten, D. M., Vana, L., Devidze, N., Yang, L., Cadelina, G., ... Ahlijanian, M. (2015). Passive immunization with phospho-tau antibodies reduces tau pathology and functional deficits in two distinct mouse tauopathy models. *PLoS ONE, 10*, e0125614.

Satake, W., Nakabayashi, Y., Mizuta, I., Hirota, Y., Ito, C., Kubo, M., ... Toda, T. (2009). Genome-wide association study identifies common variants at four loci as genetic risk factors for Parkinson's disease. *Nature Genetics, 41*, 1303–1307.

Schmidt, M. L., Zhukareva, V., Newell, K. L., Lee, V. M. Y., & Trojanowski, J. Q. (2001). Tau isoform profile and phosphorylation state in dementia pugilistica recapitulate Alzheimer's disease. *Acta Neuropathologica, 101*, 518–524.

Sherrington, R., Rogaev, E. I., Liang, Y., Rogaeva, E. A., Levesque, G., Ikeda, M., ... St George-Hyslop, P. H. (1995). Cloning of a novel gene bearing missense mutations in early-onset familial Alzheimer's disease. *Nature, 375*, 754–760.

Simón-Sánchez, J., Sculte, C., Bras, J. M., Sharma, M., Gibbs, J. R., Berg, D., ... Gasser, T. (2009). Genome-wide association study reveals genetic risk underlying Parkinson's disease. *Nature Genetics, 41*, 1308–1312.

Spillantini, M. G., Crowther, R. A., & Goedert, M. (1996). Comparison of the neurofibrillary pathology in Alzheimer's disease and familial presenile dementia with tangles. *Acta Neuropathologica, 92*, 42–48.

Spillantini, M. G., Murrell, J. R., Goedert, M., Farlow, M. R., Klug, A., & Ghetti, B. (1998). Mutation in the tau gene in familial multiple system tauopathy with presenile dementia. *Proceedings of the National Academy of Sciences of the United States of America, 95*, 7737–7741.

Spillantini, M. G., & Goedert, M. (2013). Tau pathology and neurodegeneration. *Lancet Neurology, 12*, 609–622.

Spina, S., Murrell, J. R., Yoshida, H., Ghetti, B., Bermingham, N., Sweeney, B., ... Keohane, C. (2007). The novel *Tau* mutation G335S: Clinical, neuropathological and molecular characterization. *Acta Neuropathologica, 113*, 461–470.

Stefansson, H., Helgason, A., Thorleifsson, G., Steinthorsdottir, V., Masson, G., Barnard, J., ... Stefansson, K. (2005). A common inversion under selection in Europeans. *Nature Genetics, 37*, 129–137.

Stöhr, J., Watts, J. C., Mensinger, Z. L., Oehler, A., Grillo, S. K., DeArmond, S. ,J., ... Giles, K. (2012). Purified and synthetic Alzheimer's amyloid beta (Aβ) prions. *Proceedings of the National Academy of Sciences of the United States of America, 109*, 11025–11030.

Stöhr, J., Wu, H., Nick, M., Wu, Y., Bhate, M., Condello, C., ... DeGrado, W. F. (2017). A 31-residue peptide induces aggregation of tau's microtubule-binding region in cells. *Nature Chemistry, 9*, 874–881.

Sündermann, F., Fernandez, M. P., & Morgan, T. O. (2016). An evolutionary roadmap to the microtubule-associated protein MAP Tau. *BMC Genomics, 17*, 264.

Tacik, O. P., DeTure, M., Hinkle, K. M., Lin, W. L., Sanchez-Contreras, M., Carlomagno, Y., ... Dickson, D. W. (2015). A novel tau mutation in exon 12, p. Q336H, causes hereditary Pick disease. *Journal of Neuropathology and Experimental Neurology, 74*, 1042–1052.

Togo, T., Sahara, N., Yen, S. H., Cookson, N., Ishizawa, T., Hutton, M., ... Dickson, D. W. (2002). Argyrophilic grain disease is a sporadic 4-repeat tauopathy. *Journal of Neuropathology and Experimental Neurology, 61*, 547–556.

Uchihara, T., Tsuchiya, K., Nakamura, A., & Akiyama, H. (2005). Argyrophilic grains are not always argyrophilic—distinction from neurofibrillary tangles of diffuse neurofibrillary tangles with calcification revealed by comparison between Gallyas and Campbell–Switzer methods. *Acta Neuropathologica, 110*, 158–164.

Ulrich, J., Spillantini, M. G., Goedert, M., Dukas, L., & Stähelin, H. B. (1992). Abundant neurofibrillary tangles without senile plaques in a subset of patients with senile dementia. *Neurodegeneration, 1*, 257–264.

Usenovic, M., Niroomand, S., Drolet, R. E., Yao, L., Gaspar, R. C., Hatcher, N. G., ... Parmentier-Batteur, S. (2015). Internalized tau oligomers cause neurodegeneration by inducing accumulation of pathogenic tau in human neurons derived from induced pluripotent stem cells. *Journal of Neuroscience, 35*, 14234–14250.

Valenca, G. T., Srivastava, G. P., Oliveiro-Filho, J., White, C. C., Schneider, J. A., Buchman, A. S., ... De Jager, P. L. (2016). The role of *MAPT* haplotype H2 and isoform 1N/4R in parkinsonism of older adults. *PLoS ONE, 11*, e0157452.

Verheyen, A., Diels, A., Dijkmans, J., Oyelami, T., Meneghello, G., Mertens, L., ... Cik, M. (2015). Using human iPSC-derived neurons to model TAU aggregation. *PLoS ONE, 10*, e0146127.

Von Bergen, M., Friedhoff, P., Biernat, J., Heberle, J., Mandelkow, E. M., & Mandelkow, E. (2000). Assembly of tau protein into Alzheimer paired helical filaments depends on a local sequence motif (^{306}VQIVYK^{311}) forming β structure. *Proceedings of the National Academy of Sciences of the United States of America, 97*, 5129–5134.

Von Bergen, M., Barghorn, S., Li, L., Marx, A., Biernat, J., Mandelkow, E. M., & Mandelkow, E. (2001). Mutations of tau protein in frontotemporal dementia promote aggregation of paired helical filaments by enhancing local β-structure. *Journal of Biological Chemistry, 276*, 48165–48174.

Wilhelmsen, K. C., Lynch, T., Pavlou, E., Higgins, M., & Nygaard, T. H. (1994). Localization of disinhibition–dementia–parkinsonism–amyotrophy complex to 17q21-22. *American Journal of Human Genetics, 55*, 1159–1165.

Wischik, C. M., Novak, M., Thogersen, H. C., Edwards, P. C., Runswick, M. J., Jakes, R., ... Klug, A. (1988a). Isolation of a fragment of tau derived from the core of the paired helical filament of Alzheimer disease. *Proceedings of the National Academy of Sciences of the United States of America, 85*, 4506–4510.

Wischik, C. M., Novak, M., Edwards, P. C., Klug, A., Tichelaar, W., & Crowther, R. A. (1988b). Structural characterization of the core of the paired helical filament of Alzheimer disease. *Proceedings of the National Academy of Sciences of the United States of America, 85*, 4884–4888.

Wu, J. W., Hussaini, S. A., Bastille, I. M., Rodriguez, G. A., Mrejeru, A., Rilett, K., ... Duff, K. E. (2016). Neuronal activity enhances tau propagation and tau pathology *in vivo*. *Nature Neuroscience, 19*, 1085–1092.

Xie, C., Soeda, Y., Shinzaki, Y., In, Y., Tomoo, K., Ihara, Y., & Miyasaka, T. (2015). Identification of key amino acids responsible for the distinct aggregation properties of microtubule-associated protein 2 and tau. *Journal of Neurochemistry, 135*, 19–26.

Yamada, K., Cirrito, J. R., Stewart, F. R., Jiang, H., Finn, M. B., Holmes, B. B., ... Holtzman, D. M. (2011). *In vivo* microdialysis reveals age-dependent decrease of brain interstitial fluid tau levels in P301S human tau transgenic mice. *Journal of Neuroscience, 31*, 13110–13117.

Yanamandra, K., Kfoury, N., Jiang, H., Mahan, T. E., Ma, S., Maloney, S. E., ... Holtzman, D. M. (2013). Anti-tau antibodies that block tau aggregate seeding in vitro markedly decrease pathology and improve cognition in vivo. *Neuron, 80*, 402–414.

Yetman, M. J., Lillehaug, S., Bjaalie, J. G., Leergaard, T. B., & Jankowsky, J. L. (2016). Transgene expression in the Nop-tTA driver lines not inherently restricted to the entorhinal cortex. *Brain Structure and Function, 221*, 2231–2249.

Yoshida, H., & Goedert, M. (2002). Molecular cloning and functional characterization of chicken brain tau: Isoforms with up to five tandem repeats. *Biochemistry, 41*, 15203–15211.

Yoshida, K., Hata, Y., Kinoshita, K., Takashima, S., Tanaka, K., & Nishida, N. (2017). Incipient progressive supranuclear palsy is more common than expected and may comprise clinicopathological subtypes: A forensic autopsy series. *Acta Neuropathologica, 133*, 809–823.

Zhang, B., Une, Y., Fu, X., Yan, J., Ge, F., Yao, J., ... Higuchi, K. (2008). Fecal transmission of AA amyloidosis in the cheetah contributes to high incidence of disease. *Proceedings of the National Academy of Sciences of the United States of America, 105*, 7623–7628.

Zhao, Y., Tseng, I. C., Heyser, C. J., Rockenstein, E., Mante, M., Adame, A., ... Xu, H. (2015). Appoptosin-mediated caspase cleavage of tau contributes to progressive supranuclear palsy pathogenesis. *Neuron, 87*, 963–975.

Zhong, Q., Condon, E. E., Nagaraja, H. N., & Kuret, J. (2012). Tau isoform composition influences rate and extent of filament formation. *Journal of Biological Chemistry, 287*, 20711–20719.

Further Reading

Ahmed, Z., Cooper, J., Murray, T. K., Garn, K., McNaughton, E., Clarke, H., ... O'Neill, M. J. (2014). A novel in vivo model of tau propagation with rapid and progressive neurofibrillary tangle pathology: The pattern of spread is determined by connectivity, not proximity. *Acta Neuropathologica, 127*, 667–683.

Crowther, R. A. (1991). Straight and paired helical filaments in Alzheimer disease have a common structural unit. *Proceedings of the National Academy of Sciences of the United States of America, 88*, 2288–2292.

Hu, W., Zhang, X., Tung, Y. C., Xie, S., Liu, F., & Iqbal, K. (2016). Hyperphosphorylation determines both the spread and the morphology of tau pathology. *Alzheimer's & Dementia, 12*, 1066–1077.

CHAPTER

5

Amyotrophic Lateral Sclerosis and Other TDP-43 Proteinopathies

Jorge Gomez-Deza and Christopher E. Shaw
King's College London, London, United Kingdom

OUTLINE

TDP-43 BIOLOGY

The transactivation response DNA binding protein (TDP-43) is a 414-amino-acid protein encoded by the *TARDBP* gene located in chromosome 1 (Baralle & Buratti, 2011). TDP-43 is a member of the heterogeneous nuclear ribonuclear (hnRNP) protein family. TDP-43 contains two RNA-recognition motifs (RRM), RRM1 and RRM2, that bind to DNA and RNA ad regulating the expression, editing, transport, and translation of thousands of RNA transcripts (Baralle & Buratti, 2011; Polymenidou et al., 2011). It also contains both a nuclear localization signal (NLS) and a nuclear export signal which allows it to shuttle between the nucleus and the

The Molecular and Cellular Basis of Neurodegenerative Diseases
DOI: https://doi.org/10.1016/B978-0-12-811304-2.00005-5

cytoplasm, although the protein is predominantly located in the nucleus. Additionally, it has a low complexity domain in the C-terminus that is glycine-rich and is involved in mediating many protein–protein interactions (Fig. 5.1). Given its crucial function in the processing of thousands of RNA transcripts, the levels of TDP-43 protein are tightly controlled through autoregulation, as the protein binds to and edits its own 3′ untranslated region mRNA, leading to degradation of the TDP-43 transcript. A decrease in the cellular levels of TDP-43 results in a decrease of transcript editing and an increase in translation of TDP-43 protein. Similarly, if the levels of TDP-43 are too high, TDP-43 will bind to more transcripts and induce their degradation, decreasing the levels of soluble TDP-43 (Ayala et al., 2011).

The pathobiology of TDP-43 has been the focus of intense research since it was identified as the major component of the neuronal cytoplasmic inclusions (NCIs) that characterize tau-negative frontotemporal dementia (FTD) and amyotrophic lateral sclerosis (ALS) and in myocytes in inclusion body myositis (IBM) (Neumann et al., 2006). Although it is not the dominant molecular pathology of other neurodegenerative diseases, inclusions are also present in 25%–50% of patients with Alzheimer's disease (AD) and Parkinson's disease. In this chapter, we will review the main genetic and cellular causes of ALS as well as other TDP-43 proteinopathies.

AMYOTROPHIC LATERAL SCLEROSIS

ALS is a progressive neurodegenerative disease characterized by the loss of motor neurons in the brain and spinal cord, causing progressive paralysis of limb and bulbar muscles and death due to respiratory failure within an average of 3 years from symptom onset. Cytoplasmic inclusions containing TDP-43

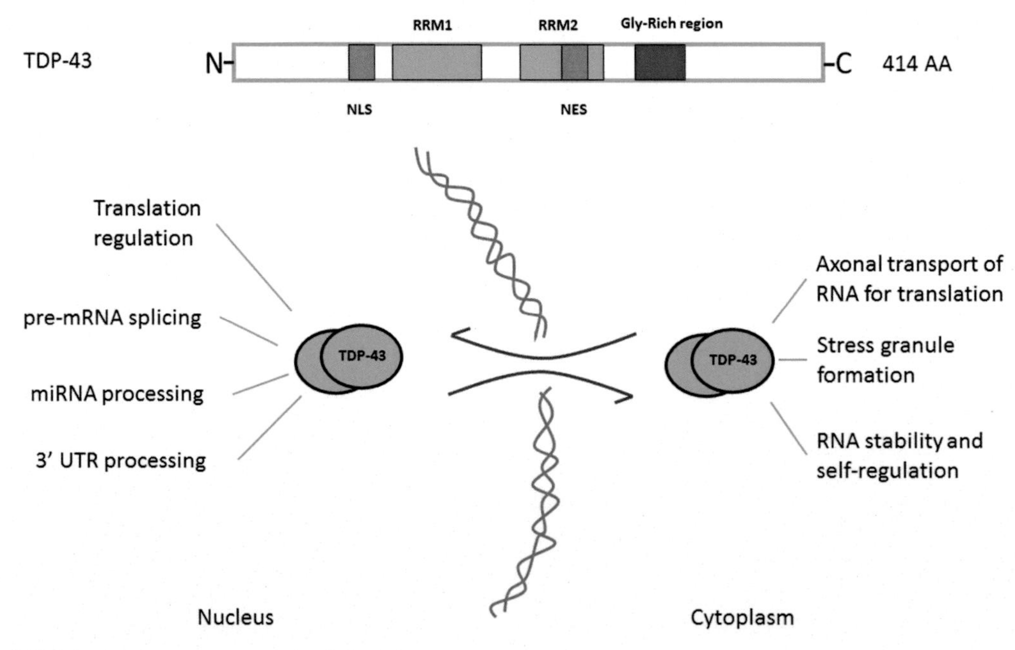

FIGURE 5.1 TDP-43 structure and cellular roles.

are the neuropathological hallmark in ~95% of ALS cases (Neumann et al., 2006).

The clinical signs of lower motor neuron degeneration include fasciculations, muscle wasting, and weakness, while signs of upper motor neuron degeneration include spasticity, hyperreflexia, and an extensor plantar response (Babinski sign). There is currently no cure, and the sole disease-modifying treatment, riluzole, has only a modest effect on survival; increasing life expectancy by only 3 months during an 18-month clinical trial (Cheah et al., 2010). The management of ALS is largely palliative and focused on reducing the impact of symptoms arising from muscle weakness in the throat, causing difficulty swallowing, or limbs, decreasing mobility and independence in the tasks of daily living. When respiratory weakness develops, noninvasive or even invasive mechanical ventilation is instituted (Mitsumoto, Brooks, & Silani, 2014). The relentless progression of disability and loss of autonomy are reasons given as to why so many ALS patients seek euthanasia. In the Netherlands, where it is legal, around 20% of ALS patients chose to die by "physician-assisted suicide" (Veldink et al., 2002).

ALS is regarded as a rare condition, with a prevalence of 3–6 per 100,000 people and an incidence of 3 per 100,000 people (Chio et al., 2013; Johnston et al., 2006). Due to improvements in healthcare and increase in life expectancy, this number is expected to rise significantly, as the incidence of ALS increases with age. ALS is mainly a late-onset disease, although juvenile onset (<25 years) and early onset (<45 years) account for ~1% and ~20% of all ALS cases, respectively (Turner et al., 2012).

There are currently no disease-specific biomarkers for ALS, and the diagnosis is based on symptoms and signs of progressive upper and lower motor neuron degeneration in the absence of radiological evidence of a structural lesion and exclusion of other conditions that may mimic ALS, which has been standardized as the "El Escorial criteria" (Brooks, 1999). The diagnostic tests include magnetic resonance imaging of the brain and spine to exclude any structural or inflammatory disorders; nerve conduction studies to exclude a peripheral neuropathy and electromyography to confirm motor neuron loss; blood tests for antiganglioside antibodies; DNA test for Kennedy's disease, and biochemical tests to exclude other rare mimics of ALS. Diagnostic accuracy exceeds 95%, but the rate of disease progression is variable, and the lack of a reliable biomarker makes it hard to objectively assess the efficacy of drugs tested in clinical trials.

To date, riluzole is the only approved therapeutic treatment for ALS in the USA and Europe. Although its mechanism of action is unknown, it is thought to reduce glutamate toxicity by inhibiting the levels presynaptic glutamate (Cheah et al., 2010). Riluzole received marketing authorization in 1995 in the USA and in 1996 in Europe. In the years that followed, over 100 small molecules have been investigated as possible treatments for ALS. Despite significant research efforts, the overwhelming majority of human clinical trials have failed to demonstrate clinical efficacy. Only very recently, oral masitinib and intravenous edaravone have been FDA approved for clinical use, but their efficacies in large cohorts and approval for clinical use in Europe still remain to be confirmed (Petrov et al., 2017).

The majority of ALS cases (~90%) appear to be sporadic, with no known family history (sporadic ALS or sALS) (Chio et al., 2013). The remaining 10% of cases have a familial history of ALS or a dementia consistent with FTD (familial ALS or fALS) as shown in Fig. 5.2. In 1993, mutations in the superoxide dismutase 1 (*SOD1*) gene were shown to cause ~20% of fALS cases (Rosen et al., 1993). It was not until recently, using next generation gene sequencing techniques, that the number of genes linked to ALS has increased dramatically. To date, mutations in over 30 different genes have been causally associated with ALS with variable degree of certainty.

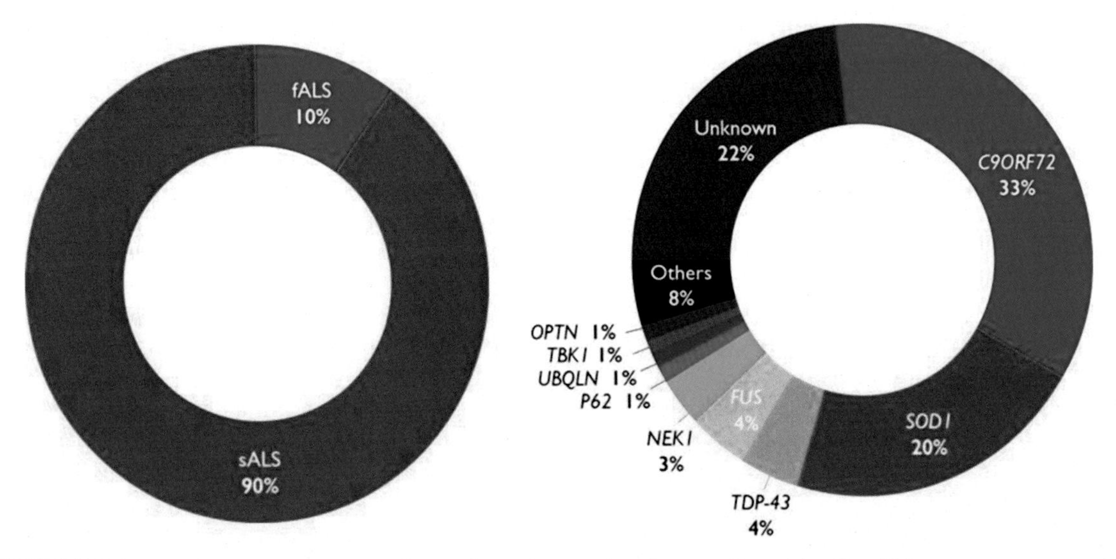

FIGURE 5.2 Diagram showing the prevalence of genetic mutations in sporadic and familiar ALS cases.

ALS–TDP-43

The *TARDBP* gene encodes for the RNA-binding protein TDP-43. Hyperphosphorylated, polyubiquitinated, insoluble cytoplasmic TDP-43 inclusions are present in the spinal cord and frontal cortex of >90% of ALS cases. TDP-43 intracellular inclusions are present in the nucleus and cytoplasm of neurons and glia (Neumann et al., 2006, 2009). TDP-43 pathology can be found in the majority of ALS patients—with the exception of those with SOD1 and fused in sarcoma (FUS) mutations—and is indistinguishable between patients with and without mutations in the *TARDBP* gene. In 2008, the first pathogenic mutations in the *TARDBP* were identified (Sreedharan et al., 2008), and since then, over 50 mutations have been reported, accounting for 4% of all fALS and 1% of sALS cases. Most mutations are clustered in the low-complexity "prion-like," glycine-rich carboxyl (C)-terminus of the protein (Ling, Polymenidou, & Cleveland, 2013; Sreedharan et al., 2008) (Fig. 5.3).

TDP-43 is a 414-amino-acid protein containing two RRM (Ling et al., 2013) that shuttles between the nucleus and the cytoplasm and has been shown to bind ~5500 different RNA transcripts, including TDP-43 mRNA, self-regulating its expression (Polymenidou et al., 2011; Tollervey et al., 2011). The presence of cytoplasmic aggregates and clearance of nuclear TDP-43 suggests that toxicity may arise though gain- or loss-of-function mechanisms, and it remains unclear which of the two is the main driver of toxicity (Diaper et al., 2013; Ratti & Buratti, 2016).

Animal models have yielded conflicting results on whether TDP-43 toxicity is due to a loss-of-function or a gain-of-function mechanism. TDP-43 knockout (KO) mice are embryonically lethal; however, heterozygous KO mice are viable and fertile, with autoregulation maintaining nearly normal TDP-43 levels (Kraemer et al., 2010). Selective removal of TDP-43 from the motor neurons in mice causes weight loss, degeneration of the motor axons, and loss of the neuromuscular junction (NMJ); however, these mice did not have reduced survival rates (Iguchi et al., 2013). Furthermore, homozygous KO of TDP-43 in zebrafish causes muscle degeneration and reduced motor neuron axonal outgrowth

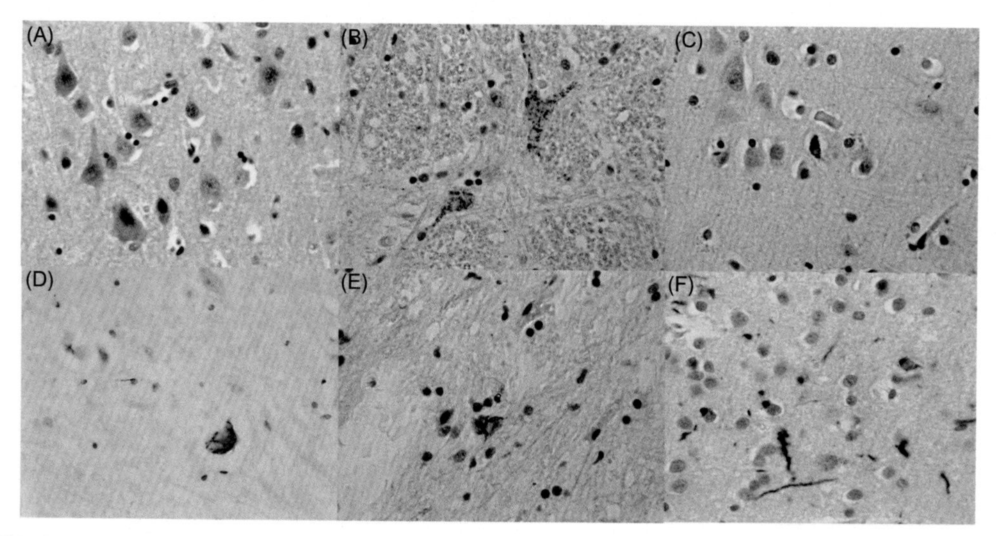

FIGURE 5.3 (A) Normal physiological expression of TDP-43 in the frontal cortex of a healthy control brain, (B) Cytoplasmic TDP-43 aggregates and nuclear clearing in the spinal cord of an ALS patient, (C) TDP-43 inclusion in the motor cortex of an ALS patient, (D) TDP-43 skein-like inclusion in the spinal cord of a TARDP mutation patient, (E) TDP-43 aggregates in the spinal cord of a ALS patient with the C9orf72 expansion, (F) TDP-43 inclusions and neurites in a patient with frontotemporal lobar degeneration (subtype C pathology).

(Schmid et al., 2013). Together, this data suggests that loss of TDP-43 may play an important role in adult motor neuron degeneration, but it may not be sufficient to cause fatal motor neuron disease.

Overexpression of TDP-43 in *Caenorhabditis elegans* causes cell death (Ash et al., 2010). This study showed that there was a correlation between endogenous TDP-43 levels and motor deficits. Moreover, the overexpression of wild-type (WT) TDP-43 and M337V TDP-43 mutant causes severe neurodegeneration and short life span in transgenic mice (Mitchell et al., 2015). These mice develop most of the observed pathology in postmortem tissue, including hyperphosphorylated, polyubiquitinated TDP-43 aggregates, and loss of upper and lower motor neurons. These studies suggest that overexpression of TDP-43 is also capable of causing neurodegeneration and reducing life span. Finally, KO and overexpression of the endogenous *tarbp* gene in zebrafish are both sufficient to cause fatal motor phenotypes in zebrafish (Kabashi et al., 2010). These results indicate that motor neuron degeneration may be driven by both the loss and overabundance of TDP-43, its accumulation in the cytoplasm, and clearance from the nucleus.

ALS–SOD1

The *SOD1* gene codes for superoxidase dismutase 1 protein which catalyzes the dismutation of superoxide radicals to molecular oxygen or hydrogen peroxide (Rosen et al., 1993). Mutations present in the *SOD1* gene were the first to be identified in fALS cases. Initially, 11 missense mutations were discovered by linkage studies; over 180 disease-causing mutations have subsequently been reported (Rosen et al., 1993; Scarrott et al., 2015). Cases with *SOD1* mutations do not show the typical TDP-43 pathology identified in 90% of ALS cases. Instead, SOD1 cytoplasmic inclusions are present in the spinal cord of *SOD1* mutation carriers (van Zundert & Brown, 2016).

Pathogenic *SOD1* mutations are predominantly single amino acid changes and account for 20% of fALS and 1% of sALS cases (White & Sreedharan, 2016). The mechanism of toxicity of *SOD1*-linked ALS is caused by misfolding and aggregation of SOD1 protein rather than by impairment of SOD1 activity (Williamson et al., 2000). Mice overexpressing G93A mutant SOD1 human protein develop progressive motor neuron loss and hind limb paralysis and are commonly used for preclinical studies. These mice however do not develop TDP-43 pathology that typifies ALS cases (Bruijn et al., 1998; Watanabe et al., 2001).

ALS-FUS

Mutations in the *FUS* gene were reported by two independent groups in 2009, having been previously linked to chromosome 16q (Abalkhail et al., 2003; Kwiatkowski et al., 2009; Ruddy et al., 2003; Sapp et al., 2003; Vance et al., 2009). FUS, like TDP-43, is another DNA and RNA binding protein. Most ALS-linked mutations reside in the extreme C-terminus of the protein and reduce its nuclear import, leading to increased cytoplasmic accumulation and aggregation (Vance et al., 2006). FUS is a 526-amino-acid nuclear protein that shuttles between the nucleus and the cytoplasm and can bind single- and double-stranded DNA as well as RNA. It has been shown to participate in multiple cellular processes such as DNA repair, axonal transport of mRNA, RNA splicing, and transcriptional regulation (Lagier-Tourenne, Polymenidou, & Cleveland, 2010).

Interestingly, like SOD1, FUS cases are also pathologically distinct, as they lack TDP-43 pathology. FUS immune-reactive inclusions are present in the spinal cord of ALS cases with *FUS* mutations and in FTD patients without FUS mutations (King et al., 2015; Vance et al., 2009). FUS-mutant ALS cases tend to have a younger age at onset, with rare cases occurring in the late teens and early 20s and more than 60% of *FUS*-mutant ALS having a disease onset before 45 years of age (Baumer et al., 2010; Shang & Huang, 2016).

Several groups have used a wide range of animal models to study the effect of expressing ALS-mutant FUS proteins. In vitro studies have shown that mutations known to cause ALS increase the levels of cytoplasmic FUS by impairing nuclear import and cause it to aggregate and sequester WT-FUS into cytoplasmic stress granules (Dormann et al., 2010; Vance et al., 2013). FUS knockdown using antisense morpholinos in zebrafish cause motor and behavioral defects and reductions in the branching and length of motor axons. Morpholino knockdown of zebrafish FUS followed by overexpression of mutant human FUS proteins also cause defects in the NMJ (Armstrong & Drapeau, 2013). Finally, murine models of FUS show that WT-FUS as well as mutant FUS overexpression cause degeneration in mice. FUS-R521C and FUS-R525L cause more severe dendritic outgrowth deficits than WT-FUS (Mitchell et al., 2013). Interestingly, transgenic mice expressing higher levels of WT-FUS also displayed early-onset motor neuron degeneration in a dose-dependent manner (Mitchell et al., 2013).

ALS-C9ORF72

In 2011, a large intronic hexanucleotide expansion GGGGCC (G4C2) in the chromosome 9 open reading frame 72 (*C9ORF72*) gene was identified and accounts for 20%–50% of all fALS cases (DeJesus-Hernandez et al., 2011; Renton et al., 2011). In ALS cases, the G4C2 repeat can be expanded between 60 and 3000 times; this contrasts with the presence of 2–20 repeats in unaffected individuals (DeJesus-Hernandez et al., 2011; Renton et al., 2011).

The TDP-43 pathology observed in *C9ORF72* cases is largely indistinguishable from other ALS cases (Mackenzie et al., 2013); however, *C9ORF72* cases also show abundant p62-positive and TDP-43-negative inclusions in non-motor regions such as the cerebellum (Al-Sarraj et al., 2011). Additionally, expanded G4C2 RNA accumulates in the nucleus of neurons and glia as RNA foci in most regions of the brain and the spinal cord (Cooper-Knock et al., 2012; DeJesus-Hernandez et al., 2011, 2017; Lee et al., 2013; Mizielinska et al., 2013). Moreover, the repeats can be translated through unconventional repeat-associated non-ATG translation of the sense and the antisense strands. Five dipeptide-repeat (DPR) proteins have been identified in postmortem tissues from C9ALS and FTD cases; poly-Gly-Pro (GP), Gly-Ala (GA), and Gly-Arg (GR) from the sense strand (Ash et al., 2013; Mori et al., 2013) and GP, Pro-Ala, and Pro-Arg (PR) from the antisense strand (Gendron et al., 2013; Mori et al., 2013). Aggregates of these five DPR proteins have been identified to varying degree in many different regions of the brain and spinal cord of *C9ORF72* expansion carriers (Ash et al., 2013; Mori et al., 2013).

The mechanisms by which the hexanucleotide repeats might exert toxicity are highly controversial. It was initially reported that the presence of the G4C2 repeats lead to a decrease in *C9ORF72* transcripts and protein, therefore suggesting toxicity could be due to haploinsufficiency (DeJesus-Hernandez et al., 2011; Renton et al., 2011). This theory subsequently was not supported, as *C9orf72*-KO mice do not develop any neurological defects or any ALS-associated pathology and instead have immune deficiency and dysfunctional macrophages and microglia (O'Rourke et al., 2016; Koppers et al., 2015). Furthermore, homozygous expansion carriers do not develop disease at an earlier age or a more rapidly progressive disease course compared to heterozygous carriers (Fratta et al., 2013), which would be expected if haploinsufficiency was the mechanism.

Sense and antisense transcripts have been identified in the major sites of neurodegeneration in *C9ORF72* expansion carriers as nuclear RNA foci that sequester RNA binding proteins (Lee et al., 2013; Mizielinska et al., 2013). We and others have shown that G4C2 RNA foci can sequester RNA binding proteins such as hnRNP-H, ADARB2, and nucleolin and increase nucleolar stress (Donnelly et al., 2013; Haeusler et al., 2014; Lee et al., 2013). Additionally, expanded transcripts can form secondary structures such as G-quadruplexes and R-loops (DNA–RNA hybrids) that bind proteins and possibly disrupt nucleocytoplasmic transport in the cell (Conlon et al., 2016; Fratta et al., 2012; Haeusler et al., 2014; Jovicic et al., 2015; Zhang et al., 2015). Finally, the presence of antisense but not sense RNA foci correlates with neurodegeneration (Cooper-Knock et al., 2015).

The expression of DPR proteins has also been strongly implicated in causing neurotoxicity. Although it is unclear which specific DPR is toxic; poly-GR and PR have been shown to be toxic in *Drosophila*, causing eye degeneration when overexpressed in the retina and reducing life span when expressed under an embryonic lethal abnormal vision (ELAV) pan-neuronal promoter (Mizielinska et al., 2014). The arginine-rich DPRs, poly-GR, and PR, may also exert toxicity through impairment of biogenesis of ribosomal RNA (Kwon et al., 2014). Additionally, poly-GA has been shown to be toxic in an aggregation-dependent manner when overexpressed in mice. Poly-GA may exert its toxicity though endoplasmic reticulum stress and caspase-3 activation (Schludi et al., 2017; Zhang et al., 2016a). However, the extremely low presence of DPR inclusions in some areas of maximal neurodegeneration, such as the spinal cord, challenges the idea that DPRs are the main source of toxicity (Gomez-Deza et al., 2015; Mackenzie et al., 2015). It is also unclear whether any of the three mechanisms are sufficient to exert toxicity on their own. Based on the current evidence, it is clear that a combination of all three processes may differentially contribute to neurodegeneration in different sites. The recent emergence of *C9ORF72* bacterial artificial construct mouse models and neurons derived from induced pluripotent stem cells (iPSCs), or iPSC-derived neurons (iPSNs), carrying the expansion opens the possibility of studying disease mechanisms in a more physiological setting (Jiang et al., 2016; Liu et al., 2016; O'Rourke et al., 2015; Peters et al., 2015).

MECHANISMS DYSREGULATED

Nucleocytoplasmic Transport

The efficacy of nucleocytoplasmic transport has been shown to deteriorate significantly during aging (D'Angelo et al., 2009), and studies using iPSNs show that it is particularly affected in neurons derived from old people compared to younger ones (Mertens et al., 2015). The presence of cytoplasmic aggregates of FUS and TDP-43 are the main pathological hallmark of 98% of ALS cases. Although TDP-43 and FUS are known to shuttle between the nucleus and the cytoplasm, they are mainly nuclear proteins. The existence of cytoplasmic aggregates therefore highlights the possibility of the nucleocytoplasmic transport of proteins may be dysregulated in ALS (Boeynaems et al., 2016; Dormann & Haass, 2011).

Most pathogenic *FUS* mutations affect its NLS and interfere with proper nuclear targeting (Dormann et al., 2010; Vance et al., 2013). Interestingly, the nuclear/cytoplasmic ratio of different FUS mutants inversely correlates with the age of onset in FUS-ALS patients (Dormann et al., 2010). Independently of the amino acid sequence, the methylation of the FUS NLS also perturbs its nuclear targeting and increases its affinity to transportin (Dormann et al., 2012; Suarez-Calvet et al., 2016).

Moreover, a large group of proteins implicated in nucleocytoplasmic transport, the karyopherins, have been shown to be misregulated in ALS and FTD (Nishimura et al., 2010), and a recent study has shown that the presence of cytoplasmic protein aggregates may interfere with the nucleocytoplasmic transport process (Woerner et al., 2016). Interestingly, DPRs translated from the G4C2 expansion, together with expanded RNA, have also been shown to affect the nucleocytoplasmic transport in several ways, which will be reviewed in greater detail later (Boeynaems et al., 2016; Jovicic et al., 2015; Zhang et al., 2015, 2016b).

Despite the importance of nucleocytoplasmic transport in aging and ALS pathology, genes regulating nucleocytoplasmic transport have not been strongly implicated in the genetics of the disease. The only exception is *GLE1*, where mutations have been linked to ALS cases, and GLE1 knockdown in zebrafish causes motor neuron deficits that can be rescued by WT but not mutant Gle1 (Kaneb et al., 2015).

Proteostasis Clearance

As with most neurodegenerative conditions, a hallmark feature of ALS is the presence of pathological protein aggregates in affected tissues. In addition, a large number of ALS mutant genes play a vital role in protein quality control, trafficking and degradation, and maintaining protein homeostasis (*VCP*, *UBQLN2*, *OPTN*, and *TBK1*). This indicates that defective proteostasis is a key pathogenic mechanism (Blokhuis et al., 2013).

The aggregation of mutant proteins in ALS is well documented. Mutant SOD1 has been shown to misfold and adopt a conformation to prevent ubiquitin-mediated degradation (Niwa et al., 2007). Mutant SOD1 forms toxic oligomers (Urushitani et al., 2002), which further accumulates as aggregates to induce a stress response in cells (Atkin et al., 2006). The oxidation of WT SOD1 also leads to misfolding and aggregation of the protein in a similar fashion to mutant SOD1. Aggregation of SOD1 has toxic effects on two major protein degradation pathways, the ubiquitin proteasome system (UPS) and autophagy.

Similarly, TDP-43 and FUS form insoluble protein aggregates in ALS and FTD but are also present in other neurodegenerative disorders such as AD, Parkinson's, and Huntington's disease (Neumann et al., 2006; Vance et al., 2009). Mutant TDP-43 and FUS are recruited into cytoplasmic stress granules upon stress. Stress granules are normally resolved by disaggregases and chaperones; however, it is thought that mutant TDP-43 and mutant FUS aggregate with stress granule components in an irreversible manner to give rise to pathological inclusions (Chen et al., 2016; Vance et al., 2013). This may be mediated by mislocalization from the nucleus to the cytoplasm of the mutant protein and the prion-like domain within the protein structure of TDP-43 and FUS (Hock & Polymenidou, 2016).

Mitochondrial Energy Production

Evidence for mitochondrial involvement in ALS was first identified when ultrastructural studies revealed an increase in the number of mitochondria in myelinated axons (Atsumi, 1981). Dysfunctional mitochondria have also been observed in transgenic mice overexpressing the G93A SOD1 mutant. The presence of mutant SOD1 aggregates correlates with mitochondrial vacuolar degeneration in G93A SOD1 mice (Dal Canto & Gurney, 1995; Wong et al., 1995). Transgenic mice expressing TDP-43 also display mitochondrial abnormalities, such as mitochondrial aggregation, and develop impaired motor function (Xu et al., 2010). Mitochondrial dysfunction may lead to apoptosis through the release of cytochrome-C into the cytoplasm.

RNA Metabolism

The first link between alterations in RNA processing and motor neuron degeneration was described over a decade ago. The expression of a neurofilament transgene was found to affect binding of a ribonucleoprotein to the neurofilament mRNA and lead to the degeneration of enteric and motor neurons in transgenic mice (Canete-Soler et al., 1999). The discovery of disease-linked mutations in multiple RNA binding proteins has highlighted dysfunctional RNA processing as a major pathogenic pathway in ALS (Kwiatkowski et al., 2009; Sreedharan et al., 2008; Vance et al., 2009). Pathological as well as in vitro and in vivo studies have provided evidence of abnormalities in multiple steps of RNA processing, including transcription, splicing, translation, and decay (Baloh, 2012). Mutations present in *TARBP*, the gene encoding for TDP-43, and *FUS*, both coding for RNA binding proteins with hundreds of target transcripts, highlight the importance of dysregulated RNA metabolism. Knockdown of TDP-43 and FUS affects the level of thousands of transcripts (Lagier-Tourenne et al., 2012).

Axonal Dysfunction

Motor neurons are characterized by a unique morphology. These cells are highly polarized, typified by axonal projections that can reach a meter in length in adult humans. This morphology requires active axonal transport of organelles, cytoskeletal, synaptic components, and trophic factors to maintain normal cellular function. Axonal transport defects have been shown to occur in presymptomatic stages of disease in animal models of motor neuron disease, and the selective loss of neuromuscular synapse subtypes are believed to precede motor neuron degeneration (Frey et al., 2000). The accumulation of neurofilaments in the soma and axons of motor neurons in the spinal cord constitutes one of the pathological hallmarks in ALS (Hirano et al., 1984). Additionally, the loss of the NMJ is common in patients and mouse models of the disease (Mitchell et al., 2015) and ALS-causing mutations in TDP-43 also affect axonal transport in iPSNs (Alami et al., 2014).

Excitotoxicity

Levels of glutamate, one of the main neurotransmitters in the central nervous system, have been shown to be elevated in the cerebrospinal fluid of ALS patients (Shaw et al., 1995). Moreover, knock down of metabolic glutamate receptor 1 in G93A SOD1 mice extends survival and increases the number of motor neurons (Milanese et al., 2014). SOD1 fALS mutations inhibit the function of excitatory amino acid transporter-2 responsible for glutamate uptake (Trotti et al., 1999). Additionally, iPSN derived from *C9ORF72* expansion carriers show an increased susceptibility to glutamate toxicity (Donnelly et al., 2013). Finally, Riluzole, the only licensed drug for ALS, reduces excitotoxicity through the inhibition of presynaptic glutamate release (Cheah et al., 2010).

Oxidative Stress

Oxidative stress arises from the imbalance in the production of reactive oxygen (ROS) and reactive nitrogen species and their clearance by the cell. ROS are important signaling molecules necessary for a range of physiological processes. However, excessive production of ROS causes damage of biomolecules interfering with their normal functions and leading ultimately to cellular dysfunction or even death.

There is strong evidence showing that oxidative stress may be a contributing factor in the pathogenesis of ALS. Spinal cord samples from ALS patients contain higher levels of oxidative damage markers and free radicals (Ihara et al., 2005; Shaw et al., 1995).

Most in vivo studies assessing the effect of oxidative stress in ALS have been carried out in the G93A SOD1 mice. The same oxidative stress makers seen to be upregulated in ALS patients have been reported in the spinal cord and cortex G93A SOD1 transgenic mice (Ferrante et al., 1997). Finally, a recent study has proposed that some DPRs translated from the G4C2 expansion may also induce oxidative stress in iPSC-derived motor neurons from *C9ORF72* expansion carriers (Lopez-Gonzalez et al., 2016).

Neuroinflammation

Another main pathological hallmark of ALS is the presence of activated microglia and infiltrated lymphocytes in the main sites of neurodegeneration (Engelhardt & Appel, 1990). Research using positron emission tomography has provided direct evidence of microglial activation in the brains of living patients with ALS. The intensity of microglial activation was correlated with the severity of disease, suggesting an active involvement of microglial activation in ALS throughout the disease process (Turner et al., 2004). Microgliosis and inflammation have been demonstrated in mutant G93A SOD1 mice and ALS cases. ALS patients contain elevated levels of proinflammatory molecules such as cytokines, ROS, chemokines, and glutamate (Komine & Yamanaka, 2015). However, it still remains unclear whether the presence of M2 protective microglia and M1 neurotoxic microglia are a response to the loss of neurons or actively contribute to neurodegeneration (Henkel et al., 2009).

Finally, it has been shown that *C9ORF72* KO mice show increased microglial activation, possibly indicating that the G4C2 expansion may be causing a decrease in the levels of C9ORF72 protein and increasing neuroinflammation (O'Rourke et al., 2016).

OTHER TDP-43 PROTEINOPATHIES

Frontotemporal Dementia

FTD is the second most common cause of presenile dementia with an onset <65 years only after AD. It is characterized by the loss of up to 70% of spindle neurons in the frontal and temporal lobes, whilst others remain unaffected. The symptoms include personality and behavioral changes, apathy, blunting of emotions, and the loss of language skills affecting men and women equally.

The symptoms of FTD are complex and heterogeneous. As a result, FTD is diagnosed as three different types, depending on the function of the frontal and temporal lobes affected. Behavioral variant FTD (BvFTD) is characterized by the patient's loss of language skills. As a result, patients diagnosed with BvFTD show changes in social behavior, impulsive traits, and loss of social awareness. Semantic dementia results from impaired word comprehension, although speech remains generally fluent. Finally, progressive nonfluent aphasia is characterized by word finding difficulty resulting in a decline in meaningful speech. Symptoms of all three forms of FTD are commonly present in the late stages of FTD.

From a pathological perspective, there are three main types for FTD subtypes; FTD-MAPT, FTD-FUS, and FTD-TDP-43. In FTD-Tau, the microtubule associated protein tau is deposited in the frontal and temporal lobes in the form of tau microfilaments in ~40% of FTD cases (Lee, Goedert, & Trojanowski, 2001). However, tau pathology can also be found in other brain areas (e.g., basal ganglia and subthalamus) depending on the specific type of FTD. Mutations in the *MAPT* gene, which encodes for tau, have been reported to cause FTD through impairing the normal splicing or increasing its aggregation propensity (Wolfe, 2012). Interestingly, Tau aggregates are present in the brains of patients with AD, which will be reviewed in greater detail in a later section. FTD-FUS is characterized by the presence of immunoreactive FUS aggregates in the frontal and temporal lobes of FTD patients. The subset of sporadic FTD cases that contain FUS aggregates lack the more common aggregate markers TDP-43 and tau (Gao, Almeida, & Lopez-Gonzalez, 2017).

Finally, FTD-TDP-43 is characterized by the presence of immunoreactive TDP-43 aggregates, common in ALS, in the frontal cortex of ~50% of FTD cases. As a result, ALS and FTD are strongly pathologically and genetically linked and are increasingly regarded as two ends of a phenotypic spectrum. In FTD-TDP-43, neuronal and glial cytoplasmic inclusions of TDP-43 are present in the brain in FTD patients but absent from the spinal cord. Like in ALS, polyubiquitinated, hyperphosphorylated TDP-43 aggregates are mainly neuronal and cytoplasmic. Biochemical analysis of insoluble protein extracts isolated from affected FTD-TDP-43 and ALS tissue has revealed a characteristic biochemical profile of TDP-43, with detection of bands at ~25, ~45 kDa, and a high molecular mass smear and the normal 43 kDa band. Further analysis demonstrated that this profile is due to N-terminal truncation, hyperphosphorylation, and ubiquitination of TDP-43 in FTD-TDP-43 and ALS (Neumann et al., 2006). The morphology and abundance of TDP-43 aggregates in FTD-TDP-43 cases is heterogeneous. As a result, FTD-TDP-43 cases have been subcategorized into four subtypes (1–4) depending on the TDP-43 inclusion morphology, laminar distribution of ubiquitin and TDP-43-positive inclusions, and relative proportion of dystrophic neurites versus NCIs (Cairns et al., 2007; Mackenzie et al., 2006, 2009). The four different subtypes are summarized in Table 5.1.

Moreover, the hexanucleotide repeat GGGGCC (G4C2) expansion in the *C9ORF72* gene has provided strong genetic evidence of the link between ALS and FTD. The expansion is present

TABLE 5.1 Pathological Classification of Frontotemporal Lobar Dementia Sybtypes According to TDP-43 Inclusions

	Type 1	Type 2	Type 3	Type 4
Pathology	Long neurites, rare or absent intranuclear inclusions	Common cytoplasmic inclusions but rare intranuclear inclusions	Common cytoplasmic inclusions but rare intranuclear inclusions and short neurites	Short neurites and common neuronal intranuclear inclusions
Glial inclusions	Rare	Frequent	Frequent	Absent
Clinical symptoms	SD	FTD, often with ALS	FTD and PNFA	IBMPFD

Classification according to Sampathu et al. (2006). *SD*, semantic dementia; *PNFA*, progressive nonfluent aphasia; *IBMPFD*, inclusion body myopathy with early-onset Paget disease and frontotemporal dementia.

in 20%–50% of fALS (Snowden et al., 2012) and 0% of sALS and FTD cases (DeJesus-Hernandez et al., 2011). Frank dementia is rare in ALS, but around 30%–50% of ALS cases present with subtle cognitive and language deficits that are typical of FTD, and it is estimated that 15% of FTD cases develop subtle signs of upper and lower motor neuron degeneration and the diagnostic criteria for ALS (Ling et al., 2013).

TDP-43 in Alzheimer's Disease and Parkinsonian Syndromes

TDP-43 pathology has been shown to be present in several other neurodegenerative conditions other than FTD and ALS. TDP-43 pathology is a highly consistent finding in most cases of ALS-Parkinsonism dementia complex of Guam (Hasegawa et al., 2007). Intracellular inclusions consisting of TDP-43 are present in up to 50% of AD cases, correlate with worsened cognition and neurodegeneration, and follow a distinct and well-established pattern of progression described in the TDP-43 in AD staging scheme (McAleese et al., 2017). Unlike in FTD, TDP-43 pathology is mostly restricted to mesial temporal regions (Neumann, 2009). TDP-43 immunoreactivity is often found in separate inclusions, or only partially colocalizes with characteristic lesions found in these diseases, such as neurofibrillary tangles. The precise role of TDP-43 in the AD pathophysiology remains unknown. Recently, TDP-43 depletion in forebrain neurons of an AD mouse model has been shown to exacerbate neurodegeneration and correlate with increased prefibrillar oligomeric Aβ and decreased Aβ plaque burden (LaClair et al., 2016). It is therefore possible that the depletion of TDP-43 from the nucleus may help to accelerate the progression of AD.

Inclusion Body Myositis

IBM are inflammatory myopathies that lead to muscle weakness and atrophy in a slowly progressive manner (Murnyák et al., 2015). Although it is the most common muscle disease affecting people over the age of 50, it is considered a rare condition affecting mostly men (Greenberg, 2011). Several proteins can be accumulated as inclusions in the cytoplasm of muscle fibers in IBM, including amyloid-β42 and its different oligomers as well as phosphorylated tau protein (Greenberg, 2011). Additionally, TDP-43 and p62 have also been identified, which are present in the inclusions found in ALS (Weihl & Pestronk, 2010). Interestingly, it is the presence of TDP-43 aggregates that differentiates IBM from other inflammatory myopathies.

Paget disease of the bone (PDB) and FTD (IBMPFD) is a special subset of IBM (Murnyák et al., 2015). It results from adult-onset muscle weakness, early-onset PDB, and premature FTD. Muscle weakness spreads involving other muscles, including those involved in breathing. In later stages, cardiac failure and cardiomyopathy are common (Murnyák et al., 2015). Early stages of FTD are characterized by lack of sleep, inability to understand simple phrases, and grammatical errors with relative preservation of memory. The inability to speak and comprehension deficits are observed in the later stages of the condition. Mean age at diagnosis for IBMPFD is 42 years; for FTD, 55 years (Gruener & Camacho, 2014). IBMPFD, FTD, and ALS have also been genetically linked. Mutations in *VCP* and *hnRNP A2/B1* and *hnRNP A1* have also been found in patients suffering from all three conditions (Kim et al., 2011, 2013). A mouse model of the disease-associated VCP R155H mutation displayed muscle weakness and brain pathology, with increased TDP-43 and ubiquitin-positive cytoplasmic inclusions (Nalbandian et al., 2013).

CONCLUSIONS

TDP-43 binds to thousands of RNA transcripts and plays an essential role in every stage of their processing. As a result, the pathological effects of its mislocalization and aggregation are likely to be profound and affect multiple cellular pathways, leading to cellular stress and degeneration. Its aggregation in ALS and FTD has not only highlighted the link between these two conditions that are now regarded as ends of a phenotypic spectrum, but it has also revealed the importance of RNA regulation in disease. TDP-43 aggregation also occurs in a proportion of patients with AD and Parkinson's disease, where it may contribute to neurodegeneration as a secondary phenomenon. These findings highlight the importance of understanding the biology of TDP-43 in health and disease potential so that we may develop new therapeutic strategies for intervention.

References

Abalkhail, H., et al. (2003). A new familial amyotrophic lateral sclerosis locus on chromosome 16q12.1–16q12.2. *The American Journal of Human Genetics*, *73*(2), 383–389.

Al-Sarraj, S., et al. (2011). p62 positive, TDP-43 negative, neuronal cytoplasmic and intranuclear inclusions in the cerebellum and hippocampus define the pathology of C9orf72-linked FTLD and MND/ALS. *Acta Neuropathologica*, *122*(6), 691–702.

Alami, N. H., et al. (2014). Axonal transport of TDP-43 mRNA granules is impaired by ALS-causing mutations. *Neuron*, *81*(3), 536–543.

Armstrong, G. A., & Drapeau, P. (2013). Loss and gain of FUS function impair neuromuscular synaptic transmission in a genetic model of ALS. *Human Molecular Genetics*, *22*(21), 4282–4292.

Ash, P. E., et al. (2010). Neurotoxic effects of TDP-43 overexpression in C. elegans. *Human Molecular Genetics*, *19*(16), 3206–3218.

Ash, P. E., et al. (2013). Unconventional translation of C9ORF72 GGGGCC expansion generates insoluble polypeptides specific to c9FTD/ALS. *Neuron*, *77*(4), 639–646.

Atkin, J. D., et al. (2006). Induction of the unfolded protein response in familial amyotrophic lateral sclerosis and association of protein-disulfide isomerase with superoxide dismutase 1. *Journal of Biological Chemistry*, *281*(40), 30152–30165.

Atsumi, T. (1981). The ultrastructure of intramuscular nerves in amyotrophic lateral sclerosis. *Acta Neuropathologica*, *55*(3), 193–198.

Ayala, Y. M., et al. (2011). TDP-43 regulates its mRNA levels through a negative feedback loop. *The EMBO Journal*, *30*(2), 277–288.

Baloh, R. H. (2012). How do the RNA-binding proteins TDP-43 and FUS relate to amyotrophic lateral sclerosis and frontotemporal degeneration, and to each other?. *Current Opinion in Neurology*, *25*(6), 701–707.

Baralle, F. E., & Buratti, E. (2011). TDP-43: Overview of the series. *The FEBS Journal*, *278*(19), 3529.

Baumer, D., et al. (2010). Juvenile ALS with basophilic inclusions is a FUS proteinopathy with FUS mutations. *Neurology*, *75*(7), 611–618.

Blokhuis, A. M., et al. (2013). Protein aggregation in amyotrophic lateral sclerosis. *Acta Neuropathologica*, *125*(6), 777–794.

Boeynaems, S., et al. (2016). Drosophila screen connects nuclear transport genes to DPR pathology in c9ALS/FTD. *Scientific Reports*, *6*, 20877.

Boeynaems, S., et al. (2016). Inside out: The role of nucleocytoplasmic transport in ALS and FTLD. *Acta Neuropathologica*, *132*(2), 159–173.

Brooks, B. R. (1999). Diagnostic dilemmas in amyotrophic lateral sclerosis. *Journal of the Neurological Sciences*, *165* (Suppl 1), S1–S9.

Bruijn, L. I., et al. (1998). Aggregation and motor neuron toxicity of an ALS-linked SOD1 mutant independent from wild-type SOD1. *Science*, *281*(5384), 1851–1854.

Cairns, N. J., et al. (2007). Neuropathologic diagnostic and nosologic criteria for frontotemporal lobar degeneration: Consensus of the Consortium for Frontotemporal Lobar Degeneration. *Acta Neuropathologica*, *114*(1), 5–22.

Canete-Soler, R., et al. (1999). Mutation in neurofilament transgene implicates RNA processing in the pathogenesis of neurodegenerative disease. *The Journal of Neuroscience*, *19*(4), 1273–1283.

Cheah, B. C., et al. (2010). Riluzole, neuroprotection and amyotrophic lateral sclerosis. *Current Medicinal Chemistry*, *17*(18), p. 1942-199.

Chen, H. J., et al. (2016). The heat shock response plays an important role in TDP-43 clearance: Evidence for dysfunction in amyotrophic lateral sclerosis. *Brain*, *139*(Pt 5), 1417–1432.

Chio, A., et al. (2013). Global epidemiology of amyotrophic lateral sclerosis: A systematic review of the published literature. *Neuroepidemiology*, *41*(2), 118–130.

Conlon, E. G., et al. (2016). The C9ORF72 GGGGCC expansion forms RNA G-quadruplex inclusions and sequesters hnRNP H to disrupt splicing in ALS patient brains. *Elife*, 5.

Cooper-Knock, J., et al. (2012). Clinico-pathological features in amyotrophic lateral sclerosis with expansions in C9ORF72. *Brain*, *135*(Pt 3), 751–764.

Cooper-Knock, J., et al. (2015). Antisense RNA foci in the motor neurons of C9ORF72-ALS patients are associated with TDP-43 proteinopathy. *Acta Neuropathologica*, *130*(1), 63–75.

D'Angelo, M. A., et al. (2009). Age-dependent deterioration of nuclear pore complexes causes a loss of nuclear integrity in postmitotic cells. *Cell*, *136*(2), 284–295.

Dal Canto, M. C., & Gurney, M. E. (1995). Neuropathological changes in two lines of mice carrying a transgene for mutant human Cu, Zn SOD, and in mice overexpressing wild type human SOD: A model of familial amyotrophic lateral sclerosis (FALS). *Brain Research*, *676*(1), 25–40.

DeJesus-Hernandez, M., et al. (2011). Expanded GGGGCC hexanucleotide repeat in noncoding region of C9ORF72 causes chromosome 9p-linked FTD and ALS. *Neuron*, *72*(2), 245–256.

DeJesus-Hernandez, M., et al. (2017). In-depth clinico-pathological examination of RNA foci in a large cohort of C9ORF72 expansion carriers. *Acta Neuropathologica*.

Diaper, D. C., et al. (2013). Drosophila TDP-43 dysfunction in glia and muscle cells cause cytological and behavioural phenotypes that characterize ALS and FTLD. *Human Molecular Genetics*, *22*(19), 3883–3893.

Donnelly, C. J., et al. (2013). RNA toxicity from the ALS/FTD C9ORF72 expansion is mitigated by antisense intervention. *Neuron*, *80*(2), 415–428.

Dormann, D., & Haass, C. (2011). TDP-43 and FUS: A nuclear affair. *Trends in Neurosciences*, *34*(7), 339–348.

Dormann, D., et al. (2010). ALS-associated fused in sarcoma (FUS) mutations disrupt Transportin-mediated nuclear import. *The EMBO Journal*, *29*(16), 2841–2857.

Dormann, D., et al. (2012). Arginine methylation next to the PY-NLS modulates Transportin binding and nuclear import of FUS. *The EMBO Journal*, *31*(22), 4258–4275.

Engelhardt, J. I., & Appel, S. H. (1990). IgG reactivity in the spinal cord and motor cortex in amyotrophic lateral sclerosis. *Archives of Neurology*, *47*(11), 1210–1216.

Ferrante, R. J., et al. (1997). Evidence of increased oxidative damage in both sporadic and familial amyotrophic lateral sclerosis. *Journal of Neurochemistry*, *69*(5), 2064–2074.

Fratta, P., et al. (2012). C9orf72 hexanucleotide repeat associated with amyotrophic lateral sclerosis and frontotemporal dementia forms RNA G-quadruplexes. *Scientific Reports*, *2*, 1016.

Fratta, P., et al. (2013). Homozygosity for the C9orf72 GGGGCC repeat expansion in frontotemporal dementia. *Acta Neuropathologica*, *126*(3), 401–409.

Frey, D., et al. (2000). Early and selective loss of neuromuscular synapse subtypes with low sprouting competence in motoneuron diseases. *The Journal of Neuroscience*, *20*(7), 2534–2542.

Gao, F. B., Almeida, S., & Lopez-Gonzalez, R. (2017). Dysregulated molecular pathways in amyotrophic lateral sclerosis-frontotemporal dementia spectrum disorder. *The EMBO Journal*.

Gendron, T. F., et al. (2013). Antisense transcripts of the expanded C9ORF72 hexanucleotide repeat form nuclear RNA foci and undergo repeat-associated non-ATG translation in c9FTD/ALS. *Acta Neuropathologica*, *126*(6), 829–844.

Gomez-Deza, J., et al. (2015). Dipeptide repeat protein inclusions are rare in the spinal cord and almost absent from motor neurons in C9ORF72 mutant amyotrophic lateral sclerosis and are unlikely to cause their degeneration. *Acta Neuropathologica Communications*, *3*, 38.

Greenberg, S. A. (2011). Inclusion body myositis. *Current Opinion in Rheumatology*, *23*(6), 574–578.

Gruener, G., & Camacho, P. (2014). Paget's disease of bone. *Handbook of Clinical Neurology*, *119*, 529–540.

Haeusler, A. R., et al. (2014). C9orf72 nucleotide repeat structures initiate molecular cascades of disease. *Nature*, *507*(7491), 195–200.

Hasegawa, M., et al. (2007). TDP-43 is deposited in the Guam parkinsonism-dementia complex brains. *Brain*, *130*(Pt 5), 1386–1394.

Henkel, J. S., et al. (2009). Microglia in ALS: The good, the bad, and the resting. *Journal of Neuroimmune Pharmacology*, *4*(4), 389–398.

Hirano, A., et al. (1984). Fine structural study of neurofibrillary changes in a family with amyotrophic lateral sclerosis. *Journal of Neuropathology & Experimental Neurology*, *43*(5), 471–480.

Hock, E. M., & Polymenidou, M. (2016). Prion-like propagation as a pathogenic principle in frontotemporal dementia. *Journal of Neurochemistry*, *138*(Suppl 1), 163–183.

Iguchi, Y., et al. (2013). Loss of TDP-43 causes age-dependent progressive motor neuron degeneration. *Brain*, *136*(Pt 5), 1371–1382.

Ihara, Y., et al. (2005). Oxidative stress and metal content in blood and cerebrospinal fluid of amyotrophic lateral sclerosis patients with and without a Cu, Zn-superoxide dismutase mutation. *Neurological Research*, *27*(1), 105–108.

Jiang, J., et al. (2016). Gain of toxicity from ALS/FTD-linked repeat expansions in C9ORF72 is alleviated by antisense oligonucleotides targeting GGGGCC-containing RNAs. *Neuron*, *90*(3), 535–550.

Johnston, C. A., et al. (2006). Amyotrophic lateral sclerosis in an urban setting: A population based study of inner city London. *Journal of Neurology*, *253*(12), 1642–1643.

Jovicic, A., et al. (2015). Modifiers of C9orf72 dipeptide repeat toxicity connect nucleocytoplasmic transport defects to FTD/ALS. *Nature Neuroscience*, *18*(9), 1226–1229.

Kabashi, E., et al. (2010). Gain and loss of function of ALS-related mutations of TARDBP (TDP-43) cause motor deficits in vivo. *Human Molecular Genetics*, *19*(4), 671–683.

Kaneb, H. M., et al. (2015). Deleterious mutations in the essential mRNA metabolism factor, hGle1, in amyotrophic lateral sclerosis. *Human Molecular Genetics*, *24*(5), 1363–1373.

Kim, E. J., et al. (2011). *Inclusion body myopathy with Paget disease of bone and frontotemporal dementia linked to VCP p. Arg155Cys in a Korean family*. *Archives of Neurology*, *68*(6), 787–796.

Kim, H. J., et al. (2013). Mutations in prion-like domains in hnRNPA2B1 and hnRNPA1 cause multisystem proteinopathy and ALS. *Nature*, *495*(7442), 467–473.

King, A., et al. (2015). ALS-FUS pathology revisited: Singleton FUS mutations and an unusual case with both a FUS and TARDBP mutation. *Acta Neuropathologica Communications*, *3*, 62.

Komine, O., & Yamanaka, K. (2015). Neuroinflammation in motor neuron disease. *Nagoya Journal of Medical Science*, *77*(4), 537–549.

Koppers, M., et al. (2015). C9orf72 ablation in mice does not cause motor neuron degeneration or motor deficits. *Annals of Neurology*, *78*(3), 426–438.

Kraemer, B. C., et al. (2010). Loss of murine TDP-43 disrupts motor function and plays an essential role in embryogenesis. *Acta Neuropathologica*, *119*(4), 409–419.

Kwiatkowski, T. J., Jr, et al. (2009). Mutations in the FUS/TLS gene on chromosome 16 cause familial amyotrophic lateral sclerosis. *Science*, *323*(5918), 1205–1208.

Kwon, I., et al. (2014). Poly-dipeptides encoded by the C9orf72 repeats bind nucleoli, impede RNA biogenesis, and kill cells. *Science*, *345*(6201), 1139–1145.

LaClair, K. D., et al. (2016). Depletion of TDP-43 decreases fibril and plaque β-amyloid and exacerbates neurodegeneration in an Alzheimer's mouse model. *Acta Neuropathologica*, *132*(6), 859–873.

Lagier-Tourenne, C., Polymenidou, M., & Cleveland, D. W. (2010). TDP-43 and FUS/TLS: Emerging roles in RNA processing and neurodegeneration. *Human Molecular Genetics*, *19*(R1), R46–R64.

Lagier-Tourenne, C., et al. (2012). Divergent roles of ALS-linked proteins FUS/TLS and TDP-43 intersect in processing long pre-mRNAs. *Nature Neuroscience*, *15*(11), 1488–1497.

Lee, V. M., Goedert, M., & Trojanowski, J. Q. (2001). Neurodegenerative tauopathies. *Annual Review of Neuroscience*, *24*, 1121–1159.

Lee, Y. B., et al. (2013). Hexanucleotide repeats in ALS/FTD form length-dependent RNA foci, sequester RNA binding proteins, and are neurotoxic. *Cell Reports*, *5*(5), 1178–1186.

Ling, S. C., Polymenidou, M., & Cleveland, D. W. (2013). Converging mechanisms in ALS and FTD: Disrupted RNA and protein homeostasis. *Neuron*, *79*(3), 416–438.

Liu, Y., et al. (2016). C9orf72 BAC mouse model with motor deficits and neurodegenerative features of ALS/FTD. *Neuron*, *90*(3), 521–534.

Lopez-Gonzalez, R., et al. (2016). Poly(GR) in C9ORF72-related ALS/FTD compromises mitochondrial function and increases oxidative stress and DNA damage in iPSC-derived motor neurons. *Neuron*, *92*(2), 383–391.

Mackenzie, I. R., et al. (2006). Heterogeneity of ubiquitin pathology in frontotemporal lobar degeneration: Classification and relation to clinical phenotype. *Acta Neuropathologica*, *112*(5), 539–549.

Mackenzie, I. R., et al. (2009). Nomenclature for neuropathologic subtypes of frontotemporal lobar degeneration: Consensus recommendations. *Acta Neuropathologica, 117*(1), 15–18.

Mackenzie, I. R., et al. (2013). Dipeptide repeat protein pathology in C9ORF72 mutation cases: Clinicopathological correlations. *Acta Neuropathologica, 126*(6), 859–879.

Mackenzie, I. R., et al. (2015). Quantitative analysis and clinico-pathological correlations of different dipeptide repeat protein pathologies in C9ORF72 mutation carriers. *Acta Neuropathologica, 130*(6), 845–861.

McAleese, K. E., et al. (2017). TDP-43 pathology in Alzheimer's disease, dementia with Lewy bodies and ageing. *Brain Pathology, 27*(4), 472–479.

Mertens, J., et al. (2015). Directly reprogrammed human neurons retain aging-associated transcriptomic signatures and reveal age-related nucleocytoplasmic defects. *Cell Stem Cell, 17*(6), 705–718.

Milanese, M., et al. (2014). Knocking down metabotropic glutamate receptor 1 improves survival and disease progression in the SOD1(G93A) mouse model of amyotrophic lateral sclerosis. *Neurobiology of Disease, 64*, 48–59.

Mitchell, J. C., et al. (2013). Overexpression of human wild-type FUS causes progressive motor neuron degeneration in an age- and dose-dependent fashion. *Acta Neuropathologica, 125*(2), 273–288.

Mitchell, J. C., et al. (2015). Wild type human TDP-43 potentiates ALS-linked mutant TDP-43 driven progressive motor and cortical neuron degeneration with pathological features of ALS. *Acta Neuropathologica Communications, 3*, 36.

Mitsumoto, H., Brooks, B. R., & Silani, V. (2014). Clinical trials in amyotrophic lateral sclerosis: Why so many negative trials and how can trials be improved? *The Lancet Neurology, 13*(11), 1127–1138.

Mizielinska, S., et al. (2013). C9orf72 frontotemporal lobar degeneration is characterised by frequent neuronal sense and antisense RNA foci. *Acta Neuropathologica, 126*(6), 845–857.

Mizielinska, S., et al. (2014). C9orf72 repeat expansions cause neurodegeneration in Drosophila through arginine-rich proteins. *Science, 345*(6201), 1192–1194.

Mori, K., et al. (2013). Bidirectional transcripts of the expanded C9orf72hexanucleotide repeat are translated into aggregating dipeptide repeat proteins. *Acta Neuropathologica, 126*(6), 881–893.

Mori, K., et al. (2013). The C9orf72GGGGCC repeat is translated into aggregating dipeptide-repeat proteins in FTLD/ALS. *Science, 339*(6125), 1335–1338.

Murnyák, B., et al. (2015). Inclusion body myositis—Pathomechanism and lessons from genetics. *Open medicine (Warsaw, Poland), 10*(1), 188–193.

Nalbandian, A., et al. (2013). A progressive translational mouse model of human valosin-containing protein disease: The VCP(R155H/ +) mouse. *Muscle & Nerve, 47*(2), 260–270.

Neumann, M. (2009). Molecular neuropathology of TDP-43 proteinopathies. *International Journal of Molecular Sciences, 10*(1), 232–246.

Neumann, M., et al. (2006). Ubiquitinated TDP-43 in frontotemporal lobar degeneration and amyotrophic lateral sclerosis. *Science, 314*(5796), 130–133.

Neumann, M., et al. (2009). Phosphorylation of S409/410 of TDP-43 is a consistent feature in all sporadic and familial forms of TDP-43 proteinopathies. *Acta Neuropathologica, 117*(2), 137–149.

Nishimura, A. L., et al. (2010). Nuclear import impairment causes cytoplasmic trans-activation response DNA-binding protein accumulation and is associated with frontotemporal lobar degeneration. *Brain, 133*(Pt 6), 1763–1771.

Niwa, J.-i, et al. (2007). Disulfide bond mediates aggregation, toxicity, and ubiquitylation of familial amyotrophic lateral sclerosis-linked mutant SOD1. *Journal of Biological Chemistry, 282*(38), 28087–28095.

O'Rourke, J. G., et al. (2015). C9orf72 BAC transgenic mice display typical pathologic features of ALS/FTD. *Neuron, 88*(5), 892–901.

O'Rourke, J. G., et al. (2016). C9orf72 is required for proper macrophage and microglial function in mice. *Science, 351*(6279), 1324–1329.

Peters, O. M., et al. (2015). Human C9ORF72 hexanucleotide expansion reproduces RNA foci and dipeptide repeat proteins but not neurodegeneration in BAC transgenic mice. *Neuron, 88*(5), 902–909.

Petrov, D., et al. (2017). ALS clinical trials review: 20 years of failure. Are we any closer to registering a new treatment? *Frontiers in Aging Neuroscience, 9*, 68.

Polymenidou, M., et al. (2011). Long pre-mRNA depletion and RNA missplicing contribute to neuronal vulnerability from loss of TDP-43. *Nature Neuroscience, 14*(4), 459–468.

Ratti, A., & Buratti, E. (2016). Physiological functions and pathobiology of TDP-43 and FUS/TLS proteins. *Journal of Neurochemistry, 138*(Suppl 1), 95–111.

Renton, A. E., et al. (2011). A hexanucleotide repeat expansion in C9ORF72 is the cause of chromosome 9p21-linked ALS-FTD. *Neuron, 72*(2), 257–268.

Rosen, D. R., et al. (1993). Mutations in Cu/Zn superoxide dismutase gene are associated with familial amyotrophic lateral sclerosis. *Nature, 362*(6415), 59–62.

Ruddy, D. M., et al. (2003). Two families with familial amyotrophic lateral sclerosis are linked to a novel locus on chromosome 16q. *The American Journal of Human Genetics, 73*(2), 390–396.

Sampathu, D. M., et al. (2006). Pathological heterogeneity of frontotemporal lobar degeneration with ubiquitin-positive inclusions delineated by ubiquitin immunohistochemistry and novel monoclonal antibodies. *The American Journal of Pathology, 169*(4), 1343–1352.

Sapp, P. C., et al. (2003). Identification of two novel loci for dominantly inherited familial amyotrophic lateral sclerosis. *The American Journal of Human Genetics, 73*(2), 397–403.

Scarrott, J. M., et al. (2015). Current developments in gene therapy for amyotrophic lateral sclerosis. *Expert Opinion on Biological Therapy, 15*(7), 935–947.

Schludi, M. H., et al. (2017). Spinal poly-GA inclusions in a C9orf72 mouse model trigger motor deficits and inflammation without neuron loss. *Acta Neuropathologica*.

Schmid, B., et al. (2013). Loss of ALS-associated TDP-43 in zebrafish causes muscle degeneration, vascular dysfunction, and reduced motor neuron axon outgrowth. *Proceedings of the National Academy of Sciences of the United States of America, 110*(13), 4986–4991.

Shang, Y., & Huang, E. J. (2016). Mechanisms of FUS mutations in familial amyotrophic lateral sclerosis. *Brain Research, 1647*, 65–78.

Shaw, P. J., et al. (1995). CSF and plasma amino acid levels in motor neuron disease: Elevation of CSF glutamate in a subset of patients. *Neurodegeneration, 4*(2), 209–216.

Shaw, P. J., et al. (1995). Oxidative damage to protein in sporadic motor neuron disease spinal cord. *Annals of Neurology, 38*(4), 691–695.

Snowden, J. S., et al. (2012). Distinct clinical and pathological characteristics of frontotemporal dementia associated with C9ORF72 mutations. *Brain, 135*(Pt 3), 693–708.

Sreedharan, J., et al. (2008). TDP-43 mutations in familial and sporadic amyotrophic lateral sclerosis. *Science, 319* (5870), 1668–1672.

Suarez-Calvet, M., et al. (2016). Monomethylated and unmethylated FUS exhibit increased binding to Transportin and distinguish FTLD-FUS from ALS-FUS. *Acta Neuropathologica, 131*(4), 587–604.

Tollervey, J. R., et al. (2011). Characterizing the RNA targets and position-dependent splicing regulation by TDP-43. *Nature Neuroscience, 14*(4), 452–458.

Trotti, D., et al. (1999). SOD1 mutants linked to amyotrophic lateral sclerosis selectively inactivate a glial glutamate transporter. *Nature Neuroscience, 2*(5), 427–433.

Turner, M. R., et al. (2004). *Evidence of widespread cerebral microglial activation in amyotrophic lateral sclerosis: An [11C](R)-PK11195 pos*itron emission tomography study. *Neurobiology of Disease, 15*(3), 601–609.

Turner, M. R., et al. (2012). Young-onset amyotrophic lateral sclerosis: Historical and other observations. *Brain, 135*(Pt 9), 2883–2891.

Urushitani, M., et al. (2002). Proteasomal inhibition by misfolded mutant superoxide dismutase 1 induces selective motor neuron death in familial amyotrophic lateral sclerosis. *Journal of Neurochemistry, 83*(5), 1030–1042.

Vance, C., et al. (2006). Familial amyotrophic lateral sclerosis with frontotemporal dementia is linked to a locus on chromosome 9p13.2–21.3. *Brain, 129*(Pt 4), 868–876.

Vance, C., et al. (2009). Mutations in FUS, an RNA processing protein, cause familial amyotrophic lateral sclerosis type 6. *Science, 323*(5918), 1208–1211.

Vance, C., et al. (2013). ALS mutant FUS disrupts nuclear localization and sequesters wild-type FUS within cytoplasmic stress granules. *Human Molecular Genetics, 22* (13), 2676–2688.

Veldink, J. H., et al. (2002). Euthanasia and physician-assisted suicide among patients with amyotrophic lateral sclerosis in the Netherlands. *The New England Journal of Medicine, 346*(21), 1638–1644.

Watanabe, M., et al. (2001). Histological evidence of protein aggregation in mutant SOD1 transgenic mice and in amyotrophic lateral sclerosis neural tissues. *Neurobiology of Disease, 8*(6), 933–941.

Weihl, C. C., & Pestronk, A. (2010). Sporadic inclusion body myositis: Possible pathogenesis inferred from biomarkers. *Current Opinion in Neurology, 23*(5), 482–488.

White, M. A., & Sreedharan, J. (2016). Amyotrophic lateral sclerosis: Recent genetic highlights. *Current Opinion in Neurology, 29*(5), 557–564.

Williamson, T. L., et al. (2000). Toxicity of ALS-linked SOD1 mutants. *Science, 288*(5465), 399.

Woerner, A. C., et al. (2016). Cytoplasmic protein aggregates interfere with nucleocytoplasmic transport of protein and RNA. *Science, 351*(6269), 173–176.

Wolfe, M. S. (2012). The role of tau in neurodegenerative diseases and its potential as a therapeutic target. *Scientifica (Cairo), 2012.*, 796024.

Wong, P. C., et al. (1995). An adverse property of a familial ALS-linked SOD1 mutation causes motor neuron disease characterized by vacuolar degeneration of mitochondria. *Neuron, 14*(6), 1105–1116.

Xu, Y. F., et al. (2010). Wild-type human TDP-43 expression causes TDP-43 phosphorylation, mitochondrial aggregation, motor deficits, and early mortality in transgenic mice. *The Journal of Neuroscience, 30*(32), 10851–10859.

Zhang, K., et al. (2015). The C9orf72 repeat expansion disrupts nucleocytoplasmic transport. *Nature, 525*(7567), 56–61.

Zhang, K., et al. (2016a). Nucleocytoplasmic transport in C9orf72-mediated ALS/FTD. *Nucleus, 7*(2), 132–137.

Zhang, Y. J., et al. (2016b). C9ORF72 poly(GA) aggregates sequester and impair HR23 and nucleocytoplasmic transport proteins. *Nature Neuroscience, 19*(5), 668–677.

van Zundert, B., & Brown, R. H., Jr. (2016). Silencing strategies for therapy of SOD1-mediated ALS. *Neuroscience Letters*.

CHAPTER

6

Parkinson's Disease and Other Synucleinopathies

Mark R. Cookson

National Institute on Aging, National Institutes of Health, Bethesda, MD, United States

OUTLINE

INTRODUCTION: THE PATHOLOGY OF PARKINSON'S DISEASE

Parkinson's disease (PD) is a relatively common worldwide, age-related neurodegenerative disorder. While the movement disorder of PD, comprising bradykinesia (slowness of movement), tremor, postural instability, and gait disturbances, is widely recognized, in practice, many people living with PD have nonmotor symptoms related to damage to multiple neuronal circuits (Langston, 2006). For example, autonomic dysfunction may occur early in the disease course (Visanji & Marras, 2015), while many people with PD develop cortical dysfunction such as dementia at later times than the classic movement problems (Aarsland, 2016). Thus, PD is a multisystem progressive disorder and, at the time of writing, is treated only symptomatically.

An important question, both in terms of understanding the clinical disease and for the neuroscience of the PD brain, is what are the

The Molecular and Cellular Basis of Neurodegenerative Diseases
DOI: https://doi.org/10.1016/B978-0-12-811304-2.00006-7

underlying molecular mechanisms of disease risk? As will be outlined later, current thinking suggests that the overall disease risk is influenced by both genetic and nongenetic mechanisms, of which the latter include age and stochastic events. However, to delineate risk requires some definition of the disease of interest. Therefore, before discussing how we think PD develops and the underlying cellular and molecular events involved in its progression, it is important to describe the underlying brain pathology, which involves protein deposition and neuronal cell death.

Protein Deposition

Aggregation of α-Synuclein in PD

Postmortem examination of the brain from a person who had lived with PD reveals intraneuronal structures that stain positive with eosin, pathology first described by Freidrich Lewy. Over time, additional ways to label these Lewy bodies have been developed, including antibodies against ubiquitin or several other protein components and lipids. However, the most reliable markers to date are antibodies against α-synuclein, which label Lewy bodies and Lewy neurites in neurons (Spillantini et al., 1997).

In contrast to the pathological accumulation in cell bodies, α-synuclein is normally concentrated at synaptic vesicles in neurons (Burré, 2015). Thus, the presence of Lewy bodies and Lewy neurites represents both a loss of normal α-synuclein at the synapse and mislocalization to other cellular compartments. Furthermore, α-synuclein in Lewy bodies has altered biochemical properties in which the protein is heavily aggregated and organized into fibrils that have strong beta-sheet structure. This is in contrast to the unfolded protein found in solution, the helical form associated with membranes (Breydo, Wu, & Uversky, 2012), or tetramers in the cellular milieu (Bartels, Choi, & Selkoe, 2011; Dettmer, Newman, Luth, Bartels, & Selkoe, 2013; Wang et al., 2011). Lewy bodies are therefore evidence of protein aggregation in PD.

α-Synuclein-positive Lewy bodies and Lewy neurites are found throughout many regions of the PD brain. Seminal work from Braak et al. (2003) identified that Lewy bodies are often found in deep brain structures and the olfactory bulb. Additionally, when Lewy bodies are found in the *substantia nigra*, part of the midbrain that projects to the striatum and is critical for initiation and termination of movement, they are usually also in deeper brain structures. Equally, if Lewy bodies are found in the cerebral cortex, they are usually also in midbrain and deeper structures. Thus, Braak et al. infer that there is a progression of Lewy body pathology from lower brain regions through the midbrain, then with progressive involvement of the cerebral cortex. This staging scheme might approximately correspond to premotor PD, motor PD, and PD with dementia. It should be noted that the cross-sectional nature of these studies precludes any direct observation of true progression from region to region in a longitudinal sense within the same patient. Nonetheless, the staging scheme is important for both classifying PD brains and in generating hypotheses about disease progression.

The Broader Set of Synucleinopathies

Given that α-synuclein is a reliable marker for PD pathology, it is reasonable to ask whether the same protein is deposited in other diseases. The most common disease with a positive answer to this question is probably diffuse Lewy body disease also known as Lewy body dementia (DLB). In DLB, α-synuclein-positive pathology is much more widespread regionally than in PD, with prominent deposition in the cerebral cortex that is associated with a specific type of dementia that includes visual hallucinations and fluctuating consciousness (Donaghy & McKeith, 2014).

Many patients with DLB also progress to a movement disorder reminiscent of PD. Therefore, the distinction between PD with dementia and DLB with parkinsonism is a matter of timing, although they might all be considered within the umbrella term of "Lewy body disorders" (Lippa et al., 2007).

α-Synuclein is also useful in labeling a related condition, multiple system atrophy (MSA), which has some aspects of a parkinsonian movement disorder. Clinically, MSA is distinguished from PD by having much more prominent autonomic dysfunction and pathologically by the presence of glial cytoplasmic inclusions (Stamelou & Bhatia, 2016). These deposits of α-synuclein are found in oligodendrocytes, cells that do not normally express α-synuclein mRNA (Chen, Mills, Halliday, & Janitz, 2015; Miller et al., 2005).

For a detailed discussion of the various pathologies related to α-synuclein, the interested reader is directed toward some more comprehensive reviews (Dickson, 2001; Halliday, Holton, Revesz, & Dickson, 2011; Jellinger, 2009). For the purposes of the discussion of cellular and molecular mechanisms, the key point here is that deposition of α-synuclein affects multiple brain systems and is associated with diverse clinical outcomes.

Changes in Cellularity

Neuronal Loss

As well as Lewy bodies and Lewy neurites, the classic neuropathological event in PD is loss of neurons in the *substantia nigra pars compacta* that use dopamine to transmit signals to medium spiny neurons in the striatum. Dopamine neurons in the *substantia nigra* contain pigment called neuromelanin, and because the loss of neurons is substantial in PD, depigmentation can be seen macroscopically in the PD brain.

The loss of dopamine is particularly important in PD for two reasons. First, one of the primary functions of the nigrostriatal pathway is to control initiation of movement. Therefore, loss of nigral neurons leads to at least some of the clinical signs of PD. Second, replacing dopamine by treating with the precursor L-DOPA remains a primary treatment for most PD patients (Fahn, 2003).

However, as might be expected given that there are nonmotor symptoms in PD, there is also evidence for neuronal damage outside of the nigrostriatal system. For example, loss of several neurotransmitters can be imaged in multiple brain regions throughout the course of PD (Brooks, 2007). How neuronal loss and accumulation of α-synuclein pathology relate to each other as the disease progresses is not always clear. However, it is important to note that the cells with Lewy pathology are not those that have been lost. Therefore, whether Lewy bodies mark cells susceptible to cell death or are instead protective is uncertain.

Reactive Gliosis

As well as loss of neurons, there is abundant evidence of changes in glial cells in PD. Most prominently, reactive microglia are often found in brain regions, including those with extensive neuronal damage and/or Lewy body pathology. Some brain regions affected in PD have particularly high numbers of microglia, including the substantia nigra pars compacta. It is generally thought that these localized reactive microglia represent an immunological reaction to damage, and it is likely that microgliosis is maladaptive for neuronal survival. Specifically, as microglia become heavily activated, they may phagocytose neurons and neuronal debris, thus contributing to disease progression (Halliday & Stevens, 2011). Microglial activation can be imaged in the living PD brain, suggesting that these cells respond to active neurodegeneration throughout the disease course (Brooks, 2007).

There is also evidence of astrocyte activation in PD, specifically of protoplasmic astrocytes in gray matter, including in regions without Lewy bodies where neuronal projections may be affected, such as the striatum and dorsal thalamus (Halliday & Stevens, 2011). However, there is evidence that astrocyte activation differs between PD and other conditions in that, in PD, astrocytes do not proliferate and transform into reactive astrocytes (Mirza, Hadberg, Thomsen, & Moos, 2000; Song et al., 2009).

One major central nervous system (CNS) cell type that appears not to be affected in PD is the myelinating oligodendrocyte. Braak et al. (2003) noted that neurons affected by α-synuclein deposition in PD generally have thin, unmyelinated axons. However, nonmyelinating oligodendrocytes can demonstrate some accumulation of α-synuclein and may contribute to neuronal damage (Halliday & Stevens, 2011).

In summary, there are two major parts to the neuropathology of PD and related disorders, namely, a protein deposition disorder that centers around α-synuclein and changes in cellularity in many brain regions, but prominently including loss of dopamine neurons in the substantia nigra and reactive gliosis. The central argument of the rest of the chapter is that these events are linked and that the identification of genetic contributors to risk of PD are critical in understanding the underlying cellular mechanisms. Therefore, I will discuss some of the critical genes involved in PD.

GENES ASSOCIATED WITH SYNUCLEINOPATHIES

SNCA Mutations and Familial PD

Point Mutations

The identification of a single point mutation in a large set of families from Italy and Greece profoundly influenced the understanding of PD etiology in which it was the first proven genetic cause of this disease. The mutation is a substitution of threonine for alanine at position 53 (A53T) in the human *SNCA* gene on chromosome 4 that encodes for α-synuclein (Polymeropoulos et al., 1997). Subsequently, additional mutations were reported in other families, including A30P (Krüger et al., 1998), E46K (Zarranz et al., 2004), H50Q (Proukakis et al., 2013), and G51D (Lesage et al., 2013). Of interest, these mutations appear to cluster in a relatively small region of the 140 amino acid full-length protein, suggesting that sequences from amino acids ~30–53 are particularly important in pathogenesis.

It is also worth noting that the phenotype and pathology of patients with point mutations in the *SNCA* gene is much more variable that "typical" PD. For example, some cases present with Lewy body dementia (Zarranz et al., 2004). Pathologically, cases with A53T mutation have extensive α-synuclein-positive neuritic pathology and tau staining across many brain regions (Duda et al., 2002; Spira, Sharpe, Halliday, Cavanagh, & Nicholson, 2001), while A30P cases have abundant Lewy bodies, again in multiple brain regions (Seidel et al., 2010). Therefore, mutations in α-synuclein tend to produce clinical and pathological phenotypes that map across multiple synucleinopathies. These observations might support the idea that these multiple diseases are etiologically related.

Multiplications

There are also families where the multiple copies of the genomic region including the *SNCA* gene are inserted in tandem in the genome. The first identified family was a large kindred with a triplication of a stretch of the genome on chromosome 4 that includes *SNCA* (Singleton et al., 2003). Subsequently, families with duplication mutations were reported (Chartier-Harlin et al., 2004; Ibáñez et al., 2004). Interestingly, there is at least one large complex family where different branches have either duplication or triplication mutations

(Fuchs et al., 2007). Additionally, there are sporadic PD cases with duplication mutations (Ahn et al., 2008; Troiano et al., 2008), which might suggest some nonpenetrant carriers, although at least one case has been shown to be a de novo acquisition of a multiplication (Brueggemann et al., 2008). Collectively, these results suggest that the *SNCA* locus is prone to rearrangements that can increase copy number of the *SNCA* gene. Although additional copies of other genes in the region are also present in these cases, examination of the regions that overlap between families suggests that extra copies of *SNCA* are sufficient to explain inheritance of PD (Ross et al., 2008).

As for point mutations, multiplications are associated with a relatively broad range of clinical and phenotypes. It has been suggested that higher expression in the triplication cases leads to a more penetrant phenotype with greater involvement of the cerebral cortex and associated phenotypes (Ahn et al., 2008; Chartier-Harlin et al., 2004; Ibáñez et al., 2004; Ross et al., 2008), although there are individuals with duplication mutations that have dementia (Uchiyama et al., 2008). The pathological phenotypes of triplication cases are also very broad and include both neuritic and, interestingly, glial pathology that might otherwise be reminiscent of MSA (Gwinn-Hardy et al., 2000).

The SNCA Locus and Sporadic Synucleinopathies

Given that *SNCA* mutations are associated with familial PD and that the protein product is deposited in sporadic PD, it is therefore reasonable to hypothesize that variation around the *SNCA* gene might contribute to risk of sporadic PD. In fact, several studies suggested that variation around both the likely promoter region and the 3′ untranslated region of *SNCA* is associated with PD risk in diverse populations (Cheng et al., 2016; Farrer et al., 2001; Maraganore et al., 2006; Mueller et al., 2005).

Further support for the idea that risk of PD relates to variation around *SNCA* came from genome-wide association studies where, in contrast to the candidate gene approach, a very large number of common variants across the genome are queried simultaneously. Initial genome-wide association studies (GWAS) nominated *SNCA* variants as the strongest risk factors across the genome (Satake et al., 2009; Simon-sanchez et al., 2010), a result that has been confirmed in subsequent meta-analyses (Lill et al., 2012; Nalls et al., 2014).

Other synucleinopathies may also be associated with variation around the same locus. A small GWAS of MSA suggested an association with SNCA variants, although with a top single nucleotide polymorphism (SNP) that is slightly different from PD (Scholz et al., 2009). Similarly, although there is an overall genetic correlation between risk of DLB and PD (Guerreiro et al., 2016), detailed resequencing of the *SNCA* locus suggests that each of DLB, PD, and PD with dementia have differing variants that drive the associations (Guella et al., 2016).

To summarize, these observations derived from pathological and genetic studies converge on the idea that α-synuclein is a critical contributor to several diseases, including PD and other related neurological conditions that affect many different brain regions and result in multiple clinical outcomes. Before discussing cellular mechanisms related to neurodegeneration in the synucleinopathies in detail, it is worth briefly discussing other genetic causes of PD.

LRRK2

Mutations in the gene encoding the leucine-rich repeat kinase 2 protein, *LRRK2*, are important for several reasons, including that they are numerous, especially in some populations (Paisán-Ruiz, Lewis, & Singleton, 2013). However, while the original mutations were found in individual families (Funayama et al.,

2005; Paisán-Ruíz et al., 2004; Zimprich et al., 2004), some amino acid variants show association with disease at the population level (Peeraully & Tan, 2012). Furthermore, variation around the *LRRK2* locus also contributes to risk of sporadic PD, as identified in metaanalysis of GWAS signals (Nalls et al., 2014). This places the *LRRK2*, like *SNCA*, in the category of a pleiomorphic risk locus (Singleton & Hardy, 2011), a genomic region which contains multiple types of genetic signals that may work by different mechanisms.

However, the effects of mutations in *LRRK2* and *SNCA* differ in important ways. While, as discussed above, *SNCA* mutations cover a clinicopathological spectrum overlapping with multiple synucleinopathies, *LRRK2* mutations are remarkably similar to typical sporadic PD (Kumari & Tan, 2009). Loss of nigral neurons is a consistent feature of LRRK2 cases but, perhaps surprisingly, the protein deposition pathology of *LRRK2* cases is highly variable (Cookson, Hardy, & Lewis, 2008). Some cases with LRRK2 have α-synuclein-positive Lewy bodies while others have tau-positive depositions, but still others lack any distinctive neuropathology (Cookson et al., 2008). Interestingly, a recent study suggests that the presence of Lewy bodies correlates with dementia, anxiety, and some autonomic features (Kalia et al., 2015). Such considerations suggest that while α-synuclein deposition is associated with variable features of PD, it is not a required event for the early and essential loss of nigral dopamine neurons.

Other Genes Associated with Synucleinopathies

Dominant Genes

In recent years, several additional mutations have been found in multiplex families as has been reviewed elsewhere (Ferreira & Massano, 2016; Olgiati et al., 2016; Singleton & Hardy, 2016; Volta, Milnerwood, & Farrer, 2015). While many of these genes are interesting, for the purposes of the current discussion, I will focus on those that have been linked to α-synucleinopathy.

Mutations in *DNAJC13*, also known as required for receptor-mediated endocytosis 8, are a relatively rare cause of PD with a single variant suggested to be associated with disease (Vilariño-Güell et al., 2014). Importantly for the discussion here, multiple members of the main family with this mutation have been shown to have Lewy bodies *postmortem* (Appel-Cresswell et al., 2014). One difficulty with *DNAJC13* is that the same family has recently been claimed to have a second mutation in a nearby gene, *TMEM230* (Deng et al., 2016). Which mutation fully accounts for disease in this family is currently unclear, but given that both *DNAJC13* and *TMEM230* are involved in similar biological processes (see below), it is likely that inferences about function are similar.

At this time, mutations in *VPS35* (Vilariño-Güell et al., 2011) have not been shown to result in a synucleinopathy in patients, although a recent cell-based study suggested that in overexpression models, mutations increase α-synuclein inclusions (Follett et al., 2016). Conversely, a mutation in *EIF4G1* was found in Lewy body-positive cases (Chartier-Harlin et al., 2011), although the pathogenicity of that gene has been challenged (e.g., Nichols et al., 2015). Finally, mutations in *CHCHD2* have been proposed to be a rare cause of dominant PD (Funayama et al., 2015), although again the proposed mutation has not been independently confirmed in additional families.

From these genetic observations, it is clear that the most useful immediate clue to genetics of synucleinopathies is *LRRK2*, although there may be useful information about the biology of synuclein from other genes, even if their status is not yet confirmed.

Recessive Genes

For the sake of completeness, it is important to include the genes for early-onset PD and more complex parkinsonian disorders (Ferreira & Massano, 2016). Many loss-of-function mutations in parkin, PINK1 and DJ-1 have been reported to be associated with early-onset parkinsonism, and the function of the encoded proteins in mitochondrial/oxidative stress pathways is well established (Han, Kim, & Son, 2014; Pickrell & Youle, 2015). However, in the majority of cases, these are not Lewy body disorders, with the possible exception of one *PINK1* case that had widespread α-synuclein deposits (Samaranch et al., 2010). Therefore, while very interesting, it is not clear whether mutations in these genes are instructive for the cell biology of synucleinopathies.

Risk Factors

Although the discussion above of GWAS hits outlined variation around the *SNCA* and *LRRK2* loci, there are an additional ~20 loci that contain risk factors for PD (Nalls et al., 2014). Although these studies have not been performed on neuropathologically confirmed cases, it is likely that these risk factors contribute to lifetime risk of synucleinopathy. Additionally, a locus on chromosome 1 that includes the gene *GBA* is associated with PD risk (Nalls et al., 2014). Mutations in *GBA* cause Gaucher's disease in the homozygous state and had previously been nominated as risk factor genes for PD when heterozygous (Sidransky et al., 2009).

Cellular Mechanisms in Synucleinopathies

The observations outlined above suggest that there are many genetic contributors to inherited PD and that at least a subset of these are relevant for sporadic PD. Furthermore, because the genes involved in PD risk include *SNCA* that encodes for the major deposited protein α-synuclein, it seems likely that the mechanisms that underlie neuropathology might be shared between familial and sporadic PD. I will therefore use the rest of this chapter to outline what some of those mechanisms might be. Because the neuropathology of synucleinopathies involves both neurons and other cell types, I will discuss mechanisms that might be restricted to neurons, followed by those that may involve nonneuronal cells.

Neuronal Events

Synuclein Accumulation

Some genetic events may immediately predict mechanisms that may be relevant to human disease, while others might be more complicated to interpret. One set of mutations that might seem straightforward are the multiplication mutations in the *SNCA* locus. We have shown that patients with the triplication have the expected doubling of α-synuclein protein in blood samples (Miller et al., 2004), demonstrating that all alleles are active. In turn, this suggests that too much α-synuclein can cause PD. It has been suggested that the risk factor variants found in GWAS might work by a similar, if subtler, manner by modestly increasing α-synuclein protein levels (Fuchs et al., 2008; McCarthy et al., 2011; Singleton & Hardy, 2011). Some nominated risk factor genes, including *GAK* (Dumitriu et al., 2011) and *GBA* (Mazzulli et al., 2011), have been proposed to modulate α-synuclein in cells. Therefore, the accumulation of the normal human wild type form of α-synuclein might be one common mechanism relevant to sporadic and familial PD. Furthermore, α-synuclein gene expression is restricted to neurons in the brain, suggesting that increased levels of this protein represent an intrinsic, cell-autonomous mode for neuropathology in PD.

As discussed earlier, Lewy bodies are composed of α-synuclein in an aggregated, fibrillar, and insoluble form. This leads to the suggestion that a critical event in the pathogenesis of synucleinopathies is the aggregation of α-synuclein. Supporting this, several mutations in α-synuclein increase aggregation rates in vitro, although the effects vary by mutation (Sahay, Ghosh, Singh, & Maji, 2016). Also, demonstrating that aggregation is directly related to toxicity has been difficult (Villar-Piqué et al., 2016), and the lack of Lewy bodies in a subset of *LRRK2* mutation cases suggests that toxicity can progress in the absence of α-synuclein deposition in humans (Cookson et al., 2008). Instead, it has been posited that aggregation into insoluble fibrils is not the toxic event, but rather that oligomeric intermediates mediate neuronal damage (Barrett & Greenamyre, 2015; Ingelsson, 2016). One might even argue that because Lewy bodies are, by definition, only found in cells that have survived at the end of PD that their presence may even be neuroprotective. At the time of writing, the relationships between α-synuclein aggregation, Lewy pathology, and cell death are therefore difficult to fully disentangle.

Further clouding the concept of toxicity mediated by α-synuclein is the observation that a Lewy body not only represents accumulation of protein but also mislocalization from its usual place at the synapse. This leads to the concept that loss of α-synuclein from adult synapses might contribute in part to neuronal loss (Benskey, Perez, & Manfredsson, 2016). Additionally, the presence of Lewy pathology in cell bodies and neurites may impede axonal transport (Volpicelli-Daley, 2016). Therefore, toxicity associated with synuclein accumulation may be a complex mixture of loss of normal function at the synapse and accumulation in other cellular compartments. I will next discuss some more specific intracellular targets that might be influenced by α-synuclein toxicity.

Vesicular Trafficking

Given that α-synuclein plays a role in synaptic dynamics (Burré, 2015), it is likely that part of the toxicity caused by altered α-synuclein might relate to the neuronal synapse. We can extend this idea to suggest that α-synuclein might damage other vesicular transport systems in cells, of which synaptic vesicles are a specialized form. This idea is supported by a variety of observations across different species (Fig. 6.1).

In yeast, screens of modifier genes for the toxicity of overexpressed α-synuclein identify multiple genes related to lipid metabolism and vesicular trafficking (Willingham, Outeiro, DeVit, Lindquist, & Muchowski, 2003). Specifically, α-synuclein can affect trafficking from the endoplasmic reticulum (ER) to the Golgi by affecting small GTPases such as Rab8a (Cooper et al., 2006; Gitler et al., 2008) and other Rabs (Soper, Kehm, Burd, Bankaitis, & Lee, 2011).

Along the same lines, results in *Caenorhabditis elegans* and *Drosophila melanogaster* highlighted the endocytic pathway (Kuwahara et al., 2008) and the endocytic protein Rab11 (Breda et al., 2015), respectively. More recent results suggest that the effects of α-synuclein on synaptic morphology and function in *Drosophila* are different from other toxic proteins (Chouhan et al., 2016), echoing similar distinctions in yeast screens (Willingham et al., 2003).

In mammalian cells, increased expression of α-synuclein inhibits neurotransmitter release and decreases steady-state levels of synapsins in vivo (Nemani et al., 2010). Similarly, mutant α-synuclein negatively regulates endocytosis in mouse neurons (Xu et al., 2016). However, deliberately targeting α-synuclein away from the synapse can result in toxicity to dopamine neurons in vivo (Burré, Sharma, & Sudhof, 2015). Because these are very different studies, it is difficult to infer the underlying molecular pathways that lead to degeneration. However, it is possible that both accumulation of α-synuclein at the synapse and loss of protein might be detrimental to neurons. Speculatively, in triplication cases, there may be an initial accumulation of α-synuclein at the synapse

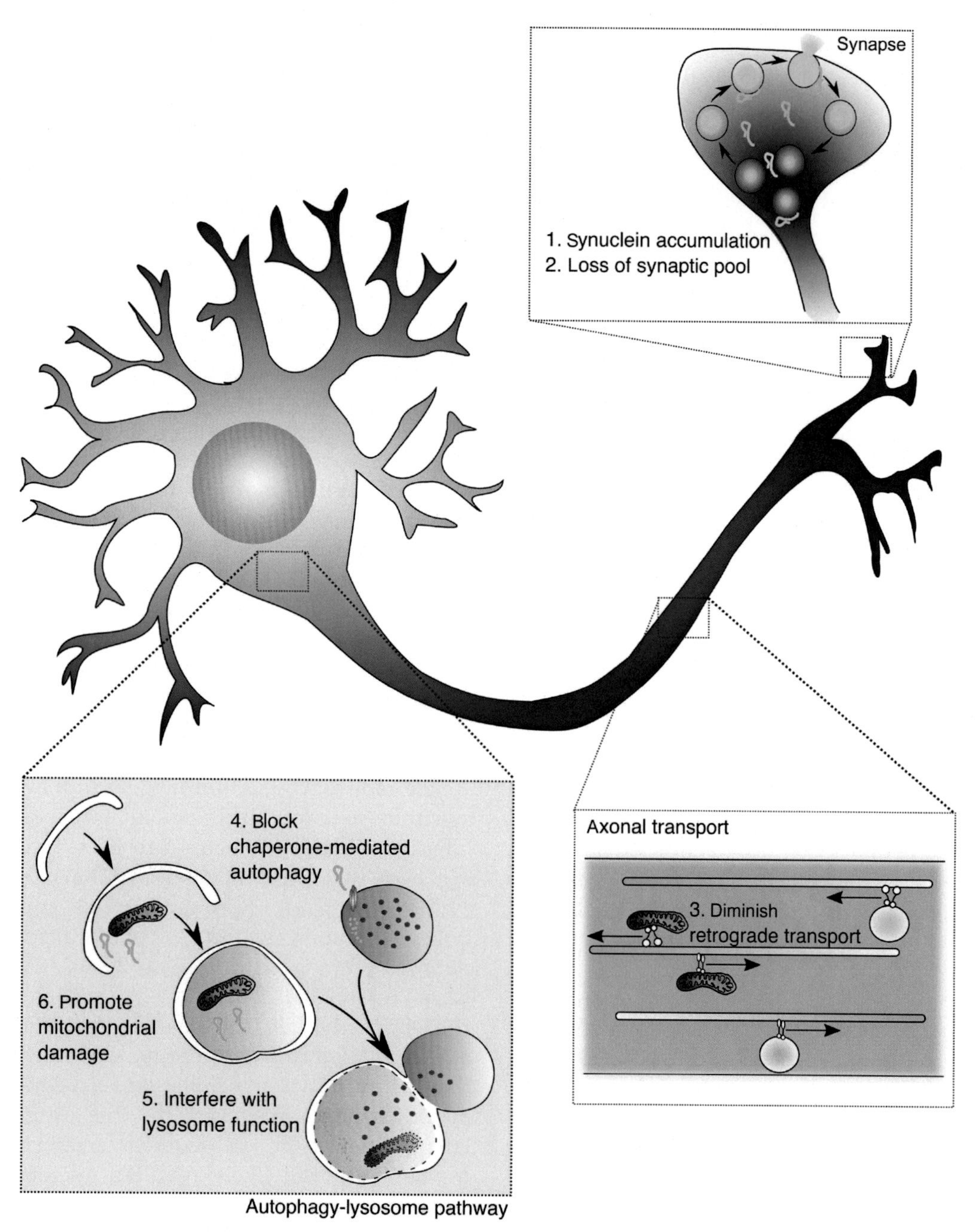

FIGURE 6.1 **Cell-autonomous mechanisms in synucleinopathies**. Schematic illustration of the many potential mechanisms by which α-synuclein mutation or duplication might affect neuronal function and, eventually health. There are, broadly, three groups of potential effects that can be separated by subcellular location (shown as inserts on the overall picture of the neuron). At the synapse, both accumulation in the presynaptic compartment *(1)* and loss of synuclein from presynaptic vesicles *(2)* have been proposed to contribute to neuronal dysfunction. Along axons, accumulating α-synuclein has been shown to block retrograde transport of organelles from the synapse back to the cell body *(3)*. Finally, in the soma of neurons, α-synuclein can inhibit aspects of the autophagy–lysosome system, including inhibition of chaperone-mediated autophagy *(4)* and lysosomal function *(5)*. Finally, α-synuclein has been proposed to cause mitochondrial damage *(6)* that would be cleared by a specialized form of autophagy called mitophagy.

that diminishes neuronal function, which could lead to increased neuronal activity in compensation. Later in the disease process, α-synuclein may accumulate in the cytoplasm, further damaging synaptic integrity. At this point, α-synuclein may have greater access to earlier stages in the secretory pathway, such as ER–Golgi transport, as first identified in the yeast models.

Several other genes associated with synucleinopathies may also impact vesicular trafficking. For example, LRRK2 phosphorylates Rab proteins (Steger et al., 2016), which in turn influence several vesicular trafficking pathways in cells (Cookson, 2016). LRRK2 also is reported to have effects at the synapse (Arranz et al., 2015; Beccano-Kelly et al., 2015; Belluzzi et al., 2016; Matta et al., 2012; Piccoli et al., 2011). Whether the effects of LRRK2 on synapses are mediated by Rabs or other substrates, and how they relate to α-synuclein is not fully resolved at the time of writing. Further supporting the concept that LRRK2 may affect vesicular trafficking, two identified protein interactors of LRRK2 (GAK and Rab7L1) are important in trafficking events in the cell (Beilina et al., 2014; Kuwahara et al., 2016; MacLeod et al., 2013). I highlight these two proteins in particular because the genes that encode them are candidate risk factor genes for PD identified in GWAS (Nalls et al., 2014; Satake et al., 2009; Simon-sanchez et al., 2010) and have been independently confirmed in several additional genetic studies (Gan-Or et al., 2012; Ma, He, & Xu, 2015; Nagle et al., 2016). These results, while needing further replication, suggest that vesicular trafficking pathways may be important for sporadic synucleinopathies.

Microtubule Function and Axonal Transport

Another neuronally restricted toxic pathway in PD relates to the axon, specifically to microtubule function. As well as vesicular proteins, LRRK2 also binds to tubulin, specifically the TUBB4 isoform that is highly expressed in the nervous system (Law et al., 2014). Furthermore, LRRK2 mutations functionally interact with microtubule transport in flies (Godena et al., 2014). This observation may link LRRK2 to another prominent risk factor gene found in GWAS, *MAPT*, which encodes the neuronal microtubule binding protein tau. In contrast to α-synuclein in Lewy bodies, tau deposition is not a typical neuropathological event in sporadic synucleinopathies, although some familial *SNCA* and *LRRK2* mutation cases do have tau pathology (Duda et al., 2002; Zimprich et al., 2004). Therefore, in sporadic PD, the effect of risk factor variants in *MAPT* is unlikely to be related to formation of tau pathology, but rather to a functional role of tau protein, which is to stabilize microtubules. Thus, both LRRK2 and MAPT point to a potential role of axonal transport as a key process affected in PD.

Finally, as discussed elsewhere, aggregated forms of α-synuclein may impair axonal transport when deposited away from the normal location of this protein at the synapse (Volpicelli-Daley, 2016).

The above discussion suggests that both synaptic function and axonal transport are good candidates for neuron-restricted cellular functions that can be affected in PD and related synucleinopathies. Furthermore, as microtubules are important for vesicular transport in neurons (Terada & Hirokawa, 2000), it would seem reasonable that these two events are related, temporally and spatially, to cell damage in PD. Extending the argument above that there is an initial accumulation of α-synuclein at the synapse followed by loss from that compartment, the subsequent accumulation in the cytoplasm likely requires microtubule-dependent transfer from synapse to cell body but may also have detrimental effects on that process (Fig. 6.1). We can therefore see that there are likely intertwined detrimental effects of α-synuclein, LRRK2, and MAPT, as well as other risk factor genes, that collectively can damage neurons in a cell-autonomous manner via vesicles and microtubules (Taymans & Cookson, 2010).

Lysosomal Dysfunction and the Autophagic Pathways

In addition to the above effects on vesicular/synaptic function and microtubule-dependent axonal transport, there are several other potential cellular functions that can be disrupted in PD. However, I will separate these out here, because they are likely important in both neuronal and nonneuronal cells. Thus, the argument that they impact PD in a cell-autonomous manner is weaker, although this still could be their major site of action. Specifically, here I will discuss a set of ultimately degradative pathways that control protein and organellar turnover via the autophagy-lysosome system.

There are several reasons to think that lysosomes might be important in PD (Dehay et al., 2013), but perhaps one of the most compelling is that some lysosomal genes contribute to PD risk. Specifically, the risk factor gene *GBA* encodes a lysosomal enzyme, beta-glucocerebrosidase, that cleaves the lipid glucocerebroside. Full deficiency of this enzyme causes Gaucher's disease, which is characterized by accumulation of lipids in a number of cell types throughout the body but particularly in macrophages. However, possessing a single mutant *GBA* allele increases risk of PD by several fold compared to nonmutation carriers (Lwin, Orvisky, Goker-Alpan, LaMarca, & Sidransky, 2004; Sidransky et al., 2009) and may also be a risk factor for DLB (Geiger et al., 2016; Mata et al., 2008; Shiner et al., 2016).

The major question that arises from these observations is how a partial deficiency in a lysosomal enzyme might affect risk of synucleinopathies. One relatively simple possibility is that α-synuclein is a substrate of lysosomal degradation either through conventional routes (Gao et al., 2008; Li, 2004) or via the more specialized process of chaperone-mediated autophagy (Cuervo, Stefanis, Fredenburg, Lansbury, & Sulzer, 2004). Consequentially, lysosomal dysfunction would be predicted to result in α-synuclein accumulation, similar to how *SNCA* risk variants work in sporadic PD.

Potentially consistent with this idea, other lysosomal genes, such as ATP13A2, cause complex forms of parkinsonism in a recessive, i.e., loss-of-function, manner (Di Fonzo et al., 2007; Ramirez et al., 2006). However, while some studies suggest that ATP13A2 mutations affect α-synuclein metabolism (Kong et al., 2014; Usenovic, Tresse, Mazzulli, Taylor, & Krainc, 2012), others have found the in vivo effects of ATP13A2 to be independent of α-synuclein (Kett et al., 2015) and vice versa (Daniel et al., 2015). Similarly, results have been mixed regarding the relationship between GBA deficiency and α-synuclein-mediated dysfunction in various model systems (Bae et al., 2015; Davis et al., 2016; Dermentzaki, Dimitriou, Xilouri, Michelakakis, & Stefanis, 2013; Mazzulli et al., 2011; Suzuki et al., 2015). Therefore, the relationship(s) between loss of lysosomal function and α-synuclein remains to be clarified.

Moreover, the association of GBA1 and other lysosomal enzymes with synucleinopathies might not be driven by simple loss of function. Notably, much of the risk of PD for GBA variants is driven by mutant alleles in the heterozygous state where at least 50% of activity is retained. Furthermore, at least one variant in GBA, E326K, is associated with risk of PD but does not cause Gaucher's disease in the homozygous state (Duran et al., 2013). Thus, this mutation is unlikely to result in a simple loss of function. Therefore, alternate hypotheses have been advanced positing a gain of GBA function, either by affecting the lysosome, influencing mitochondria, or causing ER stress (Migdalska-Richards & Schapira, 2016). None of these theories have been tested in vivo, indicating that there is a large amount of work to be done in this area.

However, focusing on lysosomes as a terminal degradative organelle may obscure the fact that they are integrated in a number of complex pathways in the cell. One cellular process that may be particularly germane to the discussion of parkinsonism is autophagy, which is so closely related to lysosomal function that the two are sometimes referred to as the autophagy-lysosomal pathway (Kenney & Benarroch, 2015) (see also Chapter 11: Neurodegenerative Diseases and Autophagy). Macroautophagy is a specific type of autophagy where a lipid membrane, the autophagosome, is formed around objects in the cytoplasm to promote their degradation via fusion with the lysosome (Suzuki, Osawa, Fujioka, & Noda, 2016). Importantly, lipids act as bidirectional mediators of function for both autophagosomes and lysosomes (Dall'Armi, Devereaux, & Di Paolo, 2013; Jaishy & Abel, 2016). Therefore, subtle lysosomal dysfunction due to mutations in a lipid metabolism enzyme like GBA might also affect autophagy pathways.

One reason to suspect that the regulation of autophagy might be affected in PD relates to mutations in LRRK2. Several laboratories have reported that LRRK2 affects autophagy markers both in cells (Alegre-Abarrategui et al., 2009; Bravo-San Pedro et al., 2013; Manzoni et al., 2013; Plowey, Cherra, Liu, & Chu, 2008; Schapansky, Nardozzi, Felizia, & Lavoie, 2014; Su, Guo, & Qi, 2015) and in vivo (Dodson, Leung, Lone, Lizzio, & Guo, 2014; Ramonet et al., 2011; Saha et al., 2015; Soukup et al., 2016; Tong et al., 2010; Tong et al., 2012). The precise mechanism(s) involved are not fully resolved at this time, although one immediate possibility is that Rab substrates (Steger et al., 2016) or interactors (Beilina et al., 2014; Kuwahara et al., 2016; MacLeod et al., 2013) are involved, given that Rabs are critical mediators of membrane trafficking (Zerial & McBride, 2001).

Another way in which autophagy may play a role in PD pathogenesis is in the more specialized process of mitophagy, a process by which mitochondria that are damaged or depolarized are removed from the cellular pool (Youle & Narendra, 2011) (see also Chapter 13: Mitochondrial Function and Neurodegenerative Diseases). Two genes involved in recessive early-onset parkinsonism, *PINK1* and *parkin*, are known to play critical roles in mitochondrial quality control (Allen, Toth, James, & Ganley, 2013; Ashrafi, Schlehe, LaVoie, & Schwarz, 2014; Clark et al., 2006; Deng, Dodson, Huang, & Guo, 2008; Gehrke et al., 2015; Kane et al., 2014; Koyano et al., 2014; Matsuda et al., 2010; McLelland, Soubannier, Chen, McBride, & Fon, 2014; Narendra et al., 2010; Park et al., 2006). Whether mitophagy plays a critical role in synucleinopathies is less certain. Most *postmortem* reports of cases with *parkin* mutations do not have Lewy bodies (Cookson et al., 2008), although a *PINK1* case has been demonstrated to have α-synuclein pathology (Samaranch et al., 2010). Therefore, the relationship of recessive parkinsonism genes to the general category of synucleinopathies has been uncertain.

However, given the long standing interest in mitochondria as contributors to PD pathogenesis (Bose & Beal, 2016), several labs have examined whether dominant PD genes can affect mitochondria. α-Synuclein can influence mitochondrial fission (Li et al., 2013) and mitophagy (Chinta, Mallajosyula, Rane, & Andersen, 2010), possibly by direct binding to mitochondrial membranes (Guardia-Laguarta et al., 2014; Nakamura et al., 2011), leading to altered mitochondrial morphology (Chen et al., 2015; Xie & Chung, 2012). LRRK2 also affects mitochondrial morphology (Cherra, Steer, Gusdon, Kiselyov, & Chu, 2013; Yue et al., 2015) and function (Cooper et al., 2012; Mortiboys, Johansen, Aasly, & Bandmann, 2010; Mortiboys et al., 2015; Su & Qi, 2013), possibly by mechanisms involving interaction with various mitochondrial proteins (Hsieh et al., 2016; Su et al., 2015; Wang et al., 2012).

In summary, there are several neuronal cell-autonomous pathways that may mediate the toxic events in synucleinopathies, principally relating to vesicular trafficking and microtubule function (Fig. 6.1). The major genes, as well as likely several risk factors and rare genes, affect these pathways directly, although they tend to have regulatory roles rather than being core components. For example, α-synuclein modulates synaptic function but is not required for neurotransmission, as readily demonstrated by the observations that many invertebrate organisms lack a homologue of this gene (Cookson, 2012) and that neurotransmission still occurs even in α/β/γ-synuclein triple knockout mice, albeit with some changes in synaptic efficiency (Anwar et al., 2011; Greten-Harrison et al., 2010). Similarly, LRRK2 modulates autophagy but does not appear to be required for the pathway to function, and PINK1/parkin are not absolutely required for mitophagy. Speculatively, this somewhat peripheral involvement of synuclein and related genes in neuronal function might partially explain why these diseases are restricted to some types of cells rather than affecting the whole body. Furthermore, perhaps disruption of regulatory events can be tolerated through early life but are harmful with aging because of failure of compensatory mechanisms, although this is a difficult concept to prove. However, another layer of complexity comes from consideration of the ways in which cell-to-cell communication may contribute to disease progression in synucleinopathies.

Noncell-Autonomous Mechanisms

Cell-to-Cell Spread

As discussed above, in MSA there is accumulation of α-synuclein protein in oligodendrocytes that do not express the associated mRNA (Miller et al., 2005). The accumulation of glial cytoplasmic inclusions in *SNCA* triplication cases (Gwinn-Hardy et al., 2000) suggests that increased expression within neurons of the normal α-synuclein protein is sufficient to replicate this effect. Although many logical possibilities can explain these observations, in my view, the most likely is that the protein moves from neurons into oligodendrocytes, i.e., α-synuclein must have spread from cell to cell.

Cell-to-cell spread is likely to be more widespread than the example of MSA, as there is evidence, particularly from studies of transplanted neurons in grafts, that α-synuclein can also be taken up into neurons (Li et al., 2008). The concept of spreading proteins that transfer pathology from brain region to region likely extends across several neurodegenerative diseases (Braak & Del Tredici, 2016).

While this topic will be reviewed in greater depth in other chapters of this volume (e.g., Chapter 8: Prion-like Propagation in Neurodegenerative Diseases), it is important to mention here as several models of the cell biology of synucleinopathies now incorporate aspects of spread or seeding (Angot & Brundin, 2009). Moreover, there are obvious therapeutic possibilities if the extracellular protein component of neuronal damage can be targeted (Vekrellis & Stefanis, 2012).

Although a full description of the mechanism(s) involved in synuclein spread has not yet been developed, it is of interest that some of the proposed events include aspects that overlap with the mechanisms of neuronal toxicity described above. Specifically, α-synuclein uptake in cells likely involves endocytosis (Oh et al., 2016; Spencer et al., 2014), and relative degradation of internalized α-synuclein compared to release involves the autophagy-lysosomal pathway (Poehler et al., 2014; Sacino et al., 2016).

Additionally, recent publications have suggested that other genes involved in PD, particularly LRRK2, may affect α-synuclein transfer between neurons (Kondo, Obitsu, & Teshima,

2011; Volpicelli-Daley et al., 2016). This reinforces the idea that there may be shared cellular mechanisms related to both neuronal cell-autonomous events and cell-to-cell transfer in synucleinopathies. It seems likely that these mechanisms relate to fundamental properties of α-synuclein, specifically its ability to bind multiple membranes in cells, both promoting toxic events and allowing access to uptake mechanisms (Fig. 6.2). What is not yet fully established is which cell-autonomous or cell-transfer mechanisms are most important in neurodegeneration in vivo.

Immunological Responses

As well as neuron-to-neuron transfer of α-synuclein, there is also strong evidence of reciprocal signaling between neurons and

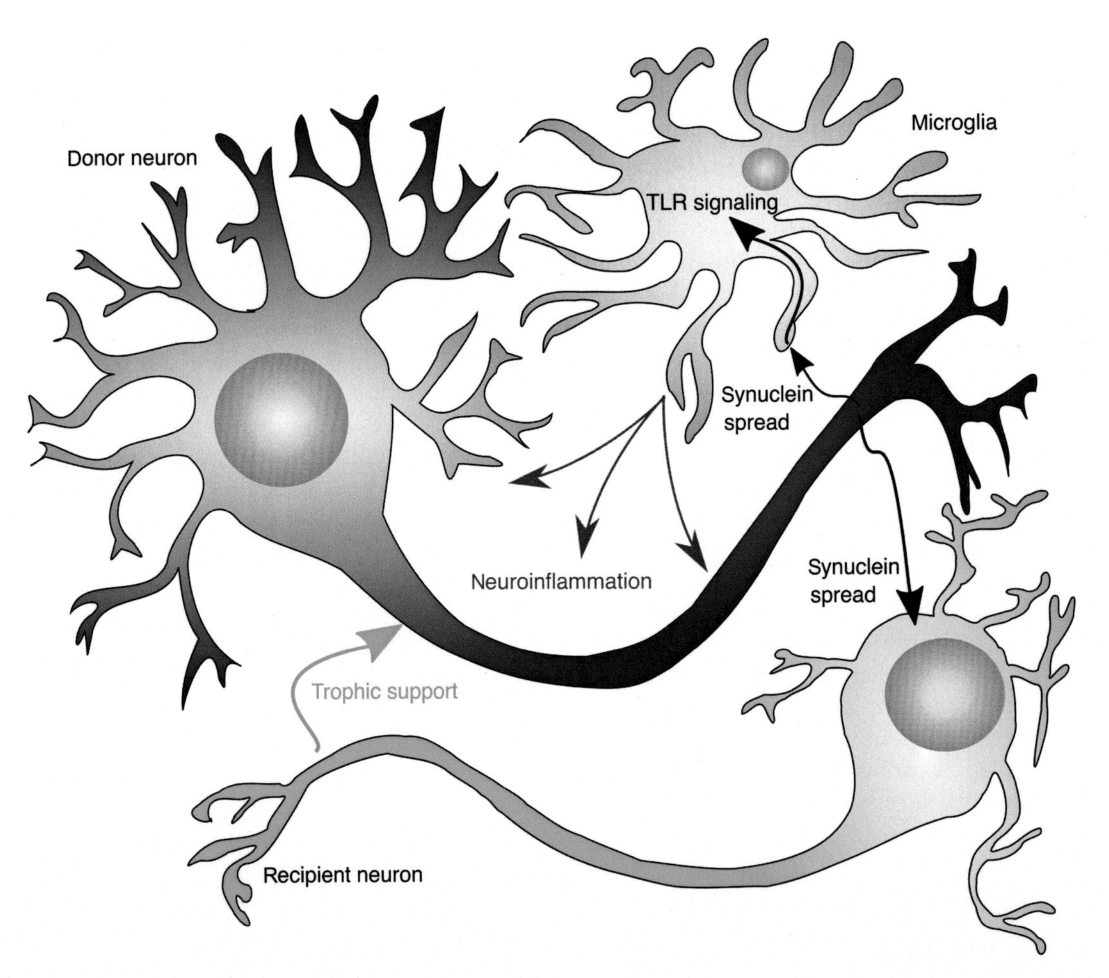

FIGURE 6.2 **Noncell autonomous mechanisms in synucleinopathies**. Schematically, this figure demonstrates two major concepts related to neuronal damage. First, α-synuclein can spread from donor neurons (shown in *blue*) into the extraneuronal space where it can be transmitted either to recipient neurons (here in *green*) where pathology can be established or to glial cells such as microglia (*brown*) triggering neuroinflammatory cascades. Second, both of these cells can have effects that might damage the original source neuron—either by loss of trophic support from other neurons or damaging inflammatory molecules from activated glia.

immune cells in PD, which is sufficiently distinct from neuronal events that it merits separate discussion (see also Chapter 14: Non-Cell Autonomous Degeneration: Role of Astrocytes in Neurodegenerative Diseases).

Several laboratories have shown that α-synuclein can bind to and activate microglia (Hoffmann et al., 2016; Kim et al., 2013; Su et al., 2008; Watson et al., 2012), likely via Toll-like receptors (TLRs), which are important mediators of innate immunity (Béraud & Maguire-Zeiss, 2012). At least a proportion of the α-synuclein that binds to microglia can be internalized and degraded, likely via lysosomes, in a manner that is inhibited by LRRK2 (Maekawa et al., 2016). LRRK2-deficient microglia show diminished inflammatory responses to α-synuclein in culture (Russo et al., 2015), which would be consistent with a higher rate of clearance. LRRK2 inhibitors have similar inhibitory effects on inflammation after stimulation of TLR4 receptors (Moehle et al., 2012).

The immune system clearly has important prosurvival roles at the organismal level, but in the context of neurological disease, activation of immune cells can be detrimental. For example, LRRK2-knockout mice are more sensitive to intestinal infectious agents (Zhang et al., 2015) but generally resistant to neurodegeneration (Daher, Volpicelli-Daley, Blackburn, Moehle, & West, 2014; Daher et al., 2015). Activation of microglia in the brain—either from local inflammatory molecules such as α-synuclein (Zhang et al., 2016) or from remote sites such as the gut due to inflammatory effects of the gut microbiome (Sampson et al., 2016)—can evoke dopaminergic cell death. Furthermore, microglial activation may secondarily affect astrocytes and also add to a toxic environment (Liddelow et al., 2017).

These considerations suggest that neuroinflammation, induced by extracellular α-synuclein and modulated by LRRK2, might contribute to noncell autonomous cell death synucleinopathies. Furthermore, because extracellular α-synuclein would be available to both neurons and nonneuronal cells, spread of disease between brain regions could be a function both of transfer of pathology to recipient neurons and inflammation at the same time. Further complicating interpretation, exposure of neurons to TLR agonists can increase endogenous α-synuclein expression and affect autophagy (Dzamko et al., 2017; Kim et al., 2015). Therefore, the spread of pathology in synucleinopathies may result from a complex mixture of protein transfer between cells and inflammatory processes. Nonetheless, these observations show how noncell autonomous mechanisms may include immune cells (Fig. 6.2).

Trophic Support

Trophic factors have attracted attention in PD research for their potential ability to protect against neuronal cell loss, even though that ability has not yet led to clinical applications (Olanow, Bartus, Volpicelli-Daley, & Kordower, 2015). However, there are a few reasons to consider that deficiencies in trophic factor signaling may be relevant to pathogenesis of synucleinopathies, probably due to noncell autonomous events (Fig. 6.2).

It was proposed many years ago in cell culture models that chronic α-synuclein expression leads to downregulation of genes associated with the dopaminergic phenotype, likely via the transcription factor Nurr1 (Baptista et al., 2003). Subsequent studies in rat models demonstrated that α-synuclein-induced changes in Nurr1 expression result in diminished signaling by glial cell line–derived neurotrophic factor (GDNF) (Decressac et al., 2012). This is a likely explanation for the lack of efficacy of GDNF against α-synuclein toxicity (Lo Bianco, Déglon, Pralong, & Aebischer, 2004), although restoration of signaling can allow GDNF to be effective (Volakakis et al., 2015). Relevant to this discussion, one source of GDNF is the postsynaptic medium spiny neurons of the striatum to which nigral dopamine neurons project (Airaksinen & Saarma, 2002).

Striatal medium spiny neurons can also express brain-derived neurotrophic factor, BDNF (Zuccato & Cattaneo, 2007), as well as LRRK2 (Mandemakers, Snellinx, O'Neill, & de Strooper, 2012), and genetic interactions between BDNF and LRRK2 variants have been reported (Liu et al., 2012). Whether part of the effects of LRRK2 mutations on dopamine neuron survival are mediated via changes in neurotrophic signaling remain to be seen, but it is certainly an appealing potential explanation for the apparent differences between where LRRK2 is expressed and which cells are affected by mutation.

SUMMARY

The discussion of cellular mechanisms in synucleinopathies presented here is grounded in observations from human disease, namely, the neuropathological outputs of the disease process and the genetic determinants of disease risk. Because there is overlap between these two lists, particularly in that α-synuclein is both a pathological marker and a risk gene for inherited and sporadic disease, it seems reasonable to assume that genetics and neuropathology are related, if imperfectly, to each other. However, the overall picture of these diseases is complex, with several candidate pathways for intracellular events that might influence neuronal dysfunction as well as contributions from nonneuronal cell types. Identifying which pathways are most amenable to intervention for therapeutics remains a major challenge for the field.

Acknowledgment

This research was supported by the Intramural Research Program of the NIH, National Institute on Aging.

References

Aarsland, D. (2016). Cognitive impairment in Parkinson's disease and dementia with Lewy bodies. *Parkinsonism & Related Disorders*, *22*(Suppl 1), S144–S148. Available from https://doi.org/10.1016/j.parkreldis.2015.09.034.

Ahn, T.-B., Kim, S. Y., Kim, J. Y., Park, S.-S., Lee, D. S., Min, H. J., … … Jeon, B. S. (2008). alpha-Synuclein gene duplication is present in sporadic Parkinson disease. *Neurology*, *70*, 43–49. Available from https://doi.org/10.1212/01.wnl.0000271080.53272.c7.

Airaksinen, M. S., & Saarma, M. (2002). The GDNF family: Signalling, biological functions and therapeutic value. *Nature Reviews Neuroscience*, *3*, 383–394. Available from https://doi.org/10.1038/nrn812.

Alegre-Abarrategui, J., Christian, H., Lufino, M. M. P., Mutihac, R., Venda, L. L., Ansorge, O., & Wade-Martins, R. (2009). LRRK2 regulates autophagic activity and localizes to specific membrane microdomains in a novel human genomic reporter cellular model. *Human Molecular Genetics*, *18*, 4022–4034. Available from https://doi.org/10.1093/hmg/ddp346.

Allen, G. F. G., Toth, R., James, J., & Ganley, I. G. (2013). Loss of iron triggers PINK1/Parkin-independent mitophagy. *EMBO Reports*, *14*, 1127–1135. Available from https://doi.org/10.1038/embor.2013.168.

Angot, E., & Brundin, P. (2009). Dissecting the potential molecular mechanisms underlying alpha-synuclein cell-to-cell transfer in Parkinson's disease. *Parkinsonism & Related Disorders*, *15*(Suppl 3), S143–S147. Available from https://doi.org/10.1016/S1353-8020(09)70802-8.

Anwar, S., Peters, O., Millership, S., Ninkina, N., Doig, N., Connor-Robson, N., … … Buchman, V. L. (2011). Functional alterations to the nigrostriatal system in mice lacking all three members of the synuclein family. *The Journal of Neuroscience: The Official Journal of the Society for Neuroscience*, *31*, 7264–7274. Available from https://doi.org/10.1523/JNEUROSCI.6194-10.2011.

Appel-Cresswell, S., Rajput, A. H., Sossi, V., Thompson, C., Silva, V., McKenzie, J., … … Rajput, A. (2014). Clinical, positron emission tomography, and pathological studies of DNAJC13p.N855S Parkinsonism. *Movement Disorders: The official Journal of the International Parkinson and Movement Disorder Society*, *29*, 1684–1687. Available from https://doi.org/10.1002/mds.26019.

Arranz, A. M., Delbroek, L., Van Kolen, K., Guimarães, M. R., Mandemakers, W., Daneels, G., … … Moechars, D. (2015). LRRK2 functions in synaptic vesicle endocytosis through a kinase-dependent mechanism. *Journal of Cell Science*, *128*, 541–552. Available from https://doi.org/10.1242/jcs.158196.

Ashrafi, G., Schlehe, J. S., LaVoie, M. J., & Schwarz, T. L. (2014). Mitophagy of damaged mitochondria occurs locally in distal neuronal axons and requires PINK1 and

Parkin. *Journal of Cell Biology, 206*, 655–670. Available from https://doi.org/10.1083/jcb.201401070.

Bae, E.-J., Yang, N. Y., Lee, C., Lee, H.-J., Kim, S., Sardi, S. P., & Lee, S.-J. (2015). Loss of glucocerebrosidase 1 activity causes lysosomal dysfunction and α-synuclein aggregation. *Experimental & Molecular Medicine, 47*, e153. Available from https://doi.org/10.1038/emm.2014.128.

Baptista, M. J., O'Farrell, C., Daya, S., Ahmad, R., Miller, D. W., Hardy, J., … … Cookson, M. R. (2003). Coordinate transcriptional regulation of dopamine synthesis genes by alpha-synuclein in human neuroblastoma cell lines. *Journal of Neurochemistry, 85*, 957–968.

Barrett, P., & Greenamyre, J. (2015). Post-translational modification of α-synuclein in Parkinson's disease. *Brain Research, 1628*, 247–253. Available from https://doi.org/10.1016/j.brainres.2015.06.002.

Bartels, T., Choi, J. G., & Selkoe, D. J. (2011). α-Synuclein occurs physiologically as a helically folded tetramer that resists aggregation. *Nature, 477*, 107–110. Available from https://doi.org/10.1038/nature10324.

Beccano-Kelly, D. A., Volta, M., Munsie, L. N., Paschall, S. A., Tatarnikov, I., Co, K., … … Milnerwood, A. J. (2015). LRRK2 overexpression alters glutamatergic presynaptic plasticity, striatal dopamine tone, postsynaptic signal transduction, motor activity and memory. *Human Molecular Genetics, 24*, 1336–1349. Available from https://doi.org/10.1093/hmg/ddu543.

Beilina, A., Rudenko, I. N., Kaganovich, A., Civiero, L., Chau, H., Kalia, S. K., … … Cookson, M. R. (2014). Unbiased screen for interactors of leucine-rich repeat kinase 2 supports a common pathway for sporadic and familial Parkinson disease. *Proceedings of the National Academy of Sciences of the United States of America*. Available from https://doi.org/10.1073/pnas.1318306111.

Belluzzi, E., Gonnelli, A., Cirnaru, M.-D., Marte, A., Plotegher, N., Russo, I., … … Greggio, E. (2016). LRRK2 phosphorylates pre-synaptic *N*-ethylmaleimide sensitive fusion (NSF) protein enhancing its ATPase activity and SNARE complex disassembling rate. *Molecular Neurodegeneration, 11*, 1. Available from https://doi.org/10.1186/s13024-015-0066-z.

Benskey, M. J., Perez, R. G., & Manfredsson, F. P. (2016). The contribution of alpha synuclein to neuronal survival and function—Implications for Parkinson's disease. *Journal of Neurochemistry, 137*, 331–359. Available from https://doi.org/10.1111/jnc.13570.

Béraud, D., & Maguire-Zeiss, K. A. (2012). Misfolded α-synuclein and Toll-like receptors: Therapeutic targets for Parkinson's disease. *Parkinsonism & Related Disorders, 18*(Suppl 1), S17–S20. Available from https://doi.org/10.1016/S1353-8020(11)70008-6.

Bose, A., & Beal, M. F. (2016). Mitochondrial dysfunction in Parkinson's disease. *Journal of Neurochemistry, 139* (Suppl 1), 216–231. Available from https://doi.org/10.1111/jnc.13731.

Braak, H., & Del Tredici, K. (2016). Potential pathways of abnormal tau and α-synuclein dissemination in sporadic Alzheimer's and Parkinson's diseases. *Cold Spring Harbor Perspectives in Biology*, 8. Available from https://doi.org/10.1101/cshperspect.a023630.

Braak, H., Del Tredici, K., Rüb, U., de Vos, R. A. I., Jansen Steur, E. N. H., & Braak, E. (2003). Staging of brain pathology related to sporadic Parkinson's disease. *Neurobiology of Aging, 24*, 197–211.

Bravo-San Pedro, J. M., Niso-Santano, M., Gómez-Sánchez, R., Pizarro-Estrella, E., Aiastui-Pujana, A., Gorostidi, A., … … González-Polo, Ra (2013). The LRRK2 G2019S mutant exacerbates basal autophagy through activation of the MEK/ERK pathway. *Cellular and Molecular Life Sciences, 70*, 121–136. Available from https://doi.org/10.1007/s00018-012-1061-y.

Breda, C., Nugent, M. L., Estranero, J. G., Kyriacou, C. P., Outeiro, T. F., Steinert, J. R., & Giorgini, F. (2015). Rab11 modulates α-synuclein-mediated defects in synaptic transmission and behaviour. *Human Molecular Genetics, 24*, 1077–1091. Available from https://doi.org/10.1093/hmg/ddu521.

Breydo, L., Wu, J. W., & Uversky, V. N. (2012). A-synuclein misfolding and Parkinson's disease. *Biochimica et Biophysica Acta, 1822*, 261–285. Available from https://doi.org/10.1016/j.bbadis.2011.10.002.

Brooks, D. J. (2007). Imaging non-dopaminergic function in Parkinson's disease. *Molecular Imaging and Biology: MIB: The Official Publication of the Academy of Molecular Imaging, 9*, 217–222. Available from https://doi.org/10.1007/s11307-007-0084-5.

Brueggemann, N., Odin, P., Gruenewald, A., Tadic, V., Hagenah, J., Seidel, G., … … Djarmati, A. (2008). Re: Alpha-synuclein gene duplication is present in sporadic Parkinson disease. *Neurology, 71*, 1294. Available from https://doi.org/10.1212/01.wnl.0000338439.00992.c7, author reply 1294.

Burré, J. (2015). The synaptic function of α-synuclein. *Journal of Parkinson's Disease, 5*, 699–713. Available from https://doi.org/10.3233/JPD-150642.

Burré, J., Sharma, M., & Sudhof, T. C. (2015). Definition of a molecular pathway mediating—synuclein neurotoxicity. *The Journal of Neuroscience, 35*, 5221–5232. Available from https://doi.org/10.1523/JNEUROSCI.4650-14.2015.

Chartier-Harlin, M.-C., Dachsel, J. C., Vilariño-Güell, C., Lincoln, S. J., Leprêtre, F., Hulihan, M. M., … … Farrer, M. J. (2011). Translation initiator EIF4G1 mutations in familial Parkinson disease. *American Journal of Human Genetics, 89*, 398–406. Available from https://doi.org/10.1016/j.ajhg.2011.08.009.

Chartier-Harlin, M.-C., Kachergus, J., Roumier, C., Mouroux, V., Douay, X., Lincoln, S., ... Destée, A. (2004). Alpha-synuclein locus duplication as a cause of familial Parkinson's disease. *The Lancet, 364*, 1167–1169. Available from https://doi.org/10.1016/S0140-6736(04)17103-1.

Chen, J., Mills, J. D., Halliday, G. M., & Janitz, M. (2015). Role of transcriptional control in multiple system atrophy. *Neurobiology of Aging, 36*, 394–400. Available from https://doi.org/10.1016/j.neurobiolaging.2014.08.015.

Chen, L., Xie, Z., Turkson, S., & Zhuang, X. (2015). A53T human α-synuclein overexpression in transgenic mice induces pervasive mitochondria macroautophagy defects preceding dopamine neuron degeneration. *The Journal of Neuroscience: The Official Journal of the Society for Neuroscience, 35*, 890–905. Available from https://doi.org/10.1523/JNEUROSCI.0089-14.2015.

Cheng, L., Wang, L., Li, N.-N., Yu, W.-J., Sun, X.-Y., Li, J.-Y., ... Peng, R. (2016). SNCA rs356182 variant increases risk of sporadic Parkinson's disease in ethnic Chinese. *Journal of the Neurological Sciences, 368*, 231–234. Available from https://doi.org/10.1016/j.jns.2016.07.032.

Cherra, S. J., Steer, E., Gusdon, A. M., Kiselyov, K., & Chu, C. T. (2013). Mutant LRRK2 elicits calcium imbalance and depletion of dendritic mitochondria in neurons. *The American Journal of Pathology, 182*, 474–484. Available from https://doi.org/10.1016/j.ajpath.2012.10.027.

Chinta, S. J., Mallajosyula, J. K., Rane, A., & Andersen, J. K. (2010). Mitochondrial α-synuclein accumulation impairs complex I function in dopaminergic neurons and results in increased mitophagy in vivo. *Neuroscience Letters, 486*, 235–239. Available from https://doi.org/10.1016/j.neulet.2010.09.061.

Chouhan, A. K., Guo, C., Hsieh, Y.-C., Ye, H., Senturk, M., Zuo, Z., ... Shulman, J. M. (2016). Uncoupling neuronal death and dysfunction in *Drosophila* models of neurodegenerative disease. *Acta Neuropathologica Communications, 4*, 62. Available from https://doi.org/10.1186/s40478-016-0333-4.

Clark, I. E., Dodson, M. W., Jiang, C., Cao, J. H., Huh, J. R., Seol, J. H., ... Guo, M. (2006). *Drosophila* pink1 is required for mitochondrial function and interacts genetically with parkin. *Nature, 441*, 1162–1166. Available from https://doi.org/10.1038/nature04779.

Cookson, M. R. (2012). Evolution of neurodegeneration. *Current Biology, 22*, R753–R761. Available from https://doi.org/10.1016/j.cub.2012.07.008.

Cookson, M. R. (2016). Cellular functions of LRRK2 implicate vesicular trafficking pathways in Parkinson's disease. *Biochemical Society Transactions, 44*, 1603–1610. Available from https://doi.org/10.1042/BST20160228.

Cookson, M. R., Hardy, J., & Lewis, P. A. (2008). Genetic neuropathology of Parkinson's disease. *International Journal of Clinical and Experimental Pathology, 1*, 217–231.

Cooper, A. A., Gitler, A. D., Cashikar, A., Haynes, C. M., Hill, K. J., Bhullar, B., ... Lindquist, S. (2006). Alpha-synuclein blocks ER–Golgi traffic and Rab1 rescues neuron loss in Parkinson's models. *Science, 313*, 324–328. Available from https://doi.org/10.1126/science.1129462.

Cooper, O., Seo, H., Andrabi, S., Guardia-Laguarta, C., Graziotto, J., Sundberg, M., ... Isacson, O. (2012). Pharmacological rescue of mitochondrial deficits in iPSC-derived neural cells from patients with familial Parkinson's disease. *Science Translational Medicine, 4*, 141ra90. Available from https://doi.org/10.1126/scitranslmed.3003985.

Cuervo, A.M., Stefanis, L., Fredenburg, R., Lansbury, P.T., Sulzer, D., 2004. Impaired degradation of mutant α-synuclein by chaperone-mediated autophagy 29831. *Science, 305* (5688), 1292–1295.

Daher, J. P. L., Abdelmotilib, H. A., Hu, X., Volpicelli-Daley, L. A., Moehle, M. S., Fraser, K. B., ... West, A. B. (2015). Leucine-rich repeat kinase 2 (LRRK2) pharmacological inhibition abates α-synuclein gene-induced neurodegeneration. *Journal of Biological Chemistry, 290*, 19433–19444. Available from https://doi.org/10.1074/jbc.M115.660001.

Daher, J. P. L., Volpicelli-Daley, L. A., Blackburn, J. P., Moehle, M. S., & West, A. B. (2014). Abrogation of α-synuclein-mediated dopaminergic neurodegeneration in LRRK2-deficient rats. *Proceedings of the National Academy of Sciences of the United States of America, 111*, 9289–9294. Available from https://doi.org/10.1073/pnas.1403215111.

Dall'Armi, C., Devereaux, K. A., & Di Paolo, G. (2013). The role of lipids in the control of autophagy. *Current Biology, 23*, R33–R45. Available from https://doi.org/10.1016/j.cub.2012.10.041.

Daniel, G., Musso, A., Tsika, E., Fiser, A., Glauser, L., Pletnikova, O., ... Moore, D. J. (2015). α-Synuclein-induced dopaminergic neurodegeneration in a rat model of Parkinson's disease occurs independent of ATP13A2 (PARK9). *Neurobiology of Disease, 73*, 229–243. Available from https://doi.org/10.1016/j.nbd.2014.10.007.

Davis, M. Y., Trinh, K., Thomas, R. E., Yu, S., Germanos, A. A., Whitley, B. N., ... Pallanck, L. J. (2016). Glucocerebrosidase deficiency in *Drosophila* results in α-synuclein-independent protein aggregation and neurodegeneration. *PLoS Genetics, 12*, e1005944. Available from https://doi.org/10.1371/journal.pgen.1005944.

Decressac, M., Kadkhodaei, B., Mattsson, B., Laguna, A., Perlmann, T., & Björklund, A. (2012). α-Synuclein-induced down-regulation of Nurr1 disrupts GDNF signaling in nigral dopamine neurons. *Science Translational Medicine*, *4*. Available from https://doi.org/10.1126/scitranslmed.3004676, 163ra156.

Dehay, B., Martinez-Vicente, M., Caldwell, G. A., Caldwell, K. A., Yue, Z., Cookson, M. R., ... Bezard, E. (2013). Lysosomal impairment in Parkinson's disease. *Movement Disorders: The official Journal of the International Parkinson and Movement Disorder Society*, *28*, 725–732. Available from https://doi.org/10.1002/mds.25462.

Deng, H., Dodson, M. W., Huang, H., & Guo, M. (2008). The Parkinson's disease genes pink1 and parkin promote mitochondrial fission and/or inhibit fusion in *Drosophila*. *Proceedings of the National Academy of Sciences of the United States of America*, *105*, 14503–14508. Available from https://doi.org/10.1073/pnas.0803998105.

Deng, H.-X., Shi, Y., Yang, Y., Ahmeti, K. B., Miller, N., Huang, C., ... Siddique, T. (2016). Identification of TMEM230 mutations in familial Parkinson's disease. *Nature Genetics*, 733–739. Available from https://doi.org/10.1038/ng.3589. (advance on).

Dermentzaki, G., Dimitriou, E., Xilouri, M., Michelakakis, H., & Stefanis, L. (2013). Loss of β-glucocerebrosidase activity does not affect alpha-synuclein levels or lysosomal function in neuronal cells. *PLoS ONE*, *8*, e60674. Available from https://doi.org/10.1371/journal.pone.0060674.

Dettmer, U., Newman, A. J., Luth, E. S., Bartels, T., & Selkoe, D. (2013). In vivo cross-linking reveals principally oligomeric forms of α-synuclein and β-synuclein in neurons and non-neural cells. *Journal of Biological Chemistry*, *288*, 6371–6385. Available from https://doi.org/10.1074/jbc.M112.403311.

Dickson, D. W. (2001). Alpha-synuclein and the Lewy body disorders. *Current Opinion in Neurology*, *14*, 423–432.

Di Fonzo, A., Chien, H. F., Socal, M., Giraudo, S., Tassorelli, C., Iliceto, G., ... Bonifati, V. (2007). ATP13A2 missense mutations in juvenile parkinsonism and young onset Parkinson disease. *Neurology*, *68*, 1557–1562. Available from https://doi.org/10.1212/01.wnl.0000260963.08711.08.

Dodson, M. W., Leung, L. K., Lone, M., Lizzio, M. A., & Guo, M. (2014). Novel ethyl methanesulfonate (EMS)-induced null alleles of the *Drosophila* homolog of LRRK2 reveal a crucial role in endolysosomal functions and autophagy in vivo. *Disease Models & Mechanisms*, *7*, 1351–1363. Available from https://doi.org/10.1242/dmm.017020.

Donaghy, P. C., & McKeith, I. G. (2014). The clinical characteristics of dementia with Lewy bodies and a consideration of prodromal diagnosis. *Alzheimer's Research & Therapy*, *6*, 46. Available from https://doi.org/10.1186/alzrt274.

Duda, J. E., Giasson, B. I., Mabon, M. E., Miller, D. C., Golbe, L. I., Lee, V. M.-Y., & Trojanowski, J. Q. (2002). Concurrence of alpha-synuclein and tau brain pathology in the Contursi kindred. *Acta Neuropathologica (Berlin)*, *104*, 7–11. Available from https://doi.org/10.1007/s00401-002-0563-3.

Dumitriu, A., Pacheco, C. D., Wilk, J. B., Strathearn, K. E., Latourelle, J. C., Goldwurm, S., ... Myers, R. H. (2011). Cyclin-G-associated kinase modifies α-synuclein expression levels and toxicity in Parkinson's disease: Results from the GenePD Study. *Human Molecular Genetics*, *20*, 1478–1487. Available from https://doi.org/10.1093/hmg/ddr026.

Duran, R., Mencacci, N. E., Angeli, A. V., Shoai, M., Deas, E., Houlden, H., ... Foltynie, T. (2013). The glucocerobrosidase E326K variant predisposes to Parkinson's disease, but does not cause Gaucher's disease. *Movement Disorders: The official Journal of the International Parkinson and Movement Disorder Society*, *28*, 232–236. Available from https://doi.org/10.1002/mds.25248.

Dzamko, N., Gysbers, A., Perera, G., Bahar, A., Shankar, A., Gao, J., ... Halliday, G. M. (2017). Toll-like receptor 2 is increased in neurons in Parkinson's disease brain and may contribute to alpha-synuclein pathology. *Acta Neuropathologica (Berlin)*, *133*, 303–319. Available from https://doi.org/10.1007/s00401-016-1648-8.

Fahn, S. (2003). Description of Parkinson's disease as a clinical syndrome. *Annals of the New York Academy of Sciences*, *991*, 1–14.

Farrer, M., Maraganore, D. M., Lockhart, P., Singleton, A., Lesnick, T. G., de Andrade, M., ... Hernandez, D. (2001). alpha-Synuclein gene haplotypes are associated with Parkinson's disease. *Human Molecular Genetics*, *10*, 1847–1851.

Ferreira, M., & Massano, J. (2016). An updated review of Parkinson's disease genetics and clinicopathological correlations. *Acta Neurologica Scandinavica*. Available from https://doi.org/10.1111/ane.12616.

Follett, J., Bugarcic, A., Yang, Z., Ariotti, N., Norwood, S. J., Collins, B. M., ... Teasdale, R. D. (2016). Parkinson disease-linked Vps35 R524W mutation impairs the endosomal association of retromer and induces α-synuclein aggregation. *Journal of Biological Chemistry*, *291*, 18283–18298. Available from https://doi.org/10.1074/jbc.M115.703157.

Fuchs, J., Nilsson, C., Kachergus, J., Munz, M., Larsson, E.-M., Schüle, B., ... Farrer, M. J. (2007). Phenotypic variation in a large Swedish pedigree due to SNCA duplication and triplication. *Neurology*, *68*, 916–922. Available from https://doi.org/10.1212/01.wnl.0000254458.17630.c5.

Fuchs, J., Tichopad, A., Golub, Y., Munz, M., Schweitzer, K. J., Wolf, B., ... Gasser, T. (2008). Genetic variability in the SNCA gene influences alpha-synuclein levels in the blood and brain. *FASEB Journal: Official Publication of the Federation of American Societies for*

Experimental Biology, *22*, 1327–1334. Available from https://doi.org/10.1096/fj.07-9348com.

Funayama, M., Hasegawa, K., Ohta, E., Kawashima, N., Komiyama, M., Kowa, H., ... Obata, F. (2005). An LRRK2 mutation as a cause for the parkinsonism in the original PARK8 family. *Annals of Neurology*, *57*, 918–921. Available from https://doi.org/10.1002/ana.20484.

Funayama, M., Ohe, K., Amo, T., Furuya, N., Yamaguchi, J., Saiki, S., ... Hattori, N. (2015). CHCHD2 mutations in autosomal dominant late-onset Parkinson's disease: A genome-wide linkage and sequencing study. *The Lancet Neurology*, *14*, 274–282. Available from https://doi.org/10.1016/S1474-4422(14)70266-2.

Gan-Or, Z., Bar-Shira, A., Dahary, D., Mirelman, A., Kedmi, M., Gurevich, T., ... Orr-Urtreger, A. (2012). Association of sequence alterations in the putative promoter of RAB7L1 with a reduced Parkinson disease risk. *Archives of Neurology*, *69*, 105–110. Available from https://doi.org/10.1001/archneurol.2011.924.

Gao, H.-M., Kotzbauer, P. T., Uryu, K., Leight, S., Trojanowski, J. Q., & Lee, V. M.-Y. (2008). Neuroinflammation and oxidation/nitration of alpha-synuclein linked to dopaminergic neurodegeneration. *The Journal of Neuroscience: The Official Journal of the Society for Neuroscience*, *28*, 7687–7698. Available from https://doi.org/10.1523/JNEUROSCI.0143-07.2008.

Gehrke, S., Wu, Z., Klinkenberg, M., Sun, Y., Auburger, G., Guo, S., & Lu, B. (2015). PINK1 and Parkin control localized translation of respiratory chain component mRNAs on mitochondria outer membrane. *Cell Metabolism*, *21*, 95–108. Available from https://doi.org/10.1016/j.cmet.2014.12.007.

Geiger, J. T., Ding, J., Crain, B., Pletnikova, O., Letson, C., Dawson, T. M., ... Scholz, S. W. (2016). Next-generation sequencing reveals substantial genetic contribution to dementia with Lewy bodies. *Neurobiology of Disease*, *94*, 55–62. Available from https://doi.org/10.1016/j.nbd.2016.06.004.

Gitler, A. D., Bevis, B. J., Shorter, J., Strathearn, K. E., Hamamichi, S., Su, L. J., ... Lindquist, S. (2008). The Parkinson's disease protein alpha-synuclein disrupts cellular Rab homeostasis. *Proceedings of the National Academy of Sciences of the United States of America*, *105*, 145–150. Available from https://doi.org/10.1073/pnas.0710685105.

Godena, V. K., Brookes-Hocking, N., Moller, A., Shaw, G., Oswald, M., Sancho, R. M., ... De Vos, K. J. (2014). Increasing microtubule acetylation rescues axonal transport and locomotor deficits caused by LRRK2 Roc-COR domain mutations. *Nature Communications*, *5*, 5245. Available from https://doi.org/10.1038/ncomms6245.

Greten-Harrison, B., Polydoro, M., Morimoto-Tomita, M., Diao, L., Williams, A. M., Nie, E. H., ... Chandra, S. S. (2010). αβγ-Synuclein triple knockout mice reveal age-dependent neuronal dysfunction. *Proceedings of the National Academy of Sciences of the United States of America*, *107*, 19573–19578. Available from https://doi.org/10.1073/pnas.1005005107.

Guardia-Laguarta, C., Area-Gomez, E., Rüb, C., Liu, Y., Magrané, J., Becker, D., ... Przedborski, S. (2014). α-Synuclein is localized to mitochondria-associated ER membranes. *The Journal of Neuroscience: The Official Journal of the Society for Neuroscience*, *34*, 249–259. Available from https://doi.org/10.1523/JNEUROSCI.2507-13.2014.

Guella, I., Evans, D. M., Szu-Tu, C., Nosova, E., Bortnick, S. F., SNCA Cognition Study Group., ... Farrer, M. J. (2016). α-Synuclein genetic variability: A biomarker for dementia in Parkinson disease. *Annals of Neurology*, *79*, 991–999. Available from https://doi.org/10.1002/ana.24664.

Guerreiro, R., Escott-Price, V., Darwent, L., Parkkinen, L., Ansorge, O., Hernandez, D. G., ... Bras, J. (2016). Genome-wide analysis of genetic correlation in dementia with Lewy bodies, Parkinson's and Alzheimer's diseases. *Neurobiology of Aging*, *38*, 214.e7–214.e10. Available from https://doi.org/10.1016/j.neurobiolaging.2015.10.028.

Gwinn-Hardy, K., Mehta, N. D., Farrer, M., Maraganore, D., Muenter, M., Yen, S. H., ... Dickson, D. W. (2000). Distinctive neuropathology revealed by alpha-synuclein antibodies in hereditary parkinsonism and dementia linked to chromosome 4p. *Acta Neuropathologica (Berlin)*, *99*, 663–672.

Halliday, G. M., Holton, J. L., Revesz, T., & Dickson, D. W. (2011). Neuropathology underlying clinical variability in patients with synucleinopathies. *Acta Neuropathologica (Berlin)*, *122*, 187–204. Available from https://doi.org/10.1007/s00401-011-0852-9.

Halliday, G. M., & Stevens, C. H. (2011). Glia: Initiators and progressors of pathology in Parkinson's disease. *Movement Disorders: The official Journal of the International Parkinson and Movement Disorder Society*, *26*, 6–17. Available from https://doi.org/10.1002/mds.23455.

Han, J.-Y., Kim, J.-S., & Son, J. H. (2014). Mitochondrial homeostasis molecules: Regulation by a trio of recessive Parkinson's disease genes. *Experimental Neurobiology*, *23*, 345–351. Available from https://doi.org/10.5607/en.2014.23.4.345.

Hoffmann, A., Ettle, B., Bruno, A., Kulinich, A., Hoffmann, A.-C., von Wittgenstein, J., ... Schlachetzki, J. C. M. (2016). Alpha-synuclein activates BV2 microglia dependent on its aggregation state. *Biochemical and Biophysical Research Communications*, *479*, 881–886. Available from https://doi.org/10.1016/j.bbrc.2016.09.109.

Hsieh, C.-H., Shaltouki, A., Gonzalez, A. E., Bettencourt da Cruz, A., Burbulla, L. F., St Lawrence, E., ... Wang, X. (2016). Functional impairment in miro degradation and mitophagy is a shared feature in familial and sporadic Parkinson's disease. *Cell Stem Cell, 19*, 709–724. Available from https://doi.org/10.1016/j.stem.2016.08.002.

Ibáñez, P., Bonnet, A.-M., Débarges, B., Lohmann, E., Tison, F., Pollak, P., ... Brice, A. (2004). Causal relation between alpha-synuclein gene duplication and familial Parkinson's disease. *The Lancet, 364*, 1169–1171. Available from https://doi.org/10.1016/S0140-6736(04)17104-3.

Ingelsson, M. (2016). Alpha-synuclein oligomers-neurotoxic molecules in Parkinson's disease and other Lewy body disorders. *Frontiers in Neuroscience, 10*, 408. Available from https://doi.org/10.3389/fnins.2016.00408.

Jaishy, B., & Abel, E. D. (2016). Lipids, lysosomes, and autophagy. *Journal of Lipid Research, 57*, 1619–1635. Available from https://doi.org/10.1194/jlr.R067520.

Jellinger, K. A. (2009). A critical evaluation of current staging of alpha-synuclein pathology in Lewy body disorders. *Biochimica et Biophysica Acta, 1792*, 730–740. Available from https://doi.org/10.1016/j.bbadis.2008.07.006.

Kalia, L. V., Lang, A. E., Hazrati, L.-N., Fujioka, S., Wszolek, Z. K., Dickson, D. W., ... Marras, C. (2015). Clinical correlations with Lewy body pathology in LRRK2-related Parkinson disease. *JAMA Neurology, 72*, 100–105. Available from https://doi.org/10.1001/jamaneurol.2014.2704.

Kane, L. A., Lazarou, M., Fogel, A. I., Li, Y., Yamano, K., Sarraf, S. A., ... Youle, R. J. (2014). PINK1 phosphorylates ubiquitin to activate Parkin E3 ubiquitin ligase activity. *Journal of Cell Biology, 205*, 143–153. Available from https://doi.org/10.1083/jcb.201402104.

Kenney, D. L., & Benarroch, E. E. (2015). The autophagy-lysosomal pathway: General concepts and clinical implications. *Neurology, 85*, 634–645. Available from https://doi.org/10.1212/WNL.0000000000001860.

Kett, L. R., Stiller, B., Bernath, M. M., Tasset, I., Blesa, J., Jackson-Lewis, V., ... Dauer, W. T. (2015). α-Synuclein-independent histopathological and motor deficits in mice lacking the endolysosomal parkinsonism protein Atp13a2. *The Journal of Neuroscience: The Official Journal of the Society for Neuroscience, 35*, 5724–5742. Available from https://doi.org/10.1523/JNEUROSCI.0632-14.2015.

Kim, C., Ho, D.-H., Suk, J.-E., You, S., Michael, S., Kang, J., ... Lee, S.-J. (2013). Neuron-released oligomeric α-synuclein is an endogenous agonist of TLR2 for paracrine activation of microglia. *Nature Communications, 4*, 1562. Available from https://doi.org/10.1038/ncomms2534.

Kim, C., Rockenstein, E., Spencer, B., Kim, H.-K., Adame, A., Trejo, M., ... Masliah, E. (2015). Antagonizing neuronal Toll-like receptor 2 prevents synucleinopathy by activating autophagy. *Cell Reports, 13*, 771–782. Available from https://doi.org/10.1016/j.celrep.2015.09.044.

Kondo, K., Obitsu, S., & Teshima, R. (2011). α-Synuclein aggregation and transmission are enhanced by leucine-rich repeat kinase 2 in human neuroblastoma SH-SY5Y cells. *Biological and Pharmaceutical Bulletin, 34*, 1078–1083.

Kong, S. M. Y., Chan, B. K. K., Park, J.-S., Hill, K. J., Aitken, J. B., Cottle, L., ... Cooper, Aa (2014). Parkinson's disease-linked human PARK9/ATP13A2 maintains zinc homeostasis and promotes α-synuclein externalization via exosomes. *Human Molecular Genetics, 23*, 2816–2833. Available from https://doi.org/10.1093/hmg/ddu099.

Koyano, F., Okatsu, K., Kosako, H., Tamura, Y., Go, E., Kimura, M., ... Matsuda, N. (2014). Ubiquitin is phosphorylated by PINK1 to activate parkin. *Nature, 510*, 162–166. Available from https://doi.org/10.1038/nature13392.

Krüger, R., Kuhn, W., Müller, T., Woitalla, D., Graeber, M., Kösel, S., ... Riess, O. (1998). Ala30Pro mutation in the gene encoding alpha-synuclein in Parkinson's disease. *Nature Genetics, 18*, 106–108. Available from https://doi.org/10.1038/ng0298-106.

Kumari, U., & Tan, E. K. (2009). LRRK2 in Parkinson's disease: Genetic and clinical studies from patients. *The FEBS Journal, 276*, 6455–6463. Available from https://doi.org/10.1111/j.1742-4658.2009.07344.x.

Kuwahara, T., Inoue, K., D'Agati, V. D., Fujimoto, T., Eguchi, T., Saha, S., ... Abeliovich, A. (2016). LRRK2 and RAB7L1 coordinately regulate axonal morphology and lysosome integrity in diverse cellular contexts. *Scientific Reports, 6*, 29945. Available from https://doi.org/10.1038/srep29945.

Kuwahara, T., Koyama, A., Koyama, S., Yoshina, S., Ren, C.-H., Kato, T., ... Iwatsubo, T. (2008). A systematic RNAi screen reveals involvement of endocytic pathway in neuronal dysfunction in alpha-synuclein transgenic *C. elegans*. *Human Molecular Genetics, 17*, 2997–3009. Available from https://doi.org/10.1093/hmg/ddn198.

Langston, J. W. (2006). The Parkinson's complex: Parkinsonism is just the tip of the iceberg. *Annals of Neurology, 59*, 591–596. Available from https://doi.org/10.1002/ana.20834.

Law, B. M. H., Spain, V. A., Leinster, V. H. L., Chia, R., Beilina, A., Cho, H. J., ... Harvey, K. (2014). A direct interaction between leucine-rich repeat kinase 2 and specific β-tubulin isoforms regulates tubulin acetylation. *Journal of Biological Chemistry, 289*, 895–908. Available from https://doi.org/10.1074/jbc.M113.507913.

Lesage, S., Anheim, M., Letournel, F., Bousset, L., Honoré, A., Rozas, N., ... French Parkinson's Disease Genetics Study Group. (2013). G51D α-synuclein mutation causes a novel parkinsonian–pyramidal syndrome.

Annals of Neurology, *73*, 459–471. Available from https://doi.org/10.1002/ana.23894.

Li, J.-Y., Englund, E., Holton, J. L., Soulet, D., Hagell, P., Lees, A. J., ... Brundin, P. (2008). Lewy bodies in grafted neurons in subjects with Parkinson's disease suggest host-to-graft disease propagation. *Nature Medicine*, *14*, 501–503. Available from https://doi.org/10.1038/nm1746.

Li, L., Nadanaciva, S., Berger, Z., Shen, W., Paumier, K., Schwartz, J., ... Hirst, W. D. (2013). Human A53T α-synuclein causes reversible deficits in mitochondrial function and dynamics in primary mouse cortical neurons. *PLoS ONE*, *8*, e85815. Available from https://doi.org/10.1371/journal.pone.0085815.

Li, W. (2004). Stabilization of α-synuclein protein with aging and familial Parkinson's disease-linked A53T mutation. *The Journal of Neuroscience*, *24*, 7400–7409. Available from https://doi.org/10.1523/JNEUROSCI.1370-04.2004.

Liddelow, S. A., Guttenplan, K. A., Clarke, L. E., Bennett, F. C., Bohlen, C. J., Schirmer, L., ... Barres, B. A. (2017). Neurotoxic reactive astrocytes are induced by activated microglia. *Nature*. Available from https://doi.org/10.1038/nature21029.

Lill, C. M., Roehr, J. T., McQueen, M. B., Kavvoura, F. K., Bagade, S., Schjeide, B.-M. M., ... Bertram, L. (2012). Comprehensive research synopsis and systematic meta-analyses in Parkinson's disease genetics: The PDGene database. *PLoS Genetics*, *8*, e1002548. Available from https://doi.org/10.1371/journal.pgen.1002548.

Lippa, C. F., Duda, J. E., Grossman, M., Hurtig, H. I., Aarsland, D., Boeve, B. F., ... DLB/PDD Working Group. (2007). DLB and PDD boundary issues: Diagnosis, treatment, molecular pathology, and biomarkers. *Neurology*, *68*, 812–819. Available from https://doi.org/10.1212/01.wnl.0000256715.13907.d3.

Liu, J., Zhou, Y., Wang, C., Wang, T., Zheng, Z., & Chan, P. (2012). Brain-derived neurotrophic factor (BDNF) genetic polymorphism greatly increases risk of leucine-rich repeat kinase 2 (LRRK2) for Parkinson's disease. *Parkinsonism & Related Disorders*, *18*, 140–143. Available from https://doi.org/10.1016/j.parkreldis.2011.09.002.

Lo Bianco, C., Déglon, N., Pralong, W., & Aebischer, P. (2004). Lentiviral nigral delivery of GDNF does not prevent neurodegeneration in a genetic rat model of Parkinson's disease. *Neurobiology of Disease*, *17*, 283–289. Available from https://doi.org/10.1016/j.nbd.2004.06.008.

Lwin, A., Orvisky, E., Goker-Alpan, O., LaMarca, M. E., & Sidransky, E. (2004). Glucocerebrosidase mutations in subjects with parkinsonism. *Molecular Genetics and Metabolism*, *81*, 70–73.

Ma, Z.-G., He, F., & Xu, J. (2015). Quantitative assessment of the association between GAK rs1564282 C/T polymorphism and the risk of Parkinson's disease. *Journal of Clinical Neuroscience: The Official Journal of the Neurosurgical Society of Australasia*, *22*, 1077–1080. Available from https://doi.org/10.1016/j.jocn.2014.12.014.

MacLeod, D. A., Rhinn, H., Kuwahara, T., Zolin, A., Di Paolo, G., McCabe, B. D., ... Abeliovich, A. (2013). RAB7L1 interacts with LRRK2 to modify intraneuronal protein sorting and Parkinson's disease risk. *Neuron*, *77*, 425–439. Available from https://doi.org/10.1016/j.neuron.2012.11.033.

Maekawa, T., Sasaoka, T., Azuma, S., Ichikawa, T., Melrose, H. L., Farrer, M. J., & Obata, F. (2016). Leucine-rich repeat kinase 2 (LRRK2) regulates α-synuclein clearance in microglia. *BMC Neuroscience*, *17*, 77. Available from https://doi.org/10.1186/s12868-016-0315-2.

Mandemakers, W., Snellinx, A., O'Neill, M. J., & de Strooper, B. (2012). LRRK2 expression is enriched in the striosomal compartment of mouse striatum. *Neurobiology of Disease*, *48*, 582–593. Available from https://doi.org/10.1016/j.nbd.2012.07.017.

Manzoni, C., Mamais, A., Dihanich, S., Abeti, R., Soutar, M. P. M., Plun-Favreau, H., ... Lewis, P. A. (2013). Inhibition of LRRK2 kinase activity stimulates macroautophagy. *Biochimica et Biophysica Acta*, *1833*, 2900–2910. Available from https://doi.org/10.1016/j.bbamcr.2013.07.020.

Maraganore, D. M., de Andrade, M., Elbaz, A., Farrer, M. J., Ioannidis, J. P., Krüger, R., ... Genetic Epidemiology of Parkinson's Disease (GEO-PD) Consortium. (2006). Collaborative analysis of alpha-synuclein gene promoter variability and Parkinson disease. *The Journal of the American Medical Association*, *296*, 661–670. Available from https://doi.org/10.1001/jama.296.6.661.

Mata, I. F., Samii, A., Schneer, S. H., Roberts, J. W., Griffith, A., Leis, B. C., ... Zabetian, C. P. (2008). Glucocerebrosidase gene mutations: A risk factor for Lewy body disorders. *Archives of Neurology*, *65*, 379–382. Available from https://doi.org/10.1001/archneurol.2007.68.

Matsuda, N., Sato, S., Shiba, K., Okatsu, K., Saisho, K., Gautier, C. A., ... Tanaka, K. (2010). PINK1 stabilized by mitochondrial depolarization recruits Parkin to damaged mitochondria and activates latent Parkin for mitophagy. *Journal of Cell Biology*, *189*, 211–221. Available from https://doi.org/10.1083/jcb.200910140.

Matta, S., Van Kolen, K., da Cunha, R., van den Bogaart, G., Mandemakers, W., Miskiewicz, K., ... Verstreken, P. (2012). LRRK2 controls an EndoA phosphorylation cycle in synaptic endocytosis. *Neuron*, *75*, 1008–1021. Available from https://doi.org/10.1016/j.neuron.2012.08.022.

Mazzulli, J. R., Xu, Y.-H., Sun, Y., Knight, A. L., McLean, P. J., Caldwell, G. A., ... Krainc, D. (2011). Gaucher disease glucocerebrosidase and α-synuclein form a bidirectional pathogenic loop in synucleinopathies. *Cell, 146*, 37–52. Available from https://doi.org/10.1016/j.cell.2011.06.001.

McCarthy, J. J., Linnertz, C., Saucier, L., Burke, J. R., Hulette, C. M., Welsh-Bohmer, K. A., & Chiba-Falek, O. (2011). The effect of SNCA 3′ region on the levels of SNCA-112 splicing variant. *Neurogenetics, 12*, 59–64. Available from https://doi.org/10.1007/s10048-010-0263-4.

McLelland, G.-L., Soubannier, V., Chen, C. X., McBride, H. M., & Fon, E. A. (2014). Parkin and PINK1 function in a vesicular trafficking pathway regulating mitochondrial quality control. *The EMBO Journal, 33*, 282–295. Available from https://doi.org/10.1002/embj.201385902.

Migdalska-Richards, A., & Schapira, A. H. V. (2016). The relationship between glucocerebrosidase mutations and Parkinson disease. *Journal of Neurochemistry, 139* (Suppl 1), 77–90. Available from https://doi.org/10.1111/jnc.13385.

Miller, D. W., Hague, S. M., Clarimon, J., Baptista, M., Gwinn-Hardy, K., Cookson, M. R., & Singleton, A. B. (2004). Alpha-synuclein in blood and brain from familial Parkinson disease with SNCA locus triplication. *Neurology, 62*, 1835–1838.

Miller, D. W., Johnson, J. M., Solano, S. M., Hollingsworth, Z. R., Standaert, D. G., & Young, A. B. (2005). Absence of alpha-synuclein mRNA expression in normal and multiple system atrophy oligodendroglia. *Journal of Neural Transmission (Vienna, Austria), 1996*(112), 1613–1624. Available from https://doi.org/10.1007/s00702-005-0378-1.

Mirza, B., Hadberg, H., Thomsen, P., & Moos, T. (2000). The absence of reactive astrocytosis is indicative of a unique inflammatory process in Parkinson's disease. *Neuroscience, 95*, 425–432.

Moehle, M. S., Webber, P. J., Tse, T., Sukar, N., Standaert, D. G., DeSilva, T. M., ... West, A. B. (2012). LRRK2 inhibition attenuates microglial inflammatory responses. *The Journal of Neuroscience: The Official Journal of the Society for Neuroscience, 32*, 1602–1611. Available from https://doi.org/10.1523/JNEUROSCI.5601-11.2012.

Mortiboys, H., Furmston, R., Bronstad, G., Aasly, J., Elliott, C., & Bandmann, O. (2015). UDCA exerts beneficial effect on mitochondrial dysfunction in LRRK2(G2019S) carriers and in vivo. *Neurology, 85*, 846–852. Available from https://doi.org/10.1212/WNL.0000000000001905.

Mortiboys, H., Johansen, K. K., Aasly, J. O., & Bandmann, O. (2010). Mitochondrial impairment in patients with Parkinson disease with the G2019S mutation in LRRK2. *Neurology, 75*, 2017–2020. Available from https://doi.org/10.1212/WNL.0b013e3181ff9685.

Mueller, J. C., Fuchs, J., Hofer, A., Zimprich, A., Lichtner, P., Illig, T., ... Gasser, T. (2005). Multiple regions of alpha-synuclein are associated with Parkinson's disease. *Annals of Neurology, 57*, 535–541. Available from https://doi.org/10.1002/ana.20438.

Nagle, M. W., Latourelle, J. C., Labadorf, A., Dumitriu, A., Hadzi, T. C., Beach, T. G., & Myers, R. H. (2016). The 4p16.3 Parkinson disease risk locus is associated with GAK expression and genes involved with the synaptic vesicle membrane. *PLoS ONE, 11*, e0160925. Available from https://doi.org/10.1371/journal.pone.0160925.

Nakamura, K., Nemani, V. M., Azarbal, F., Skibinski, G., Levy, J. M., Egami, K., ... Edwards, R. H. (2011). Direct membrane association drives mitochondrial fission by the Parkinson disease-associated protein alpha-synuclein. *Journal of Biological Chemistry, 286*, 20710–20726. Available from https://doi.org/10.1074/jbc.M110.213538.

Nalls, M. A., Pankratz, N., Lill, C. M., Do, C. B., Hernandez, D. G., Saad, M., ... Singleton, A. B. (2014). Large-scale meta-analysis of genome-wide association data identifies six new risk loci for Parkinson's disease. *Nature Genetics, 46*, 989–993. Available from https://doi.org/10.1038/ng.3043.

Narendra, D. P., Jin, S. M., Tanaka, A., Suen, D.-F., Gautier, C. A., Shen, J., ... Youle, R. J. (2010). PINK1 is selectively stabilized on impaired mitochondria to activate Parkin. *PLoS Biology, 8*, e1000298. Available from https://doi.org/10.1371/journal.pbio.1000298.

Nemani, V. M., Lu, W., Berge, V., Nakamura, K., Onoa, B., Lee, M. K., ... Edwards, R. H. (2010). Increased expression of alpha-synuclein reduces neurotransmitter release by inhibiting synaptic vesicle reclustering after endocytosis. *Neuron, 65*, 66–79. Available from https://doi.org/10.1016/j.neuron.2009.12.023.

Nichols, N., Bras, J. M., Hernandez, D. G., Jansen, I. E., Lesage, S., Lubbe, S., ... International Parkinson's Disease Genomics Consortium. (2015). EIF4G1 mutations do not cause Parkinson's disease. *Neurobiology of Aging, 36*, 2444.e1–2444.e4. Available from https://doi.org/10.1016/j.neurobiolaging.2015.04.017.

Oh, S. H., Kim, H. N., Park, H. J., Shin, J. Y., Bae, E.-J., Sunwoo, M. K., ... Lee, P. H. (2016). Mesenchymal stem cells inhibit transmission of α-synuclein by modulating clathrin-mediated endocytosis in a parkinsonian model. *Cell Reports, 14*, 835–849. Available from https://doi.org/10.1016/j.celrep.2015.12.075.

Olanow, C. W., Bartus, R. T., Volpicelli-Daley, L. A., & Kordower, J. H. (2015). Trophic factors for Parkinson's disease: To live or let die. *Movement Disorders: The official Journal of the International Parkinson and Movement Disorder Society, 30*, 1715–1724. Available from https://doi.org/10.1002/mds.26426.

Olgiati, S., Quadri, M., Fang, M., Rood, J. P. M. A., Saute, J. A., Chien, H. F., ... Bonifati, V. (2016). DNAJC6 mutations associated with early-onset Parkinson's disease. *Annals of Neurology, 79*, 244–256. Available from https://doi.org/10.1002/ana.24553.

Paisán-Ruíz, C., Jain, S., Evans, E. W., Gilks, W. P., Simón, J., van der Brug, M., ... Singleton, A. B. (2004). Cloning of the gene containing mutations that cause PARK8-linked Parkinson's disease. *Neuron, 44*, 595–600. Available from https://doi.org/10.1016/j.neuron.2004.10.023.

Paisán-Ruiz, C., Lewis, P. A., & Singleton, A. B. (2013). LRRK2: Cause, risk, and mechanism. *Journal of Parkinson's Disease, 3*, 85–103. Available from https://doi.org/10.3233/JPD-130192.

Park, J., Lee, S. B., Lee, S., Kim, Y., Song, S., Kim, S., ... Chung, J. (2006). Mitochondrial dysfunction in *Drosophila* PINK1 mutants is complemented by parkin. *Nature, 441*, 1157–1161. Available from https://doi.org/10.1038/nature04788.

Peeraully, T., & Tan, E. K. (2012). Genetic variants in sporadic Parkinson's disease: East vs West. *Parkinsonism & Related Disorders, 18*(Suppl 1), S63–S65. Available from https://doi.org/10.1016/S1353-8020(11)70021-9.

Piccoli, G., Condliffe, S. B., Bauer, M., Giesert, F., Boldt, K., De Astis, S., ... Ueffing, M. (2011). LRRK2 controls synaptic vesicle storage and mobilization within the recycling pool. *The Journal of Neuroscience: The Official Journal of the Society for Neuroscience, 31*, 2225–2237. Available from https://doi.org/10.1523/JNEUROSCI.3730-10.2011.

Pickrell, A. M., & Youle, R. J. (2015). The roles of PINK1, parkin, and mitochondrial fidelity in Parkinson's disease. *Neuron, 85*, 257–273. Available from https://doi.org/10.1016/j.neuron.2014.12.007.

Plowey, E. D., Cherra, S. J., Liu, Y.-J., & Chu, C. T. (2008). Role of autophagy in G2019S-LRRK2-associated neurite shortening in differentiated SH-SY5Y cells. *Journal of Neurochemistry, 105*, 1048–1056. Available from https://doi.org/10.1111/j.1471-4159.2008.05217.x.

Poehler, A.-M., Xiang, W., Spitzer, P., May, V. E. L., Meixner, H., Rockenstein, E., ... Klucken, J. (2014). Autophagy modulates SNCA/α-synuclein release, thereby generating a hostile microenvironment. *Autophagy, 10*, 2171–2192. Available from https://doi.org/10.4161/auto.36436.

Polymeropoulos, M. H., Lavedan, C., Leroy, E., Ide, S. E., Dehejia, A., Dutra, A., ... Nussbaum, R. L. (1997). Mutation in the alpha-synuclein gene identified in families with Parkinson's disease. *Science, 276*, 2045–2047.

Proukakis, C., Dudzik, C. G., Brier, T., MacKay, D. S., Cooper, J. M., Millhauser, G. L., ... Schapira, A. H. (2013). A novel α-synuclein missense mutation in Parkinson disease. *Neurology, 80*, 1062–1064. Available from https://doi.org/10.1212/WNL.0b013e31828727ba.

Ramirez, A., Heimbach, A., Gründemann, J., Stiller, B., Hampshire, D., Cid, L. P., ... Kubisch, C. (2006). Hereditary parkinsonism with dementia is caused by mutations in ATP13A2, encoding a lysosomal type 5 P-type ATPase. *Nature Genetics, 38*, 1184–1191. Available from https://doi.org/10.1038/ng1884.

Ramonet, D., Daher, J. P. L., Lin, B. M., Stafa, K., Kim, J., Banerjee, R., ... Moore, D. J. (2011). Dopaminergic neuronal loss, reduced neurite complexity and autophagic abnormalities in transgenic mice expressing G2019S mutant LRRK2. *PLoS ONE, 6*. Available from https://doi.org/10.1371/journal.pone.0018568.

Ross, O. A., Braithwaite, A. T., Skipper, L. M., Kachergus, J., Hulihan, M. M., Middleton, F. A., ... Farrer, M. J. (2008). Genomic investigation of alpha-synuclein multiplication and parkinsonism. *Annals of Neurology, 63*, 743–750. Available from https://doi.org/10.1002/ana.21380.

Russo, I., Berti, G., Plotegher, N., Bernardo, G., Filograna, R., Bubacco, L., & Greggio, E. (2015). Leucine-rich repeat kinase 2 positively regulates inflammation and down-regulates NF-κB p50 signaling in cultured microglia cells. *Journal of Neuroinflammation, 12*, 230. Available from https://doi.org/10.1186/s12974-015-0449-7.

Sacino, A. N., Brooks, M. M., Chakrabarty, P., Saha, K., Khoshbouei, H., Golde, T. E., & Giasson, B. I. (2016). Proteolysis of α-synuclein fibrils in the lysosomal pathway limits induction of inclusion pathology. *Journal of Neurochemistry*. Available from https://doi.org/10.1111/jnc.13743.

Saha, S., Ash, P. E. A., Gowda, V., Liu, L., Shirihai, O., & Wolozin, B. (2015). Mutations in LRRK2 potentiate age-related impairment of autophagic flux. *Molecular Neurodegeneration, 10*, 26. Available from https://doi.org/10.1186/s13024-015-0022-y.

Sahay, S., Ghosh, D., Singh, P. K., & Maji, S. K. (2016). Alteration of structure and aggregation of α-synuclein by familial Parkinson's disease associated mutations. *Current Protein & Peptide Science, 18*(112), 656–676.

Samaranch, L., Lorenzo-Betancor, O., Arbelo, J. M., Ferrer, I., Lorenzo, E., Irigoyen, J., ... Pastor, P. (2010). PINK1-linked parkinsonism is associated with Lewy body pathology. *Brain: A Journal of Neurology, 133*, 1128–1142. Available from https://doi.org/10.1093/brain/awq051.

Sampson, T. R., Debelius, J. W., Thron, T., Janssen, S., Shastri, G. G., Ilhan, Z. E., ... Mazmanian, S. K. (2016). Gut microbiota regulate motor deficits and neuroinflammation in a model of Parkinson's disease. *Cell, 167*, 1469–1480.e12. Available from https://doi.org/10.1016/j.cell.2016.11.018.

Satake, W., Nakabayashi, Y., Mizuta, I., Hirota, Y., Ito, C., Kubo, M., ... Toda, T. (2009). Genome-wide association study identifies common variants at four loci as genetic risk factors for Parkinson's disease. *Nature Genetics, 41*, 1303–1307. Available from https://doi.org/10.1038/ng.485.

Schapansky, J., Nardozzi, J. D., Felizia, F., & Lavoie, M. J. (2014). Membrane recruitment of endogenous LRRK2 precedes its potent regulation of autophagy. *Human Molecular Genetics*. Available from https://doi.org/10.1093/hmg/ddu138.

Scholz, S. W., Houlden, H., Schulte, C., Sharma, M., Li, A., Berg, D., ... Gasser, T. (2009). SNCA variants are associated with increased risk for multiple system atrophy. *Annals of Neurology, 65*, 610–614. Available from https://doi.org/10.1002/ana.21685.

Seidel, K., Schöls, L., Nuber, S., Petrasch-Parwez, E., Gierga, K., Wszolek, Z., ... Krüger, R. (2010). First appraisal of brain pathology owing to A30P mutant alpha-synuclein. *Annals of Neurology, 67*, 684–689. Available from https://doi.org/10.1002/ana.21966.

Shiner, T., Mirelman, A., Gana Weisz, M., Bar-Shira, A., Ash, E., Cialic, R., ... Giladi, N. (2016). High frequency of GBA gene mutations in dementia with Lewy bodies Among Ashkenazi Jews. *JAMA Neurology, 73*, 1448–1453. Available from https://doi.org/10.1001/jamaneurol.2016.1593.

Sidransky, E., Nalls, M. A., Aasly, J. O., Aharon-Peretz, J., Annesi, G., Barbosa, E. R., ... Ziegler, S. G. (2009). Multicenter analysis of glucocerebrosidase mutations in Parkinson's disease. *The New England Journal of Medicine, 361*, 1651–1661. Available from https://doi.org/10.1056/NEJMoa0901281.

Simon-sanchez, J., Schulte, C., Bras, J. M., Sharma, M., Gibbs, R., Berg, D., ... Foote, K. D. (2010). *Genome-wide association study reveals genetic risk underlying Parkinson's disease.* Nature Genetics, *41*, 1308–1312. Available from https://doi.org/10.1038/ng.487.Genome-Wide.

Singleton, A., & Hardy, J. (2011). A generalizable hypothesis for the genetic architecture of disease: Pleomorphic risk loci. *Human Molecular Genetics, 20*, R158–R162. Available from https://doi.org/10.1093/hmg/ddr358.

Singleton, A., & Hardy, J. (2016). The evolution of genetics: Alzheimer's and Parkinson's diseases. *Neuron, 90*, 1154–1163. Available from https://doi.org/10.1016/j.neuron.2016.05.040.

Singleton, A. B., Farrer, M., Johnson, J., Singleton, A., Hague, S., Kachergus, J., ... Gwinn-Hardy, K. (2003). α-Synuclein locus triplication causes Parkinson's disease. *Science, 302*, 841.

Song, Y. J. C., Halliday, G. M., Holton, J. L., Lashley, T., O'Sullivan, S. S., McCann, H., ... Revesz, T. R. (2009). Degeneration in different parkinsonian syndromes relates to astrocyte type and astrocyte protein expression. *Journal of Neuropathology & Experimental Neurology, 68*, 1073–1083. Available from https://doi.org/10.1097/NEN.0b013e3181b66f1b.

Soper, J. H., Kehm, V., Burd, C. G., Bankaitis, V. A., & Lee, V. M.-Y. (2011). Aggregation of α-synuclein in *S. cerevisiae* is associated with defects in endosomal trafficking and phospholipid biosynthesis. *Journal of Molecular Neuroscience, 43*, 391–405. Available from https://doi.org/10.1007/s12031-010-9455-5.

Soukup, S.-F., Kuenen, S., Vanhauwaert, R., Manetsberger, J., Hernández-Díaz, S., Swerts, J., ... Verstreken, P. (2016). A LRRK2-dependent EndophilinA phosphoswitch is critical for macroautophagy at presynaptic terminals. *Neuron, 92*, 829–844. Available from https://doi.org/10.1016/j.neuron.2016.09.037.

Spencer, B., Emadi, S., Desplats, P., Eleuteri, S., Michael, S., Kosberg, K., ... Masliah, E. (2014). ESCRT-mediated uptake and degradation of brain-targeted α-synuclein single chain antibody attenuates neuronal degeneration in vivo. *Molecular Therapy: The Journal of the American Society of Gene Therapy American Society of Gene Therapy, 22*, 1753–1767. Available from https://doi.org/10.1038/mt.2014.129.

Spillantini, M. G., Schmidt, M. L., Lee, V. M., Trojanowski, J. Q., Jakes, R., & Goedert, M. (1997). Alpha-synuclein in Lewy bodies. *Nature, 388*, 839–840. Available from https://doi.org/10.1038/42166.

Spira, P. J., Sharpe, D. M., Halliday, G., Cavanagh, J., & Nicholson, G. A. (2001). Clinical and pathological features of a Parkinsonian syndrome in a family with an Ala53Thr alpha-synuclein mutation. *Annals of Neurology, 49*, 313–319.

Stamelou, M., & Bhatia, K. P. (2016). Atypical parkinsonism—New advances. *Current Opinion in Neurology, 29*, 480–485. Available from https://doi.org/10.1097/WCO.0000000000000355.

Steger, M., Tonelli, F., Ito, G., Davies, P., Trost, M., Vetter, M., ... Mann, M. (2016). Phosphoproteomics reveals that Parkinson's disease kinase LRRK2 regulates a subset of Rab GTPases. *eLife, 5*. Available from https://doi.org/10.7554/eLife.12813.

Su, X., Maguire-Zeiss, Ka, Giuliano, R., Prifti, L., Venkatesh, K., & Federoff, H. J. (2008). Synuclein activates microglia in a model of Parkinson's disease. *Neurobiology of Aging, 29*, 1690–1701. Available from https://doi.org/10.1016/j.neurobiolaging.2007.04.006.

Su, Y.-C., Guo, X., & Qi, X. (2015). Threonine 56 phosphorylation of Bcl-2 is required for LRRK2 G2019S-induced mitochondrial depolarization and autophagy. *Biochimica et Biophysica Acta, 1852*, 12–21. Available from https://doi.org/10.1016/j.bbadis.2014.11.009.

Su, Y.-C., & Qi, X. (2013). Inhibition of excessive mitochondrial fission reduced aberrant autophagy and neuronal damage caused by LRRK2 G2019S mutation. *Human Molecular Genetics, 22*, 4545–4561. Available from https://doi.org/10.1093/hmg/ddt301.

Suzuki, H., Osawa, T., Fujioka, Y., & Noda, N. N. (2016). Structural biology of the core autophagy machinery. *Current Opinion in Structural Biology, 43*, 10–17. Available from https://doi.org/10.1016/j.sbi.2016.09.010.

Suzuki, M., Fujikake, N., Takeuchi, T., Kohyama-Koganeya, A., Nakajima, K., Hirabayashi, Y., ... Nagai, Y. (2015). Glucocerebrosidase deficiency accelerates the accumulation of proteinase K-resistant α-synuclein and aggravates neurodegeneration in a *Drosophila* model of Parkinson's disease. *Human Molecular Genetics, 24*, 6675–6686. Available from https://doi.org/10.1093/hmg/ddv372.

Taymans, J.-M., & Cookson, M. R. (2010). Mechanisms in dominant parkinsonism: The toxic triangle of LRRK2, alpha-synuclein, and tau. *BioEssays: News and Reviews in Molecular, Cellular and Developmental Biology, 32*, 227–235. Available from https://doi.org/10.1002/bies.200900163.

Terada, S., & Hirokawa, N. (2000). Moving on to the cargo problem of microtubule-dependent motors in neurons. *Current Opinion in Neurobiology, 10*, 566–573.

Tong, Y., Giaime, E., Yamaguchi, H., Ichimura, T., Liu, Y., Si, H., ... Shen, J. (2012). Loss of leucine-rich repeat kinase 2 causes age-dependent bi-phasic alterations of the autophagy pathway. *Molecular Neurodegeneration, 7*, 2. Available from https://doi.org/10.1186/1750-1326-7-2.

Tong, Y., Yamaguchi, H., Giaime, E., Boyle, S., Kopan, R., Kelleher, R. J., & Shen, J. (2010). Loss of leucine-rich repeat kinase 2 causes impairment of protein degradation pathways, accumulation of α-synuclein, and apoptotic cell death in aged mice. *Proceedings of the National Academy of Sciences of the United States of America, 107*, 9879–9884. Available from https://doi.org/10.1073/pnas.1004676107.

Troiano, A. R., Cazeneuve, C., Le Ber, I., Bonnet, A.-M., Lesage, S., & Brice, A. (2008). Re: Alpha-synuclein gene duplication is present in sporadic Parkinson disease. *Neurology, 71*, 1295. Available from https://doi.org/10.1212/01.wnl.0000338435.78120.0f, author reply 1295.

Uchiyama, T., Ikeuchi, T., Ouchi, Y., Sakamoto, M., Kasuga, K., Shiga, A., ... Ohashi, T. (2008). Prominent psychiatric symptoms and glucose hypometabolism in a family with a SNCA duplication. *Neurology, 71*, 1289–1291. Available from https://doi.org/10.1212/01.wnl.0000327607.28928.e6.

Usenovic, M., Tresse, E., Mazzulli, J. R., Taylor, J. P., & Krainc, D. (2012). Deficiency of ATP13A2 leads to lysosomal dysfunction, α-synuclein accumulation, and neurotoxicity. *The Journal of Neuroscience, 32*, 4240–4246. Available from https://doi.org/10.1523/JNEUROSCI.5575-11.2012.

Vekrellis, K., & Stefanis, L. (2012). Targeting intracellular and extracellular alpha-synuclein as a therapeutic strategy in Parkinson's disease and other synucleinopathies. *Expert Opinion on Therapeutic Targets, 16*, 421–432. Available from https://doi.org/10.1517/14728222.2012.674111.

Vilariño-Güell, C., Rajput, A., Milnerwood, A. J., Shah, B., Szu-Tu, C., Trinh, J., ... Farrer, M. J. (2014). DNAJC13 mutations in Parkinson disease. *Human Molecular Genetics, 23*, 1794–1801. Available from https://doi.org/10.1093/hmg/ddt570.

Vilariño-Güell, C., Wider, C., Ross, O. A., Dachsel, J. C., Kachergus, J. M., Lincoln, S. J., ... Farrer, M. J. (2011). VPS35 mutations in Parkinson disease. *American Journal of Human Genetics, 89*, 162–167. Available from https://doi.org/10.1016/j.ajhg.2011.06.001.

Villar-Piqué, A., Lopes da Fonseca, T., Sant'Anna, R., Szegö, É. M., Fonseca-Ornelas, L., Pinho, R., ... Outeiro, T. F. (2016). Environmental and genetic factors support the dissociation between α-synuclein aggregation and toxicity. *Proceedings of the National Academy of Sciences of the United States of America, 113*, E6506–E6515. Available from https://doi.org/10.1073/pnas.1606791113.

Visanji, N., & Marras, C. (2015). The relevance of pre-motor symptoms in Parkinson's disease. *Expert Review of Neurotherapeutics, 15*, 1205–1217. Available from https://doi.org/10.1586/14737175.2015.1083423.

Volakakis, N., Tiklova, K., Decressac, M., Papathanou, M., Mattsson, B., Gillberg, L., ... Perlmann, T. (2015). Nurr1 and retinoid X receptor ligands stimulate ret signaling in dopamine neurons and can alleviate α-synuclein disrupted gene expression. *The Journal of Neuroscience: The Official Journal of the Society for Neuroscience, 35*, 14370–14385. Available from https://doi.org/10.1523/JNEUROSCI.1155-15.2015.

Volpicelli-Daley, L. A. (2016). Effects of α-synuclein on axonal transport. *Neurobiology of Disease*. Available from https://doi.org/10.1016/j.nbd.2016.12.008.

Volpicelli-Daley, L. A., Abdelmotilib, H., Liu, Z., Stoyka, L., Daher, J. P. L., Milnerwood, A. J., ... West, A. B. (2016). G2019S-LRRK2 expression augments α-synuclein sequestration into inclusions in neurons. *The Journal of Neuroscience: The Official Journal of the Society for Neuroscience, 36*, 7415–7427. Available from https://doi.org/10.1523/JNEUROSCI.3642-15.2016.

Volta, M., Milnerwood, A. J., & Farrer, M. J. (2015). Insights from late-onset familial parkinsonism on the pathogenesis of idiopathic Parkinson's disease. *The Lancet*

Neurology, 14, 1054–1064. Available from https://doi.org/10.1016/S1474-4422(15)00186-6.

Wang, W., Perovic, I., Chittuluru, J., Kaganovich, A., Nguyen, L. T. T., Liao, J., ... Hoang, Q. Q. (2011). A soluble α-synuclein construct forms a dynamic tetramer. *Proceedings of the National Academy of Sciences of the United States of America, 108*, 17797–17802. Available from https://doi.org/10.1073/pnas.1113260108.

Wang, X., Yan, M. H., Fujioka, H., Liu, J., Wilson-Delfosse, A., Chen, S. G., ... Zhu, X. (2012). LRRK2 regulates mitochondrial dynamics and function through direct interaction with DLP1. *Human Molecular Genetics, 21*, 1931–1944. Available from https://doi.org/10.1093/hmg/dds003.

Watson, M. B., Richter, F., Lee, S. K., Gabby, L., Wu, J., Masliah, E., ... Chesselet, M.-F. (2012). Regionally-specific microglial activation in young mice over-expressing human wildtype alpha-synuclein. *Experimental Neurology, 237*, 318–334. Available from https://doi.org/10.1016/j.expneurol.2012.06.025.

Willingham, S., Outeiro, T. F., DeVit, M. J., Lindquist, S. L., & Muchowski, P. J. (2003). Yeast genes that enhance the toxicity of a mutant huntingtin fragment or alpha-synuclein. *Science, 302*, 1769–1772. Available from https://doi.org/10.1126/science.1090389.

Xie, W., & Chung, K. K. K. (2012). Alpha-synuclein impairs normal dynamics of mitochondria in cell and animal models of Parkinson's disease. *Journal of Neurochemistry, 122*, 404–414. Available from https://doi.org/10.1111/j.1471-4159.2012.07769.x.

Xu, J., Wu, X.-S., Sheng, J., Zhang, Z., Yue, H.-Y., Sun, L., ... Wu, L.-G. (2016). α-Synuclein mutation inhibits endocytosis at mammalian central nerve terminals. *The Journal of Neuroscience: The Official Journal of the Society for Neuroscience, 36*, 4408–4414. Available from https://doi.org/10.1523/JNEUROSCI.3627-15.2016.

Youle, R. J., & Narendra, D. P. (2011). Mechanisms of mitophagy. *Nature Reviews Molecular Cell Biology, 12*, 9–14. Available from https://doi.org/10.1038/nrm3028.

Yue, M., Hinkle, K. M., Davies, P., Trushina, E., Fiesel, F. C., Christenson, T. A., ... Melrose, H. L. (2015). Progressive dopaminergic alterations and mitochondrial abnormalities in LRRK2 G2019S knock-in mice. *Neurobiology of Disease, 78*, 172–195. Available from https://doi.org/10.1016/j.nbd.2015.02.031.

Zarranz, J. J., Alegre, J., Gómez-Esteban, J. C., Lezcano, E., Ros, R., Ampuero, I., ... de Yebenes, J. G. (2004). The new mutation, E46K, of alpha-synuclein causes Parkinson and Lewy body dementia. *Annals of Neurology, 55*, 164–173. Available from https://doi.org/10.1002/ana.10795.

Zerial, M., & McBride, H. (2001). Rab proteins as membrane organizers. *Nature Reviews Molecular Cell Biology, 2*, 107–117. Available from https://doi.org/10.1038/35052055.

Zhang, Q., Pan, Y., Yan, R., Zeng, B., Wang, H., Zhang, X., ... Liu, Z. (2015). Commensal bacteria direct selective cargo sorting to promote symbiosis. *Nature Immunology, 16*, 918–926. Available from https://doi.org/10.1038/ni.3233.

Zhang, W., Gao, J.-H., Yan, Z.-F., Huang, X.-Y., Guo, P., Sun, L., ... Hong, J.-S. (2016). Minimally toxic dose of lipopolysaccharide and α-synuclein oligomer elicit synergistic dopaminergic neurodegeneration: Role and mechanism of microglial NOX2 activation. *Molecular Neurobiology*. Available from https://doi.org/10.1007/s12035-016-0308-2.

Zimprich, A., Biskup, S., Leitner, P., Lichtner, P., Farrer, M., Lincoln, S., ... Gasser, T. (2004). Mutations in LRRK2 cause autosomal-dominant parkinsonism with pleomorphic pathology. *Neuron, 44*, 601–607. Available from https://doi.org/10.1016/j.neuron.2004.11.005.

Zuccato, C., & Cattaneo, E. (2007). Role of brain-derived neurotrophic factor in Huntington's disease. *Progress in Neurobiology, 81*, 294–330. Available from https://doi.org/10.1016/j.pneurobio.2007.01.003.

CHAPTER

7

Huntington's Disease and Other Polyglutamine Repeat Diseases: Molecular Mechanisms and Pathogenic Pathways

Audrey S. Dickey and Albert R. La Spada
Duke University School of Medicine, Durham, NC, United States

OUTLINE

Polyglutamine (polyQ) diseases are inherited, fatal neurodegenerative disorders caused by an expansion of the three nucleotides cytosine–adenine–guanine (CAG) repeated multiple times. CAG codes for the amino acid glutamine, so these repeats result in long stretches of glutamine residues (called polyQ tracts) in the encoded proteins. In 1991, one of us provided the first evidence that expanded CAG trinucleotide repeats can cause a human disease by discovering that the neurodegenerative disorder X-linked spinal and bulbar muscular atrophy (SBMA) results from an abnormally long CAG repeat tract in the androgen receptor (AR) gene (La Spada, Wilson, Lubahn, Harding, & Fischbeck, 1991). After this

The Molecular and Cellular Basis of Neurodegenerative Diseases
DOI: https://doi.org/10.1016/B978-0-12-811304-2.00007-9

discovery, other investigators identified the identical mutation, expansion of a CAG/polyQ repeat, as the cause of eight other inherited neurodegenerative disorders: Huntington's disease (HD), dentatorubral–pallidoluysian atrophy (DRPLA), and spinocerebellar ataxia (SCA) types 1, 2, 3, 6, 7, and 17. With the exception of SBMA, which is X-linked, the polyQ diseases all exhibit autosomal inheritance (Orr & Zoghbi, 2007), though SBMA also involves a dominantly acting mutation with females protected due to their low levels of circulating testosterone, making it a "sex-limited" disorder. Because all nine inherited neurodegenerative disorders have expanded polyQ tracts in the disease proteins, they are collectively known as the "polyQ repeat diseases" (Table 7.1).

POLYGLUTAMINE EXPANSION

Years of research suggest that disease-length CAG repeats can be further expanded during DNA replication, resulting in a phenomenon called genetic instability. This characteristic instability of the CAG repeats can be caused by mistakes that occur during DNA replication, including strand slippage, misalignment, and stalling (Mirkin, 2007). Single-strand loops or hairpins can form in repeat-containing DNA, allowing the two strands to be displaced (i.e., slippage). This can result in the addition of trinucleotides to the newly synthesized DNA strand, causing the number of CAG repeats to increase. The severity of the disease can worsen from one generation to the next as the length of the disease-causing CAG repeat expands, because with the increased repeat number, there is a worsening in disease severity and earlier disease onset. This phenomenon is called "anticipation." For example, in HD, the median age of onset decreases from 67 years for patients with 39 CAGs to 27 years for patients with 50 CAGs (Brinkman, Mezei, Theilmann, Almqvist, & Hayden, 1997).

Although the genes associated with SBMA, HD, DRPLA, and SCA 1, 2, 3, 6, 7, and 17 are

TABLE 7.1 The Polyglutamine Repeat Diseases

Polyglutamine Disorders	PolyQ Protein	Gene	Major Clinical Signs
Huntington's disease	Huntingtin	HTT (IT15)	Chorea, dementia
Dentatorubral–pallidoluysian atrophy	Atrophin 1	DRPLA	Ataxia, myoclonic epilepsy, dementia
Spinal and bulbar muscular atrophy (Kennedy's disease)	Androgen receptor	AR	Muscle weakness, lower motor neuron disease
Spinocerebellar ataxia 1	Ataxin-1	SCA1	Ataxia, eye movement abnormalities
Spinocerebellar ataxia 2	Ataxin-2	SCA2	Ataxia
Spinocerebellar ataxia 3 (Machado–Joseph disease)	Ataxin-3	SCA3/MJD	Ataxia, motor neuron disease
Spinocerebellar ataxia 6	α1A-voltage-dependent Ca^{2+} channel subunit	CACNA1A	Ataxia
Spinocerebellar ataxia 7	Ataxin-7	SCA7	Ataxia and vision loss
Spinocerebellar ataxia 17	TATA box binding protein	TBP	Ataxia, seizures

DRPLA, dentatorubral–pallidoluysian atrophy; *AR*, androgen receptor; *SCA*, spinocerebellar ataxia; *TBP*, TATA-binding protein.

structurally and functionally distinct, the polyQ diseases share certain key characteristics. The diseases are all progressive, often fatal disorders that typically begin in adulthood and worsen over a period of 10–30 years. The diseases occur only when the length of the CAG repeat tract exceeds a threshold number specific to the particular gene, typically in the low to mid 30s, with SCA6 a notable exception. Research indicates the shared expansion of polyQ tracts leads to some shared neurotoxic pathways in the different diseases. While the region of the brain that is affected differs with each disease, it appears that cell toxicity requires that toxic polyQ disease proteins misfold into assemblies variably called oligomers, fibrils, or aggregates. Nuclear and cytoplasmic aggregates are detected in all polyQ diseases and typically contain fragments of the respective disease proteins, ubiquitin, heat shock chaperones, and proteasome subunits, and several other important regulatory proteins, often transcription factors (Havel, Li, & Li, 2009). The recruitment of ubiquitin, heat shock proteins, and proteasomal subunits into these aggregates implies that protein quality control mechanisms such as the ubiquitin-proteasome system (UPS) attempt to counter polyQ disease pathogenesis, but with limited success (Chai, Koppenhafer, Bonini, & Paulson, 1999). Although the vast majority of polyQ disease proteins are ubiquitously expressed and can be detected throughout development and in adulthood, disease pathology is most prominent in neurons, with involvement of other central nervous system (CNS) cell types, such as glia and microglia (Cohen-Carmon & Meshorer, 2012). Neurons are challenged by environmental stress, physiological stress, and age-related homeostasis decompensation, and constantly require high energy levels, depending on oxidative energy metabolism which challenges this cell type with reactive oxygen species accumulation (Falkowska et al., 2015). Hence, mitochondrial dysfunction appears to be a common thread in the neurodegenerative disease cascade in polyQ disorders (Schon & Manfredi, 2003). Here, we discuss the pathobiology of the polyQ repeat expansion diseases, focusing on nuclear aggregation, intracellular protein degradation systems, proteolytic processing, mitochondrial dysfunction, nuclear trafficking/subcellular localization, and the role of posttranslational modifications, with HD serving as the prototype polyQ disease where all of these processes have been implicated. For a presentation of the clinical features of these disorders, we refer the reader to a recent review and the specific references therein (Fan, Ho, & Chi, 2014).

Aggregation

Expanded polyQ tracts tend to interact avidly with one another, so the disease-associated proteins are prone to aggregation. Aggregates were identified as intranuclear inclusions in SCA3 patients (Becher, Kotzuk, & Sharp, 1998), and then in mouse models of HD (Davies, Turmaine, & Cozens, 1997), and subsequently were confirmed in HD patients (Becher et al., 1998; DiFiglia, Sapp, & Chase, 1997). This was quickly followed by an identification of aggregates containing the polyQ-expanded disease protein in all of the other polyQ disorders. As a pathological hallmark of polyQ disease (Orr & Zoghbi, 2007), aggregation has been widely discussed as a therapeutic target, although the exact role of aggregates in the neurodegeneration observed in polyQ diseases is still under debate, and inclusions may sometimes fulfill a protective role (Arrasate, Mitra, Schweitzer, Segal, & Finkbeiner, 2004; Klement, Skinner, & Kaytor, 1998; Saudou, Finkbeiner, Devys, & Greenberg, 1998). Researchers studying polyQ proteins in vitro have learned that proteins with longer polyQ repeats aggregate more readily and promote increased cellular toxicity (Hackam, Singaraja, & Wellington, 1998). These data led to the hypothesis that the tendency of the disease proteins to aggregate is predictor of neurodegeneration. This mechanism is consistent with observations from other, more common, neurodegenerative disorders, including Alzheimer's disease (AD), Parkinson's disease (PD), and tauopathies, which are associated with the accumulation of aggregation-prone proteins.

Large fibrillar inclusions are most likely the end stage of protein nucleation and aggregation, as the initial steps in the pathogenic cascade likely begin with monomeric insoluble species that form oligomers, then protofibrils, and ultimately fibrils. However, the concentration of these intermediates may not be similar, as the conversion process may favor accumulation of certain types of misfolded species. Some may be pathological intermediates, while others may not be directly relevant to the inclusion formation seen in patients (Kodali & Wetzel, 2007). As we will discuss, proteolytic cleavage may play a major role in the kinetics of aggregation and in the mechanisms of toxicity. In the various cell models, polyQ-containing fragments or polyQ stretches themselves can form soluble oligomeric structures, which likely mediate cytotoxicity and may represent a starting point for subsequent aggregation (Iuchi, Hoffner, Verbeke, Djian, & Green, 2003; Legleiter, Mitchell, & Lotz, 2010; Takahashi et al., 2008). These oligomeric species have been identified in brain tissue from HD mouse models and HD patients (Legleiter et al., 2010; Sathasivam, Lane, & Legleiter, 2010). Assays are being developed to study monomer addition, and isolate specific aggregate species as targets for therapeutic intervention (Jayaraman, Thakur, Kar, Kodali, & Wetzel, 2011), but this work is still early on in terms of viable treatment strategies. The conversion of monomers to oligomers in HD has been described as a packing of the amino-terminal Htt segment into the oligomer core, followed by fibril elongation, and finally a third step involving the seeding of further monomer aggregation (Jayaraman et al., 2011).

Ataxin-3 also has a multistep aggregation process, where the first step involves protein aggregation independent of the polyQ tract and a second step is unique to the polyQ expansion tract, producing highly stable amyloid-like aggregates (Ellisdon, Thomas, & Bottomley, 2006). Kinetic differences between nucleation and protein folding in the nucleus versus the cytoplasm likely accounts for the observed differences seen in the nuclear aggregates in SCA1 and SCA3 and the cytoplasmic inclusions detected in SCA2 and SCA6 (Stohr, 2012). For HD, Htt protein aggregation can be influenced by various inhibitors, molecules, and interactions (Jayaraman, Mishra, & Kodali, 2012), suggesting multiple parallel pathways. Inhibiting each pathway may thus have different effects on neurotoxicity. The same was shown for ataxin-3 where different amyloid aggregates affect Ca^{2+} regulation by different mechanisms (Pellistri, Bucciantini, & Invernizzi, 2013). Altering the specific pathways of aggregation is a potential therapeutic strategy, which may not decrease the total amount of aggregation but could decrease neurotoxicity.

A widely discussed topic is how misfolded protein accumulations exert toxicity. Examination of the specific location of neuronal inclusions in patient brains revealed a disconnect between inclusion frequency and tendency to degenerate (Kuemmerle, Gutekunst, & Klein, 1999). In HD, for example, the medium spiny neurons are selectively lost but exhibit much less aggregation at the light microscope level than the large striatal interneurons (Kuemmerle et al., 1999). Hence, while initial correlations between aggregation and disease pathology suggested that aggregation directly contributed to pathology, current opinion favors a model in which large aggregates can be neuroprotective by reducing the levels of diffuse toxic mutant Htt protein in the soluble phase (Arrasate et al., 2004; Saudou et al., 1998). However, aggregates may ultimately contribute to the further demise of the neuron of CNS cell by sequestering vital proteins/enzymes (Arrasate & Finkbeiner, 2012; Todd & Lim, 2013). Studies of SCA1 reinforce this complicated relationship between aggregates and cytotoxicity, as patients with a histidine interruption in the expanded polyQ tract of ataxin-1 have a decreased amount of aggregation and do not develop disease (Sen, Dash, Pasha, & Brahmachari, 2003), indicating that the capacity to aggregate is what confers toxic potential to a polyQ disease protein. However, all aggregate species should not be lumped into one umbrella category, as aggregates visible at the light microscope level may be less relevant to the onset of disease symptoms than oligomers or so-called micro-aggregates that are only detected by sophisticated imaging or biophysical techniques (Bucciantini, Giannoni, & Chiti, 2002; Weiss, Klein, & Woodman, 2008).

One pathological feature of SCA7 that is commonly observed in neurodegenerative diseases associated with polyQ-expanded proteins is the presence of nuclear inclusions (NIs). Mutant ataxin-7 has been localized to NIs, together with a host of other proteins including ubiquitin-proteasome components (Holmberg, Duyckaerts, & Durr, 1998; Takahashi, Fujigasaki, & Zander, 2002). Several studies of SCA7 neuropathology in patients and in mice have delineated the pattern of NI formation (Einum, Townsend, Ptacek, & Fu, 2001; Garden, Libby, & Fu, 2002; Holmberg et al., 1998; La Spada, Fu, & Sopher, 2001; Takahashi et al., 2002; Yvert et al., 2001). While extensive NI formation in SCA7 is present in exquisitely vulnerable neuronal populations, such as the inferior olivary complex and basis pontis, the presence of NIs extends to other areas of the neuraxis—e.g., cerebral cortex—that are usually not subject to prominent degeneration (Holmberg et al., 1998; Lindenberg, Yvert, Muller, & Landwehrmeyer, 2000). By documenting a disconnection between NI formation and neuronal pathology, studies on SCA7 patient material contributed to an emerging view that inclusion body formation does not predict patterns of neuronal loss in polyQ diseases.

The role of aggregates as toxic or protective is relevant for the screening of large libraries of therapeutic compounds or genetic modifiers. Using aggregation as a readout is intuitive, since if any step of the aggregation pathway is toxic, then reducing the eventual product should reduce the intermediate toxic species. But, blocking the conversion of toxic oligomeric species to possible "beneficial" aggregates could actually do more harm than good (Nekooki-Machida et al., 2009). Another counter-intuitive possibility would be to increase the overall rate of aggregation to decrease the concentration of presumed toxic intermediates. Targeting the depletion of specific oligomeric or protofibrillar species with antibodies or upregulated clearance is thus a therapeutic option. Such an intervention might stem the prion-like spread of smaller fibrils and oligomers (Ren et al., 2009; Yang, Dunlap, Andrews, & Wetzel, 2002), since recent studies indicate that extracellular aggregate transmission can occur in polyQ diseases, both in vitro and in vivo, between neurons and glia, and from neuron to neuron via synaptic terminals (Babcock & Ganetzky, 2015; Pearce, Spartz, Hong, Luo, & Kopito, 2015).

Clearance of PolyQ Proteins

Because polyQ proteins form intracellular aggregates, clearance mechanisms and protein turnover pathways have been extensively investigated. The two main clearance routes for proteins in eukaryotic cells are the heat shock pathway/UPS and the autophagy (Fig. 7.1). While proteasomes predominantly degrade short-lived nuclear and cytoplasmic proteins as well as misfolded and unfolded proteins from the endoplasmic reticulum, autophagy can degrade organelles and large macromolecular complexes (Ciechanover, 2006; Yorimitsu & Klionsky, 2005).

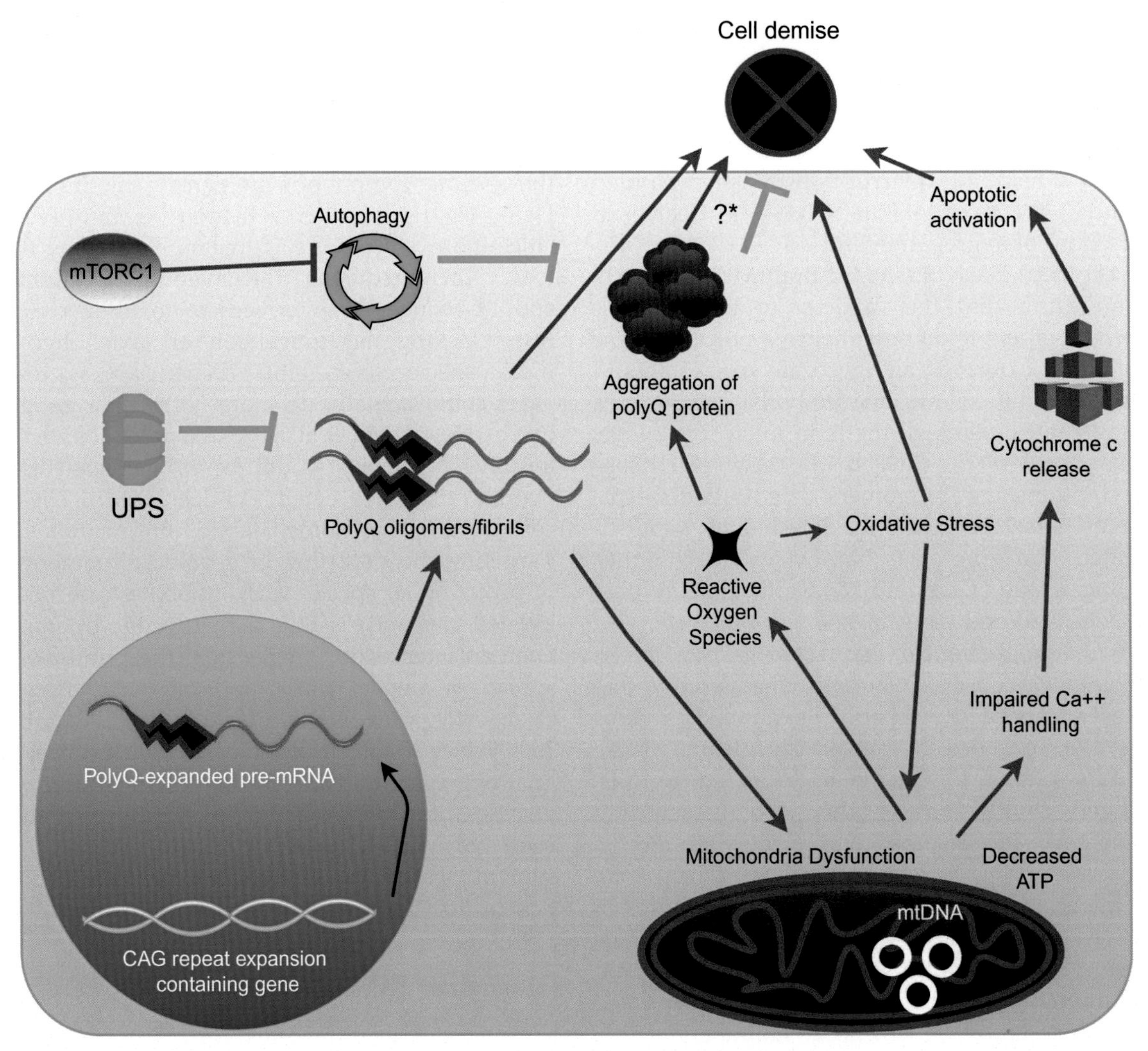

FIGURE 7.1 **Cellular pathways in polyglutamine disease pathogenesis.** PolyQ-expanded proteins cause multiple cellular pathologies (with affected organelle or process shown in red): decreased ATP production, impaired Ca^{2+} handing, and interruption of autophagy, which can increase aggregation of polyQ proteins. The UPS, autophagy, and certain types of aggregates (shown in green) can reduce levels of the toxic PolyQ-expanded proteins. *PolyQ*, polyglutamine; *UPS*, ubiquitin-proteasome system.

Impairment of the Ubiquitin-Proteasome System

The proteasome is a protein complex that degrades unneeded or damaged proteins. Proteins are targeted for proteasome-mediated degradation when ubiquitin is attached to them by enzymes called ubiquitin ligases. The coordination of heat shock proteins, chaperones, and the UPS is important for protein clearance (Wickner, Maurizi, & Gottesman, 1999). During oxidative or cellular stress, heat shock proteins are dramatically upregulated.

They bind to misfolded proteins and remodel them back to their native formation. If refolding is not possible, degradation by the proteasome is initiated. Failure in one of these systems can be compensated partially by the upregulation of the other, but prolonged failure results in protein aggregation and dysfunctional homeostasis (Kastle & Grune, 2012). Ataxin-3 (Doss-Pepe, Stenroos, Johnson, & Madura, 2003; Wang, Sawai, Kotliarova, & Kanazawa, 2000), ataxin-1 (Davidson et al., 2000; Hong, Kim, Ka, Choi, & Kang, 2002; Hong, Lee, Cho, & Kang, 2008), ataxin-7 (Matilla, Gorbea, & Einum, 2001), and huntingtin (Kalchman, Graham, & Xia, 1996) have all been shown to interact with components of the UPS under normal conditions.

Molecular chaperones and proteasome subunits are abundant within polyQ aggregates and thus may represent a cellular attempt to recruit defense mechanisms against protein misfolding and aggregation, likely in a failed attempt at degradation (Bence, Sampat, & Kopito, 2001; Jana, Zemskov, Wang, & Nukina, 2001). Involvement of the proteasome is supported by immunohistochemical studies showing that neuronal inclusions are highly ubiquitin-positive (Woulfe, 2008). A number of polyQ diseases have been associated with decreased chaperone and proteasome activity in patient samples, cell models, and animal models, with compelling data published for SCA1, SCA3, SCA7, SCA17, and HD (Diaz-Hernandez, Valera, & Moran, 2006; Jana et al., 2001; Khan et al., 2006; Park, Hong, Kim, & Kang, 2005; Seo, Sonntag, & Isacson, 2004; k; Yang, Huang, Gaertig, Li, & Li, 2014). In HD patients and animal models, ubiquitin, molecular chaperones including GRP78/BiP, HSP70, and HSP40, and the 20S, 19S, and 11S subunits of the 26S proteasome are found in aggregates alongside amino-terminal Htt (DiFiglia et al., 1997; Waelter, Boeddrich, & Lurz, 2001) (reviewed in Arrasate & Finkbeiner, 2012). Similar results have been described for SCA1 (Cummings et al., 1998), SCA3 (Schmidt, Lindenberg, & Krebs, 2002), and SCA7 (Matilla et al., 2001; Zander, Takahashi, & El Hachimi, 2001), although one study of a SCA7 knock-in mouse model yielded no significant impairment of the UPS for unknown reasons (Bowman, Yoo, Dantuma, & Zoghbi, 2005).

All of these observations suggest that polyQ-expanded proteins sequester and impair the UPS, or that polyQ expansions tracts block the proteasome, thus preventing entry of other substates (Chai et al., 1999; Paulson, Perez, & Trottier, 1997). HD transgenic mice and human HD postmortem brain do display decreased in vitro proteasome activity (Diaz-Hernandez et al., 2006). In addition, certain ataxin proteins have been shown to normally interact with components of the UPS and to serve as essential role in protein degradation. Yeast two-hybrid assays have found interactions between ataxin-7 and the S4 subunit of the 19S proteasome (Matilla et al., 2001), ataxin-1 and the ubiquitin-like protein A1Up (Davidson et al., 2000), and ataxin-3 and the ubiquitin and proteasome binding factor HHR23B (Wang et al., 2000). Ataxin-3 is well known to possess deubiquitination activity and remove attached ubiquitins from proteins targeted for degradation by the UPS (Burnett, Li, & Pittman, 2003; Chai, Berke, Cohen, & Paulson, 2004; Donaldson et al., 2003).

The carboxy terminus of the HSP70-interacting protein (CHIP) is a HSP70 cochaperone and an E3 ubiquitin ligase that protects cells from proteotoxic stress. These two roles place CHIP in a pivotal position in protein quality control (Rosser, Washburn, Muchowski, Patterson, & Cyr, 2007). CHIP directly interacts with and colocalizes to ataxin-1, ataxin-3, AR, and huntingtin aggregates (Al-Ramahi, Lam, & Chen, 2006; Williams, Knutson, Colomer Gould, & Paulson, 2009). CHIP decreases steady-state levels of mutant ataxin-1, ataxin-3, AR, and huntingtin by ubiquitination, which induces degradation. CHIP suppresses aggregation and toxicity in cell culture and *Drosophila* models of polyQ disease (Al-Ramahi et al., 2006; Williams et al., 2009). Suppression of CHIP resulted in an increased formation of microaggregates and toxicity in SCA3 transgenic mice (Williams et al., 2009). CHIP overexpression delays the disease phenotype in SBMA animal models through enhanced ubiquitylation and subsequent clearance of polyQ-expanded proteins (Adachi, Waza, & Tokui, 2007).

Another HSP70-dependent E3 ligase shown to act redundantly to CHIP on some substrates is parkin (Morishima et al., 2008). Parkin (PARK2, mutated in an autosomal recessive form of PD) mediates the targeting of proteins for proteasomal degradation and is known to interact with and modulate ataxin-2, ataxin-3, and AR, but not ataxin-1 (Choi, Ryu, & Kim, 2007; Durcan, Kontogiannea, Bedard, Wing, & Fon, 2012; Huynh, Nguyen, Pulst-Korenberg, Brice, & Pulst, 2007; Tsai, Fishman, Thakor, & Oyler, 2003). Both CHIP and Parkin colocalize to aggregates of polyQ-expanded AR (Morishima et al., 2008). Wild-type and polyQ-expanded ataxin-3 deubiquitinate parkin directly, while parkin ubiquitinates and facilitates the clearance of wild-type and mutant ataxin-2 and ataxin-3 by proteasomal degradation (Durcan et al., 2012; Huynh et al., 2007; Tsai et al., 2003). The colocalization of parkin and Htt in mouse brain as well as in patient samples has been reported (Morishima et al., 2008). No direct interaction of Htt and parkin has been described; however, it is possible that Htt and parkin indirectly interact through p97 (valosin-containing protein) (Hirabayashi, Inoue, & Tanaka, 2001). Additionally, partial suppression of parkin in HD mice worsens the neurological disease phenotype (Rubio, Rodriguez-Navarro, & Tomas-Zapico, 2009). The interaction or modulation of polyQ disease proteins by parkin may offer an explanation for occasional parkinsonian phenotypes in SCA2 and SCA3 patients. Htt binds to and is ubiquitinated by the human ubiquitin-conjugating enzyme E2–25K, which requires the polyQ domain, but is not influenced by the length of the polyQ tract (Kalchman et al., 1996). E2–25K modulates the aggregation and toxicity of mutant Htt and is recruited to aggregates in HD and SCA3 patients (Durcan et al., 2012; Kalchman et al., 1996).

Soluble oligomers, such as polyQ disease protein fragments, may target multiple cellular processes, causing "detrimental hits" that can impact cell survival (Hoffner & Djian, 2014; Kim, Hosp, & Frottin, 2016). Therefore, reducing the amount of such oligomers could be an important therapeutic strategy. This reduction might be achieved by enhancing the above-described mechanisms: chaperone-mediated refolding of polyQ proteins or degradation of misfolded proteins by the UPS. Heat shock proteins accumulate in aggregates in HD, SCA1, SCA3, and SCA7, and this has led to great interest in modulating the molecular chaperone machinery as a possible therapeutic strategy for polyQ diseases. In recent studies on HD degradation, rapid and complete clearance of polyQ-expanded Htt in neuronal cells was shown (Juenemann et al., 2013), and dynamic and reversible recruitment of proteasome subunits into inclusion bodies was observed in living cells (Schipper-Krom, Juenemann, & Jansen, 2014). Specifically targeting amino-terminal Htt to the UPS decreased its level and decreased aggregation (Juenemann et al., 2013). Furthermore, several groups demonstrated that proteasome inhibiton in mammalian cell models resulted in increased aggregation and cytotoxicity in SCA3 and HD (Wang et al., 2007; Wyttenbach, Carmichael, & Swartz, 2000), whereas overexpression of p45 (the ATPase of the 19S subunit of the proteasome) stimulates degradation of ataxin-3 (Wang, Jia, & Fei, 2007). Oral administration of a compound that induced HSP70, HSP40, and HSP90 expression markedly suppressed eye degeneration, inclusion formation, and lethality in SCA3 and HD *Drosophila* models (Fujikake et al., 2008). This compound also reduced polyQ-AR toxicity in cell and mouse models (Tokui, Adachi, & Waza, 2009; Waza, Adachi, & Katsuno, 2005). Overexpression of HSP40 suppressed ataxin-1 and ataxin-3 aggregation in vitro (Chai et al., 1999; Cummings et al., 1998), but not in Htt exon 1 overexpressing cell lines (Schipper-Krom et al., 2014). As modulation of the chaperone system through the overexpression of single proteins or combinations of different molecular chaperones has yielded inconsistent results and transient effects, combinatorial approaches are being considered.

Unfolding and remodeling of proteins is necessary for their passage through the narrow pore of the proteasome barrel, which thus likely precludes clearance of oligomers and aggregated proteins (Seo et al., 2004). Although modulating the proteasome system has been attempted, upregulation of this pathway is challenging, and thus attention has shifted to enhancing autophagy (Watson, Scholefield, Greenberg, & Wood, 2012). In polyQ diseases and likely other neurodegenerative proteinopathies, modulation of one protein quality control system often has effects on other protein quality control pathways. For example, an inhibitor of HSP90 reduced neuropathology in SCA3 mice not by affecting molecular chaperone pathway function, but instead by inducing the expression of autophagy proteins LC3-II and beclin-1 (Silva-Fernandes, Duarte-Silva, & Neves-Carvalho, 2014).

The Role of Autophagy

Autophagy is the process by which cellular components, including damaged organelles and protein aggregates, are delivered to the lysosome for degradation. Aggregation-prone polyQ proteins and fragments thereof are established substrates for autophagy-mediated clearance (Ravikumar, Duden, & Rubinsztein, 2002), and the importance of autophagy for polyQ disease toxicity is supported by a wealth of data, including detection of an HD patient age-of-onset modifier polymorphism in an essential autophagy gene (ATG7) (Metzger, Saukko, & Van Che, 2010; Metzger, Walter, & Riess, 2013). In SCA7, the truncated polyQ disease protein is degraded via autophagy in vitro (Mookerjee, Papanikolaou, & Guyenet, 2009), and in cellular models of SCA1, SCA3, SCA6, and SCA7, autophagic degradation of cytosolic aggregates in contrast to NIs has been documented (Duncan, Papanikolaou, & Ellerby, 2010; Iwata, Christianson, & Bucci, 2005; Nascimento-Ferreira, Nobrega, & Vasconcelos-Ferreira, 2013; Nascimento-Ferreira, Santos-Ferreira, & Sousa-Ferreira, 2011; Unno, Wakamori, & Koike, 2012; Yu, Ajayi, Boga, & Strom, 2012).

Defects in autophagy are directly associated with neurodegeneration, cancer, and inflammatory disease in humans (Kundu & Thompson, 2008). PolyQ tracts can regulate autophagy, so it is not surprising that in several polyQ diseases, signs of decreased autophagy efficiency have been reported (Ashkenazi, Bento, & Ricketts, 2017; Martinez-Vicente, Talloczy, & Wong, 2010). Impairment of autophagy is indicated by an increased number of autophagosomes, endosomal–lysosomal-like organelles, and multiple vesicular bodies. This has been documented in brain samples and in lymphoblasts from HD patients as well as in primary neurons and brain sections from HD mice (Davies et al., 1997; Nagata, Sawa, Ross, & Snyder, 2004; Petersen, Larsen, & Behr, 2001; Sapp, Schwarz, & Chase, 1997). Characterization of a SCA1 transgenic mouse model revealed changes in autophagic flux and a significantly altered LC3-II/-I ratio (Vig, Shao, Subramony, Lopez, & Safaya, 2009). Similar results were found in SCA7 transgenic mice where LC3 levels were significantly altered and wild-type ataxin-7 levels were stabilized by autophagy, although no stabilizing effects were found for mutant ataxin-7 (Duncan et al., 2010). Additionally, it was shown that full-length and cleaved fragments of ataxin-7 are differentially degraded. While full-length wild-type and mutant ataxin-7 are abundant in the nucleus and degraded by the UPS there, fragments of ataxin-7 in both the cytoplasm and nucleus appear to be substrates for degradation by both autophagy and the UPS (Yu et al., 2012). p62 acts as a cargo receptor for the autophagy-mediated-degradation of ubiquitinated targets (Pankiv, Clausen, & Lamark, 2007). Studies in human postmortem brain tissue from SCA3, SCA6, and HD patients revealed p62-positive cytoplasmic, axonal, and nuclear aggregates, suggesting that the autophagy pathway plays a central role involvement in the clearance of aggregated polyQ proteins (Nagaoka, Kim, & Jana, 2004; Seidel, Brunt, & de Vos, 2009; Seidel, den Dunnen, & Schultz, 2010). p62 may also contribute to the recruitment of proteasomes to nuclear aggregates of ataxin-1 and to the subsequent degradation of ataxin-1 (Pankiv et al., 2010).

Multiple strategies have been attempted to employ autophagy to degrade toxic polyQ-expanded disease proteins. Early studies of autophagy modulation demonstrated that blocking autophagy reduced cell viability and increased aggregates, while stimulating autophagy promoted clearance of wild-type Htt, mutant Htt, and a caspase-derived amino-terminal Htt proteolytic fragment (Qin, Wang, & Kegel, 2003). Pharmacological activation of autophagy has exhibited therapeutic efficacy in SCA7 (Yu, Munoz-Alarcon, & Ajayi, 2013), and beclin-1 modulation of autophagy can promote polyQ-ataxin-3 clearance and rescue motor abnormalities in a lentiviral rat model (Nascimento-Ferreira et al., 2013, 2011) and also in HD cell culture and primary neuron models (Silva-Fernandes et al., 2014; Wu, Qi, & Wang, 2012). Autophagy can also be upregulated by mammalian target of rapamycin (mTOR)-dependent and mTOR-independent pathways. Autophagy can be induced in mammalian cells by treatment with rapamycin, an inhibitor of the mTORC1 complex (Fig. 7.1). Rapamycin treatment has been shown to effectively promote the degradation of polyQ disease proteins and aggregates and exhibited therapeutic efficacy in countering polyQ neurotoxicity in HD and SCA3 in various types of model systems (Berger, Ttofi, & Michel, 2005; Menzies et al., 2010; Ravikumar, Vacher, & Berger, 2004). In terms of possible interventions that activate autophagy via an mTORC1-independent pathway, the well-known drug lithium has shown promise as a treatment for polyQ diseases. Lithium inhibits glycogen synthase kinase 3 and interferes with inositol monophosphatase function (Sarkar & Rubinsztein, 2006). Lithium treatment has yielded enhanced autophagic clearance of polyQ-Htt, polyQ-ataxin-1, and polyQ-ataxin-3, again in various types of model systems (Berger et al., 2005; Carmichael, Sugars, Bao, & Rubinsztein, 2002; Jia, Zhang, & Chen, 2013; Watase, Gatchel, & Sun, 2007; Wu, Zheng, & Huang, 2013). Combined treatment with lithium and rapamycin has also been attempted with commendable success in a *Drosophila* model of HD (Sarkar et al., 2008). Other approaches for autophagy modulation include the use of trehalose to activate autophagy in a mTORC1-independent fashion, often in combination with another small compound (Rose, Menzies, & Renna, 2010; Sarkar, Davies, Huang, Tunnacliffe, & Rubinsztein, 2007).

Transcriptional Dysregulation

One leading model for polyQ disease protein neurotoxicity is predicated upon the now well-established observation that polyQ disease proteins must translocate to the nucleus. Once localized to the nucleus, polyQ disease proteins (or fragments thereof) accumulate there, engage in aberrant interactions with nuclear proteins, and thereby interfere with the normal process of gene transcription (Zoghbi

& Orr, 2000). In strong support of this view, the vast majority of polyQ proteins (or fragments/alternative transcripts thereof) function as transcription factors, or transcription regulators. For example, the AR and TATA-binding protein (TBP) were studied as transcription factors for decades before their genes were respectively implicated as the cause of the two polyQ disorders SBMA and SCA17. For other polyQ diseases, the normal functions of proteins were initially unclear, but upon further study, such polyQ disease proteins were characterized as transcription factors. Ataxin-7 serves as an excellent case in point, as one of us, and others, discovered that ataxin-7 is a core component of a transcription coactivator complex called STAGA (Helmlinger, Hardy, & Sasorith, 2004; Palhan, Chen, & Peng, 2005). Transcription coactivator complexes are large protein complexes that mediate interactions between upstream transcription activators and the RNA polymerase II transcription complex (Blazek, Mittler, & Meisterernst, 2005; Conaway, Florens, & Sato, 2005). STAGA is the mammalian equivalent of the yeast SAGA complex, which promotes the effective interaction of classic DNA-bound transcription factors with the RNA polymerase II complex (Timmers & Tora, 2005). SAGA/STAGA contains Gcn5, which possesses intrinsic histone acetyltransferase (HAT) activity. Sgf73 in SAGA has been recognized as the yeast orthologue of mammalian ataxin-7 (Sanders, Jennings, Canutescu, Link, & Weil, 2002), as despite its limited amino acid homology, human ataxin-7 can complement Sgf73 null yeast strains, indicating that the two proteins are functionally interchangeable (McMahon, Pray-Grant, Schieltz, Yates, & Grant, 2005). Biochemical studies of the STAGA complex subsequently confirmed that ataxin-7 is indeed a core component of STAGA and the closely related TBP-free TAF containing complex (TFTC) (Helmlinger et al., 2004; Palhan et al., 2005). Both STAGA and TFTC mediate transcription activation by promoting histone acetylation; however, STAGA has been shown to interact with transcription factors linked to nuclear enzymatic activities (Martinez, Palhan, & Tjernberg, 2001). PolyQ-expanded ataxin-7 can alter the HAT activity of SAGA/STAGA/TFTC in yeast cells, retinal cells, and neuron-like cells (Helmlinger, Hardy, & Abou-Sleymane, 2006; Palhan et al., 2005; Strom, Forsgren, & Holmberg, 2005). Furthermore, polyQ-expanded ataxin-7 expressed in rod photoreceptors produces severe chromatin decondensation (Helmlinger et al., 2006). Thus, the polyQ tract expansion in ataxin-7 likely interferes with normal STAGA/TFTC function to elicit neurotoxicity in SCA7. A close homologue of atrophin-1 (the gene responsible for DRPLA) can function as a transcription corepressor in *Drosophila* (Zhang, Xu, Lee, & Xu, 2002), and human atrophin-1 interferes with CREB-binding protein (CBP)-mediated transcription (Nucifora, Sasaki, & Peters, 2001). Ataxin-1 was first suggested to modulate transcription through an interaction with the scaffolding protein PQBP-1 (Okazawa, Rich, & Chang, 2002), or through its interaction with the corepressor silencing mediator for retinoid and thyroid hormone receptors (SMRT) (Tsai, Kao, & Mitzutani, 2004); however, subsequent studies have established that PolyQ-ataxin-1 interacts with Capicua to modulate its transcription repressor function in the native complex to which both proteins belong (Lam, Bowman, & Jafar-Nejad, 2006). Ataxin-3, by binding to histone proteins, can interfere with the chromatin remodeling activities of transcription coactivators with HAT activity, in particular CBP and p300, and may therefore function as a transcription corepressor (Li, Macfarlan, Pittman, & Chakravarti, 2002). Ataxin-3 may directly interact with HDAC3 in the NCoR complex to repress transcription, suggesting that polyQ expanded-ataxin-3 may prevent proper recruitment and function of this deacetylase complex, resulting in inappropriate transcription activation (Evert, Araujo, & Vieira-Saecker, 2006).

In the case of HD, an extensive literature has established that amino-terminal truncation products of Htt are bona fide transcription factors (i.e., components of the TFIID and TFIIF complexes) and can interfere with transcription factors such as CBP, Sp1, and TAFII-130 (Dunah, Jeong, & Griffin, 2002; Nucifora et al., 2001; Zuccato, Tartari, & Crotti, 2003). Numerous studies of gene expression in various HD model systems have documented that polyQ-expanded versions of this amino-terminal Htt truncation product potently interfere with gene transcription in the early stages of the HD disease process [reviewed in (Riley & Orr, 2006)]. The full-length huntingtin protein can also regulate transcription through an interaction with repressor element-1 transcription factor in neurons, and this regulation is altered by polyQ-expanded htt in HD (Dunah et al., 2002; Nucifora et al., 2001; Zuccato et al., 2003). One important transcription dysregulation pathology that contributes to HD pathogenesis stems from the impairment of the function of peroxisome proliferator-activated receptor (PPAR) coactivator-1-alpha (PGC-1α), which is a master transcription coactivator controlling mitochondrial biogenesis, metabolism, and antioxidant defense (Handschin & Spiegelman, 2006; Kelly & Scarpulla, 2004; Puigserver et al., 1998), primarily through its coactivation of a family of nuclear receptors, known as the PPARs (α, δ, γ). Alterations in the activity of PGC-1α have been reported for HD patients and mice (Cui et al., 2006; Weydt, Pineda, & Torrence, 2006). The importance of PGC-1α for HD pathogenesis is underscored by the observation that PGC-1α overexpression in HD mice is sufficient to rescue motor phenotypes, prevent accumulation of misfolded Htt protein in CNS through increased autophagy, and reduce neurodegeneration (Tsunemi, Ashe, & Morrison, 2012). To determine the basis for PGC-1α transcription interference in HD, we performed an unbiased screen for htt-interacting proteins, and identified PPARs as candidate interactors. When we evaluated the different PPARs, we documented a physical interaction between PPARδ and htt in the cortex of BAC-HD97 transgenic mice, and confirmed that PPARδ is highly expressed in neurons of the CNS (Dickey, Pineda, & Tsunemi, 2016). Mutant htt repressed PPARδ transactivation in neurons from BAC-HD97 mice, but could be rescued by PPARδ agonist treatment or overexpression. When we examined the contribution of PPARδ to HD pathogenesis, we documented that htt physically interacts with PPARδ and represses PPARδ transactivation function to yield mitochondrial dysfunction and neurotoxicity in a variety of model systems and determined that transgenic mice expressing dominant-negative PPARδ in the striatum recapitulate HD-like behavioral, metabolic, transcriptional, and neurodegenerative phenotypes (Dickey et al., 2016). Furthermore, we observed dramatic neurological disease phenotypes in mice expressing dominant-negative PPARδ in various neural lineages, thereby identifying neurons of the CNS as a cell type where PPARδ function is absolutely essential for homeostasis. Finally, we validated a highly selective and potent PPARδ agonist, KD3010, as capable of rescuing htt neurotoxicity and engaging PPARδ targets in CNS, and performed a preclinical trial of KD3010 in HD mice, where we documented significant improvements in motor function, neurodegeneration, and survival (Dickey et al., 2016). PolyQ disease protein transcription interference may also affect the UPS, autophagy, and other cellular pathways.

Mitochondrial Dysfunction

Mitochondrial dysfunction and impaired energy metabolism contribute to the pathogenesis or progression of many neurodegenerative disorders, including AD, PD, and amyotrophic lateral sclerosis (ALS) (de Moura, dos Santos, & Van Houten, 2010; Tan, Pasinelli, & Trotti, 2014).

This can be explained by the high energy demands of neurons and other CNS cells, and their extreme reliance on mitochondrial oxidative phosphorylation to generate ATP. An extensive body of work has implicated mitochondrial dysfunction in polyQ diseases (Fig. 7.1), with much of the work focused on HD and performed more than two decades ago. In one seminal study in 1993, chronic administration of a mitochondrial toxin, 3-nitropropionic acid, yields a selective loss of medium spiny neurons in the striatum (Beal, Brouillet, & Jenkins, 1993), indicating the importance of mitochondrial function for homeostasis in the region of the basal ganglia that is selectively vulnerable in HD. This finding has been corroborated by a wide range of studies in HD cell culture models, mice, and human patients (reviewed in Lin & Beal, 2006). Directed studies on mitochondria isolated from HD patients and mice have also revealed significant defects in mitochondrial membrane potential and depolarization (Panov, Gutekunst, & Leavitt, 2002), making the case for mitochondrial dysfunction in HD quite compelling.

Oxidative Stress/ATP Production

Oxidative stress and changes in ATP production caused by altered respiratory chain complex activities indicate mitochondrial dysfunction in polyQ disease. HD patients show reduced complex II, III, and IV activities in putamen and caudate, while alterations in complex I activity have been observed in skeletal muscle (Costa & Scorrano, 2012). Dysfunctional respiratory chain complex function and increased oxidative stress have also been observed in SCA2, 3, and 12 (Dickey & Strack, 2011; Garcia-Martinez, Perez-Navarro, & Xifro, 2007; Laco, Oliveira, Paulson, & Rego, 2012; Pacheco, da Silveira, & Trott, 2013; Simon, Zheng, & Velazquez, 2007; Wang, Lee, & Lee, 2011; Yu, Kuo, Cheng, Liu, & Hsieh, 2009). Decreased complex II activity was found in lymphoblasts from SCA3 patients, in cells from transgenic mice, and in SCA3 cell models (Laco et al., 2012). In cells expressing polyQ-expanded ataxin-3, decreased activities of the antioxidant enzymes catalase, glutathione reductase, and superoxide dismutase, in combination with mitochondrial DNA damage, were detected (Yu et al., 2009). Similar findings of increased catalase levels and DNA damage were noted in SCA3 patient samples compared to healthy controls (Pacheco et al., 2013). A recent study suggests that the polyQ-containing aggregates can be reduced in SCA3 neurons by treatment with an extract of *Gardenia jasminoides*, which was found to reduce reactive oxygen species accumulation (Chang, Chen, & Wu, 2014).

Metabolism/Body Weight

Metabolic defects and weight loss at early disease stages are well-described signs of clinical disease in patinets with HD (Djousse et al., 2002; Leenders, Frackowiak, Quinn, & Marsden, 1986), SCA1 (Mahler et al., 2014), and SCA3 (Saute, Silva, & Souza, 2012), as well as in their respective disease mouse models (Hubener, Vauti, & Funke, 2011; Jafar-Nejad, Ward, Richman, Orr, & Zoghbi, 2011; Mangiarini, Sathasivam, & Seller, 1996). For HD and SCA3 patients, an inverse correlation between body mass index and CAG repeat number was reported (Aziz et al., 2008; Saute et al., 2012). In SCA1 patients, this weight loss appears despite a balance between energy intake and expenditure, though patients show increased energy expenditure and fat oxidation at resting state, which may result from altered autonomic nervous system activity and gait ataxia (Mahler et al., 2014). Advanced nuclear magnetic resonance imaging techniques can be used to study alterations in metabolite concentrations in distinct brain regions of human patients and mice. Increased lactate production was found in the cortex and basal ganglia of HD patients (Jenkins, Koroshetz, Beal, & Rosen, 1993). In SCA1 patients, cerebellum and brain stem showed decreased total *N*-acetylaspartate (NAA)

[NAA + *N*-acetylglutamate, tNAA] concentrations, as well as elevated glutamine, elevated total creatine, and elevated myoinositol concentrations (Guerrini, Lolli, & Ginestroni, 2004; Oz, Nelson, & Koski, 2010). Interestingly, levels of tNAA and myoinositol correlated with patients' ataxia scores. Similar changes in metabolite concentrations were seen in conditional SCA1 and SCA1 knock-in mouse models. Furthermore, the metabolite levels nearly returned to baseline when polyQ-ataxin-1 transgene expression was suppressed at early stages of the disease in the conditional SCA1 mouse model, and alterations in metabolite levels were observed in SCA1 knock-in mice long before any neuropathology could be detected (Emir, Brent Clark, Vollmers, Eberly, & Oz, 2013; Oz et al., 2010).

Mitochondria-Mediated Apoptotic Activation

An important role in regulating mitochondria-mediated activation of the apoptotic pathway in polyQ disease has been ascribed to the B-cell lymphoma 2 (Bcl-2) family of proteins. These proteins regulate the permeability of the outer mitochondrial membrane and release of cytochrome c and thereby control cell survival, morphology, dynamics, and membrane potential of mitochondria. Bcl-2 family members can be prosurvival or proapoptotic. The main family members inhibiting cell death are Bcl-2 and B-cell lymphoma-extra large (Bcl-xL), while the BH3-only protein Bax forms pores in the mitochondrial membrane and thus initiates apoptosis. For SCA3 and SCA7, mRNA and protein levels of Bcl-xL were downregulated in cerebellar neurons, respectively, expressing polyQ-ataxin-3 or polyQ-ataxin-7, leading to activation of caspase-3 and caspase-9, two main caspases involved in the execution of mitochondrial-induced apoptosis (Chou, Yeh, & Kuo, 2006; Wang, Yeh, & Chou, 2006). Recently, a direct interaction between ataxin-3 and Bcl-xL was reported, suggesting that ataxin-3 promotes the interaction between Bcl-xL and Bax (Wang et al., 2006). SCA3 and SCA7 mice express increased levels of Bax mRNA and protein, likely due to increased levels of activated p53, a transcription factor known to enhance Bax transcription (Chou, Lin, & Hong, 2011; Chou et al., 2006; Wang et al., 2010, 2006). Similarly, Bax levels were found to be increased in HD cells and mice (Bae, Xu, & Igarashi, 2005; Garcia-Martinez et al., 2007; Zhang, Ona, & Li, 2003), as well as in the caudate nucleus of HD patients (Vis, Schipper, & de Boer-van Huizen, 2005). Moreover, polyQ-expanded ataxin-3 was found to decrease mRNA and protein levels of the prosurvival factor Bcl-2 by affecting Bcl-2 mRNA stability (Tien, Wen, & Hsieh, 2008; Tsai, Tsai, & Hsieh, 2004). While expression of polyQ-Htt decreased Bcl-2 protein levels in cell lines and in brain samples from HD mice (Duan, Peng, & Masuda, 2008; Ju, Chen, & Lin, 2011; Ruiz, Casarejos, & Rubio, 2012), other studies did not find alterations in well-established HD mouse models, such as R6/1 (Teles et al., 2008). To understand the role of Bcl-2 in HD, R6/2 mice were crossed with transgenic mice selectively overexpressing Bcl-2 in neurons: double transgenic R6/2;Bcl-2 mice exhibited delayed onset of motor abnormalities and survived longer than singly transgenic R6/2 littermates (Zhang et al., 2003). Notwithstanding contradictory data about Bcl-2 expression in HD mouse models, this result suggests that Bcl-2 overexpression may protect neurons from toxicity elicited by polyQ-Htt.

Mitochondrial Size, Shape, and Movement

Localization of polyQ disease-causing proteins to mitochondria and their actions at the mitochondria have been subjects of intense research. For SCA3, it is known that both normal and polyQ-expanded ataxin-3 localize to mitochondria (Pozzi, Valtorta, & Tedeschi, 2008), and degradation of polyQ-expanded ataxin-3 via the UPS is promoted by a ubiquitin ligase in the outer mitochondrial membrane (Sugiura,

Yonashiro, & Fukuda, 2011). Mutant Htt was also shown to localize to the mitochondria (Orr, Li, & Wang, 2008; Rockabrand, Slepko, & Pantalone, 2007). Apart from the changes in mitochondrial bioenergetics and transcription of proteins associated with mitochondrial-mediated apoptotic activation, alterations in the shape and motility of mitochondria have repeatedly been observed in HD. Both retrograde and anterograde mitochondrial transport along axons is impaired by polyQ-Htt in cultured neurons from mice and rats (Chang, Rintoul, Pandipati, & Reynolds, 2006; Orr et al., 2008). While fragmented mitochondria have been documented in numerous cell culture models of HD and in cell lines from HD patients, more recent work has linked this observation to dysregulation of the GTPase dynamin-related protein-1 (Drp-1), which is known to promote mitochondrial fission. Calcineurin phosphorylates Drp-1, thereby increasing its activity and translocation to mitochondria, where it promotes mitochondrial fragmentation, and one study noted a higher basal activity of calcineurin in HD models (Costa, Giacomello, & Hudec, 2010). A direct physical interaction between polyQ-Htt and Drp-1 accompanied by an increase in Drp-1 enzymatic activity have been documented in brain tissue from HD mice and human patients (Shirendeb, Reddy, & Manczak, 2011). Since the balance between fission and fusion is crucial for mitochondrial function, and since neuron cell death caused by excessive mitochondrial fragmentation has been documented in other neurodegenerative disorders, such as AD and PD (de Moura et al., 2010), this pathway may offer possible therapeutic opportunities for HD patients and for patients with other polyQ diseases as well.

POSTTRANSLATIONAL MODIFICATIONS

Posttranslational modifications of polyQ disease proteins are now a well-established modifier of polyQ disease protein toxicity. Posttranslational modifications of polyQ disease proteins are diverse (Fig. 7.2), including phosphorylation, ubiquitination, SUMOylation, acetylation, and *S*-nitrosylation. The downstream consequences of altering polyQ protein posttranslational modifications are similarly wide ranging, affecting proteolytic cleavage, oligomer formation, aggregation, degradation, protein–protein interactions, and subcellular localization.

Phosphorylation

The importance of phosphorylation in SCA1 (reviewed in (Orr & Zoghbi, 2007)) and the relevance of Htt phosphorylation and kinase pathways have been illustrated in *Drosophila* and mice (Lievens, Iche, Laval, Faivre-Sarrailh, & Birman, 2008; Pardo, Colin, & Regulier, 2006; Warby, Chan, & Metzler, 2005). In SCA1, the phosphorylation of serine at position 776 in the ataxin-1 protein appears absolutely necessary for the toxicity of the mutant disease protein (Duvick, Barnes, & Ebner, 2010; Emamian, Kaytor, & Duvick, 2003). Phosphorylation at S776 and its subsequent binding and release from 14-3-3, a cytosolic protein, can stabilize ataxin-1 and modulate its localization (Lai, O'Callaghan, Zoghbi, & Orr, 2011), key events in the SCA1 pathogenic cascade (Lim, Crespo-Barreto, & Jafar-Nejad, 2008). Similarly, phosphorylation at specific serines in the Htt protein and the AR play critical roles in the pathogenesis of HD and SBMA, respectively (Mishra et al., 2012; Palazzolo, Burnett, & Young, 2007). CDK5 appears to be neuroprotective, as it may counter caspase cleavage of Htt by phosphorylating Htt at serine 434 (Luo, Vacher, Davies, & Rubinsztein, 2005; Ratovitski, Nakamura, & D'Ambola, 2007). In SCA3, dosage reduction of CDK5 by RNA interference (RNAi) in a *Drosophila* model actually enhanced polyQ-expanded ataxin-3 neurotoxicity (Liman, Deeg, & Voigt, 2014). In HD, the importance of the first 17 amino acids of Htt protein for regulating toxicity has emerged

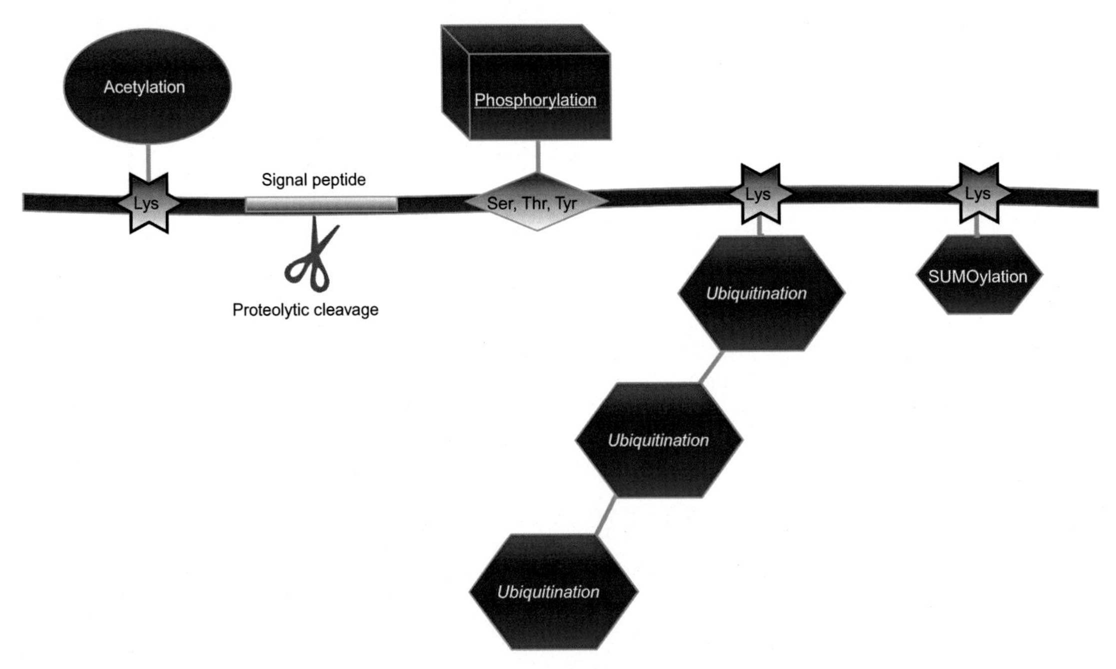

FIGURE 7.2 **Posttranslational modifications and proteolytic processing in polyglutamine disease pathogenesis.** Posttranslational modifications and proteolytic processing can modify the toxicity of PolyQ-expanded proteins. Posttranslational modifications are diverse, including phosphorylation, SUMOylation, ubiquitination, and acetylation. *PolyQ*, polyglutamine.

through in-depth investigations in cell culture models and in mice and has led to a model where phosphorylation of serines within the first 17 amino acids of Htt releases it from the endoplasmic reticulum to promote its nuclear entry and prevents its export from the nucleus during stress response (Maiuri, Woloshansky, Xia, & Truant, 2013), thereby modulating its neurotoxicity (Watkin, Arbez, & Waldron-Roby, 2014). For ataxin-3, CK-2-dependent phosphorylation of serine 340 and 352 may similarly control its nuclear entry (Mueller et al., 2009). In HD, mutation of serine 536 to an aspartic acid abolished huntingtin proteolysis at this cleavage site and reduced mutant huntingtin toxicity, further implicating phosphorylation events as key modulators of the cleavage process that drives the mutant Htt pathogenic cascade (Schilling, Gafni, & Torcassi, 2006).

Sumoylation

Enhanced immunoreactivity for SUMO1 has been reported in affected neurons of DRPLA, SCA1, SCA3, and HD patients (Ueda, Goto, & Hashida, 2002). In addition, colocalization of SUMO and polyQ protein aggregates has been found polyQ-atrophin-1-containing inclusion bodies in cell lines and material from DRPLA patients (Terashima, Kawai, Fujitani, Maeda, & Yasuda, 2002). SUMO1 immunostaining is detected in nuclear aggregates in SCA3, although this does not always overlap with polyQ-ataxin-3 (Pountney, Huang, & Burns, 2003). In contrast, HD primary neuron models exhibiting polyQ-Htt aggregation indicated an association with SUMO1 (Steffan, Agrawal, & Pallos, 2004). Similar colocalization of SUMO1 with both normal and polyQ-ataxin-1 has been observed in SCA1 (Riley, Zoghbi, & Orr, 2005).

These studies cumulatively demonstrate the presence of SUMO in neuronal inclusions in a variety of polyQ disorders and suggest a potential role for this ubiquitin-like protein in polyQ pathogenesis but, until recently, it was unclear if the polyQ disease proteins were targets for SUMOylation. Direct links to SUMOylation of disease proteins now support a possible role for this PTM in neurodegeneration for certain polyQ diseases, potentially by mediating changes in protein solubility or levels of toxic soluble oligomers. For example, in a neuronal model of DRPLA, coexpression of wild-type SUMO1 with polyQ-atrophin-1 significantly accelerated the formation of nuclear aggregates and promoted apoptosis (Terashima et al., 2002). One proposed explanation for the accelerated formation of aggregates is SUMO disruption of or competition with proteasome-mediated degradation of aggregated mutant atrophin-1. An alternate possibility is altered trafficking or sequestration of SUMO-modified proteins, resulting in their accumulation within the nucleus. With the ability of SUMO to regulate protein–protein interactions via noncovalent binding motifs (Hannich, Lewis, & Kroetz, 2005; Hecker, Rabiller, Haglund, Bayer, & Dikic, 2006; Minty, Dumont, Kaghad, & Caput, 2000; Song, Durrin, Wilkinson, Krontiris, & Chen, 2004), the recruitment of SUMO monomers or SUMOylated substrates into aggregates could further exacerbate the sequestration of cellular components critical for neuron survival. Conversely, coexpression of conjugation-deficient SUMO1 reduced, but did not abolish, the accumulation of polyQ-atrophin-1 aggregates. Expression of this conjugation-deficient SUMO promoted cell survival, as compared to the mutant polyQ protein alone or in the presence of wild-type SUMO1.

A pathogenic fragment of Htt protein has been identified as a SUMO substrate (Steffan et al., 2004). Expression of a nonhydrolyzable Htt-SUMO conjugate in a neuronal cell line resulted in a more stable protein and increased Htt-mediated transcription repression (Steffan et al., 2004). This also resulted in decreased formation of inclusions and an increase in the levels of toxic oligomers. Rhes SUMOylated polyQ-Htt, but not the normal Htt, and polyQ-Htt SUMOylation led to decreased aggregate formation, but increased cytotoxicity in vitro and in vivo (Steffan et al., 2004; Subramaniam, Sixt, Barrow, & Snyder, 2009). Such detrimental effects can be attributed to a SUMO-induced increase in toxic soluble oligomers (in HD) or by antagonizing proteasome-mediated degradation (in DRPLA). SUMO1 conjugation sites are located within the amino-terminal domain of the Htt peptide and overlap with the lysine residues targeted by ubiquitin, which may lead to competition between the two modifiers (Steffan et al., 2004). The relative contribution of ubiquitin and SUMO to HD pathogenesis has also been investigated in *Drosophila* models. Expression of mutant Htt led to progressive degeneration, which was reduced in flies heterozygous for SUMO and modestly worsened in flies with genetically reduced ubiquitination (Steffan et al., 2004). Furthermore, expression of a mutant Htt in which the conjugation lysines were mutated substantially decreased degeneration. This indicates that the availability of the target lysines is essential for neurodegeneration and that SUMOylation at these sites may produce a greater pathological response, as compared to the slight amelioration associated with altered ubiquitination. The overall dynamics and cooperation between these two modification systems may thus be more important than the conjugation of each individual modifier separately.

SUMOylation can have diverse consequences on different disease-associated polyQ proteins, sometimes having a cytotoxic rather than a cytoprotective effect. The causal protein in SBMA is the AR which, when carrying an expanded polyQ tract, results in nuclear and cytoplasmic aggregates, together with progressive neurodegeneration (Chan, Warrick, Andriola, Merry, & Bonini, 2002). Wild-type AR is a known SUMO1 substrate and SUMOylation negatively regulates its transcriptional activity (Poukka, Karvonen, Janne, & Palvimo, 2000). Disruption of SUMOylation of mutant polyQ-AR enhances its transcription function and improves SBMA disease phenotypes (Chua, Reddy, & Yu, 2015). However, expression of a catalytic-deficient mutant form of a SUMO activating enzyme (E1) greatly enhanced degeneration in a *Drosophila* model of SBMA (Chan et al., 2002). Similar observations have been made in SCA3, suggesting that down-regulation of SUMO modification of ataxin-3 is detrimental to cells. This may also involve the overlapping functions of SUMO and the UPS, as the SUMO E1 degenerative phenotype is similar to that induced by expression of a dominant-negative form of Hsp70 or an inactive proteasome β-subunit mutant (Chan et al., 2002). In contrast to the mutant proteasome, Hsp70 cannot rescue SUMO E1 mutant-induced degeneration. This is consistent with the notion that SUMOylation contributes to more than protein aggregation and degradation but may regulate other cellular pathways. This could include events involved in apoptotic pathway activation, such as SUMO1 modification of caspases (Besnault-Mascard, Leprince, & Auffredou, 2005; Hayashi, Shirakura, Uehara, & Nomura, 2006; Shirakura et al., 2005). SUMO-mediated changes in protein transport into the nucleus is yet another pathway with potential pathogenic relevance to polyQ diseases. PolyQ-ataxin-1 accumulates in SCA1 and is covalently modified by SUMO1 at five different target lysines (Riley et al., 2005). An increase in the length of the polyQ stretch negatively regulates such ataxin-1 SUMOylation, which is influenced both by phosphorylation and nuclear translocation mediated by a nuclear localization signal (NLS). This work indicates that SUMO modification may control the nuclear import/export efficiency of ataxin-1 and regulate its trafficking. This role for SUMO is consistent with findings for SUMOylated Htt (Steffan et al., 2004), where SUMO conjugation within the Htt amino-terminal domain may mask a cytoplasmic retention signal and thereby promote nuclear localization. This could explain why a SUMO-mediated increase in polyQ-Htt transcription repression has been observed.

In summary, polyQ diseases could also be considered disorders of protein trafficking and transport (Taylor, Grote, & Xia, 2006), and this cellular pathology may involve altered SUMO pathway processes for certain disorders, including especially HD. One important consequence of a nuclear shuttling defect would be to promote transcription interference, a common feature of these disorders. This makes sense, as SUMOylation is a well-defined regulator of transcription (reviewed in Gill, 2003) and several polyQ disease proteins are well-established transcription factors or transcriptional regulators (reviewed in Gatchel & Zoghbi, 2005). A direct pathological implication of SUMO in polyQ diseases is suggested by the SUMO-mediated increase in Htt transcriptional repression (Steffan et al., 2004). Many of the polyQ-interacting partners that are recruited into neuron inclusions are also SUMO substrates (Takahashi-Fujigasaki, Arai, Funata, & Fujigasaki, 2006), including PML (Yasuda, Inoue, & Hirabayashi, 1999), p53 (Steffan, Kazantsev, & Spasic-Boskovic, 2000), and *c-jun* (Yasuda et al., 1999). SUMOylation may therefore modulate their ability to interact with polyQ disease proteins or to be recruited into transcription activation complexes.

S-*Nitrosylation*

Htt is highly conserved and most of its 70 cysteine residues are conserved among vertebrates (Candiani, Pestarino, Cattaneo, & Tartari, 2007). Endogenous Htt is *S*-nitrosylated on some of these cysteine residues in multiple mouse tissues (Ni, Seth, & Fonseca, 2016). PolyQ-dependent *S*-nitrosylation is also observed in the neuron-like PC12 pheochromocytoma cell line (Westerink & Ewing, 2008). Htt undergoes both *S*-nitrosylation and *S*-acylation, and both modifications are increased by polyQ expansion (Ni et al., 2016). These posttranslational modifications would be expected to affect local features of Htt domains and might cause deleterious effects, though these aspects have not yet been studied. Their widespread distribution could have broader implications for global properties of Htt. This seems to be of general significance for polyQ-containing proteins, since parallel increases are seen for ataxin-1 (Ni et al., 2016). One indication of the higher level reorganization of ataxin-1 and Htt upon polyQ expansion is the presence of high-molecular-weight species that are *S*-nitrosylated and *S*-acylated. These species are not eliminated by reduction with β-ME, suggesting unique chemistry and possible linkage to other proteins (Ni et al., 2016). A recent report has found that disulfide formation, which may be promoted by *S*-nitrosylation (Arnelle & Stamler, 1995), can mediate Htt oligomerization (Fox, Connor, & Stiles, 2011). These changes could modulate protein conformation and the progression of disease. As expected, Htt protein aggregation increases in response to NOS (nitric oxide synthase) overexpression (Ni et al., 2016). Thus, polyQ-dependent modifications can impact the dynamics of local and perhaps global protein conformation for these proteins.

Proteolytic Cleavage

Initial studies into the common characteristics of polyQ diseases indicated that small fragments of mutant proteins containing the expanded polyQ stretch exhibited even greater cytotoxicity than their respective full-length polyQ proteins (Ikeda et al., 1996; Mangiarini et al., 1996). Proteolytic cleavage was suggested as an early or initial step in disease pathogenesis (Fig. 7.2), providing a basis for the so-called toxic fragment hypothesis (Wellington & Hayden, 1997). The presence of proteolytically derived fragments of mutant proteins has been reported for virtually all polyQ diseases. Evidence exists for the preferential nuclear accumulation of truncated polyQ-containing fragments of mutant proteins in several polyQ diseases, including HD and SCA7 (Garden et al., 2002; Li, Miwa, & Kobayashi, 1998; Paulson et al., 1997; Schilling, Becher, & Sharp, 1999; Schilling, Wood, & Duan, 1999). Several classes of endogenous proteases have been linked to the proteolysis of polyQ proteins, including caspases (Goldberg, Nicholson, & Rasper, 1996; Wellington, Ellerby, & Gutekunst, 2002; Wellington, Ellerby, & Hackam, 1998; Young, Gouw, & Propp, 2007) and calpains (Gafni & Ellerby, 2002; Gafni et al., 2004; Goffredo, Rigamonti, & Tartari, 2002; Haacke, Hartl, & Breuer, 2007; Hubener, Weber, & Richter, 2013; Kim, Yi, & Sapp, 2001). The importance of proteolytic cleavage has been thoroughly studied in HD, where protease cleavage sites have been mapped and specific proteases sought (Gafni & Ellerby, 2002; Goldberg et al., 1996; Kim et al., 2001; Wellington et al., 1998; Wellington, Singaraja, & Ellerby, 2000).

In vitro studies in HD revealed that progressive truncation of polyQ-Htt protein correlates with a propensity to aggregation and an increase in apoptotic stress (Hackam et al., 1998; Martindale, Hackam, & Wieczorek, 1998). Mice expressing the polyQ-expanded Htt exon 1 fragment showed a progressive neurological phenotype recapitulating many characteristics of human HD phenotypes, suggesting that amino-terminal polyQ-Htt is sufficient to induce neurodegeneration in vivo (Mangiarini et al., 1996). These disease fragments were also detected in human HD brain and lymphoblasts, after their recognition in SCA3 brain (Goti, Katzen, & Mez, 2004; Kim et al., 2001; Toneff, Mende-Mueller, & Wu, 2002), and were determined to be an important component of neuronal intranuclear inclusions (DiFiglia et al., 1997; Schmidt, Landwehrmeyer, &

Schmitt, 1998; Sieradzan et al., 1999). In two SCA7 mouse models expressing amino-terminal ataxin-7, fragments appeared in nuclear aggregates in correlation with onset of disease phenotypes (Garden et al., 2002; Yvert et al., 2000). The expression of an ataxin-3 fragment containing an elongated polyQ stretch induced apoptosis and cell death as well as a severe ataxia phenotype in a mouse model, showing a more rapid manifestation of SCA3-like disease when compared to mice expressing full-length polyQ-ataxin-3 (Ikeda et al., 1996). In addition, polyQ-containing ataxin-3 protein was recruited into insoluble inclusions (Haacke et al., 2006; Paulson et al., 1997). However, as with much of the current research on polyQ diseases, not all observations are in agreement. An HD mouse model expressing a polyQ-expanded Htt fragment encompassing exons 1 and 2 exhibited neither neurotoxic effects nor an HD phenotype, despite the presence of NIs (Slow, Graham, & Osmand, 2005). This illustrates that not all fragment species produce neuropathology. Another noteworthy investigation of a SCA3 gene trap mouse model showed that expression of a fusion protein comprising β-galactosidase and amino-terminal ataxin-3 without the polyQ tract led to the formation of cytoplasmic inclusions and a neurological phenotype reminiscent of SCA3 (Hubener et al., 2011). Furthermore, carboxy-terminal polyQ fragments of the 1A calcium channel, the disease protein in SCA6, showed a polyQ-independent cytotoxic nature. However, the expansion of the polyQ tract within the fragment resulted in increased resistance to proteolysis leading to an accumulation of this toxic species (Kubodera, Yokota, & Ohwada, 2003). For SCA1 and SCA2, neither an inherent cytotoxicity and aggregation propensity nor a clear impact on pathology was evident for mutant protein fragments, demanding further characterization (Huynh, Figueroa, Hoang, & Pulst, 2000; Klement et al., 1998). For ataxin-2, the disease protein in SCA2, mutant fragment constructs were shown to exhibit a pronounced aggregate formation potential in vitro (Nozaki, Onodera, Takano, & Tsuji, 2001), but further studies revealed decreased cytotoxicity of amino-terminal polyQ-ataxin-2 compared to the full-length protein (Ng, Pulst, & Huynh, 2007). Nonetheless, despite these exceptions, for the majority of polyQ diseases, there is a strong correlation between proteolytic processing of mutant proteins and more rapid disease progression. Indeed, studies of ataxin-7 in SCA7 mouse models and cell culture models support the existence of an amino-terminal truncated fragment with greater pathogenicity. An ~55 kDa ataxin-7 fragment was detected with both an amino-terminal directed ataxin-7 antibody and the 1C2 antibody in SCA7 transgenic mice and SCA7 human patients (Garden et al., 2002). Similar to Htt, ataxin-7 can be cleaved by caspase-7 (Ellerby, Hackam, & Propp, 1999), and caspase cleavage can modulate ataxin-7 cellular toxicity *in vitro* (Young et al., 2007). Caspase-7-mediated cleavage is predicted to generate a short fragment containing the amino-terminus with the polyQ tract but without the nuclear export sequence (NES), potentially resulting in greater accumulation of a polyQ-expanded ataxin-7 fragment in the nucleus. Compared to full-length mutant ataxin-7 protein, expression of a mutant ataxin-7 amino-terminal fragment similar in size to the truncation product found in vivo, exhibited enhanced nuclear localization and cellular toxicity (Taylor et al., 2006).

Caspases

The first proteases shown to cleave polyQ expanded proteins were caspases. This family of cysteine proteases is associated with apoptotic activation and inflammation but is also known to regulate other cellular functions, including cell proliferation, differentiation, and migration (Li & Yuan, 2008; McIlwain, Berger, & Mak, 2013). Caspases are involved in the execution of cell death, and an increase in caspase activation is readily detected in the course of polyQ disease progression. Presence of apoptotic cell death and caspase activation was shown in human HD brains as well as in mouse and cell models of HD (Chen, Ona, & Li, 2000; Hermel, Gafni, & Propp, 2004; Kim, Lee, & LaForet, 1999; Li, Lam, Cheng, & Li, 2000; Lunkes & Mandel, 1998; Ona, Li, & Vonsattel, 1999; Portera-Cailliau, Hedreen, Price, & Koliatsos, 1995; Zhang et al., 2003), although this is not consistent with other work that did not detect apoptotic nuclei in HD R6/2 mice (Davies et al., 1997). Cell death pathways and caspases were also reported to be switched on in other polyQ diseases, such as SCA3 (Chou et al., 2006; Ikeda et al., 1996) and SCA7 (Latouche, Lasbleiz, & Martin, 2007; Zander et al., 2001). In SCA7, activated caspase-3 was recruited into inclusions in cell culture and human SCA7 brain, and its expression was upregulated in cortical neurons (Zander et al., 2001). While caspase activation is synonymous with the execution of cell death in dividing cells and immature neurons, the process by which established adult neurons die in neurodegenerative diseases is unclear and unlikely to occur via classic apoptosis. However, activation of the apoptotic pathway likely has important implications for promoting cell stress, inflammation, and metabolic derangements that contribute to the decompensation of normal homeostasis in neurodegeneration.

The first discovery of caspase-mediated cleavage of a polyQ-disease protein was made for HD (Goldberg et al., 1996). This in vitro study indicated the specific action of caspase-3 for Htt cleavage in a polyQ-length dependent fashion. Further work identified caspase-1-dependent cleavage of Htt and confirmed caspase-3-mediated fragmentation, whereas caspase-7 and -8 appeared not to cleave full-length Htt (Wellington et al., 1998). Additionally, caspase-3 selectively processed polyQ-expanded Htt, and the resulting amino-terminal fragments formed cytoplasmic and NIs (Lunkes & Mandel, 1998). Direct evidence for caspase-mediated Htt cleavage was obtained from early-stage HD postmortem human tissue and transgenic mice. In these brain tissues, both mutant polyQ-Htt and normal wild-type Htt were substrates for caspase cleavage. Analysis of early-stage HD patient samples suggests that caspase-mediated proteolysis of polyQ-Htt may precede neurodegeneration (Wellington et al., 2002).

In general, caspase inhibition has been shown to ameliorate disease progression and reduce phenotype severity in HD mice (Chen et al., 2000; Ona et al., 1999). A broad inhibition of caspases with ZVAD-fmk in clonal striatal cells led to a reduction of specific Htt fragments and increased viability, without changing the levels of aggregated inclusions. However, treatment with the caspase-3-specific inhibitor DEVD-fmk reduced aggregates without altering cleavage or increasing cell viability (Kim et al., 1999). The generation of mouse lines expressing caspase-3- and caspase-6-resistant polyQ-expanded Htt was pursued by eliminating these specific cleavage sites, and thereby unveiled a key role of caspase-6 cleavage of Htt at amino acid position 586. Eliminating this presumed caspase-6 cleavage site, but not the caspase-3 cleavage site, was found to be sufficient to protect HD mice from neuron dysfunction and neurodegeneration in vivo (Graham, Deng, & Carroll, 2010). A further study showed that caspase-6, but not caspase-3, is activated before the onset of motor abnormalities in HD mouse brain and in HD human brain. Caspase-6 activation correlated directly with the size of the polyQ expansion and inversely with age at onset (Graham et al., 2010). Furthermore, medium spiny neurons (MSNs) expressing caspase-6-resistant Htt showed a decreased susceptibility for *N*-Methyl-D-aspartic acid (NMDA)-induced excitotoxicity, associated with lack of caspase-6 activation, when compared to MSNs expressing unmodified Htt (Graham et al., 2010; Graham, Deng, & Slow, 2006; Milnerwood, Gladding, & Pouladi, 2010). By contrast, two caspase-6 knockout HD mouse models showed that production of a 586-amino-acid-length proteolytic fragment was not prevented in brain, undermining the exclusive involvement of caspase-6 in the cleavage of polyQ-Htt (Gafni, Papanikolaou, & Degiacomo, 2012; Landles, Weiss, Franklin, Howland, & Bates, 2012).

In the case of SCA7, in vitro assays have identified caspase-7 as the responsible proteolytic enzyme for ataxin-7 fragmentation (Young et al., 2007). The mutation of two specific caspase-cleavage sites in ataxin-7 not only resulted in resistance of polyQ-expanded ataxin-7 to caspase cleavage but also attenuated cell death, aggregate formation, and transcriptional interference. Fragments of ataxin-7 corresponding to products of caspase-7 cleavage have also been found in SCA7 mice, which exhibit increased caspase-7 activation and recruitment into the nucleus by polyQ-expanded ataxin-7 (Young et al., 2007). Nonetheless, full-length polyQ-expanded ataxin-7 can form inclusions without evidence for cleavage (Zander et al., 2001). TBP, the disease protein in SCA17, was reported to show fragmentation and fragment-dependent formation of aggregates in SCA17 mice (Friedman, Wang, Li, & Li, 2008), but in vitro assays did not show a TBP substrate-specificity for caspases (Wellington et al., 1998), suggesting different proteolytic enzymes likely mediate the truncation of TBP.

Correlating with the results for Htt, caspases-1 and -3, but not caspases-7 and -8, were shown to cleave ataxin-3 in vitro producing specific fragments (Wellington et al., 1998). Caspase-1-mediated fragmentation of polyQ-expanded ataxin-3 resulted in increased aggregation, and treatment with caspase inhibitors prevented aggregation formation in vitro (Berke, Schmied, Brunt, Ellerby, & Paulson, 2004). Another in vitro study showed that mutant ataxin-3 was cleaved to a lesser extent than wild-type ataxin-3 after a common initial proteolytic step, suggesting that such mutant fragments cannot be further processed. This may result in an accumulation of aggregation-prone polyQ-expanded ataxin-3 fragments (Pozzi et al., 2008). In a *Drosophila* model, cleavage of ataxin-3 appeared to be conserved and also caspase-mediated, inducing neuronal loss, which was mitigated by mutating multiple caspase sites in ataxin-3 (Jung, Xu, Lessing, & Bonini, 2009). CDK5 has been reported to have a role in caspase-mediated ataxin-3 cleavage, as RNAi of CDK5 in a *Drosophila* model for SCA3 resulted in enhanced SCA3 toxicity (Liman et al., 2014). Complicating the issue, an in vitro study based on patient-derived induced pluripotent stem cells (iPSCs) demonstrated that upon excitotoxic stress, ataxin-3 cleavage and aggregation were not prevented by pharmacological inhibition of caspases but was abolished by inhibiting calpain activity (Koch, Breuer, & Peitz, 2011).

Calpains and Other Proteolytic Enzymes

Another group of proteolytic enzymes associated with cleavage of polyQ-expanded proteins are calpains, a class of calcium-dependent cysteine proteases. These ubiquitously expressed enzymes exhibit a multitude of regulatory cellular functions and are specialized in modulating structure, localization, and activity of their substrates (Smith & Schnellmann, 2012; Sorimachi, Hata, & Ono, 2011). In human HD tissue and in brains from HD mice, increased expression level of calpains-1, -5, -7, and -10, and elevated enzyme activity have been reported (Cowan, Fan, & Fan, 2008; Gafni & Ellerby, 2002; Gafni et al., 2004; Gladding, Sepers, & Xu, 2012). An age-dependent attenuation of calpain activity was observed in one HD mouse model, suggesting alterations in calcium signaling mechanisms with disease progression (Dau, Gladding, Sepers, & Raymond, 2014). Both wild-type and mutant Htt were identified as calpain substrates, and calpain-dependent proteolytic cleavage products of Htt were detected in murine and human HD tissue (Gafni & Ellerby, 2002; Goffredo et al., 2002; Kim et al., 1999; Landles, Sathasivam, & Weiss, 2010). Caspase-3 cleavage-derived Htt fragments undergo further proteolysis by calpains, generating smaller products and suggesting a proteolytic pathway of serial processing events (Kim et al., 2001). Additionally, calpain-derived mHtt fragments accumulate in the nucleus (Gafni et al., 2004), which correlates with cytotoxicity and aggregation in HD (Hackam et al., 1998; Martindale et al., 1998). In cell models, inhibition of calpain cleavage by mutating putative cleavage sites within the mutant Htt protein resulted in decreased proteolysis, aggregation, and toxicity (Gafni et al., 2004). Concurrent with their activation after ischemic injury, calpains were also shown to cleave full-length Htt in infarcted rat cortex and striatum, producing amino-terminal fragments (Kim, Roh, & Yoon, 2003).

Calpain-dependent proteolysis of ataxin-3, akin to observations in HD, has also been repeatedly documented (Haacke et al., 2007; Hubener et al., 2013; Koch et al., 2011). Several putative calpain cleavage sites within the ataxin-3 protein have been identified (Haacke et al., 2007; Hubener et al., 2013; Koch et al., 2011; Simoes, Goncalves, & Koeppen, 2012), accounting for the generation of a carboxy-terminal polyQ-containing and aggregation-prone fragment (Haacke et al., 2006). After activation of calpains in vitro, fragments of various sizes appear. When the endogenous calpain inhibitor calpastatin (CAST) was coexpressed in treated cells, fragment generation was suppressed and aggregation of mutant ataxin-3 was decreased (Haacke et al., 2007). Overexpression of CAST using adeno-associated virus (AAV) vectors in a lentiviral mouse model of SCA3 resulted in reduced ataxin-3 proteolysis and a marked decrease in the size and number of intranuclear inclusions of ataxin-3, while simultaneously eliciting neuroprotection (Simoes et al., 2012). In a double-mutant CAST knock-out/SCA3 mouse model, absence of the endogenous calpain inhibitor led to greater ataxin-3 fragmentation, increased aggregate load, more severe neurodegeneration, and an obviously worse behavioral phenotype (Hubener et al., 2013). In line with these observations, CAST was shown to be depleted in mouse and human SCA3 brain tissue (Simoes et al., 2012). The neuronal specificity for the molecular mechanisms underlying SC3 pathology was explored using SCA3 patient-derived iPSCs (Koch et al., 2011). After neuronal differentiation and glutamate-induced calcium influx, excitation-induced ataxin-3 cleavage and aggregation were triggered. This was observed only in neurons, not in glial cells or fibroblasts, and was abolished by calpain inhibition (Koch et al., 2011).

While caspases and calpains account for the majority of cleavage events for polyQ-expanded disease proteins, several fragmentation events cannot be explained by their proteolytic activity. Another important group of enzymes to consider is the lysosomal cathepsins, which are known to process mutant Htt. Involvement of cathepsins-D, -B, -L, and -Z (Kim, Sapp, & Cuiffo, 2006; Qin et al., 2003; Ratovitski, Chighladze, Waldron, Hirschhorn, & Ross, 2011) can produce fragments termed cp-A and cp-B (Lunkes, Lindenberg, & Ben-Haiem, 2002). For the cp-A fragment, the protease responsible for its formation displayed cathepsin-D-like properties in immortalized neuronal cell lines and gamma-secretase-like properties in primary neurons, pointing to a cell-type specific involvement of different proteolytic enzymes (Kegel, Sapp, & Alexander, 2010). A screen using 514 protease-specific short interfering RNAs (siRNAs) detected 11 enzymes, including three members of the matrix metalloproteinase (MMP) family, that cleave Htt (Miller, Holcomb, & Al-Ramahi, 2010). When knocking down the most promising candidate (MMP-10) in a striatal cell line, cleavage of mutant Htt was prevented. Examination of HD mouse models indicated that MMPs were upregulated. Loss of function of *Drosophila* MMP homologs also ameliorated mutant Htt-induced neuronal dysfunction (Miller et al., 2010). Another novel explanation for the appearance of toxic fragments of Htt is that missplicing of Htt transcripts accounts for shortened N-terminal huntingtin variants (Sathasivam, Neueder, & Gipson, 2013).

In summary, work from numerous groups indicates that proteolytic processing of polyQ-expanded proteins by a variety of enzymes represents a pivotal step in the pathogenic cascade. Thus, this offers obvious opportunities for therapeutic intervention, as inhibition of the activity of the cleavage-responsible proteases should decrease the levels of toxic fragments. One approach is to inhibit the proteolytic activity of caspases, calpains, cathepsins, or MMPs directly. Using such methods, beneficial effects were achieved for HD (Gafni et al., 2004; Kim et al., 1999; Miller et al., 2010; Ratovitski et al., 2007; Wellington et al., 2000) and SCA3 (Berke et al., 2004; Haacke et al., 2007; Jung et al., 2009; Koch et al., 2011) as noted above, but attention should be paid to potential adverse effects as well (Ratovitski et al., 2007). A similar approach is to boost the expression of endogenous calpain inhibitors, such as calpastatin, as was done in SCA3 (Haacke et al., 2007; Hubener et al., 2013; Simoes et al., 2012). A second approach is to modulate alternate pathways and achieve off-target benefits. Treating HD model R6/2 mice with a tetracycline derivative delayed disease progression and neuron cell death by reducing the levels of caspases-1 and -3 (Li & Yuan, 2008) through upstream regulation of Apaf-1 (Sancho, Herrera, & Gortat, 2011). Reducing elevated calpain activity in HD mice also had beneficial off-target benefits(Cowan et al., 2008; Gladding et al., 2012). Another option is to use a genetic approach to modulate cleavage such as induction of exon 12 skipping in Htt pre-mRNA using oligonucleotides. This modification prevented the translation of the caspase-targeted region around amino acid 586 and thereby inhibited the formation of the key amino-terminal fragment implicated in causing HD neurotoxicity (Evers, Tran, & Zalachoras, 2014).

Nuclear Trafficking and Subcellular Localization

The ability of polyQ disease proteins to shuttle back and forth between the nucleus and cytosol affects not only their normal functions but also dictates their aggregation potential and ability to cause neurodegeneration. Nuclear transport affects many aspects of protein function and metabolism, including transcription regulation, turnover and half-life, and proteolysis (Fig. 7.3). In most cases, the localization of polyQ-expanded proteins to the nucleus induces a high level of toxicity in cells, and blocking nuclear transport, as has been done in mice, confirms that subcellular localization control is a reasonable therapeutic target (Bichelmeier, Schmidt, & Hubener, 2007; Jackson, Tallaksen-Greene, Albin, & Detloff, 2003; Yang et al., 2002). Nuclear entry is controlled by the nuclear pore complex, which serves as a selective gatekeeper of entry (Ribbeck & Gorlich, 2002). The nuclear pore complex recognizes a group of proteins known as karyopherins, which carry protein cargo for entry and exit out of the nucleus. Karyopherins recognize their cargo by the presence of a specific NLS) and/or NES on proteins (reviewed by Pemberton & Paschal, 2005). The most direct way for a protein to be transported by a karyopherin is to have an identifiable NLS or NES (or combination), but secondary features such as the accessibility of this motif and its post-translational modifications, such as phosphorylation, can alter whether the localization signal elicits a biological effect.

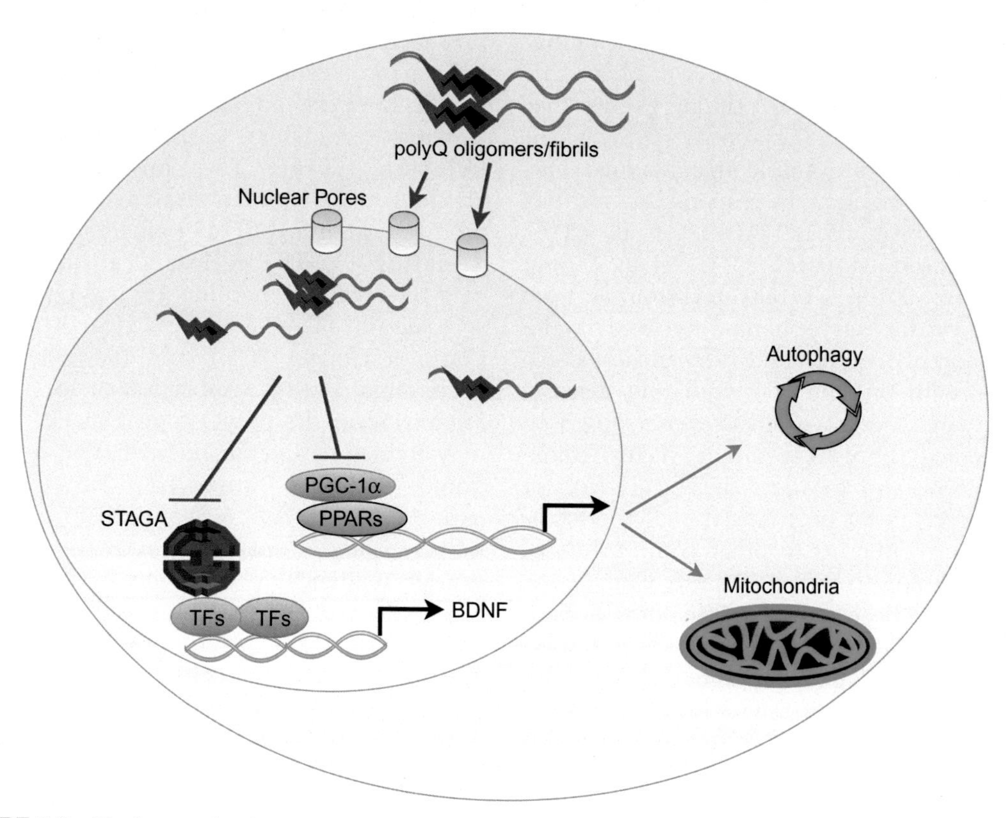

FIGURE 7.3 **Nuclear mechanisms in polyglutamine disease pathogenesis.** PolyQ-expanded proteins can shuttle into the nucleus, where they can interfere with transcription and avoid the protein clearance machinery in the cytosol. PolyQ-expanded proteins may alter nucleocytoplasmic transport at nuclear pores, as well. Transcription interference is envisioned to have multiple downstream effects, including deleterious impacts on mitochondrial function and the protein clearance machinery itself. *TFs*, transcription factors; *STAGA*, a transcription factor coactivator complex; *PGC-1α*, PPAR gamma coactivator 1 α; *PPARs*, peroxisome proliferator-activated receptors; *BDNF*, brain-derived neurotrophic factor.

NLS sequences and/or NES sequences have been identified and validated for the proteins that cause a number of polyQ diseases, including HD, SCA1, SCA3, and SCA7 (Irwin, Vandelft, & Pinchev, 2005; Kaytor et al., 1999; Klement et al., 1998; Taylor et al., 2006; Xia, Lee, Taylor, Vandelft, & Truant, 2003). In SCA1, the importance of nuclear localization for disease pathogenesis and ataxin-1 protein stability was established in now classic transgenic mouse experiments, where mutation of the NLS prevented polyQ-ataxin-1 protein from causing neurodegeneration in vivo (Klement et al., 1998). SCA3 and HD have been the thoroughly studied as well, and in particular with an eye toward possible therapeutic targeting to modify disease. Analysis of nuclear transport has involved fusing polyQ-Htt fragments to exogenous NES sequences and then showing that such machinations could prevent nuclear transport and decrease the toxicity of the fragment (Peters, Nucifora, & Kushi, 1999; Saudou et al., 1998); on the other hand, the reverse happened when such polyQ fragments were fused to an NLS (Peters et al., 1999). This work was reproduced in mice which exhibited a reduced lifespan, due to the added NLS (Schilling, Savonenko, & Klevytska, 2004). It has been difficult to tie together the cellular events that cause transport with the known pathways of nuclear entry, for example that ataxin-3 and Htt enter the nucleus in response to cellular stress and heat shock (Atwal et al., 2007; Munsie, Caron, & Atwal, 2011; Reina, Zhong, & Pittman, 2010). Current research is now focusing on the karyopherins involved in the recognition of the NLS and NES sites of these proteins with the aim of modulating disease. Cellular and oxidative stress were shown to alter the activity of Exportin1, a nuclear transport protein, and to affect the localization of polyQ proteins by posttranslational modifications of karyopherins or other components of the nuclear pore complex (Patel & Chu, 2011). Exportin1 has been shown to interact with both ataxin-3 and Htt NES sites (Chan, Tsoi, & Wu, 2011; Maiuri et al., 2013), and may be involved in the export of ataxin-7 (Taylor et al., 2006). Karyopherins B1 and B2 could regulate Htt subcellular localization by acting on a putative Htt NLS (Desmond, Atwal, Xia, & Truant, 2012). The AR is retained in the cytosol by heat shock proteins which mask its NLS but, upon binding to the androgen ligand, the NLS is exposed and AR translocates to the nucleus where it activates androgen-responsive genes (reviewed in Katsuno, Tanaka, & Adachi, 2012). The localization of the AR to the nucleus is considered necessary, but not sufficient, for disease development, as mice with a NLS deletion showed delayed onset of disease phenotype with a reduced motor deficit (Montie, Cho, & Holder, 2009).

In those polyQ disorders where an NLS or NES has not been identified, localization of the protein still appears critical for disease pathogenesis. Recent work using a polyQ-specific antibody has demonstrated that localization of ataxin-2 within the cell corresponds to the disease stages of SCA2. Cytosolic presence corresponded to early stage, while nuclear accumulation and aggregation correspond to more advanced stages of disease (Koyano, Yagishita, Kuroiwa, Tanaka, & Uchihara, 2014). The mislocalization of ataxin-2 has also been shown to be a potent modifier of ALS/TDP-43 toxicity (Elden, Kim, & Hart, 2010). The nuclear localization of the carboxy-terminal peptide of polyQ-CACNA1 is toxic in SCA6 (Kordasiewicz, Thompson, Clark, & Gomez, 2006), and the polyQ tract itself may affect subcellular localization. Expansion of the polyQ repeat in Htt protein may reduce its interaction with Tpr, a nuclear pore protein involved in nuclear export (Cornett, Cao, & Wang, 2005). Similarly, polyQ-expanded ataxin-3 and -7 exhibit a greater tendency to retention in the nucleus (Chai, Shao, Miller, Williams, & Paulson, 2002; Taylor et al., 2006). Overall, nuclear trafficking and subcellular localization are affected by many processes within the cell that are pathologically affected in polyQ disease, including posttranslational modification, aggregation, mitochondrial function, and transcription dysregulation.

OTHER MECHANISMS

The above-discussed mechanisms are not exhaustive. Other cellular processes have been implicated in the toxicity of mutant polyQ proteins. Since polyQ-expanded proteins are sticky, they can interfere with protein–protein binding and cellular signaling, organelle function (e.g., ER, mitochondria), intracellular calcium homeostasis, and axonal transport, as has been documented in various polyQ disorders (Gunawardena, Her, & Brusch, 2003; Szebenyi, Morfini, & Babcock, 2003; Tang, Slow, & Lupu, 2005; Zeron, Fernandes, & Krebs, 2004).

THERAPEUTIC OPPORTUNITIES FOR POLYQ DISEASES

As the expression of the toxic polyQ-containing protein drives all subsequent disease pathology, a very attractive therapeutic paradigm is to prevent the expression of the mutant gene product. One approach for terminating expression of a gene product is to target the messenger RNA for destruction or inhibit protein translation, a strategy known as "gene silencing." As interrupting the pathogenic cascade at its earliest step has great potential to dramatically alter disease course, interventions intended to reduce targeted gene expression are now considered among the most promising emerging therapies (Garriga-Canut, Agustin-Pavon, & Herrmann, 2012; Magen & Hornstein, 2014). There are four major approaches that are under active investigation to reduce mutant huntingtin expression: RNAi, antisense oligonucleotides (ASOs), zinc finger proteins (ZFPs), and genome editing (Fig. 7.4).

RNA Interference

In RNAi, an siRNA is introduced into cells and provides a substrate that is processed by the RNA-induced silencing complex, which is the cellular machinery that normally mediates the action of microRNAs (miRNAs)—endogenous RNAs that turn off gene expression. As delivery of naked double-stranded RNA is challenging, siRNA sequences can be encoded as short hairpin RNAs (shRNAs) that

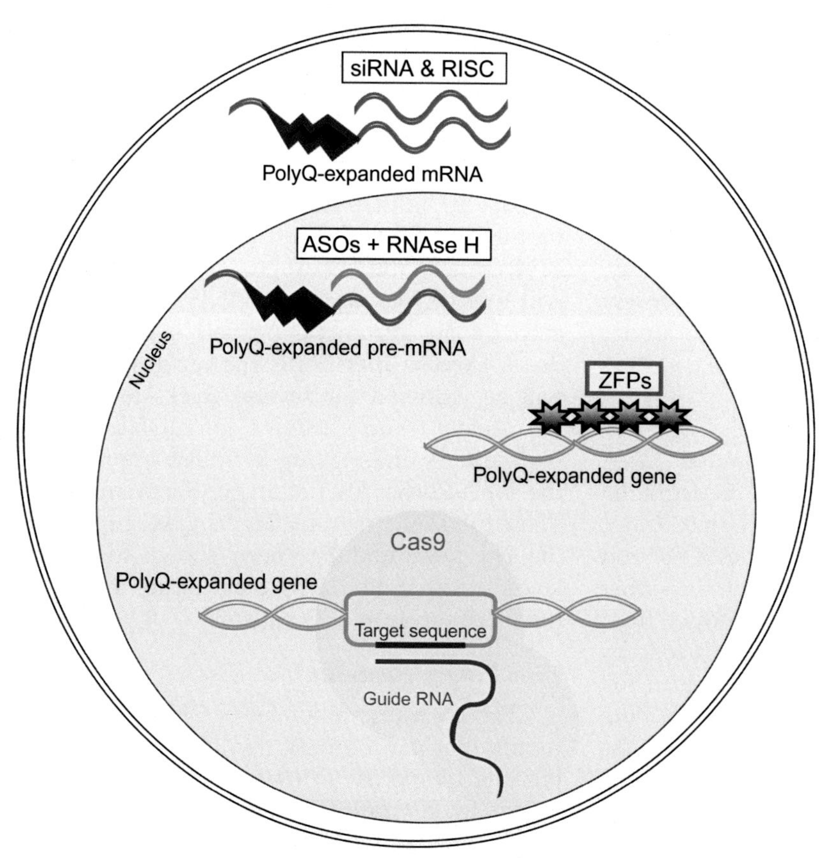

FIGURE 7.4 **Therapy approaches for polyglutamine diseases.** Here we see four different approaches that are under investigation to reduce mutant huntingtin expression: RNAi, ASOs, ZFPs, and genome editing (CRISPR/Cas9). *RNAi*, RNA interference; *ASOs*, antisense oligonucleotides; *ZFPs*, zinc finger proteins.

are expressed from viral vectors, such as AAV. While AAV delivery of shRNAs was found to be a robust and highly successful method for treating mouse models of polyQ disease (Xia, Mao, & Eliason, 2004), further studies demonstrated that high-level expression of exogenous shRNAs may cause toxicity (Grimm, Streetz, & Jopling, 2006). This prompted the development of artificial miRNA vectors, packaged into AAV, where gene-targeting shRNAs are instead embedded in the sequence of a miRNA so that the processing of the siRNA follows the full endogenous Drosher—Dicer pathway (Boudreau, Martins, & Davidson, 2009). At present, various methods for siRNA delivery are being pursued, ranging from siRNAs in nanoparticles to shRNA viral vectors to artificial miRNA viral vectors. In the case of viral vectors, delivery will require a neurosurgery procedure of stereotactic injection into specific brain regions in patient subjects, when clinical trials are initiated. In addition to viral vectors, exosomes (Alvarez-Erviti et al., 2011), cholesterol conjugation (DiFiglia, Sena-Esteves, & Chase, 2007), convection-enhanced delivery, and novel conjugates of single-stranded siRNA compounds (Lima, Prakash, & Murray, 2012; Yu, Pendergraff, & Liu, 2012) represent cutting-edge approaches that are also under development.

Antisense Oligonucleotides

ASOs are modified single-stranded DNA molecules, and ASO-bound mRNA is degraded by RNAse H (Martinez, Wright, Lopez-Fraga, Jimenez, & Paneda, 2013), or ASO-bound mRNA inhibits protein translation. In either case, the production of protein product from the target mRNA is typically reduced by greater than one-half. ASO dosage reduction can be "nonallele specific," meaning that both the disease allele and the normal allele mRNAs are targeted, or "allele-specific," meaning that the disease allele is selectively targeted, usually because of a single nucleotide polymorphism that is in linkage with the disease allele. While allele-specific targeting of the mutant gene would be ideal, numerous studies in HD mouse models have established the safety and efficacy of nonallele-specific ASOs (Boudreau et al., 2009; Kordasiewicz, Stanek, & Wancewicz, 2012). Because of dramatic successes in preclinical trial work in mice, ASO therapy is being advanced into human patients. Ionis Pharmaceuticals and Roche have initiated a randomized, double-blind, placebo-controlled Phase 1/2a clinical trial in early stage HD patients to evaluate the safety, tolerability, pharmacokinetics, and pharmacodynamics of ascending doses of an investigational ASO, IONIS-HTT_{Rx}. IONIS-HTT_{Rx} is the first therapy to enter clinical development that is designed to treat the underlying cause of this fatal disease and has been granted orphan drug designation by the FDA and the European Medicines Agency. As an ASO, IONIS-HTT_{Rx} binds to Htt RNA, promoting its destruction and decreasing the amount of Htt protein to cause toxicity. Research studies with IONIS-HTT_{Rx} in animal models of HD have been promising (Skotte, Southwell, & Ostergaard, 2014; Stanek, Yang, & Angus, 2013). Each patient will receive four doses of IONIS-HTT_{Rx} or placebo by an intrathecal (spinal) injection, with doses 4 weeks apart. After the last dose, patients will be followed for several weeks to monitor the safety and activity of IONIS-HTT_{Rx}. BioMarin is undertaking a similar approach to the Ionis-Roche ASO strategy (Sheridan, 2016). Their PRO289 program targets CAG repeats in the Htt gene and has been shown to reduce levels of expanded Htt protein in fibroblast cultures derived from HD patients.

Zinc Finger Proteins

ZFPs are transcription factor DNA-recognition motifs that can be designed to allow selective binding to specific DNA sequences, and when fused to a transcription repressor domain can be used to repress protein production by reducing transcription. They may combine the benefits of RNAi translational repression without the potential toxicity of mutant polyQ mRNA(Banez-Coronel, Porta, & Kagerbauer, 2012) or alternatively spliced genes that may lack the targeted mRNA sequence (Sathasivam et al., 2013) —alternative pathological mechanisms that have been proposed to occur in HD. Preliminary data indicates successful selective repression of mutant Htt and amelioration of motor manifestations in HD mice, but this approach still has the delivery and distribution hurdles of other viral vector methods (Garriga-Canut et al., 2012).

CRISPR

Genome editing with the recently discovered CRISPR/Cas9 system represents an alternative for contending with polyQ-expanded disorders (Cong, Ran, & Cox, 2013; Mali, Yang, & Esvelt, 2013). Treatment would involve expressing Cas9 along with a single-guide RNA (sgRNA) molecule. When coexpressed, sgRNAs bind and recruit Cas9 to a specific genomic target sequence where it mediates a double-strand DNA (dsDNA) break, activating the dsDNA break repair machinery. Targeted gene deletions by nonhomologous end joining can be made when a pair of sgRNA/Cas9 complexes binds in proximity and produce dsDNA breaks (Cong et al., 2013; Jinek et al., 2013; Ran et al., 2013). CRISPR/Cas9 gene editing in the brain reversed pathology and symptoms in a Htt 104Q-knock-in mouse model of HD (Yang, Chang, & Yang, 2017). An approach using personalized allele-specific CRISPR/Cas9 to completely inactivation of the mutant HTT allele without impacting the normal allele has also been piloted (Shin, Kim, & Chao, 2016). Such excision on the disease chromosome completely prevented the generation of mutant Htt mRNA and protein, unequivocally indicating permanent allele-specific inactivation of the HD mutant allele in primary fibroblasts, NPCs, and iPSCs from HD patients (Shin et al., 2016). The allele selectivity and broad applicability of this strategy in disorders with diverse disease haplotypes may support precision medicine through inactivation of many other gain-of-function mutations. Tests of this approach have not yet been reported in any other polyQ disorder.

Cpf1 is a more recently characterized nuclease with multiple advantages compared to Cas9 (Zetsche, Gootenberg, & Abudayyeh, 2015). Two studies show that Cpf1 displays lower off-target editing than Cas9, confirming that this protein is well suited for genome editing (Kim et al., 2016; Kleinstiver, Tsai, & Prew, 2016). Cpf1 requires only one RNA rather than the two (tracrRNA and crRNA) needed by Cas9 for cleavage (Zetsche et al., 2015). Cpf1 cleaves DNA in a staggered pattern opening up the possibility of directional gene transfer (Zetsche et al., 2015). Sticky-end-mediated gene transfer would be particularly helpful for targeting nondividing cells such as neurons, which are difficult to modify through homology-directed repair. Both Cpf1 and its guide RNAs are smaller than their SpCas9 counterparts, so they should be easier to deliver in low-capacity vectors like AAV, which are commonly used for in vivo gene delivery due to their low immunogenicity and range of serotypes allowing preferential infection of target tissues.

Concluding Remarks

PolyQ diseases share a number of key features, including an inverse correlation between CAG repeat length and age at onset/disease severity, involvement of the CNS, and an autosomal-dominant mode of inheritance due to a gain-of-function toxicity mechanism. The expanded polyQ domains in these unrelated proteins also share common pathogenic mechanisms, which offers the prospect of developing therapeutic interventions that could be efficacious for treating multiple polyQ disorders.

One appealing option is to target the polyQ-expanded disease protein for enhanced turnover by modifying posttranslational modifications. Another promising approach could prevent proteolytic cleavage and the creation of a toxic fragment, which should reduce the pathogenic cascade for pathways that involve the accumulation of oligomers, fibrils, or aggregates. Yet another option is to decrease the ability of the disease protein to reach the site of cellular toxicity by altering its translocation between the cytosol and the nucleus. Other prospects for therapy include facilitating disease protein clearance by promoting proteasomal degradation, activating the autophagy pathway, or insuring optimal lysosomal function. Enhancing mitochondrial homeostasis and function will also likely have a beneficial cascade in these diseases, as increased ATP production, better resistance to oxidative stress, and improved buffering of intracellular calcium will remedy pathological processes that always emerge in the course of disease progression. The promise of dosage reduction, either nonspecific or specific, is currently being pursued by multiple research groups, often in collaboration with biopharma. While there is currently no "cure" for any of these diseases, there are currently many possibilities being pursued to develop effective disease-modifying interventions.

Acknowledgments

Our work on polyglutamine repeat diseases is funded by the National Institutes of Health, the Muscular Dystrophy Association, and the Harrington Discovery Institute.

References

Adachi, H., Waza, M., Tokui, K., et al. (2007). CHIP overexpression reduces mutant androgen receptor protein and ameliorates phenotypes of the spinal and bulbar muscular atrophy transgenic mouse model. *The Journal of Neuroscience*, *27*, 5115–5126.

Al-Ramahi, I., Lam, Y. C., Chen, H. K., et al. (2006). CHIP protects from the neurotoxicity of expanded and wild-type ataxin-1 and promotes their ubiquitination and degradation. *The Journal of Biological Chemistry*, *281*, 26714–26724.

Alvarez-Erviti, L., Seow, Y., Yin, H., Betts, C., Lakhal, S., & Wood, M. J. (2011). Delivery of siRNA to the mouse brain by systemic injection of targeted exosomes. *Nature Biotechnology*, *29*, 341–345.

Arnelle, D. R., & Stamler, J. S. (1995). NO +, NO, and NO − donation by *S*-nitrosothiols: Implications for regulation of physiological functions by *S*-nitrosylation and acceleration of disulfide formation. *Archives of Biochemistry and Biophysics*, *318*, 279–285.

Arrasate, M., & Finkbeiner, S. (2012). Protein aggregates in Huntington's disease. *Experimental Neurology*, *238*, 1–11.

Arrasate, M., Mitra, S., Schweitzer, E. S., Segal, M. R., & Finkbeiner, S. (2004). Inclusion body formation reduces levels of mutant huntingtin and the risk of neuronal death. *Nature*, *431*, 805–810.

Ashkenazi, A., Bento, C. F., Ricketts, T., et al. (2017). Polyglutamine tracts regulate beclin 1-dependent autophagy. *Nature*, *545*, 108–111.

Atwal, R. S., Xia, J., Pinchev, D., Taylor, J., Epand, R. M., & Truant, R. (2007). Huntingtin has a membrane association signal that can modulate huntingtin aggregation, nuclear entry and toxicity. *Human Molecular Genetics*, *16*, 2600–2615.

Aziz, N. A., van der Burg, J. M., Landwehrmeyer, G. B., Brundin, P., Stijnen, T., & Roos, R. A. (2008). Weight loss in Huntington disease increases with higher CAG repeat number. *Neurology*, *71*, 1506–1513.

Babcock, D. T., & Ganetzky, B. (2015). Transcellular spreading of huntingtin aggregates in the *Drosophila* brain. *Proceedings of the National Academy of Sciences of the United States of America*, *112*, E5427–E5433.

Bae, B. I., Xu, H., Igarashi, S., et al. (2005). p53 mediates cellular dysfunction and behavioral abnormalities in Huntington's disease. *Neuron*, *47*, 29–41.

Banez-Coronel, M., Porta, S., Kagerbauer, B., et al. (2012). A pathogenic mechanism in Huntington's disease involves small CAG-repeated RNAs with neurotoxic activity. *PLoS Genetics*, *8*, e1002481.

Beal, M. F., Brouillet, E., Jenkins, B. G., et al. (1993). Neurochemical and histologic characterization of striatal excitotoxic lesions produced by the mitochondrial toxin 3-nitropropionic acid. *The Journal of Neuroscience*, *13*, 4181–4192.

Becher, M. W., Kotzuk, J. A., Sharp, A. H., et al. (1998). Intranuclear neuronal inclusions in Huntington's disease and dentatorubral and pallidoluysian atrophy: Correlation between the density of inclusions and IT15

CAG triplet repeat length. *Neurobiology of Disease, 4,* 387–397.

Bence, N. F., Sampat, R. M., & Kopito, R. R. (2001). Impairment of the ubiquitin-proteasome system by protein aggregation. *Science (New York, N.Y.), 292,* 1552–1555.

Berger, Z., Ttofi, E. K., Michel, C. H., et al. (2005). Lithium rescues toxicity of aggregate-prone proteins in *Drosophila* by perturbing Wnt pathway. *Human Molecular Genetics, 14,* 3003–3011.

Berke, S. J., Schmied, F. A., Brunt, E. R., Ellerby, L. M., & Paulson, H. L. (2004). Caspase-mediated proteolysis of the polyglutamine disease protein ataxin-3. *Journal of Neurochemistry, 89,* 908–918.

Besnault-Mascard, L., Leprince, C., Auffredou, M. T., et al. (2005). Caspase-8 sumoylation is associated with nuclear localization. *Oncogene, 24,* 3268–3273.

Bichelmeier, U., Schmidt, T., Hubener, J., et al. (2007). Nuclear localization of ataxin-3 is required for the manifestation of symptoms in SCA3: In vivo evidence. *The Journal of Neuroscience, 27,* 7418–7428.

Blazek, E., Mittler, G., & Meisterernst, M. (2005). The mediator of RNA polymerase II. *Chromosoma, 113,* 399–408.

Boudreau, R. L., Martins, I., & Davidson, B. L. (2009). Artificial microRNAs as siRNA shuttles: Improved safety as compared to shRNAs in vitro and in vivo. *Molecular Therapy: The Journal of the American Society of Gene Therapy, 17,* 169–175.

Bowman, A. B., Yoo, S. Y., Dantuma, N. P., & Zoghbi, H. Y. (2005). Neuronal dysfunction in a polyglutamine disease model occurs in the absence of ubiquitin-proteasome system impairment and inversely correlates with the degree of nuclear inclusion formation. *Human Molecular Genetics, 14,* 679–691.

Brinkman, R. R., Mezei, M. M., Theilmann, J., Almqvist, E., & Hayden, M. R. (1997). The likelihood of being affected with Huntington disease by a particular age, for a specific CAG size. *American Journal of Human Genetics, 60,* 1202–1210.

Bucciantini, M., Giannoni, E., Chiti, F., et al. (2002). Inherent toxicity of aggregates implies a common mechanism for protein misfolding diseases. *Nature, 416,* 507–511.

Burnett, B., Li, F., & Pittman, R. N. (2003). The polyglutamine neurodegenerative protein ataxin-3 binds polyubiquitylated proteins and has ubiquitin protease activity. *Human Molecular Genetics, 12,* 3195–3205.

Candiani, S., Pestarino, M., Cattaneo, E., & Tartari, M. (2007). Characterization, developmental expression and evolutionary features of the huntingtin gene in the amphioxus Branchiostoma floridae. *BMC Developmental Biology, 7,* 127.

Carmichael, J., Sugars, K. L., Bao, Y. P., & Rubinsztein, D. C. (2002). Glycogen synthase kinase-3beta inhibitors prevent cellular polyglutamine toxicity caused by the Huntington's disease mutation. *The Journal of Biological Chemistry, 277,* 33791–33798.

Chai, Y., Berke, S. S., Cohen, R. E., & Paulson, H. L. (2004). Poly-ubiquitin binding by the polyglutamine disease protein ataxin-3 links its normal function to protein surveillance pathways. *The Journal of Biological Chemistry, 279,* 3605–3611.

Chai, Y., Koppenhafer, S. L., Bonini, N. M., & Paulson, H. L. (1999). Analysis of the role of heat shock protein (Hsp) molecular chaperones in polyglutamine disease. *The Journal of Neuroscience, 19,* 10338–10347.

Chai, Y., Shao, J., Miller, V. M., Williams, A., & Paulson, H. L. (2002). Live-cell imaging reveals divergent intracellular dynamics of polyglutamine disease proteins and supports a sequestration model of pathogenesis. *Proceedings of the National Academy of Sciences of the United States of America, 99,* 9310–9315.

Chan, H. Y., Warrick, J. M., Andriola, I., Merry, D., & Bonini, N. M. (2002). Genetic modulation of polyglutamine toxicity by protein conjugation pathways in *Drosophila. Human Molecular Genetics, 11,* 2895–2904.

Chan, W. M., Tsoi, H., Wu, C. C., et al. (2011). Expanded polyglutamine domain possesses nuclear export activity which modulates subcellular localization and toxicity of polyQ disease protein via exportin-1. *Human Molecular Genetics, 20,* 1738–1750.

Chang, D. T., Rintoul, G. L., Pandipati, S., & Reynolds, I. J. (2006). Mutant huntingtin aggregates impair mitochondrial movement and trafficking in cortical neurons. *Neurobiology of Disease, 22,* 388–400.

Chang, K. H., Chen, W. L., Wu, Y. R., et al. (2014). Aqueous extract of *Gardenia jasminoides* targeting oxidative stress to reduce polyQ aggregation in cell models of spinocerebellar ataxia 3. *Neuropharmacology, 81,* 166–175.

Chen, M., Ona, V. O., Li, M., et al. (2000). Minocycline inhibits caspase-1 and caspase-3 expression and delays mortality in a transgenic mouse model of Huntington disease. *Nature Medicine, 6,* 797–801.

Choi, J. Y., Ryu, J. H., Kim, H. S., et al. (2007). Co-chaperone CHIP promotes aggregation of ataxin-1. *Molecular and Cellular Neurosciences, 34,* 69–79.

Chou, A. H., Lin, A. C., Hong, K. Y., et al. (2011). p53 activation mediates polyglutamine-expanded ataxin-3 upregulation of Bax expression in cerebellar and pontine nuclei neurons. *Neurochemistry International, 58,* 145–152.

Chou, A. H., Yeh, T. H., Kuo, Y. L., et al. (2006). Polyglutamine-expanded ataxin-3 activates mitochondrial apoptotic pathway by upregulating Bax and downregulating Bcl-xL. *Neurobiology of Disease, 21,* 333–345.

Chua, J. P., Reddy, S. L., Yu, Z., et al. (2015). Disrupting SUMOylation enhances transcriptional function and ameliorates polyglutamine androgen receptor-mediated disease. *The Journal of Clinical Investigation, 125*, 831–845.

Ciechanover, A. (2006). The ubiquitin proteolytic system: From a vague idea, through basic mechanisms, and onto human diseases and drug targeting. *Neurology, 66*, S7–S19.

Cohen-Carmon, D., & Meshorer, E. (2012). Polyglutamine (polyQ) disorders: The chromatin connection. *Nucleus, 3*, 433–441.

Conaway, J. W., Florens, L., Sato, S., et al. (2005). The mammalian Mediator complex. *FEBS Letters, 579*, 904–908.

Cong, L., Ran, F. A., Cox, D., et al. (2013). Multiplex genome engineering using CRISPR/Cas systems. *Science (New York, N.Y.), 339*, 819–823.

Cornett, J., Cao, F., Wang, C. E., et al. (2005). Polyglutamine expansion of huntingtin impairs its nuclear export. *Nature Genetics, 37*, 198–204.

Costa, V., Giacomello, M., Hudec, R., et al. (2010). Mitochondrial fission and cristae disruption increase the response of cell models of Huntington's disease to apoptotic stimuli. *EMBO Molecular Medicine, 2*, 490–503.

Costa, V., & Scorrano, L. (2012). Shaping the role of mitochondria in the pathogenesis of Huntington's disease. *The EMBO Journal, 31*, 1853–1864.

Cowan, C. M., Fan, M. M., Fan, J., et al. (2008). Polyglutamine-modulated striatal calpain activity in YAC transgenic Huntington disease mouse model: Impact on NMDA receptor function and toxicity. *The Journal of Neuroscience, 28*, 12725–12735.

Cui, L., Jeong, H., Borovecki, F., Parkhurst, C. N., Tanese, N., & Krainc, D. (2006). Transcriptional repression of PGC-1alpha by mutant huntingtin leads to mitochondrial dysfunction and neurodegeneration. *Cell, 127*, 59–69.

Cummings, C. J., Mancini, M. A., Antalffy, B., DeFranco, D. B., Orr, H. T., & Zoghbi, H. Y. (1998). Chaperone suppression of aggregation and altered subcellular proteasome localization imply protein misfolding in SCA1. *Nature Genetics, 19*, 148–154.

Dau, A., Gladding, C. M., Sepers, M. D., & Raymond, L. A. (2014). Chronic blockade of extrasynaptic NMDA receptors ameliorates synaptic dysfunction and pro-death signaling in Huntington disease transgenic mice. *Neurobiology of Disease, 62*, 533–542.

Davidson, J. D., Riley, B., Burright, E. N., Duvick, L. A., Zoghbi, H. Y., & Orr, H. T. (2000). Identification and characterization of an ataxin-1-interacting protein: A1Up, a ubiquitin-like nuclear protein. *Human Molecular Genetics, 9*, 2305–2312.

Davies, S. W., Turmaine, M., Cozens, B. A., et al. (1997). Formation of neuronal intranuclear inclusions underlies the neurological dysfunction in mice transgenic for the HD mutation. *Cell, 90*, 537–548.

Desmond, C. R., Atwal, R. S., Xia, J., & Truant, R. (2012). Identification of a karyopherin beta1/beta2 proline-tyrosine nuclear localization signal in huntingtin protein. *The Journal of Biological Chemistry, 287*, 39626–39633.

Diaz-Hernandez, M., Valera, A. G., Moran, M. A., et al. (2006). Inhibition of 26S proteasome activity by huntingtin filaments but not inclusion bodies isolated from mouse and human brain. *Journal of Neurochemistry, 98*, 1585–1596.

Dickey, A. S., Pineda, V. V., Tsunemi, T., et al. (2016). PPAR-delta is repressed in Huntington's disease, is required for normal neuronal function and can be targeted therapeutically. *Nature Medicine, 22*, 37–45.

Dickey, A. S., & Strack, S. (2011). PKA/AKAP1 and PP2A/Bbeta2 regulate neuronal morphogenesis via Drp1 phosphorylation and mitochondrial bioenergetics. *The Journal of Neuroscience, 31*, 15716–15726.

DiFiglia, M., Sapp, E., Chase, K. O., et al. (1997). Aggregation of huntingtin in neuronal intranuclear inclusions and dystrophic neurites in brain. *Science (New York, N.Y.), 277*, 1990–1993.

DiFiglia, M., Sena-Esteves, M., Chase, K., et al. (2007). Therapeutic silencing of mutant huntingtin with siRNA attenuates striatal and cortical neuropathology and behavioral deficits. *Proceedings of the National Academy of Sciences of the United States of America, 104*, 17204–17209.

Djousse, L., Knowlton, B., Cupples, L. A., Marder, K., Shoulson, I., & Myers, R. H. (2002). Weight loss in early stage of Huntington's disease. *Neurology, 59*, 1325–1330.

Donaldson, K. M., Li, W., Ching, K. A., Batalov, S., Tsai, C. C., & Joazeiro, C. A. (2003). Ubiquitin-mediated sequestration of normal cellular proteins into polyglutamine aggregates. *Proceedings of the National Academy of Sciences of the United States of America, 100*, 8892–8897.

Doss-Pepe, E. W., Stenroos, E. S., Johnson, W. G., & Madura, K. (2003). Ataxin-3 interactions with rad23 and valosin-containing protein and its associations with ubiquitin chains and the proteasome are consistent with a role in ubiquitin-mediated proteolysis. *Molecular and Cellular Biology, 23*, 6469–6483.

Duan, W., Peng, Q., Masuda, N., et al. (2008). Sertraline slows disease progression and increases neurogenesis in N171-82Q mouse model of Huntington's disease. *Neurobiology of Disease, 30*, 312–322.

Dunah, A. W., Jeong, H., Griffin, A., et al. (2002). Sp1 and TAFII130 transcriptional activity disrupted in early

Huntington's disease. *Science (New York, N.Y.), 296*, 2238–2243.

Duncan, C., Papanikolaou, T., & Ellerby, L. M. (2010). Autophagy: polyQ toxic fragment turnover. *Autophagy, 6*, 312–314.

Durcan, T. M., Kontogiannea, M., Bedard, N., Wing, S. S., & Fon, E. A. (2012). Ataxin-3 deubiquitination is coupled to Parkin ubiquitination via E2 ubiquitin-conjugating enzyme. *The Journal of Biological Chemistry, 287*, 531–541.

Duvick, L., Barnes, J., Ebner, B., et al. (2010). SCA1-like disease in mice expressing wild-type ataxin-1 with a serine to aspartic acid replacement at residue 776. *Neuron, 67*, 929–935.

Einum, D. D., Townsend, J. J., Ptacek, L. J., & Fu, Y. H. (2001). Ataxin-7 expression analysis in controls and spinocerebellar ataxia type 7 patients. *Neurogenetics, 3*, 83–90.

Elden, A. C., Kim, H. J., Hart, M. P., et al. (2010). Ataxin-2 intermediate-length polyglutamine expansions are associated with increased risk for ALS. *Nature, 466*, 1069–1075.

Ellerby, L. M., Hackam, A. S., Propp, S. S., et al. (1999). Kennedy's disease: Caspase cleavage of the androgen receptor is a crucial event in cytotoxicity. *Journal of Neurochemistry, 72*, 185–195.

Ellisdon, A. M., Thomas, B., & Bottomley, S. P. (2006). The two-stage pathway of ataxin-3 fibrillogenesis involves a polyglutamine-independent step. *The Journal of Biological Chemistry, 281*, 16888–16896.

Emamian, E. S., Kaytor, M. D., Duvick, L. A., et al. (2003). Serine 776 of ataxin-1 is critical for polyglutamine-induced disease in SCA1 transgenic mice. *Neuron, 38*, 375–387.

Emir, U. E., Brent Clark, H., Vollmers, M. L., Eberly, L. E., & Oz, G. (2013). Non-invasive detection of neurochemical changes prior to overt pathology in a mouse model of spinocerebellar ataxia type 1. *Journal of Neurochemistry, 127*, 660–668.

Evers, M. M., Tran, H. D., Zalachoras, I., et al. (2014). Preventing formation of toxic N-terminal huntingtin fragments through antisense oligonucleotide-mediated protein modification. *Nucleic Acid Therapeutics, 24*, 4–12.

Evert, B. O., Araujo, J., Vieira-Saecker, A. M., et al. (2006). Ataxin-3 represses transcription via chromatin binding, interaction with histone deacetylase 3, and histone deacetylation. *The Journal of Neuroscience, 26*, 11474–11486.

Falkowska, A., Gutowska, I., Goschorska, M., Nowacki, P., Chlubek, D., & Baranowska-Bosiacka, I. (2015). Energy metabolism of the brain, including the cooperation between astrocytes and neurons, especially in the context of glycogen metabolism. *International Journal of Molecular Sciences, 16*, 25959–25981.

Fan, H. C., Ho, L. I., Chi, C. S., et al. (2014). Polyglutamine (PolyQ) diseases: Genetics to treatments. *Cell Transplantation, 23*, 441–458.

Fox, J. H., Connor, T., Stiles, M., et al. (2011). Cysteine oxidation within N-terminal mutant huntingtin promotes oligomerization and delays clearance of soluble protein. *The Journal of Biological Chemistry, 286*, 18320–18330.

Friedman, M. J., Wang, C. E., Li, X. J., & Li, S. (2008). Polyglutamine expansion reduces the association of TATA-binding protein with DNA and induces DNA binding-independent neurotoxicity. *The Journal of Biological Chemistry, 283*, 8283–8290.

Fujikake, N., Nagai, Y., Popiel, H. A., Okamoto, Y., Yamaguchi, M., & Toda, T. (2008). Heat shock transcription factor 1-activating compounds suppress polyglutamine-induced neurodegeneration through induction of multiple molecular chaperones. *The Journal of Biological Chemistry, 283*, 26188–26197.

Gafni, J., & Ellerby, L. M. (2002). Calpain activation in Huntington's disease. *The Journal of Neuroscience, 22*, 4842–4849.

Gafni, J., Hermel, E., Young, J. E., Wellington, C. L., Hayden, M. R., & Ellerby, L. M. (2004). Inhibition of calpain cleavage of huntingtin reduces toxicity: Accumulation of calpain/caspase fragments in the nucleus. *The Journal of Biological Chemistry, 279*, 20211–20220.

Gafni, J., Papanikolaou, T., Degiacomo, F., et al. (2012). Caspase-6 activity in a BACHD mouse modulates steady-state levels of mutant huntingtin protein but is not necessary for production of a 586 amino acid proteolytic fragment. *The Journal of Neuroscience, 32*, 7454–7465.

Garcia-Martinez, J. M., Perez-Navarro, E., Xifro, X., et al. (2007). BH3-only proteins Bid and Bim(EL) are differentially involved in neuronal dysfunction in mouse models of Huntington's disease. *Journal of Neuroscience Research, 85*, 2756–2769.

Garden, G. A., Libby, R. T., Fu, Y. H., et al. (2002). Polyglutamine-expanded ataxin-7 promotes non-cell-autonomous Purkinje cell degeneration and displays proteolytic cleavage in ataxic transgenic mice. *The Journal of Neuroscience, 22*, 4897–4905.

Garriga-Canut, M., Agustin-Pavon, C., Herrmann, F., et al. (2012). Synthetic zinc finger repressors reduce mutant huntingtin expression in the brain of R6/2 mice. *Proceedings of the National Academy of Sciences of the United States of America, 109*, E3136–E3145.

Gatchel, J. R., & Zoghbi, H. Y. (2005). Diseases of unstable repeat expansion: Mechanisms and common principles. *Nature Reviews Genetics, 6*, 743–755.

Gill, G. (2003). Post-translational modification by the small ubiquitin-related modifier SUMO has big effects on transcription factor activity. *Current Opinion in Genetics & Development, 13*, 108–113.

Gladding, C. M., Sepers, M. D., Xu, J., et al. (2012). Calpain and STriatal-Enriched protein tyrosine phosphatase (STEP) activation contribute to extrasynaptic NMDA receptor localization in a Huntington's disease mouse model. *Human Molecular Genetics, 21*, 3739–3752.

Goffredo, D., Rigamonti, D., Tartari, M., et al. (2002). Calcium-dependent cleavage of endogenous wild-type huntingtin in primary cortical neurons. *The Journal of Biological Chemistry, 277*, 39594–39598.

Goldberg, Y. P., Nicholson, D. W., Rasper, D. M., et al. (1996). Cleavage of huntingtin by apopain, a proapoptotic cysteine protease, is modulated by the polyglutamine tract. *Nature Genetics, 13*, 442–449.

Goti, D., Katzen, S. M., Mez, J., et al. (2004). A mutant ataxin-3 putative-cleavage fragment in brains of Machado–Joseph disease patients and transgenic mice is cytotoxic above a critical concentration. *The Journal of Neuroscience, 24*, 10266–10279.

Graham, R. K., Deng, Y., Carroll, J., et al. (2010). Cleavage at the 586 amino acid caspase-6 site in mutant huntingtin influences caspase-6 activation in vivo. *The Journal of Neuroscience, 30*, 15019–15029.

Graham, R. K., Deng, Y., Slow, E. J., et al. (2006). Cleavage at the caspase-6 site is required for neuronal dysfunction and degeneration due to mutant huntingtin. *Cell, 125*, 1179–1191.

Grimm, D., Streetz, K. L., Jopling, C. L., et al. (2006). Fatality in mice due to oversaturation of cellular microRNA/short hairpin RNA pathways. *Nature, 441*, 537–541.

Guerrini, L., Lolli, F., Ginestroni, A., et al. (2004). Brainstem neurodegeneration correlates with clinical dysfunction in SCA1 but not in SCA2. A quantitative volumetric, diffusion and proton spectroscopy MR study. *Brain, 127*, 1785–1795.

Gunawardena, S., Her, L. S., Brusch, R. G., et al. (2003). Disruption of axonal transport by loss of huntingtin or expression of pathogenic polyQ proteins in *Drosophila*. *Neuron, 40*, 25–40.

Haacke, A., Broadley, S. A., Boteva, R., Tzvetkov, N., Hartl, F. U., & Breuer, P. (2006). Proteolytic cleavage of polyglutamine-expanded ataxin-3 is critical for aggregation and sequestration of non-expanded ataxin-3. *Human Molecular Genetics, 15*, 555–568.

Haacke, A., Hartl, F. U., & Breuer, P. (2007). Calpain inhibition is sufficient to suppress aggregation of polyglutamine-expanded ataxin-3. *The Journal of Biological Chemistry, 282*, 18851–18856.

Hackam, A. S., Singaraja, R., Wellington, C. L., et al. (1998). The influence of huntingtin protein size on nuclear localization and cellular toxicity. *The Journal of Cell Biology, 141*, 1097–1105.

Handschin, C., & Spiegelman, B. M. (2006). Peroxisome proliferator-activated receptor gamma coactivator 1 coactivators, energy homeostasis, and metabolism. *Endocrine Reviews, 27*, 728–735.

Hannich, J. T., Lewis, A., Kroetz, M. B., et al. (2005). Defining the SUMO-modified proteome by multiple approaches in Saccharomyces cerevisiae. *The Journal of Biological Chemistry, 280*, 4102–4110.

Havel, L. S., Li, S., & Li, X. J. (2009). Nuclear accumulation of polyglutamine disease proteins and neuropathology. *Molecular Brain, 2*, 21.

Hayashi, N., Shirakura, H., Uehara, T., & Nomura, Y. (2006). Relationship between SUMO-1 modification of caspase-7 and its nuclear localization in human neuronal cells. *Neuroscience Letters, 397*, 5–9.

Hecker, C. M., Rabiller, M., Haglund, K., Bayer, P., & Dikic, I. (2006). Specification of SUMO1- and SUMO2-interacting motifs. *The Journal of Biological Chemistry, 281*, 16117–16127.

Helmlinger, D., Hardy, S., Abou-Sleymane, G., et al. (2006). Glutamine-expanded ataxin-7 alters TFTC/STAGA recruitment and chromatin structure leading to photoreceptor dysfunction. *PLoS Biology, 4*, e67.

Helmlinger, D., Hardy, S., Sasorith, S., et al. (2004). Ataxin-7 is a subunit of GCN5 histone acetyltransferase-containing complexes. *Human Molecular Genetics, 13*, 1257–1265.

Hermel, E., Gafni, J., Propp, S. S., et al. (2004). Specific caspase interactions and amplification are involved in selective neuronal vulnerability in Huntington's disease. *Cell Death and Differentiation, 11*, 424–438.

Hirabayashi, M., Inoue, K., Tanaka, K., et al. (2001). VCP/p97 in abnormal protein aggregates, cytoplasmic vacuoles, and cell death, phenotypes relevant to neurodegeneration. *Cell Death and Differentiation, 8*, 977–984.

Hoffner, G., & Djian, P. (2014). Monomeric, oligomeric and polymeric proteins in Huntington disease and other diseases of polyglutamine expansion. *Brain Sciences, 4*, 91–122.

Holmberg, M., Duyckaerts, C., Durr, A., et al. (1998). Spinocerebellar ataxia type 7 (SCA7): A neurodegenerative disorder with neuronal intranuclear inclusions. *Human Molecular Genetics, 7*, 913–918.

Hong, S., Kim, S. J., Ka, S., Choi, I., & Kang, S. (2002). USP7, a ubiquitin-specific protease, interacts with ataxin-1, the SCA1 gene product. *Molecular and Cellular Neurosciences, 20*, 298–306.

Hong, S., Lee, S., Cho, S. G., & Kang, S. (2008). UbcH6 interacts with and ubiquitinates the SCA1 gene product ataxin-1. *Biochemical and Biophysical Research Communications, 371*, 256–260.

Hubener, J., Vauti, F., Funke, C., et al. (2011). N-terminal ataxin-3 causes neurological symptoms with inclusions, endoplasmic reticulum stress and ribosomal dislocation. *Brain, 134*, 1925–1942.

Hubener, J., Weber, J. J., Richter, C., et al. (2013). Calpain-mediated ataxin-3 cleavage in the molecular pathogenesis of spinocerebellar ataxia type 3 (SCA3). *Human Molecular Genetics, 22*, 508–518.

Huynh, D. P., Figueroa, K., Hoang, N., & Pulst, S. M. (2000). Nuclear localization or inclusion body formation of ataxin-2 are not necessary for SCA2 pathogenesis in mouse or human. *Nature Genetics, 26*, 44–50.

Huynh, D. P., Nguyen, D. T., Pulst-Korenberg, J. B., Brice, A., & Pulst, S. M. (2007). Parkin is an E3 ubiquitin-ligase for normal and mutant ataxin-2 and prevents ataxin-2-induced cell death. *Experimental Neurology, 203*, 531–541.

Ikeda, H., Yamaguchi, M., Sugai, S., Aze, Y., Narumiya, S., & Kakizuka, A. (1996). Expanded polyglutamine in the Machado-Joseph disease protein induces cell death in vitro and in vivo. *Nature Genetics, 13*, 196–202.

Irwin, S., Vandelft, M., Pinchev, D., et al. (2005). RNA association and nucleocytoplasmic shuttling by ataxin-1. *Journal of Cell Science, 118*, 233–242.

Iuchi, S., Hoffner, G., Verbeke, P., Djian, P., & Green, H. (2003). Oligomeric and polymeric aggregates formed by proteins containing expanded polyglutamine. *Proceedings of the National Academy of Sciences of the United States of America, 100*, 2409–2414.

Iwata, A., Christianson, J. C., Bucci, M., et al. (2005). Increased susceptibility of cytoplasmic over nuclear polyglutamine aggregates to autophagic degradation. *Proceedings of the National Academy of Sciences of the United States of America, 102*, 13135–13140.

Jackson, W. S., Tallaksen-Greene, S. J., Albin, R. L., & Detloff, P. J. (2003). Nucleocytoplasmic transport signals affect the age at onset of abnormalities in knock-in mice expressing polyglutamine within an ectopic protein context. *Human Molecular Genetics, 12*, 1621–1629.

Jafar-Nejad, P., Ward, C. S., Richman, R., Orr, H. T., & Zoghbi, H. Y. (2011). Regional rescue of spinocerebellar ataxia type 1 phenotypes by 14-3-3epsilon haploinsufficiency in mice underscores complex pathogenicity in neurodegeneration. *Proceedings of the National Academy of Sciences of the United States of America, 108*, 2142–2147.

Jana, N. R., Zemskov, E. A., Wang, G., & Nukina, N. (2001). Altered proteasomal function due to the expression of polyglutamine-expanded truncated N-terminal huntingtin induces apoptosis by caspase activation through mitochondrial cytochrome c release. *Human Molecular Genetics, 10*, 1049–1059.

Jayaraman, M., Mishra, R., Kodali, R., et al. (2012). Kinetically competing huntingtin aggregation pathways control amyloid polymorphism and properties. *Biochemistry, 51*, 2706–2716.

Jayaraman, M., Thakur, A. K., Kar, K., Kodali, R., & Wetzel, R. (2011). Assays for studying nucleated aggregation of polyglutamine proteins. *Methods (San Diego, Calif.), 53*, 246–254.

Jenkins, B. G., Koroshetz, W. J., Beal, M. F., & Rosen, B. R. (1993). Evidence for impairment of energy metabolism in vivo in Huntington's disease using localized 1H NMR spectroscopy. *Neurology, 43*, 2689–2695.

Jia, D. D., Zhang, L., Chen, Z., et al. (2013). Lithium chloride alleviates neurodegeneration partly by inhibiting activity of GSK3beta in a SCA3 *Drosophila* model. *Cerebellum (London, England), 12*, 892–901.

Jinek, M., East, A., Cheng, A., Lin, S., Ma, E., & Doudna, J. (2013). RNA-programmed genome editing in human cells. *Elife, 2*, e00471.

Ju, T. C., Chen, H. M., Lin, J. T., et al. (2011). Nuclear translocation of AMPK-alpha1 potentiates striatal neurodegeneration in Huntington's disease. *The Journal of Cell Biology, 194*, 209–227.

Juenemann, K., Schipper-Krom, S., Wiemhoefer, A., Kloss, A., Sanz Sanz, A., & Reits, E. A. (2013). Expanded polyglutamine-containing N-terminal huntingtin fragments are entirely degraded by mammalian proteasomes. *The Journal of Biological Chemistry, 288*, 27068–27084.

Jung, J., Xu, K., Lessing, D., & Bonini, N. M. (2009). Preventing Ataxin-3 protein cleavage mitigates degeneration in a *Drosophila* model of SCA3. *Human Molecular Genetics, 18*, 4843–4852.

Kalchman, M. A., Graham, R. K., Xia, G., et al. (1996). Huntingtin is ubiquitinated and interacts with a specific ubiquitin-conjugating enzyme. *The Journal of Biological Chemistry, 271*, 19385–19394.

Kastle, M., & Grune, T. (2012). Interactions of the proteasomal system with chaperones: Protein triage and protein quality control. *Progress in Molecular Biology and Translational Science, 109*, 113–160.

Katsuno, M., Tanaka, F., Adachi, H., et al. (2012). Pathogenesis and therapy of spinal and bulbar muscular atrophy (SBMA). *Progress in Neurobiology, 99*, 246–256.

Kaytor, M. D., Duvick, L. A., Skinner, P. J., Koob, M. D., Ranum, L. P., & Orr, H. T. (1999). Nuclear localization of the spinocerebellar ataxia type 7 protein, ataxin-7. *Human Molecular Genetics, 8*, 1657–1664.

Kegel, K. B., Sapp, E., Alexander, J., et al. (2010). Huntingtin cleavage product A forms in neurons and is reduced by gamma-secretase inhibitors. *Molecular Neurodegeneration, 5*, 58.

Kelly, D. P., & Scarpulla, R. C. (2004). Transcriptional regulatory circuits controlling mitochondrial biogenesis and function. *Genes & Development, 18*, 357–368.

Khan, L. A., Bauer, P. O., Miyazaki, H., Lindenberg, K. S., Landwehrmeyer, B. G., & Nukina, N. (2006). Expanded polyglutamines impair synaptic transmission and ubiquitin-proteasome system in Caenorhabditis elegans. *Journal of Neurochemistry*, *98*, 576–587.

Kim, D., Kim, J., Hur, J. K., Been, K. W., Yoon, S. H., & Kim, J. S. (2016). Genome-wide analysis reveals specificities of Cpf1 endonucleases in human cells. *Nature Biotechnology*, *34*, 863–868.

Kim, M., Lee, H. S., LaForet, G., et al. (1999). Mutant huntingtin expression in clonal striatal cells: Dissociation of inclusion formation and neuronal survival by caspase inhibition. *The Journal of Neuroscience*, *19*, 964–973.

Kim, M., Roh, J. K., Yoon, B. W., et al. (2003). Huntingtin is degraded to small fragments by calpain after ischemic injury. *Experimental Neurology*, *183*, 109–115.

Kim, Y. E., Hosp, F., Frottin, F., et al. (2016). Soluble oligomers of polyQ-expanded huntingtin target a multiplicity of key cellular factors. *Molecular Cell*, *63*, 951–964.

Kim, Y. J., Sapp, E., Cuiffo, B. G., et al. (2006). Lysosomal proteases are involved in generation of N-terminal huntingtin fragments. *Neurobiology of Disease*, *22*, 346–356.

Kim, Y. J., Yi, Y., Sapp, E., et al. (2001). Caspase 3-cleaved N-terminal fragments of wild-type and mutant huntingtin are present in normal and Huntington's disease brains, associate with membranes, and undergo calpain-dependent proteolysis. *Proceedings of the National Academy of Sciences of the United States of America*, *98*, 12784–12789.

Kleinstiver, B. P., Tsai, S. Q., Prew, M. S., et al. (2016). Genome-wide specificities of CRISPR–Cas Cpf1 nucleases in human cells. *Nature Biotechnology*, *34*, 869–874.

Klement, I. A., Skinner, P. J., Kaytor, M. D., et al. (1998). Ataxin-1 nuclear localization and aggregation: Role in polyglutamine-induced disease in SCA1 transgenic mice. *Cell*, *95*, 41–53.

Koch, P., Breuer, P., Peitz, M., et al. (2011). Excitation-induced ataxin-3 aggregation in neurons from patients with Machado–Joseph disease. *Nature*, *480*, 543–546.

Kodali, R., & Wetzel, R. (2007). Polymorphism in the intermediates and products of amyloid assembly. *Current Opinion in Structural Biology*, *17*, 48–57.

Kordasiewicz, H. B., Stanek, L. M., Wancewicz, E. V., et al. (2012). Sustained therapeutic reversal of Huntington's disease by transient repression of huntingtin synthesis. *Neuron*, *74*, 1031–1044.

Kordasiewicz, H. B., Thompson, R. M., Clark, H. B., & Gomez, C. M. (2006). C-termini of P/Q-type Ca^{2+} channel alpha1A subunits translocate to nuclei and promote polyglutamine-mediated toxicity. *Human Molecular Genetics*, *15*, 1587–1599.

Koyano, S., Yagishita, S., Kuroiwa, Y., Tanaka, F., & Uchihara, T. (2014). Neuropathological staging of spinocerebellar ataxia type 2 by semiquantitative 1C2-positive neuron typing. Nuclear translocation of cytoplasmic 1C2 underlies disease progression of spinocerebellar ataxia type 2. *Brain Pathology (Zurich, Switzerland)*, *24*, 599–606.

Kubodera, T., Yokota, T., Ohwada, K., et al. (2003). Proteolytic cleavage and cellular toxicity of the human alpha1A calcium channel in spinocerebellar ataxia type 6. *Neuroscience Letters*, *341*, 74–78.

Kuemmerle, S., Gutekunst, C. A., Klein, A. M., et al. (1999). Huntington aggregates may not predict neuronal death in Huntington's disease. *Annals of Neurology*, *46*, 842–849.

Kundu, M., & Thompson, C. B. (2008). Autophagy: Basic principles and relevance to disease. *Annual Review of Pathology*, *3*, 427–455.

Laco, M. N., Oliveira, C. R., Paulson, H. L., & Rego, A. C. (2012). Compromised mitochondrial complex II in models of Machado–Joseph disease. *Biochimica et Biophysica Acta*, *1822*, 139–149.

Lai, S., O'Callaghan, B., Zoghbi, H. Y., & Orr, H. T. (2011). 14-3-3 Binding to ataxin-1(ATXN1) regulates its dephosphorylation at Ser-776 and transport to the nucleus. *The Journal of Biological Chemistry*, *286*, 34606–34616.

Lam, Y. C., Bowman, A. B., Jafar-Nejad, P., et al. (2006). ATAXIN-1 interacts with the repressor Capicua in its native complex to cause SCA1 neuropathology. *Cell*, *127*, 1335–1347.

Landles, C., Sathasivam, K., Weiss, A., et al. (2010). Proteolysis of mutant huntingtin produces an exon 1 fragment that accumulates as an aggregated protein in neuronal nuclei in Huntington disease. *The Journal of Biological Chemistry*, *285*, 8808–8823.

Landles, C., Weiss, A., Franklin, S., Howland, D., & Bates, G. (2012). Caspase-6 does not contribute to the proteolysis of mutant huntingtin in the HdhQ150 knock-in mouse model of Huntington's disease. *PLoS Currents*, *4*, e4fd085bfc9973.

Latouche, M., Lasbleiz, C., Martin, E., et al. (2007). A conditional pan-neuronal *Drosophila* model of spinocerebellar ataxia 7 with a reversible adult phenotype suitable for identifying modifier genes. *The Journal of Neuroscience*, *27*, 2483–2492.

Leenders, K. L., Frackowiak, R. S., Quinn, N., & Marsden, C. D. (1986). Brain energy metabolism and dopaminergic function in Huntington's disease measured in vivo using positron emission tomography. *Movement Disorders: Official Journal of the Movement Disorder Society*, *1*, 69–77.

Legleiter, J., Mitchell, E., Lotz, G. P., et al. (2010). Mutant huntingtin fragments form oligomers in a polyglutamine length-dependent manner in vitro and in vivo. *The Journal of Biological Chemistry*, *285*, 14777–14790.

Li, F., Macfarlan, T., Pittman, R. N., & Chakravarti, D. (2002). Ataxin-3 is a histone-binding protein with two independent transcriptional corepressor activities. *The Journal of Biological Chemistry*, *277*, 45004–45012.

Li, J., & Yuan, J. (2008). Caspases in apoptosis and beyond. *Oncogene*, *27*, 6194–6206.

Li, M., Miwa, S., Kobayashi, Y., et al. (1998). Nuclear inclusions of the androgen receptor protein in spinal and bulbar muscular atrophy. *Annals of Neurology*, *44*, 249–254.

Li, S. H., Lam, S., Cheng, A. L., & Li, X. J. (2000). Intranuclear huntingtin increases the expression of caspase-1 and induces apoptosis. *Human Molecular Genetics*, *9*, 2859–2867.

Lievens, J. C., Iche, M., Laval, M., Faivre-Sarrailh, C., & Birman, S. (2008). AKT-sensitive or insensitive pathways of toxicity in glial cells and neurons in *Drosophila* models of Huntington's disease. *Human Molecular Genetics*, *17*, 882–894.

Lim, J., Crespo-Barreto, J., Jafar-Nejad, P., et al. (2008). Opposing effects of polyglutamine expansion on native protein complexes contribute to SCA1. *Nature*, *452*, 713–718.

Lima, W. F., Prakash, T. P., Murray, H. M., et al. (2012). Single-stranded siRNAs activate RNAi in animals. *Cell*, *150*, 883–894.

Liman, J., Deeg, S., Voigt, A., et al. (2014). CDK5 protects from caspase-induced Ataxin-3 cleavage and neurodegeneration. *Journal of Neurochemistry*, *129*, 1013–1023.

Lin, M. T., & Beal, M. F. (2006). Mitochondrial dysfunction and oxidative stress in neurodegenerative diseases. *Nature*, *443*, 787–795.

Lindenberg, K. S., Yvert, G., Muller, K., & Landwehrmeyer, G. B. (2000). Expression analysis of ataxin-7 mRNA and protein in human brain: Evidence for a widespread distribution and focal protein accumulation. *Brain Pathology (Zurich, Switzerland)*, *10*, 385–394.

Lunkes, A., Lindenberg, K. S., Ben-Haiem, L., et al. (2002). Proteases acting on mutant huntingtin generate cleaved products that differentially build up cytoplasmic and nuclear inclusions. *Molecular Cell*, *10*, 259–269.

Lunkes, A., & Mandel, J. L. (1998). A cellular model that recapitulates major pathogenic steps of Huntington's disease. *Human Molecular Genetics*, *7*, 1355–1361.

Luo, S., Vacher, C., Davies, J. E., & Rubinsztein, D. C. (2005). Cdk5 phosphorylation of huntingtin reduces its cleavage by caspases: Implications for mutant huntingtin toxicity. *The Journal of Cell Biology*, *169*, 647–656.

Magen, I., & Hornstein, E. (2014). Oligonucleotide-based therapy for neurodegenerative diseases. *Brain Research*, *1584*, 116–128.

Mahler, A., Steiniger, J., Endres, M., Paul, F., Boschmann, M., & Doss, S. (2014). Increased catabolic state in spinocerebellar ataxia type 1 patients. *Cerebellum (London, England)*, *13*, 440–446.

Maiuri, T., Woloshansky, T., Xia, J., & Truant, R. (2013). The huntingtin N17 domain is a multifunctional CRM1 and Ran-dependent nuclear and cilial export signal. *Human Molecular Genetics*, *22*, 1383–1394.

Mali, P., Yang, L., Esvelt, K. M., et al. (2013). RNA-guided human genome engineering via Cas9. *Science (New York, N.Y.)*, *339*, 823–826.

Mangiarini, L., Sathasivam, K., Seller, M., et al. (1996). Exon 1 of the HD gene with an expanded CAG repeat is sufficient to cause a progressive neurological phenotype in transgenic mice. *Cell*, *87*, 493–506.

Martindale, D., Hackam, A., Wieczorek, A., et al. (1998). Length of huntingtin and its polyglutamine tract influences localization and frequency of intracellular aggregates. *Nature Genetics*, *18*, 150–154.

Martinez, E., Palhan, V. B., Tjernberg, A., et al. (2001). Human STAGA complex is a chromatin-acetylating transcription coactivator that interacts with pre-mRNA splicing and DNA damage-binding factors in vivo. *Molecular and Cellular Biology*, *21*, 6782–6795.

Martinez, T., Wright, N., Lopez-Fraga, M., Jimenez, A. I., & Paneda, C. (2013). Silencing human genetic diseases with oligonucleotide-based therapies. *Human Genetics*, *132*, 481–493.

Martinez-Vicente, M., Talloczy, Z., Wong, E., et al. (2010). Cargo recognition failure is responsible for inefficient autophagy in Huntington's disease. *Nature Neuroscience*, *13*, 567–576.

Matilla, A., Gorbea, C., Einum, D. D., et al. (2001). Association of ataxin-7 with the proteasome subunit S4 of the 19S regulatory complex. *Human Molecular Genetics*, *10*, 2821–2831.

McIlwain, D. R., Berger, T., & Mak, T. W. (2013). Caspase functions in cell death and disease. *Cold Spring Harbor Perspectives in Biology*, *5*, a008656.

McMahon, S. J., Pray-Grant, M. G., Schieltz, D., Yates, J. R., 3rd, & Grant, P. A. (2005). Polyglutamine-expanded spinocerebellar ataxia-7 protein disrupts normal SAGA and SLIK histone acetyltransferase activity. *Proceedings of the National Academy of Sciences of the United States of America*, *102*, 8478–8482.

Menzies, F. M., Huebener, J., Renna, M., Bonin, M., Riess, O., & Rubinsztein, D. C. (2010). Autophagy induction reduces mutant ataxin-3 levels and toxicity in a mouse model of spinocerebellar ataxia type 3. *Brain*, *133*, 93–104.

Metzger, S., Saukko, M., Van Che, H., et al. (2010). Age at onset in Huntington's disease is modified by the autophagy pathway: Implication of the V471A polymorphism in Atg7. *Human Genetics*, *128*, 453–459.

Metzger, S., Walter, C., Riess, O., et al. (2013). The V471A polymorphism in autophagy-related gene ATG7 modifies age at onset specifically in Italian Huntington disease patients. *PLoS ONE*, *8*, e68951.

Miller, J. P., Holcomb, J., Al-Ramahi, I., et al. (2010). Matrix metalloproteinases are modifiers of huntingtin proteolysis and toxicity in Huntington's disease. *Neuron, 67*, 199–212.

Milnerwood, A. J., Gladding, C. M., Pouladi, M. A., et al. (2010). Early increase in extrasynaptic NMDA receptor signaling and expression contributes to phenotype onset in Huntington's disease mice. *Neuron, 65*, 178–190.

Minty, A., Dumont, X., Kaghad, M., & Caput, D. (2000). Covalent modification of p73alpha by SUMO-1. Two-hybrid screening with p73 identifies novel SUMO-1-interacting proteins and a SUMO-1 interaction motif. *The Journal of Biological Chemistry, 275*, 36316–36323.

Mirkin, S. M. (2007). Expandable DNA repeats and human disease. *Nature, 447*, 932–940.

Mishra, R., Hoop, C. L., Kodali, R., Sahoo, B., van der Wel, P. C., & Wetzel, R. (2012). Serine phosphorylation suppresses huntingtin amyloid accumulation by altering protein aggregation properties. *Journal of Molecular Biology, 424*, 1–14.

Montie, H. L., Cho, M. S., Holder, L., et al. (2009). Cytoplasmic retention of polyglutamine-expanded androgen receptor ameliorates disease via autophagy in a mouse model of spinal and bulbar muscular atrophy. *Human Molecular Genetics, 18*, 1937–1950.

Mookerjee, S., Papanikolaou, T., Guyenet, S. J., et al. (2009). Posttranslational modification of ataxin-7 at lysine 257 prevents autophagy-mediated turnover of an N-terminal caspase-7 cleavage fragment. *The Journal of Neuroscience, 29*, 15134–15144.

Morishima, Y., Wang, A. M., Yu, Z., Pratt, W. B., Osawa, Y., & Lieberman, A. P. (2008). CHIP deletion reveals functional redundancy of E3 ligases in promoting degradation of both signaling proteins and expanded glutamine proteins. *Human Molecular Genetics, 17*, 3942–3952.

de Moura, M. B., dos Santos, L. S., & Van Houten, B. (2010). Mitochondrial dysfunction in neurodegenerative diseases and cancer. *Environmental and Molecular Mutagenesis, 51*, 391–405.

Mueller, T., Breuer, P., Schmitt, I., Walter, J., Evert, B. O., & Wullner, U. (2009). CK2-dependent phosphorylation determines cellular localization and stability of ataxin-3. *Human Molecular Genetics, 18*, 3334–3343.

Munsie, L., Caron, N., Atwal, R. S., et al. (2011). Mutant huntingtin causes defective actin remodeling during stress: Defining a new role for transglutaminase 2 in neurodegenerative disease. *Human Molecular Genetics, 20*, 1937–1951.

Nagaoka, U., Kim, K., Jana, N. R., et al. (2004). Increased expression of p62 in expanded polyglutamine-expressing cells and its association with polyglutamine inclusions. *Journal of Neurochemistry, 91*, 57–68.

Nagata, E., Sawa, A., Ross, C. A., & Snyder, S. H. (2004). Autophagosome-like vacuole formation in Huntington's disease lymphoblasts. *Neuroreport, 15*, 1325–1328.

Nascimento-Ferreira, I., Nobrega, C., Vasconcelos-Ferreira, A., et al. (2013). Beclin 1 mitigates motor and neuropathological deficits in genetic mouse models of Machado–Joseph disease. *Brain, 136*, 2173–2188.

Nascimento-Ferreira, I., Santos-Ferreira, T., Sousa-Ferreira, L., et al. (2011). Overexpression of the autophagic beclin-1 protein clears mutant ataxin-3 and alleviates Machado–Joseph disease. *Brain, 134*, 1400–1415.

Nekooki-Machida, Y., Kurosawa, M., Nukina, N., Ito, K., Oda, T., & Tanaka, M. (2009). Distinct conformations of in vitro and in vivo amyloids of huntingtin-exon1 show different cytotoxicity. *Proceedings of the National Academy of Sciences of the United States of America, 106*, 9679–9684.

Ng, H., Pulst, S. M., & Huynh, D. P. (2007). Ataxin-2 mediated cell death is dependent on domains downstream of the polyQ repeat. *Experimental Neurology, 208*, 207–215.

Ni, C. L., Seth, D., Fonseca, F. V., et al. (2016). Polyglutamine tract expansion increases S-nitrosylation of huntingtin and ataxin-1. *PLoS ONE, 11*, e0163359.

Nozaki, K., Onodera, O., Takano, H., & Tsuji, S. (2001). Amino acid sequences flanking polyglutamine stretches influence their potential for aggregate formation. *Neuroreport, 12*, 3357–3364.

Nucifora, F. C., Jr., Sasaki, M., Peters, M. F., et al. (2001). Interference by huntingtin and atrophin-1 with cbp-mediated transcription leading to cellular toxicity. *Science (New York, N.Y.), 291*, 2423–2428.

Okazawa, H., Rich, T., Chang, A., et al. (2002). Interaction between mutant ataxin-1 and PQBP-1 affects transcription and cell death. *Neuron, 34*, 701–713.

Ona, V. O., Li, M., Vonsattel, J. P., et al. (1999). Inhibition of caspase-1 slows disease progression in a mouse model of Huntington's disease. *Nature, 399*, 263–267.

Orr, A. L., Li, S., Wang, C. E., et al. (2008). N-terminal mutant huntingtin associates with mitochondria and impairs mitochondrial trafficking. *The Journal of Neuroscience, 28*, 2783–2792.

Orr, H. T., & Zoghbi, H. Y. (2007). Trinucleotide repeat disorders. *Annual Review of Neuroscience, 30*, 575–621.

Oz, G., Nelson, C. D., Koski, D. M., et al. (2010). Noninvasive detection of presymptomatic and progressive neurodegeneration in a mouse model of spinocerebellar ataxia type 1. *The Journal of Neuroscience, 30*, 3831–3838.

Pacheco, L. S., da Silveira, A. F., Trott, A., et al. (2013). Association between Machado–Joseph disease and oxidative stress biomarkers. *Mutation Research, 757*, 99–103.

Palazzolo, I., Burnett, B. G., Young, J. E., et al. (2007). Akt blocks ligand binding and protects against expanded polyglutamine androgen receptor toxicity. *Human Molecular Genetics, 16*, 1593–1603.

Palhan, V. B., Chen, S., Peng, G. H., et al. (2005). Polyglutamine-expanded ataxin-7 inhibits STAGA histone acetyltransferase activity to produce retinal degeneration. *Proceedings of the National Academy of Sciences of the United States of America, 102*, 8472–8477.

Pankiv, S., Clausen, T. H., Lamark, T., et al. (2007). p62/SQSTM1 binds directly to Atg8/LC3 to facilitate degradation of ubiquitinated protein aggregates by autophagy. *The Journal of Biological Chemistry, 282*, 24131–24145.

Pankiv, S., Lamark, T., Bruun, J. A., Overvatn, A., Bjorkoy, G., & Johansen, T. (2010). Nucleocytoplasmic shuttling of p62/SQSTM1 and its role in recruitment of nuclear polyubiquitinated proteins to promyelocytic leukemia bodies. *The Journal of Biological Chemistry, 285*, 5941–5953.

Panov, A. V., Gutekunst, C. A., Leavitt, B. R., et al. (2002). Early mitochondrial calcium defects in Huntington's disease are a direct effect of polyglutamines. *Nature Neuroscience, 5*, 731–736.

Pardo, R., Colin, E., Regulier, E., et al. (2006). Inhibition of calcineurin by FK506 protects against polyglutamine-huntingtin toxicity through an increase of huntingtin phosphorylation at S421. *The Journal of Neuroscience, 26*, 1635–1645.

Park, Y., Hong, S., Kim, S. J., & Kang, S. (2005). Proteasome function is inhibited by polyglutamine-expanded ataxin-1, the SCA1 gene product. *Molecules and Cells, 19*, 23–30.

Patel, V. P., & Chu, C. T. (2011). Nuclear transport, oxidative stress, and neurodegeneration. *International Journal of Clinical and Experimental Pathology, 4*, 215–229.

Paulson, H. L., Perez, M. K., Trottier, Y., et al. (1997). Intranuclear inclusions of expanded polyglutamine protein in spinocerebellar ataxia type 3. *Neuron, 19*, 333–344.

Pearce, M. M., Spartz, E. J., Hong, W., Luo, L., & Kopito, R. R. (2015). Prion-like transmission of neuronal huntingtin aggregates to phagocytic glia in the *Drosophila* brain. *Nature Communications, 6*, 6768.

Pellistri, F., Bucciantini, M., Invernizzi, G., et al. (2013). Different ataxin-3 amyloid aggregates induce intracellular Ca(2 +) deregulation by different mechanisms in cerebellar granule cells. *Biochimica et Biophysica Acta, 1833*, 3155–3165.

Pemberton, L. F., & Paschal, B. M. (2005). Mechanisms of receptor-mediated nuclear import and nuclear export. *Traffic (Copenhagen, Denmark), 6*, 187–198.

Peters, M. F., Nucifora, F. C., Jr, Kushi, J., et al. (1999). Nuclear targeting of mutant Huntingtin increases toxicity. *Molecular and Cellular Neurosciences, 14*, 121–128.

Petersen, A., Larsen, K. E., Behr, G. G., et al. (2001). Expanded CAG repeats in exon 1 of the Huntington's disease gene stimulate dopamine-mediated striatal neuron autophagy and degeneration. *Human Molecular Genetics, 10*, 1243–1254.

Portera-Cailliau, C., Hedreen, J. C., Price, D. L., & Koliatsos, V. E. (1995). Evidence for apoptotic cell death in Huntington disease and excitotoxic animal models. *The Journal of Neuroscience, 15*, 3775–3787.

Poukka, H., Karvonen, U., Janne, O. A., & Palvimo, J. J. (2000). Covalent modification of the androgen receptor by small ubiquitin-like modifier 1 (SUMO-1). *Proceedings of the National Academy of Sciences of the United States of America, 97*, 14145–14150.

Pountney, D. L., Huang, Y., Burns, R. J., et al. (2003). SUMO-1 marks the nuclear inclusions in familial neuronal intranuclear inclusion disease. *Experimental Neurology, 184*, 436–446.

Pozzi, C., Valtorta, M., Tedeschi, G., et al. (2008). Study of subcellular localization and proteolysis of ataxin-3. *Neurobiology of Disease, 30*, 190–200.

Puigserver, P., Wu, Z., Park, C. W., Graves, R., Wright, M., & Spiegelman, B. M. (1998). A cold-inducible coactivator of nuclear receptors linked to adaptive thermogenesis. *Cell, 92*, 829–839.

Qin, Z. H., Wang, Y., Kegel, K. B., et al. (2003). Autophagy regulates the processing of amino terminal huntingtin fragments. *Human Molecular Genetics, 12*, 3231–3244.

Ran, F. A., Hsu, P. D., Wright, J., Agarwala, V., Scott, D. A., & Zhang, F. (2013). Genome engineering using the CRISPR–Cas9 system. *Nature Protocols, 8*, 2281–2308.

Ratovitski, T., Chighladze, E., Waldron, E., Hirschhorn, R. R., & Ross, C. A. (2011). Cysteine proteases bleomycin hydrolase and cathepsin Z mediate N-terminal proteolysis and toxicity of mutant huntingtin. *The Journal of Biological Chemistry, 286*, 12578–12589.

Ratovitski, T., Nakamura, M., D'Ambola, J., et al. (2007). N-terminal proteolysis of full-length mutant huntingtin in an inducible PC12 cell model of Huntington's disease. *Cell Cycle (Georgetown, Tex.), 6*, 2970–2981.

Ravikumar, B., Duden, R., & Rubinsztein, D. C. (2002). Aggregate-prone proteins with polyglutamine and polyalanine expansions are degraded by autophagy. *Human Molecular Genetics, 11*, 1107–1117.

Ravikumar, B., Vacher, C., Berger, Z., et al. (2004). Inhibition of mTOR induces autophagy and reduces toxicity of polyglutamine expansions in fly and mouse models of Huntington disease. *Nature Genetics, 36*, 585–595.

Reina, C. P., Zhong, X., & Pittman, R. N. (2010). Proteotoxic stress increases nuclear localization of ataxin-3. *Human Molecular Genetics*, *19*, 235–249.

Ren, P. H., Lauckner, J. E., Kachirskaia, I., Heuser, J. E., Melki, R., & Kopito, R. R. (2009). Cytoplasmic penetration and persistent infection of mammalian cells by polyglutamine aggregates. *Nature Cell Biology*, *11*, 219–225.

Ribbeck, K., & Gorlich, D. (2002). The permeability barrier of nuclear pore complexes appears to operate via hydrophobic exclusion. *The EMBO Journal*, *21*, 2664–2671.

Riley, B. E., & Orr, H. T. (2006). Polyglutamine neurodegenerative diseases and regulation of transcription: Assembling the puzzle. *Genes & Development*, *20*, 2183–2192.

Riley, B. E., Zoghbi, H. Y., & Orr, H. T. (2005). SUMOylation of the polyglutamine repeat protein, ataxin-1, is dependent on a functional nuclear localization signal. *The Journal of Biological Chemistry*, *280*, 21942–21948.

Rockabrand, E., Slepko, N., Pantalone, A., et al. (2007). The first 17 amino acids of Huntingtin modulate its subcellular localization, aggregation and effects on calcium homeostasis. *Human Molecular Genetics*, *16*, 61–77.

Rose, C., Menzies, F. M., Renna, M., et al. (2010). Rilmenidine attenuates toxicity of polyglutamine expansions in a mouse model of Huntington's disease. *Human Molecular Genetics*, *19*, 2144–2153.

Rosser, M. F., Washburn, E., Muchowski, P. J., Patterson, C., & Cyr, D. M. (2007). Chaperone functions of the E3 ubiquitin ligase CHIP. *The Journal of Biological Chemistry*, *282*, 22267–22277.

Rubio, I., Rodriguez-Navarro, J. A., Tomas-Zapico, C., et al. (2009). Effects of partial suppression of parkin on huntingtin mutant R6/1 mice. *Brain Research*, *1281*, 91–100.

Ruiz, C., Casarejos, M. J., Rubio, I., et al. (2012). The dopaminergic stabilizer, (–)-OSU6162, rescues striatal neurons with normal and expanded polyglutamine chains in huntingtin protein from exposure to free radicals and mitochondrial toxins. *Brain Research*, *1459*, 100–112.

Sancho, M., Herrera, A. E., Gortat, A., et al. (2011). Minocycline inhibits cell death and decreases mutant Huntingtin aggregation by targeting Apaf-1. *Human Molecular Genetics*, *20*, 3545–3553.

Sanders, S. L., Jennings, J., Canutescu, A., Link, A. J., & Weil, P. A. (2002). Proteomics of the eukaryotic transcription machinery: Identification of proteins associated with components of yeast TFIID by multidimensional mass spectrometry. *Molecular and Cellular Biology*, *22*, 4723–4738.

Sapp, E., Schwarz, C., Chase, K., et al. (1997). Huntingtin localization in brains of normal and Huntington's disease patients. *Annals of Neurology*, *42*, 604–612.

Sarkar, S., Davies, J. E., Huang, Z., Tunnacliffe, A., & Rubinsztein, D. C. (2007). Trehalose, a novel mTOR-independent autophagy enhancer, accelerates the clearance of mutant huntingtin and alpha-synuclein. *The Journal of Biological Chemistry*, *282*, 5641–5652.

Sarkar, S., Krishna, G., Imarisio, S., Saiki, S., O'Kane, C. J., & Rubinsztein, D. C. (2008). A rational mechanism for combination treatment of Huntington's disease using lithium and rapamycin. *Human Molecular Genetics*, *17*, 170–178.

Sarkar, S., & Rubinsztein, D. C. (2006). Inositol and IP3 levels regulate autophagy: Biology and therapeutic speculations. *Autophagy*, *2*, 132–134.

Sathasivam, K., Lane, A., Legleiter, J., et al. (2010). Identical oligomeric and fibrillar structures captured from the brains of R6/2 and knock-in mouse models of Huntington's disease. *Human Molecular Genetics*, *19*, 65–78.

Sathasivam, K., Neueder, A., Gipson, T. A., et al. (2013). Aberrant splicing of HTT generates the pathogenic exon 1 protein in Huntington disease. *Proceedings of the National Academy of Sciences of the United States of America*, *110*, 2366–2370.

Saudou, F., Finkbeiner, S., Devys, D., & Greenberg, M. E. (1998). Huntingtin acts in the nucleus to induce apoptosis but death does not correlate with the formation of intranuclear inclusions. *Cell*, *95*, 55–66.

Saute, J. A., Silva, A. C., Souza, G. N., et al. (2012). Body mass index is inversely correlated with the expanded CAG repeat length in SCA3/MJD patients. *Cerebellum (London, England)*, *11*, 771–774.

Schilling, B., Gafni, J., Torcassi, C., et al. (2006). Huntingtin phosphorylation sites mapped by mass spectrometry. Modulation of cleavage and toxicity. *The Journal of Biological Chemistry*, *281*, 23686–23697.

Schilling, G., Becher, M. W., Sharp, A. H., et al. (1999). Intranuclear inclusions and neuritic aggregates in transgenic mice expressing a mutant N-terminal fragment of huntingtin. *Human Molecular Genetics*, *8*, 397–407.

Schilling, G., Savonenko, A. V., Klevytska, A., et al. (2004). Nuclear-targeting of mutant huntingtin fragments produces Huntington's disease-like phenotypes in transgenic mice. *Human Molecular Genetics*, *13*, 1599–1610.

Schilling, G., Wood, J. D., Duan, K., et al. (1999). Nuclear accumulation of truncated atrophin-1 fragments in a transgenic mouse model of DRPLA. *Neuron*, *24*, 275–286.

Schipper-Krom, S., Juenemann, K., Jansen, A. H., et al. (2014). Dynamic recruitment of active proteasomes into polyglutamine initiated inclusion bodies. *FEBS Letters*, *588*, 151–159.

Schmidt, T., Landwehrmeyer, G. B., Schmitt, I., et al. (1998). An isoform of ataxin-3 accumulates in the nucleus of neuronal cells in affected brain regions of SCA3 patients. *Brain Pathology (Zurich, Switzerland)*, *8*, 669–679.

Schmidt, T., Lindenberg, K. S., Krebs, A., et al. (2002). Protein surveillance machinery in brains with spinocerebellar ataxia type 3: Redistribution and differential recruitment of 26S proteasome subunits and chaperones to neuronal intranuclear inclusions. *Annals of Neurology*, *51*, 302–310.

Schon, E. A., & Manfredi, G. (2003). Neuronal degeneration and mitochondrial dysfunction. *The Journal of Clinical Investigation*, *111*, 303–312.

Seidel, K., Brunt, E. R., de Vos, R. A., et al. (2009). The p62 antibody reveals various cytoplasmic protein aggregates in spinocerebellar ataxia type 6. *Clinical Neuropathology*, *28*(344-9).

Seidel, K., den Dunnen, W. F., Schultz, C., et al. (2010). Axonal inclusions in spinocerebellar ataxia type 3. *Acta Neuropathologica*, *120*, 449–460.

Sen, S., Dash, D., Pasha, S., & Brahmachari, S. K. (2003). Role of histidine interruption in mitigating the pathological effects of long polyglutamine stretches in SCA1: A molecular approach. *Protein Science: A publication of the Protein Society*, *12*, 953–962.

Seo, H., Sonntag, K. C., & Isacson, O. (2004). Generalized brain and skin proteasome inhibition in Huntington's disease. *Annals of Neurology*, *56*, 319–328.

Sheridan, C. (2016). Prosensa raises $30M for exon-skipping drug pipeline. *Bioworld Today*, *20*, 1.

Shin, J. W., Kim, K. H., Chao, M. J., et al. (2016). Permanent inactivation of Huntington's disease mutation by personalized allele-specific CRISPR/Cas9. *Human Molecular Genetics*, *25*, 4566–4576.

Shirakura, H., Hayashi, N., Ogino, S., Tsuruma, K., Uehara, T., & Nomura, Y. (2005). Caspase recruitment domain of procaspase-2 could be a target for SUMO-1 modification through Ubc9. *Biochemical and Biophysical Research Communications*, *331*, 1007–1015.

Shirendeb, U., Reddy, A. P., Manczak, M., et al. (2011). Abnormal mitochondrial dynamics, mitochondrial loss and mutant huntingtin oligomers in Huntington's disease: Implications for selective neuronal damage. *Human Molecular Genetics*, *20*, 1438–1455.

Sieradzan, K. A., Mechan, A. O., Jones, L., Wanker, E. E., Nukina, N., & Mann, D. M. (1999). Huntington's disease intranuclear inclusions contain truncated, ubiquitinated huntingtin protein. *Experimental Neurology*, *156*, 92–99.

Silva-Fernandes, A., Duarte-Silva, S., Neves-Carvalho, A., et al. (2014). Chronic treatment with 17-DMAG improves balance and coordination in a new mouse model of Machado–Joseph disease. *Neurotherapeutics: The Journal of the American Society for Experimental NeuroTherapeutics*, *11*, 433–449.

Simoes, A. T., Goncalves, N., Koeppen, A., et al. (2012). Calpastatin-mediated inhibition of calpains in the mouse brain prevents mutant ataxin 3 proteolysis, nuclear localization and aggregation, relieving Machado–Joseph disease. *Brain*, *135*, 2428–2439.

Simon, D. K., Zheng, K., Velazquez, L., et al. (2007). Mitochondrial complex I gene variant associated with early age at onset in spinocerebellar ataxia type 2. *Archives of Neurology*, *64*, 1042–1044.

Skotte, N. H., Southwell, A. L., Ostergaard, M. E., et al. (2014). Allele-specific suppression of mutant huntingtin using antisense oligonucleotides: Providing a therapeutic option for all Huntington disease patients. *PLoS ONE*, *9*, e107434.

Slow, E. J., Graham, R. K., Osmand, A. P., et al. (2005). Absence of behavioral abnormalities and neurodegeneration in vivo despite widespread neuronal huntingtin inclusions. *Proceedings of the National Academy of Sciences of the United States of America*, *102*, 11402–11407.

Smith, M. A., & Schnellmann, R. G. (2012). Calpains, mitochondria, and apoptosis. *Cardiovascular Research*, *96*, 32–37.

Song, J., Durrin, L. K., Wilkinson, T. A., Krontiris, T. G., & Chen, Y. (2004). Identification of a SUMO-binding motif that recognizes SUMO-modified proteins. *Proceedings of the National Academy of Sciences of the United States of America*, *101*, 14373–14378.

Sorimachi, H., Hata, S., & Ono, Y. (2011). Calpain chronicle—An enzyme family under multidisciplinary characterization. *Proceedings of the Japan Academy, Series B, Physical and Biological Sciences*, *87*, 287–327.

La Spada, A. R., Fu, Y. H., Sopher, B. L., et al. (2001). Polyglutamine-expanded ataxin-7 antagonizes CRX function and induces cone-rod dystrophy in a mouse model of SCA7. *Neuron*, *31*, 913–927.

La Spada, A. R., Wilson, E. M., Lubahn, D. B., Harding, A. E., & Fischbeck, K. H. (1991). Androgen receptor gene mutations in X-linked spinal and bulbar muscular atrophy. *Nature*, *352*, 77–79.

Stanek, L. M., Yang, W., Angus, S., et al. (2013). Antisense oligonucleotide-mediated correction of transcriptional dysregulation is correlated with behavioral benefits in the YAC128 mouse model of Huntington's disease. *Journal of Huntington's Disease*, *2*, 217–228.

Steffan, J. S., Agrawal, N., Pallos, J., et al. (2004). SUMO modification of huntingtin and Huntington's disease pathology. *Science (New York, N.Y.)*, *304*, 100–104.

Steffan, J. S., Kazantsev, A., Spasic-Boskovic, O., et al. (2000). The Huntington's disease protein interacts with p53 and CREB-binding protein and represses transcription. *Proceedings of the National Academy of Sciences of the United States of America*, *97*, 6763–6768.

Stohr, J. (2012). Prion protein aggregation and fibrillogenesis in vitro. *Sub-cellular Biochemistry, 65*, 91–108.

Strom, A. L., Forsgren, L., & Holmberg, M. (2005). A role for both wild-type and expanded ataxin-7 in transcriptional regulation. *Neurobiology of Disease, 20*, 646–655.

Subramaniam, S., Sixt, K. M., Barrow, R., & Snyder, S. H. (2009). Rhes, a striatal specific protein, mediates mutant-huntingtin cytotoxicity. *Science (New York, N.Y.), 324*, 1327–1330.

Sugiura, A., Yonashiro, R., Fukuda, T., et al. (2011). A mitochondrial ubiquitin ligase MITOL controls cell toxicity of polyglutamine-expanded protein. *Mitochondrion, 11*, 139–146.

Szebenyi, G., Morfini, G. A., Babcock, A., et al. (2003). Neuropathogenic forms of huntingtin and androgen receptor inhibit fast axonal transport. *Neuron, 40*, 41–52.

Takahashi, J., Fujigasaki, H., Zander, C., et al. (2002). Two populations of neuronal intranuclear inclusions in SCA7 differ in size and promyelocytic leukaemia protein content. *Brain, 125*, 1534–1543.

Takahashi, T., Kikuchi, S., Katada, S., Nagai, Y., Nishizawa, M., & Onodera, O. (2008). Soluble polyglutamine oligomers formed prior to inclusion body formation are cytotoxic. *Human Molecular Genetics, 17*, 345–356.

Takahashi-Fujigasaki, J., Arai, K., Funata, N., & Fujigasaki, H. (2006). SUMOylation substrates in neuronal intranuclear inclusion disease. *Neuropathology and Applied Neurobiology, 32*, 92–100.

Tan, W., Pasinelli, P., & Trotti, D. (2014). Role of mitochondria in mutant SOD1 linked amyotrophic lateral sclerosis. *Biochimica et Biophysica Acta, 1842*, 1295–1301.

Tang, T. S., Slow, E., Lupu, V., et al. (2005). Disturbed Ca^{2+} signaling and apoptosis of medium spiny neurons in Huntington's disease. *Proceedings of the National Academy of Sciences of the United States of America, 102*, 2602–2607.

Taylor, J., Grote, S. K., Xia, J., et al. (2006). Ataxin-7 can export from the nucleus via a conserved exportin-dependent signal. *The Journal of Biological Chemistry, 281*, 2730–2739.

Teles, A. V., Rosenstock, T. R., Okuno, C. S., Lopes, G. S., Bertoncini, C. R., & Smaili, S. S. (2008). Increase in bax expression and apoptosis are associated in Huntington's disease progression. *Neuroscience Letters, 438*, 59–63.

Terashima, T., Kawai, H., Fujitani, M., Maeda, K., & Yasuda, H. (2002). SUMO-1 co-localized with mutant atrophin-1 with expanded polyglutamines accelerates intranuclear aggregation and cell death. *Neuroreport, 13*, 2359–2364.

Tien, C. L., Wen, F. C., & Hsieh, M. (2008). The polyglutamine-expanded protein ataxin-3 decreases bcl-2 mRNA stability. *Biochemical and Biophysical Research Communications, 365*, 232–238.

Timmers, H. T., & Tora, L. (2005). SAGA unveiled. *Trends in Biochemical Sciences, 30*, 7–10.

Todd, T. W., & Lim, J. (2013). Aggregation formation in the polyglutamine diseases: Protection at a cost? *Molecules and Cells, 36*, 185–194.

Tokui, K., Adachi, H., Waza, M., et al. (2009). 17-DMAG ameliorates polyglutamine-mediated motor neuron degeneration through well-preserved proteasome function in an SBMA model mouse. *Human Molecular Genetics, 18*, 898–910.

Toneff, T., Mende-Mueller, L., Wu, Y., et al. (2002). Comparison of huntingtin proteolytic fragments in human lymphoblast cell lines and human brain. *Journal of Neurochemistry, 82*, 84–92.

Tsai, C. C., Kao, H. Y., Mitzutani, A., et al. (2004). Ataxin 1, a SCA1 neurodegenerative disorder protein, is functionally linked to the silencing mediator of retinoid and thyroid hormone receptors. *Proceedings of the National Academy of Sciences of the United States of America, 101*, 4047–4052.

Tsai, H. F., Tsai, H. J., & Hsieh, M. (2004). Full-length expanded ataxin-3 enhances mitochondrial-mediated cell death and decreases Bcl-2 expression in human neuroblastoma cells. *Biochemical and Biophysical Research Communications, 324*, 1274–1282.

Tsai, Y. C., Fishman, P. S., Thakor, N. V., & Oyler, G. A. (2003). Parkin facilitates the elimination of expanded polyglutamine proteins and leads to preservation of proteasome function. *The Journal of Biological Chemistry, 278*, 22044–22055.

Tsunemi, T., Ashe, T. D., Morrison, B. E., et al. (2012). PGC-1alpha rescues Huntington's disease proteotoxicity by preventing oxidative stress and promoting TFEB function. *Science Translational Medicine, 4*, 142ra97.

Ueda, H., Goto, J., Hashida, H., et al. (2002). Enhanced SUMOylation in polyglutamine diseases. *Biochemical and Biophysical Research Communications, 293*, 307–313.

Unno, T., Wakamori, M., Koike, M., et al. (2012). Development of Purkinje cell degeneration in a knockin mouse model reveals lysosomal involvement in the pathogenesis of SCA6. *Proceedings of the National Academy of Sciences of the United States of America, 109*, 17693–17698.

Vig, P. J., Shao, Q., Subramony, S. H., Lopez, M. E., & Safaya, E. (2009). Bergmann glial S100B activates myo-inositol monophosphatase 1 and co-localizes to purkinje cell vacuoles in SCA1 transgenic mice. *Cerebellum (London, England), 8*, 231–244.

Vis, J. C., Schipper, E., de Boer-van Huizen, R. T., et al. (2005). Expression pattern of apoptosis-related markers in Huntington's disease. *Acta Neuropathologica, 109*, 321–328.

Waelter, S., Boeddrich, A., Lurz, R., et al. (2001). Accumulation of mutant huntingtin fragments in aggresome-like inclusion bodies as a result of insufficient protein degradation. *Molecular Biology of the Cell, 12*, 1393–1407.

Wang, G., Sawai, N., Kotliarova, S., & Kanazawa, I. (2000). Nukina N. Ataxin-3, the MJD1 gene product, interacts with the two human homologs of yeast DNA repair protein RAD23, HHR23A and HHR23B. *Human Molecular Genetics, 9*, 1795–1803.

Wang, H., Jia, N., Fei, E., et al. (2007). p45, an ATPase subunit of the 19S proteasome, targets the polyglutamine disease protein ataxin-3 to the proteasome. *Journal of Neurochemistry, 101*, 1651–1661.

Wang, H. L., Chou, A. H., Lin, A. C., Chen, S. Y., Weng, Y. H., & Yeh, T. H. (2010). Polyglutamine-expanded ataxin-7 upregulates Bax expression by activating p53 in cerebellar and inferior olivary neurons. *Experimental Neurology, 224*, 486–494.

Wang, H. L., He, C. Y., Chou, A. H., Yeh, T. H., Chen, Y. L., & Li, A. H. (2007). Polyglutamine-expanded ataxin-7 decreases nuclear translocation of NF-kappaB p65 and impairs NF-kappaB activity by inhibiting proteasome activity of cerebellar neurons. *Cellular Signalling, 19*, 573–581.

Wang, H. L., Yeh, T. H., Chou, A. H., et al. (2006). Polyglutamine-expanded ataxin-7 activates mitochondrial apoptotic pathway of cerebellar neurons by upregulating Bax and downregulating Bcl-x(L). *Cellular Signalling, 18*, 541–552.

Wang, Y. C., Lee, C. M., Lee, L. C., et al. (2011). Mitochondrial dysfunction and oxidative stress contribute to the pathogenesis of spinocerebellar ataxia type 12 (SCA12). *The Journal of Biological Chemistry, 286*, 21742–21754.

Warby, S. C., Chan, E. Y., Metzler, M., et al. (2005). Huntingtin phosphorylation on serine 421 is significantly reduced in the striatum and by polyglutamine expansion in vivo. *Human Molecular Genetics, 14*, 1569–1577.

Watase, K., Gatchel, J. R., Sun, Y., et al. (2007). Lithium therapy improves neurological function and hippocampal dendritic arborization in a spinocerebellar ataxia type 1 mouse model. *PLoS Medicine, 4*, e182.

Watkin, E. E., Arbez, N., Waldron-Roby, E., et al. (2014). Phosphorylation of mutant huntingtin at serine 116 modulates neuronal toxicity. *PLoS ONE, 9*, e88284.

Watson, L. M., Scholefield, J., Greenberg, L. J., & Wood, M. J. (2012). Polyglutamine disease: From pathogenesis to therapy. *South African Medical Journal = Suid-Afrikaanse Tydskrif Vir Geneeskunde, 102*, 481–484.

Waza, M., Adachi, H., Katsuno, M., et al. (2005). 17-AAG, an Hsp90 inhibitor, ameliorates polyglutamine-mediated motor neuron degeneration. *Nature Medicine, 11*, 1088–1095.

Weiss, A., Klein, C., Woodman, B., et al. (2008). Sensitive biochemical aggregate detection reveals aggregation onset before symptom development in cellular and murine models of Huntington's disease. *Journal of Neurochemistry, 104*, 846–858.

Wellington, C. L., Ellerby, L. M., Gutekunst, C. A., et al. (2002). Caspase cleavage of mutant huntingtin precedes neurodegeneration in Huntington's disease. *The Journal of Neuroscience, 22*, 7862–7872.

Wellington, C. L., Ellerby, L. M., Hackam, A. S., et al. (1998). Caspase cleavage of gene products associated with triplet expansion disorders generates truncated fragments containing the polyglutamine tract. *The Journal of Biological Chemistry, 273*, 9158–9167.

Wellington, C. L., & Hayden, M. R. (1997). Of molecular interactions, mice and mechanisms: New insights into Huntington's disease. *Current Opinion in Neurology, 10*, 291–298.

Wellington, C. L., Singaraja, R., Ellerby, L., et al. (2000). Inhibiting caspase cleavage of huntingtin reduces toxicity and aggregate formation in neuronal and nonneuronal cells. *The Journal of Biological Chemistry, 275*, 19831–19838.

Westerink, R. H., & Ewing, A. G. (2008). The PC12 cell as model for neurosecretion. *Acta Physiologica (Oxford, England), 192*, 273–285.

Weydt, P., Pineda, V. V., Torrence, A. E., et al. (2006). Thermoregulatory and metabolic defects in Huntington's disease transgenic mice implicate PGC-1alpha in Huntington's disease neurodegeneration. *Cell Metabolism, 4*, 349–362.

Wickner, S., Maurizi, M. R., & Gottesman, S. (1999). Posttranslational quality control: Folding, refolding, and degrading proteins. *Science (New York, N.Y.), 286*, 1888–1893.

Williams, A. J., Knutson, T. M., Colomer Gould, V. F., & Paulson, H. L. (2009). In vivo suppression of polyglutamine neurotoxicity by C-terminus of Hsp70-interacting protein (CHIP) supports an aggregation model of pathogenesis. *Neurobiology of Disease, 33*, 342–353.

Woulfe, J. (2008). Nuclear bodies in neurodegenerative disease. *Biochimica et Biophysica Acta, 1783*, 2195–2206.

Wu, J. C., Qi, L., Wang, Y., et al. (2012). The regulation of N-terminal Huntingtin (Htt552) accumulation by Beclin1. *Acta pharmacologica Sinica, 33*, 743–751.

Wu, S., Zheng, S. D., Huang, H. L., et al. (2013). Lithium down-regulates histone deacetylase 1 (HDAC1) and induces degradation of mutant huntingtin. *The Journal of Biological Chemistry, 288*, 35500–35510.

Wyttenbach, A., Carmichael, J., Swartz, J., et al. (2000). Effects of heat shock, heat shock protein 40 (HDJ-2), and proteasome inhibition on protein aggregation in cellular models of Huntington's disease. *Proceedings of the National Academy of Sciences of the United States of America*, *97*, 2898–2903.

Xia, H., Mao, Q., Eliason, S. L., et al. (2004). RNAi suppresses polyglutamine-induced neurodegeneration in a model of spinocerebellar ataxia. *Nature Medicine*, *10*, 816–820.

Xia, J., Lee, D. H., Taylor, J., Vandelft, M., & Truant, R. (2003). Huntingtin contains a highly conserved nuclear export signal. *Human Molecular Genetics*, *12*, 1393–1403.

Yang, S., Chang, R., Yang, H., et al. (2017). CRISPR/Cas9-mediated gene editing ameliorates neurotoxicity in mouse model of Huntington's disease. *The Journal of Clinical Investigation*, *127*, 2719–2724.

Yang, S., Huang, S., Gaertig, M. A., Li, X. J., & Li, S. (2014). Age-dependent decrease in chaperone activity impairs MANF expression, leading to Purkinje cell degeneration in inducible SCA17 mice. *Neuron*, *81*, 349–365.

Yang, W., Dunlap, J. R., Andrews, R. B., & Wetzel, R. (2002). Aggregated polyglutamine peptides delivered to nuclei are toxic to mammalian cells. *Human Molecular Genetics*, *11*, 2905–2917.

Yasuda, S., Inoue, K., Hirabayashi, M., et al. (1999). Triggering of neuronal cell death by accumulation of activated SEK1 on nuclear polyglutamine aggregations in PML bodies. *Genes to Cells: Devoted to Molecular & Cellular Mechanisms*, *4*, 743–756.

Yorimitsu, T., & Klionsky, D. J. (2005). Autophagy: Molecular machinery for self-eating. *Cell Death and Differentiation*, *12*(Suppl 2), 1542–1552.

Young, J. E., Gouw, L., Propp, S., et al. (2007). Proteolytic cleavage of ataxin-7 by caspase-7 modulates cellular toxicity and transcriptional dysregulation. *The Journal of Biological Chemistry*, *282*, 30150–30160.

Yu, D., Pendergraff, H., Liu, J., et al. (2012). Single-stranded RNAs use RNAi to potently and allele-selectively inhibit mutant huntingtin expression. *Cell*, *150*, 895–908.

Yu, X., Ajayi, A., Boga, N. R., & Strom, A. L. (2012). Differential degradation of full-length and cleaved ataxin-7 fragments in a novel stable inducible SCA7 model. *Journal of Molecular Neuroscience: MN*, *47*, 219–233.

Yu, X., Munoz-Alarcon, A., Ajayi, A., et al. (2013). Inhibition of autophagy via p53-mediated disruption of ULK1 in a SCA7 polyglutamine disease model. *Journal of Molecular Neuroscience: MN*, *50*, 586–599.

Yu, Y. C., Kuo, C. L., Cheng, W. L., Liu, C. S., & Hsieh, M. (2009). Decreased antioxidant enzyme activity and increased mitochondrial DNA damage in cellular models of Machado–Joseph disease. *Journal of Neuroscience Research*, *87*, 1884–1891.

Yvert, G., Lindenberg, K. S., Devys, D., Helmlinger, D., Landwehrmeyer, G. B., & Mandel, J. L. (2001). SCA7 mouse models show selective stabilization of mutant ataxin-7 and similar cellular responses in different neuronal cell types. *Human Molecular Genetics*, *10*, 1679–1692.

Yvert, G., Lindenberg, K. S., Picaud, S., Landwehrmeyer, G. B., Sahel, J. A., & Mandel, J. L. (2000). Expanded polyglutamines induce neurodegeneration and trans-neuronal alterations in cerebellum and retina of SCA7 transgenic mice. *Human Molecular Genetics*, *9*, 2491–2506.

Zander, C., Takahashi, J., El Hachimi, K. H., et al. (2001). Similarities between spinocerebellar ataxia type 7 (SCA7) cell models and human brain: Proteins recruited in inclusions and activation of caspase-3. *Human Molecular Genetics*, *10*, 2569–2579.

Zeron, M. M., Fernandes, H. B., Krebs, C., et al. (2004). Potentiation of NMDA receptor-mediated excitotoxicity linked with intrinsic apoptotic pathway in YAC transgenic mouse model of Huntington's disease. *Molecular and Cellular Neurosciences*, *25*, 469–479.

Zetsche, B., Gootenberg, J. S., Abudayyeh, O. O., et al. (2015). Cpf1 is a single RNA-guided endonuclease of a class 2 CRISPR–Cas system. *Cell*, *163*, 759–771.

Zhang, S., Xu, L., Lee, J., & Xu, T. (2002). *Drosophila* atrophin homolog functions as a transcriptional corepressor in multiple developmental processes. *Cell*, *108*, 45–56.

Zhang, Y., Ona, V. O., Li, M., et al. (2003). Sequential activation of individual caspases, and of alterations in Bcl-2 proapoptotic signals in a mouse model of Huntington's disease. *Journal of Neurochemistry*, *87*, 1184–1192.

Zoghbi, H. Y., & Orr, H. T. (2000). Glutamine repeats and neurodegeneration. *Annual Review of Neuroscience*, *23*, 217–247.

Zuccato, C., Tartari, M., Crotti, A., et al. (2003). Huntingtin interacts with REST/NRSF to modulate the transcription of NRSE-controlled neuronal genes. *Nature Genetics*, *35*, 76–83.

CHAPTER

8

Prion-Like Propagation in Neurodegenerative Diseases

Wouter Peelaerts[1,2], *Veerle Baekelandt*[1] *and Patrik Brundin*[2]

[1]KU Leuven, Leuven, Belgium [2]Van Andel Research Institute, Grand Rapids, MI, United States

OUTLINE

The Molecular and Cellular Basis of Neurodegenerative Diseases
DOI: https://doi.org/10.1016/B978-0-12-811304-2.00008-0

INTRODUCTION

Correct protein folding is an intricate process of all living systems. Protein homeostasis and healthy cellular metabolism are crucial to prevent incorrect folding and to preserve proteins in their functional state. Proteins that adopt nonphysiological conformations can assemble into insoluble aggregates that are hallmarks of various neurodegenerative disorders. Despite the native versions of the proteins involved in these neurodegenerative diseases exhibiting differing biochemical, functional, and biological features, once they misfold, many of them appear to share an intriguing commonality. Thus, aggregated proteins involved in neurodegenerative disease can act as transmissible molecular templates, and thereby instigate and propagate disease. While the pathogenic processes underlying several progressive neurodegenerative diseases were originally considered as distinctly different, the emerging picture from the past decade of research suggests that they all share certain molecular features, which include the "prion-like" behavior of aggregation-prone proteins. This cell-to-cell transmission of assemblies of misfolded protein is suggested to significantly contribute to the progression of symptoms. Despite overwhelming experimental evidence that amyloid proteins behave as prion-like particles, their precise contributions to pathogenesis are, however, not fully understood, and this is an area of vibrant research. In this review, we will describe several neurodegenerative diseases where protein aggregation is a salient feature, and prion-like behavior of the aggregated protein has been demonstrated experimentally. We discuss the validity of the experimental models and their relevance to the clinical conditions. We describe recent advances in the understanding of molecular mechanisms controlling the prion-like propagation of disease-related proteins and identify important areas for future research. We also highlight both the commonalities and the differences between the diseases and briefly discuss how the dissection of the molecular features of prion-like neurodegenerative diseases opens perspectives for the development of novel therapies that might slow disease progression.

The prion-like concept refers to the idea that proteins can act as disease agents in a way that is similar to infectious particles such as living microbes, viruses, or prions. However, redefining the role of prion-like proteins under a common banner of infectivity has proven difficult over recent years and still yields vivid debates that are mainly about semantics. In order to frame the prion-like concept, we need to understand its historical foundations. Therefore, we will first discuss in the sociological scientific context from which it was given shape, since the prion-like concept has been the subject of a rapidly changing and evolving field of research.

A Historical Perspective on Prion-Like Aspects in Neurodegenerative Diseases

Koch's Postulates and Infectivity

When Robert Koch and Friedrich Loeffler, both German physicians, announced Koch's postulates in 1882, they established the first definitions to describe and experimentally test infectivity and the contribution of infectious living organisms in the etiology of tuberculosis (Brock, 1989; Koch, 1876). Since the late 19th century, Koch's postulates were used as an outline to define the role of a specific microbial organism in disease pathogenesis. Even though their set of rules allowed novel and important discoveries, Koch's postulates hampered subsequent identification of viruses, since viruses could not be isolated and grown in a pure culture from which they could be passaged to other organisms (Evans, 1976). Therefore, it took many years before viruses were identified, and it was recognized that infectivity not only applies to living microbes. After the discovery of viruses, Koch's postulates of infectivity were revised accordingly.

Several decades later, during late 1950s, another enigma presented itself when virologist Gajdusek and Zigas (1957) discovered that a disease could be transferred between humans during tribal rituals of cannibalism. The disease, termed Kuru, presented with variable incubation times ($\sim$4–40 years) and a short duration. It was progressive with neurological features, but fever or inflammation was not apparent (Gajdusek & Zigas, 1959). In the absence of these typical signs of bacterial or viral infections, many hypotheses to explain Kuru were formulated, but the disease remained largely enigmatic. It took about one decade to realize that Kuru shared remarkable similarities with another neurological disease, scrapie. Scrapie belongs to a group of transmissible spongiform encephalopathies (TSEs) that occurs in animals (Griffith, 1967). Previous attempts of Gajdusek to cultivate and propagate Kuru from human to small laboratory animals had failed, but because of an intriguing link with scrapie, he decided to use larger animals and perform additional transmission experiments in chimpanzee. Surprisingly, he found that Kuru could be successfully transmitted or "cultivated" from humans to chimpanzees and cause tremor and ataxias after incubation times between 1 and 3 years (Rogers, Basnight, Gibbs, & Gajdusek, 1967). His work implied that an unknown infectious agent responsible for Kuru was present in humans. Because it was possible to experimentally transmit scrapie from goat to mice, scrapie was thought to be caused by a slow virus with a very long incubation time. Further experiments showed that it was possible to isolate and preserve infectivity after filtration of diseased brain, but, strikingly, infectivity was preserved in the absence of nucleic acids (Field, Farmer, Caspary, & Joyce, 1969). After two decades of speculation and many attempts to identify the mysterious infectious agent, an amyloid protein surged to the forefront, the prion protein (PrP) (Prusiner, 1982). Only in 2010, it was finally shown that pure forms of synthetic PrP could self-propagate and transmit disease after exogenous application (Wang, Wang, Yuan, & Ma, 2010), providing final evidence for a protein-only hypothesis. The realization that the PrP can exhibit infectivity and cause disease has required additional modifications to Koch's theory of infectivity (Walker, Levine, & Jucker, 2006).

It is clear that the theory of infectivity, as initially defined by Koch and Loeffler, has changed so drastically that it is now almost a touch of precious history. Because of recent advances in neurodegenerative research, we are forced to revisit again the concept of infectivity in the context of the protein-only hypothesis. The role of proteins as putative infectious or "nucleating" agents in neurodegenerative diseases appears to apply to a much larger group of proteins than originally anticipated. Despite the fact that these pathogenic proteins have a striking resemblance to the PrP, they do show subtle differences in their mechanism of action.

Seeds of Infectivity Propagate Throughout History

Parallels between infectious diseases and various neurodegenerative disorders have long been observed. As a result, renowned and historically important scientists have proposed that infectious-like mechanisms play a role. Some of these initial hypotheses concerned clinical descriptions of more than a hundred years ago. At the beginning of the 20th century, a small group of psychiatrists described the presence of abnormal protein deposits in patients that were diagnosed with progressive memory loss. Amongst them was Alois Alzheimer, a German psychiatrist who presented his most important finding, an autopsy report on a case of presenile dementia, at the Tübingen meeting of the Southwest German Psychiatrics in 1906 (Dahm, 2006). He described clinical and histopathological examinations of a female patient, Auguste Deter, who was admitted a few years earlier at the mental asylum in Frankfurt with progressive memory loss, hallucinations, and delusions (Goedert & Ghetti, 2007; Maurer, Volk, & Gerbaldo, 1997). At postmortem examination, the brain of his patient was scarred, particularly in the cerebral cortex, with protein deposits: "In the center of an otherwise almost normal cell there stands out one or several fibrils due to their characteristic thickness and peculiar impregnability" (Alzheimer, 1906; Maurer et al., 1997). Alois Alzheimer was describing what today is known as tau neurofibrillary tangles (NFTs). In addition to NFTs, he observed miliary foci or senile plaques. The finding of extracellular plaques was however not new, because plaques were already pointed out in 1882 by the neurologists Paul Blocq and Georges Marinesco, who worked at the ward of Martin Charcot at the Salpêtrière Clinic in Paris. Both neurologists described these senile plaques as "round heaps" in the cortex of an epileptic patient. Soon after establishing his differential diagnosis of what Alzheimer thought to be a very rare disorder, he noted that more of his patients were showing striking similarities with infectious diseases and in particular syphilitic dementia (Alzheimer, Forstl, & Levy, 1991; Miklossy, 2015). As a consequence, in the beginning of the 20th century, an infectious-like disease course, such as with syphilitic dementia, was considered as one of the features required for the diagnosis of dementia.

Similar parallels were drawn for another frequently occurring neurodegenerative disease, namely, Parkinson's disease (PD). In the early 20th century, around a century after the publication of James Parkinson's assay on the shaking palsy, Fritz Jacob Heinrich Lewy, a German neurologist who worked with Alois Alzheimer at the Royal Psychiatric Clinic of Munich, examined postmortem brain samples from over 80 patients who were diagnosed with PD. In 1912, he identified peculiar lesions in various regions of the central nervous system (CNS) including the vagal nerve, the nucleus basalis of Meynert, and several nuclei in the thalamus, that all appeared to be eosinophilic and insoluble in alcohol, benzene, and chloroform, indicating that these inclusions were largely made up by proteins (Goedert, Spillantini, Del Tredici, & Braak, 2013). Already then, and in accordance with James Parkinson's hypothesis that PD was a progressive disease, Lewy hypothesized that the progression of PD pathogenesis resembled a viral spread. Lewy compared the presence and the pattern of his newly identified proteinaceous inclusions with inclusion or negri bodies that are caused by infection with Rabies virus (Lewy, 1932). A few years later, Konstantin Tretiakoff, a Russian neuropathologist who received his medical training in Paris, observed similar inclusions in the substantia nigra of PD patients (Goedert et al., 2013). In his doctoral dissertation, he reported six paralysis agitans or parkinsonian cases with severe depigmentation and neuronal loss in the substantia nigra who all had inclusions within these nigral neurons, which he named after Lewy, entitling them "corps de Lewy" or Lewy Bodies (Lees, Selikhova, Andrade, & Duyckaerts, 2008).

It is clear that Alzheimer's and Lewy's findings had important clinical and historical implications. Strikingly, their observations not only suggested that proteinaceous inclusions are important but also suggested a pattern of "spreading" of protein deposits within the CNS. This pattern prompted researchers to investigate what seemed to be a parallel between the progressive course of the clinical manifestations and the occurrence of protein deposits. In the 1990s, Braak et al. were some of the first to thoroughly investigate the spatiotemporal pattern of tau inclusions in Alzheimer's disease (AD). In a clinicopathological study of brains of demented patients who died at several different stages of the disease, the aim was to correlate one of AD pathological hallmarks with disease progression (Braak

& Braak, 1995). It became clear that NFTs, composed of hyperphosphorylated tau protein, do not distribute uniformly across the brain but start in the transentorhinal cortex located in the temporal lobe (Braak & Del Tredici, 2011). When the disease progresses, deposits are found in more regions, including limbic areas (e.g., hippocampus and insular cortex) and virtually all subdivisions of the neocortex in the most advanced cases (Braak & Braak, 1991; Braak, Alafuzoff, Arzberger, Kretzschmar, & Del Tredici, 2006). Based on this pattern, inclusions seemingly disseminate via anatomical connections between neurons. Interestingly, Alzheimer's original report mainly described NFTs with typical radiating fibrillar filaments in pyramidal neurons of the isocortex (Alzheimer, 1907). It is unclear why he did not examine associated regions, such as the allocortex, which are most affected by tau pathology (Graeber, Kosel, Grasbon-Frodl, Moller, & Mehraein, 1998). Several decades later, Braak et al. identified a spatiotemporal progression of protein deposits, which appeared to follow neural tracts and that provided a framework for disease staging correlating to clinical symptoms: early episodic memory loss that is followed by associated cognitive deficits, leading to executive dysfunction, including motor tasks, visuospatial and perceptive deficits, and semantic dysfunction until overt full blown dementia has manifested (Arnold, Hyman, Flory, Damasio, & Van Hoesen, 1991; Braak & Braak, 1991).

In comparison to tau, the progressive appearance of Aβ plaques across brain regions follows a pattern that represents almost the exact opposite direction (Arnold et al., 1991; Braak & Braak, 1991). However, the pattern is much less predictable. As a result, different staging schemes have been proposed for extracellular Aβ deposits (Braak & Braak, 1991; Thal, Rub, Orantes, & Braak, 2002). Typically, extracellular Aβ plaques are initially found in the isocortex, then in allocortical areas, such as the entorhinal cortex, hippocampus, and amygdala, followed by different subcortical nuclei in the brainstem. Interestingly, the affected brain regions are also anatomically connected, suggestive of transneuronal spread despite the fact that Aβ deposits extracellularly.

The reason why the anatomical patterns of Aβ and tau pathology in AD are so different is puzzling. Indeed, it is suggested from in vitro and in vivo experiments that Aβ and tau might influence each other's toxicity. Once their spreading patterns converge from opposite directions, they work synergistically, causing neuronal deficits and cell death (Karran, Mercken, & De Strooper, 2011; Selkoe & Hardy, 2016). Over time, the size of Aβ and tau aggregates grows, and the number of inclusions increases (Serrano-Pozo, Frosch, Masliah, & Hyman, 2011). Neuroanatomical connections and a proteinaceous spread of Aβ and tau proteins could therefore determine whether certain neurons and regions might be vulnerable to neurodegeneration in AD or not. However, since not all neuronal populations within these neuroanatomical pathways are equally affected, a spreading hypothesis of depositing toxic proteins is insufficient to explain this apparent spread of pathology. Therefore, an alternative hypothesis, which is based on selective vulnerability of susceptible neurons, might further explain why certain populations of neurons are more susceptible to toxic insults (Freer et al., 2016) (Fig. 8.1A).

Specific patterns of progressive changes in α-synuclein (αSyn) deposits have been described for PD and related synucleinopathies. Braak suggested a parallel between inclusions in different brain regions and clinical symptoms and signs. He and his colleagues proposed that the gradual involvement of more brain regions might be due to a mechanism that resembles the spreading of a neurotropic pathogen, probably virus, that can enter the body and transfer transsynaptically to the central nervous system via anatomically connected regions (Halliday,

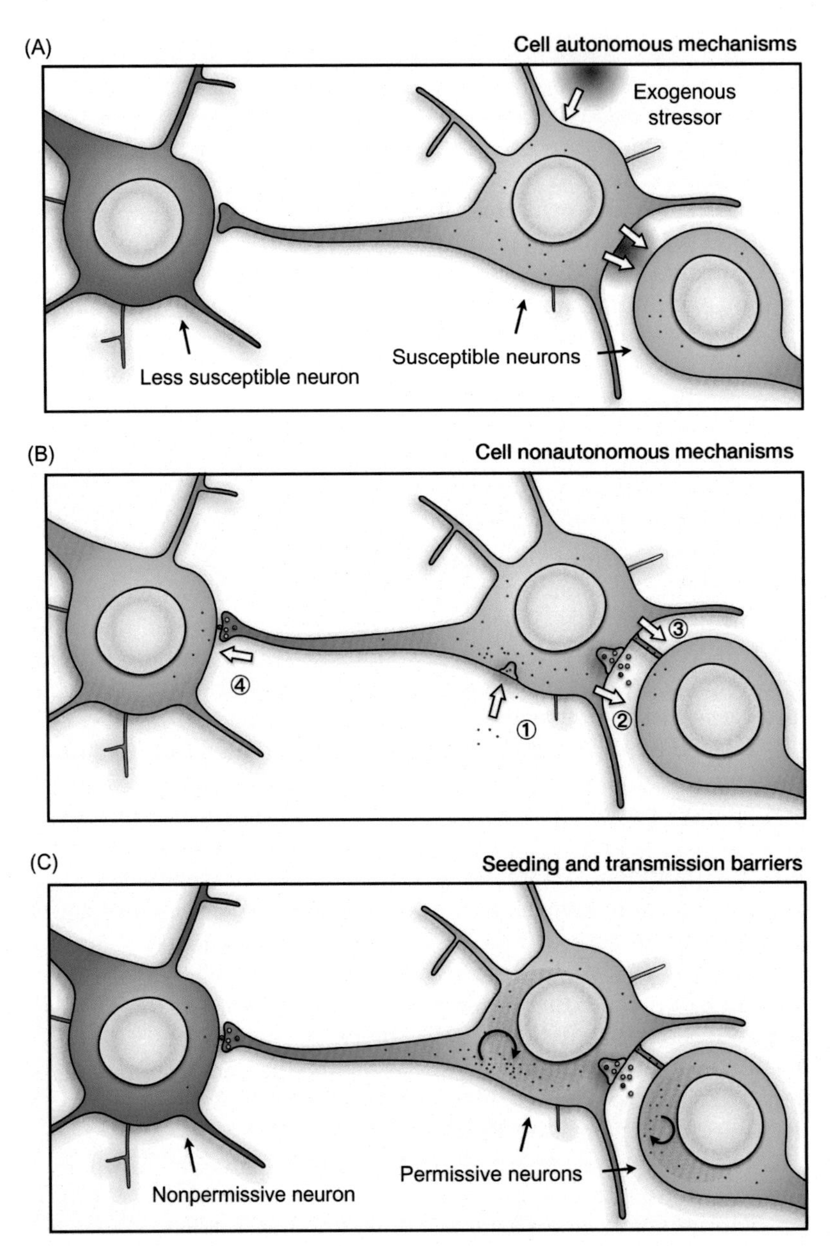

FIGURE 8.1 **Different mechanisms that contribute to the prion-like spread of disease-related proteins.** Two hypotheses currently describe the spreading pattern of misfolded proteins in prion-like disorders. (A) External stressors induce pathology or de novo protein aggregation in a selective population of susceptible neurons. Adverse stimuli can come from exogenous sources or neighboring cells. Vulnerable neurons that are poised for protein aggregation will develop pathology first, while less susceptible neurons will develop pathology during later and more advanced stages of the disease. (B) Alternatively, seeding-potent aggregates physically spread from cell to cell. They enter the cell from the outside after binding with their receptor *(1)*. After internalization, aggregates spread over the cell and transmit to second order neurons via exosomes, *(2)* tunneling nanotubes, *(3)* or the synapse *(4)*. (C) In the cytoplasm, seeding-potent assemblies can replicate via consumption of its soluble endogenous precursor. Newly generated seeds spread between neuroanatomically connected networks of susceptible or seeding-potent cells. Seeding-potent particles can also become transmitted to nonpermissive cells that are less susceptible for amplification and impose a transmission barrier. The mechanisms of pathogenic spreading and selective vulnerability are not mutually exclusive, and a combination of both mechanisms is likely to contribute to the prion-like spread of various disease-related proteins.

Hely, Reid, & Morris, 2008). Braak, Rub, Gai, and Del Tredici (2003b) even speculated, "Such a pathogen could possess unconventional prion-like properties and might consist of misfolded αSyn fragments", which was remarkable since at that time there were no reports on prion-related properties of αSyn.

In one of the first papers describing the Braak stages, it was proposed that αSyn pathology initiated in the dorsal motor nucleus of the vagal nerve (Braak et al., 2003a). However, after publication of the Braak hypothesis, other studies carefully pointed out that many PD patients could not be classified using the Braak scheme. This was mainly due to the fact that the classification system is based on the presence of vagal inclusions which are absent in half of the PD patients (Jellinger, 2008; Visanji, Brooks, Hazrati, & Lang, 2013). Therefore, a unified staging system was proposed more recently where αSyn pathology is described to initiate (1) in the olfactory bulb and (2) in the enteric neurons for which the Braak hypothesis was adapted accordingly (the "dual hit" hypothesis) (Beach et al., 2009; Kupsky, Grimes, Sweeting, Bertsch, & Cote, 1987). The enteric nervous system contains the nerves lining the digestive tract, and the presence of Lewy body lesions is accompanied with gastrointestinal problems including swallowing, bloating, nausea, and constipation (Cersosimo & Benarroch, 2012). The enteric system is innervated by the vagal nerve, which originates in the lower brainstem. This way, a direct entry route is provided from the gut to the CNS by which the disease appears to spread. From the dorsal motor nucleus of the vagal nerve, the disease can subsequently propagate to first order ascending and descending regions in the brain stem and the spinal cord, respectively. After manifesting in the brainstem, the basal ganglia and limbic circuits are further affected (Braak & Del Tredici, 2008a). These regions control essential parts of the motor system that are responsible for regulating movement and coordination. During this stage, neuronal cell death becomes more apparent, and the disease turns more severe as it shifts from peripheral-associated symptoms toward more pronounced motor-related symptoms. In other PD cases, pathology can occur exclusively in the bulbus olfactorius and bypass the midbrain region in a limbic predominant fashion (Beach et al., 2009). From the limbic system, the disease progresses slowly in a caudal-rostral pattern, affecting ascending cortical areas in a stepwise pattern. The damage inflicted to higher associative regions finally result in speech disorders, anxiety, depression, and even personality changes. An often cooccurring and debilitating clinical manifestation that involves cortical areas is dementia. This occurs in more than half of all Parkinson patients, although mostly during later stages (Aarsland, Zaccai, & Brayne, 2005; Hobson & Meara, 2004). Taken together, this stepwise pattern illustrates the gradual and prion-like nature of the disease and its typical symptomatology that translates into different clinical stages (Braak & Del Tredici, 2008b; Braak et al., 2003a). The establishment of a spreading pattern for tau and αSyn suggests that there might be cell-to-cell propagation of misfolded proteins.

The Prion-Like Hypothesis

Initial attempts of Gajdusek to determine whether AD and other neurodegenerative diseases have transmissible features were unsuccessful. In analogy with their Kuru experiments, Gajdusek et al. (1980) tested if brain isolates could transfer AD to monkeys, but they did not observe any sign of AD pathology. Notably, these studies were performed long before the anatomical substrates of AD (Aβ and tau) were identified. Consequently, for a long time, the consensus among scientists was that AD is nontransmissible. However, Prusiner (1984) and Gajdusek (1994), both pioneers in prion disease research, long speculated that under certain conditions neurological illnesses could be transferred and would exhibit certain prion-like features.

Indeed, later attempts to transmit AD pathology via brain homogenates did show that diseased AD brain had the ability to induce Aβ plaque formation and cerebral amyloid angiopathy in marmoset brain (Baker, Ridley, Duchen, Crow, & Bruton, 1994; Ridley, Baker, Windle, & Cummings, 2006). An important difference with Kuru was that Aβ pathology was only achieved after very protracted lag periods, with incubation times of 5–7 years (Baker et al., 1994; Ridley et al., 2006). Interestingly, in the early 1960s, similar observations were made for a different neurodegenerative disease, amyotrophic lateral sclerosis (ALS). Injection of spinal cord homogenate of a patient who died of ALS into monkey brain resulted in ALS-like pathology (Zilber et al., 1963). This transmissible agent could be passaged multiple times in monkeys but not in other laboratory animals (Zilber et al., 1963). The incubation time required more than 5 years. Collectively, this suggested that ALS might be of virus origin.

Eventually, more evidence that bolstered the idea of a prion-like mechanism came from a rather unexpected side: cell transplantation therapy. In PD, cell therapy aims to replace the loss of dopamine and counteract debilitating motor symptoms by grafting fetal mesencephalic cells that resupply dopamine in the striatum. Strikingly, patients who had undergone cell replacement therapy presented with Lewy bodies in their grafts at autopsy over a decade later, when their brains were examined microscopically using a range of morphological markers (Kordower, Chu, Hauser, Freeman, & Olanow, 2008; Li et al., 2008). In the two initial reports describing Lewy bodies inside grafted cells, these grafted fetal cells were 10–16 years old.

Postmortem studies on cell transplants in Huntington's disease (HD) patients also addressed the question whether aggregated protein could develop inside grafted fetal cells, which might be viewed as evidence of cell-to-cell transfer of misfolded protein. In HD, an inherited expanded polyglutamine (polyQ) repeat in the gene encoding huntingtin (Htt) results in an aggregation-prone protein and a clinical syndrome, which includes motor, cognitive, and psychiatric changes. The morphological studies on grafts in HD have yielded variable results, with the initial observations suggesting that there are no Htt aggregates inside the transplanted cells (Cicchetti et al., 2009), while a later report described Htt aggregates inside grafts of fetal striatal cells in three patients who died a decade following surgery (Cicchetti et al., 2014). HD is a monogenetic disorder, and the disease mechanism has always been considered to be cell-autonomous, albeit with a long delay of years to several decades between the first production of aggregation-prone protein (presumably starting at the birth of the neural cells) and the emergence of brain pathology and symptoms. However, the latter observations in three HD patients with grafted cells exhibiting protein aggregates, suggest that cell-to-cell spread of aggregated Htt might play a role in the progression of the disease, once the first aggregates have formed. Thus, the prion-like hypothesis is not only relevant for disorders where environmental factors are believed to trigger the disease in genetically predisposed individuals, such as PD. It might also be relevant for purely genetic disorders such as HD where one has to consider the possibility that aggregates can propagate from one cell to another and trigger neurodegeneration in a nonautonomous way.

PRION-LIKE PROTEINS IN NEURODEGENERATIVE DISEASES

Prions arise from endogenous proteins that acquire an alternative conformation and become self-sustaining. Prion disorders such as Gerstmann–Straussler–Scheinker syndrome or fatal familial insomnia (FFI) are predominantly inherited and result from gene mutations, deletions, or multiplications. In contrast, Creutzfeldt–Jacob's disease (CJD), the most common prion disorder, is mainly sporadic, and therefore it is the wild type protein that forms toxic aggregates and instigates disease. In contrast to polyQ disorders (e.g., HD, see above), that arise as a consequence of an expanded glutamine repeat in the native sequence, most of the neurodegenerative disorders with prion features are also sporadic (>90%). As with prions, different major players in these neurodegenerative disorders, such as Aβ, tau, and αSyn have the ability to acquire an alternative conformation, bind with other protein molecules, and fold into nonphysiological assemblies with a typical β-sheet conformation.

Below, we will discuss the experimental evidence that these insoluble assemblies can acquire a conformation that "seeds" subsequent aggregation reactions—resulting in an infectious amplification cycle of disease spreading and pathology. In accordance with a pathological spreading hypothesis, prions can amplify and spread, in a spatiotemporal pattern, and inflict specific disease phenotypes including different types of dementia, parkinsonian syndromes, or other movement disorders (Table 8.1).

TABLE 8.1 Overview of Prions and Neurodegenerative Disorders

Prion	Neurodegenerative Disease
Amyloid-β	AD
PrP	Kuru
	CJD
	GSS
	FFI
Tau	AD
	FTD
	FTDP-17
	Pick's disease
	PSP
	Corticobasal degeneration
	Tangle-only dementia
	White matter tauopathy with globular glial inclusions
	AGD
	CTE
αSyn	PD
	DLB
	MSA
	PAF
SOD1	ALS
TDP-43	ALS
	FTD
FUS	ALS
	FTD
Htt	HD

PrP, prion protein; *CJD*, Creutzfeldt–Jacob's disease; *GSS*, Gerstmann–Straussler–Scheinker syndrome; *FFI*, fatal familial insomnia; *AD*, Alzheimer's disease; *FTD*, frontotemporal dementia; *FTDP-17*, FTD linked to chromosome 17; *PSP*, progressive supranuclear palsy; *AGD*, argyrophilic grains disease; *CTE*, chronic traumatic encephalopathy; *PD*, Parkinson's disease; *DLB*, dementia with Lewy bodies; *αSyn*, α-Synuclein; *MSA*, multiple system atrophy; *PAF*, pure autonomic failure; *ALS*, amyotrophic lateral sclerosis; *SOD1*, superoxide dismutase 1; *TDP-43*, TAR DNA-binding protein; *FUS*, fused in sarcoma; *HD*, Huntington's disease; *Htt*, huntingtin.

Amyloid-β Prions

Small fragments of Aβ form filamentous β-sheet aggregates that deposit in the extracellular space of defined brain areas where they cluster as Aβ plaques, the typical hallmark of AD (Glenner & Wong, 1984; Masters et al., 1985). Next to plaques, Aβ can also form deposits in blood vessels walls leading to cerebral amyloid angiopathy (Glenner & Wong, 1984; Vinters, 1987). Most of the AD cases arise sporadically with several important risk factors that confer increased susceptibility to develop AD (Tanzi, 2012). Autosomal dominant forms of AD represent less than 5% of all cases. Mutations in the amyloid precursor protein (*APP*) and presenilin (*PSEN1* and *PSEN2*) genes were the first genes discovered to cause familial forms of AD (Hardy & Selkoe, 2002). PS1 and PS2 are part of the larger γ-secretase complex that cleaves APP into smaller fragments yielding Aβ peptides of different sizes (Chow, Mattson, Wong, & Gleichmann, 2010; O'Brien & Wong, 2011). Around two dozen mutations in the Aβ sequence of APP are now known to give rise to familial forms of AD (LaFerla, Green, & Oddo, 2007; Tanzi, 2012). Together, these discoveries made clear that genetic changes, which increase levels of Aβ or its tendency to aggregate, cause AD.

Although several clinical trials targeting Aβ in AD have failed, Aβ is still considered to have an important role, and based on aforementioned causative mutations, several animal models have been developed to gain more insight in the different steps that underlie AD pathogenesis. Some of these animal models have been used to test whether the prion hypothesis might be relevant to Aβ in AD. In an initial set of experiments, transgenic mice that harbor a mutation in APP and yield high levels of Aβ were injected with autopsy-derived brain extracts of AD patients (Kane et al., 2000; Walker et al., 2002). A high burden of amyloid plaques was found at the injection sites—the hippocampus and neocortex—and with extended lag times, additional deposits were observed in regions anatomically connected to these sites (Kane et al., 2000; Walker et al., 2002). These results suggested that Aβ has prion-like features. At the time, however, concerns were that human-specific viruses in the injected material, or immune-related host factors, might augment or accelerate the development of pathology in these transgenic animals. Therefore, another set of experiments was carried out using transgenic APP23 and APP-PS1 mice that produce the shorter Aβ40 and the longer Aβ42 fragments, respectively. Brain homogenates from the APP23 and APP-PS1 transgenic models were injected intracerebrally in young APP23 and APP-PS1 mice, respectively. Once again, intracerebral injections of brain homogenates accelerated the deposition of nascent Aβ amyloid (Meyer-Luehmann et al., 2006). Passive immunization using Aβ antibodies or pretreatment with formic acid neutralized the seeding response of Aβ isolates (Meyer-Luehmann et al., 2006). In line with the experiments of Gajdusek and coworkers, this work showed that a protein component in the donor brain tissue, Aβ from human patients or animal models, could induce or accelerate disease pathology in experimental animal models-akin to prions.

An important detail that remained unexplained in these initial experiments was that intracerebral injection of synthetic or pure Aβ seeds into transgenic or wild type mice failed to induce amyloid formation (Kane et al., 2000; Meyer-Luehmann et al., 2006). This was not a complete surprise given that, despite tremendous efforts, pure synthetic prions also failed to transmit disease in animals. Eventually, it was found that in vivo conversion of prions and exogenous cofactors such as lipids or RNA could alter the assembly of prions into more infectious templates and therefore enhance the infectivity of synthetic forms (Deleault et al., 2012; Gabizon, McKinley, & Prusiner, 1987; Wang et al., 2010). In order to show that the process of Aβ deposition and pathology was indeed due to Aβ, pure synthetic Aβ peptides were injected into transgenic mouse brains. With longer incubation times and higher doses, it was eventually shown that pure Aβ fibrils have nucleating properties (Stohr et al., 2012). These pure synthetic Aβ fibrils triggered plaque formation at the injection site, and later plaques were also found in anatomically connected regions. Thus, these findings are consistent with the idea that the appearance of Aβ plaques follows a stereotypic anatomical pattern due to spreading of the amyloid protein along neural tracts (Stohr et al., 2012). Most importantly, studies demonstrating that spreading of amyloid plaques in mice could be triggered by injections of pure synthetic Aβ fibrils showed that other tissue factors, which had been present in the brain-derived material used in the earlier studies, are not necessary. These landmark studies provided strong evidence that Aβ can act as a prion.

Tau Prions

Tau is a small and natively unfolded soluble protein. It is enriched in distal axons of neurons and associates with microtubules to promote their stability (Buee, Bussiere, Buee-Scherrer, Delacourte, & Hof, 2000). In various age-related disorders, tau accumulates in neurons as NFT (Goedert & Spillantini, 2006). The filaments that make up these filamentous inclusions are hyperphosphorylated and can have various morphologies (Crowther & Goedert, 2000). Intraneuronal deposits of tau are found frequently in different neurodegenerative diseases. Thus, in addition to the classical tauopathies described below, tau deposits are often present in synucleinopathies such as PD and dementia with Lewy bodies (DLB). The classical tauopathies include sporadic disorders such as frontotemporal dementia (FTD), Pick's disease (PiD), progressive supranuclear palsy (PSP), or corticobasal degeneration (CBD) (Table 8.1). Next to the presence of tau in neurons, trace amounts of tau are also found in astrocytes and oligodendrocytes (Komori, 1999). In white matter tauopathy with globular glial inclusion, argyrophilic grains disease (AGD) and to a lesser extent in other tauopathies, tau deposits are also found in glial cells (coiled bodies) in addition to neuronal pathology (tangles) (Bigio et al., 2001; Komori, 1999; Kovacs et al., 2008). Mutations in the tau gene (*MAPT*) cause familial forms of tauopathy such as FTD linked to chromosome (FTDP)-17 or other inherited forms of FTD, as well as PSP or CBD (Hutton et al., 1998; Poorkaj et al., 1998; Spillantini & Goedert, 1998; Spillantini et al., 1998c). An environmental link with progressive tau pathology is found for chronic traumatic encephalopathy that arises after repeated brain concussions or traumatic injuries in athletes or soldiers (McKee et al., 2009). In other tauopathies, intracellular inclusions of tau coexist with extracellular deposits of Aβ in AD, with intracellular deposits of fused in sarcoma (FUS) or transactivation response element DNA-binding protein (TDP-43) in FTD and with αSyn inclusions in DLB.

The fact that mutations in *MAPT* can cause disease, and that there is an abundance of tau inclusions in sporadic tauopathies strongly suggests that the development of tau aggregates is a central part of the pathogenic process. There exist several tauopathies that each exhibits a specific cellular pathology, preferred anatomical distribution of the aggregates, and distinct symptomatic profile. At the molecular level, the origin of this heterogeneity can at least in part be explained by the presence of six different tau isoforms that are the result of alternative mRNA splicing (Buee et al., 2000). Three isoforms have three repeats of the microtubule-binding region (3R tau), while the other three isoforms have four repeats (4R tau). Interestingly, these different isoforms are found to varying extents in different tauopathies (Goedert, Clavaguera, & Tolnay, 2010). In a later section describing tau strains, we discuss how these different isoforms might have different roles in disease pathology.

Stimulated by advances regarding the prion-like properties of other protein aggregates, and descriptions that progression of tau pathology follows stereotypic anatomical patterns, researchers sought to find out whether tau aggregates might also spread and behave like prions. In a first series of in vitro experiments, recombinant seeds of tau were shown to promote inclusion formation in cultured HEK293 cells (Frost, Jacks, & Diamond, 2009a; Guo & Lee, 2011; Kfoury, Holmes, Jiang, Holtzman, & Diamond, 2012). Different forms of tau, either truncated forms or full-length (4R tau), were all capable of seeding aggregates in various cellular assays (Frost et al., 2009a; Guo & Lee, 2011; Kfoury et al., 2012). Importantly, diseased human brain-derived insoluble fractions of tau also seeded tau inclusions in cultured HEK293 cells in vitro (Sanders et al., 2014). Inoculation of brain homogenate from transgenic mice that express a mutant form of tau (P301S) into the cortex or hippocampus of young transgenic mice that express full-length human tau (ALZ17) led to an acceleration of its disease phenotype and promoted deposition of tau as NFT, neuropil threads, and glial globular inclusions (Ahmed et al., 2014; Clavaguera et al., 2009). Deposits of tau pathology were also obtained in experiments where homogenized brain material from patients suffering from different types of tauopathies were injected into transgenic mice expressing full-length human 4R tau (which only show modest pathology) or into wild-type mice (Clavaguera et al., 2013). Here, injection of diseased brain material of patients with PSP, AGD, or CBD recapitulated their respective histopathological features in vivo. Injection of sonicated brain material from transgenic mice in wild-type animals triggered inclusion formation, but to a lower extent, and the inclusions were restricted to the injection site (Clavaguera et al., 2009). Similar experiments with injections of synthetic forms of full-length or truncated tau, harboring the P301S or the P301L mutation, into the brains of predisposed transgenic mice also led to earlier disease onset with variable pathologies, including neuronal dysfunction, neuronal loss, gliosis, and behavioral deficits (Iba et al., 2013; Peeraer et al., 2015; Sanders et al., 2014; Stancu et al., 2015). Pathology induction by pure recombinant tau protein further excluded the possibility that other contributing host or cofactors are necessary for pathology induction.

Not only in AD, but also in other tauopathies, tau inclusions appear to engage well-defined anatomical regions that follow stereotypic temporospatial patterns (Braak & Braak, 1991; Braak et al., 2006; Saito et al., 2004; Tolnay & Clavaguera, 2004). In AD, tau pathology can be defined according to six stages involving anatomically connected brain areas (discussed above), while for AGD, three different stages of disease progression have been proposed. Argyrophilic tau grains appear first in the ambient gyrus and then in the medial temporal lobe and cingulate gyrus. This progression of neuropathological changes correlates with a progressive cognitive decline in AGD (Saito et al., 2004). The model proposed by Braak et al., as well as other models proposed for pathology progression in specific tauopathies, have been recapitulated in different animal models. Thus, intracerebral injection of synthetic tau fibrils resulted in inclusions that spread between restricted anatomical regions in a time-dependent manner (Iba et al., 2013; Peeraer et al., 2015). This pattern of NFTs has been suggested to be more influenced by the strength of synaptic connections than the proximity of the interconnected anatomical sites (Ahmed et al., 2014). Different genetic tau mouse models of early AD with region-specific promoters also exhibit tau aggregates in nonexpressing regions; restricted transgenic tau expression in the entorhinal cortex resulted in NFTs in associated limbic regions that did not express transgenic tau (de Calignon et al., 2012; Liu et al., 2012)—in accordance with a pathogenic spreading hypothesis. Collectively, these experiments suggest that tau pathology is not a cell-autonomous mechanism and that tau also behaves as a prion.

α-Synuclein Prions

In analogy to the previously mentioned aggregation-prone proteins, αSyn is a small protein that is intrinsically disordered and lacks a well-defined tertiary structure (Fauvet et al., 2012; Theillet et al., 2016). αSyn binds membranes and is enriched at presynaptic terminals (Maroteaux, Campanelli, & Scheller, 1988; Shibayama-Imazu et al., 1993). Upon membrane association, it acquires an alternative helical conformation by wedging itself between the phospholipid head groups of the outer membrane, where it bends and tubulates membranes (Auluck, Caraveo, & Lindquist, 2010; Braun, Lacy, Ducas, Rhoades, & Sachs, 2014; Fusco et al., 2014; Ouberai et al., 2013). In addition to its role in membrane remodeling, αSyn is also considered to function as a chaperone with other vesicular proteins to regulate kinetics of endo- and exocytosis (Burre et al., 2010; Chandra, Gallardo, Fernandez-Chacon, Schluter, & Sudhof, 2005; Diao et al., 2013; Vargas et al., 2014). This remarkable functional diversity is made possible by the dynamic structural features of αSyn, which makes the protein prone to aberrant folding events. As a result, αSyn can easily form de novo aggregates, e.g., at membranes where it is highly enriched (Fares et al., 2016).

Aggregated αSyn is the main component of Lewy Bodies and Lewy neurites in PD, DLB and other rare conditions that include pure autonomic failure (Table 8.1). αSyn mainly deposits in neurons, and postmortem confirmation of Lewy pathology is essential for PD or DLB diagnosis (Berg et al., 2014; McKeith, 2006). In multiple system atrophy (MSA), αSyn inclusions are also observed in neurons, but only during later stages of the disease (Wakabayashi et al., 1998). In initial stages of MSA, αSyn deposits in oligodendrocytes and Schwann cells, which are termed glial cytoplasmic inclusions and Schwann cell cytoplasmic inclusions, respectively (Nakamura et al., 2015). MSA is a primary oligodendropathy, and its origin remains unknown. The vast majority of cases with PD, DLB, and MSA are sporadic. Rare point mutations and *SNCA* gene replications that give rise to familial forms of PD have been identified (Chartier-Harlin et al., 2004; Kruger et al., 1998; Pasanen et al., 2014; Polymeropoulos et al., 1997; Singleton et al., 2003; Zarranz et al., 2004). In cases of gene multiplications, the gene load inversely correlates with age of onset and indicates that the level of wild-type αSyn influences how rapidly the disease develops (Chartier-Harlin et al., 2004; Singleton et al., 2003).

Already two decades ago, in vitro experiments showed that αSyn aggregates have seeding properties and that they can nucleate the conversion of soluble precursor into toxic assemblies (Han, Weinreb, & Lansbury, 1995; Hashimoto et al., 1998). However, the prion-like features of αSyn and the possibility that assemblies of misfolded protein could propagate pathology from one brain region to another were triggered only a decade ago. Two papers showed that neural transplants injected into the striatum of PD patients more than 10 years prior to death contained Lewy bodies at autopsy (Kordower et al., 2008; Li et al., 2008). It was suggested that an unknown host factor triggered the formation of Lewy bodies in the young grafts in a noncell-autonomous way (Kordower et al., 2008; Li et al., 2008), and there was a mixed reception in the research field when it was suggested that this trigger might be misfolded αSyn that had undergone cell-to-cell transfer (Li et al., 2008).

Following these two clinical reports, the first experiments were initiated to test a potential spreading hypothesis. By transplanting cortical neuronal stem cells into the hippocampus of transgenic mice overexpressing human αSyn, it was shown that human αSyn could transfer to transplanted cells (Desplats et al., 2009). Alternatively, using a striatal grafting model demonstrated that overexpressed αSyn could transfer to grafted dopaminergic neurons in mice and rats and seed its soluble precursor (Angot et al., 2012; Hansen et al., 2011; Kordower et al., 2011), in accordance with previous clinical findings (Kordower et al., 2008; Li et al., 2008). Synthetic forms of aggregated αSyn were able to corrupt protein folding in HEK293 and SH-SY5Y cells and trigger Lewy-like pathology (Luk et al., 2009). These important experiments thereby confirmed that αSyn can behave in a prion-like fashion and established important mechanistic links with other prion-like neurodegenerative disorders. In subsequent studies, wild-type mice were inoculated with brain homogenates from DLB patients, and this resulted in αSyn deposition in their brain (Masuda-Suzukake et al., 2013). In monkeys, intracerebral injection of αSyn-enriched fractions derived from Lewy Bodies of PD patients also induced αSyn protein accumulation, but only with modest nigrostriatal neurodegeneration (Recasens et al., 2014). Similar experiments in αSyn transgenic animal models yielded comparable results. Injection of brain extracts from old TgM83 transgenic mice (overexpressing human A53T-mutant αSyn) into young TgM83 mice that did not yet show pathology resulted in earlier disease onset and pathology (Mougenot et al., 2012). Injection of brain material from deceased PD, DLB, and MSA patients into the striatum of transgenic mice (TgM83 and BDF1 that express mutant and wild-type αSyn, respectively) also promoted αSyn aggregate formation and accelerated synucleinopathy progression with subsequent spread of pathology to connected regions (Jones et al., 2015; Mougenot et al., 2012; Watts et al., 2013). In addition, injection of brain extracts from patients with MSA or incidental Lewy Body disease into transgenic mice that express wild-type human αSyn in the absence of endogenous αSyn (mouse αSyn has threonine at position 53 instead of alanine) was sufficient to induce hyperphosphorylated inclusions (Bernis et al., 2015). Interestingly, in experiments with autopsy-derived brain material from MSA patients, αSyn inclusions and pathology mainly propagated in neurons instead of oligodendrocytes (Prusiner et al., 2015).

To further substantiate that αSyn can behave as a prion-like causative agent in synucleinopathies, it was important to show that pure αSyn assemblies in the absence of diseased material could induce an infectious cascade of prion-like propagation. In two initial experiments, truncated forms of aggregated mouse αSyn (that lack the C-terminus) were injected into the brains of wild-type and transgenic animals (Luk et al., 2012a, 2012b; Paumier et al., 2015). These synthetic assemblies were shown to trigger a cascade of protein spreading and aggregation in a time-dependent manner (Luk et al., 2012a, 2012b; Paumier et al., 2015). αSyn aggregates made up of human recombinant protein had similar capacities to inflict synucleinopathy (Masuda-Suzukake et al., 2013; Peelaerts & Baekelandt, 2016), albeit at slower rates (Rey et al., 2016b). By further dissecting the seeding potency of different assemblies, it was shown that stable αSyn fibrillar seeds had greater nucleating capacities compared to smaller but unstable synthetic oligomers that did not have prion features in vivo (Peelaerts et al., 2015). These experiments showed that the mechanistic pathogenic features of αSyn resemble that of prions.

TDP-43 and FUS Prions

Several recent and intriguing findings have shown an unexpected overlap between ALS and FTD. Several common genetic causes and phenotypic features, such as cognitive decline and motor neuron deficits, have been linked to both neurological disorders (Lomen-Hoerth, Anderson, & Miller, 2002; Phukan, Pender, & Hardiman, 2007). Different genetic components in ALS and FTD are in distinct, but intertwined, pathways of RNA processing and protein homeostasis. Thus, it is becoming clear that disruption of any of these molecular pathways can contribute to a prion-like disease process.

FTD, also known as frontotemporal lobar degeneration (FTLD), is a well characterized tauopathy. However, in addition to tau, different histopathological features are characteristic of FTD. In FTD, tau misfolds into typical NFTs, and other ubiquitinated protein inclusions with variable morphologies that are negative for tau are also often found in the CNS of FTD patients. This has led to a FTD classification system that is based on different histopathological features, including tau or ubiquitinylated inclusions, termed FTLD-tau or FTLD-u, respectively.

The exact substrate of these ubiquitinated structures in FTD was long unknown. Recent reports described two other accumulating proteins. First, the TDP-43 was found to accumulate in the CNS of both FTD and ALS patients in patients who were negative for tau (Arai et al., 2006; Neumann et al., 2006). Shortly thereafter, various mutations in *TARBDP* (the gene that encodes TDP-43) were linked with sporadic and familial FTD and ALS (Borroni et al., 2009; Kabashi et al., 2008; Kovacs et al., 2009; Sreedharan et al., 2008). These findings established mutant TDP-43 as a causative agent and were in concert with the fact that aggregated TDP-43 is the main histopathological substrate in FTDL-u. Later, mutations in a second gene, *FUS*, were also discovered in cases of ALS and FTD. However, pathological inclusions of FUS protein in ALS are rare. This eventually led to the reclassification of FTD into different subtypes into three main neuropathological classes: FTLD-tau (45%), FTLD-TDP-43 (45%), and FTD-FUS (9%) (Mackenzie et al., 2010).

More evidence that ALS is a protein misfolding disease with prion-like features came from mechanistic links between genes that are involved in protein homeostasis and gene regulation. For instance, repeats of a noncoding GGGGCC hexanucleotide expansion in the *C9ORF72* gene that compromises nucleocytolasmic transport and promotes inclusion formation are found in both FTD and ALS (DeJesus-Hernandez et al., 2011; Patel et al., 2015; Renton et al., 2011; Zhang et al., 2015). This hexanucleotide expansion in C9orf72 is known as the most common sporadic or familial genetic abnormality in ALS (DeJesus-Hernandez et al., 2011; Renton et al., 2011). Mutations in other genes such as ubiquilin-2 (*UBQLN2*), p62/sequestome (SQSTM1), vasolin containing protein (*VCP*), and the vesicle-associated membrane-associated protein B (*VAPB*) complete the list of an entangled network of proteostasis regulators that, when perturbed, can trigger toxic protein aggregation and severe neurodegenerative deficits (Ling, Polymenidou, & Cleveland, 2013).

As most prion-like disorders, ALS is mainly a sporadic disease. Here, TDP-43 and FUS aberrant protein folding and aggregation are imperative for disease pathogenesis. TDP-43 and FUS are highly conserved proteins that bind DNA or RNA and are implicated in the regulation of several general cellular mechanisms (Ling et al., 2013). Both proteins contain a prion-like domain (similar to that of the PrP) that mediates their self-assembly into high molecular weight aggregates under native conditions (King, Gitler, & Shorter, 2012). These aggregates, that form reversible β-sheet structures, are highly enriched in cytosolic mRNA-containing stress granules where they bind to mRNA (Li, King, Shorter, & Gitler, 2013; Protter & Parker, 2016). An emerging hypothesis is therefore that failed clearance of stress granules leads to an increased probability of their assembly through their prion-like domains into toxic, irreversible amyloid (Ramaswami, Taylor, & Parker, 2013). Interestingly, different experimental models have shown that mutations in TDP-43 or FUS result in their pathogenic self-assembly in the absence of stress granules (Molliex et al., 2015; Patel et al., 2015).

TDP-43 aggregates are found in the soma of the cytosol or neurites, or in the nucleus, in motor neurons and glial cells (Neumann et al., 2006). As the disease advances, protein aggregates of TDP-43 are found in an increasing number of CNS regions, following a stepwise and progressive pattern. Recently, a classification system was proposed for FTD in an attempt to standardize the neuropathological descriptions of TDP-43 in FTD and ALS. The distribution of TDP-43 aggregates in FTD encompasses four different stages (Brettschneider et al., 2014). First, during initial stages, FTD-43 aggregates are present in the prefrontal cortex. In stage II, pathological lesions in these initial regions are more severe, and inclusions spread in a rostrocaudal and sequential manner to anatomically connected regions (Brettschneider et al., 2014). During stage III phosphorylated inclusions of TDP-43 are also found in motor neurons of the spinal cord. Only on rare occasions in patients with a protracted disease course, inclusions are also found in the occipital or visual cortex (stage IV) (Brettschneider et al., 2014).

Propagating neuronal histopathological lesions are also characteristic of ALS (Al-Chalabi et al., 2016). Neuropathology appears focally or in patches of the motor cortex and bulbar or spinal regions (stage I) (Brettschneider et al., 2013). Motor neurons show increasing burden of TDP-43 deposits and degenerate along the neuraxis in a progressive gradient from the site of disease onset toward prefrontal cortex and brainstem reticular formations (Ravits, Laurie, Fan, & Moore, 2007). During stages III and IV, there is severe involvement of postcentral cortex, limbic system, and striatum. In all stages, neuronal deposits are accompanied by inclusions in oligodendroglia that are in close proximity of affected axons of motor neurons (Brettschneider et al., 2013). In some brain areas, oligodendroglial inclusions can be more frequent than neuronal inclusions (Fatima, Tan, Halliday, & Kril, 2015). Motor deficits in patients are also radially graded and perfectly match the contiguously advancing neuronal pathology in the four different stages (Ravits et al., 2007). Collectively, these neuroanatomical patterns of aggregate spreading are consistent with a prion-like spread of pathogenic proteins along axonal projection.

Following these observations in human brains, the question was whether TDP-43 or FUS could also behave as self-propagating assemblies in experimental models. TDP-43 and FUS readily aggregate in vitro, and fibrils assembled in vitro can seed aggregation when added to cell cultures expressing TDP-43 or FUS (Furukawa, Kaneko, Watanabe, Yamanaka, & Nukina, 2011; Nomura et al., 2014; Shimonaka, Nonaka, Suzuki, Hisanaga, & Hasegawa, 2016). Autopsy-derived sarkosyl insoluble fractions from ALS and FTD patients were furthermore shown to have nucleating properties in vitro (Feiler et al., 2015; Nonaka et al., 2013; Smethurst et al., 2016). Formic acid treatment ameliorated brain-derived TDP-43 seeding capacities, while proteolysis or heat treatment did not, which suggests that a β-sheet conformation is required for cellular propagation (Nonaka et al., 2013). Taken together, these results support the notion that ADP-43 and FUS might act as prion-like proteins. However, additional experiments are warranted since "ALS aggregates," either synthetic or patient-derived, have not yet been shown to induce prion-like propagation after injection into brain or after passaging affected brain material in animals.

SOD1 Prions

Already in 1993, the first causative gene mutation in ALS was identified in superoxide dismutase 1 (SOD1) (Renton, Chio, & Traynor, 2014; Rosen et al., 1993). Also, this protein is prone to aggregation, and next to TDP-43 and FUS, proteinaceous inclusions of SOD1 complete the ensemble of histopathological substrates in ALS (Table 8.1). SOD1 is a small and stable dimer that acts as an enzyme to scavenge superoxide radicals (Polymenidou & Cleveland, 2011). Mutations in SOD1 decrease its stability and result in partial protein unfolding and its subsequent assembly into β-sheet rich structures (Tiwari & Hayward, 2005). Under physiological conditions, mutated SOD1 becomes insoluble and aggregates into fibrils that resemble Lewy body-like inclusions (Heiman-Patterson et al., 2011). These inclusions, which are found in neurons and glial cells, are eosinophilic and have a dense core with radiating filaments (Heiman-Patterson et al., 2011). Despite this strong genetic component in familial cases, abnormal protein deposits of wild-type SOD1 are also present in sporadic ALS (Polymenidou & Cleveland, 2011).

In vitro experiments showed that misfolded mutant SOD1 is self-propagating and spreads between cells (Grad et al., 2011; Munch, O'Brien, & Bertolotti, 2011). Mutant SOD1 could not only seed assemblies of mutant protein but also corrupt the folding of native SOD1 and seed the wild-type form, which is thermodynamically more stable (Grad et al., 2011). The fact that SOD1 could propagate in a cellular system was unexpected, since genetic animal models for mutant or truncated SOD1 very closely mimic several of the core features of ALS (Heiman-Patterson et al., 2011). Motor behavior in mutant SOD1 mice deteriorates in a progressive manner, and neuronal and glial SOD1 deposits accompany motor neuron loss (Ling et al., 2013). The presence of SOD1 aggregates together with the fact that these animal models recapitulate different aspects of ALS pathogenesis initially opposed the idea that ALS was a noncell-autonomous disease.

In order to assess the prion-like properties of mutant or wild-type SOD1 in vivo, spinal homogenates of transgenic SOD^{G93A} mice were used to assess whether CNS tissue containing the SOD assemblies could trigger aggregation of synthetic SOD (Chia et al., 2010). Tissue from mice expressing SOD1 mutant forms was shown to accelerate aggregation of mutant and wild-type recombinant assemblies (Chia et al., 2010). Injection of purified seeding-potent SOD1 aggregates in sciatic nerves of mutant $SOD1^{G85R}$ and $SOD1^{D90A}$ transgenic mice resulted in accelerated motor neuron pathology along restricted anatomical routes throughout the spinal cord (Ayers, Fromholt, O'Neal, Diamond, & Borchelt, 2016; Bidhendi et al., 2016). These initial experiments show that aggregates of SOD1 might corrupt native and mutant SOD1 protein and induce misfolding. Hence, the emerging picture is that the molecular mechanisms involved in diseases with SOD1 aggregation include prion-like behavior of the protein, akin to the findings reported in CJD, AD, PD, and HD (see below).

Huntington Prions

Unlike most putative prion-like disorders, HD is exclusively inherited. It belongs to a larger group of polyglutamine (polyQ) or trinucleotide disorders that are characterized by a multitude of neurological symptoms, frequently including progressive cognitive decline and movement deficits (Orr & Zoghbi, 2007). PolyQ disorders are caused by expansions of CAG repeats that increase the tendency of the affected protein to aggregate into pathological assemblies (Weber, Sowa, Binder, & Hubener, 2014). In HD, glutamine repeats in exon 1 of the *IT15* gene, that encodes the Htt protein, results in inherited HD. Normal Htt protein contains around 6 to 35 glutamine residues, and no individuals with less than 36 glutamine residues have been reported to develop HD. Incomplete penetrance occurs with 35–41 repeats, and above 41 repeats, the disease becomes fully penetrant (Rubinsztein et al., 1996; Walker, 2007). The longer the polyQ insert, the higher the tendency of the protein to aggregate (Li & Li, 1998). The age of onset is inversely correlated with the length of the repeats of the aggregating protein (Li & Li, 1998; Mahant, McCusker, Byth, Graham, & Huntington Study, 2003; Squitieri, Cannella, & Simonelli, 2002), indicting a causal relationship between corrupt protein folding and toxicity.

HD is a progressive disorder, and classical signs start at the extremities, but it is unclear how aggregate formation relates to the progression of the disease. The presence of neuronal Htt inclusions is associated with progressive neuronal death and gross atrophy of the striatum and frontoparietal cortex (Arrasate & Finkbeiner, 2012; DiFiglia et al., 1997). The inclusions are present in the cytosol of the perikaryon or inside the nucleus and are mainly composed of truncated mutant Htt (mHTT) that contains the N-terminal segment (Sieradzan et al., 1999). Although still controversial, it is suggested that inclusion bodies are sequestered toxic forms of mHTT (Arrasate, Mitra, Schweitzer, Segal, & Finkbeiner, 2004) and that nonscavenged assemblies are causal for disease pathogenesis and prion-like propagation.

In vitro experiments with mutant or aggregated Htt showed that misfolded forms transfer between cells and propagate. Large aggregates of mHtt are readily taken up in vitro and transfer via various mechanisms to postsynaptic neurons or glial cells in vivo (Jansen, Batenburg, Pecho-Vrieseling, & Reits, 2017; Pearce, Spartz, Hong, Luo, & Kopito, 2015; Pecho-Vrieseling et al., 2014). The transferred spreading particles were shown to inflict synaptic alterations and cell damage in a noncell-autonomous way (Babcock & Ganetzky, 2015; Pecho-Vrieseling et al., 2014). Brain extracts from mutant mice or HD patients seeded polyQ peptides in an in vitro assay, indicating that mHTT has propagating features (Gupta, Jie, & Colby, 2012). However, to date no study has demonstrated that aggregated Htt can recruit its wild-type counterpart. Transplantation of patient-derived fibroblasts expressing different types of mHTT into the striatum of wild-type mice resulted in mHTT spreading and disease transmission (Jeon et al., 2016). In this experimental model, transmitted mHTT induced cell loss and cognitive impairments. Aggregates of Htt have been reported to be present in neural tissue grafted to the striatum of HD patients (Cicchetti et al., 2014). Upon autopsy, these neural grafts were found to exhibit clear signs of degeneration after only a few years (Cicchetti et al., 2009). This indicates that the host environment or mHtt aggregates inflict or potentiate toxicity in tissue despite the fact that the transplant tissue does not express the mutant gene product. Accumulating evidence is thus pointing toward a prion-like mechanism in HD disease pathogenesis, but in order to establish this disease as a prion disease, many features remain to be investigated regarding HD (Table 8.2).

MECHANISTIC, FUNCTIONAL, AND PATHOGENIC PROPERTIES OF PRIONS

Bacterial, Fungal, and Mammalian Prions

Although we often associate prions with transmissible diseases, they also have normal roles that do not cause disease. Prions exist in bacteria, yeast, fungi, and mammals where a number of functional and biological properties have been described and many more probably remain undiscovered (Chiti & Dobson, 2006).

One of the first discovered yeast prions is Sup35p. The Sup35p prion is a translation termination factor that binds and sequesters its functional form in a dominant negative fashion (Teravanesyan, Dagkesamanskaya, Kushnirov, & Smirnov, 1994). The soluble and biologically active form thereby becomes insoluble and unavailable for binding with DNA, leading to its own suppression. As a consequence, the yeast cell will form novel gene products and

TABLE 8.2 Summary and Evidence for Prion Features of Different Proteins in Neurodegenerative Proteopathies

Prion	Deposition	Cellular Localization	Amplification		Spreading		Disease Transmission	
			In Cellulo	In Vivo	In Cellulo	In Vivo	Synthetic	Brain Extract
Aβ	Extracellular (Glenner & Wong, 1984; Masters et al., 1985)	–	–	Yes (Meyer-Luehmann et al., 2006)	Yes (Brahic, Bousset, Bieri, Melki, & Gitler, 2016)	Yes (Stohr et al., 2012)	Yes (Stohr et al., 2012)	Yes (Kane et al., 2000; Meyer-Luehmann et al., 2006; Walker et al., 2002)
Tau	Neuronal glial (Bigio et al., 2001; Goedert & Spillantini, 2006; Komori, 1999; Kovacs et al., 2008)	Cytoplasmic (Crowther & Goedert, 2000; Goedert & Spillantini, 2006)	Yes (Frost et al., 2009a; Guo & Lee, 2011; Kfoury et al., 2012)	Yes (Kaufman et al., 2016; Sanders et al., 2014)	Yes (Calafate et al., 2015; Wu et al., 2016)	Yes (Ahmed et al., 2014; de Calignon et al., 2012; Iba et al., 2013; Liu et al., 2012; Peeraer et al., 2015)	Yes (Iba et al., 2013; Peeraer et al., 2015; Sanders et al., 2014; Stancu et al., 2015)	Yes (Ahmed et al., 2014; Clavaguera et al., 2009, 2013)
αSyn	Neuronal glial (Berg et al., 2014; McKeith, 2006; Nakamura et al., 2015)	Cytoplasmic (Berg et al., 2014; McKeith, 2006; Nakamura et al., 2015)	Yes (Hansen et al., 2011; Luk et al., 2009)	Yes (Peelaerts et al., 2015; Prusiner et al., 2015)	Yes (Abounit et al., 2016; Hansen et al., 2011; Stuendl et al., 2016)	Yes (Kordower et al., 2011; Rey et al., 2016b; Rey, Petit, Bousset, Melki, & Brundin, 2013; Reyes et al., 2014)	Yes (Guo et al., 2013; Luk et al., 2012a, 2012b; Paumier et al., 2015; Peelaerts et al., 2015)	Yes (Bernis et al., 2015; Jones et al., 2015; Masuda-Suzukake et al., 2013; Mougenot et al., 2012; Prusiner et al., 2015; Recasens et al., 2014; Watts et al., 2013)
TDP-43	Neuronal glial (Brettschneider et al., 2013; Neumann et al., 2006)	Nuclear, cytoplasmic (Neumann et al., 2006)	Yes (Furukawa et al., 2011; Nonaka et al., 2013; Smethurst et al., 2016)	n.d.	Yes (Feiler et al., 2015; Smethurst et al., 2016)	n.d.	n.d.	n.d.
FUS	Neuronal glia (Deng et al., 2010)	Nuclear cytoplasmic (Baumer et al., 2010)	Yes (Nomura et al., 2014)	n.d.	n.d.	n.d.	n.d.	n.d.
SOD1	Neuronal glial (Heiman-Patterson et al., 2011; Polymenidou & Cleveland, 2011)	Cytoplasmic (Heiman-Patterson et al., 2011; Polymenidou & Cleveland, 2011)	Yes (Grad et al., 2011; Munch et al., 2011)	Yes (Bidhendi et al., 2016)	Yes (Grad et al., 2011; Munch et al., 2011)	Yes (Bidhendi et al., 2016)	n.d.	Yes (Ayers et al., 2016; Bidhendi et al., 2016)
Htt	Neuronal (DiFiglia et al., 1997; Landwehrmeyer et al., 1995)	Nuclear cytoplasmic (DiFiglia et al., 1997; Sieradzan et al., 1999)	n.d.	n.d.	Yes (Brahic et al., 2016; Jansen et al., 2017; Pearce et al., 2015; Pecho-Vrieseling et al., 2014)	Yes (Cicchetti et al., 2014; Jeon et al., 2016; Pecho-Vrieseling et al., 2014)	n.d.	n.d.

αSyn, α-Synuclein; *TDP-43*, TAR DNA-binding protein; *FUS*, fused in sarcoma; *SOD1*, superoxide dismutase 1; *Htt*, huntingtin.

distinctive yeast phenotypes (termed [*psi*]$^+$ or [*PSI*]$^+$) that potentially provide a survival benefit (Tanaka, Collins, Toyama, & Weissman, 2006).

Another example of a fungal prion is the heterokaryon incompatibility protein s (HET-s). HET-s is a protein-based genetic element that is important for programed cell death in fungal cells (Balguerie et al., 2003). It contains an intrinsically disordered prion sequence that is flanked by a globular domain, which is essential to prevent the prion domain from forming infectious prions (Balguerie et al., 2003). This kind of modular organization is similar to polyQ prions discussed earlier, such as Htt or ataxins, that become more infectious with increasing lengths of its polyQ tracts or after cleavage of its globular domain (Duennwald, Jagadish, Muchowski, & Lindquist, 2006). These aggregation-prone polyQ sequences, akin to other prion-like domains, are often rich in polar or hydrophobic amino acids to promote homotypic oligomerization (Gilks et al., 2004). Nucleating HET-s prions can be vertically transmitted between dividing mother and daughter cells, but not between cells after for example being released into culture medium because of inhibitory mechanisms of the HET-s protein (in *cis*) that has a slightly different sequence than HET-s (Greenwald et al., 2010). Because of sequence mismatch, this cis-acting element will be less prone for seeding and aggregation and inhibit the functional role of its prion-like state (Greenwald et al., 2010).

Importantly, nonpathogenic prion-like proteins also exist in mammalian cells. A well-studied example is the cytoplasmic polyadenylation element-binding protein or the T-cell restricted intracellular antigen 1 (TIA-1). TIA-1 has several RNA-binding domains and a glutamine-rich domain that resembles that of the PrP (Anderson & Kedersha, 2002). Upon reversible oligomerization, it undergoes liquid–liquid phase separation to yield membraneless organelles or stress granules that stall ribonucleoprotein complexes (Li, Rayman, Kandel, & Derkatch, 2014)—reminiscent of TDP-43 and FUS. In response to stress, or altered conditions that have increased expression of TIA-1, liquid TIA-1 complexes can become solid and form microaggregates that coalesce into larger cytoplasmic inclusions with a high β-sheet content. Under experimental conditions, TIA-1 assembles into insoluble seeds that propagate its soluble precursor through a cascade of protein misfolding and amplification (Gilks et al., 2004). In contrast to TDP-43 or FUS, however, that have strikingly similar functions in mammalian cells, TIA-1 is not known to cause disease.

These examples show that proteins can exist in both soluble and insoluble states with amyloid or prion-like features in various organisms. These features are required to fulfill their normal biological function, which varies greatly between the proteins.

Prions in Neurodegenerative Disorders

Nature might have selected—or not selected against—some proteins to exist in an amyloid or prion-like state under normal conditions. Other proteins, however, are selected under evolutionary pressure so that they exist very close to the limits of their own solubility (Monsellier & Chiti, 2007; Tartaglia, Pechmann, Dobson, & Vendruscolo, 2007; Watters et al., 2007). Interestingly, the expression levels of pathogenic proteins such as the PrP, tau, Aβ, or αSyn and also other nonpathogenic proteins are inversely correlated with their aggregation rate (Tartaglia et al., 2007). This means that aggregation is a natural or intricate property of proteins and not just a consequence of their sequence that might be rich in hydrophobic or polar residues—making them more or less aggregation prone. Genetic mutations or unknown environmental factors that alter the protein's expression or modify its folding state can favor the protein adopting a thermodynamically stable amyloid state. This is observed for proteins that are intrinsically disordered (Aβ, tau, or αSyn), contain a prion domain (PrP, TDP-43, or FUS) or have long repeats of glutamate residues (Htt or ataxins).

During the initial steps of misfolding and amyloid formation, amyloidogenic or prion-like proteins typically expose their hydrophobic or polar residues and cluster between these residues so that they are protected from the surrounding aqueous environment (Perutz, Johnson, Suzuki, & Finch, 1994; Uversky, Gillespie, & Fink, 2000). A resulting hydrophobic collapse will yield small, amorphous assemblies with a β-sheet conformation. This conformational change can be reversible and sometimes even functional, as in the case of stress granules (TIA-1, TDP-43, and FUS). These assemblies are not very toxic to cells since they are unstable and can easily be degraded by the cellular degradation machinery (Stefani & Dobson, 2003). However, under certain conditions, small aggregated assemblies can rearrange their conformation and form more stable assemblies that are toxic and have nucleating capacities to seed subsequent aggregation steps (Prusiner, 2013). The kinetic barriers imposed on these steps are high, and therefore aggregation in vivo will only occur after protracted periods (Knowles, Vendruscolo, & Dobson, 2014).

Despite these kinetic barriers, several factors can positively influence the tendency of proteins to form nucleating molecules. Disease-related proteins Aβ and αSyn exist in the brain at concentrations that approach their own solubility or, as in the case of αSyn, even exceed it (Baldwin et al., 2011). In addition, different isoforms of Aβ, tau, αSyn, or Htt can undergo truncation, yielding fragments with a higher tendency to aggregate (Arrasate & Finkbeiner, 2012; Li et al., 2005; Snyder et al., 1994). In other words, native assemblies of several disease-related proteins populate folding states that are thermodynamically less stable relative to their amyloid form. Cellular proteostasis mechanisms (e.g., the autophagy-lysosomal or ubiquitin-proteasome systems or disaggregase complexes) therefore need to actively engage aberrant protein folding in order to prevent the assemblies from becoming toxic and self-sustaining (Hartl, Bracher, & Hayer-Hartl, 2011; Nillegoda et al., 2015).

Under normal physiological conditions, these prion-like proteins are continuously prone to form amyloid. It is therefore not a question of *why* proteins with prion features form propagating species, but rather *when*. Once they surpass the kinetic threshold in a fluctuating energy landscape, they will form a stable template onto which soluble precursor proteins can seed. An important prerequisite for fibril growth is that soluble monomers need to exist in folding equilibrium with their seed. This also implies that not all monomers can associate with the template, which is required for homotypic seeding. Only monomers with a minimum degree of sequence similarity, and an intermediate assembly state that has a conformation similar to that of the seed, will incorporate (Knowles et al., 2014; Melki, 2017). Different proteins, such as tau and αSyn, will be incapable of direct heterotypic seeding because they lack a shared sequence. Nonetheless, if none of these amplification barriers are present, the incorporation of misfolded assemblies will yield a protofibrillar particle that can break and catalyze subsequent aggregation reactions (Knowles et al., 2009; Xue, Homans, & Radford, 2008).

MECHANISMS OF TRANSPORT AND CELL-TO-CELL PROPAGATION

Prion Propagation

How proteopathic molecules inflict toxicity and progressively cause disease are areas of intense research. The currently prevailing hypothesis is that seeding-potent particles amplify and spread from neuron to neuron to induce inclusion formation and toxicity. An alternate hypothesis posits that disease pathogenesis might be the result of a cell-autonomous mechanism that includes selective vulnerability. In that model, cellular damage would be the sum of genetic and other environmental factors that predispose vulnerable neurons to degenerate to a greater extent compared to other less susceptible populations. It has been suggested that such a selective vulnerability could explain why protein aggregates progressively appear in different interconnected brain regions, and that, e.g., an inflammatory response in one region could spread and trigger disease in an adjacent brain area.

The best known example of disease propagation via conversion of endogenous protein into toxic amyloid is the PrP^{sc} (Prusiner, 1998). Endogenous PrP^{c} turns toxic and spontaneously self-perpetuates after acquiring an alternative conformation (PrP^{sc}), thereby triggering an infectious cycle of protein aggregation and spreading. Prions in Kuru, CJD, and other TSEs instigate disease through spreading of infectious prion particles (PrP^{sc}) via neuronal connections. Different animal models have shown that, once within the brain, PrP^{sc} prions spread via neuronal contacts and not simply through diffusion along axonal projections. After peripheral administration, PrP^{sc} prions spread from neurons innervating peripheral organs (prions replicate first in the lymphoreticular system) from which they invade the spinal cord or the lower brainstem (Aguzzi, Nuvolone, & Zhu, 2013; Fraser & Dickinson, 1970; Kimberlin, Hall, & Walker, 1983; McBride et al., 2001). Intraocular inoculation of PrP^{sc} propagated infectivity through neuroanatomical connections from the retina to the contralateral superior colliculus (Liberski, Hainfellner, Sikorska, & Budka, 2012; Scott & Fraser, 1989) and intracerebral inoculation of prion-laden extracts resulted in widespread dissemination of infectious prions with brain tropism that selectively propagated along specific neuronal tracts (Collinge & Clarke, 2007). Also important is that glial cells appear to be resistant to PrP^{sc} infection, suggesting that they do not contribute to disease propagation (Prinz et al., 2004).

Although the molecular and cellular mechanisms of neuronal spreading of prions in TSEs are not well understood, it is believed that cells take up infectious particles and transfer them to their neighboring cells. PrP^{c} is synthesized in the endoplasmic reticulum and is transported via exosomes to the outside of the cellular membrane (Linden et al., 2008). Subsequently, it can become internalized and localized with endosomes, lysosomes, and multivesicular bodies. Various in vitro studies have shown that PrP^{sc} replication occurs intracellularly at lipid organelles and at the plasma membrane (Grassmann, Wolf, Hofmann, Graham, & Vorberg, 2013). Infectivity is enhanced by cofactors (Ma & Wang, 2014) or polysaccharides that bind and enrich PrP prions at lipid rafts (Diaz-Nido, Wandosell, & Avila, 2002). Successful infection does not necessarily lead to sustained replication and might be due to nonpermissive cells or nonreplicating conformers (Beringue, Vilotte, & Laude, 2008). When sustained replication does occur, infectious templates of PrP^{sc} appear to spread between cells via exosomes or tunneling nanotubes (TNTs) that act as cytoplasmic bridges between two cells (Gousset et al., 2009; Grassmann et al., 2013).

While observations in humans supporting these experimental data are lacking, these results favor a noncell-autonomous spreading hypothesis (Prusiner, 1982; Walsh & Selkoe, 2016). This, however, does not exclude that selective vulnerability plays a role in TSE, i.e., some cells, e.g., those with elaborate synaptic contacts and higher metabolic demands might be more susceptible to develop pathology.

A Hitchhiker's Guide to Infectivity

Cellular Mechanisms in Prion-Like Diseases

In light of the observations made by Braak et al., proponents of the pathogenic spreading hypothesis explored whether proteins in other neurodegenerative diseases might propagate similar to TSE prions. A direct and popular method to assess the spreading of seeding-potent particles is via cellular uptake and spreading assays. Fluorescently labeled or tagged synthetic proteins are added to cultured cells, and protein uptake as well as seeding of aggregation are assessed. From these assays, it is clear that different types of cells readily take up naked aggregates of Aβ, tau, αSyn, FTD-43, SOD1, or Htt and seed subsequent aggregation (Desplats et al., 2009; Frost et al., 2009a; Hansen et al., 2011; Holmes et al., 2013; Ren et al., 2009; Wu et al., 2013).

Several studies have shown that cells that contain aggregates can release naked or partially exposed nucleating particles through active mechanisms, or upon death (Fig. 8.1B). Uptake of these particles by neighboring cells might be mediated through cell-surface receptors. While some features of amyloid proteins are shared, the quaternary structures of PrP, Aβ, tau, αSyn, FTD-43, SOD1, and Htt nucleating agents are different, and therefore one can predict that their binding to other proteins, lipids, or polysaccharides will differ significantly. Receptor association could therefore include interactions that are aggregate specific. Notwithstanding this reasoning, various non-specific receptors such as heparan sulfate proteoglycans, integrins, or neuronexins (Holmes et al., 2013; Melki, 2017) have been identified. Furthermore, a growing list of specific aggregate receptors, that mediate internalization, includes the transmembrane receptors PrP^{c}-mGluR5 (Aβ) (Lauren, Gimbel, Nygaard, Gilbert, & Strittmatter, 2009), α7-nAchR (Aβ) (Wang, Lee, Davis, & Shank, 2000), BIN1 (tau) (Calafate, Flavin, Verstreken, & Moechars, 2016), α3-NaK-ATPase (αSyn) (Shrivastava et al., 2015), or LAG3 (αSyn) (Mao et al., 2016). The importance of these for uptake and aggregate-mediated toxicity has been demonstrated since removal or inhibition of their interaction significantly ameliorates their neurodegenerative effects (Lauren et al., 2009; Mao et al., 2016; Shrivastava et al., 2015; Wang et al., 2000).

After binding a cell-surface receptor, aggregates need to be taken up and enter the cytosol. Different endocytic mechanisms such as receptor-mediated endocytosis (tau, αSyn) or bulk or fluid endocytosis (tau, SOD1) (Guo & Lee, 2014; Lee, Desplats, Sigurdson, Tsigelny, & Masliah, 2010) play a role. Some smaller aggregate particles have been shown to diffuse passively over the cell membrane, but they fail to induce subsequent seeding steps (Haass & Selkoe, 2007). Once aggregates have entered the cell via vesicular endocytosis, it is important for nucleating assemblies to escape from their endocytic compartment in order to amplify and seed their own aggregation. How seeds escape their compartmentalization and where they instigate initial aggregation steps is now beginning to be clarified, with studies on αSyn leading the way (Flavin et al., 2017; Freeman et al., 2013). Some studies have used lipid carriers or liposomes to deliver large aggregated assemblies and found that it significantly enhances cellular uptake and release to the cytoplasm (Holmes & Diamond, 2017). One can speculate that the conversion of nascent endogenous protein into infectious particles largely depends on where it is enriched in the cell and which organelles it associates with. Soluble species of aggregated proteins such as PrP, Aβ, and αSyn are concentrated at membrane structures such as multivesicular bodies, exosomes, or synaptic vesicles and can also be found in the cytosol (Clayton & George, 1999; Takahashi et al., 2002a; Takahashi, Nam, Edgar, & Gouras, 2002b).

After translocation of protein aggregates to the cytosol via receptors and endocytosis, successful amplification depends on a permissive seeding environment. PrP prion research indicates that amplification is subject to significant barriers which hamper the interaction of the seeding template with its partially folded or unfolded monomer (Beringue et al., 2008; Collinge & Clarke, 2007). We predict that prion-like amplification also depends on a number of factors including (1) aggregate size or frangibility; (2) precursor availability or local enrichment; (3) indirect modulators including pH, salts or chaperones; (4) direct modulators such as sequence homology, mutations, or posttranslational modifications; and (5) factors that counteract protein misfolding, e.g., the protein degradation machinery. This degradation machinery is intricately linked with protein misfolding and aggregation (Wong et al., 2008). The ubiquitin-proteasomal and the endosomal–lysosomal system modulate aggregation of prion-like proteins and counteract aggregate toxicity via disaggregation (Gao et al., 2015; Jackrel & Shorter, 2015; Nillegoda et al., 2015) or by targeting nucleating particles in multivesicular bodies for lysosomal degradation (Cuervo & Wong, 2014; Settembre, Fraldi, Medina, & Ballabio, 2013). Moreover, αSyn, tau, and Htt aggregate-bearing lysosomes can be transported via TNTs that directly connect the cytoplasm of two adjacent cells (Abounit et al., 2016; Costanzo et al., 2013; Tardivel et al., 2016). Alternatively, nucleating particles are harbored inside multivesicular bodies which can form exosomes that bud from plasma membrane or directly release their content to the extracellular environment (Vella, Hill, & Cheng, 2016). TNTs and exosomes therefore provide direct transmission routes between infected and noninfected cells. It would appear that by getting rid of some of the aggregated protein and diluting the nucleating seeds, the infected cell protects itself.

By using microfluidic chambers, it is possible to study kinetics of particle spreading in neurons and to visualize particle transport. Such experiments have shown that after internalization, fibrils of Aβ, tau, αSyn, and Htt are transported from the cell body along the axon to the presynapse (Brahic et al., 2016; Freundt et al., 2012). Axonal transport occurs in both anterograde and retrograde directions (Brahic et al., 2016; Freundt et al., 2012). The kinetics of transport are relatively slow (resembling the slow component b of axonal transport) which implies that assemblies are transported naked, without being attached to membrane structures or organelles associated with faster kinetics (Freundt et al., 2012). Once seeding potent particles reach the synapse, they eventually transfer across the synapse to the adjacent neuron. This transneuronal propagation requires intact axons, synapses, and connectivity (Calafate et al., 2015; Freundt et al., 2012; Pecho-Vrieseling et al., 2014). Disruption of any of these significantly lowers transmission between first- and second-order neurons (Calafate et al., 2015; Freundt et al., 2012; Pecho-Vrieseling et al., 2014).

Many features required for endosome recycling and synaptic vesicle release appear to be hijacked by pathogenic molecules. The aforementioned observations are largely based on in vitro experiments, and it is not fully established that these mechanisms contribute to aggregate spreading in the nervous system. Most cell culture studies are performed with isolated neurons only. Also oligodendrocytes, microglial cells, and astrocytes take up prion-like assemblies; but it is not yet clear how they contribute to the spreading of nucleating particles. A recent study indicated that microglia could play a role in the uptake and subsequent release of tau via exosomes (Asai et al., 2015). In the next section, we discuss how experiments in vivo have shed light on some of these issues.

Mechanisms of Prion-Like Disease Propagation Studied In Vivo

It is challenging to address complex questions regarding how protein aggregates spread in neuronal networks, the potential modulatory role of neuronal activity, and the roles of nonneuronal cells in the brain in this context. Recent studies have provided some insight, which we summarize in this section.

Transgenic mouse models of tauopathies driven by cell-specific promoters have been used to demonstrate spreading of protein deposits to cells that lack tau expression (de Calignon et al., 2012; Liu et al., 2012). Neuronal deposits were first (after 3 months) found in the entorhinal cortex, while after longer incubation times (2 years), tau inclusions were also found in downstream regions, including the hippocampus, cingulate, and somatosensory cortices, that normally do not develop tau accumulation in this model (de Calignon et al., 2012; Liu et al., 2012). A transgenic model expressing a mutant form of tau (P301S) inoculated with purified tau brain extracts yielded network-dependent spreading patterns of protein deposits that propagated from the hippocampus to distant regions via the fimbria-fornix (Ahmed et al., 2014). Other regions in closer proximity, but lacking significant innervation from the hippocampus, did not exhibit tau pathology (Ahmed et al., 2014). Subsequent studies by other research groups showed that tau spreading was not only dependent on the density of innervating connections but also on the activity of its neuronal network (Ahmed et al., 2014; Wu et al., 2016). The release of tau was shown to be regulatable via synaptic activity in vitro (Pooler, Phillips, Lau, Noble, & Hanger, 2013). Using optogenetics to stimulate the hippocampal neuronal network in vivo led to increased tau accumulation in the hippocampus in human tau overexpressing mice (Wu et al., 2016).

Likewise, aggregation-prone Htt has been shown to spread in a transgenic model of HD (Pecho-Vrieseling et al., 2014). Employing mHTT covalently bound to a fluorescent tag (Pecho-Vrieseling et al., 2014), it has been shown that the protein can efficiently spread via the corticostriatal pathway. This spreading was dependent on active synaptic connections and could be blocked by neurotoxins that specifically inhibited synaptic vesicle fusion (Pecho-Vrieseling et al., 2014). Interestingly, Ataxin-3, a different polyQ aggregate with putative prion-like features, lacked the capacity to spread between cells (Babcock & Ganetzky, 2015). This suggests that the spreading features of polyQ proteins, or aggregated protein in general, do not solely rely on their β-sheet conformation.

Intracerebral injection of fluorescently tagged αSyn in the olfactory bulb of wild-type mice demonstrated that monomeric and oligomeric αSyn spreads rapidly, over hours to days, along olfactory pathways (Rey et al., 2013). By contrast, it was not possible to detect significant spread of tagged αSyn fibrils over this short time span (Rey et al., 2013). According to the neuropathological staging systems developed for synucleinopathies that we described earlier, the olfactory bulb is most often affected in cases of early disease and considered one of the likely initiation sites for αSyn pathology (Braak et al., 2003a). In a later study, injections of a mixture (so-called preformed fibrils, PFFs) of different sized αSyn fibrils and oligomers led to progressive propagation of αSyn aggregates over 1 year in over 40 interconnected olfactory, and related, brain structures (Rey et al., 2016b). The pathology propagated along neural tracts requiring that the aggregation prone protein transcended across 1–2 synapses and led to a progressive impairment in olfaction. The network of spreading resembled that described in the Braak staging system and was suggested to mimic that seen in prodromal PD (i.e., before the onset of motor symptoms) (Mason et al., 2016; Rey et al., 2013, 2016b). In a different study, fluorescently labeled αSyn aggregates injected in the substantia nigra spread along the nigrostriatal pathway and triggered αSyn pathology in neurons in a time-dependent manner (Peelaerts et al., 2015). Strikingly, fluorescently labeled αSyn particles that were injected intravenously were also found in the CNS indicating that systemic inoculation can be sufficient to access the CNS (Peelaerts et al., 2015). These findings are consistent with the observation that intramuscular, intraperitoneal, or intravenous injection of PFF's accelerated protein deposition in αSyn transgenic models (Ayers et al., 2017; Sacino et al., 2014).

In PD and other age-related disorders, dysfunction of autophagy, the trans-Golgi network, and endosomal trafficking are directly linked with disease pathogenesis (Abeliovich & Gitler, 2016; Gouras, 2013; Hunn, Cragg, Bolam, Spillantini, & Wade-Martins, 2015). Notably, some LRRK2 mutations which cause PD lead to deficits in these trafficking pathways (Abeliovich & Gitler, 2016; Beilina et al., 2014; MacLeod et al., 2013). Inoculation of αSyn seeds in LRRK2^{G2019S} transgenic animals results in increased deposit formation (Volpicelli-Daley et al., 2016), compared to when the same type of αSyn seeds are injected in wild-type mice, which might be explained via altered synaptic physiology or reduced degradation of the transmitting seed.

THE STRAIN HYPOTHESIS

Strains in Transmissible Spongiform Encephalopathies

Initial descriptions of strains in scrapie diseases date back to the early 1960s. At this time, scrapie was still considered to be the result of a slow virus. In an attempt to identify this virus, it was not only shown that experimental inoculation of scrapie isolates retained their capacity to passage pathogenicity from sick to healthy animals but also that they transferred disease in a strain-dependent manner (Bessen & Marsh, 1992). In passaging experiments of peripherally injected sheep isolates in mice, different incubation times and pathologies were observed (Bruce & Dickinson, 1987; Outram, Dickinson, & Fraser, 1973). This suggested that different disease-promoting agents from an unknown source had their "independent genome" and were causative of different disease phenotypes.

Before the molecular mechanisms of this strain-dependent behavior in scrapie or other TSEs could be investigated, it took many years for Alper, Cramp, Haig, and Clarke (1967) to discover that TSEs were not caused by a virus but by an infectious macromolecule that could be neutralized through hydrolysis or other denaturing protein modifications (Latarjet, Muel, Haig, Clarke, & Alper, 1970; Prusiner, 1998). This infectious particle was therefore coined a "prion," to suggest that the instigating agent was an infectious proteinaceous particle lacking nucleic acids (Prusiner, 1982). It is important to note that other factors aside from a protein substrate (such as lipids or RNA Ma & Wang, 2014) might still be required for infectivity. Moreover, after systematic and laborious biochemical characterizations of scrapie isolates, it became apparent that prions were presenting with distinctive biochemical fingerprints with a β-sheet configuration (Cohen et al., 1994; Pan et al., 1993; Telling et al., 1996). These characteristic features were invariably associated with specific disease incubation times or spongiosis with anatomical specificity within the brain (Prusiner, 1998). Proteolysis experiments on infectious samples, which were retrieved after passaging in mice or hamsters, provided more evidence that the tertiary structure of infective isolates was segregating with different disease phenotypes. These findings were subsequently confirmed in samples derived from humans. The proteolytic cleavage pattern of PrP^{Sc} from patients with FFI and familial CJD (fCJD) also yielded prion fragments of different sizes (Goldfarb et al., 1992; Medori et al., 1992; Monari et al., 1994). These proteolytic fragments could be faithfully and serially passaged to transgenic PrP mice with shorter (fCJD isolates) and longer incubation times (FFI isolates) (Goldfarb et al., 1992; Medori et al., 1992; Monari et al., 1994). These and many other follow-up studies therefore argued that different PrP^{sc} conformers, or strains, are responsible for different TSEs and that strains have a unique structural code that can transmit phenotypes.

Strains in Prion-Like Neurodegenerative Diseases

The concept of strains is now well known from the clinicopathological heterogeneity in TSEs, and researchers have speculated for a long time about the existence of strains in other neurodegenerative diseases. Intriguingly, AD, tauopathies, and synucleinopathies each comprise several distinct clinical phenotypes. If the proteins involved in these diseases acquire a certain unique pathological conformation for each of the clinical phenotypes they are associated with, it would tentatively constitute support for the existence of strains also in these disorders.

In AD, evidence is building to support the existence of conformational variants of Aβ. Depending on aggregation conditions or physicochemical properties, Aβ readily aggregates into different conformers (Karran et al., 2011; Sawaya et al., 2007). Initially, Jucker et al. tested whether different Aβ aggregates act as strains and propagate pathology (Meyer-Luehmann et al., 2006). They prepared Aβ-laden extracts from transgenic APP23 and APP-PS1 mice, which are transgenic AD mouse models that generate the shorter Aβ40 and the longer Aβ42 fragments, respectively, and both exhibit unique phenotypes (Meyer-Luehmann et al., 2006). Inoculation of brain homogenates from these two transgenic models into APP23 or APP-PS1 mice (and vice versa) triggered neuropathological features that were associated with both the Aβ inoculum and the animal model (Meyer-Luehmann et al., 2006). Furthermore, the conformations of the Aβ aggregates were maintained following passaging, i.e., when using brain extracts from infected mice to inoculate a second generation of naïve mice (Heilbronner et al., 2013). A later study showed that assemblies of recombinant Aβ40 and Aβ42 could trigger-specific types of pathology after intracerebral injection into APP transgenic mice (Stohr et al., 2014). These findings hinted that Aβ conformational variants could be the key reason why different Aβ disease phenotypes develop, and that Aβ strains might exist. Soon, studies were reported using human brain extracts from sporadic and genetic AD cases (Watts et al., 2014). Unique patterns of amyloid deposition were found after inoculating brain extracts from four patients with sporadic or genetic AD, with different mutations in APP (Arctic and Swedish mutations), into transgenic animals. These disease-specific phenotypes could be maintained following serial passaging by intracerebral injection in multiple generations in transgenic animals (Watts et al., 2014). In a different study, the structure of Aβ40 amyloid of two patients with a variable disease course and clinical presentation was assessed using solid-state NMR (ssNMR) (Lu et al., 2013). After seeded fibril growth in vitro, the resulting Aβ40 fibers from both patients showed remarkable structural differences (Lu et al., 2013). In a more comprehensive follow-up study with a larger cohort of AD patients, it was shown that different clinical manifestations (e.g., including rapidly and slowly progressing AD variants) were associated with different conformations of Aβ (Qiang, Yau, Lu, Collinge, & Tycko, 2017). In patients with slowly progressing AD or a typical disease progression, the ssNMR data of seeded Aβ40 fibrils showed one predominant fibril structure. In contrast, the spectra of Aβ40 seeded extracts from patients with a more aggressive disease exhibited greater heterogeneity with multiple fibrillar structures. When brain extracts were seeded with Aβ42 precursor, a large heterogeneity was found after amplification (Qiang et al., 2017). In a different study, a similar link was confirmed between the structural state of aggregated Aβ42 and AD clinical subtypes. Here, cases with rapidly progressing AD displayed similar numbers of Aβ42 particles as those with slower progressing disease, but the conformations of aggregated Aβ42 differed (Cohen et al., 2015).

From these studies, we know that strains of Aβ40 or Aβ42 are likely to contribute to defining the rate of neurodegeneration in AD, although it is still uncertain how and to which extent. Earlier studies have demonstrated a weak association between amyloid load and disease severity in AD (Masters & Selkoe, 2012). The existence of strains could explain this. A pattern emerges of rapidly progressing cases of AD that associate with different subpopulations of unstable Aβ42 conformers and slowly progressing cases with very low Aβ40 heterogeneity. The fact that Aβ42 amplified into a heterogeneous mixture of assemblies might reflect its increased propensity to aggregate into amyloid with energetically favorable thermodynamic stabilities. In addition, the abundance of one Aβ40 variant in typical AD patients with variable subclinical presentations and the large heterogeneity of Aβ42 seeds also suggests that other factors influence the clinical subtypes. For example, mutations in APP, *PSEN1*, *PSEN2*, polymorphisms in the *APOE* gene could contribute to disease variability without affecting strain conformation (Schellenberg & Montine, 2012). Finally, unknown environmental factors or aggregates of tau or αSyn might also contribute to the clinical variability of AD.

Tau tangles are intimately linked to another group of heterogeneous neurodegenerative diseases, with dementia as a common feature, namely, the tauopathies (Table 8.1). Tau acts as a prion and many tau isoforms with variable domain repeats give rise to discrete amyloid assemblies with different toxicities. Intriguingly, the six different tau isoforms (3R- and 4R-tau), which are the result of alternative splicing of the *MAPT* gene, are linked with specific clinical subtypes (Goedert et al., 2010). In AD, tangle-only dementia and some cases of FTDP, a mixture of 3R-and 4R-repeat tau are prominent in the pathology. PiD and cases of FTDP are characterized by neuronal inclusions of 3R-tau, whereas PSP, CBD, AGD, and white matter tauopathy with globular inclusions are associated with 4R-tau in neurons and glial cells. Since all isoforms of 3R- and 4R-tau spread similarly through functionally interconnected neuroanatomical pathways, it is tempting to speculate that structural variations in tau aggregation and templated transmission might underlie specific types of toxicities and pathologies.

In order to test if the prion strain paradigm also applies to tau, different experiments with recombinant or de novo generated aggregates of tau were performed in cell and animal models. Synthetic tau can assemble in vitro into distinct conformations that resemble strains with specific biochemical and biological properties (Frost, Ollesch, Wille, & Diamond, 2009b; Morozova, March, Robinson, & Colby, 2013). When synthetic tau aggregates encompassing the repeat domains were applied to tau-expressing cells, they induced unique cellular changes that could be passaged over subsequent cell divisions (Kaufman et al., 2016; Sanders et al., 2014). Isolation and inoculation of these putative tau strains in transgenic mice led to strain-specific tau deposits in multiple generations of animals (Kaufman et al., 2016; Sanders et al., 2014). Moreover, addition of tau-enriched brain extracts from third generation animals to tau-expressing cells recreated the original cellular changes, indicating that tau strains faithfully passaged between cells and animals (Kaufman et al., 2016; Sanders et al., 2014).

Evidence that strains might exist in humans comes from similar passaging experiments with human tissue. Injection of brain homogenate of tangle-only dementia, AD, PiD, AGD, PSP, and CBD patients in transgenic animals expressing full-length tau triggered the development of argyrophilic inclusions and other features that closely resemble the human condition (Clavaguera et al., 2013). These forms of tauopathy could be passaged between different generations of P301S transgenic and wild-type mice (Clavaguera et al., 2013). In a different experiment, tau pathology was induced after intracerebral injection of brain homogenates from CBD or AD patients in P301S transgenic mice. In a fascinating demonstration of cell-type specificity, CBD-tau inoculates induced oligodendroglial tau inclusions, whereas tau extracts derived from AD patients induced neuronal inclusions (Boluda et al., 2015). After longer incubation periods, this cell-type specific pathology spread from the injection site to other brain regions (Boluda et al., 2015). Both tau and Aβ have therefore the ability to become self-propagating, act as strains, and potentially influence disease progression in AD and tauopathies.

Patients with PD, DLB, and MSA exhibit different clinical symptoms and neuropathological profiles, but all have αSyn aggregates as a major histopathological hallmark. Similar to tau and Aβ, αSyn is known to aggregate into different recombinant strains under controlled aggregation conditions (Bousset et al., 2013; Kim et al., 2016; Ma, Hu, Zhao, Chen, & Li, 2016). Several factors appear to affect the type of αSyn fibrils that form. Thus, incubation of αSyn monomers that have different posttranslational modifications can influence how αSyn monomers assemble into fibrils. Furthermore, certain incubation conditions probably impact the fibrillization, e.g., the presence of endotoxins in the preparation or varying buffer conditions can impact the type of αSyn strains that assemble (Bousset et al., 2013; Kim et al., 2016; Ma et al., 2016). Different strains of αSyn assemblies generated from recombinant protein have distinct biochemical features and exhibit diverse toxic phenotypes. Furthermore, purified fibrils isolated from PD and MSA patients displayed straight and twisted morphologies, respectively (Spillantini & Goedert, 2000; Spillantini et al., 1998a; Spillantini, Crowther, Jakes, Hasegawa, & Goedert, 1998b).

Recently, it was shown that recombinant fibrillar variants of αSyn, which according to their appearance were termed "fibrils" and "ribbons," behave as different strains when injected into animals in vivo (Bousset et al., 2013; Peelaerts et al., 2015). Both strains of αSyn assemblies induced Lewy-like pathology after intracerebral injection in wild-type and αSyn overexpressing rats (Peelaerts et al., 2015). In rodents overexpressing A53T mutant human αSyn, the fibrils were found to be the most toxic. αSyn ribbons, on the other hand, induced abundant Lewy-like pathology, including some inclusions inside glial cells, reminiscent of MSA (Peelaerts et al., 2015). Importantly, fibrils and ribbons propagated disease-specific phenotypes and amplified in a strain-dependent manner in vivo (Peelaerts et al., 2015). Stanley Prusiner's group subsequently showed that αSyn aggregates derived from MSA brains, but not those from PD brains, propagate αSyn aggregates in transgenic M83 mice that express A53T mutant αSyn (Prusiner et al., 2015). Injections of brain extracts from MSA patients caused earlier onset of neurological deficits and resulted in intraneuronal αSyn inclusions (Prusiner et al., 2015). The reason why they did not trigger αSyn aggregation in oligodendrocytes is not clear, but one could speculate that the level of oligodendroglial αSyn expression is very low in the M83 mouse, where the transgene expression is driven by a promoter that is active in neurons.

Taken together, observations made in models of different neurodegenerative diseases demonstrate that different strains of a given misfolded protein can preferentially invade different cells and exhibit distinct anatomical propagation patterns. Most described strains are likely to represent only a fraction of possible strains, and a combination of strains might be required to develop a given phenotype. The recently discovered Aβ strains already embody a larger conformational repertoire than PrP^{sc} strains (Cohen, Appleby, & Safar, 2016). In fact, it would be surprising if not all prion-like neurodegenerative disorders, including HD and ALS, are eventually characterized by disease-specific strains. Potentially, distinct effects of different strains in, e.g., ALS are unclear, but there are emerging clues. For instance, proteolysis of TDP-43 seeded extracts of ALS patient brain yielded different proteolysis patterns (Nonaka et al., 2013). SOD1 aggregates prepared from the spinal cord of $hSOD1^{G85R}$ and $hSOD^{D90A}$ transgenic mice were shown to have different propagation rates, histopathological features, and produce different incubation times until disease onset (Bidhendi et al., 2016).

Molecular Features of Strains

The assembly of precursor protein into a strain is a complicated process. Before strain-specific assemblies are formed, soluble monomeric precursor molecules that lack structure need to fold into a stable and well-organized fibrillar aggregate. Variations in assembly conditions or the protein's physicochemical properties, such as amino acids or posttranslational modifications, will affect the protein folding and aggregation energy landscape and yield different protein assemblies (Aguzzi, Heikenwalder, & Polymenidou, 2007). For instance, protein truncation, glycation, or phosphorylation will influence the interaction between different protein molecules by lowering or strengthening intra- or intermolecular forces (Haass & Selkoe, 2007; Li & Li, 1998; Ma et al., 2016; Nelson et al., 2005; Stohr et al., 2014). In addition, complex variations that arise from the local biochemical environment (cellular compartment or test tube) likely play a role. Salts, pH, or various interactors, such as chaperones or lipids, will favor stochastic assembly of nonaggregated monomers into distinctive and unique conformations or cause strain mutation during amplification (Beringue et al., 2008; Bousset et al., 2013; Ghaemmaghami et al., 2013; Guo et al., 2013; Wang et al., 2010). Various intrinsic and extrinsic factors will change the ruggedness of the energy landscape before, during, and after strain formation and contribute to the conversion of unstructured precursor molecules or folding intermediates into complex and uniquely structured prion strains.

Strains have a uniquely structured amyloid core. It is unbranched and exhibits a typical amyloid X-ray diffraction pattern of 4.7 Å that comes from the periodic spacing of perfectly aligned polypeptide backbones along the nucleating filament (Knowles et al., 2014; Serpell, Berriman, Jakes, Goedert, & Crowther, 2000). Variations at the molecular-level arise from amino acids side chains or environmental factors that introduce alterations in the strain's β-sheet stacking. Amino acid charge or volume determines the lateral packing of intercalating β-sheets into steric zippers (Nelson et al., 2005; Sawaya et al., 2007; Tuttle et al., 2016). Steric zippers mostly involve hydrophobic interactions and form hydrophobic pockets that are void of water (Sawaya et al., 2007). In addition, a myriad of electrostatic interactions will form salt bridges or salt ladders inside the fibril its stabilizing core (Tuttle et al., 2016). A network of interactions determines the strain conformation, and under permissive aggregation conditions, the nucleus will grow several micrometers in length, but not in width. Due to conformation-specific templating, strains will seed the aggregation of endogenous protein in a strain-dependent manner and propagate unique structural information. In conjunction with their spreading capacities, strains will therefore transmit biochemical, biological, and pathological phenotypes between cells or organisms through a "mendelian" pattern of inheritance (Fig. 8.2).

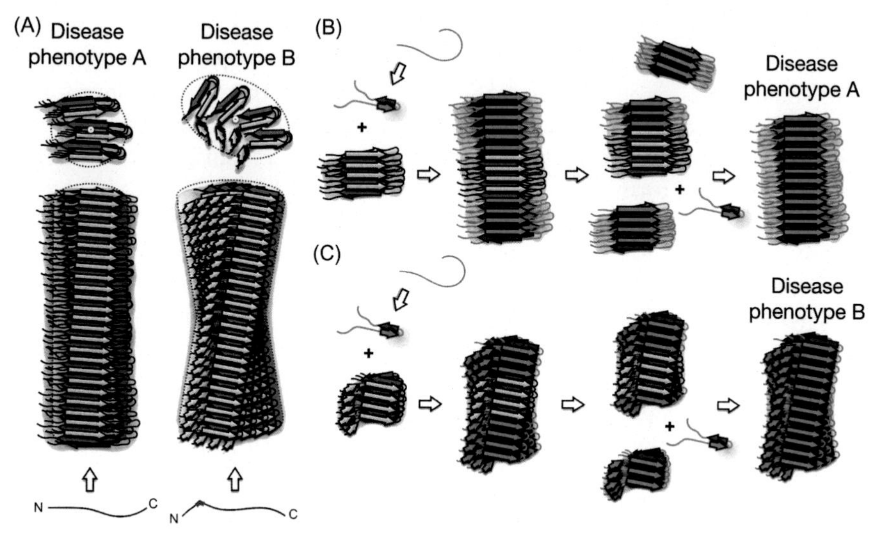

FIGURE 8.2 **How templating results in strain-specific phenotypes.** (A) Misfolding of αSyn leads to the generation of pleomorphic fibrillar assemblies. In this illustrative example, the characteristic fingerprint of cylindrical fibrils induces disease phenotype A, while a small change in the protein its primary sequence results in flat and twisted ribbon-like fibrils inducing disease phenotype B. The specific disease phenotype that develops is linked to the conformation of the strain. (B) The mature cylindrical or (C) twisted fibril can break and function as a seed. Partially folded soluble monomer will incorporate itself at the freely available ends resulting in fibril growth via faithful propagation of the fibril structural code in a strain-dependent manner. The newly assembled fibril will break and create new seeds resulting in a vicious propagating cycle of strain growth and amplification.

In a previous section, we discussed that aggregates from different disease-related proteins yield different pathologies because of structural variations, e.g., which amino acid surfaces that are exposed. Strains add another layer of structural diversity. Through their structure, strains can dictate phenotypic diversity, e.g., how they bind receptors, enter cells, amplify, propagate, and cause toxicity. From this, it is clear that by resolving strain structure and mapping the interactome of different strains of the same protein, we can start to develop more therapies and diagnostic tools.

HOW PRION STRAINS MIGHT COME TO OUR AID

Therapeutic Strategies

The prion-like concept has the potential to revolutionize the treatment of neurodegenerative diseases and provide us with novel therapeutic targets. Molecules or molecular pathways that are crucial for the spread of propagating proteins, such as lipid membranes, transmembrane receptors, exosomes, or specific endocytic pathways, could be targeted to block the uptake of misfolded proteins and prevent their subsequent spreading. The existence of strains implies that specific receptors might be involved in different closely related diseases (e.g., PD and DLB) or even in patients diagnosed with the same clinical disorder. Targeting these receptors could be possible through different strategies, e.g., LAG3-specific antibodies are currently being evaluated for cancer treatment (Nguyen & Ohashi, 2015) but could also be useful to block the spread of aggregated αSyn (since fibril binding of αSyn with LAG3 might be required for spreading (Mao et al., 2016)). The existence of different strains could therefore allow us to further fine-tune the molecular targets, in a future era of precision medicine.

Because the misfolded proteins use the extracellular space between cells, immunotherapy is an appealing therapeutic strategy. There are two principally different forms of immunotherapy. The first is *active* immunization and requires the body itself to raise antibodies to the disease-related protein. It has the advantage that only one or two injections are required to activate the patients' immune system long term, which means that it is not labor intense or costly. The challenge is to produce a vaccine that triggers a host response to conformation-specific pathogens, especially when one considers putative strains. Therefore *passive* immunization has certain advantages. By generating the antibodies in the laboratory, one could potentially select a high specificity and affinity for disease-specific aggregates. Different types of aggregates involved in different neurodegenerative disorders or belonging to a specific clinical subclass of a given disorder might then be possible to target.

The success of immunotherapy will rely on adequate preclinical models that closely mimic the human disease. Ideally, these models should involve progressive prion-like features. It has been shown that passive immunization with monoclonal antibodies ameliorated progressive neurodegenerative deficits induced by synthetic Aβ, tau, and αSyn fibrils in vivo (Lannfelt et al., 2014; Pedersen & Sigurdsson, 2015; Schenk et al., 2016; Tran et al., 2014). At first, clinical trials in AD patients that assessed the effects of different antibodies directed against soluble forms of Aβ did not meet their primary endpoints. However, a recent phase II trial in AD patients claimed a significant decline in disease progression after repeated intravenous administration of a monoclonal antibody (aducanumab) selectively targeting the β-sheet conformation of Aβ fibrils (Sevigny et al., 2016). It was reported that improvements of cognition were paralleled by reduced Aβ plaque burden and recruitment of microglia to the plaques (Sevigny et al., 2016). The positive outcome of this small phase II clinical trial has been received with caution. Nevertheless, unlike other recent trials, it used conformation-specific antibodies. Given that only about one percent of administered antibodies is estimated to reach the brain, high-affinity antibodies with a high specificity are desirable (Yu & Watts, 2013). The aducanumab antibody has an affinity that is three magnitudes greater for fibrillar Aβ relative to its soluble form (Sevigny et al., 2016). For the treatment of synucleinopathies, conformational specificity will be imperative because red blood cells are rich in soluble, monomeric αSyn (Barbour et al., 2008). Antibodies that bind with soluble proteins have the risk of becoming sequestered and neutralized. The development of antibodies that do not recognize the soluble form but only pathogenic assemblies might circumvent this problem. Several immunization trials for tau and αSyn are currently on the way and more are close to clinical testing (Lee & Lee, 2016; Pedersen & Sigurdsson, 2015).

Another way of lowering protein infectivity is by reducing the availability of precursor protein. This can be achieved by the use of small interference RNA or antisense probes (Smith et al., 2006). They can lower the expression of a toxic protein or modify the toxicity of disease-related mutants. Oligonucleotides can be administered intrathecally and will distribute throughout the CNS. This technique is undergoing a revival in ALS (Miller et al., 2013) and HD (Ionis HTT-Rx) and clinical studies are now ongoing.

Alternatively, protein aggregates, or misfolded intermediates, can be reduced by improving general protein homeostasis. Boosting autophagy via genetic or pharmacological approaches is beneficial in a wide range of experimental neurodegenerative models (Ghavami et al., 2014; Harris & Rubinsztein, 2011). By modulating the mammalian target of rapamycin (mTOR) pathway that controls autophagy, it is also possible to augment clearance mechanisms (Kim & Guan, 2015). These treatments, however, often come with a price of serious side effects (Pallet & Legendre, 2013). One way to circumvent this problem is via the use of pharmacological modulators that instead of boosting autophagy restore it. An unexpected development and a notable example in this regard come from research with antidiabetic drugs. New insulin sensitizers have been shown to have positive effects on several neurodegenerative models in vivo (Bassil et al., 2017; Ghosh et al., 2016). They have a good availability in the brain, where they restore autophagy back to its normal levels by modulating the mTOR pathway (Bassil et al., 2017; Ghosh et al., 2016). These antidiabetic drugs display mild side effects and good safety in clinical trials on diabetic and AD patients (Aviles-Olmos et al., 2013; Gejl et al., 2016; Shah et al., 2014) Future experiments will show whether they can reduce cell-to-cell transmission of protein aggregates.

Through isolation of prion strains, it is possible to perform detailed studies with ssNMR or cryo-electron microscopy (cryoEM) to solve their atomic structure. This will be important for rational design of compounds that bind the surface of prion strains. Most of the synthetic molecules that are in the pipeline today for the treatment of prion-like disorders are based on high-throughput screenings (Ross & Poirier, 2004). Some have already shown promising results in different animal models of prion-like diseases (Fellner et al., 2016; Kim & Kim, 2014; Levin et al., 2014; Ross & Poirier, 2004; Wagner et al., 2013; Wischik, Harrington, & Storey, 2014). In silico design of strain-specific compounds can lead to the identification of molecules that slow down the progression of prion-like disorders in one or more ways, e.g., via (1) blocking nucleating seeds from templating and amplification, (2) stabilization to prevent them from breaking and forming more nucleating seeds, or (3) interfering with essential sites that bind to specific receptors involved in cell-to-cell propagation. In these cases, general physiological pathways could be specifically targeted without altering their functionality. Alternatively, rational design could be used to target chaperones or disaggregase complexes that specifically interact with unique prion strains.

Diagnostic Strategies

Despite numerous efforts over recent years, virtually all the clinical trials aimed at slowing disease progression have failed. Possibly, entry of the drug into the brain has been limited, or the therapeutic target has not been engaged, or the therapeutic molecule actually does not work as desired. An important caveat is that most trials have focused on advanced disease, when it might be too late to rescue function. Therefore, diagnostic tools that pick up the disease in an early stage are desirable, and would allow treatment to be initiated earlier (Citron, 2010). In simple terms, we need to start treatment before irreversible damage has occurred. For this, we have to develop more sensitive and reliable diagnostic tools. The existence of prion strains provides a new angle to tackle this problem.

A good example of how knowledge of different prion strains could aid diagnosis is in the development of novel positron emission tomography (PET) tracers. PET ligands distribute to the brain and bind with reversible kinetics to its target (Pike, 2009). Several PET tracers bind with a good affinity to the β-sheets of extracellular Aβ plaques, which has led to their successful clinical use (Nordberg, Rinne, Kadir, & Langstrom, 2010). Some concerns, however, still exist about the binding properties of PET tracers. Because the nature of Aβ aggregates is not fully known, it is also not known exactly how the ligands bind the plaques. Certain radiotracers have low affinities for Aβ plaques in the cortex or the cerebellum but bind very strongly in other regions that carry plaques with a distinctive morphology (Cairns et al., 2009; Ikonomovic et al., 2008). Structural variations thus seem to influence the binding properties of PET ligands in the patients' brain. By solving the molecular structure of strains, hopefully it will be possible to develop imaging modalities that bind specifically, which might help to differentiate between AD subtypes and potentially other prion-like disorders. Different amyloid proteins such as tau or αSyn are also reported to bind with PET tracers in vitro with high affinity (Lockhart et al., 2007; Moghbel et al., 2012; Ye et al., 2008). Despite this, in the brain, extracellular Aβ plaques mainly retain PET ligands, which is probably due to the radiotracer's distribution and extracellular plaque availability (Lockhart et al., 2007; Moghbel et al., 2012; Ye et al., 2008).

Other diagnostic strategies could use seeding-potent particles that are present in peripheral tissues or fluids. Because of their distribution outside the brain (which we will discuss in more detail in the final section), it is possible to develop methods such as the protein misfolding cyclic amplification (PMCA) assay that can detect minute amounts of misfolded proteins. PMCA works through addition of recombinant monomer in a reaction buffer that allows seeded amplification from biological specimens. If seeds are present, soluble monomers incorporate onto the seed from which it grows. Hereafter, the amplification product is sonicated to generate more seeding-competent assemblies. This cycle is repeated several times and resembles a chain reaction of seeded growth and amplification. PMCA has been used for the detection of PrP^{sc} in CJD, and now the technique is being explored for diagnostic purposes in AD and PD (Zanusso, Monaco, Pocchiari, & Caughey, 2016). Via templated amplification of soluble Aβ, it was possible to amplify Aβ fibrils from CSF of AD patients but not from control subjects or patients that suffered from other neurodegenerative disorders (Salvadores, Shahnawaz, Scarpini, Tagliavini, & Soto, 2014). It was also found that levels of Aβ aggregates were higher in patients compared to controls (Salvadores et al., 2014). PMCA detects seeding-potent assemblies which likely are only a fraction (but highly relevant) of the total pool of assemblies. The sensitivity and specificity of Aβ-PMCA was found to be comparable to other currently used immuno-based assays in AD (Mulder et al., 2010). Similarly, a recent study showed that very low amounts of αSyn seeds can be detected in the CSF of synucleinopathy patients (Shahnawaz et al., 2017). The αSyn-PMCA assay was around 90% sensitive and specific for synucleinopathy patients compared to controls. Intriguingly, some of the control subjects that set off a positive amplification reaction were found to develop PD shortly after completion of the study, hinting that it might be possible to detect pathological assemblies before earliest disease symptoms appear.

In light of the strain hypothesis, it would also be interesting to see whether these assays could provide a seeding environment that allows faithful propagation of strains occurring in patients. Perhaps due to methodological issues, it was not possible in these two studies to discriminate between different types of AD or synucleinopathies (i.e., PD versus MSA). Because the amplification of Aβ, tau, αSyn, or other prion-like proteins is sensitive to so many variables, PMCA has been a technique that is difficult to reproduce, and it has been slow to catch on. Difficulties with optimizing and standardizing PMCA are substantial and need to be overcome before this variation-prone assay is likely to yield reproducible and clinically useful results.

Some of these difficulties have inspired the development of a slightly different detection method, the real-time quaking-induced conversion (RT-quIC) assay. This method is based on amplification in multiwell plates and shaking instead of amplification in vials that are sonicated (as in PMCA). RT-quIC is considered easier to optimize, manage, and scale up, and therefore it is more suited for standardization (Cramm et al., 2016). RT-quIC is currently being used for the diagnosis of sporadic CJD and, importantly, it has already been shown to be applicable for the detection of αSyn seeds in the CSF of synucleinopathy patients (Cramm et al., 2016; Fairfoul et al., 2016; Orru et al., 2015). It will be important to continue refining these tools so that they can be used for diagnostic purposes in several neurodegenerative disorders.

A last notable method that could benefit from the presence of conformation-specific assemblies is a strain-specific immune-based assay. Such an assay could either work by raising novel strain-specific antibodies or by taking advantage of the natural presence of autoantibodies in the blood. In the first case, strain-specific antibodies could replace standard antibodies that are used to detect soluble forms of disease-associated proteins. By taking plasma or CSF, circulating seeds could then be identified in a dose-dependent manner. If the presence of certain strains correlates with specific clinical disease features, they could also be used to stratify patients into groups. In the alternative example of immune-based strategies, autoantibodies could be used. Autoantibodies exist for Aβ and αSyn in the blood and the CSF of healthy individuals and patients (Kellner et al., 2009; Wu & Li, 2016). Tau, however, is a poor autoantigen and yields almost no autoantibodies in patients with AD, PD, HD, or ALS (Terryberry, Thor, & Peter, 1998). Autoantibodies have been considered as biomarkers for AD and PD, but various studies that compared patients and control subjects have yielded variable outcomes that are difficult to interpret (Wu & Li, 2016). Furthermore, by comparing the binding of autoantibodies with soluble precursor proteins of Aβ and αSyn, it was also not possible to detect differences between subclinical cases (Heinzel et al., 2014; Papachroni et al., 2007). Although speculative, autoantibodies could instead be implemented as a surrogate for the presence of strains. Amplification of strains from the brain of patients has been successful for Aβ and might become possible for αSyn in the near future (Qiang et al., 2017). Amplified synthetic prion strains could then be used to capture tiny amounts of autoantibodies circulating in the CSF or the plasma of disease-affected carriers and be used as a diagnostic asset.

These different assays might make up a novel discipline that could be called "strainotyping." Thereby, prion strains can come to our aide by unlocking a new branch in diagnostics via (1) detection of seeding-potent assemblies that correlate with disease severity and (2) identification of strains with a specific conformation that correlate with different clinical subtypes.

THE ISSUE OF COMMUNICABILITY OF "PRION-LIKE" DISEASES

Establishing and proving protein infectivity of prions has been an incredibly difficult task. Over more than 30 years, and despite a compelling amount of evidence, the prion hypothesis for CJD and other TSEs has been received with skepticism. It has been difficult to prove (1) that in vitro generated recombinant proteins, or de novo generated prions, are infectious in wild-type animals, and (2) that proteins could encode strain-dependent infectivity. Both these criteria were eventually met (Colby & Prusiner, 2011; Wang et al., 2010). Nucleating assemblies of αSyn have now also been shown to seed infectivity with a clear link to neurotoxicity in vivo (Luk et al., 2012a; Peelaerts et al., 2015), with infection meaning replication of the agent in the host.

Per definition, a prion is an infectious agent that acquires an alternative conformation and becomes self-propagating. To determine the minimum infective dose of a prion, factors that have to be considered include (1) the strain of the agent and (2) the genotype of the host. The protein load needs to exceed a certain threshold before prions are infectious in the host. Consequently, some TSE prions, but not all, with a high infectious titer can propagate infectivity between animals, humans, or both. Positive cases of horizontal transmission have been reported after contamination with surgical equipment, growth hormone extracts, blood transfusions, or organs transplants form infected donors giving rise to iatrogenic CJD (iCJD) (Johnson & Gibbs, 1998; Will, 2003). Most notable is "mad cow disease," i.e., bovine spongiform encephalopathy (BSE), which rapidly made the lay community, especially in the United Kingdom, aware of prions.

During the early nineties many people became exposed to BSE prions through food consumption. The entire UK population and large parts of Europe were exposed to infected meat, which led to a prion epidemic between 1986 and 1998. Remarkably, very few cases were eventually reported to develop symptoms (most might have been asymptomatic carriers). This variant of CJD (vCJD), that is different to sporadic or familial forms of CJD, has only been noted for 177 cases in the United Kingdom since the start of the observations by the National CJD Research & Surveillance Unit (Aguzzi et al., 2007). The high degree of exposure and the relatively few reported iCJD and vCJD cases further suggest that prion strains are converging toward a limited number of infectious conformations that can be maintained in a limited number of hosts (Solforosi, Milani, Mancini, Clementi, & Burioni, 2013). Certain *PRNP* polymorphisms at position 129 of the PrP crucially confer susceptibility toward BSE prions (Wadsworth et al., 2004). Without this, people do not develop vCJD. Similarly, polymorphisms in this amino acid residue in combination with a second residue located at position 178 will determine whether people might or might not develop FFI or fCJD (Goldfarb et al., 1992; Medori et al., 1992; Monari et al., 1994).

These apparent transmission or species barriers are now becoming better characterized for different TSEs, and even more stringent transmission barriers might exist for other prion-like disorders such as AD and PD. Although we cannot rule out that seeding-potent proteins from other prion-like disease are spreading between humans, it is generally believed (because of a lack of supporting data) that this does not occur. Also, other factors like exposure to endotoxins or tissue inflammation might induce aberrant protein folding and "trigger" disease in certain people. This idea is becoming increasingly popular for PD, where the idea that the disease originates in the enteric nerves innervating the gut and/or in the olfactory system (Rey, George, & Brundin, 2016a; Rey, Wesson, & Brundin, 2016c). Both the gut and the olfactory system are exposed to the external milieu, and therefore it is plausible to hypothesize that they are particularly prone to environmental toxins, pollutants, and infectious agents that promote inflammation (Klingelhoefer & Reichmann, 2015). From the gut, aggregated αSyn can spread via the vagal nerve to neurons in the brainstem (Hawkes, Del Tredici, & Braak, 2007; Jellinger, 2008), and in a similar fashion aggregated αSyn in the olfactory bulb can rapidly be transported along axons to reach several forebrain regions (Rey et al., 2016b, 2016c)

Most cases with prion-like neurodegenerative disorders are sporadic. In the experimental setting, nucleating seeds of Aβ, tau, and αSyn can transmit from the periphery to CNS after intraperitoneal, intragastric, intramuscular, or intravenous administration, reminiscent of PrP^{Sc} prions (Breid et al., 2016; Clavaguera et al., 2014; Eisele et al., 2010, 2014; Holmqvist et al., 2014; Peelaerts et al., 2015; Sacino et al., 2014; Tran et al., 2014). In transgenic mice, compared to wild-type animals, the disease process is accelerated (Breid et al., 2016; Clavaguera et al., 2014; Eisele et al., 2010, 2014; Holmqvist et al., 2014; Peelaerts et al., 2015; Sacino et al., 2014; Tran et al., 2014). Human tissue can also transmit prion-like disease to experimental animals (described in the "Prion Propagation" section). These studies demonstrate clearly that brain tissue from humans affected by neurodegenerative disease can be infectious in animals. A natural question is then whether human-to-human transmission of prion-like disease can occur. An initial study that evaluated the incidence of PD and AD in cadaveric growth hormone recipients found that there was no increase of cases with AD, FTD, or PD (Irwin et al., 2013). However, more recent examinations of individuals that underwent growth-hormone treatment and succumbed from iCJD at young age showed that an unusual high frequency of these patients exhibited gray matter and vascular Aβ pathology (Jaunmuktane et al., 2015). Similar findings were reported after transplantations with dura mater grafts from donors with amyloid pathology (Frontzek, Lutz, Aguzzi, Kovacs, & Budka, 2016; Kovacs et al., 2016; Preusser et al., 2006). Five out of seven young recipients who developed iCJD also presented with Aβ pathology (Frontzek et al., 2016; Kovacs et al., 2016; Preusser et al., 2006). Although these studies do not conclusively establish Aβ infectivity between individuals, they hint that iatrogenic spread of Aβ plaques might occur under certain circumstances. PrP^{Sc} prions are known to seed infectivity from the lymphoid tissue of the periphery after peripheral exposure (Brown et al., 1999). Peripheral replication and prion strain conformation are essential for subsequent neuroinvasion, although alternative routes, which require high infectious titers, are also described (Bett et al., 2012; Chen, Soto, & Morales, 2014). Possibly, the alleged high transmission barriers of Aβ or αSyn are due to the misfolded proteins being cleared by cellular proteostasis or immune cells in most healthy young individuals. This is suggested to be the case for Aβ, where the induction of cerebral Aβ pathology from the periphery requires three times the amount of Aβ seeds compared to intracerebral inoculation (Eisele et al., 2010, 2014). Alternatively, there is a lack of peripheral replication sites from where Aβ or αSyn seeds can effectively invade the CNS. Another possibility is that some amyloid strains are not stable during the extraction or sterilizing procedure of cadaveric growth hormone, which could explain why a horizontal transfer of prion-like disorder is difficult to document. Furthermore, with an incubation time (i.e., the lag time between exposure to the prion seed and the propagation of pathology) that could span several decades, it might be difficult to link the exposure to the seeds and the onset of disease. Finally, as alluded to above, infectious seeds might well be cleared by cellular proteostasis or immune cells in young people, but as these systems partially fail during aging, the disease pathology will propagate (Fig. 8.3). This is consistent with the observation that high age is the greatest risk factor for AD and PD.

The idea that a protein can be an infectious agent has been the subject of a contentious and vivid debate. Prions eventually became an accepted, global, and expanding idea. One of the issues with the "prion-like" concept is that it is continuously subjected to change, because scientists discover new features for different disease-related proteins every day. The prion-like hypothesis therefore embodies an elusive concept, which has made it particularly difficult to grasp. Nevertheless, different terms have been put forward such as "prion-like" or "prionoid," and these terms will remain provisional until we can decide on a set of rules that exactly delineate their meaning. Within the current framework, different disease-related proteins, as outlined in this chapter on prion-like neurodegenerative disorders, are in fact bona fide prions, as suggested by Stanley Prusiner (Prusiner et al., 2015; Prusiner, 2013).

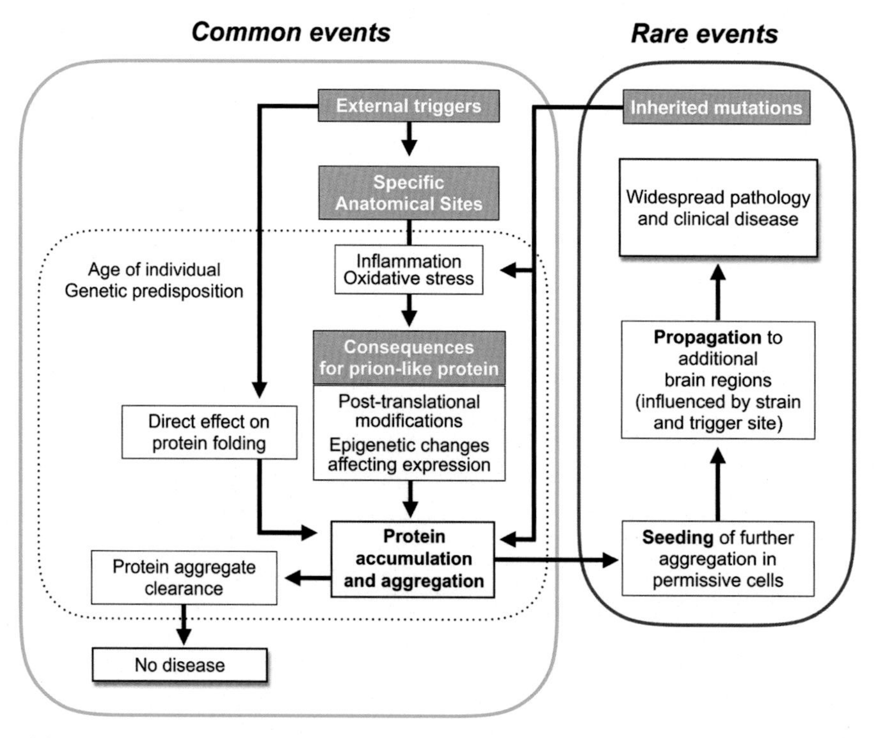

FIGURE 8.3 Schematic representation of how different triggers can lead to prion-like disorders.

This means that they become infectious within their host but does not necessarily imply that they are communicable between humans and cause disease under normal conditions. Under any circumstance, it is important that we understand the molecular mechanisms that are necessary for proteins to become infectious and the barriers that prevent them from spreading between individuals.

CONCLUSIONS

A plethora of research is now aimed at understanding the prion-like mechanisms that are causative of various progressive protein misfolding diseases. These advances are fueling a more general prion hypothesis, encompassing several neurodegenerative diseases, that awaits further validation. The emerging picture is that proteins that misfold in neurodegenerative disease are capable of seeding and amplifying pathology. It results in a progressive spread of protein inclusions and disease pathology along defined anatomical routes. Nucleating proteinaceous particles are likely to cause toxicity in cell-autonomous and noncell-autonomous ways. We need to unravel factors that inhibit seeding nucleation. By understanding the disease process and solving the structure of different prion strains, we might be able to develop novel therapeutics and diagnostic tools. This kindles hope that in the future we will be able to stop the progressive spread of seeding-potent assemblies and finally develop disease-modifying therapies for neurodegenerative disorders.

References

Aarsland, D., Zaccai, J., & Brayne, C. (2005). A systematic review of prevalence studies of dementia in Parkinson's disease. *Movement Disorders, 20*, 1255–1263.

Abeliovich, A., & Gitler, A. D. (2016). Defects in trafficking bridge Parkinson's disease pathology and genetics. *Nature, 539*, 207–216.

Abounit, S., Bousset, L., Loria, F., Zhu, S., de Chaumont, F., Pieri, L., ... Zurzolo, C. (2016). Tunneling nanotubes spread fibrillar alpha-synuclein by intercellular trafficking of lysosomes. *The EMBO Journal, 35*, 2120–2138.

Aguzzi, A., Heikenwalder, M., & Polymenidou, M. (2007). Insights into prion strains and neurotoxicity. *Nature Reviews Molecular Cell Biology, 8*, 552–561.

Aguzzi, A., Nuvolone, M., & Zhu, C. (2013). The immunobiology of prion diseases. *Nature Reviews Immunology, 13*, 888–902.

Ahmed, Z., Cooper, J., Murray, T. K., Garn, K., McNaughton, E., Clarke, H., ... Jackson, S. (2014). A novel in vivo model of tau propagation with rapid and progressive neurofibrillary tangle pathology: The pattern of spread is determined by connectivity, not proximity. *Acta Neuropathologica, 127*, 667–683.

Al-Chalabi, A., Hardiman, O., Kiernan, M. C., Chio, A., Rix-Brooks, B., & van den Berg, L. H. (2016). Amyotrophic lateral sclerosis: Moving towards a new classification system. *The Lancet Neurology, 15*, 1182–1194.

Alper, T., Cramp, W. A., Haig, D. A., & Clarke, M. C. (1967). Does the agent of scrapie replicate without nucleic acid? *Nature, 214*, 764–766.

Alzheimer, A. (1906). Über eine eigenartige schweren Erkrankungsprozeβ der Hinrinde. *Neurologisches Centralblatt*, 1129–1136.

Alzheimer, A. (1907). Uber eine eigenartige Erkrankung der Hirnrinde. *Allgemeine Zeitschrift für Psychiatrie und Psychisch-Gerichtliche Medizin*, 146–148.

Alzheimer, A., Forstl, H., & Levy, R. (1991). On certain peculiar diseases of old age. *History of Psychiatry, 2*, 71–101.

Anderson, P., & Kedersha, N. (2002). Visibly stressed: The role of eIF2, TIA-1, and stress granules in protein translation. *Cell Stress & Chaperones, 7*, 213–221.

Angot, E., Steiner, J. A., Tome, C. M. L., Ekstrom, P., Mattsson, B., Bjorklund, A., & Brundin, P. (2012). Alpha-synuclein cell-to-cell transfer and seeding in grafted dopaminergic neurons in vivo. *PLoS ONE, 7*.

Arai, T., Hasegawa, M., Akiyama, H., Ikeda, K., Nonaka, T., Mori, H., ... Hashizume, Y. (2006). TDP-43 is a component of ubiquitin-positive tau-negative inclusions in frontotemporal lobar degeneration and amyotrophic lateral sclerosis. *Biochemical and Biophysical Research Communications, 351*, 602–611.

Arnold, S. E., Hyman, B. T., Flory, J., Damasio, A. R., & Van Hoesen, G. W. (1991). The topographical and neuroanatomical distribution of neurofibrillary tangles and neuritic plaques in the cerebral cortex of patients with Alzheimer's disease. *Cerebral Cortex, 1*, 103–116.

Arrasate, M., & Finkbeiner, S. (2012). Protein aggregates in Huntington's disease. *Experimental Neurology, 238*, 1–11.

Arrasate, M., Mitra, S., Schweitzer, E. S., Segal, M. R., & Finkbeiner, S. (2004). Inclusion body formation reduces levels of mutant huntingtin and the risk of neuronal death. *Nature, 431*, 805–810.

Asai, H., Ikezu, S., Tsunoda, S., Medalla, M., Luebke, J., Haydar, T., ... Ikezu, T. (2015). Depletion of microglia and inhibition of exosome synthesis halt tau propagation. *Nature Neuroscience, 18*, 1584–1593.

Auluck, P. K., Caraveo, G., & Lindquist, S. (2010). alpha-Synuclein: Membrane interactions and toxicity in Parkinson's disease. *Annual Review of Cell and Developmental Biology, 26*, 211–233.

Aviles-Olmos, I., Dickson, J., Kefalopoulou, Z., Djamshidian, A., Ell, P., Soderlund, T., ... Lees, A. (2013). Exenatide and the treatment of patients with Parkinson's disease. *Journal of Clinical Investigation, 123*, 2730–2736.

Ayers, J. I., Brooks, M. M., Rutherford, N. J., Howard, J. K., Sorrentino, Z. A., Riffe, C. J., & Giasson, B. I. (2017). Robust central nervous system pathology in transgenic mice following peripheral injection of alpha-synuclein fibrils. *Journal of Virology, 91*.

Ayers, J. I., Fromholt, S. E., O'Neal, V. M., Diamond, J. H., & Borchelt, D. R. (2016). Prion-like propagation of mutant SOD1 misfolding and motor neuron disease spread along neuroanatomical pathways. *Acta Neuropathologica, 131*, 103–114.

Babcock, D. T., & Ganetzky, B. (2015). Transcellular spreading of huntingtin aggregates in the *Drosophila* brain. *Proceedings of the National Academy of Sciences of the United States of America, 112*, E5427–E5433.

Baker, H. F., Ridley, R. M., Duchen, L. W., Crow, T. J., & Bruton, C. J. (1994). Induction of beta (A4)-amyloid in primates by injection of Alzheimer's disease brain homogenate. Comparison with transmission of spongiform encephalopathy. *Molecular Neurobiology, 8*, 25–39.

Baldwin, A. J., Knowles, T. P., Tartaglia, G. G., Fitzpatrick, A. W., Devlin, G. L., Shammas, S. L., ... Gras, S. L. (2011). Metastability of native proteins and the phenomenon of amyloid formation. *Journal of the American Chemical Society, 133*, 14160–14163.

Balguerie, A., Dos Reis, S., Ritter, C., Chaignepain, S., Coulary-Salin, B., Forge, V., ... Riek, R. (2003). Domain organization and structure-function relationship of the HET-s prion protein of *Podospora anserina*. *The EMBO Journal, 22*, 2071–2081.

Barbour, R., Kling, K., Anderson, J. P., Banducci, K., Cole, T., Diep, L., ... Seubert, P. (2008). Red blood cells are the major source of alpha-synuclein in blood. *Neurodegenerative Diseases, 5*, 55–59.

Bassil, F., Canron, M. H., Vital, A., Bezard, E., Li, Y., Greig, N. H., ... Meissner, W. G. (2017). Insulin resistance and exendin-4 treatment for multiple system atrophy. *Brain, 140*(5), 1420–1436.

Baumer, D., Hilton, D., Paine, S. M., Turner, M. R., Lowe, J., Talbot, K., & Ansorge, O. (2010). Juvenile ALS with basophilic inclusions is a FUS proteinopathy with FUS mutations. *Neurology, 75*, 611–618.

Beach, T. G., Adler, C. H., Lue, L., Sue, L. I., Bachalakuri, J., Henry-Watson, J., ... Brooks, R. (2009). Unified staging system for Lewy body disorders: Correlation with nigrostriatal degeneration, cognitive impairment and motor dysfunction. *Acta Neuropathologica, 117*, 613–634.

Beilina, A., Rudenko, I. N., Kaganovich, A., Civiero, L., Chau, H., Kalia, S. K., ... Ndukwe, K. (2014). Unbiased screen for interactors of leucine-rich repeat kinase 2 supports a common pathway for sporadic and familial Parkinson disease. *Proceedings of the National Academy of Sciences of the United States of America, 111*, 2626–2631.

Berg, D., Postuma, R. B., Bloem, B., Chan, P., Dubois, B., Gasser, T., ... Lang, A. E. (2014). Time to redefine PD? Introductory statement of the MDS Task Force on the definition of Parkinson's disease. *Movement Disorders, 29*, 454–462.

Beringue, V., Vilotte, J. L., & Laude, H. (2008). Prion agent diversity and species barrier. *Veterinary Research, 39*, 47.

Bernis, M. E., Babila, J. T., Breid, S., Wusten, K. A., Wullner, U., & Tamguney, G. (2015). Prion-like propagation of human brain-derived alpha-synuclein in transgenic mice expressing human wild-type alpha-synuclein. *Acta Neuropathologica Communications, 3*, 75.

Bessen, R. A., & Marsh, R. F. (1992). Biochemical and physical-properties of the prion protein from 2 strains of the transmissible mink encephalopathy agent. *Journal of Virology, 66*, 2096–2101.

Bett, C., Joshi-Barr, S., Lucero, M., Trejo, M., Liberski, P., Kelly, J. W., ... Sigurdson, C. J. (2012). Biochemical properties of highly neuroinvasive prion strains. *PLoS Pathogens, 8*, e1002522.

Bidhendi, E. E., Bergh, J., Zetterstrom, P., Andersen, P. M., Marklund, S. L., & Brannstrom, T. (2016). Two superoxide dismutase prion strains transmit amyotrophic lateral sclerosis-like disease. *Journal of Clinical Investigation, 126*, 2249–2253.

Bigio, E. H., Lipton, A. M., Yen, S. H., Hutton, M. L., Baker, M., Nacharaju, P., ... Dickson, D. W. (2001). Frontal lobe dementia with novel tauopathy: Sporadic multiple system tauopathy with dementia. *Journal of Neuropathology & Experimental Neurology, 60*, 328–341.

Boluda, S., Iba, M., Zhang, B., Raible, K. M., Lee, V. M., & Trojanowski, J. Q. (2015). Differential induction and spread of tau pathology in young PS19 tau transgenic mice following intracerebral injections of pathological tau from Alzheimer's disease or corticobasal degeneration brains. *Acta Neuropathologica, 129*, 221–237.

Borroni, B., Bonvicini, C., Alberici, A., Buratti, E., Agosti, C., Archetti, S., ... Gennarelli, M. (2009). Mutation within TARDBP leads to frontotemporal dementia without motor neuron disease. *Human Mutation, 30*, E974–E983.

Bousset, L., Pieri, L., Ruiz-Arlandis, G., Gath, J., Jensen, P. H., Habenstein, B., ... Meier, B. H. (2013). Structural and functional characterization of two alpha-synuclein strains. *Nature Communications, 4*, 2575.

Braak, H., Alafuzoff, I., Arzberger, T., Kretzschmar, H., & Del Tredici, K. (2006). Staging of Alzheimer disease-associated neurofibrillary pathology using paraffin sections and immunocytochemistry. *Acta Neuropathologica, 112*, 389–404.

Braak, H., & Braak, E. (1991). Neuropathological stageing of Alzheimer-related changes. *Acta Neuropathologica, 82*, 239–259.

Braak, H., & Braak, E. (1995). Staging of Alzheimer's disease-related neurofibrillary changes. *Neurobiology of Aging, 16*, 271–278, discussion278-284.

Braak, H., & Del Tredici, K. (2008a). Cortico-basal ganglia-cortical circuitry in Parkinson's disease reconsidered. *Experimental Neurology, 212*, 226–229.

Braak, H., & Del Tredici, K. (2008b). Invited Article: Nervous system pathology in sporadic Parkinson disease. *Neurology, 70*, 1916–1925.

Braak, H., & Del Tredici, K. (2011). The pathological process underlying Alzheimer's disease in individuals under thirty. *Acta Neuropathologica, 121*, 171–181.

Braak, H., Del Tredici, K., Rub, U., de Vos, R. A., Jansen Steur, E. N., & Braak, E. (2003a). Staging of brain pathology related to sporadic Parkinson's disease. *Neurobiology of Aging, 24*, 197–211.

Braak, H., Rub, U., Gai, W. P., & Del Tredici, K. (2003b). Idiopathic Parkinson's disease: Possible routes by which vulnerable neuronal types may be subject to neuroinvasion by an unknown pathogen. *Journal of Neural Transmission, 110*, 517–536.

Brahic, M., Bousset, L., Bieri, G., Melki, R., & Gitler, A. D. (2016). Axonal transport and secretion of fibrillar forms of alpha-synuclein, Abeta42 peptide and HTTExon 1. *Acta Neuropathologica, 131*, 539–548.

Braun, A. R., Lacy, M. M., Ducas, V. C., Rhoades, E., & Sachs, J. N. (2014). Alpha-synuclein-induced membrane remodeling is driven by binding affinity, partition depth, and interleaflet order asymmetry. *Journal of the American Chemical Society, 136*, 9962–9972.

Breid, S., Bernis, M. E., Babila, J. T., Garca, M. C., Wille, H., & Tamguney, G. (2016). Neuroinvasion of alpha-synuclein prionoids after intraperitoneal and intraglossal inoculation. *Journal of Virology, 90*(20), 9182–9193.

Brettschneider, J., Del Tredici, K., Irwin, D. J., Grossman, M., Robinson, J. L., Toledo, J. B., ... Ludolph, A. C. (2014). Sequential distribution of pTDP-43 pathology in behavioral variant frontotemporal dementia (bvFTD). *Acta Neuropathologica, 127*, 423–439.

Brettschneider, J., Del Tredici, K., Toledo, J. B., Robinson, J. L., Irwin, D. J., Grossman, M., ... Baek, Y. (2013). Stages of pTDP-43 pathology in amyotrophic lateral sclerosis. *Annals of Neurology, 74*, 20–38.

Brock, T. D. (1989). Brock, Thomas D. (1988): Robert Koch. A Life in Medicine and Bacteriology: In: Scientific Revolutionaries: A Biographical Series. Science Tech. Publishers, Madison WI and J. Springer, Berlin, 364 pp. hard cover, 48—DM. *European Journal of Protistology, 25*, 85.

Brown, K. L., Stewart, K., Ritchie, D. L., Mabbott, N. A., Williams, A., Fraser, H., ... Bruce, M. E. (1999). Scrapie replication in lymphoid tissues depends on prion protein-expressing follicular dendritic cells. *Nature Medicine, 5*, 1308–1312.

Bruce, M. E., & Dickinson, A. G. (1987). Biological evidence that scrapie agent has an independent genome. *Journal of General Virology, 68*(Pt 1), 79–89.

Buee, L., Bussiere, T., Buee-Scherrer, V., Delacourte, A., & Hof, P. R. (2000). Tau protein isoforms, phosphorylation and role in neurodegenerative disorders. *Brain Research Brain Research Reviews, 33*, 95–130.

Burre, J., Sharma, M., Tsetsenis, T., Buchman, V., Etherton, M. R., & Sudhof, T. C. (2010). Alpha-synuclein promotes SNARE-complex assembly in vivo and in vitro. *Science, 329*, 1663–1667.

Cairns, N. J., Ikonomovic, M. D., Benzinger, T., Storandt, M., Fagan, A. M., Shah, A. R., ... Holtzman, D. M. (2009). Absence of Pittsburgh compound B detection of cerebral amyloid beta in a patient with clinical, cognitive, and cerebrospinal fluid markers of Alzheimer disease: A case report. *Archives of Neurology, 66*, 1557–1562.

Calafate, S., Buist, A., Miskiewicz, K., Vijayan, V., Daneels, G., de Strooper, B., ... Moechars, D. (2015). Synaptic contacts enhance cell-to-cell tau pathology propagation. *Cell Reports, 11*, 1176–1183.

Calafate, S., Flavin, W., Verstreken, P., & Moechars, D. (2016). Loss of Bin1 promotes the propagation of tau pathology. *Cell Reports, 17*, 931–940.

Cersosimo, M. G., & Benarroch, E. E. (2012). Pathological correlates of gastrointestinal dysfunction in Parkinson's disease. *Neurobiology of Disease, 46*, 559–564.

Chandra, S., Gallardo, G., Fernandez-Chacon, R., Schluter, O. M., & Sudhof, T. C. (2005). Alpha-synuclein cooperates with CSP alpha in preventing neurodegeneration. *Cell, 123*, 383–396.

Chartier-Harlin, M. C., Kachergus, J., Roumier, C., Mouroux, V., Douay, X., Lincoln, S., ... Hulihan, M. (2004). Alpha-synuclein locus duplication as a cause of familial Parkinson's disease. *The Lancet, 364*, 1167–1169.

Chen, B., Soto, C., & Morales, R. (2014). Peripherally administrated prions reach the brain at sub-infectious quantities in experimental hamsters. *FEBS Letters, 588*, 795–800.

Chia, R., Tattum, M. H., Jones, S., Collinge, J., Fisher, E. M., & Jackson, G. S. (2010). Superoxide dismutase 1 and tgSOD1 mouse spinal cord seed fibrils, suggesting a propagative cell death mechanism in amyotrophic lateral sclerosis. *PLoS ONE, 5*, e10627.

Chiti, F., & Dobson, C. M. (2006). Protein misfolding, functional amyloid, and human disease. *Annual Review of Biochemistry, 75*, 333–366.

Chow, V. W., Mattson, M. P., Wong, P. C., & Gleichmann, M. (2010). An overview of APP processing enzymes and products. *Neuromolecular Medicine, 12*, 1–12.

Cicchetti, F., Lacroix, S., Cisbani, G., Vallieres, N., Saint-Pierre, M., St-Amour, I., ... Mantovani, D. (2014). Mutant huntingtin is present in neuronal grafts in Huntington disease patients. *Annals of Neurology, 76*, 31–42.

Cicchetti, F., Saporta, S., Hauser, R. A., Parent, M., Saint-Pierre, M., Sanberg, P. R., ... Mufson, E. J. (2009). Neural transplants in patients with Huntington's disease undergo disease-like neuronal degeneration. *Proceedings of the National Academy of Sciences of the United States of America, 106*, 12483–12488.

Citron, M. (2010). Alzheimer's disease: Strategies for disease modification. *Nature Reviews Drug Discovery, 9*, 387–398.

Clavaguera, F., Akatsu, H., Fraser, G., Crowther, R. A., Frank, S., Hench, J., ... Staufenbiel, M. (2013). Brain homogenates from human tauopathies induce tau inclusions in mouse brain. *Proceedings of the National Academy of Sciences of the United States of America, 110*, 9535–9540.

Clavaguera, F., Bolmont, T., Crowther, R. A., Abramowski, D., Frank, S., Probst, A., ... Staufenbiel, M. (2009). Transmission and spreading of tauopathy in transgenic mouse brain. *Nature Cell Biology, 11*, 909–913.

Clavaguera, F., Hench, J., Lavenir, I., Schweighauser, G., Frank, S., Goedert, M., & Tolnay, M. (2014). Peripheral administration of tau aggregates triggers intracerebral tauopathy in transgenic mice. *Acta Neuropathologica, 127*, 299–301.

Clayton, D. F., & George, J. M. (1999). Synucleins in synaptic plasticity and neurodegenerative disorders. *Journal of Neuroscience Research, 58*, 120–129.

Cohen, F. E., Pan, K. M., Huang, Z., Baldwin, M., Fletterick, R. J., & Prusiner, S. B. (1994). Structural clues to prion replication. *Science, 264*, 530–531.

Cohen, M., Appleby, B., & Safar, J. G. (2016). Distinct prion-like strains of amyloid beta implicated in phenotypic diversity of Alzheimer's disease. *Prion, 10*, 9–17.

Cohen, M. L., Kim, C., Haldiman, T., ElHag, M., Mehndiratta, P., Pichet, T., … Chen, W. (2015). Rapidly progressive Alzheimer's disease features distinct structures of amyloid-beta. *Brain, 138*, 1009–1022.

Colby, D. W., & Prusiner, S. B. (2011). De novo generation of prion strains. *Nature Reviews Microbiology, 9*, 771–777.

Collinge, J., & Clarke, A. R. (2007). A general model of prion strains and their pathogenicity. *Science, 318*, 930–936.

Costanzo, M., Abounit, S., Marzo, L., Danckaert, A., Chamoun, Z., Roux, P., & Zurzolo, C. (2013). Transfer of polyglutamine aggregates in neuronal cells occurs in tunneling nanotubes. *Journal of Cell Science, 126*, 3678–3685.

Cramm, M., Schmitz, M., Karch, A., Mitrova, E., Kuhn, F., Schroeder, B., … Satoh, K. (2016). Stability and reproducibility underscore utility of RT-QuIC for diagnosis of Creutzfeldt–Jakob disease. *Molecular Neurobiology, 53*, 1896–1904.

Crowther, R. A., & Goedert, M. (2000). Abnormal tau-containing filaments in neurodegenerative diseases. *Journal of Structural Biology, 130*, 271–279.

Cuervo, A. M., & Wong, E. (2014). Chaperone-mediated autophagy: Roles in disease and aging. *Cell Research, 24*, 92–104.

Dahm, R. (2006). Alzheimer's discovery. *Current Biology: CB, 16*, R906–R910.

de Calignon, A., Polydoro, M., Suarez-Calvet, M., William, C., Adamowicz, D. H., Kopeikina, K. J., … Carlson, G. A. (2012). Propagation of tau pathology in a model of early Alzheimer's disease. *Neuron, 73*, 685–697.

DeJesus-Hernandez, M., Mackenzie, I. R., Boeve, B. F., Boxer, A. L., Baker, M., Rutherford, N. J., … Adamson, J. (2011). Expanded GGGGCC hexanucleotide repeat in noncoding region of C9ORF72 causes chromosome 9p-linked FTD and ALS. *Neuron, 72*, 245–256.

Deleault, N. R., Walsh, D. J., Piro, J. R., Wang, F., Wang, X., Ma, J., … Supattapone, S. (2012). Cofactor molecules maintain infectious conformation and restrict strain properties in purified prions. *Proceedings of the National Academy of Sciences of the United States of America, 109*, E1938–E1946.

Deng, H. X., Zhai, H., Bigio, E. H., Yan, J., Fecto, F., Ajroud, K., … Sufit, R. (2010). FUS-immunoreactive inclusions are a common feature in sporadic and non-SOD1 familial amyotrophic lateral sclerosis. *Annals of Neurology, 67*, 739–748.

Desplats, P., Lee, H. J., Bae, E. J., Patrick, C., Rockenstein, E., Crews, L., … Lee, S. J. (2009). Inclusion formation and neuronal cell death through neuron-to-neuron transmission of alpha-synuclein. *Proceedings of the National Academy of Sciences of the United States of America, 106*, 13010–13015.

Diao, J., Burre, J., Vivona, S., Cipriano, D. J., Sharma, M., Kyoung, M., … Brunger, A. T. (2013). Native alpha-synuclein induces clustering of synaptic-vesicle mimics via binding to phospholipids and synaptobrevin-2/VAMP2. *eLife, 2*, e00592.

Diaz-Nido, J., Wandosell, F., & Avila, J. (2002). Glycosaminoglycans and beta-amyloid, prion and tau peptides in neurodegenerative diseases. *Peptides, 23*, 1323–1332.

DiFiglia, M., Sapp, E., Chase, K. O., Davies, S. W., Bates, G. P., Vonsattel, J. P., & Aronin, N. (1997). Aggregation of huntingtin in neuronal intranuclear inclusions and dystrophic neurites in brain. *Science, 277*, 1990–1993.

Duennwald, M. L., Jagadish, S., Muchowski, P. J., & Lindquist, S. (2006). Flanking sequences profoundly alter polyglutamine toxicity in yeast. *Proceedings of the National Academy of Sciences of the United States of America, 103*, 11045–11050.

Eisele, Y. S., Fritschi, S. K., Hamaguchi, T., Obermuller, U., Fuger, P., Skodras, A., … Staufenbiel, M. (2014). Multiple factors contribute to the peripheral induction of cerebral beta-amyloidosis. *Journal of Neuroscience, 34*, 10264–10273.

Eisele, Y. S., Obermuller, U., Heilbronner, G., Baumann, F., Kaeser, S. A., Wolburg, H., … Jucker, M. (2010). Peripherally applied Abeta-containing inoculates induce cerebral beta-amyloidosis. *Science, 330*, 980–982.

Evans, A. S. (1976). Causation and disease: The Henle–Koch postulates revisited. *The Yale Journal of Biology and Medicine, 49*, 175–195.

Fairfoul, G., McGuire, L. I., Pal, S., Ironside, J. W., Neumann, J., Christie, S., … Rolinski, M. (2016). Alpha-synuclein RT-QuIC in the CSF of patients with alpha-synucleinopathies. *Annals of Clinical and Translational Neurology, 3*, 812–818.

Fares, M. B., Maco, B., Oueslati, A., Rockenstein, E., Ninkina, N., Buchman, V. L., … Lashuel, H. A. (2016). Induction of de novo alpha-synuclein fibrillization in a neuronal model for Parkinson's disease. *Proceedings of the National Academy of Sciences of the United States of America, 113*, E912–E921.

Fatima, M., Tan, R., Halliday, G. M., & Kril, J. J. (2015). Spread of pathology in amyotrophic lateral sclerosis: Assessment of phosphorylated TDP-43 along axonal pathways. *Acta Neuropathologica Communications, 3*, 47.

Fauvet, B., Mbefo, M. K., Fares, M. B., Desobry, C., Michael, S., Ardah, M. T., ... Lion, N. (2012). Alpha-synuclein in central nervous system and from erythrocytes, mammalian cells, and Escherichia coli exists predominantly as disordered monomer. *Journal of Biological Chemistry, 287*, 15345–15364.

Feiler, M. S., Strobel, B., Freischmidt, A., Helferich, A. M., Kappel, J., Brewer, B. M., ... Ludolph, A. C. (2015). TDP-43 is intercellularly transmitted across axon terminals. *Journal of Cell Biology, 211*, 897–911.

Fellner, L., Kuzdas-Wood, D., Levin, J., Ryazanov, S., Leonov, A., Griesinger, C., ... Stefanova, N. (2016). Anle138b partly ameliorates motor deficits despite failure of neuroprotection in a model of advanced multiple system atrophy. *Frontiers in Neuroscience, 10*, 99.

Field, E. J., Farmer, F., Caspary, E. A., & Joyce, G. (1969). Susceptibility of scrapie agent to ionizing radiation. *Nature, 222*, 90–91.

Flavin, W. P., Bousset, L., Green, Z. C., Chu, Y., Skarpathiotis, S., Chaney, M. J., ... Campbell, E. M. (2017). Endocytic vesicle rupture is a conserved mechanism of cellular invasion by amyloid proteins. *Acta Neuropathologica.*

Fraser, H., & Dickinson, A. G. (1970). Pathogenesis of scrapie in the mouse: The role of the spleen. *Nature, 226*, 462–463.

Freeman, D., Cedillos, R., Choyke, S., Lukic, Z., McGuire, K., Marvin, S., ... O'Connor, C. (2013). Alpha-synuclein induces lysosomal rupture and cathepsin dependent reactive oxygen species following endocytosis. *PLoS ONE, 8*, e62143.

Freer, R., Sormanni, P., Vecchi, G., Ciryam, P., Dobson, C. M., & Vendruscolo, M. (2016). A protein homeostasis signature in healthy brains recapitulates tissue vulnerability to Alzheimer's disease. *Science Advances, 2*, e1600947.

Freundt, E. C., Maynard, N., Clancy, E. K., Roy, S., Bousset, L., Sourigues, Y., ... Brahic, M. (2012). Neuron-to-neuron transmission of alpha-synuclein fibrils through axonal transport. *Annals of Neurology, 72*, 517–524.

Frontzek, K., Lutz, M. I., Aguzzi, A., Kovacs, G. G., & Budka, H. (2016). Amyloid-beta pathology and cerebral amyloid angiopathy are frequent in iatrogenic Creutzfeldt–Jakob disease after dural grafting. *Swiss Medical Weekly, 146*, w14287.

Frost, B., Jacks, R. L., & Diamond, M. I. (2009a). Propagation of tau misfolding from the outside to the inside of a cell. *Journal of Biological Chemistry, 284*, 12845–12852.

Frost, B., Ollesch, J., Wille, H., & Diamond, M. I. (2009b). Conformational diversity of wild-type Tau fibrils specified by templated conformation change. *Journal of Biological Chemistry, 284*, 3546–3551.

Furukawa, Y., Kaneko, K., Watanabe, S., Yamanaka, K., & Nukina, N. (2011). A seeding reaction recapitulates intracellular formation of Sarkosyl-insoluble transactivation response element (TAR) DNA-binding protein-43 inclusions. *Journal of Biological Chemistry, 286*, 18664–18672.

Fusco, G., De Simone, A., Gopinath, T., Vostrikov, V., Vendruscolo, M., Dobson, C. M., & Veglia, G. (2014). Direct observation of the three regions in alpha-synuclein that determine its membrane-bound behaviour. *Nature Communications, 5*, 3827.

Gabizon, R., McKinley, M. P., & Prusiner, S. B. (1987). Purified prion proteins and scrapie infectivity copartition into liposomes. *Proceedings of the National Academy of Sciences of the United States of America, 84*, 4017–4021.

Gajdusek, D. C. (1994). Spontaneous generation of infectious nucleating amyloids in the transmissible and nontransmissible cerebral amyloidoses. *Molecular Neurobiology, 8*, 1–13.

Gajdusek, D. C., & Zigas, V. (1957). Degenerative disease of the central nervous system in New Guinea; the endemic occurrence of kuru in the native population. *The New England Journal of Medicine, 257*, 974–978.

Gajdusek, D. C., & Zigas, V. (1959). Kuru; clinical, pathological and epidemiological study of an acute progressive degenerative disease of the central nervous system among natives of the Eastern Highlands of New Guinea. *The American Journal of Medicine, 26*, 442–469.

Gao, X., Carroni, M., Nussbaum-Krammer, C., Mogk, A., Nillegoda, N. B., Szlachcic, A., ... Bukau, B. (2015). Human Hsp70 disaggregase reverses Parkinson's-linked alpha-synuclein amyloid fibrils. *Molecular Cell, 59*, 781–793.

Gejl, M., Gjedde, A., Egefjord, L., Moller, A., Hansen, S. B., Vang, K., ... Schacht, A. (2016). In Alzheimer's disease, 6-month treatment with GLP-1 analog prevents decline of brain glucose metabolism: randomized, placebo-controlled, double-blind clinical trial. *Frontiers in Aging Neuroscience, 8*, 108.

Ghaemmaghami, S., Colby, D. W., Nguyen, H. O., Hayashi, S., Oehler, A., DeArmond, S. J., & Prusiner, S. B. (2013). Convergent replication of mouse synthetic prion strains. *The American Journal of Pathology, 182*, 866–874.

Ghavami, S., Shojaei, S., Yeganeh, B., Ande, S. R., Jangamreddy, J. R., Mehrpour, M., ... Kashani, H. H. (2014). Autophagy and apoptosis dysfunction in neurodegenerative disorders. *Progress in Neurobiology, 112*, 24–49.

Ghosh, A., Tyson, T., George, S., Hildebrandt, E. N., Steiner, J. A., Madaj, Z., ... Escobar Galvis, M. L. (2016). Mitochondrial pyruvate carrier regulates autophagy, inflammation, and neurodegeneration in experimental models of Parkinson's disease. *Science Translational Medicine, 8*, 368ra174.

Gilks, N., Kedersha, N., Ayodele, M., Shen, L., Stoecklin, G., Dember, L. M., & Anderson, P. (2004). Stress granule assembly is mediated by prion-like aggregation of TIA-1. *Molecular Biology of the Cell, 15*, 5383–5398.

Glenner, G. G., & Wong, C. W. (1984). Alzheimer's disease: Initial report of the purification and characterization of a novel cerebrovascular amyloid protein. *Biochemical and Biophysical Research Communications, 120*, 885–890.

Goedert, M., Clavaguera, F., & Tolnay, M. (2010). The propagation of prion-like protein inclusions in neurodegenerative diseases. *Trends in Neurosciences, 33*, 317–325.

Goedert, M., & Ghetti, B. (2007). Alois Alzheimer: His life and times. *Brain Pathology, 17*, 57–62.

Goedert, M., & Spillantini, M. G. (2006). A century of Alzheimer's disease. *Science, 314*, 777–781.

Goedert, M., Spillantini, M. G., Del Tredici, K., & Braak, H. (2013). 100 years of Lewy pathology. *Nature Reviews Neurology, 9*, 13–24.

Goldfarb, L. G., Petersen, R. B., Tabaton, M., Brown, P., Leblanc, A. C., Montagna, P., ... Pendelbury, W. W. (1992). Fatal familial insomnia and familial Creutzfeldt–Jakob disease—disease phenotype determined by a DNA polymorphism. *Science, 258*, 806–808.

Goudsmit, J., Morrow, C. H., Asher, D. M., Yanagihara, R. T., Masters, C. L., Gibbs, C. J., Jr., & Gajdusek, D. C. (1980). Evidence for and against the transmissibility of Alzheimer disease. *Neurology, 30*, 945–950.

Gouras, G. K. (2013). Convergence of synapses, endosomes, and prions in the biology of neurodegenerative diseases. *International Journal of Cell Biology, 2013*, 141083.

Gousset, K., Schiff, E., Langevin, C., Marijanovic, Z., Caputo, A., Browman, D. T., ... Enninga, J. (2009). Prions hijack tunnelling nanotubes for intercellular spread. *Nature Cell Biology, 11*, 328–336.

Grad, L. I., Guest, W. C., Yanai, A., Pokrishevsky, E., O'Neill, M. A., Gibbs, E., ... Plotkin, S. S. (2011). Intermolecular transmission of superoxide dismutase 1 misfolding in living cells. *Proceedings of the National Academy of Sciences of the United States of America, 108*, 16398–16403.

Graeber, M. B., Kosel, S., Grasbon-Frodl, E., Moller, H. J., & Mehraein, P. (1998). Histopathology and APOE genotype of the first Alzheimer disease patient, Auguste D. *Neurogenetics, 1*, 223–228.

Grassmann, A., Wolf, H., Hofmann, J., Graham, J., & Vorberg, I. (2013). Cellular aspects of prion replication in vitro. *Viruses, 5*, 374–405.

Greenwald, J., Buhtz, C., Ritter, C., Kwiatkowski, W., Choe, S., Maddelein, M. L., ... Leitz, D. (2010). The mechanism of prion inhibition by HET-S. *Molecular Cell, 38*, 889–899.

Griffith, J. S. (1967). Self-replication and scrapie. *Nature, 215*, 1043–1044.

Guo, J. L., Covell, D. J., Daniels, J. P., Iba, M., Stieber, A., Zhang, B., ... Trojanowski, J. Q. (2013). Distinct alpha-synuclein strains differentially promote tau inclusions in neurons. *Cell, 154*, 103–117.

Guo, J. L., & Lee, V. M. (2011). Seeding of normal Tau by pathological Tau conformers drives pathogenesis of Alzheimer-like tangles. *Journal of Biological Chemistry, 286*, 15317–15331.

Guo, J. L., & Lee, V. M. (2014). Cell-to-cell transmission of pathogenic proteins in neurodegenerative diseases. *Nature Medicine, 20*, 130–138.

Gupta, S., Jie, S., & Colby, D. W. (2012). Protein misfolding detected early in pathogenesis of transgenic mouse model of Huntington disease using amyloid seeding assay. *Journal of Biological Chemistry, 287*, 9982–9989.

Haass, C., & Selkoe, D. J. (2007). Soluble protein oligomers in neurodegeneration: Lessons from the Alzheimer's amyloid beta-peptide. *Nature Reviews Molecular Cell Biology, 8*, 101–112.

Halliday, G., Hely, M., Reid, W., & Morris, J. (2008). The progression of pathology in longitudinally followed patients with Parkinson's disease. *Acta Neuropathologica, 115*, 409–415.

Han, H., Weinreb, P. H., & Lansbury, P. T., Jr. (1995). The core Alzheimer's peptide NAC forms amyloid fibrils which seed and are seeded by beta-amyloid: Is NAC a common trigger or target in neurodegenerative disease? *Chemistry & Biology, 2*, 163–169.

Hansen, C., Angot, E., Bergstrom, A. L., Steiner, J. A., Pieri, L., Paul, G., ... Fog, K. (2011). Alpha-synuclein propagates from mouse brain to grafted dopaminergic neurons and seeds aggregation in cultured human cells. *Journal of Clinical Investigation, 121*, 715–725.

Hardy, J., & Selkoe, D. J. (2002). The amyloid hypothesis of Alzheimer's disease: Progress and problems on the road to therapeutics. *Science, 297*, 353–356.

Harris, H., & Rubinsztein, D. C. (2011). Control of autophagy as a therapy for neurodegenerative disease. *Nature Reviews Neurology, 8*, 108–117.

Hartl, F. U., Bracher, A., & Hayer-Hartl, M. (2011). Molecular chaperones in protein folding and proteostasis. *Nature, 475*, 324–332.

Hashimoto, M., Hsu, L. J., Sisk, A., Xia, Y., Takeda, A., Sundsmo, M., & Masliah, E. (1998). Human recombinant NACP/alpha-synuclein is aggregated and fibrillated in vitro: Relevance for Lewy body disease. *Brain Research, 799*, 301–306.

Hawkes, C. H., Del Tredici, K., & Braak, H. (2007). Parkinson's disease: A dual-hit hypothesis. *Neuropathology and Applied Neurobiology, 33*, 599–614.

Heilbronner, G., Eisele, Y. S., Langer, F., Kaeser, S. A., Novotny, R., Nagarathinam, A., ... Jucker, M. (2013). Seeded strain-like transmission of beta-amyloid morphotypes in APP transgenic mice. *EMBO Reports, 14*, 1017–1022.

Heiman-Patterson, T. D., Sher, R. B., Blankenhorn, E. A., Alexander, G., Deitch, J. S., Kunst, C. B., ... Cox, G. (2011). Effect of genetic background on phenotype variability in transgenic mouse models of amyotrophic lateral sclerosis: A window of opportunity in the search for genetic modifiers. *Amyotrophic Lateral Sclerosis: Official Publication of the World Federation of Neurology Research Group on Motor Neuron Diseases, 12*, 79–86.

Heinzel, S., Gold, M., Deuschle, C., Bernhard, F., Maetzler, W., Berg, D., & Dodel, R. (2014). Naturally occurring alpha-synuclein autoantibodies in Parkinson's disease: Sources of (error) variance in biomarker assays. *PLoS ONE, 9*, e114566.

Hobson, P., & Meara, J. (2004). Risk and incidence of dementia in a cohort of older subjects with Parkinson's disease in the United Kingdom. *Movement Disorders, 19*, 1043–1049.

Holmes, B. B., DeVos, S. L., Kfoury, N., Li, M., Jacks, R., Yanamandra, K., ... Bagchi, D. P. (2013). Heparan sulfate proteoglycans mediate internalization and propagation of specific proteopathic seeds. *Proceedings of the National Academy of Sciences of the United States of America, 110*, E3138–E3147.

Holmes, B. B., & Diamond, M. I. (2017). Cellular models for the study of prions. *Cold Spring Harbor Perspectives in Medicine, 7*.

Holmqvist, S., Chutna, O., Bousset, L., Aldrin-Kirk, P., Li, W., Bjorklund, T., ... Li, J. Y. (2014). Direct evidence of Parkinson pathology spread from the gastrointestinal tract to the brain in rats. *Acta Neuropathologica, 128*, 805–820.

Hunn, B. H., Cragg, S. J., Bolam, J. P., Spillantini, M. G., & Wade-Martins, R. (2015). Impaired intracellular trafficking defines early Parkinson's disease. *Trends in Neurosciences, 38*, 178–188.

Hutton, M., Lendon, C. L., Rizzu, P., Baker, M., Froelich, S., Houlden, H., ... Grover, A. (1998). Association of missense and 5′-splice-site mutations in tau with the inherited dementia FTDP-17. *Nature, 393*, 702–705.

Iba, M., Guo, J. L., McBride, J. D., Zhang, B., Trojanowski, J. Q., & Lee, V. M. (2013). Synthetic tau fibrils mediate transmission of neurofibrillary tangles in a transgenic mouse model of Alzheimer's-like tauopathy. *Journal of Neuroscience, 33*, 1024–1037.

Ikonomovic, M. D., Klunk, W. E., Abrahamson, E. E., Mathis, C. A., Price, J. C., Tsopelas, N. D., ... Paljug, W. R. (2008). Post-mortem correlates of in vivo PiB-PET amyloid imaging in a typical case of Alzheimer's disease. *Brain, 131*, 1630–1645.

Irwin, D. J., Abrams, J. Y., Schonberger, L. B., Leschek, E. W., Mills, J. L., Lee, V. M., & Trojanowski, J. Q. (2013). Evaluation of potential infectivity of Alzheimer and Parkinson disease proteins in recipients of cadaver-derived human growth hormone. *JAMA Neurology, 70*, 462–468.

Jackrel, M. E., & Shorter, J. (2015). Engineering enhanced protein disaggregases for neurodegenerative disease. *Prion, 9*, 90–109.

Jansen, A. H., Batenburg, K. L., Pecho-Vrieseling, E., & Reits, E. A. (2017). Visualization of prion-like transfer in Huntington's disease models. *Biochimica et Biophysica Acta, 1863*, 793–800.

Jaunmuktane, Z., Mead, S., Ellis, M., Wadsworth, J. D., Nicoll, A. J., Kenny, J., ... Walker, A. S. (2015). Evidence for human transmission of amyloid-beta pathology and cerebral amyloid angiopathy. *Nature, 525*, 247–250.

Jellinger, K. A. (2008). A critical reappraisal of current staging of Lewy-related pathology in human brain. *Acta Neuropathologica, 116*, 1–16.

Jeon, I., Cicchetti, F., Cisbani, G., Lee, S., Li, E., Bae, J., ... Kim, M. (2016). Human-to-mouse prion-like propagation of mutant huntingtin protein. *Acta Neuropathologica, 132*, 577–592.

Johnson, R. T., & Gibbs, C. J., Jr. (1998). Creutzfeldt–Jakob disease and related transmissible spongiform encephalopathies. *The New England Journal of Medicine, 339*, 1994–2004.

Jones, D. R., Delenclos, M., Baine, A. T., DeTure, M., Murray, M. E., Dickson, D. W., & McLean, P. J. (2015). Transmission of soluble and insoluble alpha-synuclein to mice. *Journal of Neuropathology & Experimental Neurology, 74*, 1158–1169.

Kabashi, E., Valdmanis, P. N., Dion, P., Spiegelman, D., McConkey, B. J., Vande Velde, C., ... Salachas, F. (2008). TARDBP mutations in individuals with sporadic and familial amyotrophic lateral sclerosis. *Nature Genetics, 40*, 572–574.

Kane, M. D., Lipinski, W. J., Callahan, M. J., Bian, F., Durham, R. A., Schwarz, R. D., ... Walker, L. C. (2000). Evidence for seeding of beta-amyloid by intracerebral infusion of Alzheimer brain extracts in beta-amyloid precursor protein-transgenic mice. *Journal of Neuroscience, 20*, 3606–3611.

Karran, E., Mercken, M., & De Strooper, B. (2011). The amyloid cascade hypothesis for Alzheimer's disease: An appraisal for the development of therapeutics. *Nature Reviews Drug Discovery, 10*, 698–712.

Kaufman, S. K., Sanders, D. W., Thomas, T. L., Ruchinskas, A. J., Vaquer-Alicea, J., Sharma, A. M., ... Diamond, M. I. (2016). Tau prion strains dictate patterns of cell pathology, progression rate, and regional vulnerability in vivo. *Neuron, 92*, 796–812.

Kellner, A., Matschke, J., Bernreuther, C., Moch, H., Ferrer, I., & Glatzel, M. (2009). Autoantibodies against beta-amyloid are common in Alzheimer's disease and help control plaque burden. *Annals of Neurology, 65*, 24–31.

Kfoury, N., Holmes, B. B., Jiang, H., Holtzman, D. M., & Diamond, M. I. (2012). Trans-cellular propagation of Tau aggregation by fibrillar species. *Journal of Biological Chemistry, 287*, 19440–19451.

Kim, C., Lv, G., Lee, J. S., Jung, B. C., Masuda-Suzukake, M., Hong, C. S., ... Hasegawa, M. (2016). Exposure to bacterial endotoxin generates a distinct strain of alpha-synuclein fibril. *Scientific Reports, 6*, 30891.

Kim, S., & Kim, K. T. (2014). Therapeutic Approaches for Inhibition of Protein Aggregation in Huntington's Disease. *Experimental Neurobiology, 23*, 36–44.

Kim, Y. C., & Guan, K. L. (2015). mTOR: A pharmacologic target for autophagy regulation. *Journal of Clinical Investigation, 125*, 25–32.

Kimberlin, R. H., Hall, S. M., & Walker, C. A. (1983). Pathogenesis of mouse scrapie. Evidence for direct neural spread of infection to the CNS after injection of sciatic nerve. *Journal of the Neurological Sciences, 61*, 315–325.

King, O. D., Gitler, A. D., & Shorter, J. (2012). The tip of the iceberg: RNA-binding proteins with prion-like domains in neurodegenerative disease. *Brain Research, 1462*, 61–80.

Klingelhoefer, L., & Reichmann, H. (2015). Pathogenesis of Parkinson disease-the gut-brain axis and environmental factors. *Nature Reviews Neurology, 11*, 625–636.

Knowles, T. P., Vendruscolo, M., & Dobson, C. M. (2014). The amyloid state and its association with protein misfolding diseases. *Nature Reviews Molecular Cell Biology, 15*, 384–396.

Knowles, T. P., Waudby, C. A., Devlin, G. L., Cohen, S. I., Aguzzi, A., Vendruscolo, M., ... Dobson, C. M. (2009). An analytical solution to the kinetics of breakable filament assembly. *Science, 326*, 1533–1537.

Koch, R. (1876). Untersuchungen über Bakterien: V. Die Ätiologie der Milzbrand-Krankheit, begründet auf die Entwicklungsgeschichte des Bacillus anthracis. *Beitrage zur Biology der Pflanzen*, 227–310.

Komori, T. (1999). Tau-positive glial inclusions in progressive supranuclear palsy, corticobasal degeneration and Pick's disease. *Brain Pathology, 9*, 663–679.

Kordower, J. H., Chu, Y., Hauser, R. A., Freeman, T. B., & Olanow, C. W. (2008). Lewy body-like pathology in long-term embryonic nigral transplants in Parkinson's disease. *Nature Medicine, 14*, 504–506.

Kordower, J. H., Dodiya, H. B., Kordower, A. M., Terpstra, B., Paumier, K., Madhavan, L., ... Collier, T. J. (2011). Transfer of host-derived alpha synuclein to grafted dopaminergic neurons in rat. *Neurobiology of Disease, 43*, 552–557.

Kovacs, G. G., Lutz, M. I., Ricken, G., Strobel, T., Hoftberger, R., Preusser, M., ... Fischer, P. (2016). Dura mater is a potential source of Abeta seeds. *Acta Neuropathologica, 131*, 911–923.

Kovacs, G. G., Majtenyi, K., Spina, S., Murrell, J. R., Gelpi, E., Hoftberger, R., ... Budka, H. (2008). White matter tauopathy with globular glial inclusions: A distinct sporadic frontotemporal lobar degeneration. *Journal of Neuropathology & Experimental Neurology, 67*, 963–975.

Kovacs, G. G., Murrell, J. R., Horvath, S., Haraszti, L., Majtenyi, K., Molnar, M. J., ... Spina, S. (2009). TARDBP variation associated with frontotemporal dementia, supranuclear gaze palsy, and chorea. *Movement Disorders, 24*, 1843–1847.

Kruger, R., Kuhn, W., Muller, T., Woitalla, D., Graeber, M., Kosel, S., ... Riess, O. (1998). Ala30Pro mutation in the gene encoding alpha-synuclein in Parkinson's disease. *Nature Genetics, 18*, 106–108.

Kupsky, W. J., Grimes, M. M., Sweeting, J., Bertsch, R., & Cote, L. J. (1987). Parkinson's disease and megacolon: Concentric hyaline inclusions (Lewy bodies) in enteric ganglion cells. *Neurology, 37*, 1253–1255.

LaFerla, F. M., Green, K. N., & Oddo, S. (2007). Intracellular amyloid-beta in Alzheimer's disease. *Nature Reviews Neuroscience, 8*, 499–509.

Landwehrmeyer, G. B., McNeil, S. M., Dure, L., St., Ge, P., Aizawa, H., Huang, Q., ... Bonilla, E. (1995). Huntington's disease gene: Regional and cellular expression in brain of normal and affected individuals. *Annals of Neurology, 37*, 218–230.

Lannfelt, L., Moller, C., Basun, H., Osswald, G., Sehlin, D., Satlin, A., ... Gellerfors, P. (2014). Perspectives on future Alzheimer therapies: Amyloid-beta protofibrils—A new target for immunotherapy with BAN2401 in Alzheimer's disease. *Alzheimer's Research & Therapy, 6*, 16.

Latarjet, R., Muel, B., Haig, D. A., Clarke, M. C., & Alper, T. (1970). Inactivation of the scrapie agent by near monochromatic ultraviolet light. *Nature, 227*, 1341–1343.

Lauren, J., Gimbel, D. A., Nygaard, H. B., Gilbert, J. W., & Strittmatter, S. M. (2009). Cellular prion protein mediates impairment of synaptic plasticity by amyloid-beta oligomers. *Nature, 457*, 1128–1132.

Lee, J. S., & Lee, S. J. (2016). Mechanism of Anti-alpha-Synuclein Immunotherapy. *Journal of Movement Disorders, 9*, 14–19.

Lee, S. J., Desplats, P., Sigurdson, C., Tsigelny, I., & Masliah, E. (2010). Cell-to-cell transmission of non-prion protein aggregates. *Nature Reviews Neurology, 6*, 702–706.

Lees, A. J., Selikhova, M., Andrade, L. A., & Duyckaerts, C. (2008). The black stuff and Konstantin Nikolaevich Tretiakoff. *Movement Disorders, 23*, 777–783.

Levin, J., Schmidt, F., Boehm, C., Prix, C., Botzel, K., Ryazanov, S., ... Giese, A. (2014). The oligomer modulator anle138b inhibits disease progression in a Parkinson mouse model even with treatment started after disease onset. *Acta Neuropathologica, 127*, 779–780.

Lewy, F. H. (1932). Die Entstehung der Einschlußkörper und ihre Bedeutung für die systematische Einordnung der sogenannten Viruskrankheiten. *Deutsche Zeitschrift für Nervenheilkunde, 124*, 93–110.

Li, J. Y., Englund, E., Holton, J. L., Soulet, D., Hagell, P., Lees, A. J., ... Bjorklund, A. (2008). Lewy bodies in grafted neurons in subjects with Parkinson's disease suggest host-to-graft disease propagation. *Nature Medicine, 14*, 501–503.

Li, S. H., & Li, X. J. (1998). Aggregation of N-terminal huntingtin is dependent on the length of its glutamine repeats. *Human Molecular Genetics, 7*, 777–782.

Li, W., West, N., Colla, E., Pletnikova, O., Troncoso, J. C., Marsh, L., ... Price, D. L. (2005). Aggregation promoting C-terminal truncation of alpha-synuclein is a normal cellular process and is enhanced by the familial Parkinson's disease-linked mutations. *Proceedings of the National Academy of Sciences of the United States of America, 102*, 2162–2167.

Li, X., Rayman, J. B., Kandel, E. R., & Derkatch, I. L. (2014). Functional role of Tia1/Pub1 and Sup35 prion domains: Directing protein synthesis machinery to the tubulin cytoskeleton. *Molecular Cell, 55*, 305–318.

Li, Y. R., King, O. D., Shorter, J., & Gitler, A. D. (2013). Stress granules as crucibles of ALS pathogenesis. *Journal of Cell Biology, 201*, 361–372.

Liberski, P. P., Hainfellner, J. A., Sikorska, B., & Budka, H. (2012). Prion protein (PrP) deposits in the tectum of experimental Gerstmann–Straussler–Scheinker disease following intraocular inoculation. *Folia Neuropathologica, 50*, 85–88.

Linden, R., Martins, V. R., Prado, M. A., Cammarota, M., Izquierdo, I., & Brentani, R. R. (2008). Physiology of the prion protein. *Physiological Reviews, 88*, 673–728.

Ling, S. C., Polymenidou, M., & Cleveland, D. W. (2013). Converging mechanisms in ALS and FTD: Disrupted RNA and protein homeostasis. *Neuron, 79*, 416–438.

Liu, L., Drouet, V., Wu, J. W., Witter, M. P., Small, S. A., Clelland, C., & Duff, K. (2012). Trans-synaptic spread of tau pathology in vivo. *PLoS ONE, 7*, e31302.

Lockhart, A., Lamb, J. R., Osredkar, T., Sue, L. I., Joyce, J. N., Ye, L., ... Beach, T. G. (2007). PIB is a non-specific imaging marker of amyloid-beta (Abeta) peptide-related cerebral amyloidosis. *Brain, 130*, 2607–2615.

Lomen-Hoerth, C., Anderson, T., & Miller, B. (2002). The overlap of amyotrophic lateral sclerosis and frontotemporal dementia. *Neurology, 59*, 1077–1079.

Lu, J. X., Qiang, W., Yau, W. M., Schwieters, C. D., Meredith, S. C., & Tycko, R. (2013). Molecular structure of beta-amyloid fibrils in Alzheimer's disease brain tissue. *Cell, 154*, 1257–1268.

Luk, K. C., Kehm, V., Carroll, J., Zhang, B., O'Brien, P., Trojanowski, J. Q., & Lee, V. M. (2012a). Pathological alpha-synuclein transmission initiates Parkinson-like neurodegeneration in nontransgenic mice. *Science, 338*, 949–953.

Luk, K. C., Kehm, V. M., Zhang, B., O'Brien, P., Trojanowski, J. Q., & Lee, V. M. (2012b). Intracerebral inoculation of pathological alpha-synuclein initiates a rapidly progressive neurodegenerative alpha-synucleinopathy in mice. *Journal of Experimental Medicine, 209*, 975–986.

Luk, K. C., Song, C., O'Brien, P., Stieber, A., Branch, J. R., Brunden, K. R., ... Lee, V. M. (2009). Exogenous alpha-synuclein fibrils seed the formation of Lewy body-like intracellular inclusions in cultured cells. *Proceedings of the National Academy of Sciences of the United States of America, 106*, 20051–20056.

Ma, J., & Wang, F. (2014). Prion disease and the 'protein-only hypothesis'. *Essays in Biochemistry, 56*, 181–191.

Ma, M. R., Hu, Z. W., Zhao, Y. F., Chen, Y. X., & Li, Y. M. (2016). Phosphorylation induces distinct alpha-synuclein strain formation. *Scientific Reports, 6*, 37130.

Mackenzie, I. R., Neumann, M., Bigio, E. H., Cairns, N. J., Alafuzoff, I., Kril, J., ... Holm, I. E. (2010). Nomenclature and nosology for neuropathologic subtypes of frontotemporal lobar degeneration: An update. *Acta Neuropathologica, 119*, 1–4.

MacLeod, D. A., Rhinn, H., Kuwahara, T., Zolin, A., Di Paolo, G., McCabe, B. D., ... Small, S. A. (2013). RAB7L1 interacts with LRRK2 to modify intraneuronal protein sorting and Parkinson's disease risk. *Neuron, 77*, 425–439.

Mahant, N., McCusker, E. A., Byth, K., Graham, S., & Huntington Study, G. (2003). Huntington's disease: Clinical correlates of disability and progression. *Neurology, 61*, 1085–1092.

Mao, X., Ou, M. T., Karuppagounder, S. S., Kam, T. I., Yin, X., Xiong, Y., ... Shin, J. H. (2016). Pathological alpha-synuclein transmission initiated by binding lymphocyte-activation gene 3. *Science, 353*(6307), 1513–1526.

Maroteaux, L., Campanelli, J. T., & Scheller, R. H. (1988). Synuclein: A neuron-specific protein localized to the nucleus and presynaptic nerve terminal. *Journal of Neuroscience*, *8*, 2804–2815.

Mason, D. M., Nouraei, N., Pant, D. B., Miner, K. M., Hutchison, D. F., Luk, K. C., ... Leak, R. K. (2016). Transmission of alpha-synucleinopathy from olfactory structures deep into the temporal lobe. *Molecular Neurodegeneration*, *11*, 49.

Masters, C. L., & Selkoe, D. J. (2012). Biochemistry of amyloid beta-protein and amyloid deposits in Alzheimer disease. *Cold Spring Harbor Perspectives in Medicine*, *2*, a006262.

Masters, C. L., Simms, G., Weinman, N. A., Multhaup, G., McDonald, B. L., & Beyreuther, K. (1985). Amyloid plaque core protein in Alzheimer disease and Down syndrome. *Proceedings of the National Academy of Sciences of the United States of America*, *82*, 4245–4249.

Masuda-Suzukake, M., Nonaka, T., Hosokawa, M., Oikawa, T., Arai, T., Akiyama, H., ... Hasegawa, M. (2013). Prion-like spreading of pathological alpha-synuclein in brain. *Brain*, *136*, 1128–1138.

Maurer, K., Volk, S., & Gerbaldo, H. (1997). Auguste D and Alzheimer's disease. *The Lancet*, *349*, 1546–1549.

McBride, P. A., Schulz-Schaeffer, W. J., Donaldson, M., Bruce, M., Diringer, H., Kretzschmar, H. A., & Beekes, M. (2001). Early spread of scrapie from the gastrointestinal tract to the central nervous system involves autonomic fibers of the splanchnic and vagus nerves. *Journal of Virology*, *75*, 9320–9327.

McKee, A. C., Cantu, R. C., Nowinski, C. J., Hedley-Whyte, E. T., Gavett, B. E., Budson, A. E., ... Stern, R. A. (2009). Chronic traumatic encephalopathy in athletes: Progressive tauopathy after repetitive head injury. *Journal of Neuropathology & Experimental Neurology*, *68*, 709–735.

McKeith, I. G. (2006). Consensus guidelines for the clinical and pathologic diagnosis of dementia with Lewy bodies (DLB): Report of the consortium on DLB international workshop. *Journal of Alzheimer's Disease: JAD*, *9*, 417–423.

Medori, R., Tritschler, H. J., Leblanc, A., Villare, F., Manetto, V., Chen, H. Y., ... Cortelli, P. (1992). Fatal familial insomnia, a prion disease with a mutation at codon-178 of the prion protein gene. *The New England Journal of Medicine*, *326*, 444–449.

Melki, R. (2017). How the shapes of seeds can influence pathology. *Neurobiology of Disease*.

Meyer-Luehmann, M., Coomaraswamy, J., Bolmont, T., Kaeser, S., Schaefer, C., Kilger, E., ... Jaton, A. L. (2006). Exogenous induction of cerebral beta-amyloidogenesis is governed by agent and host. *Science*, *313*, 1781–1784.

Miklossy, J. (2015). Historic evidence to support a causal relationship between spirochetal infections and Alzheimer's disease. *Frontiers in Aging Neuroscience*, *7*, 46.

Miller, T. M., Pestronk, A., David, W., Rothstein, J., Simpson, E., Appel, S. H., ... Alexander, K. (2013). An antisense oligonucleotide against SOD1 delivered intrathecally for patients with SOD1 familial amyotrophic lateral sclerosis: A phase 1, randomised, first-in-man study. *The Lancet Neurology*, *12*, 435–442.

Moghbel, M. C., Saboury, B., Basu, S., Metzler, S. D., Torigian, D. A., Langstrom, B., & Alavi, A. (2012). Amyloid-beta imaging with PET in Alzheimer's disease: Is it feasible with current radiotracers and technologies? *European Journal of Nuclear Medicine and Molecular Imaging*, *39*, 202–208.

Molliex, A., Temirov, J., Lee, J., Coughlin, M., Kanagaraj, A. P., Kim, H. J., ... Taylor, J. P. (2015). Phase separation by low complexity domains promotes stress granule assembly and drives pathological fibrillization. *Cell*, *163*, 123–133.

Monari, L., Chen, S. G., Brown, P., Parchi, P., Petersen, R. B., Mikol, J., ... Ghetti, B. (1994). Fatal familial insomnia and familial Creutzfeldt–Jakob-disease—different prion proteins determined by a DNA polymorphism. *Proceedings of the National Academy of Sciences of the United States of America*, *91*, 2839–2842.

Monsellier, E., & Chiti, F. (2007). Prevention of amyloid-like aggregation as a driving force of protein evolution. *EMBO Reports*, *8*, 737–742.

Morozova, O. A., March, Z. M., Robinson, A. S., & Colby, D. W. (2013). Conformational features of tau fibrils from Alzheimer's disease brain are faithfully propagated by unmodified recombinant protein. *Biochemistry*, *52*, 6960–6967.

Mougenot, A. L., Nicot, S., Bencsik, A., Morignat, E., Verchere, J., Lakhdar, L., ... Baron, T. (2012). Prion-like acceleration of a synucleinopathy in a transgenic mouse model. *Neurobiology of Aging*, *33*, 2225–2228.

Mulder, C., Verwey, N. A., van der Flier, W. M., Bouwman, F. H., Kok, A., van Elk, E. J., ... Blankenstein, M. A. (2010). Amyloid-beta(1–42), total tau, and phosphorylated tau as cerebrospinal fluid biomarkers for the diagnosis of Alzheimer disease. *Clinical Chemistry*, *56*, 248–253.

Munch, C., O'Brien, J., & Bertolotti, A. (2011). Prion-like propagation of mutant superoxide dismutase-1 misfolding in neuronal cells. *Proceedings of the National Academy of Sciences of the United States of America*, *108*, 3548–3553.

Nakamura, K., Mori, F., Kon, T., Tanji, K., Miki, Y., Tomiyama, M., ... Takahashi, H. (2015). Filamentous aggregations of phosphorylated alpha-synuclein in Schwann cells (Schwann cell cytoplasmic inclusions) in multiple system atrophy. *Acta Neuropathologica Communications*, *3*, 29.

Nelson, R., Sawaya, M. R., Balbirnie, M., Madsen, A. O., Riekel, C., Grothe, R., & Eisenberg, D. (2005). Structure of the cross-beta spine of amyloid-like fibrils. *Nature, 435*, 773–778.

Neumann, M., Sampathu, D. M., Kwong, L. K., Truax, A. C., Micsenyi, M. C., Chou, T. T., ... Clark, C. M. (2006). Ubiquitinated TDP-43 in frontotemporal lobar degeneration and amyotrophic lateral sclerosis. *Science, 314*, 130–133.

Nguyen, L. T., & Ohashi, P. S. (2015). Clinical blockade of PD1 and LAG3--potential mechanisms of action. *Nature Reviews Immunology, 15*, 45–56.

Nillegoda, N. B., Kirstein, J., Szlachcic, A., Berynskyy, M., Stank, A., Stengel, F., ... Aebersold, R. (2015). Crucial HSP70 co-chaperone complex unlocks metazoan protein disaggregation. *Nature, 524*, 247–251.

Nomura, T., Watanabe, S., Kaneko, K., Yamanaka, K., Nukina, N., & Furukawa, Y. (2014). Intranuclear aggregation of mutant FUS/TLS as a molecular pathomechanism of amyotrophic lateral sclerosis. *Journal of Biological Chemistry, 289*, 1192–1202.

Nonaka, T., Masuda-Suzukake, M., Arai, T., Hasegawa, Y., Akatsu, H., Obi, T., ... Akiyama, H. (2013). Prion-like properties of pathological TDP-43 aggregates from diseased brains. *Cell Reports, 4*, 124–134.

Nordberg, A., Rinne, J. O., Kadir, A., & Langstrom, B. (2010). The use of PET in Alzheimer disease. *Nature Reviews Neurology, 6*, 78–87.

O'Brien, R. J., & Wong, P. C. (2011). Amyloid precursor protein processing and Alzheimer's disease. *Annual Review of Neuroscience, 34*, 185–204.

Orr, H. T., & Zoghbi, H. Y. (2007). Trinucleotide repeat disorders. *Annual Review of Neuroscience, 30*, 575–621.

Orru, C. D., Groveman, B. R., Hughson, A. G., Zanusso, G., Coulthart, M. B., & Caughey, B. (2015). Rapid and sensitive RT-QuIC detection of human Creutzfeldt–Jakob disease using cerebrospinal fluid. *mBio, 6*.

Ouberai, M. M., Wang, J., Swann, M. J., Galvagnion, C., Guilliams, T., Dobson, C. M., & Welland, M. E. (2013). Alpha-synuclein senses lipid packing defects and induces lateral expansion of lipids leading to membrane remodeling. *Journal of Biological Chemistry, 288*, 20883–20895.

Outram, G. W., Dickinson, A. G., & Fraser, H. (1973). Developmental maturation of susceptibility to scrapie in mice. *Nature, 241*, 536–537.

Pallet, N., & Legendre, C. (2013). Adverse events associated with mTOR inhibitors. *Expert Opinion on Drug Safety, 12*, 177–186.

Pan, K. M., Baldwin, M., Nguyen, J., Gasset, M., Serban, A., Groth, D., ... Cohen, F. E. (1993). Conversion of alpha-helices into beta-sheets features in the formation of the scrapie prion proteins. *Proceedings of the National Academy of Sciences of the United States of America, 90*, 10962–10966.

Papachroni, K. K., Ninkina, N., Papapanagiotou, A., Hadjigeorgiou, G. M., Xiromerisiou, G., Papadimitriou, A., ... Buchman, V. L. (2007). Autoantibodies to alpha-synuclein in inherited Parkinson's disease. *Journal of Neurochemistry, 101*, 749–756.

Pasanen, P., Myllykangas, L., Siitonen, M., Raunio, A., Kaakkola, S., Lyytinen, J., ... Paetau, A. (2014). Novel alpha-synuclein mutation A53E associated with atypical multiple system atrophy and Parkinson's disease-type pathology. *Neurobiology of Aging, 35*(2180), e2181–e2185.

Patel, A., Lee, H. O., Jawerth, L., Maharana, S., Jahnel, M., Hein, M. Y., ... Franzmann, T. M. (2015). A liquid-to-solid phase transition of the ALS protein FUS accelerated by disease mutation. *Cell, 162*, 1066–1077.

Paumier, K. L., Luk, K. C., Manfredsson, F. P., Kanaan, N. M., Lipton, J. W., Collier, T. J., ... Schulz, E. (2015). Intrastriatal injection of pre-formed mouse alpha-synuclein fibrils into rats triggers alpha-synuclein pathology and bilateral nigrostriatal degeneration. *Neurobiology of Disease, 82*, 185–199.

Pearce, M. M., Spartz, E. J., Hong, W., Luo, L., & Kopito, R. R. (2015). Prion-like transmission of neuronal huntingtin aggregates to phagocytic glia in the *Drosophila* brain. *Nature Communications, 6*, 6768.

Pecho-Vrieseling, E., Rieker, C., Fuchs, S., Bleckmann, D., Esposito, M. S., Botta, P., ... Muller, M. (2014). Transneuronal propagation of mutant huntingtin contributes to non-cell autonomous pathology in neurons. *Nature Neuroscience, 17*, 1064–1072.

Pedersen, J. T., & Sigurdsson, E. M. (2015). Tau immunotherapy for Alzheimer's disease. *Trends in Molecular Medicine, 21*, 394–402.

Peelaerts, W., & Baekelandt, V. (2016). α-Synuclein strains and the variable pathologies of synucleinopathies. *Journal of Neurochemistry*.

Peelaerts, W., Bousset, L., Van der Perren, A., Moskalyuk, A., Pulizzi, R., Giugliano, M., ... Baekelandt, V. (2015). alpha-Synuclein strains cause distinct synucleinopathies after local and systemic administration. *Nature, 522*, 340–344.

Peeraer, E., Bottelbergs, A., Van Kolen, K., Stancu, I. C., Vasconcelos, B., Mahieu, M., ... Sluydts, E. (2015). Intracerebral injection of preformed synthetic tau fibrils initiates widespread tauopathy and neuronal loss in the brains of tau transgenic mice. *Neurobiology of Disease, 73*, 83–95.

Perutz, M. F., Johnson, T., Suzuki, M., & Finch, J. T. (1994). Glutamine repeats as polar zippers: Their possible role in inherited neurodegenerative diseases. *Proceedings of the National Academy of Sciences of the United States of America, 91*, 5355–5358.

Phukan, J., Pender, N. P., & Hardiman, O. (2007). Cognitive impairment in amyotrophic lateral sclerosis. *The Lancet Neurology, 6*, 994–1003.

Pike, V. W. (2009). PET radiotracers: Crossing the blood–brain barrier and surviving metabolism. *Trends in Pharmacological Sciences, 30*, 431–440.

Polymenidou, M., & Cleveland, D. W. (2011). The seeds of neurodegeneration: Prion-like spreading in ALS. *Cell, 147*, 498–508.

Polymeropoulos, M. H., Lavedan, C., Leroy, E., Ide, S. E., Dehejia, A., Dutra, A., ... Boyer, R. (1997). Mutation in the alpha-synuclein gene identified in families with Parkinson's disease. *Science, 276*, 2045–2047.

Pooler, A. M., Phillips, E. C., Lau, D. H., Noble, W., & Hanger, D. P. (2013). Physiological release of endogenous tau is stimulated by neuronal activity. *EMBO Reports, 14*, 389–394.

Poorkaj, P., Bird, T. D., Wijsman, E., Nemens, E., Garruto, R. M., Anderson, L., ... Schellenberg, G. D. (1998). Tau is a candidate gene for chromosome 17 frontotemporal dementia. *Annals of Neurology, 43*, 815–825.

Preusser, M., Strobel, T., Gelpi, E., Eiler, M., Broessner, G., Schmutzhard, E., & Budka, H. (2006). Alzheimer-type neuropathology in a 28 year old patient with iatrogenic Creutzfeldt–Jakob disease after dural grafting. *Journal of Neurology, Neurosurgery, and Psychiatry, 77*, 413–416.

Prinz, M., Montrasio, F., Furukawa, H., van der Haar, M. E., Schwarz, P., Rulicke, T., ... Glatzel, M. (2004). Intrinsic resistance of oligodendrocytes to prion infection. *Journal of Neuroscience, 24*, 5974–5981.

Protter, D. S., & Parker, R. (2016). Principles and properties of stress granules. *Trends in Cell Biology, 26*, 668–679.

Prusiner, S. B. (1982). Novel proteinaceous infectious particles cause scrapie. *Science, 216*, 136–144.

Prusiner, S. B. (1984). Some speculations about prions, amyloid, and Alzheimer's disease. *The New England Journal of Medicine, 310*, 661–663.

Prusiner, S. B. (1998). Prions. *Proceedings of the National Academy of Sciences of the United States of America, 95*, 13363–13383.

Prusiner, S. B. (2013). Biology and genetics of prions causing neurodegeneration. *Annual Review of Genetics, 47*, 601–623.

Prusiner, S. B., Woerman, A. L., Mordes, D. A., Watts, J. C., Rampersaud, R., Berry, D. B., ... Kravitz, S. N. (2015). Evidence for alpha-synuclein prions causing multiple system atrophy in humans with parkinsonism. *Proceedings of the National Academy of Sciences of the United States of America, 112*, E5308–E5317.

Qiang, W., Yau, W. M., Lu, J. X., Collinge, J., & Tycko, R. (2017). Structural variation in amyloid-beta fibrils from Alzheimer's disease clinical subtypes. *Nature, 541*, 217–221.

Ramaswami, M., Taylor, J. P., & Parker, R. (2013). Altered ribostasis: RNA-protein granules in degenerative disorders. *Cell, 154*, 727–736.

Ravits, J., Laurie, P., Fan, Y., & Moore, D. H. (2007). Implications of ALS focality: Rostral–caudal distribution of lower motor neuron loss postmortem. *Neurology, 68*, 1576–1582.

Recasens, A., Dehay, B., Bove, J., Carballo-Carbajal, I., Dovero, S., Perez-Villalba, A., ... Perier, C. (2014). Lewy body extracts from Parkinson disease brains trigger alpha-synuclein pathology and neurodegeneration in mice and monkeys. *Annals of Neurology, 75*, 351–362.

Ren, P. H., Lauckner, J. E., Kachirskaia, I., Heuser, J. E., Melki, R., & Kopito, R. R. (2009). Cytoplasmic penetration and persistent infection of mammalian cells by polyglutamine aggregates. *Nature Cell Biology, 11*, 219–225.

Renton, A. E., Chio, A., & Traynor, B. J. (2014). State of play in amyotrophic lateral sclerosis genetics. *Nature Neuroscience, 17*, 17–23.

Renton, A. E., Majounie, E., Waite, A., Simon-Sanchez, J., Rollinson, S., Gibbs, J. R., ... Myllykangas, L. (2011). A hexanucleotide repeat expansion in C9ORF72 is the cause of chromosome 9p21-linked ALS-FTD. *Neuron, 72*, 257–268.

Rey, N. L., George, S., & Brundin, P. (2016a). Review: Spreading the word: Precise animal models and validated methods are vital when evaluating prion-like behaviour of alpha-synuclein. *Neuropathology and Applied Neurobiology, 42*, 51–76.

Rey, N. L., Petit, G. H., Bousset, L., Melki, R., & Brundin, P. (2013). Transfer of human alpha-synuclein from the olfactory bulb to interconnected brain regions in mice. *Acta Neuropathologica, 126*, 555–573.

Rey, N. L., Steiner, J. A., Maroof, N., Luk, K. C., Madaj, Z., Trojanowski, J. Q., ... Brundin, P. (2016b). Widespread transneuronal propagation of alpha-synucleinopathy triggered in olfactory bulb mimics prodromal Parkinson's disease. *Journal of Experimental Medicine, 213*, 1759–1778.

Rey, N. L., Wesson, D. W., & Brundin, P. (2016c). The olfactory bulb as the entry site for prion-like propagation in neurodegenerative diseases. *Neurobiology of Disease*.

Reyes, J. F., Rey, N. L., Bousset, L., Melki, R., Brundin, P., & Angot, E. (2014). Alpha-synuclein transfers from neurons to oligodendrocytes. *Glia, 62*(3), 387–398.

Ridley, R. M., Baker, H. F., Windle, C. P., & Cummings, R. M. (2006). Very long term studies of the seeding of beta-amyloidosis in primates. *Journal of Neural Transmission, 113*, 1243–1251.

Rogers, N. G., Basnight, M., Gibbs, C. J., & Gajdusek, D. C. (1967). Latent viruses in chimpanzees with experimental kuru. *Nature, 216*, 446–449.

Rosen, D. R., Siddique, T., Patterson, D., Figlewicz, D. A., Sapp, P., Hentati, A., ... Deng, H. X. (1993). Mutations in Cu/Zn superoxide dismutase gene are associated with familial amyotrophic lateral sclerosis. *Nature, 362*, 59–62.

Ross, C. A., & Poirier, M. A. (2004). Protein aggregation and neurodegenerative disease. *Nature Medicine, 10* (Suppl), S10–S17.

Rubinsztein, D. C., Leggo, J., Coles, R., Almqvist, E., Biancalana, V., Cassiman, J. J., ... Curtis, A. (1996). Phenotypic characterization of individuals with 30–40 CAG repeats in the Huntington disease (HD) gene reveals HD cases with 36 repeats and apparently normal elderly individuals with 36–39 repeats. *American Journal of Human Genetics, 59*, 16–22.

Sacino, A. N., Brooks, M., Thomas, M. A., McKinney, A. B., Lee, S., Regenhardt, R. W., ... Borchelt, D. R. (2014). Intramuscular injection of alpha-synuclein induces CNS alpha-synuclein pathology and a rapid-onset motor phenotype in transgenic mice. *Proceedings of the National Academy of Sciences of the United States of America, 111*, 10732–10737.

Saito, Y., Ruberu, N. N., Sawabe, M., Arai, T., Tanaka, N., Kakuta, Y., ... Murayama, S. (2004). Staging of argyrophilic grains: An age-associated tauopathy. *Journal of Neuropathology & Experimental Neurology, 63*, 911–918.

Salvadores, N., Shahnawaz, M., Scarpini, E., Tagliavini, F., & Soto, C. (2014). Detection of misfolded Abeta oligomers for sensitive biochemical diagnosis of Alzheimer's disease. *Cell Reports, 7*, 261–268.

Sanders, D. W., Kaufman, S. K., DeVos, S. L., Sharma, A. M., Mirbaha, H., Li, A., ... Serpell, L. C. (2014). Distinct tau prion strains propagate in cells and mice and define different tauopathies. *Neuron, 82*, 1271–1288.

Sawaya, M. R., Sambashivan, S., Nelson, R., Ivanova, M. I., Sievers, S. A., Apostol, M. I., ... McFarlane, H. T. (2007). Atomic structures of amyloid cross-beta spines reveal varied steric zippers. *Nature, 447*, 453–457.

Schellenberg, G. D., & Montine, T. J. (2012). The genetics and neuropathology of Alzheimer's disease. *Acta Neuropathologica, 124*, 305–323.

Schenk, D. B., Koller, M., Ness, D. K., Griffith, S. G., Grundman, M., Zago, W., ... Kinney, G. G. (2016). First-in-human assessment of PRX002, an anti-alpha-synuclein monoclonal antibody, in healthy volunteers. *Movement Disorders.*

Scott, J. R., & Fraser, H. (1989). Transport and targeting of scrapie infectivity and pathology in the optic nerve projections following intraocular infection. *Progress in Clinical and Biological Research, 317*, 645–652.

Selkoe, D. J., & Hardy, J. (2016). The amyloid hypothesis of Alzheimer's disease at 25 years. *EMBO Molecular Medicine, 8*, 595–608.

Serpell, L. C., Berriman, J., Jakes, R., Goedert, M., & Crowther, R. A. (2000). Fiber diffraction of synthetic alpha-synuclein filaments shows amyloid-like cross-beta conformation. *Proceedings of the National Academy of Sciences of the United States of America, 97*, 4897–4902.

Serrano-Pozo, A., Frosch, M. P., Masliah, E., & Hyman, B. T. (2011). Neuropathological alterations in Alzheimer disease. *Cold Spring Harbor Perspectives in Medicine, 1*, a006189.

Settembre, C., Fraldi, A., Medina, D. L., & Ballabio, A. (2013). Signals from the lysosome: A control centre for cellular clearance and energy metabolism. *Nature Reviews Molecular Cell Biology, 14*, 283–296.

Sevigny, J., Chiao, P., Bussiere, T., Weinreb, P. H., Williams, L., Maier, M., ... Ling, Y. (2016). The antibody aducanumab reduces Abeta plaques in Alzheimer's disease. *Nature, 537*, 50–56.

Shah, R. C., Matthews, D. C., Andrews, R. D., Capuano, A. W., Fleischman, D. A., VanderLugt, J. T., & Colca, J. R. (2014). An evaluation of MSDC-0160, a prototype mTOT modulating insulin sensitizer, in patients with mild Alzheimer's disease. *Current Alzheimer Research, 11*, 564–573.

Shahnawaz, M., Tokuda, T., Waragai, M., Mendez, N., Ishii, R., Trenkwalder, C., & Soto, C. (2017). Development of a biochemical diagnosis of Parkinson disease by detection of alpha-synuclein misfolded aggregates in cerebrospinal fluid. *JAMA Neurology, 74*(2), 163–172.

Shibayama-Imazu, T., Okahashi, I., Omata, K., Nakajo, S., Ochiai, H., Nakai, Y., ... Nakaya, K. (1993). Cell and tissue distribution and developmental change of neuron specific 14 kDa protein (phosphoneuroprotein 14). *Brain Research, 622*, 17–25.

Shimonaka, S., Nonaka, T., Suzuki, G., Hisanaga, S., & Hasegawa, M. (2016). Templated Aggregation of TAR DNA-binding Protein of 43 kDa (TDP-43) by Seeding with TDP-43 Peptide Fibrils. *Journal of Biological Chemistry, 291*, 8896–8907.

Shrivastava, A. N., Redeker, V., Fritz, N., Pieri, L., Almeida, L. G., Spolidoro, M., ... Lena, C. (2015). alpha-synuclein assemblies sequester neuronal alpha3-Na^+/K^+-ATPase and impair Na^+ gradient. *The EMBO Journal, 34*, 2408–2423.

Sieradzan, K. A., Mechan, A. O., Jones, L., Wanker, E. E., Nukina, N., & Mann, D. M. (1999). Huntington's disease intranuclear inclusions contain truncated, ubiquitinated huntingtin protein. *Experimental Neurology, 156*, 92–99.

Singleton, A. B., Farrer, M., Johnson, J., Singleton, A., Hague, S., Kachergus, J., ... Nussbaum, R. (2003). alpha-Synuclein locus triplication causes Parkinson's disease. *Science, 302*, 841.

Smethurst, P., Newcombe, J., Troakes, C., Simone, R., Chen, Y. R., Patani, R., & Sidle, K. (2016). In vitro prion-like behaviour of TDP-43 in ALS. *Neurobiology of Disease, 96*, 236–247.

Smith, R. A., Miller, T. M., Yamanaka, K., Monia, B. P., Condon, T. P., Hung, G., ... Wei, H. (2006). Antisense oligonucleotide therapy for neurodegenerative disease. *Journal of Clinical Investigation, 116*, 2290–2296.

Snyder, S. W., Ladror, U. S., Wade, W. S., Wang, G. T., Barrett, L. W., Matayoshi, E. D., ... Holzman, T. F. (1994). Amyloid-beta aggregation: Selective inhibition of aggregation in mixtures of amyloid with different chain lengths. *Biophysical Journal, 67*, 1216–1228.

Solforosi, L., Milani, M., Mancini, N., Clementi, M., & Burioni, R. (2013). A closer look at prion strains: Characterization and important implications. *Prion, 7*, 99–108.

Spillantini, M. G., Crowther, R. A., Jakes, R., Cairns, N. J., Lantos, P. L., & Goedert, M. (1998a). Filamentous alpha-synuclein inclusions link multiple system atrophy with Parkinson's disease and dementia with Lewy bodies. *Neuroscience Letters, 251*, 205–208.

Spillantini, M. G., Crowther, R. A., Jakes, R., Hasegawa, M., & Goedert, M. (1998b). alpha-Synuclein in filamentous inclusions of Lewy bodies from Parkinson's disease and dementia with lewy bodies. *Proceedings of the National Academy of Sciences of the United States of America, 95*, 6469–6473.

Spillantini, M. G., & Goedert, M. (1998). Tau protein pathology in neurodegenerative diseases. *Trends in Neurosciences, 21*, 428–433.

Spillantini, M. G., & Goedert, M. (2000). The alpha-synucleinopathies: Parkinson's disease, dementia with Lewy bodies, and multiple system atrophy. *Annals of the New York Academy of Sciences, 920*, 16–27.

Spillantini, M. G., Murrell, J. R., Goedert, M., Farlow, M. R., Klug, A., & Ghetti, B. (1998c). Mutation in the tau gene in familial multiple system tauopathy with presenile dementia. *Proceedings of the National Academy of Sciences of the United States of America, 95*, 7737–7741.

Squitieri, F., Cannella, M., & Simonelli, M. (2002). CAG mutation effect on rate of progression in Huntington's disease. *Neurological Sciences, 23*(Suppl 2), S107–S108.

Sreedharan, J., Blair, I. P., Tripathi, V. B., Hu, X., Vance, C., Rogelj, B., ... Buratti, E. (2008). TDP-43 mutations in familial and sporadic amyotrophic lateral sclerosis. *Science, 319*, 1668–1672.

Stancu, I. C., Vasconcelos, B., Ris, L., Wang, P., Villers, A., Peeraer, E., ... Oyelami, T. (2015). Templated misfolding of Tau by prion-like seeding along neuronal connections impairs neuronal network function and associated behavioral outcomes in Tau transgenic mice. *Acta Neuropathologica, 129*, 875–894.

Stefani, M., & Dobson, C. M. (2003). Protein aggregation and aggregate toxicity: New insights into protein folding, misfolding diseases and biological evolution. *Journal of Molecular Medicine, 81*, 678–699.

Stohr, J., Condello, C., Watts, J. C., Bloch, L., Oehler, A., Nick, M., ... Prusiner, S. B. (2014). Distinct synthetic Abeta prion strains producing different amyloid deposits in bigenic mice. *Proceedings of the National Academy of Sciences of the United States of America, 111*, 10329–10334.

Stohr, J., Watts, J. C., Mensinger, Z. L., Oehler, A., Grillo, S. K., DeArmond, S. J., ... Giles, K. (2012). Purified and synthetic Alzheimer's amyloid beta (Abeta) prions. *Proceedings of the National Academy of Sciences of the United States of America, 109*, 11025–11030.

Stuendl, A., Kunadt, M., Kruse, N., Bartels, C., Moebius, W., Danzer, K. M., ... Schneider, A. (2016). Induction of alpha-synuclein aggregate formation by CSF exosomes from patients with Parkinson's disease and dementia with Lewy bodies. *Brain, 139*, 481–494.

Takahashi, R. H., Milner, T. A., Li, F., Nam, E. E., Edgar, M. A., Yamaguchi, H., ... Gouras, G. K. (2002a). Intraneuronal Alzheimer abeta42 accumulates in multivesicular bodies and is associated with synaptic pathology. *The American Journal of Pathology, 161*, 1869–1879.

Takahashi, R. H., Nam, E. E., Edgar, M., & Gouras, G. K. (2002b). Alzheimer beta-amyloid peptides: Normal and abnormal localization. *Histology and Histopathology, 17*, 239–246.

Tanaka, M., Collins, S. R., Toyama, B. H., & Weissman, J. S. (2006). The physical basis of how prion conformations determine strain phenotypes. *Nature, 442*, 585–589.

Tanzi, R. E. (2012). The genetics of Alzheimer disease. *Cold Spring Harbor Perspectives in Medicine, 2*, 1–10.

Tardivel, M., Begard, S., Bousset, L., Dujardin, S., Coens, A., Melki, R., ... Colin, M. (2016). Tunneling nanotube (TNT)-mediated neuron-to neuron transfer of pathological Tau protein assemblies. *Acta Neuropathologica Communications, 4*, 117.

Tartaglia, G. G., Pechmann, S., Dobson, C. M., & Vendruscolo, M. (2007). Life on the edge: A link between gene expression levels and aggregation rates of human proteins. *Trends in Biochemical Sciences, 32*, 204–206.

Telling, G. C., Parchi, P., DeArmond, S. J., Cortelli, P., Montagna, P., Gabizon, R., ... Prusiner, S. B. (1996). Evidence for the conformation of the pathologic isoform of the prion protein enciphering and propagating prion diversity. *Science, 274*, 2079–2082.

Teravanesyan, M. D., Dagkesamanskaya, A. R., Kushnirov, V. V., & Smirnov, V. N. (1994). The Sup35 omnipotent suppressor gene is involved in the maintenance of the non-mendelian determinant [Psi(+)] in the yeast *Saccharomyces cerevisiae*. *Genetics, 137*, 671–676.

Terryberry, J. W., Thor, G., & Peter, J. B. (1998). Autoantibodies in neurodegenerative diseases:

Antigen-specific frequencies and intrathecal analysis. *Neurobiology of Aging, 19*, 205–216.

Thal, D. R., Rub, U., Orantes, M., & Braak, H. (2002). Phases of A beta-deposition in the human brain and its relevance for the development of AD. *Neurology, 58*, 1791–1800.

Theillet, F. X., Binolfi, A., Bekei, B., Martorana, A., Rose, H. M., Stuiver, M., ... Goldfarb, D. (2016). Structural disorder of monomeric alpha-synuclein persists in mammalian cells. *Nature, 530*(7588), 45–50.

Tiwari, A., & Hayward, L. J. (2005). Mutant SOD1 instability: Implications for toxicity in amyotrophic lateral sclerosis. *Neurodegenerative Diseases, 2*, 115–127.

Tolnay, M., & Clavaguera, F. (2004). Argyrophilic grain disease: A late-onset dementia with distinctive features among tauopathies. *Neuropathology, 24*, 269–283.

Tran, H. T., Chung, C. H., Iba, M., Zhang, B., Trojanowski, J. Q., Luk, K. C., & Lee, V. M. (2014). Alpha-synuclein immunotherapy blocks uptake and templated propagation of misfolded alpha-synuclein and neurodegeneration. *Cell Reports, 7*, 2054–2065.

Tuttle, M. D., Comellas, G., Nieuwkoop, A. J., Covell, D. J., Berthold, D. A., Kloepper, K. D., ... Kendall, A. (2016). Solid-state NMR structure of a pathogenic fibril of full-length human alpha-synuclein. *Nature Structural & Molecular Biology, 23*, 409–415.

Uversky, V. N., Gillespie, J. R., & Fink, A. L. (2000). Why are "natively unfolded" proteins unstructured under physiologic conditions? *Proteins, 41*, 415–427.

Vargas, K. J., Makani, S., Davis, T., Westphal, C. H., Castillo, P. E., & Chandra, S. S. (2014). Synucleins regulate the kinetics of synaptic vesicle endocytosis. *Journal of Neuroscience, 34*, 9364–9376.

Vella, L. J., Hill, A. F., & Cheng, L. (2016). Focus on Extracellular Vesicles: Exosomes and Their Role in Protein Trafficking and Biomarker Potential in Alzheimer's and Parkinson's Disease. *International Journal of Molecular Sciences, 17*, 173.

Vinters, H. V. (1987). Cerebral amyloid angiopathy. A critical review. *Stroke, 18*, 311–324.

Visanji, N. P., Brooks, P. L., Hazrati, L. N., & Lang, A. E. (2013). The prion hypothesis in Parkinson's disease: Braak to the future. *Acta Neuropathologica Communications, 1*, 2.

Volpicelli-Daley, L. A., Abdelmotilib, H., Liu, Z., Stoyka, L., Daher, J. P., Milnerwood, A. J., ... Zhao, H. T. (2016). G2019S-LRRK2 Expression Augments alpha-Synuclein Sequestration into Inclusions in Neurons. *Journal of Neuroscience, 36*, 7415–7427.

Wadsworth, J. D. F., Asante, E. A., Desbruslais, M., Linehan, J. M., Joiner, S., Gowland, I., ... Hill, A. F. (2004). Human prion protein with valine 129 prevents expression of variant CJD phenotype. *Science, 306*, 1793–1796.

Wagner, J., Ryazanov, S., Leonov, A., Levin, J., Shi, S., Schmidt, F., ... Mitteregger-Kretzschmar, G. (2013). Anle138b: A novel oligomer modulator for disease-modifying therapy of neurodegenerative diseases such as prion and Parkinson's disease. *Acta Neuropathologica, 125*, 795–813.

Wakabayashi, K., Hayashi, S., Kakita, A., Yamada, M., Toyoshima, Y., Yoshimoto, M., & Takahashi, H. (1998). Accumulation of alpha-synuclein/NACP is a cytopathological feature common to Lewy body disease and multiple system atrophy. *Acta Neuropathologica, 96*, 445–452.

Walker, F. O. (2007). Huntington's disease. *The Lancet, 369*, 218–228.

Walker, L., Levine, H., & Jucker, M. (2006). Koch's postulates and infectious proteins. *Acta Neuropathologica, 112*, 1–4.

Walker, L. C., Callahan, M. J., Bian, F., Durham, R. A., Roher, A. E., & Lipinski, W. J. (2002). Exogenous induction of cerebral beta-amyloidosis in betaAPP-transgenic mice. *Peptides, 23*, 1241–1247.

Walsh, D. M., & Selkoe, D. J. (2016). A critical appraisal of the pathogenic protein spread hypothesis of neurodegeneration. *Nature Reviews Neuroscience, 17*, 251–260.

Wang, F., Wang, X., Yuan, C. G., & Ma, J. (2010). Generating a prion with bacterially expressed recombinant prion protein. *Science, 327*, 1132–1135.

Wang, H. Y., Lee, D. H., Davis, C. B., & Shank, R. P. (2000). Amyloid peptide Abeta(1–42) binds selectively and with picomolar affinity to alpha7 nicotinic acetylcholine receptors. *Journal of Neurochemistry, 75*, 1155–1161.

Watters, A. L., Deka, P., Corrent, C., Callender, D., Varani, G., Sosnick, T., & Baker, D. (2007). The highly cooperative folding of small naturally occurring proteins is likely the result of natural selection. *Cell, 128*, 613–624.

Watts, J. C., Condello, C., Stohr, J., Oehler, A., Lee, J., DeArmond, S. J., ... Prusiner, S. B. (2014). Serial propagation of distinct strains of Abeta prions from Alzheimer's disease patients. *Proceedings of the National Academy of Sciences of the United States of America, 111*, 10323–10328.

Watts, J. C., Giles, K., Oehler, A., Middleton, L., Dexter, D. T., Gentleman, S. M., ... Prusiner, S. B. (2013). Transmission of multiple system atrophy prions to transgenic mice. *Proceedings of the National Academy of Sciences of the United States of America, 110*, 19555–19560.

Weber, J. J., Sowa, A. S., Binder, T., & Hubener, J. (2014). From pathways to targets: Understanding the mechanisms behind polyglutamine disease. *BioMed Research International, 2014*, 701758.

Will, R. G. (2003). Acquired prion disease: Iatrogenic CJD, variant CJD, kuru. *British Medical Bulletin, 66*, 255–265.

Wischik, C. M., Harrington, C. R., & Storey, J. M. (2014). Tau-aggregation inhibitor therapy for Alzheimer's disease. *Biochemical Pharmacology, 88*, 529–539.

Wong, E. S., Tan, J. M., Soong, W. E., Hussein, K., Nukina, N., Dawson, V. L., ... Lim, K. L. (2008). Autophagy-mediated clearance of aggresomes is not a universal phenomenon. *Human Molecular Genetics*, *17*, 2570–2582.

Wu, J., & Li, L. (2016). Autoantibodies in Alzheimer's disease: Potential biomarkers, pathogenic roles, and therapeutic implications. *Journal of Biomedical Research*, *30*, 361–372.

Wu, J. W., Herman, M., Liu, L., Simoes, S., Acker, C. M., Figueroa, H., ... Zurzolo, C. (2013). Small misfolded Tau species are internalized via bulk endocytosis and anterogradely and retrogradely transported in neurons. *Journal of Biological Chemistry*, *288*, 1856–1870.

Wu, J. W., Hussaini, S. A., Bastille, I. M., Rodriguez, G. A., Mrejeru, A., Rilett, K., ... Boonen, R. A. (2016). Neuronal activity enhances tau propagation and tau pathology in vivo. *Nature Neuroscience*, *19*, 1085–1092.

Xue, W. F., Homans, S. W., & Radford, S. E. (2008). Systematic analysis of nucleation-dependent polymerization reveals new insights into the mechanism of amyloid self-assembly. *Proceedings of the National Academy of Sciences of the United States of America*, *105*, 8926–8931.

Ye, L., Velasco, A., Fraser, G., Beach, T. G., Sue, L., Osredkar, T., ... Lockhart, A. (2008). In vitro high affinity alpha-synuclein binding sites for the amyloid imaging agent PIB are not matched by binding to Lewy bodies in postmortem human brain. *Journal of Neurochemistry*, *105*, 1428–1437.

Yu, Y. J., & Watts, R. J. (2013). Developing therapeutic antibodies for neurodegenerative disease. *Neurotherapeutics: The Journal of the American Society for Experimental NeuroTherapeutics*, *10*, 459–472.

Zanusso, G., Monaco, S., Pocchiari, M., & Caughey, B. (2016). Advanced tests for early and accurate diagnosis of Creutzfeldt–Jakob disease. *Nature Reviews Neurology*, *12*, 427.

Zarranz, J. J., Alegre, J., Gomez-Esteban, J. C., Lezcano, E., Ros, R., Ampuero, I., ... Atares, B. (2004). The new mutation, E46K, of alpha-synuclein causes Parkinson and Lewy body dementia. *Annals of Neurology*, *55*, 164–173.

Zhang, K., Donnelly, C. J., Haeusler, A. R., Grima, J. C., Machamer, J. B., Steinwald, P., ... Vidensky, S. (2015). The C9orf72 repeat expansion disrupts nucleocytoplasmic transport. *Nature*, *525*, 56–61.

Zilber, L. A., Bajdakova, Z. L., Gardasjan, A. N., Konovalov, N. V., Bunina, T. L., & Barabadze, E. M. (1963). Study of the Etiology of Amyotrophic Lateral Sclerosis. *Bulletin of the World Health Organization*, *29*, 449–456.

CHAPTER

9

Neurodegenerative Diseases as Protein Folding Disorders

Jeremy D. Baker, Jack M. Webster, Lindsey B. Shelton, John Koren III, Vladimir N. Uversky, Laura J. Blair and Chad A. Dickey

University of South Florida, Tampa, FL, United States

OUTLINE

INTRODUCTION

This chapter focuses on neurodegeneration as a result of the misbehavior of key proteins that are aberrantly misfolded or misassembled, leading to neuronal loss producing motor or cognitive impairments. Misfolding and subsequent aggregation of proteins is linked to many

The Molecular and Cellular Basis of Neurodegenerative Diseases
DOI: https://doi.org/10.1016/B978-0-12-811304-2.00009-2

neurological disorders, including Alzheimer's disease (AD), Parkinson's disease (PD), amyotrophic lateral sclerosis (ALS), and Huntington's disease (HD), to name a few. Generally, these protein aggregates form a special structure known as amyloid, which has the distinguishing characteristic of containing β-sheet structures that can aggregate into long fibrils as well as recruit and convert protein that has not yet misfolded. How the formation of amyloid aggregates leads to neuronal dysfunction and neuron loss despite cellular machinery and processes evolved to ensure proper protein folding and regulation is an area of active research (Sorg & Grothe, 2015; Yerbury et al., 2016). Indeed, because amyloid formation results in a heterogeneous mix of many oligomeric, prefibrillar, and fibrillar structures, determining the toxic aggregated species has been difficult, and current therapeutics have failed to prevent the pathology caused by these proteins. This chapter will provide background on protein folding regulatory processes as well as illustrate certain determinants for protein misfolding. We will also describe current hypotheses on how protein aggregates cause cellular stress and neurodegeneration and provide an overview of new early detection and therapeutic strategies.

ROLES FOR PROTEIN FOLDING, MODIFICATION, AND DEGRADATION

Proteins are essential macromolecules of all living cells. Their functions are crucial and diverse, ranging from key structural components such as keratins, which form external protective layers of all land vertebrates, and tubulin, which forms the eukaryotic cytoskeleton and plays a central role in neuronal transport (Matamoros & Baas, 2016), to various enzymes catalyzing a multitude of biological reactions, such as acetylcholinesterase, a key protein within the synaptic cleft. Mammalian immune defense relies on antibody proteins, while cellular, tissue, and organ messaging depends on growth hormone proteins for communication (Steyn, Tolle, Chen, & Epelbaum, 2016). Obviously, these listed examples represent only a miniscule fraction of the immense universe of protein functions. The functional diversity of proteins arises from the multitude of polypeptide chains, whose sequences are uniquely assembled from a set of 20 major amino acids. Alternative splicing and RNA editing ensures diversification of mRNA and plays an important role in increasing the assortment of functional proteins encoded in the genome. Some proteins are folded into unique structures, whereas others exist as dynamic conformational ensembles. Biological activities of many proteins require further assembly of individual chains to large functioning protein complexes.

Furthermore, synthesized proteins can be further diversified via various chemical modifications introduced by other proteins (Walsh, Garneau-Tsodikova, & Gatto, 2005). These posttranslational modifications (PTMs) are important for regulating the wide spectrum of protein functions that play important roles in various biological processes, ranging from cell signaling to translocation of newly synthesized proteins. In fact, although DNA typically encodes 20 primary amino acids, proteins contain more than 140 different residues, because of various PTMs. As described below, abnormal PTMs can lead to neurodegeneration. Phosphorylation, the transfer of a phosphoryl group to an amino acid mediated by a variety of kinases, is the most common PTM. Phosphorylation plays a key role as a biochemical "switch" for some proteins, for instance activating or deactivating certain enzymes. A host of neurodegenerative diseases known as tauopathies may arise after the abnormal hyperphosphorylation of tau, a protein known to stabilize axonal microtubules (Kadavath et al., 2015). Acetylation, another common PTM where acetyl groups are covalently linked to an N-terminal amino group or lysine residue, generally results in changes in the stability of proteins. Histone acetylases (HATs) and histone deacetylases (HDACs) control the balance of acetylation of specific substrates. When the cellular environment changes and a neuron loses acetylation homeostasis, due to a change in HAT: HDAC equilibrium, neurodegenerative disease can result (Min et al., 2015; Saha & Pahan, 2006). Another common PTM, glycosylation, or the addition of carbohydrates to amino acid residues, is crucial to the proper folding and translocation of many cell surface and secreted proteins (Chavan & Lennarz, 2006). Traumatic brain injury often results in an altered pattern of glycosylated amyloid precursor protein (APP), which may be linked to its pathological clinical presentation in the form of cerebral plaques (Abou-Abbass et al., 2016; Hwang et al., 2010).

In addition to PTMs playing a role in proper folding of nascent proteins and regulation of protein function, a class of proteins known as molecular chaperones assist in forming the correct three-dimensional structure of proteins and help to assemble large protein complexes (Kim, Hipp, Bracher, Hayer-Hartl, & Hartl, 2013). Many molecular chaperones are activated in response to cellular stress, such as elevated temperature or changes in pH or salt concentration (Mayer & Bukau, 2005). Cellular stress, such as heat, causes a much higher incidence of misfolding, which may lead to protein aggregation. In humans, some of the more important cellular chaperones use adenosine triphosphate (ATP) as an energy source to actively support the folding or unfolding of proteins and are termed foldases or unfoldases (Mattoo & Goloubinoff, 2014; Rikhvanov, Romanova, & Chernoff, 2007; Wang & Tsou, 1998). For example, heat shock protein 70 (Hsp70), a chaperone expressed in response to causes of neuronal stress, including stroke and neurodegenerative disease, may play an important role in the suppression of aggregation and resulting toxicity of some neuropathic proteins (Chittoor-Vinod, Lee, Judge, & Notterpek, 2015; Falsone & Falsone, 2015). Some other chaperones, like the small heat shock proteins αB-crystallin and Hsp27, are ATP-independent and do not actively fold proteins. Instead, such chaperones hold proteins in certain conformations and are referred to as holdases (Yerbury et al., 2013). These small heat shock proteins have been shown to reduce toxicity caused by α-synuclein in PD (Leak, 2014). One of the more exciting recent discoveries is that of a chaperone complex of Hsp70, a DnaJ cochaperone, and the nucleotide exchange factor Hsp110, harnesses energy from ATP to physically pull apart and disassemble amyloid fibrils. This complex disassembles α-synuclein amyloid fibrils leading to a nontoxic end-product (Gao et al., 2015; Nillegoda et al., 2015). Research into harnessing the power of this human disaggregase machinery to disentangle pathological amyloids is just beginning but presents a novel therapeutic approach.

Protein homeostasis (or proteostasis) is required for cellular health and is controlled not only by the rate of synthesis of proteins, their proper folding and trafficking, PTMs and the action of molecular chaperones but also by regulated degradation (Balch, Morimoto, Dillin, & Kelly, 2008; Powers, Morimoto, Dillin, Kelly, & Balch, 2009). Degradation processes within the proteostasis network include the ubiquitin-proteasome system (UPS) and autophagy (chaperone mediated autophagy and macroautophagy) (Ciechanover & Kwon, 2015). Misfolded proteins can be marked for proteosomal degradation by covalent tagging with chains of ubiquitin protein. A downregulation of the UPS is associated with progressive onset of several neurodegenerative diseases, and it is thought that aggregated proteins can physically interfere with the proteasome machinery to reduce its activity (Bence, Sampat, & Kopito, 2001; Santos, Cardoso, Magalhaes, & Saraiva, 2007). Autophagy refers to the cellular sequestration of proteins into lysosomes, an organelle containing hydrolytic enzymes capable of degrading diverse proteins. Like the UPS, down regulation or genetic inactivation of autophagy has been linked to numerous degenerative diseases (Cuervo & Wong, 2014; Koga & Cuervo, 2011).

The remainder of this chapter will focus on the precise causes of the misfolding of proteins, how they result in toxic oligomeric and amyloid assemblies, and current explanations for how these proteins may cause disease.

WHAT IS PROTEIN MISFOLDING AND WHY DOES IT OCCUR?

Misfolding of certain proteins leads to neurodegenerative disease, due to loss of native protein function or a gain of deleterious function, the latter resulting from aggregation into toxic structures (Chiasseu et al., 2016; Gendron & Petrucelli, 2011; Winklhofer, Tatzelt, & Haass, 2008; Zempel & Mandelkow, 2015). This chapter will focus on protein aggregation as a toxic gain of function phenomenon associated with many neurological disorders.

Generally, in the case of most proteins, nascent polypeptide chains fold into properly functioning three-dimensional structures, either spontaneously or with the assistance of molecular chaperones. Resulting structures can be further tuned with the help of PTMs. Structural integrity of a folded protein is maintained via an interplay of weak conformational forces (such as hydrophobic interactions, hydrogen bonds, van der Waals interactions and electrostatic interactions) opposing the action of conformational entropy. Folded globular proteins are characterized by structures where hydrophobic residues are generally buried deep within the protein core, and amino acid side chains are engaged in multiple interactions to maintain conformation. However, because of fluctuating environmental conditions, such as changes in pH or temperature, native state proteins may partially unfold, and the resulting intermediates may begin to aggregate. Misfolding and aggregation can also be triggered by point mutations, aberrant PTMs, and many other factors. Pathological proteins undergo a series of steps from their native monomeric form to multiunit aggregates, typically rich in β-sheet structure. During this conformational change from properly folded to misfolded and aggregated forms, often involving a shift from native α-helical secondary structure to β-sheet structure, some key hydrophobic residues, originally buried within the protein core, are exposed (Ciechanover & Kwon, 2015; Díaz-Villanueva, Díaz-Molina, & García-González, 2015). These hydrophobic residues located on the protein surface cause interchain attraction manifested in multiple protein monomers coming together to form aggregates. Often, the assembly of these β-sheet-rich structures have a fibrillar morphology, and the resulting aggregated forms are termed amyloid fibrils. One should keep in mind though that, in addition to amyloid fibrils, many other types of pathological protein aggregates can be formed, ranging from various assemblies of soluble oligomers and protofibrils, to amorphous aggregates and spherulites (Kowalewski & Holtzman, 1999; Krebs, Devlin, & Donald, 2009; Roberti et al., 2012; Wetzel, Shivaprasad, & Williams, 2007).

Amyloid fibrils have been associated with more than 30 human diseases, including several neurodegenerative disorders as shown in Table 9.1 (Koo, Lansbury, & Kelly, 1999; Ross & Poirier, 2004; Wolfe & Cyr, 2011). Fibrils (or at least their core) are normally composed of polypeptides taking on a β-sheet secondary structure, which then self-associate to form long fibers (Knowles et al., 2007; Mandelkow & Mandelkow, 2012). Using atomic force microscopy and transmission electron microscopy, fibrils have been shown to consist of a cross-β-sheet secondary structure within the core of the fibrils, where the parallel chains of β-stranded peptides are arranged in an orientation perpendicular to the axis of the fiber and linked by an array of interbackbone hydrogen bonds (Fitzpatrick et al., 2013; Tycko, 2016). This structure is incredibly resistant to cellular degradation mechanics as well as experimental

TABLE 9.1 Neurodegenerative Disorders and Associated Misfolded Proteins

Disease	Misfolding or aggregating protein	Manifestation
Alzheimer's disease	Amyloid-β	Plaques
	Tau	Neurofibrillary tangles
Amyotrophic lateral sclerosis	Superoxide dismutase	Inclusion bodies
	TDP-43	
	Fused in sarcoma RNA-binding protein (FUS)	
	Dipeptide repeat proteins	
Corticobasal degeneration	Tau	Fine filamentous inclusions
		Glial plaques
Frontotemporal dementia	Tau	Pick bodies
	TDP-43	Inclusion bodies
	FUS	Inclusion bodies
Huntington's disease	Huntingtin	Inclusion bodies
Lewy body dementia	α-Synuclein	Lewy bodies
	May co-occur with (Aβ and/or tau)	
Progressive supranuclear palsy	Tau	Neurofibrillary tangles
		Tau tufted astrocytes
	May cooccur with α-synuclein	Lewy bodies
Parkinson's disease	α-Synuclein	Lewy bodies
Familial amyloid polyneuropathies	Transthyretin	Amyloid fibrils

means of protein structure disruption (Crouch et al., 2009; Gong et al., 2000; Knauer, Soreghan, Burdick, Kosmoski, & Glabe, 1992; Tsubuki, Takaki, & Saido, 2003). Amyloid fibrils may consist of hundreds and often thousands of the original monomeric protein subunits, resulting in large bodies visible by microscopy within neurons and brain tissue.

Although amyloid fibrils are the most commonly detected form of aggregation, these pathogenic proteins form a heterogeneous mixture of oligomers, protofibrils, and fibrils (Bucciantini et al., 2002; Eisenberg & Jucker, 2012). The structures and associated pathology of these assemblies is of great interest as we look for multiple toxic species responsible for the associated neurological diseases. Generally, the end-products of aggregation are insoluble fibrillar amyloid aggregates; however, there are many intermediary structures along the pathways of aggregation (Chimon et al., 2007; Moran & Zanni, 2014; Tycko, 2011, 2015). Because the aggregate structures of proteins like amyloid beta (Aβ) and tau are disordered and consist of a heterogeneous population of species, traditional methods to study structure, such as X-ray crystallography and nuclear magnetic resonance, have not yielded high-resolution structures for many smaller aggregates. However, recent advances in solid-state NMR have given us key insights into aggregate structures specifically in regard to

Aβ (Naito & Kawamura, 2007; Scheidt, Morgado, Rothemund, & Huster, 2012; Tycko, 2011).

Pathological Misfolding of Intrinsically Disordered and Mutant Proteins

Many pathological amyloidogenic proteins that cause neurodegeneration are intrinsically disordered. That is, they present not as classical globular three-dimensional structures but as a highly dynamic conformational ensembles (Uversky, 2013; Varadi, Vranken, Guharoy, & Tompa, 2015). The flexibility of intrinsically disordered proteins allows them to assume numerous conformations and increases binding affinity for a much larger range of substrates compared to fixed-conformation globular proteins. However, the flexibility and lack of a three-dimensional structure, along with an enhanced propensity to be posttranslationally modified and engaged in multiple physiological and pathological interactions makes some of these proteins more prone to aggregation and amyloid formation (Levine, Larini, LaPointe, Feinstein, & Shea, 2015; Uversky, 2010). Intrinsically disordered proteins which aggregate in neurological diseases include the microtubule binding protein tau and Aβ (Eisele et al., 2015; Knowles, Vendruscolo, & Dobson, 2014; Levine et al., 2015), both involved in AD, as well as α-synuclein, which aggregates to form Lewy bodies in PD pathology (Lashuel, Overk, Oueslati, & Masliah, 2013; Zhang, Griggs, Rochet, & Stanciu, 2013).

Meanwhile other misfolded proteins involved in neurodegeneration develop a propensity for misfolding resulting from a genetic expansion of nucleotide repeats. Expanded CAG repeats in the coding region of certain genes results in expanded polyglutamine (polyQ) tracts within the translated protein, which have a propensity to misfold into β-sheet-containing amyloids (Nagai & Popiel, 2008). For example, HD and spinocerebellar ataxias are caused by expanded polyQ tracts within the huntingtin or ataxin proteins (Takahashi, Katada, & Onodera, 2010). Additionally, expanded nucleotide repeat sequences in the noncoding region of the C9orf72 gene have been linked to frontotemporal dementia and ALS (DeJesus-Hernandez et al., 2011). This nucleotide repeat expansion becomes translated in the absence of a start codon into a mixture of five aggregating dipeptide repeat proteins translated from all six reading frames (Mori et al., 2013).

What Drives Misfolding and Amyloid Formation of Proteins?

Although it has been challenging to study the mechanisms of self-assembly of these pathological proteins by experimentation, with the help of molecular dynamic simulations and advanced biochemical techniques, a more robust picture of what drives aggregate formation is emerging. Monitoring fibril growth with a quartz crystal oscillator has revealed that the native state of these proteins is thermodynamically more stable than the amyloid state as long as a critical concentration of amyloid is not reached (Knowles et al., 2007; Knowles et al., 2014). Once a critical concentration of amyloid is present in the cell, the native state will inevitably and spontaneously be driven by thermodynamics toward the amyloid state unless cellular mechanisms prevent the transition. Experiments using small amyloid-forming fragments of several proteins have given insights into the kinetics of amyloid formation. Three kinetic stages occur in amyloid formation. First, nucleation of monomers results in the formation of a metastable critical nucleus, which can shift the protein into a growth phase. This growth phase consists of rapid assembly of oligomers into protofibrils that self-associate. Ultimately, a final dynamic stabilization phase is reached, in which mature fibrils are formed (Nasica-Labouze & Mousseau, 2012; Villar-Pique, Espargaro, Ventura, & Sabate, 2016).

The triggers that start the aggregation process vary and range from PTMs to changes in environmental factors within the cell.

Additionally, certain amino acid changes due to genetic variation within aggregation-prone proteins increase the propensity for self-assembly into amyloid (Chang, Kim, Yin, Nagaraja, & Kuret, 2008; Rizzu et al., 1999; Vogelsberg-Ragaglia et al., 2000; von Bergen, Barghorn, Biernat, Mandelkow, & Mandelkow, 2005). Although it should be noted that intrinsically ordered proteins "breathe" [i.e., they have an inherent fluctuation in conformation which may transiently expose residues normally not exposed to the cellular environment, increasing chances for aggregation (Fink, 1998; Kumar et al., 2014)], environmental stressors like pH changes, pressure changes, salt concentration changes, and temperature changes can drive partial unfolding of proteins and cause an aggregation cascade (Bhowmik et al., 2014; Fujiwara, Matsumoto, & Yonezawa, 2003; Goossens, Haelewyn, Meersman, De Ley, & Heremans, 2003; Gursky & Aleshkov, 2000; Kim, Randolph, Seefeldt, & Carpenter, 2006).

Under normal cellular conditions, environmental factors are kept relatively constant, and other factors may be predominantly responsible for triggering amyloidogenesis. A recognized trigger of amyloidogenesis of tau protein is its hyperphosphorylation, suggesting regulation via cell-signaling pathways. Tau is rich in phosphorylation sites targeted by a number of kinases. When tau is heavily phosphorylated, its association with microtubules is disrupted, and freeing tau from microtubules facilitates self-association and the formation of aggregates.

HOW DO MISFOLDED PROTEINS AND AGGREGATES CAUSE NEURODEGENERATION?

Oligomer Hypothesis

Increasing evidence supports oligomers as the causative toxic species in a number of neurodegenerative disorders, whereby fibril generation is thought to be a protective mechanism in neurons, sequestering the smaller toxic aggregates (Sengupta, Nilson, & Kayed, 2016; Spencer et al., 2016; Yang, Li, & Xu, 2017). It is well known that Aβ aberrantly accumulates in AD; however, the presence of deposited extracellular Aβ plaques do not generally correlate well with disease progression (Benilova, Karran, & De Strooper, 2012), and many healthy brains carry an Aβ plaque burden without associated cognitive deficits. This has led to the hypothesis that a smaller soluble aggregated form of Aβ is somehow pathogenic. In a mouse model of tauopathy, turning off tau expression halted neuronal loss and prevented cognitive decline, even though neurofibrillary tangles remained present and further accumulated (Santacruz et al., 2005), providing evidence for a soluble oligomeric intermediate as the toxic species. Additionally, transactive response DNA binding protein 43 kDa (TDP-43), a protein associated with ALS, has been shown to form a heterogeneous population of oligomers, and these oligomers exist in patients with TDP-43-positive frontotemporal lobar dementia (Fang et al., 2014), a form of dementia not associated with tau or Aβ. It should be pointed out again that amyloid aggregation results in the formation of a heterogeneous population of oligomers with a diverse range of structure and size. Determining the toxicity of specific oligomer assemblies is challenging, but it has been postulated that there may be an inverse correlation between oligomer size and toxicity (Sengupta et al., 2016).

Oligomers have important biochemical properties that make them more toxic than mature amyloid fibrils. Hydrophobic residues of the more inert fibrils are buried within its cross-β-sheet core; however, these residues are more readily accessible in oligomers and provide an interaction surface for other cellular structures and proteins. Oligomers, but not fibers, have been shown to interact with and disrupt membranes (Williams et al., 2011). Lastly, oligomers have been shown to seed the production of oligomers of other amyloid proteins, initiating a cascade of aberrant assembly of toxic proteins within cells (Sengupta et al., 2015). Indeed, very recent and convincing evidence has indicated that these misfolded structures of various proteins might spread from neuron to neuron in a way that resembles the pathological spread of prion proteins (Calafate et al., 2015; Pecho-Vrieseling et al., 2014; Yin, Tan, Jiang, & Yu, 2014). Precisely, how these oligomers are causing disease is just now being brought into focus, and a number of well-regulated pathways disrupted by oligomer activity may be pathogenic (Cardenas-Aguayo Mdel, Gomez-Virgilio, DeRosa, & Meraz-Rios, 2014; Ferreira & Klein, 2011). In this section, we will discuss new research illustrating just how oligomers may promote toxicity within neurons as shown in Fig. 9.1.

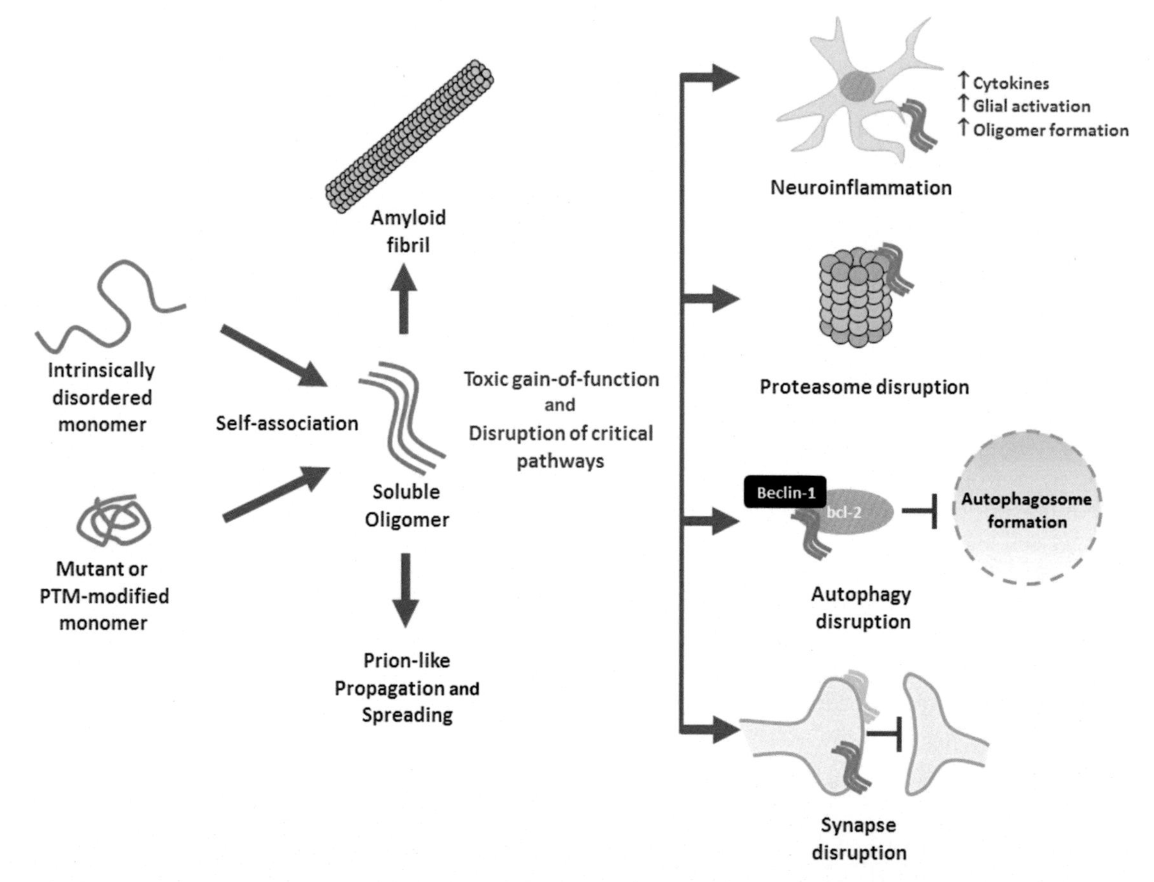

FIGURE 9.1 Some proteins can form oligomers that have deleterious effects on neurons and cause neurodegenerative disorders. Certain events like PTM or transient unfolding of proteins can lead to self-association and soluble oligomer formation. These oligomers can seed the formation of long amyloid fibers. Additionally, increasing evidence supports that misfolded proteins involved in neurodegeneration, such as tau, spread throughout the brain and propagate in a prion-like manner. Oligomers are known to directly act with key regulatory proteins throughout the brain and have been shown to increase neuroinflammation and disrupt degradation pathways, including the ubiquitin-proteasome pathway and autophagy. Lastly, oligomers can interact at the synapse to impede neurotransmission.

Neuroinflammation

Astrocytes and microglia regulate neuroinflammation through a host of cytokines, small signaling proteins that include chemokines, interferons, interleukins, and tumor necrosis factors (TNF). A number of neurodegenerative diseases are associated with increased inflammation throughout the brain, and increased activity and accumulation of microglia and astrocytes precedes deposition of insoluble amyloid, such as Aβ or tau fibrils, in AD patients (Martin, Pardo, Cork, & Price, 1994). Cytokines released by glial cells interact with neuronal receptors to promote signaling pathways that have been implicated in several neurodegenerative disorders (Sawada, Imamura, & Nagatsu, 2006; Su, Bai, & Zhang, 2016; Wang, Tan, Yu, & Tan, 2015). Chemokines like IL-1, IL-6, and TNFα, released by glial cells are significantly more abundant in AD brain tissue than healthy brains, and cytokines can upregulate pathways promoting activation of proteins associated with AD, such as protein kinase C or p38 mitogen-activating protein kinase (Salminen, Ojala, Kauppinen, Kaarniranta, & Suuronen, 2009). Prolonged, chronic activation of glial cells associates well with brain atrophy in AD (Berger et al., 2007; Lee & MacLean, 2015). Protein oligomers may be directly responsible for interacting with glial cells, causing a cascade of increasing inflammation and glial activity contributing to neuron loss (Nilson et al., 2016).

It has been proposed by the Kayed group that tau oligomers may induce a positive feedback loop by causing neuroinflammation, which in turn induces increased formation of oligomeric species. Their studies indicate that tau oligomers colocalize near astrocytes and may be responsible for astrocyte activity, leading to increased inflammation (Nilson et al., 2016). Additionally, the group found that tau oligomers colocalized with microglia, neural macrophages activated in inflammation. In contrast to astrocytes, where oligomers seem to interact near the cell surface, microglial cells apparently engulf the oligomers. Phagocytosis and subsequent release through exosomes has been reported and may play a role in neural spreading of tau (Asai et al., 2015; Medina & Avila, 2014). The authors propose that activation of microglia may result from oligomer interaction with the receptor for advanced glycation endproducts (RAGE) receptor, which leads to NFκB activation and subsequent upregulation of HMGB1, a proinflammatory cytokine. Importantly, RAGE is itself upregulated by this process, leading to chronic inflammation. Aβ can also induce release of cytokines from astrocytes. A 2011 study by the Noble group showed that Aβ treatment of astrocytes increased the amounts of cytokines CINC2α/β, IFN-γ, IL-1β, IL-1ra, IL-6, IL-13, IL-17, IP-10, and MIG (Garwood, Pooler, Atherton, Hanger, & Noble, 2011). This is important because IL-1β, IL-6, and IP-10 are all associated with plaque and tangle pathology in both human AD brains and in mouse models of AD. Further studies are needed to identify which receptors oligomers are acting on and to clarify the mechanism by which they influence inflammatory pathways.

Proteostasis Disruption

As previously mentioned, proteostasis refers to numerous pathways and cellular organelles which function to regulate protein translation, folding, localization, and degradation. Proteins abnormally accumulate in numerous neurodegenerative diseases, indicating a disruption of degradation and deaggregation processes required for neuronal survival. Oligomers of various proteins associated with neurodegenerative diseases interact with these degradation pathways and subsequently inhibit clearance of toxic proteins by the cell. We will discuss two major pathways of cellular protein degradation and highlight new areas of research showing how misassembled oligomers may be interacting with and disrupting these crucial processes.

The UPS is a two-step degradation pathway, whereby proteins targeted for degradation are first covalently conjugated to a polyubiquitin chain before being translocated to the proteasome. The proteasome is a catalytic complex responsible for the degradation of misfolded proteins. The complex consists of a 20S catalytic core along with 19S regulatory subunits on either end, which direct misfolded proteins to the catalytic core subunit. Proteolytic activity of the proteasome is reduced by proteins central to neurodegenerative diseases, including tau in tauopathies, by α-synuclein in PD, and Aβ in AD (Tseng, Green, Chan, Blurton-Jones, & LaFerla, 2008).

Oligomers of diverse neurotoxic proteins interact with and inhibit proteolytic activity of the UPS. For example, in an in vitro proteasome activity assay, oligomeric Aβ inhibited proteosomal catalytic activity in a dose-dependent manner, whereas the Aβ monomer had no effect on proteasome activity (Tseng et al., 2008). In a human cell model, Aβ interacted with and sequestered the α-subunit of the proteasome into aggregates. Further, in an in vivo mouse model overexpressing Aβ, administration of an anti-Aβ antibody results in a reduction of oligomer levels in the brain while simultaneously increasing proteasome activity (Tseng et al., 2008). In mice overexpressing a pathologically aggregating form of tau containing the P301L mutation, known to cause frontotemporal dementia, aggregates of tau directly inhibited the activity of the 26S proteasome (Myeku et al., 2016). Finally, α-synuclein oligomers directly interact and disrupt the activity of the 26S proteasome, indicating a common pathology among different oligomer forming proteins (Emmanouilidou, Stefanis, & Vekrellis, 2010).

Another major pathway through which neurons remove or degrade cellular components is through macroautophagy, whereby unneeded components are sequestered to a vacuole. This vacuole subsequently fuses to a lysosome, which contains diverse hydrolytic enzymes capable of breaking down proteins and other cellular components. Inhibition of macroautophagy leads to accumulation of toxic proteins, such as tau, Aβ, and α-synuclein oligomers (Wang et al., 2009); however, the mechanism by which neurotoxic proteins interact and hinder the macroautophagy-driven degradation pathway is unclear. Recent work suggests that Aβ has a direct conformation-specific role in modifying autophagy and subsequent apoptosis of cells. Beclin-1 is a crucial protein in the autophagy pathway, as it initiates the formation of phagosomes through recruitment of other proteins. However, if Beclin-1 interacts with bcl-2, it prevents autophagosome formation and leads to apoptosis, or neuronal death (Marquez & Xu, 2012). Interestingly, while monomers of Aβ inhibit this interaction, Aβ oligomers actually enhance the formation of Beclin-1:bcl2 complexes and cause neuronal death (Guglielmotto et al., 2014). Because oligomers of diverse toxic proteins seem to have similar effects on cellular pathways and organelles, amyloidogenic proteins other than Aβ may also directly impair autophagy.

Synaptic Dysfunction

Neuronal damage precedes fibrillary tau aggregates, suggesting that prefibrillar oligomers may be neurotoxic. Loss of neurons ultimately causes cognitive and motor impairments in animal models and in human neurodegenerative diseases. However, oligomeric forms of amyloidogenic proteins cause disruptions in synaptic activity and plasticity well before cellular degeneration. Synaptic dysfunction rather than neuron loss may be the cause of motor impairment and memory loss early in neurodegenerative disease progression (Di et al., 2016; Lacor et al., 2004). Aβ oligomers have been shown to inhibit hippocampal long-term potentiation and enhance long-term depression well before cellular degeneration (Lambert et al., 1998; Shankar et al., 2008), which suggests a potential means of reversal of neurodegenerative symptoms if diagnosed early (Dodart et al., 2002; Kotilinek et al., 2002). Soluble oligomeric amyloids associate with synaptic proteins, resulting in impairments in synaptic function. Aβ oligomers associate with postsynaptic density complexes containing NMDA receptors and cause impairment of NMDA receptor-mediated calcium entry and decreased dendritic spine density (Lacor et al., 2007; Sinnen, Bowen, Gibson, & Kennedy, 2016). Aβ oligomers have been demonstrated to cause impaired synaptic trafficking of ionotropic and metabotropic glutamate receptors (Gong et al., 2000; Lacor et al., 2004) and to induce abnormal glutamate release from hippocampal neurons, which may contribute to excitotoxic signaling (Brito-Moreira et al., 2011). Soluble amyloid oligomers also induce dysregulation of calcium sequestration and signaling through a mechanism of increased membrane permeability independent of Ca^{2+} channel activation (Demuro et al., 2005). Aβ oligomers and tau oligomers can also disrupt axonal transport mechanisms (Decker, Lo, Unger, Ferreira, & Silverman, 2010; Ward, Himmelstein, Lancia, & Binder, 2012) that are critical for the maintenance of healthy synapses. Impaired neuronal signaling resulting from oligomeric amyloid proteins should be reversible prior to neuron loss, providing hope that eventually an appropriate therapeutic combined with early detection could one day facilitate reversal of disease progression.

HOW CAN PROTEIN MISFOLDING BE TARGETED?

Biomarkers for Early Detection

Detection and diagnosis of neurodegenerative diseases continue to move away from postmortem analysis of protein aggregates in the brain toward noninvasive early detection. Many methods from genetic testing (Agrawal & Biswas, 2015), molecular and functional imaging (Risacher & Saykin, 2013), and blood and cerebrospinal fluid (CSF) biomarkers (Bernier, Kumar, Sato, & Oda, 2015), to retinal scanning (Chang et al., 2014; Ikram, Cheung, Wong, & Chen, 2012), and exhaled volatile biomarker fingerprints (Mazzatenta, Pokorski, Sartucci, Domenici, & Di Giulio, 2015) involve noninvasive tests to seek correlations between specific biomarkers and disease. Here we focus on the noninvasive detection of misfolded proteins or interacting chaperones rather than numerous other potential correlative biomarkers.

In AD, the intracellular accumulation of misfolded species of amyloidogenic proteins such as Aβ and tau is also accompanied by the secretion of these proteins from neurons. Remarkably, these proteins can be detected in the CSF and have even been detected in blood. Multiple studies have evaluated the detection of amyloidogenic proteins in the blood or CSF (Agrawal & Biswas, 2015). The focus of peripheral Aβ detection has been on the ratio of Aβ40/Aβ42 (Koyama et al., 2012), and, more recently, on smaller Aβ fragments, such as Aβ17 (Perez-Grijalba et al., 2015). Similarly, CSF and blood tau detection efforts have included both total tau and phosphorylated tau (P-tau) (Olsson et al., 2016). The extracellular chaperone clusterin has also emerged as a potential biomarker for AD (Agrawal & Biswas, 2015; Kiddle et al., 2014). Similarly, in PD, changes in the levels of α-synuclein (an intrinsically disordered protein found in pathological intraneuronal hallmarks of PD, Lewy bodies, and Lewy neurites) in various biofluids, such as CSF, plasma, blood, and saliva, are used as disease-specific and early-stage biomarkers (Atik, Stewart, & Zhang, 2016). Interestingly, the best fitting predictive models for discriminating PD patients from controls can be built using a combination of multiple CSF α-synuclein species, such as total-, oligomeric-, and phosphorylated α-synuclein, with classical AD biomarkers, such as P-tau (Majbour et al., 2016).

Nevertheless, consistency and reproducibility of using amyloidogenic proteins as blood biomarkers can be challenging. For example, variability in blood–brain barrier integrity will have a profound effect on blood detection of selected biomarkers of brain origin, leading to poor correlation with stage of disease. Additionally, quantitation of proteins with disordered regions possessing high binding potential and/or a propensity to aggregate can be challenging, as the target protein may stick to other proteins on vascular and blood cells as wells as undergo self-aggregation that would mask epitopes used for detection.

Molecular imaging of amyloidogenic protein accumulation has become a promising noninvasive detection method for correlation with neurodegenerative disease progression. Three PET imaging agents targeting Aβ accumulation have been approved by the FDA for the estimation of Aβ neuritic plaque density. However, some patients present as Aβ positive with no dementia symptoms, making the hope of a simple imaging diagnosis more complicated. Therefore, initial approval was designed to rule out AD in Aβ-negative patients with dementia (Yeo, Waddell, Khan, & Pal, 2015), but longitudinal studies of cognitively normal patients with significant Aβ burden may help to determine whether progression to AD is inevitable. A large multicenter clinical study in the United States, the Imaging Dementia-Evidence for Amyloid Scanning study, aims to evaluate how amyloid imaging affects diagnosis, patient management, and outcomes. Several tau-targeted PET imaging agents are emerging and entering clinical trials as well as becoming incorporated into clinical trials evaluating potential therapeutics (James, Doraiswamy, & Borges-Neto, 2015). The pattern of tau tracer retention appears to correlate with disease progression and with the Braak staging pattern of tau deposition in autopsy studies (Scholl et al., 2016). Many PET imaging strategies for PD focus on uptake of agents through monoamine transporters (Pavese & Brooks, 2009); however, there is an intense focus on the development of agents targeting aggregates of α-synuclein (Chu et al., 2015; Eberling, Dave, & Frasier, 2013; Zhang et al., 2014).

Therapeutics

Overall, treatment options for neurodegenerative diseases have been very disappointing, with the majority of them failing to produce any significant benefits, while others end up doing more harm than good. FDA-approved therapeutic strategies are limited to pharmacological modulation of neurotransmission to counteract a loss in synaptic connections and/or neurons (Geula & Mesulam, 1995; Reisberg et al., 2003). Future strategies for the development of therapeutics for neurodegenerative protein misfolding disorders are varied and include stem cell replacement, immunotherapy, gene therapy, targeting modulators of protein aggregation, and degradation and targeting of misfolded proteins themselves. Here we will highlight a few therapeutic strategies that are focused on modifying misfolded or aggregated protein.

Active Immunotherapy

Active immunization research through vaccination with amyloidogenic protein epitopes has shown some promise in certain animal models of neurodegeneration and in early clinical trials. Vaccination against N-terminal peptides of Aβ (Valera & Masliah, 2013; Winblad et al., 2012) or peptides and phosphopeptides representing tau protein (Kontsekova, Zilka, Kovacech, Novak, & Novak, 2014; Novak et al., 2017; Theunis et al., 2013) have been evaluated. However, active vaccination carries the risk of inducing an excessive immune response in the brain, evidenced by occurrences of meningoencephalitis in clinical trials (Kohyama & Matsumoto, 2015). Mimetics of protein epitopes are also being used for vaccination strategies designed to induce a therapeutic immune response to Aβ (Mandler, Santic, et al., 2015) and α-synuclein (Mandler, Valera, et al., 2015; Schneeberger, Mandler, Mattner, & Schmidt, 2012) in hopes of reducing the risk of autoimmunity associated with native antigens. Additionally, adjuvant-free immunotherapies, including passive and DNA immunization, may avoid undesirable immune reactions.

Passive Immunotherapy

Several antibodies developed against certain species of amyloidogenic aggregates have been extensively tested clinically as passive immunotherapy. For example, aducanumab is a human monoclonal antibody that selectively targets aggregated Aβ. Tested in patients with mild AD in a small, Phase I trial, aducanumab showed a reduction of Aβ in a dose- and time-dependent manner, as well as a clinical improvement as measured by the Clinical Dementia Rating—Sum of Boxes and Mini Mental State. Aducanumab is currently being tested in a phase III clinical trial on patients with early AD (Sevigny et al., 2016) (NCT02477800). Crenezumab, an antibody targeting multiple forms of Aβ, and gantenerumab, an antibody that specifically recognizes Aβ fibrils, are both currently being investigated in phase III clinical trials for their efficacy as AD treatments (NCT02565511, NCT02670083, NCT02051608, NCT01224106). Solanezumab was developed as a humanized monoclonal IgG1 antibody recognizing soluble Aβ, in the hopes of targeting the smaller, more toxic Aβ species. Although solanezumab has been investigated in several phase III clinical trials without producing clinically significant differences on primary outcomes (NCT00904683), it is still being tested as a combination therapy with other potential AD treatments (NCT01760005). Less specific to Aβ, a phase III clinical trial examining the role of intravenous immune globulin as an add-on treatment for AD, a part of the Gammaglobulin Alzheimer's Partnership study, failed to meet the primary outcome measures (NCT00818662).

Similarly, antibodies specific for particular species of tau are in clinical development for passive immunotherapy (Pedersen & Sigurdsson, 2015) (NCT02820896, NCT02880956, NCT02460094). Most of these antibodies recognize tau phosphorylated at specific sites (Pedersen & Sigurdsson, 2015), while others target particular peptide sequences (Yanamandra et al., 2013) or specific conformations (Jicha, Bowser, Kazam, & Davies, 1997). Certain antibodies in clinical and preclinical development specifically target only prefibrillar oligomeric forms of tau (Castillo-Carranza et al., 2014; Patterson et al., 2011; Ward et al., 2013).

The development of antibodies against α-synuclein are progressing for treatment of PD and other synucleinopathies by reducing aggregation and toxicity (Oertel & Schulz, 2016; Valera & Masliah, 2013). PRX002, a monoclonal antibody against α-synuclein, is now being investigated for its role in reducing α-synuclein levels in patients with PD (NCT02157714).

Gene Therapy

The delivery of a potentially therapeutic gene of interest into the cells of patients with disease, usually using viral vectors, is yet another therapeutic strategy. One gene therapy strategy involves the expression of genes encoding single-chain antibody fragments (intrabodies) into cells (Meli, Krako, Manca, Lecci, & Cattaneo, 2013). In this way, some of the same antibodies developed for passive immunotherapy could be reengineered for intracellular expression. Intrabodies have been developed to target Aβ (Sudol et al., 2009), α-synuclein (Emadi, Barkhordarian, Wang, Schulz, & Sierks, 2007), and regions adjacent to the polyQ tract in huntingtin (Ali, Southwell, Bugg, Ko, & Patterson, 2011; Amaro & Henderson, 2016; Messer, 2016). Similarly, an engineered protein not based on an antibody scaffold that also binds to and modulates amyloid aggregation, Aβ-binding Affibody $Z_{A\beta3}$, represents another potential gene therapy strategy (Sandberg et al., 2010). Gene therapy may also be used to deliver genome-editing technologies, such as CRISPR and zinc-finger nucleases. This has particular promise for a purely genetic disease like HD, where silencing or editing out the expanded CAG repeats at the DNA level could halt production of the aggregation-prone protein (Garriga-Canut et al., 2012; Shin et al., 2016). Additionally, gene therapy delivery may allow the expression of proteins that have a propensity to reduce the aggregation of amyloidogenic protein. Certain chaperone proteins interfere with amyloid aggregation, including αB-crystallin, clusterin, Hsp27, and Hsp70 (Abisambra et al., 2010; Narayan et al., 2011; Santhoshkumar & Sharma, 2004). Likewise, a number of nonchaperone proteins have been found to reduce amyloid aggregation, including catalase, pyruvate kinase, and the prolyl isomerase PIN1 (Liou et al., 2003; Luo, Warmlander, Graslund, & Abrahams, 2014). While natural proteins that reduce amyloid aggregation may be promising targets for gene therapy, other endogenous proteins that exacerbate amyloid aggregation may be potential therapeutic targets for inhibition.

Targeting Aggregation-Modulating Proteins

Chaperones and enzymes that associate with misfolding proteins, like tau and Aβ, are potential therapeutic targets in neurodegenerative disease. Hsp90 is a major chaperone which helps fold, refold, and stabilize various proteins. Though aberrantly misfolded, tau is normally degraded in the proteasome, Hsp90 in concert with cochaperones stabilize tau, preventing normal degradation and thereby leading to accumulation. Inhibition of Hsp90, in vitro, allows for uninterrupted tau degradation (Dickey et al., 2007). Furthermore, Hsp90 increases the formation of β-sheet-rich tau fibrils and in complex with the cochaperone FKBP51 produces neurotoxic oligomeric tau (Blair et al., 2013). Because Hsp90 helps maintain levels of kinases and other proteins required for cancer cell survival, it has become a promising target in the fight against cancer, and well-characterized drugs developed for Hsp90 inhibition may also have implications in neurodegenerative disease. Conversely, because of the importance of Hsp90 in proteostasis and the extensive list of Hsp90-interacting proteins (Echeverria, Bernthaler, Dupuis, Mayer, & Picard, 2011), targeting of Hsp90 cochaperones, like FKBP51, may prove to be an alternative therapeutic approach with less toxicity due to a smaller subset of chaperone client proteins.

Beta-secretase (BACE-1) is the enzyme that cleaves APP into the C99 fragment, which can then be cleaved by gamma-secretase into Aβ fragments. Reducing the amount of Aβ fragments using a BACE inhibitor is thought to be a good therapeutic strategy for AD. Several BACE inhibitors are currently being tested in phase III clinical trials. For example, small molecules AZD3293, CNP520, and JNJ-54861911 are all BACE-1 inhibitors being investigated for their efficacy at various stages of AD (NCT02245737, NCT02783573, NCT02565511, NCT02569398). Verubecestat is a small-molecule inhibitor of BACE-1 and BACE-2 and is currently being used in a phase III clinical trial to examine its efficacy in mild to moderate cases of AD (Scott et al., 2016) (NCT01739348, NCT01953601).

Targeting Misfolded Proteins Directly

There are several other potential therapeutics being tested for the treatment of AD, such as an inhibitor of Aβ polymerization and inflammation (ALZT-OP1) and an inhibitor of tau aggregation (TRx0237). ALZT-OP1 is a combination therapy of two FDA-approved drugs: cromolyn and ibuprofen. Both drugs have antiinflammatory mechanisms: cromolyn stabilizes mast cells and suppresses cytokine release, whereas ibuprofen is a widely used nonsteroidal antiinflammatory drug. Cromolyn has also been shown to inhibit aggregation of Aβ monomers in vitro and decrease soluble Aβ levels in vivo (Hori et al., 2015). ALZT-OP1 is currently being investigated in a phase III trial in patients with early AD. TRx0237 is a reduced form of Methylene Blue (methylthioninium chloride, MTC) that has been designed to stabilize this dye to improve its absorption, bioavailability, and tolerability (Panza, Solfrizzi, & Seripa, 2016). TRx0237 is thought to both prevent tau aggregation and dissolve existing aggregates (Panza et al., 2016). Several phase III trials have examined the effects of TRx0237 on cognition (NCT01626378, NCT01689246, and NCT01689233). Two of the clinical trials failed to meet their primary outcomes, and the results of another are not yet published. Additional phase III therapeutics are summarized in Table 9.2.

TABLE 9.2 Experimental Therapeutics Targeting Misfolded Proteins for Neurodegenerative Diseases

Therapeutic Target	Examples (Drug—Disease)	Clinical Phase
Active immunotherapy	AADvac1—AD	Phase II
	ACI-35—AD	Phase I
	CAD106—AD	Phase II
	AFFITOPE PD01A—PD	Phase I
Passive immunotherapy	Aducanumab—AD	Phase III
	Crenezumab—AD	Phase III
	Gantenerumab—AD	Phase III
	Immune globulin—AD	Phase III
	Solanezumab—AD	Phase III
	BMS-986168—tauopathies	Phase I
	C2N-8E12—tauopathies	Phase I
	PRX002—PD	Phase I
Gene therapies	Intrabodies—AD, PD, HD	
	CRISPR—HD	
Aggregation modulating proteins	AZD3293—AD	Phase III
	CNP520—AD	Phase III
	JNJ-54861911—AD	Phase III
	Verubecestat—AD	Phase III
	Hsp90 inhibitors—AD	
Targeting misfolded proteins	ALZT-OP1—AD	Phase III
	TRx0237—AD	Phase III

AD, Alzheimer's disease; *PD*, Parkinson's disease; *HD*, Huntington's disease.

CONCLUSIONS

Maintenance of cellular proteostasis is essential for cellular health and is regulated by a variety of proteins involved in proper protein folding, trafficking, and degradation pathways. Thousands of different proteins, each with multiple isoforms and PTMs, are continuously translated, properly folded, associated with binding partners, shuttled to required subcellular locations, and degraded as needed. A handful of proteins develop a propensity to misfold due to an inherently disordered structure, mutation, or PTM. In the brain, these proteins cause the appearance of large fibrillar aggregates and neurodegeneration, leading to cognitive and motor deficits, such as dementia and dystonia. However, smaller prefibrillar oligomeric aggregates also continue to accumulate throughout disease progression. These toxic oligomers can have deleterious effects on neuronal function, including synaptic dysregulation, neuroinflammation, and the disruption of protein degradation via the UPS and autophagy. Dysregulation of neuronal function may be reflected by cognitive or motor deficits in an early phase of disease progression, prior to significant neuron loss. This could represent a window of therapeutic opportunity that could be enabled by early diagnosis and oligomer-modifying drugs. Ongoing research toward an understanding of the structural determinants of toxic oligomers and the cellular pathways that modulate their formation will facilitate the development of new diagnostic and therapeutic strategies as well as the refinement of strategies currently in clinical trials.

DEDICATION

This chapter is dedicated to Dr. Chad A. Dickey, for his inspirational brilliance, creativity, and determination.

References

Abisambra, J. F., Blair, L. J., Hill, S. E., Jones, J. R., Kraft, C., Rogers, J., ... Dickey, C. A. (2010). Phosphorylation dynamics regulate Hsp27-mediated rescue of neuronal plasticity deficits in tau transgenic mice. *The Journal of Neuroscience, 30*(46), 15374–15382. Available from https://doi.org/10.1523/jneurosci.3155-10.2010.

Abou-Abbass, H., Abou-El-Hassan, H., Bahmad, H., Zibara, K., Zebian, A., Youssef, R., ... Kobeissy, F. (2016). Glycosylation and other PTMs alterations in neurodegenerative diseases: Current status and future role in neurotrauma. *Electrophoresis, 37*(11), 1549–1561. Available from https://doi.org/10.1002/elps.201500585.

Agrawal, M., & Biswas, A. (2015). Molecular diagnostics of neurodegenerative disorders. *Frontiers in Molecular Biosciences, 2*, 54. Available from https://doi.org/10.3389/fmolb.2015.00054.

Ali, K., Southwell, A. L., Bugg, C. W., Ko, J. C., & Patterson, P. H. (2011). Frontiers in neuroscience recombinant intrabodies as molecular tools and potential therapeutics for Huntington's Disease. In D. C. Lo, & R. E. Hughes (Eds.), Neurobiology of Huntington's disease: Applications to drug discovery. Boca Raton (FL): CRC Press/Taylor & Francis Llc.

Amaro, I. A., & Henderson, L. A. (2016). An intrabody drug (rAAV6-INT41) reduces the binding of N-terminal huntingtin fragment(s) to DNA to basal levels in PC12 cells and delays cognitive loss in the R6/2 animal model. *Journal of Neurodegenerative Diseases, 2016*, 10. Available from https://doi.org/10.1155/2016/7120753.

Asai, H., Ikezu, S., Tsunoda, S., Medalla, M., Luebke, J., Haydar, T., & Wolozin, B. (2015). Depletion of microglia and inhibition of exosome synthesis halt tau propagation. *Nature Neuroscience, 18*(11), 1584–1593. Available from https://doi.org/10.1038/nn.4132.

Atik, A., Stewart, T., & Zhang, J. (2016). Alpha-synuclein as a biomarker for Parkinson's disease. *Brain Pathology, 26*(3), 410–418. Available from https://doi.org/10.1111/bpa.12370.

Balch, W. E., Morimoto, R. I., Dillin, A., & Kelly, J. W. (2008). Adapting proteostasis for disease intervention. *Science, 319*(5865), 916–919. Available from https://doi.org/10.1126/science.1141448.

Bence, N. F., Sampat, R. M., & Kopito, R. R. (2001). Impairment of the ubiquitin-proteasome system by protein aggregation. *Science, 292*(5521), 1552–1555. Available from https://doi.org/10.1126/science.292.5521.1552.

Benilova, I., Karran, E., & De Strooper, B. (2012). The toxic Abeta oligomer and Alzheimer's disease: An emperor in need of clothes. *Nature Neuroscience, 15*(3), 349–357. Available from https://doi.org/10.1038/nn.3028.

Berger, Z., Roder, H., Hanna, A., Carlson, A., Rangachari, V., & Yue, M. (2007). Accumulation of pathological tau species and memory loss in a conditional model of tauopathy. *The Journal of Neuroscience, 27*(14), 3650–3662. Available from https://doi.org/10.1523/jneurosci.0587-07.2007.

Bernier, F., Kumar, P., Sato, Y., & Oda, Y. (2015). Recent progress in the identification of non-invasive biomarkers to support the diagnosis of Alzheimer's disease in clinical practice and to assist human clinical trialsIn I. Zerr (Ed.), *Alzheimer's disease – Challenges for the future*. InTech. Available from https://doi.org/10.5772/60008, Retrieved from: . Available from https://www.intechopen.com/books/alzheimer-s-disease-challenges-for-the-future/recent-progress-in-the-identification-of-non-invasive-biomarkers-to-support-the-diagnosis-of-alzheim.

Bhowmik, D., MacLaughlin, C. M., Chandrakesan, M., Ramesh, P., Venkatramani, R., Walker, G. C., & Maiti, S. (2014). pH changes the aggregation propensity of amyloid-β without altering the monomer conformation. *Physical Chemistry Chemical Physics, 16*(3), 885–889. Available from https://doi.org/10.1039/C3CP54151G.

Blair, L. J., Nordhues, B. A., Hill, S. E., Scaglione, K. M., O'Leary, J. C., Fontaine, S. N., … Dickey, C. A. (2013). Accelerated neurodegeneration through chaperone-mediated oligomerization of tau. *Journal of Clinical Investigation*, *123*(10), 4158–4169. Available from https://doi.org/10.1172/JCI69003.

Brito-Moreira, J., Paula-Lima, A. C., Bomfim, T. R., Oliveira, F. B., Sepulveda, F. J., De Mello, F. G., … Ferreira, S. T. (2011). Abeta oligomers induce glutamate release from hippocampal neurons. *Current Alzheimer Research*, *8*(5), 552–562.

Bucciantini, M., Giannoni, E., Chiti, F., Baroni, F., Formigli, L., Zurdo, J., … Stefani, M. (2002). Inherent toxicity of aggregates implies a common mechanism for protein misfolding diseases. *Nature*, *416*(6880), 507–511.

Calafate, S., Buist, A., Miskiewicz, K., Vijayan, V., Daneels, G., de Strooper, B., … Moechars, D. (2015). Synaptic contacts enhance cell-to-cell tau pathology propagation. *Cell Reports*, *11*(8), 1176–1183. Available from https://doi.org/10.1016/j.celrep.2015.04.043.

Cardenas-Aguayo Mdel, C., Gomez-Virgilio, L., DeRosa, S., & Meraz-Rios, M. A. (2014). The role of tau oligomers in the onset of Alzheimer's disease neuropathology. *ACS Chemical Neuroscience*, *5*(12), 1178–1191. Available from https://doi.org/10.1021/cn500148z.

Castillo-Carranza, D. L., Sengupta, U., Guerrero-Munoz, M. J., Lasagna-Reeves, C. A., Gerson, J. E., Singh, G., … Kayed, R. (2014). Passive immunization with Tau oligomer monoclonal antibody reverses tauopathy phenotypes without affecting hyperphosphorylated neurofibrillary tangles. *The Journal of Neuroscience*, *34*(12), 4260–4272. Available from https://doi.org/10.1523/jneurosci.3192-13.2014.

Chang, E., Kim, S., Yin, H., Nagaraja, H. N., & Kuret, J. (2008). Pathogenic missense MAPT mutations differentially modulate tau aggregation propensity at nucleation and extension steps. *Journal of Neurochemistry*, *107*(4), 1113–1123. Available from https://doi.org/10.1111/j.1471-4159.2008.05692.x.

Chang, L. Y., Lowe, J., Ardiles, A., Lim, J., Grey, A. C., Robertson, K., … Acosta, M. L. (2014). Alzheimer's disease in the human eye. Clinical tests that identify ocular and visual information processing deficit as biomarkers. *Alzheimer's & Dementia*, *10*(2), 251–261. Available from https://doi.org/10.1016/j.jalz.2013.06.004.

Chavan, M., & Lennarz, W. (2006). The molecular basis of coupling of translocation and N-glycosylation. *Trends in Biochemical Sciences*, *31*(1), 17–20. Available from https://doi.org/10.1016/j.tibs.2005.11.010.

Chiasseu, M., Cueva Vargas, J. L., Destroismaisons, L., Vande Velde, C., Leclerc, N., & Di Polo, A. (2016). Tau accumulation, altered phosphorylation, and missorting promote neurodegeneration in glaucoma. *Journal of Neuroscience*, *36*(21), 5785–5798. Available from https://doi.org/10.1523/jneurosci.3986-15.2016.

Chimon, S., Shaibat, M. A., Jones, C. R., Calero, D. C., Aizezi, B., & Ishii, Y. (2007). Evidence of fibril-like beta-sheet structures in a neurotoxic amyloid intermediate of Alzheimer's beta-amyloid. *Nature Structural & Molecular Biology*, *14*(12), 1157–1164. Available from https://doi.org/10.1038/nsmb1345.

Chittoor-Vinod, V. G., Lee, S., Judge, S. M., & Notterpek, L. (2015). Inducible HSP70 is critical in preventing the aggregation and enhancing the processing of PMP22. *ASN Neuro*, *7*(1). Available from https://doi.org/10.1177/1759091415569909, 1759091415569909.

Chu, W., Zhou, D., Gaba, V., Liu, J., Li, S., Peng, X., & Mach, R. H. (2015). Design, synthesis, and characterization of 3-(benzylidene)indolin-2-one derivatives as ligands for alpha-synuclein fibrils. *Journal of Medicinal Chemistry*, *58*(15), 6002–6017. Available from https://doi.org/10.1021/acs.jmedchem.5b00571.

Ciechanover, A., & Kwon, Y. T. (2015). Degradation of misfolded proteins in neurodegenerative diseases: Therapeutic targets and strategies. *Experimental & Molecular Medicine*, *47*, e147. Available from https://doi.org/10.1038/emm.2014.117.

Crouch, P. J., Tew, D. J., Du, T., Nguyen, D. N., Caragounis, A., Filiz, G., … White, A. R. (2009). Restored degradation of the Alzheimer's amyloid-beta peptide by targeting amyloid formation. *Journal of Neurochemistry*, *108*(5), 1198–1207. Available from https://doi.org/10.1111/j.1471-4159.2009.05870.x.

Cuervo, A. M., & Wong, E. (2014). Chaperone-mediated autophagy: Roles in disease and aging. *Cell Research*, *24*(1), 92–104. Available from https://doi.org/10.1038/cr.2013.153.

Decker, H., Lo, K. Y., Unger, S. M., Ferreira, S. T., & Silverman, M. A. (2010). Amyloid-beta peptide oligomers disrupt axonal transport through an NMDA receptor-dependent mechanism that is mediated by glycogen synthase kinase 3beta in primary cultured hippocampal neurons. *The Journal of Neuroscience*, *30*(27), 9166–9171. Available from https://doi.org/10.1523/jneurosci.1074-10.2010.

DeJesus-Hernandez, M., Mackenzie, I. R., Boeve, B. F., Boxer, A. L., Baker, M., Rutherford, N. J., … Rademakers, R. (2011). Expanded GGGGCC hexanucleotide repeat in noncoding region of C9ORF72 causes chromosome 9p-linked FTD and ALS. *Neuron*, *72*(2), 245–256. Available from https://doi.org/10.1016/j.neuron.2011.09.011.

Demuro, A., Mina, E., Kayed, R., Milton, S. C., Parker, I., & Glabe, C. G. (2005). Calcium dysregulation and membrane disruption as a ubiquitous neurotoxic mechanism of soluble amyloid oligomers. *Journal of Biological*

Chemistry, 280(17), 17294–17300. Available from https://doi.org/10.1074/jbc.M500997200.

Di, J., Cohen, L. S., Corbo, C. P., Phillips, G. R., El Idrissi, A., & Alonso, A. D. (2016). Abnormal tau induces cognitive impairment through two different mechanisms: Synaptic dysfunction and neuronal loss. *Scientific Reports, 6*, 20833. Available from https://doi.org/10.1038/srep20833.

Díaz-Villanueva, J. F., Díaz-Molina, R., & García-González, V. (2015). Protein folding and mechanisms of proteostasis. *International Journal of Molecular Sciences, 16*(8), 17193–17230. Available from https://doi.org/10.3390/ijms160817193.

Dickey, C. A., Kamal, A., Lundgren, K., Klosak, N., Bailey, R. M., Dunmore, J., ... Petrucelli, L. (2007). The high-affinity HSP90-CHIP complex recognizes and selectively degrades phosphorylated tau client proteins. *Journal of Clinical Investigation, 117*(3), 648–658. Available from https://doi.org/10.1172/jci29715.

Dodart, J. C., Bales, K. R., Gannon, K. S., Greene, S. J., DeMattos, R. B., Mathis, C., ... Paul, S. M. (2002). Immunization reverses memory deficits without reducing brain Abeta burden in Alzheimer's disease model. *Nature Neuroscience, 5*(5), 452–457. Available from https://doi.org/10.1038/nn842.

Eberling, J. L., Dave, K. D., & Frasier, M. A. (2013). alpha-synuclein imaging: A critical need for Parkinson's disease research. *Journal of Parkinson's Disease, 3*(4), 565–567. Available from https://doi.org/10.3233/jpd-130247.

Echeverria, P. C., Bernthaler, A., Dupuis, P., Mayer, B., & Picard, D. (2011). An interaction network predicted from public data as a discovery tool: Application to the Hsp90 molecular chaperone machine. *PLoS ONE, 6*(10), e26044. Available from https://doi.org/10.1371/journal.pone.0026044.

Eisele, Y. S., Monteiro, C., Fearns, C., Encalada, S. E., Wiseman, R. L., Powers, E. T., & Kelly, J. W. (2015). Targeting protein aggregation for the treatment of degenerative diseases. *Nature Reviews Drug Discovery, 14*(11), 759–780. Available from https://doi.org/10.1038/nrd4593.

Eisenberg, D., & Jucker, M. (2012). The amyloid state of proteins in human diseases. *Cell, 148*(6), 1188–1203. Available from https://doi.org/10.1016/j.cell.2012.02.022.

Emadi, S., Barkhordarian, H., Wang, M. S., Schulz, P., & Sierks, M. R. (2007). Isolation of a human single chain antibody fragment against oligomeric alpha-synuclein that inhibits aggregation and prevents alpha-synuclein-induced toxicity. *Journal of Molecular Biology, 368*(4), 1132–1144. Available from https://doi.org/10.1016/j.jmb.2007.02.089.

Emmanouilidou, E., Stefanis, L., & Vekrellis, K. (2010). Cell-produced alpha-synuclein oligomers are targeted to, and impair, the 26S proteasome. *Neurobiology of Aging, 31*(6), 953–968. Available from https://doi.org/10.1016/j.neurobiolaging.2008.07.008.

Falsone, A., & Falsone, S. F. (2015). Legal but lethal: Functional protein aggregation at the verge of toxicity. *Frontiers in Cellular Neuroscience, 9*, 45. Available from https://doi.org/10.3389/fncel.2015.00045.

Fang, Y.-S., Tsai, K.-J., Chang, Y.-J., Kao, P., Woods, R., Kuo, P.-H., ... Chen, Y.-R. (2014). Full-length TDP-43 forms toxic amyloid oligomers that are present in frontotemporal lobar dementia-TDP patients. *Nature Communications, 5*, 4824. Available from http://dx.doi.org/10.1038/ncomms5824. Available from http://www.nature.com/articles/ncomms5824#supplementary-information.

Ferreira, S. T., & Klein, W. L. (2011). The Abeta oligomer hypothesis for synapse failure and memory loss in Alzheimer's disease. *Neurobiology of Learning and Memory, 96*(4), 529–543. Available from https://doi.org/10.1016/j.nlm.2011.08.003.

Fink, A. L. (1998). Protein aggregation: Folding aggregates, inclusion bodies and amyloid. *Folding and Design, 3*(1), R9–R23. Available from https://doi.org/10.1016/S1359-0278(98)00002-9.

Fitzpatrick, A. W. P., Debelouchina, G. T., Bayro, M. J., Clare, D. K., Caporini, M. A., Bajaj, V. S., ... Dobson, C. M. (2013). Atomic structure and hierarchical assembly of a cross-β amyloid fibril. *Proceedings of the National Academy of Sciences of the United States of America, 110*(14), 5468–5473. Available from https://doi.org/10.1073/pnas.1219476110.

Fujiwara, S., Matsumoto, F., & Yonezawa, Y. (2003). Effects of salt concentration on association of the amyloid protofilaments of hen egg white lysozyme studied by time-resolved neutron scattering. *Journal of Molecular Biology, 331*(1), 21–28.

Gao, X., Carroni, M., Nussbaum-Krammer, C., Mogk, A., Nillegoda, N. B., Szlachcic, A., ... Bukau, B. (2015). Human Hsp70 disaggregase reverses Parkinson's-linked alpha-synuclein amyloid fibrils. *Molecular Cell, 59*(5), 781–793. Available from https://doi.org/10.1016/j.molcel.2015.07.012.

Garriga-Canut, M., Agustin-Pavon, C., Herrmann, F., Sanchez, A., Dierssen, M., Fillat, C., & Isalan, M. (2012). Synthetic zinc finger repressors reduce mutant huntingtin expression in the brain of R6/2 mice. *Proceedings of the National Academy of Sciences of the United States of America, 109*(45), E3136–3145. Available from https://doi.org/10.1073/pnas.1206506109.

Garwood, C. J., Pooler, A. M., Atherton, J., Hanger, D. P., & Noble, W. (2011). Astrocytes are important mediators of Aβ-induced neurotoxicity and tau phosphorylation in primary culture. *Cell Death & Disease*, *2*(6), e167. Available from https://doi.org/10.1038/cddis.2011.50.

Gendron, T. F., & Petrucelli, L. (2011). Rodent models of TDP-43 proteinopathy: Investigating the mechanisms of TDP-43-mediated neurodegeneration. *Journal of Molecular Neuroscience*, *45*(3), 486–499. Available from https://doi.org/10.1007/s12031-011-9610-7.

Geula, C., & Mesulam, M. M. (1995). Cholinesterases and the pathology of Alzheimer disease. *Alzheimer Disease and Associated Disorders*, *9*(Suppl 2), 23–28.

Gong, C. X., Lidsky, T., Wegiel, J., Zuck, L., Grundke-Iqbal, I., & Iqbal, K. (2000). Phosphorylation of microtubule-associated protein tau is regulated by protein phosphatase 2A in mammalian brain. Implications for neurofibrillary degeneration in Alzheimer's disease. *Journal of Biological Chemistry*, *275* (8), 5535–5544.

Goossens, K., Haelewyn, J., Meersman, F., De Ley, M., & Heremans, K. (2003). Pressure- and temperature-induced unfolding and aggregation of recombinant human interferon-gamma: A Fourier transform infrared spectroscopy study. *Biochemical Journal*, *370*(Pt 2), 529–535. Available from https://doi.org/10.1042/BJ20020717.

Guglielmotto, M., Monteleone, D., Piras, A., Valsecchi, V., Tropiano, M., Ariano, S., ... Tamagno, E. (2014). Aβ1-42 monomers or oligomers have different effects on autophagy and apoptosis. *Autophagy*, *10*(10), 1827–1843. Available from https://doi.org/10.4161/auto.30001.

Gursky, O., & Aleshkov, S. (2000). Temperature-dependent beta-sheet formation in beta-amyloid Abeta(1–40) peptide in water: Uncoupling beta-structure folding from aggregation. *Biochimica et Biophysica Acta*, *1476*(1), 93–102.

Hori, Y., Takeda, S., Cho, H., Wegmann, S., Shoup, T. M., Takahashi, K., ... Hudry, E. (2015). A Food and Drug Administration-approved asthma therapeutic agent impacts amyloid beta in the brain in a transgenic model of Alzheimer disease. *Journal of Biological Chemistry*, *290*(4), 1966–1978. Available from https://doi.org/10.1074/jbc.M114.586602.

Hwang, H., Zhang, J., Chung, K. A., Leverenz, J. B., Zabetian, C. P., Peskind, E. R., ... Zhang, J. (2010). Glycoproteomics in neurodegenerative diseases. *Mass Spectrometry Reviews*, *29*(1), 79–125. Available from https://doi.org/10.1002/mas.20221.

Ikram, M. K., Cheung, C. Y., Wong, T. Y., & Chen, C. P. (2012). Retinal pathology as biomarker for cognitive impairment and Alzheimer's disease. *Journal of Neurology, Neurosurgery, and Psychiatry*, *83*(9), 917–922. Available from https://doi.org/10.1136/jnnp-2011-301628.

James, O. G., Doraiswamy, P. M., & Borges-Neto, S. (2015). PET imaging of tau pathology in Alzheimer's disease and tauopathies. *Frontiers in Neurology*, *6*, 38. Available from https://doi.org/10.3389/fneur.2015.00038.

Jicha, G. A., Bowser, R., Kazam, I. G., & Davies, P. (1997). Alz-50 and MC-1, a new monoclonal antibody raised to paired helical filaments, recognize conformational epitopes on recombinant tau. *Journal of Neuroscience Research*, *48*(2), 128–132.

Kadavath, H., Hofele, R. V., Biernat, J., Kumar, S., Tepper, K., Urlaub, H., ... Zweckstetter, M. (2015). Tau stabilizes microtubules by binding at the interface between tubulin heterodimers. *Proceedings of the National Academy of Sciences of the United States of America*, *112*(24), 7501–7506. Available from https://doi.org/10.1073/pnas.1504081112.

Kiddle, S. J., Sattlecker, M., Proitsi, P., Simmons, A., Westman, E., Bazenet, C., ... Dobson, R. J. (2014). Candidate blood proteome markers of Alzheimer's disease onset and progression: A systematic review and replication study. *Journal of Alzheimer's Disease*, *38*(3), 515–531. Available from https://doi.org/10.3233/jad-130380.

Kim, Y. E., Hipp, M. S., Bracher, A., Hayer-Hartl, M., & Hartl, F. U. (2013). Molecular chaperone functions in protein folding and proteostasis. *Annual Review of Biochemistry*, *82*, 323–355. Available from https://doi.org/10.1146/annurev-biochem-060208-092442.

Kim, Y. S., Randolph, T. W., Seefeldt, M. B., & Carpenter, J. F. (2006). High-pressure studies on protein aggregates and amyloid fibrils. *Methods in Enzymology*, *413*, 237–253. Available from https://doi.org/10.1016/s0076-6879(06)13013-x.

Knauer, M. F., Soreghan, B., Burdick, D., Kosmoski, J., & Glabe, C. G. (1992). Intracellular accumulation and resistance to degradation of the Alzheimer amyloid A4/beta protein. *Proceedings of the National Academy of Sciences of the United States of America*, *89*(16), 7437–7441.

Knowles, T. P. J., Shu, W., Devlin, G. L., Meehan, S., Auer, S., Dobson, C. M., & Welland, M. E. (2007). Kinetics and thermodynamics of amyloid formation from direct measurements of fluctuations in fibril mass. *Proceedings of the National Academy of Sciences of the United States of America*, *104*(24), 10016–10021. Available from https://doi.org/10.1073/pnas.0610659104.

Knowles, T. P. J., Vendruscolo, M., & Dobson, C. M. (2014). The amyloid state and its association with protein misfolding diseases. *Nature Reviews Molecular Cell Biology*, *15*(6), 384–396. Available from https://doi.org/10.1038/nrm3810.

Koga, H., & Cuervo, A. M. (2011). Chaperone-mediated autophagy dysfunction in the pathogenesis of neurodegeneration. *Neurobiology of Disease*, *43*(1), 29–37. Available from https://doi.org/10.1016/j.nbd.2010.07.006.

Kohyama, K., & Matsumoto, Y. (2015). Alzheimer's disease and immunotherapy: What is wrong with clinical trials. *ImmunoTargets and Therapy*, *4*, 27–34. Available from https://doi.org/10.2147/itt.s49923.

Kontsekova, E., Zilka, N., Kovacech, B., Novak, P., & Novak, M. (2014). First-in-man tau vaccine targeting structural determinants essential for pathological tau-tau interaction reduces tau oligomerisation and neurofibrillary degeneration in an Alzheimer's disease model. *Alzheimer's Research & Therapy*, *6*(4), 44. Available from https://doi.org/10.1186/alzrt278.

Koo, E. H., Lansbury, P. T., Jr., & Kelly, J. W. (1999). Amyloid diseases: Abnormal protein aggregation in neurodegeneration. *Proceedings of the National Academy of Sciences of the United States of America*, *96*(18), 9989–9990.

Kotilinek, L. A., Bacskai, B., Westerman, M., Kawarabayashi, T., Younkin, L., Hyman, B. T., ... Ashe, K. H. (2002). Reversible memory loss in a mouse transgenic model of Alzheimer's disease. *The Journal of Neuroscience*, *22*(15), 6331–6335. Available from https://doi.org/20026675.

Kowalewski, T., & Holtzman, D. M. (1999). In situ atomic force microscopy study of Alzheimer's beta-amyloid peptide on different substrates: New insights into mechanism of beta-sheet formation. *Proceedings of the National Academy of Sciences of the United States of America*, *96*(7), 3688–3693.

Koyama, A., Okereke, O. I., Yang, T., Blacker, D., Selkoe, D. J., & Grodstein, F. (2012). Plasma amyloid-beta as a predictor of dementia and cognitive decline: A systematic review and meta-analysis. *Archives of Neurology*, *69* (7), 824–831. Available from https://doi.org/10.1001/archneurol.2011.1841.

Krebs, M. R. H., Devlin, G. L., & Donald, A. M. (2009). Amyloid fibril-like structure underlies the aggregate structure across the pH range for β-lactoglobulin. *Biophysical Journal*, *96*(12), 5013–5019. Available from https://doi.org/10.1016/j.bpj.2009.03.028.

Kumar, S., Tepper, K., Kaniyappan, S., Biernat, J., Wegmann, S., Mandelkow, E.-M., ... Mandelkow, E. (2014). Stages and conformations of the tau repeat domain during aggregation and its effect on neuronal toxicity. *Journal of Biological Chemistry*, *289*(29), 20318–20332. Available from https://doi.org/10.1074/jbc.M114.554725.

Lacor, P. N., Buniel, M. C., Chang, L., Fernandez, S. J., Gong, Y., Viola, K. L., ... Klein, W. L. (2004). Synaptic targeting by Alzheimer's-related amyloid beta oligomers. *The Journal of Neuroscience*, *24*(45), 10191–10200. Available from https://doi.org/10.1523/jneurosci.3432-04.2004.

Lacor, P. N., Buniel, M. C., Furlow, P. W., Clemente, A. S., Velasco, P. T., Wood, M., ... Klein, W. L. (2007). Abeta oligomer-induced aberrations in synapse composition, shape, and density provide a molecular basis for loss of connectivity in Alzheimer's disease. *The Journal of Neuroscience*, *27*(4), 796–807. Available from https://doi.org/10.1523/jneurosci.3501-06.2007.

Lambert, M. P., Barlow, A. K., Chromy, B. A., Edwards, C., Freed, R., Liosatos, M., ... Klein, W. L. (1998). Diffusible, nonfibrillar ligands derived from Abeta1-42 are potent central nervous system neurotoxins. *Proceedings of the National Academy of Sciences of the United States of America*, *95*(11), 6448–6453.

Lashuel, H. A., Overk, C. R., Oueslati, A., & Masliah, E. (2013). The many faces of α-synuclein: From structure and toxicity to therapeutic target. *Nature Reviews Neuroscience*, *14*(1), 38–48. Available from https://doi.org/10.1038/nrn3406.

Leak, R. K. (2014). Heat shock proteins in neurodegenerative disorders and aging. *Journal of Cell Communication and Signaling*, *8*(4), 293–310. Available from https://doi.org/10.1007/s12079-014-0243-9.

Lee, K. M., & MacLean, A. G. (2015). New advances on glial activation in health and disease. *World Journal of Virology*, *4*(2), 42–55. Available from https://doi.org/10.5501/wjv.v4.i2.42.

Levine, Z. A., Larini, L., LaPointe, N. E., Feinstein, S. C., & Shea, J.-E. (2015). Regulation and aggregation of intrinsically disordered peptides. *Proceedings of the National Academy of Sciences of the United States of America*, *112*(9), 2758–2763. Available from https://doi.org/10.1073/pnas.1418155112.

Liou, Y. C., Sun, A., Ryo, A., Zhou, X. Z., Yu, Z. X., Huang, H. K., ... Lu, K. P. (2003). Role of the prolyl isomerase Pin1 in protecting against age-dependent neurodegeneration. *Nature*, *424*(6948), 556–561. Available from https://doi.org/10.1038/nature01832.

Luo, J., Warmlander, S. K., Graslund, A., & Abrahams, J. P. (2014). Non-chaperone proteins can inhibit aggregation and cytotoxicity of Alzheimer amyloid beta peptide. *Journal of Biological Chemistry*, *289*(40), 27766–27775. Available from https://doi.org/10.1074/jbc.M114.574947.

Majbour, N. K., Vaikath, N. N., van Dijk, K. D., Ardah, M. T., Varghese, S., Vesterager, L. B., ... El-Agnaf, O. M. (2016). Oligomeric and phosphorylated alpha-synuclein as potential CSF biomarkers for Parkinson's disease. *Molecular Neurodegeneration*, *11*, 7. Available from https://doi.org/10.1186/s13024-016-0072-9.

Mandelkow, E.-M., & Mandelkow, E. (2012). Biochemistry and Cell Biology of Tau Protein in Neurofibrillary Degeneration. *Cold Spring Harbor Perspectives in Medicine*, *2*(7), a006247. Available from https://doi.org/10.1101/cshperspect.a006247.

Mandler, M., Santic, R., Gruber, P., Cinar, Y., Pichler, D., Funke, S. A., ... Mattner, F. (2015). Tailoring the antibody response to aggregated Ass using novel Alzheimer-vaccines. *PLoS ONE*, *10*(1), e0115237. Available from https://doi.org/10.1371/journal.pone.0115237.

Mandler, M., Valera, E., Rockenstein, E., Mante, M., Weninger, H., Patrick, C., ... Masliah, E. (2015). Active immunization against alpha-synuclein ameliorates the degenerative pathology and prevents demyelination in a model of multiple system atrophy. *Molecular Neurodegeneration*, *10*, 10. Available from https://doi.org/10.1186/s13024-015-0008-9.

Marquez, R. T., & Xu, L. (2012). Bcl-2:Beclin 1 complex: Multiple, mechanisms regulating autophagy/apoptosis toggle switch. *American Journal of Cancer Research*, *2*(2), 214–221.

Martin, L. J., Pardo, C. A., Cork, L. C., & Price, D. L. (1994). Synaptic pathology and glial responses to neuronal injury precede the formation of senile plaques and amyloid deposits in the aging cerebral cortex. *The American Journal of Pathology*, *145*(6), 1358–1381.

Matamoros, A. J., & Baas, P. W. (2016). Microtubules in health and degenerative disease of the nervous system. *Brain Research Bulletin*, *126*(Pt 3), 217–225. Available from https://doi.org/10.1016/j.brainresbull.2016.06.016.

Mattoo, R. U. H., & Goloubinoff, P. (2014). Molecular chaperones are nanomachines that catalytically unfold misfolded and alternatively folded proteins. *Cellular and Molecular Life Sciences*, *71*(17), 3311–3325. Available from https://doi.org/10.1007/s00018-014-1627-y.

Mayer, M. P., & Bukau, B. (2005). Hsp70 chaperones: Cellular functions and molecular mechanism. *Cellular and Molecular Life Sciences*, *62*(6), 670–684. Available from https://doi.org/10.1007/s00018-004-4464-6.

Mazzatenta, A., Pokorski, M., Sartucci, F., Domenici, L., & Di Giulio, C. (2015). Volatile organic compounds (VOCs) fingerprint of Alzheimer's disease. *Respiratory Physiology & Neurobiology*, *209*, 81–84. Available from https://doi.org/10.1016/j.resp.2014.10.001.

Medina, M., & Avila, J. (2014). The role of extracellular Tau in the spreading of neurofibrillary pathology. *Frontiers in Cellular Neuroscience*, *8*, 113. Available from https://doi.org/10.3389/fncel.2014.00113.

Meli, G., Krako, N., Manca, A., Lecci, A., & Cattaneo, A. (2013). Intrabodies for protein interference in Alzheimers disease. *Journal of Biological Regulators and Homeostatic Agents*, *27*(2 Suppl), 89–105.

Messer, A. (2016). Immunotherapy on experimental models for Huntington's disease. In M. Ingelsson, & L. Lannfelt (Eds.), *Immunotherapy and Biomarkers in Neurodegenerative Disorders* (pp. 139–150). New York, NY: Springer New York.

Min, S. W., Chen, X., Tracy, T. E., Li, Y., Zhou, Y., Wang, C., ... Gan, L. (2015). Critical role of acetylation in tau-mediated neurodegeneration and cognitive deficits. *Nature Medicine*, *21*(10), 1154–1162. Available from https://doi.org/10.1038/nm.3951.

Moran, S. D., & Zanni, M. T. (2014). How to get insight into amyloid structure and formation from infrared spectroscopy. *The Journal of Physical Chemistry Letters*, *5*(11), 1984–1993. Available from https://doi.org/10.1021/jz500794d.

Mori, K., Arzberger, T., Grasser, F. A., Gijselinck, I., May, S., Rentzsch, K., ... Edbauer, D. (2013). Bidirectional transcripts of the expanded C9orf72hexanucleotide repeat are translated into aggregating dipeptide repeat proteins. *Acta Neuropathologica*, *126*(6), 881–893. Available from https://doi.org/10.1007/s00401-013-1189-3.

Myeku, N., Clelland, C. L., Emrani, S., Kukushkin, N. V., Yu, W. H., & Goldberg, A. L. (2016). Tau-driven 26S proteasome impairment and cognitive dysfunction can be prevented early in disease by activating cAMP-PKA signaling. *Nature Medicine*, *22*(1), 46–53. Available from https://doi.org/10.1038/nm.4011.

Nagai, Y., & Popiel, H. A. (2008). Conformational changes and aggregation of expanded polyglutamine proteins as therapeutic targets of the polyglutamine diseases: Exposed beta-sheet hypothesis. *Current Pharmaceutical Design*, *14*(30), 3267–3279.

Naito, A., & Kawamura, I. (2007). Solid-state NMR as a method to reveal structure and membrane-interaction of amyloidogenic proteins and peptides. *Biochimica et Biophysica Acta (BBA)—Biomembranes*, *1768*(8), 1900–1912. Available from https://doi.org/10.1016/j.bbamem.2007.03.025.

Narayan, P., Orte, A., Clarke, R. W., Bolognesi, B., Hook, S., Ganzinger, K. A., ... Klenerman, D. (2011). The extracellular chaperone clusterin sequesters oligomeric forms of the amyloid-beta(1–40) peptide. *Nature Structural & Molecular Biology*, *19*(1), 79–83. Available from https://doi.org/10.1038/nsmb.2191.

Nasica-Labouze, J., & Mousseau, N. (2012). Kinetics of amyloid aggregation: a study of the GNNQQNY prion sequence. *PLOS Computational Biology*, *8*(11), e1002782. Available from https://doi.org/10.1371/journal.pcbi.1002782.

Nillegoda, N. B., Kirstein, J., Szlachcic, A., Berynskyy, M., Stank, A., Stengel, F., ... Bukau, B. (2015). Crucial HSP70 co-chaperone complex unlocks metazoan protein disaggregation. *Nature, 524*(7564), 247–251. Available from https://doi.org/10.1038/nature14884.

Nilson, A. N., English, K. C., Gerson, J. E., Barton Whittle, T., Nicolas Crain, C., Xue, J., … Kayed, R. (2016). Tau oligomers associate with inflammation in the brain and retina of tauopathy mice and in neurodegenerative diseases. *Journal of Alzheimer's Disease, 55*(3), 1083–1099. Available from https://doi.org/10.3233/JAD-160912.

Novak, P., Schmidt, R., Kontsekova, E., Zilka, N., Kovacech, B., Skrabana, R., … Novak, M. (2017). Safety and immunogenicity of the tau vaccine AADvac1 in patients with Alzheimer's disease: A randomised, double-blind, placebo-controlled, phase 1 trial. *The Lancet Neurology, 16*(2), 123–134. Available from https://doi.org/10.1016/s1474-4422(16)30331-3.

Oertel, W., & Schulz, J. B. (2016). Current and experimental treatments of Parkinson disease: A guide for neuroscientists. *Journal of Neurochemistry, 139*(Suppl 1), 325–337. Available from https://doi.org/10.1111/jnc.13750.

Olsson, B., Lautner, R., Andreasson, U., Ohrfelt, A., Portelius, E., Bjerke, M., … Zetterberg, H. (2016). CSF and blood biomarkers for the diagnosis of Alzheimer's disease: A systematic review and meta-analysis. *The Lancet Neurology, 15*(7), 673–684. Available from https://doi.org/10.1016/s1474-4422(16)00070-3.

Panza, F., Solfrizzi, V., & Seripa, D. (2016). Tau-centric targets and drugs in clinical development for the treatment of Alzheimer's disease. *BioMed Research International, 2016*, 3245935. Available from https://doi.org/10.1155/2016/3245935.

Patterson, K. R., Remmers, C., Fu, Y., Brooker, S., Kanaan, N. M., Vana, L., … Binder, L. I. (2011). Characterization of prefibrillar Tau oligomers in vitro and in Alzheimer disease. *Journal of Biological Chemistry, 286*(26), 23063–23076. Available from https://doi.org/10.1074/jbc.M111.237974.

Pavese, N., & Brooks, D. J. (2009). Imaging neurodegeneration in Parkinson's disease. *Biochimica et Biophysica Acta, 1792*(7), 722–729. Available from https://doi.org/10.1016/j.bbadis.2008.10.003.

Pecho-Vrieseling, E., Rieker, C., Fuchs, S., Bleckmann, D., Esposito, M. S., Botta, P., … Di Giorgio, F. P. (2014). Transneuronal propagation of mutant huntingtin contributes to non-cell autonomous pathology in neurons. *Nature Neuroscience, 17*(8), 1064–1072. Available from https://doi.org/10.1038/nn.3761.

Pedersen, J. T., & Sigurdsson, E. M. (2015). Tau immunotherapy for Alzheimer's disease. *Trends in Molecular Medicine, 21*(6), 394–402. Available from https://doi.org/10.1016/j.molmed.2015.03.003.

Perez-Grijalba, V., Pesini, P., Allue, J. A., Sarasa, L., Montanes, M., Lacosta, A. M., … Sarasa, M. (2015). Abeta1-17 is a major amyloid-beta fragment isoform in cerebrospinal fluid and blood with possible diagnostic value in Alzheimer's disease. *Journal of Alzheimer's Disease, 43*(1), 47–56. Available from https://doi.org/10.3233/jad-140156.

Powers, E. T., Morimoto, R. I., Dillin, A., Kelly, J. W., & Balch, W. E. (2009). Biological and chemical approaches to diseases of proteostasis deficiency. *Annual Review of Biochemistry, 78*, 959–991. Available from https://doi.org/10.1146/annurev.biochem.052308.114844.

Reisberg, B., Doody, R., Stoffler, A., Schmitt, F., Ferris, S., & Mobius, H. J. (2003). Memantine in moderate-to-severe Alzheimer's disease. *The New England Journal of Medicine, 348*(14), 1333–1341. Available from https://doi.org/10.1056/NEJMoa013128.

Rikhvanov, E. G., Romanova, N. V., & Chernoff, Y. O. (2007). Chaperone effects on prion and nonprion aggregates. *Prion, 1*(4), 217–222.

Risacher, S. L., & Saykin, A. J. (2013). Neuroimaging biomarkers of neurodegenerative diseases and dementia. *Seminars in Neurology, 33*(4), 386–416. Available from https://doi.org/10.1055/s-0033-1359312.

Rizzu, P., Van Swieten, J. C., Joosse, M., Hasegawa, M., Stevens, M., Tibben, A., … Heutink, P. (1999). High prevalence of mutations in the microtubule-associated protein tau in a population study of frontotemporal dementia in the Netherlands. *American Journal of Human Genetics, 64*(2), 414–421. Available from https://doi.org/10.1086/302256.

Roberti, M. J., Fölling, J., Celej, M. S., Bossi, M., Jovin, Thomas M., & Jares-Erijman, Elizabeth A. (2012). Imaging nanometer-sized α-synuclein aggregates by superresolution fluorescence localization microscopy. *Biophysical Journal, 102*(7), 1598–1607. Available from https://doi.org/10.1016/j.bpj.2012.03.010.

Ross, C. A., & Poirier, M. A. (2004). Protein aggregation and neurodegenerative disease. *Nature Medicine, 10*, Suppl, S10–17. Available from https://doi.org/10.1038/nm1066.

Saha, R. N., & Pahan, K. (2006). HATs and HDACs in neurodegeneration: A tale of disconcerted acetylation homeostasis. *Cell Death & Differentiation, 13*(4), 539–550. Available from https://doi.org/10.1038/sj.cdd.4401769.

Salminen, A., Ojala, J., Kauppinen, A., Kaarniranta, K., & Suuronen, T. (2009). Inflammation in Alzheimer's disease: Amyloid-β oligomers trigger innate immunity defence via pattern recognition receptors. *Progress in Neurobiology, 87*(3), 181–194. Available from https://doi.org/10.1016/j.pneurobio.2009.01.001.

Sandberg, A., Luheshi, L. M., Sollvander, S., Pereira de Barros, T., Macao, B., Knowles, T. P., … Hard, T. (2010). Stabilization of neurotoxic Alzheimer amyloid-beta oligomers by protein engineering. *Proceedings of the*

National Academy of Sciences of the United States of America, *107*(35), 15595–15600. Available from https://doi.org/10.1073/pnas.1001740107.

Santacruz, K., Lewis, J., Spires, T., Paulson, J., Kotilinek, L., Ingelsson, M., ... Ashe, K. H. (2005). Tau suppression in a neurodegenerative mouse model improves memory function. *Science*, *309*(5733), 476–481. Available from https://doi.org/10.1126/science.1113694.

Santhoshkumar, P., & Sharma, K. K. (2004). Inhibition of amyloid fibrillogenesis and toxicity by a peptide chaperone. *Molecular and Cellular Biochemistry*, *267*(1–2), 147–155.

Santos, S. D., Cardoso, I., Magalhaes, J., & Saraiva, M. J. (2007). Impairment of the ubiquitin-proteasome system associated with extracellular transthyretin aggregates in familial amyloidotic polyneuropathy. *The Journal of Pathology*, *213*(2), 200–209. Available from https://doi.org/10.1002/path.2224.

Sawada, M., Imamura, K., & Nagatsu, T. (2006). Role of cytokines in inflammatory process in Parkinson's disease. *Journal of Neural Transmission. Supplementum*, *70*, 373–381.

Scheidt, H. A., Morgado, I., Rothemund, S., & Huster, D. (2012). Dynamics of amyloid beta fibrils revealed by solid-state NMR. *Journal of Biological Chemistry*, *287*(3), 2017–2021. Available from https://doi.org/10.1074/jbc.M111.308619.

Schneeberger, A., Mandler, M., Mattner, F., & Schmidt, W. (2012). Vaccination for Parkinson's disease. *Parkinsonism & Related Disorders*, *18*(Suppl 1), S11–S13. Available from https://doi.org/10.1016/s1353-8020(11)70006-2.

Scholl, M., Lockhart, S. N., Schonhaut, D. R., O'Neil, J. P., Janabi, M., Ossenkoppele, R., ... Jagust, W. J. (2016). PET imaging of tau deposition in the aging human brain. *Neuron*, *89*(5), 971–982. Available from https://doi.org/10.1016/j.neuron.2016.01.028.

Scott, J. D., Li, S. W., Brunskill, A. P., Chen, X., Cox, K., Cumming, J. N., ... Stamford, A. W. (2016). Discovery of the 3-imino-1,2,4-thiadiazinane 1,1-dioxide derivative verubecestat (MK-8931)-A beta-site amyloid precursor protein cleaving enzyme 1 inhibitor for the treatment of Alzheimer's disease. *Journal of Medicinal Chemistry*, *59*(23), 10435–10450. Available from https://doi.org/10.1021/acs.jmedchem.6b00307.

Sengupta, U., Guerrero-Munoz, M. J., Castillo-Carranza, D. L., Lasagna-Reeves, C. A., Gerson, J. E., Paulucci-Holthauzen, A. A., ... Kayed, R. (2015). Pathological interface between oligomeric alpha-synuclein and tau in synucleinopathies. *Biological Psychiatry*, *78*(10), 672–683. Available from https://doi.org/10.1016/j.biopsych.2014.12.019.

Sengupta, U., Nilson, A. N., & Kayed, R. (2016). The role of amyloid-β oligomers in toxicity, propagation, and immunotherapy. *EBioMedicine*, *6*, 42–49. Available from https://doi.org/10.1016/j.ebiom.2016.03.035.

Sevigny, J., Chiao, P., Bussiere, T., Weinreb, P. H., Williams, L., Maier, M., ... Sandrock, A. (2016). The antibody aducanumab reduces Abeta plaques in Alzheimer's disease. *Nature*, *537*(7618), 50–56. Available from https://doi.org/10.1038/nature19323.

Shankar, G. M., Li, S., Mehta, T. H., Garcia-Munoz, A., Shepardson, N. E., Smith, I., ... Selkoe, D. J. (2008). Amyloid-beta protein dimers isolated directly from Alzheimer's brains impair synaptic plasticity and memory. *Nature Medicine*, *14*(8), 837–842. Available from https://doi.org/10.1038/nm1782.

Shin, J. W., Kim, K. H., Chao, M. J., Atwal, R. S., Gillis, T., MacDonald, M. E., ... Lee, J. M. (2016). Permanent inactivation of Huntington's disease mutation by personalized allele-specific CRISPR/Cas9. *Human Molecular Genetics*. Available from https://doi.org/10.1093/hmg/ddw286.

Sinnen, B. L., Bowen, A. B., Gibson, E. S., & Kennedy, M. J. (2016). Local and use-dependent effects of beta-amyloid oligomers on NMDA receptor function revealed by optical quantal analysis. *The Journal of Neuroscience*, *36*(45), 11532–11543. Available from https://doi.org/10.1523/jneurosci.1603-16.2016.

Sorg, C., & Grothe, M. J. (2015). The complex link between amyloid and neuronal dysfunction in Alzheimer's disease. *Brain*, *138*(Pt 12), 3472–3475. Available from https://doi.org/10.1093/brain/awv302.

Spencer, B., Desplats, P. A., Overk, C. R., Valera-Martin, E., Rissman, R. A., Wu, C., ... Masliah, E. (2016). Reducing endogenous alpha-synuclein mitigates the degeneration of selective neuronal populations in an Alzheimer's disease transgenic mouse model. *The Journal of Neuroscience*, *36*(30), 7971–7984. Available from https://doi.org/10.1523/jneurosci.0775-16.2016.

Steyn, F. J., Tolle, V., Chen, C., & Epelbaum, J. (2016). Neuroendocrine regulation of growth hormone secretion. *Comprehensive Physiology*, *6*(2), 687–735. Available from https://doi.org/10.1002/cphy.c150002.

Su, F., Bai, F., & Zhang, Z. (2016). Inflammatory cytokines and Alzheimer's disease: A review from the perspective of genetic polymorphisms. *Neuroscience Bulletin*, *32*(5), 469–480. Available from https://doi.org/10.1007/s12264-016-0055-4.

Sudol, K. L., Mastrangelo, M. A., Narrow, W. C., Frazer, M. E., Levites, Y. R., Golde, T. E., ... Bowers, W. J. (2009). Generating differentially targeted amyloid-β specific intrabodies as a passive vaccination strategy for Alzheimer's disease. *Molecular Therapy*, *17*(12), 2031–2040. Available from https://doi.org/10.1038/mt.2009.174.

Takahashi, T., Katada, S., & Onodera, O. (2010). Polyglutamine diseases: Where does toxicity come from? What is toxicity? Where are we going? *Journal of Molecular Cell Biology*, *2*(4), 180–191. Available from https://doi.org/10.1093/jmcb/mjq005.

Theunis, C., Crespo-Biel, N., Gafner, V., Pihlgren, M., Lopez-Deber, M. P., Reis, P., … Muhs, A. (2013). Efficacy and safety of a liposome-based vaccine against protein Tau, assessed in tau. P301L mice that model tauopathy. *PLoS ONE*, *8*(8), e72301. Available from https://doi.org/10.1371/journal.pone.0072301.

Tseng, B. P., Green, K. N., Chan, J. L., Blurton-Jones, M., & LaFerla, F. M. (2008). Aβ inhibits the proteasome and enhances amyloid and tau accumulation. *Neurobiology of Aging*, *29*(11), 1607–1618. Available from https://doi.org/10.1016/j.neurobiolaging.2007.04.014.

Tsubuki, S., Takaki, Y., & Saido, T. C. (2003). Dutch, Flemish, Italian, and Arctic mutations of APP and resistance of Abeta to physiologically relevant proteolytic degradation. *The Lancet*, *361*(9373), 1957–1958.

Tycko, R. (2011). Solid state NMR studies of amyloid fibril structure. *Annual Review of Physical Chemistry*, *62*, 279–299. Available from https://doi.org/10.1146/annurev-physchem-032210-103539.

Tycko, R. (2015). Amyloid polymorphism: Structural basis and neurobiological relevance. *Neuron*, *86*(3), 632–645. Available from https://doi.org/10.1016/j.neuron.2015.03.017.

Tycko, R. (2016). Molecular structure of aggregated amyloid-β: Insights from solid state nuclear magnetic resonance. *Cold Spring Harbor Perspectives in Medicine*, *6* (8). Available from https://doi.org/10.1101/cshperspect.a024083, a024083.

Uversky, V. N. (2010). Targeting intrinsically disordered proteins in neurodegenerative and protein dysfunction diseases: Another illustration of the D(2) concept. *Expert Review of Proteomics*, *7*(4), 543–564. Available from https://doi.org/10.1586/epr.10.36.

Uversky, V. N. (2013). A decade and a half of protein intrinsic disorder: Biology still waits for physics. *Protein Science: A Publication of the Protein Society*, *22*(6), 693–724. Available from https://doi.org/10.1002/pro.2261.

Valera, E., & Masliah, E. (2013). Immunotherapy for neurodegenerative diseases: Focus on alpha-synucleinopathies. *Pharmacology & Therapeutics*, *138*(3), 311–322. Available from https://doi.org/10.1016/j.pharmthera.2013.01.013.

Varadi, M., Vranken, W., Guharoy, M., & Tompa, P. (2015). Computational approaches for inferring the functions of intrinsically disordered proteins. *Frontiers in Molecular Biosciences*, *2*, 45. Available from https://doi.org/10.3389/fmolb.2015.00045.

Villar-Pique, A., Espargaro, A., Ventura, S., & Sabate, R. (2016). In vivo amyloid aggregation kinetics tracked by time-lapse confocal microscopy in real-time. *Biotechnology Journal*, *11*(1), 172–177. Available from https://doi.org/10.1002/biot.201500252.

Vogelsberg-Ragaglia, V., Bruce, J., Richter-Landsberg, C., Zhang, B., Hong, M., Trojanowski, J. Q., & Lee, V. M. Y. (2000). Distinct FTDP-17 missense mutations in tau produce tau aggregates and other pathological phenotypes in transfected CHO cells. *Molecular Biology of the Cell*, *11* (12), 4093–4104.

von Bergen, M., Barghorn, S., Biernat, J., Mandelkow, E.-M., & Mandelkow, E. (2005). Tau aggregation is driven by a transition from random coil to beta sheet structure. *Biochimica et Biophysica Acta (BBA)—Molecular Basis of Disease*, *1739*(2–3), 158–166. Available from https://doi.org/10.1016/j.bbadis.2004.09.010.

Walsh, C. T., Garneau-Tsodikova, S., & Gatto, G. J., Jr. (2005). Protein posttranslational modifications: The chemistry of proteome diversifications. *Angewandte Chemie International Edition in English*, *44*(45), 7342–7372. Available from https://doi.org/10.1002/anie.200501023.

Wang, C. C., & Tsou, C. L. (1998). Enzymes as chaperones and chaperones as enzymes. *FEBS Letters*, *425*(3), 382–384.

Wang, W.-Y., Tan, M.-S., Yu, J.-T., & Tan, L. (2015). Role of pro-inflammatory cytokines released from microglia in Alzheimer's disease. *Annals of Translational Medicine*, *3*(10), 136. Available from https://doi.org/10.3978/j.issn.2305-5839.2015.03.49.

Wang, Y., Martinez-Vicente, M., Kruger, U., Kaushik, S., Wong, E., Mandelkow, E. M., … Mandelkow, E. (2009). Tau fragmentation, aggregation and clearance: The dual role of lysosomal processing. *Human Molecular Genetics*, *18*(21), 4153–4170. Available from https://doi.org/10.1093/hmg/ddp367.

Ward, S. M., Himmelstein, D. S., Lancia, J. K., & Binder, L. I. (2012). Tau oligomers and tau toxicity in neurodegenerative disease. *Biochemical Society Transactions*, *40*(4), 667–671. Available from https://doi.org/10.1042/bst20120134.

Ward, S. M., Himmelstein, D. S., Lancia, J. K., Fu, Y., Patterson, K. R., & Binder, L. I. (2013). TOC1: Characterization of a selective oligomeric tau antibody. *Journal of Alzheimer's Disease*, *37*(3), 593–602. Available from https://doi.org/10.3233/jad-131235.

Wetzel, R., Shivaprasad, S., & Williams, A. D. (2007). Plasticity of amyloid fibrils. *Biochemistry*, *46*(1), 1–10. Available from https://doi.org/10.1021/bi0620959.

Williams, Thomas L., Johnson, Benjamin R. G., Urbanc, B., Jenkins, A., Toby, A., Connell, Simon D. A., & Serpell, Louise C. (2011). Aβ42 oligomers, but not fibrils, simultaneously bind to and cause damage to ganglioside-containing lipid membranes. *Biochemical Journal*, *439*(1), 67–77. Available from https://doi.org/10.1042/bj20110750.

Winblad, B., Andreasen, N., Minthon, L., Floesser, A., Imbert, G., Dumortier, T., … Graf, A. (2012). Safety, tolerability, and antibody response of active Abeta immunotherapy with CAD106 in patients with Alzheimer's disease: Randomised, double-blind, placebo-controlled, first-in-human study. *The Lancet Neurology, 11*(7), 597–604. Available from https://doi.org/10.1016/s1474-4422(12)70140-0.

Winklhofer, K. F., Tatzelt, J., & Haass, C. (2008). The two faces of protein misfolding: Gain- and loss-of-function in neurodegenerative diseases. *The EMBO Journal, 27*(2), 336–349. Available from https://doi.org/10.1038/sj.emboj.7601930.

Wolfe, K. J., & Cyr, D. M. (2011). Amyloid in neurodegenerative diseases: Friend or foe? *Seminars in Cell & Developmental Biology, 22*(5), 476–481. Available from https://doi.org/10.1016/j.semcdb.2011.03.011.

Yanamandra, K., Kfoury, N., Jiang, H., Mahan, T. E., Ma, S., Maloney, S. E., ... Holtzman, D. M. (2013). Anti-tau antibodies that block tau aggregate seeding in vitro markedly decrease pathology and improve cognition in vivo. *Neuron, 80*(2), 402–414. Available from https://doi.org/10.1016/j.neuron.2013.07.046.

Yang, T., Li, S., & Xu, H. (2017). Large soluble oligomers of amyloid β-protein from Alzheimer brain are far less neuroactive than the smaller oligomers to which they dissociate. *The Journal of Neuroscience, 37*(1), 152–163. Available from https://doi.org/10.1523/jneurosci.1698-16.2017.

Yeo, J. M., Waddell, B., Khan, Z., & Pal, S. (2015). A systematic review and meta-analysis of (18)F-labeled amyloid imaging in Alzheimer's disease. *Alzheimer's & Dementia (Amsterdam, The Netherlands), 1*(1), 5–13. Available from https://doi.org/10.1016/j.dadm.2014.11.004.

Yerbury, J. J., Gower, D., Vanags, L., Roberts, K., Lee, J. A., & Ecroyd, H. (2013). The small heat shock proteins αB-crystallin and Hsp27 suppress SOD1 aggregation in vitro. *Cell Stress & Chaperones, 18*(2), 251–257. Available from https://doi.org/10.1007/s12192-012-0371-1.

Yerbury, J. J., Ooi, L., Dillin, A., Saunders, D. N., Hatters, D. M., Beart, P. M., ... Ecroyd, H. (2016). Walking the tightrope: Proteostasis and neurodegenerative disease. *Journal of Neurochemistry, 137*(4), 489–505. Available from https://doi.org/10.1111/jnc.13575.

Yin, R. H., Tan, L., Jiang, T., & Yu, J. T. (2014). Prion-like mechanisms in Alzheimer's disease. *Current Alzheimer Research, 11*(8), 755–764.

Zempel, H., & Mandelkow, E. M. (2015). Tau missorting and spastin-induced microtubule disruption in neurodegeneration: Alzheimer disease and hereditary spastic paraplegia. *Molecular Neurodegeneration, 10*, 68. Available from https://doi.org/10.1186/s13024-015-0064-1.

Zhang, H., Griggs, A., Rochet, J.-C., & Stanciu, Lia A. (2013). In vitro study of α-synuclein protofibrils by cryo-EM suggests a Cu(2+)-dependent aggregation pathway. *Biophysical Journal, 104*(12), 2706–2713. Available from https://doi.org/10.1016/j.bpj.2013.04.050.

Zhang, X., Jin, H., Padakanti, P. K., Li, J., Yang, H., Fan, J., ... Tu, Z. (2014). Radiosynthesis and in vivo evaluation of two PET radioligands for imaging alpha-synuclein. *Applied Sciences (Basel), 4*(1), 66–78. Available from https://doi.org/10.3390/app4010066.

CHAPTER

10

Heat Shock Proteins and Protein Quality Control in Alzheimer's Disease

Fred W. van Leeuwen[1] and Harm H. Kampinga[2]

[1]University of Maastricht, Maastricht, The Netherlands [2]University of Groningen, Groningen, The Netherlands

OUTLINE

GENERAL INTRODUCTION

From gene transcription to posttranslational modification, several quality checks are essential before a protein is ready for biological action, locally or distantly via axonal transport in neurons. The accurate folding and control over levels of many proteins must be tightly regulated both spatially and temporally. To achieve this, the cell possesses a network of different protein quality control (PQC) systems for cotranslational protein folding via molecular

DOI: https://doi.org/10.1016/B978-0-12-811304-2.00010-9

chaperones, as the first line of defense against protein misfolding and aggregation (Balchin, Hayer-Hartl, & Hartl, 2016). Subsequently, accurate protein degradation, encompassing the ubiquitin–proteasome system (UPS) and the autophagosomal–lysosomal system, is the second line of defense. Recent data also suggest an additional PQC pathway in which misfolded proteins are excreted actively after encapsulation at the endoplasmic reticulum, a process dubbed MAPS (Misfolded Associated Protein Secretion; Lee, Takahama, Zhang, Tomarev, & Ye, 2016b).

In each of the PQC systems, an array of heat shock proteins (HSPs) are required to chaperone and direct substrates toward their correct fate and to avoid off-pathway reactions that might lead to harmful accumulations of protein aggregates (Balchin et al., 2016). In the present review, we will focus on the cross-talk between the UPS and the HSPs, concepts that are in line with the current idea that Alzheimer's disease (AD) is in an asymptomatic cellular phase long before it can be translated into clinically relevant observation and interventions (De Strooper & Karran, 2016).

Most age-related neurodegenerative diseases are hallmarked by the existence of pathological protein depositions in the brain, for example, extracellular plaques and neurofibrillary tangles in AD (Soto, 2003). Furthermore, aging and neurodegeneration in AD are often accompanied by a functionally impaired UPS (Cecarini et al., 2016) that may be both a cause and one of the detrimental consequences of the aggregation (Balch, Morimoto, Dillin, & Kelly, 2008). Synaptic plasticity is affected in AD and involves metabolism of at least 5%–8% of brain proteins each day (Dennissen, Kholod, & van Leeuwen, 2012; Rosenberg et al., 2014).

As stated, central to PQC are HSPs that interact with mis- or unfolded proteins. They generally bind to exposed hydrophobic, aggregation-prone regions, preventing their self-association and aggregation. Whilst the best-known outcome of such interaction is that the chaperone next assists in the (re)folding of its bound client protein, HSPs also can assist in supporting its clients proteasomal or autophagosomal degradation. In addition, HSPs can assist in the disassembly of protein complexes and provide the driving force for translocation of proteins across membranes (Kampinga & Craig, 2010).

Human cells contain over 100 different HSP (depending on how defined), divided over multiple HSP families (Hageman & Kampinga, 2009; Kampinga & Bergink, 2016; Kampinga et al., 2009; Li, Soroka, & Buchner, 2012) (Table 10.1). These include the main chaperones that are directly capable of substrate binding: HSPA1-9 (Hsp70s), HSPB1-10 (small HSPs), HSPC1-4 (Hsp90s), HSPD1 (Hsp60), HSPH1-4 (Hsp110s), CCT1-9 (chaperonins), DNAJA1-4, DNAJB1-12, and DNAJC1-30 (Hsp40s). In addition, these chaperones can act together and have additional coregulators such as (1) HSPE1 (Hsp10) for HSPD1; (2) DNAJs, HSPHs, BAGs, HSPBP1s, Hop, and CHIP for HSPA; or (3) AHSA1 (Aha1), PTGES3 (p23), several peptidyl-prolyl-isomerases, and many more for HSPCs.

Chaperonins, Hsp70 and Hsp90, machines are ATP-fueled, whereas small HSPs are referred to as ATP-independent chaperones that cannot actively drive substrate release after binding (often referred to as "holdases"). Several cochaperones may also have holdase activity. These holdases also are often found to be crucial to the sequestration of misfolded proteins into membrane-free compartments (Alberti & Hyman, 2016) for temporal storage during stress, until conditions restore and client processing can proceed (Kaganovich, Kopito, & Frydman, 2008; Mogk & Bukau, 2017; Nollen et al., 2001). Release from the holdases is generally thought to require assistance of the ATP-fueled chaperones in activities that than often are referred to as "foldases" or "disaggregases," although these terms are

TABLE 10.1 Overview on Different Hsp Machines and Their Cochaperone or Coregulators

	Central Component		Cochaperone/Coregulators		
Main Chaperone	**Common (nick) Name**	**NCBI Gene Name**	**Common (nick) Name**	**NCBI Gene Name**	**Effects on Main Chaperon, Other Remarks**
Small HSP	Small HSP, crystallins	HSPB1–10			ATP-independent; clients often thought to be transferred to Hsp70 machines
Chaperonins	Hsp60	HSPD1	Hsp10	HSPE1	
	TriC	CCT 1–9			
HSP70	Hsp70, Hsc70	HSPA 1–9	Hsp40	DNAJA 1–4	Substrate delivery, ATPase stimulation
				DNAJB 1–12	
				DNAJC 1–30	
			NEF	HSPBP1	ADP/ATP exchange
				BAP BAG1-6	ADP/ATP exchange + hub
				HSPH1-4	ADP/ATP exchange + cobinding to clients
			Hip		Stabilizing ADP state
			CHIP	STUB1	E3 ubiquitin ligase
			Hop		Does not regulate Hsp70 cycle but is hub between Hsp70 and Hsp90
Hsp90	Hsp90	HSPC 1–4	Hop		Does not regulate Hsp90 cycle but is hub between Hsp70 and Hsp90
			Aha1	AHSA1	ATPase stimulation, Hsp90 conformational changes
			p23	PTGES3	ATPase inhibition + cobinding to clients
			Cdc37	CDC37	ATPase inhibition + cobinding to kinase clients
			Cbp1/Melusin	CHORDC1	Cobinding to NLR receptor clients and Sgt1
			NubC	NUDC	Cobinding to specific clients

(*Continued*)

TABLE 10.1 (Continued)

	Central Component		Cochaperone/Coregulators		
Main Chaperone	Common (nick) Name	NCBI Gene Name	Common (nick) Name	NCBI Gene Name	Effects on Main Chaperon, Other Remarks
			Fkbp52	FKBP4	Peptidyl-prolyl-isomerases, maturation of certain clients
			Fkbp51	FKBP5	
			Cyp40	PPID	
			AIP	AIP	Maturation of certain clients
			CHIP	STUB1	E3 ubiquitin ligase
			PP5	PPP5C	Phosphatase
			Tpr2	DNAJC7	Also binds and regulates Hsp70
			Sgt1	SUGT1	Cobinding to NLR receptor clients and CHORDC1
			Unc45	UNC45B	Maturation of myosin fibers
			Ttc4	TTC4	Maturation of nuclear transport clients
			Tah1	RPAP3	Complex with piH1 and Hsp90

Each HSP family has a number of family members (indicated by the numbers after their names) and has members that are constitutively expressed as well as members that are strictly regulated by HSF-1, the main transcription factor driving the heat shock response (largely based on Kampinga et al., 2009; Li et al., 2012).

not exclusive or definitive for the action of a specific HSP in terms of determining the fate of the client.

In the present review, we will focus on the role of specific HSPs in protein folding and in the interaction with the UPS for protein degradation. In particular, we will discuss what is known about the role of HSPs with respect to the two aggregate-forming proteins in AD, Aß peptides and tau, how HSPs may be targets for prevention of their aggregation, and which HSP (or regulator) may be rate limiting for such manipulations.

DE NOVO FOLDING, REFOLDING AND DEGRADATION: TRIAGING

In order to understand the role of HSPs in the handling of protein clients, it must first be stated that there is a clear difference between the de novo cotranslational folding of proteins and handling folding of stress-denatured proteins. Triaging and PQC at the ribosome is therefore dependent on different factors than when it concerns dealing with partially unfolded mature proteins. This includes the requirement for different chaperones for nascent chain folding and refolding after stress (Albanèse, Yam, Baughman, Parnot, & Frydman, 2006; Dekker, Kampinga, & Bergink, 2015; for more extensive discussions on this matter).

Cotranslational folding of proteins largely occurs vectorially and localized, which a priori requires chaperones assembled at or close to the ribosomes (Balchin et al., 2016; Brandman & Hegde, 2016). This will not be further discussed in the present review. In contrast, handling mature proteins is more chaotic and can occur anywhere, without preassembled chaperone machineries present. For degradation of mature proteins, we also further need to distinguish regulated degradation (via controlled posttranscriptional modifications, such as ubiquitylation or ubiquitin-like tagging) from triaging of stress-induced unfolding or oxidation of proteins. In addition, when focusing on Aß peptides and tau that aggregate in AD, we must realize that (1) Aß peptides are not unfolded proteins and primarily form aggregates extracellularly and (2) that Tau aggregation is not initiated by intrinsic unfolding but triggered by posttranslational modifications (e.g., phosphorylation). This means that they both do not share characteristics of canonical HSP clients, bringing yet another challenge to the PQC system. It also implies that for searches toward HSPs as potential targets in AD, a strategy should be considered that does not try to improve HSP-dependent folding. Rather, such strategies should be aimed at understanding and boosting of HSP-dependent degradation of Aß peptides and tau either before or after they aggregate. Alternatively, HSP-directed strategies could be directed at protecting from the downstream effects of the aggregated species, e.g., alleviating secondary collapse of PQC due to trapping of its components within the aggregates or by binding to the aggregates to neutralize them.

In the next two sections, we will provide a general overview of the role of HSPs in proteasomal and autophagosomal degradation of mature proteins (Fig. 10.1), before more specifically addressing what is known about HSPs in AD.

HSP AND PROTEASOMAL DEGRADATION

Degradation of intact cellular proteins is largely orchestrated by ATP-dependent ubiquitination and subsequent degradation by the 26S proteasome. Ubiquitin-dependent degradation is sequentially mediated by ubiquitin-activating enzymes (E1: two human genes), ubiquitin-conjugating enzymes (E2: at least 40 genes), and over 600 different E3 ubiquitin ligases that are considered to convey specificity to the system (Bett, 2016). The polyubiquitinated substrates are next recognized by the 19S proteasomal

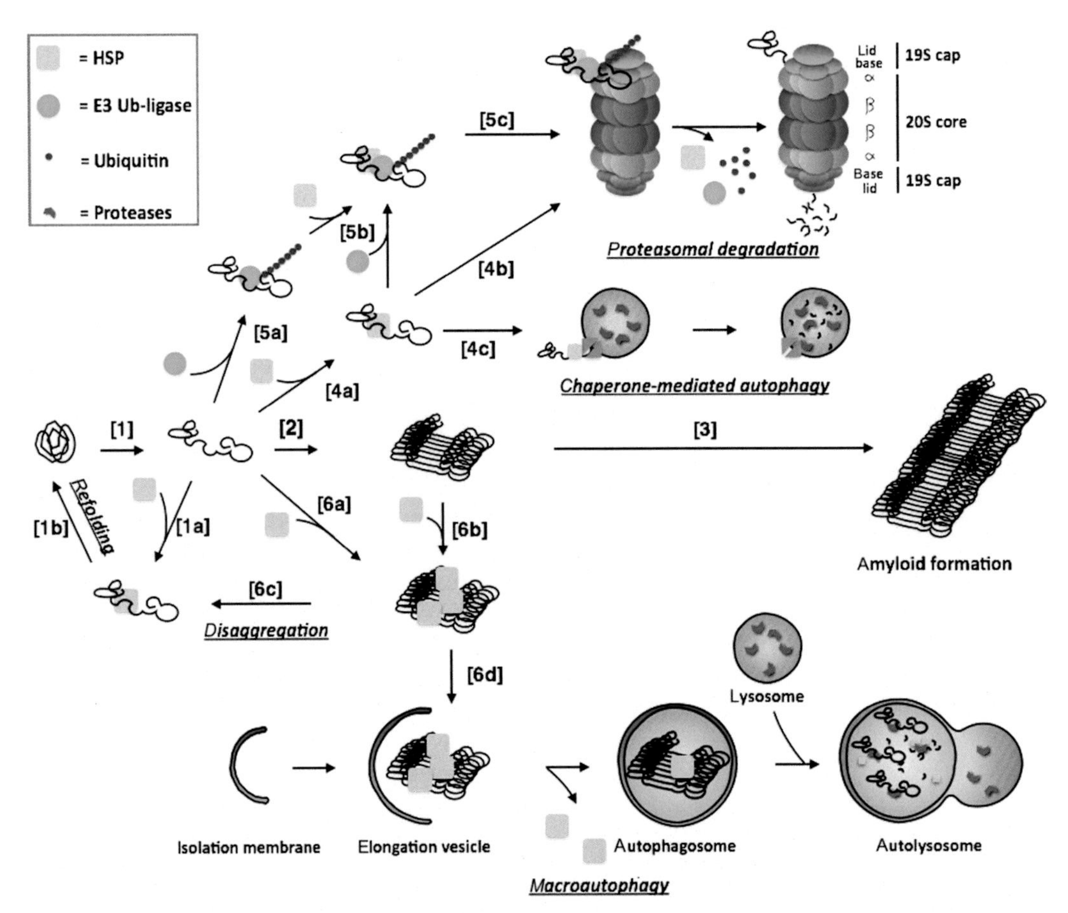

FIGURE 10.1 **Schematic overview of the roles that different heat shock proteins play in protein quality control.** When folded proteins, for whatever reason, unfold [1], they are at risk to form aggregates [2] that can result in amyloid formation [3]. The canonical role of chaperones is to bind to the unfolded intermediates [1a] and assist in their refolding [1b]. If refolding is not feasible, the same chaperone-unfolded protein interaction [4a] also can lead to their ubiquitin-independent [4b] or -dependent [5b,c] proteasomal degradation or to their degradation via chaperone-mediated autophagy [4c]. HSP may also play a role after ubiquitination of the un- or misfolded substrate recognized first by E3 ligases [5a,b] and followed by proteasomal degradation [5c]. Alternatively, HSP can bind to intermediate aggregates [6b] and disaggregate them [6c] for refolding [1a] or degradation (not indicated) or bind to intermediate aggregates [6b] for delivery to autophagosomes [6d] for autophagosomal–lysosomal degradation. A final possibility is that aggregates/amyloids are disposed via "exospheres," but the role of chaperones in this process is yet understood, and therefore this is not further indicated. The precise triaging of all these processes is not yet clearly understood, but will likely be dictated by the folding- or aggregation stage of the client and the different members of the (co)chaperones that recognize the exposed sequences or domains (see text for further details).

subunit, where they undergo deubiquitination by deubiquitinating enzymes [DUBs, at least 102 (Bett, 2016)], followed by unfolding and translocation into the proteolytic chamber of the 20S proteasomal subunit, where the actual proteolysis occurs (Pickart & Cohen, 2004). In addition to ubiquitin-dependent degradation, there are several ubiquitination- and/or ATP hydrolysis–independent pathways in which substrates can be targeted to the either the 26S proteasome with the canonical 19S subunit, to a different regulatory 11S cap (or PA28), or even directly into the 20S subunit. Structurally unstable, abnormal, misfolded, or highly oxidized proteins, some of which also can undergo ubiquitin-dependent aggregation when not misfolded, are mostly subject to ubiquitin-independent and 19S-independent degradation under conditions of cellular stress (Erales & Coffino, 2014; Zhou, 2006).

The existence of an interplay between the HSP network and the degradation machineries has been long known, but how it actually works still remains a mystery. Given that HSPs can both assist in the folding and degradation of even the same client suggests that, in part, the triaging or degradation depends on the folding state of the client itself upon its release from the chaperone. In other words, here folding or degradation would be dependent on a stochastic process, where in the repeated binding and release of the client to and from the HSP, the chaperones merely function to maintain the client in a nonaggregated, folding- or degradation-competent form (Kampinga & Craig, 2010) (Fig. 10.1).

In other cases, however, specific HSP coregulators may help to specifically target the client of their main chaperone. This is illustrated by the actions of DNAJB2, an Hsp70 cochaperone, that has two ubiquitin-interacting motifs and belongs to the family of ubiquitin-binding proteins (Andersen, Hofmann, & Hartmann-Petersen, 2005). DNAJB2 binds to ubiquitylated proteins which are targeted for degradation and which involves collaboration with a specific set of other Hsp70 coregulators, BAG1 and the E3-ligase CHIP (Lüders, Demand, & Höhfeld, 2000; Westhoff, Chapple, van der Spuy, Höhfeld, & Cheetham, 2005). Another possibility for chaperone-directed triaging is based on differential cochaperone binding to folding-competent and folding-incompetent states of the client. This is exemplified by work on the ER-resident DNAJs, where ERdj3 (DNAJB11) seems to favor HSPA5 (Bip)-dependent protein folding, whilst ERdj4 (DNAJB9) and ERdj5 (DNAJC10) target the same substrates for degradation (Behnke, Mann, Scruggs, Feige, & Hendershot, 2016; Ushioda et al., 2008). A large peptide library screen by the Henderson group recently related this to differences in recognition sequences for ERdj3 and ERdj4 that were either hidden or exposed (Behnke et al., 2016). Both the cytosolic and ER-related proteasomal pathways concern ubiquitin-dependent degradation in which ubiquitylation and chaperoning may either occur concomitantly or sequentially.

Yet another possibility is that specific cochaperones that present the clients to the main chaperones (input) already also bind to and recruit specific other regulators that control client release (output) and that directly interact with the proteasomes. An example is DNAJB2, in partnership with BAG1, which besides binding to and regulating Hsp70 also binds to the proteasome via its UBL domain (Westhoff et al., 2005). In this way, targeting clients to the proteasome may not per se require client ubiquitylation. Such may also be speculated for DNAJB6, another cochaperone of the Hsp70 machine, that was found to bind to structurally unstable proteins or fragments thereof that are not or are poorly ubiquitylated [e.g., the polyglutamine-containing huntingtin protein (Hipp et al., 2012)] and lead to their enhanced proteasomal degradation (Hageman et al., 2010).

HSP AND AUTOPHAGOSOMAL DEGRADATION

Autophagy refers to those pathways that lead to the elimination of cytoplasmic components by delivering them into mammalian lysosomes. The complex regulation of the various autophagic routes have been reviewed elsewhere extensively (Kaushik & Cuervo, 2012; Reggiori &

Klionsky, 2002; Rubinsztein, Bento, & Deretic, 2015), and cargo can comprise all kinds of cytoplasmic materials, including entire organelles. For the purpose of this review, we will only shortly introduce the basic concepts of autophagic routes and the roles of HSPs in directing proteins or protein aggregates as cargo for lysosomal degradation. For an extensive discussion of autophagy and its role in neurodegenerative diseases, see Chapter 11, Neurodegenerative Diseases and Autophagy by Fleming et al.

Chaperone-Mediated Autophagy

Chaperone-mediated autophagy (CMA) is a selective form of autophagy that targets cytosolic proteins carrying the KFERQ pentapeptide, a sequence found in approximately 30% of cytosolic proteins (Chiang & Dice, 1988; Dice, 1990). This pentapeptide, which functions as a degron, is recognized by cytosolic HSPA8 (hsc70) and is next delivered to the surface of the lysosomes, where it binds to LAMP-2A. This single-spanning membrane protein and a luminal Hsp70 isoform are the major components of the cargo translocation complex. Although it is assumed that for its involvement in CMA, HSPA8 is assisted by cochaperones (Agarraberes & Dice, 2001; Majeski & Dice, 2004), mechanistic details and evidence of cochaperone requirement are lacking. Wild-type Tau also can be degraded by CMA; however, the fragmented, aggregation-prone Tau that is involved in the pathogenesis of AD cannot be handled by this pathway (Wang et al., 2009).

Macroautophagy

Microautophagy involves the uptake of cytoplasm at the lysosomal surface, a process that has not been well characterized. In contrast, macroautophagy involves membrane engulfment around cargo and fusion of the final double-membrane structures with the lysosomes, where the cargo is degraded (Reggiori & Klionsky, 2002). In contrast to the UPS and CMA, autophagy mediates primarily nonselective and bulk degradation of many intracellular proteins or protein aggregates in one swoop, and information of the many regulators of this process, the ATG proteins, can be found elsewhere (see e.g., Nakatogawa, Suzuki, Kamada, & Ohsumi, 2009 and Chapter 11, Neurodegenerative Diseases and Autophagy of this volume).

So far, there is no evidence for the involvement of HSPs in the formation of autophagosomes or their fusion with lysosomes. However, there are some lines of evidence suggesting that targeting cargo (proteins, protein aggregates) to macroautophagy is assisted by chaperones (Fig. 10.1). The first evidence for chaperone involvement in macroautophagy was presented by the group of Landry (Carra, Seguin, Lambert, & Landry, 2008), who found that a multichaperone complex consisting of BAG3-HSPB8-HSP70 increases the autophagic flux and accelerated the disposal of polyglutamine proteins by macroautophagy. Later, other proteins were found to be cleared by this pathway, including ALS-related mutant SOD-1 (Crippa et al., 2010; Gamerdinger, Carra, & Behl, 2011). In fact, BAG3-HSPB8 are coupregulated under different conditions that impair or overwhelm the UPS (Gamerdinger et al., 2011; Minoia et al., 2014; Nivon et al., 2012), such as during aging (Gamerdinger et al., 2009). Whereas the cause for the increase in autophagy after upregulation of BAG3-HSPB8 has remained an enigma, it is thought that BAG3-HSPB8-HSP70-bound clients are transferred to the macroautophagy receptor protein SQSTM1/p62 for autophagic degradation. This process, often referred to as chaperone-assisted selective autophagy (Arndt et al., 2010), however, is not truly selective. Rather, it reroutes all Hsp70-bound proteasomal clients toward autophagosomal degradation under conditions where the UPS is compromised, which is why the process was coined as BIPASS "BAG-instructed proteasomal to autophagosomal switch and sorting" (Minoia et al., 2014).

The other chaperone involved in macroautophagy also involves a member of the small Hsp family, HSPB7. HSPB7 was picked up in a screen for suppressors of polyQ aggregation and was actually the most effective HSPB family member in this regard, an effect that also was seen in polyQ fly models (Vos et al., 2010). HSPB7 did not alter the autophagic flux but lost its effectiveness to suppress aggregation when autophagy was pharmacologically blocked or genetically impaired. Interestingly, HSPB7 was the strongest upregulated hit induced by a series of drugs that could rescue polyQ aggregation and toxicity induced in flies (Jimenez-Sanchez et al., 2015). HSPB7 localizes to growing puncta of polyQ, and the data suggest that it thereby decelerates aggregate growth, allowing the material to be engulfed in autophagosomes.

HSP, UPS, AND AD

The most commonly occurring form of AD is sporadic late-onset AD. Only about 1%–2% of all AD cases are due to autosomal-dominant mutations in amyloid precursor protein (APP), presenilin-1 (PSEN1), or presenilin 2 (PSEN2) (Gentier & van Leeuwen, 2015; Guerreiro, Gustafson, & Hardy, 2012). Although intracellular neurofibrillary tangles, consisting of hyperphosphorylated tau, and extracellular Aβ plaques are present in both sporadic and monogenic AD, it is unclear how neurodegeneration in AD starts and proceeds. It is disputed as to whether Aβ aggregation leads to cellular stress and results in tau hyperphosphorylation and aggregation (described as the amyloid cascade hypothesis), or whether tau hyperphosphorylation and aggregation precede Aβ accumulation (described as the tau axis hypothesis) (Götz, Eckert, Matamales, Ittner, & Liu, 2011). For the purpose of the present review, this issue will not be specifically discussed. Rather, we will separately discuss interrelations of HSP and the UPS with Aβ aggregation or tau aggregation.

Many changes in components of the PQC system have been observed in brains from AD patients. Below, we will shortly describe the phenotypical changes that have been described. In the "HSP: preventing neurodegenerative effects of Aβ and tau" section, we will address if and how HSP and the UPS can deal with the two pathogenic key players in AD, i.e., Aβ and tau, to delay disease onset or progression.

Phenotypical Changes in UPS in AD

Several lines of evidence point to a reduced UPS function in AD (Keller, Hanni, & Markesbery, 2000, 2002). As both Aβ and tau are UPS substrates (David et al., 2002; Lopez Salon, Pasquini, Besio Moreno, Pasquini, & Soto, 2003), a causal relation is easily assumed. In line, pooled genome-wide association studies and pathway analysis identified protein ubiquitination as one of the prime targets for AD [International Genomics of Alzheimer's Disease Consortium (IGAP), 2015]. Also, an ingenuity pathway analysis of a comparison of the proteomes of prefrontal cortex, hippocampus, and cerebellum in AD patients demonstrated 31 proteins to be significantly altered; these proteins had a strong interaction with the ubiquitin C signaling pathway. Although these data were interpreted as a dysfunctional UPS being a causative factor in AD (Manavalan et al., 2013), they also may reflect an early consequence of the disease. Perhaps the most direct evidence for an impaired UPS in AD pathogenesis comes from the discoveries of the accumulation of mutant ubiquitin B^{+1} (UBB^{+1}) protein in both sporadic and autosomal-dominant AD cases as well as other tauopathies. UBB^{+1} is a frameshift mutant of ubiquitin, caused by molecular misreading during transcription of the ubiquitin B (UBB) gene, resulting in the deletion of two nucleotides (van Leeuwen et al., 1998). Usually after mistranscription, nonsense mRNA decay is activated, but this requires a downstream intron that is not present in the UBB gene, allowing the escape of the UBB^{+1} transcript and resulting in its translation. Transcription errors are not rare and are seen in simple monotonic repeats such as GAGAG and can induce proteotoxic stress (e.g., Vermulst et al., 2015).

UBB^{+1} was first identified in neurons of patients with sporadic AD and Down's syndrome (who also exhibit AD-like brain pathology when middle-aged) (van Leeuwen et al., 1998) and also in brains of elderly individuals with mild cognitive impairment of sporadic AD cases, and later in autosomal forms of AD (van Leeuwen et al., 2000, 2006). The UBB^{+1} protein has a 19-amino acid residue C-terminal extension and is unable to ubiquitinate degradable targets. The UBB^{+1} protein itself cannot be deubiquitinated and, when present at high levels, causes a "dominant-negative" inhibition of the 26S proteasome both in cells and in transgenic mice (Fischer et al., 2009; Lam et al., 2000; Lindsten et al., 2002, 2003), culminating in downstream events, including impaired ER-associated degradation (Schipanski et al., 2014), mitochondrial dysfunction (Braun et al., 2015), and induction of HSPs expression (Hope et al., 2003). UBB^{+1} transgenic mice show a deficit in contextual memory in both water maze and fear conditioning paradigms (Fischer et al., 2009).

However, UBB^{+1} accumulation is not completely specific to AD, as it is also seen in other neurodegenerative conditions such as polyglutamine diseases and other tauopathies, although not in synucleinopathies (Fischer et al., 2003; de Pril et al., 2004). Nor is UBB^{+1} accumulation a likely initiator of the disease as, at least in polyQ diseases, it is observed after initiation of other protein aggregation (Seidel et al., 2012a, b). This idea is further substantiated by the findings that transgenic UBB^{+1} mice have no overt neuropathology (Fischer et al., 2009). Importantly, crossbreeding a transgenic mouse line with postnatal UBB^{+1} overexpression and moderate proteasomal inhibition with an AD mouse line (APPSwe/PS1Δexon9) (Garcia-Alloza et al., 2006) revealed a significant drop in the number of Aβ plaques and levels of soluble Aβ in the crossbreed (van Tijn et al., 2012). This implies that for initiation of extracellular Aβ plaque accumulation, full proteasomal activity is required. Together the data suggest that UBB^{+1} accumulation may occur as an early consequential event of proteasomal impairment in AD pathogenesis or in parallel as aging phenomenon. As such, it may however significantly contribute to disease pathogenesis by exacerbating a general overload of the PQC system and a self-perpetuating collapse of the cellular protein homeostasis. This opens possibilities for UBB^{+1} as a therapeutic target, as—different from modulation of other UPS proteins (e.g., E3 ligases, DUBs)—it is not involved in normal UPS functioning.

Apart from the cell biological consequences of UBB^{+1} expression, there is increasing evidence that the location where AD starts is different from the transentorhinal cortex (as also noted by Braak & Del Tredici, 2016). Indeed, in a phenotypical screen by the German Mouse Clinic, a respiratory dysfunction was seen in the UBB^{+1} transgenic mouse that is not due to lung insufficiency (Irmler et al., 2012). Therefore, we reexamined this transgenic mouse line and found UBB^{+1} immunoreactivity in the brainstem in two nuclei governing respiration. This unexpected finding was also

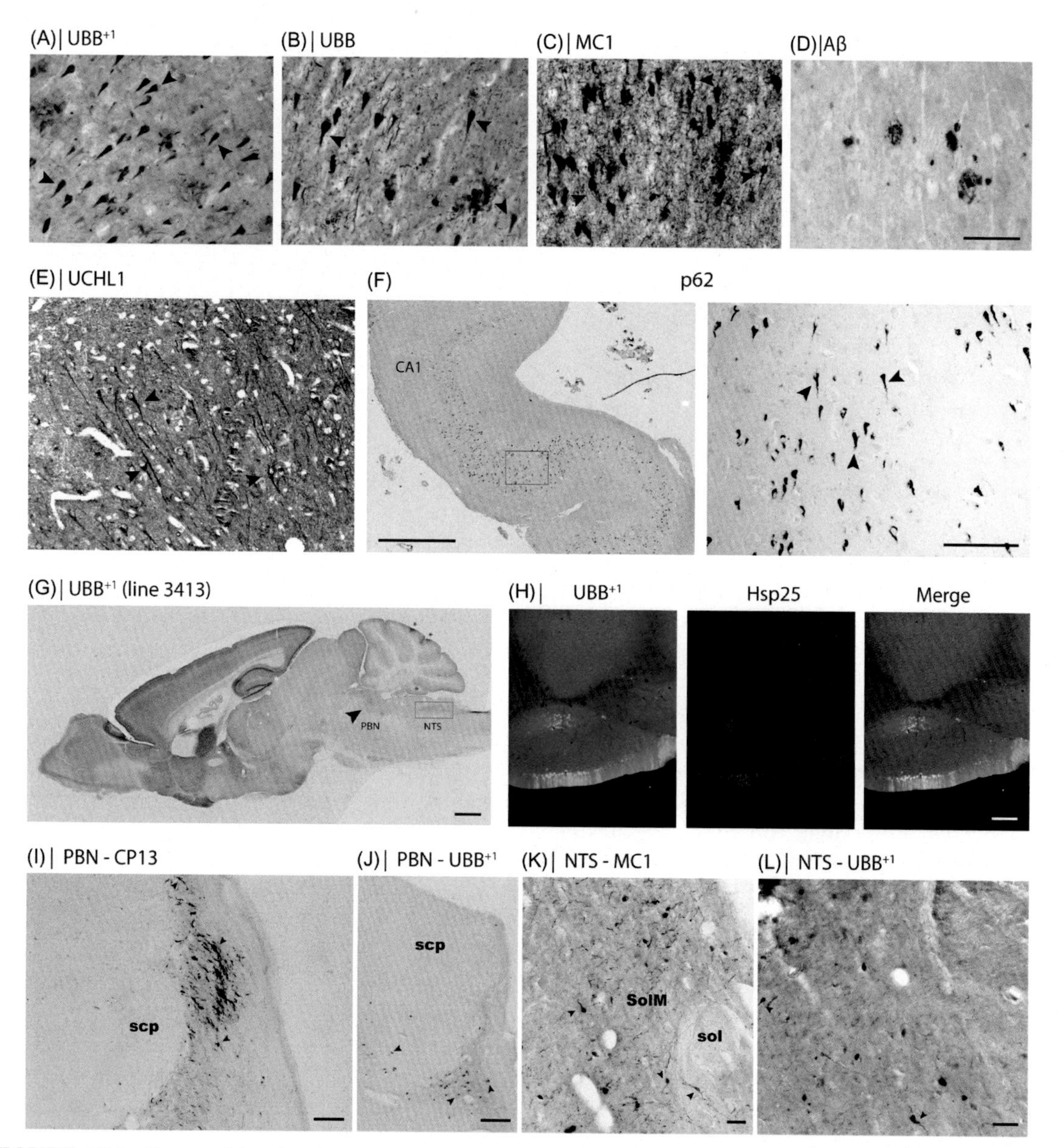

FIGURE 10.2 **Immunohistochemical representation of UPS deficits in AD, translation into transgenic mice and its validation in AD.** In AD patient (section (autopsy# 96115, Braak 5, ApoE3/3)) neurofibrillary tangles (A–C) immunoreactive for misframed ubiquitin (UBB^{+1}), wt-ubiquitin, and aberrant Tau (MC1) were shown as well as Aβ plaques (D) in 50 μm vibratome sections of the hippocampal CA1 area. Magnification bar = 100 μm. In panels E and F, UCHL1 and P62 immunoreactivity was shown in the hippocampus (CA1) of AD patient (#88028, Braak 6, ApoE3/3). Magnification bar = 100 and 400 μm for lower power, Paraffin sections of 8 μm. In panels G and H, 50 μm vibratome sections of transgenic mouse line 3413 (JAX# 008833) shows UBB^{+1} expression, not only in the forebrain, but also in the boxed area (Nucleus tractus solitarius nuclei, NTS), the lateral parabrachial nucleus (PBN, ▲) and cerebellum (*). In H colocalization of UBB^{+1} and HSP-25 was shown in Zebrin-like staining of Purkinje cells of the cerebellum; bar = 300 μm (Verheijen, Gentier, Hermes, van Leeuwen, & Hopkins, 2016). In I-L 50 μm vibratome sections of the brainstem of AD patients (# 5 and 10, Braak stage 6, van 0–6) (for details, see Gentier et al., 2015; Irmler et al., 2012). In I and J p-Tau (CP13) and UBB^{+1} staining of PBN and in NTS nuclei in K and L, MC1 and UBB^{+1} staining was shown in neurofibrillary tangles. Magnification bars in I and J 200 μm, in K 100 μm and L 25 μm. *scp*, superior cerebellar peduncle.

seen in the human brainstem of AD postmortem tissue (Irmler et al., 2012) (Fig. 10.2).

Phenotypical Changes in HSP in AD

AD brains are also characterized by a number of changes in HSP expression and distributions. (Muchowski & Wacker, 2005; Kakkar, Meister-Broekema, Minoia, Carra, & Kampinga, 2014). Strikingly, several of the intracellularly expressed HSPs, like Hsp27 (HSPB1), αB-crystalline (HSPB5), Hsp70 (HSPA1A), Grp78 (HSPA5), and HSP90, are found in extracellular amyloid plaques. How intracellular HSPs become associated with extracellular plaques still remains an enigma. Hsp27 and Hsp90 were also found to be associated with intracellular tangles in postmortem brain, which might be related to the function of these two chaperones in cytoskeletal organization and tau binding (see below).

Interestingly, significant changes were also seen in the expression of small HSPs (HSPBs) in astrocytes (Renkawek, Bosman, & Gaestel, 1993, 1994; Seidel et al., 2012b; Shinohara, Inaguma, Goto, Inagaki, & Kato, 1993; Wilhelmus et al., 2006a) and some DNAJs (Durrenberger et al., 2009) that are generally associated with astrogliosis in AD and several other neurodegenerative diseases. The precise relevance of this for AD progression is still unclear. On the one hand, this increased expression may be required for the cytoskeletal rearrangements that astrocytes undergo during the process of reactive gliosis, as HSPBs are important for cytoskeletal function and motility (Wettstein, Bellaye, Micheau, & Bonniaud, 2012). On the other hand, the increase could support functions of astrocytes in the uptake of extracellular debris and aggregates and thus could play a role in the protection against interneuronal transmission of e.g., Aβ plaques (Morales, Callegari, & Soto, 2015; Sulistio & Heese, 2016) (see also at the "Cell nonautonomous effects of HSPs" section).

HSP: PREVENTING NEURODEGENERATIVE EFFECTS OF Aβ AND TAU

As stated above, the two key aggregating polypeptides in AD are hyperphosphorylated tau and Aβ peptides. Below, we will briefly describe the aggregation characteristics of tau and Aβ peptides, discuss how some HSPs interfere with their aggregation process, and summarize the evidence on how HSP manipulations can ameliorate toxicity in cell and organismal models of AD.

HSPs and Aß Peptides

Aβ peptides are the result of APP cleavage via one of two pathways: a nonamyloidogenic pathway that leads to the generation of the most common isoform, Aβ40, or an amyloidogenic pathway that results in the generation of Aβ42 or Aβ43. AD-related mutations in APP usually affect the ratio or properties of these different Aβ species. Similarly, mutations in PSEN1 and PSEN2, which are catalytic components of the γ-secretase complex that produces Aβ, result in increased generation of the more fibrillogenic Aβ42 (De Strooper & Karran, 2016; Gentier & van Leeuwen, 2015; Götz et al., 2011; Guerreiro et al., 2012).

Increasing evidence suggests that the toxicity of Aβ is largely associated with low molecular weight oligomeric species rather than its monomeric forms or amyloid fibrils (Walsh et al., 2002) (Fig. 10.2D). This observation indicates that therapeutic strategies to reduce the toxicity associated with the aggregation process should not just aim at a generic arrest of the fibril formation but especially at the specific targeting of low oligomeric toxic species (Glabe, 2008).

However, mature fibrils may be not be inert, as they may be a main source of the smaller toxic species via fragmentation (Tipping, van Oosten-Hawle, Hewitt, & Radford, 2015). This would imply that fibril stabilization or removal may also delay the disease onset.

HSP Effects on Aβ Peptide Aggregation In Vitro: Work With Purified Proteins

In vitro, Aβ42 fibril formation involves a nucleated polymerization reaction, which proceeds from soluble monomeric peptides through nonfibrillar oligomeric species to cross β-sheet fibrils. This process is initiated by a double nucleation mechanism (primary and secondary nucleations) followed by an elongation phase (fibril growth) (Cohen et al., 2013).

A number of different HSPs have been shown to delay the amyloidogenesis of Aβ42 peptides in vitro. This includes HSPA1, HSPA5, HSPC1, and HSPA1/DNAJB1, as well as HSPB1, HSPB5, HSPB6, and HSPB8 (Kakkar et al., 2014). In particular, HSPB5 (αB-crystalline) affects the elongation phase of Aβ fibril growth (Arosio et al., 2016; Ecroyd & Carver, 2009; Mannini et al., 2012; Narayan et al., 2012; Shammas et al., 2011), most likely via binding to the elongating fibrils. In this way, it may not only prevent fibril growth but also stabilize fibrils, thus preventing shedding and secondary nucleations. Another Hsp70 cochaperone of the DNAJ family, DNAJB6, was recently found to be extremely efficient in suppressing the fibrillization of Aβ42 (Arosio et al., 2016; Månsson et al., 2014a). DNAJB6 was originally discovered as an extremely efficient suppressor of polyQ aggregation, toxicity, and neurodegeneration in cells and animal models (Hageman et al., 2010; Kakkar et al., 2016). Subsequently, it was shown to directly act on the polyQ tract, whereby it inhibited primary and secondary nucleations (Kakkar et al., 2016; Månsson et al., 2014b), most likely via interfering with H-bond formation and β-hairpin-mediated nucleation that drive polyQ aggregation (Hoop et al., 2016; Kar et al., 2013). In contrast, chaperones like HSPA1A (Hsp70) or DNAJB1 could not do so (Månsson et al., 2014b). The particular activity of DNAJB6 was recently ascribed to an S/T-rich sequence that expose hydroxyl groups well suited to interfere with such H-bonding (Kakkar et al., 2016). For Aβ peptides, H-bond formation and β-hairpin-mediated nucleation are likewise considered driving forces for amyloidogenesis (Abelein et al., 2014). In line with this, kinetic analyses of Aβ aggregation show DNAJB6 to have strong effects on primary and secondary nucleations (Arosio et al., 2016; Månsson et al., 2014a). Thus, DNAJB6 impedes formation of primary Aβ nuclei and their subsequent propagation. Compared to HSPB5, DNAJB6 is more active, with effects at much lower stoichiometric ratios. Importantly, DNAJB6 not only delays fibril formation but also binds to small Aβ42 oligomers, which are considered to be the most toxic species. Another member of the DNAJB family, DNAJB1, was also tested for its ability to inhibit Aβ42 fibril formation, and although it enhanced the ability of Hsp70 to suppress fibril formation, it was rather ineffective by itself (Evans, Wisén, & Gestwicki, 2006). This is in line with our findings that DNAJB1, unlike DNAJB6, cannot inhibit the aggregation induced by polyQ peptides, neither in vitro (Månsson et al., 2014b) nor in cells (Gillis et al., 2013). The largest dissimilarity between DNAJB1 and DNAJB6 is in their C-termini, which could also argue for an important role of the S/T-rich region in DNAJB6 in preventing Aβ42 aggregation. Indeed, the S/T region of DNAJB6 was crucial for DNAJB6 to function (see thesis Mansson 2015 @ http://www.flak.lu.se).

HSP Effects on Aβ Peptide Aggregation In Vitro: Work With Cell Models

In the classical amyloid cascade hypothesis, the accumulation of Aβ peptides into extracellular plaques is suggested as a central cause of AD (Selkoe & Hardy, 2016). Although several HSPs are also found extracellularly, specificity is unclear, as is the question of whether the abundance of extracellular HSPs is sufficient to exert chaperone effects. Regardless, extracellular HSPs (either via physiological routes or administrated exogenously) may alleviate toxicity of extracellular Aβ peptides and thus have therapeutic potential. To test this, some investigators exposed cells to purified, extracellularly added Aβ, with or without coincubation of purified HSPs, followed by evaluation of cell toxicity. Coincubation with DNAJB1, HSPB1, HSPB5, or HSPB8 protected cells against Aβ toxicity (Carnini et al., 2012; Wilhelmus et al., 2006c), which suggests that binding of these chaperones to preformed fibrils may either shield their toxic interaction with cells or/and stabilized the fibers, thus preventing shedding and generation of more toxic species.

Some reports also suggest a crucial role for intracellular Aβ in the pathogenesis of this disease and that the extracellular and intracellular pools are connected (Gouras, Almeida, & Takahashi, 2005; LaFerla, Green, & Oddo, 2007; Takahashi, Nagao, & Gouras, 2017; Wirths & Bayer, 2012). APP and its proteolytic enzymes β and γ secretases are not only localized on the plasma membrane but also in other intracellular membranes, such as the Golgi network, endosomes, and even on mitochondrial membranes. Compromised degradation of Aβ42 would then lead to accumulation of intracellular Aβ42 (Agholme & Hallbeck, 2014; Agholme, Hallbeck, Benedikz, Marcusson, & Kågedal, 2012; Domert et al., 2014). Interestingly, accumulation of intracellular Aβ was detected before the appearance of extracellular amyloid plaques in transgenic AD mice, where it was associated with the onset of cognitive impairment (Billings, Oddo, Green, McGaugh, & LaFerla, 2005; Knobloch, Konietzko, Krebs, & Nitsch, 2007). Also, in patients with AD and Down syndrome, intracellular Aβ was detected in postmortem brain (Cataldo et al., 2004). Furthermore, intracellular Aβ was suggested to be more neurotoxic than extracellular Aβ (Kienlen-Campard, Miolet, Tasiaux, & Octave, 2002), although the relevance of this must be interpreted with care, as neuronal shrinkage and dysfunction rather than cell death and neuronal loss might be the prime and earliest event linked to disease onset in AD (Gentier & van Leeuwen, 2015). Some findings even suggest that Aβ plaques are a late consequence of the intracellular Aβ peptide generation and export (Gouras et al., 2005; Gyure, Durham, Stewart, Smialek, & Troncoso, 2001; Takahashi et al., 2017), that is, intracellular Aβ somehow initiates extracellular plaques formation (Cataldo, Hamilton, & Nixon, 1994; Cataldo et al., 2004; D'Andrea, Nagele, Wang, Peterson, & Lee, 2001). If intracellular Aβ and its aggregation are indeed important in disease initiation, elevated intracellular expression of HSPs may thus also have therapeutic potential in AD.

Appropriate cellular models to evaluate HSPs on intracellular Aβ peptide aggregation and toxicity are, however, still lacking, as the free intracellular peptides seem to be very rapidly degraded. Results using Aβ-GFP fusion constructs as a model to study Aβ peptides aggregation and to discover new inhibitors for Aβ aggregation (Caine et al., 2007; Chakrabortee et al., 2012; Park, Kim, Kang, & Rhim, 2009) should be interpreted with caution, as Aβ42-GFP aggregation does not seem to reflect Aβ42 peptide aggregation. This is because (1) Aβ42-GFP, unlike the free peptide, does not form SDS-insoluble aggregates, perhaps due to GFP interfering with the required β-hairpin formation; (2) Aβ42-GFP aggregation is not seeded by exogenous Aβ42 protofibrils and fibrils; (3) suppressors like DNAJB6 that are effective in preventing required β-hairpin-driven aggregation by Aβ42 peptides via their S/T rich region did not prevent formation of amorphic Aβ42-GFP aggregates; inversely, HSPs like HSPA1A and HSPB1 that were only marginally effective in preventing aggregation by Aβ42 peptide aggregation, were effective in preventing Aβ42-GFP aggregation (our unpublished observations). Thus, solid data on whether and how intracellular Aβ aggregation may be affected by intracellular HSPs are not available. A recently developed three-dimensional human neural cell culture model of AD (Choi et al., 2014) may potentially serve as a suitable system to better evaluate this.

HSP Effects on Aβ Peptide Aggregation In Vivo: Work With Organismal Models

Only a few reports have evaluated transgenic overexpression of HSPs in in vivo Aβ models. Depletion of insulin growth factor receptor 2 (daf2 in *Caenorhabditis elegans*) as well as two of its downstream targets, leading to upregulation of HSP expression (daf16 and Hsf-1 in *C. elegans*), can ameliorate Aβ-related mobility and lifespan effects, supporting a role for intracellular events and possible intervention by chaperones in AD (Cohen, Bieschke, Perciavalle, Kelly, & Dillin, 2006). The mode of action for this could include effects on APP processing, modification of free intracellular Aβ, or its intracellular aggregation, all of which may lead to lowering of toxic oligomers. Intriguingly, one of the downstream targets of daf16, hsp16.2 (a small HSP related to the human HSPBs) alone also partially reduced amyloidogenesis and Aβ-related mobility changes (Fonte et al., 2008). Moreover, in mice containing the human Swedish double mutated APP and human presenilin-1 ΔE9 (Jankowsky et al., 2004), reduced IGF signaling protects from AD-like symptoms, and affected the type of Aβ aggregates formed (Cohen et al., 2009). Direct injection of HSF-1 [the major, stress-regulated transcription factor downstream of IGF-R1 that elevates HSP expression (Akerfelt, Morimoto, & Sistonen, 2010)] into an APP rat model (Jiang et al., 2013), genetic overexpression of HSF-1 in APP mice (Pierce et al., 2013), as well as treatment of APP mice with the HSF-1 activator celastrol (Paris et al., 2010), ameliorated disease-like effects. Finally, single overexpression of the HSF-1-regulated HSPA1 (Hoshino et al., 2011) or HSPB5 (Tóth et al., 2013) also relieve some of the AD-related features. However, these studies provide no clear evidence for whether Aβ aggregation or APP processing were directly affected by these manipulations or whether the chaperone-related protective effects were due to alleviation of secondary proteotoxic effects resulting from the primary amyloidogenic process.

In conclusion, several chaperones can interfere with the Aβ amyloidogenic process, either by preventing initiation or elongation or enhancing fiber stability. Also, protective effects in in vivo Aβ models have been observed. However, a direct connection between the two phenomena has yet to be established. A role in intracellular Aβ peptides in AD is still debated, as is a role for extracellular HSPs. The in vivo effects of HSP on AD progression may also be related to amelioration of intracellular downstream consequences of extracellular amyloid plaques, such as the disturbances of intracellular protein homeostasis, in particular effects on the UPS (see below).

HSPs and Tau

Tau is an unstructured and dynamic protein that is normally involved in stabilization of microtubules. Tau can become hyperphosphorylated by a series of kinase events that tend to reduce its affinity for microtubules, which can be restored through dephosphorylation. This reversible cycle of association and dissociation is a normal cellular process that may regulate axonal transport and other tau-dependent reactions (see Miyata, Koren, Kiray,

Dickey, & Gestwicki, 2011 for review). In AD, tau becomes modified by hyperphosphorylation and other posttranslational modifications and aggregates into filaments. Whether and how these events are causal is still a matter of debate. Aggregation can be further enhanced by proteolytic processing of tau (Götz et al., 2011; Mandelkow & Mandelkow, 2012; Miyata et al., 2011).

In postmortem brains tissues, several HSPs—small HSP in particular—have been found to be associated with neurofibrillary tangles of phosphorylated tau (Björkdahl et al., 2008), although the relevance of this is still unclear. Below, we provide a summary of the literature on HSP effects on tau aggregation and toxicity. Further details can be found in other reviews (Abisambra et al., 2011; Kakkar et al., 2014; Kosik & Shimura, 2005; Miyata et al., 2011).

HSP Effects on Tau Aggregation In Vitro: Work With Purified Proteins

Only limited in vitro work has been reported on HSP and tau binding or aggregation. HSP90 (Karagöz et al., 2014), Hsp70s (Fontaine et al., 2015; Voss, Combs, Patterson, Binder, & Gamblin, 2012), and Hsp27/HSPB1 (Shimura, Miura-Shimura, & Kosik, 2004) can bind to unstructured or hyperphosphorylated tau in vitro. In particular, Hsp70 binds to the repeat domain of tau (Fontaine et al., 2015; Voss et al., 2012) and can inhibit the aggregation of each tau isoform without affecting the normal function of tau in microtubule assembly in vitro (Voss et al., 2012).

HSP Effects on Tau Aggregation In Vitro: Work With Cell Models

In line with the cell-free data, cell models have repeatedly suggested that Hsp70 is a possible target for inhibiting tau aggregation, although its mode of action seems complex. On one hand, Hsp70 apparently can promote or stabilize tau (re)binding to microtubules (Jinwal et al., 2010a). On the other hand, and maybe in parallel, Hsp70 can assist in degrading excess tau via the UPS in combination with the E3-ligase CHIP (Petrucelli et al., 2004; Shimura, Schwartz, Gygi, & Kosik, 2004), which also was shown to be a Hsp70 cochaperone (Ballinger et al., 1999). More detailed analyses, in cells and brain-slice models of tau, showed that chemical inhibitors of the Hsp70 ATPase activity accelerate tau degradation; in contrast, activators of the Hsp70 ATPase activity stabilized tau, (Jinwal et al., 2009, 2010b; Miyata et al., 2013; Young et al., 2016). This suggest that the rate of the Hsp70 ATP cycle and "dwelling time" during which tau is associated with Hsp70 may determine the possible handling of tau (Miyata et al., 2011). However, the situation may be more complex, as increased expression of the members of the DNAJ family of Hsp70 cochaperones, DNAJA1 and DNAJB2, which stimulate the Hsp70 ATPase activity, did not decrease but rather enhanced the clearance rate of tau (Abisambra et al., 2012, 2013).

Several lines of evidence suggest Hsp90 also has potential as a target in tauopathies. Hsp90 can indirectly promote phosphorylation of tau (Tortosa et al., 2009), which could lead to an increase in pathogenic tau species and its aggregation. However, several lines of evidence support the idea that Hsp90 can, like hsp70, enhance tau clearance via CHIP-mediated, proteasomal degradation (Dickey et al., 2007). As with Hsp70, inhibition of the ATP cycle of Hsp90 may be a key to HSP90-dependent accelerated degradation of tau. For example, knockdown of p23 (Dickey et al., 2007) and Aha1 (Jinwal, Koren, & Dickey, 2013), both Hsp90 ATPase stimulators, decreases the levels of tau in cells. Also, several Hsp90/HSPC-directed drugs were found to lead to accelerated tau degradation (Luo et al., 2007; Opattova, Filipcik, Cente, & Novak, 2013). Interpretation of the latter data, however, also requires some caution, as these drugs not only inhibit Hsp90, but subsequently activate HSF-1, thus increasing levels of all stress-inducible HSPs. Which HSP is actually responsible for the effects of Hsp90 inhibitors can therefore not be discerned. Moreover, HSP90 has many additional clients beyond tau and HSF-1 from these experiments (Li et al., 2012; Picard, 2002), one of which is Akt, which also can facilitate tau degradation (Dickey et al., 2008). Targeting Hsp90 cochaperones, rather than Hsp90 itself, has also been proposed as an option (Blair, Sabbagh, & Dickey, 2014). No specific inhibitors of Aha1 have yet been identified, but chemical inhibition of p23 or Cdc37, yet another Hsp90 cofactor that regulates tau stability (Jinwal et al., 2011) is apparently feasible and could have potential in tauopathies (Blair et al., 2014). However, the in vivo specificity, effectivity, and toxicity of these drugs remain to be tested.

Finally, HspB1 (Hsp27), associated with UPS and Ubb^{+1} (Fig. 10.2), has also been suggested as a potential target for tau regulation (Shimura, Miura-Shimura, & Kosik, 2004). In this study, soluble paired helical filament (PHF) tau from AD brains was added to HCN2A cells extracellularly, and HSPB1 promoted PHF tau degradation and alleviated tau-mediated cell death. Although somewhat artificial, these studies further support the hypothesis that several chaperones delay tau aggregation and toxicity by facilitating tau degradation.

HSP Effects on Tau Aggregation In Vivo: Work With Organismal Models

Based on the cellular data, several Hsp90 inhibitors were tested for in vivo alleviation of tau aggregation and toxicity. In *Drosophila* larvae expressing human tau, Hsp90 inhibitors caused a dose-dependent reduction tau; however, these treatments failed to rescue the tau-induced larval locomotion deficits (Sinadinos et al., 2013). In a frontotemporal dementia transgenic mouse model (JNPL3, MAPT-P301L), Hsp90 inhibitors likewise reduced tau levels (Luo et al., 2007), but in this case, no functional data were reported. Intriguingly, using Tg4510 tau mice with the same mutation (MAPT-P301L) (Santacruz et al., 2005), methylene blue, among other things an Hsp70 ATPase inhibitor, not only resulted in lowering tau levels but also reduction in the rate of cognitive decline (O'Leary et al., 2010). Similarly, adeno-associated delivery of HSPB1 in Tg4510 tau mice reduced neuronal tau levels and rescued long-term potentiation deficits (Abisambra et al., 2010).

In conclusion, boosting several HSPs (HSPB1, Hsp70, Hsp90), all likely working in conjunction with the activity of the E3-ligase CHIP, could be a feasible strategy to lower tau aggregation and tangle formation and thereby delay the onset or progression of AD. Mechanistically, there are still multiple questions, e.g., how the various strategies lead to accelerated tau degradation. Despite the client promiscuity of all these chaperones and some of the drugs used, in particular of the Hsp90 inhibitors (see above), approaches such as those with the Hsp70 ATPase inhibitors could be promising (O'Leary et al., 2010; Young et al., 2016, 2017).

Cell Nonautonomous Effects of HSPs

Besides cell autonomous effects of HSPs in neurons, cell nonautonomous effects involving HSPs may affect neurodegeneration in AD. Many of these suggestions yet lack substantial direct proof, especially in mammals in vivo. Therefore, we will only briefly address these below.

Hsp's in Astrocytes/Glial

Histological studies have shown that along with degeneration of neurons, activated microglial and astrocytosis are also hallmarks AD (Osborn, Kamphuis, Wadman, & Hol, 2016) and many other neurodegenerative diseases involving protein aggregation (Pekny, Wilhelmsson, & Pekna, 2014). Astrocytes are required for neuronal survival, and abnormal astrocyte function can be a primary cause of neurodegeneration (e.g., Alexander disease) (Quinlan, Brenner, Goldman, & Messing, 2007). There is likely a delicate balance between effects of reactive (positive) and hyperreactive (negative) astrocytes on the outcome for neurodegeneration (Osborn et al., 2016; Pekny et al., 2014), but this issue is beyond the scope of this review. Here, we will only highlight a number of findings that could implicate HSP in these putative positive effects.

In Alexander disease and reactive astrocytosis in neurodegeneration in general, a striking upregulation of the expression of small HSPs in astrocytes is observed (Quinlan et al., 2007; Renkawek et al., 1993, 1994; Seidel et al., 2012b). The activated, small HSP-positive astrocytes are found at the site of Aβ deposits, and astrocytes can internalize and thereby neutralize Aβ (Nagele et al., 2004; Osborn et al., 2016; Pekny et al., 2014; Phatnani & Maniatis, 2015). Given these observations, one could speculate that the small HSPs may play a role in neuroprotection in AD by improving clearance of Aβ plaques. Some of the small HSPs, as "cytoskeletal chaperones," might be required for the drastic reorganization of the cytoskeleton (Boncoraglio, Minoia, & Carra, 2012; Carra et al., 2012; Gusev, Bogatcheva, & Marston, 2002) that occurs during astrocytosis (Osborn et al., 2016). Other chaperones or chaperone complexes that are upregulated in reactive astrocytes, e.g., HSPB8-BAG3 (Seidel et al., 2012b), a complex involved in autophagic aggregate clearance (Carra et al., 2008, 2012), may assist in uptake or autophagic clearance of plaques. Other upregulated HSPs in activated astrocytes, like the antiamyloidogenic DNAJB6 (Durrenberger et al., 2009), might serve to protect against the toxicity of internalized aggregates. Via these actions, HSPs in astrocytes may be crucial to avoiding toxic interactions of protein aggregates with neurons and/or mitigating a prion-like neuron-to-neuron transmission of Aβ plaques or extracellular tau.

Extracellular HSP—Paracrine Effects

The prime function of HSPs is maintaining intracellular protein homeostasis. However, HSPs are also found extracellularly, where they have been suggested to play a role in immunity, with relevance to cancer, autoimmune disease, and transplantation (Borges, Lang, Lopes, & Bonorino, 2016; Calderwood, Gong, & Murshid, 2016; van Eden, 1991; Shevtsov & Multhoff, 2016). In the AD brain, HSPs are found in extracellular plaques (see the "Phenotypical changes in HSP in AD" section), where they may reduce immunogenicity, prevent shedding of toxic intermediates from them, neutralize the toxic species by binding to them, or facilitate uptake by astrocytes. In addition, they may be transferred from cell to cell in a paracrine manner, providing intracellular protection for the recipient cells.

How HSPs are released into the extracellular space is unclear. Aside from cell lysis, there is evidence that HSPs may be released via exosomes (De Maio, 2014; Tytell, Lasek, & Gainer, 2016). Hsp70 can be released by glia (Hightower & Guidon, 1989), but whether this is sufficient to yield any of the abovementioned protective effects is still somewhat unclear. Nevertheless, some supportive data are available. For instance, HSPA1A and HSPB5 were found to bind amyloidogenic oligomers with consequent shielding of their reactive surfaces. Their toxicity was thereby reduced when applied extracellularly to cells (Mannini et al., 2012). Hsp70 fused with a signal peptide for secretion suppressed Aβ42 neurotoxicity and extended lifespan in *Drosophila* (Fernandez-Funez et al., 2016), although in this case, the mode of action was not revealed.

Evidence for paracrine effects have been suggested from work with polyQ cell models, where exogenous Hsp70 was shown to reduce intracellular aggregation (Guzhova et al., 2001). Intercellular chaperone transmission was also suggested from work with both cell and *Drosophila* polyglutamine models: here Hsp40 (DNAJB1), as well as Hsp70 and Hsp90, were found to be physiologically secreted from cells via exosomes, and that even the secretion itself was dependent on a function Hsp40/70 interaction (Takeuchi et al., 2015). Addition of Hsp40/Hsp70-containing exosomes suppressed intracellular aggregation, indicating exosome-mediated transmission can enhance the protein-folding environment in recipient cells (Takeuchi et al., 2015).

Finally, some HSPs may be released via more classical ER-routing under certain conditions. Intriguing data from the Wiseman lab showed that activation of the unfolded protein response (UPR) reduces extracellular aggregation of the amyloidogenic immunoglobulin light chain. This was associated with secretion of an ER-resident DNAJ protein, ERdj3 (Cooley et al., 2014; Genereux et al., 2015). Moreover, small molecules that reprogram the UPR, in particular by activating one of its branches (ATF6), also can reduce extracellular protein aggregation (Plate et al., 2016).

None of the abovementioned concepts, however, have yet been tested in models for Aβ or tau.

Cell Nonautonomous Regulation of Chaperones—Endocrine Effects

Beyond the abovementioned paracrine effects, research primarily in invertebrates has revealed that some of the main transcriptional pathways that regulate the PQC network, including molecular chaperones, can be controlled in an endocrine manner. Such transcellular signaling affects insulin and IGF-1 like signaling, including regulation of their downstream transcription factors, like DAF16/FOXO and HSF-1, as well as the different UPR networks in the ER and mitochondria. This not only pertains to signaling from the neuronal systems to distant tissues but also between intestinal, muscle, and other distal tissues back to neurons (for excellent reviews on this matter see Taylor, Berendzen, & Dillin, 2014; van Oosten-Hawle & Morimoto, 2014). If translated to mammals, these findings could lead to better understanding of disease susceptibility. Moreover, it could open new preventive or treatment avenues, by activating protective mechanisms in neurons through treatment of distal organs that are more pharmacologically accessible than the brain. Alternatively, identification of the endocrine factors specifically regulating these distal activations of the PQC network may have clinical impact.

HSPs, PQC and Extracellular Release of Aβ or Tau

As stated before (the "General Introduction" section), recent data suggest an additional PQC pathway in which misfolded proteins are excreted actively after encapsulation at the endoplasmic reticulum, a process dubbed MAPS (Lee et al., 2016b). This process was dependent on an ER-resident ubiquitin-specific protease, USP19, which originally was shown to function in ER-associated proteasomal degradation (ERAD) (Hassink et al., 2009). The catalytic domain of USP19 was found to possess an unprecedented chaperone activity, allowing recruitment of misfolded protein to the ER surface. Deubiquitinated cargos are then encapsulated into ER-associated late endosomes and secreted to the cell exterior. The action of USP19 was found to be independent of its interaction with HSP90, with whom it normally partners in the process of ERAD (Lee, Kim, Gygi, & Ye, 2014). In a study from Chad Dickey's group (Fontaine et al., 2016), extracellular release of neurodegenerative-associated proteins was also found to be a regulated process, operating likely under conditions of compromised intracellular PQC. The latter studies

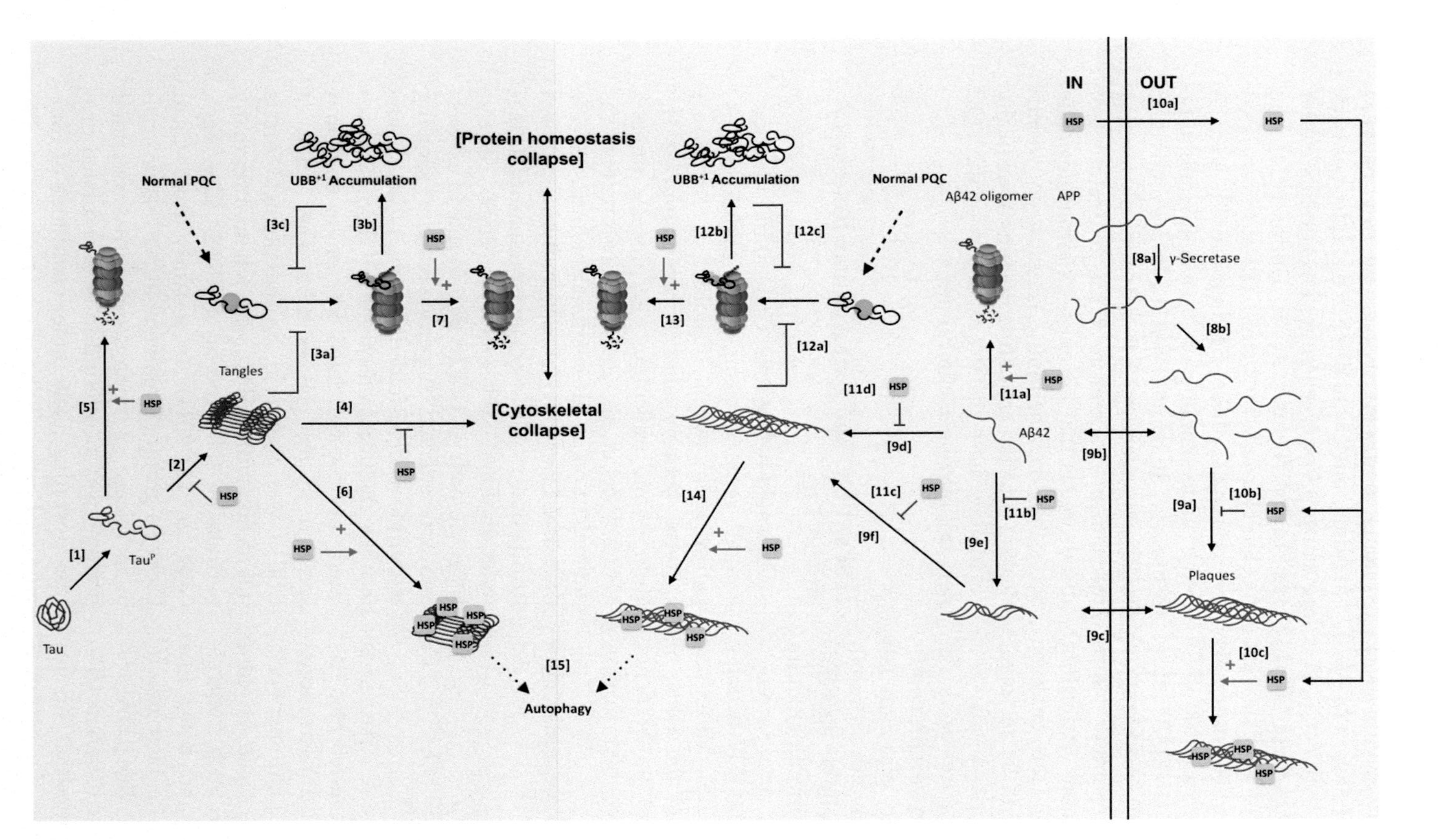

FIGURE 10.3 **Schematic overview of possible effects of HSP on tau pathology during proteasomal stress (left, red background) and Aβ pathology (right, blue background).** Hyperphosphorylation or other modifications of tau [1] can cause aggregate formation (neurofibrillary tangles; [2]). This may lead to saturation of proteasomal degradation [3a] and accumulation of misfolded proteins (e.g., UBB^{+1}) [3b] that perpetuates into a progressive functional decline of UPS systems [3c]. At the same time, the tau cascade may cause a cytoskeletal collapse, resulting in kinky and curly fibers [4], and this may further enhance the collapse of protein homeostasis. HSP may interfere in this cascade by preventing tangle formation [2], by facilitating degradation of phospho-tau via the proteasome [5], preventing the cytoskeletal collapse [4] or by binding to the tangles to make them inert [6]. Finally, increases in HSP may prevent the progressive decline of the UPS system [7]. Cleavage of the APP by the γ-secretase complex [8a] generates toxic monomeric Aβ peptide species intracellularly (not indicated) or extracellularly [8b]. These can lead to extracellular plaques [9a] or intracellular oligomers via different routes [9b–f]. How and which extracellular Aβ species are toxic is unclear. Intracellular HSPs may be released from cells [10a] to prevent monomers from oligomerizing [10b] or to bind to Aβ plaques [10c], stabilizing release of toxic species and making the plaques inert. Inside cells, Aβ oligomers may have similar effects as tau aggregates, causing a progressive decline in the protein homeostasis [12a–c]. In turn, HSP may interfere in this cascade and prevent aggregate formation [11a–d], by facilitating degradation of Aβ monomers or oligomers [11a], by preventing the progressive decline of the UPS system caused by the Aβ oligomers [13], or by binding to the Aβ oligomers to make them inert [14]. Finally, increases in HSP may assist in autophagic degradation of tau tangles and Aβ oligomers [15], although there is no experimental data yet to support a role of HSP is this process. The role of autophagy in AD is further discussed in the chapter by Rubinsztein et al. (Chapter 11: Neurodegenerative Diseases and Autophagy).

showed that tau release was fully dependent on the Hsp70 cochaperone DNAJC5 (also known as cysteine string protein), a neuronal cochaperone crucial for synaptic functioning (Donnelier & Braun, 2014). Finally, a recent study in *C. elegans* showed that adult neurons extrude large membrane-surrounded vesicles, termed exospheres. This effect was likewise augmented upon decline in intercellular PQC capacity as well as by compromised mitochondrial functioning (Melentijevic et al., 2017). Although each of these export mechanisms may provide a level of cytoprotection, this type of secretion may also contribute to the prion-like propagation of Aβ or tau (Watts & Prusiner, 2017) and therefore may be a target for inhibition to halt disease progression. These studies also further underscore the importance of maintaining an optimal intracellular PQC to prevent neurodegenerative diseases and brain aging.

PERSPECTIVES

In summary, protein homeostasis is crucial for the proper function of cells, and maintaining an optimally functioning PQC network is therefore essential to prevent neuronal degeneration. Most facets of PQC erode during aging, and this may be a central factor in the onset or progression of AD. Conversely, potentiating PQC in various ways may help to delay the disease onset, either directly, via various means of chaperoning Aβ or tau monomers or aggregates, or indirectly, by maintaining normal PQC and preventing aggregate release and propagation.

In Fig. 10.3, a summary of possible actions of HSP in the pathogenesis of AD is provided. In the same line, improving UPS efficiency, via DUBs (Lee et al., 2016a) or autophagic activities (see Chapter 11: Neurodegenerative Diseases and Autophagy by Rubinsztein et al.), is an option. Alternatively, reducing the burden on the PQC, e.g., by preventing accumulation of other misfolded proteins in general or even specific ones like UBB^{+1}, could also help to delay a disruption in protein homeostasis. However, as many of these processes are promiscuous, interference with their global activity may come with too many negative trade-offs. They may also only be effective transiently due to autoregulatory feedback loops or epigenetic changes, as was observed when trying to boost HSF-1 activity by Hsp90 inhibition (Labbadia et al., 2011). Therefore, specific targeting of cochaperones or UBB^{+1} or their transcripts may be an option.

Acknowledgments

We apologize to the authors of many excellent pieces of work which could not be discussed due to space restraints. We are greatly indebted to the critical remarks on subchapters by Drs. Eckhard Mandelkow and Yi Peng (DZNE, Bonn, Germany), Pascal Kienlen-Campard (UCL Brussels, Belgium), and Charles Glabe (UCI, Irvine, USA) and to Dr. Bert M. Verheijen (University of Utrecht, The Netherlands) and Joop J. van Heerikhuize (NHI, Amsterdam, The Netherlands) for their help to achieve this review. We would also like to acknowledge the grant support over the past years from the Prinses Beatrix Foundation (PBF) and the Dutch Brain Foundation (WAR09-23 & WAR11-31), the Samenwerkingsverband Noord-Nederland (SNN: Transitie II & Pieken, ChaperoneAge), and the Graduate School Medical Science Groningen.

References

Abelein, A., Abrahams, J. P., Danielsson, J., Gräslund, A., Jarvet, J., Luo, J., ... Wärmländer, S. K. T. S. (2014). The hairpin conformation of the amyloid β peptide is an important structural motif along the aggregation pathway. *Journal of Biological Inorganic Chemistry: JBIC: A Publication of the Society of Biological Inorganic Chemistry, 19*, 623–634.

Abisambra, J., Jinwal, U. K., Miyata, Y., Rogers, J., Blair, L., Li, X., ... Bacon, J. (2013). Allosteric heat shock protein 70 inhibitors rapidly rescue synaptic plasticity deficits by reducing aberrant tau. *Biological Psychiatry, 74*, 367–374.

Abisambra, J. F., Blair, L. J., Hill, S. E., Jones, J. R., Kraft, C., Rogers, J., ... Johnson, A. G. (2010). Phosphorylation dynamics regulate Hsp27-mediated rescue of neuronal plasticity deficits in tau transgenic mice. *The Journal of Neuroscience: The Official Journal of the Society for Neuroscience, 30*, 15374–15382.

Abisambra, J. F., Jinwal, U. K., Jones, J. R., Blair, L. J., Koren, J., & Dickey, C. A. (2011). Exploiting the diversity of the heat-shock protein family for primary and secondary tauopathy therapeutics. *Current Neuropharmacology, 9*, 623–631.

Abisambra, J. F., Jinwal, U. K., Suntharalingam, A., Arulselvam, K., Brady, S., Cockman, M., ... Dickey, C. A. (2012). DnaJA1 antagonizes constitutive Hsp70-mediated stabilization of tau. *Journal of Molecular Biology, 421*, 653–661.

Agarraberes, F. A., & Dice, J. F. (2001). A molecular chaperone complex at the lysosomal membrane is required for protein translocation. *Journal of Cell Science, 114*, 2491–2499.

Agholme, L., & Hallbeck, M. (2014). Getting rid of intracellular Aβ-loss of cellular degradation leads to transfer between connected neurons. *Current Pharmaceutical Design, 20*, 2458–2468.

Agholme, L., Hallbeck, M., Benedikz, E., Marcusson, J., & Kågedal, K. (2012). Amyloid-β secretion, generation, and lysosomal sequestration in response to proteasome inhibition: Involvement of autophagy. *Journal of Alzheimer's Disease: JAD, 31*, 343–358.

Akerfelt, M., Morimoto, R. I., & Sistonen, L. (2010). Heat shock factors: Integrators of cell stress, development and lifespan. *Nature Reviews Molecular Cell Biology, 11*, 545–555.

Albanèse, V., Yam, A. Y.-W., Baughman, J., Parnot, C., & Frydman, J. (2006). Systems analyses reveal two chaperone networks with distinct functions in eukaryotic cells. *Cell, 124*, 75–88.

Alberti, S., & Hyman, A. A. (2016). Are aberrant phase transitions a driver of cellular aging? *BioEssays: News and Reviews in Molecular, Cellular and Developmental Biology, 38*, 959–968.

Andersen, K. M., Hofmann, K., & Hartmann-Petersen, R. (2005). Ubiquitin-binding proteins: Similar, but different. *Essays in Biochemistry, 41*, 49–67.

Arndt, V., Dick, N., Tawo, R., Dreiseidler, M., Wenzel, D., Hesse, M., ... Fleischmann, B. K. (2010). Chaperone-assisted selective autophagy is essential for muscle maintenance. *Current Biology: CB, 20*, 143–148.

Arosio, P., Michaels, T. C. T., Linse, S., Månsson, C., Emanuelsson, C., Presto, J., ... Knowles, T. P. J. (2016). Kinetic analysis reveals the diversity of microscopic mechanisms through which molecular chaperones suppress amyloid formation. *Nature Communications, 7*, 10948.

Balch, W. E., Morimoto, R. I., Dillin, A., & Kelly, J. W. (2008). Adapting proteostasis for disease intervention. *Science, 319*, 916–919.

Balchin, D., Hayer-Hartl, M., & Hartl, F. U. (2016). In vivo aspects of protein folding and quality control. *Science, 353*, 43–54.

Ballinger, C. A., Connell, P., Wu, Y., Hu, Z., Thompson, L. J., Yin, L. Y., ... Patterson, C. (1999). Identification of CHIP, a novel tetratricopeptide repeat-containing protein that interacts with heat shock proteins and negatively regulates chaperone functions. *Molecular and Cellular Biology, 19*, 4535–4545.

Behnke, J., Mann, M. J., Scruggs, F.-L., Feige, M. J., & Hendershot, L. M. (2016). Members of the Hsp70 family recognize distinct types of sequences to execute ER quality control. *Molecular Cell, 63*, 739–752.

Bett, J. S. (2016). Proteostasis regulation by the ubiquitin system. *Essays in Biochemistry, 60*, 143–151.

Billings, L. M., Oddo, S., Green, K. N., McGaugh, J. L., & LaFerla, F. M. (2005). Intraneuronal Abeta causes the onset of early Alzheimer's disease-related cognitive deficits in transgenic mice. *Neuron, 45*, 675–688.

Björkdahl, C., Sjögren, M. J., Zhou, X., Concha, H., Avila, J., Winblad, B., ... Pei, J.-J. (2008). Small heat shock proteins Hsp27 or alphaB-crystallin and the protein components of neurofibrillary tangles: Tau and neurofilaments. *Journal of Neuroscience Research, 86*, 1343–1352.

Blair, L. J., Sabbagh, J. J., & Dickey, C. A. (2014). Targeting Hsp90 and its co-chaperones to treat Alzheimer's disease. *Expert Opinion on Therapeutic Targets, 18*, 1219–1232.

Boncoraglio, A., Minoia, M., & Carra, S. (2012). The family of mammalian small heat shock proteins (HSPBs): Implications in protein deposit diseases and motor neuropathies. *The International Journal of Biochemistry & Cell Biology, 44*, 1657–1669.

Borges, T. J., Lang, B. J., Lopes, R. L., & Bonorino, C. (2016). Modulation of alloimmunity by heat shock proteins. *Frontiers in Immunology, 7*, 303.

Braak, H., & Del Tredici, K. (2016). Potential pathways of abnormal tau and α-synuclein dissemination in sporadic Alzheimer's and Parkinson's diseases. *Cold Spring Harbor Perspectives in Biology, 8*, a023630.

Brandman, O., & Hegde, R. S. (2016). Ribosome-associated protein quality control. *Nature Structural & Molecular Biology, 23*, 7–15.

Braun, R. J., Sommer, C., Leibiger, C., Gentier, R. J. G., Dumit, V. I., Paduch, K., ... Magnes, C. (2015). Accumulation of basic amino acids at mitochondria dictates the cytotoxicity of aberrant ubiquitin. *Cell Reports, 10*, 1557–1571.

Caine, J., Sankovich, S., Antony, H., Waddington, L., Macreadie, P., Varghese, J., & Macreadie, I. (2007). Alzheimer's Abeta fused to green fluorescent protein induces growth stress and a heat shock response. *FEMS Yeast Research, 7*, 1230–1236.

Calderwood, S. K., Gong, J., & Murshid, A. (2016). Extracellular HSPs: The complicated roles of extracellular HSPs in immunity. *Frontiers in Immunology*, *7*, 159.

Carnini, A., Scott, L. O. M., Ahrendt, E., Proft, J., Winkfein, R. J., Kim, S.-W., ... Braun, J. E. A. (2012). Cell line specific modulation of extracellular aβ42 by Hsp40. *PLoS ONE*, *7*, e37755.

Carra, S., Seguin, S. J., Lambert, H., & Landry, J. (2008). HspB8 chaperone activity toward poly(Q)-containing proteins depends on its association with Bag3, a stimulator of macroautophagy. *Journal of Biological Chemistry*, *283*, 1437–1444.

Carra, S., Crippa, V., Rusmini, P., Boncoraglio, A., Minoia, M., Giorgetti, E., ... Poletti, A. (2012). Alteration of protein folding and degradation in motor neuron diseases: Implications and protective functions of small heat shock proteins. *Progress in Neurobiology*, *97*, 83–100.

Cataldo, A. M., Hamilton, D. J., & Nixon, R. A. (1994). Lysosomal abnormalities in degenerating neurons link neuronal compromise to senile plaque development in Alzheimer disease. *Brain Research*, *640*, 68–80.

Cataldo, A. M., Petanceska, S., Terio, N. B., Peterhoff, C. M., Durham, R., Mercken, M., ... Nixon, R. A. (2004). Abeta localization in abnormal endosomes: Association with earliest Abeta elevations in AD and Down syndrome. *Neurobiology of Aging*, *25*, 1263–1272.

Cecarini, V., Bonfili, L., Cuccioloni, M., Mozzicafreddo, M., Angeletti, M., Keller, J. N., & Eleuteri, A. M. (2016). The fine-tuning of proteolytic pathways in Alzheimer's disease. *Cellular and Molecular Life Sciences: CMLS*, *73*, 3433–3451.

Chakrabortee, S., Liu, Y., Zhang, L., Matthews, H. R., Zhang, H., Pan, N., ... Huang, Z. (2012). Macromolecular and small-molecule modulation of intracellular Aβ42 aggregation and associated toxicity. *Biochemical Journal*, *442*, 507–515.

Chiang, H. L., & Dice, J. F. (1988). Peptide sequences that target proteins for enhanced degradation during serum withdrawal. *Journal of Biological Chemistry*, *263*, 6797–6805.

Choi, S. H., Kim, Y. H., Hebisch, M., Sliwinski, C., Lee, S., D'Avanzo, C., ... Muffat, J. (2014). A three-dimensional human neural cell culture model of Alzheimer's disease. *Nature*, *515*, 274–278.

Cohen, E., Bieschke, J., Perciavalle, R. M., Kelly, J. W., & Dillin, A. (2006). Opposing activities protect against age-onset proteotoxicity. *Science*, *313*, 1604–1610.

Cohen, E., Paulsson, J. F., Blinder, P., Burstyn-Cohen, T., Du, D., Estepa, G., ... Kelly, J. W. (2009). Reduced IGF-1 signaling delays age-associated proteotoxicity in mice. *Cell*, *139*, 1157–1169.

Cohen, S. I. A., Linse, S., Luheshi, L. M., Hellstrand, E., White, D. A., Rajah, L., ... Knowles, T. P. J. (2013). Proliferation of amyloid-β42 aggregates occurs through a secondary nucleation mechanism. *Proceedings of the National Academy of Sciences of the United States of America*, *110*, 9758–9763.

Cooley, C. B., Ryno, L. M., Plate, L., Morgan, G. J., Hulleman, J. D., Kelly, J. W., ... Wiseman, R. L. (2014). Unfolded protein response activation reduces secretion and extracellular aggregation of amyloidogenic immunoglobulin light chain. *Proceedings of the National Academy of Sciences of the United States of America*, *111*, 13046–13051.

Crippa, V., Sau, D., Rusmini, P., Boncoraglio, A., Onesto, E., Bolzoni, E., ... Carra, S. (2010). The small heat shock protein B8 (HspB8) promotes autophagic removal of misfolded proteins involved in amyotrophic lateral sclerosis (ALS). *Human Molecular Genetics*, *19*, 3440–3456.

D'Andrea, M. R., Nagele, R. G., Wang, H. Y., Peterson, P. A., & Lee, D. H. (2001). Evidence that neurones accumulating amyloid can undergo lysis to form amyloid plaques in Alzheimer's disease. *Histopathology*, *38*, 120–134.

David, D. C., Layfield, R., Serpell, L., Narain, Y., Goedert, M., & Spillantini, M. G. (2002). Proteasomal degradation of tau protein. *Journal of Neurochemistry*, *83*, 176–185.

De Maio, A. (2014). Extracellular Hsp70: Export and function. *Current Protein & Peptide Science*, *15*, 225–231.

De Strooper, B., & Karran, E. (2016). The cellular phase of Alzheimer's disease. *Cell*, *164*, 603–615.

Dekker, S. L., Kampinga, H. H., & Bergink, S. (2015). DNAJs: More than substrate delivery to HSPA. *Frontiers in Molecular Biosciences*, *2*, 35.

Dennissen, F. J. A., Kholod, N., & van Leeuwen, F. W. (2012). The ubiquitin–proteasome system in neurodegenerative diseases: Culprit, accomplice or victim? *Progress in Neurobiology*, *96*, 190–207.

Dice, J. F. (1990). Peptide sequences that target cytosolic proteins for lysosomal proteolysis. *Trends in Biochemical Sciences*, *15*, 305–309.

Dickey, C. A., Kamal, A., Lundgren, K., Klosak, N., Bailey, R. M., Dunmore, J., ... Eckman, C. B. (2007). The high-affinity HSP90-CHIP complex recognizes and selectively degrades phosphorylated tau client proteins. *Journal of Clinical Investigation*, *117*, 648–658.

Dickey, C. A., Koren, J., Zhang, Y.-J., Xu, Y.-F., Jinwal, U. K., Birnbaum, M. J., ... Patterson, C. (2008). Akt and CHIP coregulate tau degradation through coordinated interactions. *Proceedings of the National Academy of Sciences of the United States of America*, *105*, 3622–3627.

Domert, J., Rao, S. B., Agholme, L., Brorsson, A.-C., Marcusson, J., Hallbeck, M., & Nath, S. (2014). Spreading of amyloid-β peptides via neuritic cell-to-cell transfer is dependent on insufficient cellular clearance. *Neurobiology of Disease*, *65*, 82–92.

Donnelier, J., & Braun, J. E. A. (2014). CSPα-chaperoning presynaptic proteins. *Frontiers in Cellular Neuroscience, 8*, 116.

Durrenberger, P. F., Filiou, M. D., Moran, L. B., Michael, G. J., Novoselov, S., Cheetham, M. E., ... Graeber, M. B. (2009). DnaJB6 is present in the core of Lewy bodies and is highly up-regulated in parkinsonian astrocytes. *Journal of Neuroscience Research, 87*, 238–245.

Ecroyd, H., & Carver, J. A. (2009). Crystallin proteins and amyloid fibrils. *Cellular and Molecular Life Sciences: CMLS, 66*, 62–81.

van Eden, W. (1991). Heat-shock proteins as immunogenic bacterial antigens with the potential to induce and regulate autoimmune arthritis. *Immunological Reviews, 121*, 5–28.

Erales, J., & Coffino, P. (2014). Ubiquitin-independent proteasomal degradation. *Biochimica et Biophysica Acta, 1843*, 216–221.

Evans, C. G., Wisén, S., & Gestwicki, J. E. (2006). Heat shock proteins 70 and 90 inhibit early stages of amyloid beta-(1-42) aggregation in vitro. *Journal of Biological Chemistry, 281*, 33182–33191.

Fernandez-Funez, P., Sanchez-Garcia, J., de Mena, L., Zhang, Y., Levites, Y., Khare, S., ... Rincon-Limas, D. E. (2016). Holdase activity of secreted Hsp70 masks amyloid-β42 neurotoxicity in *Drosophila*. *Proceedings of the National Academy of Sciences of the United States of America, 113*, E5212–5221.

Fischer, D. F., De Vos, R. A. I., Van Dijk, R., De Vrij, F. M. S., Proper, E. A., Sonnemans, M. A. F., ... Zouambia, M. (2003). Disease-specific accumulation of mutant ubiquitin as a marker for proteasomal dysfunction in the brain. *The FASEB Journal: Official Publication of the Federation of American Societies for Experimental Biology, 17*, 2014–2024.

Fischer, D. F., van Dijk, R., van Tijn, P., Hobo, B., Verhage, M. C., van der Schors, R. C., ... van Leeuwen, F. W. (2009). Long-term proteasome dysfunction in the mouse brain by expression of aberrant ubiquitin. *Neurobiology of Aging, 30*, 847–863.

Fontaine, S. N., Martin, M. D., Akoury, E., Assimon, V. A., Borysov, S., Nordhues, B. A., ... Zweckstetter, M. (2015). The active Hsc70/tau complex can be exploited to enhance tau turnover without damaging microtubule dynamics. *Human Molecular Genetics, 24*, 3971–3981.

Fontaine, S. N., Zheng, D., Sabbagh, J. J., Martin, M. D., Chaput, D., Darling, A., ... Lussier, A. (2016). DnaJ/Hsc70 chaperone complexes control the extracellular release of neurodegenerative-associated proteins. *The EMBO Journal, 35*, 1537–1549.

Fonte, V., Kipp, D. R., Yerg, J., Merin, D., Forrestal, M., Wagner, E., ... Link, C. D. (2008). Suppression of in vivo beta-amyloid peptide toxicity by overexpression of the HSP-16.2 small chaperone protein. *Journal of Biological Chemistry, 283*, 784–791.

Gamerdinger, M., Hajieva, P., Kaya, A. M., Wolfrum, U., Hartl, F. U., & Behl, C. (2009). Protein quality control during aging involves recruitment of the macroautophagy pathway by BAG3. *The EMBO Journal, 28*, 889–901.

Gamerdinger, M., Carra, S., & Behl, C. (2011). Emerging roles of molecular chaperones and co-chaperones in selective autophagy: Focus on BAG proteins. *Journal of Molecular Medicine (Berlin, Germany), 89*, 1175–1182.

Garcia-Alloza, M., Robbins, E. M., Zhang-Nunes, S. X., Purcell, S. M., Betensky, R. A., Raju, S., ... Frosch, M. P. (2006). Characterization of amyloid deposition in the APPswe/PS1dE9 mouse model of Alzheimer disease. *Neurobiology of Disease, 24*, 516–524.

Genereux, J. C., Qu, S., Zhou, M., Ryno, L. M., Wang, S., Shoulders, M. D., ... Wiseman, R. L. (2015). Unfolded protein response-induced ERdj3 secretion links ER stress to extracellular proteostasis. *The EMBO Journal, 34*, 4–19.

Gentier, R. J., & van Leeuwen, F. W. (2015). Misframed ubiquitin and impaired protein quality control: An early event in Alzheimer's disease. *Frontiers in Molecular Neuroscience, 8*, 47.

Gentier, R. J. G., Verheijen, B. M., Zamboni, M., Stroeken, M. M. A., Hermes, D. J. H. P., Küsters, B., ... Van Leeuwen, F. W. (2015). Localization of mutant ubiquitin in the brain of a transgenic mouse line with proteasomal inhibition and its validation at specific sites in Alzheimer's disease. *Frontiers in Neuroanatomy, 9*, 26.

Gillis, J., Schipper-Krom, S., Juenemann, K., Gruber, A., Coolen, S., van den Nieuwendijk, R., ... Kampinga, H. H. (2013). The DNAJB6 and DNAJB8 protein chaperones prevent intracellular aggregation of polyglutamine peptides. *Journal of Biological Chemistry, 288*, 17225–17237.

Glabe, C. G. (2008). Structural classification of toxic amyloid oligomers. *Journal of Biological Chemistry, 283*, 29639–29643.

Götz, J., Eckert, A., Matamales, M., Ittner, L. M., & Liu, X. (2011). Modes of Aβ toxicity in Alzheimer's disease. *Cellular and Molecular Life Sciences: CMLS, 68*, 3359–3375.

Gouras, G. K., Almeida, C. G., & Takahashi, R. H. (2005). Intraneuronal Abeta accumulation and origin of plaques in Alzheimer's disease. *Neurobiology of Aging, 26*, 1235–1244.

Guerreiro, R. J., Gustafson, D. R., & Hardy, J. (2012). The genetic architecture of Alzheimer's disease: Beyond APP, PSENs and APOE. *Neurobiology of Aging, 33*, 437–456.

Gusev, N. B., Bogatcheva, N. V., & Marston, S. B. (2002). Structure and properties of small heat shock proteins (sHsp) and their interaction with cytoskeleton proteins. *Biochemistry. Biokhimiia*, *67*, 511–519.

Guzhova, I., Kislyakova, K., Moskaliova, O., Fridlanskaya, I., Tytell, M., Cheetham, M., ... Margulis, B. (2001). In vitro studies show that Hsp70 can be released by glia and that exogenous Hsp70 can enhance neuronal stress tolerance. *Brain Research*, *914*, 66–73.

Gyure, K. A., Durham, R., Stewart, W. F., Smialek, J. E., & Troncoso, J. C. (2001). Intraneuronal abeta-amyloid precedes development of amyloid plaques in Down syndrome. *Archives of Pathology & Laboratory Medicine*, *125*, 489–492.

Hageman, J., & Kampinga, H. H. (2009). Computational analysis of the human HSPH/HSPA/DNAJ family and cloning of a human HSPH/HSPA/DNAJ expression library. *Cell Stress and Chaperones*, *14*, 1–21.

Hageman, J., Rujano, M. A., van Waarde, M. A. W. H., Kakkar, V., Dirks, R. P., Govorukhina, N., ... Kampinga, H. H. (2010). A DNAJB chaperone subfamily with HDAC-dependent activities suppresses toxic protein aggregation. *Molecular Cell*, *37*, 355–369.

Hassink, G. C., Zhao, B., Sompallae, R., Altun, M., Gastaldello, S., Zinin, N. V., ... Lindsten, K. (2009). The ER-resident ubiquitin-specific protease 19 participates in the UPR and rescues ERAD substrates. *EMBO Reports*, *10*, 755–761.

Hightower, L. E., & Guidon, P. T. (1989). Selective release from cultured mammalian cells of heat-shock (stress) proteins that resemble glia-axon transfer proteins. *Journal of Cellular Physiology*, *138*, 257–266.

Hipp, M. S., Patel, C. N., Bersuker, K., Riley, B. E., Kaiser, S. E., Shaler, T. A., ... Kopito, R. R. (2012). Indirect inhibition of 26S proteasome activity in a cellular model of Huntington's disease. *Journal of Cell Biology*, *196*, 573–587.

Hoop, C. L., Lin, H.-K., Kar, K., Magyarfalvi, G., Lamley, J. M., Boatz, J. C., ... van der Wel, P. C. A. (2016). Huntingtin exon 1 fibrils feature an interdigitated β-hairpin-based polyglutamine core. *Proceedings of the National Academy of Sciences of the United States of America*, *113*, 1546–1551.

Hope, A. D., de Silva, R., Fischer, D. F., Hol, E. M., van Leeuwen, F. W., & Lees, A. J. (2003). Alzheimer's associated variant ubiquitin causes inhibition of the 26S proteasome and chaperone expression. *Journal of Neurochemistry*, *86*, 394–404.

Hoshino, T., Murao, N., Namba, T., Takehara, M., Adachi, H., Katsuno, M., ... Mizushima, T. (2011). Suppression of Alzheimer's disease-related phenotypes by expression of heat shock protein 70 in mice. *The Journal of Neuroscience: The Official Journal of the Society for Neuroscience*, *31*, 5225–5234.

International Genomics of Alzheimer's Disease Consortium (IGAP). (2015). Convergent genetic and expression data implicate immunity in Alzheimer's disease. *Alzheimer's & Dementia: The Journal of the Alzheimer's Association*, *11*, 658–671.

Irmler, M., Gentier, R. J. G., Dennissen, F. J. A., Schulz, H., Bolle, I., Hölter, S. M., ... Rozman, J. (2012). Long-term proteasomal inhibition in transgenic mice by UBB(+1) expression results in dysfunction of central respiration control reminiscent of brainstem neuropathology in Alzheimer patients. *Acta Neuropathologica (Berlin)*, *124*, 187–197.

Jankowsky, J. L., Fadale, D. J., Anderson, J., Xu, G. M., Gonzales, V., Jenkins, N. A., ... Wagner, S. L. (2004). Mutant presenilins specifically elevate the levels of the 42 residue beta-amyloid peptide in vivo: Evidence for augmentation of a 42-specific gamma secretase. *Human Molecular Genetics*, *13*, 159–170.

Jiang, Y.-Q., Wang, X.-L., Cao, X.-H., Ye, Z.-Y., Li, L., & Cai, W.-Q. (2013). Increased heat shock transcription factor 1 in the cerebellum reverses the deficiency of Purkinje cells in Alzheimer's disease. *Brain Research*, *1519*, 105–111.

Jimenez-Sanchez, M., Lam, W., Hannus, M., Sönnichsen, B., Imarisio, S., Fleming, A., ... Xu, C. (2015). siRNA screen identifies QPCT as a druggable target for Huntington's disease. *Nature Chemical Biology*, *11*, 347–354.

Jinwal, U. K., Miyata, Y., Koren, J., Jones, J. R., Trotter, J. H., Chang, L., ... Shults, C. L. (2009). Chemical manipulation of hsp70 ATPase activity regulates tau stability. *The Journal of Neuroscience: The Official Journal of the Society for Neuroscience*, *29*, 12079–12088.

Jinwal, U. K., O'Leary, J. C., Borysov, S. I., Jones, J. R., Li, Q., Koren, J., ... Johnson, A. G. (2010a). Hsc70 rapidly engages tau after microtubule destabilization. *Journal of Biological Chemistry*, *285*, 16798–16805.

Jinwal, U. K., Koren, J., O'Leary, J. C., Jones, J. R., Abisambra, J. F., & Dickey, C. A. (2010b). Hsp70 ATPase modulators as therapeutics for Alzheimer's and other neurodegenerative diseases. *Molecular and Cellular Pharmacology*, *2*, 43–46.

Jinwal, U. K., Trotter, J. H., Abisambra, J. F., Koren, J., Lawson, L. Y., Vestal, G. D., ... Jones, J. R. (2011). The Hsp90 kinase co-chaperone Cdc37 regulates tau stability and phosphorylation dynamics. *Journal of Biological Chemistry*, *286*, 16976–16983.

Jinwal, U. K., Koren, J., & Dickey, C. A. (2013). Reconstructing the Hsp90/Tau machine. *Current Enzyme Inhibition*, *9*, 41–45.

Kaganovich, D., Kopito, R., & Frydman, J. (2008). Misfolded proteins partition between two distinct quality control compartments. *Nature*, *454*, 1088–1095.

Kakkar, V., Meister-Broekema, M., Minoia, M., Carra, S., & Kampinga, H. H. (2014). Barcoding heat shock proteins to human diseases: Looking beyond the heat shock response. *Disease Models & Mechanisms*, *7*, 421–434.

Kakkar, V., Månsson, C., de Mattos, E. P., Bergink, S., van der Zwaag, M., van Waarde, M. A. W. H., ... Al-Karadaghi, S. (2016). The S/T-rich motif in the DNAJB6 chaperone delays polyglutamine aggregation and the onset of disease in a mouse model. *Molecular Cell*, *62*, 272–283.

Kampinga, H. H., & Bergink, S. (2016). Heat shock proteins as potential targets for protective strategies in neurodegeneration. *Lancet Neurology*, *15*, 748–759.

Kampinga, H. H., & Craig, E. A. (2010). The HSP70 chaperone machinery: J proteins as drivers of functional specificity. *Nature Reviews Molecular Cell Biology*, *11*, 579–592.

Kampinga, H. H., Hageman, J., Vos, M. J., Kubota, H., Tanguay, R. M., Bruford, E. A., ... Hightower, L. E. (2009). Guidelines for the nomenclature of the human heat shock proteins. *Cell Stress and Chaperones*, *14*, 105–111.

Kar, K., Hoop, C. L., Drombosky, K. W., Baker, M. A., Kodali, R., Arduini, I., ... Wetzel, R. (2013). β-hairpin-mediated nucleation of polyglutamine amyloid formation. *Journal of Molecular Biology*, *425*, 1183–1197.

Karagöz, G. E., Duarte, A. M. S., Akoury, E., Ippel, H., Biernat, J., Morán Luengo, T., ... Veprintsev, D. B. (2014). Hsp90-Tau complex reveals molecular basis for specificity in chaperone action. *Cell*, *156*, 963–974.

Kaushik, S., & Cuervo, A. M. (2012). Chaperone-mediated autophagy: A unique way to enter the lysosome world. *Trends in Cell Biology*, *22*, 407–417.

Keller, J. N., Hanni, K. B., & Markesbery, W. R. (2000). Impaired proteasome function in Alzheimer's disease. *Journal of Neurochemistry*, *75*, 436–439.

Keller, J. N., Gee, J., & Ding, Q. (2002). The proteasome in brain aging. *Ageing Research Reviews*, *1*, 279–293.

Kienlen-Campard, P., Miolet, S., Tasiaux, B., & Octave, J.-N. (2002). Intracellular amyloid-beta 1-42, but not extracellular soluble amyloid-beta peptides, induces neuronal apoptosis. *Journal of Biological Chemistry*, *277*, 15666–15670.

Knobloch, M., Konietzko, U., Krebs, D. C., & Nitsch, R. M. (2007). Intracellular Abeta and cognitive deficits precede beta-amyloid deposition in transgenic arcAbeta mice. *Neurobiology of Aging*, *28*, 1297–1306.

Kosik, K. S., & Shimura, H. (2005). Phosphorylated tau and the neurodegenerative foldopathies. *Biochimica et Biophysica Acta*, *1739*, 298–310.

Labbadia, J., Cunliffe, H., Weiss, A., Katsyuba, E., Sathasivam, K., Seredenina, T., ... Luthi-Carter, R. (2011). Altered chromatin architecture underlies progressive impairment of the heat shock response in mouse models of Huntington disease. *Journal of Clinical Investigation*, *121*, 3306–3319.

LaFerla, F. M., Green, K. N., & Oddo, S. (2007). Intracellular amyloid-beta in Alzheimer's disease. *Nature Reviews Neuroscience*, *8*, 499–509.

Lam, Y. A., Pickart, C. M., Alban, A., Landon, M., Jamieson, C., Ramage, R., ... Layfield, R. (2000). Inhibition of the ubiquitin–proteasome system in Alzheimer's disease. *Proceedings of the National Academy of Sciences of the United States of America*, *97*, 9902–9906.

Lee, B.-H., Lu, Y., Prado, M. A., Shi, Y., Tian, G., Sun, S., ... Finley, D. (2016a). USP14 deubiquitinates proteasome-bound substrates that are ubiquitinated at multiple sites. *Nature*, *532*, 398–401.

Lee, J.-G., Kim, W., Gygi, S., & Ye, Y. (2014). Characterization of the deubiquitinating activity of USP19 and its role in endoplasmic reticulum-associated degradation. *Journal of Biological Chemistry*, *289*, 3510–3517.

Lee, J.-G., Takahama, S., Zhang, G., Tomarev, S. I., & Ye, Y. (2016b). Unconventional secretion of misfolded proteins promotes adaptation to proteasome dysfunction in mammalian cells. *Nature Cell Biology*, *18*, 765–776.

van Leeuwen, F. W., de Kleijn, D. P., van den Hurk, H. H., Neubauer, A., Sonnemans, M. A., Sluijs, J. A., ... Martens, G. J. (1998). Frameshift mutants of beta amyloid precursor protein and ubiquitin-B in Alzheimer's and Down patients. *Science*, *279*, 242–247.

van Leeuwen, F. W., Fischer, D. F., Kamel, D., Sluijs, J. A., Sonnemans, M. A., Benne, R., ... Hol, E. M. (2000). Molecular misreading: A new type of transcript mutation expressed during aging. *Neurobiology of Aging*, *21*, 879–891.

van Leeuwen, F. W., Hol, E. M., & Fischer, D. F. (2006). Frameshift proteins in Alzheimer's disease and in other conformational disorders: Time for the ubiquitin–proteasome system. *Journal of Alzheimer's Disease: JAD*, *9*, 319–325.

Li, J., Soroka, J., & Buchner, J. (2012). The Hsp90 chaperone machinery: Conformational dynamics and regulation by co-chaperones. *Biochimica et Biophysica Acta*, *1823*, 624–635.

Lindsten, K., de Vrij, F. M. S., Verhoef, L. G. G. C., Fischer, D. F., van Leeuwen, F. W., Hol, E. M., ... Dantuma, N. P. (2002). Mutant ubiquitin found in neurodegenerative disorders is a ubiquitin fusion degradation substrate that blocks proteasomal degradation. *Journal of Cell Biology*, *157*, 417–427.

Lindsten, K., Menéndez-Benito, V., Masucci, M. G., & Dantuma, N. P. (2003). A transgenic mouse model of the ubiquitin/proteasome system. *Nature Biotechnology*, *21*, 897–902.

Lopez Salon, M., Pasquini, L., Besio Moreno, M., Pasquini, J. M., & Soto, E. (2003). Relationship between beta-amyloid degradation and the 26S proteasome in neural cells. *Experimental Neurology, 180*, 131–143.

Lüders, J., Demand, J., & Höhfeld, J. (2000). The ubiquitin-related BAG-1 provides a link between the molecular chaperones Hsc70/Hsp70 and the proteasome. *Journal of Biological Chemistry, 275*, 4613–4617.

Luo, W., Dou, F., Rodina, A., Chip, S., Kim, J., Zhao, Q., ... Greengard, P. (2007). Roles of heat-shock protein 90 in maintaining and facilitating the neurodegenerative phenotype in tauopathies. *Proceedings of the National Academy of Sciences of the United States of America, 104*, 9511–9516.

Majeski, A. E., & Dice, J. F. (2004). Mechanisms of chaperone-mediated autophagy. *The International Journal of Biochemistry & Cell Biology, 36*, 2435–2444.

Manavalan, A., Mishra, M., Feng, L., Sze, S. K., Akatsu, H., & Heese, K. (2013). Brain site-specific proteome changes in aging-related dementia. *Experimental & Molecular Medicine, 45*, e39.

Mandelkow, E.-M., & Mandelkow, E. (2012). Biochemistry and cell biology of tau protein in neurofibrillary degeneration. *Cold Spring Harbor Perspectives in Medicine, 2*, a006247.

Mannini, B., Cascella, R., Zampagni, M., van Waarde-Verhagen, M., Meehan, S., Roodveldt, C., ... Relini, A. (2012). Molecular mechanisms used by chaperones to reduce the toxicity of aberrant protein oligomers. *Proceedings of the National Academy of Sciences of the United States of America, 109*, 12479–12484.

Månsson, C., Arosio, P., Hussein, R., Kampinga, H. H., Hashem, R. M., Boelens, W. C., ... Emanuelsson, C. (2014a). Interaction of the molecular chaperone DNAJB6 with growing amyloid-beta 42 (Aβ42) aggregates leads to sub-stoichiometric inhibition of amyloid formation. *Journal of Biological Chemistry, 289*, 31066–31076.

Månsson, C., Kakkar, V., Monsellier, E., Sourigues, Y., Härmark, J., Kampinga, H. H., ... Emanuelsson, C. (2014b). DNAJB6 is a peptide-binding chaperone which can suppress amyloid fibrillation of polyglutamine peptides at substoichiometric molar ratios. *Cell Stress and Chaperones, 19*, 227–239.

Melentijevic, I., Toth, M. L., Arnold, M. L., Guasp, R. J., Harinath, G., Nguyen, K. C., ... Gabel, C. V. (2017). *C. elegans* neurons jettison protein aggregates and mitochondria under neurotoxic stress. *Nature, 542*, 367–371.

Minoia, M., Boncoraglio, A., Vinet, J., Morelli, F. F., Brunsting, J. F., Poletti, A., ... Carra, S. (2014). BAG3 induces the sequestration of proteasomal clients into cytoplasmic puncta: Implications for a proteasome-to-autophagy switch. *Autophagy, 10*, 1603–1621.

Miyata, Y., Koren, J., Kiray, J., Dickey, C. A., & Gestwicki, J. E. (2011). Molecular chaperones and regulation of tau quality control: Strategies for drug discovery in tauopathies. *Future Medicinal Chemistry, 3*, 1523–1537.

Miyata, Y., Li, X., Lee, H.-F., Jinwal, U. K., Srinivasan, S. R., Seguin, S. P., ... Sun, D. (2013). Synthesis and initial evaluation of YM-08, a blood-brain barrier permeable derivative of the heat shock protein 70 (Hsp70) inhibitor MKT-077, which reduces tau levels. *ACS Chemical Neuroscience, 4*, 930–939.

Mogk, A., & Bukau, B. (2017). Role of sHsps in organizing cytosolic protein aggregation and disaggregation. *Cell Stress and Chaperones, 22*, 493–502.

Morales, R., Callegari, K., & Soto, C. (2015). Prion-like features of misfolded Aβ and tau aggregates. *Virus Research, 207*, 106–112.

Muchowski, P. J., & Wacker, J. L. (2005). Modulation of neurodegeneration by molecular chaperones. *Nature Reviews Neuroscience, 6*, 11–22.

Nagele, R. G., Wegiel, J., Venkataraman, V., Imaki, H., Wang, K.-C., & Wegiel, J. (2004). Contribution of glial cells to the development of amyloid plaques in Alzheimer's disease. *Neurobiology of Aging, 25*, 663–674.

Nakatogawa, H., Suzuki, K., Kamada, Y., & Ohsumi, Y. (2009). Dynamics and diversity in autophagy mechanisms: Lessons from yeast. *Nature Reviews Molecular Cell Biology, 10*, 458–467.

Narayan, P., Meehan, S., Carver, J. A., Wilson, M. R., Dobson, C. M., & Klenerman, D. (2012). Amyloid-β oligomers are sequestered by both intracellular and extracellular chaperones. *Biochemistry (Moscow), 51*, 9270–9276.

Nivon, M., Abou-Samra, M., Richet, E., Guyot, B., Arrigo, A.-P., & Kretz-Remy, C. (2012). NF-κB regulates protein quality control after heat stress through modulation of the BAG3-HspB8 complex. *Journal of Cell Science, 125*, 1141–1151.

Nollen, E. A., Salomons, F. A., Brunsting, J. F., van der Want, J. J., Sibon, O. C., & Kampinga, H. H. (2001). Dynamic changes in the localization of thermally unfolded nuclear proteins associated with chaperone-dependent protection. *Proceedings of the National Academy of Sciences of the United States of America, 98*, 12038–12043.

O'Leary, J. C., Li, Q., Marinec, P., Blair, L. J., Congdon, E. E., Johnson, A. G., ... Dickey, C. A. (2010). Phenothiazine-mediated rescue of cognition in tau transgenic mice requires neuroprotection and reduced soluble tau burden. *Molecular Neurodegeneration, 5*, 45–56.

van Oosten-Hawle, P., & Morimoto, R. I. (2014). Organismal proteostasis: Role of cell-nonautonomous regulation and transcellular chaperone signaling. *Genes & Development*, *28*, 1533–1543.

Opattova, A., Filipcik, P., Cente, M., & Novak, M. (2013). Intracellular degradation of misfolded tau protein induced by geldanamycin is associated with activation of proteasome. *Journal of Alzheimer's Disease: JAD*, *33*, 339–348.

Osborn, L. M., Kamphuis, W., Wadman, W. J., & Hol, E. M. (2016). Astrogliosis: An integral player in the pathogenesis of Alzheimer's disease. *Progress in Neurobiology*, *144*, 121–141.

Paris, D., Ganey, N. J., Laporte, V., Patel, N. S., Beaulieu-Abdelahad, D., Bachmeier, C., ... Mullan, M. J. (2010). Reduction of beta-amyloid pathology by celastrol in a transgenic mouse model of Alzheimer's disease. *Journal of Neuroinflammation*, *7*, 17.

Park, H.-J., Kim, S.-S., Kang, S., & Rhim, H. (2009). Intracellular Abeta and C99 aggregates induce mitochondria-dependent cell death in human neuroglioma H4 cells through recruitment of the 20S proteasome subunits. *Brain Research*, *1273*, 1–8.

Pekny, M., Wilhelmsson, U., & Pekna, M. (2014). The dual role of astrocyte activation and reactive gliosis. *Neuroscience Letters*, *565*, 30–38.

Petrucelli, L., Dickson, D., Kehoe, K., Taylor, J., Snyder, H., Grover, A., ... Prihar, G. (2004). CHIP and Hsp70 regulate tau ubiquitination, degradation and aggregation. *Human Molecular Genetics*, *13*, 703–714.

Phatnani, H., & Maniatis, T. (2015). Astrocytes in neurodegenerative disease. *Cold Spring Harbor Perspectives in Biology*, *7*, a020628.

Picard, D. (2002). Heat-shock protein 90, a chaperone for folding and regulation. *Cellular and Molecular Life Sciences: CMLS*, *59*, 1640–1648.

Pickart, C. M., & Cohen, R. E. (2004). Proteasomes and their kin: Proteases in the machine age. *Nature Reviews Molecular Cell Biology*, *5*, 177–187.

Pierce, A., Podlutskaya, N., Halloran, J. J., Hussong, S. A., Lin, P.-Y., Burbank, R., ... Galvan, V. (2013). Overexpression of heat shock factor 1 phenocopies the effect of chronic inhibition of TOR by rapamycin and is sufficient to ameliorate Alzheimer's-like deficits in mice modeling the disease. *Journal of Neurochemistry*, *124*, 880–893.

Plate, L., Cooley, C. B., Chen, J. J., Paxman, R. J., Gallagher, C. M., Madoux, F., ... Spicer, T. P. (2016). Small molecule proteostasis regulators that reprogram the ER to reduce extracellular protein aggregation. *eLife*, *5*, e15550.

de Pril, R., Fischer, D. F., Maat-Schieman, M. L. C., Hobo, B., de Vos, R. A. I., Brunt, E. R., ... van Leeuwen, F. W. (2004). Accumulation of aberrant ubiquitin induces aggregate formation and cell death in polyglutamine diseases. *Human Molecular Genetics*, *13*, 1803–1813.

Quinlan, R. A., Brenner, M., Goldman, J. E., & Messing, A. (2007). GFAP and its role in Alexander disease. *Experimental Cell Research*, *313*, 2077–2087.

Reggiori, F., & Klionsky, D. J. (2002). Autophagy in the eukaryotic cell. *Eukaryotic Cell*, *1*, 11–21.

Renkawek, K., Bosman, G. J., & Gaestel, M. (1993). Increased expression of heat-shock protein 27 kDa in Alzheimer disease: A preliminary study. *Neuroreport*, *5*, 14–16.

Renkawek, K., Bosman, G. J., & de Jong, W. W. (1994). Expression of small heat-shock protein hsp 27 in reactive gliosis in Alzheimer disease and other types of dementia. *Acta Neuropathologica (Berlin)*, *87*, 511–519.

Rosenberg, T., Gal-Ben-Ari, S., Dieterich, D. C., Kreutz, M. R., Ziv, N. E., Gundelfinger, E. D., & Rosenblum, K. (2014). The roles of protein expression in synaptic plasticity and memory consolidation. *Frontiers in Molecular Neuroscience*, *7*, 86.

Rubinsztein, D. C., Bento, C. F., & Deretic, V. (2015). Therapeutic targeting of autophagy in neurodegenerative and infectious diseases. *Journal of Experimental Medicine*, *212*, 979–990.

Santacruz, K., Lewis, J., Spires, T., Paulson, J., Kotilinek, L., Ingelsson, M., ... McGowan, E. (2005). Tau suppression in a neurodegenerative mouse model improves memory function. *Science*, *309*, 476–481.

Schipanski, A., Oberhauser, F., Neumann, M., Lange, S., Szalay, B., Krasemann, S., ... Glatzel, M. (2014). Lectin OS-9 delivers mutant neuroserpin to endoplasmic reticulum associated degradation in familial encephalopathy with neuroserpin inclusion bodies. *Neurobiology of Aging*, *35*, 2394–2403.

Seidel, K., Meister, M., Dugbartey, G. J., Zijlstra, M. P., Vinet, J., Brunt, E. R. P., ... den Dunnen, W. F. A. (2012a). Cellular protein quality control and the evolution of aggregates in spinocerebellar ataxia type 3 (SCA3). *Neuropathology and Applied Neurobiology*, *38*, 548–558.

Seidel, K., Vinet, J., Dunnen, W. F. A., den, Brunt, E. R., Meister, M., Boncoraglio, A., ... Kampinga, H. H. (2012b). The HSPB8-BAG3 chaperone complex is upregulated in astrocytes in the human brain affected by protein aggregation diseases. *Neuropathology and Applied Neurobiology*, *38*, 39–53.

Seidel, K., Siswanto, S., Fredrich, M., Bouzrou, M., Brunt, E. R., van Leeuwen, F. W., ... den Dunnen, W. F. A. (2016). Polyglutamine aggregation in Huntington's disease and spinocerebellar ataxia type 3: Similar mechanisms in aggregate formation. *Neuropathology and Applied Neurobiology*, *42*, 153–166.

Selkoe, D. J., & Hardy, J. (2016). The amyloid hypothesis of Alzheimer's disease at 25 years. *EMBO Molecular Medicine*, *8*, 595–608.

Shammas, S. L., Waudby, C. A., Wang, S., Buell, A. K., Knowles, T. P. J., Ecroyd, H., ... Meehan, S. (2011). Binding of the molecular chaperone αB-crystallin to Aβ amyloid fibrils inhibits fibril elongation. *Biophysical Journal, 101*, 1681–1689.

Shevtsov, M., & Multhoff, G. (2016). Heat shock protein-peptide and HSP-based immunotherapies for the treatment of cancer. *Frontiers in Immunology, 7*, 171.

Shimura, H., Miura-Shimura, Y., & Kosik, K. S. (2004). Binding of tau to heat shock protein 27 leads to decreased concentration of hyperphosphorylated tau and enhanced cell survival. *Journal of Biological Chemistry, 279*, 17957–17962.

Shimura, H., Schwartz, D., Gygi, S. P., & Kosik, K. S. (2004). CHIP-Hsc70 complex ubiquitinates phosphorylated tau and enhances cell survival. *Journal of Biological Chemistry, 279*, 4869–4876.

Shinohara, H., Inaguma, Y., Goto, S., Inagaki, T., & Kato, K. (1993). Alpha B crystallin and HSP28 are enhanced in the cerebral cortex of patients with Alzheimer's disease. *Journal of the Neurological Sciences, 119*, 203–208.

Sinadinos, C., Quraishe, S., Sealey, M., Samson, P. B., Mudher, A., & Wyttenbach, A. (2013). Low endogenous and chemical induced heat shock protein induction in a 0N3Rtau-expressing *Drosophila* larval model of Alzheimer's disease. *Journal of Alzheimer's Disease: JAD, 33*, 1117–1133.

Soto, C. (2003). Unfolding the role of protein misfolding in neurodegenerative diseases. *Nature Reviews Neuroscience, 4*, 49–60.

Sulistio, Y. A., & Heese, K. (2016). The ubiquitin–proteasome system and molecular chaperone deregulation in Alzheimer's disease. *Molecular Neurobiology, 53*, 905–931.

Takahashi, R. H., Nagao, T., & Gouras, G. K. (2017). Plaque formation and the intraneuronal accumulation of β-amyloid in Alzheimer's disease. *Pathology International, 67*, 185–193.

Takeuchi, T., Suzuki, M., Fujikake, N., Popiel, H. A., Kikuchi, H., Futaki, S., ... Nagai, Y. (2015). Intercellular chaperone transmission via exosomes contributes to maintenance of protein homeostasis at the organismal level. *Proceedings of the National Academy of Sciences of the United States of America, 112*, E2497–E2506.

Taylor, R. C., Berendzen, K. M., & Dillin, A. (2014). Systemic stress signalling: Understanding the cell non-autonomous control of proteostasis. *Nature Reviews Molecular Cell Biology, 15*, 211–217.

van Tijn, P., Dennissen, F. J. A., Gentier, R. J. G., Hobo, B., Hermes, D., Steinbusch, H. W. M., ... Fischer, D. F. (2012). Mutant ubiquitin decreases amyloid β plaque formation in a transgenic mouse model of Alzheimer's disease. *Neurochemistry International, 61*, 739–748.

Tipping, K. W., van Oosten-Hawle, P., Hewitt, E. W., & Radford, S. E. (2015). Amyloid fibres: Inert end-stage aggregates or key players in disease?. *Trends in Biochemical Sciences, 40*, 719–727.

Tortosa, E., Santa-Maria, I., Moreno, F., Lim, F., Perez, M., & Avila, J. (2009). Binding of Hsp90 to tau promotes a conformational change and aggregation of tau protein. *Journal of Alzheimer's Disease: JAD, 17*, 319–325.

Tóth, M. E., Szegedi, V., Varga, E., Juhász, G., Horváth, J., Borbély, E., ... Penke, B. (2013). Overexpression of Hsp27 ameliorates symptoms of Alzheimer's disease in APP/PS1 mice. *Cell Stress and Chaperones, 18*, 759–771.

Tytell, M., Lasek, R. J., & Gainer, H. (2016). Axonal maintenance, glia, exosomes, and heat shock proteins. *F1000Research*, 5(F1000 Faculty Rev), 205.

Ushioda, R., Hoseki, J., Araki, K., Jansen, G., Thomas, D. Y., & Nagata, K. (2008). ERdj5 is required as a disulfide reductase for degradation of misfolded proteins in the ER. *Science, 321*, 569–572.

Verheijen, B. M., Gentier, R. J. G., Hermes, D. J. H. P., van Leeuwen, F. W., & Hopkins, D. A. (2016). Selective transgenic expression of mutant ubiquitin in Purkinje cell stripes in the cerebellum. *Cerebellum (London, England)*.

Vermulst, M., Denney, A. S., Lang, M. J., Hung, C.-W., Moore, S., Moseley, M. A., ... Madden, V. (2015). Transcription errors induce proteotoxic stress and shorten cellular lifespan. *Nature Communications, 6*, 8065.

Vos, M. J., Zijlstra, M. P., Kanon, B., van Waarde-Verhagen, M. A. W. H., Brunt, E. R. P., Oosterveld-Hut, H. M. J., ... Kampinga, H. H. (2010). HSPB7 is the most potent polyQ aggregation suppressor within the HSPB family of molecular chaperones. *Human Molecular Genetics, 19*, 4677–4693.

Voss, K., Combs, B., Patterson, K. R., Binder, L. I., & Gamblin, T. C. (2012). Hsp70 alters tau function and aggregation in an isoform specific manner. *Biochemistry (Moscow), 51*, 888–898.

Walsh, D. T., Montero, R. M., Bresciani, L. G., Jen, A. Y. T., Leclercq, P. D., Saunders, D., ... Jen, L.-S. (2002). Amyloid-beta peptide is toxic to neurons in vivo via indirect mechanisms. *Neurobiology of Disease, 10*, 20–27.

Wang, Y., Martinez-Vicente, M., Krüger, U., Kaushik, S., Wong, E., Mandelkow, E.-M., ... Mandelkow, E. (2009). Tau fragmentation, aggregation and clearance: The dual role of lysosomal processing. *Human Molecular Genetics, 18*, 4153.

Watts, J. C., & Prusiner, S. B. (2017). β-Amyloid prions and the pathobiology of Alzheimer's disease. *Cold Spring Harbor Perspectives in Medicine*.

Westhoff, B., Chapple, J. P., van der Spuy, J., Höhfeld, J., & Cheetham, M. E. (2005). HSJ1 is a neuronal shuttling factor for the sorting of chaperone clients to the proteasome. *Current Biology: CB*, *15*, 1058–1064.

Wettstein, G., Bellaye, P. S., Micheau, O., & Bonniaud, P. (2012). Small heat shock proteins and the cytoskeleton: An essential interplay for cell integrity? *The International Journal of Biochemistry & Cell Biology*, *44*, 1680–1686.

Wilhelmus, M. M. M., Boelens, W. C., Otte-Höller, I., Kamps, B., Kusters, B., Maat-Schieman, M. L. C., … Verbeek, M. M. (2006a). Small heat shock protein HspB8: Its distribution in Alzheimer's disease brains and its inhibition of amyloid-beta protein aggregation and cerebrovascular amyloid-beta toxicity. *Acta Neuropathologica (Berlin)*, *111*, 139–149.

Wilhelmus, M. M. M., Otte-Höller, I., Wesseling, P., de Waal, R. M. W., Boelens, W. C., … Verbeek, M. M. (2006b). Specific association of small heat shock proteins with the pathological hallmarks of Alzheimer's disease brains. *Neuropathology and Applied Neurobiology*, *32*, 119–130.

Wilhelmus, M. M. M., Boelens, W. C., Otte-Höller, I., Kamps, B., de Waal, R. M. W., … Verbeek, M. M. (2006c). Small heat shock proteins inhibit amyloid-beta protein aggregation and cerebrovascular amyloid-beta protein toxicity. *Brain Research*, *1089*, 67–78.

Wirths, O., & Bayer, T. A. (2012). Intraneuronal Aβ accumulation and neurodegeneration: Lessons from transgenic models. *Life Sciences*, *91*, 1148–1152.

Young, Z. T., Rauch, J. N., Assimon, V. A., Jinwal, U. K., Ahn, M., Li, X., … Srinivasan, S. R. (2016). Stabilizing the Hsp70-tau complex promotes turnover in models of tauopathy. *Cell Chemical Biology*, *23*, 992–1001.

Young, Z. T., Mok, S. A., & Gestwicki, J. E. (2017). Therapeutic strategies for restoring tau homeostasis. *Cold Spring Harbor Perspectives in Medicine*.

Zhou, P. (2006). REGgamma: A shortcut to destruction. *Cell*, *124*, 256–257.

CHAPTER

11

Neurodegenerative Diseases and Autophagy

Angeleen Fleming[1,*], Mariella Vicinanza[1,*], Maurizio Renna[1,*], Claudia Puri[1,*], Thomas Ricketts[1,*], Jens Füllgrabe[1,*], Ana Lopez[1,*], Sarah M. de Jager[1,*], Avraham Ashkenazi[1,*], Mariana Pavel[1,*], Floriana Licitra[1,*], Andrea Caricasole[2,*], Stephen P. Andrews[2,*], John Skidmore[2,*] and David C. Rubinsztein[1,3]

[1]Cambridge Institute for Medical Research, University of Cambridge, Cambridge Biomedical Campus, Cambridge, United Kingdom [2]Alzheimer's Research UK Cambridge Drug Discovery Institute, University of Cambridge, Cambridge, United Kingdom [3]UK Dementia Research Institute, Cambridge Biomedical Campus, Cambridge, United Kingdom

OUTLINE

* Joint first authors.

The Molecular and Cellular Basis of Neurodegenerative Diseases
DOI: https://doi.org/10.1016/B978-0-12-811304-2.00011-0

AUTOPHAGY CELL BIOLOGY

Key Autophagy Machinery

Autophagy (macroautophagy) is a degradation process that delivers cytoplasmic materials to lysosomes. By doing so, autophagy sustains cellular renovation and homeostasis by recycling molecular building blocks (such as amino acids or fatty acids) for anabolic processes. The first morphologically recognizable autophagic precursor is a flat, double-membraned, sac-like structure (called a phagophore), whose edges elongate and fuse while engulfing a portion of the cytoplasm. The resulting structure is a spherical double-membrane organelle, called the autophagosome. The formation of autophagosomes requires several steps (nucleation, elongation, and closure) governed by conserved proteins termed ATGs (AuTophaGy-related proteins) (Mizushima, Yoshimori, & Ohsumi, 2011). Autophagy initiation and autophagosome formation require multiple interactions between different individual proteins and protein complexes. For simplicity, these are referred to by their abbreviated names in the following sections and are described in full in Table 11.1.

During autophagosome formation, the ATG8 ubiquitin-like family proteins are conjugated to the lipid phosphatidylethanolamine (PE) in autophagosomal membranes. Mammalian cells have six ATG8 orthologues; the MAP1-LC3 (LC3) and GABARAP subfamilies. Lipidated ATG8 proteins have been used to distinguish autophagosomes from other cellular membranes (Itakura & Mizushima, 2010). Measuring the LC3 lipidation, scoring the number of LC3 vesicles, and detecting the degradation of long-lived proteins or damaged organelles are the mainstay methods used for monitoring autophagy (Itakura & Mizushima, 2010). However, this requires careful interpretation since immune receptors engaged by phagocytosed cargoes can also enable LC3 recruitment to single-membrane phagosomes in a process called LC3-associated phagocytosis (Sanjuan et al., 2007).

LC3/GABARAP lipidation requires a protease and two ubiquitin-like conjugation systems (Ichimura et al., 2000; Mizushima et al., 1998) as illustrated in Fig. 11.1. The first reaction involves the conjugation of the proteins ATG12 to ATG5 in a reaction requiring the enzymatic activities of ATG7 and ATG10. The ATG5-ATG12 conjugate forms a complex with ATG16L1. The cysteine protease ATG4 cleaves the C-terminus of LC3 exposing a glycine residue (LC3-I), which is activated by the ATG7 enzyme, initiating events for the second conjugation reaction. In this second reaction, the ATG12–ATG5–ATG16L1 complex, through interaction with the ATG3, acts as the E3-like ligase that determines the site of LC3 lipidation and assists the transfer of LC3-I to PE to form LC3-II (Ichimura et al., 2000). ATG8/LC3 proteins may assist in the expansion and closure of autophagosomal membranes (Nakatogawa, Ichimura, & Ohsumi, 2007) as well as in autophagosome–lysosome fusion and inner autophagosomal membrane degradation (Nguyen et al., 2016; Tsuboyama et al., 2016).

TABLE 11.1 List of Abbreviations of Proteins, Complexes and Cellular Structures Involved in Autophagy

Abbreviation	Full Name	Function of Protein or Complex in Autophagy
AKT	Protein kinase B	Serine/threonine kinase
AMBRA-1	Activating molecule in Beclin-1-regulated autophagy	Part of VPS34 complex
AMPK	AMP-activated protein kinase	Protein kinase complex
APP	Amyloid precursor protein	Membrane protein cleaved by secretases to form Aβ peptide
ATGs	AuTophaGy-related proteins	
ATG3	AuTophaGy-related protein 3	E2-ligase-like enzymatic activity
ATG4	AuTophaGy-related protein 4	Cysteine protease
ATG7	AuTophaGy-related protein 7	E1-ligase-like enzymatic activity
ATG9	AuTophaGy-related protein 9	Organization of the preautophagosomal structure/phagophore assembly site
ATG10	AuTophaGy-related protein 10	E2-ligase-like enzymatic activity
ATG12	AuTophaGy-related protein 12	Ubiquitin-like protein
ATG14	AuTophaGy-related protein 14	Part of VPS34 complex; regulates localization of the complex
ATG16L1 complex	Complex comprising ATG5, ATG12, ATG16L1	E3-ligase-like enzymatic activity
Beclin-1	Homologue of BEC-1 (*C. elegans*) and ATG6 (yeast)	Part of VPS34 complex
COPI	Coatomer protein I	Coat protein complex required for ER-Golgi transport and early endosome formation
CREB	cAMP response element-binding protein	Transcription factor
DFCP1	Double FYVE-containing protein 1	Interact with phospholipids
E2F1	E2F transcription factor 1	Transcription factor
ER	Endoplasmic reticulum	
ERES	ER-exit sites	
ESCRT	Endosomal sorting complexes required for transport	Membrane remodeling
FAM134	Family with sequence similarity 134	Selective autophagy cargo receptor
FIP200	Focal adhesion kinase family interacting protein of 200 kDa	Part of ULK1 complex
FOXO	Forkhead box	Transcription factor
FXR	Farnesoid X receptor	Nuclear receptor
FYVE	Zinc-finger domain	Binds and inserts into PI3P membranes

(Continued)

TABLE 11.1 (Continued)

Abbreviation	Full Name	Function of Protein or Complex in Autophagy
GABARAP	γ-Aminobutyric acid receptor-associated protein	ATG8 homologue
LYNUS	Lysosome nutrient-sensing	
MAM	Mitochondria-ER-associated membranes	
MAP1-LC3 or LC3	Microtubule associated proteins 1A/1B light chain 3	ATG8 homologue
mTOR	Mechanistic target of rapamycin	Protein kinase
mTORC1	mTOR complex 1; complex comprising mTOR, RAPTOR, GβL	Kinase complex
NBR1	Neighbor Of BRCA1 gene 1	Selective autophagy cargo receptor
NCOA4	Nuclear receptor coactivator 4	Selective autophagy cargo receptor
NF-κB	Nuclear factor kappa-light-chain-enhancer of activated B cells	Transcription factor (protein complex)
NDP52/CALCOCO2	Nuclear domain 10 Protein 52/calcium binding and coiled-coil domain 2	Selective autophagy cargo receptor
OPTN	Optineurin	Selective autophagy cargo receptor
P53/TP53	Tumor protein 53	Transcription factor (tumor suppressor)
p62/SQSTM1	Ubiquitin-binding protein P62/sequestosome1	Selective autophagy cargo receptor
PARKIN	Parkinson's disease protein 2 (PARK2)	E3-Ubiquitin Protein Ligase
PE	Phosphatidylethanolamine	Phospholipids found in membranes
PI3P	Phosphatidylinositol 3-phosphate	Phospholipids found in membranes
PINK1	PTEN-induced putative kinase 1 (PARK6)	mitochondrial serine/threonine-protein kinase
PPARα	Peroxisome proliferation factor-activated receptor α	Transcription factor
RAPTOR	Regulatory-associated protein of mTOR	Part of mTOR Complex 1
RE	Recycling endosome	
RPN10	26S proteasome regulatory subunit RPN10	Proteasome component
SNARE	SNAP (soluble NSF attachment protein) Receptor	Mediate vesicle fusion
SNX18	Sorting nexin 18	Membrane remodeling
TAX1BP1	Tax1 (human T-cell leukemia virus type I) binding protein 1	Selective autophagy cargo receptor
TBC1D14	TBC1 domain family member 14	Membrane remodeling
TFEB	Transcription factor EB	Transcription factor

(Continued)

TABLE 11.1 (Continued)

Abbreviation	Full Name	Function of Protein or Complex in Autophagy
TIP60	60 kDa Tat-interactive protein (K-acetyltransferase 5, KAT5)	Acetyl transferase
TOLLIP	Toll interacting protein	Selective autophagy cargo receptor
TRAF6	TNF receptor-associated factor 6	E3-ubiquitin protein ligase
TRIM	Tripartite motif	Pathogen recognition
UBD	Ubiquitin-binding domain	Recognize and bind to ubiquitin
ULK1 and ULK2	Mammalian homologs of the *C. elegans* uncoordinated 51 kinase	serine/threonine kinase
ULK1/2-complex	Complex comprising ULK1/2, ATG13, ATG101, and FIP200	Regulation of autophagosome biogenesis
VAMPs	Vesicle-associated membrane proteins	Mediate vesicle fusion
VPS	Vacuolar protein sorting-associated protein	Vesicular transport
VPS34 complex	Complex comprising Beclin-1–ATG14–VPS15–VPS34	Class III phosphoinositide 3-kinase (PI3K)
VTI1B	Vesicle transport through interaction with t-SNAREs homolog 1B	Mediate vesicle fusion
WASH	WASP and Scar homologue	Vesicle trafficking
WDR45	WD repeat domain phosphoinositide-interacting protein 45	Interacts with PI_3
WIPIs	WD-repeat protein interacting with phosphoinositides (homologues of ATG18)	Interact with phospholipids
ZKSCAN3	Zinc-finger protein with KRAB and SCAN domains 3	Transcription factor

Autophagy initiation and autophagosome formation requires multiple interactions between different individual proteins and protein complexes. For simplicity, these are referred to by their abbreviated names in the text. These proteins and complexes have many diverse cellular functions other than those involved in autophagy; only their main role in autophagy is described here.

ATGs are Organized in Signaling Modules Upstream of LC3 Conjugation

The ULK1/2-complex is one of the most upstream signaling units in autophagosome formation. This complex includes the ULK1/2 homologues, ATG13, ATG101, and FIP200. AMP-activated protein kinase (AMPK) phosphorylates ATG13 and FIP200 (components of the ULK1/2-complex) and AMBRA-1 and Beclin-1 (components of the VPS34 complex), thereby targeting these two complexes to the preautophagosomal membrane (see Fig. 11.2) (Di Bartolomeo et al., 2010; Egan et al., 2015; Itakura & Mizushima, 2010; Jung et al., 2009; Park, Jung et al., 2016; Russell et al., 2013).

The generation of the lipid phosphatidylinositol 3-phosphate (PI3P) by the VPS34 complex at the phagophore initiation site aids the recruitment of PI3P-binding ATGs, such as DFCP1 and ATG18/WIPIs family proteins (Proikas-Cezanne, Takacs, Donnes, & Kohlbacher, 2015). WIPI2 is crucial for the localization of ATG16L1 complex

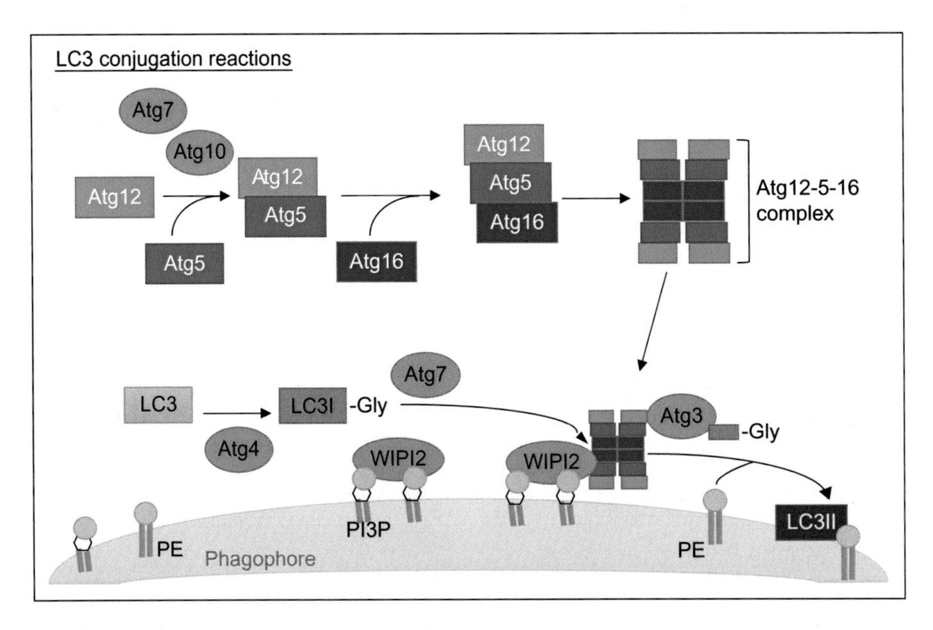

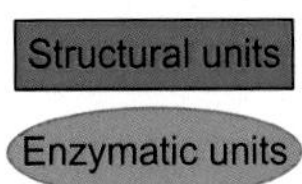

FIGURE 11.1 Conjugation steps involved in the regulation of autophagy. Two conjugation systems are required for phagophore/autophagosome membrane elongation resulting in the formation of the ATG12-5-16 complex and LC3-II. Structural components are represented as squares, and enzymatic components are represented as circles. Initial conjugation of ATG12 and ATG5 is mediated by the enzymatic activity of ATG7 and ATG10. ATG5 interacts with ATG16, generating the ATG12-5-16 conjugate, which then homo-oligomerizes to form a tetrameric structure called ATG12-5-16 complex. LC3 is cleaved by ATG4 exposing a glycine residue and generating LC3-I, which is activated by ATG7. The ATG12-5-16 complex through ATG3 interaction covalently conjugates LC3-I to PE to form LC3-II. WIPI 2 modulates the localization of the ATG12-5-16 complex through PI3P, facilitating the interaction of LC3-I with PE and its lipidation in the autophagosomal membrane. *PE*, phosphatidylethanolamine; *PI3P*, phosphatidylinositol 3-phosphate.

and dictates where the LC3 lipidation occurs (Dooley et al., 2014). Together with ULK1, WIPI2 may influence the localization of both mATG9 and ATG2A/B. mATG9, the only transmembrane protein among the core ATGs, is considered one of the suppliers of lipid bilayers to the initiation membrane during elongation (Orsi et al., 2012; Papinski et al., 2014), while ATG2A/B regulates autophagosome closure through fission/scission-type events (Knorr, Lipowsky, & Dimova, 2015; Velikkakath, Nishimura, Oita, Ishihara, & Mizushima, 2012).

Autophagosome Membrane Trafficking Events

The membrane source for autophagosome formation remains a key open question in the field. Membranes from different organelles are believed to contribute to autophagosome formation by meeting in a particular subcellular compartment representing the autophagosome platform or "isolation membrane." The isolation membrane is a compartment in proximity to mitochondria-endoplasmic reticulum

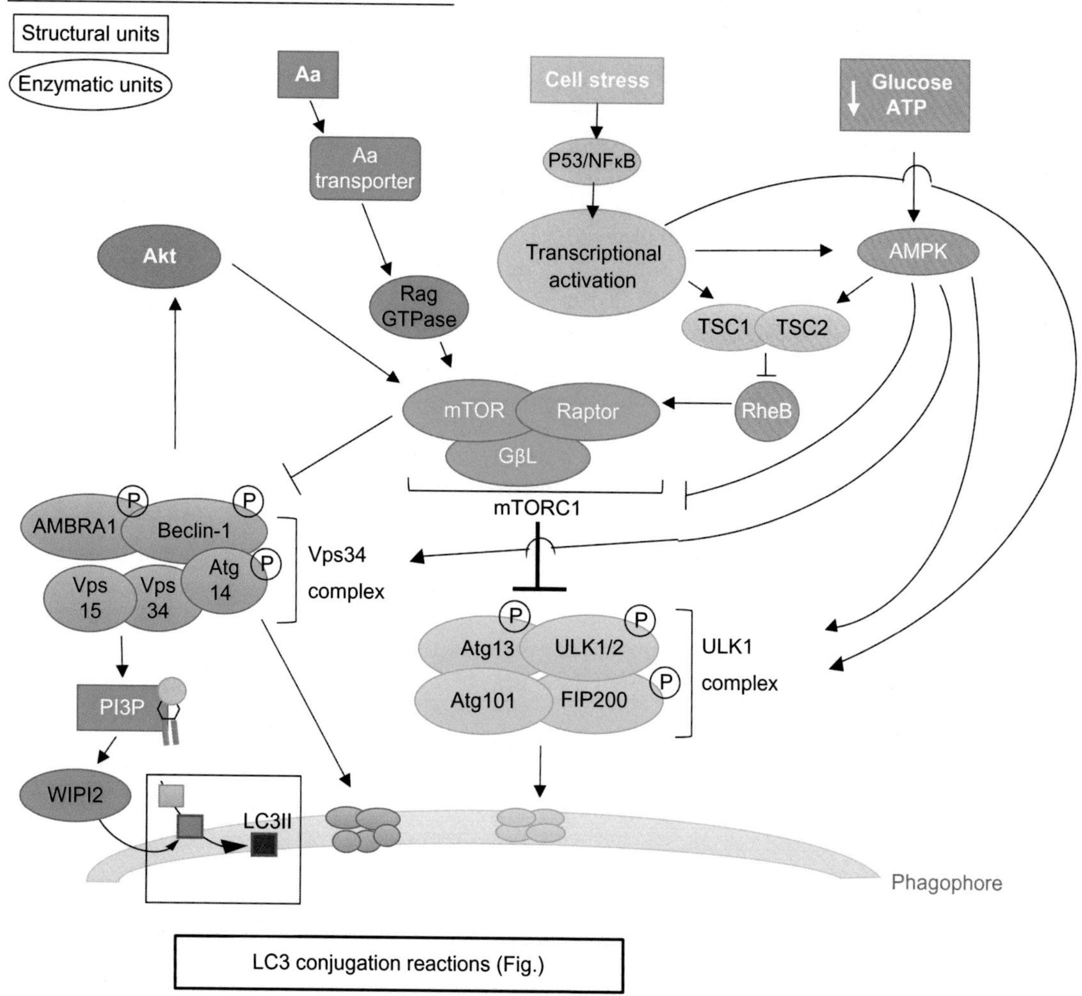

FIGURE 11.2 Initiation complexes controlling the initiation of autophagy. Many different cellular signals such as level of nutrients, ATP, or cellular stress control autophagy by modulating mTORC1 activity. The activation of the Akt and the presence of amino acids detected by Rag GTPases induce mTORC1 which in turns inhibits both Vps34 and ULK1/2 complexes and hence inhibiting autophagy. Conversely, deprivation of glucose or ATP promotes autophagy by activating AMPK, which directly phosphorylates and induces VPS34 and ULK1/2 complexes. Moreover, AMPK inactivates mTORC1 either directly or through TSC1/TSC2, *removing inhibition from ULK1/2-complex*. Cell stress, modulated by factors like p53 or NFκB, transcriptionally activates proautophagy genes such as TSC1/TSC2 and ULK1/2 and, therefore, inducing autophagy. The formation of the phagophore as a preautophagosomal structure requires the serial recruitment of ULK1/2-complex first and VPS34 complex after, which generates PI3P, crucial for autophagosome formation. *mTORC1*, mTOR complex; *AMPK*, AMP-activated protein kinase; *PI3P*, phosphatidylinositol 3-phosphate.

(ER)-associated membranes (MAMs) (Hamasaki et al., 2013; Hayashi-Nishino et al., 2009) and is labeled by ATG14, DFCP1, and WIPI2 (Axe et al., 2008; Proikas-Cezanne et al., 2015). Contact sites between the isolation membrane and surrounding organelles might contribute to the completion of the autophagosome (Biazik, Yla-Anttila, Vihinen, Jokitalo, & Eskelinen, 2015). Different sources have been proposed as donor membranes including the ER, MAM, ER-exit sites, the ER-Golgi intermediate compartment, recycling endosomes (REs), Golgi, and plasma-membrane (Axe et al., 2008; Biazik et al., 2015; Ge, Melville, Zhang, & Schekman, 2013; Ge, Zhang, & Schekman, 2014; Graef, Friedman, Graham, Babu, & Nunnari, 2013; Itakura & Mizushima, 2010; Karanasios et al., 2016; Knaevelsrud et al., 2013; Longatti et al., 2012; Park, Jung et al., 2016; Puri, Renna, Bento, Moreau, & Rubinsztein, 2013; Ravikumar, Moreau, Jahreiss, Puri, & Rubinsztein, 2010; Shibutani & Yoshimori, 2014; Tan et al., 2013; Yla-Anttila, Vihinen, Jokitalo, & Eskelinen, 2009).

The endocytic compartment is believed to play a primary role in autophagosome formation. The ATG16L1 complex and mATG9 travel on independent clathrin-coated vesicles, and these vesicles fuse in the REs in a VAMP3-dependent manner. The trafficking of these proteins from plasma-membrane to RE trafficking and subsequent vesicle fusion are essential for autophagy (Moreau, Ravikumar, Renna, Puri, & Rubinsztein, 2011; Puri et al., 2013; Ravikumar et al., 2010). The contribution of REs to autophagosome formation is supported by studies showing that autophagic proteins (e.g., ULK1 and LC3) localize on the RE and that the overexpression of RE-resident proteins (e.g., TBC1D14 and SNX18) interfere with the trafficking of mATG9 and ATG16L1 (Knaevelsrud et al., 2013; Lamb et al., 2016; Longatti et al., 2012).

Key Signaling Pathways

Low cellular energy and nutrient states signal to the autophagy pathway by posttranslationally modifying autophagy-initiating complexes or by regulating the transcription of core autophagy genes. The energy-sensing AMPK and the growth factor-regulated and nutrient-sensing kinase mammalian target of rapamycin (mTOR) oppositely regulate the ULK1/2 and VPS34 complexes (Kim et al., 2013; Kim, Kundu, Viollet, & Guan, 2011; Yuan, Russell, & Guan, 2013), and thereby autophagy, through a series of phosphorylation events.

AMPK is activated in response to low nutrients (glucose) and low energy (ATP) (Kim et al., 2013). AMPK allosteric activation by AMP binding and phosphorylation of a conserved threonine residue (Thr172) promotes autophagy by directly activating ULK1 through phosphorylation of Ser 317 and Ser 777 (under glucose starvation) (Kim et al., 2011) or Ser 555 (under nutrient starvation and mitophagy) (Yuan et al., 2013). AMPK activates the proautophagy VPS34 complex by phosphorylating Beclin-1 at Ser91/Ser94 (Orsi et al., 2012).

The mTOR complex 1 (mTORC1) is activated by nutrients (amino acids sensed by the Rag GTPases) and growth factors (signaling by receptor tyrosine kinases and the PI3K/Akt pathway) (Laplante & Sabatini, 2012). Under nutrient sufficiency, active mTORC1 inhibits autophagy by binding the ULK1/2-complex, (via raptor-ULK1/2 association) and phosphorylating ATG13 and ULK1 at Ser 757 (i.e., a different site from that phosphorylated by AMPK). This suppresses ULK1/2 kinase activity and prevents the ULK1–AMPK interaction (Ganley et al., 2009; Hosokawa et al., 2009; Jung et al., 2009; Kim et al., 2011). Under nutrient starvation, mTORC1 dissociates from the ULK1/2-complex, resulting in ULK1 activation, autophosphorylation, and phosphorylation of ATG13 and FIP200, leading to the autophagosome formation (Ganley et al., 2009; Jung et al., 2009; Kim et al., 2011). mTORC1 also inhibits the phosphoinositide 3-kinase activity of the proautophagy VPS34 complex by phosphorylating ATG14 (Velikkakath et al., 2012). Additional posttranslational modifications that increase ULK1 activity after different inducing stimuli include ubiquitination, through the E3-ligase TRAF6 (Nazio et al., 2013), and acetylation, by the acetyltransferase TIP60 (Lin et al., 2012).

A plethora of transcription factors integrate a wide range of cellular stimuli to induce core autophagy-related genes expression (Fullgrabe, Klionsky, & Joseph, 2014). Among the autophagy-associated transcription factors, the transcription factor EB (TFEB) (Sardiello et al., 2009; Settembre et al., 2011) has a prominent role, as its overexpression alone is sufficient to induce autophagy and to ameliorate the phenotype of neurodegenerative diseases (Decressac et al., 2013; Tsunemi et al., 2012) and lysosomal storage disorders (LSDs) in vivo (Medina et al., 2011; Spampanato et al., 2013). TFEB connects the lysosome nutrient-sensing machinery to the transactivation of autophagy-related genes (Settembre et al., 2012). Under fed conditions, TFEB is phosphorylated by mTORC1, which leads to its retention in the cytosol. When autophagy is induced, through the inhibition of mTORC1, TFEB becomes dephosphorylated and translocates to the nucleus (Roczniak-Ferguson et al., 2012).

Other well established transcriptional regulators of autophagy include the FOXO family, p53, E2F1, and NF-κB (Hayashi-Nishino et al., 2009). More recently, the farnesoid X receptor (FXR)/peroxisome proliferation factor-activated receptor α (PPARα)/cyclic adenosine monophosphate (cAMP) response element-binding protein (CREB) axis was discovered as a regulator of a plethora of core autophagy-related genes. Under fed conditions, the nuclear receptor FXR acts as a transcriptional repressor, while autophagy induction by starvation leads to the activation of CREB and PPARα (Lee et al., 2014; Seok et al., 2014).

Selective Autophagy

Stress-induced autophagy is thought to be nonselective, leading to bulk degradation of cytoplasm. However, autophagy contributes to intracellular homeostasis in fed conditions by selectively degrading long-lived proteins or damaged organelles, recognized by specific autophagy receptors (Shaid, Brandts, Serve, & Dikic, 2013; Svenning & Johansen, 2013).

Selective autophagic pathways are generally named after the cargo destined for degradation and include aggrephagy (protein aggregates), mitophagy (mitochondria), xenophagy (pathogens) (Deretic, Saitoh, & Akira, 2013; Melser, Lavie, & Benard, 2015; Randow & Youle, 2014; Rogov, Dotsch, Johansen, & Kirkin, 2014; Sorbara & Girardin, 2015), ER-phagy (ER) (Khaminets et al., 2015; Mochida et al., 2015), ferritinophagy (ferritin) (Dowdle et al., 2014; Mancias, Wang, Gygi, Harper, & Kimmelman, 2014), pexophagy (peroxisomes) (Kim, Hailey, Mullen, & Lippincott-Schwartz, 2008), ribophagy (ribosomes) (Kraft, Deplazes, Sohrmann, & Peter, 2008), and lipophagy (lipid droplets) (Singh et al., 2009). Autophagy receptors are generally considered as either ubiquitin-dependent or ubiquitin-independent (Khaminets, Behl, & Dikic, 2016).

Aberrantly folded or unused proteins are ubiquitinated, aggregated, and sequestered by proteins containing an ubiquitin-binding domain (i.e., p62/SQSTM1, NBR1, OPTN, TAX1BP1, NDP52/CALCOCO2, TOLLIP, and RPN10) that deliver them to lysosomes via autophagy (Bjorkoy et al., 2005; Kirkin, Lamark, Johansen, & Dikic, 2009; Lu, Psakhye, & Jentsch, 2014; Marshall, Li, Gemperline, Book, & Vierstra, 2015; Newman et al., 2012; Pankiv et al., 2007; Thurston, Ryzhakov, Bloor, von Muhlinen, & Randow, 2009; Wild et al., 2011).

Mitophagy is responsible for disposal of dysfunctional mitochondria. Multiple signals trigger mitophagy, including hypoxia and erythroid differentiation. PTEN-induced putative kinase 1 (PINK1) and PARKIN, proteins encoded by two genes that are mutated in autosomal recessive Parkinson's disease (PD), enable forms of mitophagy. This pathway is activated by nonhypoxic mitochondrial damage and linked to neurodegenerative diseases, such as PD and amyotrophic lateral sclerosis (ALS) (Hamacher-Brady & Brady, 2015; Liu et al., 2012; Melser et al., 2015; Pickrell et al., 2015; Sandoval et al., 2008). In response to mitochondrial damage, PINK1 is stabilized on the mitochondrial outer membrane and phosphorylates both cytoplasmic PARKIN and ubiquitin on mitochondria (Cunningham et al., 2015; Kane et al., 2014; Kazlauskaite et al., 2014; Kondapalli et al., 2012; Koyano et al., 2014; Ordureau et al., 2014). p62, OPTN, TAX1BP1, and NDP52 work as ubiquitylated mitochondrial protein receptors (Heo, Ordureau, Paulo, Rinehart, & Harper, 2015; Lazarou et al., 2015; Wong & Holzbaur, 2014).

Cytosolic bacteria can also be ubiquitylated and degraded by autophagy as a part of the innate immune response. Several E3 ligases attach ubiquitin chains to intracellular pathogens, for example, PARKIN on *Mycobacterium tuberculosis* (Manzanillo et al., 2013) and Lrsm1 on *Salmonella enterica* (Huett et al., 2012). Furthermore, *Salmonella*-containing endosomes can undergo ubiquitination, which recruits the autophagic machinery and ultimately incorporates them into autophagosomes (Fujita et al., 2013). Similarly, different members of the tripartite motif protein family have been linked to xenophagy (Kimura et al., 2015; Mandell et al., 2014).

A growing class of ubiquitin-independent selective autophagy pathways have been described (Khaminets et al., 2016), such as NCOA4-mediated ferritinophagy (Dowdle et al., 2014; Mancias et al., 2014) and FAM134-dependent ER-phagy (Khaminets et al., 2015), that appear specialized for degradation of one substrate cargo.

Lysosomes

Whilst the signaling pathways and machinery for the generation and trafficking of autophagosomes are important, arguably the key organelle in this process is the lysosome, since it plays a crucial role in maintaining the balance of cellular metabolism and growth by continuously mediating anabolic and catabolic processes. First described by Christian de Duve, the lysosome is a cellular organelle made of a single-lipid bilayer membrane and an acidic lumen (de Duve, 2005), which contains a complex machinery of hydrolases that are responsible for the catabolism of a vast range of substrates (Luzio, Pryor, & Bright, 2007; Saftig & Klumperman, 2009; Schroder, Wrocklage, Hasilik, & Saftig, 2010). The majority of extracellular substrates are transported to the lysosome via the endocytic pathway, and in particular through the fusion of the lysosome with late endosomes (Conner & Schmid, 2003; Huotari & Helenius, 2011), whereas intracellular substrates reach the lysosome via fusion of autophagosomes with lysosomes along the autophagic pathway. The delivery of engulfed cytosolic components is tightly coordinated so that only fully formed autophagosomes fuse with the endocytic system and deliver their contents to the lysosome. In mammalian cells, SNAREs, including VAMP7, VAMP8, and VTI1B, mediate the lysosomal fusion of autophagosomes (Fader, Sanchez, Mestre, & Colombo, 2009; Furuta, Fujita, Noda, Yoshimori, & Amano, 2010). SNAREs are membrane-anchored proteins localized on opposing membrane compartments that can interact with each other to form a highly energetically favorable complex. In order to drive membrane fusion, SNAREs must form a *trans*-SNARE complex consisting of one R-SNARE on the donor membrane and three Q-SNAREs on the acceptor membrane (Jahn & Scheller, 2006). As these SNAREs are common to other intracellular trafficking pathways, it can be difficult to assess their specific roles in endocytosis and membrane trafficking versus autophagy. The Q-SNAREs syntaxin-7, syntaxin-8, and VTI1B, along with the R-SNAREs VAMP7 and VAMP8, have been linked to the fusion of late endosomes and lysosomes and thus play indirect roles in autophagosome maturation (Pryor et al., 2004). However, these SNAREs also function in autophagosomal fusion. Indeed, VAMP8 and VTI1B, but not VAMP7, syntaxin-7 or syntaxin-8, were shown to be involved in autophagosomal fusion during clearance of intracellular bacteria (Furuta et al., 2010).

Autophagy in Neuronal Physiology

In the mouse, ubiquitous deletion of core autophagy genes results in neonatal death (Komatsu et al., 2005; Kuma et al., 2004; Sou et al., 2008); however, conditional rescue of autophagy in the nervous system can rescue this lethality (Yoshii et al., 2016). The essential role of autophagy in maintaining normal neuronal physiology has been elucidated through disruption of core autophagy genes using conditional knock-out approaches. Neuronal depletion of ATG5, ATG7, FIP200, or WIPI4 results in progressive neurodegenerative phenotypes, an accumulation of ubiquitinated and often p62/SQSTM1-positive protein inclusions, cell death, and reduced survival of the mice (Hara et al., 2006; Komatsu et al., 2006; Komatsu, Waguri, Koike et al., 2007; Liang, Wang, Peng, Gan, & Guan, 2010; Zhao et al., 2015)[2-6]. However, there are differences in the nature of the behavioral and pathological changes observed depending on the autophagy gene and cell type targeted.

Many of the original nervous system models generated did not disrupt autophagy exclusively in neurons. Depending on the promoter used, such as Nestin-Cre, nonneuronal support cells and neuronal stem cells were also targeted. Hence, there has been continued focus on more precise cellular autophagy disruption to delineate its role within specific nervous system cell types. This includes targeting subsets of neuronal cell types (Chen et al., 2013; Kaushik et al., 2011; Komatsu et al., 2006; Komatsu, Wang et al., 2007; Nishiyama, Miura, Mizushima, Watanabe, & Yuzaki, 2007; Zhou, Doggett, Sene, Apte, & Ferguson, 2015) as well as nonneuronal (e.g., Schwann) cells (Gomez-Sanchez et al., 2015; Jang et al., 2016; Jang et al., 2015). Studies disrupting core autophagy genes in neural progenitor cells also provide an emerging area of focus, with autophagy disruption resulting in reduced neural stem cell survival and neuronal maturation (Lu et al., 2014; Wang, Liang, Bian, Zhu, & Guan, 2013; Wu et al., 2016; Xi et al., 2016; Yazdankhah, Farioli-Vecchioli, Tonchev, Stoykova, & Cecconi, 2014).

In addition to knock-out studies, enhanced tools are being developed to image and monitor autophagy in vivo in normal and disease conditions, for example, GFP-LC3 reporter mice and more recently GFP-RFP-LC3 mice (Castillo, Valenzuela et al., 2013; Mizushima, Yamamoto, Matsui, Yoshimori, & Ohsumi, 2004; Pavel et al., 2016). Furthermore, mouse models have been generated to study mitophagy in vivo (McWilliams et al., 2016; Sun et al., 2015) in specific cell types as well as under varied genetic conditions such as *Atg5* or *Atg7* disruption. These tools will enable measurement of how mitophagy is modulated in broad genetic and pharmacological conditions. They may be particularly useful for PD, where there is extensive support for a central role for autophagy/mitophagy disruption, mitochondrial damage, and an established importance of the PINK1–PARKIN pathway, as reviewed (Pickrell et al., 2015).

AUTOPHAGY IN NEURODEGENERATIVE DISEASES

Although the most common neurodegenerative diseases are largely sporadic, mutations that give rise to rare familial forms and genes identified in GWAS studies have highlighted how perturbation of autophagic processes contribute to these diseases. Below, we discuss the evidence for purturbed autophagy in the pathogenesis of various neurodegenerative diseases and the genetic factors that have been identified which may affect the efficiency of autophagic processes in affected patients (summarized in Fig. 11.3).

Alzheimer's Disease

Alzheimer's disease (AD), the most common neurodegenerative disorder, is characterized by the accumulation of intraneuronal tau tangles and extracellular amyloid-beta (Aβ) deposits called plaques. Whereas abundant Aβ deposits are specific to AD, tau inclusions are also characteristic of other neurodegenerative diseases named tauopathies (Goedert & Spillantini, 2006). Aβ peptides are cleavage products of the amyloid precursor protein (APP). In sporadic cases, the accumulation of plaques appears to result from impaired Aβ clearance, in contrast with the overproduction of Aβ species in the rare dominantly inherited forms of AD

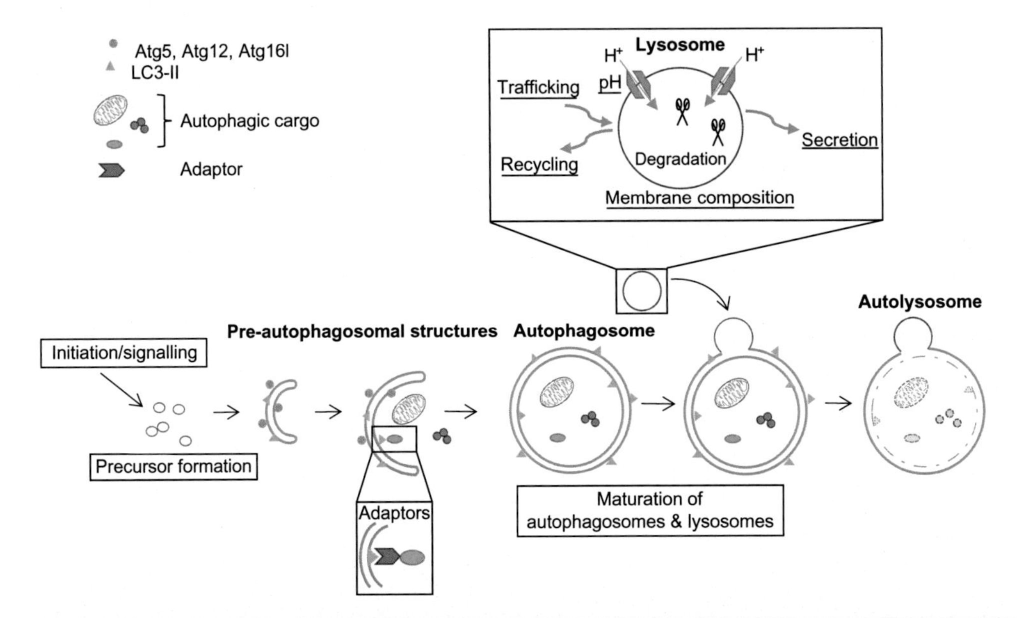

Initiation & signalling	Precursor formation	Adaptor proteins	Maturation	Autolysosome formation	Lysosome function	Trafficking	Secretion	Mitophagy
C9ORF72	WIPI4	P62	EPG5	VPS11	SPG11	Tau	Aβ	Parkin
Atg5	PICALM	optineurin	PICALM		GBA	VPS35	SYT11	PINK1
SPG15	VPS35	Htt	SigR1		TFEB	LRRK2	TFEB	SPG59
	α-syn		CHMPB2		ATP13A2	ALS2		
	malin		SPG49		PS-1	CHMPB2		
	Laforin				Tau	DYNC1H1		
					CCT5	Dynactin		
					SNX14	Dynamin2		
					NPC1			

FIGURE 11.3 Schematic diagram of autophagosome formation and degradation. Genes that are known to play a role in specific steps of the pathway and for which there are known clinical mutations are presented in the table below.

(e.g., APP mutations) (Mawuenyega et al., 2010; Potter et al., 2013). Downregulation of autophagy can enhance AD pathogenesis in model systems and similarly upregulation of the autophagic system has been shown to reduce Aβ levels in a number of models (Boland et al., 2008; Ravikumar, Sarkar, & Rubinsztein, 2008; Spilman et al., 2010; Tian, Bustos, Flajolet, & Greengard, 2011; Vingtdeux et al., 2011).

Neurons from AD patients show an abnormal accumulation of autolysosomes/lysosomes (Nixon et al., 2005; Yu et al., 2005), and autophagic failure by impaired autophagosome formation or defective lysosomal clearance appear to contribute to the increase of Aβ/tau deposits and the development of AD features in fibroblasts and brains from patients and mouse models (Li, Zhang, & Le, 2010; Nixon & Yang, 2011; Zhang, Chen, Huang, & Le, 2013). Wild-type APP has been found in autophagosomes, suggesting that autophagy might also contribute to Aβ formation (Boland et al., 2008; Yu et al., 2005). Mutations in Presenilin-1 (*PS-1*) change the way APP protein is processed into Aβ and are one of the main causes of familial AD (Citron et al., 1997). However, PS-1 is also necessary for the v-ATPase-dependent acidification of the lysosome, and PS-1 mutations are associated with elevated lysosomal pH that would be expected to affect its catabolic activity (Coffey, Beckel, Laties, & Mitchell, 2014; Lee et al., 2010; Wolfe et al., 2013). The lysosomal endopeptidase Cathepsin D (CatD) has also been implicated in the clearance of Aβ and tau peptides through the autophagy-lysosomal system. CatD has been found to colocalize with senile plaques of AD patients (Cataldo, Hamilton, & Nixon, 1994; Cataldo & Nixon, 1990; Khurana et al., 2010) and is involved in the processing APP and apolipoprotein E, both important factors in AD pathogenesis (Vidoni, Follo, Savino, Melone, & Isidoro, 2016; Zhou, Scott, Shelton, & Crutcher, 2006).

While the role of autophagy in degrading Aβ has been extensively studied in model systems (reviewed in Zare-Shahabadi, Masliah, Johnson, & Rezaei, 2015), autophagy-dependent secretion of Aβ into the extracellular space has been also reported in APP transgenic mice in which ATG7 is knocked out, suggesting that autophagy might regulate plaque formation (Nilsson et al., 2013). Indeed, the autophagic markers LC3, ATG5, and ATG12 have been associated with Aβ plaques and tau tangles in human brains (Ma, Huang, Chen, & Halliday, 2010), and both APP and Aβ peptides can be found within autophagosomes in AD mouse models (Lunemann et al., 2007).

The clathrin adapter protein PICALM (Phosphatidylinositol Binding Clathrin Assembly Protein) has been shown to interact with LC3 and target APP into autophagosomes (Tian, Chang, Fan, Flajolet, & Greengard, 2013). Polymorphisms in the *PICALM* gene are associated with increased risk of AD. PICALM is a key component of clathrin-mediated endocytosis, and loss of function of this protein inhibits autophagy at both the levels of autophagosome formation and degradation and affects tau clearance in model systems (Moreau et al., 2014). Recently, it has been also reported that levels of this protein are decreased in AD brains (Ando et al., 2013, 2016).

Depletion or downregulation of Beclin-1, an autophagy initiator, promotes accumulation and deposition of Aβ, leading to marked neurodegeneration in both cell culture and mouse models (Jaeger et al., 2010; Pickford et al., 2008). Reduced levels of Beclin-1 have been reported in the brains of AD patients (Pickford et al., 2008; Small et al., 2005), and these have been correlated with caspase 3 activation, which in turn cleaves Beclin-1 to a fragment that becomes localized to plaque regions and blood vessels in AD brains (Rohn et al., 2011).

Tauopathies

Besides AD, the accumulation of tau protein into intracellular tangles is the main feature of many other neuronal disorders termed tauopathies, including progressive supranuclear palsy (PSP), corticobasal degeneration (CBD), or frontotemporal dementias (FTDs) (Lee, Goedert, & Trojanowski, 2001). Degradation of soluble tau depends on both the ubiquitin-proteasome system and autophagy (Chesser, Pritchard, & Johnson, 2013), whereas oligomers and aggregates are mainly degraded by autophagy (Boland et al., 2008; Lee, Lee, & Rubinsztein, 2013). Hyperphosphorylated tau colocalizes with the autophagosomal marker LC3 and the autophagy cargo receptor p62/SQSTM1 in tauopathy patients with CBD or PSP (Piras, Collin, Gruninger, Graff, & Ronnback, 2016). Some pathogenic tau mutations, like P301L or A152T, impair proteasome activity leading to the accumulation of ubiquitinated proteins and small peptides in animal models (Myeku et al., 2016) that can be ameliorated by inducing autophagy (Lopez et al., 2017). Autophagy induction ameliorates the pathological consequences of aberrant tau in diverse experimental systems including primary neurons, *Drosophila*, zebrafish, and mice (Berger et al., 2006; Caccamo et al., 2013; Kruger, Wang, Kumar, & Mandelkow, 2012; Lopez et al., 2017; Moreau et al., 2014; Schaeffer et al., 2012).

Tau protein binds to and stabilizes microtubules, which are the basic components involved in axonal vesicle transport (Bernhardt & Matus, 1984; Goedert, Wischik, Crowther, Walker, & Klug, 1988; Millecamps & Julien, 2013). Mutations and/or hyperphosphorylation of tau have been reported to impair the dynein–dynactin complex, leading to disruption of axonal transport and increasing the number of autophagosomes in FTDs and AD (Butzlaff et al., 2015; Kimura, Noda, & Yoshimori, 2008; Lacovich et al., 2017; Majid et al., 2014). Small tau fibrils can interact with lysosomal membranes in vitro (Wang et al., 2009) and disrupt their permeability, resulting in lysosomal damage in a mouse model of AD (Piras et al., 2016) and in AD patients (Perez et al., 2015). Moreover, increased levels of lysosomal proteins LAMP1 and CatD have been recently reported in CBD and PSP patients (Piras et al., 2016).

Parkinson's Disease

PD results in the progressive loss of dopaminergic neurons in the substantia nigra pars compacta and is associated with the presence of Lewy bodies, alpha-synuclein (α-syn)-positive intracellular inclusions. Increased α-syn levels, as a result of multiplication of the *SNCA* gene encoding it, is sufficient to cause PD. Overexpression of α-syn in cells and mice impairs autophagy and results in mislocalization of mATG9 (Winslow et al., 2010). Accumulation of α-syn has also been described for PD-associated mutations of VPS35 (vacuolar protein sorting-associated protein 35). VPS35 is a component of the retromer complex that recruits actin nucleation-promoting WASP and Scar homologue complex to endosomes. The D620N mutation in VPS35 causes autosomal-dominant PD and, in transfected cells, has been shown to impair autophagy, mislocalize mATG9 (Zavodszky et al., 2014), and affect the trafficking of the lysosomal protein LAMP2A, leading to α-syn accumulation (Tang et al., 2015).

Heterozygous mutation of *GBA1* is the most common known genetic risk factor for PD. *GBA1* encodes glucocerebrosidase (GCase), a lysosomal enzyme that cleaves the β-glucosyl linkage of glucosylceramide (GlcCer). Deficiency in GCase activity leads to accumulation of its substrate in the lysosome and compromised lysosomal activity. Loss of GCase activity, with resultant increased GlcCer levels, leads to an increase in α-syn levels in cultured neurons, mouse, and human brain. Increased α-syn in turn inhibits lysosomal maturation and GCase activity, resulting in additional GlcCer accumulation and further α-syn accumulation (Mazzulli et al., 2011). Even in sporadic PD without *GBA1* mutations, decreased lysosomal GCase activity was directly related to reduced lysosomal chaperone-mediated autophagy and increased α-syn in the early stages of PD (Murphy et al., 2014). Treatment of iPS-derived dopaminergic (DA) neurons from PD patients with a modulator of GCase enhanced GCase activity and increased clearance of α-syn and reversed lysosomal dysfunction (Aflaki, Borger et al., 2016; Mazzulli et al., 2016). In the Thy1-SNCA mouse model of PD, where mice express mutant human A53T α-syn, ectopic expression of GCase in the striatum led to a decrease in the levels of α-syn and delayed the progression of synucleinopathy (Rockenstein et al., 2016). In addition to inhibition of the autophagy-lysosomal pathway, GBA mutations activate the unfolded protein response and lead to ER stress (in fibroblasts and iPSC-derived DA neurons from PD patients), which can be reversed by small molecule chaperones in cells and *Drosophila*, by improving GCase trafficking to the lysosomes (Fernandes et al., 2016; Sanchez-Martinez et al., 2016). Homozygous GBA mutation causes Gaucher disease, an LSD, and chaperone treatment of Gaucher patient macrophages induces autophagy, restoring autophagosome maturation, and the fusion of lysosomes with autophagosomes (Aflaki, Moaven et al., 2016). The mechanism by which GCase deficiency leads to reduced autophagy and accumulation of α-syn has been proposed to be as a result of altered lysosomal recycling, via the process of autophagy-lysosome reformation, resulting in the accumulation of defective lysosomes (Magalhaes et al., 2016).

Mutation in *ATP13A2*, which encodes a lysosomal P5-type ATPase that facilitates cation transport, results in autosomal recessive early-onset PD. Depletion of ATP13A2, through modulation of synaptotagmin 11 levels, impairs lysosomal function and the degradation of lysosomal substrates, resulting in accumulation of α-syn (Bento, Ashkenazi, Jimenez-Sanchez, & Rubinsztein, 2016; Dehay et al., 2012). Although Atp13a2-deficient mice do not exhibit degeneration of DA neurons or significant accumulation of α-syn, subunit c of mitochondrial ATP synthase accumulated in abnormal lysosomes as a result of lysosomal dysfunction and suggests that clearance of damaged mitochondria may be impaired (Sato et al., 2016). Additionally, trafficking to the lysosome of CatD, which is known to degrade α-syn, was decreased (Kett & Dauer, 2016). Studies of patient-derived cells show Zn^{2+} dyshomeostasis, mitochondrial dysfunction, and glycolytic dysfunction as a result of loss of ATP13A2 (Park, Koentjoro, Davis, & Sue, 2016).

Familial early-onset forms of PD are also caused by recessive mutations in PARK2/PARKIN (Kitada et al., 1998), encoding an E3-ubiquitin ligase, and PINK1 (Valente et al., 2004), encoding a serine-threonine kinase. As described above, both proteins control mitophagy. Both PINK1 and PARKIN patient-derived mDA neurons showed accumulation of α-syn, increased susceptibility to mitochondrial toxins, mitochondrial dysfunction, and increased intracellular DA levels (Chung et al., 2016).

Mutations in the gene encoding leucine-rich repeat kinase 2 are responsible for the majority of inherited forms of PD as well contributing to some cases of sporadic PD. Whilst the exact function of the wild-type protein is not fully understood, loss-of-function experimental models suggest an important role in intracellular vesicle trafficking (reviewed in Roosen & Cookson, 2016). However, clinical mutations are spread throughout this multidomain protein, and it is unclear how these individual mutations may affect protein function. For example, the kinetics of autophagosome formation and autophagosome–lysosome fusion has been shown to be disrupted in G2019S iPSC-derived human neurons, resulting in delayed mitophagy (Hsieh et al., 2016), and increased α-syn levels were found in iPSC-derived DA neurons from PD patients with the G2019S mutation, but not the R1441G mutation. However, in the latter example, impaired canonical NF-κB signaling was proposed to be the mechanism, rather than autophagy (Lopez de Maturana et al., 2016). In G2019S knock-in mice, with abnormally elevated excitatory synaptic activity and altered postsynaptic morphology (Matikainen-Ankney et al., 2016), no changes in α-syn were observed, but increased levels of LC3-II were reported (Yue et al., 2015). In contrast, R1441G knock-in mice showed no difference in α-syn, LC3B, or Beclin-1 expression, only perturbed DA homeostasis (Liu, Lu et al., 2014).

Polyglutamine Disorders

Nine polyglutamine (polyQ) diseases are caused by a mutation in the polyQ domain in different proteins that result in the expansion of the polyQ tract. Examples include mutant huntingtin in Huntington's disease (HD), mutant ataxin-3 in spinocerebellar ataxia type 3 (SCA3), other mutant proteins in other spinocerebellar ataxias, and mutant androgen receptor in spinal and bulbar muscular atrophy (SBMA) (Gatchel & Zoghbi, 2005). Earlier age of disease onset is often correlated with increasing length of the polyQ mutation (Andrew et al., 1993). The expansion of the polyQ domain (usually more than 35 glutamines) results in accumulation of the mutant proteins in oligomeric forms and aggregates in neurons that are found in distinct regions in the brain (e.g., striatum, cerebral cortex, and cerebellum) (Andrew et al., 1993; Gatchel & Zoghbi, 2005; Rubinsztein, 2006). While neuronal toxicity is linked with aggregate formation, toxicity is also observed in neurons without aggregates (Arrasate, Mitra, Schweitzer, Segal, & Finkbeiner, 2004).

Studies have described autophagy perturbation in several polyQ diseases, often by mechanisms that alter upstream signals required for autophagy induction. For example, the Ras homolog enriched in striatum (Rhes) binds the autophagy initiation protein, Beclin-1, and thus prevents the autophagy inhibitory interaction of bcl-2 with Beclin-1 (Mealer, Murray, Shahani, Subramaniam, & Snyder, 2014). Mutant huntingtin interacts with Rhes and reduces the beneficial effect of Rhes on Beclin-1 activation and autophagy (Mealer et al., 2014). Wild-type huntingtin serves as a scaffold or adaptor for selective autophagy that is induced by cellular stresses (Rui et al., 2015); therefore, mutant huntingtin might also decrease the efficient recruitment of autophagic cargo (Martinez-Vicente et al., 2010). Furthermore, sequestration of Beclin-1 into polyQ aggregates is seen in models of HD, SCA3, and SCA7 (Alves et al., 2014; Nascimento-Ferreira et al., 2011; Shibata et al., 2006) and might impair Beclin-1-autophagic activity. Indeed, reduced Beclin-1 levels have been observed in fibroblasts derived from SCA3 patients (Onofre et al., 2016). Interestingly, while Beclin-1 sequestration into aggregates is likely to inhibit autophagy, the formation of aggregates can also help neurons to cope with polyQ toxicity (Ravikumar et al., 2004). This is evident in studies observing the sequestration of mTOR into aggregates in HD and SCA7 mouse models (Alves et al., 2014; Ravikumar et al., 2004), which is likely to provide signals that induce autophagy to some extent. However, the net effects of autophagy-inducing versus -inhibitory signals are difficult to discern at present and may vary at different stages of disease. Since some autophagy perturbations are influenced by the soluble fraction of the mutant protein and some by its aggregated forms, it also seems that the ratio between the two fractions can affect the outcome of autophagy in these diseases.

Recent data suggest that wild-type ataxin-3 is a positive regulator of autophagy by acting as a deubiquitinase for the core autophagy protein Beclin-1, thereby protecting it from proteasome-mediated degradation. The ataxin-3-Beclin-1 interaction is enabled by the normal polyQ stretch in ataxin-3. When this tract is enlarged by the SCA3 mutation, the ataxin-3-Beclin-1 interaction is strengthened, but the deubiquitinase activity is decreased. This may result in a dominant-negative effect which lowers Beclin-1 levels and impairs autophagy in SCA3 (Ashkenazi et al., 2017). Interestingly, with other polyQ expansion diseases (e.g., HD), at least in cell lines and an animal model, there appears to be competition *in trans* of the disease-causing polyQ tract in the soluble protein with the interaction of ataxin-3 with Beclin-1. This results in a modest impairment of starvation-induced autophagy in these model systems, which may contribute to these diseases (Ashkenazi et al., 2017).

Another autophagy-associated protein that is regulated by mutant ataxin-3 is the E3-ubiquitin ligase PARKIN (Durcan et al., 2011). PARKIN recruits damaged mitochondria for degradation by autophagy (Narendra, Tanaka, Suen, & Youle, 2008) and reduced levels of parkin in the brain of a transgenic mouse model of SCA3 might be linked to disease pathogenesis (Durcan et al., 2011). Some mutant polyQ proteins also perturb transcriptional events that are important for autophagy induction. For example, Sirtuin-1 deacetylates several genes necessary for autophagy induction (Huang et al., 2015; Lee et al., 2008), and lower levels of Sirtuin-1 are observed in SCA3 mouse model (Cunha-Santos et al., 2016). Finally, mutant androgen receptor, which causes SBMA, directly interacts with TFEB, a transcription factor that coordinately regulates expression of genes involved in lysosomal biogenesis and key autophagy genes (Cortes et al., 2014; Settembre et al., 2011). As a consequence of this interaction, TFEB transactivation is abrogated and autophagy is impaired (Cortes et al., 2014).

Amyotrophic Lateral Sclerosis

ALS is predominantly sporadic, although a growing number of genes have been identified in familial forms. ALS-associated mutations in *TDP-43* (TAR DNA-binding protein 43), *SOD1* (superoxide dismutase 1), *FUS* (fused in sarcoma/translocated in sarcoma), and *C9ORF72* (Farg et al., 2014; Fecto & Siddique, 2011; Watabe et al., 2014) result in protein misfolding and the accumulation of aggregates. These intracellular aggregates correlate with an accumulation of autophagosomes and decreased proteasome activity in neurons of the spinal cord and brains in ALS patients (Chen, Zhang, Song, & Le, 2012; Cheroni et al., 2009). Growing evidence correlates defects in the autophagy system with the pathogenesis of ALS. The accumulation of the ALS-related protein FUS has been positively correlated with impaired autophagic flux (Watabe et al., 2014). The list of novel mutations in the Rab5 activator ALS2/Alsin that are associated with motor disorders is expanding (Daud et al., 2016; Siddiqi et al., 2014), and loss of ALS2 has been associated with impaired endosomal trafficking, decreased lysosome protein degradation and neurodegeneration in mouse models (Gautam et al., 2016; Hadano et al., 2010).

Recognition of misfolded ubiquitinated proteins by autophagy receptors enables selective autophagic sequestration of ubiquitinated substrates. This mechanism is relevant to many different proteins that are mutated in ALS. Mutations in the autophagy cargo receptor p62/SQSTM1 have been associated with disrupted degradation of mutant SOD1 and TDP-43 due to defective recognition of LC3-II in cell and mouse models and also in patients with ALS (Gal et al., 2009; Goode et al., 2016; Mizuno et al., 2006; Ramesh Babu et al., 2008; Teyssou et al., 2013). Similarly, clinical mutations in the LC3-binding region of p62/SQSTM1, such as L341V, also lead to the impairment of its recruitment into autophagosomes (Goode et al., 2016). Clinical mutations in the receptor ubiquilin-2 also promote abnormal protein accumulation (Williams et al., 2012; Zhang, Yang, Warraich, & Blair, 2014) in ubiquitin-positive inclusions together with mutant p62/SQSTM1 (Deng et al., 2011; Williams et al., 2012; Zhang, Yang et al., 2014). Similar results were seen for OPTN, another autophagy receptor, where familial ALS-related mutations compromise autophagosome maturation, interaction with p62/SQSTM1, and protein clearance (Maruyama et al., 2010; Shen, Li, Chen, Chern, & Tu, 2015). Most of ALS-associated mutations in OPTN are located in its myosin VI-binding domain and alter autophagosome trafficking (Shen et al., 2015; Sundaramoorthy et al., 2015; Tumbarello et al., 2012). The recent identification of ALS-associated mutations in TANK-binding kinase 1 (Cirulli et al., 2015; Freischmidt et al., 2015) identifies a link between two other familial ALS-associated proteins, since this kinase phosphorylates OPTN (Moore & Holzbaur, 2016) and p62/SQSTM1 (Pilli et al., 2012). Various proteins required for endocytic trafficking have also been implicated in ALS and FTD. Alterations in *C9ORF72* gene are the most common cause of ALS and FTD and are linked by common pathological features (DeJesus-Hernandez et al., 2011; Winklhofer, Tatzelt, & Haass, 2008). The wild-type protein is involved in the regulation of endocytic transport and colocalizes with the autophagic proteins Rab7 and Rab11 in human motor neurons (Farg et al., 2014). C9ORF72 can interact with the autophagy receptors p62/SQSTM1 and OPTN via Rab8 and Rab39, and affects autophagosome formation (Sellier et al., 2016). In addition, C9ORF72 mediates the translocation of ULK1, a kinase controlling autophagosome formation, to the phagophore via Rab1a (Webster et al., 2016). It has been recently reported that loss of C9ORF72 induces autophagy via mTOR and TFEB signaling (Ugolino et al., 2016). In addition to the coding sequence mutations in multiple genes associated with ALS/FTD, epigenetic mechanisms could play a key role in initiating ALS and FTD, especially for sporadic cases (Belzil, Katzman, & Petrucelli, 2016). For example, hypermethylation of the *C9ORF72* promoter may have a protective function against repeat length expansion that is associated with pathology (Liu, Russ et al., 2014).

Mutations in CHMP2B [charged multivesicular body (MVB) protein 2B] and the sigma nonopioid intracellular receptor 1 (SIGMAR1) have been associated with both FTD and ALS (Al-Saif, Al-Mohanna, & Bohlega, 2011; Krasniak & Ahmad, 2016; Luty et al., 2010), and both proteins play a role in vesicle trafficking. CHMP2B is essential for autophagosome–endosome fusion and endolysosomal trafficking via endosomal sorting complexes required for transport (ESCRT), and defective function of this protein impairs autophagosome degradation (Filimonenko et al., 2010; Krasniak & Ahmad, 2016; West, Lu, Marie, Gao, & Sweeney, 2015). Knockdown of SIGMAR1 impairs vesicle trafficking from the ER to the Golgi leading to reduced fusion of autophagosomes with lysosomes and consequently reduced degradation of autophagy substrates (Vollrath et al., 2014).

Hereditary Spastic Paraplegias

Hereditary spastic paraplegia (HSP) describes a heterogeneous group of inherited neurodegenerative disorders pathologically characterized by length-dependent axonal degeneration of corticospinal tracts resulting in progressive spasticity and weakness in the lower limbs. The type of HSP is designated by the loci it is associated with—to date, 50 spastic paraplegia genes (SPGs) and more than 70 distinct loci (SPG1-72) have been identified (Fink, 2013).

SPG11, encoding spatacsin, is the most commonly mutated gene in autosomal recessive HSPs. *Spg11*$^{-/-}$ mice have compromised autophagic lysosome reformation in neurons and a dramatic reduction in the number of lysosomes in the Purkinje cells (Varga et al., 2015), which result in impaired autophagosome clearance. Similarly, loss or mutation of *SPG15* (encoding spastizin) causes defects in lysosomal biogenesis and autophagosome maturation (Chang, Lee, & Blackstone, 2014; Renvoise et al., 2014). Spastizin also interacts with the Beclin-1–UVRAG–Rubicon multiprotein complex required for autophagosome maturation (Vantaggiato et al., 2013). Thus, the loss of spastizin in fibroblasts of HSP15 patients or neuronal cells leads to compromised autophagy flux due to the accumulation of immature autophagosomes (Vantaggiato et al., 2013). Additionally, depletion of the ESCRT components, such as VPS37A (encoded by *SPG53*), is also known to reduce the autophagy flux (Ganley, Wong, Gammoh, & Jiang, 2011; Rusten & Stenmark, 2009; Sahu et al., 2011).

To date, genes mutated in other SPGs have not been shown to be directly involved in regulating the overall autophagy flux or autophagosome maturation, being required only for specific types of autophagy. For example, the deubiquitinating enzyme USP8 (*SPG59*) is directly involved in regulating PARKIN deubiquitination, which is required for its efficient recruitment to damaged mitochondria. Consequently, loss of USP8 results in compromised parkin-mediated mitophagy (Durcan et al., 2014).

A rare form of HSP is caused by recessive mutations in *SPG49*, encoding TECPR2 (tectonin β-propeller containing protein 2). TECPR2 is an ATG8-binding protein that cooperates with lipidated LC3C to efficiently regulate the ER-exit sites and ER export required for the formation of early autophagosome structures (Stadel et al., 2015). Loss of TECPR2 results in reduced levels and lipidation of LC3, TECPR2 being a positive regulator of autophagy (Oz-Levi et al., 2012). Patient fibroblasts with TECPR2 mutations show compromised ER-exit and reduced autophagosome biogenesis (Stadel et al., 2015).

A very rare form of HSP (characterized by mutilating sensory neuropathy) was identified in four patients from a consanguineous Moroccan family, being caused by a loss-of-function mutation in *CCT5* (encoding the epsilon subunit of the cytosolic chaperonin CCT/TRiC) (Bouhouche, Benomar, Bouslam, Chkili, & Yahyaoui, 2006). Recently, it has been shown that CCT5 depletion in primary mouse cortical neurons and *Drosophila* causes accumulation of immature autophagosomes and undegraded autophagic cargo, such as p62 and other aggregate-prone proteins (Pavel et al., 2016). This block in autophagy flux caused by CCT5 depletion is mainly due to reduced lysosomal functioning through compromised actin cytoskeleton dynamics required for trafficking of lysosomal enzymes and V-ATPase into the lysosomes (Pavel et al., 2016).

Lafora Disease

Lafora disease is an autosomal recessive disorder characterized by neurodegeneration and abnormal accumulation of Lafora bodies (LBs), which are polyglucosan bodies comprising a long, hyperphosphorylated, and insoluble form of glycogen. Lafora disease is caused by mutations in the genes encoding either laforin (Minassian et al., 1998), a dual specificity phosphatase that dephosphorylates complex carbohydrates, or malin (Chan et al., 2003), an E3-ubiquitin ligase. Both proteins are involved in the regulation of glycogen biosynthesis. Several studies have revealed an important role for autophagy in the modulation of Lafora disease. Wild-type laforin can induce autophagy (Aguado et al., 2010), and both laforin and malin knock-out mice show impaired autophagy (Aguado et al., 2010; Criado et al., 2012), with an mTOR-dependent mechanism in laforin knock-out mice.

As several reports demonstrated that reducing glycogen synthesis resulted in prevention of formation of LBs and neurodegeneration, it was unclear whether alteration of autophagy was a cause or a consequence of neurodegeneration. Duran, Gruart, Garcia-

Rocha, Delgado-Garcia, and Guinovart (2014) used an elegant in vivo approach to shed light on this matter, demonstrating that autophagy was not upregulated in malin knock-out mice unable to synthesize glycogen in the brain. These results suggest that the autophagy defect seen upon malin knockout is a consequence of glycogen accumulation (Duran et al., 2014).

Dynein and Dynamin Mutations

As described here, many of the common neurodegenerative diseases are characterized by dysfunctional vesicle trafficking. Cytoplasmic dynein is the molecular motor that drives the retrograde transport of cargoes in cells, acting in concert with its activator dynactin. In the context of autophagy, the dynein complex enables retrograde trafficking of autophagosomes to the part of the cell where lysosomes are clustered, enabling autophagosome–lysosome fusion. Mutations in dynein heavy chain (*DYNC1H1*) cause a wide range of neuromuscular degenerative diseases, including spinal muscular atrophy and axonal Charcot–Marie–Tooth disease (Harms et al., 2012; Weedon et al., 2011). Inhibition of dynein activity results in decreased autophagosome–lysosome fusion and interferes with clearance of aggregate-prone proteins in cells, *Drosophila*, and mice (Ravikumar et al., 2005). Similar results were obtained in human glioma cells, where chemical inhibition of dynein led to an accumulation of autophagosomes indicative of impaired autophagosome–lysosome fusion (Yamamoto, Suzuki, & Himeno, 2010). Missense mutations in a dynactin gene (*DCTN1*), encoding the $p150^{Glued}$ subunit of dynactin, have been reported in familial and sporadic ALS (Puls et al., 2003). Mutations in *DCTN1* have been shown to cause Perry syndrome, a progressive neurodegenerative disorder that can present Parkinsonism and psychiatric changes (Farrer et al., 2009), suggesting a potential role for autophagic dysfunction in this disease.

Dynamin 2 mutations cause dominant centronuclear myopathy (Bitoun et al., 2005) or Charcot–Marie–Tooth disease (Zuchner et al., 2005). Dynamin 2 is a GTPase required for endocytosis and is responsible for the pinching off of nascent vesicles from intracellular membranes. The *Drosophila* orthologue of dynamin 2, *Shi*, was shown to have a direct role in autophagy, as it is required for lysosomal/autolysosomal acidification (Fang et al., 2016). Similar results are reported in knock-in mice expressing the most frequent human dynamin 2 mutation. Embryonic fibroblasts from homozygous knock-in mice showed an accumulation of immature autophagosomes, probably due to a defect in acidification (Durieux et al., 2012).

Diseases Resulting from Mutations in Core Autophagy Genes

Recent advances in next-generation sequencing have led to the identification of a growing number of autophagy-related genes showing point mutations in neurodegenerative disease patients. The majority of mutations linked to late-onset neurodegenerative disease are found in genes that are not part of the core autophagy machinery but could rather be termed "autophagy accessory" genes. However, recently, some rare congenital neurodegenerative diseases have been linked to mutations in genes that play a central role in the autophagic process (Menzies, Fleming, & Rubinsztein, 2015).

The first link between a core autophagy-related gene was established in 2012 when de novo mutations in WD-repeat domain 45 (*WDR45*) were associated to beta-propeller protein-associated neurodegeneration (BPAN), also referred to as static encephalopathy of childhood with neurodegeneration in adulthood or neurodegeneration with brain iron accumulation-5 (Haack et al., 2012; Saitsu et al.,

2013). BPAN patients have global developmental delay in early childhood and progressive dystonia, Parkinsonism, and dementia as young adults. *WDR45* encodes WIPI4, one of the four mammalian homologs of the yeast Atg18 protein (Lu et al., 2011). WIPI4 acts as an autophagy-specific PI3P-binding effector and is essential for autophagosome formation (Lu et al., 2011). The BPAN-associated mutations in *WDR45* result in lower WIPI4 protein levels, due to instability of the protein and therefore accumulation of early autophagosomal membranes and reduced autophagic flux (Saitsu et al., 2013). Noteworthy, a central nervous system (CNS)-specific knock-out of *Wdr45* in a mouse model replicated several aspects of BPAN, including poor motor coordination, impaired learning and memory, as well as autophagy defects (Zhao et al., 2015).

While the link between Vici syndrome and ectopic P-granules autophagy protein 5 (EPG5) was discovered several years ago (Cullup et al., 2013), the molecular function of EPG5 in autophagy has only been characterized recently (Wang et al., 2016). Vici syndrome is a rare autosomal recessive congenital multisystem disorder characterized by agenesis of the corpus callosum, bilateral cataracts, cutaneous hypopigmentation, progressive cardiomyopathy, and variable immunodeficiency. The affected individuals show psychomotor retardation and hypotonia (del Campo et al., 1999; Dionisi Vici et al., 1988). Most patients with Vici syndrome have truncating, splice site, or missense mutations in the *EPG5* gene with no clear mutational hotspot (Byrne et al., 2016). EPG5 acts as a Rab7 effector that mediates the fusion of autophagosomes with late endosomes/lysosomes. Various tissues from Vici patients as well as *Epg5* knock-out mice display accumulation of nondegradative autophagic vacuoles (Wang et al., 2016; Zhao, Zhao, Sun, & Zhang, 2013).

Recently, a mutation in *ATG5* was identified in two siblings with congenital ataxia with developmental delay and mental retardation. This E122D mutation results in defects of the conjugation of ATG5 to ATG12 and therefore decreased autophagy. Introduction of the corresponding mutation in a yeast system resulted in a 30%–50% reduction of induced autophagy. A corresponding *Drosophila* model replicated the ataxia phenotype, indicating a causal link of the ATG5 mutation to the ataxia phenotype (Kim et al., 2016).

Lysosomal Disorders

The lysosome has a central role in cellular catabolism and in intracellular trafficking by being at the crossroad between endocytic and autophagic pathways. Given this, disruption of this pathway at multiple points leads to impaired lysosomal function and hence defective autophagosome turnover. This can result not only from the accumulation of undegraded autophagic components in the lysosome but also accumulation of autophagosomes that may fail to fuse with the lysosome. For example, disruption of early endosomal function by depletion of COPI leads to the accumulation of autophagic structures that fail to reach the lysosome (Razi, Chan, & Tooze, 2009). Likewise, loss of function of the ESCRT complex that is required for generation of intraluminal vesicles in later endosomal structures known as MVBs not only blocks endocytic degradation but also leads to the accumulation of autophagosomes (Rusten & Stenmark, 2009). Thus, abnormal autophagic flux can be consequence of primary endocytic and lysosomal defects.

LSDs are rare, inherited disorders with variable phenotypes. They represent the most common cause of neurodegeneration in childhood but can also result in neurological impairment in adults (Poupetova et al., 2010; Wraith, 2002). Most LSDs are caused by loss-of-function of specific lysosomal hydrolases, leading to the accumulation of the substrates of these enzymes and accumulation of general autophagic substrates due to impaired autophagosome–lysosome fusion (Ballabio & Gieselmann, 2009; Platt, Boland, & van der Spoel, 2012; Settembre et al., 2008). There is increasing evidence that changes in membrane lipid composition as a result of lysosomal dysfunction contribute to lysosome fusion defects. This process appears to be relevant to LSDs, since LSD-associated membrane cholesterol abnormalities have been shown to lead to lysosomal accumulation of several substrates, lysosomal dysfunction, and impairment of endocytic membrane trafficking (Fraldi et al., 2010). In mouse models of mucopolysaccharidosis type III (Sanfilippo syndrome), defects in the breakdown of heparin sulfate cause an altered membrane lipid composition, with SNARE protein redistribution resulting in impaired autophagosome–lysosome fusion and a block in autophagy (Fraldi et al., 2010; Settembre et al., 2008).

In Krabbe disease, a defect of β-galactocerebrosidase causes the accumulation of the glycosphingolipid psychosine, which in turn alters the lipid composition of cellular membranes (Hawkins-Salsbury et al., 2013). In Niemann–Pick type A disease, where mutations in the gene encoding acid sphingomyelinase cause accumulation of sphingomyelin, defects in mATG9 trafficking and autophagosome closure have also been observed (Corcelle-Termeau et al., 2016). Niemann–Pick type C disease is a sphingolipid storage disorder that results from inherited deficiencies in intracellular lipid-trafficking proteins and is characterized by an abnormal intracellular accumulation of cholesterol and glycosphingolipids (Lloyd-Evans & Platt, 2010). Similarly, GM1 gangliosidosis and infantile neuronal ceroid lipofuscinoses have been shown to be associated with the occurrence of chronic ER stress (Sano et al., 2009; Wei et al., 2008). Sphingomyelin storage also leads to lysosomal membrane permeabilization, thereby liberating cathepsins into the cytosol (Serrano-Puebla & Boya, 2015).

Defects in posttranslational modifications or impaired trafficking of lysosomal enzymes or defective acidification can also result in lysosomal dysfunction (Colacurcio & Nixon, 2016; Hirst et al., 2015; Kyttala, Yliannala, Schu, Jalanko, & Luzio, 2005; Morimoto et al., 1989; Tiede et al., 2005). One example is represented by multiple sulfatase deficiencies, in which a failed posttranslational modification of sulfatases by an ER-resident enzyme abrogates the function of many lysosomal hydrolases (Dierks et al., 2009). Furthermore, impaired lysosomal structure, regeneration, fusion, and signaling also contribute to lysosomal malfunction (Blanz et al., 2010; Chang et al., 2014; Cortes et al., 2014; Endo, Furuta, & Nishino, 2015; Yu et al., 2010). For example, Niemann–Pick disease type C, caused by loss of NPC1 function, leads to impaired Ca^{2+} homeostasis and incorrect cholesterol trafficking with accumulation of nonesterified cholesterol and glycosphingolipids in late endosomes and lysosomes, disrupting their fusion (Lloyd-Evans et al., 2008; Lloyd-Evans & Platt, 2010; Pacheco & Lieberman, 2008). More recently, mutations in *SNX14*, encoding a sorting nexin phosphoinositol-binding protein localized on late endosome and lysosomal membranes and involved in cargo sorting upon endocytosis, were found in patients with hereditary cerebellar ataxia. Autophagosome clearance was slowed in patient cells, suggesting lysosome–autophagosome dysfunction (Akizu et al., 2015). A specific role of the autophagosome–lysosome fusion for lysosomal disease pathology is further underlined by the discovery of a homozygous missense mutation in *VPS11* in patients with a rare form of leukoencephalopathy (Zhang et al., 2016). VPS11 is a member of the homotypic fusion and protein sorting and class C core vacuole/endosome tethering complexes, and mutations lead to impaired autophagy. In zebrafish with a null mutation of *vps11*, reduction in CNS myelination and extensive neuronal death were observed in the hindbrain and midbrain (Zhang et al., 2016).

Finally, several studies have linked autophagy to the regulated secretion of the contents of secretory granules or lysosomes in specialized cells or tissues (Cadwell et al., 2008; DeSelm et al., 2011; Michaud et al., 2011; Ushio et al., 2011). Examples of such events at the plasma-membrane include secretion of lysozyme from Paneth cells, which is implicated in Crohn's disease (Cadwell et al., 2008), secretion of ATP under certain conditions (Michaud et al., 2011), secretion of cathepsin K by osteoclasts during bone resorption (DeSelm et al., 2011), and the secretion of cytotoxic proteins stored in the lytic granules of activated cytotoxic T lymphocytes (CTLs), which is a key event in killing target cells (Ushio et al., 2011). Most likely, the fusion events of autophagic vesicles with the plasma-membrane (autophagosome exocytosis) are mediated by SNAREs, in a manner similar to lysosome and lysosome-related organelles, such as lytic granules and melanosomes. Indeed, this has been demonstrated for the secretion of lytic granules: patients with familial hemophagocytic lymphohistiocytosis type 4 who have mutations in the SNARE protein syntaxin-11 have impaired degranulation of CTLs (Bryceson et al., 2007; Hong, 2005). Moreover, recent findings suggest a role for the SNARE proteins VAMP8 and VTI1B (SNAREs that regulate autophagosome−lysosome fusion) in the exocytosis of toxic proteins stored in the lytic granules of CTLs (Dressel, Elsner, Novota, Kanwar, & Fischer von Mollard, 2010). During lysosomal exocytosis, a Ca^{2+}-regulated process (mediated by the activation of the lysosomal Ca^{2+} channel MCOLN1) enables lysosomes to dock to the cell surface and fuse with the plasma-membrane, emptying their contents outside the cell. This process has an important role in secretion and plasma-membrane repair. Interestingly, lysosomal exocytosis is transcriptionally regulated by TFEB, a master regulator that coordinates lysosomal biogenesis and autophagy (Sardiello et al., 2009; Settembre et al., 2011), which promotes cellular clearance in models of lysosomal storage diseases (Medina et al., 2011). Therefore, it is tempting to speculate that cells may use this mechanism to coordinate autophagosome−lysosome degradative activity and autophagy-mediated secretory functions in response to specific stimuli, suggesting that promoting lysosomal exocytosis may represent an alternative strategy to treat disorders due to intracellular storage, such as LSDs.

AUTOPHAGY UPREGULATION

While autophagy is essential for nervous system function, data also supports it decreasing with age across diverse organisms. A range of approaches have been used to assess the potential benefits of constitutive autophagy upregulation in normal health and aging as well as neurodegenerative disease contexts, and these approaches support a clear association of upregulation with beneficial effects. Approaches within invertebrate model organisms show that induction of autophagy with spermidine extends lifespan in *Drosophila* and *Caenorhabditis elegans* (Eisenberg et al., 2009). Depletion of acetyl-coenzyme A through knocking down its synthetase specifically in the *Drosophila* brain also results in prolonged lifespan with enhanced autophagy-mediated protein clearance, which is likely the result of decreasing inhibitory acetylation of autophagy regulators (Eisenberg et al., 2014). Autophagy is induced and lifespan extended by caloric restriction in *C. elegans* (Hansen et al., 2008) as well as by pan-neuronal overexpression of Atg8a or AMPK in *Drosophila* (Simonsen et al., 2008; Ulgherait, Rana, Rera, Graniel, & Walker, 2014).

When assessing the mechanisms of how autophagy upregulation enhances survival, studies in *C. elegans* support the benefit acting through diverse pathways, which include clearance and turnover of mitochondria as well as ER homeostasis and mitotic alterations (Ghavidel et al., 2015; Palikaras, Lionaki, & Tavernarakis, 2015).

In mice, constitutive upregulation of autophagy by ubiquitous overexpression of ATG5 results in extended lifespan and reduced aging phenotypes. These include enhanced motor performance, leanness and, at a primary cellular level, enhanced resistance to oxidative damage (Pyo et al., 2013). Likewise, deletion of the polyQ tract from mouse huntingtin increases autophagy and results in lifespan extension (Zheng et al., 2010).

In cells, autophagy upregulation enhances the clearance of toxic aggregate-prone proteins, such as mutant huntingtin, alpha-synuclein, and tau (Berger et al., 2006; Ravikumar, Duden, & Rubinsztein, 2002; Webb, Ravikumar, Atkins, Skepper, & Rubinsztein, 2003). The process of autophagic engulfment enables removal of oligomeric species, which are inaccessible to the proteasome. While the protective role of autophagy has been widely replicated in animal models of various diseases caused by intracytoplasmic aggregate-prone proteins (Menzies, Fleming et al., 2015), most studies in vivo have used chemical tools that likely have autophagy-independent effects. However, overexpression of Atg5 in zebrafish induces autophagy and ameliorates tau toxicity (Lopez et al., 2017). Likewise, genetic inhibition of calpain protects against tau and huntingtin toxicity in vivo in an autophagy-dependent manner (Menzies, Garcia-Arencibia et al., 2015).

Since upregulation of autophagy has beneficial effects, a number of pharmacological autophagy stimulators have been identified and tested in preclinical models of neurodegenerative diseases (Harris & Rubinsztein, 2011; Levine, Packer, & Codogno, 2015; Sarkar & Rubinsztein, 2008). The most extensively employed preclinical pharmacological tools to probe autophagy modulation in neurodegeneration are trehalose and rapamycin.

Trehalose

Trehalose is a stable disaccharide assembled from two molecules of D-glucose (Fig. 11.4). Many organisms, notably fungi, insects, and other invertebrates, biosynthesize trehalose (Richards et al., 2002). Humans, like other mammals, do not synthesize trehalose, but significant quantities may be ingested as part of a normal diet, and some can be absorbed (Kamiya, Hirata, Matsumoto, Arai, & Yoshizane, 2004; Murray, Coupland, Smith, Ansell, & Long, 2000; Richards et al., 2002).

Trehalose has been shown to have activity in a variety of cellular and animal models of neurodegeneration (Emanuele, 2014). Trehalose treatment leads to decreased huntingtin aggregation and reduced toxicity in vitro as well as reduced aggregation and increased lifespan in the R6/2 transgenic mouse model of HD (Tanaka et al., 2004). Trehalose has also been shown to reduce toxicity and formation of mutant PABPN1 aggregates in a cellular model of oculopharyngeal muscular dystrophy (OPMD) and reduced aggregates and cell death and improved disease-associated muscle weakness in a mouse model of this disease (Davies, Sarkar, & Rubinsztein, 2006). More recently, trehalose has also been shown to decrease the processing of both APP and of C-terminal fragments derived from it, to reduce the secretion of Aβ (Tien, Karaca, Tamboli, & Walter, 2016), and to rescue impaired cognition and decrease Aβ deposition in the APP/PS-1 mouse model (Du, Liang, Xu, Sun, & Wang, 2013). Other studies have shown effects in cell and mouse models of tauopathy (Kruger et al., 2012; Schaeffer et al., 2012), tauopathy with Parkin deletion (Rodriguez-Navarro et al., 2010), SCA17 (Chen et al., 2015), and ALS (Castillo, Nassif et al.,

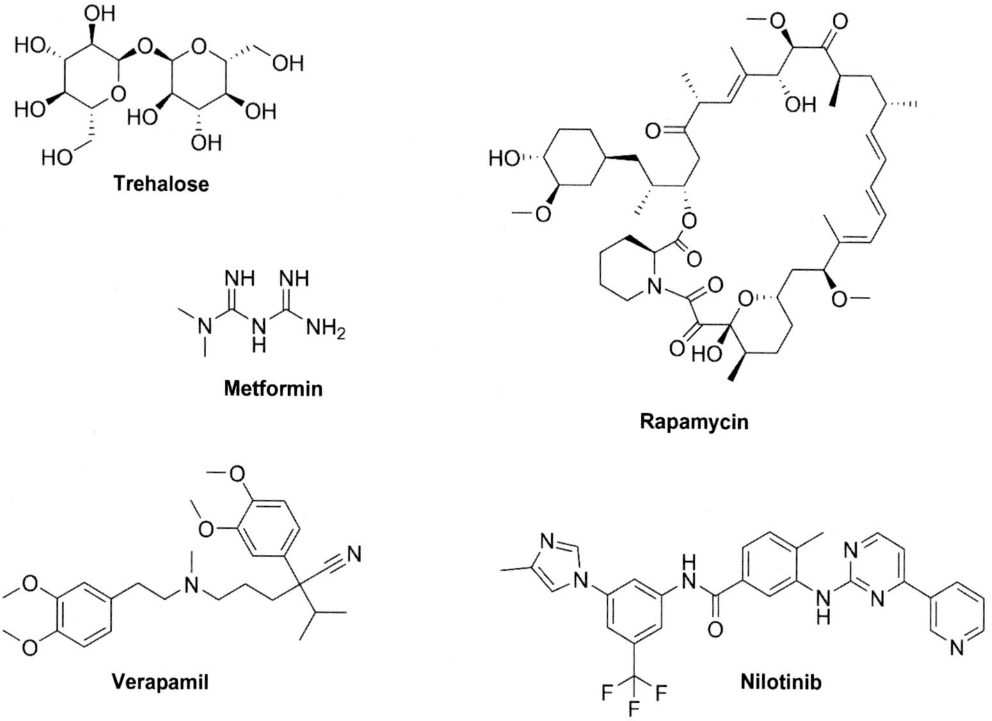

FIGURE 11.4 The chemical structures of selected autophagy enhancers with demonstrated in vivo efficacy in models of neurodegeneration.

2013; Li et al., 2015; Zhang, Chen et al., 2014). Mixed data has been reported in cell and animal models of Lewy body disease (Tanji et al., 2015) and prion disease (Aguib et al., 2009).

Mechanistically, trehalose has been identified as an mTOR-independent enhancer of autophagy (Sarkar, Davies, Huang, Tunnacliffe, & Rubinsztein, 2007) signaling via AMPK. Recently, DeBosch has proposed that trehalose enhances autophagy through inhibition of SLC2A (GLUT) transporters (DeBosch et al., 2016). IC_{50} values against GLUT transporters were reported as between 17.3 and 126 mM; these values seem inconsistent with the early reports of effects in cell assays at submillimolar concentrations. A further study by this group reports that GLUT8 functions as a trehalose-transporter responsible for trehalose entry into mammalian cells (Mayer et al., 2016).

An intravenous formulation of trehalose, known as Cabaletta, has been developed by Bioblast Pharma for the treatment of PolyA and PolyQ diseases. Cabaletta is in Phase III for OPMD and Phase II for SCA3. Preliminary results from a 24-week open-label Phase II trial have shown that Cabaletta improves dysphagia and muscle function in OPMD (Argov, Gliko-Kabir, Brais, Caraco, & Megiddo, 2016).

Rapamycin

Rapamycin, a macrolide antifungal and immunosuppressant, was first discovered and characterized as an autophagy activator over 20 years ago (Blommaart, Luiken, Blommaart, van Woerkom, & Meijer, 1995; Heitman, Movva, & Hall, 1991; Noda & Ohsumi, 1998). The mechanism of action of rapamycin involves binding to FKBP12, and the resulting complex interacts with mTORC1, inhibiting its kinase activity through an allosteric mechanism (Benjamin, Colombi, Moroni, & Hall, 2011; Heitman et al., 1991). Functionally, mTOR forms two complexes termed mTORC1 and mTORC2, with rapamycin's inhibitory action on mTOR showing selectivity toward the first (Schreiber et al., 2015). A significant effort has been devoted toward the evolution of rapamycin analogs (called "rapalogues") as well as toward the targeted development of more potent, ATP-competitive inhibitors of mTOR, such as Torin (Benjamin et al., 2011), these latter in particular as anticancer agents. However, Torin-type inhibitors inhibit both mTORC1 and mTORC2 complexes and display poor selectivity typically associated with ATP-competitive kinase inhibitors, rendering them less suitable for chronic indications.

Rapalogues (including rapamycin) have been employed to probe the therapeutic benefit of autophagy activation in a number of animal models of neurodegeneration. In the triple-transgenic AD mouse model, rapamycin as well as mTOR genetic reduction was shown to rescue cognitive deficits and decrease Aβ and tau pathology through autophagy stimulation (Caccamo, De Pinto, Messina, Branca, & Oddo, 2014; Caccamo, Majumder, Richardson, Strong, & Oddo, 2010). Prevention of AD-like cognitive deficits and lowered levels of amyloid peptide by rapamycin were reported in another AD mouse model, the PDAPP mouse (Spilman et al., 2010). In a model of prion disease, Cortes et al. reported that rapamycin delayed disease onset, increased survival, and ameliorated disease severity (Cortes, Qin, Cook, Solanki, & Mastrianni, 2012). In the P301S TAU mouse model, temsirolimus (a rapalogue) was shown to reduce tau hyperphosphorylation and improve spatial cognitive behavior (Jiang et al., 2014). Similar effects were shown for rapamycin in the same tau mouse model (Ozcelik et al., 2013). Temsirolimus ameliorated motor performance in a transgenic mouse model of spinocerebellar ataxia type 3, inducing autophagy and reducing the number of aggregates and protein levels of mutant ataxin-3 (Menzies et al., 2010), and ameliorated behavioral performance and decreased aggregate formation in the HD-N171-N82Q mouse model of HD (Ravikumar et al., 2004). In a PD mouse model (Parkin Q311X mice), rapamycin decreased dopaminergic neuronal loss, reduced synuclein levels, improved motor function, and induced autophagic markers (Siddiqui et al., 2015).

While studies in many models of neurodegenerative diseases have converged on the therapeutic efficacy of rapalogues, investigations in ALS models have provided contradictory results. In some studies, detrimental effects were reported (Zhang et al., 2011) but beneficial effects were subsequently observed after uncoupling of rapamycin's immunomodulatory function (Staats et al., 2013). The possibility that the immunomodulatory role of rapalogues may contribute to the phenotypic amelioration (or indeed worsen the phenotype) observed in different animal models of neurodegeneration remains a concern. However, the rationale that autophagy activation is therapeutically relevant in neurodegenerative diseases is strongly supported by the amelioration observed with autophagy activators with different mechanisms of action.

Repurposing of FDA-Approved Drugs as Autophagy Upregulators

There have been many reports of using repurposed FDA-approved drugs for the treatment of neurodegenerative disorders via upregulation of autophagy (Fleming, Noda, Yoshimori, & Rubinsztein, 2011; Levine et al., 2015; Rubinsztein, Bento, & Deretic, 2015). Selected examples are presented here.

Metformin is a first-line antidiabetes drug that has been demonstrated to upregulate autophagy in several preclinical animal models of neurodegenerative diseases but with mixed results on the outcome of disease progression. This FDA-approved drug is CNS-penetrant (Labuzek et al., 2010) and is an activator of AMPK. Treatment of transgenic mice expressing a mutant huntingtin with CAG repeat length of 136–151 (HD R6/2) with metformin significantly increased brain AMPK phosphorylation levels (Ma et al., 2007). Whilst decreased hind limb clasping and increased lifespan was observed in males, no effects were observed in female mice. In vivo electrophysiology demonstrated a significant increase in survival of motor units in the $SOD1^{G93A}$ mouse model of ALS, although beneficial effects were seen in different muscles of male and female mice (Kaneb, Sharp, Rahmani-Kondori, & Wells, 2011). This effect was observed with treated with 2 mg/mL metformin in drinking water; however, doses of 0.5, 2, and 5 mg/mL metformin did not significantly alter neurological scores and did not increase survival or delay disease onset in male mice. Indeed, metformin dose dependently worsened survival rates for females, possibly owing to inhibition of estrogen production (Kaneb et al., 2011). In transgenic mice overexpressing human AβPP 695 (Tg6799), metformin stimulated autophagosome accumulation and promoted the generation of Aβ through increased β- and γ-secretase activity (Son et al., 2016). The results of this study suggest that metformin heightens the progression of AD by increasing Aβ synthesis within autophagosomes.

The Bcr-Abl tyrosine kinase inhibitor nilotinib is approved for the treatment of chronic myelogenous leukemia. The compound is known to be CNS-penetrant but has a low concentration in cerebrospinal fluid (CSF) (Reinwald et al., 2014). Nilotinib has been shown to induce autophagy in vivo (Yu et al., 2013) and increase amyloid clearance in an Aβ transgenic mouse model (Lonskaya, Hebron, Desforges, Schachter, & Moussa, 2014; Lonskaya, Hebron, Selby, Turner, & Moussa, 2015).

A number of compounds that act on the $cAMP/IP_3$ pathway have shown benefits in animal models of neurodegeneration, including verapamil, clonidine, rilmenidine, and minoxidil (Rose et al., 2010; Williams et al., 2008). Verapamil is a calcium channel blocker prescribed for the treatment of hypertension, angina pectoris, and cardiac arrhythmia and is known to penetrate the CNS (Narang, Blumhardt, Doran, & Pickar, 1988). In a *Drosophila* model of HD, verapamil slowed the progression of the HD phenotype, and in a zebrafish transgenic model of HD where expression is driven in the rod photoreceptors, verapamil reduced the number of aggregates compared to untreated fish, leading to a reduction of photoreceptor degeneration (Williams et al., 2008).

Acknowledgments

We are grateful for funding from the Tau consortium, Alzheimer's Research UK, Wellcome Trust (Principal Research Fellowship to DCR 095317/Z/11/Z), the UK Dementia Research Institute (funded by the MRC, Alzheimer's Research UK and the Alzheimer's Society) (DCR) a Wellcome Trust Strategic Grant to Cambridge Institute for Medical Research (100140/Z/12/Z), NIHR Biomedical Research Unit in Dementia at Addenbrooke's Hospital, The Rosetrees Trust, FEBS (long-term fellowships to A.A. and J.F.), and Addenbrooke's Charitable Trust.

References

Aflaki, E., Borger, D. K., Moaven, N., Stubblefield, B. K., Rogers, S. A., Patnaik, S., ... Sidransky, E. (2016). A new glucocerebrosidase chaperone reduces alpha-synuclein and glycolipid levels in iPSC-derived dopaminergic neurons from patients with gaucher disease and parkinsonism. *The Journal of Neuroscience*, *36*(28), 7441–7452.

Aflaki, E., Moaven, N., Borger, D. K., Lopez, G., Westbroek, W., Chae, J. J., ... Sidransky, E. (2016). Lysosomal storage and impaired autophagy lead to inflammasome activation in Gaucher macrophages. *Aging Cell*, *15*(1), 77–88.

Aguado, C., Sarkar, S., Korolchuk, V. I., Criado, O., Vernia, S., Boya, P., ... Rubinsztein, D. C. (2010). Laforin, the most common protein mutated in Lafora disease, regulates autophagy. *Human Molecular Genetics, 19*(14), 2867–2876.

Aguib, Y., Heiseke, A., Gilch, S., Riemer, C., Baier, M., Schatzl, H. M., & Ertmer, A. (2009). Autophagy induction by trehalose counteracts cellular prion infection. *Autophagy, 5*(3), 361–369.

Akizu, N., Cantagrel, V., Zaki, M. S., Al-Gazali, L., Wang, X., Rosti, R. O., … Gleeson, J. G. (2015). Biallelic mutations in SNX14 cause a syndromic form of cerebellar atrophy and lysosome–autophagosome dysfunction. *Nature Genetics, 47*(5), 528–534.

Al-Saif, A., Al-Mohanna, F., & Bohlega, S. (2011). A mutation in sigma-1 receptor causes juvenile amyotrophic lateral sclerosis. *Annals of Neurology, 70*(6), 913–919.

Alves, S., Cormier-Dequaire, F., Marinello, M., Marais, T., Muriel, M. P., Beaumatin, F., … Sittler, A. (2014). The autophagy/lysosome pathway is impaired in SCA7 patients and SCA7 knock-in mice. *Acta Neuropathologica, 128*(5), 705–722.

Ando, K., Brion, J. P., Stygelbout, V., Suain, V., Authelet, M., Dedecker, R., … Duyckaerts, C. (2013). Clathrin adaptor CALM/PICALM is associated with neurofibrillary tangles and is cleaved in Alzheimer's brains. *Acta Neuropathologica, 125*(6), 861–878.

Ando, K., Tomimura, K., Sazdovitch, V., Suain, V., Yilmaz, Z., Authelet, M., … Brion, J. P. (2016). Level of PICALM, a key component of clathrin-mediated endocytosis, is correlated with levels of phosphotau and autophagy-related proteins and is associated with tau inclusions in AD, PSP and Pick disease. *Neurobiology of Disease, 94*, 32–43.

Andrew, S. E., Goldberg, Y. P., Kremer, B., Telenius, H., Theilmann, J., Adam, S., et al. (1993). The relationship between trinucleotide (CAG) repeat length and clinical features of Huntington's disease. *Nature Genetics, 4*(4), 398–403.

Argov, Z., Gliko-Kabir, I., Brais, B., Caraco, Y., & Megiddo, D. (2016). Intravenous trehalose improves dysphagia and muscle function in oculopharyngeal muscular dystrophy (OPMD): Preliminary results of 24 weeks open label phase 2 trial (I4.007). *Neurology, 86*(16), Supplement I4.007.

Arrasate, M., Mitra, S., Schweitzer, E. S., Segal, M. R., & Finkbeiner, S. (2004). Inclusion body formation reduces levels of mutant huntingtin and the risk of neuronal death. *Nature, 431*(7010), 805–810.

Ashkenazi, A., Bento, C. F., Ricketts, T., Vicinanza, M., Siddiqi, F., Pavel, M., … Rubinsztein, D. C. (2017). Polyglutamine tracts regulate beclin 1-dependent autophagy. *Nature, 545*(7652), 108–111.

Axe, E. L., Walker, S. A., Manifava, M., Chandra, P., Roderick, H. L., Habermann, A., … Ktistakis, N. T. (2008). Autophagosome formation from membrane compartments enriched in phosphatidylinositol 3-phosphate and dynamically connected to the endoplasmic reticulum. *The Journal of Cell Biology, 182*(4), 685–701.

Ballabio, A., & Gieselmann, V. (2009). Lysosomal disorders: From storage to cellular damage. *Biochimica et Biophysica Acta, 1793*(4), 684–696.

Belzil, V. V., Katzman, R. B., & Petrucelli, L. (2016). ALS and FTD: An epigenetic perspective. *Acta Neuropathologica, 132*(4), 487–502.

Benjamin, D., Colombi, M., Moroni, C., & Hall, M. N. (2011). Rapamycin passes the torch: A new generation of mTOR inhibitors. *Nature Reviews Drug Discovery, 10* (11), 868–880.

Bento, C. F., Ashkenazi, A., Jimenez-Sanchez, M., & Rubinsztein, D. C. (2016). The Parkinson's disease-associated genes ATP13A2 and SYT11 regulate autophagy via a common pathway. *Nature Communications, 7*, 11803.

Berger, Z., Ravikumar, B., Menzies, F. M., Oroz, L. G., Underwood, B. R., Pangalos, M. N., … Rubinsztein, D. C. (2006). Rapamycin alleviates toxicity of different aggregate-prone proteins. *Human Molecular Genetics, 15* (3), 433–442.

Bernhardt, R., & Matus, A. (1984). Light and electron microscopic studies of the distribution of microtubule-associated protein 2 in rat brain: A difference between dendritic and axonal cytoskeletons. *The Journal of Comparative Neurology, 226*(2), 203–221.

Biazik, J., Yla-Anttila, P., Vihinen, H., Jokitalo, E., & Eskelinen, E. L. (2015). Ultrastructural relationship of the phagophore with surrounding organelles. *Autophagy, 11*(3), 439–451.

Bitoun, M., Maugenre, S., Jeannet, P. Y., Lacene, E., Ferrer, X., Laforet, P., & Guicheney, P. (2005). Mutations in dynamin 2 cause dominant centronuclear myopathy. *Nature Genetics, 37*(11), 1207–1209.

Bjorkoy, G., Lamark, T., Brech, A., Outzen, H., Perander, M., Overvatn, A., & Johansen, T. (2005). p62/SQSTM1 forms protein aggregates degraded by autophagy and has a protective effect on huntingtin-induced cell death. *The Journal of Cell Biology, 171*(4), 603–614.

Blanz, J., Groth, J., Zachos, C., Wehling, C., Saftig, P., & Schwake, M. (2010). Disease-causing mutations within the lysosomal integral membrane protein type 2 (LIMP-2) reveal the nature of binding to its ligand beta-glucocerebrosidase. *Human Molecular Genetics, 19* (4), 563–572.

Blommaart, E. F., Luiken, J. J., Blommaart, P. J., van Woerkom, G. M., & Meijer, A. J. (1995). Phosphorylation of ribosomal protein S6 is inhibitory for autophagy in isolated rat hepatocytes. *The Journal of Biological Chemistry, 270*(5), 2320–2326.

Boland, B., Kumar, A., Lee, S., Platt, F., Wegiel, J., Yu, W., & Nixon, R. (2008). Autophagy induction and autophagosome clearance in neurons: Relationship to autophagic pathology in Alzheimer's disease. *The Journal of Neuroscience, 28*, 6926–6937.

Bouhouche, A., Benomar, A., Bouslam, N., Chkili, T., & Yahyaoui, M. (2006). Mutation in the epsilon subunit of the cytosolic chaperonin-containing t-complex peptide-1 (Cct5) gene causes autosomal recessive mutilating sensory neuropathy with spastic paraplegia. *Journal of Medical Genetics, 43*(5), 441–443.

Bryceson, Y. T., Rudd, E., Zheng, C., Edner, J., Ma, D., Wood, S. M., & Ljunggren, H. G. (2007). Defective cytotoxic lymphocyte degranulation in syntaxin-11 deficient familial hemophagocytic lymphohistiocytosis 4 (FHL4) patients. *Blood, 110*(6), 1906–1915.

Butzlaff, M., Hannan, S. B., Karsten, P., Lenz, S., Ng, J., Vossfeldt, H., & Voigt, A. (2015). Impaired retrograde transport by the dynein/dynactin complex contributes to Tau-induced toxicity. *Human Molecular Genetics, 24* (13), 3623–3637.

Byrne, S., Jansen, L., U-King-Im, J. M., Siddiqui, A., Lidov, H. G., Bodi, I., & Jungbluth, H. (2016). EPG5-related Vici syndrome: A paradigm of neurodevelopmental disorders with defective autophagy. *Brain, 139*(Pt 3), 765–781.

Caccamo, A., De Pinto, V., Messina, A., Branca, C., & Oddo, S. (2014). Genetic reduction of mammalian target of rapamycin ameliorates Alzheimer's disease-like cognitive and pathological deficits by restoring hippocampal gene expression signature. *The Journal of Neuroscience, 34*(23), 7988–7998.

Caccamo, A., Magri, A., Medina, D. X., Wisely, E. V., Lopez-Aranda, M. F., Silva, A. J., & Oddo, S. (2013). mTOR regulates tau phosphorylation and degradation: Implications for Alzheimer's disease and other tauopathies. *Aging Cell, 12*(3), 370–380.

Caccamo, A., Majumder, S., Richardson, A., Strong, R., & Oddo, S. (2010). Molecular interplay between mammalian target of rapamycin (mTOR), amyloid-beta, and Tau: Effects on cognitive impairments. *The Journal of Biological Chemistry, 285*(17), 13107–13120.

Cadwell, K., Liu, J. Y., Brown, S. L., Miyoshi, H., Loh, J., Lennerz, J. K., & Virgin, H. W. (2008). A key role for autophagy and the autophagy gene Atg16l1 in mouse and human intestinal Paneth cells. *Nature, 456*(7219), 259–263.

Castillo, K., Nassif, M., Valenzuela, V., Rojas, F., Matus, S., Mercado, G., & Hetz, C. (2013). Trehalose delays the progression of amyotrophic lateral sclerosis by enhancing autophagy in motoneurons. *Autophagy, 9*(9), 1308–1320.

Castillo, K., Valenzuela, V., Matus, S., Nassif, M., Onate, M., Fuentealba, Y., & Hetz, C. (2013). Measurement of autophagy flux in the nervous system in vivo. *Cell Death & Disease, 4*, e917.

Cataldo, A. M., Hamilton, D. J., & Nixon, R. A. (1994). Lysosomal abnormalities in degenerating neurons link neuronal compromise to senile plaque development in Alzheimer disease. *Brain Research, 640*(1–2), 68–80.

Cataldo, A. M., & Nixon, R. A. (1990). Enzymatically active lysosomal proteases are associated with amyloid deposits in Alzheimer brain. *Proceedings of the National Academy of Sciences of the United States of America, 87*(10), 3861–3865.

Chan, E. M., Young, E. J., Ianzano, L., Munteanu, I., Zhao, X., Christopoulos, C. C., & Scherer, S. W. (2003). Mutations in NHLRC1 cause progressive myoclonus epilepsy. *Nature Genetics, 35*(2), 125–127.

Chang, J., Lee, S., & Blackstone, C. (2014). Spastic paraplegia proteins spastizin and spatacsin mediate autophagic lysosome reformation. *The Journal of Clinical Investigation, 124*(12), 5249–5262.

Chen, S., Zhang, X., Song, L., & Le, W. (2012). Autophagy dysregulation in amyotrophic lateral sclerosis. *Brain Pathology (Zurich, Switzerland), 22*(1), 110–116.

Chen, Y., Sawada, O., Kohno, H., Le, Y. Z., Subauste, C., Maeda, T., & Maeda, A. (2013). Autophagy protects the retina from light-induced degeneration. *The Journal of Biological Chemistry, 288*(11), 7506–7518.

Chen, Z. Z., Wang, C. M., Lee, G. C., Hsu, H. C., Wu, T. L., Lin, C. W., & Hsieh-Li, H. M. (2015). Trehalose attenuates the gait ataxia and gliosis of spinocerebellar ataxia type 17 mice. *Neurochemical Research, 40*(4), 800–810.

Cheroni, C., Marino, M., Tortarolo, M., Veglianese, P., De Biasi, S., Fontana, E., & Bendotti, C. (2009). Functional alterations of the ubiquitin-proteasome system in motor neurons of a mouse model of familial amyotrophic lateral sclerosis. *Human Molecular Genetics, 18* (1), 82–96.

Chesser, A. S., Pritchard, S. M., & Johnson, G. V. (2013). Tau clearance mechanisms and their possible role in the pathogenesis of Alzheimer disease. *Frontiers in Neurology, 4*, 122.

Chung, S. Y., Kishinevsky, S., Mazzulli, J. R., Graziotto, J., Mrejeru, A., Mosharov, E. V., & Shim, J. W. (2016). Parkin and PINK1 patient iPSC-derived midbrain dopamine neurons exhibit mitochondrial dysfunction and alpha-synuclein accumulation. *Stem Cell Reports, 7* (4), 664–677.

Cirulli, E. T., Lasseigne, B. N., Petrovski, S., Sapp, P. C., Dion, P. A., Leblond, C. S., ... Goldstein, D. B. (2015). Exome sequencing in amyotrophic lateral sclerosis identifies risk genes and pathways. *Science, 347*(6229), 1436–1441.

Citron, M., Westaway, D., Xia, W., Carlson, G., Diehl, T., Levesque, G., & Selkoe, D. J. (1997). Mutant presenilins of Alzheimer's disease increase production of 42-residue amyloid beta-protein in both transfected cells and transgenic mice. *Nature Medicine, 3*(1), 67–72.

Coffey, E. E., Beckel, J. M., Laties, A. M., & Mitchell, C. H. (2014). Lysosomal alkalization and dysfunction in human fibroblasts with the Alzheimer's disease-linked presenilin 1 A246E mutation can be reversed with cAMP. *Neuroscience, 263*, 111–124.

Colacurcio, D. J., & Nixon, R. A. (2016). Disorders of lysosomal acidification-the emerging role of v-ATPase in aging and neurodegenerative disease. *Ageing Research Reviews, 32*, 75–88.

Conner, S. D., & Schmid, S. L. (2003). Regulated portals of entry into the cell. *Nature, 422*(6927), 37–44.

Corcelle-Termeau, E., Vindelov, S. D., Hamalisto, S., Mograbi, B., Keldsbo, A., Brasen, J. H., ... Jaattela, M. (2016). Excess sphingomyelin disturbs ATG9A trafficking and autophagosome closure. *Autophagy, 12*(5), 833–849.

Cortes, C. J., Miranda, H. C., Frankowski, H., Batlevi, Y., Young, J. E., Le, A., & La Spada, A. R. (2014). Polyglutamine-expanded androgen receptor interferes with TFEB to elicit autophagy defects in SBMA. *Nature Neuroscience, 17*(9), 1180–1189.

Cortes, C. J., Qin, K., Cook, J., Solanki, A., & Mastrianni, J. A. (2012). Rapamycin delays disease onset and prevents PrP plaque deposition in a mouse model of Gerstmann–Straussler–Scheinker disease. *The Journal of Neuroscience, 32*(36), 12396–12405.

Criado, O., Aguado, C., Gayarre, J., Duran-Trio, L., Garcia-Cabrero, A. M., Vernia, S., & Rodriguez de Cordoba, S. (2012). Lafora bodies and neurological defects in malin-deficient mice correlate with impaired autophagy. *Human Molecular Genetics, 21*(7), 1521–1533.

Cullup, T., Kho, A. L., Dionisi-Vici, C., Brandmeier, B., Smith, F., Urry, Z., & Jungbluth, H. (2013). Recessive mutations in EPG5 cause Vici syndrome, a multisystem disorder with defective autophagy. *Nature Genetics, 45*(1), 83–87.

Cunha-Santos, J., Duarte-Neves, J., Carmona, V., Guarente, L., Pereira de Almeida, L., & Cavadas, C. (2016). Caloric restriction blocks neuropathology and motor deficits in Machado-Joseph disease mouse models through SIRT1 pathway. *Nature Communications, 7*, 11445.

Cunningham, C. N., Baughman, J. M., Phu, L., Tea, J. S., Yu, C., Coons, M., & Corn, J. E. (2015). USP30 and parkin homeostatically regulate atypical ubiquitin chains on mitochondria. *Nature Cell Biology, 17*(2), 160–169.

Daud, S., Kakar, N., Goebel, I., Hashmi, A. S., Yaqub, T., Nurnberg, G., & Borck, G. (2016). Identification of two novel ALS2 mutations in infantile-onset ascending hereditary spastic paraplegia. *Amyotroph Lateral Scler Frontotemporal Degener, 17*(3–4), 260–265.

Davies, J. E., Sarkar, S., & Rubinsztein, D. C. (2006). Trehalose reduces aggregate formation and delays pathology in a transgenic mouse model of oculopharyngeal muscular dystrophy. *Human Molecular Genetics, 15*(1), 23–31.

DeBosch, B. J., Heitmeier, M. R., Mayer, A. L., Higgins, C. B., Crowley, J. R., Kraft, T. E., ... Moley, K. H. (2016). Trehalose inhibits solute carrier 2A (SLC2A) proteins to induce autophagy and prevent hepatic steatosis. *Science Signaling, 9*(416), ra21.

Decressac, M., Mattsson, B., Weikop, P., Lundblad, M., Jakobsson, J., & Bjorklund, A. (2013). TFEB-mediated autophagy rescues midbrain dopamine neurons from alpha-synuclein toxicity. *Proceedings of the National Academy of Sciences of the United States of America, 110*(19), E1817–1826.

de Duve, C. (2005). The lysosome turns fifty. *Nature Cell Biology, 7*(9), 847–849.

Dehay, B., Ramirez, A., Martinez-Vicente, M., Perier, C., Canron, M. H., Doudnikoff, E., & Bezard, E. (2012). Loss of P-type ATPase ATP13A2/PARK9 function induces general lysosomal deficiency and leads to Parkinson disease neurodegeneration. *Proceedings of the National Academy of Sciences of the United States of America, 109*(24), 9611–9616.

DeJesus-Hernandez, M., Mackenzie, I. R., Boeve, B. F., Boxer, A. L., Baker, M., Rutherford, N. J., & Rademakers, R. (2011). Expanded GGGGCC hexanucleotide repeat in noncoding region of C9ORF72 causes chromosome 9p-linked FTD and ALS. *Neuron, 72*(2), 245–256.

del Campo, M., Hall, B. D., Aeby, A., Nassogne, M. C., Verloes, A., Roche, C., & Quero, J. (1999). Albinism and agenesis of the corpus callosum with profound developmental delay: Vici syndrome, evidence for autosomal recessive inheritance. *American Journal of Medical Genetics, 85*(5), 479–485.

Deng, H. X., Chen, W., Hong, S. T., Boycott, K. M., Gorrie, G. H., Siddique, N., & Siddique, T. (2011). Mutations in UBQLN2 cause dominant X-linked juvenile and adult-onset ALS and ALS/dementia. *Nature, 477*(7363), 211–215.

Deretic, V., Saitoh, T., & Akira, S. (2013). Autophagy in infection, inflammation and immunity. *Nature Reviews Immunology, 13*(10), 722–737.

DeSelm, C. J., Miller, B. C., Zou, W., Beatty, W. L., van Meel, E., Takahata, Y., & Virgin, H. W. (2011). Autophagy proteins regulate the secretory component of osteoclastic bone resorption. *Developmental Cell, 21*(5), 966–974.

Di Bartolomeo, S., Corazzari, M., Nazio, F., Oliverio, S., Lisi, G., Antonioli, M., & Fimia, G. M. (2010). The dynamic interaction of AMBRA1 with the dynein motor complex regulates mammalian autophagy. *The Journal of Cell Biology*, *191*(1), 155–168.

Dierks, T., Schlotawa, L., Frese, M. A., Radhakrishnan, K., von Figura, K., & Schmidt, B. (2009). Molecular basis of multiple sulfatase deficiency, mucolipidosis II/III and Niemann–Pick C1 disease—Lysosomal storage disorders caused by defects of non-lysosomal proteins. *Biochimica et Biophysica Acta*, *1793*(4), 710–725.

Dionisi Vici, C., Sabetta, G., Gambarara, M., Vigevano, F., Bertini, E., Boldrini, R., & Fiorilli, M. (1988). Agenesis of the corpus callosum, combined immunodeficiency, bilateral cataract, and hypopigmentation in two brothers. *American Journal of Medical Genetics*, *29*(1), 1–8.

Dooley, H. C., Razi, M., Polson, H. E., Girardin, S. E., Wilson, M. I., & Tooze, S. A. (2014). WIPI2 links LC3 conjugation with PI3P, autophagosome formation, and pathogen clearance by recruiting Atg12-5-16L1. *Molecular Cell*, *55*(2), 238–252.

Dowdle, W. E., Nyfeler, B., Nagel, J., Elling, R. A., Liu, S., Triantafellow, E., ... Murphy, L. O. (2014). Selective VPS34 inhibitor blocks autophagy and uncovers a role for NCOA4 in ferritin degradation and iron homeostasis in vivo. *Nature Cell Biology*, *16*(11), 1069–1079.

Dressel, R., Elsner, L., Novota, P., Kanwar, N., & Fischer von Mollard, G. (2010). The exocytosis of lytic granules is impaired in Vti1b- or Vamp8-deficient CTL leading to a reduced cytotoxic activity following antigen-specific activation. *Journal of Immunology*, *185*(2), 1005–1014.

Du, J., Liang, Y., Xu, F., Sun, B., & Wang, Z. (2013). Trehalose rescues Alzheimer's disease phenotypes in APP/PS1 transgenic mice. *The Journal of Pharmacy and Pharmacology*, *65*(12), 1753–1756.

Duran, J., Gruart, A., Garcia-Rocha, M., Delgado-Garcia, J. M., & Guinovart, J. J. (2014). Glycogen accumulation underlies neurodegeneration and autophagy impairment in Lafora disease. *Human Molecular Genetics*, *23*(12), 3147–3156.

Durcan, T. M., Kontogiannea, M., Thorarinsdottir, T., Fallon, L., Williams, A. J., Djarmati, A., ... Fon, E. A. (2011). The Machado–Joseph disease-associated mutant form of ataxin-3 regulates parkin ubiquitination and stability. *Human Molecular Genetics*, *20*(1), 141–154.

Durcan, T. M., Tang, M. Y., Perusse, J. R., Dashti, E. A., Aguileta, M. A., McLelland, G. L., ... Fon, E. A. (2014). USP8 regulates mitophagy by removing K6-linked ubiquitin conjugates from parkin. *The EMBO Journal*, *33*(21), 2473–2491.

Durieux, A. C., Vassilopoulos, S., Laine, J., Fraysse, B., Brinas, L., Prudhon, B., ... Bitoun, M. (2012). A centronuclear myopathy—dynamin 2 mutation impairs autophagy in mice. *Traffic (Copenhagen, Denmark)*, *13*(6), 869–879.

Egan, D. F., Chun, M. G., Vamos, M., Zou, H., Rong, J., Miller, C. J., ... Shaw, R. J. (2015). Small molecule inhibition of the autophagy kinase ULK1 and identification of ULK1 Substrates. *Molecular Cell*, *59*(2), 285–297.

Eisenberg, T., Knauer, H., Schauer, A., Buttner, S., Ruckenstuhl, C., Carmona-Gutierrez, D., ... Madeo, F. (2009). Induction of autophagy by spermidine promotes longevity. *Nature Cell Biology*, *11*(11), 1305–1314.

Eisenberg, T., Schroeder, S., Andryushkova, A., Pendl, T., Kuttner, V., Bhukel, A., ... Madeo, F. (2014). Nucleocytosolic depletion of the energy metabolite acetyl-coenzyme A stimulates autophagy and prolongs lifespan. *Cell Metabolism*, *19*(3), 431–444.

Emanuele, E. (2014). Can trehalose prevent neurodegeneration? Insights from experimental studies. *Current Drug Targets*, *15*(5), 551–557.

Endo, Y., Furuta, A., & Nishino, I. (2015). Danon disease: A phenotypic expression of LAMP-2 deficiency. *Acta Neuropathologica*, *129*(3), 391–398.

Fader, C. M., Sanchez, D. G., Mestre, M. B., & Colombo, M. I. (2009). TI-VAMP/VAMP7 and VAMP3/cellubrevin: Two v-SNARE proteins involved in specific steps of the autophagy/multivesicular body pathways. *Biochimica et Biophysica Acta*, *1793*(12), 1901–1916.

Fang, X., Zhou, J., Liu, W., Duan, X., Gala, U., Sandoval, H., ... Tong, C. (2016). Dynamin Regulates Autophagy by Modulating Lysosomal Function. *Journal of Genetics and Genomics = Yi Chuan Xue Bao*, *43*(2), 77–86.

Farg, M. A., Sundaramoorthy, V., Sultana, J. M., Yang, S., Atkinson, R. A., Levina, V., ... Atkin, J. D. (2014). C9ORF72, implicated in amytrophic lateral sclerosis and frontotemporal dementia, regulates endosomal trafficking. *Human Molecular Genetics*, *23*(13), 3579–3595.

Farrer, M. J., Hulihan, M. M., Kachergus, J. M., Dachsel, J. C., Stoessl, A. J., Grantier, L. L., ... Wszolek, Z. K. (2009). DCTN1 mutations in Perry syndrome. *Nature Genetics*, *41*(2), 163–165.

Fecto, F., & Siddique, T. (2011). Making connections: Pathology and genetics link amyotrophic lateral sclerosis with frontotemporal lobe dementia. *Journal of Molecular Neuroscience: MN*, *45*(3), 663–675.

Fernandes, H. J., Hartfield, E. M., Christian, H. C., Emmanoulidou, E., Zheng, Y., Booth, H., ... Wade-Martins, R. (2016). ER stress and autophagic perturbations lead to elevated extracellular alpha-synuclein in GBA-N370S Parkinson's iPSC-derived dopamine neurons. *Stem Cell Reports*, *6*(3), 342–356.

Filimonenko, M., Isakson, P., Finley, K. D., Anderson, M., Jeong, H., Melia, T. J., ... Yamamoto, A. (2010). The selective macroautophagic degradation of aggregated proteins requires the PI3P-binding protein Alfy. *Molecular Cell*, *38*(2), 265–279.

Fink, J. K. (2013). Hereditary spastic paraplegia: Clinicopathologic features and emerging molecular mechanisms. *Acta Neuropathologica*, *126*(3), 307–328.

Fleming, A., Noda, T., Yoshimori, T., & Rubinsztein, D. C. (2011). Chemical modulators of autophagy as biological probes and potential therapeutics. *Nature Chemical Biology*, *7*(1), 9–17.

Fraldi, A., Annunziata, F., Lombardi, A., Kaiser, H. J., Medina, D. L., Spampanato, C., ... Ballabio, A. (2010). Lysosomal fusion and SNARE function are impaired by cholesterol accumulation in lysosomal storage disorders. *The EMBO Journal*, *29*(21), 3607–3620.

Freischmidt, A., Wieland, T., Richter, B., Ruf, W., Schaeffer, V., Muller, K., ... Weishaupt, J. H. (2015). Haploinsufficiency of TBK1 causes familial ALS and fronto-temporal dementia. *Nature Neuroscience*, *18*(5), 631–636.

Fujita, N., Morita, E., Itoh, T., Tanaka, A., Nakaoka, M., Osada, Y., ... Yoshimori, T. (2013). Recruitment of the autophagic machinery to endosomes during infection is mediated by ubiquitin. *The Journal of Cell Biology*, *203*(1), 115–128.

Fullgrabe, J., Klionsky, D. J., & Joseph, B. (2014). The return of the nucleus: Transcriptional and epigenetic control of autophagy. *Nature Reviews Molecular Cell Biology*, *15*(1), 65–74.

Furuta, N., Fujita, N., Noda, T., Yoshimori, T., & Amano, A. (2010). Combinational soluble *N*-ethylmaleimide-sensitive factor attachment protein receptor proteins VAMP8 and Vti1b mediate fusion of antimicrobial and canonical autophagosomes with lysosomes. *Molecular Biology of the Cell*, *21*(6), 1001–1010.

Gal, J., Strom, A. L., Kwinter, D. M., Kilty, R., Zhang, J., Shi, P., ... Zhu, H. (2009). Sequestosome 1/p62 links familial ALS mutant SOD1 to LC3 via an ubiquitin-independent mechanism. *Journal of Neurochemistry*, *111*(4), 1062–1073.

Ganley, I. G., Lam du, H., Wang, J., Ding, X., Chen, S., & Jiang, X. (2009). ULK1.ATG13.FIP200 complex mediates mTOR signaling and is essential for autophagy. *The Journal of Biological Chemistry*, *284*(18), 12297–12305.

Ganley, I. G., Wong, P. M., Gammoh, N., & Jiang, X. (2011). Distinct autophagosomal–lysosomal fusion mechanism revealed by thapsigargin-induced autophagy arrest. *Molecular Cell*, *42*(6), 731–743.

Gatchel, J. R., & Zoghbi, H. Y. (2005). Diseases of unstable repeat expansion: Mechanisms and common principles. *Nature Reviews Genetics*, *6*(10), 743–755.

Gautam, M., Jara, J. H., Sekerkova, G., Yasvoina, M. V., Martina, M., & Ozdinler, P. H. (2016). Absence of alsin function leads to corticospinal motor neuron vulnerability via novel disease mechanisms. *Human Molecular Genetics*, *25*(6), 1074–1087.

Ge, L., Melville, D., Zhang, M., & Schekman, R. (2013). The ER–Golgi intermediate compartment is a key membrane source for the LC3 lipidation step of autophagosome biogenesis. *Elife*, *2*, e00947.

Ge, L., Zhang, M., & Schekman, R. (2014). Phosphatidylinositol 3-kinase and COPII generate LC3 lipidation vesicles from the ER-Golgi intermediate compartment. *Elife*, *3*, e04135.

Ghavidel, A., Baxi, K., Ignatchenko, V., Prusinkiewicz, M., Arnason, T. G., Kislinger, T., ... Harkness, T. A. (2015). A genome scale screen for mutants with delayed exit from mitosis: Ire1-independent induction of autophagy integrates ER homeostasis into mitotic lifespan. *PLoS Genetics*, *11*(8), e1005429.

Goedert, M., & Spillantini, M. G. (2006). A century of Alzheimer's disease. *Science (New York, N.Y.)*, *314*(5800), 777–781.

Goedert, M., Wischik, C. M., Crowther, R. A., Walker, J. E., & Klug, A. (1988). Cloning and sequencing of the cDNA encoding a core protein of the paired helical filament of Alzheimer disease: Identification as the microtubule-associated protein tau. *Proceedings of the National Academy of Sciences of the United States of America*, *85*(11), 4051–4055.

Gomez-Sanchez, J. A., Carty, L., Iruarrizaga-Lejarreta, M., Palomo-Irigoyen, M., Varela-Rey, M., Griffith, M., & Jessen, K. R. (2015). Schwann cell autophagy, myelinophagy, initiates myelin clearance from injured nerves. *The Journal of Cell Biology*, *210*(1), 153–168.

Goode, A., Butler, K., Long, J., Cavey, J., Scott, D., Shaw, B., ... Layfield, R. (2016). Defective recognition of LC3B by mutant SQSTM1/p62 implicates impairment of autophagy as a pathogenic mechanism in ALS-FTLD. *Autophagy*, *12*(7), 1–11.

Graef, M., Friedman, J. R., Graham, C., Babu, M., & Nunnari, J. (2013). ER exit sites are physical and functional core autophagosome biogenesis components. *Molecular Biology of the Cell*, *24*(18), 2918–2931.

Haack, T. B., Hogarth, P., Kruer, M. C., Gregory, A., Wieland, T., Schwarzmayr, T., ... Hayflick, S. J. (2012). Exome sequencing reveals de novo WDR45 mutations causing a phenotypically distinct, X-linked dominant form of NBIA. *American Journal of Human Genetics*, *91*(6), 1144–1149.

Hadano, S., Otomo, A., Kunita, R., Suzuki-Utsunomiya, K., Akatsuka, A., Koike, M., ... Ikeda, J. E. (2010). Loss of ALS2/Alsin exacerbates motor dysfunction in a SOD1-expressing mouse ALS model by disturbing endolysosomal trafficking. *PLoS ONE*, *5*(3), e9805.

Hamacher-Brady, A., & Brady, N. R. (2015). Bax/Bak-dependent, Drp1-independent targeting of X-linked inhibitor of apoptosis protein (XIAP) into inner mitochondrial compartments counteracts smac/DIABLO-dependent effector caspase activation. *The Journal of Biological Chemistry*, *290*(36), 22005–22018.

Hamasaki, M., Furuta, N., Matsuda, A., Nezu, A., Yamamoto, A., Fujita, N., ... Yoshimori, T. (2013). Autophagosomes form at ER-mitochondria contact sites. *Nature, 495*(7441), 389–393.

Hansen, M., Chandra, A., Mitic, L. L., Onken, B., Driscoll, M., & Kenyon, C. (2008). A role for autophagy in the extension of lifespan by dietary restriction in *C. elegans*. *PLoS Genetics, 4*(2), e24.

Hara, T., Nakamura, K., Matsui, M., Yamamoto, A., Nakahara, Y., Suzuki-Migishima, R., ... Mizushima, N. (2006). Suppression of basal autophagy in neural cells causes neurodegenerative disease in mice. *Nature, 441* (7095), 885–889.

Harms, M. B., Ori-McKenney, K. M., Scoto, M., Tuck, E. P., Bell, S., Ma, D., ... Baloh, R. H. (2012). Mutations in the tail domain of DYNC1H1 cause dominant spinal muscular atrophy. *Neurology, 78*(22), 1714–1720.

Harris, H., & Rubinsztein, D. C. (2011). Control of autophagy as a therapy for neurodegenerative disease. *Nature Reviews Neurology, 8*(2), 108–117.

Hawkins-Salsbury, J. A., Parameswar, A. R., Jiang, X., Schlesinger, P. H., Bongarzone, E., Ory, D. S., ... Sands, M. S. (2013). Psychosine, the cytotoxic sphingolipid that accumulates in globoid cell leukodystrophy, alters membrane architecture. *Journal of Lipid Research, 54*(12), 3303–3311.

Hayashi-Nishino, M., Fujita, N., Noda, T., Yamaguchi, A., Yoshimori, T., & Yamamoto, A. (2009). A subdomain of the endoplasmic reticulum forms a cradle for autophagosome formation. *Nature Cell Biology, 11*(12), 1433–1437.

Heitman, J., Movva, N. R., & Hall, M. N. (1991). Targets for cell cycle arrest by the immunosuppressant rapamycin in yeast. *Science (New York, N.Y.), 253*(5022), 905–909.

Heo, J. M., Ordureau, A., Paulo, J. A., Rinehart, J., & Harper, J. W. (2015). The PINK1-PARKIN Mitochondrial Ubiquitylation Pathway Drives a Program of OPTN/NDP52 Recruitment and TBK1 Activation to Promote Mitophagy. *Molecular Cell, 60*(1), 7–20.

Hirst, J., Edgar, J. R., Esteves, T., Darios, F., Madeo, M., Chang, J., ... Robinson, M. S. (2015). Loss of AP-5 results in accumulation of aberrant endolysosomes: Defining a new type of lysosomal storage disease. *Human Molecular Genetics, 24*(17), 4984–4996.

Hong, W. (2005). Cytotoxic T lymphocyte exocytosis: Bring on the SNAREs!. *Trends in Cell Biology, 15*(12), 644–650.

Hosokawa, N., Hara, T., Kaizuka, T., Kishi, C., Takamura, A., Miura, Y., ... Mizushima, N. (2009). Nutrient-dependent mTORC1 association with the ULK1-Atg13-FIP200 complex required for autophagy. *Molecular Biology of the Cell, 20*(7), 1981–1991.

Hsieh, C. H., Shaltouki, A., Gonzalez, A. E., Bettencourt da Cruz, A., Burbulla, L. F., St Lawrence, E., ... Wang, X. (2016). Functional impairment in Miro degradation and mitophagy is a shared feature in familial and sporadic Parkinson's disease. *Cell Stem Cell, 19*(6), 709–724.

Huang, R., Xu, Y., Wan, W., Shou, X., Qian, J., You, Z., ... Liu, W. (2015). Deacetylation of nuclear LC3 drives autophagy initiation under starvation. *Molecular Cell, 57* (3), 456–466.

Huett, A., Heath, R. J., Begun, J., Sassi, S. O., Baxt, L. A., Vyas, J. M., ... Xavier, R. J. (2012). The LRR and RING domain protein LRSAM1 is an E3 ligase crucial for ubiquitin-dependent autophagy of intracellular Salmonella Typhimurium. *Cell Host & Microbe, 12*(6), 778–790.

Huotari, J., & Helenius, A. (2011). Endosome maturation. *The EMBO Journal, 30*(17), 3481–3500.

Ichimura, Y., Kirisako, T., Takao, T., Satomi, Y., Shimonishi, Y., Ishihara, N., ... Ohsumi, Y. (2000). A ubiquitin-like system mediates protein lipidation. *Nature, 408*(6811), 488–492.

Itakura, E., & Mizushima, N. (2010). Characterization of autophagosome formation site by a hierarchical analysis of mammalian Atg proteins. *Autophagy, 6*(6), 764–776.

Jaeger, P. A., Pickford, F., Sun, C. H., Lucin, K. M., Masliah, E., & Wyss-Coray, T. (2010). Regulation of amyloid precursor protein processing by the Beclin 1 complex. *PLoS ONE, 5*(6), e11102.

Jahn, R., & Scheller, R. H. (2006). SNAREs—Engines for membrane fusion. *Nature Reviews Molecular Cell Biology, 7*(9), 631–643.

Jang, S. Y., Shin, Y. K., Park, S. Y., Park, J. Y., Lee, H. J., Yoo, Y. H., ... Park, H. T. (2016). Autophagic myelin destruction by schwann cells during wallerian degeneration and segmental demyelination. *Glia, 64*(5), 730–742.

Jang, S. Y., Shin, Y. K., Park, S. Y., Park, J. Y., Rha, S. H., Kim, J. K., ... Park, H. T. (2015). Autophagy is involved in the reduction of myelinating Schwann cell cytoplasm during myelin maturation of the peripheral nerve. *PLoS ONE, 10*(1), e0116624.

Jiang, T., Yu, J. T., Zhu, X. C., Zhang, Q. Q., Cao, L., Wang, H. F., ... Tan, L. (2014). Temsirolimus attenuates tauopathy in vitro and in vivo by targeting tau hyperphosphorylation and autophagic clearance. *Neuropharmacology, 85*, 121–130.

Jung, C. H., Jun, C. B., Ro, S. H., Kim, Y. M., Otto, N. M., Cao, J., ... Kim, D. H. (2009). ULK-Atg13-FIP200 complexes mediate mTOR signaling to the autophagy machinery. *Molecular Biology of the Cell, 20*(7), 1992–2003.

Kamiya, T., Hirata, K., Matsumoto, S., Arai, C., & Yoshizane, C. (2004). Targeted disruption of the trehalase gene: Determination of the digestion and absorption of trehalose in trehalase-deficient mice. *Nutrition Research, 24*(2), 185–196.

Kane, L. A., Lazarou, M., Fogel, A. I., Li, Y., Yamano, K., Sarraf, S. A., ... Youle, R. J. (2014). PINK1 phosphorylates ubiquitin to activate Parkin E3 ubiquitin ligase activity. *The Journal of Cell Biology*, *205*(2), 143–153.

Kaneb, H. M., Sharp, P. S., Rahmani-Kondori, N., & Wells, D. J. (2011). Metformin treatment has no beneficial effect in a dose–response survival study in the SOD1(G93A) mouse model of ALS and is harmful in female mice. *PLoS ONE*, *6*(9), e24189.

Karanasios, E., Walker, S. A., Okkenhaug, H., Manifava, M., Hummel, E., Zimmermann, H., ... Ktistakis, N. T. (2016). Autophagy initiation by ULK complex assembly on ER tubulovesicular regions marked by ATG9 vesicles. *Nature Communications*, *7*, 12420.

Kaushik, S., Rodriguez-Navarro, J. A., Arias, E., Kiffin, R., Sahu, S., Schwartz, G. J., ... Singh, R. (2011). Autophagy in hypothalamic AgRP neurons regulates food intake and energy balance. *Cell Metabolism*, *14*(2), 173–183.

Kazlauskaite, A., Kondapalli, C., Gourlay, R., Campbell, D. G., Ritorto, M. S., Hofmann, K., ... Muqit, M. M. (2014). Parkin is activated by PINK1-dependent phosphorylation of ubiquitin at Ser65. *The Biochemical Journal*, *460*(1), 127–139.

Kett, L. R., & Dauer, W. T. (2016). Endolysosomal dysfunction in Parkinson's disease: Recent developments and future challenges. *Movement Disorders: Official Journal of the Movement Disorder Society*, *31*(10), 1433–1443.

Khaminets, A., Behl, C., & Dikic, I. (2016). Ubiquitin-Dependent And Independent Signals In Selective Autophagy. *Trends in Cell Biology*, *26*(1), 6–16.

Khaminets, A., Heinrich, T., Mari, M., Grumati, P., Huebner, A. K., Akutsu, M., ... Dikic, I. (2015). Regulation of endoplasmic reticulum turnover by selective autophagy. *Nature*, *522*(7556), 354–358.

Khurana, V., Elson-Schwab, I., Fulga, T. A., Sharp, K. A., Loewen, C. A., Mulkearns, E., ... Feany, M. B. (2010). Lysosomal dysfunction promotes cleavage and neurotoxicity of tau in vivo. *PLoS Genetics*, *6*(7), e1001026.

Kim, J., Kim, Y. C., Fang, C., Russell, R. C., Kim, J. H., Fan, W., ... Guan, K. L. (2013). Differential regulation of distinct Vps34 complexes by AMPK in nutrient stress and autophagy. *Cell*, *152*(1–2), 290–303.

Kim, J., Kundu, M., Viollet, B., & Guan, K. L. (2011). AMPK and mTOR regulate autophagy through direct phosphorylation of Ulk1. *Nature Cell Biology*, *13*(2), 132–141.

Kim, M., Sandford, E., Gatica, D., Qiu, Y., Liu, X., Zheng, Y., ... Burmeister, M. (2016). Mutation in ATG5 reduces autophagy and leads to ataxia with developmental delay. *Elife*, *5*, e12245.

Kim, P. K., Hailey, D. W., Mullen, R. T., & Lippincott-Schwartz, J. (2008). Ubiquitin signals autophagic degradation of cytosolic proteins and peroxisomes. *Proceedings of the National Academy of Sciences of the United States of America*, *105*(52), 20567–20574.

Kimura, S., Noda, T., & Yoshimori, T. (2008). Dynein-dependent movement of autophagosomes mediates efficient encounters with lysosomes. *Cell Structure and Function*, *33*(1), 109–122.

Kimura, T., Jain, A., Choi, S. W., Mandell, M. A., Schroder, K., Johansen, T., & Deretic, V. (2015). TRIM-mediated precision autophagy targets cytoplasmic regulators of innate immunity. *The Journal of Cell Biology*, *210*(6), 973–989.

Kirkin, V., Lamark, T., Johansen, T., & Dikic, I. (2009). NBR1 cooperates with p62 in selective autophagy of ubiquitinated targets. *Autophagy*, *5*(5), 732–733.

Kitada, T., Asakawa, S., Hattori, N., Matsumine, H., Yamamura, Y., Minoshima, S., ... Shimizu, N. (1998). Mutations in the parkin gene cause autosomal recessive juvenile parkinsonism. *Nature*, *392*(6676), 605–608.

Knaevelsrud, H., Soreng, K., Raiborg, C., Haberg, K., Rasmuson, F., Brech, A., ... Simonsen, A. (2013). Membrane remodeling by the PX-BAR protein SNX18 promotes autophagosome formation. *The Journal of Cell Biology*, *202*(2), 331–349.

Knorr, R. L., Lipowsky, R., & Dimova, R. (2015). Autophagosome closure requires membrane scission. *Autophagy*, *11*(11), 2134–2137.

Komatsu, M., Waguri, S., Chiba, T., Murata, S., Iwata, J., Tanida, I., ... Tanaka, K. (2006). Loss of autophagy in the central nervous system causes neurodegeneration in mice. *Nature*, *441*(7095), 880–884.

Komatsu, M., Waguri, S., Koike, M., Sou, Y. S., Ueno, T., Hara, T., ... Tanaka, K. (2007). Homeostatic levels of p62 control cytoplasmic inclusion body formation in autophagy-deficient mice. *Cell*, *131*(6), 1149–1163.

Komatsu, M., Waguri, S., Ueno, T., Iwata, J., Murata, S., Tanida, I., ... Chiba, T. (2005). Impairment of starvation-induced and constitutive autophagy in Atg7-deficient mice. *The Journal of Cell Biology*, *169*(3), 425–434.

Komatsu, M., Wang, Q. J., Holstein, G. R., Friedrich, V. L., Jr., Iwata, J., Kominami, E., ... Yue, Z. (2007). Essential role for autophagy protein Atg7 in the maintenance of axonal homeostasis and the prevention of axonal degeneration. *Proceedings of the National Academy of Sciences of the United States of America*, *104*(36), 14489–14494.

Kondapalli, C., Kazlauskaite, A., Zhang, N., Woodroof, H. I., Campbell, D. G., Gourlay, R., ... Muqit, M. M. (2012). PINK1 is activated by mitochondrial membrane potential depolarization and stimulates Parkin E3 ligase activity by phosphorylating Serine 65. *Open Biology*, *2*(5), 120080.

Koyano, F., Okatsu, K., Kosako, H., Tamura, Y., Go, E., Kimura, M., ... Matsuda, N. (2014). Ubiquitin is phosphorylated by PINK1 to activate parkin. *Nature*, *510*(7503), 162–166.

Kraft, C., Deplazes, A., Sohrmann, M., & Peter, M. (2008). Mature ribosomes are selectively degraded upon starvation by an autophagy pathway requiring the Ubp3p/Bre5p ubiquitin protease. *Nature Cell Biology*, *10*(5), 602–610.

Krasniak, C. S., & Ahmad, S. T. (2016). The role of CHMP2BIntron5 in autophagy and frontotemporal dementia. *Brain Research*, *1649*(Pt B), 151–157.

Kruger, U., Wang, Y., Kumar, S., & Mandelkow, E. M. (2012). Autophagic degradation of tau in primary neurons and its enhancement by trehalose. *Neurobiology of Aging*, *33*(10), 2291–2305.

Kuma, A., Hatano, M., Matsui, M., Yamamoto, A., Nakaya, H., Yoshimori, T., ... Mizushima, N. (2004). The role of autophagy during the early neonatal starvation period. *Nature*, *432*(7020), 1032–1036.

Kyttala, A., Yliannala, K., Schu, P., Jalanko, A., & Luzio, J. P. (2005). AP-1 and AP-3 facilitate lysosomal targeting of Batten disease protein CLN3 via its dileucine motif. *The Journal of Biological Chemistry*, *280*(11), 10277–10283.

Labuzek, K., Suchy, D., Gabryel, B., Bielecka, A., Liber, S., & Okopien, B. (2010). Quantification of metformin by the HPLC method in brain regions, cerebrospinal fluid and plasma of rats treated with lipopolysaccharide. *Pharmacological Reports: PR*, *62*(5), 956–965.

Lacovich, V., Espindola, S. L., Alloatti, M., Pozo Devoto, V., Cromberg, L., Carna, M., ... Falzone, T. L. (2017). Tau isoforms imbalance impairs the axonal transport of the amyloid precursor protein in human neurons. *The Journal of Neuroscience*, *37*(1), 58–69.

Lamb, C. A., Nuhlen, S., Judith, D., Frith, D., Snijders, A. P., Behrends, C., & Tooze, S. A. (2016). TBC1D14 regulates autophagy via the TRAPP complex and ATG9 traffic. *The EMBO Journal*, *35*(3), 281–301.

Laplante, M., & Sabatini, D. M. (2012). mTOR signaling in growth control and disease. *Cell*, *149*(2), 274–293.

Lazarou, M., Sliter, D. A., Kane, L. A., Sarraf, S. A., Wang, C., Burman, J. L., ... Youle, R. J. (2015). The ubiquitin kinase PINK1 recruits autophagy receptors to induce mitophagy. *Nature*, *524*(7565), 309–314.

Lee, I. H., Cao, L., Mostoslavsky, R., Lombard, D. B., Liu, J., Bruns, N. E., ... Finkel, T. (2008). A role for the NAD-dependent deacetylase Sirt1 in the regulation of autophagy. *Proceedings of the National Academy of Sciences of the United States of America*, *105*(9), 3374–3379.

Lee, J. H., Yu, W. H., Kumar, A., Lee, S., Mohan, P. S., Peterhoff, C. M., ... Nixon, R. A. (2010). Lysosomal proteolysis and autophagy require presenilin 1 and are disrupted by Alzheimer-related PS1 mutations. *Cell*, *141*(7), 1146–1158.

Lee, J. M., Wagner, M., Xiao, R., Kim, K. H., Feng, D., Lazar, M. A., & Moore, D. D. (2014). Nutrient-sensing nuclear receptors coordinate autophagy. *Nature*, *516*(7529), 112–115.

Lee, M. J., Lee, J. H., & Rubinsztein, D. C. (2013). Tau degradation: The ubiquitin-proteasome system versus the autophagy-lysosome system. *Progress in Neurobiology*, *105*, 49–59.

Lee, V. M., Goedert, M., & Trojanowski, J. Q. (2001). Neurodegenerative tauopathies. *Annual Review of Neuroscience*, *24*, 1121–1159.

Levine, B., Packer, M., & Codogno, P. (2015). Development of autophagy inducers in clinical medicine. *The Journal of Clinical Investigation*, *125*(1), 14–24.

Li, L., Zhang, X., & Le, W. (2010). Autophagy dysfunction in Alzheimer's disease. *Neuro-Degenerative Diseases*, *7*(4), 265–271.

Li, Y., Guo, Y., Wang, X., Yu, X., Duan, W., Hong, K., ... Li, C. (2015). Trehalose decreases mutant SOD1 expression and alleviates motor deficiency in early but not end-stage amyotrophic lateral sclerosis in a SOD1-G93A mouse model. *Neuroscience*, *298*, 12–25.

Liang, C. C., Wang, C., Peng, X., Gan, B., & Guan, J. L. (2010). Neural-specific deletion of FIP200 leads to cerebellar degeneration caused by increased neuronal death and axon degeneration. *The Journal of Biological Chemistry*, *285*(5), 3499–3509.

Lin, S. Y., Li, T. Y., Liu, Q., Zhang, C., Li, X., Chen, Y., ... Lin, S. C. (2012). GSK3-TIP60-ULK1 signaling pathway links growth factor deprivation to autophagy. *Science (New York, N.Y.)*, *336*(6080), 477–481.

Liu, E. Y., Russ, J., Wu, K., Neal, D., Suh, E., McNally, A. G., ... Lee, E. B. (2014). C9orf72 hypermethylation protects against repeat expansion-associated pathology in ALS/FTD. *Acta Neuropathologica*, *128*(4), 525–541.

Liu, H. F., Lu, S., Ho, P. W., Tse, H. M., Pang, S. Y., Kung, M. H., ... Ho, S. L. (2014). LRRK2 R1441G mice are more liable to dopamine depletion and locomotor inactivity. *Annals of Clinical and Translational Neurology*, *1*(3), 199–208.

Liu, L., Feng, D., Chen, G., Chen, M., Zheng, Q., Song, P., ... Chen, Q. (2012). Mitochondrial outer-membrane protein FUNDC1 mediates hypoxia-induced mitophagy in mammalian cells. *Nature Cell Biology*, *14*(2), 177–185.

Lloyd-Evans, E., Morgan, A. J., He, X., Smith, D. A., Elliot-Smith, E., Sillence, D. J., ... Platt, F. M. (2008). Niemann–Pick disease type C1 is a sphingosine storage disease that causes deregulation of lysosomal calcium. *Nature Medicine*, *14*(11), 1247–1255.

Lloyd-Evans, E., & Platt, F. M. (2010). Lipids on trial: The search for the offending metabolite in Niemann–Pick type C disease. *Traffic (Copenhagen, Denmark)*, *11*(4), 419–428.

Longatti, A., Lamb, C. A., Razi, M., Yoshimura, S., Barr, F. A., & Tooze, S. A. (2012). TBC1D14 regulates autophagosome formation via Rab11- and ULK1-positive recycling endosomes. *The Journal of Cell Biology*, *197*(5), 659–675.

Lonskaya, I., Hebron, M. L., Desforges, N. M., Schachter, J. B., & Moussa, C. E. (2014). Nilotinib-induced autophagic changes increase endogenous parkin level and ubiquitination, leading to amyloid clearance. *Journal of Molecular Medicine (Berlin, Germany)*, *92*(4), 373–386.

Lonskaya, I., Hebron, M. L., Selby, S. T., Turner, R. S., & Moussa, C. E. (2015). Nilotinib and bosutinib modulate pre-plaque alterations of blood immune markers and neuro-inflammation in Alzheimer's disease models. *Neuroscience*, *304*, 316–327.

Lopez, A., Lee, S. E., Wojta, K., Ramos, E. M., Klein, E., Chen, J., ... Rubinsztein, D. C. (2017). A152T tau allele causes neurodegeneration which can be ameliorated in a zebrafish model by autophagy induction. *Brain*, *140*(4), 1128–1146.

Lopez de Maturana, R., Lang, V., Zubiarrain, A., Sousa, A., Vazquez, N., Gorostidi, A., ... Sanchez-Pernaute, R. (2016). Mutations in LRRK2 impair NF-kappaB pathway in iPSC-derived neurons. *Journal of Neuroinflammation*, *13*(1), 295.

Lu, K., Psakhye, I., & Jentsch, S. (2014). Autophagic clearance of polyQ proteins mediated by ubiquitin-Atg8 adaptors of the conserved CUET protein family. *Cell*, *158*(3), 549–563.

Lu, Q., Yang, P., Huang, X., Hu, W., Guo, B., Wu, F., ... Zhang, H. (2011). The WD40 repeat PtdIns(3)P-binding protein EPG-6 regulates progression of omegasomes to autophagosomes. *Developmental Cell*, *21*(2), 343–357.

Lunemann, J. D., Schmidt, J., Schmid, D., Barthel, K., Wrede, A., Dalakas, M. C., & Munz, C. (2007). Beta-amyloid is a substrate of autophagy in sporadic inclusion body myositis. *Annals of Neurology*, *61*(5), 476–483.

Luty, A. A., Kwok, J. B., Dobson-Stone, C., Loy, C. T., Coupland, K. G., Karlstrom, H., ... Schofield, P. R. (2010). Sigma nonopioid intracellular receptor 1 mutations cause frontotemporal lobar degeneration-motor neuron disease. *Annals of Neurology*, *68*(5), 639–649.

Luzio, J. P., Pryor, P. R., & Bright, N. A. (2007). Lysosomes: Fusion and function. *Nature Reviews Molecular Cell Biology*, *8*(8), 622–632.

Ma, J. F., Huang, Y., Chen, S. D., & Halliday, G. (2010). Immunohistochemical evidence for macroautophagy in neurones and endothelial cells in Alzheimer's disease. *Neuropathology and Applied Neurobiology*, *36*(4), 312–319.

Ma, T. C., Buescher, J. L., Oatis, B., Funk, J. A., Nash, A. J., Carrier, R. L., & Hoyt, K. R. (2007). Metformin therapy in a transgenic mouse model of Huntington's disease. *Neuroscience Letters*, *411*(2), 98–103.

Magalhaes, J., Gegg, M. E., Migdalska-Richards, A., Doherty, M. K., Whitfield, P. D., & Schapira, A. H. (2016). Autophagic lysosome reformation dysfunction in glucocerebrosidase deficient cells: Relevance to Parkinson disease. *Human Molecular Genetics*, *25*(16), 3432–3445.

Majid, T., Ali, Y. O., Venkitaramani, D. V., Jang, M. K., Lu, H. C., & Pautler, R. G. (2014). In vivo axonal transport deficits in a mouse model of fronto-temporal dementia. *Neuroimage Clin*, *4*, 711–717.

Mancias, J. D., Wang, X., Gygi, S. P., Harper, J. W., & Kimmelman, A. C. (2014). Quantitative proteomics identifies NCOA4 as the cargo receptor mediating ferritinophagy. *Nature*, *509*(7498), 105–109.

Mandell, M. A., Jain, A., Arko-Mensah, J., Chauhan, S., Kimura, T., Dinkins, C., ... Deretic, V. (2014). TRIM proteins regulate autophagy and can target autophagic substrates by direct recognition. *Developmental Cell*, *30*(4), 394–409.

Manzanillo, P. S., Ayres, J. S., Watson, R. O., Collins, A. C., Souza, G., Rae, C. S., ... Cox, J. S. (2013). The ubiquitin ligase parkin mediates resistance to intracellular pathogens. *Nature*, *501*(7468), 512–516.

Marshall, R. S., Li, F., Gemperline, D. C., Book, A. J., & Vierstra, R. D. (2015). Autophagic degradation of the 26S proteasome is mediated by the dual ATG8/ubiquitin receptor RPN10 in arabidopsis. *Molecular Cell*, *58*(6), 1053–1066.

Martinez-Vicente, M., Talloczy, Z., Wong, E., Tang, G., Koga, H., Kaushik, S., ... Cuervo, A. M. (2010). Cargo recognition failure is responsible for inefficient autophagy in Huntington's disease. *Nature Neuroscience*, *13*(5), 567–576.

Maruyama, H., Morino, H., Ito, H., Izumi, Y., Kato, H., Watanabe, Y., ... Kawakami, H. (2010). Mutations of optineurin in amyotrophic lateral sclerosis. *Nature*, *465*(7295), 223–226.

Matikainen-Ankney, B. A., Kezunovic, N., Mesias, R. E., Tian, Y., Williams, F. M., Huntley, G. W., & Benson, D. L. (2016). Altered development of synapse structure and function in striatum caused by Parkinson's disease-linked LRRK2-G2019S mutation. *The Journal of Neuroscience*, *36*(27), 7128–7141.

Mawuenyega, K. G., Sigurdson, W., Ovod, V., Munsell, L., Kasten, T., Morris, J. C., ... Bateman, R. J. (2010). Decreased clearance of CNS beta-amyloid in Alzheimer's disease. *Science (New York, N.Y.)*, *330*(6012), 1774.

Mayer, A. L., Higgins, C. B., Heitmeier, M. R., Kraft, T. E., Qian, X., Crowley, J. R., ... DeBosch, B. J. (2016). SLC2A8 (GLUT8) is a mammalian trehalose transporter required for trehalose-induced autophagy. *Scientific Reports*, *6*, 38586.

Mazzulli, J. R., Xu, Y. H., Sun, Y., Knight, A. L., McLean, P. J., Caldwell, G. A., ... Krainc, D. (2011). Gaucher disease glucocerebrosidase and alpha-synuclein form a bidirectional pathogenic loop in synucleinopathies. *Cell*, *146*(1), 37–52.

Mazzulli, J. R., Zunke, F., Tsunemi, T., Toker, N. J., Jeon, S., Burbulla, L. F., ... Krainc, D. (2016). Activation of beta-glucocerebrosidase reduces pathological alpha-synuclein and restores lysosomal function in Parkinson's patient midbrain neurons. *The Journal of Neuroscience*, *36*(29), 7693–7706.

McWilliams, T. G., Prescott, A. R., Allen, G. F., Tamjar, J., Munson, M. J., Thomson, C., ... Ganley, I. G. (2016). mito-QC illuminates mitophagy and mitochondrial architecture in vivo. *The Journal of Cell Biology*, *214*(3), 333–345.

Mealer, R. G., Murray, A. J., Shahani, N., Subramaniam, S., & Snyder, S. H. (2014). Rhes, a striatal-selective protein implicated in Huntington disease, binds beclin-1 and activates autophagy. *The Journal of Biological Chemistry*, *289*(6), 3547–3554.

Medina, D. L., Fraldi, A., Bouche, V., Annunziata, F., Mansueto, G., Spampanato, C., ... Ballabio, A. (2011). Transcriptional activation of lysosomal exocytosis promotes cellular clearance. *Developmental Cell*, *21*(3), 421–430.

Melser, S., Lavie, J., & Benard, G. (2015). Mitochondrial degradation and energy metabolism. *Biochimica et Biophysica Acta*, *1853*(10 Pt B), 2812–2821.

Menzies, F. M., Fleming, A., & Rubinsztein, D. C. (2015). Compromised autophagy and neurodegenerative diseases. *Nature Reviews Neuroscience*, *16*(6), 345–357.

Menzies, F. M., Garcia-Arencibia, M., Imarisio, S., O'Sullivan, N. C., Ricketts, T., Kent, B. A., ... Rubinsztein, D. C. (2015). Calpain inhibition mediates autophagy-dependent protection against polyglutamine toxicity. *Cell Death and Differentiation*, *22*(3), 433–444.

Menzies, F. M., Huebener, J., Renna, M., Bonin, M., Riess, O., & Rubinsztein, D. C. (2010). Autophagy induction reduces mutant ataxin-3 levels and toxicity in a mouse model of spinocerebellar ataxia type 3. *Brain*, *133*(Pt 1), 93–104.

Michaud, M., Martins, I., Sukkurwala, A. Q., Adjemian, S., Ma, Y., Pellegatti, P., ... Kroemer, G. (2011). Autophagy-dependent anticancer immune responses induced by chemotherapeutic agents in mice. *Science (New York, N.Y.)*, *334*(6062), 1573–1577.

Millecamps, S., & Julien, J. P. (2013). Axonal transport deficits and neurodegenerative diseases. *Nature Reviews Neuroscience*, *14*(3), 161–176.

Minassian, B. A., Lee, J. R., Herbrick, J. A., Huizenga, J., Soder, S., Mungall, A. J., ... Scherer, S. W. (1998). Mutations in a gene encoding a novel protein tyrosine phosphatase cause progressive myoclonus epilepsy. *Nature Genetics*, *20*(2), 171–174.

Mizuno, Y., Amari, M., Takatama, M., Aizawa, H., Mihara, B., & Okamoto, K. (2006). Immunoreactivities ofp62, an ubiqutin-binding protein, in the spinal anterior horn cells of patients with amyotrophic lateral sclerosis. *Journal of the Neurological Sciences*, *249*(1), 13–18.

Mizushima, N., Noda, T., Yoshimori, T., Tanaka, Y., Ishii, T., George, M. D., ... Ohsumi, Y. (1998). A protein conjugation system essential for autophagy. *Nature*, *395* (6700), 395–398.

Mizushima, N., Yamamoto, A., Matsui, M., Yoshimori, T., & Ohsumi, Y. (2004). In vivo analysis of autophagy in response to nutrient starvation using transgenic mice expressing a fluorescent autophagosome marker. *Molecular Biology of the Cell*, *15*(3), 1101–1111.

Mizushima, N., Yoshimori, T., & Ohsumi, Y. (2011). The role of Atg proteins in autophagosome formation. *Annual Review of Cell and Developmental Biology*, *27*, 107–132.

Mochida, K., Oikawa, Y., Kimura, Y., Kirisako, H., Hirano, H., Ohsumi, Y., & Nakatogawa, H. (2015). Receptor-mediated selective autophagy degrades the endoplasmic reticulum and the nucleus. *Nature*, *522*(7556), 359–362.

Moore, A. S., & Holzbaur, E. L. (2016). Dynamic recruitment and activation of ALS-associated TBK1 with its target optineurin are required for efficient mitophagy. *Proceedings of the National Academy of Sciences of the United States of America*, *113*(24), E3349–3358.

Moreau, K., Fleming, A., Imarisio, S., Lopez Ramirez, A., Mercer, J. L., Jimenez-Sanchez, M., ... Rubinsztein, D. C. (2014). PICALM modulates autophagy activity and tau accumulation. *Nature Communications*, *5*, 4998.

Moreau, K., Ravikumar, B., Renna, M., Puri, C., & Rubinsztein, D. C. (2011). Autophagosome precursor maturation requires homotypic fusion. *Cell*, *146*(2), 303–317.

Morimoto, S., Martin, B. M., Yamamoto, Y., Kretz, K. A., O'Brien, J. S., & Kishimoto, Y. (1989). Saposin A: Second cerebrosidase activator protein. *Proceedings of the National Academy of Sciences of the United States of America*, *86*(9), 3389–3393.

Murphy, K. E., Gysbers, A. M., Abbott, S. K., Tayebi, N., Kim, W. S., Sidransky, E., ... Halliday, G. M. (2014). Reduced glucocerebrosidase is associated with increased alpha-synuclein in sporadic Parkinson's disease. *Brain*, *137*(Pt 3), 834–848.

Murray, I. A., Coupland, K., Smith, J. A., Ansell, I. D., & Long, R. G. (2000). Intestinal trehalase activity in a UK population: Establishing a normal range and the effect of disease. *The British Journal of Nutrition*, *83*(3), 241–245.

Myeku, N., Clelland, C. L., Emrani, S., Kukushkin, N. V., Yu, W. H., Goldberg, A. L., & Duff, K. E. (2016). Tau-driven 26S proteasome impairment and cognitive dysfunction can be prevented early in disease by activating cAMP-PKA signaling. *Nature Medicine*, *22*(1), 46–53.

Nakatogawa, H., Ichimura, Y., & Ohsumi, Y. (2007). Atg8, a ubiquitin-like protein required for autophagosome formation, mediates membrane tethering and hemifusion. *Cell, 130*(1), 165–178.

Narang, P. K., Blumhardt, C. L., Doran, A. R., & Pickar, D. (1988). Steady-state cerebrospinal fluid transfer of verapamil and metabolites in patients with schizophrenia. *Clinical Pharmacology and Therapeutics, 44*(5), 550–557.

Narendra, D., Tanaka, A., Suen, D. F., & Youle, R. J. (2008). Parkin is recruited selectively to impaired mitochondria and promotes their autophagy. *The Journal of Cell Biology, 183*(5), 795–803.

Nascimento-Ferreira, I., Santos-Ferreira, T., Sousa-Ferreira, L., Auregan, G., Onofre, I., Alves, S., ... Pereira., & de Almeida, L. (2011). Overexpression of the autophagic beclin-1 protein clears mutant ataxin-3 and alleviates Machado–Joseph disease. *Brain, 134*(Pt 5), 1400–1415.

Nazio, F., Strappazzon, F., Antonioli, M., Bielli, P., Cianfanelli, V., Bordi, M., ... Cecconi, F. (2013). mTOR inhibits autophagy by controlling ULK1 ubiquitylation, self-association and function through AMBRA1 and TRAF6. *Nature Cell Biology, 15*(4), 406–416.

Newman, A. C., Scholefield, C. L., Kemp, A. J., Newman, M., McIver, E. G., Kamal, A., & Wilkinson, S. (2012). TBK1 kinase addiction in lung cancer cells is mediated via autophagy of Tax1bp1/Ndp52 and non-canonical NF-kappaB signalling. *PLoS ONE, 7*(11), e50672.

Nguyen, T. N., Padman, B. S., Usher, J., Oorschot, V., Ramm, G., & Lazarou, M. (2016). Atg8 family LC3/GABARAP proteins are crucial for autophagosome–lysosome fusion but not autophagosome formation during PINK1/Parkin mitophagy and starvation. *The Journal of Cell Biology, 215*(6), 857–874.

Nilsson, P., Loganathan, K., Sekiguchi, M., Matsuba, Y., Hui, K., Tsubuki, S., ... Saido, T. C. (2013). Aβ secretion and plaque formation depend on autophagy. *Cell Rep, 5*(1), 61–69.

Nishiyama, J., Miura, E., Mizushima, N., Watanabe, M., & Yuzaki, M. (2007). Aberrant membranes and double-membrane structures accumulate in the axons of Atg5-null Purkinje cells before neuronal death. *Autophagy, 3*(6), 591–596.

Nixon, R. A., Wegiel, J., Kumar, A., Yu, W. H., Peterhoff, C., Cataldo, A., & Cuervo, A. M. (2005). Extensive involvement of autophagy in Alzheimer disease: An immuno-electron microscopy study. *Journal of Neuropathology and Experimental Neurology, 64*(2), 113–122.

Nixon, R. A., & Yang, D. S. (2011). Autophagy failure in Alzheimer's disease—Locating the primary defect. *Neurobiology of Disease, 43*(1), 38–45.

Noda, T., & Ohsumi, Y. (1998). Tor, a phosphatidylinositol kinase homologue, controls autophagy in yeast. *The Journal of Biological Chemistry, 273*(7), 3963–3966.

Onofre, I., Mendonca, N., Lopes, S., Nobre, R., de Melo, J. B., Carreira, I. M., ... de Almeida, L. P. (2016). Fibroblasts of Machado–Joseph disease patients reveal autophagy impairment. *Scientific Reports, 6*, 28220.

Ordureau, A., Sarraf, S. A., Duda, D. M., Heo, J. M., Jedrychowski, M. P., Sviderskiy, V. O., ... Harper, J. W. (2014). Quantitative proteomics reveal a feedforward mechanism for mitochondrial PARKIN translocation and ubiquitin chain synthesis. *Molecular Cell, 56*(3), 360–375.

Orsi, A., Razi, M., Dooley, H. C., Robinson, D., Weston, A. E., Collinson, L. M., & Tooze, S. A. (2012). Dynamic and transient interactions of Atg9 with autophagosomes, but not membrane integration, are required for autophagy. *Molecular Biology of the Cell, 23*(10), 1860–1873.

Ozcelik, S., Fraser, G., Castets, P., Schaeffer, V., Skachokova, Z., Breu, K., ... Winkler, D. T. (2013). Rapamycin attenuates the progression of tau pathology in P301S tau transgenic mice. *PLoS ONE, 8*(5), e62459.

Oz-Levi, D., Ben-Zeev, B., Ruzzo, E. K., Hitomi, Y., Gelman, A., Pelak, K., ... Lancet, D. (2012). Mutation in TECPR2 reveals a role for autophagy in hereditary spastic paraparesis. *American Journal of Human Genetics, 91*(6), 1065–1072.

Pacheco, C. D., & Lieberman, A. P. (2008). The pathogenesis of Niemann-Pick type C disease: A role for autophagy? *Expert Reviews in Molecular Medicine, 10*, e26.

Palikaras, K., Lionaki, E., & Tavernarakis, N. (2015). Coordination of mitophagy and mitochondrial biogenesis during ageing in *C. elegans*. *Nature, 521*(7553), 525–528.

Pankiv, S., Clausen, T. H., Lamark, T., Brech, A., Bruun, J. A., Outzen, H., ... Johansen, T. (2007). p62/SQSTM1 binds directly to Atg8/LC3 to facilitate degradation of ubiquitinated protein aggregates by autophagy. *The Journal of Biological Chemistry, 282*(33), 24131–24145.

Papinski, D., Schuschnig, M., Reiter, W., Wilhelm, L., Barnes, C. A., Maiolica, A., ... Kraft, C. (2014). Early steps in autophagy depend on direct phosphorylation of Atg9 by the Atg1 kinase. *Molecular Cell, 53*(3), 471–483.

Park, J. M., Jung, C. H., Seo, M., Otto, N. M., Grunwald, D., Kim, K. H., ... Kim, D. H. (2016). The ULK1 complex mediates MTORC1 signaling to the autophagy initiation machinery via binding and phosphorylating ATG14. *Autophagy, 12*(3), 547–564.

Park, J. S., Koentjoro, B., Davis, R. L., & Sue, C. M. (2016). Loss of ATP13A2 impairs glycolytic function in Kufor-Rakeb syndrome patient-derived cell models. *Parkinsonism & Related Disorders, 27*, 67–73.

Pavel, M., Imarisio, S., Menzies, F. M., Jimenez-Sanchez, M., Siddiqi, F. H., Wu, X., ... Rubinsztein, D. C. (2016).

CCT complex restricts neuropathogenic protein aggregation via autophagy. *Nature Communications, 7*, 13821.

Perez, S. E., He, B., Nadeem, M., Wuu, J., Ginsberg, S. D., Ikonomovic, M. D., & Mufson, E. J. (2015). Hippocampal endosomal, lysosomal, and autophagic dysregulation in mild cognitive impairment: Correlation with a beta and tau pathology. *Journal of Neuropathology and Experimental Neurology, 74*(4), 345–358.

Pickford, F., Masliah, E., Britschgi, M., Lucin, K., Narasimhan, R., Jaeger, P. A., ... Wyss-Coray, T. (2008). The autophagy-related protein beclin 1 shows reduced expression in early Alzheimer disease and regulates amyloid beta accumulation in mice. *The Journal of Clinical Investigation, 118*(6), 2190–2199.

Pickrell, A. M., Huang, C. H., Kennedy, S. R., Ordureau, A., Sideris, D. P., Hoekstra, J. G., ... Youle, R. J. (2015). Endogenous parkin preserves dopaminergic substantia nigral neurons following mitochondrial DNA mutagenic stress. *Neuron, 87*(2), 371–381.

Pilli, M., Arko-Mensah, J., Ponpuak, M., Roberts, E., Master, S., Mandell, M. A., ... Deretic, V. (2012). TBK-1 promotes autophagy-mediated antimicrobial defense by controlling autophagosome maturation. *Immunity, 37*(2), 223–234.

Piras, A., Collin, L., Gruninger, F., Graff, C., & Ronnback, A. (2016). Autophagic and lysosomal defects in human tauopathies: Analysis of post-mortem brain from patients with familial Alzheimer disease, corticobasal degeneration and progressive supranuclear palsy. *Acta Neuropathologica Communications, 4*, 22.

Platt, F. M., Boland, B., & van der Spoel, A. C. (2012). The cell biology of disease: Lysosomal storage disorders: The cellular impact of lysosomal dysfunction. *The Journal of Cell Biology, 199*(5), 723–734.

Potter, R., Patterson, B. W., Elbert, D. L., Ovod, V., Kasten, T., Sigurdson, W., & Bateman, R. J. (2013). Increased in vivo amyloid-beta42 production, exchange, and loss in presenilin mutation carriers. *Science Translational Medicine, 5*(189), 189ra177.

Poupetova, H., Ledvinova, J., Berna, L., Dvorakova, L., Kozich, V., & Elleder, M. (2010). The birth prevalence of lysosomal storage disorders in the Czech Republic: Comparison with data in different populations. *Journal of Inherited Metabolic Disease, 33*(4), 387–396.

Proikas-Cezanne, T., Takacs, Z., Donnes, P., & Kohlbacher, O. (2015). WIPI proteins: Essential PtdIns3P effectors at the nascent autophagosome. *Journal of Cell Science, 128*(2), 207–217.

Pryor, P. R., Mullock, B. M., Bright, N. A., Lindsay, M. R., Gray, S. R., Richardson, S. C., ... Luzio, J. P. (2004). Combinatorial SNARE complexes with VAMP7 or VAMP8 define different late endocytic fusion events. *EMBO Reports, 5*(6), 590–595.

Puls, I., Jonnakuty, C., LaMonte, B. H., Holzbaur, E. L., Tokito, M., Mann, E., ... Fischbeck, K. H. (2003). Mutant dynactin in motor neuron disease. *Nature Genetics, 33*(4), 455–456.

Puri, C., Renna, M., Bento, C. F., Moreau, K., & Rubinsztein, D. C. (2013). Diverse autophagosome membrane sources coalesce in recycling endosomes. *Cell, 154*(6), 1285–1299.

Pyo, J. O., Yoo, S. M., Ahn, H. H., Nah, J., Hong, S. H., Kam, T. I., ... Jung, Y. K. (2013). Overexpression of Atg5 in mice activates autophagy and extends lifespan. *Nature Communications, 4*, 2300.

Ramesh Babu, J., Lamar Seibenhener, M., Peng, J., Strom, A. L., Kemppainen, R., Cox, N., ... Wooten, M. W. (2008). Genetic inactivation of p62 leads to accumulation of hyperphosphorylated tau and neurodegeneration. *Journal of Neurochemistry, 106*(1), 107–120.

Randow, F., & Youle, R. J. (2014). Self and nonself: How autophagy targets mitochondria and bacteria. *Cell Host & Microbe, 15*(4), 403–411.

Ravikumar, B., Acevedo-Arozena, A., Imarisio, S., Berger, Z., Vacher, C., O'Kane, C. J., ... Rubinsztein, D. C. (2005). Dynein mutations impair autophagic clearance of aggregate-prone proteins. *Nature Genetics, 37*(7), 771–776.

Ravikumar, B., Duden, R., & Rubinsztein, D. C. (2002). Aggregate-prone proteins with polyglutamine and polyalanine expansions are degraded by autophagy. *Human Molecular Genetics, 11*(9), 1107–1117.

Ravikumar, B., Moreau, K., Jahreiss, L., Puri, C., & Rubinsztein, D. C. (2010). Plasma membrane contributes to the formation of pre-autophagosomal structures. *Nature Cell Biology, 12*(8), 747–757.

Ravikumar, B., Sarkar, S., & Rubinsztein, D. C. (2008). Clearance of mutant aggregate-prone proteins by autophagy. *Methods in Molecular Biology, 445*, 195–211.

Ravikumar, B., Vacher, C., Berger, Z., Davies, J. E., Luo, S., Oroz, L. G., ... Rubinsztein, D. C. (2004). Inhibition of mTOR induces autophagy and reduces toxicity of polyglutamine expansions in fly and mouse models of Huntington disease. *Nature Genetics, 36*(6), 585–595.

Razi, M., Chan, E. Y., & Tooze, S. A. (2009). Early endosomes and endosomal coatomer are required for autophagy. *The Journal of Cell Biology, 185*(2), 305–321.

Reinwald, M., Schleyer, E., Kiewe, P., Blau, I. W., Burmeister, T., Pursche, S., ... Bender, H. U. (2014). Efficacy and pharmacologic data of second-generation tyrosine kinase inhibitor nilotinib in BCR–ABL-positive leukemia patients with central nervous system relapse after allogeneic stem cell transplantation. *BioMed Research International, 2014*, 637059.

Renvoise, B., Chang, J., Singh, R., Yonekawa, S., FitzGibbon, E. J., Mankodi, A., ... Pierson, T. M. (2014).

Lysosomal abnormalities in hereditary spastic paraplegia types SPG15 and SPG11. *Annals of Clinical and Translational Neurology*, *1*(6), 379–389.

Richards, A. B., Krakowka, S., Dexter, L. B., Schmid, H., Wolterbeek, A. P., Waalkens-Berendsen, D. H., ... Kurimoto, M. (2002). Trehalose: A review of properties, history of use and human tolerance, and results of multiple safety studies. *Food and Chemical Toxicology: An International Journal Published for the British Industrial Biological Research Association*, *40*(7), 871–898.

Rockenstein, E., Clarke, J., Viel, C., Panarello, N., Treleaven, C. M., Kim, C., ... Sardi, S. P. (2016). Glucocerebrosidase modulates cognitive and motor activities in murine models of Parkinson's disease. *Human Molecular Genetics*, *25*(13), 2645–2660.

Roczniak-Ferguson, A., Petit, C. S., Froehlich, F., Qian, S., Ky, J., Angarola, B., ... Ferguson, S. M. (2012). The transcription factor TFEB links mTORC1 signaling to transcriptional control of lysosome homeostasis. *Science Signaling*, *5*(228), ra42.

Rodriguez-Navarro, J. A., Rodriguez, L., Casarejos, M. J., Solano, R. M., Gomez, A., Perucho, J., ... Mena, M. A. (2010). Trehalose ameliorates dopaminergic and tau pathology in parkin deleted/tau overexpressing mice through autophagy activation. *Neurobiology of Disease*, *39*(3), 423–438.

Rogov, V., Dotsch, V., Johansen, T., & Kirkin, V. (2014). Interactions between autophagy receptors and ubiquitin-like proteins form the molecular basis for selective autophagy. *Molecular Cell*, *53*(2), 167–178.

Rohn, T. T., Wirawan, E., Brown, R. J., Harris, J. R., Masliah, E., & Vandenabeele, P. (2011). Depletion of Beclin-1 due to proteolytic cleavage by caspases in the Alzheimer's disease brain. *Neurobiology of Disease*, *43*(1), 68–78.

Roosen, D. A., & Cookson, M. R. (2016). LRRK2 at the interface of autophagosomes, endosomes and lysosomes. *Molecular Neurodegeneration*, *11*(1), 73.

Rose, C., Menzies, F. M., Renna, M., Acevedo-Arozena, A., Corrochano, S., Sadiq, O., ... Rubinsztein, D. C. (2010). Rilmenidine attenuates toxicity of polyglutamine expansions in a mouse model of Huntington's disease. *Human Molecular Genetics*, *19*(11), 2144–2153.

Rubinsztein, D. C. (2006). The roles of intracellular protein-degradation pathways in neurodegeneration. *Nature*, *443*(7113), 780–786.

Rubinsztein, D. C., Bento, C. F., & Deretic, V. (2015). Therapeutic targeting of autophagy in neurodegenerative and infectious diseases. *The Journal of Experimental Medicine*, *212*(7), 979–990.

Rui, Y. N., Xu, Z., Patel, B., Chen, Z., Chen, D., Tito, A., ... Zhang, S. (2015). Huntingtin functions as a scaffold for selective macroautophagy. *Nature Cell Biology*, *17*(3), 262–275.

Russell, R. C., Tian, Y., Yuan, H., Park, H. W., Chang, Y. Y., Kim, J., ... Guan, K. L. (2013). ULK1 induces autophagy by phosphorylating Beclin-1 and activating VPS34 lipid kinase. *Nature Cell Biology*, *15*(7), 741–750.

Rusten, T. E., & Stenmark, H. (2009). How do ESCRT proteins control autophagy? *Journal of Cell Science*, *122*(Pt 13), 2179–2183.

Saftig, P., & Klumperman, J. (2009). Lysosome biogenesis and lysosomal membrane proteins: Trafficking meets function. *Nature Reviews Molecular Cell Biology*, *10*(9), 623–635.

Sahu, R., Kaushik, S., Clement, C. C., Cannizzo, E. S., Scharf, B., Follenzi, A., ... Santambrogio, L. (2011). Microautophagy of cytosolic proteins by late endosomes. *Developmental Cell*, *20*(1), 131–139.

Saitsu, H., Nishimura, T., Muramatsu, K., Kodera, H., Kumada, S., Sugai, K., ... Matsumoto, N. (2013). De novo mutations in the autophagy gene WDR45 cause static encephalopathy of childhood with neurodegeneration in adulthood. *Nature Genetics*, *45*(4), 445–449, 449e441.

Sanchez-Martinez, A., Beavan, M., Gegg, M. E., Chau, K. Y., Whitworth, A. J., & Schapira, A. H. (2016). Parkinson disease-linked GBA mutation effects reversed by molecular chaperones in human cell and fly models. *Scientific Reports*, *6*, 31380.

Sandoval, H., Thiagarajan, P., Dasgupta, S. K., Schumacher, A., Prchal, J. T., Chen, M., & Wang, J. (2008). Essential role for Nix in autophagic maturation of erythroid cells. *Nature*, *454*(7201), 232–235.

Sanjuan, M. A., Dillon, C. P., Tait, S. W., Moshiach, S., Dorsey, F., Connell, S., ... Green, D. R. (2007). Toll-like receptor signalling in macrophages links the autophagy pathway to phagocytosis. *Nature*, *450*(7173), 1253–1257.

Sano, R., Annunziata, I., Patterson, A., Moshiach, S., Gomero, E., Opferman, J., ... d'Azzo, A. (2009). GM1-ganglioside accumulation at the mitochondria-associated ER membranes links ER stress to Ca(2 +)-dependent mitochondrial apoptosis. *Molecular Cell*, *36*(3), 500–511.

Sardiello, M., Palmieri, M., di Ronza, A., Medina, D. L., Valenza, M., Gennarino, V. A., ... Ballabio, A. (2009). A gene network regulating lysosomal biogenesis and function. *Science (New York, N.Y.)*, *325*(5939), 473–477.

Sarkar, S., Davies, J. E., Huang, Z., Tunnacliffe, A., & Rubinsztein, D. C. (2007). Trehalose, a novel mTOR-independent autophagy enhancer, accelerates the clearance of mutant huntingtin and alpha-synuclein. *The Journal of Biological Chemistry*, *282*(8), 5641–5652.

Sarkar, S., & Rubinsztein, D. C. (2008). Small molecule enhancers of autophagy for neurodegenerative diseases. *Molecular bioSystems*, *4*(9), 895–901.

Sato, S., Koike, M., Funayama, M., Ezaki, J., Fukuda, T., Ueno, T., ... Hattori, N. (2016). Lysosomal storage of subunit c of mitochondrial ATP synthase in brain-specific Atp13a2-deficient mice. *The American Journal of Pathology*, *186*(12), 3074–3082.

Schaeffer, V., Lavenir, I., Ozcelik, S., Tolnay, M., Winkler, D. T., & Goedert, M. (2012). Stimulation of autophagy reduces neurodegeneration in a mouse model of human tauopathy. *Brain*, *135*(Pt 7), 2169–2177.

Schreiber, K. H., Ortiz, D., Academia, E. C., Anies, A. C., Liao, C. Y., & Kennedy, B. K. (2015). Rapamycin-mediated mTORC2 inhibition is determined by the relative expression of FK506-binding proteins. *Aging Cell*, *14*(2), 265–273.

Schroder, B. A., Wrocklage, C., Hasilik, A., & Saftig, P. (2010). The proteome of lysosomes. *Proteomics*, *10*(22), 4053–4076.

Sellier, C., Campanari, M. L., Julie Corbier, C., Gaucherot, A., Kolb-Cheynel, I., Oulad-Abdelghani, M., ... Charlet-Berguerand, N. (2016). Loss of C9ORF72 impairs autophagy and synergizes with polyQ Ataxin-2 to induce motor neuron dysfunction and cell death. *The EMBO Journal*, *35*(12), 1276–1297.

Seok, S., Fu, T., Choi, S. E., Li, Y., Zhu, R., Kumar, S., ... Kemper, J. K. (2014). Transcriptional regulation of autophagy by an FXR-CREB axis. *Nature*, *516*(7529), 108–111.

Serrano-Puebla, A., & Boya, P. (2015). Lysosomal membrane permeabilization in cell death: New evidence and implications for health and disease. *Annals of the New York Academy of Sciences*, *1371*(1), 30–44.

Settembre, C., Di Malta, C., Polito, V. A., Garcia Arencibia, M., Vetrini, F., Erdin, S., ... Ballabio, A. (2011). TFEB links autophagy to lysosomal biogenesis. *Science (New York, N.Y.)*, *332*(6036), 1429–1433.

Settembre, C., Fraldi, A., Jahreiss, L., Spampanato, C., Venturi, C., Medina, D., ... Ballabio, A. (2008). A block of autophagy in lysosomal storage disorders. *Human Molecular Genetics*, *17*(1), 119–129.

Settembre, C., Zoncu, R., Medina, D. L., Vetrini, F., Erdin, S., Erdin, S., ... Ballabio, A. (2012). A lysosome-to-nucleus signalling mechanism senses and regulates the lysosome via mTOR and TFEB. *The EMBO Journal*, *31* (5), 1095–1108.

Shaid, S., Brandts, C. H., Serve, H., & Dikic, I. (2013). Ubiquitination and selective autophagy. *Cell Death and Differentiation*, *20*(1), 21–30.

Shen, W. C., Li, H. Y., Chen, G. C., Chern, Y., & Tu, P. H. (2015). Mutations in the ubiquitin-binding domain of OPTN/optineurin interfere with autophagy-mediated degradation of misfolded proteins by a dominant-negative mechanism. *Autophagy*, *11*(4), 685–700.

Shibata, M., Lu, T., Furuya, T., Degterev, A., Mizushima, N., Yoshimori, T., ... Yuan, J. (2006). Regulation of intracellular accumulation of mutant Huntingtin by Beclin 1. *The Journal of Biological Chemistry*, *281*(20), 14474–14485.

Shibutani, S. T., & Yoshimori, T. (2014). A current perspective of autophagosome biogenesis. *Cell Research*, *24*(1), 58–68.

Siddiqi, S., Foo, J. N., Vu, A., Azim, S., Silver, D. L., Mansoor, A., ... Khor, C. C. (2014). A novel splice-site mutation in ALS2 establishes the diagnosis of juvenile amyotrophic lateral sclerosis in a family with early onset anarthria and generalized dystonias. *PLoS ONE*, *9* (12), e113258.

Siddiqui, A., Bhaumik, D., Chinta, S. J., Rane, A., Rajagopalan, S., Lieu, C. A., ... Andersen, J. K. (2015). Mitochondrial quality control via the PGC1alpha–TFEB signaling pathway is compromised by parkin Q311X mutation but independently restored by rapamycin. *The Journal of Neuroscience*, *35*(37), 12833–12844.

Simonsen, A., Cumming, R. C., Brech, A., Isakson, P., Schubert, D. R., & Finley, K. D. (2008). Promoting basal levels of autophagy in the nervous system enhances longevity and oxidant resistance in adult *Drosophila*. *Autophagy*, *4*(2), 176–184.

Singh, R., Kaushik, S., Wang, Y., Xiang, Y., Novak, I., Komatsu, M., ... Czaja, M. J. (2009). Autophagy regulates lipid metabolism. *Nature*, *458*(7242), 1131–1135.

Small, S. A., Kent, K., Pierce, A., Leung, C., Kang, M. S., Okada, H., ... Kim, T. W. (2005). Model-guided microarray implicates the retromer complex in Alzheimer's disease. *Annals of Neurology*, *58*(6), 909–919.

Son, S. M., Shin, H. J., Byun, J., Kook, S. Y., Moon, M., Chang, Y. J., & Mook-Jung, I. (2016). Metformin facilitates amyloid-beta generation by beta- and gamma-secretases via autophagy activation. *Journal of Alzheimer's Disease: JAD*, *51*(4), 1197–1208.

Sorbara, M. T., & Girardin, S. E. (2015). Emerging themes in bacterial autophagy. *Current Opinion in Microbiology*, *23*, 163–170.

Sou, Y. S., Waguri, S., Iwata, J., Ueno, T., Fujimura, T., Hara, T., ... Komatsu, M. (2008). The Atg8 conjugation system is indispensable for proper development of autophagic isolation membranes in mice. *Molecular Biology of the Cell*, *19*(11), 4762–4775.

Spampanato, C., Feeney, E., Li, L., Cardone, M., Lim, J. A., Annunziata, F., ... Raben, N. (2013). Transcription factor EB (TFEB) is a new therapeutic target for Pompe disease. *EMBO Molecular Medicine*, *5*(5), 691–706.

Spilman, P., Podlutskaya, N., Hart, M. J., Debnath, J., Gorostiza, O., Bredesen, D., ... Galvan, V. (2010). Inhibition of mTOR by rapamycin abolishes cognitive deficits and reduces amyloid-beta levels in a mouse model of Alzheimer's disease. *PLoS ONE, 5*(4), e9979.

Staats, K. A., Hernandez, S., Schonefeldt, S., Bento-Abreu, A., Dooley, J., Van Damme, P., ... Van Den Bosch, L. (2013). Rapamycin increases survival in ALS mice lacking mature lymphocytes. *Molecular Neurodegeneration, 8*, 31.

Stadel, D., Millarte, V., Tillmann, K. D., Huber, J., Tamin-Yecheskel, B. C., Akutsu, M., ... Behrends, C. (2015). TECPR2 cooperates with LC3C to regulate COPII-dependent ER export. *Molecular Cell, 60*(1), 89–104.

Sun, N., Yun, J., Liu, J., Malide, D., Liu, C., Rovira, I. I., ... Finkel, T. (2015). Measuring in vivo mitophagy. *Molecular Cell, 60*(4), 685–696.

Sundaramoorthy, V., Walker, A. K., Tan, V., Fifita, J. A., McCann, E. P., Williams, K. L., ... Atkin, J. D. (2015). Defects in optineurin- and myosin VI-mediated cellular trafficking in amyotrophic lateral sclerosis. *Human Molecular Genetics, 24*(13), 3830–3846.

Svenning, S., & Johansen, T. (2013). Selective autophagy. *Essays in Biochemistry, 55*, 79–92.

Tan, D., Cai, Y., Wang, J., Zhang, J., Menon, S., Chou, H. T., ... Walz, T. (2013). The EM structure of the TRAPPIII complex leads to the identification of a requirement for COPII vesicles on the macroautophagy pathway. *Proceedings of the National Academy of Sciences of the United States of America, 110*(48), 19432–19437.

Tanaka, M., Machida, Y., Niu, S., Ikeda, T., Jana, N. R., Doi, H., ... Nukina, N. (2004). Trehalose alleviates polyglutamine-mediated pathology in a mouse model of Huntington disease. *Nature Medicine, 10*(2), 148–154.

Tang, F. L., Erion, J. R., Tian, Y., Liu, W., Yin, D. M., Ye, J., ... Xiong, W. C. (2015). VPS35 in dopamine neurons is required for endosome-to-Golgi retrieval of Lamp2a, a receptor of chaperone-mediated autophagy that is critical for alpha-synuclein degradation and prevention of pathogenesis of Parkinson's disease. *The Journal of Neuroscience, 35*(29), 10613–10628.

Tanji, K., Miki, Y., Maruyama, A., Mimura, J., Matsumiya, T., Mori, F., ... Wakabayashi, K. (2015). Trehalose intake induces chaperone molecules along with autophagy in a mouse model of Lewy body disease. *Biochemical and Biophysical Research Communications, 465*(4), 746–752.

Teyssou, E., Takeda, T., Lebon, V., Boillee, S., Doukoure, B., Bataillon, G., ... Millecamps, S. (2013). Mutations in SQSTM1 encoding p62 in amyotrophic lateral sclerosis: Genetics and neuropathology. *Acta Neuropathologica, 125*(4), 511–522.

Thurston, T. L., Ryzhakov, G., Bloor, S., von Muhlinen, N., & Randow, F. (2009). The TBK1 adaptor and autophagy receptor NDP52 restricts the proliferation of ubiquitin-coated bacteria. *Nature Immunology, 10*(11), 1215–1221.

Tian, Y., Bustos, V., Flajolet, M., & Greengard, P. (2011). A small-molecule enhancer of autophagy decreases levels of Abeta and APP-CTF via Atg5-dependent autophagy pathway. *The FASEB Journal: Official Publication of the Federation of American Societies for Experimental Biology, 25*(6), 1934–1942.

Tian, Y., Chang, J. C., Fan, E. Y., Flajolet, M., & Greengard, P. (2013). Adaptor complex AP2/PICALM, through interaction with LC3, targets Alzheimer's APP-CTF for terminal degradation via autophagy. *Proceedings of the National Academy of Sciences of the United States of America, 110*(42), 17071–17076.

Tiede, S., Storch, S., Lubke, T., Henrissat, B., Bargal, R., Raas-Rothschild, A., & Braulke, T. (2005). Mucolipidosis II is caused by mutations in GNPTA encoding the alpha/beta GlcNAc-1-phosphotransferase. *Nature Medicine, 11*(10), 1109–1112.

Tien, N. T., Karaca, I., Tamboli, I. Y., & Walter, J. (2016). Trehalose alters subcellular trafficking and the metabolism of the Alzheimer-associated amyloid precursor protein. *The Journal of Biological Chemistry, 291*(20), 10528–10540.

Tsuboyama, K., Koyama-Honda, I., Sakamaki, Y., Koike, M., Morishita, H., & Mizushima, N. (2016). The ATG conjugation systems are important for degradation of the inner autophagosomal membrane. *Science (New York, N.Y.), 354*(6315), 1036–1041.

Tsunemi, T., Ashe, T. D., Morrison, B. E., Soriano, K. R., Au, J., Roque, R. A., ... La Spada, A. R. (2012). PGC-1alpha rescues Huntington's disease proteotoxicity by preventing oxidative stress and promoting TFEB function. *Science Translational Medicine, 4*(142), 142ra197.

Tumbarello, D. A., Waxse, B. J., Arden, S. D., Bright, N. A., Kendrick-Jones, J., & Buss, F. (2012). Autophagy receptors link myosin VI to autophagosomes to mediate Tom1-dependent autophagosome maturation and fusion with the lysosome. *Nature Cell Biology, 14*(10), 1024–1035.

Ugolino, J., Ji, Y. J., Conchina, K., Chu, J., Nirujogi, R. S., Pandey, A., ... Wang, J. (2016). Loss of C9orf72 enhances autophagic activity via deregulated mTOR and TFEB signaling. *PLoS Genetics, 12*(11), e1006443.

Ulgherait, M., Rana, A., Rera, M., Graniel, J., & Walker, D. W. (2014). AMPK modulates tissue and organismal aging in a non-cell-autonomous manner. *Cell Reports, 8*(6), 1767–1780.

Ushio, H., Ueno, T., Kojima, Y., Komatsu, M., Tanaka, S., Yamamoto, A., ... Nakano, H. (2011). Crucial role for autophagy in degranulation of mast cells. *The Journal of Allergy and Clinical Immunology, 127*(5), 1267–1276, e1266.

Valente, E. M., Abou-Sleiman, P. M., Caputo, V., Muqit, M. M., Harvey, K., Gispert, S., ... Wood, N. W. (2004). Hereditary early-onset Parkinson's disease caused by mutations in PINK1. *Science (New York, N.Y.)*, *304*(5674), 1158–1160.

Vantaggiato, C., Crimella, C., Airoldi, G., Polishchuk, R., Bonato, S., Brighina, E., ... Bassi, M. T. (2013). Defective autophagy in spastizin mutated patients with hereditary spastic paraparesis type 15. *Brain*, *136*(Pt 10), 3119–3139.

Varga, R. E., Khundadze, M., Damme, M., Nietzsche, S., Hoffmann, B., Stauber, T., ... Hubner, C. A. (2015). In vivo evidence for lysosome depletion and impaired autophagic clearance in hereditary spastic paraplegia type SPG11. *PLoS Genetics*, *11*(8), e1005454.

Velikkakath, A. K., Nishimura, T., Oita, E., Ishihara, N., & Mizushima, N. (2012). Mammalian Atg2 proteins are essential for autophagosome formation and important for regulation of size and distribution of lipid droplets. *Molecular Biology of the Cell*, *23*(5), 896–909.

Vidoni, C., Follo, C., Savino, M., Melone, M. A., & Isidoro, C. (2016). The Role of Cathepsin D in the Pathogenesis of Human Neurodegenerative Disorders. *Medicinal Research Reviews*, *36*(5), 845–870.

Vingtdeux, V., Chandakkar, P., Zhao, H., d'Abramo, C., Davies, P., & Marambaud, P. (2011). Novel synthetic small-molecule activators of AMPK as enhancers of autophagy and amyloid-beta peptide degradation. *The FASEB Journal: Official Publication of the Federation of American Societies for Experimental Biology*, *25*(1), 219–231.

Vollrath, J. T., Sechi, A., Dreser, A., Katona, I., Wiemuth, D., Vervoorts, J., ... Goswami, A. (2014). Loss of function of the ALS protein SigR1 leads to ER pathology associated with defective autophagy and lipid raft disturbances. *Cell Death & Disease*, *5*, e1290.

Wang, C., Liang, C. C., Bian, Z. C., Zhu, Y., & Guan, J. L. (2013). FIP200 is required for maintenance and differentiation of postnatal neural stem cells. *Nature Neuroscience*, *16*(5), 532–542.

Wang, Y., Martinez-Vicente, M., Kruger, U., Kaushik, S., Wong, E., Mandelkow, E. M., ... Mandelkow, E. (2009). Tau fragmentation, aggregation and clearance: The dual role of lysosomal processing. *Human Molecular Genetics*, *18*(21), 4153–4170.

Wang, Z., Miao, G., Xue, X., Guo, X., Yuan, C., Zhang, G., ... Zhang, H. (2016). The vici syndrome protein EPG5 is a Rab7 effector that determines the fusion specificity of autophagosomes with late endosomes/lysosomes. *Molecular Cell*, *63*(5), 781–795.

Watabe, K., Akiyama, K., Kawakami, E., Ishii, T., Endo, K., Yanagisawa, H., ... Tsukamoto, M. (2014). Adenoviral expression of TDP-43 and FUS genes and shRNAs for protein degradation pathways in rodent motoneurons in vitro and in vivo. *Neuropathology: Official Journal of the Japanese Society of Neuropathology*, *34*(1), 83–98.

Webb, J. L., Ravikumar, B., Atkins, J., Skepper, J. N., & Rubinsztein, D. C. (2003). Alpha-Synuclein is degraded by both autophagy and the proteasome. *The Journal of Biological Chemistry*, *278*(27), 25009–25013.

Webster, C. P., Smith, E. F., Bauer, C. S., Moller, A., Hautbergue, G. M., Ferraiuolo, L., ... De Vos, K. J. (2016). The C9orf72 protein interacts with Rab1a and the ULK1 complex to regulate initiation of autophagy. *The EMBO Journal*, *35*(15), 1656–1676.

Weedon, M. N., Hastings, R., Caswell, R., Xie, W., Paszkiewicz, K., Antoniadi, T., ... Ellard, S. (2011). Exome sequencing identifies a DYNC1H1 mutation in a large pedigree with dominant axonal Charcot–Marie–Tooth disease. *American Journal of Human Genetics*, *89*(2), 308–312.

Wei, H., Kim, S. J., Zhang, Z., Tsai, P. C., Wisniewski, K. E., & Mukherjee, A. B. (2008). ER and oxidative stresses are common mediators of apoptosis in both neurodegenerative and non-neurodegenerative lysosomal storage disorders and are alleviated by chemical chaperones. *Human Molecular Genetics*, *17*(4), 469–477.

West, R. J., Lu, Y., Marie, B., Gao, F. B., & Sweeney, S. T. (2015). Rab8, POSH, and TAK1 regulate synaptic growth in a *Drosophila* model of frontotemporal dementia. *The Journal of Cell Biology*, *208*(7), 931–947.

Wild, P., Farhan, H., McEwan, D. G., Wagner, S., Rogov, V. V., Brady, N. R., ... Dikic, I. (2011). Phosphorylation of the autophagy receptor optineurin restricts Salmonella growth. *Science (New York, N.Y.)*, *333*(6039), 228–233.

Williams, A., Sarkar, S., Cuddon, P., Ttofi, E. K., Saiki, S., Siddiqi, F. H., ... Rubinsztein, D. C. (2008). Novel targets for Huntington's disease in an mTOR-independent autophagy pathway. *Nature Chemical Biology*, *4*(5), 295–305.

Williams, K. L., Warraich, S. T., Yang, S., Solski, J. A., Fernando, R., Rouleau, G. A., ... Blair, I. P. (2012). UBQLN2/ubiquilin 2 mutation and pathology in familial amyotrophic lateral sclerosis. *Neurobiology of Aging*, *33*(10), 2527.e3–2527.e10.

Winklhofer, K. F., Tatzelt, J., & Haass, C. (2008). The two faces of protein misfolding: Gain- and loss-of-function in neurodegenerative diseases. *The EMBO Journal*, *27*(2), 336–349.

Winslow, A. R., Chen, C. W., Corrochano, S., Acevedo-Arozena, A., Gordon, D. E., Peden, A. A., ... Rubinsztein, D. C. (2010). alpha-Synuclein impairs macroautophagy: Implications for Parkinson's disease. *The Journal of Cell Biology*, *190*(6), 1023–1037.

Wolfe, D. M., Lee, J. H., Kumar, A., Lee, S., Orenstein, S. J., & Nixon, R. A. (2013). Autophagy failure in Alzheimer's disease and the role of defective lysosomal acidification. *The European Journal of Neuroscience*, *37*(12), 1949–1961.

Wong, Y. C., & Holzbaur, E. L. (2014). Optineurin is an autophagy receptor for damaged mitochondria in parkin-mediated mitophagy that is disrupted by an ALS-linked mutation. *Proceedings of the National Academy of Sciences of the United States of America, 111* (42), E4439–4448.

Wraith, J. E. (2002). Lysosomal disorders. *Seminars in Neonatology: SN, 7*(1), 75–83.

Wu, X., Fleming, A., Ricketts, T., Pavel, M., Virgin, H., Menzies, F. M., & Rubinsztein, D. C. (2016). Autophagy regulates Notch degradation and modulates stem cell development and neurogenesis. *Nature Communications, 7*, 10533.

Xi, Y., Dhaliwal, J. S., Ceizar, M., Vaculik, M., Kumar, K. L., & Lagace, D. C. (2016). Knock-out of Atg5 delays the maturation and reduces the survival of adult-generated neurons in the hippocampus. *Cell Death & Disease, 7*, e2127.

Yamamoto, M., Suzuki, S. O., & Himeno, M. (2010). The effects of dynein inhibition on the autophagic pathway in glioma cells. *Neuropathology: Official Journal of the Japanese Society of Neuropathology, 30*(1), 1–6.

Yazdankhah, M., Farioli-Vecchioli, S., Tonchev, A. B., Stoykova, A., & Cecconi, F. (2014). The autophagy regulators Ambra1 and Beclin 1 are required for adult neurogenesis in the brain subventricular zone. *Cell Death & Disease, 5*, e1403.

Yla-Anttila, P., Vihinen, H., Jokitalo, E., & Eskelinen, E. L. (2009). 3D tomography reveals connections between the phagophore and endoplasmic reticulum. *Autophagy, 5* (8), 1180–1185.

Yoshii, S. R., Kuma, A., Akashi, T., Hara, T., Yamamoto, A., Kurikawa, Y., ... Mizushima, N. (2016). Systemic analysis of Atg5-null mice rescued from neonatal lethality by transgenic ATG5 expression in neurons. *Developmental Cell, 39*(1), 116–130.

Yu, H. C., Lin, C. S., Tai, W. T., Liu, C. Y., Shiau, C. W., & Chen, K. F. (2013). Nilotinib induces autophagy in hepatocellular carcinoma through AMPK activation. *The Journal of Biological Chemistry, 288*(25), 18249–18259.

Yu, L., McPhee, C. K., Zheng, L., Mardones, G. A., Rong, Y., Peng, J., ... Lenardo, M. J. (2010). Termination of autophagy and reformation of lysosomes regulated by mTOR. *Nature, 465*(7300), 942–946.

Yu, W., Cuervo, A., Kumar, A., Peterhoff, C., Schmidt, S., Lee, J., ... Tjernberg, L. (2005). Macroautophagy—A novel beta-amyloid peptide-generating pathway activated in Alzheimer's disease. *The Journal of Cell Biology, 171*, 87–98.

Yuan, H. X., Russell, R. C., & Guan, K. L. (2013). Regulation of PIK3C3/VPS34 complexes by MTOR in nutrient stress-induced autophagy. *Autophagy, 9*(12), 1983–1995.

Yue, M., Hinkle, K. M., Davies, P., Trushina, E., Fiesel, F. C., Christenson, T. A., ... Melrose, H. L. (2015). Progressive dopaminergic alterations and mitochondrial abnormalities in LRRK2 G2019S knock-in mice. *Neurobiology of Disease, 78*, 172–195.

Zare-Shahabadi, A., Masliah, E., Johnson, G. V., & Rezaei, N. (2015). Autophagy in Alzheimer's disease. *Reviews in the Neurosciences, 26*(4), 385–395.

Zavodszky, E., Seaman, M. N., Moreau, K., Jimenez-Sanchez, M., Breusegem, S. Y., Harbour, M. E., & Rubinsztein, D. C. (2014). Mutation in VPS35 associated with Parkinson's disease impairs WASH complex association and inhibits autophagy. *Nature Communications, 5*, 3828.

Zhang, J., Lachance, V., Schaffner, A., Li, X., Fedick, A., Kaye, L. E., ... Edelmann, L. (2016). A founder mutation in VPS11 causes an autosomal recessive leukoencephalopathy linked to autophagic defects. *PLoS Genetics, 12* (4), e1005848.

Zhang, K. Y., Yang, S., Warraich, S. T., & Blair, I. P. (2014). Ubiquilin 2: A component of the ubiquitin-proteasome system with an emerging role in neurodegeneration. *The International Journal of Biochemistry & Cell Biology, 50*, 123–126.

Zhang, X., Chen, S., Song, L., Tang, Y., Shen, Y., Jia, L., & Le, W. (2014). MTOR-independent, autophagic enhancer trehalose prolongs motor neuron survival and ameliorates the autophagic flux defect in a mouse model of amyotrophic lateral sclerosis. *Autophagy, 10*(4), 588–602.

Zhang, X., Li, L., Chen, S., Yang, D., Wang, Y., Zhang, X., ... Le, W. (2011). Rapamycin treatment augments motor neuron degeneration in SOD1(G93A) mouse model of amyotrophic lateral sclerosis. *Autophagy, 7*(4), 412–425.

Zhang, X. J., Chen, S., Huang, K. X., & Le, W. D. (2013). Why should autophagic flux be assessed? *Acta Pharmacologica Sinica, 34*(5), 595–599.

Zhao, Y. G., Sun, L., Miao, G., Ji, C., Zhao, H., Sun, H., ... Zhang, H. (2015). The autophagy gene Wdr45/Wipi4 regulates learning and memory function and axonal homeostasis. *Autophagy, 11*(6), 881–890.

Zhao, Y. G., Zhao, H., Sun, H., & Zhang, H. (2013). Role of Epg5 in selective neurodegeneration and Vici syndrome. *Autophagy, 9*(8), 1258–1262.

Zheng, S., Clabough, E. B., Sarkar, S., Futter, M., Rubinsztein, D. C., & Zeitlin, S. O. (2010). Deletion of the huntingtin polyglutamine stretch enhances neuronal autophagy and longevity in mice. *PLoS Genetics, 6*(2), e1000838.

Zhou, W., Scott, S. A., Shelton, S. B., & Crutcher, K. A. (2006). Cathepsin D-mediated proteolysis of apolipoprotein E: Possible role in Alzheimer's disease. *Neuroscience, 143*(3), 689–701.

Zhou, Z., Doggett, T. A., Sene, A., Apte, R. S., & Ferguson, T. A. (2015). Autophagy supports survival and phototransduction protein levels in rod photoreceptors. *Cell Death and Differentiation*, *22*(3), 488–498.

Zuchner, S., Noureddine, M., Kennerson, M., Verhoeven, K., Claeys, K., De Jonghe, P., ... Vance, J. M. (2005). Mutations in the pleckstrin homology domain of dynamin 2 cause dominant intermediate Charcot–Marie–Tooth disease. *Nature Genetics*, *37*(3), 289–294.

Further Reading

Argov, Z., Vornovitsky, H., Blumen, S., & Caraco, Y. (2015). First human use of high dose IV trehalose: Safety, tolerability and pharmacokinetic results from the oculopharyngeal muscular dystrophy (OPMD) therapy trial (P7.068). *Neurology*, *84*(14), Supplement P7.068.

Lv, X., Jiang, H., Li, B., Liang, Q., Wang, S., Zhao, Q., & Jiao, J. (2014). The crucial role of Atg5 in cortical neurogenesis during early brain development. *Scientific Reports*, *4*, 6010.

CHAPTER

12

Neurodegenerative Diseases and Axonal Transport

Lawrence S. Goldstein and Utpal Das

University of California San Diego, School of Medicine, San Diego, CA, United States

OUTLINE

INTRODUCTION TO AXONAL TRANSPORT

Intracellular transport of macromolecules and organelles is a fundamental requirement for all cell types in mammalian organ systems. Proteins, one of the critical macromolecules essential for cell survival and integrity, are largely synthesized in association with the endoplasmic reticulum (ER), posttranslationally modified in the ER and Golgi, and eventually sorted to vesicles targeted to specific cellular locations. While every cell type shares common mechanisms for transporting proteins, neurons follow a highly elaborated mechanism to target proteins to different domains. Neurons are highly polarized cells with a cell body resembling a nonneuronal cell, with multiple dendrites projecting from the cell body. Neurons also generally have one long axon traversing thousands of microns, which in some cases such as human motor neurons, can exceed 1 m. Despite the extreme length,

The Molecular and Cellular Basis of Neurodegenerative Diseases
DOI: https://doi.org/10.1016/B978-0-12-811304-2.00012-2

axons are largely dependent on the cell body for supply of essential proteins, axonal components, and other critical molecules required for synaptic transmission (Fig. 12.1). Thus, the geometry of axons requires axonal transport to be particularly well-orchestrated, with any interference in axonal transport leading to a pathogenic condition or disease.

Inside an axon, there is two-way traffic. In anterograde transport, axonal components synthesized in the cell body are generally packaged into vesicles and translocated from the cell body into the axonal processes and toward the synapses (Fig. 12.1). These vesicles can be derived from the Golgi apparatus or from the endocytic system. In the counter (retrograde) mechanism, recycled proteins, organelles, and signaling complexes are transported from the tip of the axon toward the cell body. A significant fraction of retrograde axonal traffic may be derived from endocytosis at the presynaptic domain. Both anterograde and retrograde transport are primarily long-range microtubule-based transport that is dependent on the molecular "motors" and the microtubule "tracks." Microtubules are key components of the axonal cytoskeleton. They are polarized in axons with a faster growing "plus" end pointing toward the axon tip and a slower growing "minus" end facing the cell body. Structurally, they are tubular, 25 nm in diameter and composed of α- and β-tubulin heterodimers. Tau protein stabilizes some microtubule tracks. In the axon, motors largely fall into two categories—the kinesin superfamily and the dyneins. Kinesins primarily generate unidirectional movement of cargoes toward the microtubule plus end, thus mediating mostly anterograde transport. Complementing anterograde movement, dynein motors move toward the microtubule minus end, carrying the

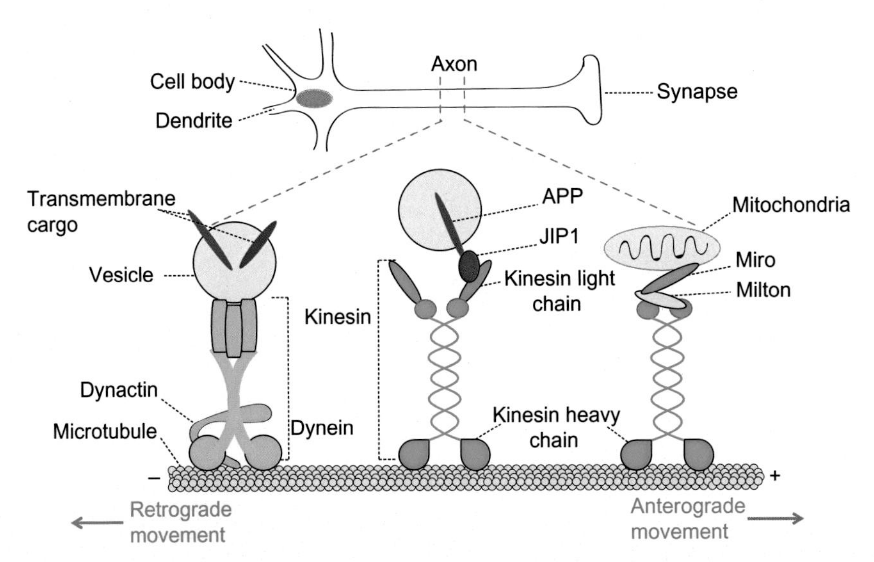

FIGURE 12.1 **Schematic showing the fast axonal transport machinery in a neuron.** After biosynthesis in the cell body, proteins are transported toward different neuronal processes. In axon, kinesin and dynein motors govern the axonal transport of membranous cargoes such as vesicles carrying membrane proteins, receptors, or mitochondria. Kinesin motor is composed of heavy and light chains. With the help of adaptor proteins such as JIP1, Miro, and Milton, kinesins carry cargoes in the anterograde direction, toward the synaptic terminals along the microtubule plus end tracks. On the other hand, proteins recycled at the synaptic terminal or destined for degradation are transported in the retrograde direction, toward the cell body along the microtubule minus ends using the dynein motor, which is composed of multiple heavy, light, and intermediate chains, along with other proteins such as dynactin.

cargoes toward the cell body and mediating retrograde transport.

Based on the speed of cargo movement, axonal transport is divided into fast and slow axonal transport. Membrane-bound vesicular cargoes and mitochondria move in the fast axonal transport component at a speed of 100–400 mm/day (1–5 μm/s). On the other hand, cytoskeletal components and soluble cytosolic proteins move in the slow axonal transport component at a speed of 0.2–5 mm/day (0.0002–0.05 μm/s). Recent studies have reported that soluble proteins organize into transport-competent multiprotein complexes and move into the axon; this slow movement is dependent on fast axonal transport of the vesicular proteins. However, at this time, specific molecular motors are not yet fully assigned to slow axonal transport, and the identity of slow transport-competent complexes are not well defined and need further experimentation. Fast axonal transport is better characterized and largely carried out by different kinesins and dynein motors.

Kinesin-1, the key molecular motor involved in anterograde fast axonal transport is a heterotetramer that consists of two kinesin heavy chains (KHCs) and two kinesin light chains (KLC) (Fig. 12.1). Structurally, each KHC molecule is divided into multiple domains, and each domain performs a specific function. The N-terminal globular motor domain binds to microtubules and ATP; the α-helical coiled-coil neck domain facilitates dimerization, and the C-terminal tail domain interacts with KLC to secure cargo binding (reviewed in Encalada & Goldstein, 2014; Gauger & Goldstein, 1993; Hirokawa, Niwa, & Tanaka, 2010; Hirokawa & Takemura, 2005; Verhey et al., 1998; Yang, Laymon, & Goldstein, 1989). When off-loaded, the C-terminal tail domain, along with the KLC, interacts with the N-terminal motor domain, inhibiting its motor activity. However, upon cargo loading, structural refolding leads to the dissociation of these domains, activating the motor function (Lawrence et al., 2004; Verhey et al., 1998). Several adaptor proteins, such as c-Jun N-terminal kinase (JNK)-interacting protein 1 (JIP1) modulate cargo binding to kinesin-1 (Bowman et al., 2000; Cavalli, Kujala, Klumperman, & Goldstein, 2005; Verhey et al., 2001).

Cytoplasmic dynein, the motor involved in retrograde fast axonal transport, is a multisubunit complex that consists of two heavy chains, two intermediate chains, four light intermediate chains, and multiple light chains (Eschbach & Dupuis, 2011) (Fig. 12.1). The globular motor domain of dynein heavy chains binds to microtubules, providing the energy required for cargo movement. The other chains are associated with cargo regulation and cargo binding, which is modulated by dynactin. Dynactin is a multiprotein complex that consists of dynactin subunit 1 (DCTN1 or p150glued), dynactin 2 (DCTN2 or dynamitin), ARP 1, CapZα/β, p24, and p27 (Pfister et al., 2006). Although dynactin is argued to mediate binding of cargo to dynein (Muresan et al., 2001), one study argues that dynein binding to vesicles is independent of dynactin and that dynactin regulates movement dynamics (Haghnia et al., 2007). Dynein-mediated transport is further regulated by Rab6 and Rab7, small GTPases that are associated with distinct membrane compartments (Deneka, Neeft, & van der Sluijs, 2003; Matanis et al., 2002).

Although kinesin-1 is associated with anterograde axonal transport and dynein with retrograde transport, recent studies provide evidence that the movement and directionality of vesicular cargoes could be an outcome of a tug-of-war between kinesin and dynein motors (Akhmanova & Hammer, 2010; Muller, Klumpp, & Lipowsky, 2008; Shubeita et al., 2008; Soppina, Rai, Ramaiya, Barak, & Mallik, 2009; Verhey, Kaul, & Soppina, 2011). A recent study by Encalada et al. reported that the motor composition on vesicles carrying cellular prion proteins (PrPc) in axons is the same irrespective of their directionalities. The study indicated that retrograde transport of the PrPc vesicles was dependent on the association of the vesicles to both kinesin-1 and KLC1 subunits (Encalada, Szpankowski, Xia, & Goldstein, 2011). Intriguingly, a recent report suggests that the kinesin-1 motor drives the slow axonal transport of dynein toward the axon tip, from which dynein starts carrying cargoes retrogradely toward the cell body (Twelvetrees et al., 2016).

AXONAL TRANSPORT AND NEURODEGENERATIVE DISEASE

Disruption in any of the three core modules of axonal transport: the cargoes, the motors, and the tracks can lead to pathological changes in the neuron, and eventually neurodegeneration. Ever since it was first postulated by Gajdusek in the mid-1980s (Gajdusek, 1985), several lines of evidence suggest axonal accumulation of proteins, and organelles can lead to multiple neurodegenerative diseases.

ALZHEIMER'S DISEASE

Alzheimer's disease (AD), the predominant form of neurodegenerative disease worldwide, is pathologically characterized by extracellular deposition of amyloid plaques made up of amyloid beta (Aβ) peptide, and intracellular accumulation of neurofibrillary tangles consisting of phosphorylated Tau. Aβ is generated by sequential cleavage of amyloid precursor protein (APP) by beta-site-APP cleavage enzyme-1 (BACE1) and presenilin, the enzymatic component of the gamma-secretase complex. On the other hand, Tau is an intracellular microtubule-binding protein that undergoes hyperphosphorylation to form paired helical filaments which eventually form filamentous accumulations within neurons and their processes (Lee, Daughenbaugh, & Trojanowski, 1994). Though AD is largely sporadic (~95%), mutations and duplication of APP or mutations in the Presenilin genes (PS1 and PS2) cause familial forms of AD (FAD) (Hardy, 2006).

Emerging evidence implicates several AD-associated proteins in carrying out and/or regulating axonal transport. APP is transported in axons through fast axonal transport (Koo et al., 1990) by the kinesin-1 motor (Ferreira, Caceres, & Kosik, 1993; Kamal, Stokin, Yang, Xia, & Goldstein, 2000). Although the physiological role of APP is still debated, an imbalance in the amount of APP within the axon (increase or decrease) has been demonstrated to lead to detrimental effects on axonal transport. A surfeit of wild-type human or Swedish-APP (APP^{swe}) expressing transgenic mice displayed phenotypes typical of axonal transport disturbance (Stokin et al., 2005), disrupted axonal transport of nerve growth factor (NGF) (Salehi et al., 2006), and caused vesicular axonal accumulation in *Drosophila* larvae (Gunawardena & Goldstein, 2001; Rusu et al., 2007; Torroja, Chu, Kotovsky, & White, 1999). These findings were further supported by in vivo MEMRI (manganese-enhanced MRI) of Tg2576 mice (Smith, Kallhoff, Zheng, & Pautler, 2007). On the other hand, loss of APP is also reported to induce axonal transport deficit (Kamal, Almenar-Queralt, LeBlanc, Roberts, & Goldstein, 2001; Kamal et al., 2000), and recent MEMRI detected significant reduction in axonal transport to the septal nuclei, amygdala, and contralateral hippocampus in the APP knockout mice (Gallagher, Zhang, Ziomek, Jacobs, & Bearer, 2012). The extent of transport deficit was further detected in the visual pathway of those animals. Ablation of the APP homologue in *Drosophila*, APPL, causes a defect in axonal transport similar to a phenotype observed in the kinesin and dynein mutants (Gunawardena & Goldstein, 2001).

Toward understanding the underlying mechanism of the APP-induced axonal transport deficit, it was first reported that APP interacts with KLC1 and mediates vesicular axonal transport (Kamal et al., 2000, 2001). This finding was supported by other studies (Cottrell et al., 2005; Fu & Holzbaur, 2013; Horiuchi, Barkus, Pilling, Gassman, & Saxton, 2005; Inomata et al., 2003; Matsuda, Matsuda, & D'Adamio, 2003; Muresan & Muresan, 2005; Satpute-Krishnan, DeGiorgis, Conley, Jang, & Bearer, 2006) which broadened understanding and led to the finding that in vivo the kinesin-1 complex interacts with APP in a ternary complex including JIP1. Additional studies also provide evidence that APP is cotransported with its cleaving enzymes BACE1 and Presenilins in axons (Almenar-Queralt et al., 2014; Gunawardena & Goldstein, 2001; Kamal et al., 2001). Recent experiments on murine primary hippocampal neurons coexpressing APP-GFP and BACE1-mCherry validates that both these proteins can be cotransported in axons and provides evidence that the transport carrier is a Golgi-derived vesicle (Das et al., 2016). These reports specify that dysregulation of cargoes carrying APP or its cleaving enzymes could lead to disruption in fast axonal transport, an increase in APP processing and Aβ generation, and eventual neurodegeneration (Gunawardena & Goldstein, 2004). Indeed, disruption to anterograde axonal transport by KLC1 depletion in the transgenic APP^{swe} mice showed axonal blockade and axonal swelling with increased Aβ production (Stokin et al., 2005). These studies also suggest that altered axonal transport can reroute APP and its cleaving enzymes to an endosomal compartment which could provide an acidic milieu optimal for the enzymes to act upon APP. However, this idea needs further experimental validation.

Further study of FAD mutations in human induced pluripotent stem cell (hIPSC)-derived neurons recently led to the surprising finding that FAD mutations $PS1^{\Delta E9}$ as well as APP^{swe} and APP^{V717F} slowed transcytosis of APP and lipoproteins from soma to axon, which leads to reduced lipoproteins and APP in the axonal transport system (Fig. 12.2) (Woodruff et al., 2016). These findings extend previous work that suggested that APP and metabolites that enter the axon to participate in axonal transport are not completely supplied by the Golgi apparatus, but instead may undergo trafficking from the Golgi apparatus to the neuronal somatodendritic membrane followed by endocytosis (Niederst, Reyna, & Goldstein, 2015). Endocytic compartments containing APP then undergo incomplete amyloidogenic processing by BACE1 and γ-secretase prior to entering the axon in an overall process referred to as transcytosis, since the proteins journey from one compartment of the neuron, the somatodendritic compartment, to the restricted axonal compartment. Most Aβ (40−50%) secreted via the axon goes through this pathway (Niederst et al., 2015). These steps are mediated in part by Rab11 and are reduced in rate by FAD mutations. Strikingly, lipoprotein transcytosis from the somatodendritic domain to the axonal domain is also reduced by FAD mutations, potentially linking the allelic differences among Apolipoprotein E (APOE) alleles to cholesterol transport to the axon. This transport may be altered by FAD mutations during axonal growth, maintenance, or repair (Woodruff et al., 2016).

Human Aβ can also impair fast axonal transport; several different possibilities for mechanism have been suggested. A dominant hypothesis is that Aβ induces glycogen synthase kinase-3β (GSK3β), which then phosphorylates microtubule-binding protein Tau. The hyperphosphorylated Tau is argued to inhibit fast axonal transport (Fig. 12.2) (Takashima et al., 1996). Other studies on cultured neurons demonstrated a similar pathway involved in inhibiting mitochondrial transport (Calkins & Reddy, 2011; Rui, Tiwari, Xie, & Zheng, 2006; Takashima et al., 1996; Vossel et al., 2010, 2015) and dense core vesicle

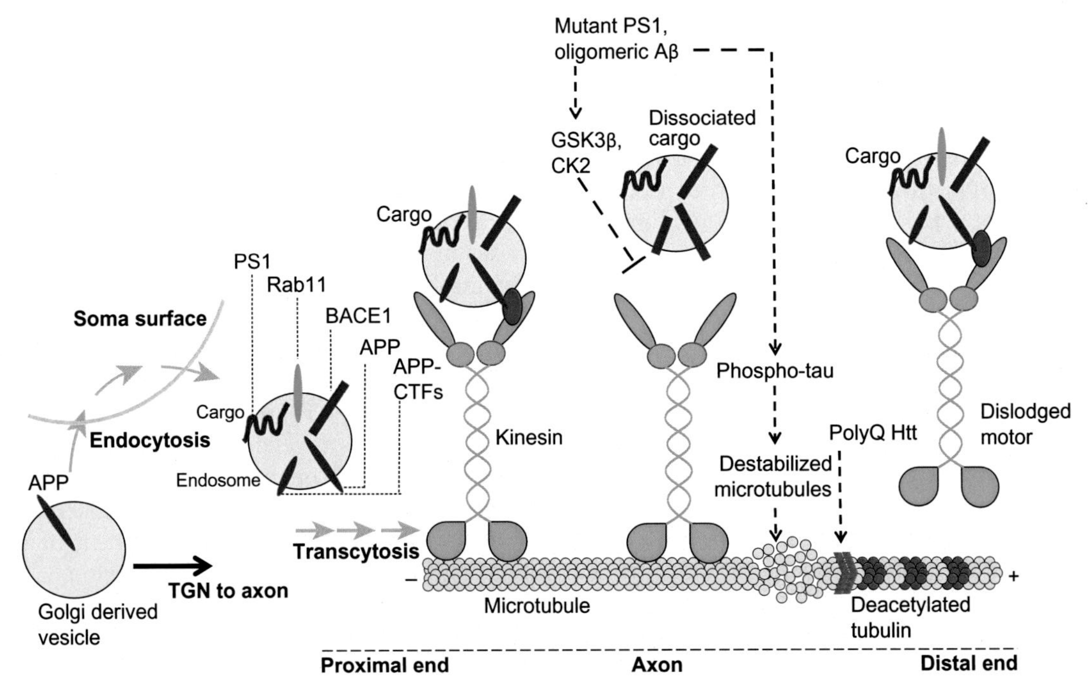

FIGURE 12.2 **Distinct routes of axonal entry and molecular determinants of axonal transport in pathogenic conditions.** Membranous cargoes can enter an axon directly from TGN (black arrow). However, they can follow an indirect pathway, where proteins from TGN first undergo endocytosis along the plasma membrane in the somatodendritic compartment and re-enter into an endosomal compartment. Endosomes carrying the proteins then enter into the axon and traffic toward the presynaptic terminals. This indirect pathway is termed transcytosis (brown repeat-arrow) and highly regulated by small GTPases such as Rab11 and mutant PS1 or APP. Various molecules can interrupt axonal transport at different junctions during protein translocation causing pathogenic conditions. Mutant PS1 or toxic oligomeric Aβ can dissociate cargoes from their respective motors by modulating GSK3β or casein kinase 2 (CK2) signaling pathways. Hyperphosphorylated Tau can destabilize microtubule tracks, thereby dislodging motors. Further, mutant proteins such as Huntingtin can aggregate and bind to microtubule tracks or deacetylate tubulin, initiating dissociation of motors from the tracks.

transport (Decker, Lo, Unger, Ferreira, & Silverman, 2010). Human Aβ applied in vivo inhibits fast axonal transport of APP, vesicular acetylcholinesterase transporter, and synaptophysin in rat sciatic nerve (Kasa et al., 2000).

A current concept is that oligomeric forms of Aβ are neurotoxic (Cleary et al., 2005; Shankar et al., 2008; Walsh et al., 2005). Structurally, a hydrophobic amino acid patch from 17 to 20 of Aβ interacts with another hydrophobic patch 31–35 and forms β-sheets and eventually oligomers (Barrow & Zagorski, 1991; Chaney, Webster, Kuo, & Roher, 1998; Huang et al., 2000; Lee et al., 1995; Li, Darden, Bartolotti, Kominos, & Pedersen, 1999). These oligomers are suggested to form pores on the neuronal membrane (Furukawa, Abe, & Akaike, 1994; Kawahara & Kuroda, 2000; Sanderson, Butler, & Ingram, 1997) and activate the intracellular p38-MAPK signaling pathway to initiate actin polymerization and subsequent actin aggregation (Song, Perides, Wang, & Liu, 2002). Aggregated actin eventually disrupts the intracellular actin network (Davis, Cribbs, Cotman, & Van Nostrand, 1999), thus impairing axonal transport

(Hiruma, Katakura, Takahashi, Ichikawa, & Kawakami, 2003; Tang et al., 2012). Perfusion of soluble oligomeric Aβ into the squid giant axon leads to inhibition of bidirectional axonal transport of vesicles by casein kinase-mediated phosphorylation of KLCs (Pigino et al., 2009). A recent report further strengthens the mechanistic link between Aβ and Tau in AD. Generating a bigenic transgenic mouse—Tg-Aβ + Tau by crossing Tg2576 (Tg-APP) and rTg4510 (Tg-Tau), Sherman et al. have shown that before cognitive decline is observed, Aβ trimers increase in the mouse brain. These trimers of Aβ then might induce conformational changes in Tau and reduce KLC1 protein levels, thus inhibiting axonal transport, including of APP (Sherman et al., 2016).

Presenilin has also been reported to be involved in fast axonal transport. Similar to Aβ, several studies suggest that Presenilin interacts with GSK3β and induces KLC phosphorylation, dislodging cargoes from the motors (Fig. 12.2) (Cai et al., 2003; Lazarov et al., 2007; Morfini, Szebenyi, Elluru, Ratner, & Brady, 2002; Pigino et al., 2003; Stokin et al., 2008; Takashima et al., 1998). Intriguingly, chemical inhibition of gamma-secretase activity or a reduction in Presenilin protein led to an increase in APP velocities in the axon (Gunawardena, Yang, & Goldstein, 2013). These changes in cargo velocity were specific to APP, indicating another role of Presenilin in AD besides gamma-cleavage (Gunawardena et al., 2013). Though the extent of Presenilin-induced axonal transport deficit is still not clear, these reports can open up new research in AD pathology. Finally, the role of ApoE4, the most prominent allelic form associated with sporadic AD, is also intriguing. Transgenic mice expressing human ApoE4 exhibit disruption in axonal transport (Tesseur et al., 2000). It was further suggested that the ApoE4 receptor binds JIP1/2 that mediates APP-kinesin interaction, an essential step for APP trafficking (Fu & Holzbaur, 2013; Verhey et al., 2001). Though preliminary, we suggest that the ApoE4 allele may reduce transcytosis of APOE-HDL (high-density lipoprotein) relative to APOE3 in axonal transport, which could be critical for sporadic AD. ApoE4 reduces the recycling of ApoE receptor at the plasma membrane along with APP and other receptors such as NMDA and AMPA. The reduced endocytic recycling of APP could lead to a decrease in the transcytosed pool of APP into the axons altering the proteolytic processing.

HUNTINGTON'S AND OTHER POLYGLUTAMINE DISEASES

Huntington's disease (HD), a progressive autosomal-dominant neurodegenerative disorder is triggered by mutations in the huntingtin (Htt) gene that translates an expansion of a polyglutamine repeat in the Htt protein. When the length of the extra glutamines exceeds 36 residues, the soluble Htt forms aggregates within the striatal neurons, leading to atrophy and degeneration. However, the underlying molecular mechanism of degeneration by the expanded repeats is still unclear. Within the neuron, soluble Htt associates with vesicles and is transported in the fast axonal component. Early studies indicated interaction of Htt with the transport component Dynactin p150^{glued} via Huntingtin-Associated Protein 1 (Engelender et al., 1997; Li, Gutekunst, Hersch, & Li, 1998), thus establishing a physiological role of Htt in axonal transport. Colin et al. assigned a regulatory role to Htt in determining the directionality of vesicular cargoes in the axon. Using brain-derived neurotrophic factor (BDNF) as a signature cargo, they provided evidence that Htt is phosphorylated at residue S421 by AKT kinase, recruiting kinesin-1 on the cargo and initiating anterograde transport. Upon dephosphorylation, Htt detaches from kinesin-1 and undergoes retrograde transport (Colin et al., 2008). Similarly, phosphorylation/dephosphorylation of residues S1181/S1201 dictates axonal transport of vesicles

(Ben M'Barek et al., 2013; Humbert et al., 2002). Therefore, it is evident that changes in Htt can lead to alteration in axonal transport. Indeed, when polyglutamine expansion mutation was introduced into the endogenous *Htt* locus, change in BDNF transport was observed, resulting in a loss of neurotrophic function and neuronal toxicity (Gauthier et al., 2004; Her & Goldstein, 2008; Trushina et al., 2004). In addition, other experimental models expressing mutant Htt with expanded polyglutamine repeats displayed axonal aggregates and impaired axonal transport in cultured neuron (Li, Li, Yu, Shelbourne, & Li, 2001), *Drosophila* (Gunawardena et al., 2003; Lee, Yoshihara, & Littleton, 2004), squid giant axon (Szebenyi et al., 2003), and mice (Trushina et al., 2004).

Another potential mechanism by which the mutant Htt could impair axonal transport is by altering alpha-tubulin acetylation at its residue K40. Acetylation at K40 is essential for kinesin-1 and microtubule association, thus promoting axonal transport (Fig. 12.2) (Reed et al., 2006). Also, a reduction in alpha-tubulin acetylation is seen in HD brain. Based on this, Dompierre et al. (2007) has shown that inhibiting histone deacetylase-6 induces tubulin acetylation, thus rescuing the vesicular axonal transport. While this is an interesting observation, further validation is essential before targeting this pathway for potential therapeutic intervention in HD. Mutant Htt can also exert its deleterious effect on transport machinery via activating the JNK3-pathway. A recent report on an HD mouse model showed JNK3-mediated phosphorylation of kinesin-1 dislodging the motor from the microtubule tracks (Morfini et al., 2009).

SPINAL AND BULBAR MUSCULAR ATROPHY

Spinal and bulbar muscular atrophy (SBMA) is an X-linked recessive hereditary disorder characterized by progressive proximal muscular atrophy, instigated by degeneration of motor neurons in the brain stem and spinal cord. At the molecular level, SBMA is caused by a polyglutamine repeat expansion of androgen receptor (AR) protein. AR is a transcription factor and typically targeted to the nucleus upon androgen binding. However, a polyglutamine expansion of AR (polyQAR) induces an abnormal distribution of kinesin, leading to disrupted fast axonal transport in squid axoplasm (Piccioni et al., 2002; Szebenyi et al., 2003). PolyQAR activates JNK3-Stress activated protein kinase 1β that phosphorylates KHCs, thereby inhibiting KHC association with microtubules (Morfini et al., 2006). Transgenic mice expressing human AR with polyQ repeats showed defective retrograde axonal transport of tetanus toxin, used as a vesicle tracker (Chevalier-Larsen et al., 2004). Recent data suggest a noncell autonomous disease mechanism instigating retrograde transport defects in motor neurons (Halievski, Kemp, Breedlove, Miller, & Jordan, 2016).

HEREDITARY SPASTIC PARAPLEGIA

Hereditary spastic paraplegias (HSPs) are a group of clinically diverse motor neuron disorders characterized by advancing weakness and spasticity (stiffness) in the lower extremities, usually in the legs. Nearly 70 genes have been identified as associating with various HSPs. Interestingly, the encoded proteins by many of these genes are related to axonal pathfinding, axonal transport, and intracellular protein trafficking (Fink, 2006; Lo Giudice, Lombardi, Santorelli, Kawarai, & Orlacchio, 2014; Novarino et al., 2014).

Severing and bundling of microtubules are essential phenomena for axonal transport on microtubules. Spastin or SPG4, an ATPase of the AAA protein family, is responsible for microtubule severing and bundling, and thus regulates the number, motility, and the dynamic distribution of microtubule plus-ends in the axon (Baas, Vidya Nadar, & Myers, 2006; Errico, Ballabio, & Rugarli, 2002; Evans, Gomes, Reisenweber, Gundersen, & Lauring, 2005; Roll-Mecak & Vale, 2005; Salinas et al., 2005). Structurally, Spastin has one N-terminal microtubule-interacting and trafficking domain (MIT + AAA) and one C-terminal ATPase domain with various cellular functions. In addition to the structural uniqueness, two Kozak's sequences are present in the gene, translating two isoforms of Spastin simultaneously, one long (68 kDa MI-Spastin) and one short (60 kDa M87-Spastin). Large deletion or single point mutations in Spastin lead to HSP. Mutant Spastin lacks or has reduced microtubule severing activity, leading to mislocalized microtubules and mitochondria in the axon (McDermott et al., 2003; Roll-Mecak & Vale, 2008; Salinas et al., 2005). Mice expressing mutant Spastin developed progressive axonopathy with axonal swelling, intraaxonal accumulation of organelles and intermediate filaments, and impaired retrograde transport (Tarrade et al., 2006). Similar axonal accumulation of membranous material has been identified in the sural nerve, indicating axonal transport deficits in one of the clinically complicated forms of HSPs. This type of HSP causes cognitive dysfunction and is triggered by mutation in the SPG11 (Spatacsin) gene (Hehr et al., 2007; Stevanin et al., 2008). A mouse model with a mutation in the SPG7 gene that encodes mitochondrial Paraplegin causes distal axonopathy of spinal and peripheral axons, with a characterized axonal swelling and organelle accumulation, indicating axonal transport defects (Ferreirinha et al., 2004).

Mutation in the KHC isoform 5a (KIF5A) (SPG10) gene, encoding a kinesin component, has also been reported to cause HSP. In in vitro settings, missense mutations at R280C, N256S, K253N in the microtubule-binding domain in the KIF5A protein exhibited a reduction in binding affinity with the microtubules and decreases transport velocity (Ebbing et al., 2008; Fichera et al., 2004). Although in vitro, these observations can be linked to the observation that postnatal conditional knockout KIF5A mice showed a reduction in neurofilament (NF) axonal transport (Xia et al., 2003), indicating a larger role of kinesin mutations in different neurodegenerative diseases.

AMYOTROPHIC LATERAL SCLEROSIS

Amyotrophic lateral sclerosis (ALS) or Lou Gehrig's disease is a motor neuron disease characterized by progressive muscle weakening, leading to paralysis, respiratory failure, and death. While largely sporadic, around 10% ALS cases are familial, with nearly 20 genes associated with it (Boylan, 2015). Although the precise pathogenic mechanism is still unclear, various factors such as protein aggregation, defective axonal transport, uncontrolled glutamate reuptake leading to excitotoxicity, and mitochondrial dysfunction have been implicated to ALS. Early microscopic studies uncovered accumulation of vesicular organelles and intermediate filaments in the proximal segments of motor axons in ALS patients (Corbo & Hays, 1992; Sasaki & Maruyama, 1992), suggesting an involvement of axonopathy and axonal transport deficit.

Mutation in the superoxide dismutase-1 (SOD1) gene, which encodes Cu/Zn SOD1 enzyme, causes both familial and sporadic ALS. SOD1 enzyme, present mainly in the cytoplasm and intermembrane space of mitochondria, is a 32-kDa protein responsible for catalyzing toxic superoxide into hydrogen peroxide and molecular oxygen. Transgenic mice expressing mutant SOD1 (G93A, G37R, or G85R) develop ALS pathology with an axonal accumulation of membrane-bound vacuoles (Bruijn et al., 1997; Gurney et al., 1994; Wong et al., 1995). Interestingly, mice expressing the SOD1 G37R and G85R mutants had reduced transport of NF isoforms [Neurofilament-light (NF-L), NF-medium (NF-M), NF-heavy (NF-H)] and Tubulin—two proteins moving in the slow axonal transport component. However, the vesicular protein movement was unaltered, and these phenotypes were prominent nearly 6 months before the disease onset (Williamson & Cleveland, 1999). The mutant mSOD1s may inhibit axonal transport by activating the p38 and CDK-p25 kinase pathways: 1. mSOD1 can activate p38, which then hyperphosphorylates Kinesin-1 leading to Kinesin-1 inhibition. Inhibited Kinesin-1 impairs axonal transport of vesicles. 2. Similarly, mSOD1 activates cdk5 which in return hyperphosphorylates Tau. This hyperphosphorulated Tau destabilizes microtubules and disrupts axonal transport (Nguyen, Lariviere, & Julien, 2001; Tortarolo et al., 2003). In isolated squid axoplasm, the oxidized form of WT-SOD1 inhibited conventional kinesin-based fast axonal transport in a p38 kinase-dependent pathway (Bosco et al., 2010). This report also suggests the association of SOD1 protein modification to sporadic ALS.

Other pathways including a mutant SOD1 can induce axonal transport deficits in ALS. These may act by damaging mitochondria, thereby reducing energy supply to the molecular motors and altering motor/cargo phosphorylation, thus modulating their association. Mutant SOD1 can exert its effect on mitochondria by two different ways—(1) SOD1G93A can undergo misfolding, form aggregates, and bind to mitochondria. Binding of misfolded SOD1 can lead to the reduction of mitochondrial transport in axon (Vande Velde et al., 2011); (2) Mutant SOD1 hyperactivates p38 that phosphorylates KLCs, dissociating mitochondrial cargoes from the motor proteins (De Vos et al., 2000). Indeed, reduction in the anterograde axonal transport and an increase in retrograde transport of mitochondria have been observed in cultured motor neurons from mice expressing SOD1G93A (De Vos et al., 2007). This may cause ATP depletion in the axon, which is required for axonal transport and synaptic transmission. Similarly, a reduction in the number of mobile mitochondria, and an increase in the stationary pool have also been reported in axons of cultured neurons from SOD1G93A mice (Bilsland et al., 2010). However, Marinkovic et al. (2012) have shown that defective organelle transport in a mouse model of ALS is not essential for motor neuron degeneration, suggesting a compensatory mechanism. This study further showed that a defect in mitochondrial transport could occur well in advance of motor neuron degeneration.

The point mutation G59S in the p150 subunit of the motor protein dynactin has been identified in patients with a slowly progressive autosomal-dominant form of ALS (Munch et al., 2004; Puls et al., 2003). This mutation induces p150 aggregation ($p150^{glued}$) and lowers its the affinity toward microtubules and EB1 resulting in the loss of dynactin function. As the dynein-based retrograde axonal transport is dependent on a functional dynactin, this mutation eventually leads to a deficit in retrograde transport (De Vos, Grierson, Ackerley, & Miller, 2008; Lai et al., 2007; Levy et al., 2006).

Phosphorylation status of NF is responsible for its axonal transport (Shea, Jung, & Pant, 2003). NF-H and NF-M subunits have multiple phosphorylation sites, and their interaction with each other and other cytoskeletal proteins are dependent on the phosphorylation status of these residues. Within the axons, hyperphosphorylated NF-H move slower than hypophosphorylated ones (Jung, Yabe, Lee, & Shea, 2000; Lewis & Nixon, 1988). Interestingly, elevated phospho-NF-H has been reported in the CSF and blood of ALS mouse models and patients with ALS (Boylan et al., 2009; Brettschneider, Petzold, Sussmuth, Ludolph, & Tumani, 2006; Ganesalingam, An, Bowser, Andersen, & Shaw, 2013; Lu et al., 2015; Mendonca et al., 2011). This evidence suggests a potential axonal transport deficit of NFs in ALS. Strengthening this concept further, Collard, Cote, and Julien (1995) have reported NF accumulation and axonal transport defects in an ALS mouse model. Moreover, a transgenic mouse overexpressing peripherin protein that disrupts NF assembly presents defective axonal transport, axonal protein aggregation, and ALS pathology during aging (Millecamps, Robertson, Lariviere, Mallet, & Julien, 2006).

Mutation in Alsin/ALS2—a guanine nucleotide exchange factor associated with the Rab5-GTPase—has been reported in some forms of ALS. Alsin is essential for endolysosomal trafficking through the fusion of autophagosomes and endosomes. Mutation in Alsin causes accumulation of immature vesicles and aggregated proteins in spinal axons (Hadano et al., 2010; Otomo et al., 2003). Alsin knockout mice exhibit disrupted endosomal trafficking associated with motor neuron abnormalities (Devon et al., 2006).

CHARCOT–MARIE–TOOTH DISEASE

Charcot–Marie–Tooth (CMT) is a clinically heterogeneous group of hereditary sensory motor neuropathies and can be autosomal-dominant or autosomal recessive. Among the autosomal-dominant forms, demyelinating CMT (CMT1) and axonal CMT (CMT2) are the most common forms of CMTs.

Eleven genes have been identified to be associated with CMT2. However, mutations in Mitofusin 2 (MFN2) are the most prominent and account for 8%–30% of the disease (Bombelli et al., 2014; Kijima et al., 2005a; Lawson, Graham, & Flanigan, 2005; Zuchner et al., 2004). MFN2, a multidomain protein essential for successful mitochondrial fusion, is localized in the mitochondrial outer membrane. A transgenic mouse expressing human MFN2 R94Q mutation, adjacent to the GTPase domain, showed CMT phenotype with locomotor impairments. Neurons from the R94Q mutant had increased mitochondria numbers in their axons, with evidence of mitochondrial aggregation (Cartoni et al., 2010). Overexpression of disease-associated MFN2 mutants L76P, R94Q, P251A, and R280H in the dorsal root ganglion neurons lead to diminished axonal transport of mitochondria in both antero- and retrograde directions, increased mitochondria numbers, and axonal clustering (Baloh, Schmidt, Pestronk, & Milbrandt, 2007; Misko, Jiang, Wegorzewska, Milbrandt, & Baloh, 2010). A possible reason for the disrupted axonal transport of mitochondria is dissociation of MFN2 from the Miro/Milton protein complex essential for mitochondrial cargo/kinesin motor association (Misko et al., 2010) (Fig. 12.1). Another study indicated the involvement of a motor protein in the defective axonal transport in CMT2A patients. Point mutation Q98L in the KIF1Bβ motor protein, implicated in the transport of synaptic vesicle precursors, disrupted the motor function of KIF1Bβ. $KIF1B^{+/-}$ mice, with a reduction in the KIF1Bβ protein, caused CMT2A-type axonopathy, suggesting axonal transport deficits in CMT2A (Zhao et al., 2001). However, later studies performed in other CMT2A pedigrees identified no mutation in KIF1Bβ (Bissar-Tadmouri et al., 2004; Kijima et al., 2005b). Instead, mutations in MFN2 were detected (Kijima et al., 2005b). Recently, G235Emutation in KIF5A was reported to be associated with CMT2 patient (Crimella et al., 2012).

Other genes involved in CMT2 disease associated with axonal transport phenotypes are HSPB1, NFL, and DYNC1H1. HSPB1 encodes a small heat-shock protein and participates in protein folding. Disease-associated mutation in HSPB1 leads to misfolding of NF and dynactin (Zhai, Lin, Julien, & Schlaepfer, 2007). The R127W, S135F, and R136W HSPB1 mutants, with higher chaperone activity compared to the wild type, displayed enhanced interaction with tubulin and microtubules, leading to microtubule stabilization and axonal transport defects (Almeida-Souza et al., 2011). Extending this, d'Ydewalle et al. (2011), generated transgenic mice expressing human HSPB1 with disease-associated mutations that developed CMT phenotypes with axonal transport defects. Likewise, a mutation in the retrograde motor DYNC1H1 impaired axonal transport a with CMT2B phenotype (Weedon et al., 2011).

PARKINSON'S DISEASE AND RELATED SYNUCLEINOPATHIES

Parkinson's disease (PD) is a progressive neurodegenerative movement disorder with a hallmark deposition of Lewy body inclusions affecting the dopaminergic neurons in the substantia nigra, with rigidity and shaking limbs. Lewy body diseases are a class of heterogeneous disorders including PD and Dementia with Lewy Body (McKeith, 2006). The principal constituent of the Lewy body inclusions is the protein alpha-synuclein, a small ~15 kDa presynaptic protein (Iwai et al., 1995; Jakes, Spillantini, & Goedert, 1994; Withers, George, Banker, & Clayton, 1997). Alpha-synuclein associates with synaptic vesicles at the presynapse and regulates neurotransmitter release and synaptic plasticity (Burre et al., 2010; Kahle et al., 2000; Lee, Jeon, & Kandror, 2008; Scott et al., 2010). More recently, several reports have suggested that physiologically stable tetramers of alpha-synuclein are responsible for its physiological function (Bartels, Choi, & Selkoe, 2011; Wang et al., 2014). Mutations in the alpha-synuclein gene (SNCA) lead to familial forms of Parkinson's. The most common mutations in the human alpha-synuclein gene are A53T, A30P, and E46K (Kruger et al., 1998; Polymeropoulos et al., 1997; Zarranz et al., 2004). Besides mutations, multiplication of SNCA gene dosages also causes familial PD (Chartier-Harlin et al., 2004; Hardy, Cai, Cookson, Gwinn-Hardy, & Singleton, 2006). The mechanism of neuritic aggregation of alpha-synuclein is still unclear. However, a study by Saha et al. (2004) in cultured neurons demonstrated that overexpressing human PD mutants—A30P, A53T, or phosphomimetic S129D disrupted alpha-synuclein transport and alpha-synuclein eventually accumulated in axons. Overexpressed wild-type as well as mutant alpha-synucleins reduced the number of axonal synaptophysin and disrupted the velocity of synaptophysin in axons of cultured rat primary midbrain neurons (Koch et al., 2015). Further, alpha-synuclein interacts with Tau protein in axons, and the synergistic effect of alpha-synuclein/Tau leads to axonal transport defects in a *Drosophila* model of PD (Roy & Jackson, 2014). Alpha-synuclein is transported in the slow axonal transport component (Roy, Winton, Black, Trojanowski, & Lee, 2007). Though bidirectional, the net movement is biased toward anterograde. According to recent work, anterogradely transported alpha-synuclein can be transmitted to and internalized by a second neuron, thus suggesting a prion-like propagation in PD (Brahic, Bousset, Bieri, Melki, & Gitler, 2016; Freundt et al., 2012). However, further experimental validation of this route of transmission is needed. A recent study shows a reduction in the levels of conventional kinesins and DYN in the nigrostriatal neurons of PD patients, along with a reduction in axonal transport mechanism (Chu et al., 2012).

Aberrant mitochondrial trafficking has been implicated in familial forms of PD. Four genes and their encoded protein products—Miro, Milton, PTEN-induced kinase 1 (PINK1), and Parkin—play a coordinated role in mitochondrial axonal transport. Milton, an adapter protein, connects Miro, a Rho GTPase facing the cytosol in the outer mitochondrial membrane, to microtubules via kinesin (Fig. 12.1). This association is essential for anterograde transport of mitochondria (Fransson, Ruusala, & Aspenstrom, 2006; Glater, Megeath, Stowers, & Schwarz, 2006; Guo et al., 2005; Stowers, Megeath, Gorska-Andrzejak, Meinertzhagen, & Schwarz, 2002). PINK1 and Parkin proteins are associated with the Miro/Milton complex (Wang et al., 2011; Weihofen, Thomas, Ostaszewski, Cookson, & Selkoe, 2009). Upon phosphorylation of Miro by PINK1, Parkin ubiquitinates Miro, inducing its proteasomal degradation and disruption of Miro/Milton complex. This disruption blocks mitochondrial transport. PINK1 is also implicated in mitochondrial fission/fusion dynamics. Overexpression of PINK1 or PARKIN leads to the inhibition of mitochondrial transport in both mammalian and *Drosophila* neurons (Chan et al., 2011; Wang et al., 2011). On the other hand, PINK1 gene knockdown in *Drosophila* larval motor neurons resulted in enhanced anterograde transport (Liu et al., 2012). Thus, dysfunction of the Miro/Milton/PINK1 pathway could contribute to the loss of dopaminergic neurons in PD.

FRONTOTEMPORAL DEMENTIA (FTD) AND RELATED TAUOPATHIES

Frontotemporal dementia (FTD) is a group of progressive neurodegenerative disorders involving the frontal and the temporal lobes in the brain, causing deterioration in behavior and personality and alteration in motor function. Two proteins—Tau and TDP43—have been reported to be responsible for FTD. Pathologically, intracellular accumulation of filamentous Tau aggregates is observed in FTD with tauopathies. Mutation in Tau also can lead to an inherited autosomal-dominant form of tauopathy or FTD with parkinsonism linked to chromosome 17 (FTDP-17) (Clark et al., 1998; Dumanchin et al., 1998; Lee, Goedert, & Trojanowski, 2001; Spillantini & Goedert, 1998). However, the underlying mechanism is still unclear. Tau is transported in the slow axonal transport component (Mercken, Fischer, Kosik, & Nixon, 1995). A transgenic mouse expressing FTDP-17-linked Tau mutation R406W led to a reduction in Tau axonal transport (Zhang et al., 2004). These mice also expressed more Tau in somato dendritic compartments compared to axons. As Tau stabilizes axonal microtubules, the reduction in Tau transport may be attributed to diminished tau binding to microtubules in the axonal compartment. Ittner et al. expressed FTD-linked Tau mutation K369I in a transgenic mouse model and observed early-onset of memory impairment and amyotrophy similar to Parkinsonism; however, L-dopa did not rescue the effect, indicating similarities to FTD with Parkinsonism. Substantia nigra neurons in these mice had impaired vesicular axonal transport (Ittner et al., 2008).

CONCLUSION

From the studies discussed here, it is now evident that axonal transport defects might evolve through distinct trajectories that may function synergistically and lead to neurodegenerative phenotypes. The essential contributors in this process could be primary, such as genetic mutations in key proteins associated with axonal transport, or secondary cellular events, such as protein aggregation. However, their quantitative contributions remain unknown in different pathologic conditions and need further experimentation. In AD, mutations in APP and Presenilin one genes causes axonal transport deficit (Kamal et al., 2001; Pigino et al., 2009; Salehi et al., 2006; Stokin et al., 2005). In addition to this, Aβ also impairs axonal transport either by activating different signaling cascades or by modulating neuronal cytoskeleton (Calkins & Reddy, 2011; Davis et al., 1999; Decker et al., 2010; Hiruma et al., 2003; Kasa et al., 2000; Pigino et al., 2009; Rui et al., 2006; Song et al., 2002; Takashima et al., 1996; Vossel et al., 2010, 2015). Similarly, in HD, polyglutamine expansion mutation of Htt alters axonal transport, resulting in loss of neurotrophic function without any axonal protein aggregates, suggesting a primary role axonal transport deficits in the neuronal toxicity of this mutant protein (Gauthier et al., 2004; Her & Goldstein, 2008; Trushina et al., 2004). However, other experimental models expressing mutant Htt with expanded polyglutamine repeats demonstrated axonal aggregates and impaired axonal transport (Gunawardena et al., 2003; Lee et al., 2004; Li et al., 2001; Szebenyi et al., 2003; Trushina et al., 2004), suggesting a potential secondary role of protein aggregation in axonal transport induced-neurodegeneration. In the case of ALS, mutant SOD1 causes axonal transport deficits by reducing the mitochondrial energy supply to motor proteins. However, SOD1 can do this by two different ways—(1) by activating p38 and dissociating motor proteins from their mitochondrial cargoes or, (2) by undergoing misfolding and forming axonal aggregates that bind to mitochondria (De Vos et al., 2000; Vande Velde et al., 2011). In certain cases, different mutations in the same gene can lead to distinct neurodegenerative diseases. Mutation in Dynactin subunit one can cause either motor neuron disease or Perry syndrome (Farrer et al., 2009; Puls et al., 2003). In either event, changes in axonal transport can lead to neurodegeneration. Going forward, to have a deeper understanding of the underlying mechanisms and to resolve apparent inconsistencies and controversies, transport studies should be performed on the most disease-relevant cell type, which is the neuron. Specifically, dynamic studies of bona fide human neurons derived from hIPSC reprogrammed from patient-derived fibroblasts (Israel et al., 2012) or introduction of defined mutations into isogenic genetic backgrounds could lead to discoveries in axonal transport defects and neurodegeneration and identification of potential therapeutic targets.

References

Akhmanova, A., & Hammer, J. A., III. (2010). Linking molecular motors to membrane cargo. *Current Opinion in Cell Biology*, *22*, 479–487.

Almeida-Souza, L., Asselbergh, B., d'Ydewalle, C., Moonens, K., Goethals, S., de Winter, V., … Gevaert, K. (2011). Small heat-shock protein HSPB1 mutants stabilize microtubules in Charcot–Marie–Tooth neuropathy. *The Journal of Neuroscience*, *31*, 15320–15328.

Almenar-Queralt, A., Falzone, T. L., Shen, Z., Lillo, C., Killian, R. L., Arreola, A. S., … Briggs, S. P. (2014). UV irradiation accelerates amyloid precursor protein (APP) processing and disrupts APP axonal transport. *The Journal of Neuroscience*, *34*, 3320–3339.

Baas, P. W., Vidya Nadar, C., & Myers, K. A. (2006). Axonal transport of microtubules: The long and short of it. *Traffic*, *7*, 490–498.

Baloh, R. H., Schmidt, R. E., Pestronk, A., & Milbrandt, J. (2007). Altered axonal mitochondrial transport in the pathogenesis of Charcot–Marie–Tooth disease from mitofusin 2 mutations. *The Journal of Neuroscience*, *27*, 422–430.

Barrow, C. J., & Zagorski, M. G. (1991). Solution structures of beta peptide and its constituent fragments: Relation to amyloid deposition. *Science*, *253*, 179–182.

Bartels, T., Choi, J. G., & Selkoe, D. J. (2011). Alpha-synuclein occurs physiologically as a helically folded tetramer that resists aggregation. *Nature*, *477*, 107–110.

Ben M'Barek, K., Pla, P., Orvoen, S., Benstaali, C., Godin, J. D., Gardier, A. M., … Humbert, S. (2013). Huntingtin mediates anxiety/depression-related behaviors and hippocampal neurogenesis. *The Journal of Neuroscience*, *33*, 8608–8620.

Bilsland, L. G., Sahai, E., Kelly, G., Golding, M., Greensmith, L., & Schiavo, G. (2010). Deficits in axonal transport precede ALS symptoms in vivo. *Proceedings of the National Academy of Sciences of the United States of America*, *107*, 20523–20528.

Bissar-Tadmouri, N., Nelis, E., Zuchner, S., Parman, Y., Deymeer, F., Serdaroglu, P., … Battaloglu, E. (2004). Absence of KIF1B mutation in a large Turkish CMT2A family suggests involvement of a second gene. *Neurology*, *62*, 1522–1525.

Bombelli, F., Stojkovic, T., Dubourg, O., Echaniz-Laguna, A., Tardieu, S., Larcher, K., … Cazeneuve, C. (2014). Charcot–Marie–Tooth disease type 2A: From typical to rare phenotypic and genotypic features. *JAMA Neurology*, *71*, 1036–1042.

Bosco, D. A., Morfini, G., Karabacak, N. M., Song, Y., Gros-Louis, F., Pasinelli, P., … McKenna-Yasek, D. (2010). Wild-type and mutant SOD1 share an aberrant conformation and a common pathogenic pathway in ALS. *Nature Neuroscience*, *13*, 1396–1403.

Bowman, A. B., Kamal, A., Ritchings, B. W., Philp, A. V., McGrail, M., Gindhart, J. G., & Goldstein, L. S. (2000). Kinesin-dependent axonal transport is mediated by the Sunday driver (SYD) protein. *Cell, 103*, 583–594.

Boylan, K. (2015). Familial amyotrophic lateral sclerosis. *Neurologic Clinics, 33*, 807–830.

Boylan, K., Yang, C., Crook, J., Overstreet, K., Heckman, M., Wang, Y., ... Shaw, G. (2009). Immunoreactivity of the phosphorylated axonal neurofilament H subunit (pNF-H) in blood of ALS model rodents and ALS patients: Evaluation of blood pNF-H as a potential ALS biomarker. *Journal of Neurochemistry, 111*, 1182–1191.

Brahic, M., Bousset, L., Bieri, G., Melki, R., & Gitler, A. D. (2016). Axonal transport and secretion of fibrillar forms of alpha-synuclein, Abeta42 peptide and HTTExon 1. *Acta Neuropathologica, 131*, 539–548.

Brettschneider, J., Petzold, A., Sussmuth, S. D., Ludolph, A. C., & Tumani, H. (2006). Axonal damage markers in cerebrospinal fluid are increased in ALS. *Neurology, 66*, 852–856.

Bruijn, L. I., Becher, M. W., Lee, M. K., Anderson, K. L., Jenkins, N. A., Copeland, N. G., ... Cleveland, D. W. (1997). ALS-linked SOD1 mutant G85R mediates damage to astrocytes and promotes rapidly progressive disease with SOD1-containing inclusions. *Neuron, 18*, 327–338.

Burre, J., Sharma, M., Tsetsenis, T., Buchman, V., Etherton, M. R., & Sudhof, T. C. (2010). Alpha-synuclein promotes SNARE-complex assembly in vivo and in vitro. *Science, 329*, 1663–1667.

Cai, D., Leem, J. Y., Greenfield, J. P., Wang, P., Kim, B. S., Wang, R., ... Greengard, P. (2003). Presenilin-1 regulates intracellular trafficking and cell surface delivery of beta-amyloid precursor protein. *Journal of Biological Chemistry, 278*, 3446–3454.

Calkins, M. J., & Reddy, P. H. (2011). Amyloid beta impairs mitochondrial anterograde transport and degenerates synapses in Alzheimer's disease neurons. *Biochimica et Biophysica Acta, 1812*, 507–513.

Cartoni, R., Arnaud, E., Medard, J. J., Poirot, O., Courvoisier, D. S., Chrast, R., & Martinou, J. C. (2010). Expression of mitofusin 2(R94Q) in a transgenic mouse leads to Charcot–Marie–Tooth neuropathy type 2A. *Brain, 133*, 1460–1469.

Cavalli, V., Kujala, P., Klumperman, J., & Goldstein, L. S. (2005). Sunday driver links axonal transport to damage signaling. *Journal of Cell Biology, 168*, 775–787.

Chan, N. C., Salazar, A. M., Pham, A. H., Sweredoski, M. J., Kolawa, N. J., Graham, R. L., ... Chan, D. C. (2011). Broad activation of the ubiquitin-proteasome system by Parkin is critical for mitophagy. *Human Molecular Genetics, 20*, 1726–1737.

Chaney, M. O., Webster, S. D., Kuo, Y. M., & Roher, A. E. (1998). Molecular modeling of the Abeta1-42 peptide from Alzheimer's disease. *Protein Engineering, 11*, 761–767.

Chartier-Harlin, M. C., Kachergus, J., Roumier, C., Mouroux, V., Douay, X., Lincoln, S., ... Hulihan, M. (2004). Alpha-synuclein locus duplication as a cause of familial Parkinson's disease. *Lancet, 364*, 1167–1169.

Chevalier-Larsen, E. S., O'Brien, C. J., Wang, H., Jenkins, S. C., Holder, L., Lieberman, A. P., & Merry, D. E. (2004). Castration restores function and neurofilament alterations of aged symptomatic males in a transgenic mouse model of spinal and bulbar muscular atrophy. *The Journal of Neuroscience, 24*, 4778–4786.

Chu, Y., Morfini, G. A., Langhamer, L. B., He, Y., Brady, S. T., & Kordower, J. H. (2012). Alterations in axonal transport motor proteins in sporadic and experimental Parkinson's disease. *Brain, 135*, 2058–2073.

Clark, L. N., Poorkaj, P., Wszolek, Z., Geschwind, D. H., Nasreddine, Z. S., Miller, B., ... Markopoulou, K. (1998). Pathogenic implications of mutations in the tau gene in pallido-ponto-nigral degeneration and related neurodegenerative disorders linked to chromosome 17. *Proceedings of the National Academy of Sciences of the United States of America, 95*, 13103–13107.

Cleary, J. P., Walsh, D. M., Hofmeister, J. J., Shankar, G. M., Kuskowski, M. A., Selkoe, D. J., & Ashe, K. H. (2005). Natural oligomers of the amyloid-beta protein specifically disrupt cognitive function. *Nature Neuroscience, 8*, 79–84.

Colin, E., Zala, D., Liot, G., Rangone, H., Borrell-Pages, M., Li, X. J., ... Humbert, S. (2008). Huntingtin phosphorylation acts as a molecular switch for anterograde/retrograde transport in neurons. *The EMBO Journal, 27*, 2124–2134.

Collard, J. F., Cote, F., & Julien, J. P. (1995). Defective axonal transport in a transgenic mouse model of amyotrophic lateral sclerosis. *Nature, 375*, 61–64.

Corbo, M., & Hays, A. P. (1992). Peripherin and neurofilament protein coexist in spinal spheroids of motor neuron disease. *Journal of Neuropathology & Experimental Neurology, 51*, 531–537.

Cottrell, B. A., Galvan, V., Banwait, S., Gorostiza, O., Lombardo, C. R., Williams, T., ... Koo, E. H. (2005). A pilot proteomic study of amyloid precursor interactors in Alzheimer's disease. *Annals of Neurology, 58*, 277–289.

Crimella, C., Baschirotto, C., Arnoldi, A., Tonelli, A., Tenderini, E., Airoldi, G., ... Scarlato, M. (2012). Mutations in the motor and stalk domains of KIF5A in spastic paraplegia type 10 and in axonal Charcot–Marie–Tooth type 2. *Clinical Genetics, 82*, 157–164.

d'Ydewalle, C., Krishnan, J., Chiheb, D. M., Van Damme, P., Irobi, J., Kozikowski, A. P., ... Van Den Bosch, L. (2011). HDAC6 inhibitors reverse axonal loss in a mouse model of mutant HSPB1-induced Charcot–Marie–Tooth disease. *Nature Medicine, 17*, 968–974.

Das, U., Wang, L., Ganguly, A., Saikia, J. M., Wagner, S. L., Koo, E. H., & Roy, S. (2016). Visualizing APP and BACE-1 approximation in neurons yields insight into the amyloidogenic pathway. *Nature Neuroscience, 19*, 55–64.

Davis, J., Cribbs, D. H., Cotman, C. W., & Van Nostrand, W. E. (1999). Pathogenic amyloid beta-protein induces apoptosis in cultured human cerebrovascular smooth muscle cells. *Amyloid, 6*, 157–164.

De Vos, K., Severin, F., Van Herreweghe, F., Vancompernolle, K., Goossens, V., Hyman, A., & Grooten, J. (2000). Tumor necrosis factor induces hyperphosphorylation of kinesin light chain and inhibits kinesin-mediated transport of mitochondria. *Journal of Cell Biology, 149*, 1207–1214.

De Vos, K. J., Chapman, A. L., Tennant, M. E., Manser, C., Tudor, E. L., Lau, K. F., ... McLoughlin, D. M. (2007). Familial amyotrophic lateral sclerosis-linked SOD1 mutants perturb fast axonal transport to reduce axonal mitochondria content. *Human Molecular Genetics, 16*, 2720–2728.

De Vos, K. J., Grierson, A. J., Ackerley, S., & Miller, C. C. (2008). Role of axonal transport in neurodegenerative diseases. *Annual Review of Neuroscience, 31*, 151–173.

Decker, H., Lo, K. Y., Unger, S. M., Ferreira, S. T., & Silverman, M. A. (2010). Amyloid-beta peptide oligomers disrupt axonal transport through an NMDA receptor-dependent mechanism that is mediated by glycogen synthase kinase 3beta in primary cultured hippocampal neurons. *The Journal of Neuroscience, 30*, 9166–9171.

Deneka, M., Neeft, M., & van der Sluijs, P. (2003). Regulation of membrane transport by rab GTPases. *Critical Reviews in Biochemistry and Molecular Biology, 38*, 121–142.

Devon, R. S., Orban, P. C., Gerrow, K., Barbieri, M. A., Schwab, C., Cao, L. P., ... Davidson, T. L. (2006). Als2-deficient mice exhibit disturbances in endosome trafficking associated with motor behavioral abnormalities. *Proceedings of the National Academy of Sciences of the United States of America, 103*, 9595–9600.

Dompierre, J. P., Godin, J. D., Charrin, B. C., Cordelieres, F. P., King, S. J., Humbert, S., & Saudou, F. (2007). Histone deacetylase 6 inhibition compensates for the transport deficit in Huntington's disease by increasing tubulin acetylation. *The Journal of Neuroscience, 27*, 3571–3583.

Dumanchin, C., Camuzat, A., Campion, D., Verpillat, P., Hannequin, D., Dubois, B., ... Charbonnier, F. (1998). Segregation of a missense mutation in the microtubule-associated protein tau gene with familial frontotemporal dementia and parkinsonism. *Human Molecular Genetics, 7*, 1825–1829.

Ebbing, B., Mann, K., Starosta, A., Jaud, J., Schols, L., Schule, R., & Woehlke, G. (2008). Effect of spastic paraplegia mutations in KIF5A kinesin on transport activity. *Human Molecular Genetics, 17*, 1245–1252.

Encalada, S. E., & Goldstein, L. S. (2014). Biophysical challenges to axonal transport: Motor-cargo deficiencies and neurodegeneration. *Annual Review of Biophysics, 43*, 141–169.

Encalada, S. E., Szpankowski, L., Xia, C. H., & Goldstein, L. S. (2011). Stable kinesin and dynein assemblies drive the axonal transport of mammalian prion protein vesicles. *Cell, 144*, 551–565.

Engelender, S., Sharp, A. H., Colomer, V., Tokito, M. K., Lanahan, A., Worley, P., ... Ross, C. A. (1997). Huntingtin-associated protein 1 (HAP1) interacts with the p150Glued subunit of dynactin. *Human Molecular Genetics, 6*, 2205–2212.

Errico, A., Ballabio, A., & Rugarli, E. I. (2002). Spastin, the protein mutated in autosomal dominant hereditary spastic paraplegia, is involved in microtubule dynamics. *Human Molecular Genetics, 11*, 153–163.

Eschbach, J., & Dupuis, L. (2011). Cytoplasmic dynein in neurodegeneration. *Pharmacology & Therapeutics, 130*, 348–363.

Evans, K. J., Gomes, E. R., Reisenweber, S. M., Gundersen, G. G., & Lauring, B. P. (2005). Linking axonal degeneration to microtubule remodeling by Spastin-mediated microtubule severing. *Journal of Cell Biology, 168*, 599–606.

Farrer, M. J., Hulihan, M. M., Kachergus, J. M., Dachsel, J. C., Stoessl, A. J., Grantier, L. L., ... Chapon, F. (2009). DCTN1 mutations in Perry syndrome. *Nature Genetics, 41*, 163–165.

Ferreira, A., Caceres, A., & Kosik, K. S. (1993). Intraneuronal compartments of the amyloid precursor protein. *The Journal of Neuroscience, 13*, 3112–3123.

Ferreirinha, F., Quattrini, A., Pirozzi, M., Valsecchi, V., Dina, G., Broccoli, V., ... Gaeta, L. (2004). Axonal degeneration in paraplegin-deficient mice is associated with abnormal mitochondria and impairment of axonal transport. *Journal of Clinical Investigation, 113*, 231–242.

Fichera, M., Lo Giudice, M., Falco, M., Sturnio, M., Amata, S., Calabrese, O., ... Neri, M. (2004). Evidence of kinesin heavy chain (KIF5A) involvement in pure hereditary spastic paraplegia. *Neurology, 63*, 1108–1110.

Fink, J. K. (2006). Hereditary spastic paraplegia. *Current Neurology and Neuroscience Reports, 6*, 65–76.

Fransson, S., Ruusala, A., & Aspenstrom, P. (2006). The atypical Rho GTPases Miro-1 and Miro-2 have essential roles in mitochondrial trafficking. *Biochemical and Biophysical Research Communications, 344*, 500–510.

Freundt, E. C., Maynard, N., Clancy, E. K., Roy, S., Bousset, L., Sourigues, Y., ... Brahic, M. (2012). Neuron-to-neuron transmission of alpha-synuclein fibrils through axonal transport. *Annals of Neurology, 72*, 517–524.

Fu, M. M., & Holzbaur, E. L. (2013). JIP1 regulates the directionality of APP axonal transport by coordinating kinesin and dynein motors. *Journal of Cell Biology, 202*, 495–508.

Furukawa, K., Abe, Y., & Akaike, N. (1994). Amyloid beta protein-induced irreversible current in rat cortical neurones. *Neuroreport, 5*, 2016–2018.

Gajdusek, D. C. (1985). Hypothesis: Interference with axonal transport of neurofilament as a common pathogenetic mechanism in certain diseases of the central nervous system. *The New England Journal of Medicine, 312*, 714–719.

Gallagher, J. J., Zhang, X., Ziomek, G. J., Jacobs, R. E., & Bearer, E. L. (2012). Deficits in axonal transport in hippocampal-based circuitry and the visual pathway in APP knock-out animals witnessed by manganese enhanced MRI. *Neuroimage, 60*, 1856–1866.

Ganesalingam, J., An, J., Bowser, R., Andersen, P. M., & Shaw, C. E. (2013). pNfH is a promising biomarker for ALS. *Amyotrophic Lateral Sclerosis and Frontotemporal Degeneration, 14*, 146–149.

Gauger, A. K., & Goldstein, L. S. (1993). The *Drosophila* kinesin light chain. Primary structure and interaction with kinesin heavy chain. *Journal of Biological Chemistry, 268*, 13657–13666.

Gauthier, L. R., Charrin, B. C., Borrell-Pages, M., Dompierre, J. P., Rangone, H., Cordelieres, F. P., ... Saudou, F. (2004). Huntingtin controls neurotrophic support and survival of neurons by enhancing BDNF vesicular transport along microtubules. *Cell, 118*, 127–138.

Glater, E. E., Megeath, L. J., Stowers, R. S., & Schwarz, T. L. (2006). Axonal transport of mitochondria requires milton to recruit kinesin heavy chain and is light chain independent. *Journal of Cell Biology, 173*, 545–557.

Gunawardena, S., & Goldstein, L. S. (2001). Disruption of axonal transport and neuronal viability by amyloid precursor protein mutations in *Drosophila*. *Neuron, 32*, 389–401.

Gunawardena, S., & Goldstein, L. S. (2004). Cargo-carrying motor vehicles on the neuronal highway: Transport pathways and neurodegenerative disease. *Journal of Neurobiology, 58*, 258–271.

Gunawardena, S., Her, L. S., Brusch, R. G., Laymon, R. A., Niesman, I. R., Gordesky-Gold, B., ... Goldstein, L. S. (2003). Disruption of axonal transport by loss of huntingtin or expression of pathogenic polyQ proteins in *Drosophila*. *Neuron, 40*, 25–40.

Gunawardena, S., Yang, G., & Goldstein, L. S. (2013). Presenilin controls kinesin-1 and dynein function during APP-vesicle transport in vivo. *Human Molecular Genetics, 22*, 3828–3843.

Guo, X., Macleod, G. T., Wellington, A., Hu, F., Panchumarthi, S., Schoenfield, M., ... Zinsmaier, K. E. (2005). The GTPase dMiro is required for axonal transport of mitochondria to *Drosophila* synapses. *Neuron, 47*, 379–393.

Gurney, M. E., Pu, H., Chiu, A. Y., Dal Canto, M. C., Polchow, C. Y., Alexander, D. D., ... Deng, H. X. (1994). Motor neuron degeneration in mice that express a human Cu, Zn superoxide dismutase mutation. *Science, 264*, 1772–1775.

Hadano, S., Otomo, A., Kunita, R., Suzuki-Utsunomiya, K., Akatsuka, A., Koike, M., ... Ikeda, J. E. (2010). Loss of ALS2/Alsin exacerbates motor dysfunction in a SOD1-expressing mouse ALS model by disturbing endolysosomal trafficking. *PLoS ONE, 5*, e9805.

Haghnia, M., Cavalli, V., Shah, S. B., Schimmelpfeng, K., Brusch, R., Yang, G., ... Goldstein, L. S. (2007). Dynactin is required for coordinated bidirectional motility, but not for dynein membrane attachment. *Molecular Biology of the Cell, 18*, 2081–2089.

Halievski, K., Kemp, M. Q., Breedlove, S. M., Miller, K. E., & Jordan, C. L. (2016). Non-cell-autonomous regulation of retrograde motoneuronal axonal transport in an SBMA mouse model. *eNeuro, 3*, 1–12.

Hardy, J. (2006). A hundred years of Alzheimer's disease research. *Neuron, 52*, 3–13.

Hardy, J., Cai, H., Cookson, M. R., Gwinn-Hardy, K., & Singleton, A. (2006). Genetics of Parkinson's disease and parkinsonism. *Annals of Neurology, 60*, 389–398.

Hehr, U., Bauer, P., Winner, B., Schule, R., Olmez, A., Koehler, W., ... Seibel, A. (2007). Long-term course and mutational spectrum of spatacsin-linked spastic paraplegia. *Annals of Neurology, 62*, 656–665.

Her, L. S., & Goldstein, L. S. (2008). Enhanced sensitivity of striatal neurons to axonal transport defects induced by mutant huntingtin. *The Journal of Neuroscience, 28*, 13662–13672.

Hirokawa, N., Niwa, S., & Tanaka, Y. (2010). Molecular motors in neurons: Transport mechanisms and roles in brain function, development, and disease. *Neuron, 68*, 610–638.

Hirokawa, N., & Takemura, R. (2005). Molecular motors and mechanisms of directional transport in neurons. *Nature Reviews Neuroscience, 6*, 201–214.

Hiruma, H., Katakura, T., Takahashi, S., Ichikawa, T., & Kawakami, T. (2003). Glutamate and amyloid beta-protein rapidly inhibit fast axonal transport in cultured rat hippocampal neurons by different mechanisms. *The Journal of Neuroscience, 23*, 8967–8977.

Horiuchi, D., Barkus, R. V., Pilling, A. D., Gassman, A., & Saxton, W. M. (2005). APLIP1, a kinesin binding JIP-1/JNK scaffold protein, influences the axonal transport of both vesicles and mitochondria in *Drosophila*. *Current Biology, 15*, 2137–2141.

Huang, T. H., Yang, D. S., Plaskos, N. P., Go, S., Yip, C. M., Fraser, P. E., & Chakrabartty, A. (2000). Structural studies of soluble oligomers of the Alzheimer beta-amyloid peptide. *Journal of Molecular Biology, 297*, 73–87.

Humbert, S., Bryson, E. A., Cordelieres, F. P., Connors, N. C., Datta, S. R., Finkbeiner, S., ... Saudou, F. (2002). The IGF-1/Akt pathway is neuroprotective in Huntington's disease and involves huntingtin phosphorylation by Akt. *Developmental Cell, 2*, 831–837.

Inomata, H., Nakamura, Y., Hayakawa, A., Takata, H., Suzuki, T., Miyazawa, K., & Kitamura, N. (2003). A scaffold protein JIP-1b enhances amyloid precursor protein phosphorylation by JNK and its association with kinesin light chain 1. *Journal of Biological Chemistry, 278*, 22946–22955.

Israel, M. A., Yuan, S. H., Bardy, C., Reyna, S. M., Mu, Y., Herrera, C., ... Boscolo, F. S. (2012). Probing sporadic and familial Alzheimer's disease using induced pluripotent stem cells. *Nature, 482*, 216–220.

Ittner, L. M., Fath, T., Ke, Y. D., Bi, M., van Eersel, J., Li, K. M., ... Gotz, J. (2008). Parkinsonism and impaired axonal transport in a mouse model of frontotemporal dementia. *Proceedings of the National Academy of Sciences of the United States of America, 105*, 15997–16002.

Iwai, A., Masliah, E., Yoshimoto, M., Ge, N., Flanagan, L., de Silva, H. A., ... Saitoh, T. (1995). The precursor protein of non-A beta component of Alzheimer's disease amyloid is a presynaptic protein of the central nervous system. *Neuron, 14*, 467–475.

Jakes, R., Spillantini, M. G., & Goedert, M. (1994). Identification of two distinct synucleins from human brain. *FEBS Letters, 345*, 27–32.

Jung, C., Yabe, J. T., Lee, S., & Shea, T. B. (2000). Hypophosphorylated neurofilament subunits undergo axonal transport more rapidly than more extensively phosphorylated subunits in situ. *Cell Motility and the Cytoskeleton, 47*, 120–129.

Kahle, P. J., Neumann, M., Ozmen, L., Muller, V., Jacobsen, H., Schindzielorz, A., ... Probst, A. (2000). Subcellular localization of wild-type and Parkinson's disease-associated mutant alpha-synuclein in human and transgenic mouse brain. *The Journal of Neuroscience, 20*, 6365–6373.

Kamal, A., Almenar-Queralt, A., LeBlanc, J. F., Roberts, E. A., & Goldstein, L. S. (2001). Kinesin-mediated axonal transport of a membrane compartment containing beta-secretase and presenilin-1 requires APP. *Nature, 414*, 643–648.

Kamal, A., Stokin, G. B., Yang, Z., Xia, C. H., & Goldstein, L. S. (2000). Axonal transport of amyloid precursor protein is mediated by direct binding to the kinesin light chain subunit of kinesin-I. *Neuron, 28*, 449–459.

Kasa, P., Papp, H., Kovacs, I., Forgon, M., Penke, B., & Yamaguchi, H. (2000). Human amyloid-beta1-42 applied in vivo inhibits the fast axonal transport of proteins in the sciatic nerve of rat. *Neuroscience Letters, 278*, 117–119.

Kawahara, M., & Kuroda, Y. (2000). Molecular mechanism of neurodegeneration induced by Alzheimer's beta-amyloid protein: Channel formation and disruption of calcium homeostasis. *Brain Research Bulletin, 53*, 389–397.

Kijima, K., Numakura, C., Goto, T., Takahashi, T., Otagiri, T., Umetsu, K., & Hayasaka, K. (2005a). Small heat shock protein 27 mutation in a Japanese patient with distal hereditary motor neuropathy. *Journal of Human Genetics, 50*, 473–476.

Kijima, K., Numakura, C., Izumino, H., Umetsu, K., Nezu, A., Shiiki, T., ... Hayasaka, K. (2005b). Mitochondrial GTPase mitofusin 2 mutation in Charcot–Marie–Tooth neuropathy type 2A. *Human Genetics, 116*, 23–27.

Koch, J. C., Bitow, F., Haack, J., d'Hedouville, Z., Zhang, J. N., Tonges, L., ... Liman, J. (2015). Alpha-synuclein affects neurite morphology, autophagy, vesicle transport and axonal degeneration in CNS neurons. *Cell Death & Disease, 6*, e1811.

Koo, E. H., Sisodia, S. S., Archer, D. R., Martin, L. J., Weidemann, A., Beyreuther, K., ... Price, D. L. (1990). Precursor of amyloid protein in Alzheimer disease undergoes fast anterograde axonal transport. *Proceedings of the National Academy of Sciences of the United States of America, 87*, 1561–1565.

Kruger, R., Kuhn, W., Muller, T., Woitalla, D., Graeber, M., Kosel, S., ... Riess, O. (1998). Ala30Pro mutation in the gene encoding alpha-synuclein in Parkinson's disease. *Nature Genetics, 18*, 106–108.

Lai, C., Lin, X., Chandran, J., Shim, H., Yang, W. J., & Cai, H. (2007). The G59S mutation inp150(glued) causes dysfunction of dynactin in mice. *The Journal of Neuroscience, 27*, 13982–13990.

Lawrence, C. J., Dawe, R. K., Christie, K. R., Cleveland, D. W., Dawson, S. C., Endow, S. A., ... Howard, J. (2004). A standardized kinesin nomenclature. *Journal of Cell Biology, 167*, 19–22.

Lawson, V. H., Graham, B. V., & Flanigan, K. M. (2005). Clinical and electrophysiologic features of CMT2A with mutations in the mitofusin 2 gene. *Neurology, 65*, 197–204.

Lazarov, O., Morfini, G. A., Pigino, G., Gadadhar, A., Chen, X., Robinson, J., ... Sisodia, S. S. (2007). Impairments in fast axonal transport and motor neuron deficits in transgenic mice expressing familial Alzheimer's disease-linked mutant presenilin 1. *The Journal of Neuroscience, 27*, 7011–7020.

Lee, J. P., Stimson, E. R., Ghilardi, J. R., Mantyh, P. W., Lu, Y. A., Felix, A. M., ... Van Criekinge, M. (1995). 1H NMR of A beta amyloid peptide congeners in water solution. Conformational changes correlate with plaque competence. *Biochemistry, 34*, 5191–5200.

Lee, S. J., Jeon, H., & Kandror, K. V. (2008). Alpha-synuclein is localized in a subpopulation of rat brain synaptic vesicles. *Acta Neurobiologiae Experimentalis (Warsaw), 68*, 509–515.

Lee, V. M., Daughenbaugh, R., & Trojanowski, J. Q. (1994). Microtubule stabilizing drugs for the treatment of Alzheimer's disease. *Neurobiology of Aging, 15*(Suppl 2), S87–S89.

Lee, V. M., Goedert, M., & Trojanowski, J. Q. (2001). Neurodegenerative tauopathies. *Annual Review of Neuroscience, 24*, 1121–1159.

Lee, W. C., Yoshihara, M., & Littleton, J. T. (2004). Cytoplasmic aggregates trap polyglutamine-containing proteins and block axonal transport in a *Drosophila* model of Huntington's disease. *Proceedings of the National Academy of Sciences of the United States of America, 101*, 3224–3229.

Levy, J. R., Sumner, C. J., Caviston, J. P., Tokito, M. K., Ranganathan, S., Ligon, L. A., ... Puls, I. (2006). A motor neuron disease-associated mutation in p150Glued perturbs dynactin function and induces protein aggregation. *Journal of Cell Biology, 172*, 733–745.

Lewis, S. E., & Nixon, R. A. (1988). Multiple phosphorylated variants of the high molecular mass subunit of neurofilaments in axons of retinal cell neurons: Characterization and evidence for their differential association with stationary and moving neurofilaments. *Journal of Cell Biology, 107*, 2689–2701.

Li, H., Li, S. H., Yu, Z. X., Shelbourne, P., & Li, X. J. (2001). Huntingtin aggregate-associated axonal degeneration is an early pathological event in Huntington's disease mice. *The Journal of Neuroscience, 21*, 8473–8481.

Li, L., Darden, T. A., Bartolotti, L., Kominos, D., & Pedersen, L. G. (1999). An atomic model for the pleated beta-sheet structure of Abeta amyloid protofilaments. *Biophysical Journal, 76*, 2871–2878.

Li, S. H., Gutekunst, C. A., Hersch, S. M., & Li, X. J. (1998). Interaction of huntingtin-associated protein with dynactin P150Glued. *The Journal of Neuroscience, 18*, 1261–1269.

Liu, S., Sawada, T., Lee, S., Yu, W., Silverio, G., Alapatt, P., ... Kanao, T. (2012). Parkinson's disease-associated kinase PINK1 regulates Miro protein level and axonal transport of mitochondria. *PLoS Genetics, 8*, e1002537.

Lo Giudice, T., Lombardi, F., Santorelli, F. M., Kawarai, T., & Orlacchio, A. (2014). Hereditary spastic paraplegia: Clinical-genetic characteristics and evolving molecular mechanisms. *Experimental Neurology, 261*, 518–539.

Lu, C. H., Petzold, A., Topping, J., Allen, K., Macdonald-Wallis, C., Clarke, J., ... Fratta, P. (2015). Plasma neurofilament heavy chain levels and disease progression in amyotrophic lateral sclerosis: Insights from a longitudinal study. *Journal of Neurology, Neurosurgery, and Psychiatry, 86*, 565–573.

Marinkovic, P., Reuter, M. S., Brill, M. S., Godinho, L., Kerschensteiner, M., & Misgeld, T. (2012). Axonal transport deficits and degeneration can evolve independently in mouse models of amyotrophic lateral sclerosis. *Proceedings of the National Academy of Sciences of the United States of America, 109*, 4296–4301.

Matanis, T., Akhmanova, A., Wulf, P., Del Nery, E., Weide, T., Stepanova, T., ... De Zeeuw, C. I. (2002). Bicaudal-D regulates COPI-independent Golgi-ER transport by recruiting the dynein–dynactin motor complex. *Nature Cell Biology, 4*, 986–992.

Matsuda, S., Matsuda, Y., & D'Adamio, L. (2003). Amyloid beta protein precursor (AbetaPP), but not AbetaPP-like protein 2, is bridged to the kinesin light chain by the scaffold protein JNK-interacting protein 1. *Journal of Biological Chemistry, 278*, 38601–38606.

McDermott, C. J., Grierson, A. J., Wood, J. D., Bingley, M., Wharton, S. B., Bushby, K. M., & Shaw, P. J. (2003). Hereditary spastic paraparesis: Disrupted intracellular transport associated with spastin mutation. *Annals of Neurology, 54*, 748–759.

McKeith, I. G. (2006). Consensus guidelines for the clinical and pathologic diagnosis of dementia with Lewy bodies (DLB): Report of the Consortium on DLB International Workshop. *Journal of Alzheimer's Disease, 9*, 417–423.

Mendonca, D. M., Martins, S. C., Higashi, R., Muscara, M. N., Neto, V. M., Chimelli, L., & Martinez, A. M. (2011). Neurofilament heavy subunit in cerebrospinal fluid: A biomarker of amyotrophic lateral sclerosis? *Amyotrophic Lateral Sclerosis, 12*, 144–147.

Mercken, M., Fischer, I., Kosik, K. S., & Nixon, R. A. (1995). Three distinct axonal transport rates for tau, tubulin, and other microtubule-associated proteins: Evidence for dynamic interactions of tau with microtubules in vivo. *The Journal of Neuroscience, 15*, 8259–8267.

Millecamps, S., Robertson, J., Lariviere, R., Mallet, J., & Julien, J. P. (2006). Defective axonal transport of neurofilament proteins in neurons overexpressing peripherin. *Journal of Neurochemistry, 98*, 926–938.

Misko, A., Jiang, S., Wegorzewska, I., Milbrandt, J., & Baloh, R. H. (2010). Mitofusin 2 is necessary for transport of axonal mitochondria and interacts with the Miro/Milton complex. *The Journal of Neuroscience, 30*, 4232–4240.

Morfini, G., Pigino, G., Szebenyi, G., You, Y., Pollema, S., & Brady, S. T. (2006). JNK mediates pathogenic effects of polyglutamine-expanded androgen receptor on fast axonal transport. *Nature Neuroscience, 9*, 907–916.

Morfini, G., Szebenyi, G., Elluru, R., Ratner, N., & Brady, S. T. (2002). Glycogen synthase kinase 3 phosphorylates kinesin light chains and negatively regulates kinesin-based motility. *The EMBO Journal, 21*, 281–293.

Morfini, G. A., You, Y. M., Pollema, S. L., Kaminska, A., Liu, K., Yoshioka, K., … Han, D. (2009). Pathogenic huntingtin inhibits fast axonal transport by activating JNK3 and phosphorylating kinesin. *Nature Neuroscience, 12*, 864–871.

Muller, M. J., Klumpp, S., & Lipowsky, R. (2008). Tug-of-war as a cooperative mechanism for bidirectional cargo transport by molecular motors. *Proceedings of the National Academy of Sciences of the United States of America, 105*, 4609–4614.

Munch, C., Sedlmeier, R., Meyer, T., Homberg, V., Sperfeld, A. D., Kurt, A., … Ludolph, A. C. (2004). Point mutations of the p150 subunit of dynactin (DCTN1) gene in ALS. *Neurology, 63*, 724–726.

Muresan, V., Stankewich, M. C., Steffen, W., Morrow, J. S., Holzbaur, E. L., & Schnapp, B. J. (2001). Dynactin-dependent, dynein-driven vesicle transport in the absence of membrane proteins: A role for spectrin and acidic phospholipids. *Molecular Cell, 7*, 173–183.

Muresan, Z., & Muresan, V. (2005). Coordinated transport of phosphorylated amyloid-beta precursor protein and c-Jun NH2-terminal kinase-interacting protein-1. *Journal of Cell Biology, 171*, 615–625.

Nguyen, M. D., Lariviere, R. C., & Julien, J. P. (2001). Deregulation of Cdk5 in a mouse model of ALS: Toxicity alleviated by perikaryal neurofilament inclusions. *Neuron, 30*, 135–147.

Niederst, E. D., Reyna, S. M., & Goldstein, L. S. (2015). Axonal amyloid precursor protein and its fragments undergo somatodendritic endocytosis and processing. *Molecular Biology of the Cell, 26*, 205–217.

Novarino, G., Fenstermaker, A. G., Zaki, M. S., Hofree, M., Silhavy, J. L., Heiberg, A. D., … Mansour, L. (2014). Exome sequencing links corticospinal motor neuron disease to common neurodegenerative disorders. *Science, 343*, 506–511.

Otomo, A., Hadano, S., Okada, T., Mizumura, H., Kunita, R., Nishijima, H., … Suga, E. (2003). ALS2, a novel guanine nucleotide exchange factor for the small GTPase Rab5, is implicated in endosomal dynamics. *Human Molecular Genetics, 12*, 1671–1687.

Pfister, K. K., Shah, P. R., Hummerich, H., Russ, A., Cotton, J., Annuar, A. A., … Fisher, E. M. (2006). Genetic analysis of the cytoplasmic dynein subunit families. *PLoS Genetics, 2*, e1.

Piccioni, F., Pinton, P., Simeoni, S., Pozzi, P., Fascio, U., Vismara, G., … Poletti, A. (2002). Androgen receptor with elongated polyglutamine tract forms aggregates that alter axonal trafficking and mitochondrial distribution in motor neuronal processes. *The FASEB Journal, 16*, 1418–1420.

Pigino, G., Morfini, G., Atagi, Y., Deshpande, A., Yu, C., Jungbauer, L., … Brady, S. (2009). Disruption of fast axonal transport is a pathogenic mechanism for intraneuronal amyloid beta. *Proceedings of the National Academy of Sciences of the United States of America, 106*, 5907–5912.

Pigino, G., Morfini, G., Pelsman, A., Mattson, M. P., Brady, S. T., & Busciglio, J. (2003). Alzheimer's presenilin 1 mutations impair kinesin-based axonal transport. *The Journal of Neuroscience, 23*, 4499–4508.

Polymeropoulos, M. H., Lavedan, C., Leroy, E., Ide, S. E., Dehejia, A., Dutra, A., … Boyer, R. (1997). Mutation in the alpha-synuclein gene identified in families with Parkinson's disease. *Science, 276*, 2045–2047.

Puls, I., Jonnakuty, C., LaMonte, B. H., Holzbaur, E. L., Tokito, M., Mann, E., … Oh, S. J. (2003). Mutant dynactin in motor neuron disease. *Nature Genetics, 33*, 455–456.

Reed, N. A., Cai, D., Blasius, T. L., Jih, G. T., Meyhofer, E., Gaertig, J., & Verhey, K. J. (2006). Microtubule acetylation promotes kinesin-1 binding and transport. *Current Biology, 16*, 2166–2172.

Roll-Mecak, A., & Vale, R. D. (2005). The *Drosophila* homologue of the hereditary spastic paraplegia protein, spastin, severs and disassembles microtubules. *Current Biology, 15*, 650–655.

Roll-Mecak, A., & Vale, R. D. (2008). Structural basis of microtubule severing by the hereditary spastic paraplegia protein spastin. *Nature, 451*, 363–367.

Roy, B., & Jackson, G. R. (2014). Interactions between tau and alpha-synuclein augment neurotoxicity in a *Drosophila* model of Parkinson's disease. *Human Molecular Genetics, 23*, 3008–3023.

Roy, S., Winton, M. J., Black, M. M., Trojanowski, J. Q., & Lee, V. M. (2007). Rapid and intermittent cotransport of slow component-b proteins. *The Journal of Neuroscience, 27*, 3131–3138.

Rui, Y., Tiwari, P., Xie, Z., & Zheng, J. Q. (2006). Acute impairment of mitochondrial trafficking by beta-amyloid peptides in hippocampal neurons. *The Journal of Neuroscience, 26*, 10480–10487.

Rusu, P., Jansen, A., Soba, P., Kirsch, J., Lower, A., Merdes, G., ... Kins, S. (2007). Axonal accumulation of synaptic markers in APP transgenic *Drosophila* depends on the NPTY motif and is paralleled by defects in synaptic plasticity. *European Journal of Neuroscience, 25*, 1079–1086.

Saha, A. R., Hill, J., Utton, M. A., Asuni, A. A., Ackerley, S., Grierson, A. J., ... Hanger, D. P. (2004). Parkinson's disease alpha-synuclein mutations exhibit defective axonal transport in cultured neurons. *Journal of Cell Science, 117*, 1017–1024.

Salehi, A., Delcroix, J. D., Belichenko, P. V., Zhan, K., Wu, C., Valletta, J. S., ... Chung, P. P. (2006). Increased App expression in a mouse model of Down's syndrome disrupts NGF transport and causes cholinergic neuron degeneration. *Neuron, 51*, 29–42.

Salinas, S., Carazo-Salas, R. E., Proukakis, C., Cooper, J. M., Weston, A. E., Schiavo, G., & Warner, T. T. (2005). Human spastin has multiple microtubule-related functions. *Journal of Neurochemistry, 95*, 1411–1420.

Sanderson, K. L., Butler, L., & Ingram, V. M. (1997). Aggregates of a beta-amyloid peptide are required to induce calcium currents in neuron-like human teratocarcinoma cells: Relation to Alzheimer's disease. *Brain Research, 744*, 7–14.

Sasaki, S., & Maruyama, S. (1992). Increase in diameter of the axonal initial segment is an early change in amyotrophic lateral sclerosis. *Journal of the Neurological Sciences, 110*, 114–120.

Satpute-Krishnan, P., DeGiorgis, J. A., Conley, M. P., Jang, M., & Bearer, E. L. (2006). A peptide zipcode sufficient for anterograde transport within amyloid precursor protein. *Proceedings of the National Academy of Sciences of the United States of America, 103*, 16532–16537.

Scott, D. A., Tabarean, I., Tang, Y., Cartier, A., Masliah, E., & Roy, S. (2010). A pathologic cascade leading to synaptic dysfunction in alpha-synuclein-induced neurodegeneration. *The Journal of Neuroscience, 30*, 8083–8095.

Shankar, G. M., Li, S., Mehta, T. H., Garcia-Munoz, A., Shepardson, N. E., Smith, I., ... Lemere, C. A. (2008). Amyloid-beta protein dimers isolated directly from Alzheimer's brains impair synaptic plasticity and memory. *Nature Medicine, 14*, 837–842.

Shea, T. B., Jung, C., & Pant, H. C. (2003). Does neurofilament phosphorylation regulate axonal transport? *Trends in Neurosciences, 26*, 397–400.

Sherman, M. A., LaCroix, M., Amar, F., Larson, M. E., Forster, C., Aguzzi, A., ... Lesne, S. E. (2016). Soluble conformers of abeta and tau alter selective proteins governing axonal transport. *The Journal of Neuroscience, 36*, 9647–9658.

Shubeita, G. T., Tran, S. L., Xu, J., Vershinin, M., Cermelli, S., Cotton, S. L., ... Gross, S. P. (2008). Consequences of motor copy number on the intracellular transport of kinesin-1-driven lipid droplets. *Cell, 135*, 1098–1107.

Smith, K. D., Kallhoff, V., Zheng, H., & Pautler, R. G. (2007). In vivo axonal transport rates decrease in a mouse model of Alzheimer's disease. *Neuroimage, 35*, 1401–1408.

Song, C., Perides, G., Wang, D., & Liu, Y. F. (2002). Beta-amyloid peptide induces formation of actin stress fibers through p38 mitogen-activated protein kinase. *Journal of Neurochemistry, 83*, 828–836.

Soppina, V., Rai, A. K., Ramaiya, A. J., Barak, P., & Mallik, R. (2009). Tug-of-war between dissimilar teams of microtubule motors regulates transport and fission of endosomes. *Proceedings of the National Academy of Sciences of the United States of America, 106*, 19381–19386.

Spillantini, M. G., & Goedert, M. (1998). Tau protein pathology in neurodegenerative diseases. *Trends in Neurosciences, 21*, 428–433.

Stevanin, G., Azzedine, H., Denora, P., Boukhris, A., Tazir, M., Lossos, A., ... Alegria, P. (2008). Mutations in SPG11 are frequent in autosomal recessive spastic paraplegia with thin corpus callosum, cognitive decline and lower motor neuron degeneration. *Brain, 131*, 772–784.

Stokin, G. B., Almenar-Queralt, A., Gunawardena, S., Rodrigues, E. M., Falzone, T., Kim, J., ... McGowan, E. (2008). Amyloid precursor protein-induced axonopathies are independent of amyloid-beta peptides. *Human Molecular Genetics, 17*, 3474–3486.

Stokin, G. B., Lillo, C., Falzone, T. L., Brusch, R. G., Rockenstein, E., Mount, S. L., ... Goldstein, L. S. (2005). Axonopathy and transport deficits early in the pathogenesis of Alzheimer's disease. *Science, 307*, 1282–1288.

Stowers, R. S., Megeath, L. J., Gorska-Andrzejak, J., Meinertzhagen, I. A., & Schwarz, T. L. (2002). Axonal transport of mitochondria to synapses depends on milton, a novel *Drosophila* protein. *Neuron, 36*, 1063–1077.

Szebenyi, G., Morfini, G. A., Babcock, A., Gould, M., Selkoe, K., Stenoien, D. L., ... Brady, S. T. (2003). Neuropathogenic forms of huntingtin and androgen receptor inhibit fast axonal transport. *Neuron, 40*, 41–52.

Takashima, A., Murayama, M., Murayama, O., Kohno, T., Honda, T., Yasutake, K., ... Wolozin, B. (1998). Presenilin 1 associates with glycogen synthase kinase-3beta and its substrate tau. *Proceedings of the National Academy of Sciences of the United States of America, 95*, 9637–9641.

Takashima, A., Noguchi, K., Michel, G., Mercken, M., Hoshi, M., Ishiguro, K., & Imahori, K. (1996). Exposure of rat hippocampal neurons to amyloid beta peptide (25–35) induces the inactivation of phosphatidyl inositol-3 kinase and the activation of tau protein kinase I/glycogen synthase kinase-3 beta. *Neuroscience Letters, 203*, 33–36.

Tang, Y., Scott, D. A., Das, U., Edland, S. D., Radomski, K., Koo, E. H., & Roy, S. (2012). Early and selective impairments in axonal transport kinetics of synaptic cargoes induced by soluble amyloid beta-protein oligomers. *Traffic, 13*, 681–693.

Tarrade, A., Fassier, C., Courageot, S., Charvin, D., Vitte, J., Peris, L., … Roblot, N. (2006). A mutation of spastin is responsible for swellings and impairment of transport in a region of axon characterized by changes in microtubule composition. *Human Molecular Genetics, 15*, 3544–3558.

Tesseur, I., Van Dorpe, J., Bruynseels, K., Bronfman, F., Sciot, R., Van Lommel, A., & Van Leuven, F. (2000). Prominent axonopathy and disruption of axonal transport in transgenic mice expressing human apolipoprotein E4 in neurons of brain and spinal cord. *The American Journal of Pathology, 157*, 1495–1510.

Torroja, L., Chu, H., Kotovsky, I., & White, K. (1999). Neuronal overexpression of APPL, the *Drosophila* homologue of the amyloid precursor protein (APP), disrupts axonal transport. *Current Biology, 9*, 489–492.

Tortarolo, M., Veglianese, P., Calvaresi, N., Botturi, A., Rossi, C., Giorgini, A., … Bendotti, C. (2003). Persistent activation of p38 mitogen-activated protein kinase in a mouse model of familial amyotrophic lateral sclerosis correlates with disease progression. *Molecular and Cellular Neuroscience, 23*, 180–192.

Trushina, E., Dyer, R. B., Badger, J. D., 2nd, Ure, D., Eide, L., Tran, D. D., … Mandavilli, B. S. (2004). Mutant huntingtin impairs axonal trafficking in mammalian neurons in vivo and in vitro. *Molecular and Cellular Biology, 24*, 8195–8209.

Twelvetrees, A. E., Pernigo, S., Sanger, A., Guedes-Dias, P., Schiavo, G., Steiner, R. A., … Holzbaur, E. L. (2016). The dynamic localization of cytoplasmic dynein in neurons is driven by kinesin-1. *Neuron, 90*, 1000–1015.

Vande Velde, C., McDonald, K. K., Boukhedimi, Y., McAlonis-Downes, M., Lobsiger, C. S., Bel Hadj, S., … Cleveland, D. W. (2011). Misfolded SOD1 associated with motor neuron mitochondria alters mitochondrial shape and distribution prior to clinical onset. *PLoS ONE, 6*, e22031.

Verhey, K. J., Kaul, N., & Soppina, V. (2011). Kinesin assembly and movement in cells. *Annual Review of Biophysics, 40*, 267–288.

Verhey, K. J., Lizotte, D. L., Abramson, T., Barenboim, L., Schnapp, B. J., & Rapoport, T. A. (1998). Light chain-dependent regulation of Kinesin's interaction with microtubules. *Journal of Cell Biology, 143*, 1053–1066.

Verhey, K. J., Meyer, D., Deehan, R., Blenis, J., Schnapp, B. J., Rapoport, T. A., & Margolis, B. (2001). Cargo of kinesin identified as JIP scaffolding proteins and associated signaling molecules. *Journal of Cell Biology, 152*, 959–970.

Vossel, K. A., Xu, J. C., Fomenko, V., Miyamoto, T., Suberbielle, E., Knox, J. A., … Mucke, L. (2015). Tau reduction prevents Abeta-induced axonal transport deficits by blocking activation of GSK3beta. *Journal of Cell Biology, 209*, 419–433.

Vossel, K. A., Zhang, K., Brodbeck, J., Daub, A. C., Sharma, P., Finkbeiner, S., … Mucke, L. (2010). Tau reduction prevents Abeta-induced defects in axonal transport. *Science, 330*, 198.

Walsh, D. M., Klyubin, I., Shankar, G. M., Townsend, M., Fadeeva, J. V., Betts, V., … Selkoe, D. J. (2005). The role of cell-derived oligomers of Abeta in Alzheimer's disease and avenues for therapeutic intervention. *Biochemical Society Transactions, 33*, 1087–1090.

Wang, L., Das, U., Scott, D. A., Tang, Y., McLean, P. J., & Roy, S. (2014). Alpha-synuclein multimers cluster synaptic vesicles and attenuate recycling. *Current Biology, 24*, 2319–2326.

Wang, X., Winter, D., Ashrafi, G., Schlehe, J., Wong, Y. L., Selkoe, D., … Schwarz, T. L. (2011). PINK1 and Parkin target Miro for phosphorylation and degradation to arrest mitochondrial motility. *Cell, 147*, 893–906.

Weedon, M. N., Hastings, R., Caswell, R., Xie, W., Paszkiewicz, K., Antoniadi, T., … Ellard, S. (2011). Exome sequencing identifies a DYNC1H1 mutation in a large pedigree with dominant axonal Charcot–Marie–Tooth disease. *American Journal of Human Genetics, 89*, 308–312.

Weihofen, A., Thomas, K. J., Ostaszewski, B. L., Cookson, M. R., & Selkoe, D. J. (2009). Pink1 forms a multiprotein complex with Miro and Milton, linking Pink1 function to mitochondrial trafficking. *Biochemistry, 48*, 2045–2052.

Williamson, T. L., & Cleveland, D. W. (1999). Slowing of axonal transport is a very early event in the toxicity of ALS-linked SOD1 mutants to motor neurons. *Nature Neuroscience, 2*, 50–56.

Withers, G. S., George, J. M., Banker, G. A., & Clayton, D. F. (1997). Delayed localization of synelfin (synuclein, NACP) to presynaptic terminals in cultured rat hippocampal neurons. *Brain Research. Developmental Brain Research, 99*, 87–94.

Wong, P. C., Pardo, C. A., Borchelt, D. R., Lee, M. K., Copeland, N. G., Jenkins, N. A., … Price, D. L. (1995). An adverse property of a familial ALS-linked SOD1 mutation causes motor neuron disease characterized by vacuolar degeneration of mitochondria. *Neuron, 14,* 1105–1116.

Woodruff, G., Reyna, S. M., Dunlap, M., Van Der Kant, R., Callender, J. A., Young, J. E., ... Goldstein, L. S. (2016). Defective transcytosis of APP and lipoproteins in human iPSC-derived neurons with familial Alzheimer's disease mutations. *Cell Rep, 17,* 759–773.

Xia, C. H., Roberts, E. A., Her, L. S., Liu, X., Williams, D. S., Cleveland, D. W., & Goldstein, L. S. (2003). Abnormal neurofilament transport caused by targeted disruption of neuronal kinesin heavy chain KIF5A. *Journal of Cell Biology, 161,* 55–66.

Yang, J. T., Laymon, R. A., & Goldstein, L. S. (1989). A three-domain structure of kinesin heavy chain revealed by DNA sequence and microtubule binding analyses. *Cell, 56,* 879–889.

Zarranz, J. J., Alegre, J., Gomez-Esteban, J. C., Lezcano, E., Ros, R., Ampuero, I., ... Atares, B. (2004). The new mutation, E46K, of alpha-synuclein causes Parkinson and Lewy body dementia. *Annals of Neurology, 55,* 164–173.

Zhai, J., Lin, H., Julien, J. P., & Schlaepfer, W. W. (2007). Disruption of neurofilament network with aggregation of light neurofilament protein: A common pathway leading to motor neuron degeneration due to Charcot–Marie–Tooth disease-linked mutations in NFL and HSPB1. *Human Molecular Genetics, 16,* 3103–3116.

Zhang, B., Higuchi, M., Yoshiyama, Y., Ishihara, T., Forman, M. S., Martinez, D., ... Lee, V. M. (2004). Retarded axonal transport of R406W mutant tau in transgenic mice with a neurodegenerative tauopathy. *The Journal of Neuroscience, 24,* 4657–4667.

Zhao, C., Takita, J., Tanaka, Y., Setou, M., Nakagawa, T., Takeda, S., ... Takei, Y. (2001). Charcot–Marie–Tooth disease type 2A caused by mutation in a microtubule motor KIF1Bbeta. *Cell, 105,* 587–597.

Zuchner, S., Mersiyanova, I. V., Muglia, M., Bissar-Tadmouri, N., Rochelle, J., Dadali, E. L., ... Senderek, J. (2004). Mutations in the mitochondrial GTPase mitofusin 2 cause Charcot–Marie–Tooth neuropathy type 2A. *Nature Genetics, 36,* 449–451.

CHAPTER

13

Mitochondrial Function and Neurodegenerative Diseases

Heather M. Wilkins, Ian Weidling, Scott Koppel, Xiaowan Wang, Alex von Schulze and Russell H. Swerdlow

University of Kansas School of Medicine, Kansas City, KS, United States

OUTLINE

INTRODUCTION

Biologists first described mitochondria over 100 years ago (Lewis & Lewis, 1914). Since then, considerable insight into their structure and function has emerged. For example, the chemiosmotic hypothesis explains the process through which mitochondria produce energy (Mitchell, 1961;

The Molecular and Cellular Basis of Neurodegenerative Diseases
DOI: https://doi.org/10.1016/B978-0-12-811304-2.00013-4

Mitchell & Moyle, 1967), and unique among nonnuclear organelles mitochondria contain their own genome (Nass & Nass, 1963a, 1963b).

Initial efforts defined the mitochondria's essential role in cell bioenergetics, while subsequent studies offered a fuller appreciation of how mitochondria integrate into the greater cell milieu. It now appears that although cells regulate mitochondria, mitochondria also regulate their host cells. Signals pass bidirectionally between mitochondria and the nucleus (Finkel, 2001, 2003; Liu & Butow, 2006; Schieke & Finkel, 2006) and facilitate a molecular and physiologic coordination that affects cell growth, survival, and death.

A growing realization of what mitochondria do and how they do it increasingly implicate mitochondria in health and disease. Many believe mitochondria play an important role in aging and specifically brain aging (Harman, 1972; Lopez-Otin et al., 2013; Muller et al., 2007), and both hypothesis-driven and serendipitous discoveries show mitochondrial dysfunction occurs in multiple neurodegenerative diseases. Sometimes, mitochondrial defects mediate disease pathology, and in some cases, mitochondrial defects actually drive the disease. Regardless, understanding the causes and consequences of mitochondrial dysfunction provides insight into neurodegeneration and reveals therapeutic targets that could lead to new treatments. Accordingly, this chapter provides an overview of mitochondrial structure and function, catalogs different mitochondrial lesions observed in specific neurodegenerative disorders, and discusses therapeutic implications.

MITOCHONDRIA AND BIOENERGETICS

Section Overview

Mitochondria contribute to ATP production, mediate respiratory fluxes, and facilitate other energy metabolism pathways. They influence or mediate oxidative stress levels, redox states, and, when necessary, programed cell death. These phenomena, individually or in combination, may contribute to neurodegeneration. This section provides an overview of basic mitochondrial biology and function, which provides a necessary background for understanding their potential role in neurodegenerative diseases.

Mitochondrial Structure

Mitochondria contain inner and outer membranes separated by an intermembrane space. The inner mitochondrial membrane (IMM) encloses the matrix (Palade, 1953; Sjostrand, 1953) and contains enzymes that mediate oxidative phosphorylation (OXPHOS). The IMM increases its surface by folding into cristae, which accommodates more OXPHOS proteins and facilitates ATP production (Mannella, 2006).

Similar to other eukaryotic membranes, phosphatidylcholine and phosphatidyl-ethanolamine phospholipids comprise the outer mitochondrial membrane (OMM). Cardiolipin enrichment distinguishes the IMM (Horvath & Daum, 2013). Cardiolipin interacts with OXPHOS enzymes (Fry & Green, 1981; Robinson, 1993). Compositional IMM and OMM features define permeability differences. Most small molecules and ions freely cross the OMM but not the IMM. Outer and inner membrane translocases help import exogenous proteins to the intermembrane space, IMM, and matrix (Pfanner & Meijer, 1997).

Mitochondrial morphology is highly dynamic and differs between cell types (Amchenkova et al., 1988; Collins et al., 2002; De Giorgi, Lartigue, & Ichas, 2000). Within a cell, the mitochondrial pool transitions between tubular networks and fragmented structures. Physical subtypes include small globules, swollen globules, straight tubules, twisted tubules, branched tubules, and loops

(Peng et al., 2011). Fission and fusion events (discussed below) additionally regulate structure (Bereiter-Hahn, 1990).

Mitochondrial morphologies vary with cell cycle and health (Frank et al., 2001; Mitra et al., 2009; Twig et al., 2008). Exposing mitochondria to excess ADP, which stimulates respiration, changes IMM folding and matrix volume (Hackenbrock, 1966). Altering glycolysis to respiration ratios alters tubule diameter and branching (Rossignol et al., 2004).

Mitochondrial DNA

Even before the discovery of the double helix, studies found cells contained nonnuclear nucleic acid accumulations (Ephrussi, Hottinger, & Chimenes, 1949). Investigators called these nonnuclear nucleic acid pools ρ DNA until the 1960s, when Nass and Nass (1963a, 1963b) showed they resided in mitochondria.

Mitochondrial DNA (mtDNA) plasmids, typically illustrated as DNA circles, sit within the matrix and physically associate with the IMM. mtDNA does not interact with histones, but instead with transcription factor A of the mitochondria (TFAM), which influences transcription and replication (Scarpulla, 2008). In 1981, the entire human 16,569 base pair mtDNA "Cambridge sequence" was published (Anderson et al., 1981), which subsequently underwent minor revisions (Andrews et al., 1999).

The mtDNA consists of "light" and "heavy" strands, with the heavy strand containing a higher proportion of guanine and adenine nucleotides. The Cambridge sequence reports the light strand. RNA sequences transcribed from the heavy strand are therefore sense to the Cambridge sequence, while RNA sequences transcribed from the light strand are antisense.

All but 1 (ND6) of the 13 protein-coding "structural" mtDNA genes template from the heavy strand. In addition to these 13 protein-coding structural genes, the mtDNA also contains 24 "synthetic" genes, whose products service structural gene translation. They include 22 tRNA and 2 rRNA genes (Anderson et al., 1981).

A 1.1-kB control region, also called the "noncoding region", lacks genes and consists of promoter sites and a displacement loop (the "D-loop") (Fig. 13.1). At the D-loop, the heavy strand can separate from the light strand to let an independent ~650 nucleotide piece of DNA called the 7S DNA, which matches the heavy strand sequence, anneal to the light strand (Clayton, 2000; Nicholls & Minczuk, 2014). This third DNA strand helps initiate new mtDNA replication. In addition to the displacement region, the control region contains hypervariable sequence sections, conserved sequence block stretches, the termination-associated sequence region, a light strand promoter (LSP) that initiates translation from the light strand (producing RNA that is sense to the heavy strand), and heavy strand promoters (HSPs) that initiate translation from the heavy strand (producing RNA that is sense to the light strand). It also contains the site of origin for new heavy strand synthesis, called the origin of the heavy strand or OH, which corresponds to the 5′end of the 7S DNA. A ~200 nucleotide piece of RNA (the 7S RNA) produced at the LSP primes production of the 7S DNA.

The origin of light strand synthesis, or OL, does not reside in the control region. When nascent heavy strand replication reaches approximately two-thirds of its full length, covering a stretch called the "major arc," displacement of the parent light and heavy strands opens the OL to begin synthesis of a new light strand, using the parent heavy strand as its template.

Transcription begins at the LSP or HSP sites and proceeds in opposite directions, dependent on the promoter, around the DNA circle. This generates polycistronic RNA transcripts. mtDNA lacks introns, so splicing of heterogeneous transcripts is not required, but tRNA genes within the polycistronic transcripts facilitate processing into individual mRNA, rRNA, and tRNA species. The mRNAs receive a

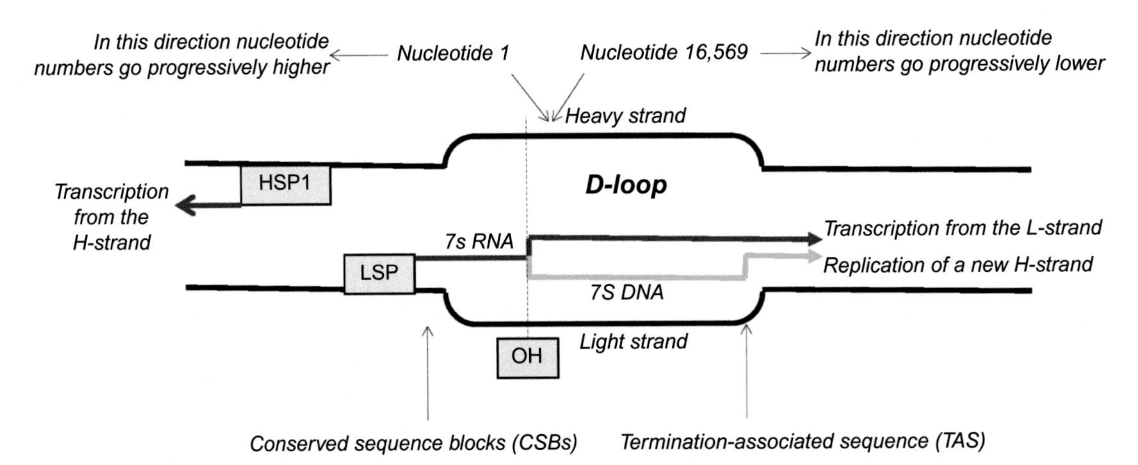

FIGURE 13.1 **The mtDNA control region.** The ~1.1-kB control region spans mtDNA nucleotides 576–16,024 and contains the D-loop and neighboring promoter sequences. The heavy strand promoter (HSP) encompasses nucleotides 545–567 and the LSP encompasses nucleotides 392–445. The origin of the heavy strand (OH) begins around nucleotides 190–200. Two hypervariable regions (HSV1 between nucleotides 16,024 and 16,383, and HSV2 between nucleotides 57 and 372) are not marked in the figure. The three CSBs reside within HSV2, in the region between the LSP and the start of the D-loop. The TAS sits near nucleotide 16,172. The parent mtDNA heavy and light strands are shown in black, new RNA synthesis is shown in red, and new DNA synthesis is shown in green. *mtDNA*, mitochondrial DNA; *HSP*, heavy strand promoter; *CSB*, conserved sequence block; *LSP*, light strand promoter; *TAS*, termination-associated sequence.

polyadenylation tail prior to translation at mitochondrial ribosomes that contain the mtDNA-encoded 12S and 16S rRNA products. At these ribosomes, the codon dictionary occasionally diverges from the nuclear codon dictionary. Similar to bacteria, mtDNA-encoded peptides start with a formylmethionine residue.

Cells contain many mitochondria, and most mitochondria maintain multiple mtDNA copies. Different cell types can therefore contain quite different amounts of mtDNA, from less than 100 to over 100,000 copies (Shoubridge & Wai, 2007). The term homoplasmy describes the presence of only one mtDNA sequence within a cell, while heteroplasmy refers to a state in which mtDNA wild-type and mutated sequences coexist within a cell. Whether a particular mtDNA mutation confers a functional consequence depends on mutation pathogenicity and the percentage of mutated mtDNA plasmids. These factors determine whether the mutation burden surpasses a functional "threshold." Relatively pathogenic mutations surpass their thresholds at lower heteroplasmic burdens than relatively benign mutations.

Dividing cells partition their mitochondria and, by extension, their mtDNA. If a heteroplasmic cell divides in a way that segregates its different mtDNA sequences, the daughter cells will inherit different mtDNA sequences and possibly different functional characteristics. The terms "mitotic segregation" and "replicative segregation" refer to this phenomenon.

During sexual reproduction, organisms receive their mitochondria and mtDNA from the oocyte; under normal conditions sperm do not noticeably contribute (Giles et al., 1980). Due to this "maternal inheritance," mothers contribute more than fathers to mitochondrial function. Women with oocyte mtDNA mutations can pass them to their offspring, while men with sperm mtDNA mutations cannot. For this reason, disease-causing mtDNA mutations sometimes show a recognizable maternal inheritance pattern. However, due to heteroplasmy, threshold, and mitotic segregation mtDNA mutation-related diseases also present in a sporadic fashion (Parker, 1990).

The Respiratory Chain

The 13 mtDNA structural genes encode catalytically important OXPHOS subunits. These include 7 NADH:ubiquinone oxidoreductase (complex I), 1 coenzyme Q:cytochrome *c* oxidoreductase (cytochrome bc1 complex; complex III), 3 cytochrome *c* oxidase (COX; complex IV), and 2 ATP synthase (complex V) subunits. The nucleus contains over 70 genes that also produce OXPHOS subunits, as well as proteins that generate nonpeptide OXPHOS components and that assemble the respiratory holoenzymes.

The chemiosmotic theory of Mitchell describes the mechanisms through which OXPHOS produces energy (Mitchell, 1961; Mitchell & Moyle, 1967). OXPHOS takes advantage of basic thermodynamic principles, as molecules that contain high-energy electrons in their reduced state, especially nicotinamide adenine dinucleotide (NAD) and flavin adenine dinucleotide (FAD), readily transfer their high-energy electrons to oxygen.

Since the transfer of high-energy electrons from reduced NAD (NADH) and reduced FAD (FADH2) to oxygen releases considerable energy, cells evolved respiratory chains that allow for a gradual, as opposed to abrupt, energy release. After electrons from NADH enter the respiratory chain at complex I (Sazanov, 2007), and electrons from FADH2 enter at complex II, they pass through a series of electron carriers with increasingly higher electron affinities (Fig. 13.2). Based on this escalating affinity gradient, as the electrons progressively release energy and pass to lower energy states, they advance through the respiratory chain. After initially navigating the iron–sulfur clusters of complexes I and II, the electrons reach ubiquinone (coenzyme Q), which reduces to ubiquinol and shuttles the electrons to complex III. Within complex III, coenzyme Q, an iron–sulfur cluster, and cytochromes pass the electrons to cytochrome *c*, which shuttles them to complex IV. Complex IV stabilizes oxygen in such a way that four electrons reach two oxygen atoms, and four protons access the reduced oxygen to generate two molecules of water.

Energy released by electrons as they pass through the respiratory chain drives the transfer of protons from the mitochondrial matrix to the intermembrane space. This creates an electrochemical gradient, $\Delta\psi$, with a major component being the voltage gradient that arises from the

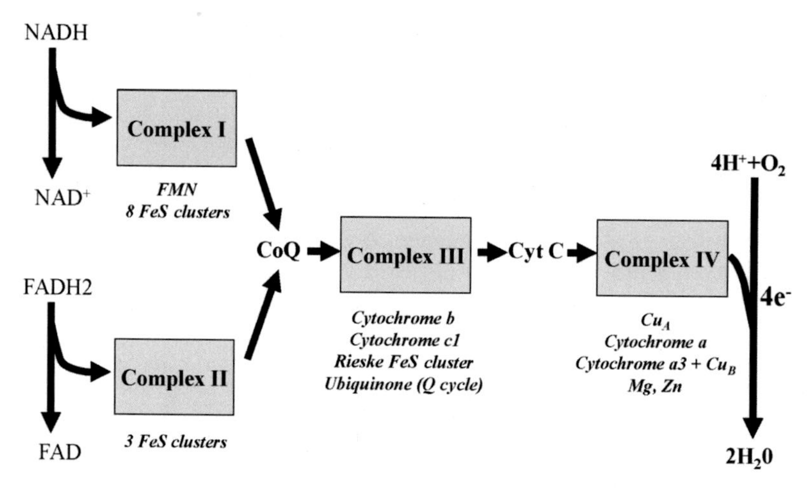

FIGURE 13.2 **The ETC.** The respiratory chain includes complexes I–V, whereas the ETC only includes complexes I–IV. The cartoon depicts the passage of electrons through the ETC, as they travel from higher to lower energy states and pass through carriers with increasingly higher electron affinities. The electron carriers present in each complex are indicated below the name of the complex. *Cu*, copper; *FeS*, iron–sulfur complex; *FMN*, flavin mononucleotide; *Mg*, magnesium; *Zn*, zinc; *ETC*, electron transport chain.

separation of positive and negative charges across the IMM (ΔV), and a minor component that reflects the resulting pH gradient (ΔpH). The charge separation also defines a mitochondrial membrane potential; increasing this potential induces hyperpolarization, and decreasing it induces depolarization.

This electrochemical gradient-determined "proton motive force" favors the return of protons to the matrix, but the IMM creates a physical barrier that maintains the separation. Complex V, however, lets the intermembrane space protons reaccess the matrix. As this happens, complex V harvests energy from the proton flux and uses this energy to phosphorylate ADP to ATP. When a high proportion of protons pumped due to the passage of electrons end up supporting ATP synthesis, a high state of respiratory "coupling" exists. Under some physiologic and nonphysiologic conditions, though, protons may reaccess the matrix independently of complex V and support the production of heat instead of the production of ATP. Increasing heat-to-ATP ratios define progressively greater states of "uncoupled" respiration.

Mitochondrial Biogenesis

Mitochondrial biogenesis involves the synthesis of new mtDNA, protein, and membrane, although new mitochondria are themselves generated from preexisting mitochondria through mitochondrial fission and do not form de novo (Jornayvaz & Shulman, 2010). In many cell types, PPARγ coactivator 1 alpha (PGC-1α) regulates new mitochondria synthesis. First recognized as a regulator of mitochondrial biogenesis in brown adipose tissue and skeletal muscle, PGC-1α increases following exposure to cold and exercise (Puigserver et al., 1998). A similar molecule, PGC-1β, also stimulates mitochondrial biogenesis but does not respond to cold or exercise (Meirhaeghe et al., 2003).

Under resting conditions, PGC-1α primarily resides in the cytosol. Upon activation, it translocates to the nucleus, where it coactivates various energy metabolism-relevant transcription factors. Chief among these are the nuclear respiratory factors (NRFs) 1 and 2, estrogen-related receptors (ERRs), and peroxisome proliferator-activated receptors (PPARs) (Ventura-Clapier, Garnier, & Veksler, 2008). Increased NRF activation induces TFAM transcription, which facilitates mtDNA synthesis (Virbasius & Scarpulla, 1994). Induction of ERR pathways promotes mtDNA synthesis and increased glucose utilization (Giguere, 2008). PPAR activation upregulates infrastructure that mediates fatty acid β-oxidation (Narkar et al., 2008).

PGC-1α activators include 5′-AMP-activated protein kinase (AMPK), calcineurin, p38-mitogen-activated protein kinase, sirtuin 1, transducer of regulated CREB, and cyclic GMP (cGMP) generated from nitric oxide signaling (Akimoto et al., 2005; Bergeron et al., 2001; Nisoli et al., 2003; Rodgers et al., 2005; Ryder et al., 2003; Wu et al., 2006). Mitochondrial stress, including oxidative stress, can initiate mitochondrial mass upregulation.

Fusion/Fission

Lewis and Lewis (1914) noted mitochondria appeared as small-to-large granules, short-to-long rods, or thread-like networks. We now recognize that fusion and fission events partly determine this heterogeneity (Fig. 13.3). Mitochondrial fusion and fission help cells adapt to or implement a variety of processes including apoptosis, cell division, cell movement, and bioenergetic demand (Westermann, 2012).

GTPase proteins mediate mitochondrial fusion. Optic atrophy 1 (OPA-1) accomplishes IMM fusion, and two mitofusin proteins (Mfn1 and Mfn2) accomplish OMM fusion (Alexander et al., 2000; Delettre et al., 2000;

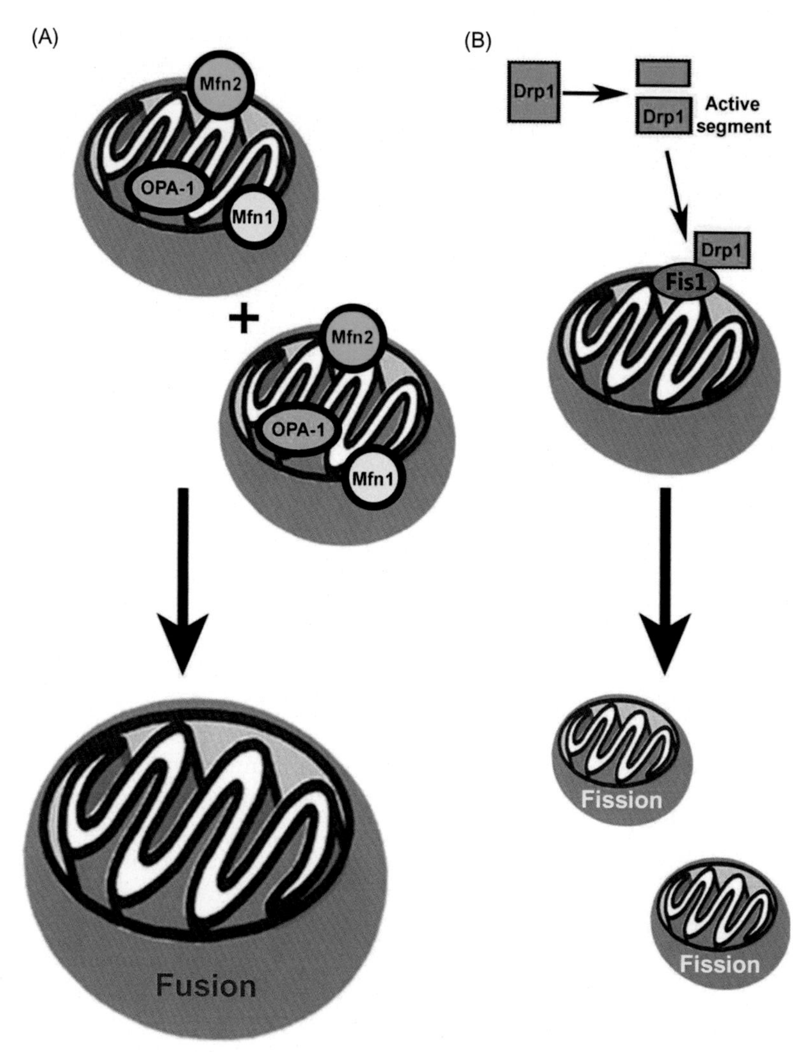

FIGURE 13.3 **Mitochondrial fusion and fission.** (A) Between two fusing mitochondria, OPA-1 mediates inner membrane fusion and the Mfn1 and Mfn2 mediate outer membrane fusion. (B) To enable mitochondrial fission, the phosphorylation and cleavage of inactive DRP1 produces an active fragment that accesses mitochondria, where it complexes with Fis1 to mediate outer and inner membrane fission. *OPA-1*, optic atrophy 1; *Mfn*, mitofusin.

Eura et al., 2003; Santel et al., 2003). OPA-1 localizes to the intermembrane space, and Mfn1 and Mfn2 reside at the OMM. OPA-1 function requires Mfn1, but not Mfn2 (Cipolat et al., 2004). Mfn2 also contributes to mitochondria-endoplasmic reticulum interactions (de Brito & Scorrano, 2008) and mevalonate pathway function, as Mfn2 knock-out causes coenzymeQ deficiency (Silva Ramos, Larsson, & Mourier, 2016).

Mitochondrial membrane depolarization inhibits fusion (Ishihara et al., 2003; Legros et al., 2002; Olichon et al., 2003). Loss of OPA-1 function dissipates the membrane potential, leading to cytochrome *c* release and cell death (Olichon et al., 2003). Genetic ablation of both Mfn1 and Mfn2, or of OPA-1, completely blocks fusion, reduces respiration, and increases mitochondrial membrane potential variation (Chen, Chomyn, & Chan, 2005; Cogliati et al., 2013). Conversely, fusion facilitates ATP synthesis by enabling ATP synthase dimerization and the formation of electrically coupled mitochondrial networks (Amchenkova et al., 1988; Cogliati et al., 2013; Skulachev, 2001; Westermann, 2012). Energy substrate deprivation enhances fusion and network formation. Fusion-mediated elongation may increase overall mitochondrial mass by precluding phagophore engulfment and mitophagy (Gomes, Di Benedetto, & Scorrano, 2011; Rambold et al., 2011).

The dynamin-related protein 1 (Drp1) GTPase, with assistance from the protein Fis1, executes both inner and outer membrane fission (Horbay & Bilyy, 2016; James et al., 2003; Smirnova et al., 2001). Drp1 cycles between the OMM and cytosol. Fission, like fusion, is regulated by bioenergetic demand and may provide a mechanism for pinching off damaged mitochondrial material from a larger mitochondrial network.

Drp1 activation requires dephosphorylation by calcineurin, which promotes association of Drp1 with the OMM (Cereghetti et al., 2008). Conversely, phosphorylation by protein kinase A, a cyclic AMP (cAMP) dependent kinase, inhibits Drp1. Other posttranslational modifications influence Drp1 activity (Chang & Blackstone, 2010; Elgass et al., 2013), as do mitochondrial membrane potential and cytosolic Ca^{2+} concentrations (Cereghetti et al., 2008).

Autophagy and Mitophagy

Applemans et al. described lysosomes in 1955, identified their association with acid phosphatase (Appelmans, Wattiaux, & De Duve, 1955), and characterized their role in intracellular digestion. These investigators subsequently noted intracellular digestion also featured "autophagosomes," which lacked digestive enzymes and acid hydrolase (De Duve & Wattiaux, 1966). Autophagosomes and lysosomes facilitate autophagy, which encompasses three subtypes: macroautophagy, microautophagy, and the mechanistically unrelated process of chaperone-mediated autophagy (Feng et al., 2014).

Macroautophagy represents the classic form and begins with the formation of a double-membraned autophagosome that engulfs cytoplasmic contents. The fully formed autophagosome fuses with a lysosome, and delivers its contents to the lysosome lumen. Degradation ensues, and the degradation products reaccess the cytoplasm for reuse. The process requires a set of proteins encoded by the autophagy-related genes (*ATG*), initially characterized in *Saccharomyces cerevisiae*, and later extended to their eukaryotic orthologues (Cebollero & Reggiori, 2009).

During nutrient deprivation, autophagy maintains homeostasis by scavenging substrates for anabolic processes. The mammalian target of rapamycin (mTOR) monitors catabolic versus anabolic signals, such that starvation inhibits mTOR activity and activates autophagy (Jung et al., 2010).

Mitochondrial function influences autophagy flux. Mitochondria-generated superoxide induces autophagy through unclear mechanisms (Ding & Yin, 2012). AMPK, which monitors cell energy supplies, stimulates autophagy, and energy deficiency caused by mitochondrial dysfunction can activate autophagy (Kim et al., 2011; Zhao et al., 2016).

Selective autophagy pathways facilitate the degradation of mitochondria, peroxisomes, unfolded proteins, bacteria, viruses, and endoplasmic reticulum (Johansen & Lamark, 2011). Work performed in cultured rat hepatocytes provided initial evidence of selective mitochondrial autophagy, or "mitophagy." In that study, serum starvation in the presence of glucagon triggered an increased association of mitochondria with lysosomes and autophagosomes (Rodriguez-Enriquez et al., 2006). A subsequent study demonstrated hypoxia also induces mitophagy (Zhang et al., 2008).

Evidence suggests mitophagy preferentially removes damaged mitochondria. Failing mitochondria depolarize, which prevents the normal translocation of PTEN-induced putative kinase 1 (PINK1) protein to the matrix. PINK1 translocation arrest causes it to project out from the mitochondrion and into the cytoplasm, where it recruits parkin, a ubiquitin ligase. This initiates autophagosome engulfment (Narendra et al., 2008) via a process that involves ubiquitination of mitochondrial proteins, binding of adaptor proteins to the ubiquitinated proteins, and adaptor-mediated transport to autophagosomes (Fig. 13.4) (Ding & Yin, 2012).

The Mitochondrial Unfolded Protein Response

The accumulation of misfolded protein within mitochondria initiates a mitochondrial unfolded protein response (UPR^{mt}) (Yoneda et al., 2004; Zhao et al., 2002). The UPR^{mt} arises in response to an array of mitochondrial

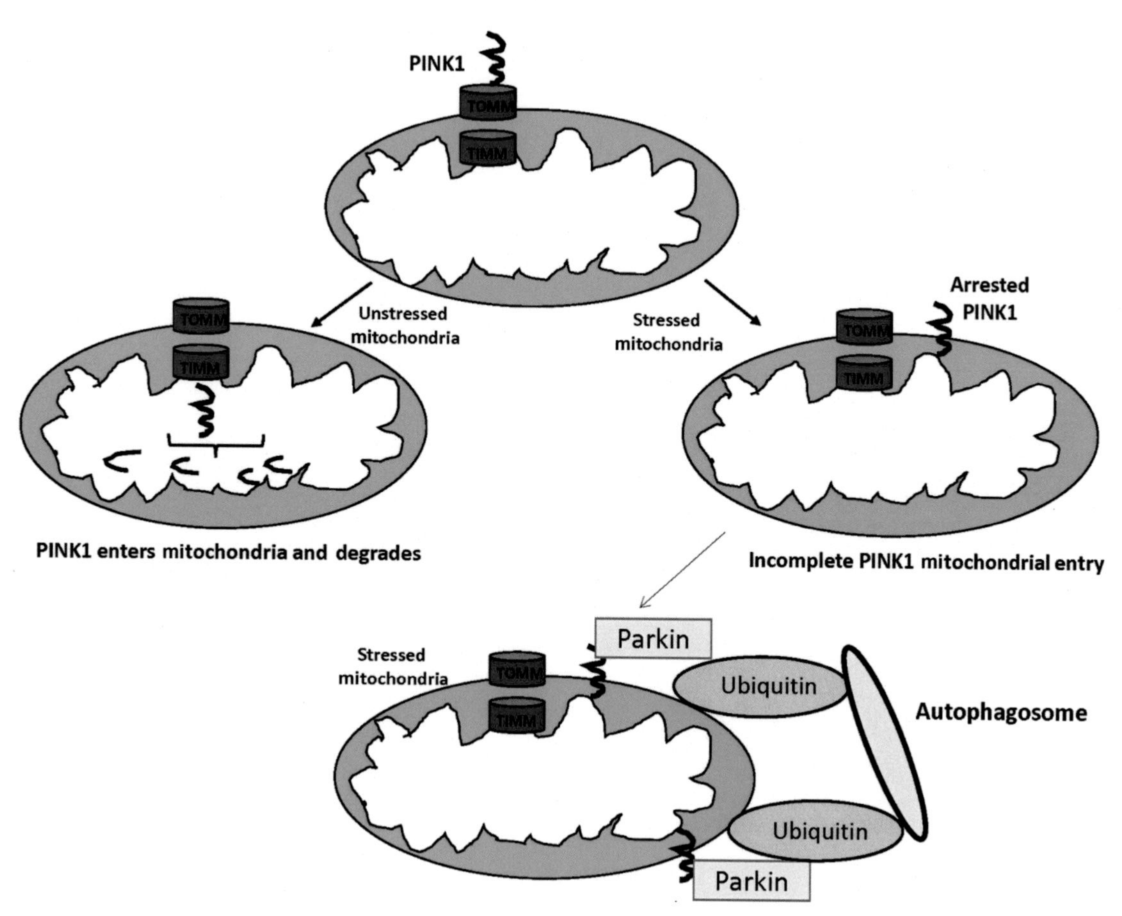

FIGURE 13.4 **PINK1-parkin mediated autophagy.** Unstressed mitochondria with normal membrane potentials import PINK1 and degrade it. Under stress/depolarization conditions, PINK1 import arrests to reveal a cytoplasmic extension. This cytoplasmic extension recruits parkin, which facilitates mitochondrial ubiquitination, autophagosome activation, and mitophagy. *PINK1*, PTEN-induced putative kinase 1.

insults, including mtDNA damage, reactive oxygen species (ROS) induction, electron transport chain (ETC) dysfunction, and mitochondrial protease inhibition (Martinus et al., 1996; Nargund et al., 2012; Yoneda et al., 2004; Zhao et al., 2002). Regardless of the precise inciting event, the transcriptional response appears well conserved, as well as designed to preserve mitochondrial function (Fig. 13.5). It includes increased transcription of genes known to enhance protein folding, mitochondrial fission, mitochondrial autophagy, ROS responses, and glycolytic flux. Another critical component includes the reduced production of mitochondria-targeted proteins. If the UPR signaling persists or mitochondrial dysfunction is severe, the UPR^{mt} signaling pathway can also promote apoptosis (Lin & Haynes, 2016).

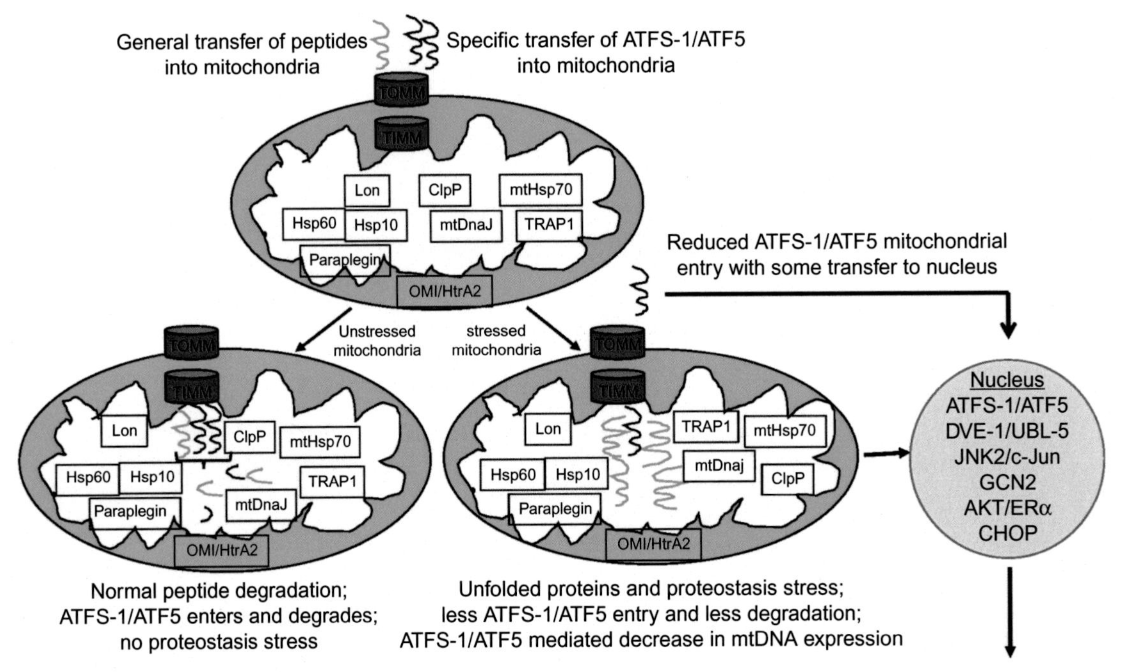

FIGURE 13.5 **The UPR^{mt}.** Stressed mitochondria may not degrade internal peptides fast enough, leading to intramitochondrial proteostatic stress. This gives rise to retrograde signaling events from the mitochondria to the nucleus, including translocation of some ATFS-1 (in *C. elegans*) or possibly ATF5 (in mammals) to the nucleus, and implementation of JNK2/c-Jun, GCN2, AKT/ERα, DVE-1/UBL-5, and CHOP-C/EPBβ mediated signals to and in the nucleus. Cumulatively, these responses increase the expression of proteins that directly degrade or refold peptides within mitochondria, as well as the expression of proteins (such as CHOP) that further promote the expression of peptidases and chaperones. The transcription and translation of other proteins that could enter the mitochondria and further increase protein stress is also reduced. In the stressed state some ATFS-1/ATF5 still continues to access the mitochondria, but does not degrade, and inhibits mtDNA gene expression. *JNK2*, c-Jun-N-terminal kinase 2; *ERα*, estrogen receptor α; *CHOP*, C/EBPβ homologous protein; *mtDNA*, mitochondrial DNA.

Ongoing research will better define UPR^{mt} underlying mechanisms, but one proposed early event involves a depolarization-induced activation of c-Jun-N-terminal kinase 2 (JNK2). JNK2 promotes c-Jun binding to the AP-1 promoter element, which leads to the transcription of CCAAT/enhancer-binding protein beta (C/EBPβ) and C/EBPβ homologous protein (CHOP) (Pellegrino, Nargund, & Haynes, 2013; Weiss et al., 2003). CHOP and C/EBPβ dimerize, bind CHOP promoters, and increase heat shock protein (Hsp) transcription. Relevant heat shock chaperones include mtHsp70, Hsp60, and Hsp10 (Aldridge, Horibe, & Hoogenraad, 2007; Cheng et al., 1989; Herrmann et al., 1994; Horibe & Hoogenraad, 2007).

Through a parallel action, the UPR^{mt} exhibits a general control nonderepressable 2 mediated phosphorylation of eukaryotic initiation factor 2 alpha, which inhibits protein translation (Baker et al., 2012; Hinnebusch, 1994). This reduces the synthesis of new proteins that could access mitochondria and exacerbate an existing protein overload state.

Unfolded intermembrane space proteins induce AKT phosphorylation and activate estrogen receptor α (ERα). ERα, in turn, increases transcription of the promitochondrial biogenesis NRF1 transcription factor and the HtrA2 mitochondrial protease (Papa & Germain, 2011). Accumulation of unfolded proteins within the matrix also triggers the UPR^{mt}, as does inhibition of the matrix proteases Lon and ClpP (Haynes et al., 2007; Martinus et al., 1996).

Caenorhabditis elegans studies inform much of our current UPR^{mt} understanding. In *C. elegans*, failing mitochondrial protein import through the translocase of the OMM (TOMM) or translocase of the IMM (TIMM) apparatus triggers the response (Chacinska et al., 2009; Hill et al., 1998). The response itself relies on ATFS-1, a basic leucine zipper protein that normally undergoes mitochondrial import and degradation. In the setting of mitochondrial depolarization or other states of mitochondrial dysfunction, import does not occur, ATFS-1 redistributes to the nucleus, and this initiates UPR^{mt}-related gene transcription (Nargund et al., 2012, 2015). Recent data suggest the protein ATF5 represents the mammalian homolog of ATFS-1 (Fiorese et al., 2016).

In addition to maintaining their own protein homeostasis infrastructure and mounting a distinct UPR, mitochondria appear to play an important role in whole-cell protein homeostasis. As is the case for nuclear-encoded proteins with specific mitochondrial functions, proteins without recognized mitochondrial activities similarly access the matrix via the TIMM and TOMM translocases, where peptidase-mediated degradation then ensues (Ruan et al., 2017).

Some proteins, it seems, form OMM-associated aggregates. These peptide aggregates may target specific mitochondria for particular fates (Zhou et al., 2014). Alternatively, the aggregates can undergo chaperone-mediated solubilization to peptide monomers, with mitochondrial import of the monomers and subsequent monomer proteolysis (Ruan et al., 2017).

Apoptosis and Cell Death

Apoptosis efficiently removes damaged, infected, or unnecessary cells. Two main apoptotic pathways exist, the extrinsic and intrinsic pathways (Fig. 13.6A). The extrinsic apoptotic pathway initiates from extracellular signals and converges with the intrinsic pathway (Desagher & Martinou, 2000; Kantari & Walczak, 2011). Mitochondria themselves activate the intrinsic pathway, with critical levels of mitochondrial dysfunction serving as the trigger (Czabotar et al., 2014; Desagher & Martinou, 2000).

In the extrinsic pathway, activation and dimerization of death receptors by death ligands such as Fas-associated death domain, tumor necrosis factor (TNF), and TNF-related apoptosis-inducing ligand initiate the process

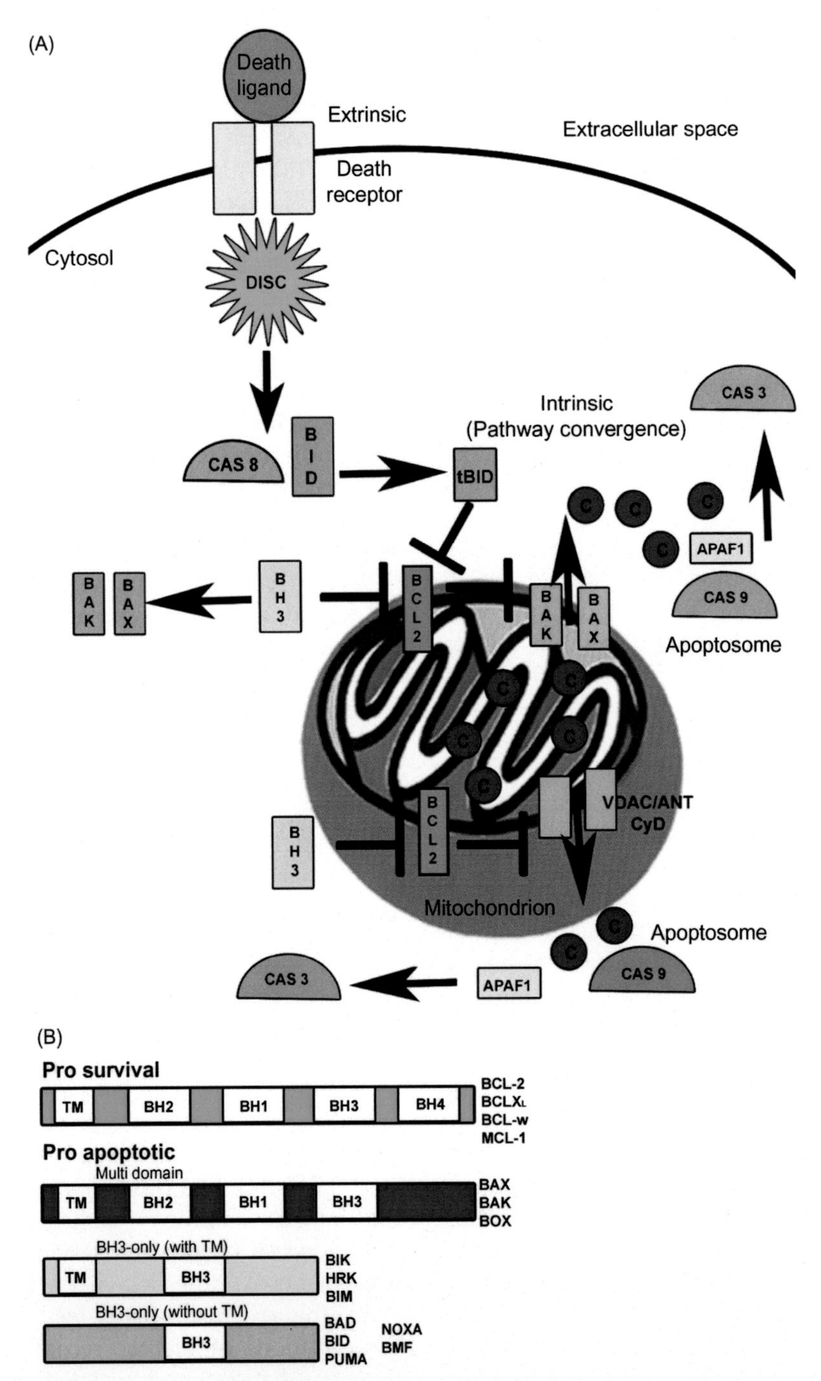

FIGURE 13.6 **Apoptosis.** (A) A schematic overview of the extrinsic and intrinsic apoptotic pathways. Other abbreviations are defined in the text. (B) Structure-defined apoptosis molecule categories with members from each category listed. *CAS*, caspase; *c*, cytochrome *c*; *CyD*, cyclophilin D; *TM*, transmembrane domain.

(Cordeiro & Bidere, 2013; Donepudi et al., 2003; Elmore, 2007; Scott et al., 2009; Walczak & Haas, 2008). Following death receptor activation, the death-inducing signaling complex activates cysteine-aspartic protease (caspase) 8 (Donepudi et al., 2003), which cleaves and activates the BCL-2 homology (BH)3-only protein, BH3-interacting death domain agonist (BID), to truncated BID (t-BID) (Kantari & Walczak, 2011; Li et al., 1998; Schug et al., 2011). t-BID activates caspase 3, the executioner caspase (Kantari & Walczak, 2011; Li et al., 1998).

The BCL-2 protein family controls intrinsic apoptosis. These proteins share a common BH domain and include activators (BH3-only peptides) or inhibitors (BCL-2) of apoptosis (Fig. 13.6B) (Burlacu, 2003; Czabotar et al., 2014). The BCL-2 pathway acts as a "tripartite apoptotic switch" (Czabotar et al., 2014; Green, Galluzzi, & Kroemer, 2014), such that BH3-only proteins inhibit antiapoptotic proteins (such as BCL-2) and activate proapoptotic effectors (BAX and BAK). For example, the BH3-only proteins BIM, PUMA, and t-BID directly activate BAX and BAK, and the BH3-only protein BAD inhibits BCL-2 (Cheng et al., 2001; Kim et al., 2006; Kuwana et al., 2005; Vela et al., 2013).

Antiapoptotic proteins such as BCL-2, on the other hand, normally sequester and inhibit the proapoptotic BAK and BAX (Kim et al., 2006). BH3-only protein activation inhibits BCL-2, which allows BAX and BAK to oligomerize on the OMM, form pores that facilitate mitochondrial outer membrane permeabilization, and allow the release of proapoptotic factors such as cytochrome *c* (Antonsson et al., 2000; Czabotar et al., 2014; Kim et al., 2006; Ow et al., 2008; Suzuki, Youle, & Tjandra, 2000; von Ahsen et al., 2000; Vela et al., 2013). In the cytoplasm, cytochrome *c* forms an "apoptosome" complex with caspase 9 and apoptosis protease-activating factor 1 (Adrain & Martin, 2001; Jiang & Wang, 2000; Riedl & Salvesen, 2007; Zou et al., 1999). The apoptosome activates additional caspases and degrades other cytosolic proteins (Adrain & Martin, 2001).

Mitochondrial permeability transition pore (mPTP) activation also enables the release of mitochondrial prodeath proteins. mPTP formation begins with mitochondrial swelling, which typically associates with excess mitochondrial Ca^{2+} uptake with subsequent reduction of the mitochondrial membrane potential (Crompton, 1999). Classic descriptions of the mPTP emphasized structural or functional contributions from the adenine nucleotide transporter (ANT), the voltage-dependent anion channel (VDAC), and cyclophilin D. Uncertainty still exists, however, over the exact composition of the mPTP (Crompton, Virji, & Ward, 1998; Shimizu, Narita, & Tsujimoto, 1999; Tsujimoto & Shimizu, 2002). Interactions between BAX/BAK and either the VDAC or ANT are reported (Marzo et al., 1998), which suggests interactions may also occur between proapoptotic BCL-2 family proteins and the mPTP.

Proteins that facilitate apoptosis frequently interact with metabolic enzymes (Green & Kroemer, 2004; Green et al., 2014; Ow et al., 2008). For example, VDAC activates hexokinase II, which increases glycolytic ATP production (Majewski et al., 2004; Pastorino & Hoek, 2008), and both apoptosis and OXPHOS feature cytochrome *c* (Ow et al., 2008). Metabolic fluxes also modulate cell death pathways, as cells can prevent apoptosis by upregulating glyceraldehyde-3-phosphate dehydrogenase, which increases glycolysis and helps cells maintain their mitochondrial pool (Colell et al., 2007; Tait et al., 2010). Frank energy failure leads to the release of apoptotic initiating factor from mitochondria and initiates necroptosis, a caspase-independent cell death process (Green et al., 2014). Nutrient availability (in the form of glucose, lipids, and amino acids), levels of energy pathway intermediates (such as acetyl coA), and redox states (NADP + /NADPH and NAD + /NADH ratios) influence cell death pathways (Green et al., 2014).

Free Radical and Redox Biology

Free radicals are highly reactive molecules that oxidize proteins, lipids, and nucleic acids (Murphy, 2009). Free radical-related oxidative modifications increase with advancing age and were previously postulated to at least partly drive aging (Balaban, Nemoto, & Finkel, 2005), although this remains unproven.

The transfer of high-energy electrons to molecular oxygen produces ROS, which can have both beneficial and harmful effects. A majority of cell ROS are generated at complexes I and III (Cadenas et al., 1977; Lambert & Brand, 2004). When the mitochondrial membrane hyperpolarizes, NAD+ /NADH ratio falls, or if the available pool of coenzyme Q is low, high-energy electrons are increasingly likely to escape the ETC (Lambert, Buckingham, & Brand, 2008). A portion of these electrons then react with molecular oxygen to form the highly reactive superoxide anion (O_2^-) (Lambert & Brand, 2004).

At low rates of ROS generation, mitochondria can detoxify superoxide by first using a superoxide dismutase to process it to hydrogen peroxide (Weisiger & Fridovich, 1973), followed by a glutathione peroxidase- or catalase-mediated conversion to water (Day, 2009). However, when these defense mechanisms are overwhelmed, excess ROS can accumulate and cause lipid peroxidation, protein oxidation, and DNA damage.

Two periredoxins, thioredoxin 2 and glutaredoxin, can reverse protein oxidation (Zhang et al., 2014). Their reducing ability partly depends on the activity of glutathione reductase, glutathione levels, and the availability of pentose phosphate shunt-generated NADPH (Lillig & Holmgren, 2007). Recognition of protein oxidation reversibility led to the prediction that low levels of protein oxidation facilitate physiologic intracellular signaling events (D'Autreaux & Toledano, 2007).

Increased ROS production can drive the formation of reactive nitrogen species (RNS), most notably nitric oxide and peroxynitrite (Pacher, Beckman, & Liaudet, 2007). RNS modify proteins via cysteine nitrosylation and tyrosine nitration (Wiseman & Thurmond, 2012). Similar to ROS-mediated posttranslational modifications, RNS modifications transmit physiologic signals and perhaps contribute to disease.

Separate from their role in free radical biology, electrons also define pairs of reduced and oxidized molecules, such as the reduced (NADH) and oxidized (NAD+) forms of NAD. "Redox" ratios defined in this way impact cell biology by gating metabolic fluxes. For example, high mitochondrial NAD+ / NADH ratios facilitate catabolism of carbon metabolites through the pyruvate dehydrogenase complex and Krebs cycle enzymes, which increases the amount of high-energy NADH electrons that feed into OXPHOS. Low NAD+ /NADH ratios, on the other hand, impede Krebs cycle flux, promoting the mitochondrial export of malate for gluconeogenesis, or of citrate for lipid synthesis. NAD+ / NADH ratios influence glycolysis flux, with elevated ratios promoting forward motion. Enzymes of the sirtuin family monitor and respond to NAD + /NADH ratios (Finkel, Deng, & Mostoslavsky, 2009; Guarente, 2007; Haigis & Guarente, 2006; Lin & Guarente, 2003; Revollo, Grimm, & Imai, 2004; Yang & Sauve, 2006), which mark and reflect a cell's bioenergetic state.

MITOCHONDRIA IN NEURODEGENERATIVE DISEASES

Section Overview

Neurodegenerative diseases typically feature insidious onset and progressive decline. Signs and symptoms arise from nervous system dysfunction and ultimately neuron loss. Anatomically distinct patterns of neurodysfunction and neurodegeneration produce definable and ideally recognizable clinical syndromes. In practice though, syndromic and neuroanatomic overlaps occur, which limits diagnostic accuracy.

Clinical, gross pathology, and histopathology features originally defined the different neurodegenerative diseases. Recent diagnostic schemes, though, tend to emphasize molecular and genetic parameters. This occasionally affects issues of disease "lumping" and "splitting," as mutations in different genes can cause very similar clinical disorders, while a specific mutation can produce different phenotypes.

Substantial clinical and neuroanatomic diversity exists. Some diseases solely affect the central nervous system, some the peripheral nervous system, and some affect both. Age of onset varies greatly between disorders, from early-onset disorders that strike infants (e.g., Leigh's disease) to those that primarily affect geriatric populations. Nuclear gene and mtDNA mutations both cause specific neurodegenerative diseases, although many affected individuals lack specific discernable mutations, and for some diseases, sporadic cases vastly outnumber apparent Mendelian and maternally inherited cases.

Symptomatic treatments exist for some conditions and sometimes confer a substantial clinical benefit [e.g., L-DOPA in Parkinson's disease (PD)]. Disease-modifying treatments that meaningfully alter the course of an underlying disease, though, remain quite rare. For many disorders, no effective treatment exists. In the case of the known genetic diseases, lack of therapeutic progress reflects an incomplete understanding of how a particular mutation mediates dysfunction at the molecular level, how to remedy that dysfunction, or both. In the case of the sporadic diseases, uncertainty over what pathologies warrant therapeutic targeting complicates the picture. For any given neurodegenerative disease, many parameters change, and we frequently do not know whether particular changes represent true lesions, compensatory adaptations, or epiphenomena. In an attempt to circumvent this uncertainty, investigators occasionally extrapolate insights generated from studies of Mendelian presentations to sporadic presentations. Unfortunately, we generally do not know how rigorously the Mendelian variants we study model the sporadic variants we seek to better understand.

Commonalities between groups of neurodegenerative diseases may also yield insight into their underlying causes. Protein aggregations represent one such commonality, leading some to classify these disorders as specific proteinopathies. Mitochondrial perturbations constitute another common theme, and for this reason, others label these disorders "neurodegenerative mitochondriopathies" (Swerdlow, 2009). This latter perspective informs the following sections, which provide a focused overview of the neurodegenerative mitochondriopathies.

Diseases Arising Exclusively from mtDNA Mutations

Holt, Harding, and Morgan-Hughes (1988), Wallace et al. (1988), and Zeviani et al. (1988) first demonstrated in 1988 that mtDNA mutations cause human disease. The Holt et al. (1988) and Zeviani et al. (1988) work applied to myopathies that featured clear-cut mitochondrial dysfunction. The Wallace et al. work revealed a G11778A mutation in the mtDNA ND4 gene causes Leber's hereditary optic neuropathy (LHON), a neurodegenerative disorder of the optic nerve that manifests as an acquired blindness syndrome (Howell, 1998; Newman, 1993; Parker, Oley, & Parks, 1989; Wallace et al., 1988). In LHON, recognition of occasional multigenerational matrilineal kindreds directed the mutation search to mtDNA.

Subsequent studies revealed associations between various maternally inherited syndromes and mtDNA nucleotide substitutions. Many of these syndromes involve nervous system and muscle defects, and pathological studies demonstrated neuron loss (Filosto et al., 2007; Swerdlow, 2009). These "mitochondrial encephalomyopathies" include a series of descriptively named conditions including mitochondrial encephalopathy, lactic acidosis, and stroke-like episodes (MELAS) syndrome (Chomyn et al., 1992; Flierl, Reichmann, & Seibel, 1997; Goto, Nonaka, & Horai, 1990; Ihara et al., 1989; King et al., 1992); myoclonic epilepsy and ragged red fiber (MERRF) syndrome (Masucci et al., 1995; Shoffner et al., 1990; Takeda et al., 1988); and neuropathy, ataxia, and retinitis pigmentosa syndrome (Rojo et al., 2006). Other maternally inherited, mtDNA mutation-associated encephalomyopathies include maternally inherited Leigh syndrome and Kearns–Sayre syndrome (Lestienne & Ponsot, 1988; Rojo et al., 2006; Zeviani et al., 1988).

Despite the fact that mtDNA mutations cause these conditions (Dimauro, 2013), other genetic and lifestyle–environmental factors still influence them (D'Aurelio et al., 2010). For example, LHON affects more men than women, which could reflect the presence of nuclear gene or physiologic modifiers (Newman, 1993), and alcohol consumption can precipitate LHON onset.

In some cases, age of onset and degree of penetrance greatly vary. For some mutations, the likelihood of manifesting a partial phenotype exceeds the chance of manifesting a full phenotype. The A3243G MELAS mutation, for instance, more commonly presents within the context of uncomplicated diabetes mellitus than it does as the complete MELAS syndrome (Maassen, Janssen, & Hart, 2005).

The frequent association between mtDNA mutations and nervous system dysfunction begs the question of why these mutations preferentially affect the brain and nerves (Filosto et al., 2007). Proposed explanations include a unique sensitivity of neurons to energy shortages, as well as the presence of extended projections over which mitochondria must migrate. Neuronal differentiation and limited replacement potential may also contribute to tissue susceptibility. For heteroplasmic mutations, a time-related drift toward the mutant species may occur, causing mutation burdens to eventually surpass functional thresholds.

Different mutation types produce different histologic characteristics. In muscle, mtDNA synthetic gene mutations, which occur in MELAS and MERRF, give rise to subsarcolemmal mitochondrial proliferations called ragged red fibers. Structural gene mutations do not produce a similar degree of mitochondrial proliferation. This may reflect the fact that synthetic gene mutations cause deficient synthesis of respiratory chain holoenzymes (Enriquez, Chomyn, & Attardi, 1995), while structural gene mutations cause the assembly of defective respiratory chain holoenzymes.

Diseases Arising Exclusively from Nuclear DNA Mutations

For some Mendelian neurodegenerative diseases, the product of the responsible gene normally localizes to mitochondria and mediates a required mitochondrial function. Examples include Friedreich's ataxia, Wilson's disease, dominant optic atrophy, Mohr–Tranebjærg syndrome (MTS), several Charcot–Marie–Tooth (CMT) subtypes, and several hereditary spastic paraparesis subtypes.

Friedreich's ataxia arises from recessive mutations of the frataxin gene, which encodes a protein (frataxin) that enters the mitochondrial matrix and participates in mitochondrial iron homeostasis (Pandolfo, 2008). Presumably, loss-of-function frataxin mutations impair the assembly of iron–sulfur clusters and perhaps heme, thereby causing respiratory chain enzyme defects (Rotig et al., 1997; Koutnikova et al., 1997). Clinical consequences include sensory loss, weakness, and dyscoordination with onset typically occurring during the second decade.

Recessive mutations in the ATPase copper transporting beta polypeptide gene (ATP7B) cause Wilson's disease (Das & Ray, 2006). ATP7B localizes to the mitochondrial matrix and plays a role in mitochondrial copper handling (Bull et al., 1993; Lutsenko & Cooper, 1998; Tanzi et al., 1993). Tissues from affected individuals show impaired respiratory chain function, including reduced COX activity (Gu et al., 2000; Rossi et al., 2004). In addition to developing liver failure, patients not treated with copper chelation therapy acquire a hypokinetic movement disorder that reflects an underlying basal ganglia degeneration.

MTS phenotypes vary, but the classic syndrome presents as a deafness-dystonia syndrome that can also include progressive vision loss and dementia (Binder et al., 2003; Swerdlow & Wooten, 2001). Causal mutations occur in the TIMM 8A subunit, which helps mediate the mitochondrial import of nuclear-encoded mitochondrial proteins (Koehler et al., 1999; Roesch et al., 2002). A number of dominant optic atrophy subtypes exist, the most common of which involves mutation of the OPA-1 protein that facilitates mitochondrial fusion (Chun & Rizzo, 2016). The main symptom is progressive vision loss. One of the CMT disorders, CMT2A, presents as a progressive disorder of the peripheral nerves and involves mutations in the gene that encodes mitofusin 2 (Zuchner et al., 2004). Mutations in a gene called ganglioside-induced differentiation-associated protein 1 (GDAP1), a glutathione transferase, cause a different form of CMT called CMT4A (Niemann et al., 2005; Pedrola et al., 2005). GDAP1 mostly localizes to the OMM, where it appears to participate in regulating oxidative stress as well as mitochondrial fission-fusion events. Some hereditary spastic paraparesis gene mutations, including those in the Hsp60 and paraplegin genes, alter proteins that function within mitochondria and specifically help manage organelle proteostasis (Casari et al., 1998; Hansen et al., 2002).

For other Mendelian neurodegenerative diseases, mutant protein-mitochondrial associations are less clear, but studies of affected tissues or disease models consistently reveal mitochondrial dysfunction. Huntington's disease (HD) represents a good example of this. HD, an autosomal-dominant, hyperkinetic movement disorder arises from a polyglutamine repeat expansion of the huntingtin (HTT) gene (MacDonald et al., 1993; Vonsattel & DiFiglia, 1998). Pervasive ETC defects occur across multiple tissues of affected patients (Arenas et al., 1998; Browne & Beal, 2004; Gu et al., 1996; Parker et al., 1990). Further, the complex II inhibitors malonate and 3-nitroproprionic acid generate HD-like damage in rodents (Beal et al., 1993; Greene et al., 1993). The mutant HTT gene product, polyglutamine-expanded HTT, associates with the OMM and interferes with mitochondrial calcium handling (Panov et al., 2002). Mutant HTT also reportedly interferes with PGC-1α, and through this interaction impairs mitochondrial biogenesis (Cui et al., 2006).

Alzheimer's Disease

Alzheimer's disease (AD) presents as a progressive dementia and represents the most common neurodegenerative disease (Swerdlow, 2007b). Classically emphasized AD histopathologies include extracellular fibril aggregates enriched in β-amyloid (Aβ) protein (plaques), and intracellular aggregates consisting of tau microtubule-associated protein (tangles).

The vast majority of affected individuals, over 99%, present in an ostensibly sporadic fashion and without a clearly demonstrable Mendelian inheritance pattern (Swerdlow, 2007c). Despite this, genes do influence sporadic AD risk (Guerreiro et al., 2013; Harold et al., 2009; Jonsson et al., 2013; Jun et al., 2010; Naj et al., 2011; Rogaeva et al., 2007; Seshadri et al., 2010). Overall though, advancing age represents the greatest risk factor, and prevalence and incidence rise sharply with advancing age (Hebert et al., 2003; Jorm & Jolley, 1998; Polvikoski et al., 2001; Yaffe et al., 2011). The term "late onset AD" (LOAD) generally applies to non-Mendelian cases.

Very rare instances of Mendelian, autosomal-dominant, "familial AD" (FAD) also occur (Ryman et al., 2014). Documented autosomal-dominant families number in hundreds. Mutations in three known genes, amyloid precursor protein (APP), presenilin 1 (PSEN1), and presenilin 2, determine autosomal-dominant FAD (Goate et al., 1991; Levy-Lahad et al., 1995; Sherrington et al., 1995). Age of onset varies depending on the specific mutation, although most manifest symptoms in the fifth or sixth decade.

Fluorodeoxyglucose positron emission tomography (FDG PET) and mitochondria studies demonstrate perturbed AD energy metabolism (Swerdlow, 2012). FDG PET reveals focal reductions in brain glucose utilization with advancing age (De Santi et al., 1995; Marano et al., 2013; Nugent et al., 2014; Willis et al., 2002), which extend in magnitude and distribution with AD (de Leon et al., 1983; Ferris et al., 1980; Foster et al., 1983; Friedland et al., 1983; Mosconi et al., 2009a). Asymptomatic individuals with an APOE4 allele or AD-affected mother, who have an increased AD risk, show AD-like glucose utilization patterns (Mosconi et al., 2007; Reiman et al., 1996; Small et al., 1995). Neuronal loss, synaptic degradation, or bioenergetic changes could contribute to reduced AD brain glucose utilization, although changes in pre- or asymptomatic subjects suggest bioenergetic changes are indeed a contributor. In support of this notion, AD subject brain homogenates show less glucose consumption than control homogenates (Swerdlow et al., 1994), even though homogenization disrupts synapses and neutralizes their potential impact.

O_2 PET and homogenate studies show altered AD brain oxygen consumption, which likely reflects compromised mitochondrial capacity or integrity (Frackowiak et al., 1981; Fukuyama et al., 1994; Sims et al., 1987). Electron microscopy reveals perturbed mitochondrial structures (Baloyannis, 2006; Hirai et al., 2001; Johnson & Blum, 1970; Wisniewski, Terry, & Hirano, 1970) and mitochondrial fission increases (Manczak, Calkins, & Reddy, 2011; Wang et al., 2009). Several mitochondria-localized enzymes, including pyruvate dehydrogenase complex (Sorbi, Bird, & Blass, 1983), α-ketoglutarate dehydrogenase complex (Gibson et al., 1988), and COX, demonstrate reduced activities (Bosetti et al., 2002; Cardoso et al., 2004a; Curti et al., 1997; Fisar et al., 2016; Kish et al., 1992; Mancuso et al., 2003; Maurer, Zierz, & Moller, 2000; Mutisya, Bowling, & Beal, 1994; Parker, Filley, & Parks, 1990; Parker et al., 1994a, 1994b; Parker & Parks, 1995; Simonian & Hyman, 1993; Valla, Berndt, & Gonzalez-Lima, 2001; Valla et al., 2006; Wong-Riley et al., 1997). The AD COX defect, interestingly, also exists outside the brain (Bosetti et al., 2002; Cardoso et al., 2004a; Curti et al., 1997; Fisar et al., 2016; Mancuso et al., 2003; Parker et al., 1990, 1994a; Valla et al., 2006) and at least partly depends on mtDNA (Swerdlow, 2012; Swerdlow et al., 1997). mtDNA alterations include reduced PCR-amplifiable levels, despite an apparent increase in autophagosome-deposited mtDNA (de la Monte et al., 2000; Hirai et al., 2001). Increased oxidative modifications (Lovell & Markesbery, 2007; Bradley-Whitman et al., 2014; Mecocci, MacGarvey, & Beal, 1994), increased deletions (Corral-Debrinski et al., 1994; Hamblet & Castora, 1997; Krishnan et al., 2012; Phillips, Simpkins, & Roby, 2014), and possibly increased somatic point mutations (Chang et al., 2000; Coskun, Beal, & Wallace, 2004; Coskun et al., 2010) also occur. Inherited mtDNA signatures reportedly associate with and influence AD risk (Fesahat et al., 2007; Lakatos et al., 2010; Maruszak et al., 2009;

Santoro et al., 2010; van der Walt et al., 2004; Wang & Brinton, 2016), while epidemiologic and biomarker-based endophenotype studies show maternal inheritance bias (Andrawis et al., 2012; Berti et al., 2011; Debette et al., 2009; Duara et al., 1993; Edland et al., 1996; Honea et al., 2010, 2011, 2012; Liu et al., 2013; Mosconi et al., 2007, 2009b, 2010a, 2010b, 2010c, 2011; Okonkwo et al., 2014; Reiter et al., 2012).

Studies report links between mitochondria and AD-associated proteins. The most studied LOAD risk gene, APOE, encodes the apolipoprotein E protein, which primarily plays a role in lipid transport (Corder et al., 1993; Strittmatter et al., 1993). One APOE isoform, APOE4, appears to move up the age of AD onset and thereby increase lifetime risk (Blacker & Tanzi, 1998). Among the hypotheses that attempt to explain this association, one emphasizes that apolipoprotein E4 protein undergoes a unique proteolytic cut that reveals a mitochondrial targeting sequence. The resulting peptide derivative localizes to mitochondria and interferes with mitochondrial function (Chang et al., 2005; Chen et al., 2011; Mahley & Huang, 2012). Variants in an APOE-adjacent gene, the TOMM40 gene, also associate with AD risk (Roses, 2010; Roses et al., 2010). While mitochondrial dysfunction in AD provides an attractive rationale for how this gene and its product could influence risk, the exact contribution of TOMM40 remains uncertain, since various TOMM40 and APOE variants exist in linkage disequilibrium (Roses et al., 2013).

Studies report the tau and PSEN1 proteins both localize to mitochondria (Choi et al., 2014; Hansson et al., 2004). APP and Aβ protein colocalize with mitochondria and can disrupt their function (Anandatheerthavarada et al., 2003; Anandatheerthavarada & Devi, 2007; Caspersen et al., 2005; Devi et al., 2006; Hansson Petersen et al., 2008; Lustbader et al., 2004; Manczak et al., 2006).

Conversely, mitochondrial function and cell bioenergetics influence APP processing and Aβ production (Gabuzda et al., 1994; Gasparini et al., 1997; Khan et al., 2000; Onyango et al., 2010; Webster et al., 1998). Rodent studies suggest increased synaptic activity, which consumes energy supplies and increases mitochondrial respiration, can increase brain Aβ levels (Bero et al., 2011; Kang et al., 2009; Roh et al., 2014; Yamamoto et al., 2015). Decreasing respiratory infrastructure, on the other hand, reduces brain Aβ levels (Fukui et al., 2007; Pinto et al., 2013). In mice that express a human mutant APP transgene, introducing mtDNA sequence changes profoundly affects brain amyloidosis (Scheffler et al., 2012). In APP transgenic mice, acquisition of mtDNA mutations accelerates Aβ plaque deposition (Kukreja et al., 2014). Mitochondrial function also influences tau protein status (Blass et al., 1990, 1991; Escobar-Khondiker et al., 2007; Hoglinger et al., 2005; Rottscholl et al., 2016; Szabados et al., 2004; Yamada et al., 2014).

Exactly where mitochondria sit on the AD pathology spectrum and timeline remains unclear. An epiphenomenal or functionally inconsequential role seems less and less likely as investigators increasingly link mitochondria and bioenergetics to the classic AD pathologies and AD-implicated proteins.

The fact that some AD-associated mitochondria and energy metabolism changes also manifest to lesser degrees in normal aging, and are not brain-limited, suggest mitochondria may play a relatively upstream role in this disease (Swerdlow et al., 2017). Specifically, AD patient platelets and fibroblasts show reduced COX activity, which suggests a possible inherited genetic basis (Bosetti et al., 2002; Cardoso et al., 2004a; Curti et al., 1997; Fisar et al., 2016; Mancuso et al., 2003; Parker et al., 1990, 1994a; Valla et al., 2006).

Because COX contains mtDNA-encoded subunits, investigators previously used a cytoplasmic hybrid approach to test whether mtDNA might contribute to low AD COX activity (Swerdlow, 2012). In these experiments, investigators first removed the endogenous mtDNA from neuronal cell lines to create neuronal "ρ0" cells. Next, they acquired platelet mitochondria from individuals with AD and from age-matched non-AD subjects. Transferring the platelet mitochondria to the ρ0 cells generated cytoplasmic hybrid cells, or cybrid cells, that after a prolonged selection phase differed only in the origin of their mtDNA (Fig. 13.7).

Since cybrid cells created by this approach contain identical nuclei, biochemical differences between cybrid lines generated from different mtDNA donors presumably reflect differences in their mtDNA sequence (Swerdlow, 2007a). In these experiments, relative to cybrid lines that contained mtDNA from control subjects, cybrid lines containing mtDNA from AD subjects showed lower COX activity (Swerdlow et al., 1997). Cybrid studies further demonstrated other molecular phenomena that recapitulate pathologies observed in AD brains, including oxidative stress, abnormal mitochondrial morphologies and movement, reduced oxygen consumption and glucose utilization, reduced mitochondrial membrane potential, and Aβ accumulations (Bijur, Davis, & Jope, 1999; Cardoso et al., 2004b; Costa et al., 2012; Cassarino et al., 1998; Davis et al., 1997; De Sarno et al., 2000; Gan et al., 2014a, 2014b; Ghosh et al., 1999; Jeong et al., 2015; Khan et al., 2000; Onyango, Tuttle, & Bennett, 2005b; Onyango, Bennett, & Tuttle, 2005d; Silva et al., 2013a; Onyango et al., 2010;

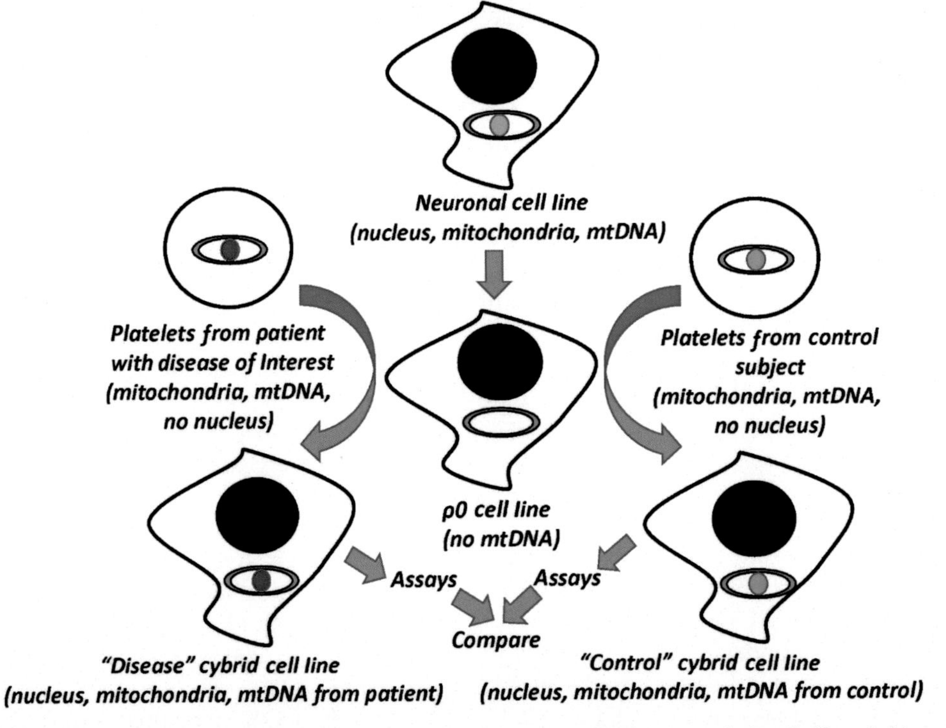

FIGURE 13.7 **Cybrids.** A neuronal cell line has its endogenous mtDNA removed (small *blue circle*) to create a respiration-incompetent ρ0 cell line. Next, study subjects provide platelets, which lack nuclei but contain mitochondria and mtDNA. Mixing these platelet "cytoplasts" with ρ0 cells allows for the transfer of platelet mitochondria and mtDNA to the ρ0 cells, which restores their mtDNA and respiration ability. The various cybrid lines thus formed differ only in the origin and content of their mtDNA, so functional differences between the cell lines reflect differences in their mtDNA sequence. *mtDNA*, mitochondrial DNA.

Sheehan et al., 1997b; Silva et al., 2013b; Swerdlow et al., 1997; Thiffault & Bennett, 2005; Trimmer et al., 2000, 2004a; Trimmer & Borland, 2005; Yu et al., 2016; Zhang et al., 2011). Partly based on cybrid experiments, Swerdlow and Khan proposed the sporadic AD "mitochondrial cascade hypothesis" (Swerdlow, Burns, & Khan, 2010, 2014; Swerdlow & Khan, 2004, 2009), in which mitochondrial dysfunction represents the primary molecular event that initiates and perhaps drives AD neurodysfunction, neurodegeneration, and histopathology changes.

Parkinson's Disease

PD neurodegeneration initially features a loss of dopaminergic substantia nigra neurons (Swerdlow, 2011). As this neuron population innervates the striatum to help regulate movement, the disease presents clinically as a progressive hypokinetic movement disorder. The age of onset varies, but advancing age nevertheless represents a major risk factor. The classic histologic change includes intracellular protein aggregations called Lewy bodies, which primarily consist of insoluble, fibrillar α-synuclein protein.

Similar to AD, PD presents within Mendelian and sporadic contexts (Tanner & Goldman, 1996). The prevalence of the more common sporadic forms increases with advancing age. The Mendelian versions demonstrate substantial genetic heterogeneity, include both autosomal-dominant and recessive variants, and occasionally feature a relatively early age of onset.

In the 1980s, a chemical contaminant of a synthetic opioid was found to destroy substantia nigra dopaminergic neurons and produce a subacute PD-like syndrome (Langston et al., 1983). Following ingestion, this contaminant, 1-methyl-4-phenyl-1,2,3,6-tetrahydropyridine, undergoes conversion to 1-methyl-4-phenylpyridinium, which concentrates within mitochondria and inhibits complex I of the respiratory chain (Nicklas, Vyas, & Heikkila, 1985). Rats systemically treated with rotenone, another complex I inhibitor, also experience a nigrostriatal-focused loss of dopamine neurons and develop Lewy body accumulations (Betarbet et al., 2000). For unclear reasons, therefore, substantia nigra dopaminergic cells are particularly sensitive to complex I inhibition (Parker & Swerdlow, 1998). Potential explanations for this should presumably reflect characteristics that uniquely define this cell type, such as their exceptionally high baseline and age-related accumulation of mtDNA deletions and point mutations (Kraytsberg et al., 2006; Simon et al., 2004).

Working from a hypothesis that proposed mtDNA may contribute to sporadic neurodegenerative diseases by affecting respiratory chain function, Parker et al. measured ETC enzyme activities in platelet mitochondria from PD and control individuals and found a selective reduction in complex I activity in the PD subjects (Parker, Boyson, & Parks, 1989). In that same year, 1989, investigators reported similar complex I activity or protein subunit reductions in PD subject brains and muscle (Bindoff et al., 1989; Mizuno et al., 1989; Schapira et al., 1989). Other studies confirmed these observations and extended them to PD subject lymphocytes and fibroblasts (Barroso et al., 1993; Benecke, Strumper, & Weiss, 1993; Bindoff et al., 1991; Blin et al., 1994; Cardellach et al., 1993; Haas et al., 1995; Krige et al., 1992; Mytilineou et al., 1994; Nakagawa-Hattori et al., 1992; Parker, Parks, & Swerdlow, 2008; Shoffner et al., 1991; Yoshino et al., 1992).

Investigators subsequently used a cybrid approach to test whether mtDNA contributes to this apparently systemic complex I defect (Swerdlow et al., 1996). Multiple studies comparing cybrid cell lines generated from PD subjects with cybrid cell lines generated from control subjects report biochemical and molecular differences. Relative to control cybrid lines, PD cybrid lines show reduced complex I activity, a decreased maximum respiratory capacity, increased oxidative stress, perturbed calcium homeostasis, mitochondrial membrane potential depolarization, altered mitochondrial morphologies, decreased mitochondrial movement, and α-synuclein aggregations (Borland et al., 2009; Cassarino et al., 1997, 2000; Esteves et al., 2008a, 2008b, 2010a, 2010b, 2010c; Ghosh et al., 1999; Gu et al., 1998; Keeney et al., 2009; Onyango, Tuttle, & Bennett, 2005a, 2005c; Sheehan et al., 1997a; Shults & Miller, 1998; Swerdlow et al., 1996, 1998; Trimmer & Bennett, 2009; Trimmer et al., 2000, 2004b, 2009; Veech et al., 2000).

The results of these cybrid studies collectively implicate a role for mtDNA in PD. In support of this, association studies report mtDNA haplogroups influence PD risk (Autere et al., 2004; Ghezzi et al., 2005; Pyle et al., 2005; van der Walt et al., 2003). Some studies claim associations between single nucleotide mtDNA variants, although replication across cohorts does not uniformly occur (Kirchner et al., 2000; Mayr-Wohlfart, Rodel, & Henneberg, 1997; Ross et al., 2003; Shoffner et al., 1993; Simon et al., 2000, 2010; Swerdlow et al., 2006; van der Walt et al., 2003). Neurons from PD patients also show, relative to control subjects, an excess presence of heteroplasmic mtDNA deletion and point mutations that could represent either somatic or inherited variants (Bender et al., 2006b; Lin et al., 2012; Smigrodzki, Parks, & Parker, 2004; Parker & Parks, 2005).

The Mendelian forms also demonstrate mitochondrial links (Tan & Skipper, 2007). The first PD-implicated nuclear gene, SNCA, encodes α-synuclein, which colocalizes with mitochondria and can perturb mitochondrial function (Bir et al., 2014; Liu et al., 2009; Luth et al., 2014; Martin et al., 2014; Mullin & Schapira, 2013; Reeve et al., 2015; Sarafian et al., 2013; Zhang et al., 2008). Mutations in the leucine-rich repeat kinase 2 (LRRK2) gene cause an autosomal-dominant PD variant that occurs in particularly high frequencies in Ashkenazi Jewish and some North African PD cohorts and may further function as a sporadic PD susceptibility gene (Esteves, Swerdlow, & Cardoso, 2014; Lesage et al., 2005; Ozelius et al., 2006; Paisan-Ruiz et al., 2004; Satake et al., 2009; Simon-Sanchez et al., 2009; Zabetian et al., 2006; Zimprich et al., 2004). Its gene product, LRRK2 protein, colocalizes with the OMM. The mechanisms that mediate the association remain uncertain, and could involve changes in the protein's kinase function, other functions such as its GTPase capacity, or protein levels (Esteves et al., 2014). Reports indicate mutations influence mitochondria-related autophagy or apoptosis physiology (Iaccarino et al., 2007; Saez-Atienzar et al., 2014). Mutations in the gene that encodes the DJ-1 protein cause autosomal-recessive PD (Bonifati et al., 2003). The underlying mechanisms remain unclear, although studies suggest a mitochondrial localization and perhaps an antioxidant function (Bonifati et al., 2003; Canet-Aviles et al., 2004; Taira et al., 2004). Some claim mutations in the gene that encodes the mtDNA polymerase gamma may predispose to PD, presumably by accelerating an accumulation of somatic mtDNA mutations (Luoma et al., 2007).

Mutations in the genes for PINK1 and parkin appear to disrupt mitophagy and through this cause autosomal-recessive PD (Kitada et al., 1998; Valente et al., 2004). The PINK1 gene product, PINK1 protein, targets mitochondria. In the setting of a normal mitochondrial membrane potential, PINK1 enters the matrix, where internal peptidases degrade it. Membrane potential depolarization interferes with this mitochondrial internalization, so that a PINK1 tail projects from the depolarized mitochondrion and into the cytoplasm. This PINK1 tail recruits parkin, which tags the mitochondrion for mitophagy-mediated disposal (see Fig. 13.4) (Vives-Bauza et al., 2010). Mutations in either the PINK1 or parkin genes, therefore, impair mitochondrial quality control, which over time leads to PD (Tan & Skipper, 2007; Youle & Narendra, 2011).

Well-recognized relationships between mitochondria and Mendelian PD variants, especially those revealed through studies of parkin and PINK1, support the view that mitochondria and their dysfunction feature prominently in PD. The Mendelian forms further provide insight into the sporadic cases. Specifically, impaired mitochondria accumulate when their number exceeds the neuron's capacity to remove them. This occurs due to a breakdown of the infrastructure that removes impaired mitochondria, or if the neuron mitochondrial pool fails to such a degree that this infrastructure is simply overwhelmed. These two scenarios could overlap, with pathologies in the mitochondrial pool inducing pathologies of the Mendelian PD-associated proteins. For example, studies show complex I dysfunction promotes α-synuclein aggregation (Lee et al., 2002; Sherer et al., 2002, 2003; Trimmer et al., 2004b).

Amyotrophic Lateral Sclerosis

Amyotrophic lateral sclerosis (ALS), a progressive disorder of weakness that arises from the loss of central and peripheral motor neurons, presents within familial and sporadic contexts (Swerdlow et al., 2000). The less common familial types comprise 5%–10% of the total cases, with approximately 20% of these resulting from mutations in the superoxide dismutase 1 (SOD1) gene. One extensively studied SOD1 mutation, the G93A mutation, causes the predominantly cytosolic protein to colocalize with mitochondria, where it causes profound mitochondrial dysfunction (Higgins, Jung, & Xu, 2003). In G93A SOD1 ALS rodent models, mitochondrial pathologies precede motor decline. Noted consequences include mitochondrial membrane potential depolarization, organelle swelling, reduced ATP production, disrupted calcium homeostasis, ETC impairments, BCL-2 and VDAC dysfunction, spinal cord mitochondrial vacuolization, reduced axon mitochondria transport rates, and overall reduced axon mitochondria content (Carri et al., 1997; De Vos et al., 2007; Ferri et al., 2006; Israelson et al., 2010; Jaarsma et al., 2000; Jung, Higgins, & Xu, 2002a; Jung, Higgins, & Xu, 2002b; Kirkinezos et al., 2005; Kruman et al., 1999; Pasinelli et al., 2004; Sasaki et al., 2004; Wong et al., 1995).

Mutations in the TARDBP gene that encodes the transactive response DNA-binding protein 43 kDa (TAR DNA-binding protein 43; TDP43) also cause familial ALS. Interestingly, wild-type and A315T-mutant TDP43 both localize to mitochondria and influence mitochondrial function (Lu et al., 2012; Wang et al., 2013). A315T TDP43 transgenic mice show deficient axon mitochondrial transport and altered mitochondrial morphology (Magrane et al., 2014).

In sporadic ALS, dorsal root ganglion cells contain increased numbers of abnormal mitochondrial inclusions, spinal cord anterior horn cells exhibit excessive mitochondrial aggregation and variable morphologies, and motor neurons demonstrate reduced COX activity (Borthwick et al., 1999; Sasaki, Horie, & Iwata, 2007; Sasaki & Iwata, 2007). ALS spinal cord shows less overall mtDNA, increased mtDNA mutations, and ETC lesions (Wiedemann et al., 2002). Muscle biopsies from ALS subjects also manifest mitochondrial dysfunction in the form of increased COX-negative fibers, increased mtDNA deletions, an overall reduced mtDNA content, ETC enzyme defects, and altered mitochondrial morphology (Afifi et al., 1966; Chung & Suh, 2002; Crugnola et al., 2010; Krasnianski et al., 2005; Vielhaber et al., 2000; Wiedemann et al., 1998).

Following exposure to a chemical uncoupling agent, which induces a maximum respiratory rate in mitochondria, lymphocytes from ALS subjects increase their oxygen consumption to a lesser extent than lymphocytes from control subjects (Curti et al., 1996). Hepatocytes from ALS subjects contain excessive amounts of abnormal mitochondrial structures, including swollen mitochondria and intramitochondrial inclusions (Nakano, Hirayama, & Terao, 1987). Platelets from ALS subjects show relatively depolarized mitochondria and increased apoptosis activity (Shrivastava et al., 2011). mtDNA haplogroups may also influence ALS risk (Mancuso et al., 2004). Observations such as these suggest systemic mitochondrial pathology occur in sporadic ALS patients.

In further support of this possibility, mitochondrial dysfunction appears to perpetuate in cultured cells that receive platelet mtDNA from ALS subjects. In one study of cybrid cell lines generated from ALS and control subjects, the cybrid lines from the ALS subjects demonstrated a trend toward lower complex III and COX activities, a significant reduction in complex I activity, evidence of increased oxidative stress, perturbed calcium homeostasis, and an increased proportion of morphologically altered mitochondria (Swerdlow et al., 1998).

THERAPEUTIC TARGETING OF MITOCHONDRIA

For mitochondrial disorders that arise from mitochondrial gene mutations, mtDNA represents a particularly attractive target. Mitochondrial gene therapy presents unique challenges but also opportunities. One potentially achievable objective includes the manipulation of heteroplasmic ratios. To this end, investigators previously explored using peptide nucleic acids to selectively block replication of a heteroplasmic mutant mtDNA species, while sparing replication of the wild-type sequence (Taylor et al., 1997). Transgenic expression of mitochondria-targeted restriction enzymes that selectively degrade particular mutation-containing mtDNA molecules, or through the introduction of transcription activator-like effector nucleases, offer other options (Alexeyev et al., 2008; Bacman et al., 2010, 2012; Reddy et al., 2015). Less specific strategies for addressing heteroplasmic mutations include leveraging rapamycin to promote mitophagy, which lets cells use their intrinsic mitochondrial quality control infrastructure to preferentially remove damaged mitochondria and enrich for wild-type mtDNA (Dai et al., 2014).

From a practical standpoint, mtDNA gene therapy still faces considerable hurdles. mtDNA manipulations designed not to treat, but rather to prevent the transmission of mtDNA disorders, though, seem poised to reach the clinic. New mitochondrial replacement therapy approaches, also referred to as "mitochondrial donation," allow the transfer of a nucleus from an oocyte or zygote that contains mutant mtDNA to an oocyte that does not (Herbert & Turnbull, 2017; Wolf, Mitalipov, & Mitalipov, 2015). Children born through these procedures will contain genetic material from three individuals, including two who provide nuclear genes and one who provides the mitochondrial genes.

Preclinical therapeutic development focuses on an array of fundamental mitochondrial processes. Examples include the use of drugs that influence mitochondrial fission and fusion (Gan et al., 2014; Reddy, 2014), mitochondria-targeted antioxidants (Kelso et al., 2001), drugs that promote mitochondrial biogenesis (Ghosh et al., 2007; Tadaishi et al., 2011; Wenz et al., 2008, 2010), drugs that alter redox couple ratios (Bai et al., 2011; Wilkins et al., 2014, 2016; Yang & Sauve, 2016), modifying proteins that monitor redox ratios (Lagouge et al., 2006; Pearson et al., 2008), and drugs that inhibit specific protein–mitochondria interactions (Yao et al., 2011b).

Initial attempts at clinical mitochondrial medicine focused on energy production and oxidative stress in patients with mitochondrial encephalomyopathies. Interventions included the administration of mature electron carriers such as coenzyme Q or its water-soluble analog, idebenone (Ihara et al., 1989; Ikejiri et al., 1996); electron coenzyme precursors including riboflavin and nicotinamide (Penn et al., 1992); electron donors such as vitamins C and K3 (Eleff et al., 1984); dichloroacetate, which helps activate pyruvate dehydrogenase complex (Kaufmann et al., 2006; Saitoh et al., 1998; Stacpoole et al., 2008); and creatine, which stores energy in the form of high-energy phosphate bonds (Komura et al., 2003). While occasional anecdotal reports or small case series reported benefits, controlled clinical trials, when performed, did not support such claims (Kaufmann et al., 2006). Despite this, clinical data point to occasional, if minor, demonstrable benefits in other conditions. These include a study that assessed the value of riboflavin in migraine prevention (Schoenen, Jacquy, & Lenaerts, 1998), and studies that evaluated the impact of idebenone on Friedreich's ataxia (Di Prospero et al., 2007; Mariotti et al., 2003).

Manipulations of mitochondria or mitochondria-related phenomena in other neurodegenerative disorders similarly include antioxidants (Bozik et al., 2014; Cudkowicz et al., 2011; Parkinson Study Group, 1993), ETC coenzyme supplements (Cooper & Schapira, 2007; Feigin et al., 1996; Mancuso et al., 2010; Shults et al., 2002; Storch et al., 2007; Stamelou et al., 2008), supplements to help store high-energy phosphates (Bender et al., 2006a; Shefner et al., 2004; Verbessem et al., 2003), supplements to increase levels of respiration substrates (Henderson et al., 2009; Reger et al., 2004), and drugs that promote autophagy (Gordon et al., 2007). Reported results include no effect, slightly accelerated decline, and perhaps slight improvements or slowing of decline.

In AD, two studies suggested vitamin E in high doses might slightly slow decline rates (Dysken et al., 2014; Sano et al., 1997). Other AD clinical trials that intentionally targeted energy metabolism or related phenomena evaluated coenzyme Q and idebenone (either to enhance electron transfer through the ETC or reduce oxidative stress) (Galasko et al., 2012; Gutzmann & Hadler, 1998; Thal et al., 2003; Weyer et al., 1997), caprylic triglyceride (to promote ketone body synthesis) (Henderson et al., 2009; Reger et al., 2004), rosiglitazone (to promote mitochondrial biogenesis) (Gold et al., 2010; Risner et al., 2006), and latrepirdine (to "stabilize" mitochondrial membranes) (Bachurin et al., 2003; Bezprozvanny, 2010; Zhang et al., 2010). These interventions showed little or no benefit, or their major trials were inconclusive, which could reflect no physiologic engagement of their intended targets (coenzyme Q, idebenone), concentrations or exposure durations insufficient to robustly alter or engage their targets (rosiglitazone, caprylic triglyceride), or a profoundly limited understanding of their mitochondrial effects (latrepirdine). This experience underscores why rational design of mitochondrial medicine must consider mechanistic data and pursue new mechanistic insight (Swerdlow, 2014).

Currently, investigators are working to develop a new generation of respiration-promoting compounds with distinct pharmacokinetic and pharmacodynamic properties (Ehinger et al., 2016; Rodan et al., 2015; Zhang et al., 2015). A relatively new strategy, "bioenergetic medicine" (Swerdlow, 2014, 2016), uses redox manipulations and the law of mass action to promote bioenergetic fluxes, and bioenergetic medicine therapies are currently entering early stage clinical trials (Swerdlow et al., 2016).

In preclinical studies, two nonpharmacologic interventions, exercise and diet, alter brain mitochondria (Bough et al., 2006; Dorsey et al., 2013; Kukreja et al., 2014; Mohamed et al., 2010; Navarro et al., 2004; Noh et al., 2004; Selfridge et al., 2015; Sleiman et al., 2016; Steiner et al., 2011; Yao et al., 2011a). Changes to insulin levels, insulin sensitivity, brain respiration substrates, neurotrophic factors, or inflammatory cytokines could mediate these effects. To date, human clinical trials of exercise or bioenergetics-altering diets, such as the ketogenic diet, suggest these interventions could confer benefits and further studies to address this possibility are underway (Castellano et al., 2017; Krikorian et al., 2012; Morris et al., 2017; Vanitallie et al., 2005; Witte et al., 2009). For these nonpharmacologic interventions, even small benefits would provide critical proof-of-principle support for mitochondrial therapies, and justify the development of more potent and convenient pharmacologic mimetics.

CONCLUSIONS

Mitochondria increasingly appear to play a role in a variety, and perhaps majority, of neurodegenerative diseases (Swerdlow, 2009). A limited nervous system ability to tolerate or adapt to mitochondrial dysfunction could account for this. Extensive energy demands, limited bioenergetic flux options for different cell types, the complexities of coordinating bioenergetic interplay among different cell types, physical structures that include extended projections, and prolonged cell life spans that facilitate an accumulation of damage could all enhance susceptibility.

For some neurodegenerative diseases mitochondria clearly initiate pathology, for some they clearly mediate pathology induced by more upstream factors, and for some their hierarchical position remains unclear. When inherited mtDNA mutations segregate with a disorder, a causal position seems most likely. Associations with nuclear gene mutations that encode critical mitochondrial proteins also suggest an upstream position.

Many disorders, though, are not straightforward. Some neurodegenerative disease-implicated nuclear genes encode proteins that perform a variety of cell functions. Mutations in these genes can generate unique and sometimes harmful mitochondrial interactions, and changes in mitochondrial function or quality may themselves create new mitochondria-protein relationships.

For the latter situation, age-related mitochondrial changes may prove particularly relevant. The fact that age-related mitochondrial changes do in fact occur (Swerdlow et al., 2017) raises the question of whether combinations of mtDNA and nuclear DNA signatures interact to define an individual's lifetime risk of developing a particular disease, the age at which it manifests, and its rate of progression. This principle may particularly apply to the late-onset, sporadic neurodegenerative diseases including sporadic AD, PD, and ALS.

Regardless of whether mitochondria initiate, mediate, or exacerbate neuronal dysfunction or degeneration, mitochondria represent valid therapeutic targets. What types of manipulations are necessary, and how to achieve those manipulations remains unclear, and different diseases may require quite different interventions. Moving forward, we especially need to keep in mind that only some of the biochemical and molecular changes we observe in the neurodegenerative diseases represent true pathologies. Most changes probably reflect compensatory adaptations or attempts at adaptation. An advancing appreciation of fundamental mitochondrial biology and physiology will ideally address how to distinguish adaptation from true pathology, indicate the exact parameters we need to manipulate, and how to achieve the necessary manipulations.

Acknowledgment

The study is supported in part by the University of Kansas Alzheimer's Disease Center (P30 AG035982).

References

Adrain, C., & Martin, S. J. (2001). The mitochondrial apoptosome: A killer unleashed by the cytochrome seas. *Trends in Biochemical Sciences*, *26*(6), 390–397.

Afifi, A. K., et al. (1966). Ultrastructure of atrophic muscle in amyotrophic lateral sclerosis. *Neurology*, *16*(5), 475–481.

Akimoto, T., et al. (2005). Exercise stimulates Pgc-1alpha transcription in skeletal muscle through activation of the p38 MAPK pathway. *Journal of Biological Chemistry*, *280*(20), 19587–19593.

Aldridge, J. E., Horibe, T., & Hoogenraad, N. J. (2007). Discovery of genes activated by the mitochondrial unfolded protein response (mtUPR) and cognate promoter elements. *PLoS ONE*, *2*(9), e874.

Alexander, C., et al. (2000). OPA1, encoding a dynamin-related GTPase, is mutated in autosomal dominant optic atrophy linked to chromosome 3q28. *Nature Genetics*, *26*(2), 211–215.

Alexeyev, M. F., et al. (2008). Selective elimination of mutant mitochondrial genomes as therapeutic strategy for the treatment of NARP and MILS syndromes. *Gene Therapy*, *15*(7), 516–523.

Amchenkova, A. A., et al. (1988). Coupling membranes as energy-transmitting cables. I. Filamentous mitochondria in fibroblasts and mitochondrial clusters in cardiomyocytes. *The Journal of Cell Biology*, *107*(2), 481–495.

Anandatheerthavarada, H. K., & Devi, L. (2007). Amyloid precursor protein and mitochondrial dysfunction in Alzheimer's disease. *Neuroscientist*, *13*(6), 626–638.

Anandatheerthavarada, H. K., et al. (2003). Mitochondrial targeting and a novel transmembrane arrest of Alzheimer's amyloid precursor protein impairs mitochondrial function in neuronal cells. *Journal of Cell Biology*, *161*(1), 41–54.

Anderson, S., et al. (1981). Sequence and organization of the human mitochondrial genome. *Nature*, *290*(5806), 457–465.

Andrawis, J. P., et al. (2012). Effects of ApoE4 and maternal history of dementia on hippocampal atrophy. *Neurobiology of Aging*, *33*(5), 856–866.

Andrews, R. M., et al. (1999). Reanalysis and revision of the Cambridge reference sequence for human mitochondrial DNA. *Nature Genetics*, *23*(2), 147.

Antonsson, B., et al. (2000). Bax oligomerization is required for channel-forming activity in liposomes and to trigger cytochrome c release from mitochondria. *Biochemical Journal*, *345*(Pt 2), 271–278.

Appelmans, F., Wattiaux, R., & De Duve, C. (1955). Tissue fractionation studies. 5. The association of acid phosphatase with a special class of cytoplasmic granules in rat liver. *Biochemical Journal*, *59*(3), 438–445.

Arenas, J., et al. (1998). Complex I defect in muscle from patients with Huntington's disease. *Annals of Neurology*, *43*(3), 397–400.

Autere, J., et al. (2004). Mitochondrial DNA polymorphisms as risk factors for Parkinson's disease and Parkinson's disease dementia. *Human Genetics*, *115*(1), 29–35.

Bachurin, S. O., et al. (2003). Mitochondria as a target for neurotoxins and neuroprotective agents. *Annals of the New York Academy of Sciences*, *993*, 334–344, discussion345-9.

Bacman, S. R., et al. (2010). Organ-specific shifts in mtDNA heteroplasmy following systemic delivery of a mitochondria-targeted restriction endonuclease. *Gene Therapy*, *17*(6), 713–720.

Bacman, S. R., et al. (2012). Manipulation of mtDNA heteroplasmy in all striated muscles of newborn mice by AAV9-mediated delivery of a mitochondria-targeted restriction endonuclease. *Gene Therapy*, *19*(11), 1101–1106.

Bai, P., et al. (2011). PARP-1 inhibition increases mitochondrial metabolism through SIRT1 activation. *Cell Metabolism*, *13*(4), 461–468.

Baker, B. M., et al. (2012). Protective coupling of mitochondrial function and protein synthesis via the eIF2alpha kinase GCN-2. *PLoS Genetics*, *8*(6), e1002760.

Balaban, R. S., Nemoto, S., & Finkel, T. (2005). Mitochondria, oxidants, and aging. *Cell*, *120*(4), 483–495.

Baloyannis, S. J. (2006). Mitochondrial alterations in Alzheimer's disease. *Journal of Alzheimer's Disease*, *9*(2), 119–126.

Barroso, N., et al. (1993). Respiratory chain enzyme activities in lymphocytes from untreated patients with Parkinson disease. *Clinical Chemistry*, *39*(4), 667–669.

Beal, M. F., et al. (1993). Neurochemical and histologic characterization of striatal excitotoxic lesions produced by the mitochondrial toxin 3-nitropropionic acid. *The Journal of Neuroscience*, *13*(10), 4181–4192.

Bender, A., et al. (2006a). Creatine supplementation in Parkinson disease: A placebo-controlled randomized pilot trial. *Neurology*, *67*(7), 1262–1264.

Bender, A., et al. (2006b). High levels of mitochondrial DNA deletions in substantia nigra neurons in aging and Parkinson disease. *Nature Genetics*, *38*(5), 515–517.

Benecke, R., Strumper, P., & Weiss, H. (1993). Electron transfer complexes I and IV of platelets are abnormal in

Parkinson's disease but normal in Parkinson-plus syndromes. *Brain, 116*(Pt 6), 1451–1463.
Bereiter-Hahn, J. (1990). Behavior of mitochondria in the living cell. *International Review of Cytology, 122*, 1–63.
Bergeron, R., et al. (2001). Chronic activation of AMP kinase results in NRF-1 activation and mitochondrial biogenesis. *American Journal of Physiology. Endocrinology and Metabolism, 281*(6), E1340–E1346.
Bero, A. W., et al. (2011). Neuronal activity regulates the regional vulnerability to amyloid-beta deposition. *Nature Neuroscience, 14*(6), 750–756.
Berti, V., et al. (2011). Structural brain changes in normal individuals with a maternal history of Alzheimer's. *Neurobiology of Aging, 32*, 2325.e17–2325.e26.
Betarbet, R., et al. (2000). Chronic systemic pesticide exposure reproduces features of Parkinson's disease. *Nature Neuroscience, 3*(12), 1301–1306.
Bezprozvanny, I. (2010). The rise and fall of Dimebon. *Drug News & Perspectives, 23*(8), 518–523.
Bijur, G. N., Davis, R. E., & Jope, R. S. (1999). *Rapid activation of heat shock factor-1 DNA binding by H_2O_2 and modulation by glutathione in human neuroblastoma and Alzheimer's disease cybrid cells. Brain Research. Molecular Brain Research, 71*(1), 69–77.
Binder, J., et al. (2003). Clinical and molecular findings in a patient with a novel mutation in the deafness-dystonia peptide (DDP1) gene. *Brain, 126*(Pt 8), 1814–1820.
Bindoff, L. A., et al. (1989). Mitochondrial function in Parkinson's disease. *The Lancet, 2*(8653), 49.
Bindoff, L. A., et al. (1991). Respiratory chain abnormalities in skeletal muscle from patients with Parkinson's disease. *Journal of the Neurological Sciences, 104*(2), 203–208.
Bir, A., et al. (2014). alpha-Synuclein-induced mitochondrial dysfunction in isolated preparation and intact cells: Implications in the pathogenesis of Parkinson's disease. *Journal of Neurochemistry, 131*(6), 868–877.
Blacker, D., & Tanzi, R. E. (1998). The genetics of Alzheimer disease: Current status and future prospects. *Archives of Neurology, 55*(3), 294–296.
Blass, J. P., et al. (1990). Induction of Alzheimer antigens by an uncoupler of oxidative phosphorylation. *Archives of Neurology, 47*(8), 864–869.
Blass, J. P., et al. (1991). Expression of 'Alzheimer antigens' in cultured skin fibroblasts. *Archives of Neurology, 48*(7), 709–717.
Blin, O., et al. (1994). Mitochondrial respiratory failure in skeletal muscle from patients with Parkinson's disease and multiple system atrophy. *Journal of the Neurological Sciences, 125*(1), 95–101.
Bonifati, V., et al. (2003). Mutations in the DJ-1 gene associated with autosomal recessive early-onset parkinsonism. *Science, 299*(5604), 256–259.
Borland, M. K., et al. (2009). Relationships among molecular genetic and respiratory properties of Parkinson's disease cybrid cells show similarities to Parkinson's brain tissues. *Biochimica et Biophysica Acta, 1792*(1), 68–74.
Borthwick, G. M., et al. (1999). Mitochondrial enzyme activity in amyotrophic lateral sclerosis: Implications for the role of mitochondria in neuronal cell death. *Annals of Neurology, 46*(5), 787–790.
Bosetti, F., et al. (2002). Cytochrome c oxidase and mitochondrial F1F0-ATPase (ATP synthase) activities in platelets and brain from patients with Alzheimer's disease. *Neurobiology of Aging, 23*(3), 371–376.
Bough, K. J., et al. (2006). Mitochondrial biogenesis in the anticonvulsant mechanism of the ketogenic diet. *Annals of Neurology, 60*(2), 223–235.
Bozik, M. E., et al. (2014). A post hoc analysis of subgroup outcomes and creatinine in the phase III clinical trial (EMPOWER) of dexpramipexole in ALS. *Amyotrophic Lateral Sclerosis and Frontotemporal Degeneration, 15*(5–6), 406–413.
Bradley-Whitman, M. A., et al. (2014). Nucleic acid oxidation: An early feature of Alzheimer's disease. *Journal of Neurochemistry, 128*(2), 294–304.
Browne, S. E., & Beal, M. F. (2004). The energetics of Huntington's disease. *Neurochemical Research, 29*(3), 531–546.
Bull, P. C., et al. (1993). The Wilson disease gene is a putative copper transporting P-type ATPase similar to the Menkes gene. *Nature Genetics, 5*(4), 327–337.
Burlacu, A. (2003). Regulation of apoptosis by Bcl-2 family proteins. *Journal of Cellular and Molecular Medicine, 7*(3), 249–257.
Cadenas, E., et al. (1977). Production of superoxide radicals and hydrogen peroxide by NADH-ubiquinone reductase and ubiquinol-cytochrome c reductase from beef-heart mitochondria. *Archives of Biochemistry and Biophysics, 180*(2), 248–257.
Canet-Aviles, R. M., et al. (2004). The Parkinson's disease protein DJ-1 is neuroprotective due to cysteine-sulfinic acid-driven mitochondrial localization. *Proceedings of the National Academy of Sciences of the United States of America, 101*(24), 9103–9108.
Cardellach, F., et al. (1993). Mitochondrial respiratory chain activity in skeletal muscle from patients with Parkinson's disease. *Neurology, 43*(11), 2258–2262.
Cardoso, S. M., et al. (2004a). Cytochrome c oxidase is decreased in Alzheimer's disease platelets. *Neurobiology of Aging, 25*(1), 105–110.
Cardoso, S. M., et al. (2004). Mitochondria dysfunction of Alzheimer's disease cybrids enhances Abeta toxicity. *Journal of Neurochemistry, 89*(6), 1417–1426.

Carri, M. T., et al. (1997). Expression of a Cu, Zn superoxide dismutase typical of familial amyotrophic lateral sclerosis induces mitochondrial alteration and increase of cytosolic Ca2 + concentration in transfected neuroblastoma SH-SY5Y cells. *FEBS Letters, 414*(2), 365–368.

Casari, G., et al. (1998). Spastic paraplegia and OXPHOS impairment caused by mutations in paraplegin, a nuclear-encoded mitochondrial metalloprotease. *Cell, 93*(6), 973–983.

Caspersen, C., et al. (2005). Mitochondrial Abeta: A potential focal point for neuronal metabolic dysfunction in Alzheimer's disease. *The FASEB Journal, 19*(14), 2040–2041.

Cassarino, D. S., et al. (1997). Elevated reactive oxygen species and antioxidant enzyme activities in animal and cellular models of Parkinson's disease. *Biochimica et Biophysica Acta, 1362*(1), 77–86.

Cassarino, D. S., et al. (1998). Cyclosporin A increases resting mitochondrial membrane potential in SY5Y cells and reverses the depressed mitochondrial membrane potential of Alzheimer's disease cybrids. *Biochemical and Biophysical Research Communications, 248*(1), 168–173.

Cassarino, D. S., et al. (2000). Interaction among mitochondria, mitogen-activated protein kinases, and nuclear factor-kappaB in cellular models of Parkinson's disease. *Journal of Neurochemistry, 74*(4), 1384–1392.

Castellano, C. A., et al. (2017). A 3-month aerobic training program improves brain energy metabolism in mild Alzheimer's disease: Preliminary results from a neuroimaging study. *Journal of Alzheimer's Disease, 56*(4), 1459–1468.

Cebollero, E., & Reggiori, F. (2009). Regulation of autophagy in yeast Saccharomyces cerevisiae. *Biochimica et Biophysica Acta, 1793*(9), 1413–1421.

Cereghetti, G. M., et al. (2008). Dephosphorylation by calcineurin regulates translocation of Drp1 to mitochondria. *Proceedings of the National Academy of Sciences of the United States of America, 105*(41), 15803–15808.

Chacinska, A., et al. (2009). Importing mitochondrial proteins: Machineries and mechanisms. *Cell, 138*(4), 628–644.

Chang, C. R., & Blackstone, C. (2010). Dynamic regulation of mitochondrial fission through modification of the dynamin-related protein Drp1. *Annals of the New York Academy of Sciences, 1201*, 34–39.

Chang, S., et al. (2005). Lipid- and receptor-binding regions of apolipoprotein E4 fragments act in concert to cause mitochondrial dysfunction and neurotoxicity. *Proceedings of the National Academy of Sciences of the United States of America, 102*(51), 18694–18699.

Chang, S. W., et al. (2000). The frequency of point mutations in mitochondrial DNA is elevated in the Alzheimer's brain. *Biochemical and Biophysical Research Communications, 273*(1), 203–208.

Chen, H., Chomyn, A., & Chan, D. C. (2005). Disruption of fusion results in mitochondrial heterogeneity and dysfunction. *Journal of Biological Chemistry, 280*(28), 26185–26192.

Chen, H. K., et al. (2011). Apolipoprotein E4 domain interaction mediates detrimental effects on mitochondria and is a potential therapeutic target for Alzheimer disease. *Journal of Biological Chemistry, 286*(7), 5215–5221.

Cheng, E. H., et al. (2001). BCL-2, BCL-X(L) sequester BH3 domain-only molecules preventing BAX- and BAK-mediated mitochondrial apoptosis. *Molecular Cell, 8*(3), 705–711.

Cheng, M. Y., et al. (1989). Mitochondrial heat-shock protein hsp60 is essential for assembly of proteins imported into yeast mitochondria. *Nature, 337*(6208), 620–625.

Choi, J., et al. (2014). Brain diabetic neurodegeneration segregates with low intrinsic aerobic capacity. *Annals of Clinical and Translational Neurology, 1*(8), 589–604.

Chomyn, A., et al. (1992). MELAS mutation in mtDNA binding site for transcription termination factor causes defects in protein synthesis and in respiration but no change in levels of upstream and downstream mature transcripts. *Proceedings of the National Academy of Sciences of the United States of America, 89*(10), 4221–4225.

Chun, B. Y., & Rizzo, J. F., III (2016). Dominant optic atrophy: Updates on the pathophysiology and clinical manifestations of the optic atrophy 1 mutation. *Current Opinion in Ophthalmology, 27*(6), 475–480.

Chung, M. J., & Suh, Y. L. (2002). Ultrastructural changes of mitochondria in the skeletal muscle of patients with amyotrophic lateral sclerosis. *Ultrastructural Pathology, 26*(1), 3–7.

Cipolat, S., et al. (2004). OPA1 requires mitofusin 1 to promote mitochondrial fusion. *Proceedings of the National Academy of Sciences of the United States of America, 101*(45), 15927–15932.

Clayton, D. A. (2000). Transcription and replication of mitochondrial DNA. *Human Reproduction, 15*(Suppl 2), 11–17.

Cogliati, S., et al. (2013). Mitochondrial cristae shape determines respiratory chain supercomplexes assembly and respiratory efficiency. *Cell, 155*(1), 160–171.

Colell, A., et al. (2007). GAPDH and autophagy preserve survival after apoptotic cytochrome c release in the absence of caspase activation. *Cell, 129*(5), 983–997.

Collins, T. J., et al. (2002). Mitochondria are morphologically and functionally heterogeneous within cells. *The EMBO Journal, 21*(7), 1616–1627.

Cooper, J. M., & Schapira, A. H. (2007). Friedreich's ataxia: Coenzyme Q10 and vitamin E therapy. *Mitochondrion*, *7* (Suppl), S127–S135.

Cordeiro, N., & Bidere, N. (2013). Visualization of fas-mediated death-inducing signaling complex formation by immunoprecipitation. *Methods in Molecular Biology*, *979*, 43–49.

Corder, E. H., et al. (1993). Gene dose of apolipoprotein E type 4 allele and the risk of Alzheimer's disease in late onset families. *Science*, *261*(5123), 921–923.

Corral-Debrinski, M., et al. (1994). Marked changes in mitochondrial DNA deletion levels in Alzheimer brains. *Genomics*, *23*(2), 471–476.

Coskun, P. E., et al. (2010). Systemic mitochondrial dysfunction and the etiology of Alzheimer's disease and down syndrome dementia. *Journal of Alzheimer's Disease*, *20*(Suppl 2), S293–S310.

Coskun, P. E., Beal, M. F., & Wallace, D. C. (2004). Alzheimer's brains harbor somatic mtDNA control-region mutations that suppress mitochondrial transcription and replication. *Proceedings of the National Academy of Sciences of the United States of America*, *101*(29), 10726–10731.

Costa, R. O., et al. (2012). Amyloid beta-induced ER stress is enhanced under mitochondrial dysfunction conditions. *Neurobiology of Aging*, *33*(4), 824.e5–824.e16.

Crompton, M. (1999). The mitochondrial permeability transition pore and its role in cell death. *Biochemical Journal*, *341*(Pt 2), 233–249.

Crompton, M., Virji, S., & Ward, J. M. (1998). Cyclophilin-D binds strongly to complexes of the voltage-dependent anion channel and the adenine nucleotide translocase to form the permeability transition pore. *European Journal of Biochemistry*, *258*(2), 729–735.

Crugnola, V., et al. (2010). Mitochondrial respiratory chain dysfunction in muscle from patients with amyotrophic lateral sclerosis. *Archives of Neurology*, *67*(7), 849–854.

Cudkowicz, M., et al. (2011). The effects of dexpramipexole (KNS-760704) in individuals with amyotrophic lateral sclerosis. *Nature Medicine*, *17*(12), 1652–1656.

Cui, L., et al. (2006). Transcriptional repression of PGC-1alpha by mutant huntingtin leads to mitochondrial dysfunction and neurodegeneration. *Cell*, *127*(1), 59–69.

Curti, D., et al. (1996). Amyotrophic lateral sclerosis: Oxidative energy metabolism and calcium homeostasis in peripheral blood lymphocytes. *Neurology*, *47*(4), 1060–1064.

Curti, D., et al. (1997). Oxidative metabolism in cultured fibroblasts derived from sporadic Alzheimer's disease (AD) patients. *Neuroscience Letters*, *236*(1), 13–16.

Czabotar, P. E., et al. (2014). Control of apoptosis by the BCL-2 protein family: Implications for physiology and therapy. *Nature Reviews Molecular Cell Biology*, *15*(1), 49–63.

Dai, Y., et al. (2014). Rapamycin drives selection against a pathogenic heteroplasmic mitochondrial DNA mutation. *Human Molecular Genetics*, *23*(3), 637–647.

Das, S. K., & Ray, K. (2006). Wilson's disease: An update. *Nature Clinical Practice Neurology*, *2*(9), 482–493.

D'Aurelio, M., et al. (2010). Mitochondrial DNA background modifies the bioenergetics of NARP/MILS ATP6 mutant cells. *Human Molecular Genetics*, *19*(2), 374–386.

D'Autreaux, B., & Toledano, M. B. (2007). ROS as signalling molecules: Mechanisms that generate specificity in ROS homeostasis. *Nature Reviews Molecular Cell Biology*, *8*(10), 813–824.

Davis, R. E., et al. (1997). Mutations in mitochondrial cytochrome c oxidase genes segregate with late-onset Alzheimer disease. *Proceedings of the National Academy of Sciences of the United States of America*, *94*(9), 4526–4531.

Day, B. J. (2009). Catalase and glutathione peroxidase mimics. *Biochemical Pharmacology*, *77*(3), 285–296.

de Brito, O. M., & Scorrano, L. (2008). Mitofusin 2 tethers endoplasmic reticulum to mitochondria. *Nature*, *456* (7222), 605–610.

De Duve, C., & Wattiaux, R. (1966). Functions of lysosomes. *Annual Review of Physiology*, *28*, 435–492.

De Giorgi, F., Lartigue, L., & Ichas, F. (2000). Electrical coupling and plasticity of the mitochondrial network. *Cell Calcium*, *28*(5), 365–370.

de la Monte, S. M., et al. (2000). Mitochondrial DNA damage as a mechanism of cell loss in Alzheimer's disease. *Laboratory Investigation*, *80*(8), 1323–1335.

de Leon, M. J., et al. (1983). Positron emission tomographic studies of aging and Alzheimer disease. *AJNR. American Journal of Neuroradiology*, *4*(3), 568–571.

De Santi, S., et al. (1995). Age-related changes in brain: II. Positron emission tomography of frontal and temporal lobe glucose metabolism in normal subjects. *Psychiatric Quarterly*, *66*(4), 357–370.

De Sarno, P., et al. (2000). Alterations in muscarinic receptor-coupled phosphoinositide hydrolysis and AP-1 activation in Alzheimer's disease cybrid cells. *Neurobiology of Aging*, *21*(1), 31–38.

De Vos, K. J., et al. (2007). Familial amyotrophic lateral sclerosis-linked SOD1 mutants perturb fast axonal transport to reduce axonal mitochondria content. *Human Molecular Genetics*, *16*(22), 2720–2728.

Debette, S., et al. (2009). Association of parental dementia with cognitive and brain MRI measures in middle-aged adults. *Neurology*, *73*(24), 2071–2078.

Delettre, C., et al. (2000). Nuclear gene OPA1, encoding a mitochondrial dynamin-related protein, is mutated in dominant optic atrophy. *Nature Genetics*, *26*(2), 207–210.

Desagher, S., & Martinou, J. C. (2000). Mitochondria as the central control point of apoptosis. *Trends in Cell Biology*, *10*(9), 369–377.

Devi, L., et al. (2006). Accumulation of amyloid precursor protein in the mitochondrial import channels of human Alzheimer's disease brain is associated with mitochondrial dysfunction. *The Journal of Neuroscience, 26*(35), 9057–9068.

Di Prospero, N. A., et al. (2007). Neurological effects of high-dose idebenone in patients with Friedreich's ataxia: A randomised, placebo-controlled trial. *The Lancet Neurology, 6*(10), 878–886.

Dimauro, S. (2013). Mitochondrial encephalomyopathies—Fifty years on: The Robert Wartenberg Lecture. *Neurology, 81*(3), 281–291.

Ding, W. X., & Yin, X. M. (2012). Mitophagy: Mechanisms, pathophysiological roles, and analysis. *Biological Chemistry, 393*(7), 547–564.

Donepudi, M., et al. (2003). Insights into the regulatory mechanism for caspase-8 activation. *Molecular Cell, 11*(2), 543–549.

Dorsey, E. R., et al. (2013). A randomized, double-blind, placebo-controlled study of latrepirdine in patients with mild to moderate Huntington disease. *JAMA Neurology, 70*(1), 25–33.

Duara, R., et al. (1993). A comparison of familial and sporadic Alzheimer's disease. *Neurology, 43*(7), 1377–1384.

Dysken, M. W., et al. (2014). Effect of vitamin E and memantine on functional decline in Alzheimer disease: The TEAM-AD VA cooperative randomized trial. *The Journal of the American Medical Association, 311*(1), 33–44.

Edland, S. D., et al. (1996). Increased risk of dementia in mothers of Alzheimer's disease cases: Evidence for maternal inheritance. *Neurology, 47*(1), 254–256.

Ehinger, J. K., et al. (2016). Cell-permeable succinate prodrugs bypass mitochondrial complex I deficiency. *Nature Communications, 7*, 12317.

Eleff, S., et al. (1984). 31P NMR study of improvement in oxidative phosphorylation by vitamins K3 and C in a patient with a defect in electron transport at complex III in skeletal muscle. *Proceedings of the National Academy of Sciences of the United States of America, 81*(11), 3529–3533.

Elgass, K., et al. (2013). Recent advances into the understanding of mitochondrial fission. *Biochimica et Biophysica Acta, 1833*(1), 150–161.

Elmore, S. (2007). Apoptosis: A review of programmed cell death. *Toxicologic Pathology, 35*(4), 495–516.

Enriquez, J. A., Chomyn, A., & Attardi, G. (1995). MtDNA mutation in MERRF syndrome causes defective aminoacylation of tRNA(Lys) and premature translation termination. *Nature Genetics, 10*(1), 47–55.

Ephrussi, B., Hottinger, H., & Chimenes, A. (1949). Action de l'acriflavine sur les levures, I: la mutation "petite clonie.". *Annales De l'Institut Pasteur, 76*, 531.

Escobar-Khondiker, M., et al. (2007). Annonacin, a natural mitochondrial complex I inhibitor, causes tau pathology in cultured neurons. *The Journal of Neuroscience, 27*(29), 7827–7837.

Esteves, A. R., et al. (2008a). Mitochondrial function in Parkinson's disease cybrids containing an nt2 neuron-like nuclear background. *Mitochondrion, 8*(3), 219–228.

Esteves, A. R., et al. (2010a). Dysfunctional mitochondria uphold calpain activation: Contribution to Parkinson's disease pathology. *Neurobiology of Disease, 37*(3), 723–730.

Esteves, A. R., et al. (2010b). Microtubule depolymerization potentiates alpha-synuclein oligomerization. *Frontiers in Aging Neuroscience, 1*, 5.

Esteves, A. R., et al. (2010c). Mitochondrial respiration and respiration-associated proteins in cell lines created through Parkinson's subject mitochondrial transfer. *Journal of Neurochemistry, 113*(3), 674–682.

Esteves, A. R., Swerdlow, R. H., & Cardoso, S. M. (2014). LRRK2, a puzzling protein: Insights into Parkinson's disease pathogenesis. *Experimental Neurology, 261*, 206–216.

Esteves, A. R., et al. (2008b). Oxidative stress involvement in alpha-synuclein oligomerization in Parkinsons disease cybrids. *Antioxidants & Redox Signaling*.

Eura, Y., et al. (2003). Two mitofusin proteins, mammalian homologues of FZO, with distinct functions are both required for mitochondrial fusion. *Journal of Biochemistry, 134*(3), 333–344.

Feigin, A., et al. (1996). Assessment of coenzyme Q10 tolerability in Huntington's disease. *Movement Disorders, 11*(3), 321–323.

Feng, Y., et al. (2014). The machinery of macroautophagy. *Cell Research, 24*(1), 24–41.

Ferri, A., et al. (2006). Familial ALS-superoxide dismutases associate with mitochondria and shift their redox potentials. *Proceedings of the National Academy of Sciences of the United States of America, 103*(37), 13860–13865.

Ferris, S. H., et al. (1980). Positron emission tomography in the study of aging and senile dementia. *Neurobiology of Aging, 1*(2), 127–131.

Fesahat, F., et al. (2007). Do haplogroups H and U act to increase the penetrance of Alzheimer's disease? *Cellular and Molecular Neurobiology, 27*(3), 329–334.

Filosto, M., et al. (2007). Neuropathology of mitochondrial diseases. *Bioscience Reports, 27*(1–3), 23–30.

Finkel, T. (2001). Reactive oxygen species and signal transduction. *International Union of Biochemistry and Molecular Biology Life, 52*(1–2), 3–6.

Finkel, T. (2003). Oxidant signals and oxidative stress. *Current Opinion in Cell Biology, 15*(2), 247–254.

Finkel, T., Deng, C. X., & Mostoslavsky, R. (2009). Recent progress in the biology and physiology of sirtuins. *Nature*, *460*(7255), 587–591.

Fiorese, C. J., et al. (2016). The transcription factor ATF5 mediates a mammalian mitochondrial UPR. *Current Biology*, *26*(15), 2037–2043.

Fisar, Z., et al. (2016). Mitochondrial respiration in the platelets of patients with Alzheimer's disease. *Current Alzheimer Research*, *13*(8), 930–941.

Flierl, A., Reichmann, H., & Seibel, P. (1997). Pathophysiology of the MELAS 3243 transition mutation. *Journal of Biological Chemistry*, *272*(43), 27189–27196.

Foster, N. L., et al. (1983). Alzheimer's disease: Focal cortical changes shown by positron emission tomography. *Neurology*, *33*(8), 961–965.

Frackowiak, R. S., et al. (1981). Regional cerebral oxygen supply and utilization in dementia. A clinical and physiological study with oxygen-15 and positron tomography. *Brain*, *104*(Pt 4), 753–778.

Frank, S., et al. (2001). The role of dynamin-related protein 1, a mediator of mitochondrial fission, in apoptosis. *Developmental Cell*, *1*(4), 515–525.

Friedland, R. P., et al. (1983). *Regional cerebral metabolic alterations in dementia of the Alzheimer type: Positron emission tomo*graphy with [18F]fluorodeoxyglucose. *Journal of Computer Assisted Tomography*, *7*(4), 590–598.

Fry, M., & Green, D. E. (1981). Cardiolipin requirement for electron transfer in complex I and III of the mitochondrial respiratory chain. *Journal of Biological Chemistry*, *256*(4), 1874–1880.

Fukui, H., et al. (2007). Cytochrome c oxidase deficiency in neurons decreases both oxidative stress and amyloid formation in a mouse model of Alzheimer's disease. *Proceedings of the National Academy of Sciences of the United States of America*, *104*(35), 14163–14168.

Fukuyama, H., et al. (1994). Altered cerebral energy metabolism in Alzheimer's disease: A PET study. *The Journal of Nuclear Medicine*, *35*(1), 1–6.

Gabuzda, D., et al. (1994). Inhibition of energy metabolism alters the processing of amyloid precursor protein and induces a potentially amyloidogenic derivative. *Journal of Biological Chemistry*, *269*(18), 13623–13628.

Galasko, D. R., et al. (2012). Antioxidants for Alzheimer disease: A randomized clinical trial with cerebrospinal fluid biomarker measures. *Archives of Neurology*, *69*(7), 836–841.

Gan, X., et al. (2014a). Inhibition of ERK-DLP1 signaling and mitochondrial division alleviates mitochondrial dysfunction in Alzheimer's disease cybrid cell. *Biochimica et Biophysica Acta*, *1842*(2), 220–231.

Gan, X., et al. (2014b). Oxidative stress-mediated activation of extracellular signal-regulated kinase contributes to mild cognitive impairment-related mitochondrial dysfunction. *Free Radical Biology and Medicine*, *75*, 230–240.

Gasparini, L., et al. (1997). Effect of energy shortage and oxidative stress on amyloid precursor protein metabolism in COS cells. *Neuroscience Letters*, *231*(2), 113–117.

Ghezzi, D., et al. (2005). Mitochondrial DNA haplogroup K is associated with a lower risk of Parkinson's disease in Italians. *European Journal of Human Genetics*, *13*(6), 748–752.

Ghosh, S., et al. (2007). The thiazolidinedione pioglitazone alters mitochondrial function in human neuron-like cells. *Molecular Pharmacology*, *71*(6), 1695–1702.

Ghosh, S. S., et al. (1999). Use of cytoplasmic hybrid cell lines for elucidating the role of mitochondrial dysfunction in Alzheimer's disease and Parkinson's disease. *Annals of the New York Academy of Sciences*, *893*, 176–191.

Gibson, G. E., et al. (1988). Reduced activities of thiamine-dependent enzymes in the brains and peripheral tissues of patients with Alzheimer's disease. *Archives of Neurology*, *45*(8), 836–840.

Giguere, V. (2008). Transcriptional control of energy homeostasis by the estrogen-related receptors. *Endocrine Reviews*, *29*(6), 677–696.

Giles, R. E., et al. (1980). Maternal inheritance of human mitochondrial DNA. *Proceedings of the National Academy of Sciences of the United States of America*, *77*(11), 6715–6719.

Goate, A., et al. (1991). Segregation of a missense mutation in the amyloid precursor protein gene with familial Alzheimer's disease. *Nature*, *349*(6311), 704–706.

Gold, M., et al. (2010). Rosiglitazone monotherapy in mild-to-moderate Alzheimer's disease: Results from a randomized, double-blind, placebo-controlled phase III study. *Dementia and Geriatric Cognitive Disorders*, *30*(2), 131–146.

Gomes, L. C., Di Benedetto, G., & Scorrano, L. (2011). During autophagy mitochondria elongate, are spared from degradation and sustain cell viability. *Nature Cell Biology*, *13*(5), 589–598.

Gordon, P. H., et al. (2007). Efficacy of minocycline in patients with amyotrophic lateral sclerosis: A phase III randomised trial. *The Lancet Neurology*, *6*(12), 1045–1053.

Goto, Y., Nonaka, I., & Horai, S. (1990). A mutation in the tRNA(Leu)(UUR) gene associated with the MELAS subgroup of mitochondrial encephalomyopathies. *Nature*, *348*(6302), 651–653.

Green, D. R., & Kroemer, G. (2004). The pathophysiology of mitochondrial cell death. *Science*, *305*(5684), 626–629.

Green, D. R., Galluzzi, L., & Kroemer, G. (2014). Cell biology. Metabolic control of cell death. *Science*, *345*(6203), 1250256.

Greene, J. G., et al. (1993). Inhibition of succinate dehydrogenase by malonic acid produces an "excitotoxic" lesion in rat striatum. *Journal of Neurochemistry, 61*(3), 1151–1154.

Gu, M., et al. (1996). Mitochondrial defect in Huntington's disease caudate nucleus. *Annals of Neurology, 39*(3), 385–389.

Gu, M., et al. (1998). Mitochondrial DNA transmission of the mitochondrial defect in Parkinson's disease. *Annals of Neurology, 44*(2), 177–186.

Gu, M., et al. (2000). Oxidative-phosphorylation defects in liver of patients with Wilson's disease. *The Lancet, 356* (9228), 469–474.

Guarente, L. (2007). Sirtuins in aging and disease. *Cold Spring Harbor Symposia on Quantitative Biology, 72*, 483–488.

Guerreiro, R., et al. (2013). TREM2 variants in Alzheimer's disease. *The New England Journal of Medicine, 368*(2), 117–127.

Gutzmann, H., & Hadler, D. (1998). Sustained efficacy and safety of idebenone in the treatment of Alzheimer's disease: Update on a 2-year double-blind multicentre study. *Journal of Neural Transmission. Supplementa, 54*, 301–310.

Haas, R. H., et al. (1995). Low platelet mitochondrial complex I and complex II/III activity in early untreated Parkinson's disease. *Annals of Neurology, 37*(6), 714–722.

Hackenbrock, C. R. (1966). Ultrastructural bases for metabolically linked mechanical activity in mitochondria. I. Reversible ultrastructural changes with change in metabolic steady state in isolated liver mitochondria. *Journal of Cell Biology, 30*(2), 269–297.

Haigis, M. C., & Guarente, L. P. (2006). Mammalian sirtuins—Emerging roles in physiology, aging, and calorie restriction. *Genes & Development, 20*(21), 2913–2921.

Hamblet, N. S., & Castora, F. J. (1997). Elevated levels of the Kearns–Sayre syndrome mitochondrial DNA deletion in temporal cortex of Alzheimer's patients. *Mutation Research, 379*(2), 253–262.

Hansen, J. J., et al. (2002). Hereditary spastic paraplegia SPG13 is associated with a mutation in the gene encoding the mitochondrial chaperonin Hsp60. *American Journal of Human Genetics, 70*(5), 1328–1332.

Hansson, C. A., et al. (2004). Nicastrin, presenilin, APH-1, and PEN-2 form active gamma-secretase complexes in mitochondria. *Journal of Biological Chemistry, 279*(49), 51654–51660.

Hansson Petersen, C. A., et al. (2008). The amyloid beta-peptide is imported into mitochondria via the TOM import machinery and localized to mitochondrial cristae. *Proceedings of the National Academy of Sciences of the United States of America, 105*(35), 13145–13150.

Harman, D. (1972). The biologic clock: The mitochondria? *Journal of the American Geriatrics Society, 20*(4), 145–147.

Harold, D., et al. (2009). Genome-wide association study identifies variants at CLU and PICALM associated with Alzheimer's disease. *Nature Genetics, 41*(10), 1088–1093.

Haynes, C. M., et al. (2007). ClpP mediates activation of a mitochondrial unfolded protein response in C. elegans. *Developmental Cell, 13*(4), 467–480.

Hebert, L. E., et al. (2003). Alzheimer disease in the US population: Prevalence estimates using the 2000 census. *Archives of Neurology, 60*(8), 1119–1122.

Henderson, S. T., et al. (2009). Study of the ketogenic agent AC-1202 in mild to moderate Alzheimer's disease: A randomized, double-blind, placebo-controlled, multicenter trial. *Nutrition & Metabolism (London), 6*, 31.

Herbert, M., & Turnbull, D. (2017). Mitochondrial donation—Clearing the final regulatory hurdle in the United Kingdom. *The New England Journal of Medicine, 376*(2), 171–173.

Herrmann, J. M., et al. (1994). Mitochondrial heat shock protein 70, a molecular chaperone for proteins encoded by mitochondrial DNA. *Journal of Cell Biology, 127*(4), 893–902.

Higgins, C. M., Jung, C., & Xu, Z. (2003). ALS-associated mutant SOD1G93A causes mitochondrial vacuolation by expansion of the intermembrane space and by involvement of SOD1 aggregation and peroxisomes. *BMC Neuroscience, 4*, 16.

Hill, K., et al. (1998). Tom40 forms the hydrophilic channel of the mitochondrial import pore for preproteins [see comment]. *Nature, 395*(6701), 516–521.

Hinnebusch, A. G. (1994). The eIF-2 alpha kinases: Regulators of protein synthesis in starvation and stress. *Seminars in Cell Biology, 5*(6), 417–426.

Hirai, K., et al. (2001). Mitochondrial abnormalities in Alzheimer's disease. *The Journal of Neuroscience, 21*(9), 3017–3023.

Hoglinger, G. U., et al. (2005). The mitochondrial complex I inhibitor rotenone triggers a cerebral tauopathy. *Journal of Neurochemistry, 95*(4), 930–939.

Holt, I. J., Harding, A. E., & Morgan-Hughes, J. A. (1988). Deletions of muscle mitochondrial DNA in patients with mitochondrial myopathies. *Nature, 331*(6158), 717–719.

Honea, R. A., et al. (2011). Progressive regional atrophy in normal adults with a maternal history of Alzheimer disease. *Neurology, 76*, 822–829.

Honea, R. A., et al. (2012). Maternal family history is associated with Alzheimer's disease biomarkers. *Journal of Alzheimer's Disease, 31*(3), 659–668.

Honea, R. A., et al. (2010). Reduced gray matter volume in normal adults with a maternal family history of Alzheimer disease. *Neurology, 74*(2), 113–120.

Horbay, R., & Bilyy, R. (2016). Mitochondrial dynamics during cell cycling. *Apoptosis, 21*(12), 1327–1335.

Horibe, T., & Hoogenraad, N. J. (2007). The chop gene contains an element for the positive regulation of the mitochondrial unfolded protein response. *PLoS ONE*, *2*(9), e835.

Horvath, S. E., & Daum, G. (2013). Lipids of mitochondria. *Progress in Lipid Research*, *52*(4), 590–614.

Howell, N. (1998). Leber hereditary optic neuropathy: Respiratory chain dysfunction and degeneration of the optic nerve. *Vision Research*, *38*(10), 1495–1504.

Iaccarino, C., et al. (2007). Apoptotic mechanisms in mutant LRRK2-mediated cell death. *Human Molecular Genetics*, *16*(11), 1319–1326.

Ihara, Y., et al. (1989). Mitochondrial encephalomyopathy (MELAS): Pathological study and successful therapy with coenzyme Q10 and idebenone. *Journal of the Neurological Sciences*, *90*(3), 263–271.

Ikejiri, Y., et al. (1996). Idebenone improves cerebral mitochondrial oxidative metabolism in a patient with MELAS. *Neurology*, *47*(2), 583–585.

Ishihara, N., et al. (2003). Regulation of mitochondrial morphology by membrane potential, and DRP1-dependent division and FZO1-dependent fusion reaction in mammalian cells. *Biochemical and Biophysical Research Communications*, *301*(4), 891–898.

Israelson, A., et al. (2010). Misfolded mutant SOD1 directly inhibits VDAC1 conductance in a mouse model of inherited ALS. *Neuron*, *67*(4), 575–587.

Jaarsma, D., et al. (2000). Human Cu/Zn superoxide dismutase (SOD1) overexpression in mice causes mitochondrial vacuolization, axonal degeneration, and premature motoneuron death and accelerates motoneuron disease in mice expressing a familial amyotrophic lateral sclerosis mutant SOD1. *Neurobiology of Disease*, *7* (6 Pt B), 623–643.

James, D. I., et al. (2003). hFis1, a novel component of the mammalian mitochondrial fission machinery. *Journal of Biological Chemistry*, *278*(38), 36373–36379.

Jeong, J. H., et al. (2015). Dose-specific effect of simvastatin on hypoxia-induced HIF-1alpha and BACE expression in Alzheimer's disease cybrid cells. *BMC Neurology*, *15*, 127.

Jiang, X., & Wang, X. (2000). Cytochrome c promotes caspase-9 activation by inducing nucleotide binding to Apaf-1. *Journal of Biological Chemistry*, *275*(40), 31199–31203.

Johansen, T., & Lamark, T. (2011). Selective autophagy mediated by autophagic adapter proteins. *Autophagy*, *7* (3), 279–296.

Johnson, A. B., & Blum, N. R. (1970). Nucleoside phosphatase activities associated with the tangles and plaques of Alzheimer's disease: A histochemical study of natural and experimental neurofibrillary tangles. *Journal of Neuropathology & Experimental Neurology*, *29*(3), 463–478.

Jonsson, T., et al. (2013). Variant of TREM2 associated with the risk of Alzheimer's disease. *The New England Journal of Medicine*, *368*(2), 107–116.

Jorm, A. F., & Jolley, D. (1998). The incidence of dementia: A meta-analysis. *Neurology*, *51*(3), 728–733.

Jornayvaz, F. R., & Shulman, G. I. (2010). Regulation of mitochondrial biogenesis. *Essays in Biochemistry*, *47*, 69–84. Available from https://doi.org/10.1042/bse0470069.

Jun, G., et al. (2010). Meta-analysis confirms CR1, CLU, and PICALM as Alzheimer disease risk loci and reveals interactions with APOE genotypes. *Archives of Neurology*, *67*(12), 1473–1484.

Jung, C., Higgins, C. M., & Xu, Z. (2002a). A quantitative histochemical assay for activities of mitochondrial electron transport chain complexes in mouse spinal cord sections. *Journal of Neuroscience Methods*, *114*(2), 165–172.

Jung, C., Higgins, C. M., & Xu, Z. (2002b). Mitochondrial electron transport chain complex dysfunction in a transgenic mouse model for amyotrophic lateral sclerosis. *Journal of Neurochemistry*, *83*(3), 535–545.

Jung, C. H., et al. (2010). mTOR regulation of autophagy. *FEBS Letters*, *584*(7), 1287–1295.

Kang, J. E., et al. (2009). Amyloid-beta dynamics are regulated by orexin and the sleep-wake cycle. *Science*, *326* (5955), 1005–1007.

Kantari, C., & Walczak, H. (2011). Caspase-8 and bid: Caught in the act between death receptors and mitochondria. *Biochimica et Biophysica Acta*, *1813*(4), 558–563.

Kaufmann, P., et al. (2006). Dichloroacetate causes toxic neuropathy in MELAS: A randomized, controlled clinical trial. *Neurology*, *66*(3), 324–330.

Keeney, P. M., et al. (2009). Cybrid models of Parkinson's disease show variable mitochondrial biogenesis and genotype-respiration relationships. *Experimental Neurology*, *220*(2), 374–382.

Kelso, G. F., et al. (2001). Selective targeting of a redox-active ubiquinone to mitochondria within cells: Antioxidant and antiapoptotic properties. *Journal of Biological Chemistry*, *276*(7), 4588–4596.

Khan, S. M., et al. (2000). Alzheimer's disease cybrids replicate beta-amyloid abnormalities through cell death pathways. *Annals of Neurology*, *48*(2), 148–155.

Kim, H., et al. (2006). Hierarchical regulation of mitochondrion-dependent apoptosis by BCL-2 subfamilies. *Nature Cell Biology*, *8*(12), 1348–1358.

Kim, J., et al. (2011). AMPK and mTOR regulate autophagy through direct phosphorylation of Ulk1. *Nature Cell Biology*, *13*(2), 132–141.

King, M. P., et al. (1992). Defects in mitochondrial protein synthesis and respiratory chain activity segregate with the tRNA(Leu(UUR)) mutation associated with mitochondrial myopathy, encephalopathy, lactic acidosis, and strokelike episodes. *Molecular and Cellular Biology, 12*(2), 480–490.

Kirchner, S. C., et al. (2000). Mitochondrial ND1 sequence analysis and association of the T4216C mutation with Parkinson's disease. *Neurotoxicology, 21*(4), 441–445.

Kirkinezos, I. G., et al. (2005). Cytochrome c association with the inner mitochondrial membrane is impaired in the CNS of G93A-SOD1 mice. *The Journal of Neuroscience, 25*(1), 164–172.

Kish, S. J., et al. (1992). Brain cytochrome oxidase in Alzheimer's disease. *Journal of Neurochemistry, 59*(2), 776–779.

Kitada, T., et al. (1998). Mutations in the parkin gene cause autosomal recessive juvenile parkinsonism. *Nature, 392* (6676), 605–608.

Koehler, C. M., et al. (1999). Human deafness dystonia syndrome is a mitochondrial disease. *Proceedings of the National Academy of Sciences of the United States of America, 96*(5), 2141–2146.

Komura, K., et al. (2003). Effectiveness of creatine monohydrate in mitochondrial encephalomyopathies. *Pediatric Neurology, 28*(1), 53–58.

Koutnikova, H., et al. (1997). Studies of human, mouse and yeast homologues indicate a mitochondrial function for frataxin. *Nature Genetics, 16*(4), 345–351.

Krasnianski, A., et al. (2005). Mitochondrial changes in skeletal muscle in amyotrophic lateral sclerosis and other neurogenic atrophies. *Brain, 128*(Pt 8), 1870–1876.

Kraytsberg, Y., et al. (2006). Mitochondrial DNA deletions are abundant and cause functional impairment in aged human substantia nigra neurons. *Nature Genetics, 38*(5), 518–520.

Krige, D., et al. (1992). Platelet mitochondrial function in Parkinson's disease. The Royal Kings and Queens Parkinson Disease Research Group. *Annals of Neurology, 32*(6), 782–788.

Krikorian, R., et al. (2012). Dietary ketosis enhances memory in mild cognitive impairment. *Neurobiology of Aging, 33*(2), 425.e19–425.e27.

Krishnan, K. J., et al. (2012). Mitochondrial DNA deletions cause the biochemical defect observed in Alzheimer's disease. *Neurobiology of Aging, 33*(9), 2210–2214.

Kruman, I. I., et al. (1999). ALS-linked Cu/Zn-SOD mutation increases vulnerability of motor neurons to excitotoxicity by a mechanism involving increased oxidative stress and perturbed calcium homeostasis. *Experimental Neurology, 160*(1), 28–39.

Kukreja, L., et al. (2014). Increased mtDNA mutations with aging promotes amyloid accumulation and brain atrophy in the APP/Ld transgenic mouse model of Alzheimer's disease. *Molecular Neurodegeneration, 9*, 16.

Kuwana, T., et al. (2005). BH3 domains of BH3-only proteins differentially regulate Bax-mediated mitochondrial membrane permeabilization both directly and indirectly. *Molecular Cell, 17*(4), 525–535.

Lagouge, M., et al. (2006). Resveratrol improves mitochondrial function and protects against metabolic disease by activating SIRT1 and PGC-1alpha. *Cell, 127*(6), 1109–1122.

Lakatos, A., et al. (2010). Association between mitochondrial DNA variations and Alzheimer's disease in the ADNI cohort. *Neurobiology of Aging, 31*(8), 1355–1363.

Lambert, A. J., & Brand, M. D. (2004). Superoxide production by NADH: Ubiquinone oxidoreductase (complex I) depends on the pH gradient across the mitochondrial inner membrane. *Biochemical Journal, 382*(Pt 2), 511–517.

Lambert, A. J., Buckingham, J. A., & Brand, M. D. (2008). Dissociation of superoxide production by mitochondrial complex I from NAD(P)H redox state. *FEBS Letters, 582* (12), 1711–1714.

Langston, J. W., et al. (1983). Chronic Parkinsonism in humans due to a product of meperidine-analog synthesis. *Science, 219*(4587), 979–980.

Lee, H. J., et al. (2002). Formation and removal of alpha-synuclein aggregates in cells exposed to mitochondrial inhibitors. *Journal of Biological Chemistry, 277*(7), 5411–5417.

Legros, F., et al. (2002). Mitochondrial fusion in human cells is efficient, requires the inner membrane potential, and is mediated by mitofusins. *Molecular Biology of the Cell, 13*(12), 4343–4354.

Lesage, S., et al. (2005). LRRK2 haplotype analyses in European and North African families with Parkinson disease: A common founder for the G2019S mutation dating from the 13th century. *American Journal of Human Genetics, 77*(2), 330–332.

Lestienne, P., & Ponsot, G. (1988). Kearns–Sayre syndrome with muscle mitochondrial DNA deletion. *The Lancet, 1* (8590), 885.

Levy-Lahad, E., et al. (1995). Candidate gene for the chromosome 1 familial Alzheimer's disease locus. *Science, 269*(5226), 973–977.

Lewis, M. R., & Lewis, W. H. (1914). Mitochondria in tissue culture. *Science, 39*(1000), 330–333.

Li, H., et al. (1998). Cleavage of BID by caspase 8 mediates the mitochondrial damage in the Fas pathway of apoptosis. *Cell, 94*(4), 491–501.

Lillig, C. H., & Holmgren, A. (2007). Thioredoxin and related molecules—From biology to health and disease. *Antioxidants & Redox Signaling*, *9*(1), 25–47.

Lin, M. T., et al. (2012). Somatic mitochondrial DNA mutations in early Parkinson and incidental Lewy body disease. *Annals of Neurology*, *71*(6), 850–854.

Lin, S. J., & Guarente, L. (2003). Nicotinamide adenine dinucleotide, a metabolic regulator of transcription, longevity and disease. *Current Opinion in Cell Biology*, *15*(2), 241–246.

Lin, Y. F., & Haynes, C. M. (2016). Metabolism and the UPR(mt). *Molecular Cell*, *61*(5), 677–682.

Liu, G., et al. (2009). alpha-Synuclein is differentially expressed in mitochondria from different rat brain regions and dose-dependently down-regulates complex I activity. *Neuroscience Letters*, *454*(3), 187–192.

Liu, Z., & Butow, R. A. (2006). Mitochondrial retrograde signaling. *Annual Review of Genetics*, *40*, 159–185.

Liu, Z., et al. (2013). A cross-sectional study on cerebrospinal fluid biomarker levels in cognitively normal elderly subjects with or without a family history of Alzheimer's disease. *CNS Neuroscience & Therapeutics*, *19*(1), 38–42.

Lopez-Otin, C., et al. (2013). The hallmarks of aging. *Cell*, *153*(6), 1194–1217.

Lovell, M. A., & Markesbery, W. R. (2007). Oxidative DNA damage in mild cognitive impairment and late-stage Alzheimer's disease. *Nucleic Acids Research*, *35*(22), 7497–7504.

Lu, J., et al. (2012). Mitochondrial dysfunction in human TDP-43 transfected NSC34 cell lines and the protective effect of dimethoxy curcumin. *Brain Research Bulletin*, *89* (5-6), 185–190.

Luoma, P. T., et al. (2007). Mitochondrial DNA polymerase gamma variants in idiopathic sporadic Parkinson disease. *Neurology*, *69*(11), 1152–1159.

Lustbader, J. W., et al. (2004). ABAD directly links Abeta to mitochondrial toxicity in Alzheimer's disease. *Science*, *304*(5669), 448–452.

Luth, E. S., et al. (2014). *Soluble, prefibrillar alpha-synuclein oligomers promote complex I-dependent, Ca^{2+}-induced mitochondrial dysfunction. Journal of Biological Chemistry*, *289* (31), 21490–21507.

Lutsenko, S., & Cooper, M. J. (1998). Localization of the Wilson's disease protein product to mitochondria. *Proceedings of the National Academy of Sciences of the United States of America*, *95*(11), 6004–6009.

Maassen, J. A., Janssen, G. M., & Hart, L. M. T. (2005). Molecular mechanisms of mitochondrial diabetes (MIDD). *Annals of Medicine*, *37*(3), 213–221.

MacDonald, Marcy E., Ambrose, Christine M., Duyao, Mabel P., Myers, Richard H., Lin, Carol, Srinidhi, Lakshmi, … … Lehrach, Hans (1993). A novel gene containing a trinucleotide repeat that is expanded and unstable on Huntington's disease chromosomes. The Huntington's Disease Collaborative Research Group. *Cell*, *72*(6), 971–983.

Magrane, J., et al. (2014). Abnormal mitochondrial transport and morphology are common pathological denominators in SOD1 and TDP43 ALS mouse models. *Human Molecular Genetics*, *23*(6), 1413–1424.

Mahley, R. W., & Huang, Y. (2012). Apolipoprotein e sets the stage: Response to injury triggers neuropathology. *Neuron*, *76*(5), 871–885.

Majewski, N., et al. (2004). Akt inhibits apoptosis downstream of BID cleavage via a glucose-dependent mechanism involving mitochondrial hexokinases. *Molecular and Cellular Biology*, *24*(2), 730–740.

Mancuso, M., et al. (2003). Decreased platelet cytochrome c oxidase activity is accompanied by increased blood lactate concentration during exercise in patients with Alzheimer disease. *Experimental Neurology*, *182*(2), 421–426.

Mancuso, M., et al. (2004). Could mitochondrial haplogroups play a role in sporadic amyotrophic lateral sclerosis? *Neuroscience Letters*, *371*(2–3), 158–162.

Mancuso, M., et al. (2010). Coenzyme Q10 in neuromuscular and neurodegenerative disorders. *Current Drug Targets*, *11*(1), 111–121.

Manczak, M., et al. (2006). Mitochondria are a direct site of A beta accumulation in Alzheimer's disease neurons: Implications for free radical generation and oxidative damage in disease progression. *Human Molecular Genetics*, *15*(9), 1437–1449.

Manczak, M., Calkins, M. J., & Reddy, P. H. (2011). Impaired mitochondrial dynamics and abnormal interaction of amyloid beta with mitochondrial protein Drp1 in neurons from patients with Alzheimer's disease: Implications for neuronal damage. *Human Molecular Genetics*, *20*(13), 2495–2509.

Mannella, C. A. (2006). Structure and dynamics of the mitochondrial inner membrane cristae. *Biochimica et Biophysica Acta (BBA)—Molecular Cell Research*, *1763* (5–6), 542–548.

Marano, C. M., et al. (2013). Longitudinal studies of cerebral glucose metabolism in late-life depression and normal aging. *International Journal of Geriatric Psychiatry*, *28* (4), 417–423.

Mariotti, C., et al. (2003). Idebenone treatment in Friedreich patients: One-year-long randomized placebo-controlled trial. *Neurology*, *60*(10), 1676–1679.

Martin, L. J., et al. (2014). Mitochondrial permeability transition pore regulates Parkinson's disease development in mutant alpha-synuclein transgenic mice. *Neurobiology of Aging*, *35*(5), 1132–1152.

Martinus, R. D., et al. (1996). Selective induction of mitochondrial chaperones in response to loss of the mitochondrial genome. *European Journal of Biochemistry, 240*(1), 98–103.

Maruszak, A., et al. (2009). Mitochondrial haplogroup H and Alzheimer's disease—Is there a connection? *Neurobiology of Aging, 30*(11), 1749–1755.

Marzo, I., et al. (1998). Bax and adenine nucleotide translocator cooperate in the mitochondrial control of apoptosis. *Science, 281*(5385), 2027–2031.

Masucci, J. P., et al. (1995). In vitro analysis of mutations causing myoclonus epilepsy with ragged-red fibers in the mitochondrial tRNA(Lys)gene: Two genotypes produce similar phenotypes. *Molecular and Cellular Biology, 15*(5), 2872–2881.

Maurer, I., Zierz, S., & Moller, H. J. (2000). A selective defect of cytochrome c oxidase is present in brain of Alzheimer disease patients. *Neurobiology of Aging, 21*(3), 455–462.

Mayr-Wohlfart, U., Rodel, G., & Henneberg, A. (1997). Mitochondrial tRNA(Gln) and tRNA(Thr) gene variants in Parkinson's disease. *European Journal of Medical Research, 2*(3), 111–113.

Mecocci, P., MacGarvey, U., & Beal, M. F. (1994). Oxidative damage to mitochondrial DNA is increased in Alzheimer's disease. *Annals of Neurology, 36*(5), 747–751.

Meirhaeghe, A., et al. (2003). Characterization of the human, mouse and rat PGC1 beta (peroxisome-proliferator-activated receptor-gamma co-activator 1 beta) gene in vitro and in vivo. *Biochemical Journal, 373*(Pt 1), 155–165.

Mitchell, P. (1961). Coupling of phosphorylation to electron and hydrogen transfer by a chemi-osmotic type of mechanism. *Nature, 191*, 144–148.

Mitchell, P., & Moyle, J. (1967). Chemiosmotic hypothesis of oxidative phosphorylation. *Nature, 213*(5072), 137–139.

Mitra, K., et al. (2009). A hyperfused mitochondrial state achieved at G1-S regulates cyclin E buildup and entry into S phase. *Proceedings of the National Academy of Sciences of the United States of America, 106*(29), 11960–11965.

Mizuno, Y., et al. (1989). Deficiencies in complex I subunits of the respiratory chain in Parkinson's disease. *Biochemical and Biophysical Research Communications, 163*(3), 1450–1455.

Mohamed, H. E., et al. (2010). Biochemical effect of a ketogenic diet on the brains of obese adult rats. *Journal of Clinical Neuroscience, 17*(7), 899–904.

Morris, J. K., et al. (2017). Aerobic exercise for Alzheimer's disease: A randomized controlled pilot trial. *PLoS ONE, 12*(2), e0170547.

Mosconi, L., et al. (2007). Maternal family history of Alzheimer's disease predisposes to reduced brain glucose metabolism. *Proceedings of the National Academy of Sciences of the United States of America, 104*(48), 19067–19072.

Mosconi, L., et al. (2009a). FDG-PET changes in brain glucose metabolism from normal cognition to pathologically verified Alzheimer's disease. *European Journal of Nuclear Medicine and Molecular Imaging, 36*(5), 811–822.

Mosconi, L., et al. (2009b). Declining brain glucose metabolism in normal individuals with a maternal history of Alzheimer disease. *Neurology, 72*(6), 513–520.

Mosconi, L., et al. (2010a). Increased fibrillar amyloid-{beta} burden in normal individuals with a family history of late-onset Alzheimer's. *Proceedings of the National Academy of Sciences of the United States of America, 107*(13), 5949–5954.

Mosconi, L., et al. (2010b). Maternal transmission of Alzheimer's disease: Prodromal metabolic phenotype and the search for genes. *Human Genomics, 3*(4), 170–193.

Mosconi, L., et al. (2010c). Oxidative stress and amyloid-beta pathology in normal individuals with a maternal history of Alzheimer's. *Biological Psychiatry, 68*(10), 913–921.

Mosconi, L., et al. (2011). Reduced mitochondria cytochrome oxidase activity in adult children of mothers with Alzheimer's disease. *Journal of Alzheimer's Disease, 27*(3), 483–490.

Muller, F. L., et al. (2007). Trends in oxidative aging theories. *Free Radical Biology and Medicine, 43*(4), 477–503.

Mullin, S., & Schapira, A. (2013). alpha-Synuclein and mitochondrial dysfunction in Parkinson's disease. *Molecular Neurobiology, 47*(2), 587–597.

Murphy, Michael P. (2009). How mitochondria produce reactive oxygen species. *Biochemical Journal, 417*(Pt 1), 1–13.

Mutisya, E. M., Bowling, A. C., & Beal, M. F. (1994). Cortical cytochrome oxidase activity is reduced in Alzheimer's disease. *Journal of Neurochemistry, 63*(6), 2179–2184.

Mytilineou, C., et al. (1994). Impaired oxidative decarboxylation of pyruvate in fibroblasts from patients with Parkinson's disease. *Journal of Neural Transmission. Parkinson's Disease and Dementia Section, 8*(3), 223–228.

Naj, A. C., et al. (2011). Common variants at MS4A4/MS4A6E, CD2AP, CD33 and EPHA1 are associated with late-onset Alzheimer's disease. *Nature Genetics, 43*(5), 436–441.

Nakagawa-Hattori, Y., et al. (1992). Is Parkinson's disease a mitochondrial disorder? *Journal of the Neurological Sciences, 107*(1), 29–33.

Nakano, Y., Hirayama, K., & Terao, K. (1987). Hepatic ultrastructural changes and liver dysfunction in amyotrophic lateral sclerosis. *Archives of Neurology*, *44*(1), 103–106.

Narendra, D., et al. (2008). Parkin is recruited selectively to impaired mitochondria and promotes their autophagy. *Journal of Cell Biology*, *183*(5), 795–803.

Nargund, A. M., et al. (2012). Mitochondrial import efficiency of ATFS-1 regulates mitochondrial UPR activation. *Science*, *337*(6094), 587–590.

Nargund, A. M., et al. (2015). Mitochondrial and nuclear accumulation of the transcription factor ATFS-1 promotes OXPHOS recovery during the UPR(mt). *Molecular Cell*, *58*(1), 123–133.

Narkar, V. A., et al. (2008). AMPK and PPARdelta agonists are exercise mimetics. *Cell*, *134*(3), 405–415.

Nass, M. M., & Nass, S. (1963a). Intramitochondrial fibers with DNA characteristics. I. fixation and electron staining reactions. *Journal of Cell Biology*, *19*, 593–611.

Nass, S., & Nass, M. M. (1963b). Intramitochondrial fibers with DNA characteristics. II. enzymatic and other hydrolytic treatments. *Journal of Cell Biology*, *19*, 613–629.

Navarro, A., et al. (2004). Beneficial effects of moderate exercise on mice aging: Survival, behavior, oxidative stress, and mitochondrial electron transfer. *American Journal of Physiology. Regulatory, Integrative and Comparative Physiology*, *286*(3), R505–R511.

Newman, N. J. (1993). Leber's hereditary optic neuropathy. New genetic considerations. *Archives of Neurology*, *50*(5), 540–548.

Nicholls, T. J., & Minczuk, M. (2014). In D-loop: 40 years of mitochondrial 7S DNA. *Experimental Gerontology*, *56*, 175–181.

Nicklas, W. J., Vyas, I., & Heikkila, R. E. (1985). Inhibition of NADH-linked oxidation in brain mitochondria by 1-methyl-4-phenyl-pyridine, a metabolite of the neurotoxin, 1-methyl-4-phenyl-1,2,5,6-tetrahydropyridine. *Life Sciences*, *36*(26), 2503–2508.

Niemann, A., et al. (2005). Ganglioside-induced differentiation associated protein 1 is a regulator of the mitochondrial network: New implications for Charcot–Marie–Tooth disease. *Journal of Cell Biology*, *170*(7), 1067–1078.

Nisoli, E., et al. (2003). Mitochondrial biogenesis in mammals: The role of endogenous nitric oxide. *Science*, *299*(5608), 896–899.

Noh, H. S., et al. (2004). A cDNA microarray analysis of gene expression profiles in rat hippocampus following a ketogenic diet. *Brain Research. Molecular Brain Research*, *129*(1–2), 80–87.

Nugent, S., et al. (2014). Brain glucose and acetoacetate metabolism: A comparison of young and older adults. *Neurobiology of Aging*, *35*(6), 1386–1395.

Okonkwo, O. C., et al. (2014). Cerebral blood flow is diminished in asymptomatic middle-aged adults with maternal history of Alzheimer's disease. *Cerebral Cortex*, *24*(4), 978–988.

Olichon, A., et al. (2003). Loss of OPA1 perturbates the mitochondrial inner membrane structure and integrity, leading to cytochrome c release and apoptosis. *Journal of Biological Chemistry*, *278*(10), 7743–7746.

Onyango, I. G., et al. (2010). Nerve growth factor attenuates oxidant-induced beta-amyloid neurotoxicity in sporadic Alzheimer's disease cybrids. *Journal of Neurochemistry*, *114*(6), 1605–1618.

Onyango, I. G., Tuttle, J. B., & Bennett, J. P., Jr. (2005a). Activation of p38 and N-acetylcysteine-sensitive c-Jun NH2-terminal kinase signaling cascades is required for induction of apoptosis in Parkinson's disease cybrids. *Molecular and Cellular Neuroscience*, *28*(3), 452–461.

Onyango, I. G., Tuttle, J. B., & Bennett, J. P., Jr. (2005b). Altered intracellular signaling and reduced viability of Alzheimer's disease neuronal cybrids is reproduced by beta-amyloid peptide acting through receptor for advanced glycation end products (RAGE). *Molecular and Cellular Neuroscience*, *29*(2), 333–343.

Onyango, I. G., Tuttle, J. B., & Bennett, J. P., Jr. (2005c). Brain-derived growth factor and glial cell line-derived growth factor use distinct intracellular signaling pathways to protect PD cybrids from H2O2-induced neuronal death. *Neurobiology of Disease*, *20*(1), 141–154.

Onyango, I. G., Bennett, J. P., Jr., & Tuttle, J. B. (2005d). Endogenous oxidative stress in sporadic Alzheimer's disease neuronal cybrids reduces viability by increasing apoptosis through pro-death signaling pathways and is mimicked by oxidant exposure of control cybrids. *Neurobiology of Disease*, *19*(1–2), 312–322.

Ow, Y. P., et al. (2008). Cytochrome c: Functions beyond respiration. *Nature Reviews Molecular Cell Biology*, *9*(7), 532–542.

Ozelius, L. J., et al. (2006). LRRK2 G2019S as a cause of Parkinson's disease in Ashkenazi Jews. *The New England Journal of Medicine*, *354*(4), 424–425.

Pacher, P., Beckman, J. S., & Liaudet, L. (2007). Nitric oxide and peroxynitrite in health and disease. *Physiological Reviews*, *87*(1), 315.

Paisan-Ruiz, C., et al. (2004). Cloning of the gene containing mutations that cause PARK8-linked Parkinson's disease. *Neuron*, *44*(4), 595–600.

Palade, G. E. (1953). An electron microscope study of the mitochondrial structure. *Journal of Histochemistry & Cytochemistry*, *1*(4), 188–211.

Pandolfo, M. (2008). Friedreich ataxia. *Archives of Neurology*, *65*(10), 1296–1303.

Panov, A. V., et al. (2002). Early mitochondrial calcium defects in Huntington's disease are a direct effect of polyglutamines. *Nature Neuroscience*, *5*(8), 731–736.

Papa, L., & Germain, D. (2011). Estrogen receptor mediates a distinct mitochondrial unfolded protein response. *Journal of Cell Science*, *124*(Pt 9), 1396–1402.

Parker, W. D., Jr., & Parks, J. K. (2005). Mitochondrial ND5 mutations in idiopathic Parkinson's disease. *Biochemical and Biophysical Research Communications*, *326*(3), 667–669.

Parker, W. D., Jr., & Parks, J. K. (1995). Cytochrome c oxidase in Alzheimer's disease brain: Purification and characterization. *Neurology*, *45*(3 Pt 1), 482–486.

Parker, W. D., Jr., & Swerdlow, R. H. (1998). Mitochondrial dysfunction in idiopathic Parkinson disease. *American Journal of Human Genetics*, *62*(4), 758–762.

Parker, W. D., Jr., Oley, C. A., & Parks, J. K. (1989). A defect in mitochondrial electron-transport activity (NADH-coenzyme Q oxidoreductase) in Leber's hereditary optic neuropathy. *The New England Journal of Medicine*, *320* (20), 1331–1333.

Parker, W. D., Jr., Filley, C. M., & Parks, J. K. (1990). Cytochrome oxidase deficiency in Alzheimer's disease. *Neurology*, *40*(8), 1302–1303.

Parker, W. D., Jr, et al. (1990). Evidence for a defect in NADH: Ubiquinone oxidoreductase (complex I) in Huntington's disease. *Neurology*, *40*(8), 1231–1234.

Parker, W. D., Jr., et al. (1994a). Reduced platelet cytochrome c oxidase activity in Alzheimer's disease. *Neurology*, *44*(6), 1086–1090.

Parker, W. D., Jr., et al. (1994b). Electron transport chain defects in Alzheimer's disease brain. *Neurology*, *44*(6), 1090–1096.

Parker, W. D., Jr., Parks, J. K., & Swerdlow, R. H. (2008). Complex I deficiency in Parkinson's disease frontal cortex. *Brain Research*, *1189*, 215–218.

Parker, W. D., Jr, Boyson, S. J., & Parks, J. K. (1989). Abnormalities of the electron transport chain in idiopathic Parkinson's disease. *Annals of Neurology*, *26*(6), 719–723.

Parker, W. D. (1990). Sporadic neurologic disease and the electron transport chain: A hypothesis. In R. M. Pascuzzi (Ed.), *Proceedings of the 1989 scientific meeting of the American Society for Neurological Investigation: New developments in neuromuscular disease*. Bloomington, Indiana: Indiana University Printing Services.

Parkinson Study Group. (1993). Effects of tocopherol and deprenyl on the progression of disability in early Parkinson's disease. *The New England Journal of Medicine*, *328*(3), 176–183.

Pasinelli, P., et al. (2004). Amyotrophic lateral sclerosis-associated SOD1 mutant proteins bind and aggregate with Bcl-2 in spinal cord mitochondria. *Neuron*, *43*(1), 19–30.

Pastorino, J. G., & Hoek, J. B. (2008). Regulation of hexokinase binding to VDAC. *Journal of Bioenergetics and Biomembranes*, *40*(3), 171–182.

Pearson, K. J., et al. (2008). Resveratrol delays age-related deterioration and mimics transcriptional aspects of dietary restriction without extending life span. *Cell Metabolism*, *8*(2), 157–168.

Pedrola, L., et al. (2005). GDAP1, the protein causing Charcot–Marie–Tooth disease type 4A, is expressed in neurons and is associated with mitochondria. *Human Molecular Genetics*, *14*(8), 1087–1094.

Pellegrino, M. W., Nargund, A. M., & Haynes, C. M. (2013). Signaling the mitochondrial unfolded protein response. *Biochimica et Biophysica Acta*, *1833*(2), 410–416.

Peng, J. Y., et al. (2011). Automatic morphological subtyping reveals new roles of caspases in mitochondrial dynamics. *PLoS Computational Biology*, *7*(10), e1002212.

Penn, A. M., et al. (1992). MELAS syndrome with mitochondrial tRNA(Leu)(UUR) mutation: Correlation of clinical state, nerve conduction, and muscle 31P magnetic resonance spectroscopy during treatment with nicotinamide and riboflavin. *Neurology*, *42*(11), 2147–2152.

Pfanner, N., & Meijer, M. (1997). Mitochondrial biogenesis: The Tom and Tim machine. *Current Biology*, *7*(2), R100–R103.

Phillips, N. R., Simpkins, J. W., & Roby, R. K. (2014). Mitochondrial DNA deletions in Alzheimer's brains: A review. *Alzheimer's & Dementia*, *10*(3), 393–400.

Pinto, M., et al. (2013). Mitochondrial DNA damage in a mouse model of Alzheimer's disease decreases amyloid beta plaque formation. *Neurobiology of Aging*, *34*(10), 2399–2407.

Polvikoski, T., et al. (2001). Prevalence of Alzheimer's disease in very elderly people: A prospective neuropathological study. *Neurology*, *56*(12), 1690–1696.

Puigserver, P., et al. (1998). A cold-inducible coactivator of nuclear receptors linked to adaptive thermogenesis. *Cell*, *92*(6), 829–839.

Pyle, A., et al. (2005). Mitochondrial DNA haplogroup cluster UKJT reduces the risk of PD. *Annals of Neurology*, *57* (4), 564–567.

Rambold, A. S., et al. (2011). Tubular network formation protects mitochondria from autophagosomal degradation during nutrient starvation. *Proceedings of the National Academy of Sciences of the United States of America*, *108*(25), 10190–10195.

Reddy, P., et al. (2015). Selective elimination of mitochondrial mutations in the germline by genome editing. *Cell*, *161*(3), 459–469.

Reddy, P. H. (2014). Inhibitors of mitochondrial fission as a therapeutic strategy for diseases with oxidative stress and mitochondrial dysfunction. *Journal of Alzheimer's Disease*, *40*(2), 245–256.

Reeve, A. K., et al. (2015). Aggregated alpha-synuclein and complex I deficiency: Exploration of their relationship in differentiated neurons. *Cell Death & Disease*, *6*, e1820.

Reger, M. A., et al. (2004). Effects of beta-hydroxybutyrate on cognition in memory-impaired adults. *Neurobiology of Aging*, *25*(3), 311–314.

Reiman, E. M., et al. (1996). Preclinical evidence of Alzheimer's disease in persons homozygous for the epsilon 4 allele for apolipoprotein E. *The New England Journal of Medicine*, *334*(12), 752–758.

Reiter, K., et al. (2012). Cognitively normal individuals with AD parents may be at risk for developing aging-related cortical thinning patterns characteristic of AD. *Neuroimage*, *61*(3), 525–532.

Revollo, J. R., Grimm, A. A., & Imai, S. (2004). The NAD biosynthesis pathway mediated by nicotinamide phosphoribosyltransferase regulates Sir2 activity in mammalian cells. *Journal of Biological Chemistry*, *279*(49), 50754–50763.

Riedl, S. J., & Salvesen, G. S. (2007). The apoptosome: Signalling platform of cell death. *Nature Reviews Molecular Cell Biology*, *8*(5), 405–413.

Risner, M. E., et al. (2006). Efficacy of rosiglitazone in a genetically defined population with mild-to-moderate Alzheimer's disease. *The Pharmacogenomics Journal*, *6*(4), 246–254.

Robinson, N. C. (1993). Functional binding of cardiolipin to cytochrome c oxidase. *Journal of Bioenergetics and Biomembranes*, *25*(2), 153–163.

Rodan, L. H., et al. (2015). L-Arginine affects aerobic capacity and muscle metabolism in MELAS (mitochondrial encephalomyopathy, lactic acidosis and stroke-like episodes) syndrome. *PLoS ONE*, *10*(5), e0127066.

Rodgers, J. T., et al. (2005). Nutrient control of glucose homeostasis through a complex of PGC-1alpha and SIRT1. *Nature*, *434*(7029), 113–118.

Rodriguez-Enriquez, S., et al. (2006). Tracker dyes to probe mitochondrial autophagy (mitophagy) in rat hepatocytes. *Autophagy*, *2*(1), 39–46.

Roesch, K., et al. (2002). Human deafness dystonia syndrome is caused by a defect in assembly of the DDP1/TIMM8a-TIMM13 complex. *Human Molecular Genetics*, *11*(5), 477–486.

Rogaeva, E., et al. (2007). The neuronal sortilin-related receptor SORL1 is genetically associated with Alzheimer disease. *Nature Genetics*, *39*(2), 168–177.

Roh, J. H., et al. (2014). Potential role of orexin and sleep modulation in the pathogenesis of Alzheimer's disease. *Journal of Experimental Medicine*, *211*(13), 2487–2496.

Rojo, A., et al. (2006). NARP–MILS syndrome caused by 8993 T>G mitochondrial DNA mutation: A clinical, genetic and neuropathological study. *Acta Neuropathologica*, *111*(6), 610–616.

Roses, A. D. (2010). An inherited variable poly-T repeat genotype in TOMM40 in Alzheimer disease. *Archives of Neurology*, *67*(5), 536–541.

Roses, A. D., et al. (2010). A TOMM40 variable-length polymorphism predicts the age of late-onset Alzheimer's disease. *The Pharmacogenomics Journal*, *10*(5), 375–384.

Roses, A. D., et al. (2013). TOMM40 and APOE: Requirements for replication studies of association with age of disease onset and enrichment of a clinical trial. *Alzheimer's & Dementia*, *9*(2), 132–136.

Ross, O. A., et al. (2003). mt4216C variant in linkage with the mtDNA TJ cluster may confer a susceptibility to mitochondrial dysfunction resulting in an increased risk of Parkinson's disease in the Irish. *Experimental Gerontology*, *38*(4), 397–405.

Rossi, L., et al. (2004). Mitochondrial dysfunction in neurodegenerative diseases associated with copper imbalance. *Neurochemical Research*, *29*(3), 493–504.

Rossignol, R., et al. (2004). Energy substrate modulates mitochondrial structure and oxidative capacity in cancer cells. *Cancer Research*, *64*(3), 985–993.

Rotig, A., et al. (1997). Aconitase and mitochondrial iron–sulphur protein deficiency in Friedreich ataxia. *Nature Genetics*, *17*(2), 215–217.

Rottscholl, R., et al. (2016). Chronic consumption of Annona muricata juice triggers and aggravates cerebral tau phosphorylation in wild-type and MAPT transgenic mice. *Journal of Neurochemistry*.

Ruan, L., et al. (2017). Cytosolic proteostasis through importing of misfolded proteins into mitochondria. *Nature*, *543*(7645), 443–446.

Ryder, J. W., et al. (2003). Skeletal muscle reprogramming by activation of calcineurin improves insulin action on metabolic pathways. *Journal of Biological Chemistry*, *278*(45), 44298–44304.

Ryman, D. C., et al. (2014). Symptom onset in autosomal dominant Alzheimer disease: A systematic review and meta-analysis. *Neurology*, *83*(3), 253–260.

Saez-Atienzar, S., et al. (2014). The LRRK2 inhibitor GSK2578215A induces protective autophagy in SH-SY5Y cells: Involvement of Drp-1-mediated mitochondrial fission and mitochondrial-derived ROS signaling. *Cell Death & Disease*, *5*, e1368.

Saitoh, S., et al. (1998). Effects of dichloroacetate in three patients with MELAS. *Neurology*, *50*(2), 531–534.

Sano, M., et al. (1997). A controlled trial of selegiline, alpha-tocopherol, or both as treatment for Alzheimer's disease. The Alzheimer's disease cooperative study. *The New England Journal of Medicine*, *336*(17), 1216–1222.

Santel, A., et al. (2003). Mitofusin-1 protein is a generally expressed mediator of mitochondrial fusion in mammalian cells. *Journal of Cell Science*, *116*(Pt 13), 2763–2774.

Santoro, A., et al. (2010). Evidence for sub-haplogroup h5 of mitochondrial DNA as a risk factor for late onset Alzheimer's disease. *PLoS ONE, 5*(8), e12037.

Sarafian, T. A., et al. (2013). Impairment of mitochondria in adult mouse brain overexpressing predominantly full-length, N-terminally acetylated human alpha-synuclein. *PLoS ONE, 8*(5), e63557.

Sasaki, S., & Iwata, M. (2007). Mitochondrial alterations in the spinal cord of patients with sporadic amyotrophic lateral sclerosis. *Journal of Neuropathology & Experimental Neurology, 66*(1), 10–16.

Sasaki, S., et al. (2004). Ultrastructural study of mitochondria in the spinal cord of transgenic mice with a G93A mutant SOD1 gene. *Acta Neuropathologica, 107*(5), 461–474.

Sasaki, S., Horie, Y., & Iwata, M. (2007). Mitochondrial alterations in dorsal root ganglion cells in sporadic amyotrophic lateral sclerosis. *Acta Neuropathologica, 114*(6), 633–639.

Satake, W., et al. (2009). Genome-wide association study identifies common variants at four loci as genetic risk factors for Parkinson's disease. *Nature Genetics, 41*(12), 1303–1307.

Sazanov, L. A. (2007). Respiratory complex I: Mechanistic and structural insights provided by the crystal structure of the hydrophilic domain. *Biochemistry, 46*(9), 2275–2288.

Scarpulla, R. C. (2008). Transcriptional paradigms in mammalian mitochondrial biogenesis and function. *Physiological Reviews, 88*(2), 611–638.

Schapira, A. H., et al. (1989). Mitochondrial complex I deficiency in Parkinson's disease. *The Lancet, 1*(8649), 1269.

Scheffler, K., et al. (2012). Mitochondrial DNA polymorphisms specifically modify cerebral beta-amyloid proteostasis. *Acta Neuropathologica, 124*(2), 199–208.

Schieke, S. M., & Finkel, T. (2006). Mitochondrial signaling, TOR, and life span. *Biological Chemistry, 387*(10–11), 1357–1361.

Schoenen, J., Jacquy, J., & Lenaerts, M. (1998). Effectiveness of high-dose riboflavin in migraine prophylaxis. A randomized controlled trial. *Neurology, 50*(2), 466–470.

Schug, Z. T., et al. (2011). BID is cleaved by caspase-8 within a native complex on the mitochondrial membrane. *Cell Death & Differentiation, 18*(3), 538–548.

Scott, F. L., et al. (2009). The Fas-FADD death domain complex structure unravels signalling by receptor clustering. *Nature, 457*(7232), 1019–1022.

Selfridge, J. E., et al. (2015). Effect of one month duration ketogenic and non-ketogenic high fat diets on mouse brain bioenergetic infrastructure. *Journal of Bioenergetics and Biomembranes, 47*(1–2), 1–11.

Seshadri, S., et al. (2010). Genome-wide analysis of genetic loci associated with Alzheimer disease. *The Journal of the American Medical Association, 303*(18), 1832–1840.

Sheehan, J. P., et al. (1997a). Altered calcium homeostasis in cells transformed by mitochondria from individuals with Parkinson's disease. *Journal of Neurochemistry, 68*(3), 1221–1233.

Sheehan, J. P., et al. (1997b). Calcium homeostasis and reactive oxygen species production in cells transformed by mitochondria from individuals with sporadic Alzheimer's disease. *The Journal of Neuroscience, 17*(12), 4612–4622.

Shefner, J. M., et al. (2004). A clinical trial of creatine in ALS. *Neurology, 63*(9), 1656–1661.

Sherer, T. B., et al. (2002). An in vitro model of Parkinson's disease: Linking mitochondrial impairment to altered alpha-synuclein metabolism and oxidative damage. *The Journal of Neuroscience, 22*(16), 7006–7015.

Sherer, T. B., et al. (2003). Subcutaneous rotenone exposure causes highly selective dopaminergic degeneration and alpha-synuclein aggregation. *Experimental Neurology, 179*(1), 9–16.

Sherrington, R., et al. (1995). Cloning of a gene bearing missense mutations in early-onset familial Alzheimer's disease. *Nature, 375*(6534), 754–760.

Shimizu, S., Narita, M., & Tsujimoto, Y. (1999). Bcl-2 family proteins regulate the release of apoptogenic cytochrome c by the mitochondrial channel VDAC. *Nature, 399*(6735), 483–487.

Shoffner, J. M., et al. (1990). Myoclonic epilepsy and ragged-red fiber disease (MERRF) is associated with a mitochondrial DNA tRNA(Lys) mutation. *Cell, 61*(6), 931–937.

Shoffner, J. M., et al. (1991). Mitochondrial oxidative phosphorylation defects in Parkinson's disease. *Annals of Neurology, 30*(3), 332–339.

Shoffner, J. M., et al. (1993). Mitochondrial DNA variants observed in Alzheimer disease and Parkinson disease patients. *Genomics, 17*(1), 171–184.

Shoubridge, E. A., & Wai, T. (2007). Mitochondrial DNA and the mammalian oocyte. *Current Topics in Developmental Biology, 77*, 87–111.

Shrivastava, M., et al. (2011). Mitochondrial perturbance and execution of apoptosis in platelet mitochondria of patients with amyotrophic lateral sclerosis. *International Journal of Neuroscience, 121*(3), 149–158.

Shults, C. W., & Miller, S. W. (1998). Reduced complex I activity in parkinsonian cybrids. *Movement Disorders, 13*(suppl. 2), 217.

Shults, C. W., et al. (2002). Effects of coenzyme Q10 in early Parkinson disease: Evidence of slowing of the functional decline. *Archives of Neurology, 59*(10), 1541–1550.

Silva Ramos, E., Larsson, N. G., & Mourier, A. (2016). Bioenergetic roles of mitochondrial fusion. *Biochimica et Biophysica Acta*.

Silva, D. F., et al. (2013a). Bioenergetic flux, mitochondrial mass and mitochondrial morphology dynamics in AD and MCI cybrid cell lines. *Human Molecular Genetics, 22*(19), 3931–3946.

Silva, D. F., et al. (2013b). Prodromal metabolic phenotype in MCI cybrids: Implications for Alzheimer's disease. *Current Alzheimer Research, 10*(2), 180–190.

Simon, D. K., et al. (2000). *Mitochondrial DNA mutations in complex I and tRNA genes in Parkinson's disease. Neurology, 54*(3), 703–709.

Simon, D. K., et al. (2010). Maternal inheritance and mitochondrial DNA variants in familial Parkinson's disease. *BMC Medical Genetics, 11*, 53.

Simon, D. K., et al. (2004). Somatic mitochondrial DNA mutations in cortex and substantia nigra in aging and Parkinson's disease. *Neurobiology of Aging, 25*(1), 71–81.

Simonian, N. A., & Hyman, B. T. (1993). Functional alterations in Alzheimer's disease: Diminution of cytochrome oxidase in the hippocampal formation. *Journal of Neuropathology & Experimental Neurology, 52*(6), 580–585.

Simon-Sanchez, J., et al. (2009). Genome-wide association study reveals genetic risk underlying Parkinson's disease. *Nature Genetics, 41*(12), 1308–1312.

Sims, N. R., et al. (1987). Mitochondrial function in brain tissue in primary degenerative dementia. *Brain Research, 436*(1), 30–38.

Sjostrand, F. S. (1953). Electron microscopy of mitochondria and cytoplasmic double membranes. *Nature, 171*(4340), 30–32.

Skulachev, V. P. (2001). Mitochondrial filaments and clusters as intracellular power-transmitting cables. *Trends in Biochemical Sciences, 26*(1), 23–29.

Sleiman, S. F., et al. (2016). Exercise promotes the expression of brain derived neurotrophic factor (BDNF) through the action of the ketone body beta-hydroxybutyrate. *Elife, 5*.

Small, G. W., et al. (1995). Apolipoprotein E type 4 allele and cerebral glucose metabolism in relatives at risk for familial Alzheimer disease. *The Journal of the American Medical Association, 273*(12), 942–947.

Smigrodzki, R., Parks, J., & Parker, W. D. (2004). High frequency of mitochondrial complex I mutations in Parkinson's disease and aging. *Neurobiology of Aging, 25*(10), 1273–1281.

Smirnova, E., et al. (2001). Dynamin-related protein Drp1 is required for mitochondrial division in mammalian cells. *Molecular Biology of the Cell, 12*(8), 2245–2256.

Sorbi, S., Bird, E. D., & Blass, J. P. (1983). Decreased pyruvate dehydrogenase complex activity in Huntington and Alzheimer brain. *Annals of Neurology, 13*(1), 72–78.

Stacpoole, P. W., et al. (2008). Role of dichloroacetate in the treatment of genetic mitochondrial diseases. *Advanced Drug Delivery Reviews, 60*(13–14), 1478–1487.

Stamelou, M., et al. (2008). Short-term effects of coenzyme Q10 in progressive supranuclear palsy: A randomized, placebo-controlled trial. *Movement Disorders, 23*(7), 942–949.

Steiner, J. L., et al. (2011). Exercise training increases mitochondrial biogenesis in the brain. *Journal of Applied Physiology, 111*(4), 1066–1071.

Storch, A., et al. (2007). Randomized, double-blind, placebo-controlled trial on symptomatic effects of coenzyme Q(10) in Parkinson disease. *Archives of Neurology, 64*(7), 938–944.

Strittmatter, W. J., et al. (1993). Apolipoprotein E: High-avidity binding to beta-amyloid and increased frequency of type 4 allele in late-onset familial Alzheimer disease. *Proceedings of the National Academy of Sciences of the United States of America, 90*(5), 1977–1981.

Suzuki, M., Youle, R. J., & Tjandra, N. (2000). Structure of Bax: Coregulation of dimer formation and intracellular localization. *Cell, 103*(4), 645–654.

Swerdlow, R., et al. (1994). Brain glucose metabolism in Alzheimer's disease. *The American Journal of the Medical Sciences, 308*(3), 141–144.

Swerdlow, R. H. (2012). Mitochondria and cell bioenergetics: Increasingly recognized components and a possible etiologic cause of Alzheimer's disease. *Antioxidants & Redox Signaling, 16*(12), 1434–1455.

Swerdlow, R. H. (2007a). Mitochondria in cybrids containing mtDNA from persons with mitochondriopathies. *Journal of Neuroscience Research, 85*(15), 3416–3428.

Swerdlow, R. H. (2007b). Pathogenesis of Alzheimer's disease. *Clinical Interventions in Aging, 2*(3), 347–359.

Swerdlow, R. H. (2007c). Is aging part of Alzheimer's disease, or is Alzheimer's disease part of aging? *Neurobiology of Aging, 28*(10), 1465–1480.

Swerdlow, R. H. (2009). The neurodegenerative mitochondriopathies. *Journal of Alzheimer's Disease, 17*(4), 737–751.

Swerdlow, R. H. (2011). Does mitochondrial DNA play a role in Parkinson's disease? A review of cybrid and other supportive evidence. *Antioxidants & Redox Signaling, 16*(9), 950–964.

Swerdlow, R. H. (2016). Bioenergetics and metabolism: A bench to bedside perspective. *Journal of Neurochemistry, 139*(Suppl 2), 126–135.

Swerdlow, R. H. (2014). Bioenergetic medicine. *British Journal of Pharmacology, 171*(8), 1854–1869.

Swerdlow, R. H., & Wooten, G. F. (2001). A novel deafness/dystonia peptide gene mutation that causes dystonia in female carriers of Mohr–Tranebjaerg syndrome. *Annals of Neurology, 50*(4), 537–540.

Swerdlow, R. H., & Khan, S. M. (2004). A "mitochondrial cascade hypothesis" for sporadic Alzheimer's disease. *Medical Hypotheses, 63*(1), 8–20.

Swerdlow, R. H., & Khan, S. M. (2009). The Alzheimer's disease mitochondrial cascade hypothesis: An update. *Experimental Neurology, 218*(2), 308–315.

Swerdlow, R. H., et al. (1996). Origin and functional consequences of the complex I defect in Parkinson's disease. *Annals of Neurology, 40*(4), 663–671.

Swerdlow, R. H., et al. (1997). Cybrids in Alzheimer's disease: A cellular model of the disease? *Neurology, 49*(4), 918–925.

Swerdlow, R. H., et al. (1998). Matrilineal inheritance of complex I dysfunction in a multigenerational Parkinson's disease family. *Annals of Neurology, 44*(6), 873–881.

Swerdlow, R. H., et al. (2017). Mitochondria, cybrids, aging, and Alzheimer's disease. *Progress in Molecular Biology and Translational Science, 146*, 259–302.

Swerdlow, R. H., et al. (1998). Mitochondria in sporadic amyotrophic lateral sclerosis. *Experimental Neurology, 153*(1), 135–142.

Swerdlow, R. H., et al. (2000). Role of mitochondria in amyotrophic lateral sclerosis. *Amyotrophic Lateral Sclerosis and Other Motor Neuron Disorders, 1*(3), 185–190.

Swerdlow, R. H., et al. (2006). Complex I polymorphisms, bigenomic heterogeneity, and family history in Virginians with Parkinson's disease. *Journal of the Neurological Sciences, 247*(2), 224–230.

Swerdlow, R. H., et al. (2016). Tolerability and pharmacokinetics of oxaloacetate 100 mg capsules in Alzheimer's subjects. *BBA Clinical, 5*, 120–123.

Swerdlow, R. H., Burns, J. M., & Khan, S. M. (2010). The Alzheimer's disease mitochondrial cascade hypothesis. *Journal of Alzheimer's Disease, 20*(Suppl 2), S265–S279.

Swerdlow, R. H., Burns, J. M., & Khan, S. M. (2014). The Alzheimer's disease mitochondrial cascade hypothesis: Progress and perspectives. *Biochimica et Biophysica Acta, 1842*(8), 1219–1231.

Szabados, T., et al. (2004). A chronic Alzheimer's model evoked by mitochondrial poison sodium azide for pharmacological investigations. *Behavioural Brain Research, 154*(1), 31–40.

Tadaishi, M., et al. (2011). Effect of exercise intensity and AICAR on isoform-specific expressions of murine skeletal muscle PGC-1alpha mRNA: A role of beta(2)-adrenergic receptor activation. *American Journal of Physiology. Endocrinology and Metabolism, 300*(2), E341–E349.

Taira, T., et al. (2004). DJ-1 has a role in antioxidative stress to prevent cell death. *EMBO Reports, 5*(2), 213–218.

Tait, S. W., et al. (2010). Resistance to caspase-independent cell death requires persistence of intact mitochondria. *Developmental Cell, 18*(5), 802–813.

Takeda, S., et al. (1988). Neuropathology of myoclonus epilepsy associated with ragged-red fibers (Fukuhara's disease). *Acta Neuropathologica, 75*(5), 433–440.

Tan, E. K., & Skipper, L. M. (2007). *Pathogenic mutations in Parkinson disease. Human Mutation, 28*(7), 641–653.

Tanner, C. M., & Goldman, S. M. (1996). Epidemiology of Parkinson's disease. *Neurologic Clinics, 14*(2), 317–335.

Tanzi, R. E., et al. (1993). The Wilson disease gene is a copper transporting ATPase with homology to the Menkes disease gene. *Nature Genetics, 5*(4), 344–350.

Taylor, R. W., et al. (1997). Selective inhibition of mutant human mitochondrial DNA replication in vitro by peptide nucleic acids. *Nature Genetics, 15*(2), 212–215.

Thal, L. J., et al. (2003). Idebenone treatment fails to slow cognitive decline in Alzheimer's disease. *Neurology, 61* (11), 1498–1502.

Thiffault, C., & Bennett, J. P., Jr. (2005). *Cyclical mitochondrial deltapsiM fluctuations linked to electron transport, F0F1 ATP-synthase and mitochondrial Na^+/Ca^{+2} exchange are reduced in Alzheimer's disease cybrids. Mitochondrion, 5* (2), 109–119.

Trimmer, P. A., & Bennett, J. P., Jr. (2009). The cybrid model of sporadic Parkinson's disease. *Experimental Neurology, 218*(2), 320–325.

Trimmer, P. A., & Borland, M. K. (2005). Differentiated Alzheimer's disease transmitochondrial cybrid cell lines exhibit reduced organelle movement. *Antioxidants & Redox Signaling, 7*(9–10), 1101–1109.

Trimmer, P. A., et al. (2000). Abnormal mitochondrial morphology in sporadic Parkinson's and Alzheimer's disease cybrid cell lines. *Experimental Neurology, 162*(1), 37–50.

Trimmer, P. A., et al. (2004a). Mitochondrial abnormalities in cybrid cell models of sporadic Alzheimer's disease worsen with passage in culture. *Neurobiology of Disease, 15*(1), 29–39.

Trimmer, P. A., et al. (2004b). Parkinson's disease transgenic mitochondrial cybrids generate Lewy inclusion bodies. *Journal of Neurochemistry, 88*(4), 800–812.

Trimmer, P. A., et al. (2009). Reduced axonal transport in Parkinson's disease cybrid neurites is restored by light therapy. *Molecular Neurodegeneration, 4*, 26.

Tsujimoto, Y., & Shimizu, S. (2002). The voltage-dependent anion channel: An essential player in apoptosis. *Biochimie, 84*(2–3), 187–193.

Twig, G., et al. (2008). Fission and selective fusion govern mitochondrial segregation and elimination by autophagy. *The EMBO Journal, 27*(2), 433–446.

Valente, E. M., et al. (2004). Hereditary early-onset Parkinson's disease caused by mutations in PINK1. *Science, 304*(5674), 1158–1160.

Valla, J., et al. (2006). Impaired platelet mitochondrial activity in Alzheimer's disease and mild cognitive impairment. *Mitochondrion, 6*(6), 323–330.

Valla, J., Berndt, J. D., & Gonzalez-Lima, F. (2001). Energy hypometabolism in posterior cingulate cortex of Alzheimer's patients: Superficial laminar cytochrome oxidase associated with disease duration. *The Journal of Neuroscience, 21*(13), 4923–4930.

van der Walt, J. M., et al. (2003). Mitochondrial polymorphisms significantly reduce the risk of Parkinson disease. *American Journal of Human Genetics, 72*(4), 804–811.

van der Walt, J. M., et al. (2004). Analysis of European mitochondrial haplogroups with Alzheimer disease risk. *Neuroscience Letters, 365*(1), 28–32.

Vanitallie, T. B., et al. (2005). Treatment of Parkinson disease with diet-induced hyperketonemia: A feasibility study. *Neurology, 64*(4), 728–730.

Veech, G. A., et al. (2000). Disrupted mitochondrial electron transport function increases expression of anti-apoptotic bcl-2 and bcl-X(L) proteins in SH-SY5Y neuroblastoma and in Parkinson disease cybrid cells through oxidative stress. *Journal of Neuroscience Research, 61*(6), 693–700.

Vela, L., et al. (2013). Direct interaction of Bax and Bak proteins with Bcl-2 homology domain 3 (BH3)-only proteins in living cells revealed by fluorescence complementation. *Journal of Biological Chemistry, 288*(7), 4935–4946.

Ventura-Clapier, R., Garnier, A., & Veksler, V. (2008). Transcriptional control of mitochondrial biogenesis: The central role of PGC-1alpha. *Cardiovascular Research, 79* (2), 208–217.

Verbessem, P., et al. (2003). Creatine supplementation in Huntington's disease: A placebo-controlled pilot trial. *Neurology, 61*(7), 925–930.

Vielhaber, S., et al. (2000). Mitochondrial DNA abnormalities in skeletal muscle of patients with sporadic amyotrophic lateral sclerosis. *Brain, 123*(Pt 7), 1339–1348.

Virbasius, J. V., & Scarpulla, R. C. (1994). Activation of the human mitochondrial transcription factor A gene by nuclear respiratory factors: A potential regulatory link between nuclear and mitochondrial gene expression in organelle biogenesis. *Proceedings of the National Academy of Sciences of the United States of America, 91*(4), 1309–1313.

Vives-Bauza, C., et al. (2010). PINK1-dependent recruitment of Parkin to mitochondria in mitophagy. *Proceedings of the National Academy of Sciences of the United States of America, 107*(1), 378–383.

von Ahsen, O., et al. (2000). Preservation of mitochondrial structure and function after Bid- or Bax-mediated cytochrome c release. *Journal of Cell Biology, 150*(5), 1027–1036.

Vonsattel, J. P., & DiFiglia, M. (1998). Huntington disease. *Journal of Neuropathology & Experimental Neurology, 57* (5), 369–384.

Walczak, H., & Haas, T. L. (2008). Biochemical analysis of the native TRAIL death-inducing signaling complex. *Methods in Molecular Biology, 414*, 221–239.

Wallace, D. C., et al. (1988). Mitochondrial DNA mutation associated with Leber's hereditary optic neuropathy. *Science, 242*(4884), 1427–1430.

Wang, W., et al. (2013). The ALS disease-associated mutant TDP-43 impairs mitochondrial dynamics and function in motor neurons. *Human Molecular Genetics, 22*(23), 4706–4719.

Wang, X., et al. (2009). Impaired balance of mitochondrial fission and fusion in Alzheimer's disease. *The Journal of Neuroscience, 29*(28), 9090–9103.

Wang, Y., & Brinton, R. D. (2016). Triad of risk for late onset Alzheimer's: Mitochondrial haplotype, APOE genotype and chromosomal sex. *Frontiers in Aging Neuroscience, 8*, 232.

Webster, M. T., et al. (1998). The effects of perturbed energy metabolism on the processing of amyloid precursor protein in PC12 cells. *Journal of Neural Transmission, 105*(8-9), 839–853.

Weisiger, R. A., & Fridovich, I. (1973). Superoxide dismutase. Organelle specificity. *Journal of Biological Chemistry, 248*(10), 3582–3592.

Weiss, C., et al. (2003). JNK phosphorylation relieves HDAC3-dependent suppression of the transcriptional activity of c-Jun. *The EMBO Journal, 22*(14), 3686–3695.

Wenz, T., et al. (2008). Activation of the PPAR/PGC-1alpha pathway prevents a bioenergetic deficit and effectively improves a mitochondrial myopathy phenotype. *Cell Metabolism, 8*(3), 249–256.

Wenz, T., et al. (2010). A metabolic shift induced by a PPAR panagonist markedly reduces the effects of pathogenic mitochondrial tRNA mutations. *Journal of Cellular and Molecular Medicine.*

Westermann, B. (2012). Bioenergetic role of mitochondrial fusion and fission. *Biochimica et Biophysica Acta, 1817* (10), 1833–1838.

Weyer, G., et al. (1997). A controlled study of 2 doses of idebenone in the treatment of Alzheimer's disease. *Neuropsychobiology, 36*(2), 73–82.

Wiedemann, F. R., et al. (1998). Impairment of mitochondrial function in skeletal muscle of patients with amyotrophic lateral sclerosis. *Journal of the Neurological Sciences, 156*(1), 65–72.

Wiedemann, F. R., et al. (2002). Mitochondrial DNA and respiratory chain function in spinal cords of ALS patients. *Journal of Neurochemistry, 80*(4), 616–625.

Wilkins, H. M., et al. (2014). Oxaloacetate activates brain mitochondrial biogenesis, enhances the insulin pathway, reduces inflammation and stimulates neurogenesis. *Human Molecular Genetics, 23*(24), 6528–6541.

Wilkins, H. M., et al. (2016). Oxaloacetate enhances neuronal cell bioenergetic fluxes and infrastructure. *Journal of Neurochemistry, 137*(1), 76–87.

Willis, M. W., et al. (2002). Age, sex and laterality effects on cerebral glucose metabolism in healthy adults. *Psychiatry Research, 114*(1), 23–37.

Wiseman, D. A., & Thurmond, D. C. (2012). The good and bad effects of cysteine S-nitrosylation and tyrosine nitration upon insulin exocytosis: A balancing act. *Current Diabetes Reviews, 8*(4), 303–315.

Wisniewski, H., Terry, R. D., & Hirano, A. (1970). Neurofibrillary pathology. *Journal of Neuropathology & Experimental Neurology, 29*(2), 163–176.

Witte, A. V., et al. (2009). Caloric restriction improves memory in elderly humans. *Proceedings of the National Academy of Sciences of the United States of America, 106*(4), 1255–1260.

Wolf, D. P., Mitalipov, N., & Mitalipov, S. (2015). Mitochondrial replacement therapy in reproductive medicine. *Trends in Molecular Medicine, 21*(2), 68–76.

Wong, P. C., et al. (1995). An adverse property of a familial ALS-linked SOD1 mutation causes motor neuron disease characterized by vacuolar degeneration of mitochondria. *Neuron, 14*(6), 1105–1116.

Wong-Riley, M., et al. (1997). Cytochrome oxidase in Alzheimer's disease: Biochemical, histochemical, and immunohistochemical analyses of the visual and other systems. *Vision Research, 37*(24), 3593–3608.

Wu, Z., et al. (2006). Transducer of regulated CREB-binding proteins (TORCs) induce PGC-1alpha transcription and mitochondrial biogenesis in muscle cells. *Proceedings of the National Academy of Sciences of the United States of America, 103*(39), 14379–14384.

Yaffe, K., et al. (2011). Mild cognitive impairment, dementia, and their subtypes in oldest old women. *Archives of Neurology, 68*(5), 631–636.

Yamada, E. S., et al. (2014). Annonacin, a natural lipophilic mitochondrial complex I inhibitor, increases phosphorylation of tau in the brain of FTDP-17 transgenic mice. *Experimental Neurology, 253*, 113–125.

Yamamoto, K., et al. (2015). Chronic optogenetic activation augments abeta pathology in a mouse model of Alzheimer disease. *Cell Reports, 11*(6), 859–865.

Yang, T., & Sauve, A. A. (2006). NAD metabolism and sirtuins: Metabolic regulation of protein deacetylation in stress and toxicity. *The AAPS Journal, 8*(4), E632–E643.

Yang, Y., & Sauve, A. A. (2016). NAD+ metabolism: Bioenergetics, signaling and manipulation for therapy. *Biochimica et Biophysica Acta, 1864*(12), 1787–1800.

Yao, J., et al. (2011a). 2-Deoxy-D-glucose treatment induces ketogenesis, sustains mitochondrial function, and reduces pathology in female mouse model of Alzheimer's disease. *PLoS ONE, 6*(7), e21788.

Yao, J., et al. (2011b). Inhibition of amyloid-beta (Abeta) peptide-binding alcohol dehydrogenase-Abeta interaction reduces Abeta accumulation and improves mitochondrial function in a mouse model of Alzheimer's disease. *The Journal of Neuroscience, 31*(6), 2313–2320.

Yoneda, T., et al. (2004). Compartment-specific perturbation of protein handling activates genes encoding mitochondrial chaperones. *Journal of Cell Science, 117*(Pt 18), 4055–4066.

Yoshino, H., et al. (1992). Mitochondrial complex I and II activities of lymphocytes and platelets in Parkinson's disease. *Journal of Neural Transmission. Parkinson's Disease and Dementia Section, 4*(1), 27–34.

Youle, R. J., & Narendra, D. P. (2011). Mechanisms of mitophagy. *Nature Reviews Molecular Cell Biology, 12*(1), 9–14.

Yu, Q., et al. (2016). Antioxidants rescue mitochondrial transport in differentiated Alzheimer's disease transmitochondrial cybrid cells. *Journal of Alzheimer's Disease, 54*(2), 679–690.

Zabetian, C. P., et al. (2006). LRRK2 G2019S in families with Parkinson disease who originated from Europe and the Middle East: Evidence of two distinct founding events beginning two millennia ago. *American Journal of Human Genetics, 79*(4), 752–758.

Zeviani, M., et al. (1988). Deletions of mitochondrial DNA in Kearns–Sayre syndrome. *Neurology, 38*(9), 1339–1346.

Zhang, H., et al. (2008). Mitochondrial autophagy is an HIF-1-dependent adaptive metabolic response to hypoxia. *Journal of Biological Chemistry, 283*(16), 10892–10903.

Zhang, H., et al. (2011). Puerarin protects Alzheimer's disease neuronal cybrids from oxidant-stress induced apoptosis by inhibiting pro-death signaling pathways. *Experimental Gerontology, 46*(1), 30–37.

Zhang, H., et al. (2014). Glutaredoxin 2 reduces both thioredoxin 2 and thioredoxin 1 and protects cells from apoptosis induced by auranofin and 4-hydroxynonenal. *Antioxidants & Redox Signaling, 21*(5), 669–681.

Zhang, L., et al. (2008). Semi-quantitative analysis of alpha-synuclein in subcellular pools of rat brain neurons: An immunogold electron microscopic study using a C-terminal specific monoclonal antibody. *Brain Research, 1244*, 40–52.

Zhang, L., et al. (2015). Modulation of mitochondrial complex I activity averts cognitive decline in multiple animal models of familial Alzheimer's Disease. *EBioMedicine, 2*(4), 294–305.

Zhang, S., et al. (2010). Dimebon (latrepirdine) enhances mitochondrial function and protects neuronal cells from death. *Journal of Alzheimer's Disease, 21*(2), 389–402.

Zhao, B., et al. (2016). Mitochondrial dysfunction activates the AMPK signaling and autophagy to promote cell survival. *Genes & Diseases, 3*(1), 82–87.

Zhao, Q., et al. (2002). A mitochondrial specific stress response in mammalian cells. *The EMBO Journal*, *21*(17), 4411–4419.

Zhou, C., et al. (2014). Organelle-based aggregation and retention of damaged proteins in asymmetrically dividing cells. *Cell*, *159*(3), 530–542.

Zimprich, A., et al. (2004). Mutations in LRRK2 cause autosomal-dominant parkinsonism with pleomorphic pathology. *Neuron*, *44*(4), 601–607.

Zou, H., et al. (1999). An APAF-1. cytochrome c multimeric complex is a functional apoptosome that activates procaspase-9. *Journal of Biological Chemistry*, *274*(17), 11549–11556.

Zuchner, S., et al. (2004). Mutations in the mitochondrial GTPase mitofusin 2 cause Charcot–Marie–Tooth neuropathy type 2A. *Nature Genetics*, *36*(5), 449–451.

CHAPTER

14

Non-cell Autonomous Degeneration: Role of Astrocytes in Neurodegenerative Diseases

Sarah E. Smith[1] *and Azad Bonni*[2]

[1]Medical Scientist Training Program, Washington University School of Medicine, St. Louis, MO, United States [2]Department of Neuroscience, Washington University School of Medicine, St. Louis, MO, United States

OUTLINE

INTRODUCTION

Besides neurons, glial cells represent the major cell type in the central nervous system (CNS). Glial cells are classified into the two broad categories: macroglia and microglia. Comprised of astrocytes, oligodendrocytes, and ependymal cells, macroglial cells are derived from neuroepithelial precursor cells and thus share their origin with neurons (Gotz & Huttner, 2005). Microglia, however, are related to macrophages, and are thought to represent resident CNS immune cells (Ginhoux, Lim, Hoeffel, Low, & Huber, 2013). Among the glial cells, astrocytes are arguably the least understood cell type, and although we have learned a great deal in recent decades about their functions during development and in the mature CNS,

The Molecular and Cellular Basis of Neurodegenerative Diseases
DOI: https://doi.org/10.1016/B978-0-12-811304-2.00014-6

many important roles for astrocytes likely remain to be discovered.

Gray matter astroglia form a pattern of non-overlapping, regularly spaced zones, each occupied by a single astrocyte (Oberheim et al., 2009). The neuropil is composed of highly ramified astrocytic processes, which make contact with synapses, blood vessels, and other astrocytes (Oberheim et al., 2009; Oberheim, Goldman, & Nedergaard, 2012). Although they are broadly categorized as gray matter protoplasmic and white matter fibrous astrocytes, regional and developmental variability lead to further classification based on cytoarchitecture as well as protein expression and function (Emsley & Macklis, 2006; Miller & Raff, 1984). Astrocytes have been implicated in diverse biological processes in the developing and mature CNS, including synapse development and pruning, modulation of the blood–brain barrier and control of cerebral blood flow, ionic buffering of the extracellular environment, neuronal metabolism, information processing, regulation of inflammation, and oxidative stress (Fig. 14.1) (Allaman, Belanger, & Magistretti, 2011; Barres, 2008; Belanger, Allaman, & Magistretti, 2011; Oberheim et al., 2012).

Signals from astrocytes are essential for the formation of functional synapses. Without astrocytes, cultured neurons exhibit reduced synapse formation and synapse immaturity (Pfrieger & Barres, 1997; Ullian, Sapperstein, Christopherson, & Barres, 2001). Gliogenesis and synaptogenesis both occur together after neurogenesis, suggesting that astrocytes may promote synapse formation (Ullian et al., 2001). In vivo evidence has revealed a role for astrocytes in synapse formation and function (Clarke & Barres, 2013). For example, although ocular dominance columns are thought to be stable after the critical period, young astrocytes transplanted into the mature visual cortex promote de novo synapse formation and plasticity (Muller & Best, 1989). Several secreted factors mediate the synaptogenic effects of astrocytes. Immature astrocytes secrete the extracellular matrix glycoprotein thrombospondin to induce the formation of synapses in neurons, although these synapses are postsynaptically silent without the presence of other factors (Christopherson et al., 2005). Other astrocyte-secreted factors, such as glypicans, are required for the formation of postsynaptically functional synapses (Allen et al., 2012). Additionally, astrocytes secrete cholesterol complexed with apolipoprotein E (ApoE), which serves to assist synapse formation and function (Clarke & Barres, 2013; Mauch et al., 2001). Astrocytes are also critical for synapse elimination. Immature astrocytes promote tagging of unwanted synapses by the complement protein C1q, which leads to phagocytosis of synapses by microglia (Stevens et al., 2007). Astrocytes also directly phagocytose synapses during development and adulthood (Chung et al., 2013).

A major function of mature astrocytes is creating an extracellular environment that is supportive for neurons. The importance of astrocytes in conditioning the extracellular environment has been observed in vitro, where neurons are significantly less viable without astrocytes or astrocyte-conditioned medium present (Banker, 1980). Astrocytes take up glutamate from the synaptic space via the excitatory amino acid transporters (EAATs): glutamate transporter 1 (GLT -1) and glutamate aspartate transporter (GLAST). Then, astrocytes convert glutamate to glutamine, and shuttle glutamine back to neurons (Anderson & Swanson, 2000). Glutamate transport by astrocytes prevents neuronal excitotoxicity and facilitates neurotransmission by supplying neurons with glutamine, the precursor to glutamate (Anderson & Swanson, 2000). Astrocytes are also responsible for absorbing potassium released by neurons during depolarization, a process known as potassium spatial buffering (Kofuji & Newman, 2004). This process is critical because neurons are highly sensitive to changes in extracellular potassium (Kofuji & Newman, 2004). Astrocytic uptake of potassium after

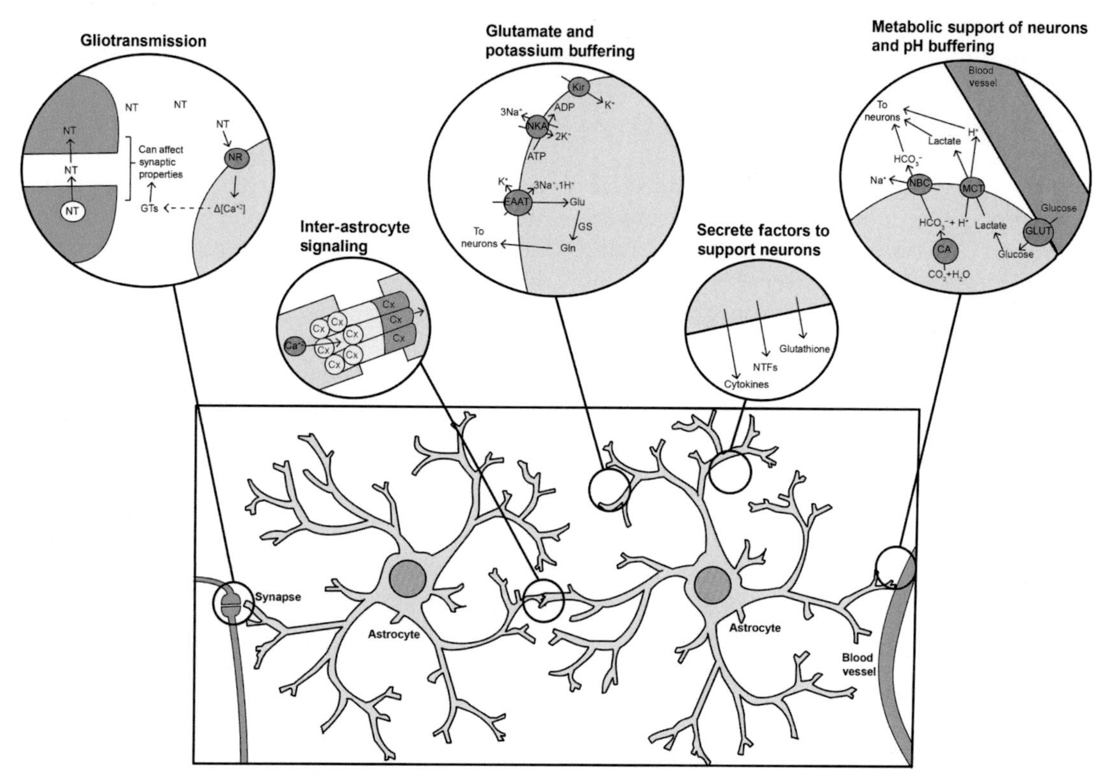

FIGURE 14.1 Physiological functions of astrocytes. For further detail, see text or reviews of astrocyte-neuron physiology (Allaman et al., 2011; Barres, 2008; Belanger et al., 2011; Oberheim et al., 2012). *Gliotransmission:* In the tripartite synapse, astrocytic processes ensheath the synapse, where they respond to and influence synaptic transmission. In response to NT release and binding to NTRs on astrocytes, astrocytic Ca^{2+} transients occur. Calcium signaling then indirectly (dashed line) leads to release of GTs such as adenosine, ATP, and D-serine via secretory lysosomes or hemichannels onto neurons (Pascual et al., 2005; Perea et al., 2009). *Interastrocyte signaling:* Astrocytes signal to other astrocytes via passage of calcium and other small molecules through gap junctions, which are comprised of 6 transmembrane Cx proteins (Houades et al., 2006; Kuchibhotla et al., 2009; Srinivasan et al., 2016; Tang et al., 2015). *Glu and potassium buffering:* Astrocytes take up Glu via the EAAT. Astrocytes then convert Glu to Gln via GS, and finally shuttle Gln back to neurons. EAATs are antiporters which import one molecule of Glu, one molecule of hydrogen and three molecules of sodium, and export one molecule of potassium (Jensen, Fahlke, Bjorn-Yoshimoto, & Bunch, 2015). Astrocytes also buffer potassium primarily by potassium inward rectifying channels (Kir) and NKA (Kofuji & Newman, 2004). *Secretion of molecules to neurons:* Astrocytes release a variety of molecules to support neuronal survival and function, including antiinflammatory cytokines, the antioxidant glutathione, and a host of NTFs including CNTF, GDNF, BDNF, and VEGF (Allaman et al., 2011; Belanger et al., 2011; Oberheim et al., 2012). *Metabolic support of neurons and pH buffering:* In the neuron-astrocyte lactate shuttle, astrocytes extract glucose from the blood and metabolize it to lactate, which is then secreted and taken up by neurons during periods of intense activity (Allaman et al., 2011). The neuron-astrocyte lactate shuttle is coupled to neuronal activity via Glu uptake by the astrocyte. Astrocytes also buffer pH. One mechanism of pH buffering is schematized here: astrocytic CA converts carbon dioxide and water into bicarbonate (HCO_3^-) and hydrogen ions. Hydrogen ions are cotransported out of the astrocyte with lactate via MCTs and bicarbonate is exported via sodium-bicarbonate cotransporters (NBCs) (Sofroniew, 2009). *NT*, neurotransmitter; *NTR*, NT receptor; *GT*, gliotransmitter; *Cx*, connexin; *Glu*, glutamate; *EAAT*, excitatory amino acid transporter; *Gln*, glutamine; *GS*, Gln synthetase; *NKA*, Na/K ATPases; *NTF*, neurotrophic factor; *CA*, carbonic anhydrase; *MCT*, monocarboxylate transporter; *NBC*, sodium-bicarbonate cotransporter.

neuronal depolarization is primarily accomplished by potassium inward rectifying channels, especially Kir4.1, whereas Na/K ATPases actively transport potassium into the cell under baseline conditions (Kofuji & Newman, 2004). Astrocytes also perform pH buffering of the extracellular solution (Allaman et al., 2011).

Neurons rely heavily upon astrocytic secretion of supportive molecules, including metabolites and neurotrophic factors. Growing evidence suggests that astrocytes share the metabolic cost of maintaining ion gradients to support synaptic and action potentials in neurons (Belanger et al., 2011). In the neuron-astrocyte lactate shuttle, astrocytes extract glucose from the blood and metabolize it to lactate, which is then secreted and taken up by neurons during periods of intense activity (Allaman et al., 2011). Astrocytes also regulate local blood flow to match demand, a process mediated by release of vasoactive compounds from arteriole-adjacent astrocytic end-feet (Belanger et al., 2011). In states of neuroinflammation, astrocytes modulate permeability of the blood–brain barrier via these end-feet (Barres, 2008). However, even with proper ionic balance and metabolites in solution, neurons do not thrive without astrocytes. In addition to regulating ion gradients and metabolism, astrocytes release a variety of molecules to support neuronal survival and function, including antiinflammatory cytokines, the antioxidant glutathione, a host of neurotrophic factors including ciliary neurotrophic factor (CNTF), glial cell-line derived neurotrophic factor (GDNF), brain-derived neurotrophic factor (BDNF) and vascular endothelial growth factor (VEGF) and likely yet to be identified molecules (Allaman et al., 2011; Belanger et al., 2011; Oberheim et al., 2012).

Finally, a role for astrocytes in information processing is beginning to be characterized. Although not electrically excitable per se, astrocytes display spontaneous fluctuations in calcium levels (Schummers, Yu, & Sur, 2008). Calcium signals propagate as waves between neighboring astrocytes, respond to neurotransmitters, and correlate with neuronal activity or behavior (Srinivasan et al., 2016; Tang et al., 2015). The spread of calcium waves between astrocytes occurs via Connexin-43-containing gap junctions within a glial network or circuit, and the degree of interglial connectivity is modulated by region, age, and disease (Houades et al., 2006; Kuchibhotla, Lattarulo, Hyman, & Bacskai, 2009). What is the function of astrocytic calcium fluctuations? In what is referred to as the tripartite synapse, astrocytic processes ensheath the presynaptic and postsynaptic regions, where they respond to and influence synaptic transmission (Perea, Navarrete, & Araque, 2009). Gliotransmitters such as adenosine, adenosine triphosphate (ATP), and D-serine are released via secretory lysosomes or hemichannels onto neurons (Pascual et al., 2005; Perea et al., 2009). Our understanding of these biological processes is in its infancy, but exciting new studies suggest that tasks such as olfactory-induced chemotaxis and touch-induced startle in *Drosophila* require information flow through astrocytes (Ma, Stork, Bergles, & Freeman, 2016).

In light of the complex ways in which astrocytes contribute to essential neuronal functions, it is not surprising that astrocytic dysfunction can contribute to disease pathogenesis. Impairment of astrocyte function nearly always accompanies neurological disease, although the effects on glia may not be readily disambiguated from secondary glial responses. This is because in response to neuronal dysfunction or death, astrocytes undergo a highly heterogeneous, gradual phenotypic, and functional change known as reactive astrocytosis or astrogliosis (Sofroniew, 2009). Reactive astrocytes exhibit a complex combination of beneficial and detrimental effects on gene expression that varies depending upon the nature of the pathogenic mechanisms in neurons (Sofroniew, 2009). For instance, in response to CNS injury or ischemia, astrocytes in the surrounding parenchyma upregulate the intermediate filament glial fibrillary acidic protein (GFAP) and other cytoskeletal proteins (Yang & Wang, 2015). Upregulation of GFAP is associated with astrocyte hypertrophy and hyperplasia, which forms, in severe cases, a glial scar (Yang & Wang, 2015). Reactive astrocytosis also functions to reseal the blood–brain barrier, preventing infection and regulating the inflammatory response (Bush et al., 1999). However, reactive astrocytes also cause adverse consequences, including disruption of glutamate handling, alteration of calcium dynamics, activation of microglia, and upregulation of inflammatory gene expression (Pekny & Nilsson, 2005; Pekny & Pekna, 2014; Sofroniew, 2009). In this chapter, we will discuss the complex and often overlooked roles of astrocyte reactivity and dysfunction in neurodegenerative diseases.

ASTROCYTES IN AMYOTROPHIC LATERAL SCLEROSIS

Amyotrophic lateral sclerosis (ALS) is the most common motor neuron disease, affecting 1 in 25,000 Americans (Mehta et al., 2014). Characterized by relentlessly progressive paralysis, patients typically die of ventilatory failure within 5 years after onset of symptoms (Shaw & Wood-Allum, 2010; Worms, 2001). Riluzole, the only drug approved for treatment of ALS, prolongs survival about 2–3 months, so the development of new treatments is critical (Miller, Mitchell, & Moore, 2012). Pathologically, the disease is characterized by motor neuron degeneration, with loss and atrophy of lower motor neurons from the anterior horn of the spinal cord and brainstem motor nuclei (Cirulli et al., 2015). The mechanisms underlying selective motor neuron cell death are not well understood (Vucic, Rothstein, & Kiernan, 2014).

Approximately 5%–10% of ALS cases are familial (Byrne et al., 2011). Of the known genetic mutations in familial ALS, the best characterized are dominant, gain-of-toxic-function mutations in the gene encoding copper–zinc superoxide dismutase 1 (SOD1), which are responsible for approximately 20% of familial ALS (Pasinelli & Brown, 2006; Rosen et al., 1993). SOD1 is an enzyme normally responsible for the processing of oxidative free radicals, but how mutant SOD1 confers toxicity to motor neurons remains to be elucidated (Gurney et al., 1994). Transgenic expression of mutant human SOD1 in mice is the most commonly used preclinical model of ALS. High copy number transgenic expression of human SOD1 with a G93A mutation ($SOD1^{G93A}$) leads to motor neuron degeneration, astrogliosis, paralysis, and death by 6 months of age (Gurney et al., 1994; Tu et al., 1996). Other mutations in human SOD1, including G37R and G85R, have also been modeled in transgenic mice and recapitulate this phenotype (Bruijn et al., 1997; Wong et al., 1995).

Although ALS is primarily a disease of motor neurons, glial cells play a crucial role in disease pathogenesis, in particular in mediating progression following symptom onset. Germline expression of mutant SOD1 causes a severe ALS phenotype in mice, but expression of $SOD1^{G93A}$ in motor neurons alone is insufficient to cause disease (Fig. 14.2) (Lino, Schneider, & Caroni, 2002; Pramatarova, Laganiere, Roussel, Brisebois, & Rouleau, 2001). In chimeric mice that harbor both wild-type (WT) and $SOD1^{G93A}$-expressing cells, motor neurons undergo degeneration in the vicinity of mutant SOD1 glial cells, even if motor neurons are wild type (Clement et al., 2003).

Cell-autonomous and non-cell autonomous contributions to motor neuron degeneration in $SOD1^{G93A}$ mice have been further dissected using Cre/loxP targeting of specific cell types (Sauer, 1998). Cre-mediated disruption of $SOD1^{G93A}$ in spinal motor neurons results in mice with substantially later onset of disease and prolongation of the early phase (Fig. 14.2) (Boillee et al., 2006). By contrast, inactivation of

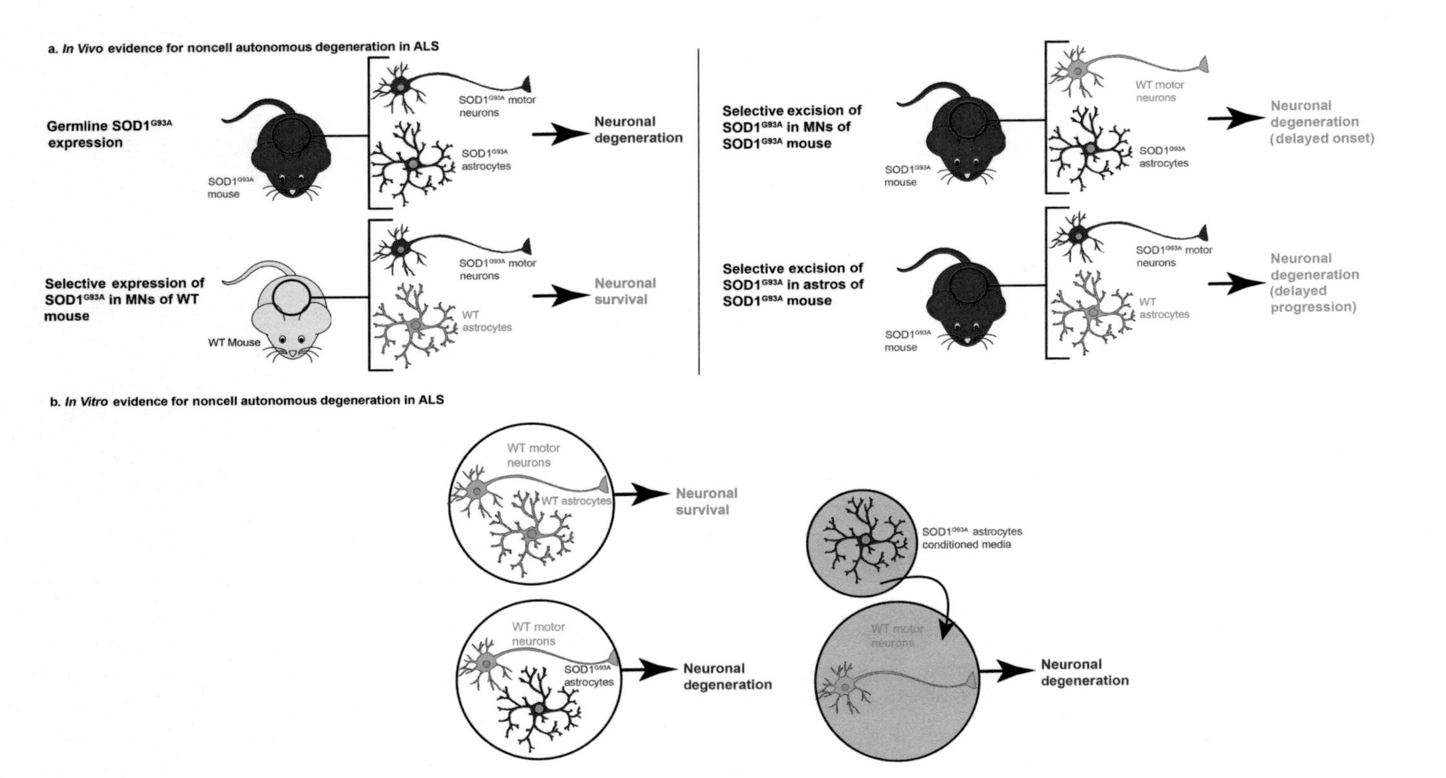

FIGURE 14.2 Evidence for non-cell autonomous degeneration in ALS. (A) In vivo, germline expression of mutant SOD1 causes spinal motor neuron death, but expression of $SOD1^{G93A}$ in motor neurons alone is insufficient to cause neurodegeneration (Lino et al., 2002; Pramatarova et al., 2001). Cre-mediated disruption of $SOD1^{G93A}$ in spinal motor neurons of $SOD1^{G93A}$ mice results in motor neuron degeneration and ALS symptomatology, although symptoms are delayed in onset relative to non-Cre expressing $SOD1^{G93A}$ mice (Boillee et al., 2006). By contrast, inactivation of $SOD1^{G93A}$ specifically in astrocytes does not affect the timing of onset or the early phase, but substantially prolongs the symptomatic phase or disease progression relative to non-Cre expressing $SOD1^{G93A}$ mice (Yamanaka et al., 2008). (B) In vitro, motor neurons are viable when cocultured with WT astrocytes, but degenerate when cocultured with $SOD1^{G93A}$ astrocytes (Di Giorgio et al., 2007; Nagai et al., 2007). Addition of conditioned medium from primary cultures of $SOD1^{G93A}$ astrocytes is toxic to motor neurons (Nagai et al., 2007). *ALS*, amyotrophic lateral sclerosis; *SOD1*, superoxide dismutase 1.

$SOD1^{G93A}$ specifically in astrocytes does not affect the timing of onset or the early phase, but substantially prolongs the symptomatic phase (Yamanaka et al., 2008). In mouse models of other ALS-associated SOD1 mutations, elimination of mutant SOD1 expression specifically in astrocytes delays onset (Wang, Gutmann, & Roos, 2011). Cell-type specific knockouts of mutant SOD1 in motor neurons and astrocytes have similar effects on overall survival (Clement et al., 2003; Yamanaka et al., 2008). These studies reveal that astrocytes aggravate degeneration and hasten cell death once motor neurons have begun to die. $SOD1^{G93A}$ astrocytes transplanted into wild-type mouse spinal cord induce motor neuron death and muscle dysfunction (Papadeas, Kraig, O'Banion, Lepore, & Maragakis, 2011). In a rat model of familial ALS, wild-type astrocytes transplanted into the spinal cord extend survival, reduce paralysis, and slow motor neuron degeneration (Lepore et al., 2008). Together, these studies, combined with findings of astrogliosis in ALS patients and mouse models (Sugiyama et al., 2013; Yang et al., 2011), suggest astrocytes play a critical role in motor neuron degeneration.

Coculture assays of glia and neurons have also been employed to study non-cell autonomous motor neuron degeneration in ALS. When cocultured, astrocytes expressing mutant SOD1 (G93A, G37R, and G85R) are toxic to wild-type motor neurons (Fig. 14.2) (Di Giorgio, Carrasco, Siao, Maniatis, & Eggan, 2007; Nagai et al., 2007). Motor neurons plated on astrocytes from mutant SOD1 mice undergo cell death earlier than neurons plated on astrocytes from wild type mice (Nagai et al., 2007). Furthermore, the toxic effect of $SOD1^{G93A}$ astrocytes is specific to motor neurons. $SOD1^{G93A}$ astrocytes do not induce cell death in other neuron subtypes, including GABAergic neurons, dorsal root ganglion neurons, and dorsal spinal cord interneurons (Nagai et al., 2007). Likewise, astrocytes cultured directly from ALS patients or converted from patient fibroblasts trigger degeneration of cocultured WT motor neurons (Haidet-Phillips et al., 2011; Meyer et al., 2014). SOD^{G93A} neurons show more pronounced degeneration than wild-type neurons when cocultured with $SOD1^{G93A}$ glia (Di Giorgio et al., 2007). Remarkably, addition of just the conditioned medium from primary cultures of $SOD1^{G93A}$ astrocytes is similarly toxic to motor neurons as the $SOD1^{G93A}$ astrocytes in the coculture paradigm (Fig. 14.2) (Nagai et al., 2007).

Until recently, the cell-intrinsic mechanism by which mutant SOD1 astrocytes trigger non-cell autonomous degeneration of motor neurons remained largely unknown. Gallardo et al. discovered a protein complex, comprised of the actin-binding protein α-adducin and ion pump α2-Na/K ATPase, that triggers $SOD1^{G93A}$ astrocyte toxicity (Fig. 14.3) (Gallardo et al., 2014). α-Adducin and α2-Na/K ATPase are upregulated in $SOD1^{G93A}$ cultured astrocytes, $SOD1^{G93A}$ mouse spinal cord astrocytes, and, importantly, in spinal cord lysates from patients with familial and sporadic ALS (Gallardo et al., 2014). Knockdown of α-adducin or α2-Na/K ATPase by RNAi in cultured $SOD1^{G93A}$ astrocytes prevents degeneration of cocultured motor neurons (Gallardo et al., 2014). In vivo knockdown of α-adducin or α2-Na/K ATPase in the spinal cord of $SOD1^{G93A}$ mice by lentiviral-mediated RNAi reduces death of ventral horn motor neurons (Gallardo et al., 2014). Remarkably, inactivation of one allele of the α2-Na/K ATPase gene prolongs longevity of $SOD1^{G93A}$ mice, increasing average lifespan by 20 days (Gallardo et al., 2014). $SOD1^{G93A}$ mice heterozygous for the α2-Na/K ATPase allele are much more mobile than $SOD1^{G93A}$ littermates, suggesting that inactivation of α2-Na/K ATPase prolongs the healthspan of $SOD1^{G93A}$ mice (Gallardo et al., 2014). The effect of heterozygous inactivation of the α2-Na/K ATPase gene on the survival and health of $SOD1^{G93A}$ mice appears to result from inhibition of disease progression (Gallardo et al., 2014). Together with the observation that α2-Na/K

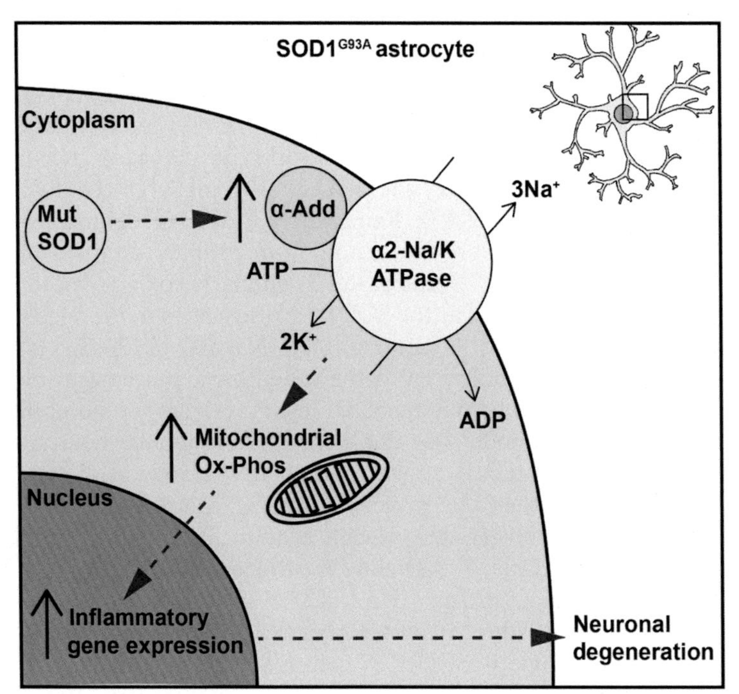

FIGURE 14.3 Role of the α-adducin/α2-Na/K ATPase complex in ALS. $SOD1^{G93A}$ astrocytes express mutant SOD1 protein, which leads to upregulation of the α-adducin/α2-Na/K ATPase complex. Presumably due to depletion of cellular ATP stores, upregulation of the α-adducin/α2-Na/K ATPase complex stimulates mitochondrial respiration (Mitochondrial Oxidative Phosphorylation or Ox-Phos). Upregulation of the α-adducin/α2-Na/K ATPase complex in $SOD1^{G93A}$ astrocytes also triggers expression of a program of genes encoding inflammatory cytokines. Inflammatory gene expression leads to neuronal degeneration through secretion of yet-unidentified toxic factors (Gallardo et al., 2014). *ALS*, amyotrophic lateral sclerosis; *SOD1*, superoxide dismutase 1.

ATPase is selectively, though not exclusively, expressed in astrocytes in the CNS (McGrall, Phillips, & Sweadner, 1991), these findings are consistent with the interpretation that α2-Na/K ATPase activity in astrocytes triggers non-cell autonomous degeneration of motor neurons in $SOD1^{G93A}$ mice. However, the cellular site of action of α2-Na/K ATPase in ALS disease progression in vivo remains to be established.

How might upregulation of α2-Na/K ATPase activity in astrocytes trigger their toxic effects on motor neurons in ALS? Gallardo et al. (2014) have provided clues on consequences of α2-Na/K ATPase activation in astrocyte (Fig. 14.3). Because the enzymatic activity of Na/K ATPase requires substantial ATP consumption in the cell (Wayne Albers & Siegel, 1999), upregulation of α2-Na/K ATPase in $SOD1^{G93A}$ astrocytes is anticipated to increase demand for ATP. Consistent with this prediction, mitochondrial respiration is stimulated in $SOD1^{G93A}$ astrocytes, an effect that is diminished by inactivation of a single copy of the α2-Na/K ATPase gene (Gallardo et al., 2014). In addition to stimulation of mitochondrial respiration, upregulation of α2-Na/K ATPase in $SOD1^{G93A}$ astrocytes triggers the expression of a program of genes encoding inflammatory cytokines (Gallardo et al., 2014). Although the role of specific cytokines induced in $SOD1^{G93A}$ astrocytes in an α2-Na/K ATPase-dependent manner in ALS pathogenesis remains to be elucidated, these observations support the conclusion that α2-Na/K ATPase triggers the expression and/or secretion of factors that mediate non-cell autonomous degeneration of motor neurons. Consistent with this observation, conditioned medium from $SOD1^{G93A}$ astrocytes, but not from $SOD1^{G93A}$ astrocytes with heterozygous inactivation of the α2-Na/K ATPase gene, causes degeneration of cocultured motor neurons (Gallardo et al., 2014).

The identification of the α-adducin/α2-Na/K ATPase complex as a major cell-intrinsic mechanism that triggers non-cell autonomous degeneration in ALS may provide the opportunity for novel treatments in ALS. Remarkably, Na/K ATPase is the target of cardiac glycosides such as digoxin, which has been widely used in the treatment of congestive heart failure (Hauptman & Kelly, 1999). In coculture experiments, digoxin robustly protects motor neurons from SOD1^{G93A} astrocyte-induced degeneration (Gallardo et al., 2014). These findings raise the prospect that inhibitors of α2-Na/K ATPase may be useful in the treatment of ALS.

What are the neuronal downstream effectors of mutant SOD1-astrocyte-induced motor neuron degeneration? Recently, Song et al. (2016) have identified a role for the major histocompatibility complex class I (MHCI) proteins in astrocyte-mediated toxicity. The expression of MHCI proteins is reduced in motor neurons of end-stage SOD1^{G93A} mice and human ALS patients (Song et al., 2016). Interestingly, MHCI downregulation on motor neurons is a non-cell autonomous effect; SOD1^{G93A} astrocytes cause reduced expression of MHCI in cocultured motor neurons (Song et al., 2016). Lentiviral expression of MHCI H2^k protects motor neurons from SOD1^{G93A} astrocyte toxicity in vitro (Song et al., 2016). In vivo, lentiviral injection of MHCI H2^k in spinal cord motor neurons of SOD1^{G93A} mice improves motor function and significantly prolongs lifespan by 21 days (Song et al., 2016). The signals by which SOD1^{G93A} astrocytes trigger downregulation of MHCI expression in motor neurons remain unknown, although these signals might be linked to endoplasmic reticulum stress. Thapsigargin, which triggers endoplasmic reticulum stress in motor neurons (Nishitoh et al., 2008), recapitulated the loss of MHCI in motor neurons induced by SOD1^{G93A} astrocytes (Song et al., 2016).

Finally, the factors secreted by mutant SOD1 astrocytes that are toxic to motor neurons remain to be identified. In vivo mechanistic insights are lacking in this area, but in vitro studies suggest possible players. Exocytosis of mutant SOD1 protein has been suggested as one possible mediator of astrocyte-induced neuronal degeneration (Basso et al., 2013). Chromogranin, a protein enriched in secretory granules, is colocalized with mutant SOD1 protein in reactive astrocytes in the spinal cord of SOD1^{G37R} mice (Urushitani et al., 2006). SOD1^{G93A} protein added to culture medium induces microgliosis and death of motor neurons (Urushitani et al., 2006). Proteomic analyses of SOD1^{G93A}-astrocyte-conditioned medium reveal elevated levels of SOD1^{G93A} protein (Basso et al., 2013). Exosomes isolated from SOD1^{G93A} astrocytes contain SOD1^{G93A} protein and, by transferring mutant protein, contribute to neuron death (Basso et al., 2013).

Reactive astrocytes may secrete inflammatory molecules that act directly on neurons or alter microglial reactivity and thus indirectly affect neurons (Sofroniew, 2009). Levels of several common cytokines and chemokines are altered in SOD1^{G93A}-astrocyte-conditioned medium (Nagai et al., 2007; Pehar et al., 2004). In response to cerebrospinal fluid from ALS patients, astrocytes become reactive and increase production of inflammatory cytokines, including IL-6 and TNF-α, and reduce production of antiinflammatory cytokines, including IL-10 (Mishra et al., 2016). SOD1^{G93A} astrocytes release TGFβ1, which damages neurons both directly and indirectly via microglia and T cells (Endo et al., 2015). In vivo pharmacological inhibition of TGFβ1 prolongs survival of SOD1^{G93A} mice, suggesting its potential as a therapeutic target (Endo et al., 2015).

Damage of motor neurons via glutamate excitotoxicity is another possible mechanism of glial-dependent toxicity of SOD1^{G93A} astrocytes. Defects in astrocyte glutamate transport are associated with ALS. For example, levels of the glutamate transporter GLT-1 are decreased in the motor cortex and spinal cord of ALS patients, and glutamate levels in the cerebrospinal fluid are increased (Rothstein, Martin, & Kuncl, 1992; Rothstein, Van Kammen, Levey,

Martin, & Kuncl, 1995; Spreux-Varoquaux et al., 2002). Moreover, GLT-1 downregulation occurs prior to motor neuron degeneration in SOD^{G93A} rats, supporting the possibility that loss of GLT-1 may contribute to neuronal death (Howland et al., 2002). Finally, downregulation of GLT-1 has functional significance in vivo. $SOD1^{G93A}$ mice heterozygous for the GLT-1 allele have more severe disease course than $SOD1^{G93A}$ mice (Pardo et al., 2006).

Neuron-astrocyte metabolic cooperation is also altered in ALS (Allaman et al., 2011). Lactate efflux transporters are impaired in $SOD1^{G93A}$ astrocytes, resulting in decreased spinal cord lactate levels, and increasing the concentration of lactate in conditioned medium from $SOD1^{G93A}$ astrocytes partially rescues degeneration of motor neurons (Ferraiuolo et al., 2011). Additionally, neurons rely on astrocytes for trophic support. Secretion of neurotrophic factors, including GDNF and CNTF, by $SOD1^{G93A}$ astrocytes is altered, and neurotrophic factor supplementation can improve health of motor neurons cocultured with $SOD1^{G93A}$ astrocytes (Mishra et al., 2016; Nagai et al., 2007; Pun, Santos, Saxena, Xu, & Caroni, 2006). Finally, mutant SOD1 astrocytes have derangements in oxidative stress handling, mitochondrial function, calcium signaling, interastrocyte connectivity, and apoptotic pathways, which damage motor neurons in vitro (Apps & Garwicz, 2005; Gallardo et al., 2014; Kawamata et al., 2014; Pasinelli, Houseweart, Brown, & Cleveland, 2000; Rojas et al., 2015; Fritz et al., 2013; Takeuchi et al., 2011). Alterations in astrocytic support have not been yet demonstrated to trigger motor neuron degeneration in vivo.

In sum, non-cell autonomous degeneration of motor neurons by astrocytes plays a critical role in disease progression in ALS. Degeneration of motor neurons in response to ALS-astrocyte-conditioned medium demonstrates cell death is at least partially a result of secreted factors. ALS astrocytes upregulate an α-adducin/α2-Na/K ATPase complex that results in inflammatory gene transcription and degeneration of motor neurons via secreted factors (Gallardo et al., 2014). Motor neurons downregulate MHCI proteins in response to ALS astrocyte-secreted factors, which results in degeneration and death of motor neurons (Song et al., 2016). Although the role of astrocytes in non-cell autonomous degeneration of motor neurons in ALS is beginning to be understood, much more study is required to elucidate these mechanisms and their relevance in ALS disease pathogenesis.

ASTROCYTES IN ALZHEIMER'S DISEASE

Alzheimer's disease (AD) is a neurodegenerative disease characterized by extracellular amyloid-β (Aβ) plaques deposition and intracellular neurofibrillary tangles of hyperphosphorylated tau. These features represent the pathological correlates of progressive decline in memory and other cognitive functions, causing disability and eventual death (Ballard et al., 2011; Huang & Mucke, 2012; Scheltens et al., 2016; Small & Duff, 2008). The burden of AD, both on individual patients and society as a whole, cannot be overstated. AD is the leading cause of dementia, affecting 11% of people over 65 years and 32% of those over 85 years in age (Meyer et al., 2014). The complex pathophysiology of this disease is reviewed elsewhere (Ballard et al., 2011; Huang & Mucke, 2012; Scheltens et al., 2016; Small & Duff, 2008). Recent advances in preclinical mouse models of AD, molecular tools, and imaging technologies have accelerated the pace of research, including in the area of non-cell autonomous degeneration of neurons (Huang & Mucke, 2012; Webster, Bachstetter, Nelson, Schmitt, & Van Eldik, 2014). The roles of astrocytes in AD process are beginning to be understood.

The reactive phenotype of astrocytes in AD has been observed in human pathological specimens (Akiyama et al., 1996; Akiyama et al., 1999; Funato et al., 1998; Kurt, Davies, & Kidd, 1999). Aβ plaques are generated by peptides cleaved from amyloid precursor protein (APP) by the action of secretases. APP is a neuronal protein, so until recently, the role of glial cells in the buildup of Aβ has been overlooked. However, Aβ is frequently colocalized with astrocytes in postmortem brain from AD patients (Akiyama et al., 1996, 1999; Funato et al., 1998; Kurt et al., 1999). In addition to microglia, astrocytes phagocytose Aβ for degradation (Sokolowski & Mandell, 2011). Synaptic proteins of neuronal origin are found within Aβ-containing astrocytes, suggesting astrocytes take up pathological Aβ from degenerating neurons and synapses (Nagele, D'Andrea, Lee, Venkataraman, & Wang, 2003). Astrocytes also phagocytose synapses during physiological synapse pruning (Chung et al., 2013). Reactive astrocytes take up Aβ, although it is unclear from pathology whether astrocytosis occurs prior to or as a consequence of Aβ phagocytosis. Astrocytes containing Aβ express more GFAP than quiescent astrocytes and have alterations in shape and size (Nagele et al., 2003; Simpson et al., 2010).

The development of preclinical models of AD has accelerated the pace of research on disease mechanisms. Mouse models of AD have mutations in APP, Presenilin-1, and/or tau, which recapitulate many aspects of human disease, including amyloid and tau pathology, gliosis, cognitive deficits, and neuronal loss (Webster et al., 2014). Mice with mutant human APP transgenic expression, including the PDAPP model with mutant APP expressed under the PDGFβ promotor and Tg2576 lines, show deposition of plaques in the brain starting at 6 to 9 months of age and cognitive deficits thereafter (Lee & Han, 2013). APP/PS1 mice express double-mutant APP as well as mutant Presenilin-1 and show plaque deposition slightly earlier, around 6 months, but no neuronal loss (Lee & Han, 2013). A drawback to all of these mouse models is the lack of tau pathology in the brain. The $3 \times$ Tg line, which harbors mutants in APP, Presenilin-1, and tau, develops both amyloid-β plaques and neurofibrillary tangles (Lee & Han, 2013).

Preclinical studies using these mouse lines reveal that astrocytes internalize Aβ. Just as in human pathological studies, astrogliosis surrounds and phagocytose plaques in the $3 \times$ Tg mouse brain (Olabarria, Noristani, Verkhratsky, & Rodriguez, 2010). In situ, wild-type cultured astrocytes plated on brain slices from aged APP/PS1 mice reduce the quantity of Aβ (Koistinaho et al., 2004; Pihlaja et al., 2008; Wyss-Coray et al., 2003). Likewise, astrocytes have the capacity to degrade Aβ in vivo (Pihlaja et al., 2008, 2011). Wild-type astrocytes injected into the hippocampus in APP/PS1 mice reduce hippocampal Aβ burden (Pihlaja et al., 2008, 2011). The mechanism of phagocytosis is currently unknown but appears to involve the AD-risk factor protein ApoE. While wild-type cultured astrocytes plated on hippocampal slices from PDAPP mice significantly reduces Aβ levels in the slice, astrocytes from ApoE knockout mice do not affect Aβ abundance (Koistinaho et al., 2004).

Although astrocytes promote clearance of Aβ peptides, Aβ-containing astrocytes have detrimental effects on neuronal health. Primary neurons undergo cell death upon addition of Aβ to culture medium, and coculture with wild-type astrocytes protects neurons from cell death. However, neurons cultured with Aβ-exposed astrocytes die (Garwood, Pooler, Atherton, Hanger, & Noble, 2011; Paradisi, Sacchetti, Balduzzi, Gaudi, & Malchiodi-Albedi, 2004). Interestingly, conditioned medium from Aβ-exposed astrocytes triggers neuronal death, demonstrating that these "sick" astrocytes may release soluble toxic factors (Garwood et al., 2011). Exposure to Aβ leads cultured astrocytes to increase expression of GFAP, Connexin-43 and S100β, reduce expression of GLT-1 and increase their surface area (Mei, Ezan, Giaume, & Koulakoff, 2010; Mori et al., 2010; Olabarria et al., 2010; Orellana et al., 2011a, 2011b; Simpson et al., 2010; Tong et al., 2014). These reactive alterations appear to have functional implications in vivo, as overexpression of S100β in Tg2576 mice worsens astrogliosis and results in earlier and more severe Aβ pathology (Mori et al., 2010).

How Aβ activates astrocytes has been the subject of significant inquiry. Aβ influences cellular homeostasis in primary astrocytes, leading to mitochondrial dysfunction, altered calcium signaling, increased glucose metabolism, and oxidative stress (Abarmov, Canevari, & Duchen, 2003; Abramov, Canevari, & Duchen, 2004; Allanman et al., 2010). The consequence of these changes in astrocytes on neurons is not fully understood. High levels of oxidative stress in astrocytes may impair their capacity to buffer neuronal oxidative stress, possibly through Aβ-induced depletion of astrocytic antioxidant glutathione (Abarmov et al., 2003). Aβ also activates NADPH oxidase in cultured astrocytes, leading to generation of reactive oxygen species. Inhibiting NADPH oxidase in Aβ-exposed astrocytes is neuroprotective (Abramov et al., 2004). Inflammatory signals including NF-κB and TNF-α are also activated in Aβ-exposed cultured astrocytes, and blockade of these signals reduces neurotoxicity (Jana & Pahan, 2010; Orellana et al., 2011a).

Finally, in vivo evidence suggests that glial-dependent alterations in tonic levels of neurotransmitters may contribute to neuronal death. In microdialysate from hippocampi of APP/PS1 mice, extracellular GABA concentration is significantly increased (Jo et al., 2014). Blocking synthesis or release of excess astrocytic GABA in vivo partially rescues cognitive and memory deficits in APP/PS1 mice (Jo et al., 2014). Astrocytic production of excess GABA reduces spike probability and alters synaptic plasticity in the dentate gyrus, one plausible contributor to memory loss in AD patients (Jo et al., 2014). In addition to releasing excess GABA, as with other causes of reactive astrogliosis, cultured astrocytes exposed to Aβ release excess glutamate (Orellana et al., 2011a; Simpson et al., 2010)

Although astrocytes may exert both beneficial and detrimental effects in AD, separating these effects has proven challenging. For example, neurons likely benefit from astrocytic uptake of Aβ. However, uptake of Aβ also triggers pathogenic alterations in these astroglia, converting them into drivers of disease progression. Further study, particularly addressing cellular mechanisms in vivo, is required to dissect the role of astrocytes in AD.

ASTROCYTES IN PARKINSON'S DISEASE

Parkinson's disease (PD) is a slowly progressive neurodegenerative disorder with early, prominent death of dopaminergic neurons in the substantia nigra pars compacta (SNpc), a mesencephalic structure projecting to the striatum (Kalia & Lang, 2015). Loss of dopaminergic input to the basal ganglia leads to the cardinal movement features of PD, including bradykinesia, rigidity, tremor, and postural imbalance (Kalia & Lang, 2015). PD is complex and heterogeneous in presentation, and patients may present with nonmotor symptoms including cognitive or mood impairment (Bachoo et al., 2004). Likewise, histopathological features of PD are complex. Despite recent advances in diagnostic imaging modalities utilizing radioactive tracers, postmortem identification of Lewy bodies and Lewy neurites, abnormal aggregates of misfolded α-synuclein in the soma or neuronal processes respectively, remains the gold standard for PD diagnosis (Kalia & Lang, 2015). Lewy body pathology and neurodegeneration affect nondopaminergic neurons outside the basal ganglia (Kalia &

Lang, 2015). Activated astrocytes and microglia may be also found in pathological examination of the brain in PD (Kalia & Lang, 2015). We focus here on the influence of astrocytes in the onset and progression of PD.

Astrocytes, like neurons, accumulate ubiquitinated inclusions of misfolded α-synuclein in PD (Braak, Sastre, & Del Tredici, 2007). The anatomic distribution of α-synuclein-containing astrocytes in human pathological specimens parallels that of Lewy body pathology in neurons (Braak et al., 2007). In particular, protoplasmic astrocytes, but not fibrous astrocytes, appear to accumulate α-synuclein (Song et al., 2009). Astrocytes in PD do not appear to adopt the morphology of classic reactive astroglia, at least not early in disease (Song et al., 2009). However, the protective functions of astrocytes may be impaired in PD.

The identification of monogenic causes of PD has been crucial for understanding molecular mechanisms of disease pathogenesis. Mutations in the gene encoding α-synuclein (*SNCA*) represent a monogenic cause of PD (Kalia & Lang, 2015). Transgenic mice overexpressing wild-type human α-synuclein develop loss of dopaminergic terminals in the basal ganglia and mild motor symptoms (Masliah et al., 2000). Mice expressing A53T α-synuclein specifically in neurons (Thy1-hSYN^{A53T}) display a severe phenotype consisting of debilitating motor impairment, paralysis, and death around 8 months (Chandra, Gallardo, Fernandez-Chacon, Schluter, & Sudhof, 2005). However, expression of A53T α-synuclein in astrocytes (GFAP-hSYN^{A53T}) leads to a more severe phenotype, with loss of midbrain dopaminergic neurons and spinal motor neurons, astrogliosis, paralysis, and death by 4 months or earlier (Gu et al., 2010). Although this phenotype does not precisely mimic the features of PD, these results suggest that mutant α-synuclein may influence degeneration of SNpc neurons non-cell autonomously. Reactive astrogliosis occurs in other synucleinopathies, suggesting a possible role for non-cell autonomous degeneration in these disorders as well (Song et al., 2009).

How α-synuclein in astrocytes damages neurons remains unclear. As in PD patients, astrocytes in GFAP-hSYN^{A53T} mice develop α-synuclein inclusions (Gu et al., 2010). Even in Thy1-hSYN^{A53T} mice, in which astrocytes do not express mutant α-synuclein, astrocytes nevertheless accumulate mutant α-synuclein protein likely via transcellular spread (Lee et al., 2010). Astrocytes that develop inclusions exhibit reactive changes including inflammatory gene transcription and downregulation of GLT-1 (Gu et al., 2010; Lee et al., 2010). The precise mechanism by which mutant α-synuclein leads to reactive astrogliosis remains unknown, but it may involve oxidative stress. The redox-sensitive transcription factor Nrf2, when overexpressed in astrocytes in Thy1-hSYN^{A53T} mice, extends lifespan, increases neuron survival, reduces gliosis, and decreases α-synuclein aggregation (Gan, Vargas, Johnson, & Johnson, 2012). Consistent with its proposed role in alleviating oxidative stress, Nrf2 has been implicated in protection against oxidative stress in other neurodegenerative disease models as well (Johnson et al., 2008). As in other neurodegenerative diseases with misfolded and aggregated proteins, α-synuclein may trigger oxidative stress via activation of the unfolded protein response (UPR).

Oxidative stress is also implicated in other causes of PD. Loss-of-function mutation of the gene encoding the antioxidant protein DJ-1 causes PD. Knockout of DJ-1 in astrocytes triggers toxicity in cocultured neurons, and DJ-1 overexpression in astrocytes reduces toxicity (Gan, Johnson, & Johnson, 2010; Mullett & Hinkle, 2009). Mutations in the E3 ubiquitin ligase Parkin also cause PD, via accumulation of toxic proteasome substrates (Kalia & Lang, 2015). Parkin knockout mice have alterations in dopamine signaling and reduction of dopamine transporters, although they fail to show degeneration of nigrostriatal neurons (Itier et al., 2003). Although Parkin has higher baseline expression in neurons, Parkin is upregulated in astrocytes in response to oxidative stress (Ledesma et al., 2002). Cell proliferation is impaired and apoptosis is increased in glial cells in Parkin knockout mice (Solano et al., 2008). Parkin knockout specifically leads to a greater susceptibility to oxidative stress and neurotoxins. Conditioned medium from Parkin knockout glial cells is less neuroprotective, an effect thought to be mediated by defective oxidative stress signaling via glutathione peroxidase in astrocytes (Damier, Hirsch, Zhang, Agid, & Javoy-Agid, 1993; Solano et al., 2008).

A major limitation to research on PD pathophysiology is the lack of genetic mouse models that develop progressive and selective loss of dopaminergic neurons and mimic patient symptomology. Mice expressing mutant α-synuclein develop motor neuron degeneration, with paralysis and spinal cord pathology. Mice with mutations in Parkin likewise lack nigrostriatal neurodegeneration. One exception is the drug-induced model of PD, caused by administration of the neurotoxin 1-methyl-4-phenyl-1,2,3,6-tetrahydropyridine (MPTP). In both humans and mice, MPTP causes severe, selective death of dopaminergic neurons in the striatum and parkinsonian motor deficits, including reduction of voluntary movement (Meredith & Rademacher, 2011). Importantly, astrocytes are the major source of the enzyme monoanime oxidase B (MAO-B), which converts MPTP into its toxic isoform in the brain. This model represents a predominately non-cell autonomous model of nigrostriatal degeneration (Meredith & Rademacher, 2011). Although the exact mechanisms of degeneration are unclear, MPTP toxicity has been linked to neuroinflammation, oxidative stress, and the UPR (Du et al., 2001; Hashida et al., 2012). Further studies in the MPTP model will provide clues to the selective nature of dopaminergic neuron loss in PD, as will the development of genetic mouse models of PD with selective loss of nigrostriatal dopaminergic neurons.

ASTROCYTES IN HUNTINGTON'S DISEASE

Huntington's disease (HD) is an autosomal-dominant neurodegenerative disease caused by expanded repeats of the DNA bases cytosine, adenine, and guanine (CAG repeats) in the gene encoding Huntingtin (Walker, 2007). Symptoms typically develop in middle age and consist mainly of progressively worsening motor skills and executive function (Walker, 2007). Atrophy of the caudate is typically seen, and the putamen also shows atrophic change (Waldvogel, Kim, Tippett, Vonsattel, & Faull, 2015). Striatal medium spiny neurons are most vulnerable, but cortical pyramidal neurons and Purkinje cells may eventually be affected (Waldvogel et al., 2015). The histopathological hallmark is nuclear inclusions containing glutamine-expanded Huntingtin protein (Waldvogel et al., 2015). Molecular mechanisms of neuronal death in HD are poorly understood, and ongoing studies are reviewed elsewhere (Ross & Tabrizi, 2011). Although the mechanisms by which astrocytes are toxic to neurons in HD are less well defined than in ALS, non-cell autonomous neurodegeneration is thought to represent a major contributor to disease pathogenesis.

As in other neurodegenerative disorders, there is progressive astrocytosis in HD (Faideau et al., 2010). Astrocytes display the typical morphologic reactive changes in HD brain as well as upregulation of Connexin-43 and GFAP and downregulation of GLT-1 (Faideau et al., 2010; Vis et al., 1998). Astrogliosis in the striatum occurs prior to cell death, suggesting a possible role for glia in the early disease stage (Faideau et al., 2010). Mutant Huntingtin inclusions are also present in astrocytes prior to neurodegeneration, albeit at lower levels than in neurons (Faideau et al., 2010).

Mouse models of HD that express mutant human Huntingtin recapitulate aspects of patient pathology and symptomatology. In humans, 41 CAG repeats (41Q) cause fully penetrant disease, but transgenic mice require expression of 80Q or more to be symptomatic (Menalled et al., 2009; Walker, 2007). In the most widely used mouse model of HD (R6/2 mouse), in which truncated human Huntingtin with 120Q repeats is expressed, abnormal movements, weight loss, and other neurological abnormalities begin around 10–15 weeks (Menalled et al., 2009). R6/2 mice develop loss of striatal neurons and neuronal inclusions (Mangiarini et al., 1996). Surprisingly, although neuronal loss in R6/2 occurs predominantly in the striatum, expression of mutant Huntingtin exclusively in striatal neurons or cortical pyramidal neurons does not lead to locomotor deficits (Gu et al., 2005, 2007). These results suggest that cell-autonomous damage is insufficient to cause disease. However, motor deficits and cell death occur if mutant Huntingtin is expressed throughout the brain using the Nestin-Cre driver (Bradford et al., 2009; Gu et al., 2005). Expression of 160Q Huntingtin under the astrocyte *gfa2* promotor results in severe motor impairment, gliosis, and reduced lifespan (Bradford et al., 2009). Non-cell autonomous molecular pathology is thus required for motor phenotypes. Nonetheless, striatal pathology is potentiated by cell-autonomous effects of mutant Huntingtin. Transgenic mice that express mutant Huntingtin in both neurons and astrocytes die sooner and develop earlier more severe symptoms than those with mutant Huntingtin in neurons or astrocytes only (Bradford et al., 2010).

How do mutant Huntingtin-expressing astrocytes cause dysfunction of striatal medium spiny neurons? This question remains to be addressed in part because modeling astrocyte toxicity in HD using in vitro techniques has been challenging. Astrocytes infected with 130Q mutant Huntingtin-expressing adenovirus are highly toxic to cocultured cortical neurons (Shin et al., 2005). However, astrocytes from R6/2 mice are not overtly toxic to neurons in vitro (Shin et al., 2005). Striatal and cortical neurons plated on R6/2 astrocytes die in greater numbers when challenged with toxic levels of glutamate than neurons on wild-type astrocytes, and conditioned medium from R6/2 astrocytes is less supportive of neurite growth and neuronal migration than medium from wild type astrocytes (Chou et al., 2008; Shin et al., 2005).

Some mechanistic insights have been gained from in vivo studies. Recently, Tong et al. (2014) discovered that loss of astrocytic potassium uptake channel Kir4.1 is critical to striatal neuron dysfunction. Prior to the onset of astrogliosis, R6/2 mice have reduced expression of Kir4.1 in striatal astrocytes (Tong et al., 2014). In vivo microdialysis from the striatum of R6/2 mice showed elevated levels of potassium, and striatal neurons are hyperexcitable (Tong et al., 2014). Viral rescue of Kir4.1 corrects striatal potassium concentration and neuronal hyperexcitability and improves motor function and lifespan of R6/2 mice (Tong et al., 2014). Kir4.1 loss also has downstream effects that may compound the neuronal hyperexcitability. The glutamate transporter GLT-1 is downregulated in R6/2 striatal astrocytes, and expression is restored with viral rescue of Kir4.1 (Tong et al., 2014). The increase in extracellular glutamate and potassium leads to significant alterations in striatal astrocytic calcium activity in vivo, which has potential unexplored effects of its own on neuronal health (Hashida et al., 2012). Further application of in vivo calcium imaging is needed to define how these signaling abnormalities originate and the nature of their downstream effects.

R6/2 astrocytes have further dysfunction. In electrophysiological analyses of R6/2 striatal slices, release of tonic GABA by astrocytes appears to be increased (Wojtowicz, Dvorzhak,

Semtner, & Grantyn, 2013). R6/2 astrocytes also appear to have deficits in neurotrophic support and cholesterol transport, although the relevance of these to neuronal health still is unclear (Arregui, Benitez, Razgado, Vergara, & Segovia, 2011; Battaglia et al., 2011; Valenza et al., 2010). Mechanistic advances in this field will require in vitro models that better model non-cell autonomous degeneration and in vivo cell-type specific genetic studies.

ASTROCYTES IN SPINOCEREBELLAR ATAXIA TYPE 7

Spinocerebellar ataxia type 7 (SCA7) is an autosomal-dominant cerebellar ataxia and a prototypical example of a primarily non-cell autonomous neurodegenerative disease (Lobsiger & Cleveland, 2007). Like HD, SCA7 is caused by expanded CAG repeats leading to pathological protein misfolding (Schöls, Bauer, Schmidt, Schulte, & Riess, 2004). The mutant protein in SCA7 is ataxin-7, a protein of unknown function (Schöls et al., 2004). Expanded polyglutamine at the N-terminus of ataxin-7 causes protein aggregation (Schöls et al., 2004), leading to the symptoms of ataxia with visual loss (Taroni & DiDonato, 2004).

Interestingly, Purkinje cell degeneration in SCA7 occurs via predominantly non-cell autonomous effects. In transgenic mice expressing 92Q ataxin-7 in all neurons except Purkinje cells, Purkinje cells nevertheless exhibit more severe degeneration than other neurons, leading to ataxia and death prior to 1 year (Garden et al., 2002). By contrast, selective expression of 90Q ataxin-7 exclusively in Purkinje cells fails to induce their degeneration until 16 months of age, despite the presence of ubiquitinated ataxin inclusions prior to that (Yvert et al., 2000). Importantly, Purkinje neuron degeneration occurs even if ataxin-7 is selectively expressed in Bergmann glia, a cerebellar-specific subtype of astrocyte. Expression of ataxin-7 under the astrocyte-specific *gfa2* promotor results in ataxia and Purkinje cell degeneration by 12 months (Custer et al., 2006). Glutamate uptake deficits may partially explain the mechanism whereby ataxin-7-expressing Bergmann glia induce degeneration of Purkinje cells. Bergmann glia shows impaired glutamate uptake and downregulation of GLAST (Custer et al., 2006). Purkinje cells are also particularly vulnerable to injury resulting from glutamate imbalance because of the enormous size of the Purkinje cell dendritic arbor and their numerous glutamatergic synapses (Bellamy, 2006).

NON-CELL AUTONOMOUS ROLES OF ASTROCYTES IN OTHER DISEASES

Astrocytic non-cell autonomous toxicity to neurons is not unique to the field of neurodegenerative disease. In Rett syndrome, an X-linked disorder caused by loss of function in methyl-CpG-binding protein 2 (MeCP2), restoration of MeCP2 function selectively in astrocytes partially rescues hypoactivity and anxiety behaviors as well as reduced dendritic complexity and shrunken somas (Lioy et al., 2011). These results suggest that loss of MeCP2 leads to behavioral deficits in a non-cell autonomous fashion. Similarly, in the lysosomal storage disorder, multiple sulfatase deficiency, deletion of the causative allele in only astrocytes causes degeneration of cortical neurons in vitro and in vivo (Di Malta, Fryer, Settembre, & Ballabio, 2012). These studies suggest that pathogenic mechanisms and therapeutic targets of glial-dependent neuron death may be generalizable beyond neurodegenerative disease.

PERSPECTIVES

Our understanding of the role of astrocytes in neurodegenerative diseases has evolved dramatically over the past two decades. Although astrocytes become reactive in response to loss of neurons, particularly in late stages of disease, astrocytic dysfunction is now thought to play critical roles at earlier stages of neurodegenerative diseases. Genetic studies in mice have demonstrated that activation of cell-autonomous pathogenic mechanisms are often insufficient to cause disease. In studies of ALS, selective expression of mutant SOD1 in motor neurons fails to induce the clinical and pathological features of motor neuron degeneration (Lino et al., 2002). Conversely, transgenic mutant SOD1 mice phenocopy ALS even upon inactivation of mutant SOD1 in motor neurons (Boillee et al., 2006). In studies of HD, expression of mutant Huntingtin in striatal neurons does not trigger striatal neurodegeneration (Gu et al., 2007). In some disorders, non-cell autonomous mechanisms appear to be primarily responsible for neurodegeneration. For example, expression of mutant ataxin-7 in astrocytes leads to Purkinje cell degeneration in SCA7 (Custer et al., 2006), whereas expression of mutant ataxin-7 in Purkinje cells fails to trigger neurodegeneration until much later (Yvert et al., 2000).

A major goal of studies of glia-dependent neurodegeneration is identification of the underlying mechanisms. Many of these studies have focused on astrocyte-dependent motor neuron degeneration in ALS. At least three biological processes have been implicated: glial-cell intrinsic mechanisms, astrocyte-derived secreted factors that act on motor neurons, and signaling pathways operating in neurons in response to pathogenic glial cues. Until recently, the glial cell–intrinsic mechanisms operating in ALS remained unknown. The identification of α-adducin/α2-Na/K ATPase as a critical pathogenic mechanism in ALS astrocytes advances our understanding of ALS and provides a starting point for greater insights into glial-specific neurodegenerative signals (Gallardo et al., 2014). Astrocyte-derived secreted factors that induce motor neuron degeneration remain to be identified. In motor neurons, the identification of glial-dependent downregulation of neuronal MHCI molecules represents an important step toward understanding the neuronal response to pathogenic glial signals (Song et al., 2016).

Although the pathways by which astrocytic dysfunction triggers neurodegeneration may vary among different diseases, common mechanisms may be at play. Astrocyte-induced excitotoxicity resulting from impaired potassium or glutamate uptake is thought to contribute to neurodegeneration in several neurodegenerative diseases. In the R6/2 mouse model of HD, downregulation of the potassium uptake channel Kir4.1 occurs in striatal astrocytes prior to reactive astrocytosis, and viral expression of Kir4.1 in astrocytes ameliorates the resultant hyperexcitability of striatal neurons and motor dysfunction (Tong et al., 2014). In SCA7, expression of mutant ataxin-7 in astrocytes results in downregulation of the glutamate transporter GLAST and thereby leads to excitotoxic death of Purkinje cells (Custer et al., 2006). In ALS, hyperexcitability of motor neurons occurs early in the course of disease (Fritz et al., 2013; Vucic & Kiernan, 2006). Astrocytes in ALS patients may have impaired glutamate uptake secondary to downregulation of GLT-1 (Rothstein et al., 1995).

Another common theme in glial-dependent neurodegeneration is the accumulation of misfolded mutant proteins in astrocytes, which occurs in AD, PD, HD, and ALS (Braak et al., 2007; Faideau et al., 2010; Forseberg, Andersen, Marklund, & Brannstrom, 2011; Nagele et al., 2003). The effect of misfolded proteins on cell-autonomous pathways of neurodegeneration has been studied intensively, though these mechanisms remain poorly understood (Soto, 2003). The mechanisms by which misfolded proteins induce astrocytic dysfunction are even more poorly understood. Misfolded proteins trigger distinct proinflammatory, reactive and oxidative stress signaling cascades in astrocytes that lead to toxic effects in neurons.

In many cases, the in vivo relevance of mechanisms of non-cell autonomous degeneration remain to be determined. Rigorous genetic studies in mice are required. These studies should employ mouse models that faithfully model the clinical features, cell-specific pathology, and presumably molecular mechanisms of human disease. Better models are also critical to identifying therapeutic targets with the best chance of success. Conditional knockout mouse genetic strategies should facilitate dissection of non-cell autonomous contributions to neurodegeneration. New Cre lines that more selectively and completely target astrocytes, such as the Aldh1L1-ERT Cre line (Srinivasan et al., 2016), should facilitate targeting of astrocyte-specific disease mediators.

In vitro models derived from human patients, such as fully humanized primary cocultures and iPS-cell derived models (Juopperi et al., 2012; Re et al., 2014; Sances et al., 2016), represents an important direction for future studies. Induced pluripotent stem cells (iPSCs), derived by reprogramming patient fibroblasts, have the potential to self-renew and differentiate into virtually any cell type (Avior, Sagi, & Benvenisty, 2016). To study non-cell autonomous mechanisms of disease in a particular patient, iPSC-derived human neurons and astrocytes may be generated. Although in vitro studies have limitations, iPSC culture models also permit high-throughput screens, modeling numerous genetic mutations at once, and the potential for personalized medicine (Avior et al., 2016).

Recent advances in our understanding of glial-dependent mechanisms of neurodegeneration provide the prospect for novel therapeutic approaches in the treatment of neurodegenerative diseases. Development of therapies aimed at salvaging neurons by mitigating cell-autonomous mechanisms in neurodegenerative disease has thus far been met with repeated failures. The identification of glial mechanisms offers a novel alternate approach for treatment of neurodegenerative disorders. In that regard, the crucial role of α2-Na/K ATPase in glial-dependent motor neuron degeneration in ALS suggests that inhibition of α2-Na/K ATPase may prove beneficial in the treatment of ALS and potentially other neurodegenerative diseases (Gallardo et al., 2014). Remarkably, the Na/K ATPase inhibitor, digoxin, which has been widely used in the treatment of congestive heart failure (Gheorghiade, van Veldhuisen, & Colucci, 2006), robustly protects motor neurons from cocultured $SOD1^{G93A}$ astrocytes (Gallardo et al., 2014). However, because digoxin blocks both α1 and α2-Na/K ATPase and has a narrow therapeutic index (Hauptman & Kelly, 1999), digoxin is unlikely to represent a potential therapeutic agent in ALS. Therefore, it will be important to identify specific α2-Na/K ATPase inhibitors for potential treatment of ALS.

Further research into the mechanisms by which astrocytes become reactive may yield additional potential therapeutic targets. Zamanian et al. (2012) used transcriptomic analyses to classify reactive astrocytes into A1 and A2 subtypes. Genes that promote further inflammation are upregulated in A1 astrocytes, whereas neurotrophic factors are upregulated in A2 astrocytes (Zamanian et al., 2012). Liddelow et al. have recently discovered that A1 reactive astrocytes may be the subtype predominately formed in neurodegenerative diseases. A1 astrocytes lose the ability to promote synapse formation and synapse elimination, and they promote neuronal death via a toxic secreted factor that remains to be identified (Liddelow et al., 2017). Interestingly, nonreactive astrocytes can be converted to reactive astrocytes upon treatment with the cytokines

IL-1α and TNF and the complement protein C1q (Liddelow et al., 2017). Targeting these inducers of A1 astrocytes may therefore prove useful in the treatment of neurodegenerative diseases.

In sum, neurodegeneration occurs as a result of convergence of cell-autonomous and non-cell autonomous effects of toxic mutant protein. Further research into the fundamental mechanisms of astrocytic dysfunction and how astrocytic dysfunction triggers harmful effects in neurons will be critical to our ability to understand and treat neurodegenerative disease.

Acknowledgments

We thank the members of the Bonni Laboratory for helpful advice and critical reading of the manuscript. Supported by the Mathers Foundation (A.B.) and NINDS Grant F30 NS100217–01 (to S.E.S.). We apologize to authors whose work was not included due to space constraints.

References

Abarmov, A. Y., Canevari, L., & Duchen, M. R. (2003). Changes in intracellular calcium and glutathione in astrocytes as the primary mechanism of amyloid neurotoxicity. *The Journal of Neuroscience, 23*(12), 5088–5095.

Abramov, A. Y., Canevari, L., & Duchen, M. R. (2004). Beta-amyloid peptides induce mitochondrial dysfunction and oxidative stress in astrocytes and death of neurons through activation of NADPH oxidase. *The Journal of Neuroscience, 24*(2), 565–575.

Akiyama, H., Mori, H., Saido, T., Kondo, H., Ikeda, K., & McGeer, P. L. (1999). Occurence of the diffuse amyloid β-protein (Aβ) deposits with numberous Aβ-containing glial cells in the cerebral cortex of patients with Alzheimer's disease. *Glia, 25*, 324–331.

Akiyama, H., Schwab, C., Kondo, H., Mori, H., Kametani, F., Kenji, I., … McGeer, P. L. (1996). Granules in glial cells of patients with Alzheimer's disease are immunopositive for C-terminal sequences of β-amyloid protein. *Neuroscience Letters, 206*, 169–172.

Albers, R. W., & Siegel, G. J. (1999). The ATP-dependent Na + ,K + pump. In G. J. Siegel, B. W. Agranoff, & R. W. Albers (Eds.), Basic Neurochemistry: Molecular, Cellular and Medical Aspects (6th ed). Philadelphia: Lippincott-Raven.

Allaman, I., Belanger, M., & Magistretti, P. J. (2011). Astrocyte-neuron metabolic relationships: For better and for worse. *Trends in Neurosciences, 34*(2), 76–87.

Allanman, I., Gavillet, M., Belanger, M., Laroche, T., Viertl, D., Lashuel, H. A., & Magistretti, P. J. (2010). Amyloid-beta aggregates cause alterations of astrocytic metabolic phenotype: Impact on neuronal viability. *The Journal of Neuroscience, 30*(9), 3326–3338.

Allen, N. J., Bennett, M. L., Foo, L. C., Wang, G. X., Chakraborty, C., Smith, S. J., & Barres, B. A. (2012). Astrocyte glypicans 4 and 6 promote formation of excitatory synapses via GluA1 AMPA receptors. *Nature, 486*(7403), 410–414.

Anderson, C. M., & Swanson, R. A. (2000). Astrocyte glutamate transport: Review of properties, regulation, and physiological functions. *Glia, 32*, 1–14.

Apps, R., & Garwicz, M. (2005). Anatomical and physiological foundations of cerebellar information processing. *Nature Reviews Neuroscience, 6*(4), 297–311.

Arregui, L., Benitez, J. A., Razgado, L. F., Vergara, P., & Segovia, J. (2011). Adenoviral astrocyte-specific expression of BDNF in the striata of mice transgenic for Huntington's disease delays the onset of the motor phenotype. *Cellular and Molecular Neurobiology, 31*(8), 1229–1243.

Avior, Y., Sagi, I., & Benvenisty, N. (2016). Pluripotent stem cells in disease modelling and drug discovery. *Nature Reviews Molecular Cell Biology, 17*(3), 170–182.

Bachoo, R. M., Kim, R. S., Ligon, K. L., Maher, E. A., Brennan, C., Billings, N., … DePinho, R. A. (2004). Molecular diversity of astrocytes with implications for neurological disorders. *Proceedings of the National Academy of Sciences of the United States of America, 101*(22), 8384–8389.

Ballard, C., Gauthier, S., Corbett, A., Brayne, C., Aarsland, D., & Jones, E. (2011). Alzheimer's disease. *The Lancet, 377*(9770), 1019–1031.

Banker, G. A. (1980). Trophic interactions between astroglial cells and hippocampal neurons in culture. *Science, 209*(4458), 809–810.

Barres, B. A. (2008). The mystery and magic of glia: A perspective on their roles in health and disease. *Neuron, 60*(3), 430–440.

Basso, M., Pozzi, S., Tortarolo, M., Fiordaliso, F., Bisighini, C., Pasetto, L., … Bonetto, V. (2013). Mutant copper–zinc superoxide dismutase (SOD1) induces protein secretion pathway alterations and exosome release in astrocytes: Implications for disease spreading and motor neuron pathology in amyotrophic lateral sclerosis. *Journal of Biological Chemistry, 288*(22), 15699–15711.

Battaglia, G., Cannella, M., Riozzi, B., Orobello, S., Maat-Schieman, M. L., Aronica, E., … Squitieri, F. (2011). Early defect of transforming growth factor beta1 formation in Huntington's disease. *Journal of Cellular and Molecular Medicine, 15*(3), 555–571.

Belanger, M., Allaman, I., & Magistretti, P. J. (2011). Brain energy metabolism: Focus on astrocyte-neuron metabolic cooperation. *Cell Metabolism, 14*(6), 724–738.

Bellamy, T. C. (2006). Interactions between Purkinje neurones and Bergmann glia. *Cerebellum, 5*(2), 116–126.

Boillee, S., Yamanaka, K., Lobsiger, C. S., Copeland, N. G., Jenkins, N. A., Kassiotis, G., ... Cleveland, D. W. (2006). Onset and progression in inherited ALS determined by motor neurons and microglia. *Science, 312*(5778), 1389–1392.

Braak, H., Sastre, M., & Del Tredici, K. (2007). Development of alpha-synuclein immunoreactive astrocytes in the forebrain parallels stages of intraneuronal pathology in sporadic Parkinson's disease. *Acta Neuropathologica, 114*(3), 231–241.

Bradford, J., Shin, J., Roberts, M., Wang, C., Li, X., & Li, S. (2009). Expression of mutant huntingtin in mouse brain astrocytes causes age-dependent neurological symptoms. *Proceedings of the National Academy of Sciences of the United States of America, 106*(52), 22480–22485.

Bradford, J., Shin, J. Y., Roberts, M., Wang, C. E., Sheng, G., Li, S., & Li, X. J. (2010). Mutant huntingtin in glial cells exacerbates neurological symptoms of Huntington disease mice. *Journal of Biological Chemistry, 285*(14), 10653–10661.

Bruijn, L. I., Becher, M. W., Lee, M. K., Anderson, K. L., Jenkins, N. A., Copeland, N. G., ... Cleveland, D. W. (1997). ALS-linked SOD1 mutant G85R mediates damage to astrocytes and promotes rapidly progressive disease with SOD1-containing inclusions. *Neuron, 18*, 327–338.

Bush, T. G., Puvanachandra, N., Horner, C. H., Polito, A., Ostenfeld, T., Svendsen, C. N., ... Sofroniew, M. V. (1999). Leukocyte infiltration, neuronal degeneration, and neurite outgrowth after ablation of scar-forming, reactive astrocytes in adult transgenic mice. *Neuron, 23*, 297–308.

Byrne, S., Walsh, C., Lynch, C., Bede, P., Elamin, M., Kenna, K., ... Hardiman, O. (2011). Rate of familial amyotrophic lateral sclerosis: A systematic review and meta-analysis. *Journal of Neurology, Neurosurgery, and Psychiatry, 82*(6), 623–627.

Chandra, S., Gallardo, G., Fernandez-Chacon, R., Schluter, O. M., & Sudhof, T. C. (2005). Alpha-synuclein cooperates with CSPalpha in preventing neurodegeneration. *Cell, 123*(3), 383–396.

Chou, S. Y., Weng, J. Y., Lai, H. L., Liao, F., Sun, S. H., Tu, P. H., ... Chern, Y. (2008). Expanded-polyglutamine huntingtin protein suppresses the secretion and production of a chemokine (CCL5/RANTES) by astrocytes. *The Journal of Neuroscience, 28*(13), 3277–3290.

Christopherson, K. S., Ullian, E. M., Stokes, C. C., Mullowney, C. E., Hell, J. W., Agah, A., ... Barres, B. A. (2005). Thrombospondins are astrocyte-secreted proteins that promote CNS synaptogenesis. *Cell, 120*(3), 421–433.

Chung, W. S., Clarke, L. E., Wang, G. X., Stafford, B. K., Sher, A., Chakraborty, C., ... Barres, B. A. (2013). Astrocytes mediate synapse elimination through MEGF10 and MERTK pathways. *Nature, 504*(7480), 394–400.

Cirulli, E. T., Lasseigne, B. N., Petrovski, S., Sapp, P. C., Dion, P. A., Leblond, C. S., ... Goldstein, D. B. (2015). Exome sequencing in amyotrophic lateral sclerosis identifies risk genes and pathways. *Science, 347*(6229), 1436–1441.

Clarke, L. E., & Barres, B. A. (2013). Emerging roles of astrocytes in neural circuit development. *Nature Reviews Neuroscience, 14*(5), 311–321.

Clement, A. M., Nguyen, M. D., Roberts, E. A., Garcia, M. L., Bolliee, S., Rule, M., ... Cleveland, D. W. (2003). Wild-type nonneuronal cells extend survival of SOD1 mutant motor neurons in ALS mice. *Science, 302*, 113–117.

Custer, S. K., Garden, G. A., Gill, N., Rueb, U., Libby, R. T., Schultz, C., ... La Spada, A. R. (2006). Bergmann glia expression of polyglutamine-expanded ataxin-7 produces neurodegeneration by impairing glutamate transport. *Nature Neuroscience, 9*(10), 1302–1311.

Damier, P., Hirsch, E. C., Zhang, P., Agid, Y., & Javoy-Agid, F. (1993). Glutathione peroxidase, glial cells and Parkinson's disease. *Neuroscience, 52*(1), 1–6.

Di Giorgio, F. P., Carrasco, M. A., Siao, M. C., Maniatis, T., & Eggan, K. (2007). Non-cell autonomous effect of glia on motor neurons in an embryonic stem cell-based ALS model. *Nature Neuroscience, 10*(5), 608–614.

Di Malta, C., Fryer, J. D., Settembre, C., & Ballabio, A. (2012). Astrocyte dysfunction triggers neurodegeneration in a lysosomal storage disorder. *Proceedings of the National Academy of Sciences of the United States of America, 109*(35), E2334–E2342.

Du, Y., Ma, Z., Lin, S., Dodel, R. C., Gao, F., Bales, K. R., ... Paul, S. M. (2001). Minocycline prevents nigrostriatal dopaminergic neurodegeneration in the MPTP model of Parkinson's disease. *Proceedings of the National Academy of Sciences of the United States of America, 98*(25), 14669–14674.

Emsley, J. G., & Macklis, J. D. (2006). Astroglial heterogeneity closely reflects the neuronal-defined anatomy of the adult murine CNS. *Neuron Glia Biology, 2*(3), 175–186.

Endo, F., Komine, O., Fujimori-Tonou, N., Katsuno, M., Jin, S., Watanabe, S., ... Yamanaka, K. (2015). Astrocyte-derived TGF-beta1 accelerates disease progression in ALS mice by interfering with the neuroprotective functions of microglia and T cells. *Cell Reports, 11*(4), 592–604.

Faideau, M., Kim, J., Cormier, K., Gilmore, R., Welch, M., Auregan, G., ... Bonvento, G. (2010). In vivo expression of polyglutamine-expanded huntingtin by mouse striatal astrocytes impairs glutamate transport: A correlation with Huntington's disease subjects. *Human Molecular Genetics*, *19*(15), 3053–3067.

Ferraiuolo, L., Higginbottom, A., Heath, P. R., Barber, S., Greenald, D., Kirby, J., & Shaw, P. J. (2011). Dysregulation of astrocyte-motoneuron cross-talk in mutant superoxide dismutase 1-related amyotrophic lateral sclerosis. *Brain*, *134*(Pt 9), 2627–2641.

Forseberg, K., Andersen, P. M., Marklund, S., & Brannstrom, T. (2011). Glial nuclear aggregates of superoxide dismutase-1 are regularly present in patients with amyotrophic lateral sclerosis. *Acta Neuropathologica*, *121*, 623–634.

Fritz, E., Izaurieta, P., Weiss, A., Mir, F. R., Rojas, P., Gonzalez, D., ... van Zundert, B. (2013). Mutant SOD1-expressing astrocytes release toxic factors that trigger motoneuron death by inducing hyperexcitability. *Journal of Neurophysiology*, *109*(11), 2803–2814.

Funato, H., Yoshimura, M., Yamazaki, T., Saido, T. C., Ito, Y., Yokofujita, J., ... Ihara, Y. (1998). Astrocytes containing amyloid β-protein (Aβ)-positive granules are associated with Aβ40-positive diffuse plaques in the aged human brain. *The American Journal of Pathology*, *152*(4), 983–992.

Gallardo, G., Barowski, J., Ravits, J., Siddique, T., Lingrel, J. B., Robertson, J., ... Bonni, A. (2014). An alpha2-Na/K ATPase/alpha-adducin complex in astrocytes triggers non-cell autonomous neurodegeneration. *Nature Neuroscience*, *17*(12), 1710–1719.

Gan, L., Johnson, D. A., & Johnson, J. A. (2010). Keap1-Nrf2 activation in the presence and absence of DJ-1. *European Journal of Neuroscience*, *31*(6), 967–977.

Gan, L., Vargas, M. R., Johnson, D. A., & Johnson, J. A. (2012). Astrocyte-specific overexpression of Nrf2 delays motor pathology and synuclein aggregation throughout the CNS in the alpha-synuclein mutant (A53T) mouse model. *The Journal of Neuroscience*, *32*(49), 17775–17787.

Garden, G. A., Libby, R. T., Fu, Y., Kinoshita, Y., Huang, J., Possin, D. E., ... La Spada, A. R. (2002). Polyglutamine-expanded ataxin-7 promotes non cell-autonomous purkinje cell degeneration and displays proteolytic cleavage in ataxic transgenic mice. *The Journal of Neuroscience*, *22*(12), 4897–4905.

Garwood, C. J., Pooler, A. M., Atherton, J., Hanger, D. P., & Noble, W. (2011). Astrocytes are important mediators of Abeta-induced neurotoxicity and tau phosphorylation in primary culture. *Cell Death & Disease*, *2*, e167.

Gheorghiade, M., van Veldhuisen, D. J., & Colucci, W. S. (2006). Contemporary use of digoxin in the management of cardiovascular disorders. *Circulation*, *113*(21), 2556–2564.

Ginhoux, F., Lim, S., Hoeffel, G., Low, D., & Huber, T. (2013). Origin and differentiation of microglia. *Frontiers in Cellular Neuroscience*, *7*, 45.

Gotz, M., & Huttner, W. B. (2005). The cell biology of neurogenesis. *Nature Reviews Molecular Cell Biology*, *6*(10), 777–788.

Gu, X., Andre, V. M., Cepeda, C., Li, S. H., Li, X. J., Levine, M. S., ... Yang, X. W. (2007). Pathological cell–cell interactions are necessary for striatal pathogenesis in a conditional mouse model of Huntington's disease. *Molecular Neurodegeneration*, *2*, 8.

Gu, X., Li, C., Wei, W., Lo, V., Gong, S., Li, S. H., ... Yang, X. W. (2005). Pathological cell–cell interactions elicited by a neuropathogenic form of mutant Huntingtin contribute to cortical pathogenesis in HD mice. *Neuron*, *46*(3), 433–444.

Gu, X. L., Long, C. X., Sun, L., Xie, C., Lin, X., & Cai, H. (2010). Astrocytic expression of Parkinson's disease-related A53T alpha-synuclein causes neurodegeneration in mice. *Molecular Brain Research*, *3*, 12.

Gurney, M. E., Pu, H., Chiu, A. Y., Dal Canto, M. C., Polchow, C. Y., Alexander, D. D., ... Siddique, T. (1994). Motor neuron degeneration in mice that express a human Cu,Zn superoxide dismutase mutation. *Science*, *264*, 1771–1775.

Haidet-Phillips, A. M., Hester, M. E., Miranda, C. J., Meyer, K., Braun, L., Frakes, A., ... Kaspar, B. K. (2011). Astrocytes from familial and sporadic ALS patients are toxic to motor neurons. *Nature Biotechnology*, *29*(9), 824–830.

Hashida, K., Kitao, Y., Sudo, H., Awa, Y., Maeda, S., Mori, K., ... Hori, O. (2012). ATF6alpha promotes astroglial activation and neuronal survival in a chronic mouse model of Parkinson's disease. *PLoS ONE*, *7*(10), e47950.

Hauptman, P. J., & Kelly, R. A. (1999). Digitalis. *Cardiovascular Drugs*, *99*, 1265–1270.

Houades, V., Rouach, N., Ezan, P., Kirchhoff, F., Koulakoff, A., & Giaume, C. (2006). Shapes of astrocyte networks in the juvenile brain. *Neuron Glia Biology*, *2*(1), 3–14.

Howland, D. S., Liu, J., She, Y., Goad, B., Maragakis, N. J., Kim, B., ... Rothstein, J. D. (2002). Focal loss of the glutamate transporter EAAT2 in a transgenic rat model of SOD1 mutant-mediated amyotrophic lateral sclerosis (ALS). *Proceedings of the National Academy of Sciences of the United States of America*, *99*(3), 1604–1609.

Huang, Y., & Mucke, L. (2012). Alzheimer mechanisms and therapeutic strategies. *Cell*, *148*(6), 1204–1222.

Itier, J. M., Ibanez, P., Mena, M. A., Abbas, N., Cohen-Salmon, C., Bohme, G. A., ... Garcia de Yebenes, J. (2003). Parkin gene inactivation alters behaviour and dopamine neurotransmission in the mouse. *Human Molecular Genetics*, *12*(18), 2277–2291.

Jana, A., & Pahan, K. (2010). Fibrillar amyloid-beta-activated human astroglia kill primary human neurons via neutral sphingomyelinase: Implications for Alzheimer's disease. *The Journal of Neuroscience, 30*(38), 12676–12689.

Jensen, A. A., Fahlke, C., Bjorn-Yoshimoto, W. E., & Bunch, L. (2015). Excitatory amino acid transporters: Recent insights into molecular mechanisms, novel modes of modulation and new therapeutic possibilities. *Current Opinion in Pharmacology, 20*, 116–123.

Jo, S., Yarishkin, O., Hwang, Y. J., Chun, Y. E., Park, M., Woo, D. H., ... Lee, C. J. (2014). GABA from reactive astrocytes impairs memory in mouse models of Alzheimer's disease. *Nature Medicine, 20*(8), 886–896.

Johnson, J. A., Johnson, D. A., Kraft, A. D., Calkins, M. J., Jakel, R. J., Vargas, M. R., ... Chen, P. C. (2008). The Nrf2-ARE pathway: An indicator and modulator of oxidative stress in neurodegeneration. *Annals of the New York Academy of Sciences, 1147*, 61–69.

Juopperi, T. A., Kim, W. R., Chiang, C. H., Yu, H., Margolis, R. L., Ross, C. A., ... Song, H. (2012). Astrocytes generated from patient induced pluripotent stem cells recapitulate features of Huntington's disease patient cells. *Molecular Brain Research, 5*, 17.

Kalia, L. V., & Lang, A. E. (2015). Parkinson's disease. *The Lancet, 386*(9996), 896–912.

Kawamata, H., Ng, S. K., Diaz, N., Burstein, S., Morel, L., Osgood, A., ... Yang, Y. (2014). Abnormal intracellular calcium signaling and SNARE-dependent exocytosis contributes to SOD1G93A astrocyte-mediated toxicity in amyotrophic lateral sclerosis. *The Journal of Neuroscience, 34*(6), 2331–2348.

Kofuji, P., & Newman, E. A. (2004). Potassium buffering in the central nervous system. *Neuroscience, 129*(4), 1045–1056.

Koistinaho, M., Lin, S., Wu, X., Esterman, M., Koger, D., Hanson, J., ... Paul, S. M. (2004). Apolipoprotein E promotes astrocyte colocalization and degradation of deposited amyloid-beta peptides. *Nature Medicine, 10*(7), 719–726.

Kuchibhotla, K. V., Lattarulo, C. R., Hyman, B. T., & Bacskai, B. J. (2009). Synchronous hyperactivity and intercellular calcium waves in astrocytes in Alzheimer mice. *Science, 323*(5918), 1211–1215.

Kurt, M. A., Davies, D. C., & Kidd, M. (1999). β-Amyloid immunoreactivity in astrocytes in Alzheimer's disease brain biopsies: An electron microscope study. *Experimental Neurology, 158*, 221–228.

Ledesma, M. D., Galvan, C., Hellias, B., Dotti, C., & Jensen, P. H. (2002). Astrocytic but not neuronal increased expression and redistribution of parkin during unfolded protein stress. *Journal of Neurochemistry, 83*, 1431–1440.

Lee, H. J., Suk, J. E., Patrick, C., Bae, E. J., Cho, J. H., Rho, S., ... Lee, S. J. (2010). Direct transfer of alpha-synuclein from neuron to astroglia causes inflammatory responses in synucleinopathies. *Journal of Biological Chemistry, 285*(12), 9262–9272.

Lee, J. E., & Han, P. L. (2013). An update of animal models of Alzheimer disease with a reevaluation of plaque depositions. *Experimental Neurobiology, 22*(2), 84–95.

Lepore, A. C., Rauck, B., Dejea, C., Pardo, A. C., Rao, M. S., Rothstein, J. D., ... Maragakis, N. J. (2008). Focal transplantation-based astrocyte replacement is neuroprotective in a model of motor neuron disease. *Nature Neuroscience, 11*(11), 1294–1301.

Liddelow, S. A., Guttenplan, K. A., Clarke, L. E., Bennett, F. C., Bohlen, C. J., Schirmer, L., ... Barres, B. A. (2017). Neurotoxic reactive astrocytes are induced by activated microglia. *Nature, 541*(7638), 481–487.

Lino, M. M., Schneider, C., & Caroni, P. (2002). Accumulation of SOD1 mutants in postnatal motoneurons does not cause motoneuron pathology or motoneuron Disease. *The Journal of Neuroscience, 22*(12), 4825–4832.

Lioy, D. T., Garg, S. K., Monaghan, C. E., Raber, J., Foust, K. D., Kaspar, B. K., ... Mandel, G. (2011). A role for glia in the progression of Rett's syndrome. *Nature, 475*(7357), 497–500.

Lobsiger, C. S., & Cleveland, D. W. (2007). Glial cells as intrinsic components of non-cell-autonomous neurodegenerative disease. *Nature Neuroscience, 10*(11), 1355–1360.

Ma, Z., Stork, T., Bergles, D. E., & Freeman, M. R. (2016). Neuromodulators signal through astrocytes to alter neural circuit activity and behaviour. *Nature, 539*(7629), 428–432.

Mangiarini, L., Sathasivam, K., Seller, M., Cozens, B., Harper, A., Hetherington, C., ... Bates, G. P. (1996). Exon 1 of the HD gene with an expanded CAG repeat is sufficient to cause a progressive neurological phenotype in transgenic mice. *Cell, 87*, 493–506.

Masliah, E., Rockenstein, E., Veinbergs, I., Mallory, M., Hashimoto, M., Takeda, A., ... Mucke, L. (2000). Dopaminergic loss and inclusion body formation in α-synuclein mice: Implications for neurodegenerative disorders. *Science, 287*(5456), 1265–1269.

Mauch, D. H., Nagler, K., Schumacher, S., Goritz, C., Muller, E., Otto, A., ... Pfrieger, F. W. (2001). CNS synaptogenesis promoted by glia-derived cholesterol. *Science, 294*(5545), 1354–1357.

McGrall, K. M., Phillips, J. M., & Sweadner, K. J. (1991). Immunofluorescent localization of three Na,K-ATPase isozymes in the rat central nervous system: Both neurons and glia can express more than one Na,K-ATPase. *The Journal of Neuroscience, 11*(2), 361–391.

Mehta, P., Antao, V., Kaye, W., Sanchez, M., Williamson, D., Bryan, L., ... Horton, K. (2014). Prevalence of amyotrophic lateral sclerosis—United States, 2010–2011. *Morbidity and Mortality Weekly Report, 63*(7), 1–14.

Mei, X., Ezan, P., Giaume, C., & Koulakoff, A. (2010). Astroglial connexin immunoreactivity is specifically altered at beta-amyloid plaques in beta-amyloid precursor protein/presenilin1 mice. *Neuroscience, 171*(1), 92–105.

Menalled, L., El-Khodor, B. F., Patry, M., Suarez-Farinas, M., Orenstein, S. J., Zahasky, B., ... Brunner, D. (2009). Systematic behavioral evaluation of Huntington's disease transgenic and knock-in mouse models. *Neurobiology of Disease, 35*(3), 319–336.

Meredith, G. E., & Rademacher, D. J. (2011). MPTP mouse models of Parkinson's disease: An update. *Journal of Parkinson's Disease, 1*(1), 19–33.

Meyer, K., Ferraiuolo, L., Miranda, C. J., Likhite, S., McElroy, S., Renusch, S., ... Kaspar, B. K. (2014). Direct conversion of patient fibroblasts demonstrates non-cell autonomous toxicity of astrocytes to motor neurons in familial and sporadic ALS. *Proceedings of the National Academy of Sciences of the United States of America, 111*(2), 829–832.

Miller, R.G., Mitchell, J.D., & Moore, D.H. (2012). Riluzole for amyotrophic lateral sclerosis (ALS)/motor neuron disease (MND). The Cochrane Database of Systematic Reviews, *3*, CD001447.

Miller, R. H., & Raff, M. C. (1984). Fibrous and protoplasmic astrocytes are biochemically and developmentally distinct. *The Journal of Neuroscience, 4*(2), 585–592.

Mishra, P. S., Dhull, D. K., Nalini, A., Vijayalakshmi, K., Sathyaprabha, T. N., Alladi, P. A., ... Raju, T. R. (2016). Astroglia acquires a toxic neuroinflammatory role in response to the cerebrospinal fluid from amyotrophic lateral sclerosis patients. *Journal of Neuroinflammation, 13*(1), 212.

Mori, T., Koyama, N., Arendash, G. W., Horikoshi-Sakuraba, Y., Tan, J., & Town, T. (2010). Overexpression of human S100B exacerbates cerebral amyloidosis and gliosis in the Tg2576 mouse model of Alzheimer's disease. *Glia, 58*(3), 300–314.

Muller, C. M., & Best, J. (1989). Ocular dominance plasticity in adult cat visual cortex after transplantation of cultured astrocytes. *Nature, 342*, 427–430.

Mullett, S. J., & Hinkle, D. A. (2009). DJ-1 knock-down in astrocytes impairs astrocyte-mediated neuroprotection against rotenone. *Neurobiology of Disease, 33*(1), 28–36.

Nagai, M., Re, D. B., Nagata, T., Chalazonitis, A., Jessell, T. M., Wichterle, H., & Przedborski, S. (2007). Astrocytes expressing ALS-linked mutated SOD1 release factors selectively toxic to motor neurons. *Nature Neuroscience, 10*(5), 615–622.

Nagele, R. G., D'Andrea, M. R., Lee, H., Venkataraman, V., & Wang, H.-Y. (2003). Astrocytes accumulate Aβ42 and give rise to astrocytic amyloid plaques in Alzheimer disease brains. *Brain Research, 971*(2), 197–209.

Nishitoh, H., Kadowaki, H., Nagai, A., Maruyama, T., Yokota, T., Fukutomi, H., ... Ichijo, H. (2008). ALS-linked mutant SOD1 induces ER stress- and ASK1-dependent motor neuron death by targeting Derlin-1. *Genes & Development, 22*(11), 1451–1464.

Oberheim, N. A., Goldman, S. A., & Nedergaard, M. (2012). Heterogeneity of astrocytic form and function. *Methods in Molecular Biology, 814*, 23–45.

Oberheim, N. A., Takano, T., Han, X., He, W., Lin, J. H., Wang, F., ... Nedergaard, M. (2009). Uniquely hominid features of adult human astrocytes. *The Journal of Neuroscience, 29*(10), 3276–3287.

Olabarria, M., Noristani, H. N., Verkhratsky, A., & Rodriguez, J. J. (2010). Concomitant astroglial atrophy and astrogliosis in a triple transgenic animal model of Alzheimer's disease. *Glia, 58*(7), 831–838.

Orellana, J. A., Froger, N., Ezan, P., Jiang, J. X., Bennett, M. V., Naus, C. C., ... Saez, J. C. (2011a). ATP and glutamate released via astroglial connexin 43 hemichannels mediate neuronal death through activation of pannexin 1 hemichannels. *Journal of Neurochemistry, 118*(5), 826–840.

Orellana, J. A., Shoji, K. F., Abudara, V., Ezan, P., Amigou, E., Saez, P. J., ... Giaume, C. (2011b). Amyloid beta-induced death in neurons involves glial and neuronal hemichannels. *The Journal of Neuroscience, 31*(13), 4962–4977.

Papadeas, S. T., Kraig, S. E., O'Banion, C., Lepore, A. C., & Maragakis, N. J. (2011). Astrocytes carrying the superoxide dismutase 1 (SOD1G93A) mutation induce wild-type motor neuron degeneration in vivo. *Proceedings of the National Academy of Sciences of the United States of America, 108*(43), 17803–17808.

Paradisi, S., Sacchetti, B., Balduzzi, M., Gaudi, S., & Malchiodi-Albedi, F. (2004). Astrocyte modulation of in vitro beta-amyloid neurotoxicity. *Glia, 46*(3), 252–260.

Pardo, A. C., Wong, V., Benson, L. M., Dykes, M., Tanaka, K., Rothstein, J. D., ... Maragakis, N. J. (2006). Loss of the astrocyte glutamate transporter GLT1 modifies disease in SOD1(G93A) mice. *Experimental Neurology, 201*(1), 120–130.

Pascual, O., Casper, K. B., Kubera, C., Zhang, J., Revilla-Sanchez, R., Sul, J., ... Haydon, P. G. (2005). Astrocytes purinergic signaling coordinates synaptic networks. *Science, 310*, 113–116.

Pasinelli, P., & Brown, R. H. (2006). Molecular biology of amyotrophic lateral sclerosis: Insights from genetics. *Nature Reviews Neuroscience, 7*(9), 710–723.

Pasinelli, P., Houseweart, M. K., Brown, R. H., & Cleveland, D. W. (2000). Caspase-1 and -3 are sequentially activated in motor neuron death in Cu,Zn superoxide dismutase-mediated familial amyotrophic lateral sclerosis. *Proceedings of the National Academy of Sciences of the United States of America*, *97*(25), 13901–13906.

Pehar, M., Cassina, P., Vargas, M. R., Castellanos, R., Viera, L., Beckman, J. S., ... Barbeito, L. (2004). Astrocytic production of nerve growth factor in motor neuron apoptosis: Implications for amyotrophic lateral sclerosis. *Journal of Neurochemistry*, *89*(2), 464–473.

Pekny, M., & Nilsson, M. (2005). Astrocyte activation and reactive gliosis. *Glia*, *50*(4), 427–434.

Pekny, M., & Pekna, M. (2014). Astrocyte reactivity and reactive astrogliosis: Costs and benefits. *Physiological reviews*, *94*(4), 1077–1098.

Perea, G., Navarrete, M., & Araque, A. (2009). Tripartite synapses: Astrocytes process and control synaptic information. *Trends in Neurosciences*, *32*(8), 421–431.

Pfrieger, F. W., & Barres, B. A. (1997). Synaptic efficacy enhanced by Glial cells in vitro. *Science*, *277*(5332), 1684–1687.

Pihlaja, R., Koistinaho, J., Kauppinen, R., Sandholm, J., Tanila, H., & Koistinaho, M. (2011). Multiple cellular and molecular mechanisms are involved in human Abeta clearance by transplanted adult astrocytes. *Glia*, *59*(11), 1643–1657.

Pihlaja, R., Koistinaho, J., Malm, T., Sikkila, H., Vainio, S., & Koistinaho, M. (2008). Transplanted astrocytes internalize deposited beta-amyloid peptides in a transgenic mouse model of Alzheimer's disease. *Glia*, *56*(2), 154–163.

Pramatarova, A., Laganiere, J., Roussel, J., Brisebois, K., & Rouleau, G. A. (2001). Neuron-specific expression of mutant superoxide dismutase 1 in transgenic mice does not lead to motor impairment. *The Journal of Neuroscience*, *21*(10), 3369–3374.

Pun, S., Santos, A. F., Saxena, S., Xu, L., & Caroni, P. (2006). Selective vulnerability and pruning of phasic motoneuron axons in motoneuron disease alleviated by CNTF. *Nature Neuroscience*, *9*(3), 408–419.

Re, D. B., Le Verche, V., Yu, C., Amoroso, M. W., Politi, K. A., Phani, S., ... Przedborski, S. (2014). Necroptosis drives motor neuron death in models of both sporadic and familial ALS. *Neuron*, *81*(5), 1001–1008.

Rojas, F., Gonzalez, D., Cortes, N., Ampuero, E., Hernandez, D. E., Fritz, E., ... van Zundert, B. (2015). Reactive oxygen species trigger motoneuron death in non-cell-autonomous models of ALS through activation of c-Abl signaling. *Frontiers in Cellular Neuroscience*, *9*, 203.

Rosen, D. R., Siddique, T., Patterson, D., Figlewicz, D. A., Sapp, P., Hentati, A., ... Deng, H. X. (1993). Mutations in Cu/Zn superoxide dismutase gene are associated with familial amyotrophic lateral sclerosis. *Nature*, *362* (6415), 59–62.

Ross, C. A., & Tabrizi, S. J. (2011). Huntington's disease: From molecular pathogenesis to clinical treatment. *The Lancet Neurology*, *10*(1), 83–98.

Rothstein, J. D., Martin, L. J., & Kuncl, R. W. (1992). Decreased Glutamate Transport by the Brain and Spinal Cord in Amyotrophic Lateral Sclerosis. *The New England Journal of Medicine*, *326*, 1464–1468.

Rothstein, J. D., Van Kammen, M. V., Levey, A. I., Martin, L. J., & Kuncl, R. W. (1995). Selective loss of glial glutamate transporter GLT-1 in amyotrophic lateral sclerosis. *Annals of Neurology*, *38*(1), 73–84.

Sances, S., Bruijn, L. I., Chandran, S., Eggan, K., Ho, R., Klim, J. R., ... Svendsen, C. N. (2016). Modeling ALS with motor neurons derived from human induced pluripotent stem cells. *Nature Neuroscience*, *19*(4), 542–553.

Sauer, B. (1998). Inducible gene targeting in mice using the Cre/lox system. *METHODS: A Companion to Methods in Enzymology*, *14*, 381–392.

Scheltens, P., Blennow, K., Breteler, M. M. B., de Strooper, B., Frisoni, G. B., Salloway, S., ... Van der Flier, W. M. (2016). Alzheimer's disease. *The Lancet*, *388*(10043), 505–517.

Schöls, L., Bauer, P., Schmidt, T., Schulte, T., & Riess, O. (2004). Autosomal dominant cerebellar ataxias: Clinical features, genetics, and pathogenesis. *The Lancet Neurology*, *3*(5), 291–304.

Schummers, J., Yu, H., & Sur, M. (2008). Tuned responses of astrocytes and their influence on hemodynamic signals in the visual cortex. *Science*, *320*(5883), 1638–1643.

Shaw, P. J., & Wood-Allum, C. (2010). Motor neurone disease: A practical update on diagnosis and management. *Clinical Medicine*, *10*(3), 252–258.

Shin, J. Y., Fang, Z. H., Yu, Z. X., Wang, C. E., Li, S. H., & Li, X. J. (2005). Expression of mutant huntingtin in glial cells contributes to neuronal excitotoxicity. *Journal of Cell Biology*, *171*(6), 1001–1012.

Simpson, J. E., Ince, P. G., Lace, G., Forster, G., Shaw, P. J., Matthews, F., ... Function, MRC Cognitive Function, & Ageing Neuropathology Study Group. (2010). Astrocyte phenotype in relation to Alzheimer-type pathology in the ageing brain. *Neurobiology of Aging*, *31*(4), 578–590.

Small, S. A., & Duff, K. (2008). Linking Abeta and tau in late-onset Alzheimer's disease: A dual pathway hypothesis. *Neuron*, *60*(4), 534–542.

Sofroniew, M. V. (2009). Molecular dissection of reactive astrogliosis and glial scar formation. *Trends in Neurosciences*, *32*(12), 638–647.

Sokolowski, J. D., & Mandell, J. W. (2011). Phagocytic clearance in neurodegeneration. *The American Journal of Pathology*, *178*(4), 1416–1428.

Solano, R. M., Casarejos, M. J., Menendez-Cuervo, J., Rodriguez-Navarro, J. A., Garcia de Yebenes, J., & Mena, M. A. (2008). Glial dysfunction in parkin null mice: Effects of aging. *The Journal of Neuroscience*, *28*(3), 598–611.

Song, S., Miranda, C. J., Braun, L., Meyer, K., Frakes, A. E., Ferraiuolo, L., ... Kaspar, B. K. (2016). Major histocompatibility complex class I molecules protect motor neurons from astrocyte-induced toxicity in amyotrophic lateral sclerosis. *Nature Medicine*, *22*(4), 397–403.

Song, Y. J. C., Halliday, G. M., Holton, J. L., Lashley, T., O'Sullivan, S. S., McCann, H., ... Revesz, T. R. (2009). Degeneration in different parkinsonian syndromes relates to astrocyte type and astrocyte protein expression. *Journal of Neuropathology & Experimental Neurology*, *68*(10), 1073–1083.

Soto, C. (2003). Unfolding the role of protein misfolding in neurodegenerative diseases. *Nature Reviews Neuroscience*, *4*(1), 49–60.

Spreux-Varoquaux, O., Bensimon, G., Lacomblez, L., Salachas, F., Pradat, P. F., Le Forestier, N., ... Meininger, V. (2002). Glutamate levels in cerebrospinal fluid in amyotrophic lateral sclerosis: A reappraisal using a new HPLC method with coulometric detection in a large cohort of patients. *Journal of the Neurological Sciences*, *193*, 73–78.

Srinivasan, R., Lu, T. Y., Chai, H., Xu, J., Huang, B. S., Golshani, P., ... Khakh, B. S. (2016). New transgenic mouse lines for selectively targeting astrocytes and studying calcium signals in astrocyte processes in situ and in vivo. *Neuron*, *92*(6), 1181–1195.

Stevens, B., Allen, N. J., Vazquez, L. E., Howell, G. R., Christopherson, K. S., Nouri, N., ... Barres, B. A. (2007). The classical complement cascade mediates CNS synapse elimination. *Cell*, *131*(6), 1164–1178.

Sugiyama, M., Takao, M., Hatsuta, H., Funabe, S., Ito, S., Obi, T., ... Murayama, S. (2013). Increased number of astrocytes and macrophages/microglial cells in the corpus callosum in amyotrophic lateral sclerosis. *Neuropathology*, *33*(6), 591–599.

Takeuchi, H., Mizoguchi, H., Doi, Y., Jin, S., Noda, M., Liang, J., ... Suzumura, A. (2011). Blockade of gap junction hemichannel suppresses disease progression in mouse models of amyotrophic lateral sclerosis and Alzheimer's disease. *PLoS ONE*, *6*(6), e21108.

Tang, W., Szokol, K., Jensen, V., Enger, R., Trivedi, C. A., Hvalby, O., ... Nagelhus, E. A. (2015). Stimulation-evoked Ca2 + signals in astrocytic processes at hippocampal CA3-CA1 synapses of adult mice are modulated by glutamate and ATP. *The Journal of Neuroscience*, *35*(7), 3016–3021.

Taroni, F., & DiDonato, S. (2004). Pathways to motor incoordination: The inherited ataxias. *Nature Reviews Neuroscience*, *5*(8), 641–655.

Tong, X., Ao, Y., Faas, G. C., Nwaobi, S. E., Xu, J., Haustein, M. D., ... Khakh, B. S. (2014). Astrocyte Kir4.1 ion channel deficits contribute to neuronal dysfunction in Huntington's disease model mice. *Nature Neuroscience*, *17*(5), 694–703.

Tu, P. H., Raju, P., Robinson, K. A., Gurnery, M. E., Trojanowski, J. Q., & Lee, V. M. (1996). Transgenic mice carrying a human mutant superoxide dismutase transgene develop neuronal cytoskeletal pathology resembling human amyotrophic lateral sclerosis lesions. *Proceedings of the National Academy of Sciences of the United States of America*, *93*, 3155–3160.

Ullian, E. M., Sapperstein, S. K., Christopherson, K. S., & Barres, B. A. (2001). Control of synapse number by glia. *Science*, *291*(5504), 657–661.

Urushitani, M., Sik, A., Sakurai, T., Nukina, N., Takahashi, R., & Julien, J. P. (2006). Chromogranin-mediated secretion of mutant superoxide dismutase proteins linked to amyotrophic lateral sclerosis. *Nature Neuroscience*, *9*(1), 108–118.

Valenza, M., Leoni, V., Karasinska, J. M., Petricca, L., Fan, J., Carroll, J., ... Cattaneo, E. (2010). Cholesterol defect is marked across multiple rodent models of Huntington's disease and is manifest in astrocytes. *The Journal of Neuroscience*, *30*(32), 10844–10850.

Vis, J. C., Nicholson, L. F. B., Faull, R. L. M., Evans, W. H., Severs, N. J., & Green, C. R. (1998). Connexin expression in Huntington's diseased human brain. *Cell Biology International*, *22*(11112), 837–847.

Vucic, S., & Kiernan, M. C. (2006). Novel threshold tracking techniques suggest that cortical hyperexcitability is an early feature of motor neuron disease. *Brain*, *129*(9), 2436–2446.

Vucic, S., Rothstein, J. D., & Kiernan, M. C. (2014). Advances in treating amyotrophic lateral sclerosis: Insights from pathophysiological studies. *Trends in Neurosciences*, *37*(8), 433–442.

Waldvogel, H. J., Kim, E. H., Tippett, L. J., Vonsattel, J. P., & Faull, R. L. (2015). The neuropathology of Huntington's disease. *Current Topics in Behavioral Neurosciences*, *22*, 33–80.

Walker, F. O. (2007). Huntington's disease. *The Lancet*, *369* (9557), 218–228.

Wang, L., Gutmann, D. H., & Roos, R. P. (2011). Astrocyte loss of mutant SOD1 delays ALS disease onset and progression in G85R transgenic mice. *Human Molecular Genetics*, *20*(2), 286–293.

Webster, S. J., Bachstetter, A. D., Nelson, P. T., Schmitt, F. A., & Van Eldik, L. J. (2014). Using mice to model Alzheimer's dementia: An overview of the clinical disease and the preclinical behavioral changes in 10 mouse models. *Frontiers in Genetics*, *5*(88), 1–23.

Wojtowicz, A. M., Dvorzhak, A., Semtner, M., & Grantyn, R. (2013). Reduced tonic inhibition in striatal output neurons from Huntington mice due to loss of astrocytic GABA release through GAT-3. *Frontiers in Neural Circuits*, *7*, 188.

Wong, P. C., Pardo, C. A., Borchelt, D. R., Lee, M. K., Copeland, N. G., Jenkins, N. A., ... Price, D. L. (1995). An adverse property of a familial ALS-linked SOD1 mutation causes motor neuron disease characterized by vacuolar degeneration of mitochondria. *Neuron*, *14*, 1105–1116.

Worms, P. M. (2001). The epidemiology of motor neuron diseases: A review of recent studies. *Journal of the Neurological Sciences*, *191*, 3–9.

Wyss-Coray, T., Loike, J. D., Brionne, T. C., Lu, E., Anankov, R., Yan, F., ... Husemann, J. (2003). Adult mouse astrocytes degrade amyloid-beta in vitro and in situ. *Nature Medicine*, *9*(4), 453–457.

Yamanaka, K., Chun, S. J., Boillee, S., Fujimori-Tonou, N., Yamashita, H., Gutmann, D. H., ... Cleveland, D. W. (2008). Astrocytes as determinants of disease progression in inherited amyotrophic lateral sclerosis. *Nature Neuroscience*, *11*(3), 251–253.

Yang, W. W., Sidman, R. L., Taksir, T. V., Treleaven, C. M., Fidler, J. A., Cheng, S. H., ... Shihabuddin, L. S. (2011). Relationship between neuropathology and disease progression in the SOD1(G93A) ALS mouse. *Experimental Neurology*, *227*(2), 287–295.

Yang, Z., & Wang, K. K. (2015). Glial fibrillary acidic protein: From intermediate filament assembly and gliosis to neurobiomarker. *Trends in Neurosciences*, *38*(6), 364–374.

Yvert, G., Lindenberg, K. S., Picaud, S., Landwehrmeyer, G. B., Sahel, J., & Mandel, J. (2000). Expanded polyglutamines induce neurodegeneration and trans-neuronal alterations in cerebellum and retina of SCA7 transgenic mice. *Human Molecular Genetics*, *9*(17), 2491–2506.

Zamanian, J. L., Xu, L., Foo, L. C., Nouri, N., Zhou, L., Giffard, R. G., & Barres, B. A. (2012). Genomic analysis of reactive astrogliosis. *The Journal of Neuroscience*, *32*(18), 6391–6410.

CHAPTER

15

Neurodegenerative Diseases and RNA-Mediated Toxicity

Tiffany W. Todd and Leonard Petrucelli

Mayo Clinic, Jacksonville, FL, United States

OUTLINE

THE IDENTIFICATION OF RNA-MEDIATED TOXICITY: THE MYOTONIC DYSTROPHIES AND CTG REPEATS

In 1992, three studies identified the genetic cause of an autosomal-dominant myotonic dystrophy (DM) as a CTG repeat expansion in the 3′ untranslated region (UTR) of the gene *dystrophia myotonica protein kinase* (*DMPK*) (Brook et al., 1992; Fu et al., 1992; Mahadevan et al., 1992). This multisystemic disease is characterized by myotonia, muscular dystrophy, cardiac defects, cataracts, and endocrine

The Molecular and Cellular Basis of Neurodegenerative Diseases
DOI: https://doi.org/10.1016/B978-0-12-811304-2.00015-8

disorders (Meola & Cardani, 2015). Initially it was hypothesized that the repeat affected *DMPK* expression, and early studies did detect a decrease in the mRNA levels of both *DMPK* and a flanking gene, *SIX5*, in patients (Fu et al., 1993; Klesert, Otten, Bird, & Tapscott, 1997; Thornton, Griggs, & Moxley, 1994). Multiple groups then produced *DMPK* and *SIX5* knockout mice to see if this disease could be explained by a loss-of-function mechanism. *DMPK* knockout mice show some muscle and cardiac defects (Berul et al., 1999; Jansen et al., 1996; Reddy et al., 1996), and *SIX5* knockout mice show an increased level of cataracts (Klesert et al., 2000; Sarkar et al., 2000). But neither of these models can fully recapitulate the human disease, suggesting that there must be another mechanism. Furthermore, shortly after the genetic cause of this disorder was reported, additional studies described a second form of DM that occurred in the absence of this repeat expansion (Meola, Sansone, Radice, Skradski, & Ptacek, 1996; Ricker et al., 1994; Thornton et al., 1994; Udd, Krahe, Wallgren-Pettersson, Falck, & Kalimo, 1997). This second disorder arises instead from a CCTG repeat in the first intron of *zinc finger protein 9* (*ZFN9*), also known as *nucleic acid binding protein* (Liquori et al., 2001). Studies have also argued against a loss of ZFN9 protein expression in this case (Margolis, Schoser, Moseley, Day, & Ranum, 2006), although *ZFN9* heterozygous and knockout mice do display phenotypes that are reminiscent of DM (Chen et al., 2007), suggesting that loss of function may still play a role. Nevertheless, the two diseases, now known as DM type 1 (DM1) and DM type 2 (DM2), respectively, are both characterized by similar clinical features and are therefore believed to arise from a common pathogenic mechanism (Meola & Cardani, 2015). As the repeat expansions in DM1 and 2 are noncoding, it seems unlikely that this common mechanism depends on the effects of a mutant protein. Yet in both cases, the inheritance is autosomal-dominant, leading to the hypothesis that pathogenicity may be manifesting at the RNA level. And in fact, early studies suggested that the repeat-expanded RNA is not properly metabolized (Wang et al., 1995), but is instead retained in the nucleus in DM1 (Davis, McCurrach, Taneja, Singer, & Housman, 1997; Hamshere, Newman, Alwazzan, Athwal, & Brook, 1997), where it accumulates into inclusions called "RNA foci" (Davis et al., 1997; Taneja, McCurrach, Schalling, Housman, & Singer, 1995). Later studies verified that RNA foci can be detected in DM2 as well (Mankodi et al., 2001). Furthermore, the expression of a CTG repeat in the 3′UTR of the unrelated *human skeletal actin* transgene induces DM-like phenotypes in mice (Mankodi et al., 2000), arguing that the repeat RNA alone is sufficient to result in disease. These findings and additional follow-up studies mark DM as the first established RNA-mediated disease.

The Sequestration Hypothesis

As research on the DMs began to focus on the RNA, the next question was to determine how these noncoding repeat transcripts could result in toxicity. One hypothesis was that the repeat expansion may act as a sink for RNA-binding proteins (RBPs) that recognize the $(CUG)_n$ and $(CCUG)_n$ sequences. In this model, the repeat-containing RNA would act as a competitive inhibitor, sequestering the RBPs into RNA foci and ultimately disrupting their native functions, resulting in cellular defects and disease (Fig. 15.1). In order to identify which proteins are sequestered by RNA foci in DM1, Timchenko et al. carried out a band-shifting assay to identify proteins that can interact with CUG repeats. This study led to the identification of CUG-binding protein 1 (CUGBP1) (Timchenko et al., 1996a; Timchenko, Timchenko, Caskey, & Roberts, 1996b), a member of the CELF family of proteins that plays a role in RNA splicing, processing,

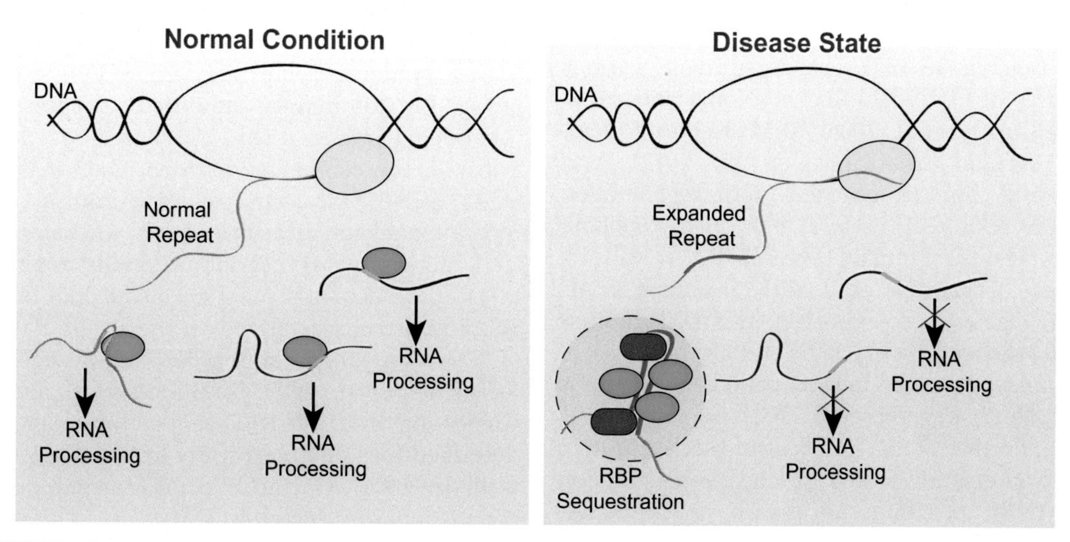

FIGURE 15.1 **The RBP sequestration hypothesis.** In many neurodegenerative diseases, RNA containing expanded repeats accumulates into RNA foci in patient cells. A common hypothesis is that the expanded repeat interacts with numerous RBPs, sequestering them into foci. This can include proteins that normally recognize the repeat sequence, as well as proteins that preferentially interact with the expanded repeat. The sequestration of these proteins prevents them from carrying out their native functions, resulting in widespread RNA processing defects that contribute to disease. *RBPs*, RNA-binding proteins.

translation, and stability (Dasgupta & Ladd, 2012). Subsequent studies have established that CUGBP1 is hyperphosphorylated in DM1, stabilizing the protein (Kuyumcu-Martinez, Wang, & Cooper, 2007) and accounting for the upregulation observed in multiple studies (Cardani et al., 2013; Dansithong, Paul, Comai, & Reddy, 2005; Savkur, Philips, & Cooper, 2001). This is believed to contribute to changes in RNA splicing and translation in DM1 (Huichalaf et al., 2010; Paul et al., 2006; Philips, Timchenko, & Cooper, 1998; Savkur et al., 2001), and overexpression of CUGBP1 in mice results in splicing changes that are also observed in patients (Ho, Bundman, Armstrong, & Cooper, 2005). Overexpression of CUGBP1 enhances CUG repeat-induced toxicity in *Drosophila*, further suggesting that increases in CUGBP1 expression can contribute to disease (de Haro et al., 2006).

CUGBP1 levels are not elevated in DM2 (Cardani et al., 2013), however, suggesting that changes in the expression of this protein cannot account for all of the phenotypes associated with DM. Furthermore, the upregulation of this protein is not consistent with the initial hypothesis that some RBPs may be sequestered and inhibited by the repeat-containing RNA and, indeed, CUGBP1 is not sequestered into RNA foci in DM1 (Michalowski et al., 1999). This prompted Miller et al. to investigate whether additional proteins could interact with the $(CUG)_n$ sequence, and their screen identified a family of proteins homologous to *Drosophila* muscleblind (Miller et al., 2000). Subsequent analyses revealed that the three human orthologs of muscleblind, muscleblind-like protein 1 (MNBL1), MNBL2, and MNBL3, can all colocalize with RNA foci in both DM1 and DM2 (Fardaei et al., 2002; Holt et al., 2009). MNBL proteins are RBPs that regulate alternative splicing at least in part by acting as CELF protein antagonists (Pascual, Vicente, Monferrer, & Artero, 2006).

Consistent with the hypothesis that the sequestration of these three proteins is important for pathogenesis, loss of MNBL protein function in mice can recapitulate various features of DM, including myotonia and cataracts (Kanadia et al., 2003), REM sleep and memory deficits (Charizanis et al., 2012), and impaired muscle function and regeneration (Poulos et al., 2013). *Mnbl1 and Mnbl2* knockout mice also accumulate several RNA splicing defects (Charizanis et al., 2012; Kanadia et al., 2003) that are also observed in a DM1 mouse model (Osborne et al., 2009) and are reminiscent of changes observed in DM1 patients (Charizanis et al., 2012; Kanadia et al., 2003; Suenaga et al., 2012). Furthermore, while loss of MNBL protein function can mimic disease, the overexpression of MNBL proteins can rescue repeat RNA-induced toxicity in both *Drosophila* (de Haro et al., 2006) and mouse models of DM1 (Kanadia et al., 2006), strongly supporting a role for these proteins in DM. Overall, DM1 and DM2 are now largely believed to be due to defects in alternative splicing that arise as a result of the sequestration of MNBL proteins into RNA foci and a disruption in the expression of other RBPs including CUGBP1, hnRNP H, and Staufen1 (Meola & Cardani, 2015). These seminal findings were also crucial for the study of other neurodegenerative diseases that are associated with repeat expansions and are now believed to be caused by mutant RNA.

Additional CTG Repeat-Related Diseases: Variations on a Theme

In addition to DM1, three other diseases have been attributed to the expansion of a CTG repeat (Table 15.1), and it is reasonable to predict that these diseases may also be due to the sequestration of MNBL proteins and other RBPs. Several studies support this possibility, and these findings are summarized below. Understanding why these diseases manifest with different symptoms is an interesting avenue of future research.

Spinocerebellar Ataxia (SCA) Type 8

The spinocerebellar ataxias (SCAs) are a group of dominantly inherited, progressive neurodegenerative disorders characterized by a loss of the cerebellar neurons. One of these SCAs, SCA8, was found to arise from a CTG repeat expansion in a previously uncharacterized, noncoding transcript originally referred to as *SCA8* (Koob et al., 1999). And like DM1, SCA8 is also characterized by the accumulation of CUG repeat-containing RNA into nuclear foci (Chen et al., 2009; Daughters et al., 2009), suggesting that the RBP sequestration model described for DM1 may apply to this disease as well. In 2004, Mutsaddi et al. carried out a genetic modifier screen in *Drosophila* to identify factors that modulate SCA8-associated phenotypes. This study identified three RBPs that, when mutated, enhanced toxicity, suggesting the function of these proteins may be negatively impacted by the expression of the repeat. These three RBPs were *split ends*, *staufen*, and, unsurprisingly, *muscleblind* (Mutsuddi, Marshall, Benzow, Koob, & Rebay, 2004). A later study by Daughters et al. found that human MNBL1 does indeed colocalize with RNA foci in the brains of SCA8 patients and that there is a dysregulation of MNBL1 and CUGBP1-dependent alternative splicing in SCA8 bacterial artificial chromosome (BAC) mice and in postmortem SCA8 brains. This dysregulation could be rescued in repeat-expressing SK-N-SH cells by overexpression of MNBL1, and, when the SCA8 mice were mated to *Mnbl1* knockout mice, the loss of MNBL1 expression enhanced SCA8-associated motor deficits (Daughters et al., 2009). Overall, SCA8 is believed to arise at least in part from an RNA-mediated mechanism that closely resembles the mechanisms discovered in DM1.

Huntington's Disease-Like 2

Huntington's disease (HD)-like 2 (HDL2) is a disease that resembles HD in both its clinical

TABLE 15.1 RNA-mediated Diseases

				Pathomechanisms currently associated with this disease:		
Disease	**Affected Gene**	**Repeat**	**Repeat Location**	**RNA foci, RBP sequestration**	**Bidirectional Transcription**	**RAN Translation**
DM1 (Myotonic Dystrophy Type 1)	*DMPK*	CTG	non-coding (3'UTR)	Yes		Yes
DM2 (Myotonic Dystrophy Type 2)	*ZFN9*	CCTG	non-coding (intron)	Yes		
SCA8 (Spinocerebellar Ataxia Type 8)	*ATXN8OS*	CTG	non-coding (3'UTR)	Yes	Yes	Yes
HDL2 (Huntington's Disease Like 2)	*JPH3*	CTG	non-coding (3'UTR, intron)	Yes	Possibly	
FECD (Fuchs Endothelial Corneal Dystrophy)	*TCF4*	CTG	non-coding (intron)	Yes		
FXTAS (Fragile X Tremor/Ataxia Syndrome)	*FMR1*	CGG	non-coding (5'UTR)	Yes	Yes	Yes
FXPOI (Fragile X Primary Ovarian Insufficiency)	*FMR1*	CGG	non-coding (5'UTR)	Yes	Yes	Yes
SCA10 (Spinocerebellar Ataxia Type 10)	*ATXN10*	ATTCT	non-coding (intron)	Yes		
SCA31 (Spinocerebellar Ataxia Type 31)	*TK2* and *BEAN*	TGGAA, TAGAA, TAAAA, and TAAAATAGAA	non-coding (intron)	Yes	Yes	
SCA36 (Spinocerebellar Ataxia Type 36)	*NOP56*	TGGGCC	non-coding (intron)	Yes		
c9FTD/ALS (C9orf72-associated Frontotemporal Dementia and Amyotrophic Lateral Sclerosis)	*C9orf72*	GGGGCC	non-coding (intron, promotor)	Yes	Yes	Yes
SCA12 (Spinocerebellar Ataxia Type 12)	*PPP2R2B*	CAG	non-coding (intron) or coding (exon)			
HD (Huntington's Disease)	*HTT*	CAG	coding (exon)	RBP binding	Yes	Yes
SBMA, DRPLA, SCA1, 2, 3,6,7,17 (Other Polyglutamine Disorders)	[various]	CAG	coding (exon)	RBP binding	Yes	

This table summarizes diseases currently associated with an RNA-mediated toxicity mechanism, as described throughout this chapter. The affected gene and relative location of the expanded repeat is noted, as well as a list of which of the major mechanisms discussed are currently associated with each disorder. It is worth noting that this list is a snapshot of the current state of the literature on these diseases and is likely to change in the future as a greater understanding of the pathomechanisms underlying each disease is achieved. *DM1 and DM2*, myotonic dystrophy types 1 and 2; *SCA8, 10, 12, 31, and 36*, spinocerebellar ataxia types 8, 10, 12, 31, and 36; *HDL2*, Huntington's disease-like 2; *FECD*, Fuchs endothelial corneal dystrophy; *FXTAS*, fragile X-associated tremor/ataxia syndrome; *FXPOI*, fragile X primary ovarian insufficiency; *c9FTD/ALS*, c9orf72-associated frontotemporal dementia and amyotrophic lateral sclerosis; *HD*, Huntington's disease.

presentation and neuropathology. Both are autosomal-dominant, late-onset disorders associated with a loss of striatal neurons (Rudnicki, Pletnikova, Vonsattel, Ross, & Margolis, 2008). Unlike HD, however, HDL2 is not a classic polyglutamine disorder but instead arises from the expansion of a CTG repeat in exon 2A of the gene *junctophilin-3* (*JPH3*) (Holmes et al., 2001; Margolis et al., 2001). This exon is alternatively spliced so that the repeat is absent from the transcript that encodes the full-length JPH3 protein, but is present in four splice variants where it can either encode polyalanine or polyleucine, or exist in the 3′UTR. JPH3 is a brain-specific protein that is thought to play a role in connecting the endoplasmic reticulum to the plasma membrane and in regulating calcium signaling (Takeshima, Komazaki, Nishi, Iino, & Kangawa, 2000). There is evidence that JPH3 levels are reduced in HDL2 patients (Seixas et al., 2012), and *JPH3* knockout mice show a progressive motor phenotype (Nishi et al., 2002; Seixas et al., 2012), suggesting HDL2 may arise in part from a loss-of-function mechanism. At the same time, however, the expanded CTG repeat is transcribed into a CUG repeat-containing mRNA that accumulates into RNA foci in BAC transgenic mice (Wilburn et al., 2011), as well as in transfected cells and in the striatum and frontal cortex of HDL2 patients (Rudnicki et al., 2007). These foci also colocalize with MBNL1 (Rudnicki et al., 2007), suggesting that HDL2, like SCA8, may also arise from a pathogenic mechanism that is similar to that described for DM1.

Fuchs Endothelial Corneal Dystrophy

Recently, another disease was added to the list of disorders associated with CTG repeat expansions and RNA-mediated toxicity. But unlike the other diseases described in this book, this disease is not a rare neurodegenerative disorder, but a very common eye disease that results in the degeneration of the endothelial cells on the internal surface of the cornea. Fuchs endothelial corneal dystrophy (FECD) is a late-onset, inherited disorder characterized by the formation of guttae: collagenous deposits in-between the corneal endothelial cells. In severe cases, this can lead to a degeneration of the endothelial cells and vision loss that can only be treated by corneal replacement (Vedana, Villarreal, & Jun, 2016). FECD is the most common cause of corneal dystrophy, and affects nearly 5% of Caucasian individuals over 40 in the United States, making it far more common than all other microsatellite expansion disorders worldwide (Du et al., 2015). The genetic basis of FECD is heterogeneous, but in 2014, Wieben et al. identified a CTG repeat in *transcription factor 4* that results in FECD when expanded to over 50 repeats (Wieben et al., 2012), and this finding has been replicated in other FECD populations (Mootha, Gong, Ku, & Xing, 2014). The expanded repeat forms nuclear RNA foci in fibroblasts derived from patient skin biopsies, as well as in the corneal endothelial cells of FECD patients (Du et al., 2015; Mootha et al., 2015). And once again, similar to what is known for other CTG repeat-associated diseases, MBNL1 colocalizes with RNA foci in patient samples. RNA-Seq analysis revealed widespread changes in RNA splicing in patient corneal endothelial cells compared to controls, including multiple changes that have also been reported in DM1 and are known to be sensitive to changes in MNBL1 activity (Du et al., 2015; Wieben et al., 2017). These intriguing findings suggest that the RNA-mediated toxicity associated with neurodegenerative diseases need not be restricted to neurons and in fact could be a more widespread disease mechanism capable of affecting various cell types. Additional studies on the pathogenic mechanisms in FECD will shed light on this possibility and perhaps help to explain the occurrence of nonneurological symptoms in some neurodegenerative diseases.

RNA FOCI AND THE SEQUESTRATION HYPOTHESIS IN OTHER REPEAT-ASSOCIATED DISEASES

The sequestration hypothesis described for DM is not restricted to CTG repeat-associated diseases but can apply to other expanded repeat-associated diseases as well. In these diseases, MNBL proteins may not play a prominent role, but the expanded RNA still forms RNA foci and is believed to sequester and inhibit other RBPs. The role of RNA foci in these diseases is summarized below.

Fragile X-Associated Tremor/Ataxia Syndrome

Fragile X syndrome (FXS) is a common form of mental retardation associated with the expansion of a CGG repeat in the 5′UTR of the gene *fragile X mental retardation 1* (*FMR1*) (Verkerk et al., 1991). FXS patients harbor repeats over 200 units in length, resulting in translational impairment and a subsequent loss of the fragile X mental retardation protein (FMRP) (Feng et al., 1995). FXS is therefore believed to be caused by loss-of-function mechanisms. In contrast, individuals harboring between 50 and 200 repeats, referred to as "premutation carriers," can show an increase in *FMR1* mRNA levels without a dramatic change in FMRP protein expression (Hagerman et al., 2001; Tassone et al., 2000). These premutation carriers are at risk to develop fragile X-associated tremor/ataxia syndrome (FXTAS), a neurodegenerative disorder characterized by gait ataxia, action tremors, and cognitive deficits (Hagerman et al., 2001). Women carrying the premutation are also at risk for early menopause and premature ovarian failure, a condition referred to as fragile X-associated primary ovarian insufficiency (FXPOI) (Allingham-Hawkins et al., 1999; Buijsen et al., 2016; Sherman, 2000). The fact that the *FMR1* RNA is increased in FXTAS and FXPOI, while FMRP levels remain relatively unchanged, has led to the conclusion that the toxicity in these diseases arises from an RNA-mediated mechanism (Galloway & Nelson, 2009; Hagerman et al., 2001; Li & Jin, 2012).

Models of FXTAS Support an RNA-Mediated Mechanism

The CGG repeat–containing *FMR1* mRNA accumulates into nuclear aggregates throughout the cerebrum and brainstem in FXTAS patients. These inclusions are similar to the RNA foci observed in CTG repeat-associated diseases, although they also bear resemblance to protein inclusions (Greco et al., 2006, 2002; Iwahashi et al., 2006; Tassone et al., 2000). These inclusions can be recapitulated in some, but not all cultured cell types, including in neuronal and ovarian cell lines (Sellier et al., 2010). In many of these models, inclusion formation correlates with a reduction in cell viability and other defects (Arocena et al., 2005; Handa et al., 2005; Hoem et al., 2011; Sellier et al., 2010), and this toxicity is dosage- and repeat length-dependent (Hoem et al., 2011). Of note, toxicity was not observed when cells were transfected with a repeat-containing reporter construct that lacked a promoter (Arocena et al., 2005), suggesting that transcription is important for cell death and supporting a role for the repeat-containing RNA in toxicity. Interestingly, expression of the repeat RNA in HEK293T cells was associated with changes in global gene expression (Handa et al., 2005), suggesting that the repeat-containing RNA may negatively impact gene transcription. Further evidence to support a role for the repeat-containing RNA in FXTAS comes from studies using animal models. In 2003, Jin et al. subcloned a *FMR1* fragment from a premutation carrier into a *Drosophila* expression vector to generate flies expressing 90 CGG repeats in the 5′UTR of an EGFP reporter. This nontranslated repeat formed inclusions and induced a progressive toxicity that was dosage- and repeat length-dependent (Jin et al., 2003). Two different knock-in mouse models of FXTAS have also been developed, and both models recapitulate the elevated *FMR1* levels and nuclear inclusions characteristic of FXTAS. Both models show behavioral abnormalities, although the exact presentation differs between the two mice and neither completely recapitulates the human disease (Entezam et al., 2007; Willemsen, 2003). Of interest, ovarian abnormalities have also been reported in one of these knock-in models, suggesting that it is a model of FXPOI as well (Hoffman et al., 2012). Importantly, two groups have also developed inducible FXTAS mouse models in which the expression of an expanded CGG repeat RNA alone, in the absence of the *FMR1* locus, is sufficient to induce inclusion formation and behavioral deficits (Hashem et al., 2009; Hukema et al., 2015).

Identifying the RBPs Sequestered by CGG Repeats

Based on the sequestration hypothesis for RNA-mediated toxicity in CTG-related diseases, it was predicted that the CGG repeat RNA could also bind to and inhibit RBPs and that this could contribute to FXTAS. Studies have focused on identifying which RBPs are affected in this disease, and a few key candidates have been described. In 2007, two papers published in *Neuron* identified Pur α and hnRNP A2/B1 as RBPs that bind to the expanded CGG repeat (Jin et al., 2007; Sofola, Jin, Qin et al., 2007). Overexpression of these proteins was able to rescue repeat-related toxicity in a FXTAS *Drosophila* model, suggesting that their function may be inhibited by the expression of the CGG repeat. Pur α knockout mice develop tremors and spontaneous seizures associated with a loss of hippocampal and cerebellar neurons, suggesting it plays an important role in neurons and neurodevelopment (Khalili et al., 2003). Within flies, Pur α interacts with the *Drosophila* ortholog of the p68 RNA helicase, Rm62, and expression of the CGG repeat leads to a posttranscriptional decrease in Rm62 expression, resulting in the nuclear accumulation of various mRNAs (Qurashi, Li, Zhou, Peng, & Jin, 2011). CUGBP1 was also able to rescue CGG repeat–induced toxicity in flies, and it was noted that this protein is recruited to RNA foci through interactions with hnRNP A2/B1 (Sofola, Jin, Qin et al., 2007). Similarly, TAR DNA-binding protein 43 (TDP-43) also suppresses FXTAS-mediated toxicity in flies through its interaction with hnRNP A2/B1 (He, Krans, Freibaum, Taylor, & Todd, 2014), while its knockdown enhances toxicity (Galloway et al., 2014). TDP-43 is not localized to RNA foci in this disease (He et al., 2014), however, but instead shows reduced expression at the mRNA level in FXTAS brains (Galloway et al., 2014). Together, these findings suggest a putative model in which CGG repeat–containing RNA foci sequester hnRNP A2/B1 and inhibit its activity, resulting in RNA splicing defects and pathology. Expression of the repeat RNA also induces a decrease in TDP-43 expression, resulting in additional splicing defects and toxicity. Overexpression of TDP-43 in *Drosophila* is able to rescue repeat-induced toxicity by reducing the negative effects of TDP-43 loss, but also by promoting the binding of TDP-43 to hnRNP A2/B1, which prevents its accumulation into foci and allows for a regain of its activity. Importantly, both hnRNP A2/B1 (Iwahashi et al., 2006) and Pur α (Jin et al., 2007) have been reported to colocalize with inclusions in FXTAS patient brain.

In 2010, Sellier et al. also carried out a RNA pull-down assay to identify proteins capable of binding the CGG repeat RNA using COS-7 cells and knock-in mouse brain lysates. This study identified hnRNP G and MNBL1 as proteins that colocalize with the CGG RNA foci (Sellier et al., 2010). MNBL1 was also reported to colocalize with inclusions in FXTAS patient brains (Iwahashi et al., 2006), although its overexpression failed to suppress the retinal degeneration seen in CGG-repeat-expressing flies (Sofola, Jin, Qin et al., 2007). Interestingly, in cells, the incorporation of these two RBPs into foci increased over time, suggesting that they were actively recruited into the inclusions (Sellier et al., 2010). Further analysis revealed that the RBP Src associated in mitosis 68 (Sam68) was required for the recruitment of both hnRNP G and MNBL1 into the aggregates. Sam68 appears to be sequestered into $(CGG)_n$ RNA foci in cells and in FXTAS patient tissue, reducing its activity and contributing to alternative splicing defects (Sellier et al., 2010). Sam68 knockout mice show phenotypes reminiscent of FXTAS, specifically ataxia-like motor coordination deficits (Lukong & Richard, 2008). Knockdown of Sam68 also results in a decrease in hippocampal dendritic spines in both mice and hippocampal neuronal cultures (Klein, Younts, Castillo, & Jordan, 2013). Given that $(CGG)_n$ RNA foci are frequently found in the hippocampus of FXTAS patients (Greco et al., 2006), and that FXTAS induced pluripotent stem cell (iPSC)-derived neurons show reduced synaptic signaling (Liu, Koscielska et al., 2012), these studies on Sam68 suggest it may be an important mediator of FXTAS pathogenesis.

Finally, in 2013, Sellier et al. performed another RNA pull-down screen and identified 10 proteins that preferentially interacted with the expanded CGG repeat as opposed to nonpathogenic controls. One of these proteins was DiGeorge syndrome critical region 8 (DGCR8) (Sellier et al., 2013), a major component of the miRNA biogenesis pathway. DGCR8 binds pri-miRNA transcripts and recruits the RNAase III enzyme Drosha. The pri-miRNA is then cleaved to form a pre-miRNA transcript, which is exported from the nucleus and subsequently processed by Dicer to form a mature miRNA (Petri, Malmevik, Fasching, Akerblom, & Jakobsson, 2014). Expression of the CGG repeat in COS-7 cells caused DGCR8 and Drosha to switch from a diffuse staining pattern to intense localization into the RNA inclusions. The authors did not detect any evidence that Drosha acts to cleave the repeat RNA, but the expression of the repeat did reduce the cleavage of specific pri-miRNAs (Sellier et al., 2013), suggesting that DGCR8 and Drosha are sequestered into aggregates and that the repeat RNA can negatively impact miRNA biogenesis. Consistent with this idea, analysis of FXTAS brains reveals a decrease in mature miRNA levels (Alvarez-Mora et al., 2013; Sellier et al., 2013) that is also observed in *Drosophila* (Tan, Poidevin, Li, Chen, & Jin, 2012), and knock-in mouse (Zongaro et al., 2013) models of FXTAS. Data from a *Caenorhabditis elegans* model of FXTAS also supports a role for the miRNA pathway in repeat-associated toxicity (Juang et al., 2014). Consistent with this hypothesis, overexpression of DGCR8 rescued repeat-induced cellular toxicity and dendritic branching defects in primary mouse cortical neurons (Sellier et al., 2013).

Protein-Coding CAG Repeats in Huntington's Disease (HD) and Other Polyglutamine Disorders

The polyglutamine diseases are a family of nine diseases that arise from the expansion of a CAG repeat within the coding region of a gene, encoding a span of glutamines within the host protein and leading to protein aggregation and a toxic gain-of-function (Todd & Lim, 2013). These diseases have long been considered protein-mediated conditions, and much of the research to date has focused on this mechanistic aspect. Nevertheless, there is increasing evidence that the repeat-containing RNA may also represent a toxic entity in HD and other polyglutamine disorders (Chan, 2014; Nalavade, Griesche, Ryan, Hildebrand, & Krauss, 2013). Early evidence that the CAG repeat RNA may play a role in polyglutamine pathogenesis comes from a study using a *Drosophila* model of SCA3. While expression of a pure CAG repeat is toxic in these flies, the authors noted that this toxicity was mitigated when the repeat was interrupted with CAA codons. Both of these repeats result in similar levels of polyglutamine protein, suggesting that the CAG repeat RNA is responsible for at least some of the toxicity observed in these flies (Li, Yu, Teng, & Bonini, 2008). This is perhaps not surprising, given that SCA1 and SCA2, for instance, are known to be caused by pure CAG repeats in patients, while the repeat is interspersed with CAA or CAT codons in nondisease controls. These interspersed codons are predicted to prevent the formation of RNA secondary structures in the repeat region, while the pure expanded CAG repeat can form stable hairpins (de Mezer, Wojciechowska, Napierala, Sobczak, & Krzyzosiak, 2011; Nalavade et al., 2013). Pure CAG repeats are also known to be toxic even in the absence of translation: placing a pure expanded CAG repeat in the 3′UTR of a reporter construct leads to neuronal toxicity and behavioral defects in *C. elegans* (Wang et al., 2011), *Drosophila* (Li et al., 2008), and mice (Hsu et al., 2011). Moreover, these pure CAG repeats accumulated into RNA foci in these models (Hsu et al., 2011; Wang et al., 2011), although foci were rare in *Drosophila* (Li et al., 2008). Finally, SCA12 is caused by an expanded CAG repeat that, depending on the isoform, is located either in or flanking exon 7 of *PPP2R2B*, a gene that encodes a regulatory subunit of protein phosphatase 2A. Although RNA foci have not been noted in this disease, and the repeat expansion is known to influence the expression of the PPP2R2B protein, it remains possible that this disease also arises from the toxic effects of a CAG repeat-containing RNA in the absence of a polyglutamine protein (Cohen & Margolis, 2016).

Given the potential for CAG repeat RNA to form foci and elicit disease phenotypes, it is possible that this RNA-mediated toxicity is also caused by a sequestration of RBPs. Consistent with this idea, a screen for modifiers of SCA1 phenotypes in flies identified multiple RBPs, suggesting that alterations in RNA processing plays a role in this disease. However, the authors also note that Ataxin1, the protein associated with SCA1, may be an RBP itself (Fernandez-Funez et al., 2000), and this may suggest that altered RNA processing in this disease could be ascribed to protein gain-of-function as well. Interestingly, several lines of evidence suggest that MBNL1 may also play a role in polyglutamine diseases, although its exact role is unclear. In an early study on a *Drosophila* model of SCA3, Li et al. (2008) found that overexpression of either muscleblind or human MNBL1 enhances polyglutamine-mediated toxicity. This is in contrast to what has been found for DM1, where overexpression of MNBL1 compensates for the loss of its function and is thereby beneficial. Nonetheless, MNBL1 has been reported to bind both CAG repeats and CUG repeats with comparable affinities (Yuan et al., 2007), and it therefore could theoretically be sequestered by the CAG repeat-containing RNA as well. Indeed, muscleblind was found to colocalize with CAG RNA foci in *C. elegans*, and in this model, its overexpression did partially reverse the lethality and locomotor defects associated with the overexpression of pure CAG repeat RNA (Wang et al., 2011). In a study by de Mezer et al. (2011), the CAG repeat-expanded *huntingtin* (*HTT*) and *androgen receptor* transcripts, associated with HD and spinal and bulbar muscular atrophy, respectively, did not form foci in cells but did colocalize with MNBL1 in the nucleus. Nuclear accumulations of CAG repeat RNA were also observed in HD and SCA3 patient fibroblasts, and these also colocalized with MNBL1 (Mykowska, Sobczak, Wojciechowska, Kozlowski, & Krzyzosiak, 2011). Furthermore, expression of pure expanded CAG repeats and of expanded *ATXN3* and *HTT* led to defects in alternative splicing that mirror what is seen in DM1 (Mykowska et al., 2011). Taken together, these studies suggest that the sequestration of MNBL1 and a loss of its function in RNA processing could contribute to polyglutamine diseases via a mechanism that is analogous to what is seen in DM1. At the same time, MNBL1 may not be sequestered in some polyglutamine models, and in these contexts it may have a different effect on disease phenotypes.

In addition to MNBL1, CAG repeats have also been shown to bind to and perhaps sequester nucleolin (NCL), a key protein in the nucleolus (Tsoi & Chan, 2014). In a series of studies, Tsoi et al. first identified an abnormal accumulation of CAG repeat RNA in the nucleus in *Drosophila* and in mice. This accumulation is attributed to a defect in nuclear export caused by the interaction of the repeat-containing RNA with the NXF1/U2AF65 RNA export pathway (Tsoi, Lau, Lau, & Chan, 2011). In a subsequent study, the authors determined that the nuclear CAG repeat RNA was localized to the nucleolus, where it binds NCL and prevents the binding of NCL to the upstream control element of the ribosomal RNA (rRNA) promotor. This results in a decrease in rRNA transcription and nucleolar stress that can trigger apoptosis (Tsoi & Chan, 2013; Tsoi, Lau, Tsang, Lau, & Chan, 2012).

Pentanucleotide and Hexanucleotide Repeats in Inherited SCAs

There are 40 different SCAs identified to date. Like those already described, all of these autosomal-dominant diseases are characterized primarily by a progressive cerebellar ataxia associated with the degeneration and loss of the cerebellar Purkinje cells. Of these 40, the genetic cause has been identified for 28 disorders, and, of these, 11 are associated with an expanded repeat (Sun, Lu, & Wu, 2016). The expansions associated with SCA1, 2, 3, 6, 7, and 17 are located within the coding region of the affected gene and give rise to polyglutamine expansions in the host protein (Todd & Lim, 2013). In contrast, SCA8, 10, 12, 31, and 36 are associated with repeat expansions in noncoding regions, and RNA foci have been identified in patient samples and/or models of these SCAs, with the exception of the little-understood SCA12 (Cohen & Margolis, 2016). In this section, we describe the limited but interesting findings associated with SCA10, 31, and 36.

Spinocerebellar Ataxia Type 10

SCA10 is a rare disorder described primarily in patients in Latin America, although the disease was also recently identified in China (Wang et al., 2015). In patients of Mexican descent, it is also associated with epilepsy, while patients in Brazil show a pure ataxia phenotype (Teive et al., 2011). In 2000, the causative mutation for SCA10 was identified as an ATTCT pentanucleotide expansion in an intron of a previously uncharacterized gene originally designated *E46*, but now called *Ataxin10* (*ATXN10*). While nonaffected individuals harbor a range of 10–22 repeats, this repeat is expanded by up to 2.5 kb in SCA10 patients (Matsuura et al., 2000). While there is some evidence suggesting that the ATXN10 protein is important for the survival of cerebellar neurons (Marz et al., 2004), the repeat expansion does not appear to affect the expression or processing of the *ATXN10* gene in patient-derived cells (Wakamiya et al., 2006). *ATXN10* knockout mice have also been generated, and while homozygous knockout of this gene is embryonic lethal, heterozygous mice, which show a 50% decrease in ATXN10 expression, fail to develop any SCA10-like phenotypes (Wakamiya et al., 2006). These studies argue against a pure loss-of-function mechanism for SCA10, and instead, an RNA gain-of-function mechanism is favored.

Support for this model comes from studies using patient-derived cells and transgenic $(ATTCT)_{500}$ mice. RNA fluorescence in situ hybridization analysis revealed that expanded ATTCT repeats formed RNA foci in both the nucleus and cytoplasm of patient-derived fibroblasts, as well as in cells transfected with a luciferase reporter construct that harbors a $(ATTCT)_{500}$ repeat within the rabbit *beta-globin* intron. These foci were also detected in the brains of mice expressing this same reporter construct (White et al., 2010), and well as in a subsequent mouse model that expressed the $(ATTCT)_{500}$ repeat in the 3′UTR of the luciferase reporter (White et al., 2012). While the phenotype of the first mouse model was not described, mice expressing the expanded repeat in the 3′UTR show locomotor dysfunction, increased susceptibility to seizures, and neuronal degeneration in the hippocampus and cortex but paradoxically do not show any cerebellar degeneration (White et al., 2012).

RNA pull-down experiments using brain extracts from the mice expressing the intronic repeat identified hnRNP K as a consistent interactor. This finding could be recapitulated using patient fibroblasts, and furthermore, hnRNP K colocalized with foci in cells and in brains from the two transgenic mouse models (White et al., 2010, 2012). Patient-derived cells also showed defects in the alternative splicing of *β-tropomyosin*, as would be consistent with an impairment in hnRNP K activity (White et al., 2010). The current model suggests that the binding of the repeat to hnRNP K disrupts its activity and also inhibits its binding to protein kinase C delta (PKCδ). This triggers the translocation of PKCδ to the mitochondria and the activation of caspase-3 and apoptosis (White et al., 2010). The accumulation of PKCδ in the mitochondria of transgenic mice (White et al., 2010, 2012) and patient fibroblasts (White et al., 2010) supports this model, and studies suggest that this effect can be reversed by overexpressing hnRNP K or knocking down the repeat-containing transcript (White et al., 2010). While compelling, it remains to be seen if this model holds true in SCA10 patients, where the presence of RNA foci and the relative localization of hnRNP K and PKCδ has yet to be determined.

Spinocerebellar Ataxia Type 31

SCA31 is characterized by a late onset, pure cerebellar ataxia with Purkinje cell degeneration. This degeneration is uniquely associated with the formation of an amorphous "halo" around the degenerating cells that appears to be comprised of neuritic sprouts emanating from the Purkinje cells and synaptophysin-positive granules presumably generated from the presynaptic terminals of interacting neurons (Ishikawa & Mizusawa, 2010). SCA31 is associated with a 2.5–3.8-kb insertion into human chromosome 16 that may have originated from paracentronic heterochromatin (i.e., tightly packed DNA located near centromeres) (Sato et al., 2009). This insert contains pentanucleotide repeats of $(TGGAA)_n$, $(TAGAA)_n$, and $(TAAAA)_n$, as well as regions of $(TAAAATAGAA)_n$ (Niimi et al., 2013; Sato et al., 2009). It is only rarely observed in control individuals, and the length of the insert is inversely correlated with the age of onset in SCA31 (Sato et al., 2009). The inserted repeats reside in introns of two genes transcribed in opposite directions—*thymidine kinase 2* and *brain-expressed, associated with Nedd4*—but it does not appear to affect the expression or splicing of either of these genes (Sato et al., 2009). Instead, the repeats form RNA foci in the Purkinje Cells of SCA31 patients (Niimi et al., 2013; Sato et al., 2009). Although there is still much to be learned about how these foci contribute to SCA31 pathogenesis, the expression of the repeat was toxic in HEK293T and PC12 cells (Niimi et al., 2013), and the RNA can bind the RBPs serine/arginine-rich splicing factor 1 (SRSF1) and SRSF9 in vitro (Sato et al., 2009), perhaps supporting a role for these repeats in sequestering RBPs and disrupting RNA processing.

Spinocerebellar Ataxia Type 36

SCA36 is a rare disorder that has primarily been identified in patients in the Asida river region of Japan (Kobayashi et al., 2011) and Galacia, Spain (Garcia-Murias et al., 2012), although it has also been identified in patients of French, Portuguese, and Chinese descent (Obayashi et al., 2015; Zeng et al., 2016). In addition to the characteristic cerebellar ataxia, SCA36 patients also present with motor neuron symptoms including fasciculations and muscle atrophy of the tongue and, particularly in Japanese cases, skeletal muscle (Garcia-Murias et al., 2012; Kobayashi et al., 2011). In some cases, mild cerebral involvement has also been reported (Abe et al., 2012; Garcia-Murias et al., 2012). Little is known about the mechanism underlying toxicity in SCA36, but this rare disorder is intriguingly associated with an intronic hexanucleotide repeat that differs from the amyotrophic lateral sclerosis (ALS) and frontotemporal dementia (FTD)-associated repeat in C9orf72 (described below) by only one nucleotide. Specifically, SCA36 is associated with a TGGGCC repeat located in the first intron of nucleolar protein 56 (*NOP56*) (Garcia-Murias et al., 2012; Kobayashi et al., 2011). The expression of *NOP56* RNA was not altered in SCA36 lymphoblastoid cell lines (LCLs), arguing against haploinsufficiency as a mechanism in SCA36. There was, however, a decrease in the expression of a nearby gene encoding the miRNA miR1292 (Kobayashi et al., 2011). The importance of this miRNA to SCA36 is still unknown. Intriguingly, initial studies revealed that the expanded TGGGCC repeat could accumulate into nuclear RNA foci in patient LCLs (Kobayashi et al., 2011), as well as in the brain of a SCA36 patient, where they were found to vary in size and number (Liu et al., 2014). Within LCLs, these foci colocalized with SRSF2, suggesting that SCA36 may also arise from the sequestration of RBPs. Further studies are needed to better understand the molecular mechanisms involved in this rare disorder.

Hexanucleotide GGGGCC repeats in C9orf72-Associated ALS and FTD (c9FTD/ALS)

Since the identification of the hexanucleotide expansion in *C9orf72* as a genetic cause of both ALS and FTD (DeJesus-Hernandez et al., 2011; Renton et al., 2011), several studies have focused on understanding how this repeat leads to pathogenesis. We recently reviewed the potential mechanisms associated with c9orf72-associated FTD and ALS (c9FTD/ALS) (Todd & Petrucelli, 2016) but have included in this chapter an updated summary of the literature as it pertains to RNA-mediated toxicity. The C9orf72 repeat RNA is known to form G-quadruplex structures as well as hairpins (Fratta et al., 2012; Haeusler et al., 2014; Reddy, Zamiri, Stanley, Macgregor, & Pearson, 2013; Su et al., 2014). Early observations demonstrated that the repeat RNA accumulates into RNA foci in patient neurons (DeJesus-Hernandez et al., 2011), and it has recently been argued that these foci may be comprised of a single RNA transcript (Liu et al., 2017). Given that RNA foci are not observed in all cells in c9FTD/ALS patients, this finding would suggest that a few RNA molecules and their interactors can have widespread effects. Nonetheless, work from our lab and others have demonstrated that there are significant defects in alternative splicing and gene transcription in c9FTD/ALS patient samples (Cooper-Knock, Kirby, Highley, & Shaw, 2015; Prudencio et al., 2015), fibroblasts (Lagier-Tourenne et al., 2013), and iPSC-derived neurons (Donnelly et al., 2013; Sareen et al., 2013), suggesting a dysregulation of RNA processing is an important factor in c9FTD/ALS. This would be consistent with a model in which RNA foci sequester RBPs in this disease. Based on this hypothesis, several groups have tried to identify the RBPs that interact with the repeat RNA and/or colocalize with RNA foci in c9FTD/ALS patients and models. Overall these studies have resulted in a mix of different candidate RBPs, including Pur α (Jin et al., 2007; Rossi et al., 2015; Sareen et al., 2013; Xu et al., 2013), adenosine deaminase RNA-specific B2 (Donnelly et al., 2013), hnRNP A3 (Mori, Lammich et al., 2013), and hnRNP U, hnRNP F, and hnRNP K (Haeusler et al., 2014). Unfortunately, few studies have focused on looking at the functional output of the potential sequestration of these proteins, and for this reason, it remains unclear which, if any, of these RBPs are chiefly responsible for degeneration in these diseases. If c9ALS/FTD is in part mediated by the sequestration of an RBP, however, perhaps one of the strongest candidates is hnRNP H. Two studies identified this protein as an RBP capable of binding the *C9orf72* GGGGCC repeat and colocalizing with RNA foci in patient tissue (Lee et al., 2013; Rossi et al., 2015). Importantly, one study also found that the alternative splicing of *TARBP2* RNA, a known hnRNP H target, was altered in cells expressing the expanded repeat, suggesting that there is a loss of hnRNP H function in this model (Lee et al., 2013). This is consistent with work from our lab in which the hnRNP H binding motif was found to be enriched in alternatively spliced cassette exons and their flanking intronic regions in the frontal cortex and cerebellum of C9ALS patients. Interestingly, the same effect was seen in sporadic ALS cases (Prudencio et al., 2015), suggesting that a loss of hnRNP H function may be a more general feature of ALS. Furthermore, a recent study by Conlon et al. used UV crosslinking to identify proteins that bind the *C9orf72* repeat RNA in vitro, and the major hit from this analysis was also hnRNP H. Interestingly, this study also noted splicing defects in c9ALS cerebellum versus controls that would correspond to a defect in hnRNP H activity, and these changes correlated with disease duration. In other words, in their sample set, patients with the shortest disease duration had the highest degree of aberrant exon inclusion (Conlon et al., 2016). The identification of misspliced hnRNP H targets in patient tissue in two separate studies strongly supports a role for this RBP in c9FTD/ALS pathogenesis.

Similar to what has been reported for CAG RNA repeats, GGGGCC RNA repeats were also found to interact with NCL (Haeusler et al., 2014), and this protein occasionally overlaps with *C9orf72* RNA foci (Cooper-Knock, Higginbottom et al., 2015; Haeusler et al., 2014). In addition, a disruption in the typical pattern of the nucleolus was observed in c9FTD/ALS patient B lymphocytes, fibroblasts, and one iPSC-derived neuronal model (Haeusler et al., 2014), although it was not observed in a second iPSC model (Almeida et al., 2013). This altered pattern suggests that the expanded repeat may result in nucleolar stress in c9FTD/ALS, and indeed, a decrease in rRNA maturation was observed in patient tissues (Haeusler et al., 2014). In support of this hypothesis, an altered nucleolar pattern was observed in BAC transgenic c9FTD/ALS mouse model (O'Rourke et al., 2015) as well as in a subset of neurons in the motor cortex of c9FTD/ALS patients (Cooper-Knock, Higginbottom et al., 2015; Haeusler et al., 2014). Recently, Celona et al. carried out another RNA pull-down assay to identified RBPs capable of binding the *C9orf72* repeat and identified zinc finger protein 106 (Zfp106) as an additional interactor. This protein normally localizes to the nucleolus, raising the possibility that if the interaction of the GGGGCC repeat with Zfp106 disrupts its normal function, it could contribute to nucleolar stress and toxicity. Consistent with the idea, when Zfp106 was overexpressed in a c9FTD/ALS *Drosophila* model, it was able to rescue repeat-induced toxicity. Interestingly, *Zfp106* knockout mice show a progressive neuromuscular phenotype characterized by the loss of spinal motor neurons and associated neurogenic muscle loss. Zfp106 also interacts with other ALS-associated proteins, including TDP-43 and fused in sarcoma (FUS), suggesting it may be part of a common pathway that, when disrupted, results in motor neuron degeneration (Celona et al., 2017).

BIDIRECTIONAL TRANSCRIPTION

It has become increasingly apparent that much of the human genome is transcribed in both directions, leading to the generation of antisense transcripts that complement known protein-coding and noncoding genes. These natural antisense transcripts (NATs) have been shown to regulate gene expression through a variety of mechanisms (Faghihi & Wahlestedt, 2009; Magistri, Faghihi, St Laurent, & Wahlestedt, 2012). NATs corresponding to various disease-associated genes have been identified and suggested to play a role in chromatin modulation, gene expression, and pathogenesis, including, for instance, the antisense transcript of beta-secretase 1 in Alzheimer's disease (Faghihi et al., 2008), the antisense transcript of PINK1 in Parkinson's disease (Scheele et al., 2007), an antisense transcript from the *Six5* region in DM1 (Cho et al., 2005), *ATXN2-AS* in SCA2 (Li et al., 2016), and the noncoding RNA *SCAANT1* in SCA7 (Sopher et al., 2011). Bidirectional transcription has also been reported for FXTAS (Ladd et al., 2007) and c9FTD/ALS (Gendron et al., 2013; Mori, Arzberger et al., 2013; Zu et al., 2013).

Perhaps one of the most interesting and compelling cases for the role of bidirectional transcription in disease is that of SCA8 (Fig. 15.2). As explained above, SCA8 was first identified as an RNA-mediated disease similar to DM1, as the CTG repeat in the untranslated gene originally named "*SCA8*" results in RNA foci that colocalize with MBNL1 and result in RNA processing defects (Daughters et al., 2009; Koob et al., 1999). This repeat runs in the "sense" direction but is antisense to the gene *Kelch-Like 1* (*KLHL1*) (Nemes, Benzow, Moseley, Ranum, & Koob, 2000). It is possible that "*SCA8*" negatively regulates the expression of *KLHL1*, although it remains to be verified in SCA8 patients. *KLHL1* knockout mice show cerebellar defects, suggesting its loss could conceivably play a role in SCA8 pathogenesis (He et al., 2006). In 2006, a BAC mouse model of SCA8 was developed that expresses the CTG repeat-containing transcript and shows neurological abnormalities upon behavioral testing. There was no neuronal loss detected in these mice, but histological

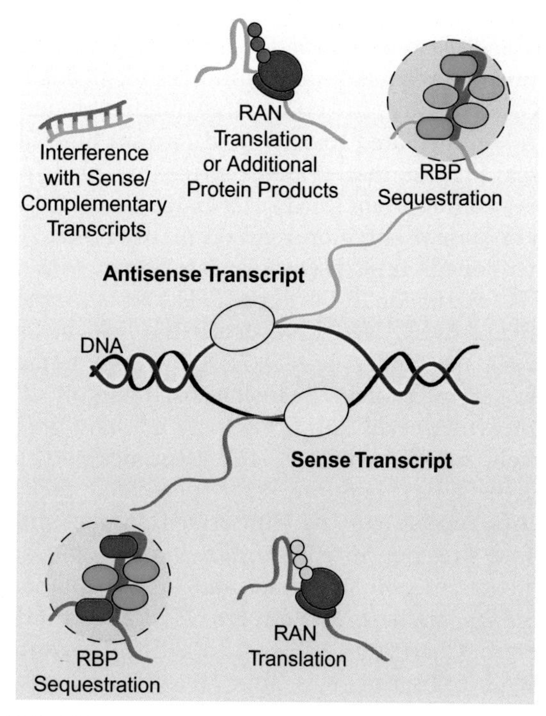

FIGURE 15.2 **Bidirectional transcription.** There is increasing evidence that several regions of the genome are bidirectionally transcribed, producing transcripts that are antisense to known coding and noncoding genes. These transcripts are thought to regulate gene expression in a number of ways. In neurodegenerative disease, these antisense transcripts can also form RNA foci, sequestering RBPs and inhibiting their function. The transcripts can also encode additional repeat proteins that can be produced either through canonical or RAN translation, depending on the gene context. *RBPs*, RNA-binding proteins.

examination revealed ubiquitin-positive inclusions that stained positive for the antibody 1C2 (Moseley et al., 2006). This antibody was raised against the polyglutamine repeat of TATA-binding protein (TBP) and is often used as a marker of polyglutamine pathology (Trottier et al., 1995). Further investigation revealed that this pure polyglutamine protein could also be detected in SCA8 patients and is produced from a CAG repeat transcribed antisense to the CTG repeat. The new transcript was named *Ataxin8* (*ATXN8*), and the original CTG repeat-containing transcript was renamed *ATXN8 opposite strand* (Moseley et al., 2006). Thus, bidirectional transcription at this locus in SCA8 results in both RNA-mediated and protein-mediated toxicity, and current hypotheses suggest that both of these mechanisms may be important for disease pathogenesis.

A similar situation may occur in HDL2. As mentioned, this disease is associated with a CTG expansion in the gene *JPH3* and is characterized by $(CUG)_n$ RNA foci (Rudnicki et al., 2007). Yet, similar to HD, HDL2 is also characterized by protein inclusions that largely do not overlap with the RNA foci and stain positively for the antibody 1C2 (Rudnicki et al., 2008). It has therefore been speculated that a CAG repeat may be transcribed in the antisense direction in HDL2, allowing for the production of polyglutamine and an explanation for the similarities between HDL2 and HD (Batra, Charizanis, & Swanson, 2010). This antisense transcript was first detected in mice expressing a BAC spanning the *JPH3* locus and harboring 120 CTG repeats (Wilburn et al., 2011). Antisense transcripts of a nonexpanded repeat were later detected in human samples, but expanded repeat transcripts have remained difficult to detect in HDL2 patients (Seixas et al., 2012). Furthermore, Seixas et al. were unable to detect any evidence of expanded polyglutamine protein in HDL2 patient tissue, suggesting that, if this protein is produced, it is at much lower levels than those seen in other polyglutamine disorders like HD. The 1C2 antibody has also been suggested to cross-react with polyalanine (Sugaya, Matsubara, Miyamoto, Kawata, & Hayashi, 2003) and polyleucine (Dorsman et al., 2002) in cellular models. While both of these proteins can be produced from the CTG repeat in *JBH3*, Seixas et al. (2012) also failed to detect expanded versions of these proteins in patient tissue. This leaves the speculation that the 1C2-positive staining of the HDL2 aggregates may simply be a red herring explained by an earlier report that TBP is itself sequestered into these aggregates (Rudnicki et al., 2008). Based on these findings, HDL2 may primarily arise from the effects of the toxic $(CUG)_n$ RNA and a loss of *JPH3* expression (Seixas et al., 2012), although the presence of aggregates in patient brains still suggests the possibility of protein-mediated toxicity as well.

Bidirectional transcription has also been implicated in SCA2, a polyglutamine disease mediated by the expansion of a CAG repeat in the gene *Ataxin2*. A recent study identified a CUG RNA transcript generated from a gene antisense to *Ataxin2*, named *ATXN-AS* (Li et al., 2016). These CUG repeat-containing transcripts were detected in SCA2 fibroblasts, iPSCs, neural stem cells, LCLs, and patient postmortem brains. Expression of the repeat in SK-N-MC neuroblastoma cells and mouse primary cortical neurons results in a significant increase in cellular toxicity, suggesting that expanded *ATXN2-AS* may play a role in disease. In support of this hypothesis, the repeat-containing transcript accumulated into RNA foci that colocalized with MBNL1 in SK-N-MC cells. Foci were also detected in the Purkinje cells in SCA2 patient brains, where they may contribute to RNA splicing abnormalities (Li et al., 2016).

Additional evidence supporting a role for bidirectional transcription in neurodegenerative disease pathogenesis comes from two independent studies carried out in *Drosophila*. Both Lawlor et al. and Yu et al. found that coexpression of expanded CTG and antisense CAG repeats enhanced retinal degeneration phenotypes compared to the expression of either repeat alone. Further analyses revealed that these repeat RNAs formed small 21-nucleotide duplexes that were processed by Dicer and Argonaute-2 (Lawlor et al., 2011; Yu, Teng, & Bonini, 2011). In one of these studies, this bidirectional transcription led to an overall disruption in the endogenous miRNA profile of *Drosophila* neurons, presumably due to competition from the double-stranded RNA produced from the repeats (Lawlor et al., 2011). The second study reported that the coexpression of the two repeats ultimately leads to a decrease in the expression of CAG repeat-containing genes through RNA interference (Yu et al., 2011). This is consistent with an earlier report suggesting that repeat RNA can be processed by Dicer to produce short interfering RNAs that target endogenous genes with complementary repeats (Krol et al., 2007). A later report also described the generation of 21-nucleotide CAG repeat RNA transcripts in HD models and noted that these transcripts resulted in the silencing of CTG repeat-containing genes (Banez-Coronel et al., 2012). When a similar experiment was carried out in *Drosophila* expressing FXTAS-associated CCG and antisense CGG repeats, however, an opposite effect was observed. While the expression of either repeat alone resulted in a retinal degeneration phenotype, coexpression of both repeats suppressed this toxicity. This effect was also dependent upon Dicer and Argonaute-2, suggesting that, similar to the CTG/CAG repeats, these CCG/CGG repeats also formed small duplex RNAs that could be processed by the RNA interference machinery, and, in this case, feedback to inhibit the expression of each repeat-containing transcript (Sofola, Jin, Botas, & Nelson, 2007). It is worth noting, however, that the *FMR1* transcript is often upregulated in premutation carriers (Hagerman et al., 2001; Tassone et al., 2000), suggesting that the expression of the endogenous antisense repeat is not sufficient to reduce expression levels in the presence of the premutation expansion. Furthermore, it remains to be seen if these RNA interference mechanisms play a role in disease outside of *Drosophila* and cellular models.

Finally, bidirectional transcription has also been suggested to contribute to repeat instability across expanded repeats. This conclusion is based on studies using human fibroblasts that were engineered to express single-copy genomic integrations of 800 CTG repeats with a CMG/CBA promotor to drive sense expression, a ROSA26 promotor to drive antisense expression, or both promotors to allow for bidirectional transcription. Repeat instability was increased with transcription in either direction and further enhanced by bidirectional transcription (Nakamori, Pearson, & Thornton, 2010). Bidirectional transcription also leads to the formation of R-loops, nucleic acid structures formed from the hybridization of the nascent RNA to the DNA template (Reddy et al., 2010). These structures, while important for gene regulation, can also promote genomic instability by leaving the exposed single-stranded DNA vulnerable to damage (Skourti-Stathaki & Proudfoot, 2014).

REPEAT-ASSOCIATED NON-ATG (RAN) TRANSLATION: WHEN RNA TOXICITY RESULTS IN PROTEOTOXICITY

In 2011, Laura Ranum's group mutated the ATG start site upstream of an ATXN8 minigene with the goal of investigating the role of the encoded polyglutamine protein in SCA8 (Zu et al., 2011). To their surprise, disrupting this canonical translational start site did not prevent the expression of the repeat protein. To confirm this result, they subcloned the repeat downstream of six STOP codons and included three downstream protein tags in each reading frame. The expression of the protein tags in transfected cells confirmed that, once again, polyglutamine proteins could be produced in the absence of a start codon and, furthermore, this non-ATG-dependent translation could occur in all three reading frames. The expression of these repeat proteins depended upon the length of the repeat, with longer repeats resulting in protein expression. This phenomenon was therefore named repeat-associated non-ATG translation, or RAN translation for short (Zu et al., 2011). This pivotal finding triggered a series of studies aimed at identifying RAN-translated products in various repeat-associated diseases. RAN translation was first confirmed in human postmortem samples from SCA8 and DM1 patients, where it results in the accumulation of polyalanine and polyglutamine, respectively (Zu et al., 2011). RAN-translated products have since been identified in patients with c9FTD/ALS (Ash et al., 2013; Mori, Weng et al., 2013), FXTAS (Todd et al., 2013), FXPOI (Buijsen et al., 2016), and HD (Banez-Coronel et al., 2015). Furthermore, in association with the role of bidirectional transcription in neurodegenerative diseases, RAN translation of the antisense repeat strand can also occur, resulting in the formation and accumulation of additional repeat proteins. Antisense RAN products have been identified in vitro and in vivo for the repeats associated with c9FTD/ALS (Gendron et al., 2013; Mori, Arzberger et al., 2013; Zu et al., 2013), HD (Banez-Coronel et al., 2015), and FXTAS (Krans, Kearse, & Todd, 2016). Fig. 15.3 summarizes the RAN translation products that can theoretically be produced in each disease, although not all of these repeat proteins have been identified in patient samples. For instance, although all three potential repeat proteins encoded by the SCA8 CTG repeat were identified in vitro in the aforementioned study, only polyalanine and polyglutamine were investigated in vivo (Zu et al., 2011).

RAN-Translated Proteins as Mediators of Disease

Since the discovery of RAN-translated proteins in patient postmortem tissue, several studies have focused on determining whether these proteins play any role in disease pathogenesis. Some of these findings are summarized below.

The c9-RAN Proteins

Of the diseases characterized by RAN translation, perhaps the most focus has been on c9FTD/ALS. It had previously been reported that in repeat-positive patient tissue, there are ubiquitin- and p62-positive inclusions that are negative for TDP-43 (Al-Sarraj et al., 2011). With the identification of RAN translation, it

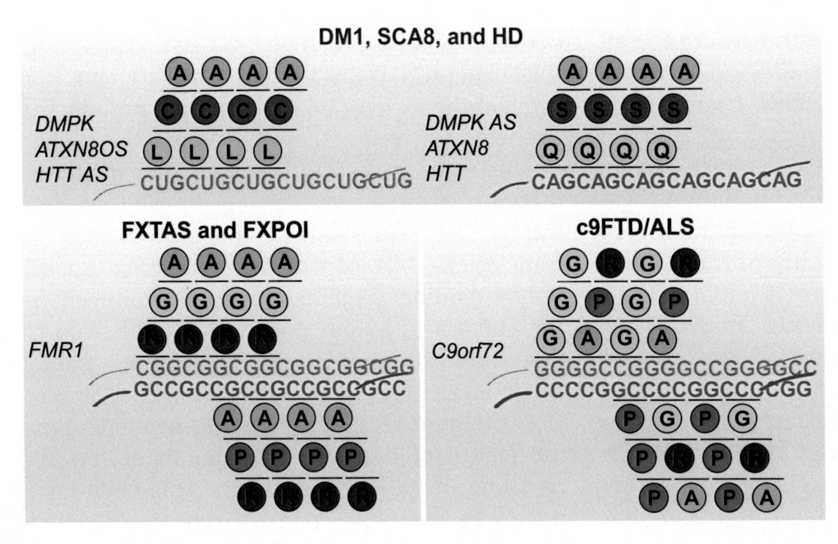

FIGURE 15.3 **RAN translation.** In 2011, work from Laura Ranum's group identified a novel form of translation that occurs across expanded repeats in the absence of a canonical start codon. RAN translation results in the production of repetitive proteins that accumulate in cells and contribute to toxicity. Different repeats encode for different repeat proteins, as outlined in this figure.

was speculated that these inclusions may contain protein products generated from the RAN translation of the GGGGCC hexanucleotide repeat. With bidirectional transcription, this repeat can result in up to five different dipeptide repeat "c9-RAN proteins": poly-glycine-proline [poly(GP)], poly-glycine-alanine [poly (GA)], and poly-glycine-arginine [poly(GR)] in the sense direction and poly-proline-alanine [poly(PA)], poly-proline-arginine [poly(PR)] and, again, poly(GP) in the antisense direction. Several studies have confirmed that all five of these c9-RAN proteins can be detected in repeat-positive patient tissue, albeit at different regional abundancies (Ash et al., 2013; Davidson et al., 2014; Gendron et al., 2013; Gomez-Deza et al., 2015; Mackenzie et al., 2015; Mori, Arzberger et al., 2013; Zu et al., 2013). Studies using *Drosophila* as a model have argued that these c9-RAN proteins are crucial for the development of *C9orf72*-related toxicity (Mizielinska et al., 2014; Tran et al., 2015). And it was recently reported that c9-RAN proteins can be transmitted between cells in neuronal cultures (Chang, Jeng, Chiang, Hwang, & Chen, 2016; Westergard et al., 2016), a phenomenon that may result a spreading of pathology in patient brains. The exact contribution of these different proteins to FTD and ALS is an active field of investigation, and several studies suggest that the different c9-RAN proteins may affect neuronal function through diverse mechanisms, all of which could conceivably contribute to disease. We've chosen to focus on more recent developments here, but a more in-depth discussion on potential pathomechanisms in c9FTD/ALS can found in Todd & Petrucelli, 2016.

Of the five potential c9-RAN proteins, toxicity has chiefly been associated with only three: the arginine-containing c9-RAN proteins poly (PR) and poly(GR), and the highly aggregation-prone poly(GA). Poly(PA), which is rare in patients (Mackenzie et al., 2015), has not resulted in toxicity in any models generated to date (Boeynaems et al., 2016; Mizielinska et al., 2014; Tao et al., 2015; Wen et al., 2014), with the exception of a recent report that suggests that the expression of particularly long poly(PA) peptides may adversely affect viability and activity in cultured cells (Callister, Ryan, Sim, Rollinson, & Pickering-Brown, 2016). Poly(GP), the only c9-RAN protein encoded by both the sense and antisense strand, has resulted in toxicity when expressed in HEK293T cells (Zu et al., 2013) but was later found to be nontoxic in other models (Freibaum et al., 2015; Tao et al., 2015; Wen et al., 2014). Regardless of its exact role in pathogenesis, however, this fifth c9-RAN protein can be detected in patient cerebral spinal fluid and has the potential to be an important biomarker in c9FTD/ALS clinical studies (Su et al., 2014).

The arginine-containing c9-RAN proteins are highly toxic when expressed in *Drosophila* (Boeynaems et al., 2016; Freibaum et al., 2015; Lee et al., 2016; Mizielinska et al., 2014; Wen et al., 2014; Yang et al., 2015) and when expressed in various cell and neuronal models (Kwon et al., 2014; Tao et al., 2015; Wen et al., 2014; Yamakawa et al., 2015; Zu et al., 2013). In attempts to explain the toxicity associated with these proteins, recent studies have employed coimmunoprecipitation techniques to screen for poly(PR) and/or poly(GR) interactors in cultured cells. Kanekura et al. (2016) screened for proteins capable of interacting with poly (PR) and noted an enrichment in RBPs and proteins involved in mRNA translation, prompting them to hypothesize that the toxicity induced by poly(PR) in cells is due to a disruption in protein translation. Lopez-Gonzalez et al. (2016) carried out an interactome analysis for poly(GR) in HEK293T cells and also noted an abundance of RBPs, as well as ribosomal proteins. In this study, poly(GR), but not poly (GA), was found to impair mitochondrial function and induce DNA damage and oxidative stress in iPSCs. The interaction of this c9-RAN protein with mitochondrial ribosomal proteins may contribute to this downstream effect (Lopez-Gonzalez et al., 2016). Finally, Lee et al. (2016) and Lin et al. (2016) published complementary papers in *Cell* that also identified RBPs as interactors of poly(PR), as well as proteins associated with RNA granules, spliceosomes, intermediate filaments, the nuclear pore, and nucleoli. Many of these interactors also coimmunoprecipitate with poly(GR) and can modulate poly(GR)-induced toxicity in *Drosophila* (Lee et al., 2016), suggesting that both of the arginine-containing c9-RAN proteins may impact the same cellular pathways. Interestingly, several of the proteins that interact with the arginine-containing c9-RAN proteins contain low-complexity domains (LCDs) (Lee et al., 2016; Lin et al., 2016). This finding implicates the arginine-containing c9-RAN proteins into the broader hypothesis that the disruption of LCD-mediated granule formation is a key mechanism in neurodegenerative disease. LCDs, also known as intrinsically disordered regions or prion-like domains, are regions of proteins that are often repetitive and generally unstructured. They are commonly found in proteins that are components of RNA granules and other membrane-less organelles. Kato et al. (2012) found that the LCD domains of both FUS and hnRNP A2 can undergo a reversible phase separation in vitro to form hydrogels that are composed of amyloid-like cross-beta polymers, leading to the hypothesis that RNA granules form through a reversible phase shift mediated by the polymerization of LCDs. In 2015, two studies argued that RNA granules actually form via a dynamic liquid–liquid phase separation (LLPS) that is regulated by both LCD–LCD interactions and by the interaction of the proteins with RNA (Lin, Protter, Rosen, & Parker, 2015; Molliex et al., 2015). This LLPS model for RNA granule formation is consistent with studies showing that these membrane-less organelles are dynamic and "liquid-like" in nature, forming, fusing, and dissolving in response to the cellular environment (Zhu & Brangwynne, 2015). Chemical footprinting reveals that LCDs adopt a similar structure in both liquid droplets and hydrogels, and that this same structure can be detected in membrane-less organelles in cells (Xiang et al., 2015). These findings favor a model in which granules form on a continuum, with LCD-containing proteins transitioning from a soluble state, to a dynamic liquid droplet, to a more static fibril-containing state (Lin et al., 2015). Hydrogel formation is comparable to a late step on a continuum of phase separation that occurs as liquid droplets "age" and become less dynamic (Alberti & Hyman, 2016). Or, alternatively, hydrogels may be reminiscent of the more stable "cores" that were recently observed by super resolution microscopy of stress granules (Jain et al., 2016). Interestingly, several neurodegenerative disease-associated proteins contain LCDs, including FUS, TDP-43, hnRNP A1/2, and ataxin2, and disease-causing mutations in these proteins may regulate their ability to undergo LLPS (Conicella, Zerze, Mittal, & Fawzi, 2016; Murakami et al., 2015; Patel et al., 2015; Schmidt & Rohatgi, 2016) or promote the conversion of the droplet protein into pathogenic fibrils (Molliex et al., 2015). The association of the c9-RAN proteins with LCDs further implicates changes in RNA granule dynamics in c9FTD/ALS.

One of the LCD-containing proteins that interacted with both arginine-containing c9-RAN proteins was nucleophosmin 1 (NPM1) (Lee et al., 2013), a component of the granular component (GC) of nucleoli. Earlier studies noted that, unlike the other c9-RAN proteins, poly(PR) and poly(GR) often localize to the nucleus in cells (Zu et al., 2013), where they have been reported to accumulate in the nucleolus (Kwon et al., 2014; Tao et al., 2015; Wen et al., 2014; Yamakawa et al., 2015)—specifically in the GC (Lee et al., 2016). These findings led to the hypothesis that these proteins result in toxicity by inducing nucleolar stress, and indeed there is evidence for nucleolar stress in several c9FTD/ALS models (Haeusler et al., 2014; Kwon et al., 2014; O'Rourke et al., 2015; Tao et al., 2015; Wen et al., 2014; Yang et al., 2015) and in a subset of neurons in patients (Cooper-Knock, Higginbottom et al., 2015; Haeusler et al., 2014). NPM1 helps maintain the liquid-like properties of the nucleolar GC through a process that depends on its three LCDs (Mitrea et al., 2016). Evidence suggests that, at stoichiometric ratios, poly(PR) and poly (GR) can bridge interactions between the NPM1 molecules and facilitate the induction of LLPS. When poly(PR) and poly(GR) are in excess, however, they can outcompete known NPM1 interactors and ultimately reduce and inhibit droplet formation. These findings may explain the observation that the expression of these proteins alters the dynamics of nucleoli in living cells. This effect is not limited to the nucleolus, but instead, the arginine-containing c9-RAN proteins seem to affect the biophysical properties of many LCD-containing proteins, resulting in changes in the dynamics of many membraneless organelles, including stress granules, nuclear speckles, and Cajal bodies.

Another family of proteins to emerge from the screens for poly(PR) protein interactors were proteins associated with the nuclear pore, particularly nucleoporins. These proteins contain regions of phenylalanine-glycine (FG) repeats that make up the central channel of the nuclear pore. A follow-up study by Shi et al. (2017) demonstrated that poly(PR) binds to cross-beta polymers generated from these FG repeats and appears to stabilize the nucleoporin proteins in their polymeric state, thereby impeding nuclear transport. This is consistent with earlier reports that nucleocytoplasmic transport factors are modulators of toxicity in both yeast (Jovicic et al., 2015) and flies (Boeynaems et al., 2016) expressing only poly (PR), and with a prevailing hypothesis that the disruption of the nucleocytoplasmic transport pathway is a major pathogenic mechanism in c9FTD/ALS and perhaps other diseases (Freibaum et al., 2015; Jovicic et al., 2015; Zhang et al., 2015). It is unclear, however, if the arginine-containing c9-RAN proteins are the main driver of these transport defects, or if other pathogenic entities are also at fault. In fact, recent studies also implicate poly(GA) in this process. Our lab found that poly(GA) aggregates can sequester components of the nucleocytoplasmic transport pathway in mice (Zhang et al., 2016). Similarly, another group recently found that expression of poly(GA) in HeLa cells led to a inhibition of the importin-α/β nuclear import pathway, as measured by a RFP reporter (Khosravi et al., 2017). This reporter used the nuclear localization sequence from TDP-43, supporting the possibility that poly(GA) contributes to the mislocalization of TDP-43 in c9FTD/ALS. Indeed, the authors found that expression of poly(GA) and poly (GR) led to an increase in the cytoplasmic localization of endogenous TDP-43 in hippocampal neurons, although neither was sufficient to induce its aggregation in this system (Khosravi et al., 2017). This is consistent with a separate report that nucleocytoplasmic transport is generally affected by the presence of cytoplasmic protein aggregates (Woerner et al., 2015), and indeed, targeting poly(GA) to the nucleus reduced this defective TDP-43 localization (Khosravi et al., 2017). Poly(GA)-expressing mice show only rare phospho-TDP-43 inclusions (Zhang et al., 2016), however, suggesting that it is not the only factor contributing to the formation of TDP-43 pathology in c9FTD/ALS.

While poly(GA) does not induce the same potent toxicity observed with poly(PR) and poly(GR) in *Drosophila* (Freibaum et al., 2015; Mizielinska et al., 2014; Yang et al., 2015), its expression does result in an increase in adult fly lethality (Mizielinska et al., 2014) and is toxic in cellular (Chang et al., 2016; May et al., 2014; Zhang et al., 2014), zebrafish (Ohki et al., 2017), and mouse models (Zhang et al., 2014).

Early studies suggested that it results in toxicity by impairing the ubiquitin-proteasome system (UPS) and inducing endoplasmic reticulum stress (May et al., 2014; Zhang et al., 2014). Support for this hypothesis also comes from our lab's recently published poly(GA) mouse model. These mice demonstrate that the expression of poly(GA) alone results in progressive neuronal loss, astrogliosis, and brain atrophy, as well as behavioral defects reminiscent of ALS/FTD, including motor and cognitive deficits, hyperactivity, and anxiety-like behavior. These phenotypes are dependent upon the formation of poly(GA) aggregates which sequester the UPS-related HR23 proteins, inhibiting their function (Zhang et al., 2016). Overexpression of HR23B in primary neurons expressing poly (GA) restores HR23 protein function, reduces poly(GA) inclusion numbers, and decreases poly(GA)-induced caspase-3 activation, suggesting that this loss of HR23 protein function is important for poly(GA) aggregation and neurotoxicity. Interestingly, this sequestration appears to be specific to poly(GA) compared to the other c9-RAN proteins, and colocalization of HR23 proteins with poly(GA) aggregates can be detected in c9FTD/ALS patient brain (Zhang et al., 2016).

FMRpolyG and FMRpolyA in FXTAS

A pathological hallmark of FXTAS is the presence of nuclear inclusions in both neurons and astrocytes, as detected by immunohistochemistry in patient tissue (Greco et al., 2006, 2002; Tassone, 2004). These inclusions contain the *FMR1* RNA and RBPs (Iwahashi et al., 2006; Sellier et al., 2010), likening them to RNA foci, as described earlier. Yet unlike typical RNA foci, the inclusions seen in FXTAS patients and models are large and ubiquitin-positive and contain chaperone proteins and components or regulators of the proteasome (Hukema et al., 2015; Iwahashi et al., 2006; Jin et al., 2003; Willemsen, 2003), features that are reminiscent of protein inclusions. Furthermore, when a repeat-containing *FMR1* 5′UTR is fused to EGFP and expressed in *Drosophila*, ubiquitin-positive EGFP inclusions form (Jin et al., 2003), but only a fraction contain the *FMR1* RNA (Todd et al., 2013). Based on these observations, Todd et al. predicted that these inclusions may contain RAN translation products. The CGG repeat could potentially encode polypeptides comprised of glycine, alanine, and arginine, and indeed both polyglycine and polyalanine peptides were detected in cellular and *Drosophila* models of FXTAS. These two proteins were named FMRpolyG and FMRpolyA, respectively. Importantly, an antibody raised against FMRpolyG labeled the inclusions present in FXTAS patient brain. Interestingly, polyarginine has not been detected, suggesting that RAN translation may not occur in this third reading frame (Kearse et al., 2016; Todd et al., 2013). This polyarginine protein would be in-frame with the FMRP protein, which is also not reported to be aggregated in patients. The *FMR1* gene can also be transcribed in the antisense direction, resulting in a CCG repeat in *ASFMR1* isoform a (Ladd et al., 2007). This repeat would be predicted to encode repeats of proline, alanine, and, once again, arginine, and, indeed, when the repeat was cloned into a luciferase reporter vector in all three reading frames, all three luciferase-fusion proteins were detected in COS-7 cells, even in the absence of an ATG start site. Furthermore, although the authors were unable to generate a polyarginine-specific antibody, antibodies raised against polyproline and polyalanine were used to demonstrate that both of these proteins are present in ubiquitin-positive inclusions in FXTAS patient brains (Krans et al., 2016). Finally, FMRpolyG was also recently identified in ubiquitin-positive inclusions in the ovaries of a woman with FXPOI (Buijsen et al., 2016), suggesting that RAN translation may play a role in this disease as well. Further studies are needed to establish this possibility, but it is worth noting that inclusions were also found in the ovaries of aged CGG repeat knock-in mice, and these mice had less ovulations with less fresh corpora lutea compared to wild-type mice (Buijsen et al., 2016).

The exact role of the RAN-translated products in FXTAS is still unclear, but some studies have begun to address this question. In one study, the expression of an expanded CGG repeat–containing mCherry reporter in HeLa cells led to an impairment of the UPS (Oh et al., 2015). Overexpression of FMRpolyG, but not FMRpolyA, exacerbated this phenotype. Furthermore, inhibiting the proteasome in a *Drosophila* model of FXTAS enhanced the retinal degeneration phenotype, while overexpression of the chaperone HSP70 suppressed the repeat-induced toxicity. These effects were also dependent upon the expression of FMRpolyG (Oh et al., 2015). Taken together, these studies suggest that RAN-translated FMRpolyG may negatively impact protein degradation pathways in FXTAS, and further in vivo studies could help establish this possibility.

RAN Translation in Huntington's Disease

RAN translation has primarily been described in diseases caused by repeat expansions located in noncoding regions, yet in 2015, Bañez-Coronel et al. discovered that RAN translation could also occur across an open reading frame—specifically the expanded CAG repeat located in exon 1 of HTT. In addition to polyglutamine, this CAG repeat can also encode polypeptides of alanine and serine. All three polypeptides were detected in HEK293T cells expressing a HTT exon 1 construct that lacks an ATG start site, and they expressed in a repeat length-dependent manner. More importantly, immunohistochemistry and western blots using antibodies raised against the novel RAN proteins revealed that both polyalanine and polyserine accumulate in HD brains, and furthermore, polyleucine and polycystine proteins are also generated from the antisense strand in vivo in patients. These proteins accumulate specifically in regions associated with degeneration and neuroinflammation in HD, as well as in the cerebellum in juvenile-onset cases. The proteins were even detected in white matter regions that are reported to be affected in HD but have not been shown to express the polyglutamine protein, suggesting that these RAN-translated proteins could help explain aspects of the disease that were previously not well understood. In support of this hypothesis, the expression of these polypeptides in SH-SY5Y and T98 glial cells resulted in increased cell death (Banez-Coronel et al., 2015).

The Mechanism of RAN Translation

There is still much to be learned about how expanded repeats result in noncanonical protein translation. Initial studies suggested that RAN translation depended upon the formation of RNA hairpins, based on the finding that RAN translation could occur in vitro with hairpin-forming repeats such as $(CAG)_n$ and $(CTG)_n$, but not with a nonhairpin-forming CAA repeat (Zu et al., 2011). Studies have also confirmed that the repeat in *C9orf72* is capable of forming both hairpins and length-dependent G-quadruplex RNA secondary structures (Fratta et al., 2012; Haeusler et al., 2014; Reddy et al., 2013; Su et al., 2014). These secondary structures may contribute to ribosomal stalling, allowing RAN translation to occur (Green, Linsalata, & Todd, 2016). It is also clear that repeat length is important for RAN translation, although the exact repeat length required may depend on the nature of the repeat and the surrounding context. For instance, in the case of the *C9orf72* repeat, RAN translation has been detected in *Drosophila* expressing pure, UAS promotor-driven repeats as short as 28 (Freibaum et al., 2015) or 36 repeats (Mizielinska et al., 2014). In contrast, when 160 repeats were expressed in the context of a *C9orf72* minigene, RAN translation was not detected at standard temperatures and was only slightly detected when the flies were raised at high temperatures (Tran et al., 2015). Similarly, in transfected cells, RAN-translated products could be detected in cells expressing 30–40 repeats directly downstream of a series of STOP codons (Zu et al., 2013), but longer repeats appeared to be required when expressed downstream of a portion of the *C9orf72* intron (Mori, Weng et al., 2013; Su et al., 2014). Although these variations may depend upon methods of detection and/or expression across different labs, the ability of the surrounding regions to influence RAN translation efficiency has also been reported for other disease-associated repeats. In the original studies on RAN translation of the ATXN8 repeat, it was noted that placing STOP codons immediately upstream of the CAG repeat was able to influence the expression of the different polypeptides (Zu et al., 2011). A similar phenomenon was observed for the CGG repeat in the *FMR1* 5′UTR. In this case, FMRpolyG expression was inhibited when a STOP codon was placed at either 6 or 12 base pairs from the start of the repeat, but not when the STOP codon was located 21 or more base pairs away. This suggests that in this context, RAN translation initiates upstream of the repeat, and further investigation has indicated that the presence of at least one AUG-like codon is required (Todd et al., 2013), although this may no longer be a requirement when the repeat extends beyond a given threshold. At longer repeat lengths, some RAN translation may initiate within the repeat, at least when measured in vitro (Kearse et al., 2016). This finding is further supported by the fact that FMRpolyG can be detected in the Dutch knock-in mouse model (Willemsen, 2003) but not in the NIH knock-in mouse (Entezam et al., 2007; Todd et al., 2013). The NIH mouse model retains a STOP codon present in the mouse *fmr1* gene that is absent in the human gene. As this STOP is 18 base pairs upstream of the repeat and in-frame with polyglycine, it can presumably prevent RAN translation of the repeat (Todd et al., 2013). In contrast, placing STOP codons as close as 6 base pairs upstream of the repeat did not prevent the production of FMRpolyA in cells or in vitro (Kearse et al., 2016; Todd et al., 2013), suggesting that RAN translation initiation occurs within the repeat for this reading frame.

Recently, Kearse et al. expanded their studies on the RAN translation of *FMR1* to better understand the mechanism underlying the production of FMRpolyG and FMRpolyA. They generated CGG luciferase reporter constructs to reliably detect repeat length-dependent FMRpolyG-Luciferase and FMRpolyA-luciferase fusion proteins in rat primary cortical neurons and HeLa cells, as well as in an in vitro mammalian translation system. Using these reporters, the authors were able to demonstrate that RAN translation of the *FMR1* repeat is less efficient than canonical translation but also requires an m^{7}G cap and eIF4E for translation initiation. The eIF4A helicase was also required, suggesting that ribosomal scanning is necessary for RAN translation at this repeat. It will be interesting to see if these requirements hold true for other repeat expansions. FMRpolyG in particular may be unique in the sense that it is initiated upstream of the repeat and translation in this frame was even independent of the repeat when analyzed in vitro (Kearse et al., 2016).

CONCLUSIONS AND THERAPEUTIC DIRECTIONS

In conclusion, substantial evidence now supports a role for the mutant RNA in repeat-associated neurodegenerative diseases. While this has primarily been studied in the context of noncoding repeats, studies from the polyglutamine disorders suggests that even traditionally protein-associated diseases may be mediated in part by RNA-mediated mechanisms as well. This includes the aggregation of mutant RNA transcripts into RNA foci, which leads to the sequestration and inhibition of RBPs (Fig. 15.1), the bidirectional transcription of the repeat loci (Fig. 15.2), and the accumulation of toxic proteins generated by RAN translation of sense and antisense repeats (Fig. 15.3). As more evidence suggests a role for the mutant RNA in neurodegenerative disease, the need for therapeutics that target this pathogenic agent is warranted. Multiple approaches are currently being used to target the mutant RNA in various disease models. A promising approach that is actively being investigated in several systems is the use of antisense oligonucleotides (ASOs) to target the mutant RNA. ASOs are short, single-stranded nucleic acids that bind to target RNA through canonical Watson-Crick base pairing and influence its processing and translation. For RNA-mediated diseases, one of most advantageous applications of ASOs is to promote the degradation of the mutant mRNA transcript, thereby inhibiting disease pathogenesis at the source (Southwell, Skotte, Bennett, & Hayden, 2012). While the use of ASOs appears to be a promising way to target repeat-containing RNA, these approaches largely target only the sense strand, leaving the antisense strand intact. An alternative therapeutic approach may be to target the transcription elongation factor Spt4. This factor is specifically important for the translation of extended repeats, and it has been found that knocking down Spt4 can reduce mutant RNA levels in HD (Liu, Chang et al., 2012) and c9ALS/FTD (Kramer et al., 2016) models. Importantly, both the sense and antisense transcripts were reduced in the c9ALS/FTD study, and this led to a decrease in RNA foci formation, RAN translation, and repeat-induced toxicity in animal models (Kramer et al., 2016). As research continues to explain the mechanisms underlying RNA-mediated toxicity, it is likely that these and other therapeutic approaches will successfully target the expanded RNA and mitigate its toxicity in neurodegenerative disease.

References

Abe, K., Ikeda, Y., Kurata, T., Ohta, Y., Manabe, Y., Okamoto, M., et al. (2012). Cognitive and affective impairments of a novel SCA/MND crossroad mutation Asidan. *European Journal of Neurology, 19*(8), 1070–1078.

Alberti, S., & Hyman, A. A. (2016). Are aberrant phase transitions a driver of cellular aging? *BioEssays: News and Reviews in Molecular, Cellular and Developmental Biology, 38*(10), 959–968.

Allingham-Hawkins, D. J., Babul-Hirji, R., Chitayat, D., Holden, J. J., Yang, K. T., Lee, C., et al. (1999). Fragile X premutation is a significant risk factor for premature ovarian failure: The International Collaborative POF in Fragile X study—Preliminary data. *American Journal of Medical Genetics, 83*(4), 322–325.

Almeida, S., Gascon, E., Tran, H., Chou, H. J., Gendron, T. F., Degroot, S., et al. (2013). Modeling key pathological features of frontotemporal dementia with C9ORF72 repeat expansion in iPSC-derived human neurons. *Acta Neuropathologica, 126*(3), 385–399.

Al-Sarraj, S., King, A., Troakes, C., Smith, B., Maekawa, S., Bodi, I., et al. (2011). p62 positive, TDP-43 negative, neuronal cytoplasmic and intranuclear inclusions in the cerebellum and hippocampus define the pathology of C9orf72-linked FTLD and MND/ALS. *Acta Neuropathologica, 122*(6), 691–702.

Alvarez-Mora, M. I., Rodriguez-Revenga, L., Madrigal, I., Torres-Silva, F., Mateu-Huertas, E., Lizano, E., et al. (2013). MicroRNA expression profiling in blood from fragile X-associated tremor/ataxia syndrome patients. *Genes, Brain, and Behavior, 12*(6), 595–603.

Arocena, D. G., Iwahashi, C. K., Won, N., Beilina, A., Ludwig, A. L., Tassone, F., et al. (2005). Induction of inclusion formation and disruption of lamin A/C structure by premutation CGG-repeat RNA in human cultured neural cells. *Human Molecular Genetics, 14*(23), 3661–3671.

Ash, P. E., Bieniek, K. F., Gendron, T. F., Caulfield, T., Lin, W. L., Dejesus-Hernandez, M., et al. (2013). Unconventional translation of C9ORF72 GGGGCC expansion generates insoluble polypeptides specific to c9FTD/ALS. *Neuron, 77*(4), 639–646.

Banez-Coronel, M., Ayhan, F., Tarabochia, A. D., Zu, T., Perez, B. A., Tusi, S. K., et al. (2015). RAN Translation in Huntington Disease. *Neuron, 88*(4), 667–677.

Banez-Coronel, M., Porta, S., Kagerbauer, B., Mateu-Huertas, E., Pantano, L., Ferrer, I., et al. (2012). A pathogenic mechanism in Huntington's disease involves small CAG-repeated RNAs with neurotoxic activity. *PLoS Genetics*, *8*(2), e1002481.

Batra, R., Charizanis, K., & Swanson, M. S. (2010). Partners in crime: Bidirectional transcription in unstable microsatellite disease. *Human Molecular Genetics*, *19*(R1), R77–R82.

Bennion Callister, J., Ryan, S., Sim, J., Rollinson, S., & Pickering-Brown, S. M. (2016). Modelling C9orf72 dipeptide repeat proteins of a physiologically relevant size. *Human Molecular Genetics*, *25*, 5069–5082.

Berul, C. I., Maguire, C. T., Aronovitz, M. J., Greenwood, J., Miller, C., Gehrmann, J., et al. (1999). DMPK dosage alterations result in atrioventricular conduction abnormalities in a mouse myotonic dystrophy model. *The Journal of Clinical Investigation*, *103*(4), R1–7.

Boeynaems, S., Bogaert, E., Michiels, E., Gijselinck, I., Sieben, A., Jovicic, A., et al. (2016). *Drosophila* screen connects nuclear transport genes to DPR pathology in c9ALS/FTD. *Scientific Reports*, *6*, 20877.

Brook, J. D., McCurrach, M. E., Harley, H. G., Buckler, A. J., Church, D., Aburatani, H., et al. (1992). Molecular basis of myotonic dystrophy: Expansion of a trinucleotide (CTG) repeat at the 3′ end of a transcript encoding a protein kinase family member. *Cell*, *69*(2), 385.

Buijsen, R. A., Visser, J. A., Kramer, P., Severijnen, E. A., Gearing, M., Charlet-Berguerand, N., et al. (2016). Presence of inclusions positive for polyglycine containing protein, FMRpolyG, indicates that repeat-associated non-AUG translation plays a role in fragile X-associated primary ovarian insufficiency. *Human Reproduction*, *31*(1), 158–168.

Cardani, R., Bugiardini, E., Renna, L. V., Rossi, G., Colombo, G., Valaperta, R., et al. (2013). Overexpression of CUGBP1 in skeletal muscle from adult classic myotonic dystrophy type 1 but not from myotonic dystrophy type 2. *PLoS ONE*, *8*(12), e83777.

Celona, B., Dollen, J. V., Vatsavayai, S. C., Kashima, R., Johnson, J. R., Tang, A. A., et al. (2017). Suppression of C9orf72 RNA repeat-induced neurotoxicity by the ALS-associated RNA-binding protein Zfp106. *eLife*, *6*, pii: e19032.

Chan, H. Y. (2014). RNA-mediated pathogenic mechanisms in polyglutamine diseases and amyotrophic lateral sclerosis. *Frontiers in Cellular Neuroscience*, *8*, 431.

Chang, Y. J., Jeng, U. S., Chiang, Y. L., Hwang, I. S., & Chen, Y. R. (2016). The glycine-alanine dipeptide repeat from C9orf72 hexanucleotide expansions forms toxic amyloids possessing cell-to-cell transmission properties. *The Journal of Biological Chemistry*, *291*(10), 4903–4911.

Charizanis, K., Lee, K. Y., Batra, R., Goodwin, M., Zhang, C., Yuan, Y., et al. (2012). Muscleblind-like 2-mediated alternative splicing in the developing brain and dysregulation in myotonic dystrophy. *Neuron*, *75*(3), 437–450.

Chen, I. C., Lin, H. Y., Lee, G. C., Kao, S. H., Chen, C. M., Wu, Y. R., et al. (2009). Spinocerebellar ataxia type 8 larger triplet expansion alters histone modification and induces RNA foci. *BMC Molecular Biology*, *10*, 9.

Chen, W., Wang, Y., Abe, Y., Cheney, L., Udd, B., & Li, Y. P. (2007). Haploinsufficiency for Znf9 in Znf9 + /- mice is associated with multiorgan abnormalities resembling myotonic dystrophy. *Journal of Molecular Biology*, *368*(1), 8–17.

Cho, D. H., Thienes, C. P., Mahoney, S. E., Analau, E., Filippova, G. N., & Tapscott, S. J. (2005). Antisense transcription and heterochromatin at the DM1 CTG repeats are constrained by CTCF. *Molecular Cell*, *20*(3), 483–489.

Cohen, R. L., & Margolis, R. L. (2016). Spinocerebellar ataxia type 12: Clues to pathogenesis. *Current Opinion in Neurology*, *29*(6), 735–742.

Conicella, A. E., Zerze, G. H., Mittal, J., & Fawzi, N. L. (2016). ALS mutations disrupt phase separation mediated by alpha-helical structure in the TDP-43 low-complexity C-terminal domain. *Structure*, *24*(9), 1537–1549.

Conlon, E. G., Lu, L., Sharma, A., Yamazaki, T., Tang, T., Shneider, N. A., et al. (2016). The C9ORF72 GGGGCC expansion forms RNA G-quadruplex inclusions and sequesters hnRNP H to disrupt splicing in ALS brains. *eLife*, *5*, pii: e17820.

Cooper-Knock, J., Higginbottom, A., Stopford, M. J., Highley, J. R., Ince, P. G., Wharton, S. B., et al. (2015). Antisense RNA foci in the motor neurons of C9ORF72-ALS patients are associated with TDP-43 proteinopathy. *Acta Neuropathologica*, *130*(1), 63–75.

Cooper-Knock, J., Kirby, J., Highley, R., & Shaw, P. J. (2015). The spectrum of C9orf72-mediated neurodegeneration and amyotrophic lateral sclerosis. *Neurotherapeutics: The Journal of the American Society for Experimental NeuroTherapeutics*, *12*(2), 326–339.

Dansithong, W., Paul, S., Comai, L., & Reddy, S. (2005). MBNL1 is the primary determinant of focus formation and aberrant insulin receptor splicing in DM1. *The Journal of Biological Chemistry*, *280*(7), 5773–5780.

Dasgupta, T., & Ladd, A. N. (2012). The importance of CELF control: Molecular and biological roles of the CUG-BP, Elav-like family of RNA-binding proteins. *Wiley Interdisciplinary Reviews. RNA*, *3*(1), 104–121.

Daughters, R. S., Tuttle, D. L., Gao, W., Ikeda, Y., Moseley, M. L., Ebner, T. J., et al. (2009). RNA gain-of-function in spinocerebellar ataxia type 8. *PLoS Genetics*, *5*(8), e1000600.

Davidson, Y. S., Barker, H., Robinson, A. C., Thompson, J. C., Harris, J., Troakes, C., et al. (2014). Brain distribution of dipeptide repeat proteins in frontotemporal lobar degeneration and motor neurone disease associated with expansions in C9ORF72. *Acta Neuropathologica Communications*, *2*, 70.

Davis, B. M., McCurrach, M. E., Taneja, K. L., Singer, R. H., & Housman, D. E. (1997). Expansion of a CUG trinucleotide repeat in the 3′ untranslated region of myotonic dystrophy protein kinase transcripts results in nuclear retention of transcripts. *Proceedings of the National Academy of Sciences of the United States of America, 94*(14), 7388–7393.

de Haro, M., Al-Ramahi, I., De Gouyon, B., Ukani, L., Rosa, A., Faustino, N. A., et al. (2006). MBNL1 and CUGBP1 modify expanded CUG-induced toxicity in a *Drosophila* model of myotonic dystrophy type 1. *Human Molecular Genetics, 15*(13), 2138–2145.

DeJesus-Hernandez, M., Mackenzie, I. R., Boeve, B. F., Boxer, A. L., Baker, M., Rutherford, N. J., et al. (2011). Expanded GGGGCC hexanucleotide repeat in noncoding region of C9ORF72 causes chromosome 9p-linked FTD and ALS. *Neuron, 72*(2), 245–256.

de Mezer, M., Wojciechowska, M., Napierala, M., Sobczak, K., & Krzyzosiak, W. J. (2011). Mutant CAG repeats of Huntingtin transcript fold into hairpins, form nuclear foci and are targets for RNA interference. *Nucleic Acids Research, 39*(9), 3852–3863.

Donnelly, C. J., Zhang, P. W., Pham, J. T., Haeusler, A. R., Mistry, N. A., Vidensky, S., et al. (2013). RNA toxicity from the ALS/FTD C9ORF72 expansion is mitigated by antisense intervention. *Neuron, 80*(2), 415–428.

Dorsman, J. C., Pepers, B., Langenberg, D., Kerkdijk, H., Ijszenga, M., den Dunnen, J. T., et al. (2002). Strong aggregation and increased toxicity of polyleucine over polyglutamine stretches in mammalian cells. *Human Molecular Genetics, 11*(13), 1487–1496.

Du, J., Aleff, R. A., Soragni, E., Kalari, K., Nie, J., Tang, X., et al. (2015). RNA toxicity and missplicing in the common eye disease Fuchs endothelial corneal dystrophy. *The Journal of Biological Chemistry, 290*(10), 5979–5990.

Entezam, A., Biacsi, R., Orrison, B., Saha, T., Hoffman, G. E., Grabczyk, E., et al. (2007). Regional FMRP deficits and large repeat expansions into the full mutation range in a new Fragile X premutation mouse model. *Gene, 395*(1–2), 125–134.

Faghihi, M. A., Modarresi, F., Khalil, A. M., Wood, D. E., Sahagan, B. G., Morgan, T. E., et al. (2008). Expression of a noncoding RNA is elevated in Alzheimer's disease and drives rapid feed-forward regulation of beta-secretase. *Nature Medicine, 14*(7), 723–730.

Faghihi, M. A., & Wahlestedt, C. (2009). Regulatory roles of natural antisense transcripts. *Nature Reviews Molecular Cell Biology, 10*(9), 637–643.

Fardaei, M., Rogers, M. T., Thorpe, H. M., Larkin, K., Hamshere, M. G., Harper, P. S., et al. (2002). Three proteins, MBNL, MBLL and MBXL, co-localize in vivo with nuclear foci of expanded-repeat transcripts in DM1 and DM2 cells. *Human Molecular Genetics, 11*(7), 805–814.

Feng, Y., Zhang, F., Lokey, L. K., Chastain, J. L., Lakkis, L., Eberhart, D., et al. (1995). Translational suppression by trinucleotide repeat expansion at FMR1. *Science, 268*(5211), 731–734.

Fernandez-Funez, P., Nino-Rosales, M. L., de Gouyon, B., She, W. C., Luchak, J. M., Martinez, P., et al. (2000). Identification of genes that modify ataxin-1-induced neurodegeneration. *Nature, 408*(6808), 101–106.

Fratta, P., Mizielinska, S., Nicoll, A. J., Zloh, M., Fisher, E. M., Parkinson, G., et al. (2012). C9orf72 hexanucleotide repeat associated with amyotrophic lateral sclerosis and frontotemporal dementia forms RNA G-quadruplexes. *Scientific Reports, 2*, 1016.

Freibaum, B. D., Lu, Y., Lopez-Gonzalez, R., Kim, N. C., Almeida, S., Lee, K. H., et al. (2015). GGGGCC repeat expansion in C9orf72 compromises nucleocytoplasmic transport. *Nature, 525*(7567), 129–133.

Fu, Y. H., Friedman, D. L., Richards, S., Pearlman, J. A., Gibbs, R. A., Pizzuti, A., et al. (1993). Decreased expression of myotonin-protein kinase messenger RNA and protein in adult form of myotonic dystrophy. *Science, 260*(5105), 235–238.

Fu, Y. H., Pizzuti, A., Fenwick, R. G., Jr., King, J., Rajnarayan, S., Dunne, P. W., et al. (1992). An unstable triplet repeat in a gene related to myotonic muscular dystrophy. *Science, 255*(5049), 1256–1258.

Galloway, J. N., & Nelson, D. L. (2009). Evidence for RNA-mediated toxicity in the fragile X-associated tremor/ataxia syndrome. *Future Neurology, 4*(6), 785.

Galloway, J. N., Shaw, C., Yu, P., Parghi, D., Poidevin, M., Jin, P., et al. (2014). CGG repeats in RNA modulate expression of TDP-43 in mouse and fly models of fragile X tremor ataxia syndrome. *Human Molecular Genetics, 23*(22), 5906–5915.

Garcia-Murias, M., Quintans, B., Arias, M., Seixas, A. I., Cacheiro, P., Tarrio, R., et al. (2012). 'Costa da Morte' ataxia is spinocerebellar ataxia 36: Clinical and genetic characterization. *Brain: A Journal of Neurology, 135*(Pt 5), 1423–1435.

Gendron, T. F., Bieniek, K. F., Zhang, Y. J., Jansen-West, K., Ash, P. E., Caulfield, T., et al. (2013). Antisense transcripts of the expanded C9ORF72 hexanucleotide repeat form nuclear RNA foci and undergo repeat-associated non-ATG translation in c9FTD/ALS. *Acta Neuropathologica, 126*(6), 829–844.

Gomez-Deza, J., Lee, Y. B., Troakes, C., Nolan, M., Al-Sarraj, S., Gallo, J. M., et al. (2015). Dipeptide repeat protein inclusions are rare in the spinal cord and almost absent from motor neurons in C9ORF72 mutant amyotrophic lateral sclerosis and are unlikely to cause their degeneration. *Acta Neuropathologica Communications, 3*(1), 38.

Greco, C. M., Berman, R. F., Martin, R. M., Tassone, F., Schwartz, P. H., Chang, A., et al. (2006). Neuropathology of fragile X-associated tremor/ataxia syndrome (FXTAS). *Brain: A Journal of Neurology, 129*(Pt 1), 243–255.

Greco, C. M., Hagerman, R. J., Tassone, F., Chudley, A. E., Del Bigio, M. R., Jacquemont, S., et al. (2002). Neuronal intranuclear inclusions in a new cerebellar tremor/ataxia syndrome among fragile X carriers. *Brain: A Journal of Neurology, 125*(Pt 8), 1760–1771.

Green, K. M., Linsalata, A. E., & Todd, P. K. (2016). RAN translation-What makes it run?. *Brain Research, 1647*, 30–42.

Haeusler, A. R., Donnelly, C. J., Periz, G., Simko, E. A., Shaw, P. G., Kim, M. S., et al. (2014). C9orf72 nucleotide repeat structures initiate molecular cascades of disease. *Nature, 507*(7491), 195–200.

Hagerman, R. J., Leehey, M., Heinrichs, W., Tassone, F., Wilson, R., Hills, J., et al. (2001). Intention tremor, parkinsonism, and generalized brain atrophy in male carriers of fragile X. *Neurology, 57*(1), 127–130.

Hamshere, M. G., Newman, E. E., Alwazzan, M., Athwal, B. S., & Brook, J. D. (1997). Transcriptional abnormality in myotonic dystrophy affects DMPK but not neighboring genes. *Proceedings of the National Academy of Sciences of the United States of America, 94*(14), 7394–7399.

Handa, V., Goldwater, D., Stiles, D., Cam, M., Poy, G., Kumari, D., et al. (2005). Long CGG-repeat tracts are toxic to human cells: Implications for carriers of Fragile X premutation alleles. *FEBS Letters, 579*(12), 2702–2708.

Hashem, V., Galloway, J. N., Mori, M., Willemsen, R., Oostra, B. A., Paylor, R., et al. (2009). Ectopic expression of CGG containing mRNA is neurotoxic in mammals. *Human Molecular Genetics, 18*(13), 2443–2451.

He, F., Krans, A., Freibaum, B. D., Taylor, J. P., & Todd, P. K. (2014). TDP-43 suppresses CGG repeat-induced neurotoxicity through interactions with HnRNP A2/B1. *Human Molecular Genetics, 23*(19), 5036–5051.

He, Y., Zu, T., Benzow, K. A., Orr, H. T., Clark, H. B., & Koob, M. D. (2006). Targeted deletion of a single Sca8 ataxia locus allele in mice causes abnormal gait, progressive loss of motor coordination, and Purkinje cell dendritic deficits. *The Journal of Neuroscience: The Official Journal of the Society for Neuroscience, 26*(39), 9975–9982.

Ho, T. H., Bundman, D., Armstrong, D. L., & Cooper, T. A. (2005). Transgenic mice expressing CUG-BP1 reproduce splicing mis-regulation observed in myotonic dystrophy. *Human Molecular Genetics, 14*(11), 1539–1547.

Hoem, G., Raske, C. R., Garcia-Arocena, D., Tassone, F., Sanchez, E., Ludwig, A. L., et al. (2011). CGG-repeat length threshold for FMR1 RNA pathogenesis in a cellular model for FXTAS. *Human Molecular Genetics, 20*(11), 2161–2170.

Hoffman, G. E., Le, W. W., Entezam, A., Otsuka, N., Tong, Z. B., Nelson, L., et al. (2012). Ovarian abnormalities in a mouse model of fragile X primary ovarian insufficiency. *The Journal of Histochemistry and Cytochemistry: Official Journal of the Histochemistry Society, 60*(6), 439–456.

Holmes, S. E., O'Hearn, E., Rosenblatt, A., Callahan, C., Hwang, H. S., Ingersoll-Ashworth, R. G., et al. (2001). A repeat expansion in the gene encoding junctophilin-3 is associated with Huntington disease-like 2. *Nature Genetics, 29*(4), 377–378.

Holt, I., Jacquemin, V., Fardaei, M., Sewry, C. A., Butler-Browne, G. S., Furling, D., et al. (2009). Muscleblind-like proteins: Similarities and differences in normal and myotonic dystrophy muscle. *The American Journal of Pathology, 174*(1), 216–227.

Hsu, R. J., Hsiao, K. M., Lin, M. J., Li, C. Y., Wang, L. C., Chen, L. K., et al. (2011). Long tract of untranslated CAG repeats is deleterious in transgenic mice. *PLoS ONE, 6*(1), e16417.

Huichalaf, C., Sakai, K., Jin, B., Jones, K., Wang, G. L., Schoser, B., et al. (2010). Expansion of CUG RNA repeats causes stress and inhibition of translation in myotonic dystrophy 1 (DM1) cells. *FASEB Journal: Official Publication of the Federation of American Societies for Experimental Biology, 24*(10), 3706–3719.

Hukema, R. K., Buijsen, R. A., Schonewille, M., Raske, C., Severijnen, L. A., Nieuwenhuizen-Bakker, I., et al. (2015). Reversibility of neuropathology and motor deficits in an inducible mouse model for FXTAS. *Human Molecular Genetics, 24*(17), 4948–4957.

Ishikawa, K., & Mizusawa, H. (2010). The chromosome 16q-linked autosomal dominant cerebellar ataxia (16q-ADCA): A newly identified degenerative ataxia in Japan showing peculiar morphological changes of the Purkinje cell: The 50th Anniversary of Japanese Society of Neuropathology. *Neuropathology: Official Journal of the Japanese Society of Neuropathology, 30*(5), 490–494.

Iwahashi, C. K., Yasui, D. H., An, H. J., Greco, C. M., Tassone, F., Nannen, K., et al. (2006). Protein composition of the intranuclear inclusions of FXTAS. *Brain: A Journal of Neurology, 129*(Pt 1), 256–271.

Jain, S., Wheeler, J. R., Walters, R. W., Agrawal, A., Barsic, A., & Parker, R. (2016). ATPase-modulated stress granules contain a diverse proteome and substructure. *Cell, 164*(3), 487–498.

Jansen, G., Groenen, P. J., Bachner, D., Jap, P. H., Coerwinkel, M., Oerlemans, F., et al. (1996). Abnormal myotonic dystrophy protein kinase levels produce only mild myopathy in mice. *Nature Genetics, 13*(3), 316–324.

Jin, P., Duan, R., Qurashi, A., Qin, Y., Tian, D., Rosser, T. C., et al. (2007). Pur alpha binds to rCGG repeats and modulates repeat-mediated neurodegeneration in a *Drosophila* model of fragile X tremor/ataxia syndrome. *Neuron, 55*(4), 556–564.

Jin, P., Zarnescu, D. C., Zhang, F., Pearson, C. E., Lucchesi, J. C., Moses, K., et al. (2003). RNA-Mediated Neurodegeneration Caused by the Fragile X Premutation rCGG Repeats in *Drosophila. Neuron, 39*(5), 739–747.

Jovicic, A., Mertens, J., Boeynaems, S., Bogaert, E., Chai, N., Yamada, S. B., et al. (2015). Modifiers of C9orf72 dipeptide repeat toxicity connect nucleocytoplasmic transport defects to FTD/ALS. *Nature Neuroscience, 18*(9), 1226–1229.

Juang, B. T., Ludwig, A. L., Benedetti, K. L., Gu, C., Collins, K., Morales, C., et al. (2014). Expression of an expanded CGG-repeat RNA in a single pair of primary sensory neurons impairs olfactory adaptation in *Caenorhabditis elegans. Human Molecular Genetics, 23*(18), 4945–4959.

Kanadia, R. N., Johnstone, K. A., Mankodi, A., Lungu, C., Thornton, C. A., Esson, D., et al. (2003). A muscleblind knockout model for myotonic dystrophy. *Science, 302* (5652), 1978–1980.

Kanadia, R. N., Shin, J., Yuan, Y., Beattie, S. G., Wheeler, T. M., Thornton, C. A., et al. (2006). Reversal of RNA missplicing and myotonia after muscleblind overexpression in a mouse poly(CUG) model for myotonic dystrophy. *Proceedings of the National Academy of Sciences of the United States of America, 103*(31), 11748–11753.

Kanekura, K., Yagi, T., Cammack, A. J., Mahadevan, J., Kuroda, M., Harms, M. B., et al. (2016). Poly-dipeptides encoded by the C9ORF72 repeats block global protein translation. *Human Molecular Genetics, 25*(9), 1803–1813.

Kato, M., Han, T. W., Xie, S., Shi, K., Du, X., Wu, L. C., et al. (2012). Cell-free formation of RNA granules: Low complexity sequence domains form dynamic fibers within hydrogels. *Cell, 149*(4), 753–767.

Kearse, M. G., Green, K. M., Krans, A., Rodriguez, C. M., Linsalata, A. E., Goldstrohm, A. C., et al. (2016). CGG repeat-associated non-AUG translation utilizes a cap-dependent scanning mechanism of initiation to produce toxic proteins. *Molecular Cell, 62*(2), 314–322.

Khalili, K., Del Valle, L., Muralidharan, V., Gault, W. J., Darbinian, N., Otte, J., et al. (2003). Pur is essential for postnatal brain development and developmentally coupled cellular proliferation as revealed by genetic inactivation in the mouse. *Molecular and Cellular Biology, 23* (19), 6857–6875.

Khosravi, B., Hartmann, H., May, S., Mohl, C., Ederle, H., Michaelsen, M., et al. (2017). Cytoplasmic poly-GA aggregates impair nuclear import of TDP-43 in C9orf72 ALS/FTLD. *Human Molecular Genetics, 26*, 790–800.

Klein, M. E., Younts, T. J., Castillo, P. E., & Jordan, B. A. (2013). RNA-binding protein Sam68 controls synapse number and local beta-actin mRNA metabolism in dendrites. *Proceedings of the National Academy of Sciences of the United States of America, 110*(8), 3125–3130.

Klesert, T. R., Cho, D. H., Clark, J. I., Maylie, J., Adelman, J., Snider, L., et al. (2000). Mice deficient in Six5 develop cataracts: Implications for myotonic dystrophy. *Nature Genetics, 25*(1), 105–109.

Klesert, T. R., Otten, A. D., Bird, T. D., & Tapscott, S. J. (1997). Trinucleotide repeat expansion at the myotonic dystrophy locus reduces expression of DMAHP. *Nature Genetics, 16*(4), 402–406.

Kobayashi, H., Abe, K., Matsuura, T., Ikeda, Y., Hitomi, T., Akechi, Y., et al. (2011). Expansion of intronic GGCCTG hexanucleotide repeat in NOP56 causes SCA36, a type of spinocerebellar ataxia accompanied by motor neuron involvement. *American Journal of Human Genetics, 89*(1), 121–130.

Koob, M. D., Moseley, M. L., Schut, L. J., Benzow, K. A., Bird, T. D., Day, J. W., et al. (1999). An untranslated CTG expansion causes a novel form of spinocerebellar ataxia (SCA8). *Nature Genetics, 21*(4), 379–384.

Kramer, N. J., Carlomagno, Y., Zhang, Y. J., Almeida, S., Cook, C. N., Gendron, T. F., et al. (2016). Spt4 selectively regulates the expression of C9orf72 sense and antisense mutant transcripts. *Science, 353*(6300), 708–712.

Krans, A., Kearse, M. G., & Todd, P. K. (2016). Repeat-associated non-AUG translation from antisense CCG repeats in fragile X tremor/ataxia syndrome. *Annals of Neurology, 80*(6), 871–881.

Krol, J., Fiszer, A., Mykowska, A., Sobczak, K., de Mezer, M., & Krzyzosiak, W. J. (2007). Ribonuclease dicer cleaves triplet repeat hairpins into shorter repeats that silence specific targets. *Molecular Cell, 25*(4), 575–586.

Kuyumcu-Martinez, N. M., Wang, G. S., & Cooper, T. A. (2007). Increased steady-state levels of CUGBP1 in myotonic dystrophy 1 are due to PKC-mediated hyperphosphorylation. *Molecular Cell, 28*(1), 68–78.

Kwon, I., Xiang, S., Kato, M., Wu, L., Theodoropoulos, P., Wang, T., et al. (2014). Poly-dipeptides encoded by the C9orf72 repeats bind nucleoli, impede RNA biogenesis, and kill cells. *Science, 345*(6201), 1139–1145.

Ladd, P. D., Smith, L. E., Rabaia, N. A., Moore, J. M., Georges, S. A., Hansen, R. S., et al. (2007). An antisense transcript spanning the CGG repeat region of FMR1 is upregulated in premutation carriers but silenced in full mutation individuals. *Human Molecular Genetics, 16*(24), 3174–3187.

Lagier-Tourenne, C., Baughn, M., Rigo, F., Sun, S., Liu, P., Li, H. R., et al. (2013). Targeted degradation of sense and antisense C9orf72 RNA foci as therapy for ALS and frontotemporal degeneration. *Proceedings of the National Academy of Sciences of the United States of America, 110* (47), E4530–4539.

Lawlor, K. T., O'Keefe, L. V., Samaraweera, S. E., van Eyk, C. L., McLeod, C. J., Maloney, C. A., et al. (2011). Double-stranded RNA is pathogenic in *Drosophila* models of expanded repeat neurodegenerative diseases. *Human Molecular Genetics*, *20*(19), 3757–3768.

Lee, K. H., Zhang, P., Kim, H. J., Mitrea, D. M., Sarkar, M., Freibaum, B. D., et al. (2016). C9orf72 dipeptide repeats impair the assembly, dynamics, and function of membrane-less organelles. *Cell*, *167*(3), 774–788.e717.

Lee, Y. B., Chen, H. J., Peres, J. N., Gomez-Deza, J., Attig, J., Stalekar, M., et al. (2013). Hexanucleotide repeats in ALS/FTD form length-dependent RNA foci, sequester RNA binding proteins, and are neurotoxic. *Cell Reports*, *5*(5), 1178–1186.

Li, L. B., Yu, Z., Teng, X., & Bonini, N. M. (2008). RNA toxicity is a component of ataxin-3 degeneration in *Drosophila*. *Nature*, *453*(7198), 1107–1111.

Li, P. P., Sun, X., Xia, G., Arbez, N., Paul, S., Zhu, S., et al. (2016). ATXN2-AS, a gene antisense to ATXN2, is associated with spinocerebellar ataxia type 2 and amyotrophic lateral sclerosis. *Annals of Neurology*, *80*(4), 600–615.

Li, Y., & Jin, P. (2012). RNA-mediated neurodegeneration in fragile X-associated tremor/ataxia syndrome. *Brain Research*, *1462*, 112–117.

Lin, Y., Mori, E., Kato, M., Xiang, S., Wu, L., Kwon, I., et al. (2016). Toxic PR poly-dipeptides encoded by the C9orf72 repeat expansion target LC domain polymers. *Cell*, *167*(3), 789–802.e712.

Lin, Y., Protter, D. S., Rosen, M. K., & Parker, R. (2015). Formation and maturation of phase-separated liquid droplets by RNA-binding proteins. *Molecular Cell*, *60*(2), 208–219.

Liquori, C. L., Ricker, K., Moseley, M. L., Jacobsen, J. F., Kress, W., Naylor, S. L., et al. (2001). Myotonic dystrophy type 2 caused by a CCTG expansion in intron 1 of ZNF9. *Science*, *293*(5531), 864–867.

Liu, C. R., Chang, C. R., Chern, Y., Wang, T. H., Hsieh, W. C., Shen, W. C., et al. (2012). Spt4 is selectively required for transcription of extended trinucleotide repeats. *Cell*, *148*(4), 690–701.

Liu, J., Hu, J., Ludlow, A. T., Pham, J. T., Shay, J. W., Rothstein, J. D., et al. (2017). c9orf72 disease-related foci are each composed of one mutant expanded repeat RNA. *Cell Chemical Biology*, *24*(2), 141–148.

Liu, J., Koscielska, K. A., Cao, Z., Hulsizer, S., Grace, N., Mitchell, G., et al. (2012). Signaling defects in iPSC-derived fragile X premutation neurons. *Human Molecular Genetics*, *21*(17), 3795–3805.

Liu, W., Ikeda, Y., Hishikawa, N., Yamashita, T., Deguchi, K., & Abe, K. (2014). Characteristic RNA foci of the abnormal hexanucleotide GGCCUG repeat expansion in spinocerebellar ataxia type 36 (Asidan). *European Journal of Neurology*, *21*(11), 1377–1386.

Lopez-Gonzalez, R., Lu, Y., Gendron, T. F., Karydas, A., Tran, H., Yang, D., et al. (2016). Poly(GR) in C9ORF72-related ALS/FTD compromises mitochondrial function and increases oxidative stress and DNA damage in iPSC-derived motor neurons. *Neuron*, *92*(2), 383–391.

Lukong, K. E., & Richard, S. (2008). Motor coordination defects in mice deficient for the Sam68 RNA-binding protein. *Behavioural Brain Research*, *189*(2), 357–363.

Mackenzie, I. R., Frick, P., Grasser, F. A., Gendron, T. F., Petrucelli, L., Cashman, N. R., et al. (2015). Quantitative analysis and clinico-pathological correlations of different dipeptide repeat protein pathologies in C9ORF72 mutation carriers. *Acta Neuropathologica*, *130*(6), 845–861.

Magistri, M., Faghihi, M. A., St Laurent, G., 3rd, & Wahlestedt, C. (2012). Regulation of chromatin structure by long noncoding RNAs: Focus on natural antisense transcripts. *Trends in Genetics: TIG*, *28*(8), 389–396.

Mahadevan, M., Tsilfidis, C., Sabourin, L., Shutler, G., Amemiya, C., Jansen, G., et al. (1992). Myotonic dystrophy mutation: An unstable CTG repeat in the 3′ untranslated region of the gene. *Science*, *255*(5049), 1253–1255.

Mankodi, A., Logigian, E., Callahan, L., McClain, C., White, R., Henderson, D., et al. (2000). Myotonic dystrophy in transgenic mice expressing an expanded CUG repeat. *Science*, *289*(5485), 1769–1773.

Mankodi, A., Urbinati, C. R., Yuan, Q. P., Moxley, R. T., Sansone, V., Krym, M., et al. (2001). Muscleblind localizes to nuclear foci of aberrant RNA in myotonic dystrophy types 1 and 2. *Human Molecular Genetics*, *10*(19), 2165–2170.

Margolis, J. M., Schoser, B. G., Moseley, M. L., Day, J. W., & Ranum, L. P. (2006). DM2 intronic expansions: Evidence for CCUG accumulation without flanking sequence or effects on ZNF9 mRNA processing or protein expression. *Human Molecular Genetics*, *15*(11), 1808–1815.

Margolis, R. L., O'Hearn, E., Rosenblatt, A., Willour, V., Holmes, S. E., Franz, M. L., et al. (2001). A disorder similar to Huntington's disease is associated with a novel CAG repeat expansion. *Annals of Neurology*, *50*(3), 373–380.

Marz, P., Probst, A., Lang, S., Schwager, M., Rose-John, S., Otten, U., et al. (2004). Ataxin-10, the spinocerebellar ataxia type 10 neurodegenerative disorder protein, is essential for survival of cerebellar neurons. *The Journal of Biological Chemistry*, *279*(34), 35542–35550.

Matsuura, T., Yamagata, T., Burgess, D. L., Rasmussen, A., Grewal, R. P., Watase, K., et al. (2000). Large expansion of the ATTCT pentanucleotide repeat in spinocerebellar ataxia type 10. *Nature Genetics, 26*(2), 191–194.

May, S., Hornburg, D., Schludi, M. H., Arzberger, T., Rentzsch, K., Schwenk, B. M., et al. (2014). C9orf72 FTLD/ALS-associated Gly-Ala dipeptide repeat proteins cause neuronal toxicity and Unc119 sequestration. *Acta Neuropathologica, 128*(4), 485–503.

Meola, G., & Cardani, R. (2015). Myotonic dystrophies: An update on clinical aspects, genetic, pathology, and molecular pathomechanisms. *Biochimica et Biophysica Acta, 1852*(4), 594–606.

Meola, G., Sansone, V., Radice, S., Skradski, S., & Ptacek, L. (1996). A family with an unusual myotonic and myopathic phenotype and no CTG expansion (proximal myotonic myopathy syndrome): A challenge for future molecular studies. *Neuromuscular Disorders: NMD, 6*(3), 143–150.

Michalowski, S., Miller, J. W., Urbinati, C. R., Paliouras, M., Swanson, M. S., & Griffith, J. (1999). Visualization of double-stranded RNAs from the myotonic dystrophy protein kinase gene and interactions with CUG-binding protein. *Nucleic Acids Research, 27*(17), 3534–3542.

Miller, J. W., Urbinati, C. R., Teng-Umnuay, P., Stenberg, M. G., Byrne, B. J., Thornton, C. A., et al. (2000). Recruitment of human muscleblind proteins to (CUG)(*n*) expansions associated with myotonic dystrophy. *The EMBO Journal, 19*(17), 4439–4448.

Mitrea, D. M., Cika, J. A., Guy, C. S., Ban, D., Banerjee, P. R., Stanley, C. B., et al. (2016). Nucleophosmin integrates within the nucleolus via multi-modal interactions with proteins displaying R-rich linear motifs and rRNA. *eLife, 5*.

Mizielinska, S., Gronke, S., Niccoli, T., Ridler, C. E., Clayton, E. L., Devoy, A., et al. (2014). C9orf72 repeat expansions cause neurodegeneration in *Drosophila* through arginine-rich proteins. *Science, 345*(6201), 1192–1194.

Molliex, A., Temirov, J., Lee, J., Coughlin, M., Kanagaraj, A. P., Kim, H. J., et al. (2015). Phase separation by low complexity domains promotes stress granule assembly and drives pathological fibrillization. *Cell, 163*(1), 123–133.

Mootha, V. V., Gong, X., Ku, H. C., & Xing, C. (2014). Association and familial segregation of CTG18.1 trinucleotide repeat expansion of TCF4 gene in Fuchs' endothelial corneal dystrophy. *Investigative Ophthalmology & Visual Science, 55*(1), 33–42.

Mootha, V. V., Hussain, I., Cunnusamy, K., Graham, E., Gong, X., Neelam, S., et al. (2015). TCF4 triplet repeat expansion and nuclear RNA foci in Fuchs' endothelial corneal dystrophy. *Investigative Ophthalmology & Visual Science, 56*(3), 2003–2011.

Mori, K., Arzberger, T., Grasser, F. A., Gijselinck, I., May, S., Rentzsch, K., et al. (2013). Bidirectional transcripts of the expanded C9orf72 hexanucleotide repeat are translated into aggregating dipeptide repeat proteins. *Acta Neuropathologica, 126*(6), 881–893.

Mori, K., Lammich, S., Mackenzie, I. R., Forne, I., Zilow, S., Kretzschmar, H., et al. (2013). hnRNP A3 binds to GGGGCC repeats and is a constituent of p62-positive/TDP43-negative inclusions in the hippocampus of patients with C9orf72 mutations. *Acta Neuropathologica, 125*(3), 413–423.

Mori, K., Weng, S. M., Arzberger, T., May, S., Rentzsch, K., Kremmer, E., et al. (2013). The C9orf72 GGGGCC repeat is translated into aggregating dipeptide-repeat proteins in FTLD/ALS. *Science, 339*(6125), 1335–1338.

Moseley, M. L., Zu, T., Ikeda, Y., Gao, W., Mosemiller, A. K., Daughters, R. S., et al. (2006). Bidirectional expression of CUG and CAG expansion transcripts and intranuclear polyglutamine inclusions in spinocerebellar ataxia type 8. *Nature Genetics, 38*(7), 758–769.

Murakami, T., Qamar, S., Lin, J. Q., Schierle, G. S., Rees, E., Miyashita, A., et al. (2015). ALS/FTD mutation-induced phase transition of FUS liquid droplets and reversible hydrogels into irreversible hydrogels impairs RNP granule function. *Neuron, 88*(4), 678–690.

Mutsuddi, M., Marshall, C. M., Benzow, K. A., Koob, M. D., & Rebay, I. (2004). The spinocerebellar ataxia 8 noncoding RNA causes neurodegeneration and associates with staufen in *Drosophila*. *Current Biology: CB, 14* (4), 302–308.

Mykowska, A., Sobczak, K., Wojciechowska, M., Kozlowski, P., & Krzyzosiak, W. J. (2011). CAG repeats mimic CUG repeats in the misregulation of alternative splicing. *Nucleic Acids Research, 39*(20), 8938–8951.

Nakamori, M., Pearson, C. E., & Thornton, C. A. (2010). Bidirectional transcription stimulates expansion and contraction of expanded (CTG)*(CAG) repeats. *Human Molecular Genetics, 20*(3), 580–588.

Nalavade, R., Griesche, N., Ryan, D. P., Hildebrand, S., & Krauss, S. (2013). Mechanisms of RNA-induced toxicity in CAG repeat disorders. *Cell Death & Disease, 4*, e752.

Nemes, J. P., Benzow, K. A., Moseley, M. L., Ranum, L. P., & Koob, M. D. (2000). The SCA8 transcript is an antisense RNA to a brain-specific transcript encoding a novel actin-binding protein (KLHL1). *Human Molecular Genetics, 9* (10), 1543–1551.

Niimi, Y., Takahashi, M., Sugawara, E., Umeda, S., Obayashi, M., Sato, N., et al. (2013). Abnormal RNA structures (RNA foci) containing a penta-nucleotide repeat (UGGAA)*n* in the Purkinje cell nucleus is associated with spinocerebellar ataxia type 31 pathogenesis. *Neuropathology: Official Journal of the Japanese Society of Neuropathology, 33*(6), 600–611.

Nishi, M., Hashimoto, K., Kuriyama, K., Komazaki, S., Kano, M., Shibata, S., et al. (2002). Motor discoordination in mutant mice lacking junctophilin type 3. *Biochemical and Biophysical Research Communications, 292*(2), 318–324.

Obayashi, M., Stevanin, G., Synofzik, M., Monin, M. L., Duyckaerts, C., Sato, N., et al. (2015). Spinocerebellar ataxia type 36 exists in diverse populations and can be caused by a short hexanucleotide GGCCTG repeat expansion. *Journal of Neurology, Neurosurgery, and Psychiatry, 86*(9), 986–995.

Oh, S. Y., He, F., Krans, A., Frazer, M., Taylor, J. P., Paulson, H. L., et al. (2015). RAN translation at CGG repeats induces ubiquitin proteasome system impairment in models of fragile X-associated tremor ataxia syndrome. *Human Molecular Genetics, 24*(15), 4317–4326.

Ohki, Y., Wenninger-Weinzierl, A., Hruscha, A., Asakawa, K., Kawakami, K., Haass, C., et al. (2017). Glycine-alanine dipeptide repeat protein contributes to toxicity in a zebrafish model of C9orf72 associated neurodegeneration. *Molecular Neurodegeneration, 12*(1), 6.

O'Rourke, Jacqueline G., Bogdanik, L., Muhammad, A. K. M. G., Gendron, Tania F., Kim, Kevin J., Austin, A., et al. (2015). C9orf72 BAC transgenic mice display typical pathologic features of ALS/FTD. *Neuron, 88*(5), 892–901.

Osborne, R. J., Lin, X., Welle, S., Sobczak, K., O'Rourke, J. R., Swanson, M. S., et al. (2009). Transcriptional and post-transcriptional impact of toxic RNA in myotonic dystrophy. *Human Molecular Genetics, 18*(8), 1471–1481.

Pascual, M., Vicente, M., Monferrer, L., & Artero, R. (2006). The Muscleblind family of proteins: An emerging class of regulators of developmentally programmed alternative splicing. *Differentiation; Research in Biological Diversity, 74*(2–3), 65–80.

Patel, A., Lee, H. O., Jawerth, L., Maharana, S., Jahnel, M., Hein, M. Y., et al. (2015). A liquid-to-solid phase transition of the ALS protein FUS accelerated by disease mutation. *Cell, 162*(5), 1066–1077.

Paul, S., Dansithong, W., Kim, D., Rossi, J., Webster, N. J., Comai, L., et al. (2006). Interaction of muscleblind, CUG-BP1 and hnRNP H proteins in DM1-associated aberrant IR splicing. *The EMBO Journal, 25*(18), 4271–4283.

Petri, R., Malmevik, J., Fasching, L., Akerblom, M., & Jakobsson, J. (2014). miRNAs in brain development. *Experimental Cell Research, 321*(1), 84–89.

Philips, A. V., Timchenko, L. T., & Cooper, T. A. (1998). Disruption of splicing regulated by a CUG-binding protein in myotonic dystrophy. *Science, 280*(5364), 737–741.

Poulos, M. G., Batra, R., Li, M., Yuan, Y., Zhang, C., Darnell, R. B., et al. (2013). Progressive impairment of muscle regeneration in muscleblind-like 3 isoform knockout mice. *Human Molecular Genetics, 22*(17), 3547–3558.

Prudencio, M., Belzil, V. V., Batra, R., Ross, C. A., Gendron, T. F., Pregent, L. J., et al. (2015). Distinct brain transcriptome profiles in C9orf72-associated and sporadic ALS. *Nature Neuroscience, 18*(8), 1175–1182.

Qurashi, A., Li, W., Zhou, J. Y., Peng, J., & Jin, P. (2011). Nuclear accumulation of stress response mRNAs contributes to the neurodegeneration caused by fragile X premutation rCGG repeats. *PLoS Genetics, 7*(6), e1002102.

Reddy, K., Tam, M., Bowater, R. P., Barber, M., Tomlinson, M., Nichol Edamura, K., et al. (2010). Determinants of R-loop formation at convergent bidirectionally transcribed trinucleotide repeats. *Nucleic Acids Research, 39*(5), 1749–1762.

Reddy, K., Zamiri, B., Stanley, S. Y., Macgregor, R. B., Jr., & Pearson, C. E. (2013). The disease-associated r (GGGGCC)*n* repeat from the C9orf72 gene forms tract length-dependent uni- and multimolecular RNA G-quadruplex structures. *The Journal of Biological Chemistry, 288*(14), 9860–9866.

Reddy, S., Smith, D. B., Rich, M. M., Leferovich, J. M., Reilly, P., Davis, B. M., et al. (1996). Mice lacking the myotonic dystrophy protein kinase develop a late onset progressive myopathy. *Nature Genetics, 13*(3), 325–335.

Renton, A. E., Majounie, E., Waite, A., Simon-Sanchez, J., Rollinson, S., Gibbs, J. R., et al. (2011). A hexanucleotide repeat expansion in C9ORF72 is the cause of chromosome 9p21-linked ALS-FTD. *Neuron, 72*(2), 257–268.

Ricker, K., Koch, M. C., Lehmann-Horn, F., Pongratz, D., Otto, M., Heine, R., et al. (1994). Proximal myotonic myopathy: A new dominant disorder with myotonia, muscle weakness, and cataracts. *Neurology, 44*(8), 1448–1452.

Rossi, S., Serrano, A., Gerbino, V., Giorgi, A., Di Francesco, L., Nencini, M., et al. (2015). Nuclear accumulation of mRNAs underlies G4C2-repeat-induced translational repression in a cellular model of C9orf72 ALS. *Journal of Cell Science, 128*(9), 1787–1799.

Rudnicki, D. D., Holmes, S. E., Lin, M. W., Thornton, C. A., Ross, C. A., & Margolis, R. L. (2007). Huntington's disease—Like 2 is associated with CUG repeat-containing RNA foci. *Annals of Neurology, 61*(3), 272–282.

Rudnicki, D. D., Pletnikova, O., Vonsattel, J. P., Ross, C. A., & Margolis, R. L. (2008). A comparison of huntington disease and huntington disease-like 2 neuropathology. *Journal of Neuropathology and Experimental Neurology, 67*(4), 366–374.

Sareen, D., O'Rourke, J. G., Meera, P., Muhammad, A. K., Grant, S., Simpkinson, M., et al. (2013). Targeting RNA foci in iPSC-derived motor neurons from ALS patients with a C9ORF72 repeat expansion. *Science Translational Medicine*, *5*(208), 208ra149.

Sarkar, P. S., Appukuttan, B., Han, J., Ito, Y., Ai, C., Tsai, W., et al. (2000). Heterozygous loss of Six5 in mice is sufficient to cause ocular cataracts. *Nature Genetics*, *25* (1), 110–114.

Sato, N., Amino, T., Kobayashi, K., Asakawa, S., Ishiguro, T., Tsunemi, T., et al. (2009). Spinocerebellar ataxia type 31 is associated with "inserted" penta-nucleotide repeats containing (TGGAA)*n*. *American Journal of Human Genetics*, *85*(5), 544–557.

Savkur, R. S., Philips, A. V., & Cooper, T. A. (2001). Aberrant regulation of insulin receptor alternative splicing is associated with insulin resistance in myotonic dystrophy. *Nature Genetics*, *29*(1), 40–47.

Scheele, C., Petrovic, N., Faghihi, M. A., Lassmann, T., Fredriksson, K., Rooyackers, O., et al. (2007). The human PINK1 locus is regulated in vivo by a non-coding natural antisense RNA during modulation of mitochondrial function. *BMC Genomics*, *8*, 74.

Schmidt, H. B., & Rohatgi, R. (2016). In vivo formation of vacuolated multi-phase compartments lacking membranes. *Cell Reports*, *16*(5), 1228–1236.

Seixas, A. I., Holmes, S. E., Takeshima, H., Pavlovich, A., Sachs, N., Pruitt, J. L., et al. (2012). Loss of junctophilin-3 contributes to huntington disease-like 2 pathogenesis. *Annals of Neurology*, *71*(2), 245–257.

Sellier, C., Freyermuth, F., Tabet, R., Tran, T., He, F., Ruffenach, F., et al. (2013). Sequestration of DROSHA and DGCR8 by expanded CGG RNA repeats alters microRNA processing in fragile X-associated tremor/ataxia syndrome. *Cell Reports*, *3*(3), 869–880.

Sellier, C., Rau, F., Liu, Y., Tassone, F., Hukema, R. K., Gattoni, R., et al. (2010). Sam68 sequestration and partial loss of function are associated with splicing alterations in FXTAS patients. *The EMBO Journal*, *29*(7), 1248–1261.

Sherman, S. L. (2000). Premature ovarian failure in the fragile X syndrome. *American Journal of Medical Genetics*, *97* (3), 189–194.

Shi, K. Y., Mori, E., Nizami, Z. F., Lin, Y., Kato, M., Xiang, S., et al. (2017). Toxic PRn poly-dipeptides encoded by the C9orf72 repeat expansion block nuclear import and export. *Proceedings of the National Academy of Sciences of the United States of America*.

Skourti-Stathaki, K., & Proudfoot, N. J. (2014). A double-edged sword: R loops as threats to genome integrity and powerful regulators of gene expression. *Genes & Development*, *28*(13), 1384–1396.

Sofola, O. A., Jin, P., Botas, J., & Nelson, D. L. (2007). Argonaute-2-dependent rescue of a *Drosophila* model of FXTAS by FRAXE premutation repeat. *Human Molecular Genetics*, *16*(19), 2326–2332.

Sofola, O. A., Jin, P., Qin, Y., Duan, R., Liu, H., de Haro, M., et al. (2007). RNA-binding proteins hnRNP A2/B1 and CUGBP1 suppress fragile X CGG premutation repeat-induced neurodegeneration in a *Drosophila* model of FXTAS. *Neuron*, *55*(4), 565–571.

Sopher, B. L., Ladd, P. D., Pineda, V. V., Libby, R. T., Sunkin, S. M., Hurley, J. B., et al. (2011). CTCF regulates ataxin-7 expression through promotion of a convergently transcribed, antisense noncoding RNA. *Neuron*, *70*(6), 1071–1084.

Southwell, A. L., Skotte, N. H., Bennett, C. F., & Hayden, M. R. (2012). Antisense oligonucleotide therapeutics for inherited neurodegenerative diseases. *Trends in Molecular Medicine*, *18*(11), 634–643.

Su, Z., Zhang, Y., Gendron, T. F., Bauer, P. O., Chew, J., Yang, W. Y., et al. (2014). Discovery of a biomarker and lead small molecules to target r(GGGGCC)-associated defects in c9FTD/ALS. *Neuron*, *83*(5), 1043–1050.

Suenaga, K., Lee, K. Y., Nakamori, M., Tatsumi, Y., Takahashi, M. P., Fujimura, H., et al. (2012). Muscleblind-like 1 knockout mice reveal novel splicing defects in the myotonic dystrophy brain. *PLoS ONE*, *7* (3), e33218.

Sugaya, K., Matsubara, S., Miyamoto, K., Kawata, A., & Hayashi, H. (2003). An aggregate-prone conformational epitope in trinucleotide repeat diseases. *Neuroreport*, *14* (18), 2331–2335.

Sun, Y. M., Lu, C., & Wu, Z. Y. (2016). Spinocerebellar ataxia: Relationship between phenotype and genotype-a review. *Clinical Genetics*, *90*(4), 305–314.

Takeshima, H., Komazaki, S., Nishi, M., Iino, M., & Kangawa, K. (2000). Junctophilins: A novel family of junctional membrane complex proteins. *Molecular Cell*, *6* (1), 11–22.

Tan, H., Poidevin, M., Li, H., Chen, D., & Jin, P. (2012). MicroRNA-277 modulates the neurodegeneration caused by fragile X premutation rCGG repeats. *PLoS Genetics*, *8*(5), e1002681.

Taneja, K. L., McCurrach, M., Schalling, M., Housman, D., & Singer, R. H. (1995). Foci of trinucleotide repeat transcripts in nuclei of myotonic dystrophy cells and tissues. *The Journal of Cell Biology*, *128*(6), 995–1002.

Tao, Z., Wang, H., Xia, Q., Li, K., Jiang, X., Xu, G., et al. (2015). Nucleolar stress and impaired stress granule formation contribute to C9orf72 RAN translation-induced cytotoxicity. *Human Molecular Genetics*, *24*(9), 2426–2441.

Tassone, F. (2004). Intranuclear inclusions in neural cells with premutation alleles in fragile X associated tremor/ataxia syndrome. *Journal of Medical Genetics*, *41*(4), e43.

Tassone, F., Hagerman, R. J., Taylor, A. K., Gane, L. W., Godfrey, T. E., & Hagerman, P. J. (2000). Elevated levels of FMR1 mRNA in carrier males: A new mechanism of involvement in the fragile-X syndrome. *American Journal of Human Genetics, 66*(1), 6–15.

Teive, H. A., Munhoz, R. P., Arruda, W. O., Raskin, S., Werneck, L. C., & Ashizawa, T. (2011). Spinocerebellar ataxia type 10—A review. *Parkinsonism & Related Disorders, 17*(9), 655–661.

Thornton, C. A., Griggs, R. C., & Moxley, R. T., III. (1994). Myotonic dystrophy with no trinucleotide repeat expansion. *Annals of Neurology, 35*(3), 269–272.

Timchenko, L. T., Miller, J. W., Timchenko, N. A., DeVore, D. R., Datar, K. V., Lin, L., et al. (1996a). Identification of a (CUG)*n* triplet repeat RNA-binding protein and its expression in myotonic dystrophy. *Nucleic Acids Research, 24*(22), 4407–4414.

Timchenko, L. T., Timchenko, N. A., Caskey, C. T., & Roberts, R. (1996b). Novel proteins with binding specificity for DNA CTG repeats and RNA CUG repeats: Implications for myotonic dystrophy. *Human Molecular Genetics, 5*(1), 115–121.

Todd, P. K., Oh, S. Y., Krans, A., He, F., Sellier, C., Frazer, M., et al. (2013). CGG repeat-associated translation mediates neurodegeneration in fragile X tremor ataxia syndrome. *Neuron, 78*(3), 440–455.

Todd, T. W., & Lim, J. (2013). Aggregation formation in the polyglutamine diseases: Protection at a cost? *Molecules and Cells, 36*(3), 185–194.

Todd, T. W., & Petrucelli, L. (2016). Insights into the pathogenic mechanisms of Chromosome 9 open reading frame 72 (C9orf72) repeat expansions. *Journal of Neurochemistry, 138*(Suppl 1), 145–162.

Tran, H., Almeida, S., Moore, J., Gendron, T. F., Chalasani, U., Lu, Y., et al. (2015). Differential toxicity of nuclear RNA foci versus dipeptide repeat proteins in a *Drosophila* model of C9ORF72 FTD/ALS. *Neuron, 87*(6), 1207–1214.

Trottier, Y., Lutz, Y., Stevanin, G., Imbert, G., Devys, D., Cancel, G., et al. (1995). Polyglutamine expansion as a pathological epitope in Huntington's disease and four dominant cerebellar ataxias. *Nature, 378*(6555), 403–406.

Tsoi, H., & Chan, H. Y. (2013). Expression of expanded CAG transcripts triggers nucleolar stress in Huntington's disease. *Cerebellum, 12*(3), 310–312.

Tsoi, H., & Chan, H. Y. (2014). Roles of the nucleolus in the CAG RNA-mediated toxicity. *Biochimica et Biophysica Acta, 1842*(6), 779–784.

Tsoi, H., Lau, C. K., Lau, K. F., & Chan, H. Y. (2011). Perturbation of U2AF65/NXF1-mediated RNA nuclear export enhances RNA toxicity in polyQ diseases. *Human Molecular Genetics, 20*(19), 3787–3797.

Tsoi, H., Lau, T. C., Tsang, S. Y., Lau, K. F., & Chan, H. Y. (2012). CAG expansion induces nucleolar stress in polyglutamine diseases. *Proceedings of the National Academy of Sciences of the United States of America, 109*(33), 13428–13433.

Udd, B., Krahe, R., Wallgren-Pettersson, C., Falck, B., & Kalimo, H. (1997). Proximal myotonic dystrophy—A family with autosomal dominant muscular dystrophy, cataracts, hearing loss and hypogonadism: Heterogeneity of proximal myotonic syndromes? *Neuromuscular Disorders: NMD, 7*(4), 217–228.

Vedana, G., Villarreal, G., Jr., & Jun, A. S. (2016). Fuchs endothelial corneal dystrophy: Current perspectives. *Clinical Ophthalmology, 10*, 321–330.

Verkerk, A. J., Pieretti, M., Sutcliffe, J. S., Fu, Y. H., Kuhl, D. P., Pizzuti, A., et al. (1991). Identification of a gene (FMR-1) containing a CGG repeat coincident with a breakpoint cluster region exhibiting length variation in fragile X syndrome. *Cell, 65*(5), 905–914.

Wakamiya, M., Matsuura, T., Liu, Y., Schuster, G. C., Gao, R., Xu, W., et al. (2006). The role of ataxin 10 in the pathogenesis of spinocerebellar ataxia type 10. *Neurology, 67*(4), 607–613.

Wang, J., Pegoraro, E., Menegazzo, E., Gennarelli, M., Hoop, R. C., Angelini, C., et al. (1995). Myotonic dystrophy: Evidence for a possible dominant-negative RNA mutation. *Human Molecular Genetics, 4*(4), 599–606.

Wang, K., McFarland, K. N., Liu, J., Zeng, D., Landrian, I., Xia, G., et al. (2015). Spinocerebellar ataxia type 10 in Chinese Han. *Neurology Genetics, 1*(3), e26.

Wang, L. C., Chen, K. Y., Pan, H., Wu, C. C., Chen, P. H., Liao, Y. T., et al. (2011). Muscleblind participates in RNA toxicity of expanded CAG and CUG repeats in *Caenorhabditis elegans*. *Cellular and Molecular Life Sciences: CMLS, 68*(7), 1255–1267.

Wen, X., Tan, W., Westergard, T., Krishnamurthy, K., Markandaiah, S. S., Shi, Y., et al. (2014). Antisense proline-arginine RAN dipeptides linked to C9ORF72-ALS/FTD form toxic nuclear aggregates that initiate in vitro and in vivo neuronal death. *Neuron, 84*(6), 1213–1225.

Westergard, T., Jensen, B. K., Wen, X., Cai, J., Kropf, E., Iacovitti, L., et al. (2016). Cell-to-cell transmission of dipeptide repeat proteins linked to C9orf72-ALS/FTD. *Cell Reports, 17*(3), 645–652.

White, M., Xia, G., Gao, R., Wakamiya, M., Sarkar, P. S., McFarland, K., et al. (2012). Transgenic mice with SCA10 pentanucleotide repeats show motor phenotype and susceptibility to seizure: A toxic RNA gain-of-function model. *Journal of Neuroscience Research, 90*(3), 706–714.

White, M. C., Gao, R., Xu, W., Mandal, S. M., Lim, J. G., Hazra, T. K., et al. (2010). Inactivation of hnRNP K by expanded intronic AUUCU repeat induces apoptosis via translocation of PKCdelta to mitochondria in spinocerebellar ataxia 10. *PLoS Genetics, 6*(6), e1000984.

Wieben, E. D., Aleff, R. A., Tang, X., Butz, M. L., Kalari, K. R., Highsmith, E. W., et al. (2017). Trinucleotide repeat expansion in the transcription factor 4 (TCF4) gene leads to widespread mRNA splicing changes in Fuchs' endothelial corneal dystrophy. *Investigative Ophthalmology & Visual Science, 58*(1), 343–352.

Wieben, E. D., Aleff, R. A., Tosakulwong, N., Butz, M. L., Highsmith, W. E., Edwards, A. O., et al. (2012). A common trinucleotide repeat expansion within the transcription factor 4 (TCF4, E2-2) gene predicts Fuchs corneal dystrophy. *PLoS ONE, 7*(11), e49083.

Wilburn, B., Rudnicki, D. D., Zhao, J., Weitz, T. M., Cheng, Y., Gu, X., et al. (2011). An antisense CAG repeat transcript at JPH3 locus mediates expanded polyglutamine protein toxicity in Huntington's disease-like 2 mice. *Neuron, 70*(3), 427–440.

Willemsen, R. (2003). The FMR1 CGG repeat mouse displays ubiquitin-positive intranuclear neuronal inclusions; implications for the cerebellar tremor/ataxia syndrome. *Human Molecular Genetics, 12*(9), 949–959.

Woerner, A. C., Frottin, F., Hornburg, D., Feng, L. R., Meissner, F., Patra, M., et al. (2015). Cytoplasmic protein aggregates interfere with nucleo-cytoplasmic transport of protein and RNA. *Science, 351*(6269), 173–176.

Xiang, S., Kato, M., Wu, L. C., Lin, Y., Ding, M., Zhang, Y., et al. (2015). The LC domain of hnRNPA2 adopts similar conformations in hydrogel polymers, liquid-like droplets, and nuclei. *Cell, 163*(4), 829–839.

Xu, Z., Poidevin, M., Li, X., Li, Y., Shu, L., Nelson, D. L., et al. (2013). Expanded GGGGCC repeat RNA associated with amyotrophic lateral sclerosis and frontotemporal dementia causes neurodegeneration. *Proceedings of the National Academy of Sciences of the United States of America, 110*(19), 7778–7783.

Yamakawa, M., Ito, D., Honda, T., Kubo, K., Noda, M., Nakajima, K., et al. (2015). Characterization of the dipeptide repeat protein in the molecular pathogenesis of c9FTD/ALS. *Human Molecular Genetics, 24*(6), 1630–1645.

Yang, D., Abdallah, A., Li, Z., Lu, Y., Almeida, S., & Gao, F. B. (2015). FTD/ALS-associated poly(GR) protein impairs the Notch pathway and is recruited by poly (GA) into cytoplasmic inclusions. *Acta Neuropathologica, 130*(4), 525–535.

Yu, Z., Teng, X., & Bonini, N. M. (2011). Triplet repeat-derived siRNAs enhance RNA-mediated toxicity in a *Drosophila* model for myotonic dystrophy. *PLoS Genetics, 7*(3), e1001340.

Yuan, Y., Compton, S. A., Sobczak, K., Stenberg, M. G., Thornton, C. A., Griffith, J. D., et al. (2007). Muscleblind-like 1 interacts with RNA hairpins in splicing target and pathogenic RNAs. *Nucleic Acids Research, 35*(16), 5474–5486.

Zeng, S., Zeng, J., He, M., Zeng, X., Zhou, Y., Liu, Z., et al. (2016). Genetic and clinical analysis of spinocerebellar ataxia type 36 in Mainland China. *Clinical Genetics, 90*(2), 141–148.

Zhang, K., Donnelly, C. J., Haeusler, A. R., Grima, J. C., Machamer, J. B., Steinwald, P., et al. (2015). The C9orf72 repeat expansion disrupts nucleocytoplasmic transport. *Nature, 525*(7567), 56–61.

Zhang, Y. J., Gendron, T. F., Grima, J. C., Sasaguri, H., Jansen-West, K., Xu, Y. F., et al. (2016). C9ORF72 poly (GA) aggregates sequester and impair HR23 and nucleocytoplasmic transport proteins. *Nature Neuroscience, 19*(5), 668–677.

Zhang, Y. J., Jansen-West, K., Xu, Y. F., Gendron, T. F., Bieniek, K. F., Lin, W. L., et al. (2014). Aggregation-prone c9FTD/ALS poly(GA) RAN-translated proteins cause neurotoxicity by inducing ER stress. *Acta Neuropathologica, 128*(4), 505–524.

Zhu, L., & Brangwynne, C. P. (2015). Nuclear bodies: The emerging biophysics of nucleoplasmic phases. *Current Opinion in Cell Biology, 34*, 23–30.

Zongaro, S., Hukema, R., D'Antoni, S., Davidovic, L., Barbry, P., Catania, M. V., et al. (2013). The 3′ UTR of FMR1 mRNA is a target of miR-101, miR-129-5p and miR-221: Implications for the molecular pathology of FXTAS at the synapse. *Human Molecular Genetics, 22*(10), 1971–1982.

Zu, T., Gibbens, B., Doty, N. S., Gomes-Pereira, M., Huguet, A., Stone, M. D., et al. (2011). Non-ATG-initiated translation directed by microsatellite expansions. *Proceedings of the National Academy of Sciences of the United States of America, 108*(1), 260–265.

Zu, T., Liu, Y., Banez-Coronel, M., Reid, T., Pletnikova, O., Lewis, J., et al. (2013). RAN proteins and RNA foci from antisense transcripts in C9ORF72 ALS and frontotemporal dementia. *Proceedings of the National Academy of Sciences of the United States of America, 110*(51), E4968–4977.

CHAPTER

16

Neuroinflammation in Age-Related Neurodegenerative Diseases

Kathryn P. MacPherson, Maria E. de Sousa Rodrigues, Amarallys F. Cintron and Malú G. Tansey

Emory University School of Medicine, Atlanta, GA, United States

OUTLINE

PERIPHERAL AND BRAIN-IMMUNE MEDIATORS IN BRAIN HEALTH AND DISEASE

Once thought to be immunologically inert, the central nervous system (CNS) is now recognized as a highly regulated and immune-competent system. It is now accepted that the CNS parenchyma is immune-specialized and has a unique relationship with the peripheral immune system. Initiation of antigen-driven proinflammatory adaptive immune responses is tightly regulated, and the kinetics of innate immune responses are modulated in health and disease. The blood–brain barrier (BBB) acts as a physical barrier that separates the brain from cells and solutes that mediate peripheral immune responses; however, this does not prevent interaction with the peripheral immune system or immune cell extravasation into the CNS (Ransohoff & Engelhardt, 2012). Under homeostatic conditions,

The Molecular and Cellular Basis of Neurodegenerative Diseases
DOI: https://doi.org/10.1016/B978-0-12-811304-2.00016-X

inflammatory responses in the brain are influenced by BBB permeability, cytokine and chemokine signaling, neuron–microglia cross talk, microglial activation, regulation of T-cell responses, and entry of peripheral immune cells into the CNS (Carson, Doose, Melchior, Schmid, & Ploix, 2006). Together, this close regulation allows for immune surveillance innate and adaptive responses to invasion of foreign pathogens that can be quickly resolved while minimizing negative outcomes on neuronal function and survival.

Evidence supports a role for inflammatory mechanisms in the regulation of brain function and health, including an especially tight relationship with neuroinflammatory processes. Increasing evidence supports a role of neuroinflammation in the regulation of synaptic organization and development of mature circuits (Schafer et al., 2012). Cellular and signaling contributions of the peripheral immune system, including macrophages and neutrophils of the innate arm and T cells and B cells of adaptive arm, have increasingly been shown to play a role not only in autoimmune disease of the nervous system in conjunction with brain-resident microglia but also in other neuroinflammatory diseases (Ransohoff, Schafer, Vincent, Blachere, & Bar-Or, 2015). How these populations of peripheral immune cells work with or against the brain-resident immune cells to influence health and risk or progression of disease is a topic of intense investigation. Conditions associated with low-grade chronic peripheral inflammation, such as metabolic syndrome, insulin resistance, diabetes, and obesity, commonly associated with unhealthy diet, have all been linked to increased risk for AD (Ojo & Brooke, 2015; Walker & Harrison, 2015). Peripheral inflammation may accelerate neurodegenerative disease by increasing neuroinflammation, altering BBB integrity, and altering peripheral immune cell trafficking to the CNS (Gonzalez, Elgueta, Montoya, & Pacheco, 2014). Knowledge of how peripheral and CNS-resident immune cells contribute to brain and BBB health and dysfunction is an essential part of understanding their role in neuroinflammatory pathology and neurodegenerative diseases and will reveal opportunities for development of novel immunomodulatory interventions to prevent, delay, or arrest these diseases.

Contributors to the CNS Inflammatory Environment

Microglia

Once thought to merely provide the "glue" to support neuronal network, glial cells are now known to be an integral part of maintaining a healthy brain. New studies have identified the origin of the majority of microglia, originally thought to derive from monocytes, to be the embryonic yolk sac (Ginhoux, Lim, Hoeffel, Low, & Huber, 2013; Tay, Hagemeyer, & Prinz, 2016). However, newly developed techniques and models have led investigators to conclude that non–yolk sac contributions to the adult microglial pool cannot be ruled out (reviewed in Tay et al., 2016). Yolk sac–derived microglia populate the brain early in embryogenesis, dependent on the development of the circulatory system, and they persist in the brain into adulthood (Ginhoux et al., 2010). In contrast, macrophages present early in development, also initially derived from the yolk sac during embryonic development, are no longer found in blood in adulthood, having been replaced by bone-marrow derived macrophages (Ginhoux et al., 2010). Development of yolk sac microglia is dependent on colony stimulating factor-1 receptor (CSF-1R) expression and its ligand interleukin-34 (IL-34). CSF-1R was shown to play an important role in microglia homeostasis and is highly expressed by cortical and hippocampal neurons, but less so in the brainstem and cerebellum (Greter et al., 2012). Thus, environmental cues from the developing brain contribute to microglial longevity as well as region-specific microglial maintenance.

Microglia serve as the primary mediator of immunological defense in the brain, capable of cytokine and chemokine production, phagocytosis, and antigen presentation. Microglia are known to play diverse roles in maintaining brain health and function. Microglia have been shown to rapidly respond to localized brain injury, specifically to released ATP (Davalos et al., 2005), a response that is impaired following systemic inflammation (Gyoneva et al., 2014). Recently a role for microglia in the formation of mature neuronal circuit organization through complement-directed synaptic pruning has been reported (Schafer et al., 2012). These same mechanisms of complement-directed phagocytosis have also been implicated in Alzheimer's disease (AD) pathogenesis. Increased levels of complement proteins have recently been shown to precede early synapse loss, to correlate with levels of soluble amyloid beta (Aβ) oligomers, and to be necessary for Aβ oligomer-induced synaptic loss and Aβ oligomer-induced long-term potentiation impairment (Hong et al., 2016). The complement protein C3 and its receptor CR3 have been shown to play a role in microglial phagocytosis of Aβ fibrils (Fu et al., 2012). Increased levels of C1qa and CD68 implicate microglia as the main source of complement and the mediators of synaptic loss in a manner similar to that of developmental synaptic pruning, which could be mediated via cytokine signaling or microglial activation (Hong et al., 2016). Complement component C1q has also been shown to activate microglia to produce soluble tumor necrosis factor (TNF), a proinflammatory cytokine that can activate pro-apoptotic signaling pathways (Farber et al., 2009).

Within the CNS, microglia are a major source of cytokines such as TNF, IL-1, IL-6, and IFN-γ, as well as a source of neurotropic factors, including NGF and BDNF. The nature of the activating stimulus, the specific microglia response, as well as the chronic duration of activation have implicated cytokine signaling as a potential mechanism of neurodegenerative disease (Smith, Das, Ray, & Banik, 2012). Regulation of microglia activation through neuronal "off" signals such as, TGF-β, CD22, CD200, and CX3CL1, and "on" signals, such as TREM2 and ATP, protects the brain from potentially harmful immune responses, and evidence suggests this regulation is disrupted in neurodegenerative diseases (Biber, Neumann, Inoue, & Boddeke, 2007). CD200, expressed on the neuronal membrane, and its receptor CD200R, expressed on microglia, have been shown to play an inhibitory function in brain inflammation: blocking antibodies worsened neuroinflammatory disease models and macrophage infiltration (Biber et al., 2007). CD200R is also expressed on T cells (Rijkers et al., 2008), suggesting T cells infiltration into the brain may also be regulated via neuronal input. Fractalkine (CX3CL1), expressed in neurons, binds to microglial CX3CR1 and has been shown to play a role in regulation of events mediated by microglial–neuronal interaction throughout life, including neurotoxicity and cytokine production (reviewed in Limatola & Ransohoff, 2014). Microglia activation and IL-6 and TNF levels were increased with the loss of CX3CR1 and further increased with the addition of AD-like pathology. While CX3CR1 loss did not alter Aβ accumulation, it was shown to downregulate fractalkine signaling, and cognitive deficits were enhanced in an AD model with loss of CX3CR1 (Cho et al., 2011). Dysfunction of these regulatory mechanisms of microglial activation have been linked to impaired phagocytosis of Aβ and increased risk for development of AD, as reviewed in Meyer-Luehmann & Prinz (2015). Such regulatory mechanisms may suggest potential therapeutic targets as our understanding of the microglial contribution to neurodegenerative disease increases.

Macrophages

Macrophages are bone marrow–derived monocytes that traffic to the brain and differentiate into antigen-presenting macrophages (Hickey & Kimura, 1988). Recently, subtypes of macrophages have been identified in distinct locations of the brain; perivascular macrophages, meningeal macrophages, and choroid plexus macrophages (Prinz, Priller, Sisodia, & Ransohoff, 2011). These macrophage populations can sample the contents of the cerebrospinal fluid (CSF) and present antigen via major histocompatibility complex II (MHC II), to contribute to immune surveillance of the brain, and play an important role in response to neuroinflammatory disease, such as AD (Meyer-Luehmann & Prinz, 2015). Perivascular CXC3R1$^+$ macrophages extend dendritic arms into CNS vessel lumen, to present CNS antigens to circulating T cells under homeostatic conditions and with increasing frequency with neuroinflammatory disease (Barkauskas et al., 2013). Understanding the role of these distinct populations and their role in mediating immune regulation via their cross talk with microglia may be a key to understanding neuroinflammation disease. However, studying these populations of macrophages in the brain as well as macrophages that have trafficked into the parenchyma is challenging. In the periphery, macrophages can be identified by the presence of markers CD11b, CD45, and lack of Ly6G; however, these same markers also identify microglia in the brain. To distinguish between microglia and peripheral monocytes that have infiltrated the brain and differentiated into macrophages, markers CCR2 and Ly6C have been used to identify macrophages, while Cx3CR1 has been used to identify microglia (Saederup et al., 2010). Importantly, the traditional markers for microglia (CD45, Iba1, and CD68) will also label macrophage populations. The markers of CCR2 and Ly6C, however, are downregulated by macrophages a short while after residing in inflamed tissue; thus, these markers (CCR2 and Ly6C) are able to identify recently infiltrated macrophage populations (Greter, Lelios, & Croxford, 2015). A majority of the peripheral macrophages that infiltrate the CNS will be CD45^{high} as compared to homeostatic/resting CD45^{low} microglia. Using relative levels of CD45 expression can thus work as a way to distinguish between macrophages and microglia among CD11b$^+$ populations. However, upon activation, microglia are able to express CD45 at higher levels (Sedgwick et al., 1991). Thus, CD45 is not a specific marker that can distinguish between microglia and macrophage populations.

Studies that assess RNA at the cellular level in conjunction with lineage tracking are viewed as key towards advancing the field, because they can reveal novel markers to distinguish between microglia and macrophages in the CNS. A recent study by Goldmann and colleagues assessed microglia, subdural meninges macrophages (1ba1$^+$, Cx3CR1$^+$, and close to fibroblasts expressing ER-TR7), perivascular macrophages (Iba1$^+$, CX3CR1$^+$, and close to the abluminal side of CD31$^+$ endothelial cells), and choroid plexus macrophages (in the stroma and epithelial layer of the choroid plexus). They found evidence suggesting that not all brain macrophage populations are derived from bone marrow populations. Fate-mapping and chimeric studies revealed that microglia, perivascular macrophages, and meningeal macrophages are all embryonic derived cells that under homeostatic conditions are stable populations not replaced by bone marrow–derived macrophages after birth; in contrast, choroid plexus macrophages are short-lived and repopulated by bone marrow–derived macrophages (Goldmann et al., 2016). In addition, unbiased quantitative single-cell RNA sequencing revealed that although perivascular macrophages are transcriptionally similar to microglia, they are distinguishable by *Mrc1* (encoding CD206), CD36 expression, and higher *Ptprc* (encoding CD45) expression, whereas microglia are distinguishable by higher *P2ry12* and *Hexb* gene expression (Goldmann et al., 2016). These distinct populations were also shown to display unique motility: meningeal macrophages were more ameboid in shape and—overall as a population—less mobile than microglia, whereas perivascular macrophages rarely moved their cell body and constantly extended and retracted their protrusions along the blood vessel wall (Goldmann et al., 2016). Together, the evidence from this study refines our understanding of the origin of these macrophage populations and their potential role in neuroinflammatory disease.

As noted above, only perivascular macrophages are replaced by peripheral blood monocytes under homeostatic conditions. This is also supported by studies of chimeric animals obtained by parabiosis, a model, which does not compromise the BBB through lethal irradiation, that have shown microglia are maintained independently of bone marrow–derived progenitors, even in neuroinflammatory conditions (Ajami, Bennett, Krieger, Tetzlaff, & Rossi, 2007). However, parabiosis generally yields lower chimeric levels than irradiation. A model independent of irradiation and parabiosis, utilizing busulfan to ablate bone marrow cells while preserving BBB integrity, found minimal myeloid cell recruitment in homeostatic conditions, but this recruitment was somewhat enhanced during neurodegeneration (Kierdorf, Katzmarski, Haas, & Prinz, 2013). Recent studies suggest that, under conditions of inflammation, macrophages can traffic into the parenchyma and play a necessary role in plaque clearance. In an AD preclinical mouse model, microglia depletion induced monocyte infiltration into the brain, and over time these cells adopted feature similar to microglia. However, unexpectedly, they did not alter amyloid plaque deposition (Varvel et al., 2015), suggesting that macrophages alone are not sufficient to clear plaques. Indeed, macrophages may be contributing to progression of pathology. Specifically, TREM2-positive, $CD45^{high}/Ly6C^{+}$ myeloid cells, but not P2RY12 positive microglia, have been shown to contribute to AD pathogenesis (Jay et al., 2015). Together these studies highlight the importance of elucidating the role of peripheral macrophages in AD.

T cells

T cells of the adaptive immune system are increasingly shown to play a role in regulation of brain immunity and mediating neuroinflammatory disease (Kipnis, 2016). CNS-specific $CD4^{+}$ T cells have been shown to support cognition and brain health, whereas the choroid plexus plays an important role in the accumulation and recruitment of T cells as the neuroimmunological interface between the brain and peripheral immune system (Baruch & Schwartz, 2013). In the healthy brain, T cells patrol the CSF and become activated once they encounter their specific antigen presented to them by brain antigen-presenting cells (APCs), such as microglia, macrophages, and even astrocytes under certain conditions (Engelhardt & Ransohoff, 2005). Once active, T cells are able to enter the brain parenchyma, and other inflamed tissues, and carry out their effector functions (Engelhardt & Ransohoff, 2012). Following termination of the inflammatory event, the effector T-cell population contracts, and memory T cells remain that are able to rapidly mount recall responses to antigens. Effector memory T cells migrate to tissue and display immediate effector functions, whereas central memory T cells home to lymphoid organs and proliferate, increasing the effector population (Sallusto, Geginat, & Lanzavecchia, 2004).

Naive $CD4^{+}$ T cells become activated when encountering an antigen that is specific to their T-cell receptor and presented on MHC II expressed by APCs and upon receiving costimulatory input. Once activated, naive T cells will differentiate into effector T cells influenced by the inflammatory and cytokine milieu. There are many subsets of $CD4^+$ T cells that each have distinct effector functions. Th1 cells are IFN-γ-producing T cells that play an essential role in phagocytic macrophage activation. Th2 cells produce IL-10, among other cytokines, and play a role in terminating immune responses. Th17 cells lead to the induction of proinflammatory cytokines and chemokines at sites of inflammation. Regulatory $CD4^+$ T cells (Treg) also produce IL-10 and play a role in suppression of the immune response (Luckheeram, Zhou, Verma, & Xia, 2012). Tregs, which suppress effector immune responses, have been shown to be elevated in elderly populations. Although not further elevated with AD pathology, Tregs displayed increased suppressive activity (Rosenkranz et al., 2007). Removal of the Treg suppressive effect on the immune system via transient depletion or inhibition of Tregs in an AD mouse model has been shown to increase leukocyte trafficking at the choroid plexus, increase brain $CD45^{high}CD11b^+$ and Treg populations, as well as decrease plaque burden, lower proinflammatory cytokine mRNA expression, and improve on cognitive function (Baruch et al., 2015).

Naive $CD8^+$ T cells are activated when encountering their T-cell receptor–specific antigen presented on MHC I, expressed on essentially all nucleated cells, expanding into populations of effector $CD8^+$ T cells producing proinflammatory cytokines such as IFN-γ and TNF. Effector $CD8^+$ T cells are cytotoxic and capable of inducing apoptosis, through release of granule enzymes and the pore-forming molecule perforin, and upregulating programmed cell death triggers (CD95L) in target cells (Harty, Tvinnereim, & White, 2000). MHC I and $CD8^+$ T cells have been shown to play a role in brain function and neurodegenerative diseases (reviewed in Cebrian et al., 2014). MHC I, expressed by neurons and present at synapses, limits synapse density by forming a macromolecular complex with the insulin receptor, inhibiting synapse-promoting insulin receptor signaling (Dixon-Salazar et al., 2014). Expression of neuronal MHC I is increased in postmortem human samples in regions with neurodegenerative disease pathology and in murine primary neuronal cultures in response to IFN-γ and activated microglia (Cebrian et al., 2014). In experimental autoimmune encephalomyelitis (EAE; see below) models, $CD8^+$ T cells become differentiated via MHC I presentation, the expression of which is induced by IFN-γ, and they directly injure neuronal axons through granzyme-B-dependent mechanisms (Sauer, Schmalstieg, & Howe, 2013). Together, this evidence supports a role of MHC I and $CD8^+$ T cell–induced changes in neuronal function and death that may play a role in mediating cognitive symptoms of neurodegenerative disease. These roles may be better revealed with further investigation.

Work in the EAE model, a commonly used mouse model of multiple sclerosis, has begun to reveal the role T cells play in neurodegenerative diseases and the mechanisms of T-cell recruitment (Engelhardt & Ransohoff, 2005). Inflammatory T cells have been shown to initially enter at the choroid plexus in a CCR6-dependent manner. However, following initiation of inflammation in the EAE model, a second wave to trafficking T cells are also able to enter in a CCR6-independent manner (Axtell & Steinman, 2009). Thus, understanding the mechanisms of trafficking, as well as phenotypic changes in trafficking leukocytes, under both homeostatic and inflammatory conditions, is crucial to understanding how the peripheral immune system contributes to neurodegenerative disease.

Astrocytes

Astrocytes are a heterogeneous population of glia cells, derived from radial glia cells in the ventricular zone during development in a regionally and temporally distinct manner (Bayraktar, Fuentealba, Alvarez-Buylla, & Rowitch, 2015).

Astrocytes are the most abundant glial cell in the brain and participate in many aspects of brain health, including modulation of synaptic formation and removal, synaptic function, and neuronal survival (Allen, 2014). Astrocytes also play a supportive role for neuroinflammatory responses in the CNS and divide in response to inflammatory stimuli, such as TNF (Barna, Estes, Jacobs, Hudson, & Ransohoff, 1990). Astrocytes not only form an integral part the BBB that blocks potential pathogens from entering the brain (Goldstein, 1988) but also function as APCs (Fontana, Fierz, & Wekerle, 1984). APCs, in both the brain and periphery, are able to communicate with antigen-specific T cells of the adaptive immune system through peptide presentation on the MHC. Astrocyte function can be regulated through immunomodulatory signaling produced by astrocytes in an autocrine manner as well as by other cells in the brain. Astrocytes are able to produce many immune factors (CXCL12, CCL2, CD40, IL15, CCL8, and CXCL1) in response to stimulation, and they can upregulate receptors to immune signaling molecules such as IL-1β, IL-6, INF-γ, TNF, TGF-β, and lipopolysaccharide (LPS). These factors are known to modulate astrocytes with regard to cell morphology, cell growth, cell proliferation, glutamate regulation, antigen presentation, and pro- and antiinflammatory cytokine production (Sofroniew, 2014).

Recent work has shown that activation of astrocytes leads to a decrease in their protective function and can also lead to neuronal death by soluble factors. LUHMES-IMA (immortalized postmitotic human neuron cell line–immortalized mouse astrocyte) co-cultures protected LUHMES neurons from MPP+ and NO-induced death. However, pretreatment for 6 days with either TNF or IL-1β or both (but not IFN-γ), which selectively activated IMA but not LUHMES cells, led to significant neuronal degeneration without the need for additional toxicants (Efremova et al., 2017). Similar responses were seen with cultures of human astrocytes after 4 days of stimulation: evidence of morphological changes as well as neurodegeneration was observed, indicating that these are not cross-species effects. In addition, it was determined that stimulated IMA conditioned media (CM) induced LUHMES cell death. This was not due to increased extracellular glutamate, suggesting that other soluble factors are responsible for neuronal death in response to astrocyte activation (Efremova et al., 2017).

Astrocyte-enriched human cultures treated with IFN-γ, IL-1β, TNF-α, but not TGF-β showed increased MHC II and MHC I as well as the cellular adhesion molecules, intercellular adhesion molecule 1 (ICAM) and vascular cell adhesion protein 1 (VCAM) (Aloisi et al., 1995). This upregulation in MHC enables astrocytes to present antigen to $CD4^+$ and $CD8^+$ T cells and is known to play a role in models of neurodegeneration (Cebrian et al., 2014; Engelhardt & Ransohoff, 2012). Astrocyte MHC II expression is downregulated through direct functional neuronal contact and acts to prevent the astrocytes from inducing T-cell activation (Tontsch & Rott, 1993). Upregulation in ICAM and VCAM suggest that trafficking of peripheral immune cells, such as macrophages and T cells, into the parenchyma may be facilitated during astrocyte stimulation. Astrocyte–macrophage cross talk alters astrocyte proliferation and inflammatory responses. CM from proinflammatory M1 macrophage cultures, but not antiinflammatory homeostatic M2 macrophage cultures, induced astrocyte proliferation (Haan, Zhu, Wang, Wei, & Song, 2015). History of exposure to either pro- or antiinflammatory peripheral macrophages can alter the secretory profile from astrocytes and can lead to either an enhanced or diminished response (Haan et al., 2015). CM from astrocytes incubated with M2 CM decreased proliferation and TNF production in M1 macrophages, whereas CM from astrocytes incubated with M1 CM increased TNF production in M1 macrophages (Haan et al., 2015). These results suggest that history of exposure can be a contributing factor in how astrocytes contribute to the local inflammatory environment and may play an important role in chronic neuroinflammation. In many human neuroinflammatory diseases, peripheral inflammation is present that may further alter the inflammatory profile of infiltrating macrophages into the CNS, and these macrophages may participate in cross talk with astrocytes and thereby contribute to chronic neuroinflammation.

Together these nonneuronal cell types mediate and regulate the inflammatory environment of the brain and contribute to overall brain health and function. When regulatory mechanisms are altered due to either acute or chronic inflammation, these cells work together to mediate neuroinflammatory disease. Understanding how each population is altered by inflammatory disease conditions can help to elucidate the broader network of immune cross talk between these cell types that potentially mediate disease. Conditions that result in chronic peripheral inflammation have been shown to increase risk for AD (Cunningham et al., 2009; Perry, Cunningham, & Holmes, 2007; Walker & Harrison, 2015) and investigation into how these conditions impact these immune cell populations as well as CNS disease states may yield strategies for therapeutic treatment.

Regulation of Neuroinflammation

As discussed in the previous section, the immune system is an integral player both in brain health and function and actively communicates with the peripheral immune system. The inflammatory response, especially microglial activation, is tightly regulated both through neuroimmune mechanisms, discussed above, as well as peripheral input: regulation of peripheral immune cell traffic at the BBB, surveillance at the deep cervical lymph nodes (DCLNs), as well as regulating of inflammatory signaling from inflamed tissue. To avoid immunodeficiency and chronic systemic inflammation, precise regulation of inflammatory responses is achieved in part through brain-derived immunoregulatory output via the autonomic nervous system and the vagus nerve, which serves as a bidirectional communication between the peripheral immune system and the brain (Pavlov & Tracey, 2015). Stimulation of the vagus nerve above the celiac ganglion, where neural cell bodies project axons to the splenic nerve, innervates the spleen and inhibits cytokine release. This inflammatory reflex is dependent on acetylcholine-producing T cells residing in the spleen. The released acetylcholine inhibits cytokine-producing macrophages via $\alpha7$ nicotinic acetylcholine receptors, limiting systemic levels of cytokines such as TNF (Rosas-Ballina et al., 2011). Vagal input to the brain as well as cytokine signaling across the BBB and leukocyte trafficking through the lymphatic systems all play an integral role in the regulation of neuroinflammation.

Lymphatics and DCLNs

To provide immunological support for the CNS, the peripheral immune system must be able to survey for foreign antigens, and this occurs via cellular surveillance of the CSF and perivascular spaces and through antigen drainage to the DCLNs. Brain-derived antigens drain via the CSF through the nasal mucosa and into the DCLNs. Thus, the CSF is thought to act as the functional equivalent of lymph (Ransohoff & Engelhardt, 2012). To show that specific antibody-forming cells expand in the lymphoid tissues following inoculation in the brain, Walter et al. used an fluorescein isothiocyanate (FITC)-labeled antigen to track antigen movement as well and anti-FITC antibody to allow cell localization. Two hours after antigen injection into the subarachnoid space, FITC is observed in the nasal mucosa, subcapsular sinus, and DCLNs. Four days after injection of antigen, specific antibody-forming cells were found in the DCLNs, fewer in superficial lymph nodes, but none within axillary lymph nodes, which peaked 6–8 days after injection (Walter, Valera, Takahashi, Matsuno, & Ushiki, 2006).

Solutes from the interstitial fluid are eliminated from the brain along basement membranes, through the wall of capillaries and arteries. Dextran (3-kDa), injected into the striatum, diffused through the parenchyma, and was located within blood vessels and capillary walls after 5 minutes, and 24 hours later within phagocytic myeloid cells. Fluorospheres are also phagocytosed by perivascular macrophages located in the perivascular space. Intraventricularly injected 0.02 and 1 μm fluorospheres, but not dextran, were found within cervical lymph nodes (Carare et al., 2008). While it is clear from these data that solutes are able to drain from the brain to the DCLNs, the authors suggest there is little evidence to support the drainage of cells along this perivascular route. However, evidence supports that cells can drain via the subarachnoid space to the nasal mucosa and cervical lymph nodes, which have been shown to play a crucial role in mediating neuroinflammatory disease (reviewed in Laman & Weller, 2013).

When considering what drains to the DCLNs, the inflammatory state of the CNS may be important, as context matters. Although microglia are highly motile cells that can traffic to the site of injury within the brain (Davalos et al., 2005; Gyoneva et al., 2014), there is no evidence that microglia leave the CNS to enter the periphery. Under homeostatic conditions, in which there is minimal peripheral immune cell traffic to the brain, there may be little cellular drainage to the DCLNs. However, under chronic inflammatory conditions, such as neurodegenerative disease, where peripheral immune cell trafficking into the CNS has been implicated, the cellular drainage into DCLNs may be increased or altered. T cells as well as myeloid cells have been found to drain to the DCLNs when injected into the parenchyma (Goldmann et al., 2006; Hatterer et al., 2006), suggesting these cell types are capable of trafficking into the brain parenchyma in response to inflammation and then trafficking to the DCLNs, where they may play a role in mediating neuroinflammatory disease. Recently, an alternative route, the lymphatic system in the brain, which drains CSF to the DCLNs and contains $CD3^+$ T cells as well as $CD11c^+$ and $B220^+$ cells, has been visualized in the rodent, localized in the dura matter at the base of the skull (Aspelund et al., 2015; Louveau et al., 2015). However, the presence of this tissue is not yet confirmed in humans. Dye placed within the parenchyma has been shown to drain to the DCLNs but not to the superficial lymph nodes (Aspelund et al., 2015). This ground-breaking finding offers a new route through which CSF as well as APCs can drain to the lymphatic system and mediate immune surveillance of the brain.

Blood–Brain Barrier

The BBB comprised tight junctions and adherens junctions between the brain endothelial cells, and its overall function is to limit the entry of plasma components, red blood cells, and leukocytes into the brain (Zlokovic, 2008). The changes in the expression level of junctional proteins, such as occludins, ZO-1, claudin-5, VE-cadherin, and β-catenin, have been implicated in increased BBB permeability and neuroinflammatory disease (Zlokovic, 2008). The outer wall of the BBB can be crossed by immune cells that then remain in the CSF-drained perivascular spaces, separated from the CNS parenchyma by the glia limitans, where immune surveillance is carried out (Engelhardt & Ransohoff, 2012). Different sets of adhesion molecules and cytokine signals are used to regulate traffic across the BBB and the glia limitans (Engelhardt, 2008). CCL2 induces recruitment of immune cells across the BBB, but induction of MMP-2 and MMP-9, upregulated by TNF, are also required to cross the glia limitans to access the parenchyma (Engelhardt & Ransohoff, 2012). TNF may be a key mediator of BBB permeability during disease, as an elevated plasma level of TNF is associated with increased risk for dementia and acceleration of AD pathology (Engelhart et al., 2004; Holmes et al., 2009). Evidence supports a role for TNF as an inflammatory mediator of BBB permeability that enables peripheral immune cells to more readily cross into the CNS (Rezai-Zadeh, Gate, & Town, 2009).

Changes in BBB transport systems have been strongly implicated in cognitive dysfunction and neurodegenerative disease (reviewed in Erickson & Banks, 2013; Zlokovic, 2008). While the choroid plexus, a highly permeable region of the BBB, plays a role in mediating neuroinflammatory disease. As discussed in the sections above, recent studies have shown that the population of macrophages that reside in the choroid plexus are derived from blood populations (Goldmann et al., 2016). This alternative origin may assist these more permeable regions in combating pathogens and/or make these regions more susceptible to neuroinflammation. Dysfunction of the choroid plexus plays a role in mediating AD-like pathology. One month after transplantation of choroid plexus epithelial cells, into the brain of 9-month-old APP/PS1 mice, Aβ accumulation, astrogliosis, and cognitive deterioration were significantly attenuated (Bolos et al., 2014). The authors suggest a role of secretion of Aβ degradation enzymes to account for this improvement in AD-like pathology (Bolos et al., 2014); an additional mechanism may be the renewed regulation of immune cell trafficking at the choroid plexus. Indeed, $CD4^{+}$ T cells have been shown to play a role at the choroid plexus in mediating cognition and AD-like pathology (Baruch et al., 2015). Changes in leukocyte trafficking across the BBB may have disease-mediating consequences as well as suggest potential therapeutic targets to modulate disease progression.

Cytokine signaling—TNF as a model

The cytokine and chemokine signals that mediate recruitment of peripheral immune to the CNS have become popular targets of investigation. New models and pharmacological tools based on these signals provide us with new ways to study modulation of immune cell recruitment to the CNS to investigate how peripheral immune populations impact brain health and disease. TNF is produced by both microglia and macrophages and is a key modulator of proinflammatory cytokine cascades and immune cell activation (MacEwan, 2002). TNF exists in two functional forms: soluble TNF (sTNF) and transmembrane TNF (tmTNF), each with distinct functional bioactivities. sTNF signaling, via TNFR1, results in proinflammatory actions, whereas tmTNF signaling, via TNFR2, results in pro-survival signaling and plays an important role in immune system development (McCoy & Tansey, 2008). Although the exact role of TNF in brain function is not well understood, elevations in systemic sTNF are associated with accelerated cognitive decline and increased conversion of mild cognitive impairment (MCI) to AD in elderly populations (Holmes et al., 2009).

A recent study demonstrated that a single systemic TNF challenge induced a transient working memory deficit and significant increase in hippocampal and hypothalamic inflammation in prion-injected mice as compared to control; however, this manipulation alone was unable to exacerbate brain pathology (Hennessy et al., 2016). These data suggest that, because there are no acute effects on pathology, TNF levels need to be chronically rather than acutely elevated or TNF is not the sole mediator of brain pathology in that model. This idea is supported by the evidence that acute systemic LPS, but not TNF, is able to increase the number of apoptotic cells in the hippocampus independent of sTNF in a prion model (Hennessy et al., 2016). Peripheral administration of LPS induces TNF expression in the CNS, and such an increase requires TNF receptor signaling and can be long-lasting, because it can be detectable 10 months after the initial LPS challenge (Qin et al., 2007). However, TNF is not the only inflammatory factor elicited by LPS treatment, and chronic and acute treatment of LPS produce different cytokine profiles in the serum and brain. TNF levels are significantly increased in the brain following repeated LPS administration as compared to a single administration (Erickson & Banks, 2011), suggesting that chronic peripheral inflammation rather than acute inflammation increased brain TNF levels. TNF is at the apex of a wider network of proinflammatory mediators that have been implicated in neuroinflammatory disease (Steinman, 2013). Therefore, understanding how TNF acts to modulate the proinflammatory environment and how it contributes to AD-like pathology will help us understand the CNS global network of inflammatory changes that occur with AD pathology.

Both TNF and LPS act at the BBB to modulate its permeability. Peripheral LPS increases TNF receptor expression at the BBB, particularly the expression of TNFR1, leading to upregulation of factors that increase BBB permeability and allowing infiltration of cells and other factors normally excluded from the brain parenchyma (Mackay, Loetscher, Stueber, Gehr, & Lesslauer, 1993; Nadeau & Rivest, 1999). Evidence of increased BBB permeability is present in brains of AD patients: increased CSF to serum albumin ratios, microvascular pathology, and upregulation of factors that regulate immune cell infiltration across the BBB have been reported (Sharma, Castellani, Smith, & Sharma, 2012). Within the CNS, TNF can also contribute to neuroinflammation through activation of resident immune cells and recruitment of peripheral immune cells (Qin et al., 2007). Specific regulation of peripheral immune cell traffic across the BBB is crucial to maintain the highly controlled immunological responses of the CNS.

Examining changes in these regulatory structures and pathways in response to central or peripheral inflammation can shed light on how regulation and activation of infiltrating populations is changing over the course of disease. TNF plays an active role in modulation of the BBB and contributes to activation of peripheral immune populations. Thus, TNF may represent a valid target for modulation of immune cell traffic to the CNS, to ameliorate chronic neuroinflammation in neurodegenerative diseases. However, signaling modulation at barrier structures may renew regulation of trafficking populations and boost the brain's natural ability to combat neuroinflammatory disease (Fig. 16.1).

THE ROLE OF INFLAMMATION IN AGE-RELATED NEURODEGENERATIVE DISEASE: A PARADIGM SHIFT

Numerous research studies and review articles have been published on the potential role and regulation of cellular and molecular mechanisms of neuroinflammation implicated in a number of neurodegenerative diseases (Blandini, 2013; Hooten, Beers, Zhao, & Appel, 2015; McAlpine & Tansey, 2008; McManus & Heneka, 2017; Mena & Garcia de Yebenes, 2008; Sanchez-Guajardo, Barnum, Tansey, & Romero-Ramos, 2013; Tufekci, Meuwissen, Genc, & Genc, 2012). Therefore, rather than repeat and review their content here, we will focus on how the field of neurodegeneration has interpreted the data and how new technological advances have contributed to the paradigm shift on the role of central and peripheral inflammation in neurodegenerative disease.

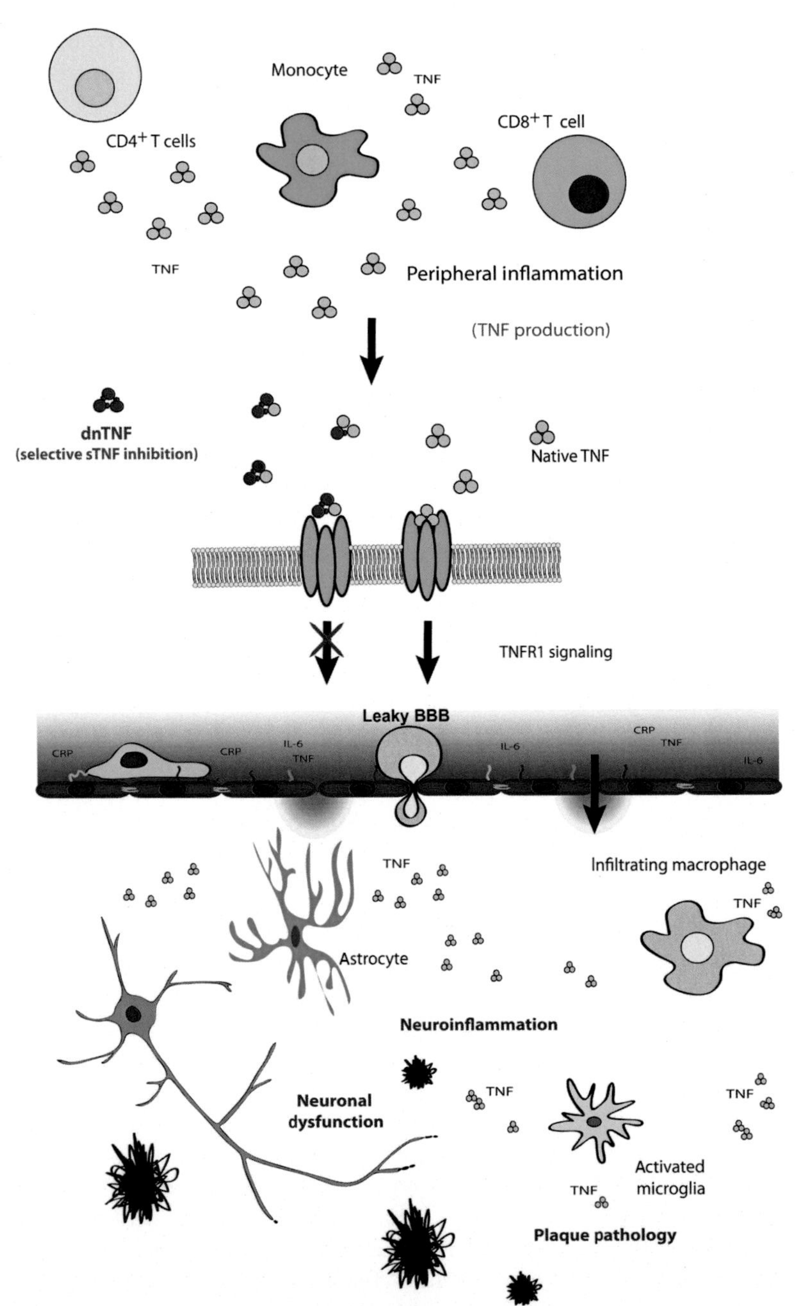

FIGURE 16.1 **Modeling the effects of peripheral inflammation and sTNF signaling on peripheral immune cell traffic to the brain and AD-like pathology.** Chronic peripheral inflammation leads to systemic increase in cytokines and other inflammatory factors, such as sTNF. Systemic sTNF, signaling via the TNFR1 receptor, leads to BBB breakdown, altered peripheral immune cell trafficking to the brain, and neuroinflammation that together contribute to neuronal dysfunction and the progression of AD-like pathology. Selective inhibition of sTNF signaling, via XPro1595, will enable the BBB to reestablish the regulation of immune cell traffic to the brain as well as dampen neuroinflammation, restore synaptic function, and slow the progression of AD-like pathology.

One of the earliest reports of the presence of activated microglia, the hallmark of neuroinflammation, in human postmortem brain analysis came from Foix and Nicolesco in 1925, when they reported that "neuroglia" was present at the site of Lewy bodies and neuromelanin-containing extracellular inclusions in Parkinson's disease (PD) brain (Foix,

1925). Yet this observation did not receive a great deal of attention in light of the prevailing idea at the time that the brain was an immune-privileged organ into which peripheral immune cells rarely entered (Engelhardt & Coisne, 2011), a belief that persisted until only a few years ago. Nevertheless, Edith and Patrick McGeer should be credited with reviving interest into the role of neuroinflammation in age-related neurodegenerative diseases such as AD and PD in the late 1980s (McGeer, Itagaki, Boyes, & McGeer, 1988; McGeer, Itagaki, Tago, & McGeer, 1988). They confirmed Foix's findings and raised the possibility that activated microglia in postmortem brains of people with these diseases might not have been the result of neuronal cell death. Rather, microglial activation might have begun early in the course of these diseases, raising the interesting possibility that neuroinflammation could play an active role in neurodegeneration. This innovative idea was supported decades later by multiple epidemiological studies noting that individuals who consumed large doses of non-steroidal antiinflammatory drugs (NSAIDs) chronically displayed lower incidence of PD or AD (Chen et al., 2005; Gao, Chen, Schwarzschild, & Ascherio, 2011; McGeer, Rogers, & McGeer, 2016). However, not all NSAID studies demonstrate neuroprotection in PD or other neurodegenerative diseases (Fondell et al., 2012; Samii, Etminan, Wiens, & Jafari, 2009). One possible reason may be related to the timing of those regimens in relation to the age of the individuals who consumed them, and their genetic make-up also likely influences the degree of protection that can be derived. Finally, as far as we know, the extent to which NSAID action is mediated through antiinflammatory mechanisms is unclear. Interestingly, a recent bioinformatics analysis of such epidemiological studies suggested that NSAIDs may afford neuroprotection via actions on oxidative phosphorylation or ribosomal pathways (Nevado-Holgado & Lovestone, 2017).

In the last decade, the concept of an immune-privileged brain was challenged and revisited. Specifically, several studies in animals demonstrated that the BBB was not a physical but a biochemical barrier, the permeability of which could be regulated (Franzen et al., 2003; Ransohoff & Perry, 2009). These studies were followed by the unequivocal demonstration by two independent laboratories that the brain possesses a system of lymphatic vessels through which innate and adaptive immune cells that have entered the brain from periphery drain from the brain into DCLNs (Aspelund et al., 2015; Louveau et al., 2015). Thus, the concept of CNS immune privilege has begun to be revisited in detail (Aspelund et al., 2015; Louveau, Harris, & Kipnis, 2015) as it has important implications for understanding neurodegenerative disease pathogenesis, prevention, and development of therapeutic approaches to combat disease.

Another important revelation that has contributed to the paradigm shift on the role of inflammation and immune responses is the discovery in genome-wide association studies (GWAS) of common variants of genes expressed primarily or exclusively in immune cells or of genes in inflammatory pathways that modulate risk for age-related diseases such as AD and PD (Ahmed et al., 2012; Guo et al., 2011; Hollingworth et al., 2011; Jones et al., 2010; Lambert et al., 2010; Morgan, 2011; Nalls et al., 2011; Pankratz et al., 2012; Raj et al., 2012; Sun et al., 2012). Compelling evidence suggests that inflammatory triggers such as LPS or TNF can kill vulnerable neurons when administered to animals acutely or chronically. However, there is little such evidence in humans, except for the unusual association of parkinsonism in patients who survived a flu outbreak after the Spanish War and those who contracted von Economo's encephalitis during WWI (Henry, Smeyne, Jang, Miller, & Okun, 2010). No clear and compelling evidence supports neuroinflammation as a primary trigger for age-related neurodegenerative disease. Nevertheless, we and others have proposed that peripheral inflammation and/or infection can exacerbate neuronal dysfunction and enhance oxidative stress in a way that hastens cell death (Frank-Cannon, Alto, McAlpine, & Tansey, 2009; Tansey & Goldberg, 2010), triggering a chronic feed-forward vicious cycle of neuroinflammation and neurodegeneration. In addition to the epidemiological, genetic, and fluid biomarker evidence, neuroimaging data (Gerhard, 2016; Gerhard et al., 2006; Schain & Kreisl, 2017) now strongly suggest that immune and inflammatory pathways are activated early in PD.

The interaction of genes and environment continues to be a pervasive issue in neurodegeneration. One such environmental risk is pesticide exposure. Long-term exposure to pesticides and herbicides is linked to increased risk of developing PD (Brown, Rumsby, Capleton, Rushton, & Levy, 2006). Pesticide exposure also results in a proinflammatory phenotype, whereby those chronically exposed to pesticides have an increased likelihood of developing autoimmune disorders (Parks et al., 2011). In a more recent study, the interaction between a specific SNP, rs3192882, which was associated with altered risk for PD, and environmental exposure to pesticides demonstrated a significant increased risk for PD. Those with the high-risk form of the SNP displayed an increased baseline and inducible MHC II expression in APCs (Kannarkat et al., 2015). These studies suggest an important balance between immunological processes and environmental exposures in determining an individual's susceptibility to PD. We posit that immune responses are activated early (perhaps in the prodromal stages of disease) in response to environmental triggers that synergize with genetic predisposition (or protective) variants to determine an individual's life-long risk of developing neurodegeneration, an idea supported by a work from our group (Kannarkat et al., 2015). The accumulation of epigenetic modifications from different sources and levels of exposures could be a link between genetic predisposition and environmental determinants of neurological disease (Lill, 2016; Sankowski, Mader, & Valdes-Ferrer, 2015). These epigenetic changes can be important components in the progression of conditions such as AD and PD (Maloney & Lahiri, 2016). When those exposures can promote biochemical changes to pathological levels is still poorly understood. Nevertheless, it is well accepted that environmental and genetic interactions can affect neuroimmune responses and inflammation in a dynamic fashion (Musaelyan et al., 2014).

THE ROLE OF PERIPHERAL INFLAMMATION IN NEURODEGENERATIVE DISEASE

Evidences suggests that genetic background can explain only a small percentage of chronic modern diseases and that the remaining causes of common disorders appear to emerge from environmental influences (Miller & Spencer, 2014; Sears & Genuis, 2012; Wohleb & Delpech, 2017). Conditions of increased peripheral inflammation, including acute infections and chronic inflammatory disease, are associated with development of amyloidosis, dementia, and delirium in elderly patients, and have been implicated as risk factors for development of AD (Biessels, Staekenborg, Brunner, Brayne, & Scheltens, 2006; Holmes, El-Okl, & Williams, 2003; Kamer et al., 2008; Miklossy & McGeer, 2016; Misiak, Leszek, & Kiejna, 2012; Murray et al., 2012). Epidemiological data suggest that the chronic use of NSAIDs reduces risk for AD. However, clinical and animals studies have not found promising results for the use of NSAIDs to treat AD (McGeer & McGeer, 2007), suggesting that a more targeted antiinflammatory treatment is warranted. The peripheral immune system plays an active role in surveying the CNS for immunological threats, and evidence suggests it may also play a role in neurodegenerative disease, as discussed in the sections above. However, it is still unclear how changes in peripheral immune cell populations over the course of AD pathology affect disease outcome, and how chronic peripheral inflammation modulates these mechanisms.

Inflammatory Risk Factors for AD

Mutations within the APP, PSEN1, and PSEN2 genes are known to confer early-onset AD (Tanzi, 2012). However, genetic variants also associate with sporadic late-onset AD, most strongly APOE, to increase disease risk, and many of these identified genes are expressed in the immune system and play a role in immune function, such as *CR1*, *CLU*, and *EPHA1* (reviewed in Tanzi, 2012). The *CD33* allele SNP re3865444 has been associated with protection against AD and reduction in both CD33 protein expression and insoluble Aβ42 (Griciuc et al., 2013). CD33 plays a role in the regulation of the innate immune system (Crocker, 2005) and has been shown to inhibit microglial clearance of Aβ42 in vitro (Griciuc et al., 2013). Moreover, certain variants in *TREM2* are increased in AD populations and increase AD susceptibility (Guerreiro et al., 2013). Increased levels of TREM2 were associated with increased levels of Aβ, and in mouse models, TREM2 is increased in myeloid cells surrounding Aβ plaques (Guerreiro et al., 2013). Recent meta-analysis of four GWAS samples of European ancestry revealed five newly associated genomic regions: HLA-DRB5-HLA-DRB1, PTK2B, SORL1, SLC24A4-RIN3, and DSG2 (Lambert et al., 2013). Although associations with AD and CD33 were not replicated in these studies, of these newly identified risk genes, HLA-DRB5-HLA-DRB1 is associated with immunocompetence and histocompatibility. Together these data indicate that changes in regulation and function of the immune system increase risk for AD.

Alterations in Trafficking of Peripheral Immune Cells to Brain

Increased number of T cells have been found in the brain in AD patients as well as animal models of AD-like pathology (Togo et al., 2002). Moreover, markers of myeloid activation are increased, suggestive of increased microglial activation and peripheral macrophage infiltration with aging or AD-like pathology (Martin, Boucher, Fontaine, & Delarasse, 2017). Together, these data support a role for peripheral immune cells in AD progression. Although some studies suggest these populations meditate disease, others suggest they are not significant contributors (Prokop et al., 2015; Varvel et al., 2015), and even play a protective role. TREM2 expression was increased on $CD11b^+$/ $CD45^{high}$ myeloid cells that are associated with Aβ deposits in an age-dependent manner, and these cells were found also to express Ly6C, suggesting they were peripherally derived macrophages. Loss of TREM2 significantly reduced this population of myeloid cells surrounding plaques and lowered Aβ accumulation, while not reducing microglia populations, suggesting that TREM2 and peripheral macrophages may be contributing to AD-like pathology (Jay et al., 2015). However, studies in a model of brain microglial depletion followed by rapid repopulation by peripheral myeloid cells have shown that peripheral myeloid cells were not recruited to Aβ plaques and did not alter Aβ pathology, even following treatment with Aβ-specific antibodies (Prokop et al., 2015). These studies suggest that the effects of peripheral macrophages on disease may only be relevant in the context of interactions with brain-resident microglia. However, studies examining overexpression of CCL2, a chemokine that recruits peripheral macrophages to the CNS, or the loss of its receptor CCR2, have shown that loss of CCL2–CCR2 signaling accelerated memory impairments, synaptic impairment, and Aβ deposition, supporting a role for peripheral macrophage recruitment in suppression of AD-like pathology (El Khoury et al., 2007; Kiyota et al., 2009). Combined together, these studies suggest that a certain degree of peripheral macrophage infiltration is optimal to maintain brain health, and deviations from this mean may contribute to AD-like pathology.

Macrophages are not the only peripheral immune cells shown to provide protective effects against AD-like pathology. Mice lacking function B and T cells have increased levels of Aβ and neuroinflammation that is reduced upon replacement of missing B- and T-cell populations (Marsh et al., 2016). However, some reports indicate that not all T-cell subsets are protective under every circumstance. As discussed above, Treg populations have been shown to have a suppressive effect on peripheral immune cell recruitment that when removed mitigated AD-like pathology (Baruch et al., 2015). Increased levels of T cells have also been found in postmortem AD brains as well as animal models of AD (Ferretti et al., 2016; Togo et al., 2002). Thus, optimal T-cell trafficking, and activation state, may have to be paired with optimal macrophage trafficking and activation state to maintain brain health, especially during the conditions of neuroinflammatory disease. Timing of immune cell trafficking to the brain may play a role in the determination of whether the contributions are protective or pathogenic and may be altered by chronic peripheral inflammation.

Chronic Peripheral Inflammation as a Risk Factor for Neurodegenerative Disease

Chronic stress is associated with neuroinflammation, impaired neurogenesis, microglia activation, and changes in neural plasticity that are frequently present in the progression of dementia and neurologic disease (Bollinger, Bergeon Burns, & Wellman, 2016; Wohleb & Delpech, 2017; Zhao et al., 2016). Evidence suggests that stress can worsen the progression of AD and PD. Animal models of stress/glucocorticoids exposure present increased production of Aβ. Additionally, under stress conditions, corticotropin-releasing factor receptor signaling increases hippocampal tau phosphorylation (Rissman et al., 2012). Chronic stress exaggerates motor deficits and neuroinflammation in a 1-methyl-4-phenyl-1,2,3,6-tetrahydropyridine (MPTP) mouse model of PD and affects mesolimbic and mesocortical dopamine system (Lauretti, Di Meco, Merali, & Pratico, 2016; Smith, Jadavji, Colwell, Katrina Perehudoff, & Metz, 2008). High-fat and high-carbohydrate consumption is associated with activation of the toll-like receptor-4 pathway, endoplasmatic reticulum (ER) stress, and insulin resistance (Balakumar et al., 2016). ER stress and mitochondrial impairment frequently play a role in neurologic disease characterized by misfolded protein accumulation, such as AD and PD (Chaudhari, Talwar, Parimisetty, Lefebvre d'Hellencourt, & Ravanan, 2014). Recently, obesity-induced hypothalamic inflammation was found to promote an imbalance in proteostasis and increases in ubiquitin–proteasome degradation system dysfunction (Cavadas, Aveleira, Souza, & Velloso, 2016). The cumulative effect of chronic stress and unhealthy diet can increase the risk and susceptibility to neurological disorders by multiple neuroinflammatory mechanisms (Machado et al., 2014; Ricci, Fuso, Ippoliti, & Businaro, 2012). Animal and human studies show that metabolic and inflammatory changes associated with the above exposures can be sustained even after removal of the initial stimuli. Thus, the detrimental outcomes resulted from chronic stress and high-caloric diet may be due to a cumulative effect during the life span (Deak et al., 2015; de Sousa Rodrigues et al., 2017).

A recent study demonstrated that psychological stress and high-fat and high-fructose diet interplay to promote an intricate net of metabolic and inflammatory changes in brain, gut, and liver. The interaction between those two exposures increased plasma cholesterol and insulin resistance in a context of peripheral inflammation (de Sousa Rodrigues et al., 2017). Growing literature demonstrates that changes in insulin resistance and brain–gut axis can play a role in PD, AD, and other neurological disorders (Rao & Gershon, 2016). Insulin affects neuronal function, synaptic plasticity, and cell survival (Bassil, Fernagut, Bezard, & Meissner, 2014; Lee, Zabolotny, Huang, Lee, & Kim, 2016). The systemic inflammation present in obesity represents an additional challenge to the energetic supply of the brain. More specifically, insulin actions are mediated by interactions with Akt/protein kinase B and the mitogen-activated protein kinase signaling pathways (Chami, Steel, De La Monte, & Sutherland, 2016). Animal studies show Aβ-triggered microglial release of proinflammatory cytokines in a context of brain neuronal elevation of insulin receptor-1 serine phosphorylation with insulin resistance (Chami et al., 2016; Talbot & Wang, 2014). More specifically, chronic inflammation, mitochondrial impairment, and oxidative stress may connect AD and Type 2 diabetes, a risk factor for vascular dementia and AD (Verdile et al., 2015; Xu et al., 2011). Obesity comorbidities such as insulin resistance, hypercholesterolemia, and hypertension alone or in conjunction to chronic inflammation can trigger neuroinflammation and neurodegeneration (Cai, 2013; Lee & Mattson, 2014; McGuire & Ishii, 2016; van Dijk et al., 2015). Cholesterol and other lipids have an important role in prion-like diseases, such as AD, and can impact Aβ generation by modulating secretase activities (Di Paolo & Kim, 2011; Hannaoui, Shim, Cheng, Corda, & Gilch, 2014). In obese states, excess of food-related molecules and gut-bacterial products can affect adipose tissue, gut, liver, and other structures that play important roles in the intricate immune responses that ultimately can affect brain functions (Yiu, Dorweiler, & Woo, 2016). Some adipokines secreted by adipose tissue are deregulated in obese status, directly or indirectly affecting neuroinflammatory and neurodegenerative processes. Adiponectin acts in the brain to alter AMPK and inflammatory and insulin pathways, and was recently implicated in AD-like pathogenesis (Ng et al., 2016).

There is strong evidence that some of the above-described processes might link metabolic inflammation to gut and BBB disruption (Pan et al., 2011). In obese status, circulating bacterial and food compounds, mitochondrial dysfunction, and oxidative stress increase peripheral inflammation that may have an impact on BBB and gut structures (de Sousa Rodrigues et al., 2017; Pan et al., 2011). Animal and human studies indicate that gut alterations can precede the motor symptoms of PD. Increased gut permeability was found as an important component of PD and may play a role in the disease progression (Houser & Tansey, 2017).

Multiple inflammatory and metabolic mechanisms can affect brain functions and effectively have an impact on neurological conditions (Fig. 16.2). The investigation of the impact of lifestyle, exercise, and healthy diet in the prevention of neuroinflammatory conditions has become an area of intense focus. Animal models reveal multiple mechanisms associated with physical activity that potentially may reduce risk factors for AD and PD. Decreasing circulating LPS, modulation of immune system towards an antiinflammatory profile, and downregulation of TLR-4 are some of the frequent benefits associated with exercising (Stewart et al., 2005). Additionally, physical activity can improve glucose and lipid profiles and reduce hypercholestolemia (Spielman, Little, & Klegeris, 2016). Studies using an MPTP animal model of PD show that exercise is associated with reduction of α-synuclein, microglia activation, and inflammation (Jang et al., 2017; Sung et al., 2012). Improving metabolic and immune profiles during the life span may have protective effects in age-related neurodegenerative conditions.

The Permissive Environment for Neurodegeneration

Immune senescence in aging human populations may be a contributing factor to the development of neurodegenerative disease. Age is the number one risk factor for AD. As we age, our immune system no longer functions with the same vitality, leading to immunodeficiency as well as autoimmune responses, as specialized immune cells begin failing to recognize "self" antigens (Goronzy, Li, Yang, & Weyand, 2013). In AD patients, peripheral macrophages have decreased phagocytic functioning compared to healthy age-matched controls (Fiala et al., 2005), suggesting that there is disease-related impairment in addition to immunosenescence. Recent work in a mouse model of AD has shown age-dependent changes in peripheral immune cell trafficking patterns to the brain as well as age-dependent, genotype-independent changes in DCLN immune cell populations (MacPherson et al., 2017). These findings suggest that age and pathology may interact to alter the immune response to AD-like pathology.

The brain's resident microglia become activated during normal aging. This process is associated with PD, AD, and other neurodegenerative disease and is believed to drive a cascade of toxic events that lead to the release of proinflammatory cytokines and loss of protective mediators (Aarsland et al., 2001; Akiyama et al., 2000; Bartels & Leenders, 2007; Block & Hong, 2005; Cribbs et al., 2012; Griffin et al., 1989; Hobson & Meara, 2004; Hughes et al., 2000; Swardfager et al., 2010; Whitton, 2007). Evidence supports the hypothesis that neuroinflammation resulting from chronic microglial activation, due to aging and/or the deposition of toxic proteins, leads to a disruption of equilibrium, thereby permitting the neurodegenerative process (Block & Hong, 2005; Colton & Wilcock, 2010; Smith et al., 2012). These activated microglia are present,

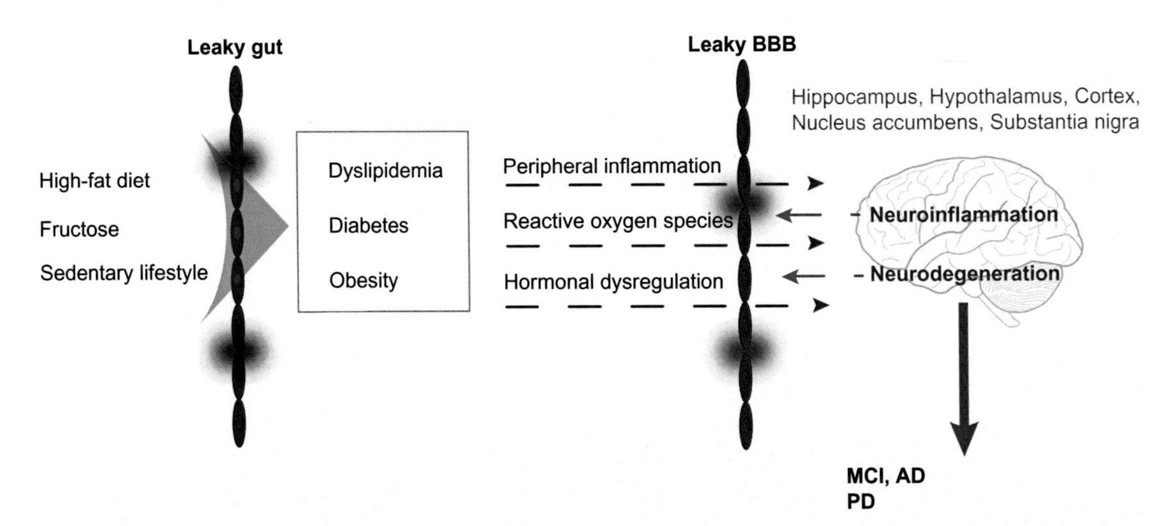

FIGURE 16.2 **Various factors can contribute to the development and maintenance of central neuroinflammation.** The energetic imbalance associated with a high-caloric diet can affect the permeability and microbiome of the gut, resulting in immune system activation, obesity, and its comorbidities. The signaling disruption that occurs with metabolic inflammation is frequently associated with mitochondrial impairment and ER stress that enhances excess production of ROS. The complex interplay between these factors can potentially affect BBB permeability and vulnerable brain regions, ultimately leading to development of neurological diseases.

years before the onset of neuropathological changes, which further supports the idea that inflammation creates gradual shift to a nonequilibrium state (Cagnin, Kassiou, Meikle, & Banati, 2006; Gerhard et al., 2006; Imamura et al., 2003).

In neurodegenerative diseases, growing evidence implicates inflammation as an important feature that is closely involved in the disease process. The permissive environment caused by inflammation can promote the aggregation of proteins. Inflammation has been shown to interfere in the normal trafficking of proteins, which consequently affects the protein function, interaction, and levels (Correia, Patel, Dutta, & Julien, 2015). There is now little doubt that protein misfolding and aggregation are crucial events in many neurodegenerative diseases, including AD. Much insight can be gained into AD and other proteopathies by investigating the self-assembly, transport, and spread of these proteopathic seeds in the misfolded protein propagation model. Knowing the conditions under which proteins misfold and how they serve as templates for the conversion of benign proteins into toxic ones could eventually yield therapeutic interventions that can prevent the domino effect seen in these proteopathies. The impact of investigation into protein misfolding and propagation goes far beyond AD. In recent years, the list of pathogenic proteins that appear to emerge, proliferate, and cause disease by a prion-like process has grown dramatically and includes Aβ, α-synuclein, TDP-43, tau, and many others (Aguzzi & Rajendran, 2009; Braak et al., 2003; Gavett, Stern, & McKee, 2011; Goedert, Falcon, Clavaguera, & Tolnay, 2014; Grad, Fernando, & Cashman, 2015; Jucker & Walker, 2011; Munch & Bertolotti, 2010; Polymenidou & Cleveland, 2012; Walker & Jucker, 2015; Westermark & Westermark, 2010). Evidence for prion-like inducibility even has been found in some cancers, cystic fibrosis, and systemic amyloidoses (Dobson, 1999; Silva, De Moura Gallo, Costa, & Rangel, 2014; Westermark & Westermark, 2010).

How Does the Prion-Like Cascade of Protein Aggregation Begin?

Chronic traumatic encephalopathy (CTE), a progressive neurodegenerative disease associated with a prolonged history of repetitive head impacts, is characterized by the accumulation of hyperphosphorylated tau in neurons, astrocytes, and cellular processes. Recent evidence has shown that high levels of microglial activation (as measured by CD68 expression) is positively correlated with the degree of tau pathology in CTE (Cherry et al., 2016). In patients with amyotrophic lateral sclerosis and frontotemporal dementia, TDP-43 typically found in the nucleus is mislocalized to the cytoplasm, whereby insoluble aggregates of this protein are formed in degenerating neurons (Arai et al., 2006; Neumann et al., 2006). In a recent study, chronic LPS treatment enhanced the cytoplasmic mislocalization and aggregation of TDP-43 in neurons in the spinal cord of transgenic mice (Correia et al., 2015). In PD, some genes that cause familial PD or increase the risk of developing PD are related to immune function (LRRK2 and HLA-DR) (Kannarkat et al., 2015). Positron emission tomography imaging of neuroinflammatory markers in PD shows early involvement and cerebral propagation in microglia. This is further supported by the postmortem analysis of PD brains, which are characterized by activated microglia (McGeer, Itagaki, Boyes, et al., 1988), higher density of astrocytes (Damier, Hirsch, Zhang, Agid, & Javoy-Agid, 1993), and infiltrating T cells (Brochard et al., 2009). Higher levels of inflammatory cytokines and immune cell dysfunction are features shared among PD, AD, and other neurodegenerative diseases. Systemic inflammation, denoted by increased microglial activation and CNS inflammation, can be caused by chronic diseases (Holmes et al., 2009), obesity, and Type 2 diabetes (Takeda et al., 2010), all of which are associated with increased risk of developing

AD. It was once thought that in neurodegenerative diseases inflammation was the result of disease pathology, but increasing evidence supports inflammation as an instigator to a permissive environment that fuels accumulation of toxic proteins and disrupts cellular processes.

Is There a Path Forward for Therapeutic Intervention?

Modulation of Peripheral Immune Cell Activation and/or Their Trafficking May Be a Potential Therapeutic Strategy

Our group has recently demonstrated that AD-like pathology, inhibition of sTNF, as well as chronic peripheral inflammation modulates peripheral immune cell trafficking to the brain. Our data suggest that modulation of trafficking patterns via inhibition of sTNF is protective, whereas modulation via chronic peripheral inflammation may lead to dysregulation of the neuroinflammatory response (MacPherson et al., 2017). In addition, our data indicate that these changes in trafficking patterns occur in conjunction with Aβ accumulation but prior to significant neurodegeneration (MacPherson et al., 2017; Oakley et al., 2006). Moreover, our group has not found significant cognitive deficits at these early time points (unpublished observations). Thus, modulation of peripheral immune cell trafficking to the CNS is supported as an early intervention prior to detectable cognitive deficits. However, further investigation into the direct effects of more specific subsets of trafficking populations is warranted. A recently published study found that the use of fingolimod, a drug that targets the sphingosine-1 receptor and acts by sequestering $CCR7^+$ T cells populations within lymph nodes (Blaho et al., 2015), significantly reduced markers of AD-like pathology in the 5xFAD mouse model (Aytan et al., 2016). While fingolimod has been shown to have direct effects on microglia (Blaho & Hla, 2014), this report also supports a role for the modulation of T-cell trafficking to the brain to treat AD pathology. Further support for this idea comes from the modulation of the timing of Treg trafficking, in that temporary inhibition of Tregs improved AD-like pathology (Baruch et al., 2015). Although sequestering subsets of immune cells may lead to potential negative side effects of suppressing overall immune function, modulation of trafficking patterns at the BBB may lead to specific regulation of immune cell trafficking into the brain. Approaches to such modulation include inhibition of TNF signaling with etanercept (Enbrel) or XPro1595 (a dominant-negative TNF that selectively neutralizes sTNF) or modulation of adhesion molecules that regulate BBB permeability directly. Future studies on the effect of XPro1595 as compared to etanercept in mouse models of neurodegeneration will help to elucidate the direct versus indirect effects of sTNF signaling in modulating BBB permeability and immune cell traffic into the CNS. Further work investigating the BBB as a whole or region-specific changes—such as at the choroid plexus, where the BBB is highly permeable—will enable investigators to determine the extent to which more specific BBB-targeted approaches can help mitigate the effects of chronic peripheral inflammation on AD-like pathology. Such work may also reveal whether modulation of trafficking patterns of specific immune cell subsets (i.e., increase entry of protective populations, while minimizing entry of cells that exacerbate inflammation and tissue damage) is also a plausible path forward to ameliorate disease progression.

Second-Hit Model of Progression of AD-like Pathology

Investigation of immune cell regulation in mouse models of neurodegenerative diseases that replicates observed neuroinflammatory changes is crucial to understand how disease progression leads to neuroinflammatory pathology in humans. In the reverse direction, however, understanding how various chronic inflammatory diseases increase risk for AD in human populations is also an unmet need. Identifying environmental conditions that modulate the traffic of immune cell populations into the brain may provide a more relevant model to study the dynamic changes in the peripheral circulation and brain–immune system network that occur during the course of neuroinflammatory pathology. This second-hit model, in which chronic peripheral inflammation affects AD- or PD-like pathology in mouse models, is a starting point to help us elucidate the mechanisms by which changes in peripheral inflammation may disrupt peripheral immune cell populations to increase risk for AD. With this second-hit model, one can assess the contribution of genetic (e.g., FAD or AD-risk genes) as well as environmental (e.g., diet-induced peripheral inflammation) factors and their interaction to begin formulating hypotheses about how peripheral immune cell trafficking changes in at-risk human populations contribute to the development of AD. For example, the ε4 allele of apolipoprotein E (ApoE), important in lipid transport and metabolism, has been linked to increased risk for AD (Tanzi, 2012). Investigation into how lipid handling is altered in high-caloric diet models, in conjunction with AD-associated transgenes, to modulate AD-like pathology may reveal additional therapeutic targets beyond peripheral immune cell trafficking. Such targets may be more relevant to certain subpopulations of AD patients and could help us stratify subjects for clinical trials by enriching cohorts for those with the highest levels of central and peripheral inflammation and dysregulated immune cell phenotypes.

Assessing Changes in Peripheral Immune Cell Populations in At-Risk Human Populations

The number of risk factors for sporadic AD and PD suggests there may be several pathogenic routes to a single diagnosis, and identification of subpopulations of patients with greater immune dysfunction may reveal such routes. Although modulation of peripheral immune cell trafficking by chronic peripheral inflammatory conditions may work in an animal model, this has yet to be investigated in human populations. Reports suggest that peripheral macrophage function is impaired in AD patients (Fiala et al., 2005). However, deep immunophenotyping by flow cytometry in human populations at risk or clinically diagnosed for MCI or AD needs to be explored, to follow up on studies done in mice. Although epidemiological data suggest that chronic NSAID use is protective against AD, clinical trials were unable to provide support for their use as an AD therapy (McGeer & McGeer, 2007). These negative results may be due to the timing of the intervention (e.g., dosing too late in the course of AD). They may have also been due in part to inclusion of a heterogeneous group of AD patients, some of which may have had less immune dysregulation or no underlying inflammatory diseases that we posit can contribute to AD pathophysiology. To move forward with clinical trials of immune therapies that can modulate peripheral immune trafficking to the brain and neuroinflammation, we must first establish that populations of AD patients and especially individuals at risk for AD show dysregulation in peripheral T-cell and monocyte populations that are potentially trafficking to the brain. To this end, our research group is in the early stages of investigating how subpopulations of myeloid cells and T cells are altered within the blood and CSF in populations at risk for AD because of mid-life hypertension, metabolic syndrome or obesity, and parental history of AD. Furthermore, it is critical to assess how peripheral immune cell populations change within the blood and CSF of populations with decreased risk for AD, either genetic (ApoE ε2 status) (Tanzi, 2012) or through drug interventions with chronic NSAIDs earlier in life (McGeer & McGeer, 2007) or with anti-hypertension medications (Forette et al., 2002). Such findings will further increase our understanding of the central and peripheral mechanisms that influence peripheral immune cell phenotype and trafficking patterns to the brain to promote chronic neuroinflammation and affect the course of neuroinflammatory disease.

References

Aarsland, D., Andersen, K., Larsen, J. P., Lolk, A., Nielsen, H., & Kragh-Sorensen, P. (2001). Risk of dementia in Parkinson's disease: A community-based, prospective study. *Neurology, 56*, 730–736.

Aguzzi, A., & Rajendran, L. (2009). The transcellular spread of cytosolic amyloids, prions, and prionoid. *Neuron, 64*, 783–790.

Ahmed, I., Tamouza, R., Delord, M., Krishnamoorthy, R., Tzourio, C., Mulot, C., ... Elbaz, A. (2012). Association between Parkinson's disease and the HLA-DRB1 locus. *Movement Disorders: Official Journal of the Movement Disorder Society, 27*, 1104–1110.

Ajami, B., Bennett, J. L., Krieger, C., Tetzlaff, W., & Rossi, F. M. (2007). Local self-renewal can sustain CNS microglia maintenance and function throughout adult life. *Nature Neuroscience, 10*, 1538–1543.

Akiyama, H., Barger, S., Barnum, S., Bradt, B., Bauer, J., Cole, G. M., ... Wyss-Coray, T. (2000). Inflammation and Alzheimer's disease. *Neurobiology of Aging, 21*, 383–421.

Allen, N. J. (2014). Astrocyte regulation of synaptic behavior. *Annual Review of Cell and Developmental Biology, 30*, 439–463.

Aloisi, F., Borsellino, G., Care, A., Testa, U., Gallo, P., Russo, G., ... Levi, G. (1995). Cytokine regulation of astrocyte function: In-vitro studies using cells from the human brain. *International Journal of Developmental Neuroscience: The Official Journal of the International Society for Developmental Neuroscience, 13*, 265–274.

Arai, T., Hasegawa, M., Akiyama, H., Ikeda, K., Nonaka, T., Mori, H., ... Oda, T. (2006). TDP-43 is a component of ubiquitin-positive tau-negative inclusions in frontotemporal lobar degeneration and amyotrophic lateral sclerosis. *Biochemical and Biophysical Research Communications, 351*, 602–611.

Aspelund, A., Antila, S., Proulx, S. T., Karlsen, T. V., Karaman, S., Detmar, M., ... Alitalo, K. (2015). A dural lymphatic vascular system that drains brain interstitial fluid and macromolecules. *The Journal of Experimental Medicine, 212*, 991–999.

Axtell, R. C., & Steinman, L. (2009). Gaining entry to an uninflamed brain. *Nature Immunology, 10*, 453–455.

Aytan, N., Choi, J. K., Carreras, I., Brinkmann, V., Kowall, N. W., Jenkins, B. G., & Dedeoglu, A. (2016). Fingolimod modulates multiple neuroinflammatory markers in a mouse model of Alzheimer's disease. *Sci Rep, 6*, 24939.

Balakumar, M., Raji, L., Prabhu, D., Sathishkumar, C., Prabu, P., Mohan, V., & Balasubramanyam, M. (2016). High-fructose diet is as detrimental as high-fat diet in the induction of insulin resistance and diabetes mediated by hepatic/pancreatic endoplasmic reticulum (ER) stress. *Molecular and Cellular Biochemistry, 423*, 93–104.

Barkauskas, D. S., Evans, T. A., Myers, J., Petrosiute, A., Silver, J., & Huang, A. Y. (2013). Extravascular CX3CR1+ cells extend intravascular dendritic processes into intact central nervous system vessel lumen. *Microscopy and Microanalysis: The Official Journal of Microscopy Society of America, Microbeam Analysis Society, Microscopical Society of Canada, 19*, 778–790.

Barna, B. P., Estes, M. L., Jacobs, B. S., Hudson, S., & Ransohoff, R. M. (1990). Human astrocytes proliferate in response to tumor necrosis factor alpha. *Journal of Neuroimmunology, 30*, 239–243.

Bartels, A. L., & Leenders, K. L. (2007). Neuroinflammation in the pathophysiology of Parkinson's disease: Evidence from animal models to human in vivo studies with [11C]-PK11195 PET. *Movement Disorders: Official Journal of the Movement Disorder Society, 22*, 1852–1856.

Baruch, K., Rosenzweig, N., Kertser, A., Deczkowska, A., Sharif, A. M., Spinrad, A., ... Schwartz, M. (2015). Breaking immune tolerance by targeting Foxp3(+) regulatory T cells mitigates Alzheimer's disease pathology. *Nature Communications, 6*, 7967.

Baruch, K., & Schwartz, M. (2013). CNS-specific T cells shape brain function via the choroid plexus. *Brain, Behavior, and Immunity, 34*, 11–16.

Bassil, F., Fernagut, P. O., Bezard, E., & Meissner, W. G. (2014). Insulin, IGF-1 and GLP-1 signaling in neurodegenerative disorders: Targets for disease modification?. *Progress in Neurobiology, 118*, 1–18.

Bayraktar, O. A., Fuentealba, L. C., Alvarez-Buylla, A., & Rowitch, D. H. (2015). Astrocyte development and heterogeneity. *Cold Spring Harbor Perspectives in Biology, 7*, a020362.

Biber, K., Neumann, H., Inoue, K., & Boddeke, H. W. (2007). Neuronal 'On' and 'Off' signals control microglia. *Trends in Neurosciences, 30*, 596–602.

Biessels, G., Staekenborg, S., Brunner, E., Brayne, C., & Scheltens, P. (2006). Risk of dementia in diabetes mellitus: A systematic review. *Lancet Neurology, 5*, 64–74.

Blaho, V. A., Galvani, S., Engelbrecht, E., Liu, C., Swendeman, S. L., Kono, M., ... Hla, T. (2015). HDL-bound sphingosine-1-phosphate restrains lymphopoiesis and neuroinflammation. *Nature, 523*, 342–346.

Blaho, V. A., & Hla, T. (2014). An update on the biology of sphingosine 1-phosphate receptors. *Journal of Lipid Research, 55*, 1596–1608.

Blandini, F. (2013). Neural and immune mechanisms in the pathogenesis of Parkinson's disease. *Journal of Neuroimmune Pharmacology: The Official Journal of the Society on NeuroImmune Pharmacology, 8*, 189–201.

Block, M. L., & Hong, J. S. (2005). Microglia and inflammation-mediated neurodegeneration: Multiple triggers with a common mechanism. *Progress in Neurobiology, 76*, 77–98.

Bollinger, J. L., Bergeon Burns, C. M., & Wellman, C. L. (2016). Differential effects of stress on microglial cell activation in male and female medial prefrontal cortex. *Brain, Behavior, and Immunity, 52*, 88–97.

Bolos, M., Antequera, D., Aldudo, J., Kristen, H., Bullido, M. J., & Carro, E. (2014). Choroid plexus implants rescue Alzheimer's disease-like pathologies by modulating amyloid-beta degradation. *Cellular and Molecular Life Sciences: CMLS, 71*, 2947–2955.

Braak, H., Del Tredici, K., Rub, U., de Vos, R. A., Jansen Steur, E. N., & Braak, E. (2003). Staging of brain pathology related to sporadic Parkinson's disease. *Neurobiology of Aging, 24*, 197–211.

Brochard, V., Combadiere, B., Prigent, A., Laouar, Y., Perrin, A., Beray-Berthat, V., ... Hunot, S. (2009). Infiltration of CD4 + lymphocytes into the brain contributes to neurodegeneration in a mouse model of Parkinson disease. *The Journal of Clinical Investigation, 119*, 182–192.

Brown, T. P., Rumsby, P. C., Capleton, A. C., Rushton, L., & Levy, L. S. (2006). Pesticides and Parkinson's disease – Is there a link? *Environmental Health Perspectives, 114*, 156–164.

Cagnin, A., Kassiou, M., Meikle, S. R., & Banati, R. B. (2006). In vivo evidence for microglial activation in neurodegenerative dementia. *Acta Neurologica Scandinavica. Supplementum, 185*, 107–114.

Cai, D. (2013). Neuroinflammation and neurodegeneration in overnutrition-induced diseases. *Trends in Endocrinology and Metabolism: TEM, 24*, 40–47.

Carare, R. O., Bernardes-Silva, M., Newman, T. A., Page, A. M., Nicoll, J. A., Perry, V. H., & Weller, R. O. (2008). Solutes, but not cells, drain from the brain parenchyma along basement membranes of capillaries and arteries: Significance for cerebral amyloid angiopathy and neuroimmunology. *Neuropathology and Applied Neurobiology, 34*, 131–144.

Carson, M. J., Doose, J. M., Melchior, B., Schmid, C. D., & Ploix, C. C. (2006). CNS immune privilege: Hiding in plain sight. *Immunological Reviews, 213*, 48–65.

Cavadas, C., Aveleira, C. A., Souza, G. F., & Velloso, L. A. (2016). The pathophysiology of defective proteostasis in the hypothalamus-from obesity to ageing. *Nature Reviews Endocrinology, 12*, 723–733.

Cebrian, C., Zucca, F. A., Mauri, P., Steinbeck, J. A., Studer, L., Scherzer, C. R., ... Sulzer, D. (2014). MHC-I expression renders catecholaminergic neurons susceptible to T-cell-mediated degeneration. *Nature Communications, 5*, 3633.

Chami, B., Steel, A. J., De La Monte, S. M., & Sutherland, G. T. (2016). The rise and fall of insulin signaling in Alzheimer's disease. *Metabolic Brain Disease, 31*, 497–515.

Chaudhari, N., Talwar, P., Parimisetty, A., Lefebvre d'Hellencourt, C., & Ravanan, P. (2014). A molecular web: Endoplasmic reticulum stress, inflammation, and oxidative stress. *Frontiers in Cellular Neuroscience, 8*, 213.

Chen, H., Jacobs, E., Schwarzschild, M. A., McCullough, M. L., Calle, E. E., Thun, M. J., & Ascherio, A. (2005). Nonsteroidal antiinflammatory drug use and the risk for Parkinson's disease. *Annals of Neurology, 58*, 963–967.

Cherry, J. D., Tripodis, Y., Alvarez, V. E., Huber, B., Kiernan, P. T., Daneshvar, D. H., ... Stein, T. D. (2016). Microglial neuroinflammation contributes to tau accumulation in chronic traumatic encephalopathy. *Acta Neuropathologica Communications, 4*, 112.

Cho, S.-H., Sun, B., Zhou, Y., Kauppinen, T., Halabisky, B., Wes, P., ... Gan, L. (2011). CX3CR1 protein signaling modulates microglial activation and protects against plaque-independent cognitive deficits in a mouse model of Alzheimer disease. *The Journal of Biological Chemistry, 286*, 32713–32722.

Colton, C., & Wilcock, D. M. (2010). Assessing activation states in microglia. *CNS & Neurological Disorders Drug Targets, 9*, 174–191.

Correia, A. S., Patel, P., Dutta, K., & Julien, J. P. (2015). Inflammation induces TDP-43 mislocalization and aggregation. *PLoS One, 10*, e0140248.

Cribbs, D. H., Berchtold, N. C., Perreau, V., Coleman, P. D., Rogers, J., Tenner, A. J., & Cotman, C. W. (2012). Extensive innate immune gene activation accompanies brain aging, increasing vulnerability to cognitive decline and neurodegeneration: A microarray study. *Journal of Neuroinflammation, 9*, 179.

Crocker, P. R. (2005). Siglecs in innate immunity. *Current Opinion in Pharmacology, 5*, 431–437.

Cunningham, C., Campion, S., Lunnon, K., Murray, C., Woods, J., ... Perry, V. (2009). Systemic inflammation induces acute behavioral and cognitive changes and accelerates neurodegenerative disease. *Biological Psychiatry*, *65*, 304–312.

Damier, P., Hirsch, E. C., Zhang, P., Agid, Y., & Javoy-Agid, F. (1993). Glutathione peroxidase, glial cells and Parkinson's disease. *Neuroscience*, *52*, 1–6.

Davalos, D., Grutzendler, J., Yang, G., Kim, J. V., Zuo, Y., Jung, S., ... Gan, W. B. (2005). ATP mediates rapid microglial response to local brain injury in vivo. *Nature Neuroscience*, *8*, 752–758.

Deak, T., Quinn, M., Cidlowski, J. A., Victoria, N. C., Murphy, A. Z., & Sheridan, J. F. (2015). Neuroimmune mechanisms of stress: Sex differences, developmental plasticity, and implications for pharmacotherapy of stress-related disease. *Stress (Amsterdam, Netherlands)*, *18*, 367–380.

de Sousa Rodrigues, M. E., Bekhbat, M., Houser, M. C., Chang, J., Walker, D. I., Jones, D. P., ... Tansey, M. G. (2017). Chronic psychological stress and high-fat high-fructose diet disrupt metabolic and inflammatory gene networks in the brain, liver, and gut and promote behavioral deficits in mice. *Brain, Behavior, and Immunity*, *59*, 158–172.

Di Paolo, G., & Kim, T. W. (2011). Linking lipids to Alzheimer's disease: Cholesterol and beyond. *Nature Reviews. Neuroscience*, *12*, 284–296.

Dixon-Salazar, T. J., Fourgeaud, L., Tyler, C. M., Poole, J. R., Park, J. J., & Boulanger, L. M. (2014). MHC class I limits hippocampal synapse density by inhibiting neuronal insulin receptor signaling. *The Journal of Neuroscience*, *34*, 11844–11856.

Dobson, C. M. (1999). Protein misfolding, evolution and disease. *Trends in Biochemical Sciences*, *24*, 329–332.

Efremova, L., Chovancova, P., Adam, M., Gutbier, S., Schildknecht, S., & Leist, M. (2017). Switching from astrocytic neuroprotection to neurodegeneration by cytokine stimulation. *Archives of Toxicology*, *91*(1), 231–246.

El Khoury, J., Toft, M., Hickman, S. E., Means, T. K., Terada, K., Geula, C., & Luster, A. D. (2007). Ccr2 deficiency impairs microglial accumulation and accelerates progression of Alzheimer-like disease. *Nature Medicine*, *13*, 432–438.

Engelhardt, B. (2008). Immune cell entry into the central nervous system: Involvement of adhesion molecules and chemokines. *Journal of the Neurological Sciences*, *274*, 23–26.

Engelhardt, B., & Coisne, C. (2011). Fluids and barriers of the CNS establish immune privilege by confining immune surveillance to a two-walled castle moat surrounding the CNS castle. *Fluids Barriers CNS*, *8*, 4.

Engelhardt, B., & Ransohoff, R. M. (2005). The ins and outs of T-lymphocyte trafficking to the CNS: Anatomical sites and molecular mechanisms. *Trends in Immunology*, *26*, 485–495.

Engelhardt, B., & Ransohoff, R. M. (2012). Capture, crawl, cross: The T cell code to breach the blood-brain barriers. *Trends in Immunology*, *33*, 579–589.

Engelhart, M., Geerlings, M., Meijer, J., Kiliaan, Amanda, Ruitenberg, A., van Swieten, J., ... Breteler, M. (2004). Inflammatory proteins in plasma and the risk of dementia: The rotterdam study. *Archives of Neurology*, *61*, 668–672.

Erickson, M. A., & Banks, W. A. (2011). Cytokine and chemokine responses in serum and brain after single and repeated injections of lipopolysaccharide: Multiplex quantification with path analysis. *Brain, Behavior, and Immunity*, *25*, 1637–1648.

Erickson, M. A., & Banks, W. A. (2013). Blood-brain barrier dysfunction as a cause and consequence of Alzheimer's disease. *Journal of Cerebral Blood Flow and Metabolism: Official Journal of the International Society of Cerebral Blood Flow and Metabolism*, *33*, 1500–1513.

Farber, K., Cheung, G., Mitchell, D., Wallis, R., Weihe, E., Schwaeble, W., & Kettenmann, H. (2009). C1q, the recognition subcomponent of the classical pathway of complement, drives microglial activation. *Journal of Neuroscience Research*, *87*, 644–652.

Ferretti, M. T., Merlini, M., Spani, C., Gericke, C., Schweizer, N., Enzmann, G., ... Nitsch, R. M. (2016). T-cell brain infiltration and immature antigen-presenting cells in transgenic models of Alzheimer's disease-like cerebral amyloidosis. *Brain, Behavior, And Immunity*, *54*, 211–225.

Fiala, M., Lin, J., Ringman, J., Kermani-Arab, V., Tsao, G., Patel, A., ... Bernard, G. (2005). Ineffective phagocytosis of amyloid-beta by macrophages of Alzheimer's disease patients. *Journal of Alzheimer's Disease: JAD*, *7*, 221–232, discussion 55–62.

Foix, C., & Nicolesco, J. (1925). Les noyaux gris centraux et la región Mésencéphalo-sous-optique., Suivi d'un apéndice sur l'anatomie pathologique de la maladie de Parkinson. In Masson et Cie (Ed.), *Anatomie cérébrale*. Paris: Masson et Cie.

Fondell, E., O'Reilly, E. J., Fitzgerald, K. C., Falcone, G. J., McCullough, M. L., Thun, M. J., ... Ascherio, A. (2012). Non-steroidal anti-inflammatory drugs and amyotrophic lateral sclerosis: Results from five prospective cohort studies. *Amyotrophic Lateral Sclerosis: Official Publication of the World Federation of Neurology Research Group on Motor Neuron Diseases*, *13*, 573–579.

Fontana, A., Fierz, W., & Wekerle, H. (1984). Astrocytes present myelin basic protein to encephalitogenic T-cell lines. *Nature, 307*, 273–276.

Forette, F., Seux, M. L., Staessen, J. A., Thijs, L., Babarskiene, M. R., Babeanu, S., ... Investigators Systolic Hypertension in Europe. (2002). The prevention of dementia with antihypertensive treatment: New evidence from the Systolic Hypertension in Europe (Syst-Eur) study. *Archives of Internal Medicine, 162*, 2046–2052.

Frank-Cannon, T. C., Alto, L. T., McAlpine, F. E., & Tansey, M. G. (2009). Does neuroinflammation fan the flame in neurodegenerative diseases? *Molecular Neurodegeneration, 4*, 47.

Franzen, B., Duvefelt, K., Jonsson, C., Engelhardt, B., Ottervald, J., Wickman, M., ... Schuppe-Koistinen, I. (2003). Gene and protein expression profiling of human cerebral endothelial cells activated with tumor necrosis factor-alpha. *Brain Research. Molecular Brain Research, 115*, 130–146.

Fu, H., Liu, B., Frost, J., Hong, S., Jin, M., Ostaszewski, B., ... Lemere, C. (2012). Complement component C3 and complement receptor type 3 contribute to the phagocytosis and clearance of fibrillar Aβ by microglia. *Glia, 60*, 993–1003.

Gao, X., Chen, H., Schwarzschild, M. A., & Ascherio, A. (2011). Use of ibuprofen and risk of Parkinson disease. *Neurology, 76*, 863–869.

Gavett, B. E., Stern, R. A., & McKee, A. C. (2011). Chronic traumatic encephalopathy: A potential late effect of sport-related concussive and subconcussive head trauma. *Clinics in Sports Medicine, 30*, 179–188, xi.

Gerhard, A. (2016). TSPO imaging in parkinsonian disorders. *Clinical and Translational Imaging, 4*, 183–190.

Gerhard, A., Pavese, N., Hotton, G., Turkheimer, F., Es, M., Hammers, A., ... Brooks, D. J. (2006). In vivo imaging of microglial activation with [11C](R)-PK11195 PET in idiopathic Parkinson's disease. *Neurobiology of Disease, 21*, 404–412.

Ginhoux, F., Greter, M., Leboeuf, M., Nandi, S., See, P., Gokhan, S., ... Merad, M. (2010). Fate mapping analysis reveals that adult microglia derive from primitive macrophages. *Science (New York, N.Y.), 330*, 841–845.

Ginhoux, F., Lim, S., Hoeffel, G., Low, D., & Huber, T. (2013). Origin and differentiation of microglia. *Frontiers in Cellular Neuroscience, 7*, 45.

Goedert, M., Falcon, B., Clavaguera, F., & Tolnay, M. (2014). Prion-like mechanisms in the pathogenesis of tauopathies and synucleinopathies. *Current Neurology and Neuroscience Reports, 14*, 495.

Goldmann, J., Kwidzinski, E., Brandt, C., Mahlo, J., Richter, D., & Bechmann, I. (2006). T cells traffic from brain to cervical lymph nodes via the cribroid plate and the nasal mucosa. *Journal of Leukocyte Biology, 80*, 797–801.

Goldmann, T., Wieghofer, P., Jordao, M. J., Prutek, F., Hagemeyer, N., Frenzel, K., ... Prinz, M. (2016). Origin, fate and dynamics of macrophages at central nervous system interfaces. *Nature Immunology, 17*, 797–805.

Goldstein, G. W. (1988). Endothelial cell-astrocyte interactions. A cellular model of the blood-brain barrier. *Annals of the New York Academy of Sciences, 529*, 31–39.

Gonzalez, H., Elgueta, D., Montoya, A., & Pacheco, R. (2014). Neuroimmune regulation of microglial activity involved in neuroinflammation and neurodegenerative diseases. *Journal of Neuroimmunology, 274*, 1–13.

Goronzy, J. J., Li, G., Yang, Z., & Weyand, C. M. (2013). The janus head of T cell aging—Autoimmunity and immunodeficiency. *Frontiers in Immunology, 4*, 131.

Grad, L. I., Fernando, S. M., & Cashman, N. R. (2015). From molecule to molecule and cell to cell: Prion-like mechanisms in amyotrophic lateral sclerosis. *Neurobiology of Disease, 77*, 257–265.

Greter, M., Lelios, I., & Croxford, A. L. (2015). Microglia versus myeloid cell nomenclature during brain inflammation. *Frontiers in Immunology, 6*, 249.

Greter, M., Lelios, I., Pelczar, P., Hoeffel, G., Price, J., Leboeuf, M., ... Becher, B. (2012). Stroma-derived interleukin-34 controls the development and maintenance of langerhans cells and the maintenance of microglia. *Immunity, 37*, 1050–1060.

Griciuc, A., Serrano-Pozo, A., Parrado, A. R., Lesinski, A. N., Asselin, C. N., Mullin, K., ... Tanzi, R. E. (2013). Alzheimer's disease risk gene CD33 inhibits microglial uptake of amyloid beta. *Neuron, 78*, 631–643.

Griffin, W. S., Stanley, L. C., Ling, C., White, L., MacLeod, V., Perrot, L. J., ... Araoz, C. (1989). Brain interleukin 1 and S-100 immunoreactivity are elevated in Down syndrome and Alzheimer disease. *Proceedings of the National Academy of Sciences of the United States of America, 86*, 7611–7615.

Guerreiro, R., Wojtas, A., Bras, J., Carrasquillo, M., Rogaeva, E., Majounie, E., ... Group Alzheimer Genetic Analysis. (2013). TREM2 variants in Alzheimer's disease. *The New England Journal of Medicine, 368*, 117–127.

Guo, Y., Deng, X., Zheng, W., Xu, H., Song, Z., Liang, H., ... Deng, H. (2011). HLA rs3129882 variant in Chinese Han patients with late-onset sporadic Parkinson disease. *Neuroscience Letters, 501*, 185–187.

Gyoneva, S., Davalos, D., Biswas, D., Swanger, S. A., Garnier-Amblard, E., Loth, F., ... Traynelis, S. F. (2014). Systemic inflammation regulates microglial responses to tissue damage in vivo. *Glia, 62*, 1345–1360.

Haan, N., Zhu, B., Wang, J., Wei, X., & Song, B. (2015). Crosstalk between macrophages and astrocytes affects proliferation, reactive phenotype and inflammatory response, suggesting a role during reactive gliosis following spinal cord injury. *Journal of Neuroinflammation, 12*, 109.

Hannaoui, S., Shim, S. Y., Cheng, Y. C., Corda, E., & Gilch, S. (2014). Cholesterol balance in prion diseases and Alzheimer's disease. *Viruses, 6*, 4505–4535.

Harty, J. T., Tvinnereim, A. R., & White, D. W. (2000). CD8 + T cell effector mechanisms in resistance to infection. *Annual Review of Immunology, 18*, 275–308.

Hatterer, E., Davoust, N., Didier-Bazes, M., Vuaillat, C., Malcus, C., Belin, M. F., & Nataf, S. (2006). How to drain without lymphatics? Dendritic cells migrate from the cerebrospinal fluid to the B-cell follicles of cervical lymph nodes. *Blood, 107*, 806–812.

Hennessy, E., Gormley, S., Lopez-Rodriguez, A. B., Murray, C., Murray, C., & Cunningham, C. (2016). Systemic TNF-alpha produces acute cognitive dysfunction and exaggerated sickness behavior when superimposed upon progressive neurodegeneration. *Brain, Behavior, and Immunity*.

Henry, J., Smeyne, R. J., Jang, H., Miller, B., & Okun, M. S. (2010). Parkinsonism and neurological manifestations of influenza throughout the 20th and 21st centuries. *Parkinsonism & Related Disorders, 16*, 566–571.

Hickey, W. F., & Kimura, H. (1988). Perivascular microglial cells of the CNS are bone marrow-derived and present antigen in vivo. *Science (New York, N.Y.), 239*, 290–292.

Hobson, P., & Meara, J. (2004). Risk and incidence of dementia in a cohort of older subjects with Parkinson's disease in the United Kingdom. *Movement Disorders: Official Journal of the Movement Disorder Society, 19*, 1043–1049.

Hollingworth, P., Harold, D., Sims, R., Gerrish, A., Lambert, J. C., Carrasquillo, M. M., ... Williams, J. (2011). Common variants at ABCA7, MS4A6A/MS4A4E, EPHA1, CD33 and CD2AP are associated with Alzheimer's disease. *Nature Genetics, 43*, 429–435.

Holmes, C., Cunningham, C., Zotova, E., Woolford, J., Dean, C., Kerr, S., ... Perry, V. H. (2009). Systemic inflammation and disease progression in Alzheimer disease. *Neurology, 73*, 768–774.

Holmes, C., El-Okl, M., Williams, A., Cunningham, C., Wilcockson, D., & Perry, V. H. (2003). Systemic infection, interleukin 1β, and cognitive decline in Alzheimer's disease. *Journal of Neurology, Neurosurgery, and Psychiatry, 74*(6), 788-9.

Hong, S., Beja-Glasser, V. F., Nfonoyim, B. M., Frouin, A., Li, S., Ramakrishnan, S., ... Stevens, B. (2016). Complement and microglia mediate early synapse loss in Alzheimer mouse models. *Science (New York, N.Y.), 352*, 712–716.

Hooten, K. G., Beers, D. R., Zhao, W., & Appel, S. H. (2015). Protective and toxic neuroinflammation in amyotrophic lateral sclerosis. *Neurotherapeutics: The Journal of the American Society for Experimental NeuroTherapeutics, 12*, 364–375.

Houser, Madelyn C., & Tansey, Malú G. (2017). The gut-brain axis: Is intestinal inflammation a silent driver of Parkinson's disease pathogenesis?. *NPJ Parkinson's Disease, 3*, 3.

Hughes, T. A., Ross, H. F., Musa, S., Bhattacherjee, S., Nathan, R. N., Mindham, R. H., & Spokes, E. G. (2000). A 10-year study of the incidence of and factors predicting dementia in Parkinson's disease. *Neurology, 54*, 1596–1602.

Imamura, K., Hishikawa, N., Sawada, M., Nagatsu, T., Yoshida, M., & Hashizume, Y. (2003). Distribution of major histocompatibility complex class II-positive microglia and cytokine profile of Parkinson's disease brains. *Acta Neuropathologica, 106*, 518–526.

Jang, Y., Koo, J. H., Kwon, I., Kang, E. B., Um, H. S., Soya, H., ... Cho, J. Y. (2017). Neuroprotective effects of endurance exercise against neuroinflammation in MPTP-induced Parkinson's disease mice. *Brain Research, 1655*, 186–193.

Jay, T. R., Miller, C. M., Cheng, P. J., Graham, L. C., Bemiller, S., Broihier, M. L., ... Lamb, B. T. (2015). TREM2 deficiency eliminates TREM2 + inflammatory macrophages and ameliorates pathology in Alzheimer's disease mouse models. *The Journal of Experimental Medicine, 212*, 287–295.

Jones, L., Holmans, P. A., Hamshere, M. L., Harold, D., Moskvina, V., Ivanov, D., ... Williams, J. (2010). Genetic evidence implicates the immune system and cholesterol metabolism in the aetiology of Alzheimer's disease. *PLoS One, 5*, e13950.

Jucker, M., & Walker, L. C. (2011). Pathogenic protein seeding in Alzheimer disease and other neurodegenerative disorders. *Annals of Neurology, 70*, 532–540.

Kamer, A., Dasanayake, A., Craig, R., Glodzik-Sobanska, L., Bry, M., & de Leon, M. (2008). Alzheimer's disease and peripheral infections: The possible contribution from periodontal infections, model and hypothesis. *Journal of Alzheimer's Disease: JAD, 13*, 437–449.

Kannarkat, G. T., Cook, D. A., Lee, J. K., Chang, J., Chung, J., Sandy, E., ... Tansey, M. G. (2015). Common genetic variant association with altered HLA expression, synergy with pyrethroid exposure, and risk for Parkinson's disease: An observational and case-control study. *NPJ Parkinsons Disease, 1*, pii: 15002.

Kierdorf, K., Katzmarski, N., Haas, C. A., & Prinz, M. (2013). Bone marrow cell recruitment to the brain in the absence of irradiation or parabiosis bias. *PLoS One, 8*, e58544.

Kipnis, J. (2016). Multifaceted interactions between adaptive immunity and the central nervous system. *Science (New York, N.Y.), 353*, 766–771.

Kiyota, T., Yamamoto, M., Xiong, H., Lambert, M. P., Klein, W. L., Gendelman, H. E., ... Ikezu, T. (2009). CCL2 accelerates microglia-mediated Abeta oligomer formation and progression of neurocognitive dysfunction. *PLoS One, 4*, e6197.

Laman, J. D., & Weller, R. O. (2013). Drainage of cells and soluble antigen from the CNS to regional lymph nodes. *Journal of Neuroimmune Pharmacology: The Official Journal of the Society on NeuroImmune Pharmacology, 8*, 840–856.

Lambert, J. C., Grenier-Boley, B., Chouraki, V., Heath, S., Zelenika, D., Fievet, N., ... Amouyel, P. (2010). Implication of the immune system in Alzheimer's disease: Evidence from genome-wide pathway analysis. *Journal of Alzheimer's disease: JAD, 20*, 1107–1118.

Lambert, J. C., Ibrahim-Verbaas, C. A., Harold, D., Naj, A. C., Sims, R., Bellenguez, C., ... Amouyel, P. (2013). Meta-analysis of 74,046 individuals identifies 11 new susceptibility loci for Alzheimer's disease. *Nature Genetics, 45*, 1452–1458.

Lauretti, E., Di Meco, A., Merali, S., & Pratico, D. (2016). Chronic behavioral stress exaggerates motor deficit and neuroinflammation in the MPTP mouse model of Parkinson's disease. *Translational Psychiatry, 6*, e733.

Lee, E. B., & Mattson, M. P. (2014). *The neuropathology of obesity: Insights from human disease, . Acta Neuropathol* (127, pp. 3–28).

Lee, S. H., Zabolotny, J. M., Huang, H., Lee, H., & Kim, Y. B. (2016). Insulin in the nervous system and the mind: Functions in metabolism, memory, and mood. *Molecular Metabolism, 5*, 589–601.

Lill, C. M. (2016). Genetics of Parkinson's disease. *Molecular and Cellular Probes, 30*(6), 386–396.

Limatola, C., & Ransohoff, R. M. (2014). Modulating neurotoxicity through CX3CL1/CX3CR1 signaling. *Frontiers in Cellular Neuroscience, 8*, 229.

Louveau, A., Harris, T. H., & Kipnis, J. (2015). Revisiting the mechanisms of CNS immune privilege. *Trends in Immunology, 36*, 569–577.

Louveau, A., Smirnov, I., Keyes, T. J., Eccles, J. D., Rouhani, S. J., Peske, J. D., ... Kipnis, J. (2015). Structural and functional features of central nervous system lymphatic vessels. *Nature, 523*, 337–341.

Luckheeram, R. V., Zhou, R., Verma, A. D., & Xia, B. (2012). CD4(+)T cells: Differentiation and functions. *Clinical & Developmental Immunology, 2012*, 925135.

MacEwan, D. (2002). TNF receptor subtype signalling: Differences and cellular consequences. *Cellular Signalling*, 4(6), 477-92.

Machado, A., Herrera, A. J., de Pablos, R. M., Espinosa-Oliva, A. M., Sarmiento, M., Ayala, A., ... Cano, J. (2014). Chronic stress as a risk factor for Alzheimer's disease. *Reviews in the Neurosciences, 25*, 785–804.

Mackay, F., Loetscher, H., Stueber, D., Gehr, G., & Lesslauer, W. (1993). Tumor necrosis factor alpha (TNF-alpha)-induced cell adhesion to human endothelial cells is under dominant control of one TNF receptor type, TNF-R55. *The Journal of Experimental Medicine, 177*, 1277–1286.

MacPherson, K. P., Sompol, P., Kannarkat, G. T., Chang, J., Sniffen, L., Wildner, M. E., ... Tansey, M. G. (2017). Peripheral administration of the soluble TNF inhibitor XPro1595 modifies brain immune cell profiles, decreases beta-amyloid plaque load, and rescues impaired long-term potentiation in 5xFAD mice. *Neurobiology of Disease, 102*, 81–95.

Maloney, B., & Lahiri, D. K. (2016). Epigenetics of dementia: Understanding the disease as a transformation rather than a state. *Lancet Neurology, 15*, 760–774.

Marsh, S. E., Abud, E. M., Lakatos, A., Karimzadeh, A., Yeung, S. T., Davtyan, H., ... Blurton-Jones, M. (2016). The adaptive immune system restrains Alzheimer's disease pathogenesis by modulating microglial function. *Proceedings of the National Academy of Sciences of the United States of America, 113*, E1316–E1325.

Martin, E., Boucher, C., Fontaine, B., & Delarasse, C. (2017). Distinct inflammatory phenotypes of microglia and monocyte-derived macrophages in Alzheimer's disease models: Effects of aging and amyloid pathology. *Aging Cell, 16*, 27–38.

McAlpine, F. E., & Tansey, M. G. (2008). Neuroinflammation and tumor necrosis factor signaling in the pathophysiology of Alzheimer's disease. *Journal of Inflammation Research, 1*, 29–39.

McCoy, M., & Tansey, M. (2008). TNF signaling inhibition in the CNS: Implications for normal brain function and neurodegenerative disease. *Journal of Neuroinflammation, 5*, 45.

McGeer, P., & McGeer, E. (2007). NSAIDs and Alzheimer disease: Epidemiological, animal model and clinical studies. *Neurobiology of Aging, 28*, 639–647.

McGeer, P. L., Itagaki, S., Boyes, B. E., & McGeer, E. G. (1988). Reactive microglia are positive for HLA-DR in the substantia nigra of Parkinson's and Alzheimer's disease brains. *Neurology, 38*, 1285–1291.

McGeer, P. L., Itagaki, S., Tago, H., & McGeer, E. G. (1988). Occurrence of HLA-DR reactive microglia in Alzheimer's disease. *Annals of the New York Academy of Sciences, 540*, 319–323.

McGeer, P. L., Rogers, J., & McGeer, E. G. (2016). Inflammation, antiinflammatory agents, and Alzheimer's disease: The last 22 years. *Journal of Alzheimer's Disease: JAD, 54*, 853–857.

McGuire, M. J., & Ishii, M. (2016). Leptin dysfunction and Alzheimer's disease: Evidence from cellular, animal, and human studies. *Cellular and Molecular Neurobiology, 36*, 203–217.

McManus, R. M., & Heneka, M. T. (2017). Role of neuroinflammation in neurodegeneration: New insights. *Alzheimer's Research & Therapy*, *9*, 14.

Mena, M. A., & Garcia de Yebenes, J. (2008). Glial cells as players in parkinsonism: The "good," the "bad," and the "mysterious" glia. *The Neuroscientist: A Review Journal Bringing Neurobiology, Neurology and Psychiatry*, *14*, 544–560.

Meyer-Luehmann, M., & Prinz, M. (2015). Myeloid cells in Alzheimer's disease: Culprits, victims or innocent bystanders? *Trends in Neurosciences*, *38*, 659–668.

Miklossy, J., & McGeer, P. L. (2016). Common mechanisms involved in Alzheimer's disease and type 2 diabetes: A key role of chronic bacterial infection and inflammation. *Aging (Albany NY)*, *8*, 575–588.

Miller, A. A., & Spencer, S. J. (2014). Obesity and neuroinflammation: A pathway to cognitive impairment. *Brain, Behavior, and Immunity*, *42*, 10–21.

Misiak, B., Leszek, J., & Kiejna, A. (2012). Metabolic syndrome, mild cognitive impairment and Alzheimer's disease--the emerging role of systemic low-grade inflammation and adiposity. *Brain Research Bulletin*, *89*, 144–149.

Morgan, K. (2011). The three new pathways leading to Alzheimer's disease. *Neuropathology and Applied Neurobiology*, *37*, 353–357.

Munch, C., & Bertolotti, A. (2010). Exposure of hydrophobic surfaces initiates aggregation of diverse ALS-causing superoxide dismutase-1 mutants. *Journal of Molecular Biology*, *399*, 512–525.

Murray, C., Sanderson, D., Barkus, C., Deacon, R., Rawlins, J., Bannerman, D., & Cunningham, Colm (2012). Systemic inflammation induces acute working memory deficits in the primed brain: Relevance for delirium. *Neurobiology of Aging*, *33*, 603–616000.

Musaelyan, K., Egeland, M., Fernandes, C., Pariante, C. M., Zunszain, P. A., & Thuret, S. (2014). Modulation of adult hippocampal neurogenesis by early-life environmental challenges triggering immune activation. *Neural Plasticity*, *2014*, 194396.

Nadeau, S., & Rivest, S. (1999). Effects of circulating tumor necrosis factor on the neuronal activity and expression of the genes encoding the tumor necrosis factor receptors (p55 and p75) in the rat brain: A view from the blood-brain barrier. *Neuroscience*, *93*, 1449–1464.

Nalls, M. A., Couper, D. J., Tanaka, T., van Rooij, F. J., Chen, M. H., Smith, A. V., … Ganesh, S. K. (2011). Multiple loci are associated with white blood cell phenotypes. *PLoS Genetics*, *7*, e1002113.

Neumann, M., Sampathu, D. M., Kwong, L. K., Truax, A. C., Micsenyi, M. C., Chou, T. T., … Lee, V. M. (2006). Ubiquitinated TDP-43 in frontotemporal lobar degeneration and amyotrophic lateral sclerosis. *Science (New York, N.Y.)*, *314*, 130–133.

Nevado-Holgado, A. J., & Lovestone, S. (2017). Determining the molecular pathways underlying the protective effect of non-steroidal anti-inflammatory drugs for Alzheimer's disease: A bioinformatics approach. *Computational and Structural Biotechnology Journal*, *15*, 1–7.

Ng, R. C., Cheng, O. Y., Jian, M., Kwan, J. S., Ho, P. W., Cheng, K. K., … Chan, K. H. (2016). Chronic adiponectin deficiency leads to Alzheimer's disease-like cognitive impairments and pathologies through AMPK inactivation and cerebral insulin resistance in aged mice. *Molecular Neurodegeneration*, *11*, 71.

Oakley, H., Cole, S., Logan, S., Maus, E., Shao, P., Craft, J., … Vassar, R. (2006). Intraneuronal beta-amyloid aggregates, neurodegeneration, and neuron loss in transgenic mice with five familial Alzheimer's disease mutations: Potential factors in amyloid plaque formation. *The Journal of Neuroscience: The Official Journal of the Society for Neuroscience*, *26*, 10129–10140.

Ojo, O., & Brooke, J. (2015). Evaluating the association between diabetes, cognitive decline and dementia. *International Journal of Environmental Research and Public Health*, *12*, 8281–8294.

Pan, W., Stone, K. P., Hsuchou, H., Manda, V. K., Zhang, Y., & Kastin, A. J. (2011). Cytokine signaling modulates blood-brain barrier function. *Current Pharmaceutical Design*, *17*, 3729–3740.

Pankratz, N., Beecham, G. W., DeStefano, A. L., Dawson, T. M., Doheny, K. F., Factor, S. A., … Pd Gwas Consortium. (2012). Meta-analysis of Parkinson's disease: Identification of a novel locus, RIT2. *Annals of Neurology*, *71*, 370–384.

Parks, C. D., Walitt, B. T., Pettinger, M., Chen, J. C., de Roos, A. J., Hunt, J., … Howard, B. V. (2011). Insecticide use and risk of rheumatoid arthritis and systemic lupus erythematosus in the women's health initiative observational study. *Arthritis Care & Research*, *63*, 184–194.

Pavlov, V. A., & Tracey, K. J. (2015). Neural circuitry and immunity. *Immunologic Research*, *63*, 38–57.

Perry, V., Cunningham, C., & Holmes, C. (2007). Systemic infections and inflammation affect chronic neurodegeneration. *Nature Reviews. Immunology*, *7*, 161–167.

Polymenidou, M., & Cleveland, D. W. (2012). Prion-like spread of protein aggregates in neurodegeneration. *The Journal of Experimental Medicine*, *209*, 889–893.

Prinz, M., Priller, J., Sisodia, S., & Ransohoff, R. (2011). Heterogeneity of CNS myeloid cells and their roles in neurodegeneration. *Nature Neuroscience*, *14*, 1227–1235.

Prokop, S., Miller, K. R., Drost, N., Handrick, S., Mathur, V., Luo, J., … Heppner, F. L. (2015). Impact of peripheral myeloid cells on amyloid-beta pathology in Alzheimer's disease-like mice. *The Journal of Experimental Medicine*, *212*, 1811–1818.

Qin, L., Wu, X., Block, M., Liu, Y., Breese, G., Hong, J., ... Crews, F. (2007). Systemic LPS causes chronic neuroinflammation and progressive neurodegeneration. *Glia, 55*, 453–462.

Raj, T., Shulman, J. M., Keenan, B. T., Chibnik, L. B., Evans, D. A., Bennett, D. A., ... De Jager, P. L. (2012). Alzheimer disease susceptibility loci: Evidence for a protein network under natural selection. *American Journal of Human Genetics, 90*, 720–726.

Ransohoff, R. M., & Engelhardt, B. (2012). The anatomical and cellular basis of immune surveillance in the central nervous system. *Nature Reviews. Immunology, 12*, 623–635.

Ransohoff, R. M., & Perry, V. H. (2009). Microglial physiology: Unique stimuli, specialized responses. *Annual Review of Immunology, 27*, 119–145.

Ransohoff, R. M., Schafer, D., Vincent, A., Blachere, N. E., & Bar-Or, A. (2015). Neuroinflammation: Ways in which the immune system affects the brain. *Neurotherapeutics: The Journal of the American Society for Experimental NeuroTherapeutics, 12*, 896–909.

Rao, M., & Gershon, M. D. (2016). The bowel and beyond: The enteric nervous system in neurological disorders. *Nature Reviews Gastroenterology & Hepatology, 13*, 517–528.

Rezai-Zadeh, K., Gate, D., & Town, T. (2009). CNS infiltration of peripheral immune cells: D-Day for neurodegenerative disease?. *Journal of Neuroimmune Pharmacology: The Official Journal of the Society on NeuroImmune Pharmacology, 4*, 462–475.

Ricci, S., Fuso, A., Ippoliti, F., & Businaro, R. (2012). Stress-induced cytokines and neuronal dysfunction in Alzheimer's disease. *Journal of Alzheimer's disease: JAD, 28*, 11–24.

Rijkers, E. S., de Ruiter, T., Baridi, A., Veninga, H., Hoek, R. M., & Meyaard, L. (2008). The inhibitory CD200R is differentially expressed on human and mouse T and B lymphocytes. *Molecular Immunology, 45*, 1126–1135.

Rissman, R. A., Staup, M. A., Lee, A. R., Justice, N. J., Rice, K. C., Vale, W., & Sawchenko, P. E. (2012). Corticotropin-releasing factor receptor-dependent effects of repeated stress on tau phosphorylation, solubility, and aggregation. *Proceedings of the National Academy of Sciences of the United States of America, 109*, 6277–6282.

Rosas-Ballina, M., Olofsson, P. S., Ochani, M., Valdes-Ferrer, S. I., Levine, Y. A., Reardon, C., ... Tracey, K. J. (2011). Acetylcholine-synthesizing T cells relay neural signals in a vagus nerve circuit. *Science (New York, N.Y.), 334*, 98–101.

Rosenkranz, D., Weyer, S., Tolosa, E., Gaenslen, A., Berg, D., Leyhe, T., ... Stoltze, L. (2007). Higher frequency of regulatory T cells in the elderly and increased suppressive activity in neurodegeneration. *Journal of Neuroimmunology, 188*, 117–127.

Saederup, N., Cardona, A., Croft, K., Mizutani, M., Cotleur, A., Tsou, C., ... Charo, I. (2010). Selective chemokine receptor usage by central nervous system myeloid cells in CCR2-red fluorescent protein knock-in mice. *PLoS One, 5*, e13693.

Sallusto, F., Geginat, J., & Lanzavecchia, A. (2004). Central memory and effector memory T cell subsets: Function, generation, and maintenance. *Annual Review of Immunology, 22*, 745–763.

Samii, A., Etminan, M., Wiens, M. O., & Jafari, S. (2009). NSAID use and the risk of Parkinson's disease: Systematic review and meta-analysis of observational studies. *Drugs & Aging, 26*, 769–779.

Sanchez-Guajardo, V., Barnum, C. J., Tansey, M. G., & Romero-Ramos, M. (2013). Neuroimmunological processes in Parkinson's disease and their relation to alpha-synuclein: Microglia as the referee between neuronal processes and peripheral immunity. *ASN Neuro, 5*, 113–139.

Sankowski, R., Mader, S., & Valdes-Ferrer, S. I. (2015). Systemic inflammation and the brain: Novel roles of genetic, molecular, and environmental cues as drivers of neurodegeneration. *Frontiers in Cellular Neuroscience, 9*, 28.

Sauer, B. M., Schmalstieg, W. F., & Howe, C. L. (2013). Axons are injured by antigen-specific CD8 (+) T cells through a MHC class I- and granzyme B-dependent mechanism. *Neurobiology of Disease, 59*, 194–205.

Schafer, D. P., Lehrman, E. K., Kautzman, A. G., Koyama, R., Mardinly, A. R., Yamasaki, R., ... Stevens, B. (2012). Microglia sculpt postnatal neural circuits in an activity and complement-dependent manner. *Neuron, 74*, 691–705.

Schain, M., & Kreisl, W. C. (2017). Neuroinflammation in neurodegenerative disorders—A review. *Current Neurology and Neuroscience Reports, 17*, 25.

Sears, M. E., & Genuis, S. J. (2012). Environmental determinants of chronic disease and medical approaches: Recognition, avoidance, supportive therapy, and detoxification. *Journal of Environmental Research and Public Health, 2012*, 356798.

Sedgwick, J. D., Schwender, S., Imrich, H., Dorries, R., Butcher, G. W., & ter Meulen, V. (1991). Isolation and direct characterization of resident microglial cells from the normal and inflamed central nervous system. *Proceedings of the National Academy of Sciences of the United States of America, 88*, 7438–7442.

Sharma, H., Castellani, R., Smith, M., & Sharma, A. (2012). The blood-brain barrier in Alzheimer's disease: Novel therapeutic targets and nanodrug delivery. *International Review of Neurobiology, 102*, 47–90.

Silva, J. L., De Moura Gallo, C. V., Costa, D. C., & Rangel, L. P. (2014). Prion-like aggregation of mutant p53 in cancer. *Trends in Biochemical Sciences, 39*, 260–267.

Smith, J. A., Das, A., Ray, S. K., & Banik, N. L. (2012). Role of pro-inflammatory cytokines released from microglia in neurodegenerative diseases. *Brain Research Bulletin, 87*, 10–20.

Smith, L. K., Jadavji, N. M., Colwell, K. L., Katrina Perehudoff, S., & Metz, G. A. (2008). Stress accelerates neural degeneration and exaggerates motor symptoms in a rat model of Parkinson's disease. *The European Journal of Neuroscience, 27*, 2133–2146.

Sofroniew, M. V. (2014). Multiple roles for astrocytes as effectors of cytokines and inflammatory mediators. *The Neuroscientist: A Review Journal Bringing Neurobiology, Neurology and Psychiatry, 20*, 160–172.

Spielman, L. J., Little, J. P., & Klegeris, A. (2016). Physical activity and exercise attenuate neuroinflammation in neurological diseases. *Brain Research Bulletin, 125*, 19–29.

Steinman, L. (2013). Inflammatory cytokines at the summits of pathological signal cascades in brain diseases. *Science Signaling, 6*, pe3.

Stewart, L. K., Flynn, M. G., Campbell, W. W., Craig, B. A., Robinson, J. P., McFarlin, B. K., ... Talbert, E. (2005). Influence of exercise training and age on CD14 + cell-surface expression of toll-like receptor 2 and 4. *Brain, Behavior, and Immunity, 19*, 389–397.

Sun, C., Wei, L., Luo, F., Li, Y., Li, J., Zhu, F., ... Xu, P. (2012). HLA-DRB1 alleles are associated with the susceptibility to sporadic Parkinson's disease in Chinese Han population. *PLoS One, 7*, e48594.

Sung, Y. H., Kim, S. C., Hong, H. P., Park, C. Y., Shin, M. S., Kim, C. J., ... Cho, H. J. (2012). Treadmill exercise ameliorates dopaminergic neuronal loss through suppressing microglial activation in Parkinson's disease mice. *Life Sciences, 91*, 1309–1316.

Swardfager, W., Lanctot, K., Rothenburg, L., Wong, A., Cappell, J., & Herrmann, N. (2010). A meta-analysis of cytokines in Alzheimer's disease. *Biological Psychiatry, 68*, 930–941.

Takeda, S., Sato, N., Uchio-Yamada, K., Sawada, K., Kunieda, T., Takeuchi, D., ... Morishita, R. (2010). Diabetes-accelerated memory dysfunction via cerebrovascular inflammation and Abeta deposition in an Alzheimer mouse model with diabetes. *Proceedings of the National Academy of Sciences of the United States of America, 107*, 7036–7041.

Talbot, K., & Wang, H. Y. (2014). The nature, significance, and glucagon-like peptide-1 analog treatment of brain insulin resistance in Alzheimer's disease. *Alzheimer's & Dementia: The Journal of the Alzheimer's Association, 10*, S12–S25.

Tansey, M. G., & Goldberg, M. S. (2010). Neuroinflammation in Parkinson's disease: Its role in neuronal death and implications for therapeutic intervention. *Neurobiology of Disease, 37*, 510–518.

Tanzi, R. E. (2012). The genetics of Alzheimer disease. *Cold Spring Harbor Perspectives in Medicine, 2*, pii: a006296.

Tay, T. L., Hagemeyer, N., & Prinz, M. (2016). The force awakens: Insights into the origin and formation of microglia. *Current Opinion in Neurobiology, 39*, 30–37.

Togo, T., Akiyama, H., Iseki, E., Kondo, H., Ikeda, K., Kato, M., ... Kosaka, K. (2002). Occurrence of T cells in the brain of Alzheimer's disease and other neurological diseases. *Journal of Neuroimmunology, 124*, 83–92.

Tontsch, U., & Rott, O. (1993). Cortical neurons selectively inhibit MHC class II induction in astrocytes but not in microglial cells. *International Immunology, 5*, 249–254.

Tufekci, K. U., Meuwissen, R., Genc, S., & Genc, K. (2012). Inflammation in Parkinson's disease. *Advances in Protein Chemistry and Structural Biology, 88*, 69–132.

van Dijk, G., van Heijningen, S., Reijne, A. C., Nyakas, C., van der Zee, E. A., & Eisel, U. L. (2015). Integrative neurobiology of metabolic diseases, neuroinflammation, and neurodegeneration. *Frontiers in Neuroscience, 9*, 173.

Varvel, N. H., Grathwohl, S. A., Degenhardt, K., Resch, C., Bosch, A., Jucker, M., & Neher, J. J. (2015). Replacement of brain-resident myeloid cells does not alter cerebral amyloid-beta deposition in mouse models of Alzheimer's disease. *The Journal of Experimental Medicine, 212*, 1803–1809.

Verdile, G., Keane, K. N., Cruzat, V. F., Medic, S., Sabale, M., Rowles, J., ... Newsholme, P. (2015). Inflammation and oxidative stress: The molecular connectivity between insulin resistance, obesity, and Alzheimer's disease. *Mediators of Inflammation, 2015*.

Walker, J. M., & Harrison, F. E. (2015). Shared neuropathological characteristics of obesity, type 2 diabetes and Alzheimer's disease: Impacts on cognitive decline. *Nutrients, 7*, 7332–7357.

Walker, L. C., & Jucker, M. (2015). Neurodegenerative diseases: Expanding the prion concept. *Annual Review of Neuroscience*.

Walter, B. A., Valera, V. A., Takahashi, S., Matsuno, K., & Ushiki, T. (2006). Evidence of antibody production in the rat cervical lymph nodes after antigen administration into the cerebrospinal fluid. *Archives of Histology and Cytology, 69*, 37–47.

Westermark, G. T., & Westermark, P. (2010). Prion-like aggregates: Infectious agents in human disease. *Trends in Molecular Medicine, 16*, 501–507.

Whitton, P. S. (2007). Inflammation as a causative factor in the aetiology of Parkinson's disease. *British Journal of Pharmacology, 150*, 963–976.

Wohleb, E. S., & Delpech, J. C. (2017). Dynamic cross-talk between microglia and peripheral monocytes underlies stress-induced neuroinflammation and behavioral consequences. *Progress in Neuro-Psychopharmacology & Biological Psychiatry*, *79*, 40–48.

Xu, W. L., Atti, A. R., Gatz, M., Pedersen, N. L., Johansson, B., & Fratiglioni, L. (2011). Midlife overweight and obesity increase late-life dementia risk: A population-based twin study. *Neurology*, *76*, 1568–1574.

Yiu, J. H., Dorweiler, B., & Woo, C. W. (2016). Interaction between gut microbiota and toll-like receptor: From immunity to metabolism. *Journal of Molecular Medicine (Berlin, Germany)*.

Zhao, Q., Wu, X., Yan, S., Xie, X., Fan, Y., Zhang, J., ... You, Z. (2016). The antidepressant-like effects of pioglitazone in a chronic mild stress mouse model are associated with PPARgamma-mediated alteration of microglial activation phenotypes. *Journal of Neuroinflammation*, *13*, 259.

Zlokovic, B. V. (2008). The blood-brain barrier in health and chronic neurodegenerative disorders. *Neuron*, *57*, 178–201.

CHAPTER

17

Neurodegenerative Diseases and the Aging Brain

Stephen K. Godin, Jinsoo Seo and Li-Huei Tsai

Massachusetts Institute of Technology, Cambridge, MA, United States

OUTLINE

GENERAL MECHANISMS UNDERLYING NEURONAL CELL DYSFUNCTION AND COGNITIVE DECLINE

It is natural to believe that the difficulty that older individuals have with learning and memory are due to simply having more information to process and parse, but unfortunately, studies from humans, nonhuman primates and rodents have consistently shown that age-dependent cognitive declines are caused by alterations, mostly deleterious, of multiple factors involved in brain function. Significantly more work is required to fully understand the mechanisms underlying age-related cognitive decline to develop effective therapeutics to prevent the cognitive impairments associated with neurodegeneration. Decades of research have gone into cataloging the changes that occur in every cellular system with age, and significant work exists detailing the specific defects known to occur in

The Molecular and Cellular Basis of Neurodegenerative Diseases
DOI: https://doi.org/10.1016/B978-0-12-811304-2.00017-1

the aging brain. Declines in neuronal protein quality control mechanisms, mitochondrial function, and DNA damage repair pathways have been extensively documented. Importantly, they also contribute to a cellular environment that is susceptible to neurodegeneration. It is important to note that there are crucial differences between the changes that are associated with cognitive impairment in normal aging brains compared to neurodegenerative brains. Critically, the neurodegenerating brain shows significant levels of neuron death compared to healthy older brains, suggesting that additional mechanisms underlie the severe cognitive declines in neurodegeneration. It is also important to note that much of aged population show minimal cognitive declines, with only certain individuals exhibiting significant cognitive impairment. What key factors differentiate these high-functioning aged brains from cognitively impaired, non-neurodegenerative brains remain unknown. Moving forward, the accumulated data from transcriptome analysis could help illuminate which genetic factors determine the impact of aging on cognitive function. In this chapter, we go over recent work on the different age-associated declines in the cellular quality control mechanisms and how these changes make the aging brain susceptible to neurodegenerative stimuli.

PROTEIN DEGRADATION AND SYNAPSE LOSS

One of the common hallmarks of neurodegenerative diseases is the toxic buildup of protein aggregates in the brain. Clear examples include the accrual of misfolded prions during Creutzfeldt–Jakob disease (Chapters 2 and 8), the buildup of mutant Huntington protein in Huntington's disease (HD) (Chapter 7), alpha-synuclein deposits during Parkinson's disease (PD) (Chapter 6), and the accumulation of extracellular amyloid beta plaques and intracellular neurofibrillary tangles of Tau protein in Alzheimer's disease (AD) (Chapters 3 and 4). The formation of these deposits is a slow process that takes decades and requires systemic failures of the protein folding machinery (Chapter 9), the ubiquitin–proteasome protein degradation pathway (Chapter 10), and autophagy (Chapter 11). While these protein aggregates take years to form, their deposition in the aging brain is likely accelerated due to age-associated declines in the efficacy of each pathway as discussed below.

Importantly, a wealth of literature supports the idea that these protein aggregates are toxic to local neurons, and their presence results in widespread synapse dysfunction, loss, and eventual neuronal death. Although their direct mechanisms of toxicity have been extensively studied and include aberrant activation of NMDA receptors, reactive microgliosis, and ROS production, they also exert a deleterious effect on neurons through indirect mechanisms such as simultaneously inhibiting the proteasomal degradation of damaged proteins and inducing the unfolded protein response (UPR). These defects result in endoplasmic reticulum (ER) stress, mitochondrial dysfunction and ultimately reduce protein synthesis to a level unable to support synapse function (Scheper & Hoozemans, 2015). Age-associated defects in the UPR, the ubiquitin–proteasome pathway, and autophagy are discussed below in the context of both normal aging and neurodegeneration, and how these defects result in synapse loss in neurons.

UNFOLDED PROTEIN RESPONSE

As discussed in detail in Chapter 9, one of the major ways that a cell prevents the accumulation of the kinds of misfolded proteins that characterize neurodegenerative diseases is through the UPR. To briefly summarize the UPR, under basal conditions of low stress, the protein folding chaperone BiP is bound to the sensors of unfolded protein stress, ATF6, PERK, and IRE1 in the ER (Joshi, Kornfeld, & Mochly-Rosen, 2016). However, if the cell is exposed to significant amounts of unfolded proteins, an increasing amount of BiP binds to the unfolded proteins, releasing ATF6, PERK, and IRE1 and allowing their activation and signaling to induce the UPR as detailed in Chapter 9. During normal aging, the expression and functionality of multiple branches of the UPR decline in the brain which results in higher basal levels of unfolded proteins and a less pronounced and effective UPR during stress conditions. During aging, the protein chaperone BiP decreases in the cortex and hippocampus which reduces the efficacy of basal protein folding (Taylor, 2016). Beyond sensing unfolded proteins, the levels of numerous ER protein folding chaperones, such as GRP94, calnexin, calreticulin, thiol-disulfide oxidoreductase, and disulfide isomerase all decrease with age, resulting in a reduced capacity to deal with the unfolded proteins that characterize neurodegenerative disease (Taylor, 2016). At the same time that the protein folding machinery declines, the sensors of UPR also become less effective with age. Sensors of the UPR, such as PERK, decline in the rat hippocampus during aging, while levels of GADD34, which bypass the UPR-induced inhibition of translation, increase with age. Together, these changes reduce the basal ability of the cell to correctly fold proteins and to mount a proper UPR during neurodegeneration.

During many forms of neurodegeneration, chronic activation of the UPR is observed in animal models and postmortem tissue. In AD, BiP expression is higher in postmortem brains compared to control, and other studies have found elevated phospho-PERK, phospho-IRE1, and phospho-eIF2 alpha in AD neurons which suggest a role for chronic and systemic UPR activation in AD neurodegeneration (Joshi et al., 2016). Suggesting a functional role for the UPR in AD, the level of UPR activation increases with the Braak stages of AD progression (Joshi et al., 2016). Although how the UPR contributes to neurodegeneration is an area of active research, there are several ways chronic UPR may result in neuronal dysfunction. First, activation of the UPR triggers excessive expression of BACE1, resulting in an increase in beta-amyloid production (Scheper & Hoozemans, 2015). Thus, chronic, low-level UPR activation caused by aging may result in more rapid accumulation of beta -amyloid into plaques, elevating AD progression. Additionally, prolonged activation of the UPR may result in cell death through both apoptosis and necrosis, contributing to the neuronal loss observed in these systems (Iurlaro & Munoz-Pinedo, 2016). Finally, a direct outcome of prolonged UPR activation is a global reduction in protein synthesis which has been shown to reduce the protein transport required for synapse function, eventually leading to the synapse loss that is a common feature of multiple neurodegenerative pathways (Gong, Radulovic, Figueiredo-Pereira, & Cardozo, 2016; Patrick, 2006; Zhao, Hegde, & Martin, 2003).

Beyond the direct effects of poor protein maintenance in neuronal function, the UPR results in systemic ER stress which has severe consequences for neuronal health. The ER serves as an essential calcium storage for the neuron that plays critical roles in both mitochondrial health as well as neuronal synapse transmission. One outcome of ER stress, induced by activation of the UPR, is calcium release by the ER, resulting in excess NADH production by multiple calcium-dependent dehydrogenases that in turn results in an increase in mitochondrial respiration and an increased production of reactive oxygen species (ROS) as shown in Fig. 17.1 (De Mario, Quintana-Cabrera, Martinvalet, & Giacomello, 2016). During periods of chronic stress, such as that induced by the accumulation of intracellular tau tangles in AD, the elevated ROS produced by the mitochondria further drives

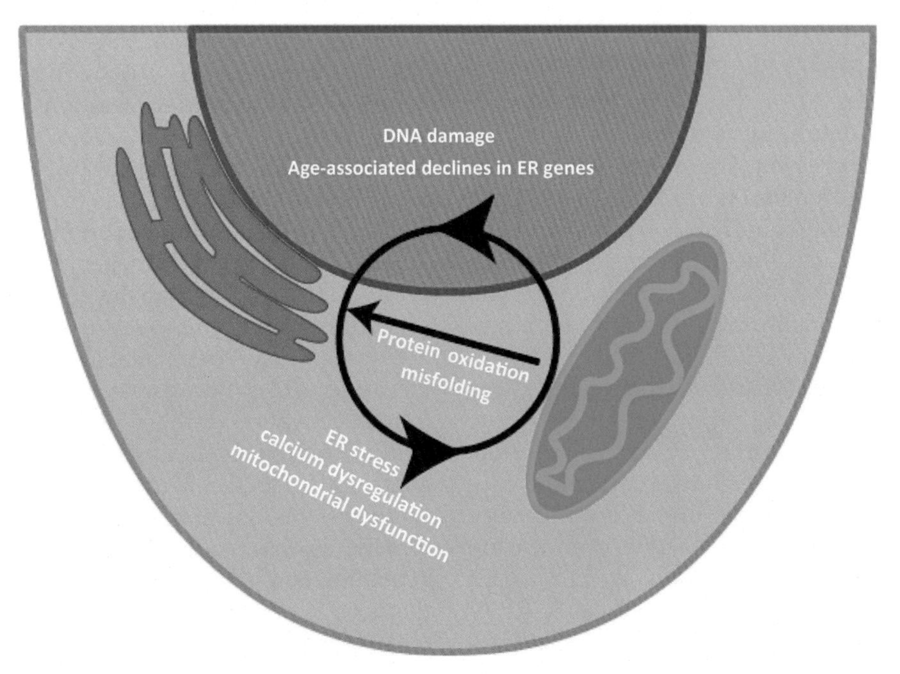

FIGURE 17.1 A schematic demonstrating how the decline in ER function can in turn drive mitochondrial stress and elevated DNA damage through elevated ROS production, forming a positive feedback loop of worsening metabolic function.

protein misfolding and damage, taxing the declining capabilities of the UPR as well as the ubiquitin–proteasome system (UPS) and autophagy, as discussed below (De Mario et al., 2016). Additionally, the elevated ROS can result in nuclear DNA damage, a hallmark of many kinds of neurodegeneration, and mitochondrial damage which is present in many PD patients (Kim, Kim, Rhie, & Yoon, 2015). Combined with the age-associated declines in the efficacy of numerous DNA damage repair pathways, as discussed below, the initiation of ROS production by the UPR and mitochondrial dysfunction leads to a further decline in cellular homeostasis.

UBIQUITIN–PROTEASOME

While the UPR prevents the production and release of poorly folded proteins, different quality control systems are used by the cell to maintain the integrity of the proteome via degradation of damaged proteins and organelles. The UPS of protein degradation (discussed in detail in Chapter 10) is an essential protein quality control mechanism in cells that serve to remove misfolded, damaged, or unnecessary proteins. In the UPS, proteins are polyubiquitinated and targeted for degradation by the 26S proteasome, thus removing damaged proteins and ensuring the functionality of most cellular systems. During aging, however, declines occur in the efficacy of the UPS in most tissues, including the brain and spine (Ciechanover & Kwon, 2015; Tydlacka, Wang, Wang, Li, & Li, 2008). Intriguingly, defects in the UPS may be exacerbated by the age-dependent accumulation of UBB + 1, which occurs in the areas of the brain that are most likely to be impacted by AD (van Leeuwen et al., 1998). UBB + 1 is caused by "molecular misreading" of ubiquitin mRNA resulting in a frameshift of ubiquitin that removes Glycine 76, essential for normal polyubiquitin chain formation, as well as adding 20 additional amino acids onto ubiquitin's C-terminal end (Chadwick, Gentle, Strachan, & Layfield, 2012). UBB + 1 accumulates in the brain with age where it acts as a seed, forming multiple polyubiquitin chains that resist 26S proteasomal degradation. These chains are able to act as inhibitors by binding to the proteasome and preventing its degradation of other substrates. The age-dependent decrease in the UPS (Tydlacka et al., 2008) corresponds to a reduced turnover of the proteome, which in turn impairs the functionality of essentially all cellular systems (Dice, 1982; Lopez Salon, Morelli, Castano, Soto, & Pasquini, 2000; Makrides, 1983; Zouambia et al., 2008). The lower efficiency of the UPS additionally places a higher burden on the UPR, discussed above, and the autophagy pathways, described below, which themselves lose efficacy with age. Combined, these defects result in a global decrease in protein quality and a positive feedback loop that can contribute to worsening of proteome quality with age.

Beyond normal aging, the role of the UPS in neurodegenerative disease has been extensively studied. Multiple groups have found that the accumulation of amyloid beta, phosphorylated tau, and other protein aggregates, all inhibit the UPS by blocking its catalytic domain without being degraded, essentially blocking or reducing the flux of proteins through the UPS (Gentier & van Leeuwen, 2015; Gregori, Fuchs, Figueiredo-Pereira, Van Nostrand, & Goldgaber, 1995; Lopez Salon, Pasquini, Besio Moreno, Pasquini, & Soto, 2003). Beyond causing general impairments in cellular homeostasis via a global decrease in cellular proteome quality, the inhibition of the UPS by cellular aggregates can initiate positive feedback loops in different neurodegenerative diseases. In AD, for instance, many members of the presenilin complex such as PS1, APH-1,

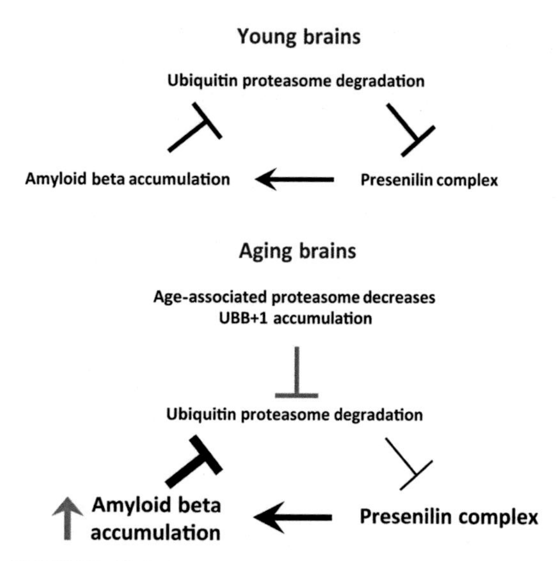

FIGURE 17.2 A model of how one element of amyloid beta accumulation, the Presenilin complex, is normally regulated in young brains, compared to the factors that cause dysregulation of the Presenilin complex and amyloid beta accumulation in the aged brain, potentially setting the stage for neurodegenerative decline to occur.

NCT, and PEN2 are all targeted for degradation in the 26S proteasome (Chadwick et al., 2012). Inhibition of the proteasome results in elevated levels of these components and a corresponding increase in amyloid beta deposition (Chadwick et al., 2012), as shown in Fig. 17.2. Furthermore, the UPS has integral roles in neuronal signaling, neurotransmitter release, and synaptic membrane receptor turnover, all of which contribute to synaptic health and plasticity (Patrick, 2006; Zhao et al., 2003).

AUTOPHAGY

The final method by which the cell can degrade defective proteins and organelles is the lysosome-dependent pathway of autophagy. The various branches of the autophagy pathway (discussed in detail in Chapter 11) have incredibly important roles in preventing the accumulation of the protein aggregates, such as amyloid beta and alpha-synuclein, which characterize neurodegenerative diseases (Xilouri & Stefanis, 2016). Additionally, autophagic degradation of mitochondria is essential for preventing the accumulation of damaged and nonfunctional mitochondria, such as those produced during persistent ER stress as discussed above (Brunk & Terman, 2002b; Masters & O'Neill, 2011; Ntsapi & Loos, 2016; Rubinsztein, Marino, & Kroemer, 2011). As with the UPR, each pathway of autophagy discussed below declines and becomes dysfunctional with age. Impaired autophagy directly contributes to neurodegeneration by failing to clear protein aggregations as well as by disrupting the homeostasis of multiple cellular functions which contribute to a positive feedback loop of increasing cellular stress and dysfunction.

In chaperone-mediated autophagy (CMA), individual proteins are targeted for degradation in the lysosome without the formation of autophagosomes. In this pathway, Hsc70 recognizes and binds a unique, KFERQ-like motif, present in up to 30% of all cytosolic proteins. Hsc70 then directs these proteins to the lysosome where they are translocated into the lysosome by LAMP-2A and degraded (Ntsapi & Loos, 2016; Xilouri & Stefanis, 2016). CMA contributes to the normal degradation of multiple neurodegeneration-associated proteins such as alpha-synuclein in PD, phosphorylated tau in AD, and Huntington protein in HD (Xilouri & Stefanis, 2016). Unfortunately, in aged tissues, lysosomes have significantly reduced translocation of CMA targets into them due to an elevated degradation of LAMP-2A in the lysosome, which could contribute to an increase in the accumulation of the protein aggregates that define different neurodegenerative disorders (Cuervo & Dice, 2000; Kiffin et al., 2007).

In another autophagy pathway, macroautophagy, proteins are sequestered into double membrane–bound vesicles called autophagosomes that are targeted for degradation in the lysosome (Ntsapi & Loos, 2016). This pathway contributes to the degradation of many long-lived cytosolic proteins as well as damaged organelles such as mitochondria during mitophagy. Like CMA, macroautophagy targets many of the proteins involved in neurodegeneration, such as amyloid beta, APP, and PS1 for degradation (Ntsapi & Loos, 2016). And, like the other protein quality control pathways, there is a wealth of literature that finds macroautophagy efficiency declines with aging (Ntsapi & Loos, 2016; Xilouri & Stefanis, 2016). For instance, numerous genes involved in macroautophagy such as Atg6, Atg7, and BCEN1 (discussed in chapter 11) are downregulated in the aged brain (Lipinski et al., 2010). Importantly, in AD postmortem brains, autophagic vacuoles are found to accumulate in dystrophic neurites (Nixon et al., 2005) suggesting that autophagy is overwhelmed and less functional in the AD brain. Unsurprisingly, the defects in macroautophagy that occur with age are also associated with an increase in amyloid beta, alpha-synuclein, or Huntington accumulation (Cherra & Chu, 2008; Ntsapi & Loos, 2016). Together, these defects would directly contribute to the elevated levels of the proteins that form the toxic aggregates of several neurodegenerative diseases

Finally, defects in the lysosomes themselves are evident in older tissues, which contribute to a global decrease in autophagic capacity regardless of the pathway. Older tissues are well known to accumulate lipofuscin, the so-called "aging pigment," formed by the incomplete degradation of cytosolic products (Ntsapi & Loos, 2016). Numerous studies have suggested a role for lipofuscin in the age-related decline in lysosomal function, which contributes to a lower turnover of damaged proteins and mitochondria in neurons (Brunk & Terman, 2002a; Nunomura & Miyagishi, 1993; Oenzil, Kishikawa, Mizuno, & Nakano, 1994;

Porta, 2002; Sulzer et al., 2008; Terman & Brunk, 1998). The inability to rapidly remove damaged mitochondria in particular may contribute to a positive feedback loop of worsening cellular stress through multiple mechanisms involving elevated ROS production. First, ROS can directly damage proteins, placing a higher burden on the already-defective protein degradation pathways, allowing a further accumulation of damaged mitochondria and protein aggregates in neurons. The excess ROS can additionally damage nuclear DNA, causing a decrease in the expression of genes involved in preventing the accumulation of neurodegenerative protein aggregates. One example is p62/SQSTM1, Sequestosome-1, which functions to recruit polyubiquitinated protein aggregates to autophagosomes (Ntsapi & Loos, 2016). In AD patients, p62 expression is dramatically decreased due to oxidative damage of its promoter, which would contribute to a general decline in the protein and mitochondrial quality control mechanisms, resulting in a worsening positive feedback loop (Babu, Geetha, & Wooten, 2005; Du, Wooten, Gearing, & Wooten, 2009; Du, Wooten, & Wooten, 2009). Coupled with the age-related decline in DNA repair mechanisms discussed below, failures in the protein and mitochondria degradation pathways have clear implications in causing a worsening cycle of cellular stress that defines neurodegeneration.

OXIDATIVE DAMAGE IN THE AGING AND NEURODEGENERATING BRAIN

The brain has one of the highest potentials for oxidative damage in the human body. Despite its small size, it utilizes approximately 20% of the total oxygen consumed, and many studies suggest that up to 1%–2% of the total oxygen consumed by the mitochondria is converted to ROS (Cadenas & Davies, 2000; Kim et al., 2015). During aging, the expression of mitochondrial genes in the brain has been shown to decline, indicating impaired mitochondrial function which may further elevate the levels of ROS production in the aged brain (Bishop, Lu, & Yankner, 2010; Blalock et al., 2004; Liang et al., 2008; Miller, Oldham, & Geschwind, 2008). The extent of oxidative damage in aging is also supported by the observation that from mice to humans, the expression of oxidative stress response genes increases with age (Bishop et al., 2010; Yankner, Lu, & Loerch, 2008). This elevated ROS production likely results in persistent DNA damage which may underlie some of the physiological changes seen in the aging brain. For example, the promoters of many of the genes known to decline with age in the brain show elevated levels of oxidative damage, suggesting that ROS-mediated DNA damage is driving their reduced expression (Bishop et al., 2010; Lu et al., 2004; Ntsapi & Loos, 2016). Reduced protein levels of numerous quality control genes could contribute to a positive feedback loop of worsening damage, leading to further impairment of neuronal function as summarized in Fig. 17.3. For instance, as detailed above, p62 has important roles in protein and mitochondrial quality control, and its expression is reduced in older brains due to oxidative damage throughout its promoter. Its reduced expression may contribute to more mitochondrial dysfunction and more ROS production, speeding decline of other genes in the aging brain. Further work is needed to determine if many of the other quality control genes whose expression declines with age are also driven by oxidative damage in their promoters.

Although the normal aging brain shows clear signs of oxidative damage, the neurodegenerating brain is further enriched for different DNA damage signatures (Hou, Song, Croteau, Akbari, & Bohr, 2016; Lu et al., 2004). Compared to age-matched controls, neurodegenerating brains have an excess of oxidative

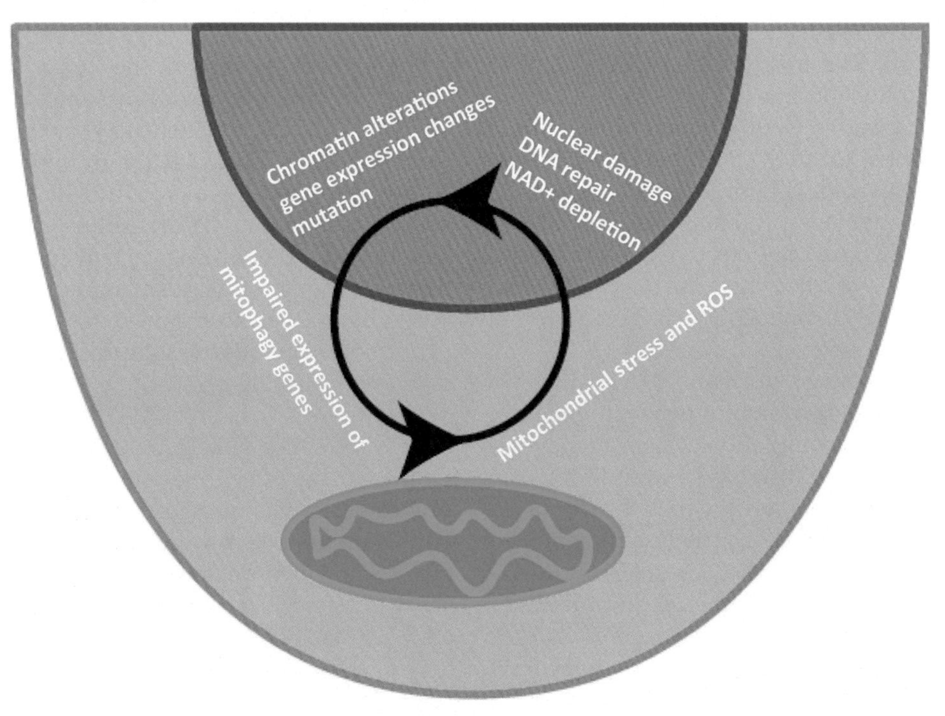

FIGURE 17.3 A model emphasizing how reduced mitochondrial function and quality control can result in DNA damage, nuclear changes, and the further deterioration of mitochondrial function with age.

lesions, such as 8-oxoguanine, 5-hydroxyuracil, and 8-hydroxyadenine in both the nuclear and mitochondrial DNA (Hou et al., 2016). The presence of elevated levels of oxidative DNA damage above age-matched controls is a conserved hallmark of patients with AD and PD, as well as amyotrophic lateral sclerosis (ALS) (Hou et al., 2016). The presence of this DNA damage in mitochondrial and nuclear DNA can result in further impairment of the mitochondria and elevated ROS production as described above for the aging brain. The neurodegenerating brain, however, also has several additional defects that further contribute to these positive feedback loops. In AD, for instance, several studies have found that oxidative stress drives amyloid beta accumulation potentially further exacerbating AD pathology (Zhao & Zhao, 2013). Critically, amyloid beta accumulation drives oxidative stress, further worsening the oxidative stress of the neurodegenerating brain compared to healthy aging brains (Kim et al., 2015; Matsuoka, Picciano, La Francois, & Duff, 2001; Zhao & Zhao, 2013). These studies are supported by in vivo mouse models of amylogenic AD such as the APP and PS1 mice, which have elevated levels of H_2O_2 and more severe peroxidation damage, indicative of oxidative stress (Kim et al., 2015; Matsuoka et al., 2001; Zhao & Zhao, 2013). Beyond amyloid beta, Tau-driven neurodegeneration models such as the P301S mice also show higher ROS levels which could contribute to the DNA damage seen in AD (David et al., 2005; Halverson, Lewis, Frausto, Hutton, & Muma, 2005; Kim et al., 2015; Yoshiyama et al., 2007). The role of ROS and mitochondrial dysfunction is just clear in PD, where studies have found both elevated levels of oxidized base damage and point mutations in the mitochondrial DNA. These mitochondrial mutations are found in Complex I proteins for the electron transport chain, which would further drive mitochondrial dysfunction and ROS production (Blesa, Trigo-Damas, Quiroga-Varela, & Jackson-Lewis, 2015; Kim et al., 2015; Schapira, 2008). Furthermore, treatment of animals with drugs that inhibit Complex I, such as 1-methyl-4-phenyl-1,2,3,4-tetrahydropyridine (MPTP) is sufficient to induce loss of dopaminergic neurons and leads to PD (Lin & Beal, 2006). Similarly, ROS production plays a critical role in many cases of ALS where autosomal dominant mutations in SOD1, a critical ROS scavenger in cells, account for 20% of familial cases of ALS (Andersen, 2006). While the exact contribution of ROS to both the normal changes in the healthy aging brain and the pathological changes in neurodegenerating brain will require much more study, there is unequivocal evidence supporting the role of mitochondrial dysfunction and ROS damage in these processes.

Beyond being exposed to a higher ROS burden, the aging brain also experiences a decline in the efficiency of the base excision repair (BER) pathway that removes the vast majority of the oxidative lesions which further contributes to their accumulation. BER is a multistep process (reviewed in detail in Dianov & Hubscher, 2013) where numerous modified bases, including the oxidative lesions discussed above, are recognized by DNA glycosylases and excised, leaving ssDNA breaks or abasic sites that are further processed by the apurinic/apyrimidinic endonuclease to ssDNA breaks. From there, these ssDNA breaks are filled in, primarily by polymerase beta, and then relegated by Ligase III to finish the repair. The aging brain sees an upregulation of many stress response genes, including several components of the BER pathway along with a concomitant increase in persistent oxidative DNA damage(Lu et al., 2004), suggesting that DNA damage repair is critical in driving normal aging. AD patients, however, exhibit clear defects in BER efficiency when compared to age-matched controls. Using postmortem tissue, up to 40% reductions in the rate of excision of oxidative lesions have been found in AD patients compared to healthy controls, along with a corresponding increase in oxidative lesion levels in AD patients (Weissman et al., 2007). Furthermore, the rate at which excised lesions are filled in by polymerases appear to be reduced in the neurodegenerating brain, likely leaving the cell with an increase in abasic sites and ssDNA breaks that can contribute to the elevated single-stranded breaks (SSBs) discussed below. Further supporting the role for DNA damage in neurodegeneration, recent work has found that introducing mutations in the BER machinery, specifically polymerase beta which fills in the excised nucleotides during BER, elevates neural death and learning defects in the AD mouse model used (Sykora et al., 2015). The exacerbation of the AD mouse model phenotypes by defects in BER inhibition suggests that proper repair of oxidative damage is critical for determining the rate at which neurodegeneration occurs. Additionally, in these studies, the use of a mouse deficient for polymerase beta likely means that the reduced BER capacity of these mice results in elevated levels of DNA SSBs, whose role in neurodegeneration is discussed in detail below. It is important to note, however, that many mouse models, where BER or DNA damage repairs are defective, do not, alone, give rise to neurodegeneration. Only when combined with an additional neurodegenerative stimulus are their phenotypes and roles in neuroprotection revealed. Due to these limitations, significantly more work is necessary to unravel the importance of DNA damage and repair in human aging and neurodegeneration.

DNA BREAK REPAIR IN THE NEURODEGENERATING BRAIN

DNA break repair has long been viewed as essential in neurodevelopment as defects in many HR (homologous recombination) and NHEJ (non-homologous end joining) genes result in embryonic lethality or debilitating developmental diseases such as Ataxia Telangiectasia (Deans, Griffin, Maconochie, & Thacker, 2000; Kerzendorfer & O'Driscoll, 2009; Savitsky et al., 1995). The role of DNA SSBs and double-stranded breaks (DSBs) in the normal brain aging, however, is an ongoing area of research. Part of the recent interest in DSB repair in the brain is due to the observation that normal neural activity directly results in the formation of DSBs, and that these DSBs are specifically induced in the promoters of immediate early response genes to allow rapid expression of genes involved in memory and learning (Madabhushi et al., 2015; Madabhushi, Pan, & Tsai, 2014; Suberbielle et al., 2013). These incredibly unexpected findings raised immediately as-yet unanswered questions about how many DSBs a neuron may purposefully induce over the course of its life, how efficiently these breaks are repaired, and if mutations and misrepaired events at these sites of induced DSBs accumulate in older brains. Part of these concerns lies in the fact that in most tissues, DSB repair by NHEJ becomes less efficient and more error-prone with age, potentially suggesting that repeated break/repair cycles during learning could eventually induce deleterious mutations in older brains (Gorbunova & Seluanov, 2016; Ju et al., 2006). Additionally, as neurons are postmitotic population, they are unable to repair breaks using the generally error-free repair mechanism of HR, placing even higher demands on the NHEJ machinery. Moving forward, significant work needs to be performed to determine if the normal aging brain (1) shows elevated levels of DSBs or reduced repair efficiency, (2) if these breaks are localized to the early response genes known to specifically induce DSBs during learning, (3) if these genes are accumulating mutations at a higher rate than other genes not exposed to induced DSBs, and, finally (4) if the ability of the neuron to induce these DSBs is impaired with aging. The next several years will likely produce several exciting studies that improve and fundamentally alter how DSBs are viewed in the aging brain.

Although relatively little is known about the role of DSBs in healthy aging, significant work exists detailing the extensive DNA damage in neurodegenerative brains compared to age-matched controls. For instance, early studies in AD postmortem tissues found that AD brains contain significantly more DSBs and SSBs than healthy brains (Adamec, Vonsattel, & Nixon, 1999; Mullaart, Boerrigter, Ravid, Swaab, & Vijg, 1990). Although these studies are correlative in nature, the fact that in many studies the appearance of DSBs either precedes pathology or worsens with pathological progression has suggested a potentially causative role (Adamec et al., 1999; Kim et al., 2008; Mullaart et al., 1990). Intriguingly, the brains of AD patients also accumulate significant amounts of chromosomal abnormalities, indicative of improper DSB repair (Hou et al., 2016). In line with these defects, AD brains have been found to exhibit defects in DSB repair by NHEJ, suggesting a possible mechanism by which these elevated gross chromosomal rearrangements may occur (Hou et al., 2016; Shackelford, 2006). It is important to note, however, that in mouse models of AD, only a select few recapitulate the DSBs seen in human AD patients (Kim et al., 2008; Suberbielle et al., 2013). While significantly more work is required to explain this discrepancy, one explanation for the lack of DSBs in mouse models is that, in humans, the brain's capacity to repair DSBs gradually erodes over decades as has been shown in numerous other tissues, allowing the formation and accumulation of these DSBs during disease progression (Gorbunova & Seluanov, 2016). Another unanswered question is are the DSBs seen in AD patients randomly distributed or localized to specific regions? One exciting hypothesis is that these DSBs may be persistently generated in the promoters of immediate early response genes and thus may interfere with normal learning and memory. This model seems plausible in that amyloid beta has been shown to artificially drive neuronal activity and lead to excitotoxicity. Persistent activation of neurons may result in chronic DSB formation, misrepair and mutagenesis, and either cell death or dysregulation in the newly mutated early response genes. Together, these results strongly implicate DSBs as having an important role in neurodegeneration, although more work is required to fully understand if these DSBs are a primary disease-causing lesion or merely a symptom of disease progression.

The role of SSBs in aging and neurodegeneration is also an important, ongoing field of research. Like with DSBs, the role of SSBs in normal brain aging is underexplored, although some evidence suggests the accumulation of SSBs in normal and diseased aging (Adamec et al., 1999). Much of the recent interest in SSBs during neurodegeneration comes from two rare, inherited forms of neurodegeneration caused by inherited mutations in SSB repair proteins: Ataxia with oculomotor apraxia-1 (AOA1) and Spinocerebellar ataxia with axonal neuropathy (SCAN1). AOA1 is caused by mutations in aprataxin (APTX) which functions in the processing of DNA ends during SSBR, while SCAN1 is caused by mutations in tyrosyl-DNA phosphodiesterase 1 (TDP1) which processes the ssDNA ends generated by topoisomerase 1 (El-Khamisy et al., 2009; Katyal et al., 2007; Madabhushi et al., 2015). Either mutation results in the misrepair or accumulation of ssDNA breaks in affected cells and, interestingly, results most prominently in neurodegeneration (El-Khamisy et al., 2005). There are several hypotheses to explain why the brain is so affected by these diseases compared to other tissues. First, as discussed above, the brain likely suffers a higher oxidative burden than most other cells due to its massive oxygen requirements. While SSBs can form from many sources, significant literature exists to suggest that the BER machinery generates SSBs during normal repair of the kinds of oxidative lesions produced by ROS as discussed above (Boiteux & Guillet, 2004). Secondly, as postmitotic cells, neurons are uniquely constrained in their repair of SSBs. Normal, mitotic cells will convert unrepaired SSBs to one-ended DSBs during replication as reviewed in Boiteux & Guillet (2004), allowing the repair of SSBs that have evaded the normal SSB repair machinery. These repair-resistant SSBs could slowly accumulate in neurons and, in the case of SCAN1 and AOA1 patients, rapidly accumulate and result in cell death and neurodegeneration (El-Khamisy et al., 2009; El-Khamisy et al., 2005; Katyal et al., 2007; Madabhushi et al., 2014). While these neurodegenerative diseases are quite rare and distinct from the more common, age-associated neurodegenerative diseases such as AD, PD, and HD, they clearly show that the accumulation of SSBs and their aberrant repair results in severe neurotoxicity. Thus, they underline the critical and as-yet unmet need to evaluate both normal aged and neurodegenerative brains for the accumulation and repair of SSBs.

DNA DAMAGE REINFORCES THE METABOLIC AND GENE EXPRESSION CHANGES IN THE AGING AND NEURODEGENERATING BRAIN

DNA damage, beyond a direct role in impaired gene expression and mutation, further contributes to the aging and neurodegenerating brain by indirectly inducing changes in chromatin structure and energy metabolism. These changes are elicited by activation of both PARP1 (Poly[ADP-ribose] polymerase 1) and SIRT1 (silent mating type information regulation 2 homolog 1) during the DNA damage response. Both SIRT1 and PARP1 have critical roles in DNA damage repair and also compete for the same pool of cellular NAD+ to carry out their functions as discussed below. Overactivation of either enzyme, for example by persistent DNA damage seen in neurodegeneration or age, can reduce the NAD+ available to the other, resulting in both metabolic challenges and impaired functionality of both enzymes. These changes, detailed below, can contribute to the worsening cycle of damage, stress, and energetic decline seen in aging and neurodegeneration.

SIRT1 has many functions, including the regulation of chromatin structure, modifications, and gene expression. SIRT1 directly modifies numerous transcription factors involved in numerous cellular processes, including mitochondrial function, biogenesis, and energetics (Canto & Auwerx, 2012; Canto, Sauve, & Bai, 2013), and silencing repetitive DNA regions in the chromatin. These activities are carried about by deacetylating numerous proteins, thus altering their localization and activity, using NAD+ as an essential co-factor. In response to DNA damage, such as the DSBs commonly present in AD as described above, SIRT1 relocalizes from around the nucleus to specific sites of DNA damage where it has critical roles in initiating the DNA damage response. Persistent DNA damage, however, permanently alters the localization of SIRT1 throughout the nucleus, causing widespread chromatin and gene expression alterations (Oberdoerffer et al., 2008). Many of these changes are induced by altering the normal levels of histone acetylation, which SIRT1 can do both directly as well as through its role in regulating specific histone deacetylases (Dobbin et al., 2013; Madabhushi et al., 2014). Notably, the changes in chromatin structure and gene expression seen in the normal aging brain are recapitulated by the same changes caused by SIRT1 relocalization following DNA damage, suggesting that mislocalization of SIRT1 could drive some of the changes in normal aging (Oberdoerffer et al., 2008).

While the role of SIRT1 in normal aging requires significantly more research, the importance of SIRT1 in neurodegeneration such as AD and ALS has been shown (Dobbin et al., 2013; Kim et al., 2007). SIRT1 directly promotes neuronal survival by promoting efficient DSB repair and reducing DSBs during neurodegeneration or exposure to toxic chemicals (Kim et al., 2007). Unsurprisingly, SIRT1 has been found to be neuroprotective in numerous AD, PD, and HD's mice models (Dobbin et al., 2013; Pasinetti et al., 2011). However, the movement of SIRT1 away from its normal chromatin localization can potentially promote a neurodegenerative environment. In AD, for example, SIRT1 has a role in promoting the expression of ADAM10, a gene which promotes the cleavage of amyloid precursor protein into non-toxic, non-amyloidogenic peptides (Herskovits & Guarente, 2014; Lee et al., 2014), and SIRT1 has roles in preventing the formation of toxic acetylation in Tau models of neurodegeneration such as the P301L mouse (Kim et al., 2007; Min et al., 2010). In this way, the persistent DNA damage presents in neurodegeneration causes indirect changes in gene expression that can further worsen disease pathology. Excitingly, pharmacological activation of SIRT1 has been found to be beneficial in many of these neurodegeneration models (Dobbin et al., 2013; Madabhushi et al., 2014). Moving forward, it will be important to determine if the changes in the expression of UPR, autophagy, the UPS, mitochondrial quality control, and DNA repair detailed above are dependent upon SIRT1 localization, and if pharmacological activation of SIRT1 can prevent some of the changes that appear to set the stage in the aging brain for neurodegeneration to occur.

SIRT1's activity is also regulated, indirectly, by the activity of PARP1, an NAD + -dependent poly[ADP-Ribose] polymerase, during DNA damage repair. PARP1 is a well-characterized protein with critical roles in SSB repair, which functions to modify various nuclear proteins with poly ADP-ribose (PAR) chains. PARP1 rapidly binds to both DNA SSBs and DSBs, where it functions to utilize NAD + to add PAR to itself and to other signaling proteins to promote recognition and repair of the break using the methods described above and reviewed in Javle & Curtin (2011). The role of PARP1 is absolutely critical for repair of the DNA SSBs and DSBs that accumulate in neurodegenerative brains, as discussed above, and unsurprisingly, PARP1 is activated in both human AD patients and numerous AD mouse models (Martire, Mosca, & d'Erme, 2015). Despite its critical role in promoting DNA damage repair, chronic activation of PARP1 has toxic effects on the cellular metabolism and energetics that can actually contribute to worsening neurodegeneration. As PARP1 utilizes NAD + to create PAR chains, PARP1 activation following DNA damage can deplete 80%–90% of the total cellular NAD + within minutes (Martire et al., 2015). This depletion of NAD + can result in impaired glycolysis and an energy deficit, a potentially dangerous situation in an energy-intensive organ like the brain (Alano et al., 2010; Martire et al., 2015). Given the elevated levels of DNA damage in the neurodegenerating brain, the chronic PARP1 activation observed likely further impairs cellular metabolism, placing a heavy demand on mitochondria which are already suffering age and neurodegenerative decline as discussed above. This further stress contributes to their dysfunction and ROS production, which plays into neurodegeneration, and can contribute directly to neuronal death by apoptosis (Alano et al., 2010; Martire et al., 2015). The importance of PARP1 activation also extends past the stress

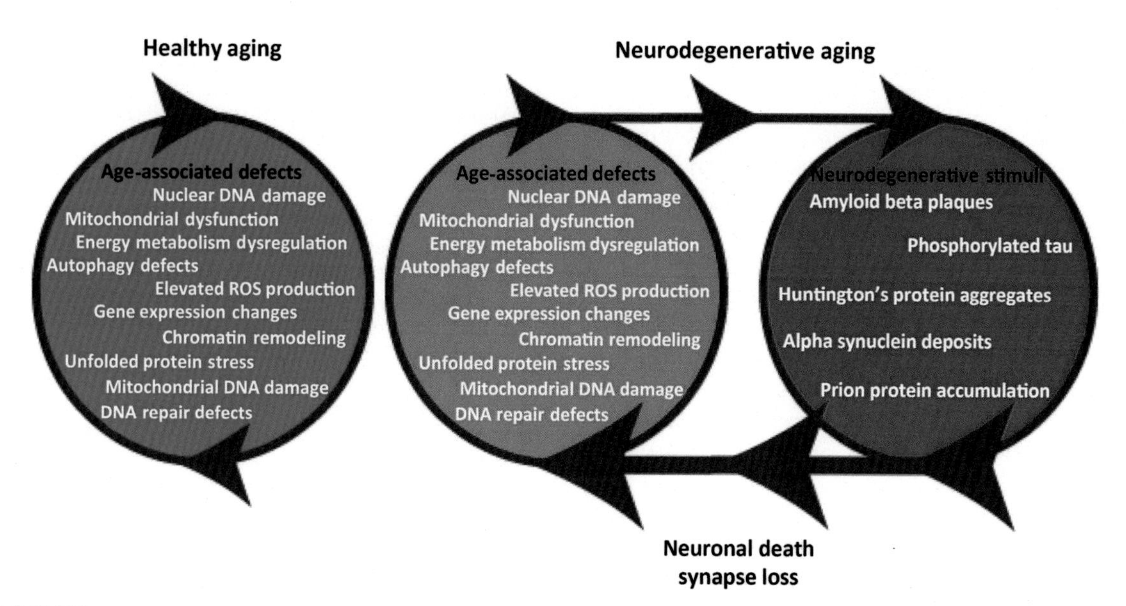

FIGURE 17.4 A model of the various forms of declines seen in the homeostatic quality controls of the normal aging brain (left) compared to how these same declines feed into, and are worsened by, additional neurodegenerative stimuli (right).

it places on an aging cell's energy metabolism by directly inhibiting SIRT1's activity, helping solidify the age-associated chromatin changes mediated by reduced SIRT1 as discussed above. SIRT1's deacetylation activity requires NAD+ as a co-factor, and, as discussed above, PARP1's activation during DNA damage utilizes nearly all of the free NAD+ in a cell. This has been shown to have an inhibitory effect on SIRT1, contributing to the chromatin changes caused by reduced SIRT1 activity as described above. Additionally, the age and DNA damage-associated chromatin and gene expression changes mediated by SIRT1 localization discussed above could be recapitulated by persistent NAD+ depletion leading to chronic SIRT1 inhibition. Taken together, the chronic activation of PARP1 and subsequent depletion of NAD+ and the relocalization of SIRT1 to sites of DNA damage can overtax the aging brain's impaired metabolism and mitochondria, further contributing to the neurodegenerative signals in Fig. 17.4.

CONCLUSIONS

Taken together, the normal aging brain suffers numerous declines in multiple quality control systems. These changes contribute to a gradual erosion of memory and learning seen in many aged individuals and make the aging brain susceptible to numerous forms of neurodegeneration. The presence of additional neurodegenerative stimuli places an extra burden on the eroded homeostatic systems in the brain (Fig. 17.4) and can result in numerous forms of neurodegeneration which exhibit unique pathologies compared to the declines seen in the normal aging brain. Understanding how to prevent or reverse the age-associated declines outlined in this chapter will be essential in generating useful therapeutics for patients.

Acknowledgment

Supported by the NIH grant R37NS051874, the Robert A. and Renee E. Belfer Family Foundation and Belfer Neurodegeneration Consortium.

References

Adamec, E., Vonsattel, J. P., & Nixon, R. A. (1999). DNA strand breaks in Alzheimer's disease. *Brain Research, 849* (1–2), 67–77.

Alano, C. C., Garnier, P., Ying, W., Higashi, Y., Kauppinen, T. M., & Swanson, R. A. (2010). NAD+ depletion is necessary and sufficient for poly(ADP-ribose) polymerase-1-mediated neuronal death. *The Journal of Neuroscience, 30*(8), 2967–2978. Available from https://doi.org/10.1523/JNEUROSCI.5552-09.2010.

Andersen, P. M. (2006). Amyotrophic lateral sclerosis associated with mutations in the CuZn superoxide dismutase gene. *Current Neurology and Neuroscience Reports, 6*(1), 37–46.

Babu, J. R., Geetha, T., & Wooten, M. W. (2005). Sequestosome 1/p62 shuttles polyubiquitinated tau for proteasomal degradation. *Journal of Neurochemistry, 94* (1), 192–203. Available from https://doi.org/10.1111/j.1471-4159.2005.03181.x.

Bishop, N. A., Lu, T., & Yankner, B. A. (2010). Neural mechanisms of ageing and cognitive decline. *Nature, 464*(7288), 529–535. Available from https://doi.org/10.1038/nature08983.

Blalock, E. M., Geddes, J. W., Chen, K. C., Porter, N. M., Markesbery, W. R., & Landfield, P. W. (2004). Incipient Alzheimer's disease: Microarray correlation analyses reveal major transcriptional and tumor suppressor responses. *Proceedings of the National Academy of Sciences of the United States of America, 101*(7), 2173–2178. Available from https://doi.org/10.1073/pnas.0308512100.

Blesa, J., Trigo-Damas, I., Quiroga-Varela, A., & Jackson-Lewis, V. R. (2015). Oxidative stress and Parkinson's disease. *Frontiers in Neuroanatomy, 9*, 91. Available from https://doi.org/10.3389/fnana.2015.00091.

Boiteux, S., & Guillet, M. (2004). Abasic sites in DNA: Repair and biological consequences in Saccharomyces cerevisiae. *DNA Repair, 3*(1), 1–12.

Brunk, U. T., & Terman, A. (2002a). Lipofuscin: Mechanisms of age-related accumulation and influence on cell function. *Free Radical Biology & Medicine, 33*(5), 611–619.

Brunk, U. T., & Terman, A. (2002b). The mitochondrial-lysosomal axis theory of aging: Accumulation of damaged mitochondria as a result of imperfect autophagocytosis. *European Journal of Biochemistry, 269*(8), 1996–2002.

Cadenas, E., & Davies, K. J. (2000). Mitochondrial free radical generation, oxidative stress, and aging. *Free Radical Biology & Medicine*, *29*(3–4), 222–230.

Canto, C., & Auwerx, J. (2012). Targeting sirtuin 1 to improve metabolism: All you need is NAD(+)?. *Pharmacological Reviews*, *64*(1), 166–187. Available from https://doi.org/10.1124/pr.110.003905.

Canto, C., Sauve, A. A., & Bai, P. (2013). Crosstalk between poly(ADP-ribose) polymerase and sirtuin enzymes. *Molecular Aspects of Medicine*, *34*(6), 1168–1201. Available from https://doi.org/10.1016/j.mam.2013.01.004.

Chadwick, L., Gentle, L., Strachan, J., & Layfield, R. (2012). Review: Unchained maladie—A reassessment of the role of Ubb(+1) -capped polyubiquitin chains in Alzheimer's disease. *Neuropathology and Applied Neurobiology*, *38*(2), 118–131. Available from https://doi.org/10.1111/j.1365-2990.2011.01236.x.

Cherra, S. J., III, & Chu, C. T. (2008). Autophagy in neuroprotection and neurodegeneration: A question of balance. *Future Neurology*, *3*(3), 309–323. Available from https://doi.org/10.2217/14796708.3.3.309.

Ciechanover, A., & Kwon, Y. T. (2015). Degradation of misfolded proteins in neurodegenerative diseases: Therapeutic targets and strategies. *Experimental & Molecular Medicine*, *47*, e147. Available from https://doi.org/10.1038/emm.2014.117.

Cuervo, A. M., & Dice, J. F. (2000). Age-related decline in chaperone-mediated autophagy. *The Journal of Biological Chemistry*, *275*(40), 31505–31513. Available from https://doi.org/10.1074/jbc.M002102200.

David, D. C., Hauptmann, S., Scherping, I., Schuessel, K., Keil, U., Rizzu, P., ... Gotz, J. (2005). Proteomic and functional analyses reveal a mitochondrial dysfunction in P301L tau transgenic mice. *The Journal of Biological Chemistry*, *280*(25), 23802–23814. Available from https://doi.org/10.1074/jbc.M500356200.

Deans, B., Griffin, C. S., Maconochie, M., & Thacker, J. (2000). Xrcc2 is required for genetic stability, embryonic neurogenesis and viability in mice. *The EMBO Journal*, *19*(24), 6675–6685. Available from https://doi.org/10.1093/emboj/19.24.6675.

De Mario, A., Quintana-Cabrera, R., Martinvalet, D., & Giacomello, M. (2016). (Neuro)degenerated Mitochondria-ER contacts. *Biochemical and Biophysical Research Communications*, *483*(4), 1096–1109. Available from https://doi.org/10.1016/j.bbrc.2016.07.056.

Dianov, G. L., & Hubscher, U. (2013). Mammalian base excision repair: The forgotten archangel. *Nucleic Acids Research*, *41*(6), 3483–3490. Available from https://doi.org/10.1093/nar/gkt076.

Dice, J. F. (1982). Altered degradation of proteins microinjected into senescent human fibroblasts. *The Journal of Biological Chemistry*, *257*(24), 14624–14627.

Dobbin, M. M., Madabhushi, R., Pan, L., Chen, Y., Kim, D., Gao, J., ... Tsai, L. H. (2013). SIRT1 collaborates with ATM and HDAC1 to maintain genomic stability in neurons. *Nature Neuroscience*, *16*(8), 1008–1015. Available from https://doi.org/10.1038/nn.3460.

Du, Y., Wooten, M. C., Gearing, M., & Wooten, M. W. (2009). Age-associated oxidative damage to the p62 promoter: Implications for Alzheimer disease. *Free Radical Biology & Medicine*, *46*(4), 492–501. Available from https://doi.org/10.1016/j.freeradbiomed.2008.11.003.

Du, Y., Wooten, M. C., & Wooten, M. W. (2009). Oxidative damage to the promoter region of SQSTM1/p62 is common to neurodegenerative disease. *Neurobiology of Disease*, *35*(2), 302–310. Available from https://doi.org/10.1016/j.nbd.2009.05.015.

El-Khamisy, S. F., Katyal, S., Patel, P., Ju, L., McKinnon, P. J., & Caldecott, K. W. (2009). Synergistic decrease of DNA single-strand break repair rates in mouse neural cells lacking both Tdp1 and aprataxin. *DNA Repair*, *8*(6), 760–766. Available from https://doi.org/10.1016/j.dnarep.2009.02.002.

El-Khamisy, S. F., Saifi, G. M., Weinfeld, M., Johansson, F., Helleday, T., Lupski, J. R., & Caldecott, K. W. (2005). Defective DNA single-strand break repair in spinocerebellar ataxia with axonal neuropathy-1. *Nature*, *434*(7029), 108–113. Available from https://doi.org/10.1038/nature03314.

Gentier, R. J., & van Leeuwen, F. W. (2015). Misframed ubiquitin and impaired protein quality control: An early event in Alzheimer's disease. *Frontiers in Molecular Neuroscience*, *8*, 47. Available from https://doi.org/10.3389/fnmol.2015.00047.

Gong, B., Radulovic, M., Figueiredo-Pereira, M. E., & Cardozo, C. (2016). The ubiquitin-proteasome system: Potential therapeutic targets for Alzheimer's disease and spinal cord injury. *Frontiers in Molecular Neuroscience*, *9*, 4. Available from https://doi.org/10.3389/fnmol.2016.00004.

Gorbunova, V., & Seluanov, A. (2016). DNA double strand break repair, aging and the chromatin connection. *Mutation Research*, *788*, 2–6. Available from https://doi.org/10.1016/j.mrfmmm.2016.02.004.

Gregori, L., Fuchs, C., Figueiredo-Pereira, M. E., Van Nostrand, W. E., & Goldgaber, D. (1995). Amyloid beta-protein inhibits ubiquitin-dependent protein degradation in vitro. *The Journal of Biological Chemistry*, *270*(34), 19702–19708.

Halverson, R. A., Lewis, J., Frausto, S., Hutton, M., & Muma, N. A. (2005). Tau protein is cross-linked by transglutaminase in P301L tau transgenic mice. *The Journal of Neuroscience*, *25*(5), 1226–1233. Available from https://doi.org/10.1523/JNEUROSCI.3263-04.2005.

Herskovits, A. Z., & Guarente, L. (2014). SIRT1 in neurodevelopment and brain senescence. *Neuron*, *81*(3), 471–483. Available from https://doi.org/10.1016/j.neuron.2014.01.028.

Hou, Y., Song, H., Croteau, D. L., Akbari, M., & Bohr, V. A. (2016). Genome instability in Alzheimer disease. *Mechanisms of Ageing and Development*, *161*(Pt A), 83–94. Available from https://doi.org/10.1016/j.mad.2016.04.005.

Iurlaro, R., & Munoz-Pinedo, C. (2016). Cell death induced by endoplasmic reticulum stress. *The FEBS Journal*, *283*(14), 2640–2652. Available from https://doi.org/10.1111/febs.13598.

Javle, M., & Curtin, N. J. (2011). The role of PARP in DNA repair and its therapeutic exploitation. *British Journal of Cancer*, *105*(8), 1114–1122. Available from https://doi.org/10.1038/bjc.2011.382.

Joshi, A. U., Kornfeld, O. S., & Mochly-Rosen, D. (2016). The entangled ER-mitochondrial axis as a potential therapeutic strategy in neurodegeneration: A tangled duo unchained. *Cell Calcium*, *60*(3), 218–234. Available from https://doi.org/10.1016/j.ceca.2016.04.010.

Ju, Y. J., Lee, K. H., Park, J. E., Yi, Y. S., Yun, M. Y., Ham, Y. H., ... Park, G. H. (2006). Decreased expression of DNA repair proteins Ku70 and Mre11 is associated with aging and may contribute to the cellular senescence. *Experimental & Molecular Medicine*, *38*(6), 686–693. Available from https://doi.org/10.1038/emm.2006.81.

Katyal, S., el-Khamisy, S. F., Russell, H. R., Li, Y., Ju, L., Caldecott, K. W., & McKinnon, P. J. (2007). TDP1 facilitates chromosomal single-strand break repair in neurons and is neuroprotective in vivo. *The EMBO Journal*, *26*(22), 4720–4731. Available from https://doi.org/10.1038/sj.emboj.7601869.

Kerzendorfer, C., & O'Driscoll, M. (2009). Human DNA damage response and repair deficiency syndromes: Linking genomic instability and cell cycle checkpoint proficiency. *DNA Repair*, *8*(9), 1139–1152. Available from https://doi.org/10.1016/j.dnarep.2009.04.018.

Kiffin, R., Kaushik, S., Zeng, M., Bandyopadhyay, U., Zhang, C., Massey, A. C., ... Cuervo, A. M. (2007). Altered dynamics of the lysosomal receptor for chaperone-mediated autophagy with age. *Journal of Cell Science*, *120*(Pt 5), 782–791. Available from https://doi.org/10.1242/jcs.001073.

Kim, D., Frank, C. L., Dobbin, M. M., Tsunemoto, R. K., Tu, W., Peng, P. L., ... Tsai, L. H. (2008). Deregulation of HDAC1 by p25/Cdk5 in neurotoxicity. *Neuron*, *60*(5), 803–817. Available from https://doi.org/10.1016/j.neuron.2008.10.015.

Kim, D., Nguyen, M. D., Dobbin, M. M., Fischer, A., Sananbenesi, F., Rodgers, J. T., ... Tsai, L. H. (2007). SIRT1 deacetylase protects against neurodegeneration in models for Alzheimer's disease and amyotrophic lateral sclerosis. *The EMBO Journal*, *26*(13), 3169–3179. Available from https://doi.org/10.1038/sj.emboj.7601758.

Kim, G. H., Kim, J. E., Rhie, S. J., & Yoon, S. (2015). The role of oxidative stress in neurodegenerative diseases. *Experimental Neurobiology*, *24*(4), 325–340. Available from https://doi.org/10.5607/en.2015.24.4.325.

Lee, H. R., Shin, H. K., Park, S. Y., Kim, H. Y., Lee, W. S., Rhim, B. Y., ... Kim, C. D. (2014). Cilostazol suppresses beta-amyloid production by activating a disintegrin and metalloproteinase 10 via the upregulation of SIRT1-coupled retinoic acid receptor-beta. *Journal of Neuroscience Research*, *92*(11), 1581–1590. Available from https://doi.org/10.1002/jnr.23421.

Liang, W. S., Reiman, E. M., Valla, J., Dunckley, T., Beach, T. G., Grover, A., ... Stephan, D. A. (2008). Alzheimer's disease is associated with reduced expression of energy metabolism genes in posterior cingulate neurons. *Proceedings of the National Academy of Sciences of the United States of America*, *105*(11), 4441–4446. Available from https://doi.org/10.1073/pnas.0709259105.

Lin, M. T., & Beal, M. F. (2006). Mitochondrial dysfunction and oxidative stress in neurodegenerative diseases. *Nature*, *443*(7113), 787–795. Available from https://doi.org/10.1038/nature05292.

Lipinski, M. M., Zheng, B., Lu, T., Yan, Z., Py, B. F., Ng, A., ... Yuan, J. (2010). Genome-wide analysis reveals mechanisms modulating autophagy in normal brain aging and in Alzheimer's disease. *Proceedings of the National Academy of Sciences of the United States of America*, *107*(32), 14164–14169. Available from https://doi.org/10.1073/pnas.1009485107.

Lopez Salon, M., Morelli, L., Castano, E. M., Soto, E. F., & Pasquini, J. M. (2000). Defective ubiquitination of cerebral proteins in Alzheimer's disease. *Journal of Neuroscience Research*, *62*(2), 302–310, doi:10.1002/1097–4547(20001015)62:2 < 302::AID-JNR15 > 3.0.CO;2-L.

Lopez Salon, M., Pasquini, L., Besio Moreno, M., Pasquini, J. M., & Soto, E. (2003). Relationship between beta-amyloid degradation and the 26S proteasome in neural cells. *Experimental Neurology*, *180*(2), 131–143.

Lu, T., Pan, Y., Kao, S. Y., Li, C., Kohane, I., Chan, J., & Yankner, B. A. (2004). Gene regulation and DNA damage in the ageing human brain. *Nature*, *429*(6994), 883–891. Available from https://doi.org/10.1038/nature02661.

Madabhushi, R., Gao, F., Pfenning, A. R., Pan, L., Yamakawa, S., Seo, J., ... Tsai, L. H. (2015). Activity-induced DNA breaks govern the expression of neuronal early-response genes. *Cell*, *161*(7), 1592–1605. Available from https://doi.org/10.1016/j.cell.2015.05.032.

Madabhushi, R., Pan, L., & Tsai, L. H. (2014). DNA damage and its links to neurodegeneration. *Neuron, 83*(2), 266–282. Available from https://doi.org/10.1016/j.neuron.2014.06.034.

Makrides, S. C. (1983). Protein synthesis and degradation during aging and senescence. *Biological Reviews of the Cambridge Philosophical Society, 58*(3), 343–422.

Martire, S., Mosca, L., & d'Erme, M. (2015). PARP-1 involvement in neurodegeneration: A focus on Alzheimer's and Parkinson's diseases. *Mechanisms of Ageing and Development, 146–148*, 53–64. Available from https://doi.org/10.1016/j.mad.2015.04.001.

Masters, S. L., & O'Neill, L. A. (2011). Disease-associated amyloid and misfolded protein aggregates activate the inflammasome. *Trends in Molecular Medicine, 17*(5), 276–282. Available from https://doi.org/10.1016/j.molmed.2011.01.005.

Matsuoka, Y., Picciano, M., La Francois, J., & Duff, K. (2001). Fibrillar beta-amyloid evokes oxidative damage in a transgenic mouse model of Alzheimer's disease. *Neuroscience, 104*(3), 609–613.

Miller, J. A., Oldham, M. C., & Geschwind, D. H. (2008). A systems level analysis of transcriptional changes in Alzheimer's disease and normal aging. *The Journal of Neuroscience, 28*(6), 1410–1420. Available from https://doi.org/10.1523/JNEUROSCI.4098-07.2008.

Min, S. W., Cho, S. H., Zhou, Y., Schroeder, S., Haroutunian, V., Seeley, W. W., ... Gan, L. (2010). Acetylation of tau inhibits its degradation and contributes to tauopathy. *Neuron, 67*(6), 953–966. Available from https://doi.org/10.1016/j.neuron.2010.08.044.

Mullaart, E., Boerrigter, M. E., Ravid, R., Swaab, D. F., & Vijg, J. (1990). Increased levels of DNA breaks in cerebral cortex of Alzheimer's disease patients. *Neurobiology of Aging, 11*(3), 169–173.

Nixon, R. A., Wegiel, J., Kumar, A., Yu, W. H., Peterhoff, C., Cataldo, A., & Cuervo, A. M. (2005). Extensive involvement of autophagy in Alzheimer disease: An immuno-electron microscopy study. *Journal of Neuropathology and Experimental Neurology, 64*(2), 113–122.

Ntsapi, C., & Loos, B. (2016). Caloric restriction and the precision-control of autophagy: A strategy for delaying neurodegenerative disease progression. *Experimental Gerontology, 83*, 97–111. Available from https://doi.org/10.1016/j.exger.2016.07.014.

Nunomura, A., & Miyagishi, T. (1993). Ultrastructural observations on neuronal lipofuscin (age pigment) and dense bodies induced by a proteinase inhibitor, leupeptin, in rat hippocampus. *Acta Neuropathologica, 86*(4), 319–328.

Oberdoerffer, P., Michan, S., McVay, M., Mostoslavsky, R., Vann, J., Park, S. K., ... Sinclair, D. A. (2008). SIRT1 redistribution on chromatin promotes genomic stability but alters gene expression during aging. *Cell, 135*(5), 907–918. Available from https://doi.org/10.1016/j.cell.2008.10.025.

Oenzil, F., Kishikawa, M., Mizuno, T., & Nakano, M. (1994). Age-related accumulation of lipofuscin in three different regions of rat brain. *Mechanisms of Ageing and Development, 76*(2–3), 157–163.

Pasinetti, G. M., Wang, J., Marambaud, P., Ferruzzi, M., Gregor, P., Knable, L. A., & Ho, L. (2011). Neuroprotective and metabolic effects of resveratrol: Therapeutic implications for Huntington's disease and other neurodegenerative disorders. *Experimental Neurology, 232*(1), 1–6. Available from https://doi.org/10.1016/j.expneurol.2011.08.014.

Patrick, G. N. (2006). Synapse formation and plasticity: Recent insights from the perspective of the ubiquitin proteasome system. *Current Opinion in Neurobiology, 16*(1), 90–94. Available from https://doi.org/10.1016/j.conb.2006.01.007.

Porta, E. A. (2002). Pigments in aging: An overview. *Annals of the New York Academy of Sciences, 959*, 57–65.

Rubinsztein, D. C., Marino, G., & Kroemer, G. (2011). Autophagy and aging. *Cell, 146*(5), 682–695. Available from https://doi.org/10.1016/j.cell.2011.07.030.

Savitsky, K., Bar-Shira, A., Gilad, S., Rotman, G., Ziv, Y., Vanagaite, L., ... Shiloh, Y. (1995). A single ataxia telangiectasia gene with a product similar to PI-3 kinase. *Science, 268*(5218), 1749–1753.

Schapira, A. H. (2008). Mitochondria in the aetiology and pathogenesis of Parkinson's disease. *Lancet Neurology, 7*(1), 97–109. Available from https://doi.org/10.1016/S1474-4422(07)70327-7.

Scheper, W., & Hoozemans, J. J. (2015). The unfolded protein response in neurodegenerative diseases: A neuropathological perspective. *Acta Neuropathologica, 130*(3), 315–331. Available from https://doi.org/10.1007/s00401-015-1462-8.

Shackelford, D. A. (2006). DNA end joining activity is reduced in Alzheimer's disease. *Neurobiology of Aging, 27*(4), 596–605. Available from https://doi.org/10.1016/j.neurobiolaging.2005.03.009.

Suberbielle, E., Sanchez, P. E., Kravitz, A. V., Wang, X., Ho, K., Eilertson, K., ... Mucke, L. (2013). Physiologic brain activity causes DNA double-strand breaks in neurons, with exacerbation by amyloid-beta. *Nature Neuroscience, 16*(5), 613–621. Available from https://doi.org/10.1038/nn.3356.

Sulzer, D., Mosharov, E., Talloczy, Z., Zucca, F. A., Simon, J. D., & Zecca, L. (2008). Neuronal pigmented autophagic vacuoles: Lipofuscin, neuromelanin, and ceroid as macroautophagic responses during aging and disease. *Journal of Neurochemistry, 106*(1), 24–36. Available from https://doi.org/10.1111/j.1471-4159.2008.05385.x.

Sykora, P., Misiak, M., Wang, Y., Ghosh, S., Leandro, G. S., Liu, D., ... Bohr, V. A. (2015). DNA polymerase beta deficiency leads to neurodegeneration and exacerbates Alzheimer disease phenotypes. *Nucleic Acids Research, 43*(2), 943–959. Available from https://doi.org/10.1093/nar/gku1356.

Taylor, R. C. (2016). Aging and the UPR(ER). *Brain Research, 1648*(Pt B), 588–593. Available from https://doi.org/10.1016/j.brainres.2016.04.017.

Terman, A., & Brunk, U. T. (1998). Ceroid/lipofuscin formation in cultured human fibroblasts: The role of oxidative stress and lysosomal proteolysis. *Mechanisms of Ageing and Development, 104*(3), 277–291.

Tydlacka, S., Wang, C. E., Wang, X., Li, S., & Li, X. J. (2008). Differential activities of the ubiquitin-proteasome system in neurons versus glia may account for the preferential accumulation of misfolded proteins in neurons. *The Journal of Neuroscience, 28*(49), 13285–13295. Available from https://doi.org/10.1523/JNEUROSCI.4393-08.2008.

van Leeuwen, F. W., de Kleijn, D. P., van den Hurk, H. H., Neubauer, A., Sonnemans, M. A., Sluijs, J. A., ... Hol, E. M. (1998). Frameshift mutants of beta amyloid precursor protein and ubiquitin-B in Alzheimer's and Down patients. *Science, 279*(5348), 242–247.

Weissman, L., Jo, D. G., Sorensen, M. M., de Souza-Pinto, N. C., Markesbery, W. R., Mattson, M. P., & Bohr, V. A. (2007). Defective DNA base excision repair in brain from individuals with Alzheimer's disease and amnestic mild cognitive impairment. *Nucleic Acids Research, 35*(16), 5545–5555. Available from https://doi.org/10.1093/nar/gkm605.

Xilouri, M., & Stefanis, L. (2016). Chaperone mediated autophagy in aging: Starve to prosper. *Ageing Research Reviews*. Available from https://doi.org/10.1016/j.arr.2016.07.001.

Yankner, B. A., Lu, T., & Loerch, P. (2008). The aging brain. *Annual Review of Pathology, 3*, 41–66. Available from https://doi.org/10.1146/annurev.pathmechdis.2.010506.092044.

Yoshiyama, Y., Higuchi, M., Zhang, B., Huang, S. M., Iwata, N., Saido, T. C., ... Lee, V. M. (2007). Synapse loss and microglial activation precede tangles in a P301S tauopathy mouse model. *Neuron, 53*(3), 337–351. Available from https://doi.org/10.1016/j.neuron.2007.01.010.

Zhao, Y., Hegde, A. N., & Martin, K. C. (2003). The ubiquitin proteasome system functions as an inhibitory constraint on synaptic strengthening. *Current Biology, 13*(11), 887–898.

Zhao, Y., & Zhao, B. (2013). Oxidative stress and the pathogenesis of Alzheimer's disease. *Oxidative Medicine and Cellular Longevity, 2013*, 316523. Available from https://doi.org/10.1155/2013/316523.

Zouambia, M., Fischer, D. F., Hobo, B., De Vos, R. A., Hol, E. M., Varndell, I. M., ... Van Leeuwen, F. W. (2008). Proteasome subunit proteins and neuropathology in tauopathies and synucleinopathies: Consequences for proteomic analyses. *Proteomics, 8*(6), 1221–1236. Available from https://doi.org/10.1002/pmic.200700679.

Index

Note: Page numbers followed by "*f*" and "*t*" refer to figures and tables, respectively.

I

O

Q

R

S

T

U

V

W

X

Z

Printed in the United States
By Bookmasters